REVIEWS in MINERALOGY and GEOCHEMISTRY

Volume 43   2001

STABLE ISOTOPE GEOCHEMISTRY

Editors:

JOHN W. VALLEY

Department of Geology & Geophysics
University of Wisconsin
Madison, Wisconsin

DAVID R. COLE

Chemical and Analytical Sciences Division
Oak Ridge National Laboratory
Oak Ridge, Tennessee

COVER: Metamorphosed magnetite in a granulite facies marble. Oxygen isotope ratios, measured by ion microprobe from 8 µm spots within each pit, show homogeneity across the core of each crystal. However, depth profiles into crystal faces reveal low $\delta^{18}O$ rims and gradients up to 9‰ per 10 µm. These results indicate slow cooling after the Grenville orogeny, explain the failure of conventional oxygen isotope thermometry for this sample, and document post-tectonic fluid infiltration. Correct interpretation of this sample would not be possible without spatially resolved microanalysis.

See Valley & Graham (1991) *Contributions to Mineralogy and Petrology* 109:38-52.

Series Editor for MSA: **Paul H. Ribbe**

Virginia Polytechnic Institute and State University
Blacksburg, Virginia

MINERALOGICAL SOCIETY of AMERICA

D1158692

REVIEWS IN MINERALOGY
AND GEOCHEMISTRY

(Formerly: REVIEWS IN MINERALOGY)

ISSN 1529-6466

Volume 43

Stable Isotope Geochemistry

ISBN 0-939950-55-3

** This volume is the fifth of a series of review volumes published jointly
under the banner of the Mineralogical Society of America and the
Geochemical Society. The newly titled *Reviews in Mineralogy and
Geochemistry* has been numbered contiguously with the previous series,
Reviews in Mineralogy.

*Additional copies of this volume as well as others in
this series may be obtained at moderate cost from:*

THE MINERALOGICAL SOCIETY OF AMERICA
1015 EIGHTEENTH STREET, NW, SUITE 601
WASHINGTON, DC 20036 U.S.A.

Dedication

Dr. William C. Luth has had a long and distinguished career in research, education and in the government. He was a leader in experimental petrology and in training graduate students at Stanford University. His efforts at Sandia National Laboratory and at the Department of Energy's headquarters resulted in the initiation and long-term support of many of the cutting edge research projects whose results form the foundations of these short courses. Bill's broad interest in understanding fundamental geochemical processes and their applications to national problems is a continuous thread through both his university and government career. He retired in 1996, but his efforts to foster excellent basic research, and to promote the development of advanced analytical capabilities gave a unique focus to the basic research portfolio in Geosciences at the Department of Energy. He has been, and continues to be, a friend and mentor to many of us. It is appropriate to celebrate his career in education and government service with this series of courses in cutting-edge geochemistry that have particular focus on Department of Energy-related science, at a time when he can still enjoy the recognition of his contributions.

FOREWORD

The scientific editors of this volume, John Valley and David Cole, organized a Short Course on *Stable Isotope Geochemistry* presented November 2-4, 2001 in conjunction with the annual meetings of the Geological Society of America in Boston, Massachusetts. The contributors to this review volume were the instructors. The Mineralogical Society of America (MSA) sponsored the course and is the publisher and distributor of this and other books in the *Reviews in Mineralogy and Geochemistry* series. Alex Speer, Executive Director of MSA, and Myrna Byer, of Printing Service Associates, Inc., are acknowledged for their considerable contributions to the success of this volume.

MSA has been in partnership with the Geochemical Society (GS) since 2000 and the publication of Volume 39, *Transformation Processes in Minerals*. As a result, the *Reviews* volumes are now covering an ever-widening range of subjects. An additional series editor, Jodi J. Rosso, has joined the team to manage those short-course publications and other books which will be sponsored by the Geochemical Society.

MSA and the editors of *Stable Isotope Geochemistry* are particularly grateful to the Geosciences Research Program (Nick Woodward, director) of the U.S. Department of Energy for financial support of this publication and student scholarships through a grant to MSA.

Paul H. Ribbe
Series Editor for MSA
Blacksburg, Virginia
September 14, 2001

PREFACE

This volume follows the 1986 *Reviews in Mineralogy* (Vol. 16) in approach but reflects significant changes in the field of Stable Isotope Geochemistry. In terms of new technology, new sub-disciplines, and numbers of researchers, the field has changed more in the past decade than in any other since that of its birth. Unlike the 1986 volume, which was restricted to high temperature fields, this book covers a wider range of disciplines. However, it would not be possible to fit a comprehensive review into a single volume. Our goal is to provide state-of-the-art reviews in chosen subjects that have emerged or advanced greatly since 1986.

The field of Stable Isotope Geochemistry was born of a good idea and nurtured by technology. In 1947, Harold Urey published his calculated values of reduced partition function for oxygen isotopes and his idea (a good one!) that the fractionation of oxygen isotopes between calcite and water might provide a means to estimate the temperatures of geologic events. Building on wartime advances in electronics, Alfred Nier then designed and built the dual-inlet, gas-source mass-spectrometer capable of making measurements of sufficient precision and accuracy. This basic instrument and the associated extraction techniques, mostly from the 1950s, are still in use in many labs today. These techniques have become "conventional" in the sense of traditional, and they provide the benchmark against which the accuracy of other techniques is compared.

The 1986 volume was based almost exclusively on natural data obtained solely from conventional techniques. Since then, revolutionary changes in sample size, accuracy, and cost have resulted from advances in continuous flow mass-spectrometry, laser heating, ion microprobes, and computer automation. The impact of new technology has differed by discipline. Some areas have benefited from vastly enlarged data sets, while others have capitalized on *in situ* analysis and/or micro- to nanogram size samples, and others have developed because formerly intractable samples can now be analyzed. Just as Stable Isotope Geochemistry is being reborn by new good ideas, it is still being nurtured by new technology.

The organization of the chapters in this book follows the didactic approach of the 2001 short course in Boston. The first three chapters present the principles and data base for equilibrium isotope fractionation and for kinetic processes of exchange. Both inorganic and biological aspects are considered. The next chapter reviews isotope compositions throughout the solar system including mass-independent fractionations that are increasingly being recognized on Earth. The fifth chapter covers the primitive compositions of the mantle and subtle variations found in basalts. This is followed by three chapters on metamorphism, isotope thermometry, fluid flow, and hydrothermal alteration. The next chapter considers water cycling in the atmosphere and the ice record. And finally, there are four chapters on the carbon cycle, the sulfur cycle, organic isotope geochemistry and extinctions in the geochemical record.

The editors thank the authors of individual chapters and also all those who generously assisted in the scientific review of these chapters: J. Alt, R. Bidigare, I. Bindeman, J. Bischoff, J. Bowman, P. Brown, D. Canfield, W. Casey, F. Corsetti, J. Eiler, J. Farquhar, J. Farver, V. Fereirra, H. Fricke, M. Gibbs, B. Hanson, R. Harmon, J. Hayes, M. Hendricks, J. Horita, C. Johnson, Y. Katzir, E. King, A. Knoll, J.S. Lackey, C. Lesher, K. Macleod, P. Meyers, W. Peck, B. Putlitz, F. Robert, G. Roselle, J. Severinghaus, Z. Sharp, M. Spicuzza, M. Thiemens, D. Tinker, and K. von Damn. Alex Speer and the MSA Business office assisted with many aspects of the Short Course in Boston, before the November 2001 Meeting of the Geological Society of America.

We thank the U.S. Department of Energy for providing publication support for this book and scholarships for students who attended the Short Course.

We are especially indebted to Paul H. Ribbe, series editor of the *Reviews in Mineralogy and Geochemistry*, formerly, *Reviews in Mineralogy*. Paul was responsible for general editing, final composition of the text, and motivating us to meet the production deadlines. Paul was also the series editor for the 1986 *Stable Isotopes* volume and has produced a remarkable total of more than forty *RiM* or *RiMG* volumes. This is a major achievement in affordable publishing of forefront science.

John W. Valley
Madison, Wisconsin

David R. Cole
Oak Ridge, Tennessee

August 15, 2001

STABLE ISOTOPE GEOCHEMISTRY
Table of Contents

3   Fractionation of Carbon and Hydrogen Isotopes in Biosynthetic Processes

John M. Hayes

4 Stable Isotope Variations in Extraterrestrial Materials
Kevin D. McKeegan, Laurie A. Leshin

5 Oxygen Isotope Variations of Basaltic Lavas and Upper Mantle Rocks
John M. Eiler

6 Stable Isotope Thermometry at High Temperatures

John W. Valley

7   Stable Isotope Transport and Contact Metamorphic Fluid Flow
Lukas P. Baumgartner, John W. Valley

8   Stable Isotopes in Seafloor Hydrothermal Systems:
Vent fluids, hydrothermal deposits, hydrothermal alteration, and microbial processes
W. C. Shanks, III

9   Oxygen- and Hydrogen-Isotopic Ratios of Water in Precipitation: Beyond Paleothermometry

Richard B. Alley, Kurt M. Cuffey

10   Isotopic Evolution of the Biogeochemical Carbon Cycle During the Precambrian

David J. Des Marais

11   Isotopic Biogeochemistry of Marine Organic Carbon
Katherine H. Freeman

12   Biogeochemistry of Sulfur Isotopes
D. E. Canfield

13   Stratigraphic Variation in Marine Carbonate Carbon Isotope Ratios

Robert L. Ripperdan

1   Equilibrium Oxygen, Hydrogen and Carbon Isotope Fractionation Factors Applicable to Geologic Systems

Thomas Chacko

Department of Earth and Atmospheric Sciences
University of Alberta
Edmonton, Alberta T6G 2E3, Canada

David R. Cole and Juske Horita

Chemical and Analytical Sciences Division
Oak Ridge National Laboratory
Oak Ridge, Tennessee 37831

INTRODUCTION

As demonstrated by the chapters in this short course, stable isotope techniques are an important tool in almost every branch of the earth sciences. Central to many of these applications is a quantitative understanding of equilibrium isotope partitioning between substances. Indeed, it was Harold Urey's (1947) thermodynamically based estimate of the temperature-dependence of $^{18}O/^{16}O$ fractionation between calcium carbonate and water, and a recognition of how this information might be used to determine the temperatures of ancient oceans, that launched the science of stable isotope geochemistry. The approach pioneered by Urey has since been used to estimate temperatures for a wide range of geological processes (e.g. Emiliani 1955; Anderson et al. 1971; Clayton 1986; Valley, this volume). In addition to their geothermometric applications, equilibrium fractionation data are also important in the study of fluid-rock interactions, including those associated with diagenetic, hydrothermal, and metamorphic processes (Baumgartner and Valley, this volume; Shanks, this volume). Finally, a knowledge of equilibrium fractionation is a necessary first step in evaluating isotopic disequilibrium, a widespread phenomenon that is increasingly being used to study temporal relationships in geological systems (Cole and Chakraborty, this volume).

In the fifty-four years since the publication of Urey's paper, equilibrium fractionation data have been reported for many minerals and fluids of geological interest. These data were derived from: (1) theoretical calculations following the methods developed by Urey (1947) and Bigeleisen and Mayer (1947); (2) direct laboratory experiments; (3) semi-empirical bond-strength models; and (4) measurement of fractionations in natural samples. Each of these methods has its advantages and disadvantages. However, the availability of a variety of methods for calibrating fractionation factors has led to a plethora of calibrations, not all of which are in agreement. In this chapter, we evaluate the major methods for determining fractionation factors. We also compile data on oxygen, hydrogen, and carbon isotope fractionation factors for geologically relevant mineral and fluid systems. Our compilation focuses primarily on experimental and natural sample calibrations of fractionations factors as large compilations of theoretical (Richet et al. 1977; Kieffer 1982) and bond-strength (e.g. Hoffbauer et al. 1994; Zheng 1999a) calibrations already exist in the literature. The chapter begins with a general overview of the theoretical basis of stable isotope fractionation, and theoretical methods for calculating fractionation factors. The reader is referred to the earlier review papers of Richet et al. (1977), Clayton (1981), O'Neil (1986) and Kyser (1987), and the recent textbook by Criss (1999) for more detailed

1529-6466/00/0043-0001$10.00

discussion of theoretical topics. Our emphasis will be on advances in the determination of fractionation factors and on our understanding of the variables that control isotopic fractionation behavior made since the publication of *Reviews in Mineralogy*, Volume 16 (Valley et al. 1986).

THEORETICAL BACKGROUND

Comparison of cation and isotope exchange reactions

The equilibrium fractionation of isotopes between substances is analogous in many ways to the partitioning of cations (such as Fe and Mg) between minerals. Both processes can be described in terms of chemical reactions in which isotopes or cations are exchanged between two coexisting phases. For example, the partitioning of Fe and Mg between orthopyroxene and olivine can be described by the reaction:

$$FeSiO_3 + 1/2\ Mg_2SiO_4 = MgSiO_3 + 1/2\ Fe_2SiO_4$$

where the equilibrium constant, K_1, for this reaction is:

$$K_1 = \frac{(a_{MgSiO_3}^{opx})(a_{Fe_2SiO_4}^{ol})^{1/2}}{(a_{FeSiO_3}^{opx})(a_{Mg_2SiO_4}^{ol})^{1/2}}$$

where a_i^p is the thermodynamic activity of component i in phase p. Assuming ideal mixing of Fe and Mg on the octahedral sites of orthopyroxene and olivine, K_1 can be recast as:

$$K_1 = \frac{(Mg/Fe)_{opx}}{[(Mg/Fe)_{ol}^2]^{1/2}} = \frac{(Mg/Fe)_{opx}}{(Mg/Fe)_{ol}}$$

Similarly, the partitioning of ^{18}O and ^{16}O between olivine and orthopyroxene is described by the reaction:

$$1/3\ MgSi^{16}O_3 + 1/4\ Mg_2Si^{18}O_4 = 1/3\ MgSi^{18}O_3 + 1/4\ Mg_2Si^{16}O_4$$

Assuming ideal mixing of oxygen isotopes among the different oxygen sites in olivine and orthopyoxene, this gives an equilibrium constant, K_2, of:

$$K_2 = \frac{[(^{18}O/^{16}O)_{opx}^3]^{1/3}}{[(^{18}O/^{16}O)_{ol}^4]^{1/4}} = \frac{(^{18}O/^{16}O)_{opx}}{(^{18}O/^{16}O)_{ol}} = \alpha_{opx-ol}$$

If the reaction is written such that one mole of ^{18}O and ^{16}O atoms are exchanged between the two minerals, K_2 is equal to α_{opx-ol}, the oxygen isotope fractionation factor between orthopyroxene and olivine.

As with all chemical reactions, the standard state Gibbs free energy change for an isotope exchange reaction at a given pressure and temperature is related to the equilibrium constant by:

$$\Delta G_R^o(T, P) = \Delta H_R^o - T\Delta S_R^o + P\Delta V_R^o = -RT\ \ln K \tag{1}$$

In principle, the free energy change and in turn the equilibrium constant for such reactions can be calculated from conventional thermodynamic data (molar enthalpy, entropy, volume data) on the end-member isotopic species denoted in the reaction. This approach, however, is generally not practicable because of the paucity of thermodynamic data on isotopically 'pure' end-members. Moreover, even if such data were widely available, the Gibbs free energy changes associated with most isotope exchange reactions are too small (typically a few tens of joules or less, compared to thousands of joules for cation exchange reactions) to permit precise calculations using classical thermodynamic methods.

Despite its limitations, the discussion above is useful for illustrating the formal similarities between cation and isotope exchange reactions. Equation (1) also shows that equilibrium constants for all exchange reactions are dependent on temperature. More specifically, ln K varies linearly with T^{-1} if ΔG^o_R is independent of temperature. This T^{-1} temperature-dependency generally applies to cation exchange reactions because values of ΔG^o_R for such reactions are approximately constant over a wide range of temperatures. In the case of isotope exchange reactions, however, ΔG^o_R varies significantly with temperature, which results in higher order temperature-dependencies (T^{-2}). The effect of pressure on ln K is determined by the volume change for the reaction. For most isotope exchange reactions, ΔV^o_R is small, resulting in correspondingly small pressure effects on ln K. However, as discussed further below, pressure effects can be significant for hydrogen isotope fractionations, particular those involving water.

QUANTUM MECHANICAL REASONS FOR ISOTOPIC FRACTIONATION

The existence of small but significant free energy changes in isotope exchange reactions implies energetic differences between chemical species differing only in their isotopic composition. These energy differences are entirely a quantum mechanical phenomenon arising from the effect of atomic mass on the vibrational energy of molecules. Consider a diatomic molecule, which can be represented by two masses, m_1 and m_2, attached by a spring (Fig. 1a). The force (F) exerted on the masses is equal to the displacement (x) of the spring from the rest position times the force constant (k) of the spring (i.e. the spring's stiffness):

$$F = -kx,$$

The potential energy, PE, of the spring is given by the equation:

$$PE = kx^2/2$$

which defines a parabola with a minimum potential energy when the spring is at the rest position (x = 0), and increasing potential energy when the spring is compressed (-x) or stretched (+x) from that position (Fig. 1b). The vibrational frequency, ν, of the spring is given by:

$$\nu = \frac{1}{2\pi}\sqrt{\frac{k}{\mu}} \tag{2}$$

where μ is the reduced mass and given by:

$$\mu = \frac{m_1 m_2}{m_1 + m_2}$$

Derivation of the equations given above can be found in most introductory physics or physical chemistry textbooks. McMillan (1985) also provides a helpful summary.

In its simplest form, the spring-chemical bond analogy is referred to as the harmonic oscillator approximation. Several facets of this analogy are useful to keep in mind. Firstly, the rest position of the spring corresponds to the optimal distance between the nuclei of the two atoms, and the minimum in the potential energy curve. If the atoms are pushed closer or pulled further away than this optimal distance, electrical forces act to

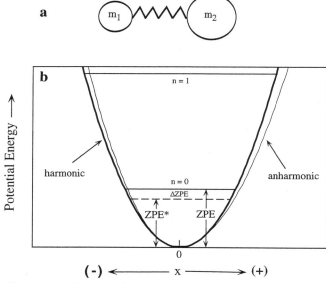

Figure 1. (a) Drawing of a spring with two masses attached (simple harmonic oscillator), which is an analogue for a diatomic molecule. (b) Schematic plot showing variation in the potential energy of harmonic (and anharmonic) oscillators as they are displaced from the rest position (x = 0). Energy levels are given by values of n. The zero point energy (ZPE) is the difference in energy between the bottom of the potential energy well and the energy of the ground vibrational state (n = 0). Note that the zero point energy of a molecule made with the heavy isotope (ZPE*) is lower than that of the molecule made with the light isotope. The magnitude of ΔZPE (ZPE-ZPE*) for a substance exerts a major control on its isotopic fractionation behavior.

restore the atoms towards the equilibrium position. Secondly, a strong chemical bond can be thought of as a stiff spring (i.e. a spring with a large force constant). It follows from Equation (2) that, all else being equal, strong bonds generally have higher vibrational frequencies than weak bonds. Thirdly, according to the Born-Oppenheimer approximation, isotopic substitution has no effect on the force constant of a bond. The magnitude of the force constant is determined by the electronic interaction between atoms, and is independent of the masses of the two nuclei. Thus, the potential energy curves of molecules comprising heavy and light isotopes of an element are identical.

Given the last statement, classical mechanics predicts no energy differences between two molecules that differ only in their isotopic composition. At a temperature of absolute zero, both molecules should have energies corresponding to the bottom of their identical potential energy wells. Quantum theory, however, indicates that the vibrational energy, E, is quantized and given by:

$$E = (n + 1/2)h\nu \tag{3}$$

where n corresponds to the energy levels 0, 1, 2, 3, etc., and h is Planck's constant. Thus, even at absolute zero, where all molecules are in the ground state (n = 0), the vibrational energy of these molecules lies some distance above the bottom of the potential energy well. The energy difference between the bottom of the potential energy well and the energy of ground vibrational state is referred to as the zero point energy or ZPE (Fig. 1b). Importantly, although the potential energy curves of molecules made up of light and heavy isotopes of an element are identical, their ZPE's are different because of the effect of mass on vibrational frequency. More specifically, it can readily be shown from Equation (2) that the ratio of the vibrational frequencies of isotopically heavy and light molecules of a particular compound is given by:

$$\frac{v^*}{v} = \sqrt{\frac{\mu}{\mu^*}} \tag{4}$$

where the asterisks denote the molecule containing the heavy isotope. Because of the inverse relationship between frequency and mass, the heavy molecule has a lower vibrational frequency, and hence a lower ZPE than the light molecule (Fig. 1b). This implies that a isotopically heavy molecule is always energetically more stable than its isotopically light counterpart.

It should be clear from the discussion above that *all* substances will be stabilized by heavy isotope substitution, and thus prefer to form bonds with the heavy isotope. The key issue for partitioning of isotopes *between* substances is the preference of one substance over another for the heavy isotope. This is determined by the degree to which a molecule's vibrational energy is lowered by heavy isotope substitution. At low temperatures, where all molecules are in their ground state, the magnitude of energy lowering is, to a good approximation, given by:

$$\Delta ZPE = ZPE - ZPE^* = 1/2\ h(v - v^*) = 1/2\ h\Delta v \tag{5}$$

In a competition for the heavy isotope, the substance with the larger ΔZPE (Fig. 1b) is more stabilized by the isotopic substitution, and therefore takes the lion's share of the heavy isotope. It should be noted that with increasing temperature, a progressively larger fraction of molecules are excited to higher energy levels (n > 0). In those cases, ΔZPE remains an important factor, but not the only factor in determining isotope fractionation behavior.

Earlier in this section, we stated that strong bonds tend to have higher vibrational frequencies than weak bonds. By rearranging terms in equation (4), it can be shown that, other things being equal, bonds with high vibrational frequency undergo larger frequency shifts (Δv) on isotope substitution than bonds with low vibrational frequency.

$$\Delta v = v - v^* = v\left(1 - \sqrt{\frac{\mu}{\mu^*}}\right) \tag{6}$$

From Equation (5), it follows that large frequency shifts lead to large ΔZPE, and consequently an affinity for the heavy isotope. The important generalization that stems from Equations (5) and (6) is that the heavy isotope favors substances with strong bonds. An example of this correlation between bond strength and heavy isotope partitioning is the sequence of ^{18}O enrichment observed in coexisting silicate minerals. Taylor and Epstein (1962) noted that minerals with abundant Si-O bonds are enriched in the heavy isotope of oxygen (^{18}O) relative to minerals with fewer Si-O bonds. This correlation reflects the high strength and therefore high vibrational frequency of Si-O bonds relative to other cation-oxygen bonds in silicates, and the effect of these parameters on ΔZPE.

The discussion above is based on the harmonic oscillator approximation, which is the simplest model for representing the energetics of a molecule. In this model, the potential energy curve is symmetrical and the energy levels are equally spaced. More realistic models, such as that of Morse (1929), have asymmetric potential energy curves, non-uniformly spaced energy levels, and numerically different values of ZPE than given in the harmonic model. Despite these differences, the general principles outlined in this section also apply to the more complex models.

CALCULATING FRACTIONATION FACTORS

Theory

The detailed calculation of isotopic fractionation factors follows the approach of Urey (1947) and Bigeleisen and Mayer (1947), and the reader is referred to those papers for further explanation of the equations given below. A summary of the nomenclature used in these equations is given in Appendix 1. The equations were originally derived for ideal gases, and require additional approximations if applied to liquids or solids. The calculation of fractionation factors involves the partition function, Q, a statistical mechanical parameter that describes all possible energy states of a substance. The equilibrium constant for an isotope exchange reaction can be expressed as a ratio of the partition functions of the two sides of the reaction. For example, consider a generalized isotope exchange reaction between substances A and B:

$$aA + bB* = aA* + bB$$

where a and b represent stoichiometric coefficients, and the asterisk, here and in all subsequent references, denotes the substance made with the heavy isotope. The equilibrium constant, K_{A-B}, for this reaction can be expressed as:

$$K_{A-B} = \frac{(Q*)_A^a (Q)_B^b}{(Q)_A^a (Q*)_B^b} = \frac{(Q*/Q)_A^a}{(Q*/Q)_B^b} \qquad (7)$$

To a good approximation, the total partition function (Q) for each species in the reaction is the product of the translational (tr), rotational (rot), and vibrational (vib) partition functions:

$$Q = Q_{tr} \times Q_{rot} \times Q_{vib} \qquad (8)$$

Taking these partition functions individually, the translational partition function is given by:

$$Q_{tr} = \frac{(2\pi M k_b T)^{3/2}}{h^3} V$$

where M is the molecular weight, k_b is Boltzmann's constant and V is the volume of the system. Fortunately, the partition function of each species in the reaction need not be evaluated in the calculation of the equilibrium constant, only the ratio of partition functions of a species and its isotopically substituted derivative (e.g. $[Q*]_A/[Q]_A$). Because all ideal gases occupy the same volume at a given pressure and temperature, all terms except molecular weight cancel in the ratio of translational partition functions:

$$(Q*/Q)_{tr} = \left(\frac{M*}{M}\right)^{3/2}$$

Similarly, most terms cancel in the calculation of the ratio of rotational partition functions. For diatomic molecules and linear polyatomic molecules, this ratio is given by:

$$(Q*/Q)_{rot} = \frac{\sigma I*}{\sigma* I}$$

where I is the moment of inertia and σ is the symmetry number, which is the number of equivalent ways of orienting a molecule in space. For example, $\sigma = 1$ for heteronuclear diatomic molecules (e.g. NO or HD), and $\sigma = 2$ for homonuclear diatomic molecules (e.g. O_2). The rotational partition function ratio for non-linear polyatomic molecules is:

$$(Q*/Q)_{rot} = \frac{\sigma}{\sigma*}\left(\frac{I_A*I_B*I_C*}{I_A I_B I_C}\right)^{1/2}$$

where I_A, I_B, and I_C are the three principal moments of inertia.

In the harmonic oscillator approximation, the vibrational partition function ratio is given by:

$$(Q*/Q)_{vib} = \prod_i \frac{e^{-U_i*/2}}{e^{-U_i/2}}\frac{1-e^{-U_i}}{1-e^{-U_i*}}$$

where $U_i = hv_i/k_bT$ and i is a running index of vibrational modes. There is only one vibrational mode for diatomic molecules (i = 1). For linear and non-linear polyatomic molecules consisting of s atoms, there are 3s-5 and 3s-6 vibrational modes, respectively, all of which must be considered in the calculation of Q*/Q.

Combining the contributions of translational, rotational and vibrational partition functions yields:

$$(Q*/Q) = \left(\frac{M*}{M}\right)^{3/2}\frac{\sigma I*}{\sigma*I}\frac{e^{-U_i*/2}}{e^{-U_i/2}}\frac{1-e^{-U_i}}{1-e^{-U_i*}}$$

for diatomic molecules and:

$$(Q*/Q) = \left(\frac{M*}{M}\right)^{3/2}\frac{\sigma}{\sigma*}\left(\frac{I_A*I_B*I_C*}{I_A I_B I_C}\right)^{1/2}\prod_i\frac{e^{-U_i*/2}}{e^{-U_i/2}}\frac{1-e^{-U_i}}{1-e^{-U_i*}}$$

for polyatomic molecules. The moments of inertia can be removed from the expressions through use of the Teller-Redlich spectroscopic theorem (Urey 1947). This yields:

$$(Q*/Q) = \left(\frac{m*}{m}\right)^{3r/2}\frac{\sigma}{\sigma*}\prod_i\frac{v_i*}{v_i}\frac{e^{-U_i*/2}}{e^{-U_i/2}}\frac{1-e^{-U_i}}{1-e^{-U_i*}} \tag{9}$$

where m* and m are the atomic weights of the isotopes being exchanged, and r is the number of atoms of the element being exchanged present in the molecule (e.g. r = 1 for oxygen exchange in CO; r = 2 for oxygen exchange in CO_2). Equation (9) forms the basis for the calculation of fractionation factors for gaseous substances.

Several features of Equation (9) are noteworthy (cf. Richet et al. 1977; Criss 1999). The first three terms on the right hand side of the equation ($[m*/m]^{3r/2}$, $[\sigma/\sigma*]$, $[v*/v]$) which take into account the effect of translation and rotation on (Q*/Q), are independent of temperature. The fourth term ($e^{-U_i*/2}/e^{-U_i/2}$) varies with temperature, but is mainly controlled by the ZPE difference of the isotopically heavy and light molecules. The last term ($[1-e^{-U_i}]/[1-e^{-U_i*}]$) relates to the spacing of energy levels. At low temperatures, where nearly all molecules are in the ground vibrational state, this term is close to unity, and therefore does not contribute appreciably to Q*/Q. The term has a progressively larger effect on Q*/Q as temperature increases. Finally, the mass term ($[m*/m]^{3r/2}$) in Equation (9) cancels in the calculation of an equilibrium constant ($[Q*/Q]_A/[Q*/Q]_B$). That is, the mass term for one molecule (A) taken to the stoichiometrically appropriate exponent equals that for the other molecule (B) involved in the exchange reaction. Thus, the mass term need not be considered for our purposes. By convention, partition function ratios with the mass term omitted are called reduced partition function ratios, and sometimes referred to by the symbol f. f is formally defined as:

$$f = \frac{Q*}{Q}\left(\frac{m}{m*}\right)^{3r/2}$$
(10)

In tabulations of partition function calculations (e.g. Richet et al. 1977), reduced partition ratios are commonly reported as $f^{1/r} = \beta$ values, or $1000 \ln \beta$ values. In such cases, the fractionation factor between two substances is simply:

$$\alpha_{(A-B)} = \beta_A/\beta_B \text{ and } 1000 \ln \alpha_{(A-B)} = 1000 \ln \beta_A - 1000 \ln \beta_B$$

The input data required in calculating fractionation factors are the vibrational frequencies of all chemical species participating in an isotope exchange reaction. In many cases, however, frequencies have only been measured for molecules made with the abundant isotope (e.g. ^{16}O); the frequencies of molecules containing the rare isotope (e.g. ^{18}O) must be calculated. The simplest way to calculate the unknown frequencies is through the harmonic oscillator approximation (Eqn. 4). More rigorous and accurate calculations of frequencies require force-field models, which are available for many common gaseous molecules (e.g. Richet et al. 1977).

An example calculation

As an example, we show the calculation of the $^{18}O/^{16}O$ fractionation factor between CO_2 and CO. Such calculations were computationally laborious in Urey's time but today can readily be done on a spreadsheet. The exchange reaction controlling oxygen isotope fractionation in the CO_2-CO system is:

$$C^{18}O + 1/2 \, C^{16}O_2 = C^{16}O + 1/2 \, C^{18}O_2$$

with an equilibrium constant given by:

$$K_{CO_2-CO} = \frac{Q_{C^{18}O_2}^{1/2} \, Q_{C^{16}O}}{Q_{C^{16}O_2}^{1/2} \, Q_{C^{18}O}} = \frac{(Q*/Q)_{CO_2}^{1/2}}{(Q*/Q)_{CO}} = \alpha_{(CO_2 - CO)}$$

For isotope exchange reactions written as above involving only isotopically pure molecules (e.g. pure $C^{16}O$ or $C^{18}O$), the symmetry number of a molecule and its isotopic derivative are identical. Therefore, $\sigma/\sigma* = 1$, and the term need not be included in the calculations. The vibrational frequencies used in our calculations are the same as those on which Urey's (1947) calculations are based. However, Urey corrected these frequencies for anharmonicity (zero-order frequencies), whereas we used observed (measured) fundamental frequencies with no anharmonicity correction (see discussions in Bottinga 1969a, p. 52; McMillan 1985, p. 15; and Polyakov and Kharlashina 1995, p. 2568). Vibrational frequencies are generally reported in wave numbers (ω), which have units of cm^{-1}. For partition function calculations, wave numbers must be converted to units of sec^{-1} by multiplying by c, the velocity of light ($\nu = c\omega$).

There is one vibrational mode for diatomic molecules such as CO, and four (3s-5) vibrational modes for linear tri-atomic molecules such as CO_2. The four modes of CO_2 correspond to different vibrational motions of the CO_2 molecule, the symmetric stretching vibration (ω_1), the asymmetric stretching vibration (ω_3), and two lower-frequency bending vibrations (ω_2) (Fig. 2). The two bending modes are referred to as degenerate because they have the same vibrational frequency. Therefore, although it is

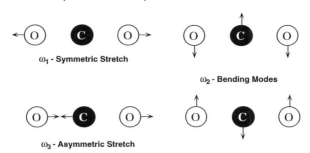

ω_1 - Symmetric Stretch

ω_2 - Bending Modes

ω_3 - Asymmetric Stretch

Figure 2. Vibrational modes of the CO_2 molecule, the symmetric stretching vibration (ω_1), the asymmetric stretching vibration (ω_3), and the two bending vibrations (ω_2). Note that ω_1 only involves movement of oxygen atoms, whereas ω_2 and ω_3 involve movement of both oxygen and carbon atoms. This results in a larger ^{18}O frequency shift (Δv) for the ω_1 vibration.

listed only once in Table 1, ω_2 must be counted twice in calculating the partition function ratio of CO_2 using Equation (9). Note also that the magnitude of frequency shifts (Δv) for the isotopically substituted CO_2 molecule varies with vibrational mode (Table 1). The largest shift is associated with the ω_1 vibration, which relates to the fact that this vibrational mode involves only movement of oxygen atoms, whereas the other three modes involve the movement of both oxygen and carbon atoms (Fig. 2). The reduced mass of the $^{12}C^{16}O_2$ molecule undergoing the ω_1, ω_2 and ω_3 vibrations is then given by:

$$\mu_{\omega1} = \frac{m_o m_o}{m_o + m_o} \qquad \mu_{\omega2,\omega3} = \frac{2m_o m_c}{2m_o + m_c}$$

(see Polyakov and Kharlashina 1995) where m_o and m_c are the masses of the ^{16}O and ^{12}C atoms, respectively. As a result of these relationships, the change in reduced mass on ^{18}O substitution ($[\mu/\mu^*]^{1/2}$), and therefore the frequency shift, is significantly greater for the ω_1 vibration than for the other vibrational modes.

Table 1 shows the contribution of individual terms to the total and reduced partition function ratios of CO and CO_2. The dominant contributor to the reduced partition function ratios of both molecules is the ZPE term, particularly at low temperatures. Thus, as noted above, isotope partitioning between substances is strongly influenced by the magnitude of ΔZPE (cf. Bigeleisen 1965). The large frequency shift associated with the ω_1 vibration of the CO_2 molecule results in a large ΔZPE, and therefore a tendency for CO_2 to concentrate the heavy isotope at low temperature. At higher temperature, the magnitudes of vibrational frequencies (high v) play an increasingly important role in isotope partitioning. In the case of the CO_2-CO system, the high vibrational frequency of the CO molecule results in a so-called crossover in fractionations between 0 and 500°C (crossover occurs at 292°C). That is, at T > 292°C, ^{18}O partitions into CO rather than into CO_2. Such crossovers are entirely consistent with equilibrium fractionation of isotopes, and, as noted by Urey (1947), Stern et al. (1968) and Spindel et al. (1970), are not uncommon in gaseous substances.

The calculations above are made on the basis of the harmonic oscillator approximation, and include no explicit corrections for anharmonicity. The effects of anharmonicity can be incorporated by using calculated zero-order frequencies rather than observed fundamental frequencies (see above), and by adding anharmonic corrections to the ZPE and energy level spacing terms of Equation (9) (Urey 1947; Bottinga 1968; Richet et al. 1977). Urey (1947) included anharmonic effects in his calculations of partition function

Table 1. Calculation of $^{18}O/^{16}O$ fractionation factors between CO_2 and CO.

Vibrat. mode	T(°C)	(1)ω (cm^{-1})	(1)ω* (cm^{-1})	(2)$(m^*/m)^{3r/2}$	ν^*/ν	$e^{-U^*/2}/e^{-U/2}$	$(1-e^{-U})/(1-e^{-U^*})$	Q^*/Q	(4)$f^{1/r}$ = β
CO		$^{12}C^{16}O$	$^{12}C^{18}O$						
ω_θ	0	2140.8	2089.1	1.193728	0.975843	1.145914	1.000004	1.334870	1.118236
ω_θ	500	2140.8	2089.1	1.193728	0.975843	1.049296	1.001920	1.224662	1.025914
CO$_2$		$^{12}C^{16}O_2$	$^{12}C^{18}O_2$						
ω_1	0	1342.5	1265.6	1.424987	0.942681	1.224661	1.000425		
(3)ω_2(2)	0	667.0	656.8	1.424987	0.984695	1.027521	1.001700		
ω_3	0	2355.0	2319.0	1.424987	0.984695	1.099579	1.000001		
	0							1.829514	1.133085
ω_1	500	1342.5	1265.6	1.424987	0.942681	1.074226	1.013986		
(3)ω_2(2)	500	667.0	656.8	1.424987	0.984695	1.009544	1.007858		
ω_3	500	2355.0	2319.0	1.424987	0.984695	1.034106	1.000879		
	500							1.496938	1.024935

K (0°C) = 1.013277   **1000 lnα (0°C) = 13.19 ‰**

K (500°C) = 0.999046   **1000 lnα (500°C) = -0.95 ‰**

(1) Observed fundamental vibrational frequencies for CO and CO_2, corresponding to zero-order frequencies (corrected for anharmonicity) given in Urey (1947). The vibrational frequencies of the $C^{18}O$ and $C^{18}O_2$ molecules (ω^*) were calculated using Equation (4) (see text).

(2) The term r in the exponent represents the number of atoms of the isotope being exchanged present in the molecule (e.g., r = 1 for CO and 2 for CO_2).

(3) ω_2 for the CO_2 molecule is degenerate. That is, two vibrational modes of the molecule have this frequency.

(4) f is the reduced partition function ratio where $f = (Q^*/Q) (m/m^*)^{3r/2}$. The equilibrium constants listed above are for the isotope exchange reaction given in the text; K (CO_2-CO) = α (CO_2-CO) = $[(Q^*/Q)^{0.5}_{CO_2}/(Q^*/Q)_{CO}]$ = β_{CO_2}/β_{CO}.

ratios for CO_2 and CO. Because we used the same vibrational frequencies as Urey, his results in the CO_2-CO system can be compared directly with those given in Table 1. For the five temperatures (273-600 K) for which Urey reported fractionation factors, our calculations neglecting anharmonicity are within 0.02 to 0.30‰ of his. Thus, the net effect of anharmonicity on fractionation factors in this system is relatively small, albeit in some cases outside measurement error. In their detailed partition function ratio calculations for gaseous molecules, Richet et al. (1977) showed that the largest anharmonic correction is to the ZPE term but that the magnitude of this correction decreases with increasing with temperature. The anharmonic correction to the energy level spacing term is much smaller but increases with increasing temperature. For most gaseous substances, the net effect of including the anharmonicity terms is to decrease the calculated β value (Urey 1947; Richet et al. 1977). Thus, in the calculation of fractionation factors from β values, the anharmonicity correction for one substance is often partly cancelled by the anharmonicity correction for the other substance in the exchange couple. More detailed discussions of anharmonicity are given in Bottinga (1968), Richet et al. (1977), Gillet et al. (1996) and Polyakov (1998).

Calculation of fractionation factors for gases, liquids and fluids

Following the original compilation by Urey (1947), several studies have calculated fractionation factors involving geologically relevant gaseous molecules. The most sophisticated and widely cited of these studies is that of Richet et al. (1977), who reported oxygen, hydrogen, carbon, sulfur, nitrogen and chlorine isotope fractionation factors for a large number of gaseous species. Their calculations used the latest (at the time) spectroscopic data and theoretical models, and included anharmonicity terms for all the gaseous species considered. In general, their calculated fractionation factors are in good agreement with experimental data.

Although the basic principles still hold, isotopic fractionation theory developed for ideal gases is not directly applicable to liquids. Indeed, it has long been known from experiments that gases fractionate isotopes relative to liquids of the same composition. For example, the equilibrium hydrogen and oxygen isotope fractionation between liquid and gaseous H_2O is 73 and 9.2‰, respectively, at 25°C (Majoube 1971b; Horita and Wesolowski 1994). These large fractionations are the result of two effects in the condensed phase (Bigeleisen 1961; Van Hook 1975). First, translational and rotational energy levels, which for free moving gaseous molecules are well represented by an energy continuum, become quantized in liquids because of interactions among molecules. Thus, whereas there is no isotope fractionation associated solely with translation or rotation in gases, there is such a fractionation in the liquid phase. This effect favors partitioning of the heavy isotope into a liquid relative to a gas of the same composition. The second effect relates to the influence of intermolecular interactions on the vibrational characteristics of individual molecules in the liquid. Thus, vibrational frequencies of substance measured in the gas phase are not identical to those of the same substance in the liquid phase. This second effect may cause isotopic fractionation in the opposite direction to the first (Bigeleisen 1961), and lead to a crossover in liquid-vapor fractionations, as has been documented for D/H fractionations in the $H_2O_{(l)}$-$H_2O_{(v)}$ system.

We are not aware of any direct calculations of partition function ratios for supercritical fluids. However, experimentally determined oxygen and carbon isotope fractionations in the CO_2-calcite system at supercritical conditions are in good agreement with fractionations derived from theoretical calculations (Chacko et al. 1991; Scheele and Hoefs 1992; Rosenbaum 1994). This suggests that supercritical CO_2 is well represented by the ideal gas approximation as the calculations were made on this basis (Rosenbaum

1997). The same conclusion does not appear to hold for H_2O (Bottinga 1968; Clayton et al. 1989; Rosenbaum 1997; Driesner 1997). For example, oxygen partition function ratios for supercritical H_2O derived empirically from mineral-H_2O exchange experiments are distinctly lower than those calculated using the ideal gas approximation (Rosenbaum 1997). As first suggested by Bottinga (1968), this discrepancy may be due to the non-ideality of H_2O under supercritical conditions, and the effect of this non-ideality on the vibrational frequencies of the H_2O molecule. Another possible reason for the discrepancy is the solubility of minerals in water at elevated pressures and temperatures. Thus, partition function ratios of H_2O derived empirically from mineral-H_2O experiments may be different from those of a pure H_2O fluid at comparable P-T conditions (Hu and Clayton, in press).

Calculation of fractionation factors for minerals

Principles. Partition function ratios can also be calculated for minerals, but these calculations are complex and require a number of approximations. The general approach is to treat a mineral as a large molecule consisting of 3s independent oscillators, where s is the number of atoms in the unit cell. For example, quartz, which contains 9 atoms in its unit cell (Si_3O_6), has a total of 27 vibrational modes. Of these, 24 (3s-3) are so-called optical modes because their vibrational frequencies are derived from optical spectroscopic techniques (e.g. IR or Raman spectroscopy). The optical modes are subdivided into internal modes, which concern the vibrational motions of individual functional groups within the mineral (e.g. Si-O in silicates and CO_3 in carbonates), and external modes, which correspond to vibrations of the mineral lattice. The remaining 3 modes are referred to as acoustic modes because they relate to sound velocity. The frequencies of the acoustic modes are typically derived from heat capacity and spectroscopic data using Debye-Einstein models (e.g. Bottinga 1968; Kawabe 1978). Using these data, the partition function ratio for the unit cell in the harmonic approximation is given by (Kieffer 1982):

$$(Q^*/Q) = \prod_{i=1}^{3s} \frac{e^{-U_i^*/2}}{e^{-U_i/2}} \frac{1-e^{-U_i}}{1-e^{-U_i^*}} \tag{11}$$

Equation (11) for minerals is similar to Equation (9) for gases but ignores translational and rotational contributions to the partition function, as such motions are restricted or absent in solids (Bottinga 1968). In detailed calculations, the three acoustic modes are treated separately from the optical modes because each acoustic mode represents a continuous spectrum of vibrations rather than a single vibrational frequency. As such, the acoustic modes are best evaluated by means of Debye functions rather than the Einstein functions typically used to treat the optical modes (see Eqn. 6 in Bottinga (1968) for the mathematical details of dealing with the acoustic modes). The total partition function ratio given by Equation (11) can be converted to a reduced partition function ratio using Equation (10) (Kieffer 1982), but r in the case of minerals is the number of atoms being exchanged in the unit cell (e.g. r = 6 for oxygen exchange in quartz).

The largest uncertainty in the calculation of partition function ratios for minerals is the magnitude of frequency shifts on isotope substitution. Because direct spectroscopic measurements of minerals made with the less abundant isotope are not widely available, these shifts must usually be calculated or estimated in some other way. The detailed approach to calculating frequency shifts employs the methods of lattice dynamics to determine force constants for each vibrational mode, from which the vibrational frequencies for a mineral and its isotopic derivative can be predicted (e.g. Bottinga 1968;

Elcombe and Hulston 1975; Kawabe 1978). Kieffer (1982) took a less detailed approach in calculating oxygen isotope partition function ratios for 11 silicate minerals, calcite and rutile. As input for her calculations, Kieffer used the measured spectra for the ^{16}O forms of minerals, divided the vibrational modes for these minerals into four groups, and then developed a set of rules for estimating the frequency shift associated with each group on ^{18}O substitution. Importantly, she applied the same rules to each mineral considered in her study, which resulted in an internally consistent set of partition function ratios. Most of Kieffer's calculated fractionation factors are in excellent agreement with experimental data (Clayton and Kieffer 1991).

Table 2. Calculation of oxygen isotope partition function ratio for quartz at 25°C.

ω (cm^{-1})	[1]shift factor	[2]g_i	[3]$f(x)^{g_i}$	Q*/Q	f	1000 lnβ
[a]102	0.96522	1	1.03675	5.3344	1.8435	101.95
[a]122	0.96522	1	1.03705			
[a]164	0.96522	1	1.03786			
128	0.93828	2	1.14019			
205.6	0.93677	1	1.07279	ln (Q*/Q)$_{optical}$ = 1.56455		
263.1	0.95173	2	1.11772	ln (Q*/Q)$_{acoustic}$ = 0.10963		
354.3	0.97883	1	1.02662	ln (Q*/Q)$_{total}$ = 1.67418		
363.5	0.97387	1	1.03332			
393.8	0.9647	1	1.04688			
401.8	0.95744	1	1.05741			
450	0.9426	2	1.17246			
463.6	0.95513	1	1.06491			
509	0.9426	1	1.08850			
697.4	0.98179	2	1.06803			
796.7	0.98795	1	1.02450			
808.6	0.98417	2	1.06652			
1066.1	0.96398	2	1.20647			
1083	0.96602	1	1.09400			
1160.6	0.95614	2	1.28109			
1231.9	0.965	1	1.11031			

(1) Frequency shift factor (ω*/ω)

(2) g^i is the degeneracy of the ith vibrational mode.

(3) $f(x)^{g_i} = [(e^{-U_i*/2}/(e^{-U_i/2})][(1-e^{-U_i})/(1-e^{-U_i*})]^{g_i}$. Frequencies and frequency shift factors were taken from the compilation of Polyakov and Kharlashina (1994), except for the shift factors of the three acoustic modes (denoted by the (a) superscript), which were calculated using the high-temperature product rule (see text).

Example calculation for quartz. We show, as an example, the calculation of a partition function ratio for quartz at 25°C (Table 2). The input data are taken from the compilation of frequencies and frequency shifts factors (ω*/ω) given in Polyakov and Kharlashina (1994), with minor modifications (see below). The shift factors in this compilation are mostly from Sato and McMillan (1987), who directly measured the spectrum of ^{18}O quartz. The degeneracy column in the table gives the number of vibrational modes with a particular frequency, and therefore the number of times that mode must be counted in the calculations. Including degeneracies, there are 27 vibrational modes.

(Q*/Q) was calculated by substituting ω and ω^* for each vibrational mode into Equation (11). For the sake of simplicity, all the vibrational modes, including the acoustic modes, were represented by Einstein functions. More rigorous calculations would treat the acoustic modes using Debye functions. Given a value for (Q^*/Q), f and β values are given by:

$$f = (Q*/Q)\left(\frac{m_{16}}{m_{18}}\right)^9 \qquad 1000\ \ln\beta = \frac{1}{6}1000\ \ln f$$

When divided into contributions of optical and acoustic modes, the three low frequency acoustic modes contribute only about 6% to the partition function ratio for quartz (Table 2). The percentage is similarly low or lower in most minerals (O'Neil 1986). Thus, imperfect information on the acoustic frequencies typically does not lead to large errors in calculated partition function ratios.

The 1000 ln β value for quartz at 25°C given in Table 2 (101.95) can be compared to values of 102.04 and 104.54 calculated by Kieffer (1982; corrected for a rounding error by Clayton et al. 1989) and Clayton and Kieffer (1991), respectively. The 2.6‰ difference in the results of these calculations is primarily due to differences in the input vibrational frequency data, and illustrates the sensitivity of theoretical calculations to these input data. It should be noted, however, that same input data yield 1000 ln β values of 11.43 (this study), 11.55 (Kieffer 1982), and 11.71 (Clayton and Kieffer 1991) at 1000 K, a range of only 0.3‰. Thus, the absolute magnitude of the discrepancy decreases markedly with increasing temperature, a consequence of the way that uncertainties in the input data propagate with temperature in Equations (9) or (11) (Richet et al. 1977; Clayton and Kieffer 1991; Farquhar 1995).

High-temperature product rule. An important consideration in partition function calculations for minerals is ensuring the proper high-temperature limiting behavior. That is, ln f should go to zero as temperature goes to infinity (see below). However, this requirement is not always met in the calculations because of rounding errors and uncertainties in frequency shift factors. The problem can be avoided through use of the high-temperature product rule (Becker 1971; Kieffer 1982; Chacko et al. 1991):

$$\prod_{i=1}^{3s}\left(\frac{\omega}{\omega^*}\right)^{g_i} = \left(\frac{m^*}{m}\right)^{3r/2}$$

where g_i is the degeneracy of a vibrational mode. If the product of the frequency shift factors for all vibrational modes does not equal the quantity on the right hand side of the equation, ln f will not go to zero at infinite temperature. To correct the problem, one or more of the frequency shift factors must be adjusted so as to satisfy the equation. Typically, it is the shift factors for the lower frequency modes that are modified (O'Neil et al. 1969; Kieffer 1982; Chacko et al. 1991). In Table 2, the shift factors for the acoustic modes of quartz were changed to fulfill the product rule. It should be emphasized that this procedure can be quantitatively important, and affects partition function ratios at both high and low temperature. In the case of the CO_2-calcite system, calculations that utilized the product rule gave results in much better agreement with experimental data than those that did not (Chacko et al. 1991).

Other theoretical methods. In addition to the procedure described above, three other theoretical methods have been developed for calculating fractionation factors involving minerals. The first is based on computer simulation of crystal structures and first principles prediction of their thermodynamic properties (Patel et al. 1991; Dove et al. 1994). Reduced partition function ratios for calcite and a number of silicate minerals calculated using this approach are within 13% of those calculated by traditional methods. Fractionation factors (at 1000 K) derived from these calculations are within 1‰ and, in many cases, within 0.5‰ of those given by experiments. These early results suggest that the *ab initio* approach to calculating fractionation factors is promising, and should be pursued. At present, however, the approach may not be sufficiently accurate to provide quantitatively reliable fractionation factors. The second of the alternative methods, which is based on thermodynamic perturbation theory, can be applied to single element substances such as graphite or diamond (Polyakov and Kharlashina 1995). The method uses only heat capacity data for the minerals of interest as input. Application of this method to the diamond-graphite and calcite-graphite systems yielded results in good agreement with more standard theoretical calculations (Bottinga 1969b), and natural sample data (Valley and O'Neil 1981; Kitchen and Valley 1995), respectively. The third method is also based on thermodynamic perturbation theory but uses Mössbauer data as input (Polyakov 1997; Polyakov and Mineev 2000). The method can be applied to two-element compounds if one of the elements (e.g. Fe) has a Mössbauer-sensitive isotope. Polyakov and Mineev (2000) used this approach to calculate iron, sulfur, and oxygen isotope reduced partition function ratios for a number of minerals. Iron isotope fractionations derived with this approach are in agreement with fractionations calculated using more traditional theoretical methods and vibrational spectroscopic data (Schauble et al. 2001).

VARIABLES INFLUENCING THE MAGNITUDE OF FRACTIONATION FACTORS

Temperature

Temperature is in many cases the single most important variable in controlling isotope fractionation behavior. Bigeleisen and Mayer (1947), Urey (1947), Stern et al. (1968), Bottinga and Javoy (1973), and Criss (1991) evaluated the temperature-dependence of fractionation factors from a theoretical perspective. The following is based largely on the lucid explanation provided in Criss' (1991) paper. Written in logarithmic form, the reduced partition function ratio of a diatomic gas (see Eqn. 9) is given by:

$$\ln f = \ln\left[\frac{\sigma}{\sigma*}\frac{v*}{v}\right] + \left[\frac{U-U*}{2}\right] + \ln\left[\frac{1-e^{-U}}{1-e^{-U*}}\right] \qquad (12)$$

As noted previously, the quantity $[(1-e^{-U})/(1-e^{-U*})]$ is approximately unity at low temperature. Thus, at low temperature, Equation (12) reduces to:

$$\ln f \cong \ln\left[\frac{\sigma}{\sigma*}\frac{v*}{v}\right] + \left[\frac{U-U*}{2}\right] \qquad (13)$$

Because $U = hv/k_bT$, this equation is of the form y = constant + slope (T^{-1}), with a slope given by $\Delta ZPE/k_b$. The equation indicates that reduced partition function ratios, and in turn fractionation factors (ln α), vary linearly with respect to T^{-1} at low temperature. The same relationship applies to polyatomic molecules but in that case both terms in Equation (13) are summations over all vibrational modes of the molecule.

At high temperature, the last term on the right-hand side of Equation (12) is significantly greater than zero and must be considered in the calculation. Criss (1991) expanded this term in a Taylor series, which, after canceling like terms, gives the following equation for diatomic molecules:

$$\ln f = \ln\left[\frac{\sigma}{\sigma*}\right] + \left[\frac{U^2 - U^{*2}}{24}\right] - \left[\frac{U^4 - U^{*4}}{2880}\right] + \left[\frac{U^6 - U^{*6}}{181,440}\right] - \left[\frac{U^8 - U^{*8}}{9,676,800}\right] + \ldots$$

The higher order terms (terms 3 and above on the right hand side) become vanishingly small at high temperature, which results in:

$$\ln f \cong \ln\left[\frac{\sigma}{\sigma*}\right] + \left[\frac{U^2 - U^{*2}}{24}\right] \quad (14)$$

This equation is of the form y = constant + slope (T^{-2}), and implies that fractionation factors vary linearly with respect to T^{-2} at high temperature. Note also that $\ln f \to 0$ as $T \to$ provided that the symmetry numbers of a molecule and its isotopic derivative ($\sigma/\sigma*$) are the same.

The theoretical considerations discussed above provide insight into the expected temperature-dependence of fractionation factors at low- and high-temperature limits. The exact temperatures at which these limits occur depend on the substance being considered. Bigeleisen and Mayer (1947) showed that the reduced partition function ratios approach the T^{-1} and T^{-2} temperature-dependencies when values of $hc\omega/k_bT$ are >20 and <2, respectively. Therefore, gaseous molecules, which commonly have vibrational frequencies in the range 2000-3000 cm^{-1}, only enter the T^{-1} region at temperatures below -53°C. Minerals, which typically have vibrational frequencies of ~1000 cm^{-1}, enter the T^{-1} region at even lower temperature. The T^{-2} temperature-dependency requires temperatures in excess of 1100 and 600°C for most gases and minerals, respectively.

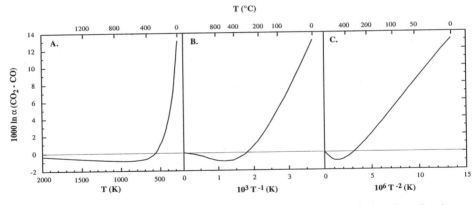

Figure 3. Calculated oxygen isotope fractionation factors between CO_2 and CO plotted as a function of (A) T, (B) 10^3T^{-1}, and (C) 10^6T^{-2}. Note that the variation in the fractionation factor with temperature cannot be represented by any one simple function of temperature. Calculations follow the procedure outlined in the text.

It should be apparent from the discussion above that geologically relevant temperatures are transitional between the high- and low-temperature limits for many substances. Thus, fractionation factors between substances cannot be represented as a simple function of temperature over a broad temperature range (Clayton 1981). This is illustrated in Figure 3, which shows calculated oxygen isotope fractionations in the CO_2-CO system as a function of T, T^{-1} and T^{-2}. With increasing temperature, the magnitude of the fractionation factor in this system first decreases, changes sign and increases, and then finally decreases and approaches zero. None of the plots are linear over the entire temperature range, although individual segments of each plot are approximately linear. Because of these complexities, equations describing the variation of ln α with temperature take a number of forms. For lower temperature fractionations (T < 100°C), one commonly finds equations of the form ln α = $A(T^{-1})$ + B or ln α = $A(T)$ + B, where A and B are constants. Higher temperature fractionations are generally represented by ln α = $A(T^{-2})$, ln α = $A(T^{-2})$ + B, or ln α = $A(T^{-2})$ + $B(T^{-1})$ + C . Simple equations such as these are useful in interpolating fractionation data but should not be used to extrapolate fractionations outside the prescribed temperature range. Numerical representation of fractionation curves over a broader temperature range requires more complex equations. For example, Clayton and Kieffer (1991) used third-order polynomial expressions in T^{-2} to represent reduced partition function ratios for minerals between 400 K and infinite temperature (see also Polyakov and Kharlashina 1995; Horita 2001).

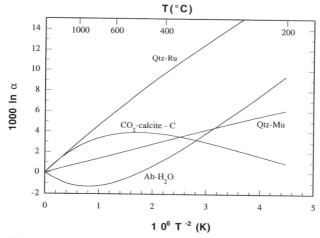

Figure 4. Temperature-dependence of oxygen or carbon isotope fractionation factors in some representative mineral-fluid and mineral-mineral systems as indicated by theoretical calculations. All curves are for oxygen isotope fractionations except for CO_2-calcite, which is for carbon isotope fractionations. Note the relative linearity of mineral-mineral fractionation curves compared to mineral-fluid fractionation curves. Reduced partition function ratios of quartz, rutile, and muscovite taken from Clayton and Kieffer (1991) and Chacko et al. (1996). Calculations for CO_2, calcite and H_2O taken from Chacko et al. (1991) and Richet et al. (1977).

It is instructive to compare the temperature-dependence of fractionation factors in some representative mineral-fluid and mineral-mineral systems (Fig. 4). The exact magnitude of fractionation factors shown on the plot may not be accurate because of uncertainties inherent to the theoretical calculations on which they are based. However, theory does place strict constraints on the temperature-dependence of fractionation factors, and hence the basic shape of fractionation curves (Clayton and Kieffer 1991). The shapes of oxygen isotope fractionation curves in most mineral-H_2O systems are broadly similar to the albite-H_2O system shown on the figure (Bottinga and Javoy 1973; Matthews et al. 1983a). At low temperatures, ^{18}O is concentrated in the mineral whereas, at higher temperatures, it partitions into the H_2O fluid. Analogous to the CO_2-CO example given above, this complex temperature-dependency results from the very high vibrational frequencies (1600-3900 cm^{-1}) but small ^{18}O frequency shifts of the water molecule relative to those of most minerals (~1000 cm^{-1}). Similarly, large differences in the vibrational characteristics of CO_2 and calcite result in a parabolically shaped fractionation curve for carbon and oxygen (not shown) fractionations in the CO_2-calcite system. Although no calculations are available, experimental data suggest that hydrogen isotope fractionation factors between hydrous minerals and water also have complex temperature-dependencies (e.g. Vennemann and O'Neil 1996). Mineral-mineral systems typically show a simpler temperature-dependency (Fig. 4). At temperatures above 600°C, oxygen isotope fractionations between anhydrous minerals (e.g. quartz-rutile) are approximately linear through the origin when plotted against T^{-2} (e.g. Bottinga and Javoy 1973). The same conclusion appears to hold for fractionations between anhydrous and hydrous minerals (quartz-muscovite), although detailed calculations have only been reported for one hydrous mineral, muscovite (Kieffer 1982; Chacko et al. 1996). At temperatures below 600°C, some mineral-mineral fractionation curves become significantly non-linear. Thus, straight-line extrapolations of these curves to lower temperature can result in the calculation of erroneous fractionation factors, especially at temperatures below 400°C.

Pressure

Oxygen isotope fractionations. Isotopic fractionation is generally thought to be independent of pressure. However, in specific cases, pressure can lead to significant changes in the magnitude of fractionation factors. From classical thermodynamics, the effect of pressure on an equilibrium constant is given by:

$$\left(\frac{\partial \ln K}{\partial P}\right)_T = -\frac{\Delta V_R}{RT}$$

The volume change associated with an isotope exchange reaction is predictably small, but is not zero because of minor differences in the molar volumes of a molecule and its isotopic derivative. These differences arise because of the anharmonicity of vibrations, which results in slightly greater mean bond lengths for the isotopically light molecule (Clayton et al. 1975). Clayton (1981) estimated a volume decrease in calcite on ^{18}O substitution of approximately 0.0025 cm^3 per mole of oxygen. Most of this volume change is cancelled by a similar magnitude change for the other phase participating in the isotope exchange reaction. Therefore, the net pressure effect on oxygen isotope fractionation factors is expected to be small, on the order of 0.003‰ per kbar (at 1000 K) in most cases (Clayton 1981). Clayton et al. (1975) directly investigated this problem through experiments in the calcite-water system at 500°C. Mineral-water systems are particularly conducive to showing pressure effects because the volume change of the water molecule on ^{18}O substitution is expected to be much smaller than that of most minerals. Therefore, ΔV_R for mineral-water reactions should be significantly larger than for many other isotope exchange reactions. Nevertheless, Clayton et al. found no measurable change of fractionation factors in the calcite-water system between 1 and 20 kbar. This result was subsequently corroborated in the quartz-water and albite-water systems (Matsuhisa et al. 1979; Matthews et al. 1983b)

Polyakov and Kharlashina (1994) took a different approach to investigating pressure effects on isotope fractionation. Using the methods of mineral physics, they examined the influence of pressure (or volume) on vibrational frequencies. This is given by:

$$\gamma_i = -\left(\frac{\partial \ln v_i}{\partial \ln V}\right)_T$$

where γ_i is the mode Grüneisen parameter of the ith vibrational mode. The magnitude of this parameter for a given substance can be obtained from molar volume, heat capacity, thermal expansion and compressibility data for that substance. The effect of pressure on β values is then given by:

$$\left(\frac{\partial \beta}{\partial P}\right)_T = -\frac{\gamma T}{B_T}\left(\frac{\partial \beta}{\partial T}\right)_V$$

where B_T is the isothermal bulk modulus of the substance. Using this approach, Polyakov and Kharlashina (1994) calculated changes in the ln β values of a number of minerals associated with a 10 kbar increase in pressure (Fig. 5). They also suggested that their calculated 10-kbar pressure effect could be extrapolated in approximately linear fashion; that is, the effect at 20 kbar should be about twice that at 10 kbar. For oxygen isotopes, the changes in ln β values are very small, except at low temperature. At temperatures of 500°C and above, ln β values of all the minerals investigated shift by less than 0.2‰, and fractionation factors between minerals by less than 0.1‰. Similar to the findings of Clayton et al. (1975), this suggests that pressure effects on oxygen isotope fractionation between minerals will be close to or within analytical uncertainty, except for changes in pressure of greater than 20-30 kbar.

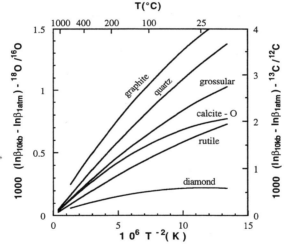

Figure 5. The effect of a 10 kbar pressure increase on the 1000 ln β values of minerals (after Polyakov and Kharlashina 1994). The curves for quartz, grossular, calcite and rutile (oxygen isotope fractionation) are calculated from the equations given in that paper. The curves for graphite and diamond (carbon isotope fractionation) are interpolated from their figure 7. Note that the effect of pressure on oxygen isotope fractionations between minerals is small except at low temperature (T < 300°C) or very high pressure (ΔP > 20 kbar). In contrast, the pressure effect on carbon isotope fractionations involving graphite is significant even at high temperature.

Carbon isotope fractionations. The same conclusion does not hold for carbon isotope fractionation factors involving graphite. Polyakov and Kharlashina (1994) noted that the β values of graphite are much more strongly affected by pressure than those of calcite or diamond (Fig. 5). As a consequence, the pressure effect on carbon isotope fractionations in the diamond-graphite and calcite-graphite systems is significant, even at high temperature. The pressure effect on graphite is, in fact, large enough to induce a fractionation crossover in the diamond-graphite system, with graphite becoming the ^{13}C-enriched phase at high pressure.

Hydrogen isotope fractionations. Several recent studies indicate that pressure is an important variable in the hydrogen isotope fractionation behavior of hydrous mineral-H_2O systems. Using spectroscopic data on high temperature and pressure H_2O as input, Driesner (1997) calculated large pressure effects on the D/H reduced partition function ratio of water. The effects were largest at the critical temperature of water (374°C), where the calculations predict a 20‰ decrease in the reduced partition function ratio from 0.2 to 2 kbar. Driesner assumed that the effect of pressure on the reduced partition function ratios of hydrous minerals would be much smaller than on those of the water molecule. Therefore, most of the calculated shifts for water should translate to similar magnitude shifts in mineral-H_2O fractionation factors. Horita et al. (1999) tested this hypothesis with experiments in the brucite-water system. At 380°C, they found a 12.4‰ increase in brucite-H_2O fractionations associated with a pressure increase from 0.15 to 8 kbar (Fig. 6). This change in fractionation factor is well outside experimental error and represents the first unequivocal demonstration of a pressure effect on D/H fractionations. Significantly, more than one half of the total change in the fractionation factor occurs from 0.15 to 0.54 kbar, which is the region of P-T space characterized by the proportionately largest increase in the density of water. The direction of the pressure effect documented by Horita et al. (1999) is the same as that indicated by Driesner's (1997) calculations, but its magnitude is markedly smaller. It must be emphasized, however, that pressure effects on brucite-water D/H fractionation are substantial, and in fact larger than temperature effects in this system over the temperature range from 200 to 600°C (Horita et al., in press) (Fig. 7).

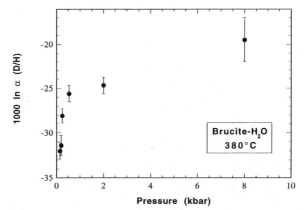

Figure 6. Pressure effect on D/H fractionations in the brucite-H_2O system at 380°C (after Horita et al. 1999). Note that increasing pressure decreases water's affinity for deuterium. The largest change in the fractionation factors occurs below 0.5 kbar, which is the region of P-T space with the proportionately largest increase in the density of water.

Although experimental details are not given, Mineev and Grinenko (1996) also report significant pressure effects on D/H fractionations in the serpentine-H_2O system at 100 and 200°C. They suggest that the large discrepancy between the experimental serpentine-water calibration of Sakai and Tatsumi (1978), and the natural sample calibration of Wenner and Taylor (1973) can be attributed to differences in pressure. Similarly, the differences in tourmaline-H_2O D/H fractionations obtained by Blamart

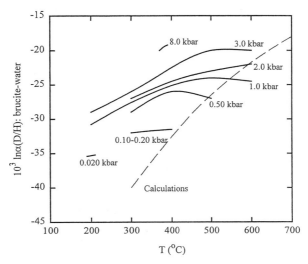

Figure 7. Experimental results of brucite-water D/H fractionation as a function of temperature and pressure. The calculated curve (dashed) is based on theoretical calculations for brucite at 0 kbar (Horita et al., in press), and water (Richet et al. 1977). The calculated curve has large errors (±20‰ at 300°C to ±15‰ at 700°C). The experiments indicate that the effect of pressure on D/H fractionations in this system is larger than the effect of temperature from 200-500°C. Modified after Horita et al. (in press).

et al. (1989) and Jibao and Yaqian (1997) can in part be attributed to differences in experimental pressures. In all the cases investigated thus far, the effect of increasing pressure is to decrease the water molecule's affinity of deuterium. Although the direction of the pressure effect seems well established from both theory and experiment, the magnitude of the effect and the relative contributions of mineral and water to that effect need to be evaluated with additional experiments and calculations.

Mineral composition

Even if isotope fractionation factors are well determined for the compositional end-member of a particular mineral, application of these data to the full range of geological samples commonly requires additional information on how compositional variation in that mineral or mineral group affects fractionation behavior. Compositional effects on fractionation factors have been investigated in a number of different ways including theoretical calculations, experiments, natural sample data, and bond-strength methods. We defer our discussion of bond-strength methods to a later section but see Zheng (1999b) for an application of this methodology to the assessment of fractionations in carbonate and sulfate minerals.

Compositional effects in carbonates. The classic experimental study of O'Neil et al. (1969), which was later refined by Kim and O'Neil (1997), systematically investigated the effect of cation substitution on the oxygen isotope fractionation behavior of carbonate minerals. Figure 8 shows experimentally determined carbonate-H_2O fractionation factors at 240°C plotted against the radius and mass of the divalent cation. Although there is an overall negative correlation between the fractionation factor and both of these variables, the correlation with mass is considerably better. This suggests that cation mass rather than radius is the major variable controlling fractionations between carbonates (O'Neil et

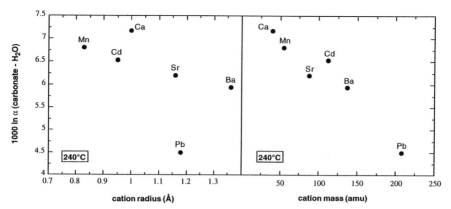

Figure 8. The effect of cation substitution on carbonate-H₂O fractionation factors at 240°C (O'Neil et al. 1969; Kim and O'Neil 1997). Note that the change in fractionation factor with cation substitution correlates with both cation radius and cation mass; however, the correlation with mass is considerably better suggesting that cation mass is dominant variable.

al. 1969; Kim and O'Neil 1997). Golyshev et al. (1981) reached the opposite conclusion on the basis of lattice dynamical calculations for a large number of carbonate minerals. It should be noted, however, that Golyshev et al.'s calculated ln β values are at least as well correlated with cation mass as with cation radius. Both O'Neil et al. and Golyshev et al. pointed out that cation radius mainly affects the internal frequencies of the CO_3 ion, whereas cation mass affects the lattice vibrations (acoustic and external optical frequencies). Notably, the frequencies of the lattice vibrations are much more strongly modified by cation substitution than are the internal frequencies.

Compositional effects in silicates. Compositional effects are also important in the oxygen isotope fractionation behavior of silicate minerals. The effects are complex, however, because of the large number of substitution mechanisms that operate in these minerals. Table 3 summarizes some of the major substitutions and their estimated effects on isotope fractionation (cf. Kohn and Valley 1998a). Of the common substitution schemes, the plagioclase substitution has the largest isotopic effect. Experimental data (O'Neil and Taylor 1967; Matsuhisa et al. 1979; Clayton et al. 1989) and theoretical calculations (Kieffer 1982) indicate a 1.05 to 1.2‰ fractionation between albite and anorthite at 1000 K. This fractionation reflects the higher Si to Al ratio of albite, and the affinity of the Si-O bond relative to the Al-O bond for ^{18}O (Taylor and Epstein 1962). In contrast to plagioclase, K ↔ Na substitution in the alkali feldspars, which does not affect the Si to Al ratio, has no measurable isotopic effect (Schwarcz 1966; O'Neil and Taylor 1967).

Like the plagioclase substitution, the jadeite ($NaAlSi_2O_6$)-diopside ($CaMgSi_2O_6$) substitution also involves Al, but Al in this case substitutes into an octahedral rather than a tetrahedral site. As demonstrated by a comparison of experimental data in the diopside-H₂O and jadeite-H₂O systems (Matthews et al. 1983a), the isotopic effect of this substitution is an enrichment in ^{18}O (Δ(jd-di) = 0.99‰ at 1000 K). Another substitution involving Al, the Tschermak substitution ($[Al^{oct}Al^{tet}]$ ↔ $[M^{2+}Si]$), can be significant in pyroxenes, amphiboles and micas. Unfortunately, there are no data with which to evaluate its isotopic effect directly. A crude estimate can be obtained by combining the measured isotopic effects of the plagioclase and jadeite substitutions. The large ^{18}O depletion associated with replacement of Si by Al in the tetrahedral site (plagioclase) should be mostly cancelled by ^{18}O enrichment associated with replacement of a divalent

Table 3. Effect of compositional substitutions on oxygen isotope fractionation in silicates.

Substitution	Example	Experimental	Theoretical	Natural
NaSi↔CaAl	plagioclase	[1]1.05, [2]1.07, [3]1.09	[4]1.2	----
NaAl↔Ca(Mg,Fe)	jadeite-diopside	[5]0.99	----	----
K↔Na	alkali feldspar	[2]0	----	[6]0
Fe↔Mg	pyroxene, garnet	[5]0.08	[7]0	----
Mn↔Ca	garnet	[8]0	----	----
(Fe,Mg)↔Ca	pyroxene	[9,10]0.49	[4]0.4	[11]0.55-0.75
(Fe,Mg)↔Ca	garnet	----	[4]0.18, [7]0.12	[11]0.5
Al^{3+}↔Fe^{3+}	garnet	[12]1.3	[4]0.5, [7]0.45	[11]0.7
F↔OH	phlogopite	[13]0.52	----	----

Notes: Substitution schemes are written such that the left-hand side has the greater affinity for ^{18}O. Experimental, theoretical and natural (sample) refer to the methodology used to evaluate the magnitude of the fractionation factor. All fractionations are reported at 1000 K and refer to the isotopic fractionation between end-members. For example, the numbers listed for the plagioclase substitution represent the fractionation between end-member albite and anorthite at 1000 K.

Sources of data: 1 = Clayton et al. (1989); 2 = O'Neil and Taylor, (1967); 3 = Matsuhisa et al. (1979); 4 = Kieffer (1982); 5 = Matthews et al. (1983a); 6 = Schwarcz (1966); 7 = Rosenbaum and Mattey (1995); 8 = Lichtenstein and Hoernes (1992); 9 = Chiba et al. (1989); 10 = Rosenbaum et al. (1994); 11 = Kohn and Valley (1998b); 12 = Taylor (1976); 13 = Chacko et al. (1996).

cation by Al in the octahedral site (jadeite). Therefore, the net effect of Tschermak substitution is a relatively small depletion in ^{18}O. Following the approach of Kohn and Valley (1998a), we estimate the magnitude of that depletion at 1000 K to be ~0.1‰ for amphiboles, ~0.2‰ for micas, and ~0.4‰ for pyroxenes per mole of Al substitution in the tetrahedral site.

Fe^{2+}-Mg^{2+} substitutions are common in many silicate minerals. Experimental data on calcic clinopyroxenes ($Ca[Fe,Mg]Si_2O_6$) at 700°C indicate no significant difference in mineral-H_2O fractionation factors between Fe and Mg end-members (Matthews et al. 1983a). Similarly, calculated reduced partition function ratios of pyrope ($Mg_3Al_2Si_3O_{12}$) and almandine ($Fe_3Al_2Si_3O_{12}$) garnet are identical (Rosenbaum and Mattey 1995). Collectively, these observations suggest that the isotopic effect of Fe-Mg substitution in silicates is negligible, at least at high temperature. The same may be true for Ca-Mn substitutions as mineral-H_2O experiments with grossular ($Ca_3Al_2Si_3O_{12}$) and spessartine ($Mn_3Al_2Si_3O_{12}$) garnets gave identical fractionations at 750°C (Lichtenstein and Hoernes 1992). In contrast to Ca-Mn, there does appear to be a small but significant isotopic effect associated with Ca-Mg and Ca-Fe^{2+} substitutions. Theoretical calculations (Kieffer 1982), and data from experiments (Chiba et al. 1989; Rosenbaum et al. 1994) and natural samples (Kohn and Valley 1998b) indicate fractionations of 0.4, 0.5 and 0.55-0.75‰, respectively, between orthopyroxene ($[Fe,Mg]_2Si_2O_6$) and calcic clinopyroxene at 1000 K. Similarly, calculations and natural sample data suggest a 0.1 to 0.5‰ fractionation between pyrope-almandine and grossular garnet at 1000 K (Kieffer 1982; Rosenbaum and Mattey 1995; Kohn and Valley 1998b).

There is also a significant isotopic effect associated with the Fe^{3+}-Al^{3+} substitution mechanism. Taylor (1976) reported a 1.7‰ fractionation between grossular and andradite ($Ca_3Fe^{3+}_2Si_3O_{12}$) garnet in hydrothermal experiments at 600°C, which translates to a 1.3‰ fractionation at 1000 K. Theoretical calculations (Kieffer 1982; Rosenbaum and Mattey 1995), and natural sample data (Kohn and Valley 1998b) indicate smaller fractionations of 0.5 to 0.8‰ at that temperature.

The generalizations that stem from the above observations are similar to the ones made long ago by Taylor and Epstein (1962). Namely, the dominant compositional variable affecting oxygen isotope fractionations between silicates is the identity of the tetrahedral cation. With the exception of Al, the octahedral and cubic (8-fold) cations are of secondary importance, although, as shown in Table 3, not insignificant in all cases. Monovalent cations have little effect on oxygen isotope fractionations.

There is good overall agreement between theory, experiment and natural sample data as regards the direction, and, in some cases, the magnitude of isotopic effects associated with compositional substitutions in silicates. This agreement is encouraging because it suggests that fractionation factors for silicate solid solutions can be predicted by combining fractionation data for end-members with the estimated isotopic effects of the various substitution mechanisms. The reader is referred to the paper of Kohn and Valley (1998a) for an elegant methodology for making such calculations.

Compositional effects on hydrogen isotope fractionation. As was the case with oxygen, hydrogen isotope fractionation factors are also influenced by mineral composition. The exact nature of the compositional dependence, however, remains unclear. From the results of their pioneering exchange experiments between micas, amphiboles and water, Suzuoki and Epstein (1976) concluded that the identity of the octahedral cation is the key compositional variable in the hydrogen isotope fractionation behavior of hydrous minerals. This observation can be rationalized by noting that the hydroxyl unit is more closely associated with the octahedral cations than with other cations in these mineral structures. Of the common octahedral cations, Al has the greatest affinity for deuterium, followed closely by Mg. Fe, on the other hand, has a strong affinity for hydrogen over deuterium. Suzuoki and Epstein (1976) suggested that these compositional effects are systematic and might be used to predict fractionation factors regardless of mineral species. Their proposed equation, which is applicable from 400 to 850°C, is:

$$1000 \ln \alpha \, (\text{mineral-}H_2O) = -22.4 \, (10^6 T^{-2}) + 28.2 + (2 \, X_{Al} - 4 \, X_{Mg} - 68 \, X_{Fe})$$

where X_{Al}, X_{Mg} and X_{Fe} are the mole fractions of each cation in the octahedral site. This equation correctly predicts the magnitude (but not the temperature-dependence) of brucite-H_2O fractionation factor (Satake and Matsuo 1984; Horita et al. 1999) to within 10‰ at 400 and 500°C. It is also supported in a general way by natural sample data, which suggest a negative correlation between the δD values of minerals and their Fe/Mg ratio (e.g. Marumo et al. 1980). There are, however, a number of complicating factors. Minerals such as boehmite, epidote and chlorite, which show a significant degree of hydrogen bonding (i.e. hydrogen exists as O-H--O units rather than O-H units), partition deuterium less strongly than predicted by the equation (Suzuoki and Epstein 1976; Graham et al. 1980, 1987). Additionally, a detailed study of hydrogen isotope fractionations between amphiboles and water does not show the compositional dependence indicated by Suzuoki and Epstein's equation, and suggests that the A-site cation in amphiboles may also play a role in hydrogen isotope partitioning (Graham et al. 1984). The issue of compositional effects on D/H fractionations remains largely unresolved and awaits further experimental studies.

Solution composition

The presence of dissolved species in the fluid phase can impact fractionation factors to a comparable or larger degree than the largest mineral composition effects. This effect applies specifically to fractionations between an aqueous fluid, and some other mineral, gas or fluid phase. The seminal work on the isotopic solution effect, or 'salt effect' as it is commonly known, was done by H. Taube in the 1950s. This and subsequent work clearly demonstrated that the effects of many dissolved salts of geochemical interest on isotopic fractionation are non-trivial at or near room temperature (Taube 1954; Sofer and Gat 1972 1975; Stewart and Friedman 1975; Bopp et al. 1977; O'Neil and Truesdell 1991; Kakiuchi 1994). However, there was little information on the salt effect at elevated temperatures, and the available data were controversial with respect to the temperature and concentration-dependency of the effect (e.g. Truesdell 1974). In an attempt to resolve the controversy, several investigators carried out studies of oxygen and hydrogen isotope salt effects at elevated temperature, using a variety of experimental techniques (Matsuhisa et al. 1979, quartz- and albite-water at 600 and 700°C; Graham and Sheppard 1980, epidote-water 250-550°C; Kendall et al. 1983, calcite-water at 275°C; Kazahaya 1986, liquid-vapor to 345°C; Zhang et al. 1989, quartz-water at 180-550°C; Zhang et al. 1994, cassiterite- and wolframite-water at 200-420°C; Poulson and Schoonen 1994, dissolved HCO_3-water at 100-300°C; Driesner 1996, Driesner and Seward 2000, liquid-vapor at 50-413°C and calcite-water at 350-500°C; Kakiuchi 2000, water-H_2 vapor). Some of these studies described a complex dependence of the isotope effects on temperature and solution composition with large uncertainties (Graham and Sheppard 1980; Kazahaya 1986; and Poulson and Schoonen 1994). Other studies observed little or no effect of dissolved salts on oxygen isotope fractionation in mineral-water systems (Matsuhisa et al. 1979; Kendall et al. 1983; Zhang et al. 1989, 1994). In perhaps the most comprehensive set of studies, Horita et al. (1993a,b; 1994, 1995a,b; 1996, 1997) investigated the effect of a number of dissolved salts, particularly NaCl, on isotopic partitioning from room temperature to 500°C.

Terminology. The magnitude of the salt effect is conventionally represented by:

$$\Gamma = \frac{\alpha_{A\text{-aqueous soln}}}{\alpha_{A\text{-pure water}}} \tag{15}$$

where A is a reference phase with which both pure water and the salt-bearing solution are exchanged in separate experiments. Water vapor is commonly used as the reference phase at lower temperatures, whereas minerals are used at higher temperature. Although Γ is formally defined as an activity-composition ratio (see Horita et al. 1993a for details), in practice, it is a measure of how much the fractionation factor between phase A and an aqueous fluid changes as material is dissolved into the fluid.

Single salt solutions. Horita et al. (1993a,b; 1995a) determined values of Γ in single salt solutions (NaCl, KCl, $MgCl_2$, $CaCl_2$, Na_2SO_4, $MgSO_4$) by means of the $H_2O_{(v)}$-$H_2O_{(l)}$ equilibration method at temperatures from 50 to 350°C. Figure 9 shows representative results from the 100°C experiments. For hydrogen isotopes, Γ_H is greater than one for all of the salt solutions studied, and increases with salt concentration. Magnitudes of the effects are in the order $CaCl_2 \geq MgCl_2 > MgSO_4 > KCl \approx NaCl > Na_2SO_4$ at the same molality. Γ_O, on the other hand, is slightly less than or very close to 1, except for KCl solutions at 50°C. The measured salt effect trends for both hydrogen and oxygen are linear with the molalities of the salt solutions, and either decrease with temperature, or are nearly constant over the temperature range 50-100°C. Salt solutions of divalent cations (Ca and Mg) exhibit much larger oxygen isotope effects than those of monovalent cations (Na and K). Magnitudes of the oxygen isotope effects in NaCl solutions, and of hydrogen isotope effects in Na_2SO_4 and $MgSO_4$, increase slightly from 50 to 100°C. The systematic changes of Γ with temperature and molality permit fitting of data for each salt to simple equations. A summary of these equations is given in Table 4. These data indicate that the identity of the cation largely controls oxygen isotope salt effects in

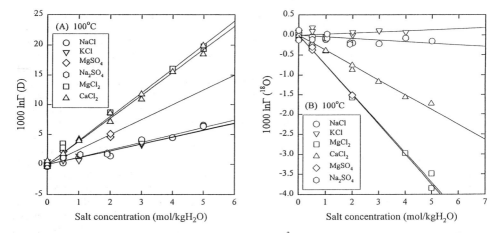

Figure 9. Experimental determined isotope salt effects (10^3 lnΓ) reported by Horita et al. (1993a) for (A) D/H and (B) $^{18}O/^{16}O$ fractionation at 100°C plotted against molality of the salt solution. The data were obtained by measuring the isotopic composition of water vapor over pure water and over salt solutions of the same isotopic composition. The solid lines are linear regressions with zero intercept through the data of each salt composition. Note that $MgCl_2$ and $CaCl_2$ have the largest (positive) D/H salt effects whereas $MgCl_2$ and $MgSO_4$ have the largest (negative) $^{18}O/^{16}O$ salt effect.

Table 4. Isotope salt effects determined by vapor-liquid equilibration.

Salt	Isotopes	Function (m=molality; T in K)	T Range (°C)
NaCl:	D/H	10^3 ln$\Gamma = m(0.01680T-13.79+3255/T)$	10-350
	$^{18}O/^{16}O$	10^3 ln$\Gamma = m(-0.033+8.93\times10^{-7}T^2-2.12\times10^{-9}T^3)$	10-350
KCl:	D/H	10^3 ln$\Gamma = m(-5.1+2278.4/T)$	20-100
	$^{18}O/^{16}O$	10^3 ln$\Gamma = m(-0.612+230.83/T)$	25-100
$MgCl_2$:	D/H	10^3 ln$\Gamma = 4.14m$	50-100
	$^{18}O/^{16}O$	10^3 ln$\Gamma = m(0.841-582.73/T)$	25-100
$CaCl_2$:	D/H	10^3 ln$\Gamma = m(0.0412T-31.38+7416.8/T)$	50-200
	$^{18}O/^{16}O$	10^3 ln$\Gamma = m(0.2447-211.09/T)$	50-200
Na_2SO_4:	D/H	10^3 ln$\Gamma = 0.86m$	50-100
	$^{18}O/^{16}O$	10^3 ln$\Gamma = -0.143m$	50-100
$MgSO_4$:	D/H	10^3 ln$\Gamma = m(8.45-2221.8/T)$	50-100
	$^{18}O/^{16}O$	10^3 ln$\Gamma = m(0.414-432.33/T)$	0-100

Regressions are based on data reported by Horita et al. (1993a,b; 1995a).
For the definition of Γ, see Equation (15).

water. This can be rationalized by noting that cations interact strongly with the negatively-charged dipole of water molecules. Sofer and Gat (1972) pointed out that the ionic charge to radius ratio (ionic potential) of cations correlates positively with the measured oxygen isotope salt effects. For the same reason, it is expected that anions control hydrogen isotope salt effects, but their relationship is more complex. For example, there is a positive correlation between the radius of alkali and halogen ions and the magnitude of hydrogen isotope salt effects (Horita et al. 1993a).

Complex salt solutions. Isotope salt effects have also been investigated in complex salt solutions consisting of mixtures of two or more salts in the system Na-K-Mg-Ca-Cl-SO_4-H_2O (Horita et al. 1993b). Some of the mixed salt solutions examined in that study were similar to natural brine compositions, such as from the Salton Sea geothermal system. The measured oxygen and hydrogen isotope salt effects in mixed salt solutions to very high ionic strengths (2-9) agree closely with calculations based on the assumption of a simple additive property of the isotope salt effects of individual salts in the solutions:

$$10^3 \ln \Gamma_{\text{mixed salt soln}} = \Sigma \ (10^3 \ln \Gamma_{\text{single salt soln}}) = \Sigma \ \{m_i(a_i + b_i/T)\}$$

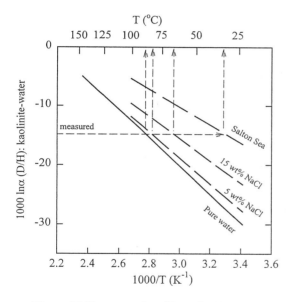

Figure 10. Calculated salt effects on D/H fractionation in the system kaolinite-water from 0 to 150°C (modified after Horita et al. 1993b). Calculations are based on the empirical equations given in Table 4. The dashed lines are the calculated curves for 1 and 3 molal NaCl solutions, and a synthetic Salton Sea brine (3.003 molal NaCl + 0.502 molal KCl + 0.990 molal $CaCl_2$). The kaolinite-pure water curve (solid line) is from Liu and Epstein (1984). Note that neglect of the salt effect could result in large errors in calculated temperatures and/or fluid compositions when mineral-water fractionation data are applied to natural samples.

Figure 10 illustrates the effect of salinity and salt composition on the kaolinite-water D/H fractionation factor, calculated assuming that the salt effect on minerals is the same as the salt effect on water vapor coexisting with the brine (Horita et al. 1993b). The effects of 1 and 3 m NaCl solutions and a representative Salton Sea brine composition (Williams and McKibben 1989) are shown as examples. The fractionation factor between kaolinite and pure water is taken from Liu and Epstein (1984). Note that the temperature of formation of a diagenetic phase calculated from the isotopic compositions of coexisting brine and mineral could be incorrect by as much as 80°C, or the calculated hydrogen isotope composition of the brine at a known temperature could be in error by as much as 15‰, if the salt effect on isotopic partitioning is ignored. Discrepancies reported in the literature between temperatures obtained from mineral-water isotope geothermometers, and temperatures derived from bore hole measurements, fluid inclusions, or other chemical geothermometers could result from neglect of isotope salt effects.

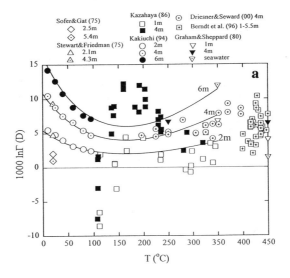

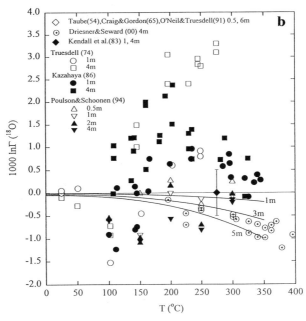

Figure 11. (a) Hydrogen isotope salt effects in NaCl solutions to 450°C. Plotted points represent data obtained from liquid-vapor equilibration, H_2-water equilibration, and mineral-water exchange experiments. (b) Oxygen isotope salt effects in NaCl solutions to 400°C. Plotted points represent data from liquid-vapor equilibration, CO_2-water equilibration, and mineral-water exchange experiments. Solid curves are based on the liquid-vapor equilibration experiments of Horita et al. (1995a).

Salt effect at high temperature. Horita et al. (1995a), Berndt et al. (1996), Shmulovich et al. (1999) and Driesner and Seward (2000) extended the results for liquid-vapor equilibration of NaCl solutions to 600°C. The results from Horita et al. (1995a) to 350°C are shown in Figure 11. The value of $10^3 \ln \Gamma(^{18}O/^{16}O)$ is always negative and its magnitude increases with increasing temperature. In contrast, the magnitude of $10^3 \ln \Gamma(D/H)$ decreases from 10°C to about 150°C, and then increases gradually to 350°C. The fractionation factors for both oxygen and hydrogen isotopes approach zero smoothly at the critical temperature of a given NaCl solution. The systematic nature of these results indicates that the complex temperature and concentration-dependencies of salt effects reported by Truesdell (1974) and Kazahaya (1986) are an experimental artifact.

Another aspect to be considered in liquid-vapor experiments is that the vapor pressure and density of water vapor in equilibrium with pure water compared to that in equilibrium with NaCl solutions are different at a given temperature. With increasing temperature, water vapor becomes more dense and non-ideal, and the formation of water clusters (dimer, trimer, etc.) becomes significant. On the basis of molecular dynamics and *ab initio* calculations, Driesner (1997) suggested that water clusters partition hydrogen isotopes much differently than the water monomer. Thus, the effect of NaCl on liquid-vapor isotope partitioning may reflect not only the isotope salt effect in liquid water, but also changes in the isotopic properties of water vapor. Furthermore, with increasing temperature, the concentration of NaCl in water vapor also increases, possibly causing an isotope salt effect in the vapor phase as well. Therefore, although the liquid-vapor equilibration method can provide the most precise results on the isotope salt effect, application of high-temperature data obtained with this technique may not be straightforward.

In an attempt to document the salt effect on D/H fractionation at elevated temperatures, Horita et al. (unpublished) conducted a study of the system brucite $(Mg[OH]_2)-H_2O\pm NaCl\pm MgCl_2$ from 200 to 500°C and salt concentrations up to 5 molal. Dissolved NaCl and $MgCl_2$ consistently increased D/H fractionation factors by a small amount at all temperatures studied (Fig. 12). These data can be compared to data obtained on the epidote-water+NaCl system by Graham and Sheppard (1980). The results of the two studies are generally consistent, although the latter show slightly larger D/H effects. In contrast to the results for hydrogen isotopes, recent studies in the system calcite-H_2O-NaCl (Horita et al. 1995a; Hu and Clayton, in press) indicate that the oxygen isotope salt effect is negligible from 300-700°C and 1 to 15 kbar.

Combined effects. Because temperature, pressure and solution composition all affect the physical properties of water, these three variables can act in concert to influence fractionation factors (Horita et al., in press; Hu and Clayton, in press). Increases in pressure and NaCl concentration both work to decrease the fluid's affinity for deuterium. The isotopic effect is most pronounced at low pressures and NaCl concentrations, which relates to the fact that the largest changes in the density of the fluid occurs over this region of pressure-composition space. It appears that the fractionation factor and the density of aqueous NaCl solutions are closely related to each other. With additional systematic experiments, empirical equations can be designed that relate the isotope salt effects and the density of aqueous solutions.

METHODS OF CALIBRATING FRACTIONATION FACTORS

As noted at the beginning of the chapter, the four main methods for calibrating fractionation factors are theoretical calculations, semi-empirical bond-strength models, natural sample data, and laboratory experiments. Theoretical methods have already been

Chacko, Cole & Horita

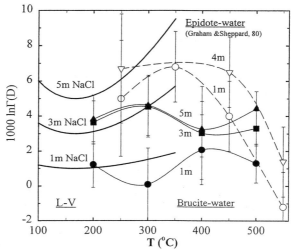

Figure 12. The effect of NaCl on the hydrogen isotope fractionation (10^3 lnΓ) obtained from experiments involving liquid-vapor equilibration (solid bold curves) compared to those obtained from brucite-water (solid curves, Horita et al., in press) and epidote-water (dashed curves, Graham and Sheppard 1980) partial exchange results. Data from Horita et al. (in press); Solid circles = 1 molal NaCl, Solid squares = 3 molal NaCl, Solid triangles = 5 molal NaCl. Data from Graham and Sheppard (1980): Open, inverted triangles = 4 molal NaCl, Open circles = 1 molal NaCl.

discussed at length. Below, we critically summarize key elements of the other three calibration methods.

Semi-empirical bond-strength calibration

This method of calibration has been applied specifically to determining oxygen isotope fractionation factors involving minerals. The method is based on the observation that the sequence of ^{18}O enrichment in minerals is correlated with average cation-oxygen bond strengths in those minerals (Taylor and Epstein 1962; Garlick 1966). As noted previously, statistical mechanical theory predicts such a correlation because bond strength relates to vibrational frequency, and in turn ZPE differences between isotopic species. Bond-strength methods, which include the original methodology of Taylor and Epstein (1962), the site potential method (Smyth and Clayton 1988; Smyth 1989), and the increment method (Schütze 1980; Richter and Hoernes 1988; Hoffbauer et al. 1994; Zheng 1999a and references cited therein), attempt to quantify the relationship between bond strength and isotopic fractionation. All of these methods involve two major steps: (1) formulate an internally consistent measure of bond strength, and (2) link bond strength to fractionation factors.

Taylor and Epstein method. Taylor and Epstein (1962) focussed on three major bond types in silicate minerals, the Si-O-Si bond (e.g. quartz), the Si-O-Al bond (e.g. anorthite) and the Si-O-M^{2+} bond (e.g. olivine). They assigned δ^{18}O values to each of these bond types on the basis of their isotopic analyses of quartz, anorthite and olivine in igneous rocks, and suggested that the δ^{18}O values of other silicates could be estimated through linear combination of these three bond types. Quartz-albite and quartz-diopside fractionation factors calculated with this method are in excellent agreement with those determined experimentally (Clayton et al. 1989; Chiba et al. 1989). Savin and Lee (1988) and Saccocia et al. (1998) applied a modified version of this method to calculating fractionation factors for phyllosilicate minerals.

Site-potential method. Smyth (1989) developed the site potential model for calculating electrostatic site energies associated with various anion sites in minerals. In essence, an anion's site potential is the energy (in electron volts) required to remove that anion from its position in the crystal. Thus, an oxygen site potential provides a convenient monitor of how strongly bound that oxygen atom is within the mineral structure. Given data on individual oxygen sites in a mineral, a mean oxygen site potential for the mineral can be calculated from a weighted average of all of the oxygen sites. Smyth used this approach to calculate mean oxygen site potentials for 165 minerals. Following Smyth and Clayton (1988), Figure 13a plots experimentally determined quartz-mineral fractionation factors versus the difference in mean quartz-mineral site potentials. With the exception of forsterite, anhydrous and hydrous silicate minerals show a good linear correlation on this plot, which suggests that the trend may be useful in predicting fractionation factors for other silicates. Calcite, apatite and oxide minerals, however, fall well off the silicate trend. The poor fit of the oxide minerals can be attributed to neglect of cation mass in the site potential model (Smyth and Clayton 1988). It is well known that cation mass affects vibrational frequencies and frequency shifts on isotopic substitution. Thus, the significantly greater mass of cations in the oxide minerals relative to those in silicates would be expected to cause differences in their isotopic

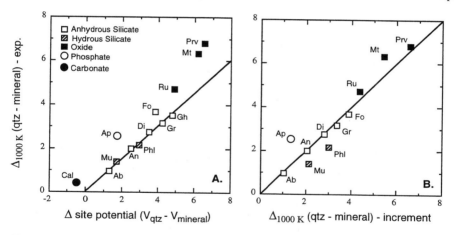

Figure 13. Comparison of quartz-mineral fractionation factors given by bond-strength methods and experiments at 1000 K (modified after Chacko et al. 1996). (A) Comparison with the oxygen site potential model of Smyth (1989) where V_{qtz} and $V_{mineral}$ are the mean oxygen site potentials of quartz and the mineral of interest, respectively. A least-squares regression through the origin fitted to all silicate data points except forsterite yields the equation: $\Delta_{1000K}(qtz\text{-}mineral) = 0.751 \ (V_{qtz} - V_{mineral})$. (B) Comparison of experimental data with the increment method calculations of Zheng (1993 1996 1997 1999a). Note that Zheng applied a low-temperature 'correction' factor (D) in his earlier papers (e.g., Zheng 1991 1993). That correction factor was not applied in some later studies (Zheng 1997 1999a). For internal consistency, the correction factor must be applied to all minerals or not at all. Following Zheng (1999a), the correction factor was not included in calculating fractionation factors for the minerals shown on the plot. The line represents 1:1 correspondence. Designations: Ab = albite; An = anorthite; Ap = apatite; Cal = calcite; Di = diopside; Fo = forsterite; Gh = gehlenite; Gr = grossular; Mt = magnetite; Mu = muscovite; Prv = CaTiO$_3$-perovskite; Ru = rutile. Sources of experimental data: Clayton et al. (1989), Chiba et al. (1989), Gautason et al. (1993), Fortier and Lüttge (1995), Rosenbaum and Mattey (1995), and Chacko et al. (1989, 1996).

fractionation behavior. The deviation of calcite and apatite from the silicate trend is likely due to the strongly covalent nature of bonding in carbonate and phosphate functional groups (Smyth and Clayton 1988). The site potential model, on the other hand, assumes that bonding in crystals is fully ionic (Smyth 1989). The site potential method also provides no direct indication of the temperature-dependence of isotopic fractionation.

Increment method. The increment method is also based on bond strengths but attempts to incorporate the effect of cation mass in its parameterization. Although they differ in detail, the various formulations of this method assign ^{18}O increment values (i_{ct-O}) for individual cation-oxygen bonds, which are then referenced to the increment value for the silicon-oxygen bond. Increment values are calculated from data on cation valence and coordination number, and cation-oxygen bond lengths. Cation mass effects are incorporated by treating the cation-oxygen pair as a diatomic molecule, and calculating the change in the reduced mass of this molecule on ^{18}O substitution (Eqn. 4). Empirically derived parameters are also included for strongly-bonded and weakly-bonded cations, and the OH^- group in hydrous silicates is treated separately. The total ^{18}O increment value for a mineral ($I-^{18}O$), which is a weighted average of the increment values of its constituent cation-oxygen bonds, is an indication of the mineral's *relative* affinity for ^{18}O. To calculate fractionation factors, this relative scale of ^{18}O enrichment must be linked to an absolute scale derived from experiments or statistical mechanical theory. The reduced partition ratios of quartz obtained from experiments or theory have generally been used to make this link. Figure 13b shows a comparison of fractionation factors at 1000 K given by the increment method calculations of Zheng (1993 1996 1997 1999a) with those indicated by experiments. There is good agreement between the two approaches for anhydrous silicates. However, hydrous silicates, apatite, and to a lesser extent, oxide minerals, deviate from the 1:1 correspondence line. Comparisons with other formulations of the increment method (e.g. Hoffbauer et al. 1994) give broadly similar results. These comparisons suggest that, although useful for anhydrous silicates, the increment method in its current form does not adequately deal with the effects of cation mass, hydroxyl groups or covalent bonding on fractionation behavior.

Summary. Bond-strength methods are in wide use because of the relative ease of determining fractionation factors for a large number of minerals with this approach. It must be emphasized, however, that these methods do not comprise a separate theory of isotopic fractionation. They are a derivative approach in which bond strengths serve as a proxy for the vibrational energies that are the root cause of fractionation. Thus, at best, bond-strength methods are only as good as the various parameterizations that go into linking bond strengths to vibrational energies, and ultimately to fractionation factors. In general, the mathematical forms of these parameterizations are not rigorously grounded in theory. The statements above are not meant to imply that bond-strength methods have no value. In cases where experiments have not been done or where theoretical calculations have not been made, these methods may provide a reasonable interim estimate of fractionation factors. Such estimates must, however, be regarded with caution until confirmed by independent methods.

Natural sample calibration

In principle, isotopic analyses of natural samples can also be used to calibrate fractionation factors. Effective use of the natural sample method, however, requires that several stringent criteria be met. (1) The phases being calibrated first attained isotopic equilibrium at some temperature, and subsequently retained their equilibrium isotopic compositions, (2) equilibration temperatures in the samples of interest are well determined by independent methods, and (3) the geothermometers used to determine temperature equilibrated (or re-equilibrated) at the same conditions as the isotopic system being calibrated. Rigorous application of these criteria significantly limits the number of samples suitable for use with this calibration method (cf. Kohn and Valley 1998b,c). Moreover, for many samples, it is difficult to demonstrate unambiguously that these criteria have been met. Despite the potential pitfalls, the natural sample method has been widely applied, and can in favorable cases provide important insights on isotope fractionation behavior. In this regard, Valley (this volume) describes how micro-analytical techniques can be used to select the optimal samples for use in natural sample calibrations.

Probably the most widely used set of natural sample calibrations is that of Bottinga and Javoy (1975). With oxygen isotope data on minerals from a large number of igneous and metamorphic rocks serving as input, these workers followed a bootstrap procedure for calibrating fractionation factors. More specifically, they estimated temperatures for individual samples with modified laboratory calibrations (Bottinga and Javoy 1973) of oxygen isotope geothermometers comprising feldspar, and one of quartz, muscovite or magnetite. They then empirically calibrated fractionation factors for other minerals (pyroxene, olivine, garnet, amphibole, biotite, and ilmenite) present in the same samples. The validity of this approach depends critically on whether all the minerals involved in the calibration preserved their equilibrium isotopic compositions (Clayton 1981). On the basis primarily of concordancy of the isotopic temperatures that they obtained, Bottinga and Javoy (1975) argued that most of the samples that they examined were in fact in isotopic equilibrium. Deines (1977) came to the opposite conclusion upon detailed statistical evaluation of the same body of isotopic data. It is also now well established from laboratory diffusion data, and from numerical modeling of natural sample data that the minerals involved in these calibrations have markedly different susceptibilities to re-equilibration, and thus are not likely to be in equilibrium in samples that have cooled slowly from high temperature (e.g. Giletti 1986; Eiler et al. 1992; Farquhar et al. 1993; Jenkin et al. 1994; Kohn and Valley 1998b,c). Therefore, purely on theoretical grounds, the approach to natural sample calibration taken by Bottinga and Javoy (1975) is suspect. Nevertheless, some of the mineral-pair fractionation factors reported in that study are in good agreement with the best available experimental calibrations. This agreement is probably fortuitous as two out of the three reference calibrations (feldspar-muscovite, feldspar-magnetite) used in the Bottinga and Javoy study are in substantial disagreement with the same set of experimental data.

The natural sample method is most likely to be successful with rocks formed at low temperatures, with rapidly cooled volcanic rocks, and with isotopically refractory minerals (e.g. garnet, graphite) in more slowly cooled rocks. Minerals formed in low-temperature environments are less susceptible to diffusional re-equilibration on cooling. They can, however, undergo changes in their isotopic compositions through recrystallization during processes such as diagenesis or deformation. Provided that this has not occurred, and that the mineral or minerals being calibrated have not become isotopically zoned during growth, the fractionations measured in such samples may yield reliable estimates of equilibrium fractionation factors. This approach to calibration also requires that the initial formation conditions of the mineral (temperature, the isotopic composition of the fluid, etc.) are well characterized. Examples of natural calibrations using low-temperature samples include the calcite-H_2O and gibbsite-H_2O oxygen isotope calibrations of Epstein et al. (1953) and Lawrence and Taylor (1971), respectively.

The very rapid cooling associated with the formation of volcanic rocks makes it likely that any phenocryst minerals present in these rocks retain their original isotopic composition. Thus, if these phenocrysts initially crystallized in isotopic equilibrium (cf.

Anderson et al. 1971), and were not mineralogically altered after crystallization, they are well suited for use in natural sample calibrations. We are not aware of any calibrations based exclusively on the analysis of minerals in volcanic rocks.

In part because of improvements in analytical techniques over the past decade, there has been an increase in natural sample calibrations involving refractory minerals found in slowly cooled metamorphic and igneous rocks. Examples include calcite-graphite carbon isotope fractionations (Valley and O'Neil 1981; Wada and Suzuki 1983; Dunn and Valley 1992; Kitchen and Valley 1995), and garnet-zircon, garnet-pyroxene, garnet-staurolite, garnet-kyanite, and quartz-kyanite oxygen isotope fractionations (Valley et al. 1994; Sharp 1995; Kohn and Valley 1998b,c). The low diffusion rates of oxygen or carbon in these minerals greatly reduce the possibility of re-equilibration effects on cooling, and in that respect make them well suited for natural sample calibration. A concern with highly refractory mineral pairs, however, is the persistence of lower-temperature isotopic compositions during prograde metamorphic evolution. A possible example of this problem is found in the calcite-graphite system. The temperature coefficient of fractionations in this system given by two independent theoretical calibrations (Chacko et al. 1991; Polyakov and Kharlashina 1995) is in excellent agreement with that given by two high-temperature (T > 650°C) natural sample calibrations (Valley and O'Neil 1981; Kitchen and Valley 1995) (Fig. 14). In contrast, three other natural sample calibrations (Wada and Suzuki 1983; Morikiyo 1984; Dunn and Valley 1992), based primarily on data from lower-temperature samples (T = 270-650°C), indicate temperature coefficients that are higher by a factor of 1.5 to 2.0. Extrapolation of the lower temperature natural sample calibrations to the high-temperature limit (ln α = 0) requires the shape of the calcite-graphite fractionation curve

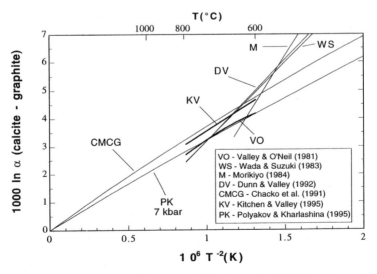

Figure 14. Comparison of theoretical (Chacko et al. 1991; Polyakov and Kharlashina 1995) and natural sample (Valley and O'Neil 1981; Wada and Suzuki 1983; Morikiyo 1984; Dunn and Valley 1992; Kitchen and Valley 1995) calibrations of the calcite-graphite fractionation factor. Note that the three lower-temperature natural sample calibrations (WS, M and DV) require a much different shape for the fractionation curve than indicated by the theoretical calculations. See text for discussion.

to be distinctly convex towards the temperature axis. This is in marked contrast to the weakly concave shape indicated by the theoretical calculations. Although the approximations required in making theoretical calculations result in significant uncertainty in the absolute magnitude of calculated fractionations, the calculations do place strict constraints on the basic shape of fractionation curves (Clayton and Kieffer 1991; Chacko et al. 1991). It is unlikely, therefore, that the shape of the fractionation curve implied by the Wada and Suzuki (1983), Morikiyo (1984), and Dunn and Valley (1992) calibrations is correct. Chacko et al. (1991) suggested that the deviation of the low-temperature natural sample data points from the calculated curve is due to the incomplete equilibration of calcite-graphite pairs at temperatures below about 650°C (see also discussion in Dunn and Valley 1992). If this interpretation is correct, calcite-graphite pairs formed at lower temperatures are unsuitable for use in natural sample calibrations, or for stable isotope thermometry.

Similar problems can occur in refractory silicate mineral pairs if the two minerals being calibrated formed at different temperatures, or if either one of the minerals formed over a range of temperatures in a prograde metamorphic sequence. In the first case, isotopic equilibrium may never have been established between the two minerals, whereas, in the second case, one or both minerals may be isotopically zoned. Sharp (1995), Kohn and Valley (1998c) and Tennie et al. (1998) noted these possibilities for natural sample calibrations involving the highly refractory minerals kyanite and garnet. Kohn and Valley (1998c) argued convincingly, however, that such problems can be overcome by carefully selecting samples with the appropriate textural and petrological characteristics. This screening process requires a detailed understanding of the metamorphic reaction history of samples.

Experimental calibration

Philosophy and methodology of experiments. Laboratory experiments are the most direct method of calibration in that they require the least number of *a priori* assumptions, and also generally permit control of the variables that may influence fractionation factors. The other calibration methods, although perhaps valid, must be regarded as tentative until confirmed by laboratory experiments. The reader is referred to the detailed reviews of experimental methods provided by O'Neil (1986) and Chacko (1993) as only the major points are summarized here.

A useful frame of reference for discussing experimental methodology is to consider the design of an ideal experiment. Such an experiment would involve direct isotopic exchange between the two substances of interest. For example, the ideal experiment for determining the oxygen isotope fractionation factor between olivine and orthopyroxene would be one in which the two minerals are intimately mixed and allowed to equilibrate at the desired temperature until isotopic equilibrium is established. To confirm the attainment of equilibrium, a companion experiment would be carried out at the same temperature consisting of olivine and orthopyroxene with the same chemical composition and structural state but with an initial isotopic fractionation on the opposite side of the equilibrium value. Obtaining the same olivine-orthopyroxene fractionation factor in both experiments would constitute a successful experimental reversal. Additional criteria for an ideal experiment include no chemical and textural changes in the starting materials during the course of the experiment. That is, isotopic exchange is accomplished exclusively through a diffusional process, which in turn is driven solely by the free energy change for the exchange reaction.

For practical reasons, most, if not all, experimental studies depart to some extent from this ideal experiment. Firstly, a direct exchange between the phases of interest is often not possible. For example, it would be impossible to obtain a clean physical separation between olivine and orthopyroxene (for isotopic analysis) after these minerals had been ground to a sufficiently fine grain size to yield reasonable amounts of isotopic exchange on laboratory time-scales. As a result, fractionation factors are often determined indirectly by exchanging phases with a common isotopic exchange medium in separate experiments, and combining the resulting data to obtain the fractionation factor of interest. The exchange media that have been used to date, H_2O, CO_2, H_2, $CaCO_3$, and $BaCO_3$ were chosen partly because they are easily separated from the phase of interest by physical or chemical techniques after the exchange experiment.

A second problem that plagues many exchange experiments is the sluggishness of exchange rates. In particular, the diffusion rates of oxygen and carbon in many minerals are slow enough to preclude a close approach to isotopic equilibrium in typical laboratory time scales through purely diffusional processes. As a consequence, many experiments in oxygen and carbon isotope systems induce exchange either by recrystallization of preexisting phases or crystallization of new phases (synthesis) during the experiment. Both procedures result in an experiment that is less than ideal. Synthesis experiments generally involve the inversion of a polymorph or crystallization of gels or oxide mixes. The free energy change associated with these processes is about 1000 times greater than those associated with isotope exchange reactions (Matthews et al. 1983b). These processes can also be very rapid and unidirectional, possibly resulting in kinetic rather than equilibrium isotopic fractionations. This was demonstrated elegantly by Matsuhisa et al. (1978) in hydrothermal experiments at 250°C, where quartz-water fractionations obtained by inverting cristobalite to quartz were about 3‰ different than those obtained by direct quartz-water exchange experiments. They interpreted these results to reflect a kinetic isotope effect associated with the crystallization of quartz from cristobalite. Similar non-equilibrium effects were noted by Matthews et al. (1983a) in crystallizing wollastonite and diopside from mixes of constituent oxides. In other cases, synthesis and direct exchange experiments appear to give comparable results (O'Neil and Taylor 1967; Matsuhisa et al. 1979; Lichtenstein and Hoernes 1992; Scheele and Hoefs 1992).

Despite the problems inherent to the mineral synthesis technique, this technique may be the only viable means of obtaining reasonable amounts of isotopic exchange in low temperature experiments (T < 250°C). With care, the technique may indeed provide reliable equilibrium fractionation factors. Criteria that have been used to suggest an approach to equilibrium fractionations in such experiments include controlled precipitation rates, an understanding of reaction pathways and the likelihood of kinetic fractionations associated with particular pathways, similar fractionations obtained in synthesis from different starting materials, and a concurrence of fractionation factors obtained with synthesis and direct exchange techniques (e.g. O'Neil and Taylor 1969; Scheele and Hoefs 1992; Bird et al. 1993; Kim and O'Neil 1997; Bao and Koch 1999). It should be emphasized, however, that even when these criteria are met, synthesis experiments cannot rigorously demonstrate the attainment of isotopic equilibrium.

Recrystallization of existing phases during the course of an experiment also involves driving forces other than solely those of the exchange reaction. However, the free energy change that drives recrystallization is approximately the same order of magnitude as that associated with isotopic exchange (Matthews et al. 1983b). As such, experiments involving recrystallization are less likely to incorporate kinetic effects than those requiring the growth of new phases. In contrast to this view, Sharp and Kirschner (1994) argued that recrystallization based experiments involving quartz do incorporate significant kinetic effects. In general, however, recrystallization experiments, although less than ideal, are clearly preferable to synthesis experiments, and often provide the best available means of calibrating fractionation factors at moderate to high temperature.

Two additional features of isotope exchange experiments are useful in the acquisition of equilibrium fractionation data. Firstly, there is a substantial increase in oxygen isotope exchange rates with increasing pressure (Clayton et al. 1975; Matthews et al. 1983a,b; Goldsmith 1991). Thus, for a given experimental run time, experiments performed at higher pressure show a closer approach to equilibrium and thereby provide more tightly constrained data. Secondly, because isotope exchange reactions run in both forward and reverse directions involve exchange of the same element between phases, it is reasonable to assume that they have nearly identical reaction rates. On the basis of this assumption, Northrop and Clayton (1966) developed the following equation for extracting equilibrium oxygen and carbon isotope fractionation factors from partially exchanged experimental data:

$$\ln \alpha^i = \ln \alpha^{eq} - 1/F \,(\ln \alpha^f - \ln \alpha^i)$$

where the superscripts *i, f* and *eq* refer to the initial, final and equilibrium fractionations, respectively, and F is the fractional approach to equilibrium during the experiment. This equation indicates that if forward and reverse directions of an isotope exchange reaction have the same exchange rates, then data from three or more companion experiments run at the same conditions, and for the same amount of time should be linear when plotted as $\ln \alpha^i$ versus $(\ln \alpha^f - \ln \alpha^i)$. The slope of this line gives the fractionation approach to equilibrium and the y intercept gives the extrapolated equilibrium fractionation factor. This method becomes progressively more reliable as F approaches unity because the intercept between the plotted line and the y-axis occurs at higher angles. The slightly modified version of this equation developed for hydrogen isotope exchange experiments is (Suzuoki and Epstein 1976):

$$(\alpha^i - 1) = (\alpha^{eq} - 1) - 1/F \,(\alpha^f - \alpha^i)$$

More recently, Criss (1999, p. 204) derived a different equation for extracting an equilibrium fractionation factor ($\alpha_{A\text{-}B}$) from a pair of exchange experiments (labeled 1 and 2) that has not proceeded completely to equilibrium:

$$(1 - H)(\alpha^{eq})^2 + [H\alpha_1^i + H\alpha_2^f - \alpha_2^i - \alpha_1^f](\alpha^{eq}) + [\alpha_1^f \alpha_2^i - H\alpha_2^f \alpha_1^i] = 0$$

where

$$H = \left(\frac{R_B^i}{R_B^f}\right)_1 \left(\frac{R_B^f}{R_B^i}\right)_2$$

and R_B is the isotope ratio (e.g. $^{18}O/^{16}O$) of phase B. The quadratic formula is used to solve the equation for α^{eq}. Criss (1999) proposed that this equation yields more exact and more correct results than the commonly used equation of Northrop and Clayton (1966). In practice, the difference in equilibrium fractionation factors obtained with the two equations is small in most cases.

Hydrothermal- and carbonate-exchange techniques. The majority of available experimental fractionation data are for oxygen isotope fractionations involving minerals. Much of the data, particularly the early data, were obtained using water as the isotopic exchange medium. These experiments were either done at ambient pressure (typically synthesis experiments), or in cold-seal pressure vessels at pressures of 1 to 3 kbar (e.g. O'Neil and Taylor 1967; O'Neil et al. 1969; Clayton et al. 1972). Later experiments were done in a piston cylinder apparatus (at ~15 kbar) to exploit the pressure enhancement of exchange rates (Clayton et al. 1975; Matsuhisa et al. 1979; Matthews et al. 1983a,b).

Even though there were some differences in experimental methodologies, the agreement between the various studies was in general excellent for some of the major mineral-water systems (e.g. albite-water, quartz- and calcite-water systems). Matthews et al. (1983a) and Matthews (1994) provide compilations of mineral-pair fractionation factors derived from the high-temperature hydrothermal experiments.

Despite the apparent success of the hydrothermal technique, there are a number of theoretical and practical drawbacks to this method of experimentation. Under hydrothermal conditions, some minerals dissolve excessively, melt, or react to form hydrous phases, making them unsuited to this type of investigation. Furthermore, the high vibrational frequencies of the water molecule results in complex temperature-dependencies for mineral-H_2O fractionations (Fig. 3), thereby making it difficult to extrapolate fractionations outside the experimentally investigated temperature range. To overcome some of these difficulties, Clayton et al. (1989) developed a technique that uses $CaCO_3$ rather than water as the common isotope exchange medium. Rosenbaum et al. (1994) used a variation of this technique with $BaCO_3$ as the exchange medium. Advantages of the so-called carbonate-exchange technique include: (1) the ability to carry out experiments at high temperatures (up to 1400°C) because of the high thermal stability of many mineral-carbonate systems, (2) the avoidance of problems associated with mineral solubility and its potential effect on fractionation factors, (3) ease of mineral separation, and (4) the relative ease and high precision of carbonate isotopic analysis. Extrapolation of experimental data is also simplified in that anhydrous mineral-carbonate fractionations should be approximately linear through the origin on fractionation plots at temperatures above 600°C (Clayton et al. 1989). There now exists a large body of oxygen isotope fractionation data for minerals acquired with the carbonate-exchange technique (summarized in Table 5).

Other experimental techniques. There have been several other recent advances in experimental techniques for determining fractionation factors for oxygen, carbon, and hydrogen isotope systems. CO_2 has successfully been used as an exchange medium for determining carbon or oxygen isotope fractionation factors for melts, glasses and carbonate, silicate and oxide minerals at both high and low pressure (Mattey et al. 1990; Chacko et al. 1991; Stolper and Epstein 1991; Scheele and Hoefs 1992; Matthews et al. 1994; Rosenbaum 1994; Palin et al. 1996; Fayek and Kyser 2000). Vennemann and O'Neil (1996) used H_2 gas as an exchange medium in low-pressure experiments to obtain hydrogen isotope fractionation factors for hydrous minerals. Horita (2001) was able to obtain tightly reversed carbon isotope fractionation data in the notoriously sluggish CO_2-CH_4 system by using transition-metal catalysts to accelerate exchange rates. Fortier et al. (1995) and Chacko et al. (1999) used the ion microprobe to obtain oxygen and hydrogen isotope analyses, respectively, of run products in magnetite-water and epidote-water exchange experiments. The latter study was novel in that it used millimeter-sized, single crystals of epidote instead finely ground epidote powder as starting material, and then analyzed the outer 0.5-2 μm of the isotopically exchanged crystals with an ion microprobe. The advantage of such an approach is that it allows D/H fractionation factors to be determined in experiments in which most or all of the isotopic exchange occurs by a diffusional process, rather than by a combination of diffusion and recrystallization (i.e. an ideal exchange experiment). The disadvantage of the approach is that the precision of hydrogen isotope analyses on the ion microprobe is currently a factor of 2 to 4 poorer than can obtained by conventional methods. That precision will no doubt improve with advances in instrumentation. Figure 15 shows Chacko et al.'s (1999) data for the epidote-H_2O system. Despite initial fractionations that were in most cases far from equilibrium, the companion experiments at each temperature were successful in bracketing the equilibrium fractionation factor within analytical error. This technique should be useful

Table 5. Coefficients for mineral-pair oxygen isotope fractionation factors.

	Cal	Ab	Mu	F-Phl	An	Phl	*Ap	Di	Gr	Gh	Fo	Ru	Mt	Prv
Qtz	0.38	0.94	1.37	1.64	1.99	2.16	2.51	2.75	3.15	3.50	3.67	4.69	6.29	6.80
Cal		0.56	0.99	1.26	1.61	1.78	2.13	2.37	2.77	3.12	3.29	4.31	5.91	6.42
Ab			0.43	0.70	1.05	1.22	1.57	1.81	2.21	2.56	2.73	3.75	5.35	5.86
Mu				0.27	0.62	0.79	1.14	1.38	1.78	2.13	2.30	3.32	4.92	5.43
F-Phl					0.35	0.52	0.87	1.11	1.51	1.86	2.03	3.05	4.65	5.16
An						0.17	0.52	0.76	1.72	1.51	1.68	2.70	4.30	4.81
Phl							0.35	0.59	0.99	1.34	1.51	2.53	4.13	4.64
Ap								0.24	0.64	0.99	1.16	2.18	3.78	4.29
Di									0.40	0.75	0.92	1.94	3.54	4.05
Gr										0.35	0.52	1.54	3.14	3.65
Gh											0.17	1.19	2.79	3.30
Fo												1.02	2.62	3.13
Ru													1.60	2.11
Mt														0.51

Coefficients for mineral-pair fractionation factors of the form $1000 \ln \alpha = A \times 10^6 / T^2 (\text{K})$, where the coefficient A is given in the table.

Equations should not be extrapolated below ~600°C. All data derived from experiments using the carbonate-exchange technique. Qtz = quartz; Cal = calcite; Ab = albite; Mu = muscovite; F-Phl = fluorophlogopite; Phl = hydroxyphlogopite; An = anorthite; Ap = apatite; Di = diopside; Gr = grossular; Gh = gehlenite; Fo = forsterite; Ru = rutile; Mt = magnetite; Prv = CaTiO3 - perovskite. Data from Clayton et al. (1989), Chiba et al. (1989); Chacko et al. (1989, 1996), Gautason et al. (1993), Fortier and Lüttge (1995), and Rosenbaum and Mattey (1995).

*The equation for apatite is not the same as that given in Fortier and Lüttge (1995). Following Chacko et al.'s (1996) suggestion for hydrous minerals, Fortier and Lüttge's 500-800°C apatite-calcite fractionation data have been regressed by a straight line through the origin on a plot of $1000 \ln \alpha$ vs. $10^6/T^2$.

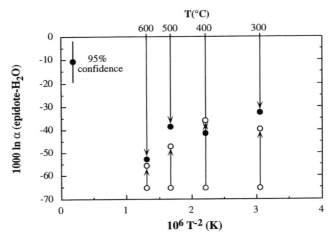

Figure 15. Experimental D/H fractionation data in the epidote-water system plotted as a function of $10^6 T^{-2}$ (modified after Chacko et al. 1999). The experiments involved exchange between water and large, single crystals of epidote. The isotopically exchanged crystals were analyzed by ion microprobe. Note that equilibrium fractionation factor was bracketed within analytical error (±6-10‰) to temperatures as low as 300°C.

for determining D/H fractionation factors in other mineral-water systems. On a more general level, micro-analytical techniques such as the ion microprobe hold great promise for isotope exchange experiments in that they open up the possibility of extending these experiments to significantly lower temperature.

SUMMARY OF FRACTIONATION FACTORS

Appendices 2-4 are an annotated list of published experimental and natural sample calibrations of oxygen, carbon and hydrogen isotope fractionation factors applicable to geological systems. The reader is referred to the Introduction section for sources of information on fractionation factors derived from theoretical and bond-strength methods. In this section, we extract from the larger tabulation some sets of fractionation data that have wide applicability. Our goal is to provide an overview of the current state of knowledge on these key fractionations, and to highlight areas where more work needs to be done. A summary of this type is necessarily subjective, and the reader is encouraged to consult the references given in this section and in the bibliography for alternative points of view.

Oxygen isotope fractionation factors

Mineral-pair fractionations. As detailed above, there currently exists two large sets of experimental data on oxygen isotope fractionation factors between minerals, one obtained using water, and the other using carbonate as the isotopic exchange medium. Mineral-pair fractionation factors derived from the two data sets are generally in good agreement except for fractionations involving quartz or calcite (Clayton et al. 1989; Chiba et al. 1989; Chacko 1993, Matthews 1994). The discrepancies are enigmatic because both hydrothermal and carbonate-exchange experiments meet many of the criteria listed above for an ideal exchange experiment. Moreover, the discrepancies cannot be attributed to inter-laboratory differences as most of both data sets were acquired at the same laboratory. Hu and Clayton (in press) provided a resolution to this paradox with an experimental investigation of the oxygen isotope salt effect on quartz-H_2O and calcite-H_2O fractionations at high pressure and temperature. They found a significant salt effect in the quartz-H_2O system but no salt effect in otherwise comparable experiments in the calcite-H_2O system. The implications of these results are profound. If salt was the only dissolved constituent in the aqueous fluid, the magnitude of the salt effect (Γ) should be identical in both sets of experiments. The fact that $\Gamma_{qtz\text{-fluid}}$ is different than $\Gamma_{calcite\text{-fluid}}$ indicates that minerals also dissolve appreciably at experimental conditions, and importantly, that the nature of the dissolved mineral species significantly affects fractionations in mineral-H_2O systems. Thus, mineral-pair fractionation factors derived from different sets of mineral-H_2O experiments may not be internally consistent because the characteristics of the fluid phase may vary with the mineral under investigation. To ensure internal consistency, two minerals must be exchanged with the same fluid. Hu and Clayton (in press) tested this idea with three-phase experiments in the quartz-calcite-H_2O and phlogopite-calcite-H_2O systems. The mineral-fluid fractionation factors derived from these three-phase experiments are different than those obtained in two-phase mineral-fluid experiments. However, the quartz-calcite and phlogopite-calcite fractionation factors derived from the three-phase experiments are the same as those obtained in the two-phase mineral-calcite experiments of Clayton et al. (1989) and Chacko et al. (1996). These results indicate that, in a choice between the two data sets, the carbonate-exchange technique provides the better mineral-pair fractionation data.

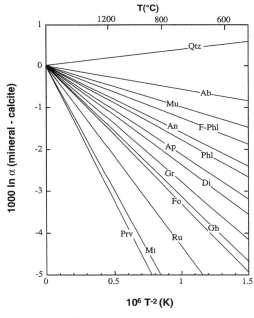

Figure 16. Summary of mineral-calcite fractionation factors given by the carbonate-exchange technique. All data have been fit by straight lines through the origin. Abbreviations and sources of data given in Table 5.

Mineral-calcite fractionation factors obtained with the carbonate-exchange technique are shown in graphical form in Figure 16. Each fractionation line shown on the plot is constrained by sets of experiments conducted at two to five different temperatures from 500 to 1300°C. These mineral-calcite fractionation data can be combined to obtain a large matrix of mineral-pair fractionation factors (Table 5). The straight-line equations given in Table 5 are adequate for calculating fractionation factors at high temperature but should not be used at temperatures of below about 600°C. Clayton and Kieffer (1991) developed a methodology for extrapolating these fractionation data to lower temperatures. With the calculated partition function ratios for calcite serving as a baseline (Chacko et al. 1991), they used the theoretical calculations (Becker 1971; Kieffer 1982) to constrain the basic shape of each fractionation curve. They then applied a correction factor to the calculated partition function ratios of individual minerals to optimize agreement between theory and experiment. It should be noted that the magnitude of the correction factor required to bring the calculations into agreement with the experiments was small for all minerals except rutile (Clayton and Kieffer 1991; Chacko et al. 1996). The advantage of this fitting procedure is that it allows experimental data, which necessarily must be obtained over a limited temperature range, to be extrapolated to lower temperature in a manner that is theoretically justifiable. Table 6 gives polynomial expressions for calculating the reduced partition function ratios for minerals derived through this fitting procedure.

Table 6. Reduced partition function ratios for minerals.

Mineral	1000 ln β
Calcite	$11.781\ x - 0.420\ x^2 + 0.0158\ x^3$
Quartz	$12.116\ x - 0.370\ x^2 + 0.0123\ x^3$
Albite	$11.134\ x - 0.326\ x^2 + 0.0104\ x^3$
Muscovite	$10.766\ x - 0.412\ x^2 + 0.0209\ x^3$
Anorthite	$9.993\ x - 0.271\ x^2 + 0.0082\ x^3$
Phlogopite	$9.969\ x - 0.382\ x^2 + 0.0194\ x^3$
F-phlogopite	$10.475\ x - 0.401\ x^2 + 0.0203\ x^3$
Diopside	$9.237\ x - 0.199\ x^2 + 0.0053x^3$
Forsterite	$8.236\ x - 0.142\ x^2 + 0.0032\ x^3$
Rutile	$7.258\ x - 0.125\ x^2 + 0.0033\ x^3$
Magnetite	$5.674\ x - 0.038\ x^2 + 0.0003\ x^3$

Polynomial expressions describing oxygen reduced partition function ratios (1000 ln β) for minerals where $x = 10^6/T^2(K)$. The equations are only applicable at temperatures above 400K. The fractionation factor between any two phases at a particular temperature is given by the algebraic difference in their 1000 lnβ values (1000 ln β_A – 1000 ln β_B). Expressions are from Clayton and Kieffer (1991) and Chacko et al. (1996).

Mineral-pair fractionation factors obtained with the carbonate-exchange technique have gained wide, but not universal, acceptance. In particular, Sharp and Kirschner (1994) questioned the validity of quartz-calcite fractionations given by this technique. Their natural sample calibration of this important system gave systematically larger fractionations than indicated by the experiments of Clayton et al. (1989). They attributed the discrepancy to a kinetic isotope effect in the experiments engendered by the rapid rate of quartz recrystallization relative to the rate of oxygen diffusion in the calcite exchange medium. We note, however, that diffusion rates of oxygen in quartz and calcite at 1000°C (Giletti and Yund 1984; Farver 1994) are rapid enough to isotopically homogenize small grains (radii < 3 μm) of these minerals by volume diffusion in 24 hours. Thus, the 1000°C quartz-calcite experiments, which used fine grained starting materials (diameter

1 to 10 μm) and were held at temperature for 24 hours, were of sufficiently long duration to establish a true diffusional equilibrium between the two minerals. The fractionation factor obtained by Clayton et al. in these high-temperature experiments is entirely consistent with their results at lower temperature (600-800°C), and does not agree with that given by the natural sample calibration. This suggests that the difference between the two calibrations has some other fundamental cause than the one proposed by Sharp and Kirschner. We believe the bulk of the evidence favors the experimental calibration. Nevertheless, given its importance, it seems prudent to attempt to re-determine fractionation factors in the quartz-calcite system with an independent method. For example, it may be possible to indirectly determine quartz-calcite fractionations through a combination of data from CO_2-quartz and CO_2-calcite experiments at high pressure and temperature.

Another mineral that bears further investigation is kyanite. Existing experimental and natural sample calibrations of quartz-kyanite oxygen isotope fractionations are widely discrepant (Sharp 1995; Tennie et al. 1998), as are three bond-strength estimates of these fractionations (Smyth and Clayton 1988; Hoffbauer et al. 1994; Zheng 1999a). Although the experimental data for kyanite were obtained with the carbonate-exchange technique (Tennie et al. 1998), these data are not entirely compatible with the rest of the carbonate-exchange data set because the experiments involved polymorphic inversion of andalusite to kyanite, rather than direct kyanite-carbonate exchange. As such, it cannot unambiguously be shown that the fractionations measured in the experiments represent true equilibrium values. The highly refractory nature of kyanite in terms of oxygen isotope exchange makes it a difficult mineral to work with from both an experimental and natural sample perspective. However, that same characteristic potentially makes it a very useful mineral for elucidating the metamorphic history of rocks. Additional studies, although difficult, should be pursued.

Mineral-fluid oxygen isotope fractionations at low temperature. Unlike the case for high-temperature fractionations between minerals, the current status of our knowledge on oxygen isotope fractionation between minerals and fluids/gases at low-temperatures (<200°C) is far from satisfactory. Although isotopic fractionation factors of many rock-forming minerals have been calculated over a wide range of temperature by theoretical and bond-strength methods discussed in the previous sections, the accuracy of these results needs to be examined by independent experimental studies. Because of extremely sluggish isotopic exchange by direct reactions (dissolution-precipitation and diffusion) between most minerals and aqueous fluids at low temperatures, virtually all experimental studies conducted to date employed various methods of synthesis by means of homogeneous precipitation from solutions and replacement of reactant minerals followed by aging. As noted previously, the attainment of isotopic equilibrium in such studies can only be inferred. Furthermore, the number of minerals amenable to low temperature synthesis on laboratory time-scales is very limited because of high activation energies of nucleation and growth.

Probably the best constrained fractionation data at this temperature range is for carbonate minerals. Oxygen isotope fractionations have been determined for a number of metal carbonates through synthesis experiments involving either precipitation of a mineral from solution or replacement of a pre-existing carbonate mineral (Fig. 17). A striking feature of these results is that all the carbonate-water fractionation curves are positioned parallel to one another in a similar but not identical order of [18]O enrichment as observed at higher temperature in the experiments of O'Neil et al. (1969) (Fig. 8). Agreement between these experimental results, and theoretical and increment method calculations by Golyshev et al. (1981) and Zheng (1999b), respectively, is not entirely

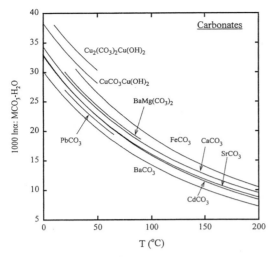

Figure 17. Plot of experimentally determined oxygen isotope fractionation factors between various metal carbonates and water at low temperatures. The carbonates were synthesized by precipitation from solution or replacement of calcite. Data sources: $CaCO_3$ (calcite), $SrCO_3$, and $BaCO_3$ – O'Neil et al. (1969), corrected in Friedman and O'Neil (1977); $CdCO_3$ – Kim and O'Neil (1997); $PbCO_3$ – Melchiorre et al. (2001); $CuCO_3Cu(OH)_2$ – Melchiorre et al. (1999); $(CuCO_3)_2Cu(OH)_2$ – Melchiorre et al. (2000); $BaMg(CO_3)_2$ – Bottcher (2000).

satisfactory. In particular, the order of ^{18}O enrichment indicated by the numerical calibration methods is not the same, with the largest discrepancies for Pb (theoretical) and Cd (increment) carbonates. Furthermore, the experimental study of Tarutani et al. (1969) indicates that aragonite is slightly enriched in ^{18}O relative to calcite at 25°C, whereas the increment method predicts a large fractionation in the opposite direction. In terms of its applicability to natural samples, the most important of these carbonate-water systems is the calcite-water system. The detailed experimental study of Kim and O'Neil (1997) examined the effect of precipitation rate and solution ionic strength on calcite-water fractionations between 0 and 40°C. They found that precipitation rate had essentially no effect on the measured fractionations whereas solutions of high ionic strength resulted in systematically larger, disequilibrium fractionations. The equation reported by Kim and O'Neil (1997) for equilibrium calcite-water fractionations at low temperature is:

$$1000 \ln \alpha = 18.03 \, (10^3 \, T^{-1}) - 32.42$$

The situation for other minerals at low temperature is less satisfactory - as in the case of the iron oxides. Several fractionation factors proposed for magnetite-water exhibit an extremely wide range (Fig. 18). Empirical-theoretical calculations (Becker and Clayton 1976; Rowe et al. 1994) and empirical calculations by the increment method (Zheng 1995) predicted very negative values for this fractionation. The curve by O'Neil and Clayton (1964) is based on an extrapolation of results from high-temperature exchange experiments to one analysis of natural magnetite sample of teeth from marine chiton. Recent results of cultured biogenic magnetite samples at low temperatures (Zhang et al. 1997; Mandernack et al. 1999) together with those of magnetite obtained from modern geothermal system (Blattner et al. 1983) point to more positive values of magnetite-water fractionation (Fig. 18).

Oxygen isotopic fractionations between various Fe(III)-oxides and water are even more poorly constrained (Fig. 19). In addition to equations based on isotopic analysis of natural hematite in low-temperature environments (Clayton and Epstein 1961; Clayton 1963), several investigators recently reported fractionation factors for hematite and goethite obtained from laboratory synthesis experiments. They show a large (>10‰) variation especially at temperatures below 40°C, depending on reaction pathways and solution compositions (e.g. pH). Mineral precipitates at low temperatures tend to be

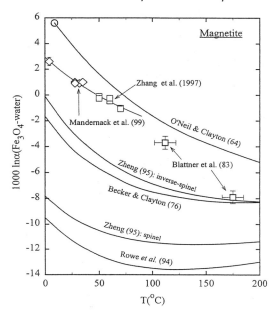

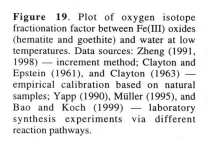

Figure 18. Plot of oxygen isotope fractionation factor between magnetite and water at low temperatures. Data sources: Becker and Clayton (1976) and Rowe et al. (1994) — empirical-theoretical calculations; Zheng (1995) — empirical increment method; O'Neil and Clayton (1964) — extrapolation of high-tem-perature experimental results to an analysis of magnetite teeth from marine chiton; Blattner et al. (1983) — natural samples from active geothermal system in New Zealand; Zhang et al. (1997) — extracellular magnetite produced by Fe(III)-reducing bacteria in culture; Mandernack et al. (1999) — magnetite produced intracellular by magnetotactic bacteria in culture.

Figure 19. Plot of oxygen isotope fractionation factor between Fe(III) oxides (hematite and goethite) and water at low temperatures. Data sources: Zheng (1991, 1998) — increment method; Clayton and Epstein (1961), and Clayton (1963) — empirical calibration based on natural samples; Yapp (1990), Müller (1995), and Bao and Koch (1999) — laboratory synthesis experiments via different reaction pathways.

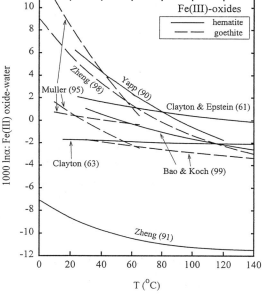

poorly crystalline and fine-grained (<100 nm) with large amounts of adsorbed water. These characteristics pose not only technical problems in handling and isotopic analysis (drying and rapid redox reactions), but also fundamental questions regarding equilibrium fractionation factors for nano-sized particles with large specific surface areas.

One way to facilitate isotopic exchange between minerals and water at low temperatures is by means of microbial and enzymatic activities. Examples are intra- and extra-cellular biological precipitation of magnetite as mentioned above (Zhang et al. 1997; Mandernack et al. 1999), and enzymatic isotopic exchange between phosphate and water (Blake et al. 1998). It is likely that biological activities result in isotopic effects different from inorganic equilibrium fractionations. Such "isotopic biosignatures" may be recognizable in the rock record, and thus serve as a tool in the search for ancient life.

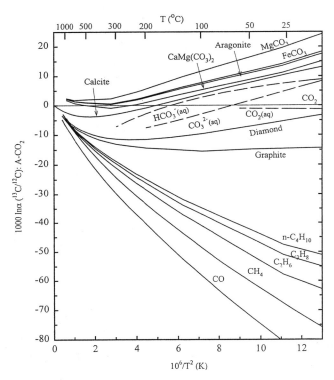

Figure 20. Plot of carbon isotope fractionation between various geologic materials and gaseous CO_2. Data sources: calcite — calculations by Chacko et al. (1991); aragonite, $CaMg(CO_3)_2$, $MgCO_3$ and $FeCO_3$, — calculations by Golyshev et al. (1981); HCO_3^-(aq) — experimental results by Malinin et al. (1967); CO_3^{2-}(aq) — calculations by Halas et al. (1997); CO_2(aq) — experimental results by Zhang et al. (1995); diamond and graphite — Bottinga (1969b); CO and CH_4 — Richet et al. (1977); hydrocarbons — Galimov (1985).

Carbon isotope fractionation factors

Figure 20 shows the general pattern of equilibrium ^{13}C enrichment among various geologic materials, exclusive of biological compounds, relative to gaseous CO_2. Because only a limited number of carbon-bearing systems have been investigated experimentally, this plot is constructed largely from theoretical calculations by Bottinga (1969b), Galimov (1985), Richet et al. (1977), Golyshev et al. (1981), Chacko et al. (1991), and Halas et al. (1997). There is a strong positive correlation between ^{13}C enrichment and the oxidation state or number of covalent bonds of carbon. Carbonate minerals, as a group, are the most enriched in ^{13}C, whereas CO and light hydrocarbons are the most depleted.

The magnitude of ^{13}C isotopic fractionations of various gaseous species (CO and hydrocarbon) relative to CO_2 decreases rapidly with increasing temperature, so that T^{-3} (K) (and even higher-order) terms are required to describe these curves numerically. There are minima in the fractionation curves of calcite, graphite and diamond, and fractionation crossovers are indicated for dissolved CO_2 species (HCO_3^- and CO_3^{2-}), calcite, and possibly other carbonate minerals.

Of the various fractionation curves shown on Figure 20, the best studied system involving a mineral is the system calcite-CO_2, which has been investigated theoretically and experimentally. Direct exchange experiments between calcite and CO_2 at high pressure (1-13 kbar) and temperature (400-1200°C) by Chacko et al. (1991), Scheele and Hoefs (1992), and Rosenbaum (1994) showed that CO_2 is enriched in ^{13}C by 2 to 3.5‰ relative to calcite over this temperature range. These experimental data are in marked disagreement with the calculations of Golyshev et al. (1981), which, over the same temperature range, indicate much smaller calcite-CO_2 fractionations, and even fractionations of the opposite sign. This casts serious doubt on the magnitude of carbonate-CO_2 fractionations given by Golyshev et al.'s calculations for other carbonate minerals (Fig. 20), although the order of ^{13}C enrichment amongst the carbonates indicated by those calculations may be correct. There is excellent agreement between the high-temperature experimental data and the theoretical calculations of Chacko et al. (1991). However, the same calculations gave fractionations that differed by 1.6-2.0‰ from those obtained in experiments involving slow, controlled precipitation of calcite from supersaturated solutions at 10-40°C (Romanek et al. 1992). In fact, the calculations for calcite are very similar to experimental aragonite-CO_2 fractionations obtained over the same temperature range (Romanek et al. 1992). Further investigations are needed to resolve this discrepancy at low temperatures in this very important system.

As originally suggested by Valley and O'Neil (1981), the calcite-graphite system is an important one, and can potentially serve as a very effective geothermometer in metamorphosed carbonate rocks. There are several theoretical and natural sample calibrations of fractionations in this system (Valley and O'Neil 1981; Wada and Suzuki 1983; Morikiyo 1984; Chacko et al. 1991; Kitchen and Valley 1995; Polyakov and Kharlashina 1995) (Fig. 14), and also an experimental calibration (Scheele and Hoefs 1992) derived from a combination of data from CO_2-graphite and CO_2-calcite experiments. In the temperature range 600-1200°C, the experimental calibration indicates markedly larger fractionations than given by the other calibrations. This discrepancy likely reflects a problem with the CO_2-graphite experiments, which, in order to overcome the problem of low exchange rates, used a very high graphite to CO_2 ratio. Although such a procedure greatly enhances exchange rates, the measured fractionations may represent those between CO_2 and graphite surface atoms rather than between CO_2 and bulk graphite (see Hamza and Broecker 1974 for a discussion of surface effects). There is good to excellent agreement between natural sample calibrations of Valley and O'Neil (1981) and Kitchen and Valley (1995), which are based on high-temperature natural sample data, and the theoretical calibration of Polyakov and Kharlashina (1995), which takes into account the significant pressure effect on fractionations in this system. Although the matter remains controversial, in our opinion, the current best estimate of equilibrium fractionations in the calcite-graphite system is given by the calibration of Polyakov and Kharlashina.

A recent experimental study on carbon isotope fractionation in the system CO_2-CH_4 (Horita 2001) showed that results of theoretical calculations by Richet et al. (1977) are accurate (to ±0.9‰) over the temperature range from 200 to 600°C. However, in the application of fractionation data for C-O-H gases, perhaps a more important question is rates and mechanisms of isotopic exchange in natural environments. Many experimental studies indicate that isotopic exchange among gaseous species of the C-O-H system is driven by a series of elemental chemical reactions rather than by simple intermolecular collisions between the gaseous species of interest. It is also known that homogeneous gaseous reactions are very sluggish, and that these reactions can be enhanced tremendously by heterogeneous reactions on the surface of solids or minerals. Gaseous species of the C-O-H system (particularly CO_2-CH_4) from sedimentary and geothermal systems commonly appear to be out of isotopic equilibrium (Ohmoto 1986; Horita 2001). Many natural materials, however, can catalyze isotopic exchange among C-O-H gases. In this context, kinetic isotopic fractionation may be more important than equilibrium fractionation in the study of C-O-H gases at low to moderate (perhaps up to 500-600°C) temperatures.

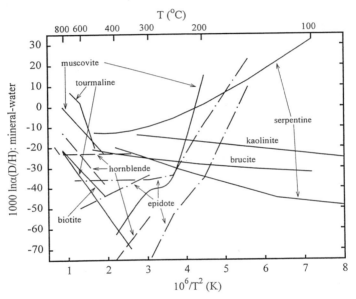

Figure 21. Plot of hydrogen isotope fractionation between various hydrous minerals and water. Note the large discrepancies of existing calibrations of epidote-water and hornblende-water fractionations. Data sources: serpentine — Wenner and Taylor (1973) and Sakai and Tsutsumi (1978); kaolinite — Gilg and Sheppard (1996); brucite — modified after Satake and Matsuo (1984); epidote — Graham et al. (1980), Vennemann and O'Neil (1996), and Chacko et al. (1999); hornblende — Suzuoki and Epstein (1976), Graham et al. (1984), and Vennemann and O'Neil (1996); muscovite — Suzuoki and Epstein (1976) and Vennemann and O'Neil (1996); biotite — Suzuoki and Epstein (1976); tourmaline — Blamart et al. (1989) and Jibao and Yaiqian (1997).

Hydrogen isotope fractionation factors

Figure 21 summarizes some of the available data on D/H fractionation factors in hydrous mineral-H_2O systems. These fractionations are mainly derived from experimental data, but also include calibrations based on natural samples. The plot shows a bewildering array of shapes for mineral-water fractionation curves. Moreover, as noted by Vennemann and O'Neil (1996), there are major, unresolved discrepancies between published experimental calibrations as regards the magnitude of fractionation factors in individual mineral-water systems. This is well illustrated by the hornblende-water and epidote-water systems, where three independent experimental determinations in each system show marked differences in the magnitude of fractionation factors, and the shape of fractionation curves. As discussed earlier, differences in experimental pressures can account for some, but not all of the differences in the measured fractionation factors (Horita et al. 1999; Horita et al., in press). Differences in the experimental procedures of the various studies undoubtedly also contribute to the discrepancies. However, a review of these procedures does not reveal an obvious 'best' calibration. The difference in hornblende-H_2O (or actinolite-H_2O) fractionations given by the Suzuoki and Epstein (1976) and Graham et al. (1984) experiments is particularly unsettling because, except for the mineral to fluid ratio in the experimental charges, these studies used broadly similar procedures.

The lack of consensus on D/H fractionations in mineral-water systems limits the confidence with which these data can be used to estimate the hydrogen isotope compositions of fluids that have interacted with rocks. To resolve these issues, additional careful studies are required that systematically investigate the effect of mineral composition, fluid composition, confining pressure, and mineral to fluid ratio on D/H fractionation factors in these systems. A complementary avenue of research would be the development of theoretical methods for calculating D/H partition function ratios for hydrous minerals. Even if such calculations could not accurately predict the magnitude of fractionations, they would provide a useful framework for fitting and extrapolating experimental data.

CONCLUSIONS

In his concluding thoughts in the 1986 MSA review volume, O'Neil anticipated that experimental developments then underway would help to resolve major discrepancies in experimental, theoretical and natural sample estimates of equilibrium oxygen isotope fractionation factors between minerals. That expectation has to a large extent been realized. There is now considerably better agreement on high-temperature oxygen isotope fractionation factors for a significant number of common rock-forming minerals. However, as noted above, there still remains controversy, or lack of sufficient data on fractionations for some important minerals including quartz and kyanite. Additional experiments and carefully designed studies of natural samples should help to resolve some of these difficulties. Important insights can also be gained from further theoretical studies. In particular, the theoretical methodology developed by Kieffer (1982) has proven to be remarkably successful in predicting oxygen isotope fractionation factors for silicates. Her work should be expanded to include calculations for the aluminosilicate polymorphs, amphiboles, oxides, high-pressure mantle minerals, and other minerals that have not been, or may not be readily amenable to experimental investigation. The new methods of theoretical calculation developed over the past decade, such as the *ab initio* (Patel et al. 1991), and perturbation theory (Polyakov and Kharlashina 1995; Polyakov and Mineev 2000) methods, show great promise, and should also be refined and extended.

In contrast to the situation described above, there is a general lack of consensus on oxygen, carbon and hydrogen isotope fractionation factors for many important mineral-fluid systems, particularly at low temperature. It is clear that many different variables may affect the magnitude of fractionation factors measured in these systems including among others mineral composition, mineral zoning, solution composition, precipitation rate, and biological activity. Although the potential significance of these variables has long been appreciated, recent developments in experimental and analytical techniques, including some of those described in this chapter, make it possible to investigate their effects on a more detailed and precise level.

ACKNOWLEDGMENTS

We respectfully dedicate this chapter to Professor Robert N. Clayton and the late Professor Julian R. Goldsmith in honor of their landmark contributions to our understanding of stable isotope fractionation in geological systems. TC thanks Neil Banerjee and Suman De for helpful comments on an earlier version of this manuscript, and Karlis Muehlenbachs for many discussions. We are very grateful to Bob Clayton, James Farquhar, Val Ferreira, Bob Luth and John Valley for timely reviews. Their comments materially improved the manuscript and caught some serious errors. Any remaining errors or omissions are entirely our responsibility. TC's research is supported by a Canadian NSERC research grant. The work of DRC and JH is sponsored by the Division of Chemical Sciences, Geosciences, and Biosciences, Office of Basic Energy Sciences, U.S. Department of Energy under contract DE-AC05-00OR22725, Oak Ridge National Laboratory, managed and operated by UT-Battelle, LLC.

REFERENCES

Addy SK, Garlick GD (1974) Oxygen isotope fractionation between rutile and water. Contrib Mineral Petrol 45:119-121

Agrinier P (1991) The natural calibration of $^{18}O/^{16}O$ geothermometers: Application to the quartz-rutile pair. Chem Geol 91:49-64

Anderson AT II, Clayton RN, Mayeda TK (1971) Oxygen isotope thermometry of mafic igneous rocks. J Geol 79:715-729

Arnason B. (1969) Equilibrium constant for the fractionation of deuterium between ice and water. J Phys Chem 73:3191-3194

Bao H, Koch PL (1999) Oxygen isotope fractionation in ferric oxide-water systems: low-temperature synthesis. Geochim Cosmochim Acta 63:599-613

Bechtel A, Hoernes S (1990) Oxygen isotope fractionation between oxygen of different sites in illite minerals: A potential single-mineral thermometer. Contrib Mineral Petrol 104:463-470

Becker RH (1971) Carbon and oxygen isotope ratios in iron-formation and associated rocks from the Hammersley Range of Western Australia and their implications. PhD Dissertation, University of Chicago

Becker RH, Clayton RN (1976) Oxygen isotope study of a Precambrian banded iron-formation. Hamersley Range, Western Australia. Geochim Cosmochim Acta 40:1153-1165

Berndt ME, Seal RR, II, Shanks WC, III, Seyfried WE Jr (1996) Hydrogen isotope systematics of phase separation in subseafloor hydrothermal systems: Experimental calibration and theoretical models. Geochim Cosmochim Acta 60:1595-1604

Bigeleisen J (1961) Statistical mechanics of isotope effects on the thermodynamic properties of condensed systems. J Chem Phys 34:1485-1493

Bigeleisen J (1965) Chemistry of isotopes. Science 147:463-471

Bigeleisen J, Mayer MG (1947) Calculation of equilibrium constants for isotopic exchange reactions. J Phys Chem 13:261-267

Bird MI, Longstaffe FJ, Fyfe WS, Bildgen P (1993) Oxygen-isotope fractionation in titanium-oxide minerals at low temperature. Geochim Cosmochim Acta 57:3083-3091

Bird MI, Longstaffe FJ, Fyfe WS, Takaki K, Chivas AR (1994) Oxygen-isotope fractionation in gibbsite: synthesis experiments versus natural samples. Geochim Cosmochim Acta 58:5267-5277

Blake RE, O'Neil JR, Garcia GA (1998) Effects of microbial activity on the $\delta^{18}O$ of dissolved inorganic phosphate and textual features of synthetic apatites. Am Mineral 83:1516-1531

Blamart D, Pichavant M, Sheppard SMF (1989) Experimental determination of the D/H isotopic fractionation between tourmaline and water at 600, 500 °C and 3 kbar. C R Acad Sci Paris 308:39-44

Blattner P (1973) Oxygen from liquids for isotopic analysis, and a new determination of αCO_2-H_2O at 25°C. Geochim Cosmochim Acta 37:2691-2693

Blattner P, Bird GW (1974) Oxygen isotope fractionation between quartz and K-feldspar at 600°C. Earth Planet Sci Lett 23:21-27

Blattner P, Braithwaite WR, Glover RB (1983) New evidence on magnetite oxygen isotope geothermometers at 175°C and 112°C in Wairakei steam pipelines (New Zealand). Isotope Geosci 1:195-204

Bopp P, Heinzinger K, Klemm A (1977) Oxygen isotope fractionation. Z Naturforschung 32a:1419-1425

Bottcher WE (1994) $^{13}C/^{12}C$ partitioning during synthesis of $Na_2Ca(CO_3)_2 2H_2O$. J Chem Soc Chem Commun, p 1485

Bottcher WE (2000) Stable isotope fractionation during experimental formation of norsethite ($BaMg[CO_3]_2$): A mineral analogue of dolomite. Aquatic Geochem 6:201-212

Bottinga Y (1968) Calculations of fractionation factors for carbon and oxygen isotopic exchange in the system calcite-carbon dioxide-water. J Phys Chem 72:800-808

Bottinga Y (1969a) Calculated fractionation factors for carbon and hydrogen isotope exchange in the system calcite-carbon dioxide-graphite-methane-hydrogen-water vapor. Geochim Cosmochim Acta 33:49-64

Bottinga Y (1969b) Carbon isotope fractionation between graphite, diamond and carbon dioxide. Earth Planet Sci Lett 5:301-307

Bottinga Y, Craig H (1969) Oxygen isotope fractionation between CO_2 and water, and the isotopic composition of marine atmospheric CO_2. Earth Planet Sci Lett 5:285-295

Bottinga Y, Javoy M (1973) Comments on oxygen isotope geothermometry. Earth Planet Sci Lett 20:250-265

Bottinga Y, Javoy M (1975) Oxygen isotope partitioning among the minerals and triplets in igneous and metamorphic rocks. Rev Geophys Space Phys 13:401-418

Brandriss ME, O'Neil JR, Edlund MB, Stoermer EF (1998) Oxygen isotope fractionation between diatomaceous silica and water. Geochim Cosmochim Acta 62:1119-1125

Brenninkmeijer CAM, Kraft P, Mook WG (1983) Oxygen isotope fractionation between CO_2 and H_2O. Isotope Geosci 1:181-190

Capuano RM (1992) The temperature-dependence of hydrogen isotope fractionation between clay minerals and water: Evidence from a geopressured system. Geochim Cosmochim Acta 56:2547-2554

Carothers WW, Adami LH, Rosenbauer RJ (1988) Experimental oxygen isotope fractionation between siderite-water and phosphoric acid liberated CO_2-siderite. Geochim Cosmochim Acta 52:2445-2450

Cerrai E, Marchetti C, Renzoni, R, Roseo L, Silvestri M, Villani S (1954) A thermal method for concentrating heavy water. Chem Engin Prog Symp Ser 50, No 11, Nuclear Engineering-Part I, p 271-280

Chacko T (1993) Experimental studies of equilibrium oxygen and carbon isotope fractionation between phases. *In* Luth RW (ed) Experiments at High Pressure and Application to the Earth's Mantle. Mineral Assoc Canada Short Course 21:357-384

Chacko T, Mayeda TK, Clayton RN (1989) Oxygen isotope fractionations in the system gehlenite-calcite. EOS Trans Am Geophys Union 70:489

Chacko T, Mayeda TK, Clayton RN, Goldsmith JR (1991) Oxygen and carbon isotope fractionations between CO_2 and calcite. Geochim Cosmochim Acta 55:2867-2882

Chacko T, Hu X, Mayeda TK, Clayton RN, Goldsmith JR (1996) Oxygen isotope fractionations in muscovite, phlogopite, and rutile. Geochim Cosmochim Acta 60:2595-2608

Chacko T, Riciputi LR, Cole DR, Horita J (1999) A new technique for determining equilibrium hydrogen isotope fractionation factors using the ion microprobe: Application to the epidote-water system. Geochim Cosmochim Acta 63:1-10

Chen C-H, Liu K-K, Shieh Y-N (1988) Geochemical and isotopic studies of bauxitization in the Tatun volcanic area, northern Taiwan. Chem Geol 68:41-56

Chiba H, Kusakabe M, Hirano S, Matsuo S, Somiya S (1981) Oxygen isotope fractionation factors between anhydrite and water from 100 to 550°C. Earth Planet Sci Lett 53:55-62

Chiba H, Chacko T, Clayton RN, Goldsmith JR (1989) Oxygen isotope fractionations involving diopside, forsterite, magnetite, and calcite: Application to geothermometry. Geochim Cosmochim Acta 53:2985-2995

Clayton RN (1963) High-temperature isotopic thermometry. *In* Nuclear Geology on Geothermal Areas. Cons Naz Ric Lab Geol Nucl, p 222-229

Clayton RN (1981) Isotopic thermometry. *In* Newton RC, Navrotsky A, Wood BJ (eds) The Thermodynamics of Minerals and Melts. Springer-Verlag, Berlin, p 85-109.

Clayton RN (1986) High temperature isotope effects in the early solar system. *In* Valley JW, Taylor HP Jr, O'Neil JR (eds) Stable Isotopes in High Temperature Geological Processes. Rev Mineral 16:129-140

Clayton RN, Epstein S (1961) The use of oxygen isotopes in high-temperature geological thermometery. J Geol 69:447-452

Clayton RN, Kieffer SW (1991) Oxygen isotope thermometer calibrations. *In* Taylor HP Jr, O'Neil JR, Kaplan IR (eds) Stable Isotope Geochemistry: A Tribute to Samuel Epstein. Geochem Soc Spec Pub 3:3-10.

Clayton RN, Jones BF, Berner RA (1968) Isotopic studies of dolomite formation under sedimentary conditions. Geochim Cosmochim Acta 32:415-432

Clayton RN, O'Neil JR, Mayeda TK (1972) Oxygen isotope exchange between quartz and water. J Geophys Res 77:3057-3067

Clayton RN, Goldsmith JR, Karel KJ, Mayeda TK, Newton RC (1975) Limits on the effect of pressure on isotopic fractionation. Geochim Cosmochim Acta 39:1197-1201

Clayton RN, Goldsmith JR, Mayeda TK (1989) Oxygen isotope fractionation in quartz, albite, anorthite, and calcite. Geochim Cosmochim Acta 53:725-733

Cole DR, Ripley EM (1999) Oxygen isotope fractionation between chlorite and water form 170 to 350°C: A preliminary assessment based on partial exchange and fluid/rock experiments. Geochim Cosmochim Acta 63:449-457

Compston W, Epstein S (1958) A method for the preparation of carbon dioxide from water vapor for oxygen isotopic analysis. EOS Trans Am Geophys Union 39:511

Craig H, Hom B (1968) Relationships of deuterium, oxygen-18, and chlorinity in the formation of sea ice. EOS Trans Am Geophys Union 216

Criss RE (1991) Temperature-dependence of isotopic fractionation factors. *In* Taylor HP Jr, O'Neil JR, Kaplan IR (eds) Stable Isotope Geochemistry: A Tribute to Samuel Epstein. Geochem Soc Spec Pub 3:11-16

Criss RE (1999) Principles of Stable Isotope Distribution. Oxford Univ Press, New York

Deines P (1977) On the oxygen isotope distribution among triplets in igneous and metamorphic rocks. Geochim Cosmochim Acta 41:1709-1730

Deuser WG, Degens ET (1967) Carbon isotope fractionation in the system CO_2(gas)- CO_2(aqueous)-HCO_3^- (aqueous). Nature 215:1033-1035

Dobson PF, Epstein S, Stolper EM (1989) Hydrogen isotope fractionation between coexisting vapor and silicate glasses and melts at low pressure. Geochim Cosmochim Acta 53:2723-2730

Dove MT, Winkler B, Leslie M, Harris MJ, Salje EKH (1992) A new interatomic model for calcite: applications to lattice dynamics studies, phase transition, and isotopic fractionation. Am Mineral 77:244-250

Downs WF, Touysinhthiphonexay Y, Deines P (1981) A direct determination of the oxygen isotope fractionation between quartz and magnetite at 600 and 800°C and 5 kbar. Geochim Cosmochim Acta 45:2065-2072

Driesner T (1996) Aspects of stable isotope fractionation in hydrothermal solutions. PhD dissertation, ETH-Zürich, 93 p

Driesner T (1997) The effect of pressure on deuterium-hydrogen fractionation in high-temperature water. Science 277:791-794

Driesner T, Seward TM (2000) Experimental and simulation study of salt effects and pressure/density effects on oxygen and hydrogen stable isotope liquid-vapor fractionation for 4 molal NaCl and KCl aqueous solutions to >400°C. Geochim Cosmochim Acta 64:1773-1784

Dugan JP, Borthwick J (1986) Carbon dioxide-water oxygen isotope fractionation factor using chlorine trifluoride and guanidine hydrochloride techniques. Analyt Chem 58:3052-3054

Dugan JP, Borthwick J, Harmon RS, Gagnier MA, Glahn JE, Kinsel EP, MacLeod S, Viglino JA (1985) Guanidine hydrochloride method for determination of water oxygen isotope ratios and the oxygen-18 fractionation between carbon dioxide and water at 25°C. Analyt Chem 57:1734-1736

Dunn SR, Valley JW (1992) Calcite-graphite isotope thermometry: a test for polymetamorphism in marble, Tudor gabbro aureole, Ontario, Canada. J Metam Geol 10:487-501

Eiler JM, Baumgartner LP, Valley JW (1992) Intercrystalline stable isotope diffusion: a fast grain boundary model. Contrib Mineral Petrol 112:543-557

Eiler JM, Kitchen N, Rahn TA (2000) Experimental constraints on the stable-isotope systematics of CO_2 ice/vapor systems and relevance to the study of Mars. Geochim Cosmochim Acta 64:733-746

Elcombe MM, Hulston JR (1975) Calculation of sulphur isotope fractionation between sphalerite and galena using lattice dynamics. Earth Planet Sci Lett 28:172-180

Emiliani C (1955) Pleistocene paleotemperatures. J Geol 63:538-578

Emrich K, Ehhalt DH, Vogel JC (1970) Carbon isotope fractionation during the precipitation of calcium carbonate. Earth Planet Sci Lett 8:363-371

Epstein S, Buchsbaum R, Lowenstam HA, Urey HC (1953) Revised carbonate-water isotopic temperature scale. Geol Soc Am Bull 64:1315-1326

Epstein S, Graf DL, Degens ET (1963) Oxygen isotope studies on the origin of dolomite. *In* Isotope and Cosmic Chemistry. H Craig et al. (eds) North-Holland, Amsterdam, p 169-180

Escande M, Decarreau A, Labeyrie L (1984) Etude experimentale de l'echangeabilite des isotopes de l'oxygene des smectites. C R Acad Sci Paris 299:707-710

Eslinger EV (1971) Mineralogy and oxygen isotope ratios of hydrothermal and low-grade metamorphic argillaceous rocks. PhD Dissertation, Case Western Reserve University, Cleveland, Ohio

Eslinger EV, Savin SM (1973) Mineralogy and oxygen isotope geochemistry of the hydrothermally altered rocks of the Ohaki-Broadlands, New Zealand geothermal area. Am J Sci 273:240-267

Eslinger EV, Savin SM, Yeh H (1979) Oxygen isotope geothermometry of diagenetically altered shales. SEPM Spec Publ 26:285-305

Farquhar J (1995) Strategies for high-temperature oxygen isotope thermometry. PhD Dissertation, University of Alberta

Farquhar J, Chacko T, Frost BR (1993) Strategies for high-temperature oxygen isotope thermometry: a worked example from the Laramie Anorthosite complex Wyoming USA. Earth Planet Sci Lett 117:407-422

Farver JR (1994) Oxygen self-diffusion in calcite: dependence on temperature and water fugacity. Earth Planet Sci Lett 121:575-587

Fayek M, Kyser TK (2000) Low temperature oxygen isotopic fractionation in the uraninite-UO_3-CO_2-H_2O system. Geochim Cosmochim Acta 64:2185-2197

Feng X, Savin SM (1993) Oxygen isotope studies of zeolites - stilbite, analcite, heulandite, and clinoptilolite; III. Oxygen isotope fractionation between stilbite and water or water vapor. Geochim Cosmochim Acta 57:4239-4247

Fontes JC, Gonfiantini R (1967) Fractionnement istopique de l'hydrogene dans l'eau de cristallisation du gypse. C R Acad Sc Paris 265:4-6

Fortier SM, Lüttge A (1995) An experimental calibration of the temperature-dependence of oxygen isotope fractionation between apatite and calcite at high temperature (350-800°C). Chem Geol 125:281-290

Fortier SM, Lüttge A, Satir M, Metz P (1994) Oxygen isotope fractionation between fluorphlogopite and calcite: An experimental investigation of temperature-dependence and F⁻/OH⁻ effects. Eur J Mineral 6:53-65

Fortier SM, Cole DR, Wesolowski DJ, Riciputi LR, Paterson BA, Valley JW, Horita J (1995) Determination of equilibrium magnetite-water oxygen isotope fractionation factor at 350°C: a comparison of ion microprobe and laser fluorination techniques. Geochim Cosmochim Acta 59:3871-3875

Frantz JD, Dubessy J, Mysen B (1993) An optical cell for Raman spectroscopic studies of supercritical fluids and its application to the study of water to 500°C and 2000 bar. Chem Geol 106:9-26

Friedman I, O'Neil JR (1977) Compilation of Stable Isotope Fractionation Factors of Geochemical Interest. U SGeol Surv Prof Paper 440-KK

Friedman I, Gleason J, Sheppard RA, Gude AJ, III (1993) Deuterium fractionation as water diffuses into silicic volcanic ash. *In* Climate Change in Continental Isotopic Records. PK Swart et al. (eds) Am Geophys Union 321-323

Fritz P, Smith DGW (1970) The isotopic composition of secondary dolomites. Geochim Cosmochim Acta 34:1161-1173

Galimov EM (1985) The Biological Fractionation of Isotopes. Academic Press, Orlando, Florida

Garlick GD (1966) Oxygen isotope fractionation in igneous rocks. Earth Planet Sci Lett 1:361-368

Gautason B, Chacko T, Muehlenbachs K (1993) Oxygen isotope partitioning among perovskite ($CaTiO_3$), cassiterite (SnO_2) and calcite ($CaCO_3$). Geol Assoc Canada/ Mineral Assoc Canada Abstr Programs, Edmonton, Alberta, p A34

Gilg HA, Sheppard SMF (1996) Hydrogen isotope fractionation between kaolinite and water revisited. Geochim Cosmochim Acta 60:529-533

Gillet P, McMillan P, Schott J, Badro J, Grzechnik A (1996) Thermodynamic properties and isotopic fractionation of calcite from vibrational spectroscopy of ${}^{18}O$-substituted calcite. Geochim Cosmochim Acta 60:3471-3485

Giletti BJ (1986) Diffusion effects on oxygen isotope temperatures of slowly cooled igneous and metamorphic rocks. Earth Planet Sci Lett 77:218-228

Giletti BJ, Yund RA (1984) Oxygen diffusion in quartz. J Geophys Res 89B:4039-4046

Girard J-P, Savin SM (1996) Intracrystalline fractionation of oxygen isotopes between hydroxyl and non-hydroxyl sites in kaolinite measured by thermal dehydroxylation and partial fluorination. Geochim Cosmochim Acta 60:469-487

Goldsmith JR (1991) Pressure-enhanced Al/Si diffusion and oxygen isotope exchange. *In* Ganguly J (ed) Diffusion, Atomic Ordering and Mass Transport. Springer-Verlag, Berlin, p 221-247

Golyshev SI, Padalko NL, Pechenkin, SA (1981) Fractionation of stable oxygen and carbon isotopes in carbonate systems. Geochem Int'l 18:85-99

Gonfiantini R, Fontes JC (1963) Oxygen isotopic fractionation in the water of crystallization of gypsum. Nature 200:644-646

Graham CM, Sheppard SMF, Heaton THE (1980) Experimental hydrogen isotope studies:hydrogen isotope fractionation in the systems epidote-H_2O, zoisite- H_2O and AlO(OH)-H_2O. Geochim Cosmochim Acta 44:353-364

Graham CM, Sheppard SMF (1980) Experimental hydrogen isotope studies, II. Fractionations in the systems epidote-NaCl-H_2O, epidote-$CaCl_2$-H_2O and epidote-seawater, and the hydrogen isotope composition of natural epidote. Earth Planet Sci Lett 48:237-251

Graham CM, Harmon RS, Sheppard SMF (1984) Experimental hydrogen isotope studies: Hydrogen isotope exchange between amphibole and water. Am Mineral 69:128-138

Graham CM, Viglino JA, Harmon RS (1987) Experimental study of hydrogen-isotope exchange between aluminous chlorite and water and of diffusion in chlorite. Am Mineral 72:566-579

Grinenko VA, Mineyev SD, Devirts AL, Lagutina Ye P (1987) Hydrogen isotope fractionation in the lizardite-water system at 100°C and 1 atm. Geochem Int'l 24:100-104.

Grootes PM, Mook WG, Vogel JC (1969) Isotopic fractionation between gaseous and condensed carbon dioxide. Z Physik 221:257-273

Grossman EL, Ku T-L (1986) Oxygen and carbon isotope fractionation in biogenic aragonite: temperature effects. Chem Geol 59:59-74

Gu Z (1980) Determination of the separation facto for isotopic exchange reaction between H_2O and CO_2 at 25°C. He Huaxue Yu Fangshe Huaxue 2:112-115 (in Chinese)

Halas S, Wolacewicz W (1982) The experimental study of oxygen isotope exchange reaction between dissolved bicarbonate and water. J Chem Phys 76:5470-5472

Halas S, Szaran J, Niezgoda H (1997) Experimental determination of carbon isotope equilibrium fractionation between dissolved carbonate and carbon dioxide. Geochim. Cosmochim. Acta 61:2691-2695

Hamza MS, Epstein S (1980) Oxygen isotopic fractionation between oxygen of different sites in hydroxyl-bearing silicate minerals. Geochim Cosmochim Acta 44:173-182

Hamza MS, Broecker WS (1974) Surface effect on the isotopic fractionation between CO_2 and some carbonate minerals. Geochim Cosmochim Acta 38:669-681

Hariya U, Tsutsumi M (1981) Hydrogen isotopic composition of MnO(OH) minerals from manganese oxide and massive sulfide (Kuroko) deposits in Japan. Contrib Mineral Petrol 77:256-261

Hoering TC (1961) The physical chemistry of isotopic substances: The effect of physical changes in isotope fractionation. Carnegie Inst Washington Yearbook 60:201-204

Hoffbauer R, Hoernes S, Fiorentini E (1994) Oxygen isotope thermometry based on a refined increment method and its application to granulite-grade rocks from Sri Lanka. Precambrian Res 66:199-220

Horibe Y, Craig H (1995) D/H fractionation in the system methane-hydrogen-water. Geochim Cosmochim Acta 59:5209-5217

Horibe Y, Shigehara K, Takakuwa Y (1973) Isotope separation factor of carbon dioxide-water system and isotopic composition of atmospheric oxygen. J Geophys Res 78:2625-2629

Horita J (1989) Stable isotope fractionation factors of water in hydrated saline minerals-brine systems. Earth Planet Sci Lett 95:173-179

Horita J (2001) Carbon isotope exchange in the system CO_2-CH_4 at elevated temperatures. Geochim Cosmochim Acta 65:1907-1919

Horita J, Wesolowski DJ (1994) Liquid-vapor fractionation of oxygen and hydrogen isotopes of water from the freezing to the critical temperature. Geochim Cosmochim Acta 58:3425-3437

Horita J, Wesolowski DJ, Cole DR (1993a) The activity-composition relationship of oxygen and hydrogen isotopes in aqueous salt solutions: I. Vapor-liquid water equilibration of single salt solutions from 50 to 100°C. Geochim Cosmochim Acta 57:2797-2817

Horita J, Cole DR, Wesolowski DJ (1993b) The activity-composition relationship of oxygen and hydrogen isotopes in aqueous salt solutions: II. Vapor-liquid water equilibration of mixed salt solutions from 50 to 100°C and geochemical implications. Geochim Cosmochim Acta 57:4703-4711

Horita J, Cole DR, Wesolowski DJ (1994) Salt effects on stable isotope partitioning and their geochemical implications for geothermal brines. In Proc 19th Workshop on Geothermal Reservoir Engineering, Stanford University, p 285-290

Horita J, Cole DR, Wesolowski DJ (1995a) The activity-composition relationship of oxygen and hydrogen isotopes in aqueous salt solutions: III. Vapor-liquid water equilibration of NaCl solutions from to 350°C. Geochim Cosmochim Acta 59:1139-1151

Horita J, Wesolowski DJ, Cole DR (1995b) D/H and $^{18}O/^{16}O$ partitioning between water liquid and vapor in the system H_2O-Na-K-Ca-Mg-Cl-SO_4 from 0 to 350°C. In Physical Chemistry of Aqueous Systems: Meeting the Needs of Industry. Proc 12th Int'l Conf on the Properties of Water and Steam. HJ White Jr et al. (eds) Begell House, p 505-510

Horita J, Cole DR, Wesolowski DJ, Fortier SM (1996) Salt effects on isotope partitioning and their geochemical implications: An overview. *In* Proc Todai Int'l Symp on Cosmochronology and Isotope Geoscience, p 33-36

Horita J, Cole DR, Wesolowski DJ (1997) Salt effects on oxygen and hydrogen isotope partitioning between aqueous salt solutions and coexisting phases at elevated temperatures. *In* Proc 5th Int'l Symp on Hydrothermal Reaction*s*. Gatlinburg, Tennessee, p 194-197

Horita J, Driesner T, Cole DR (1999) Pressure effect on hydrogen isotope fractionation between brucite and water at elevated temperatures. Science 286:1545 -1547

Horita J, Cole DR, Polyakov VB, Driesner T (in press) Experimental and theoretical study of pressure effects on hydrogen isotope fractionation in the system brucite-water at elevated temperatures. Geochim Cosmochim Acta

Hu G, Clayton RN (in press) Oxygen isotope salt effects at high pressure and high temperature, and the calibration of oxygen isotope geothermometers. Geochim Cosmochim Acta

James AT, Baker DR (1976) Oxygen isotope exchange between illite and water at 22°C. Geochim. Cosmochim. Acta 40:235-239

Javoy M, Pineau F, Iiyama I (1978) Experimental determination of the isotopic fractionation between gaseous CO_2 and carbon dissolved in tholeiitic magma. Contrib Mineral Petrol 67:35-39

Jenkin GRT, Farrow CM, Fallick AE, Higgins D (1994) Oxygen isotope exchange and closure temperatures in cooling rocks. J Metam Geol 12:221-235

Jibao G, Yaqian Q (1997) Hydrogen isotope fractionation and hydrogen diffusion in the tourmaline-water system. Geochim Cosmochim Acta 61:4679-4688

Kakiuchi M (1994) Temperature-dependence of fractionation of hydrogen isotopes in aqueous sodium chloride solutions. J Sol Chem 23:1073-1087

Kakiuchi M (2000) Distribution of isotopic molecules, H_2O, HDO, and D_2O in vapor and liquid phases in pure water and aqueous solution systems. Geochim Cosmochim Acta 64:1485-1492

Karlsson HR, Clayton RN (1990) Oxygen isotope fractionation between analcime and water: An experimental study. Geochim Cosmochim Acta 54:1359-1368

Kazahaya K (1986) Chemical and isotopic studies on hydrothermal solutions. PhD dissertation, Tokyo Inst Technology

Kawabe I (1978) Calculation of oxygen isotope fractionation in quartz-water system with special reference to the low temperature fractionation. Geochim Cosmochim Acta 42:613-621

Kendall C, Chou. I-M, Coplen TB (1983) Salt effect on oxygen isotope equilibria. EOS Trans Am Geophys Union 64:334-335

Kieffer SW (1982) Thermodynamic and Lattice vibrations of Minerals: 4.Application to phase equilibria, isotope fractionation, and high pressure thermodynamics properties. Rev Geophys Space Phys 20:827-849

Kim S-T, O'Neil JR (1997) Equilibrium and nonequilibrium oxygen isotope effects in synthetic carbonates. Geochim Cosmochim Acta 61:3461-3475

Kitchen NE, Valley JW (1995) Carbon isotope thermometry in marbles of the Adirondack Mountains, New York. J Metam Geol 13:577-594

Kita I, Taguchi S, Matsubaya O (1985) Oxygen isotope fractionation between amorphous silica and water at 34-93°C. Nature 314:83-34

Koehler G, Kyser TK (1996) The significance of hydrogen and oxygen stable isotopic fractionations between carnallite and brine at low temperature: Experimental and empirical results. Geochim Cosmochim Acta 60:2721-2726

Kohn MJ, Valley JW (1998a) Oxygen isotope geochemistry of amphiboles: isotope effects of cation substitutions in minerals. Geochim Cosmochim Acta 62:1947-1958

Kohn MJ, Valley JW (1998b) Effects of cation substitution in garnet and pyroxene on equilibrium oxygen isotope fractionations. J Metam Geol 16:625-639

Kohn MJ, Valley JW (1998c) Obtaining equilibrium oxygen isotope fractionations from rocks: theory and example. Contrib Mineral Petrol 132:209-224

Kotzer TG, Kyser TK, King RW, Kerrich R (1993) An empirical oxygen- and hydrogen-isotope geothermometer for quartz-tourmaline and tourmaline-water. Geochim Cosmochim Acta 57:3421-3426

Kulla JB (1979) Oxygen and hydrogen isotope fractionation factors determined in clay-water systems. PhD Dissertation, University of Illinois at Urbana-Champaign

Kulla JB, Anderson TF (1978) Experimental oxygen isotope fractionation between kaolinite and water. *In* Short Papers of the 4th International Congress, Geochronology, Cosmochronology, Isotope Geology. RE Zartman (ed) U S Geol Surv Open file Report 78-70, p 234-235

Kuroda Y, Hariya Y, Suzuoki T, Matsuo S (1982) D/H fractionation between water and the melts of quartz, K-feldspar, albite and anorthite at high temperature and pressure. Geochem J 16:73-78

Kusakabe M, Robinson BW (1977) Oxygen and sulfur isotope equilibria in the $BaSO_4$-HSO_4^--H_2O system from 110 to 350°C and applications. Geochim Cosmochim Acta 41:1033-1040

Kyser TK (1987) Equilibrium fractionation factors for stable isotopes. *In* Kyser TK (ed) Stable Isotope Geochemistry of Low Temperature Fluids. Mineral Assoc Canada Short Course 13:1-84

Labeyrie L (1974) New approach to surface seawater paleotemperatures using $^{18}O/^{16}O$ ratios in silica of diatom frustules. Nature 248:40-41

Lambert SJ, Epstein S (1980) Stable isotope investigations of an active geothermal system in Valles Caldera, Jemez Mountains, New Mexico. J Volcan Geotherm Res 8:111-129.

Lawrence JR, Taylor HP Jr (1971) Deuterium and oxygen-18 correlation: Clay minerals and hydroxides in Quaternary soils compared to meteoric waters. Geochim Cosmochim Acta 35:993-1003

Lawrence JR, Taylor HP Jr (1972) Hydrogen and oxygen isotope systematics in weathering profiles. Geochim Cosmochim Acta 36:1377-1393

Lecuyer C, Grandjean P, Sheppard SMF (1999) Oxygen isotope exchange between dissolved phosphate and water at temperatures 135°C: Inorganic versus biological fractionations. Geochim Cosmochim Acta 63:855-862

Lehmann M, Siegenthaler U (1991) Equilibrium oxygen- and hydrogen-isotope fractionation between ice and water. J Glaciol 37:23-26

Lesniak PM, Sakai H (1989) Carbon isotope fractionation between dissolved carbonate (CO_3^{2-}) and $CO_2(g)$ at 25° and 40°C. Earth Planet Sci Lett 95:297-301

Lichtenstein U, Hoernes S (1992) Oxygen isotope fractionation between grossular-spessartine garnet and water: an experimental investigation. Eur J Mineral 4:239-249

Liu K-K, Epstein S (1984) The hydrogen isotope fractionation between kaolinite and water. Isotope Geosci 2:335-350

Lloyd RM (1968) Oxygen isotope behavior in the sulfate-water system. J Geophys Res 73:6099-6110

Lowenstam HA (1962) Magnetite in denticle capping in recent chitons (polyplacophora). Geol Soc Am Bull 73:435

Majoube M (1971a) Fractionnement en ^{18}O entre la glace et la vapeur d'eau. J Chim Phys 68:625-636

Majoube M (1971b) Fractionnement en oxygene 18 et en deuterium entre l'eau et sa vapeur. J Chim Phys 68:1423-1436

Majzoub M (1966) Une methode d'analyse isotopique de l'oxygene sur des microquantites d'eau determination des coefficients de partage a l'equilibre de l'oxygene 18 entre H_2O et CO_2, D_2O et CO_2. J Chim Phys 63:563-568

Malinin SD, Kropotiva OI, Grinenko VA (1967) Experimental determination of equilibrium constants for carbon isotope exchange in the system $CO_2(gas)$-$HCO_3^-(sol)$ under hydrothermal conditions. Geochem Int'l 4:764-771

Mandernack KW, Bazylinski DA, Shanks WC III, Bullen TD (1999) Oxygen and iron isotope studies of magnetite produced by magnetotactic bacteria. Science 285:1892-1896

Marumo K, Nagasawa K, Kuroda Y (1980) Mineralogy and hydrogen isotope composition of clay minerals in the Ohnuma geothermal area, northeastern Japan. Earth Planet Sci Lett 47:255-262

Matsubaya O, Sakai H (1973) Oxygen and hydrogen isotopic study on the water of crystallization of gypsum from the Kuroko type mineralization. Geochem J 7:153-165

Matsuhisa Y, Matsubaya O, Sakai H (1971) BrF_5 technique for the oxygen isotopic analysis of silicates and water. Mass Spectrometer 19:124-133

Matsuhisa Y, Goldsmith JR, Clayton RN (1979) Oxygen isotopic fractionation in the system quartz-albite-anorthite-water. Geochim Cosmochim Acta 43:1131-1140

Matsuo S, Friedman I, Smith GI (1972) Studies of Quaternary saline lakes. I. Hydrogen isotope fractionation in saline minerals. Geochim Cosmochim Acta 36:427-435

Mattey DP (1991) Carbon dioxide solubility and carbon isotope fractionation in basaltic melt. Geochim Cosmochim Acta 55:3467-3473

Mattey DP, Taylor WR, Green DH, Pillinger CT (1990) Carbon isotopic fractionation between CO_2 vapour, silicate and carbonate melts: An experimental study to 30 kbar. Contrib Mineral Petrol 104:492-505

Matthews A (1994) Oxygen isotope geothermometers for metamorphic rocks. J Metam Geol 12:211-219

Matthews A, Katz A (1977) Oxygen isotope fractionation during the dolomitization of calcium carbonate. Geochim Cosmochim Acta 41:1431-1438

Matthews A, Schliestedt M (1984) Evolution of blueschist and greenschist rocks of Sifnos, Cylades, Greece. Contrib Mineral Petrol 88:150-163

Matthews A, Beckinsale RD, Durham JJ (1979) Oxygen isotope fractionation between rutile and water and geothermometry of metamorphic eclogites. Mineral Mag 43:405-413

Matthews A, Goldsmith JR, Clayton RN (1983a) Oxygen isotope fractionations involving pyroxenes: The calibration of mineral-pair geothermometers. Geochim Cosmochim Acta 47:631-644

Matthews A, Goldsmith JR, Clayton RN (1983b) On the mechanisms and kinetics of oxygen isotope exchange in quartz and feldspars at elevated temperatures and pressures. Geol Soc Am Bull 94:396-412

Matthews A, Goldsmith JR, Clayton RN (1983c) Oxygen isotope fractionation between zoisite and water. Geochim Cosmochim Acta 47:645-654

Matthews A, Palin, JM, Epstein S, Stolper EM (1994) Experimental study of $^{18}O/^{16}O$ partitioning between crystalline albite, albitic glass, and CO_2 gas. Geochim Cosmochim Acta 58:5255-5266

McCrea JM (1950) On the isotopic chemistry of carbonates and a paleotemperature scale. J Chem Phys 18:849-857

McMillan P (1985) Vibrational spectroscopy in the mineral sciences. *In* Kieffer SW, Navrotsky A (eds) Microscopic to Macroscopic—Atomic Environments to Mineral Thermodynamics. Rev Mineral 14:9-63

Melchiorre EB, Criss RE, Rose TP (1999) Oxygen and carbon isotope study of natural and synthetic malachite. Econ Geol 94:245-259

Melchiorre EB, Criss RE, Rose TP (2000) Oxygen and carbon isotope study of natural and synthetic azurite. Econ Geol 95:621-628.

Melchiorre EB, Williams PA, Bevins RE (2001) A low temperature oxygen isotope thermometer for cerussite, with applications at Broken Hill, New South Wales, Australia. Geochim Cosmochim Acta 65:2527-2533

Merlivat L, Nief G (1967) Fractionnement isotopique lors de changements d'etat solide-vapeur et liquide-vapeur de l'eau a des temperatures inferieures a 0°C. Tellus 19:122-127

Mineev SD, Grinenko VA (1996) The pressure influence on hydrogen isotopes fractionation in the serpentine-water system. V M Goldschmidt Conf Abstr 1:404

Mizutani Y, Rafter TA (1969) Oxygen isotopic composition of sulfate-Part 3. Oxygen isotopic fractionation in the bisulfate ion-water system. New Zealand J Sci 12:54-59

Mook WG, Bommerson JC, Staverman WH (1974) Carbon isotope fractionation between dissolved bicarbonate and gaseous carbon dioxide. Earth Planet Sci Lett 22:169-176

Morikiyo T (1984) Carbon isotopic study on coexisting calcite and graphite in the Ryoke metamorphic rocks, northern Kiso district. Contrib Mineral Petrol 87:251-259

Morse PM (1929) Diatomic molecules according to the wave mechanics. II. Vibrational levels. Phys Rev 34:57-64

Müller J (1995) Oxygen isotopes in iron (III) oxides: A new preparation line; mineral-water fractionation factors and paleo-environmental considerations. Isotopes Environ Health Stud 31:301-302

Nahr T, Botz R, Bohrmann G, Schmidt M (1998) Oxygen isotopic composition of low-temperature authigenic clinoptilolite. Earth Planet Sci Lett 160:369-381

Northrop DA, Clayton RN (1966) Oxygen-isotope fractionations in systems containing dolomite. J Geol 74:174-196

Noto M, Kusakabe M (1997) An experimental study of oxygen isotope fractionation between wairakite and water. Geochim Cosmochim Acta 61:2083-2093

Ohmoto H (1986) Stable isotope geochemistry of ore deposits. *In* Valley JW, Taylor HP Jr, O'Neil JR (eds) Stable Isotopes in High Temperature Geological Processes. Rev Mineral 16:491-559

O'Neil, JR (1963) Oxygen isotope fractionation studies in mineral systems. PhD Dissertation, University of Chicago

O'Neil JR (1968) Hydrogen and oxygen isotope fractionation between ice and water. J Phys Chem 723:683-3684.

O'Neil JR (1986) Theoretical and experimental aspects of isotopic fractionation. *In* Valley JW, Taylor HP Jr, O'Neil JR (eds) Stable Isotopes in High Temperature Geological Processes. Rev Mineral 16:1-40

O'Neil JR, Clayton RN (1964) Oxygen isotope geothermometry. *In* Isotope and Cosmic Chemistry (eds., H. Craig et al.) North-Holland, Amsterdam, p157-168

O'Neil JR, Epstein S (1966) A method for oxygen isotope analysis of milligram quantities of water and some of its applications. J Geophys Res 71:4955-4961

O'Neil JR, Taylor HP Jr (1967) The oxygen isotope and cation exchange chemistry of feldspars. J Geophys Res 74:6012-6022

O'Neil JR, Taylor HP Jr (1969) Oxygen isotope equilibrium between muscovite and water. Am Mineral 52:1414-1437

O'Neil JR, Barnes I (1971) C^{13} and O^{18} compositions in some fresh-water carbonates associated with ultramafic rocks and serpentines: western United States. Geochim Cosmochim Acta 35:687-697

O'Neil JR, Truesdell AH (1991) Oxygen isotope fractionation studies of solute-water interactions. *In* Taylor HP Jr, O'Neil JR, Kaplan IR (eds) Stable Isotope Geochemistry: A Tribute to Samuel Epstein, Geochem Soc Spec Pub 3:17-25

O'Neil JR, Clayton RN, Mayeda TK (1969) Oxygen isotope fractionation in divalent metal carbonates. J Chem Phys 51:5547-5558

O'Neil JR, Adami LH, Epstein S (1975) Revised value for the ^{18}O fractionation between CO_2 and water at 25°C. U S Geol Surv J Res 3:623-624

Palin JM, Epstein S, Stolper EM (1996) Oxygen isotope partitioning between rhyolitic glass/melt and CO_2: An experimental study at 550-950°C and 1 bar. Geochim Cosmochim Acta 60:1963-1973

Patel A, Price GD, Mendelssohn MJ (1991) A computer simulation approach to modelling the structure, thermodynamics and oxygen isotope equilibria of silicates. Phys Chem Minerals 17:690-699

Pineau F, Shilobreeva, S, Kadik A, Javoy M (1998) Water solubility and D/H fractionation in the system basaltic andesite-H_2O at 1250°C and between 0.5 and 3 kbars. Chem Geol 147:173-184

Polyakov VB (1997) Equilibrium fractionation of iron isotopes: estimation from Mössbauer spectroscopy data. Geochim Cosmochim Acta 61:4213-4217

Polyakov VB (1998) On anharmonic and pressure corrections to the equilibrium isotopic constants for minerals, Geochim Cosmochim Acta 62:3077-3088

Polyakov VB, Kharlashina NN (1994) Effect of pressure on equilibrium isotopic fractionation. Geochim Cosmochim Acta 58:4739-4750

Polyakov VB, Kharlashina NN (1995) The use of heat capacity data to calculate carbon dioxide fractionation between graphite, diamond, and carbon dioxide: A new approach. Geochim Cosmochim Acta 59:2561-2572

Polyakov VB, Mineev SD (2000) The use of Mössbauer spectroscopy in stable isotope geochemistry. Geochim Cosmochim Acta 64:849-865

Poulson SR, Schoonen MAA (1994) Variations of the oxygen isotope fractionation between $NaCO_3^-$ and water due to the presence of NaCl at 100-300°C. Chem Geol (Isotope Geosci Sec) 116:305-315

Pradahananga TM, Matsuo S (1985a) Intracrystalline site preference of hydrogen isotopes in borax. J Phys Chem 89:72-76

Pradahananga TM, Matsuo S (1985b) D/H fractionation in sulfate hydrate-water systems. J Phys Chem 89:1869-1872

Richet P, Bottinga Y, Javoy M (1977) A review of hydrogen, carbon, nitrogen, oxygen, sulphur, and chlorine stable isotope fractionation among gaseous molecules. Ann Rev Earth Planet Sci 5:65-110

Richet P, Roux J, Pineau F (1986) Hydrogen isotope fractionation in the system H_2O-liquid $NaAlSi_3O_8$: new data and comments on D/H fractionation in hydrothermal experiments. Earth Planet Sci Lett 78:115-120

Richter R, Hoernes S (1988) The application of the increment method in comparison with experimentally derived and calculated O-isotope fractionations. Chem Erde 48:1-18

Rolston JH, Hartog J den, Butler JP (1976) The deuterium isotope separation factor between hydrogen and liquid water. J Phys Chem 80:1064-1067

Romanek CS, Grossman EL, Morse JW (1992) Carbon isotopic fractionation in synthetic aragonite and calcite: Effects of temperature and precipitation rate. Geochim Cosmochim Acta 56:419-430

Rosenbaum JM (1993) Room temperature oxygen isotope exchange between liquid CO_2 and H_2O. Geochim Cosmochim Acta 57:3195-3198

Rosenbaum JM (1994) Stable isotope exchange between carbon dioxide and calcite at 900°C. Geochim Cosmochim Acta 58:3747-3753

Rosenbaum JM (1997) Gaseous, liquid and supercritical H_2O and CO_2: oxygen isotope fractionation behavior. Geochim Cosmochim Acta 61:4993-5003

Rosenbaum JM, Mattey DP (1995) Equilibrium garnet-calcite oxygen isotope fractionation. Geochim Cosmochim Acta 59:2839-2842

Rosenbaum JM, Kyser TK, Walker D (1994) High temperature oxygen isotope fractionation in the enstatite-olivine-$BaCO_3$ system. Geochim Cosmochim Acta 58:2653-2660

Rowe MW, Clayton RN, Mayeda TK (1994) Oxygen isotopes in separated components of CI and CM meteorites. Geochim Cosmochim Acta 58:5341-5347

Rubinson M, Clayton RN (1969) Carbon-13 fractionation between aragonite and water. Geochim Cosmochim Acta 33:997-1002

Rye RO, Stoffregen RE (1995) Jarosite-water oxygen and hydrogen isotope fractionations: Preliminary experimental data. Econ Geol 90:2336-2342

Rye RO, Bethke PM, Wasserman MD (1992) The stable isotope geochemistry of acid-sulfate alteration. Econ Geol 87:240-262

Saccocia PJ, Seewald JS, Shank WC III (1998) Hydrogen and oxygen isotope fractionation between brucite and aqueous NaCl solutions from 250-450°C. Geochim Cosmochim Acta 62:485-492

Sakai H, Tsutsumi M (1978) D/H fractionation factors between serpentine and water at 10° to 500°C and 2000 bar water pressure, and the D/H ratios of natural serpentine. Earth Planet Sci Lett 40:231-242

Satake H, Matsuo S (1984) Hydrogen isotopic fractionation factor between brucite and water in the temperature range from 100 to 510°C. Contrib Mineral Petrol 86:19-24

Sato RK, McMillan PF (1987) Infrared spectra of the isotopic species of alpha quartz. J Phys Chem 91:3494-3498

Savin SM, Lee M (1988) Isotopic studies of phyllosilicates. *In* Bailey SW (ed) Hydrous Phyllosilicates (exclusive of micas). Rev Mineral 19:189-219

Scheele N, Hoefs J (1992) Carbon isotope fractionation between calcite, graphite and CO_2: An experimental study. Contrib Mineral Petrol 112:35-45

Schauble E, Rossman GR, Taylor, HP (2001) Theoretical estimates of equilibrium Fe-isotope fractionations from vibrational spectroscopy. Geochim Cosmochim Acta 65:2487-2497

Schütze H (1980) Der Isotopenindex—eine Inkrementenmethode zur näherungsweisen Berechnung von Isotopenaustauschgleichgewichten zwischen kristallinen Substanzen. Chem Erde 39:321-334

Schwarcz HP (1966) Oxygen isotope fractionation between host and exsolved phases in perthite. Geol Soc Am Bull 77:879-882

Sharp ZD (1995) Oxygen isotope geochemistry of the Al_2SiO_5 polymorphs. Am J Sci 295:1-19

Sharp ZD, Kirschner DL (1994) Quartz-calcite oxygen isotope thermometry: a calibration based on natural isotopic variations. Geochim Cosmochim Acta 58:4491-4501

Sheppard SMF, Schwarcz HP (1970) Fractionation of carbon and oxygen isotopes and magnesium between coexisting metamorphic calcite and dolomite. Contrib Mineral Petrol 26:161-198

Sheppard SMF, Gilg HA (1996) Stable isotope geochemistry of clay minerals. Clay Minerals 31:1-24

Sheppard SMF, Nielsen RL, Taylor HP Jr (1969) Oxygen and hydrogen isotope ratios of clay minerals from porphyry copper deposits. Econ Geol 64:755-777

Shilobreyeva SN, Devirts AL, Kadik AA, Lagutina YP (1992) Distribution of hydrogen isotopes in basalt liquid-water equilibrium at 3 kbar and 1250°C. Geochem Int'l 29:130-134

Shmulovich K, Landwehr D, Simon K, Heinrich W (1999) Stable isotope fractionation between liquid and vapor in water-salt systems up to 600°C. Chem Geol 157:343-354

Smyth JR (1989) Electrostatic characterization of oxygen sites in minerals. Geochim Cosmochim Acta 53:1101-1110

Smyth JR, Clayton RN (1988) Correlation of oxygen isotope fractionations and electrostatic site potentials in silicates. EOS Trans Am Geophys Union 69:1514

Sofer Z (1978) Isotopic composition hydration water in gypsum. Geochim Cosmochim Acta 42:1141-1149

Sofer Z, Gat JR (1972) Activities and concentrations of oxygen-18 in concentrated aqueous salt solutions: Analytical and geophysical implications. Earth Planet Sci Lett 15:232-238

Sofer Z, Gat JR (1975) The isotope composition of evaporating brines: Effects of the isotopic activity ratio in saline solutions. Earth Planet Sci Lett 26:179-186

Sommer MA, Rye D (1978) Oxygen and carbon isotope internal thermometry using benthic calcite and aragonite foraminifera pairs. Short Papers, 4th Int'l Conf. Geochron Cosmochron Isotope Geol, U S Geol Surv Open File Rep. 78-701, p 408-410

Spindel W, Stern MJ, Monse EU (1970) Further study on temperature-dependences of isotope effects. J Chem Phys 52:2022-2035

Staschewski D (1964) Experimentelle bestimmung der O^{18}/O^{16}-trennfaktoren in den systemen CO_2/H_2O und CO_2/D_2O. Berichte Bunsengesell 68:454-459

Stern MJ, Spindel W, Monse EU (1968) Temperature-dependences of isotope effects. J Chem Phys 48:2908-2919

Stewart MK (1974) Hydrogen and oxygen isotope fractionation during crystallization of mirabilite and ice. Geochim Cosmochim Acta 38:167-172

Stewart MK, Friedman I (1975) Deuterium fractionation between aqueous salt solutions and water vapor. J Geophys Res 80:3812-3818

Stoffregen RE, Rye RO, Wasserman MD (1994) Experimental studies of alunite: I. ^{18}O-^{16}O and D-H fractionation factors between alunite and water at 250-450°C. Geochim Cosmochim Acta 58:903-916

Stolper E, Epstein S (1991) An experimental study of oxygen isotope partitioning between silica glass and CO_2 vapor. *In* Taylor HP Jr, O'Neil JR, Kaplan IR (eds) Stable Isotope Geochemistry: A Tribute to Samuel Epstein. Geochem Soc Spec Pub 3:35-51

Suess, VH (1949) Das gleichgewicht H_2 + HDO = HD + H_2O und die weiteren Austauschgleichgewichte im System H_2, D_2, und H_2O. Z Naturforschung 4a:328-332

Sushchevskaya TM, Ustinov VI, Nekrasov IY, Gavrilov YY, Grinenko VA (1985) The oxygen-isoope fractionation factor in cassiterite synthesis. Geochem Int'l 23:57-60

Suzuoki T, Kimura T (1973) D/H and $^{18}O/^{16}O$ fractionation in ice-water system. Mass Spectrom 21:229-233

Suzuoki T, Epstein S (1976) Hydrogen isotope fractionation between OH-bearing minerals and water. Geochim Cosmochim Acta 40:1229-1240

Szaran J (1997) Achievement of carbon isotope equilibrium in the system HCO_3^- (solution)-CO_2 (gas). Chem Geol 142:79-86

Tarutani T, Clayton RN, Mayeda TK (1969) The effect of polymorphism and magnesium substitution on oxygen isotope fractionation between calcium carbonate and water. Geochim Cosmochim Acta 33:987-996

Taube H (1954) Use of oxygen isotope effects in the study of hydration of ions. J Phys Chem 58:523-528

Taylor BE (1976) Origin and significance of C-O-H fluids in the formation of Ca-Fe-Si skarn, Osgood Mountains, Humboldt County, Nevada. PhD Dissertation, Stanford University

Taylor BE, Westrich HR (1985) Hydrogen isotope exchange and water solubility in experiments using natural rhyolite obsidian. EOS Trans Am Geophys Union 66:387.

Taylor HP Jr (1974) The application of oxygen and hydrogen isotope studies to problems of hydrothermal alteration and ore deposition. Econ Geol 69:843-883

Taylor HP Jr, Epstein S (1962) Relationship between O^{18}/O^{16} ratios in coexisting minerals of igneous and metamorphic rocks. Part I. Principles and experimental results. Geol. Soc Am Bull 73:461-480

Tennie A, Hoffbauer R, Hoernes S (1998) The oxygen isotope fractionation behaviour of kyanite in experiment and nature. Contrib Mineral Petrol 133:346-355

Truesdell AH (1974) Oxygen isotope activities and concentrations in aqueous salt solutions at elevated temperatures: Consequences for isotope geochemistry. Earth Planet Sci Lett 23:387-396

Turner JV (1982) Kinetic fractionation of carbon-13 during calcium carbonate precipitation. Geochim Cosmochim Acta 46:1183-1191

Urey HC (1947) The thermodynamic properties of isotopic substances. J Chem Soc (London), p 562-581

Usdowski E, Hoefs J (1993) Oxygen isotope exchange between carbonic acid, bicarbonate, carbonate, and water: A re-examination of the data of McCrea (1950) and an expression for the overall partitioning of oxygen isotopes between the carbonate species and water. Geochim Cosmochim Acta 57:3815-3818

Usdowski E, Michaelis J, Bottcher ME, Hoefs J (1991) Factors for the oxygen isotope equilibrium fractionation between aqueous and gaseous CO_2, carbonic acid, bicarbonate, carbonate, and water (19°C). Z Phys Chemie 170:237-249

Ustinov VI, Grinenko VA (1990) Determining isotope-equilibrium constants for mineral assemblages. Geochem Int'l 27 (10):1-9

Valley JW, O'Neil JR (1981) $^{13}C/^{12}C$ exchange between calcite and graphite: a possible thermometer in Grenville marbles. Geochim Cosmochim Acta 45:411-419

Valley JW, Taylor HP Jr, O'Neil JR (eds) (1986) Stable Isotopes in High Temperature Geological Processes. Rev Mineral, Vol 16

Valley JW, Chiarenzelli JR, McLelland JM (1994) Oxygen isotope geochemistry of zircon. Earth Planet Sci Lett 126:187-206

Van Hook WA (1975) Condensed phase isotope effects, especially vapor pressure isotope effects: aqueous solutions. In Rock PA (ed) Isotopes and Chemical Principles. Am Chem Soc Symp 11:101-130.

Vennemann TW, O'Neil JR (1996) Hydrogen isotope exchange between hydrous minerals and molecular hydrogen: I. A new approach for the determination of hydrogen isotope fractionation at moderate temperature. Geochim Cosmochim Acta 60:2437-2451

Vitali F, Longstaffe FJ, Bird MI, Caldwell WGE (2000) Oxygen-isotope fractionation between aluminum-hydroxide phases and water at <60°C: Results of decade-long synthesis experiments. Clays & Clay Minerals 48:230-237

Vitali F, Longstaffe FJ, Bird MI, Gage KL, Caldwell WGE (2001) Hydrogen-isotope fractionation in aluminum hydroxides: Synthesis products versus natural samples from bauxites. Geochim Cosmochim Acta 65:1391-1398

Vogel JC, Grootes PM, Mook WG (1970) Isotopic fractionation between gaseous and dissolved carbon dioxide. Z Phys 230:225-238

Wada H, Suzuki H (1983) Carbon isotopic thermometry calibrated by dolomite-calcite solvus temperatures. Geochim Cosmochim Acta 47:697-706.

Wendt I (1968) Fractionation of carbon isotopes and its temperature-dependence in the system CO_2-gas-CO_2 in solution and HCO_3-CO_2 in solution. Earth Planet Sci Lett 4:64-68

Wenner DB, Taylor HP Jr (1971) Temperatures of serpentinization of ultramafic rocks based on $^{18}O/^{16}O$ fractionations between coexisting minerals and magnetite. Contrib Mineral Petrol 32:165-185

Wenner DB, Taylor HP Jr (1973) Oxygen and hydrogen isotope studies of the serpentinization of ultramafic rocks in oceanic environments and continental ophiolite complexes. Am J Sci 273:207-239

Williams AE, McKibben MA (1989) A brine interface in the Salton Sea geothermal system, California: Fluid geochemical and isotopic characterization. Geochim Cosmochim Acta 53:1905-1920

Xu B-L, Zheng Y-F (1999) Experimental studies of oxygen and hydrogen isotope fractionations between precipitated brucite and water at low temperatures. Geochim Cosmochim Acta 63:2009-2018

Yapp CJ (1987) Oxygen and hydrogen isotope variations among goethites (FeOOH) and the determination of paleotemperatures. Geochim Cosmochim Acta 51:355-364.

Yapp CJ (1990) Oxygen isotopes in iron (III) oxides 1. Mineral-water fractionation factors. Chem Geol 85:329-335

Yapp CJ, Pedley MD (1985) Stable hydrogen isotopes in iron oxides. II. D/H variations among natural goethites. Geochim Cosmochim Acta 49:487-495

Yaqian Q, Jibao G (1993) Study of hydrogen isotope equilibrium and kinetic fractionation in the ilvaite-water system. Geochim Cosmochim Acta 57:3073-3082

Yeh H-W (1980) D/H ratios and late-stage dehydration of shales during burial. Geochim Cosmochim Acta 44:341-352

Yeh H-W, Savin SM (1977) The mechanism of burial metamorphism of argillaceous sediments: 3. Oxygen isotope evidence. Geol Soc Am Bull 88:1321-1330

Zhang C, Liu S, Phelps TJ, Cole DR, Horita J, Fortier SM, Elless M, Valley JW (1997) Physiochemical, mineralogical, and isotopic characterization of magnetite-rich iron oxides formed by thermophilic iron-reducing bacteria. Geochim Cosmochim Acta 61:4621-4632

Zhang CL, Horita J, Cole DR, Zhou J, Lovley DR, Phelps TJ (2001) Temperature-dependent oxygen and carbon isotope fractionations of biogenic siderite. Geochim Cosmochim Acta 65:2257-2271

Zhang L, Liu J, Zhou H, Chen Z (1989) Oxygen isotope fractionation in the quartz-water-salt system. Econ Geol 84:1643-1650

Zhang L, Liu J, Chen Z, Zhou H (1994) Experimental investigations of oxygen isotope fractionation in cassiterite and wolframite. Econ Geol 89:150-157

Zhang J, Quay PD, Wilbur DO (1995) Carbon isotope fractionation during gas-water exchange and dissolution of CO_2. Geochim Cosmochim Acta 59:107-114

Zheng Y-F (1991) Calculation of oxygen isotope fractionation in metal oxides. Geochim Cosmochim Acta 55:2299-2307

Zheng Y-F (1993) Calculation of oxygen isotope fractionation in hydroxyl-bearing silicates. Earth Planet Sci Lett 120:247-263

Zheng Y-F (1996) Oxygen isotope fractionations involving apatites: application to paleotemperature determination. Chem Geol 127:177-187

Zheng Y-F (1997) Prediction of high-temperature oxygen isotope fractionation factors between mantle minerals. Phys Chem Mineral 24:356-364

Zheng Y-F (1999a) Calculation of oxygen isotope fractionation in minerals. Episodes 22:99-106

Zheng Y-F (1999b) Oxygen isotope fractionation in carbonate and sulphate minerals. Geochem J 33:109-126

Zheng Y-F, Metz P, Satir M, Sharp ZD (1994a) An experimental calibration of oxygen isotope fractionation between calcite and forsterite in the presence of a CO_2-H_2O liquid. Chem Geol 116:17-27

Zheng Y-F, Metz P, Satir M (1994b) Oxygen isotope fractionation between calcite and tremolite: An experimental study. Contrib Mineral Petrol 118:249-255

APPENDICES

Appendix 1. Definition of commonly used terms and symbols.

Term	Symbol	Definition
Force (spring) constant	k	spring stiffness
Reduced mass	μ	$m_1 m_2/(m_1+m_2)$ diatomic molecules where m is the mass
Planck's constant	h	6.62608×10^{-34} J-sec
Boltzmann's constant	k_b	1.380658×10^{-23} J/K
Vibrational frequency	ν	$(1/2\pi)(k/\mu)^{0.5}$ (sec^{-1})
Wave number	ω	$\omega = \nu/c$ (cm^{-1})
Velocity of light (vacuum)	c	2.99792×10^{10} cm/sec
Frequency shift	Δν	$\nu - \nu^*$
Frequency shift factor	----	$\nu^*/\nu = \omega^*/\omega$
Zero Point Energy	ZPE	$h\nu/2$
Change in ZPE	ΔZPE	$ZPE - ZPE^* = h\Delta\nu/2$
Symmetry number	σ	Number of equivalent ways to orient a molecule in space
U	U	$h\nu/k_b T$
Number of atoms in a compound	s	s = number of atoms in a gaseous compound (e.g., s=3 for CO_2), or in the unit cell of crystalline solids (e.g., s = 9 in quartz - Si_3O_6)
r	r	Number of atoms of the element being exchanged in the substance of interest (e.g., r = 2 for oxygen isotope exchange in CO_2; r = 6 in Si_3O_6)
Mass of element	m, m*	Atomic mass of element being exchanged
Partition Function Ratio	Q*/Q	Equation (9) – gases Equation (11) – solids
Reduced Partition Function Ratio	f	$(Q^*/Q)(m/m^*)^{3r/2}$
β value (factor)	β	$f^{1/r}$
Fractionation Factor	α or $10^3 \ln\alpha$	$\alpha_{A-B} = \dfrac{R_A}{R_B} = \dfrac{\beta_A}{\beta_B}$ where R is an isotope ratio
Isotope salt effect	Γ	$\Gamma = \dfrac{\alpha_{A\text{-aqueous soln}}}{\alpha_{A\text{-pure water}}}$ where A is a reference phase

The asterisk in all cases denotes a substance made with the heavy isotope.

Appendix 2. Oxygen isotope fractionation factors: calibrations based on experiments or natural samples.

#	Phases	Reference	Method	$1000 \ln \alpha$	T(°C)	Comments
	Gases, Liquid, Ice:					
1	H_2O: ice-vapor	Majoube (1971a)	Ex	$11839/T-28.224$	-33.4 - 0	Slow growth of ice in a stirred water.
2	H_2O: ice-liquid	O'Neil (1968)	Ex	$+3.0$	0?	
3	H_2O: ice-liquid	Craig and Hom (1968)	Ex	$+2.65,+2.70$	0?	Freezing from 4‰ chlorinity and seawater.
4	H_2O: ice-liquid	Suzuki and Kimura (1973)	Ex	$+2.8\pm0.1$	0?	Freezing from 2.5m NaCl soln.
5	H_2O: ice-liquid	Stewart (1974)	Ex	$+2.2$	-10	Slow growth of ice in a stirred water.
6	H_2O: ice-liquid	Lehmann and Siegenthaler (1991)	Ex	$+2.91\pm0.03$	0	Slow growth of ice in a stirred water.
7	H_2O: liquid-vapor	Majoube (1971b)	Ex	$1.137\,(10^3/T^2) - 0.4156$ $(10^3/T) - 2.0667$	0-100	Slow distillation of liquid water.
8	H_2O: liquid-vapor	Horita and Wesolowski (1994), and all references therein	Ex	$-7.685 + 6.7123\,(10^3/T)$ $- 1.6664\,(10^6/T^2)$ $+ 0.35041\,(10^9/T^3)$	0-374	High precision experimental calibration using three different apparatus to cover the full temperature range. This large dataset was combined with selected earlier experimental studies to generate the calibration curve.
9	CO_2:solid-vapor	Eiler et al. (2000)	Ex	$2868/T-14.5$	130-150(K)	By sublimation-condensation and isotopic exchange.
10	CO_2:liquid-vapor	Grootes et al. (1969)	Ex	$+0.06$ to $+1.03$	220-303(K)	Liquid-vapor equilibration for >4 hr.
11	CO_2(liq)- H_2O(l)	Rosenbaum (1993)	Ex	$+41.94\pm0.27$	25.3	Direct exchange at P=100-200 bar.
12	CO_2-H_2O(l)	Compton and Epstein (1958), Staschewski (1964), Mazjoub (1966), O'Neil and Epstein (1966), Bottinga and Craig (1969), Matsuhisa et al. (1971), Blattner (1973), Horibe et al. (1973), O'Neil et al. (1975), Gu (1980), Brenninkmeijer et al. (1983), Dugan et al. (1985), and Dugan and Borthwick (1986)	Ex	$+39.89$ to $+41.53$ $10^3(\alpha-1) =17604/T - 17.93$ (Brenninkmeijer et al., 1983)	5-100	Large number of measurements at room temperature. Brenninkmeijer et al. combined their own and earlier experimental results. Fractionations adjusted such that $\alpha(CO_2$-$H_2O)$ at 25°C = 1.0412.
13	CO_2-H_2O(l)	Truesdell (1974)	Ex	$3.97\,(10^6/T^2) + 0.31$	130-350	Experiments at P=28 bars. Fractionations adjusted to $\alpha(CO_2$-$H_2O)$ at 25°C = 1.0412.
	Dissolved Species:					
14	CO_2 (aq)- H_2O	Usdowski and Hoefs (1993)	Ex	$+56.3$	19	Based on data by McCrea (1950) and Usdowski et al. (1991).
15	H_2CO_3 (aq)- H_2O	Usdowski and Hoefs (1993)	Ex	$+38.7,+38.8$	19, 25	Based on data by McCrea (1950) and Usdowski et al. (1991).
16	HCO_3^- (aq)- H_2O	Usdowski and Hoefs (1993)	Ex	$+34.0,+34.5$	19, 25	Based on data by McCrea (1950) and Usdowski et al. (1991).
17	CO_3^{2-} (aq)- H_2O	Usdowski and Hoefs (1993)	Ex	$+18.1,+18.2$	19, 25	Based on data by McCrea (1950) and Usdowski et al. (1991).
18	HCO_3^- (aq)- H_2O	Halas and Wolacewicz (1982)	Ex	$2.92\,(10^6/T^2) - 2.66$	25-45	0.03m $NaHCO_3$ soln.
19	$NaCO_3^-$ (aq)- H_2O	Poulson and Schoonen (1994)	Ex	$2.7\,(10^6/T^2) - 5.7$	100-300	0.1M Na_2CO_3+0.1M NaOH+0-4M NaCl soln.
20	HSO_4^-(aq)- H_2O	Lloyd (1968)	Ex	$3.251\,(10^6/T^2) - 5.6$	72.5-348	Direct exchange between dissolved Na_2SO_4 and water at pH=4-9.
21	HSO_4^- (aq)- H_2O	Mizutani and Rafter (1969)	Ex	$2.88\,(10^6/T^2) - 4.1$	110-200	Direct exchange between dissolved H_2SO_4 and water.
22	$H_2PO_4^-$(aq)- H_2O	Lecuyer et al. (1999)	Ex	$18.35\,(10^6/T) - 32.29$	75-135	Direct exchange between dissolved KH_2PO_4 and water at pH=5. 28-100% exchange. Extrapolated values more than 10‰ greater than results of biogenic phosphates.

#		Reference	Type	Equation	Temperature (°C)	Comments
	Hydrated Salts:					
23	carnallite(KMgCl₃·6H₂O) – H₂O	Horita (1989)	Ex	+7.5 to +8.8	10-40	By precipitation and aging 30-67 days. Determined on the activity scale of brines.
24	carnallite(KMgCl₃·6H₂O) – H₂O	Koehler and Kyser (1996)	Ex	-0.1 to +1.3	22-45	By synthesis and exchange.
25	bischofite(MgCl₂·6H₂O) – H₂O	Horita (1989)	Ex	+7.6 to +8.2	10-40	By precipitation and aging 30-67 days. Determined on the activity scale of brines.
26	tachyhydrite(CaMg₂Cl₆·12H₂O) – H₂O	Horita (1989)	Ex	+9.5	25	By precipitation and aging 30-67 days. Determined on the activity scale of brines.
27	gypsum (CaSO₄·2H₂O)- H₂O	Gonfiantini and Fontes (1963), and Matsubaya and Sakai (1973), and Sofer (1978)	Ex	+2.9 to +4.1	17-57	Slow precipitation.
28	mirabilite (Na₂SO₄·10H₂O)- H₂O	Stewart (1974)	Ex	+1.4 to +2.0	0-25	Synthesis from aqueous solutions. Determined on the composition scale of brine.
	Carbonates:					
29	aragonite/calcite-H₂O	McCrea (1950)	Ex	$\delta^{18}O=15.7 (10^3/T) - 54.2$ (Florida seawater) $\delta^{18}O=16.4 (10^3/T) - 57.6$ (Cape Cod seawater)	-1.2 to 79.8	Slow precipitation by CO₂ degassing.
30	calcite-H₂O	Epstein et al. (1953)	Mx	$2.73 (10^6/T^2) - 2.71$	7-30	Combination of data obtained from biogenic precipitation of calcite in tank experiments and analysis of natural samples. Regression line fit through data given in Epstein et al. (1953) after recalculation following the method outlined in Tarutani et al. (1969).
31	calcite-H₂O	O'Neil et al. (1969)	Ex	$2.78 (10^6/T^2) – 2.89$ corrected in Friedman and O'Neil (1977)	0-500	200-500°C experiments involved exchange between carbonate minerals and ammonium chloride solutions. P=1 kbar. Experiments at 0 and 25 involved controlled precipitation of carbonate minerals from bicarbonate solutions. P=1 atm. 100% exchange in all experiments.
32	calcite-H₂O	Clayton et al. (1975)	Ex	$1.22 - 1.33$ (500°) $0.01 - 0.10$ (700°C)	500, 700	Pressure effects investigated to 20 kbars in pure water.
33	strontianite (SrCO₃)-H₂O	O'Neil et al. (1969)	Ex	$2.69 (10^6/T^2) - 3.24$ corrected in Friedman and O'Neil (1977)	0-500	As above.
34	witherite (BaCO₃)- H₂O	O'Neil et al. (1969)	Ex	$2.57 (10^6/T^2) - 4.23$ corrected in Friedman and O'Neil (1977)	0-500	As above.
35	cerrusite (PbCO₃) - H₂O	O'Neil et al. (1969)	Ex	4.51, 6.92	240, 201	No equation given.
36	rhodochrosite (MnCO₃) - H₂O	O'Neil et al. (1969)	Ex	6.76	240	As above.
37	otavite (CdCO₃) - H₂O	O'Neil et al. (1969)	Ex	2.95, 5.97	320, 250	As above.

No.	Mineral pair	Reference	Type	Equation	T (°C)	Comments
37	calcite-H_2O	Kim and O'Neil (1997)	Ex	$18.03 (10^3/T) - 32.42$	10-40	Low-T controlled precipitation experiments.
38	witherite ($BaCO_3$) - H_2O	Kim and O'Neil (1997)	Ex	$2.63 (10^6/T^2) - 4.04$	0-500	Low-T controlled precipitation experiments combined with the high-T experiments of O'Neil et al. (1969). Revised acid fractionation factors used in recalculating the isotopic compositions of $BaCO_3$, determined in O'Neil et al. (1969).
39	otavite ($CdCO_3$) - H_2O	Kim and O'Neil (1997)	Ex	$2.76 (10^6/T^2) - 3.96$	0-500	As above.
40	siderite ($FeCO_3$)- H_2O	Carothers et al. (1988)	Ex	$3.13(10^6/T^2) - 3.50$	33-197	Slow addition of $FeCl_2$ soln to $NaHCO_3$ soln.
41	siderite ($FeCO_3$)- H_2O	Zhang et al. (2001)	Ex	$2.56 (10^6/T^2) + 1.69$	45-75	Based on microbial siderite precipitated by thermophilic Fe(III)-reducing bacteria in culture.
42	malachite ($CuCO_3Cu(OH)_2$)- H_2O	Melchiorre et al. (1999)	Ex	$2.66 (10^6/T^2) + 2.66$	0-50	Replacement of calcite with Cu^{2+}-bearing soln.
43	azurite ($(CuCO_3)_2Cu(OH)_2$)- H_2O	Melchiorre et al. (2000)	Ex	$2.67 (10^6/T^2) + 4.75$	10-45	Replacement of calcite with Cu^{2+}-bearing soln.
44	cerussite($PbCO_3$)- H_2O	Melchiorre et al. (2001)	Ex	$2.63 (10^6/T^2) - 3.58$	20-65	Replacement of calcite with Pb^{2+}-bearing soln.
45	norsethite ($BaMg(CO_3)_2$) - H_2O	Bottcher (2000)	Ex	$2.83 (10^6/T^2) - 2.85$	20-90	Formed from $BaCO_3$ and $MgCO_3,3H_2O$ in $NaHCO_3$ soln.
46	hydromagnesite($Mg_4(OH)_4(CO_3)_3$)· H_2O dolomite-H_2O	O'Neil et al. (1971)	Ex	$+31.19, +37.08$	0, 25	Slow precipitation from $Mg(HCO_3)_2$ soln.
47		Clayton et al. (1968)	N	$+34.5$	20±5	Based on sedimentary dolomite from Deep Springs Lake, California.
48	dolomite-H_2O	Northrop and Clayton (1966)	Ex	$3.20 (10^6/T^2) - 2.00$	300-510	Direct exchange. 3-50% exchange. P= 1 kbar.
49	dolomite-H_2O	Matthews and Katz (1977)	Ex	$3.06 (10^6/T^2) - 3.24$	252-295	Hydrothermal dolomitization of calcite or aragonite in the presence of Ca-Mg-Sr chloride solutions. P= 1 atm.
50	protodolomite-H_2O	Fritz and Smith (1970)	Ex	$+23.4$ to $+31.6$ $3.2 (10^6/T^2) - 2.0$	25-78.6	Precipitated from a Ca-Mg-CO_3 soln. The equation is based on extrapolation to the 25°C datum of Clayton et al. (1968).
51	CO_2-calcite	O'Neil and Epstein (1966)	Ex	$1.93 (10^6/T^2) + 3.92$	350-610	Direct exchange experiments using very large calcite to CO_2 ratio. Measured values may represent surface fractionations rather than true equilibrium fractionations. P=0.3 bars.
52	CO_2-dolomite	O'Neil and Epstein (1966)	Ex	$1.31 (10^6/T^2) + 3.62$	350-610	As above.
53	CO_2-calcite	Chacko et al. (1991)	Ex	$-0.038435 + 5.0077x - 1.0703x^2 + 0.15452x^3 - 0.014366x^4 + 0.00073624x^5 - 0.000015567x^6$, where $x=10^6/T^2$	400-800	Direct exchange. 39-94% exchange. P=10 kbar. Equation represents theoretical calculations that closely fit the experimental data. The equation reproduces the calculated fractionations from 273-4000K.
54	CO_2-calcite	Scheele and Hoefs (1992)	Ex	$5.92-2.31$	500-1200	Aragonite starting material inverted to calcite during the experiment. No equation given but fractionations generally larger than those given by Chacko et al. (1991) by -0.5‰.

#		Reference		Equation	T range	Notes
55	CO_2-calcite	Rosenbaum (1994)	Ex	3.30	900	Direct exchange. 97% exchange. P=12.5 kbar.
56	aragonite-calcite	Tarutani et al. (1969)	Ex	0.6	25	Combination of data from experiments in which calcite or aragonite were slowly precipitated from aqueous bicarbonate solutions.
57	aragonite-calcite	Grossman and Ku (1986)	N	$0.76 - 0.017T(°C)$	0-25	Based on the analysis of aragonitic foraminifer *Hogelundia* and coexisting calcitic foraminifer *Uvigerina*. Large scatter about the regression line indicates that the apparent temperature dependence is not statistically significant.
58	dolomite-calcite	Epstein et al. (1963)	Ex	+0.9	550	Dolomitization of natural calcite with $CaCl_2$ soln
59	dolomite-calcite	Northrop and Clayton (1966)	Ex	$0.50 (10^6/T^2)$	300-510	Combination of dolomite-H_2O experiments of Northrop and Clayton (1966) and calcite-H_2O experiment of O'Neil (1963).
60	dolomite-calcite	O'Neil and Epstein (1966)	Ex	$0.56 (10^6/T^2) + 0.45$	350-610	Combination of CO_2-calcite and CO_2-dolomite experiments.
61	dolomite-calcite	Sheppard and Schwarz (1970)	N	$0.45 (10^6/T^2) - 0.40$	100-650	Based on analysis of co-existing calcite-dolomite pairs in regionally metamorphosed marbles and calcareous schists. Temperatures derived from calcite-dolomite solvus thermometry.
Silica Group:						
62	quartz-calcite	Clayton et al. (1989)	Ex	$0.38 (10^6/T^2)$	600-1000	Direct exchange. 72-99% exchange. P=15 kbar.
63	quartz-calcite	Sharp and Kirschner (1994)	N	$0.87 (10^6/T^2)$	100-700	Primarily based on analyses of low-grade marbles and veins.
64	quartz-H_2O	Clayton et al. (1972)	Ex	$2.51 (10^6/T^2) - 1.96$	500-750	Direct exchange. 100% exchange. P=1 kbar.
65	quartz- H_2O	Bottinga and Javoy (1973)	Mx	$4.10 (10^6/T^2) - 3.7$	500-800	Based on selected experimental data from Clayton et al. (1972), theoretical considerations and data from natural samples.
66	quartz-H_2O	Matsuhisa et al. (1979)	Ex	$2.05 (10^6/T^2) - 1.14$	500-800	Direct exchange. 87-100% exchange. P=15 kbar.
67	quartz-H_2O	Matsuhisa et al. (1979)	Ex	$3.34 (10^6/T^2) - 3.31$	250-500	Direct exchange. 36-87% exchange. P=15 kbar.
68	quartz-H_2O	Mathews et al. (1983b)	Ex	1.44, 3.99, 9.12	600, 400, 250	Three isotope technique. No equation given but data fits equation of Matsuhisa et al (1979).
69	quartz-H_2O	Zhang et al. (1989)	Ex	$3.306 (10^6/T^2) - 2.71$	180-550	Conversion of silica gel to quartz in up to 40wt-% NaCl, NaF, and KCl. Little salt effect observed.
70	amorphous silica-H_2O	Kita et al. (1985)	Ex	$3.52 (10^6/T^2) - 4.35$	34-93	Based on the analysis of amorphous silica precipitating in geothermal power plant waters.
71	biogenic silica-H_2O	Labeyrie (1974)	N	$41.2 - 0.25T(°C)$	4-27	Based on the analysis of sponge spicules and diatoms formed under known conditions.
72	biogenic silica-water	Brandriss et al. (1998)	Ex	$15.56 (10^3/T) - 20.92$	3.6-20.0	Based on cultured fresh water diatoms. 3-4‰ smaller than Labeyrie (1974).
73	quartz-microcline	Blattner and Bird (1974)	Ex	1.8	600	Quartz and alkali feldspar directly exchanged in the presence of a common water or KCl solution.
74	quartz-cassiterite(SnO_2)	Zhang et al. (1994)	Ex	+8.5, +6.9	400, 500	Reaction between silica gel and amorphous SnO_2 in the presence of water.
Feldspars:						
75	albite-calcite	Clayton et al. (1989)	Ex	$-0.56 (10^6/T^2)$	600-800	Direct exchange. 88-100% exchange. P=11-16 kbar
76	albite-H_2O	O'Neil and Taylor (1967)	Ex	$2.91 (10^6/T^2) - 3.41$	350-800	Alkali exchange with aqueous chloride solutions. Found no difference in the fractionation behavior of Na- and K-

#	Mineral pair	Reference		Equation	T (°C)	Comments
77	albite-H_2O	Bottinga and Javoy (1973)	Mx	$3.13\ (10^6/T^2) - 3.7$	500-800	feldspar. 100% exchange. P=1 kbar. Based on experimental data from O'Neil and Taylor (1967), theoretical considerations and data from natural samples.
78	albite-H_2O	Matsuhisa et al. (1979)	Ex	$1.59\ (10^6/T^2) - 1.16$	500-700	Direct exchange. 79-100% exchange. P=7-15 kbar. P=7-12 kbar.
79	albite-H_2O	Matsuhisa et al. (1979)	Ex	$2.39\ \text{x}10^6/T^2 - 2.51$	400-500	Direct exchange. 50-79% exchange. P=12 kbar.
80	albite-H_2O	Matthews et al. (1983b)	Ex	$0.95, 1.38$	600, 500	Three isotope technique. No equation given but data fits equation of Matsuhisa et al (1979). P=15 kbar.
81	anorthite-calcite	Clayton et al. (1989)	Ex	$-1.59\ (10^6/T^2)$	600-800	Direct exchange. 95-100% exchange. P=9-12 kbar.
82	anorthite-H_2O	O'Neil and Taylor (1967)	Ex	$2.15\ (10^6/T^2) - 3.82$	500-800	Exchange of Ba feldspar or anorthite with $CaCl_2$ solutions. 96-100% exchange. P=1 kbar.
83	anorthite-H_2O	Bottinga and Javoy (1973)	Mx	$2.09\ (10^6/T^2)^2 - 3.70$	500-800	Based on experimental data from O'Neil and Taylor (1967), theoretical considerations and data from natural samples.
84	anorthite-H_2O	Matsuhisa et al. (1979)	Ex	$1.04\ (10^6/T^2) - 2.01$	500-750	Direct exchange. 100% exchange. P=4-10 kbar.
85	anorthite-H_2O	Matsuhisa et al. (1979)	Ex	$1.49\ (10^6/T^2) - 2.81$	400-500	Direct exchange. 95-100% exchange. P=2-4 kbar.
86	anorthite-H_2O	Matthews et al. (1983b)	Ex	-1.31	600	Three isotope technique. Less than 50% exchange. P=7 kbar.
Olivine:						
87	forsterite-calcite	Chiba et al. (1989)	Ex	$-3.29\ (10^6/T^2)$	700-1300	Direct exchange. 61-100% exchange. P=15-16 kbar.
88	forsterite-calcite	Zheng et al. (1994a)	Ex	$-3.17\ (10^6/T^2) - 0.44$	600-900	Exchange of forsterite and calcite in the presence of a CO_2-H_2O fluid. 60-90% exchange. P=3-12 kbar.
89	forsterite-$BaCO_3$	Rosenbaum et al. (1994)	Ex	$-1.95, -1.7, -2.12, -1.32, -1.16$	1009-1409 in 100°C increments	No equation given but all points are similar to forsterite-calcite fractionations of Chiba et al. (1989) except the 1209°C datum.
Garnet:						
90	grossular/andradite-calcite	Rosenbaum and Mattey (1995)	Ex	$-2.77\ (10^6/T^2)$	800-1200	Direct exchange between grossular$_{0.7}$ andradite$_{0.19}$ pyrope$_{0.03}$ garnet and calcite. 60-97% exchange. P=23 kbar.
91	grossular-H_2O	Taylor (1976)	Ex	-1.6	600	Hydrothermal synthesis of grossular from gel. P= 2 kbar.
92	andradite-H_2O	Taylor (1976)	Ex	-3.28	600	Hydrothermal synthesis of andradite from gel. P= 2 kbar.
93	grossular/spessartine-H_2O	Lichtenstein and Hoernes (1992)	Ex	-2.1	750	Direct exchange between grossular and spessartine rich garnet and water. Both sets of experiments yielded nearly identical fractionation factors. 56-59% exchange. P=16 kbar. Synthesis experiments at 750°C yield a spessartine-H_2O fractionation of -2.5.
94	grossular-H_2O	Matthews (1994)	Ex	$-1.77, -1.51$	700, 800	Direct exchange of grossular with 0.86M NaF solution. 90-92% exchange. P=1.6-2.0 kbar. No equation given.
95	grossular-andradite	Kohn and Valley (1998b)	N	$0.6\text{ to }0.8$	727	Based on analysis of coexisting garnet-wollastonite pairs in granulite-facies calc-silicates.
96	garnet-quartz	Bottinga and Javoy (1975)	N	$-2.88\ (10^6/T^2)$	>500	Based on quartz-garnet data on natural samples where temperatures for those samples were determined by the Bottinga and Javoy (1973) calibrations of the feldspar-quartz, feldspar-muscovite or feldspar-magnetite isotope thermometers.

			N	0		
97	garnet-zircon	Valley et al. (1994)	N		800-1000	Based on analysis of garnet-zircon pairs from the Adirondack mountains.
Aluminosilicate						
98	kyanite-calcite	Tennie et al. (1998)	Ex	$-2.62\ (10^6/T^2)$	625-725	Isotopic exchange induced by polymorphic inversion of andalusite to kyanite in the presence of calcite. P=13 kbar.
99	kyanite-quartz	Sharp (1995)	N	$-2.17\ (10^6/T^2)$	535-1300	Based on the analysis of coexisting quartz, kyanite and garnet and an assumed $\Delta(\text{qtz-grt}) = 3.1\times10^6/T^2$.
100	sillimanite-quartz	Sharp (1995)	N	$-2.36\ (10^6/T^2)$	535-1300	Based on the analysis of coexisting quartz, sillimanite and garnet and an assumed $\Delta(\text{qtz-grt}) = 3.1\times10^6/T^2$.
Sorosilicates						
101	zoisite-H_2O	Matthews et al. (1983c)	Ex	0.49, 0.20, -0.33, -0.5, -0.62	400, 450, 500, 600, 700	Direct exchange at 600 and 700°C using the three isotope method. 54-65% exchange. Synthesis from glass in 400-600°C experiments. Results from direct exchange and synthesis experiments agree at 600. No equation given.
102	epidote-quartz	Matthews and Schliestedt (1984)	Ex	$-(1.56 + 1.92\beta_{Ps})(10^6/T^2)$		Based on the experimental zosite-water and quartz-water calibrations of Matthews et al. (1983b) and Matsuhisa et al. (1979). Effect of Fe^{3+} substitution is estimated using the grossular/andradite data of Taylor (1976). β_{Ps} is the mole fraction of pistacite ($Ca_2Fe_3Si_3O_{12}OH$) component in the epidote.
103	gehlenite-calcite	Chacko et al. (1989)	Ex	$-3.12\ (10^6/T^2)$	700-1000	Direct exchange. 90-100% exchange. P=15 kbar.
Cyclosilicates:						
104	tourmaline-quartz	Kotzer et al. (1993)	N	$-1.0\ (10^6/T^2) - 0.39$	200-600	Based on the analysis of co-existing quartz, muscovite and tourmaline from several ore deposits. Temperatures based on quartz-muscovite fractionations compiled by Eslinger et al. (1979).
Pyroxenes:						
105	diopside-calcite	Chiba et al. (1989)	Ex	$-2.37\ (10^6/T^2)$	600-1200	Direct exchange. 61-100% exchange. P=15-16 kbar.
106	enstatite-$BaCO_3$	Rosenbaum et al. (1994)	Ex	-0.97, -1.20, -1.5, -1.2, -1.11	1009-1409 in 100°C increments	Direct exchange. 91-100% exchange. P= 30 kbar. No equation given. Experimental data are non-linear with respect to $1/T^2$ even at these high temperatures.
107	diopside-H_2O	Matthews et al (1983a)	Ex	-1.27, -1.08, -0.98	600, 700, 800	Direct exchange using three-isotope method. 55- 79% exchange. P=13-18 kbar. No equation given.
108	wollastonite- H_2O	Matthews et al (1983a)	Ex	-1.41, -1.37, -1.19, -1.03	500, 600, 700, 800	Direct exchange using three-isotope method. 46-100% exchange. P=9-20 kbar. No equation given.
109	hedenbergite- H_2O	Matthews et al (1983a)	Ex	-1.00	700	Direct exchange using three-isotope method. 52-73% exchange. P=13 kbar. No equation given.
110	jadeite- H_2O	Matthews et al (1983a)	Ex	0.37, 0.21	500, 600	Direct exchange using three-isotope method. 20-39% exchange. P=16-18 kbar. No equation given.
Amphiboles:						
111	tremolite-calcite	Zheng et al. (1994b)	Ex	$-3.80\ (10^6/T^2) + 1.67$	520-680	Exchange of tremolite and calcite in the presence of a CO_2-H_2O fluid. 48-81% exchange. P=3-10 kbar. Extrapolation to equilibrium fractionations at 520, 560, 580°C are suspect

No.	Mineral pair	Reference	Type	Equation	T (°C)	Comments
112	amphibole-quartz	Bottinga and Javoy (1975)	N	$-3.15\,(10^6/T^2) + 0.30$	>500	because of unequal exchange rates in companion experiments at each of those temperatures. Based on natural quartz-amphibole data where temperatures were determined by the Bottinga and Javoy (1973) calibrations of the feldspar-quartz, feldspar-muscovite or feldspar-magnetite isotope thermometers.
Micas:						
113	muscovite-calcite	Chacko et al. (1996)	Ex	$-0.99\,(10^6/T^2)$	550-650	Direct exchange. 80-100% exchange. P=15 kbar. Linear equation fit through the origin. Equation should not be extrapolated below 500°C.
114	phlogopite-calcite	Chacko et al. (1996)	Ex	$-1.78\,(10^6/T^2)$	650-800	Direct exchange. 98-100% exchange. P=15 kbar. Equation as in muscovite-calcite.
115	fluorophlogopite-calcite	Chacko et al. (1996)	Ex	$-1.26\,(10^6/T^2)$	500-800	Direct exchange. 44-90% exchange. P=15 kbar. Equation as in muscovite-calcite. Fluorophlogopite has F/(F+OH)=1
116	fluorophlogopite-calcite	Fortier et al. (1994)	Ex	$-1.84\,(10^6/T^2) + 0.43$	400-800	Direct exchange. 31-91% exchange. P=11 kbar. Equation as in muscovite-calcite. Fluorophlogopite has F/(F+OH)=0.75. Experimental data for this mixed fluoro-hydroxy phlogopite fall in between the equations for end-member hydroxy- and flurophlogopite-calcite fractionations given by Chacko et al. (1996).
117	muscovite-H₂O	O'Neil and Taylor (1969)	Ex	$2.38\,(10^6/T^2) - 3.89$	400-650	Experiments involved synthesis of muscovite from gels or reaction of paragonite or kaolinite with KCl solutions.
118	muscovite-H₂O	Bottinga and Javoy (1973)	Mx	$1.90\,(10^6/T^2) - 3.10$	500-800	Based on the experimental data of O'Neil and Taylor (1969), natural samples, and theoretical estimate of the effect of OH groups on fractionation behavior.
119	muscovite-quartz	Matthews and Schliestedt (1984)	Ex	$-1.55\,(10^6/T^2)$	500-650	Based on the experimental muscovite-water and quartz-water calibrations of O'Neil and Taylor (1969) and Matsuhisa et al. (1979) with a straight line constrained to go through the origin.
120	biotite-quartz	Bottinga and Javoy (1975)	N	$-3.69\,(10^6/T^2) + 0.60$		Based on natural quartz-biotite data where temperatures were determined by the Bottinga and Javoy (1973) calibrations of the feldspar-quartz, feldspar-muscovite or feldspar-magnetite isotope thermometers.
Other Hydrous Phyllosilicates:						
121	chlorite-H₂O	Wenner and Taylor (1971)	N	$1.56\,(10^6/T^2) - 4.70$	150-400	Based on the analysis of natural chlorites and coexisting minerals in metasediments.
122	chlorite-H₂O	Cole and Ripley (1999)	Ex	$2.693\,(10^9/T^3) - 6.342\,(10^6/T^2) + 2.969\,(10^3/T)$	150-400	Direct exchange at 350°C. 12% exchange. P=0.25 kbar. This high temperature datum is combined with results from granite-fluid experiments at 170-300°C in which chlorite is formed by alteration of biotite.
123	kaolinite-H₂O	Eslinger (1971)	Mx	$2.50\,(10^6/T^2) - 2.87$	0-350	Combination of single sample of hydrothermal kaolinite from Broadlands, New Zealand and model calculations.
124	kaolinite-H₂O	Kulla and Anderson (1978)	Ex	$2.42\,(10^6/T^2) - 4.45$	170-320	Hydrothermal synthesis of kaolinite from gels.

		Reference		Equation	T range	Comments
125	kaolinite- H_2O	Sheppard and Gilg (1996)	Mx	$2.76(10^6 T^{-2}) - 6.75$	0–350	Regression line fit through a combination of experimental data (Kulla and Anderson, 1978) and various natural occurrences.
126	smectite- H_2O	Escande et al (1984)	Ex	$3.31(10^6 T^{-2}) - 4.82$	25–95	Synthesis of Mg-rich smectite (stevensonite and saponite) under hydrothermal conditions. Savin and Lee (1988) point out that Escande et al.'s (1984) technique of analyzing smectites may have resulted in an overestimation of the smectite- H_2O fractionation factor.
127	smectite- H_2O	Savin and Lee (1988)	Mx	$2.58(10^6 T^{-2}) - 4.19$	0–350	Modification of the equation of Yeh and Savin (1977). Based on analysis of authigenic smecite formed at 1°C (Δ(sm-H_2O)=30.3 ‰), natural smectite-illite pairs, the quartz-illite curve of Eslinger and Savin (1973), and the quartz-H_2O curve of Matsuhisa et al. (1979).
128	smectite- H_2O	Sheppard and Gilg (1996)	Mx	$2.55(10^6 T^{-2}) - 4.05$	0–350	Regression line fit through a combination of experimental data (Kulla 1979) and various natural occurrences.
129	illite- H_2O	Savin and Lee (1988)	Mx	$2.39(10^6 T^{-2}) - 4.19$		Based on the natural sample quartz-illite curve of Eslinger and Savin (1973), and the quartz-H_2O curve of Matsuhisa et al. (1979).
130	illite- H_2O	James and Baker (1976)	Ex	23.4	22	Fractionation determined in experiments in which illite was suspended in 2N NaCl and 0.2N NaTPB-EDTA solution. The Northrop-Clayton extrapolation procedure indicates 140% exchange – a significant overshoot of the equilibrium fractionation. Thus, the derived fractionation factor may not be reliable.
131	illite- H_2O	Sheppard and Gilg (1996)	Mx	$2.39(10^6 T^{-2}) - 3.76$	0–350	Regression line fit through a combination of experimental data (O'Neil and Taylor, 1969) and various natural occurrences.
132	illite:framework-OH	Bechtel and Hoernes (1990)	N	$-0.076\,T(°C) + 30.42$	200–300	Determined intra-mineral fractionation by thermal dehydration and partial fluorination methods. Hamza and Epstein (1980) and Girard and Savin (1996) also attempted on many hydrous minerals, but the techniques are incomplete and the results are ambiguous.
133	serpentine- H_2O	Wenner and Taylor (1971)	N	$1.56(10^6 T^{-2}) - 4.70$	150–400	Same equation as that for chlorite- H_2O.
Oxides and Hydroxides:						
134	magnetite-calcite	Chiba et al. (1989)	Ex	$-5.91(10^6 T^{-2})$	800–1200	Direct exchange. P=15 kbar.
135	magnetite- siderite	Zhang et al. (2001)	Ex	$-1.76(10^6 T^{-2}) - 9.43$	45–75	Based on microbial siderite-water and magnetite-water fractionations established by thermophilic Fe(III)-reducing bacteria.
136	magnetite-H_2O	Lowenstam(1962)	N	5.6	9	Measurement of magnetite teeth from the species *Cryptochiton stelleri*, which grew in water at 9°C.
137	magnetite-H_2O	O'Neil (1963)	Ex	$-5.5, -4.4$	700, 800	Direct exchange. 84-94% exchange. Fractionations calculated by Matthews et al. (1983b) based on the experimental data of O'Neil (1963).

No.	Mineral pair	Reference	Type	Equation	T (°C)	Comments
138	magnetite-H_2O	Bottinga and Javoy (1973)	N	$-1.47 (10^6/T^2) - 3.70$		Based on the analysis of feldspar-magnetite pairs from mafic lavas (Anderson et al.,1971), where solidification temperatures are relatively well known.
139	magnetite-H_2O	Blattner et al. (1983)	Ex	$-3.7, -7.9$	112, 175	Based on analysis of magnetite precipitated in steam pipelines of the Wairakei geothermal power station.
140	magnetite-H_2O	Fortier et al. (1995)	Ex	-8.60	350	Magnetite grown from fine-grained hematite. P = 1 kbar. Isotopic analysis by ion microprobe.
141	magnetite-rich FeO-H_2O	Zhang et al. (1997)	Ex	$0.80 (10^6/T^2) - 7.74$	45-75	Based on Fe_3O_4-rich iron oxides precipitated by Fe(III)-reduction by thermophilic bacteria in culture.
142	magnetite-H_2O	Mandernack et al . (1999)	Ex	$0.79 (10^6/T^2) - 7.64$	4-75	Based on microbial Fe_3O_4 grown within mangetotactic bacteria in culture and those of Zhang et al. (1997).
143	magnetite-quartz	Downs et al. (1981)	Ex	$-7.8, -6.1$	600, 800	Fayalite is oxidized to quartz and magnetite under hydrothermal conditions. P = 5 kbar. No equation given but regression line through the origin fitted to these two data points is in agreement with the results of Chiba et al. (1989).
144	hematite - H_2O	Clayton and Epstein (1961)	Mx	$0.413 (10^6/T^2) - 2.56$	25-120	Based on the analysis of co-existing quartz, calcite and hematite and experimental calcite-water fractionation factor of Clayton (1961).
145	hematite (geothite)-H_2O	Yapp (1990)	Ex	$1.63 (10^6/T^2) - 12.3$	25-120	Based on the synthesis of hematite (T 62°C) or geothite (T<62°C) from $Fe(NO_3)_3$ solutions. Fractionation behavior of the two minerals was isotopically indistinguishable.
146	hematite - H_2O	Bao and Koch (1999)	Ex	$0.733 (10^6/T^2) - 6.914$	30-140	Synthesis of hematite from $FeCl_3$ solutions by the addition of $NaHCO_3$ solution. Fractionations are similar to Yapp (1990) at T>95°C but differ significantly at lower temperatures. The authors attribute the discrepancy to differences in the washing and drying protocols applied to the hematite precipitates.
147	goethite - H_2O	Müller (1995)	Ex	$1.10 (10^6/T^2) - 12.1$ (KOH) $0.3 (10^6/T^2) - 3.0$ (NaOH) $2.76 (10^6/T^2) - 23.7$ (hydrolysis)	10-65	Precipitation from $Fe(NO_3)_2$ soln by titrating KOH and NaOH solution, and by hydrolysis. The hydrolysis results differ significantly from the KOH and NaOH results.
148	goethite - H_2O	Bao and Koch (1999)	Ex	$1.907 (10^6/T^2) - 8.004$	35-140	Synthesis of geothite from $FeCl_3$ solutions by the addition of NaOH solution. See above for details.
149	akaganeite (β-FeOOH) - H_2O	Bao and Koch (1999)	Ex	$3.927 (10^6/T^2) - 12.157$	35-95	Synthesis of akaganeite by hydrolysis of $FeCl_3$ solutions.
150	rutile-calcite	Chacko et al. (1996)	Ex	$-4.31 (10^6/T^2)$	800-1000	Direct exchange. 53-90% exchange. P=15 kbar.
151	rutile-H_2O	Addy and Garlick (1974)	Ex	$-4.1 (10^6/T^2) + 0.96$	575-775	Exchange by crystallization of amorphous TiO_2 powder.
152	rutile-H_2O	Matthews et al. (1979)	Ex	$-4.72 (10^6/T^2) +1.62$	300-700	Exchange by controlled oxidation of Ti metal powder under hydrothermal conditions.
153	rutile-H_2O	Bird et al. (1993)	Ex	6.1, 3.0	22, 50	Synthesis of rutile from $TiCl_4$ solutions.
154	anatase-H_2O	Bird et al. (1993)	Ex	8.7, 4.9	22, 50	Synthesis of anatase from $TiCl_4$ solutions.
155	rutile-quartz	Matthews and Schliestedt (1984)	Ex	$-4.54 (10^6/T^2)$	500-700	Based on the experimental rutile-water and quartz-water calibrations of Matthews et al. (1979) and Matsuhisa et al. (1979).

No.	Mineral pair	Reference	Type	Equation	Temp. range (°C)	Comment
156	rutile-quartz	Agrinier (1991)	N	$-4.78\,(10^6/T^2)$	450-800	Based on the analysis of quartz, rutile pairs in metamorphic rocks, primarily eclogites. Temperatures based on the Bottinga and Javoy (1975) calibrations of the quartz-muscovite and quartz-garnet thermometers.
157	gibbsite - H_2O	Lawrence and Taylor (1971)	N	$\sim +18$	0-30	From analysis of samples from a large number of natural weathering profiles.
158	gibbsite - H_2O	Chen et al. (1988)	N	$\sim +16$	0-30	Based on the analysis of gibbsite from bauxite deposits in Taiwan.
159	gibbsite - H_2O	Bird et al. (1994)	Ex	$1.31\,(10^6/T^2) - 1.78$	8-51	By synthesis and aging for 3-56 months.
160	gibbsite - H_2O	Vitali et al. (2000)	Ex	$2.04\,(10^6/T^2) - 3.61\,(10^3/T) + 3.65$	0-60	By synthesis and aging for 3-56 months.
161	brucite-H_2O	Saccocia et al. (1998)	Ex	$9.54\,(10^6/T^2) - 35.3\,(10^3/T) + 26.58$	250-450	Direct exchange with 3.2 and 10wt% NaCl soln. P=500 bar.
162	brucite-H_2O	Xu and Zheng (1999)	Ex	$1.56\,(10^6/T^2) - 14.1$	15-120	Synthesis by hydrolysis of Mg_3N_2 and $MgCl_2$, and MgO.
163	uraninite(UO_2)-H_2O	Fayek and Kyser (2000)	Ex	$16.58\,(10^6/T^2) - 77.52\,(10^3/T) + 77.48$	100-300	Combined experimental results of UO_2-CO_2 exchange with CO_2-H_2O of Truesdell (1974).
164	UO_3-H_2O	Fayek and Kyser (2000)	Ex	$-2.21\,(10^6/T^2) + 25.06\,(10^3/T) - 49.50$	100-300	Combined experimental results of UO_3-CO_2 exchange with CO_2-H_2O of Truesdell (1974).
165	cassiterite(SnO_2)-H_2O	Zhang et al. (1994)	Ex	$10.13\,(10^6/T^2) - 26.09\,(10^3/T) + 12.58$	250-370	Synthesis from amorphous SnO_2 or $SnCl_3$ soln.
166	cassiterite(SnO_2)-H_2O	Sushchevskaya et al. (1985)	Ex	$+14.5$ (25°C)	25-450	Precipitation from $SnCl_3$ soln (25°C) and oxidation of Sn (300-450°C).
167	scheelite($CaWO_4$)-H_2O	Ustinov and Grinenko (1990)	Ex	-1.1 to 0 (300-450°C) $+10.8, +1.8$	25,100	Precipitation.
168	wolframite[(Fe,Mn)WO_4]-H_2O	Zhang et al. (1994)	Ex	$3.13\,(10^6/T^2) - 6.42\,(10^3/T) - 0.12$	200-420	Synthesis from Na_2WO_4, $FeCl_2$, and $MnCl_2$, in the presence of up to 30wt% NaCl or NaF.
169	perovskite($CaTiO_3$)-calcite	Gautason et al. (1993)	Ex	$-6.42\,(10^6/T^2)$	800-1000	Direct exchange. 67-99% exchange. P= 15 kbar.
Sulfates:						
170	anhydrite- H_2O	Lloyd (1968)	Ex	$3.878\,(10^6/T^2) - 3.4$	100-500	Direct exchange between anhydrite and water in $1N\ H_2SO_4$ solution. P=690 bars.
171	anhydrite- H_2O	Chiba et al. (1981)	Ex	$3.21\,(10^6/T^2) - 4.72$	100-550	Direct exchange between anhydrite and water in 1m NaCl, HCl or H_2SO_4 solutions. 28-98% exchange. P=1-1000 bars.
172	barite- H_2O	Kusakabe and Robinson (1977)	Ex	$2.64\,(10^6/T^2) - 5.3$ (salt-effect not corrected)	110-350	Direct exchange between barite and water in 1m NaCl or 1mNaCl-1m H_2SO_4 solution. 26-98% exchange.
173	alunite(SO_4)- H_2O	Stoffregen et al. (1994)	Ex	$3.09\,(10^6/T^2) - 2.94$	250-450	Cation exchange of natroanunite with 0.7m K_2SO_4. 8-95% exchange.
174	alunite(OH)- H_2O	Stoffregen et al. (1994)	Ex	$2.28\,(10^6/T^2) - 3.90$	250-450	Cation exchange of natroanunite with 0.7m K_2SO_4. 8-95% exchange.
175	jarosite (SO_4)- H_2O	Rye and Stoffregen (1995)	Ex	$1.43\,(10^6/T^2) + 1.86$	150-250	Cation exchange of natroanunite with H_2SO_4-K_2SO_4. 40-100% exchange.
176	jarosite(OH)- H_2O	Rye and Stoffregen (1995)	Ex	$2.1\,(10^6/T^2) - 8.77$	150-250	Cation exchange of natroanunite with H_2SO_4-K_2SO_4. 40-100% exchange.

Zeolites:						
177	analcime- H_2O	Karlsson and Clayton (1990)	Mx	$2.78 (10^6/T^2) - 2.89$	25-400	Combined results of direct exchange at 300-400°C and two natural samples.
178	analcime(channel-water)- H_2O	Karlsson and Clayton (1990)	Ex	$1.01 (10^6/T^2) - 8.87$	300-400	Direct exchange at 1.5-5.0 kbar.
179	stilbite- H_2O	Feng and Savin (1993)	Ex	$-2.4 + 2.7 (10^6/T^2)$	220-300	Direct exchange at low pressures (21 Torr).
180	wairakite- $H_2O_{(v)}$	Noto and Kusakabe (1997)	Ex	$2.46 (10^6/T^2) - 1.76$	250-400	Direct exchange at 0.5-1.5 kbar. 63-98% exchange.
181	wairakite (channel-water)- H_2O	Noto and Kusakabe (1997)	Ex	$0.79 (10^6/T^2) - 3.07$	250-400	Direct exchange at 0.5-1.5 kbar. 94-99% exchange.
182	clinoptilolite-water	Nahr et al. (1998)	N	$+31.6\pm0.2, +26.6$	25, 40	Based on analysis of authigenic clinoptilolite samples in oceanic sediments from three ODP sites.
	CO_2- **Silicate/Melt/Glass:**					
183	CO_2-silica glass	Stolper and Epstein (1991)	Ex	$+2.14$ to $+4.30$	550-950	Direct exchange between silica wool or power and CO_2 at 0.5 bar.
184	CO_2-silica glass	Matthews et al. (1994)	Ex	$+2.10$ to $+3.11$	750-950	Direct exchange between silica wool or power and CO_2 at 1 bar.
185	CO_2-albite glass	Matthews et al. (1994)	Ex	$+3.58$ to $+4.75$	750-950	Direct exchange between albite glass and CO_2 at 1 bar.
186	CO_2-albite	Matthews et al. (1994)	Ex	$+3.36$ to $+4.74$	750-950	Direct exchange between crystalline albite and CO_2 at 1 bar.
187	CO_2-rhyolitic glass/melt	Palin et al. (1996)	Ex	$+2.95$ to $+5.08$	550-950	Direct exchange between rhyolitic glass and CO_2 at 0.8-1.5 bar.

Notes: 1. Ex = experimental; N = natural sample; M = mixed experimental/natural sample calibration. 2. Equations given with temperature in Kelvin unless otherwise indicated.

Appendix 3. Carbon isotope fractionation factors: calibrations based on experiments or natural samples.

#	Phases	Reference	[1]Method	[2]$1000 \ln \alpha$	T(°C)	Comments
	CO_2 and Aqueous Species:					
1	CO_2:solid-vapor	Eiler et al. (2000)	Ex	-0.2 to +0.4	130-150(K)	By sublimation-condensation and isotopic exchange.
2	CO_2:liquid-vapor	Grootes et al. (1969)	Ex	-0.57 to -0.14	220-303(K)	Liquid-vapor equilibration for >4 hr.
3	$CO_2(aq)$-$CO_2(g)$	Wendt (1968), Vogel et al. (1970), and Zhang et al. (1995)	Ex	0.0041 T(°C)-1.18 (Vogel) 0.0049 T(°C)-1.31 (Zhang)	0-60	^{13}C is depleted and ^{18}O is enriched in the aqueous phase.
4	$HCO_3^-(aq)$-$CO_2(g)$	Deuser and Degens (1967), Wendt (1968), Emrich et al. (1970), Mook et al. (1974), Turner (1982), Lesniak and Sakai (1989), and Zhang et al. (1995)	Ex	7.92 to 8.27 at 25°C with -0.064 to -0.141‰/°C gradient	5-60	Consistent results by many investigators.
5	$HCO_3^-(aq)$-$CO_2(g)$	Szaran (1997)	Ex	-0.0954 T(°C) + 10.41	7-70	Also investigated the relaxation time.
6	$HCO_3^-(aq)$-$CO_2(g)$	Malinin et al. (1967)	Ex	-7.3 to +7.5	23-286	Cross-over at about 150°C.
7	$CO_3^{2-}(aq)$-$CO_2(g)$	Turner (1982), Lesniak and Sakai (1989), Zhang et al. (1995) and Halas et al. (1997)	Ex	-1.5 to +8.5	4-80	Cross-over at about 67°C (Halas et al., 1997).
	Carbonates-Aqueous Solution:					
8	calcite-$HCO_3^-(aq)$	Rubinson and Clayton (1969)	Ex	0.9±0.2	25	Slow precipitation by CO_2 degassing.
9	calcite-$CO_2(g)$	Emrich et al. (1970)	Ex	9.61±0.28 at 25°C	20-60	Slow precipitation by CO_2 degassing.
10	calcite-$HCO_3^-(aq)$	Turner (1982)	Ex	1.83 to 2.26	25	Slow precipitation by adding NaOH to Ca-HCO_3 soln.
11	calcite-$CO_2(g)$	Romanek et al. (1992)	Ex	11.98 - 0.12 T(°C)	10-40	Controlled seeded precipitation at constant composition.
12	aragonite-$HCO_3^-(aq)$	Rubinson and Clayton (1969)	Ex	2.7±0.2	25	Slow precipitation by CO_2 degassing.
13	aragonite-$CO_2(g)$	Romanek et al. (1992)	Ex	13.88 - 0.13 T(°C)	10-40	Controlled seeded precipitation at constant composition.
14	siderite ($FeCO_3$) - $HCO_3^-(aq)$	Carothers et al. (1988)	Ex	-4.20 to +2.86	33-197	Slow injection of $FeCl_2$ soln to $NaHCO_3$ soln.
15	gaylussite ($Na_2CO_3 \cdot CaCO_3 \cdot 5H_2O$)- $CO_3^{2-}(aq)$	Matsuo et al. (1972)	Ex	1.9 to 3.1	8-35	Formed by mixing Na_2CO_3 and $CaCl_2$ solns.
16	trona ($Na_2CO_3 \cdot NaHCO_3 \cdot 2H_2O$)- $CO_3^{2-}(aq)$	Matsuo et al. (1972)	Ex	0.7 to 1.8	18-35	Formed from Na_2CO_3 and $NaHCO_3$ solns onto seed crystal.
17	pirssonite ($Na_2 \cdot Ca(CO_3)_2 \cdot 2H_2O$)- $CO_3^{2-}(aq)$	Bottcher (1994)	Ex	2.3 and 3.3	60-90	Reaction of $CaCO_3$ with Na_2CO_3 soln.
18	CO_2-malachite ($CuCO_3 \cdot Cu(OH)_2$)	Melchiorre et al (1999)	Ex	-1.85 ($10^6/T^2$)+ 10.51	0-50	Replacement of calcite with Cu^{2+}-bearing soln.
19	norsethite ($BaMg(CO_3)_2$)- CO_2	Bottcher (2000)	Ex	1.78 ($10^6/T^2$) - 10.16	20-90	Formed from $BaCO_3$ and $MgCO_3 \cdot 3H_2O$ in $NaHCO_3$ soln.

		Reference		Equation	T (°C)	Comments
CO$_2$-Calcite/Graphite:						
20	CO$_2$-calcite	Chacko et al. (1991)	Ex	2.91 to 5.08 $-0.10028 + 5.4173x$ $-2.5076x^2 + 0.47193x^3$ $-0.049501x^4 + 0.0027046x^5$ $-0.0000059409x^6$ where $x=10^6/T^2$	400-950	Direct exchange. 19-98% exchange. P=10 kbar. Large errors due to small degrees of exchange. Equation represents theoretical calculations that fit the experimental data of Chacko et al (1991) and Rosenbaum (1994) within error. The data of Scheele and Hoefs (1992) are displaced to slightly larger fractionations. The equation reproduces the calculated fractionations from 273-4000K.
21	CO$_2$-calcite	Rosenbaum (1994)	Ex	2.70	900	Direct exchange. P=12.5 kbar.
22	CO$_2$-calcite	Scheele and Hoefs (1992)	Ex	$-3.46 (10^6/T^2) + 9.58 (10^3/T) - 2.72$	500-1200	Most experiments involved inversion of aragonite to calcite. P=1-5 kbar.
23	CO$_2$-graphite	Scheele and Hoefs (1992)	Ex	$4.53 (10^6/T^2) + 3.04$	600-1200	Direct exchange. 33-79% exchange. P=5-15 kbar. Small CO$_2$/C ratio (1/32) used in the experiments may have resulted in the incorporation of surface effects in the measured fractionations.
Carbonates-Graphite/Diamond						
24	aragonite-calcite	Sommer and Rye (1978)	N	$2.56 - 0.065T(^\circ C)$	0-25	Based on analysis of coexisting calcite and aragonite tests and shells.
25	dolomite-calcite	Sheppard and Schwarz (1970)	N	$0.18 (10^6/T^2) + 0.17$	100-650	Based on analyses of coexisting pairs of metamorphic origin. Preliminary experimental results show much larger fractionations.
26	calcite-graphite	Valley and O'Neil (1981)	N	$-0.00748 T(^\circ C) + 8.68$	610-760	Based on analyses of metamorphic samples from the Adirondack Mountains.
27	calcite-graphite	Wada and Suzuki (1983)	N	$5.6 (10^6/T^2) - 2.4$	400-680	Based on analyses of metamorphic aureole samples from central Japan.
28	calcite-graphite	Morikiyo (1984)	N	$8.9 (10^6/T^2) - 7.1$	270-650	Based on analyses of metamorphic aureole samples from central Japan.
29	calcite-graphite	Dunn and Valley (1992)	N	$5.81 (10^6/T^2) - 2.61$	400-800	Based on analyses of metamorphic samples from Tudor aureole, Canada.
30	calcite-graphite	Kitchen and Valley (1995)	N	$3.56 (10^6/T^2)$	700-800	Based on analyses of metamorphic samples from the Adirondack Mountains.
31	dolomite-graphite	Wada and Suzuki (1983)	N	$5.9 (10^6/T^2) - 1.9$	400-680	Based on analyses of metamorphic aureole samples from central Japan.
32	diamond-graphite	Hoering (1961)	Ex	0.3	>1700	Conversion of graphite to diamond at P=70 kbar.
CO$_2$ in Melts:						
33	CO$_2$-CO$_3^{2-}$ in melt	Javoy et al. (1978)	Ex	4.0 to 4.6	1120-1280	Solution of CO$_2$ into tholeiitic silicate melt at 7.0-8.4 kbar.
34	CO$_2$-CO$_3^{2-}$ in melt	Mattey et al. (1990) and Mattey (1991)	Ex	1.8 to 2.7	1200-1400	Solution of CO$_2$ into silicate (sodamelilite and basalt) and carbonate melts at P=5-39 kbar.
Gases:						
35	CO$_2$-CH$_4$	Horita (2001)	Ex	$26.70 - 49.137 (10^3/T) + 40.828 (10^6/T^2) - 7.512(10^9/T^3)$	200-600	Complete exchange catalyzed by Ni catalyst.

Notes: 1. Ex = experimental; N = natural sample; M = mixed experimental/natural sample calibration. 2. Equations given with temperature in Kelvin unless otherwise indicated.

Appendix 4. Hydrogen isotope fractionation factors: calibrations based on experiments or natural samples.

#	Phases	Reference	Method	^{2}1000 ln α	T(°C)	Comments
	Gases, Liquid, Ice					
1	H_2O: ice-vapor	Merlivat and Nief (1967)	Ex	$\log \alpha = -0.041 + 7.074/T^2$	−40 - 0	Slow growth of ice in stirred water.
2	H_2O: ice-liquid	O'Neil (1968)	Ex	+18.7±0.9	0?	Freezing from 4‰ chlorinity water and seawater.
3	H_2O: ice-liquid	Craig and Horn (1968)	Ex	+19.5, +20.3	0?	Seeded-growth.
4	H_2O: ice-liquid	Arnason (1969)	Ex	+20.6±0.7	0	
5	H_2O: ice-liquid	Suzuki and Kimura (1973)	Ex	+20.4±0.5	0?	
6	H_2O: ice-liquid	Stewart (1974)	Ex	+24	−10	Freezing from 2.5m NaCl soln.
7	H_2O: ice-liquid	Lehmann and Siegenthaler (1991)	Ex	+21.2±0.4	0	Slow growth of ice in stirred water.
8	H_2O: liquid-vapor	Majoube (1971b)	Ex	$24.844 (10^6/T^2) - 76.248(10^3/T) + 52.612$	0-100	Slow distillation of liquid water.
9	H_2O: liquid-vapor	Horita and Wesolowski (1994), and all references therein	Ex	$1158.8 (T^3/10^9) - 1620.1 (T^2/10^6) + 794.84 (T/10^3) + 2.9992 (10^9/T^3) - 161.04 + 27870/T^2$	0-374	High precision experimental calibration using three different apparatus to cover the full temperature range. This large dataset was combined with selected earlier experimental studies to generate the calibration curve.
10	CH_4-H_2	Horibe and Craig (1995)	Ex	$\alpha = 0.8994 + 183540/T^2$	200-500	Complete exchange with Ni-thoria catalyst.
11	$H_2O_{(v)}$-H_2	Suess (1949)	Ex	$\log \alpha = 203/T - 0.132$	80-200	
12	$H_2O_{(v)}$-H_2	Cerrai et al. (1954)	Ex	3.04 to 1.18 (α)	51-742	Dynamic exchange in a reactor with Pt catalyst.
13	$H_2O_{(liq)}$-H_2	Rolston et al. (1976)	Ex	$\ln \alpha = -0.2143 + 368.9/T + 27870/T^2$	7-97	Exchange catalyzed by Pt catalyst.
	Hydrated Salts:					
14	trona ($Na_2CO_3 \cdot NaHCO_3 \cdot 2H_2O$) − H_2O	Matsuo et al. (1972)	Ex	$1.420 (10^7/T^2) + 2.356 (10^4/T)$	8-35	Synthesis from aqueous solutions. Determined on the composition scale of brine.
15	gaylussite ($Na_2CO_3 \cdot CaCO_3 \cdot 5H_2O$) − H_2O	Matsuo et al. (1972)	Ex	−15 to −13	8-35	Synthesis from aqueous solutions. Determined on composition scale of brine.
16	natron ($Na_2CO_3 \cdot 10H_2O$) − H_2O	Pradhananga and Matsuo (1985b)	Ex	+15 to +17	5-10	Synthesis from aqueous solutions. Determined on the composition scale of brine.
17	borax ($Na_2B_4O_7 \cdot 10H_2O$) − H_2O	Matsuo et al. (1972)	Ex	0 to +2	8-35	Synthesis from aqueous solutions. Determined on the composition scale of brine.
		Pradhananga and Matsuo (1985a)	Ex	+2 to +5	5-25	
18	carnallite(KMgCl₃ · 6H₂O) − H_2O	Horita (1989)	Ex	−44 to −40	10-40	By precipitation and aging 30-67 days. Determined on the activity scale of brines.
19	carnallite(KMgCl₃ · 6H₂O) − H_2O	Koehler and Kyser (1996)	Ex	$-18.4 (10^6/T^2) + 162$	22-45	By synthesis and exchange.
20	bischofite(MgCl₂ · 6H₂O) − H_2O	Horita (1989)	Ex	−47 to −37	10-40	By precipitation and aging 30-67 days. Determined on the activity scale of brines.
21	tachyhydrite(CaMg₂Cl₆ · 12H₂O) − H_2O	Horita (1989)	Ex	−46 to −36	25-40	By precipitation and aging 30-67 days. Determined on the activity scale of brines.
22	gypsum (CaSO₄ · 2H₂O) − H_2O	Fontes and Gonfiantini (1967), Sofer (1978), and Matsubaya and Sakai (1973)	Ex	−15 to −20	17-57	Slow precipitation.
23	mirabilite (Na₂SO₄ · 10H₂O) − H_2O	Stewart (1974), Pradhananga and Matsuo (1985b)	Ex	+13 to +19	0-25	Synthesis from aqueous solutions. Determined on the composition scale of brine.

No.	Mineral pair	Reference	Type	1000 ln α	T (°C)	Comments
24	episomite (MgSO$_4$ 7H$_2$O)- H$_2$O	Pradhananga and Matsuo (1985b)	Ex	-1	25	Synthesis from aqueous solutions. Determined on the composition scale of brine.
	Hydroxides:					
25	boehmite- H$_2$O	Suzuoki and Epstein (1976)	Ex	-66.2	400	Direct exchange. 93% exchange. P=1 kbar (Suzuoki, pers. commun.).
26	boehmite- H$_2$O	Graham et al. (1980)	Ex	-37.4, -36.2, -53.5, -49.9	150, 200, 280, 380	Direct exchange. 77-100% exchange. P = 2 kbar.
27	diaspore- H$_2$O	Graham et al. (1980)	Ex	-78.7	380	Direct exchange. 25% exchange. P = 4 kbar.
28	gibbsite- H$_2$O	Lawrence and Taylor (1971)	N	-15	0-30	From analysis of samples from a large number of natural weathering profiles.
29	gibbsite- H$_2$O	Chen et al. (1988)	N	-8	0-30	From analysis of samples from a large number of natural weathering profiles in Taiwan.
30	gibbsite- H$_2$O	Vitali et al. (2001)	Ex	-2±6	9-60	Synthesis over 10 years.
31	gibbsite- H$_2$O	Vitali et al. (2001)	N	-8±10	Ambient	Based on analysis of 12 samples from various localities in the tropics.
32	goethite- H$_2$O$_{(v)}$	Yapp and Pedley (1985)	Ex	-75.8, -89.9	100, 145	Direct exchange between natural goethite and water vapor. 52-75% exchange. P=1 bar. Corresponding geothite-H$_2$O(l) fractionations are -103.6 at both 100 and 145°C using the liquid-vapor fractionation curve of Horita et al. (1994).
33	goethite- H$_2$O$_{(l)}$	Yapp and Pedley (1985)	N	-105	~25	Based on the analysis of natural goethite and associated modern waters.
34	goethite- H$_2$O$_{(l)}$	Yapp (1987)	Ex	~-95	25, 62	Synthesis of goethite from Fe(NO$_3$)$_3$ solutions.
35	manganite- H$_2$O$_{(l)}$	Hariya and Tsutsumi (1981)	Ex	-211 to -192	150-250	Direct exchange at 500 bar. 46-90% exchange.
36	brucite- H$_2$O	Satake and Matsuo (1984)	Ex	8.72 (10^6/T^2) - 3.86 (10^4/T) + 14.5	144-510	Direct exchange. 58-100% exchange. P=1-1060 bars.
37	brucite- H$_2$O	Saccoccia et al. (1998)	Ex	-32 to -22	250-450	Direct exchange with 3.2 and 10wt% NaCl soln. P=500 bar.
38	brucite- H$_2$O	Xu and Zheng (1999)	Ex	4.88 (10^6/T^2) - 22.54	25-90	Synthesis by hydrolysis of Mg$_3$N$_2$, and MgCl$_2$. Large discrepancy with the data of Satake and Matsuo (1984).
39	brucite- H$_2$O	Horita et al. (1999)	Ex	-31.9 to -19.5	380	Increase with pressure from 150 to 8000 bar.
	Sorosilicates:					
40	epidote- H$_2$O	Graham et al. (1980)	Ex	29.2 (10^6/T^2) - 138.8 -35.9	200-300 300-650	Direct exchange. 30-100% exchange. P=2-4 kbar. Epidote composition is pistacite (Fe/Fe+Al) = 0.29. Experiments indicate fractionation factor is independent of temperature in the 300-650°C temperature range.
41	epidote- H$_2$O	Chacko et al. (1999)	Ex	9.3 (10^6/T^2) - 61.9	300-600	Direct exchange. Experiments done on large single crystals and analyzed with the ion microprobe. P = 1.2 - 2.2 kbar. Epidote composition is pistacite (Fe/Fe+Al) = 0.33.
42	epidote- H$_2$	Vennemann and O'Neil (1996)	Ex	110.756 (10^6/T^2) + 149.98 (10^3/T) - 158.685	150-400	Direct exchange of epidote with H$_2$ gas. 71-91% exchange. P = 0.3-2 bars. At 300°C, epidote-H$_2$O fractionations derived from these results are approximately 35‰ more negative than those reported in the direct epidote-H$_2$O exchange experiments of Graham et al. (1980) and Chacko et al. (1980). Epidote composition is pistacite (Fe/Fe+Al) = 0.30.
43	zoisite- H$_2$O	Graham et al. (1980)	Ex	-15.07 (10^6/T^2) - 27.73	280-650	Direct exchange. 35-100% exchange. P= 4 kbar. Zoisite composition is pistacite (Fe/Fe+Al) = 0.03.
44	clinozoisite- H$_2$O	Graham et al. (1980)	Ex	-37.8, -37.2	300, 450	Direct exchange. 60-100% exchange. P= 2 kbar. Zoisite composition is pistacite (Fe/Fe+Al) = 0.09.

No.	Mineral pair	Reference	Type	$10^3 \ln \alpha$	T (°C)	Comments
45	ilvaite - H_2O	Yaqian and Jibao (1993)	Ex	-105 (350-550°C) -29.95 $(10^6/T^2) - 60.62$ (550-750°C)	350-750	Direct exchange at P=50-203 bar. 50-100% exchange.
	Cyclosilicates:					
46	tourmaline- H_2O	Blamart et al. (1989)	Ex	-39.3 $(10^6/T^2) + 63.4$	500-600	Direct exchange at P=3 kbar. 72-92% exchange.
47	tourmaline- H_2O	Jibao and Yaqian (1997)	Ex	-27.9 $(10^6/T^2) + 2.3$	450-800	Direct exchange at P=150-250 bar. 11-81% exchange.
48	tourmaline- H_2O	Kotzer et al. (1993)	N	-27.2 $(10^6/T^2) + 28.1$	350-600	Based on the analysis of co-existing quartz, muscovite and tourmaline from several ore deposits. Temperatures based on quartz-muscovite oxygen isotope fractionations compiled by Eslinger et al. (1979).
	Amphiboles:					
49	hornblende- H_2O	Suzuoki and Epstein (1976)	Ex	-23.9 $(10^6/T^2) + 7.9$	400-700	Direct exchange. 19-100% exchange. Pressure unspecified. Composition of the hornblende not clearly specified but appears to actinolite or actinolitic hornblende.
50	pargasite- H_2O	Graham et al. (1984)	Ex	$-21.86, -18.12$	700, 850	Direct exchange. 83-90% exchange. P = 4-6 kbar.
51	ferroan pargasitic hornblende - H_2O	Graham et al. (1984)	Ex	-31.0 $(10^6/T^2) + 1.1$ -23.1	850-950 350-700	Direct exchange. 46-100% exchange. P = 2-8 kbar. Experiments indicate fractionation factor is independent of temperature in the 350-700°C temperature range.
52	hornblende- H_2	Vennemann and O'Neil (1996)	Ex	452, 326	300, 400	Direct exchange of hornblende with H_2 gas. 71-91% exchange. P = 2 bars. Fe^{3+} content of the hornblende decreased and water content increased during the experiment indicating that isotopic exchange was accompanied by partial reduction of Fe.
53	tremolite- H_2O	Graham et al. (1984)	Ex	-31.0 $(10^6/T^2) +14.9$ -21.7	650-850 350-650	Direct exchange. 75-99% exchange. P = 2-4 kbar. Experiments indicate fractionation factor is independent of temperature in the 350-650°C temperature range.
54	actinolite- H_2O	Graham et al. (1984)	Ex	-29	400	Direct exchange. 59% exchange. P = 2 kbar.
	Micas:					
55	muscovite- H_2O	Suzuoki and Epstein (1976)	Ex	-22.1 $(10^6/T^2) + 19.1$	450-750	Direct exchange. 39-100% exchange. P=1 kbar (Suzuoki, pers. commun.).
56	muscovite- H_2	Vennemann and O'Neil (1996)	Ex	316.851 $(10^6/T^2) - 584.796$ $(10^3/T) + 515.684$	200-400	Direct exchange of muscovite with H_2 gas. 13-80% exchange. P =0.4-2 bars.
57	lepidolite- H_2O	Suzuoki and Epstein (1976)	Ex	$+15.4$	650	Direct exchange. 45% exchange. P=1 kbar (Suzuoki, pers. commun.).
58	phlogopite- H_2O	Suzuoki and Epstein (1976)	Ex	$-14.9, -9.9$	575, 650	Direct exchange. 36-81% exchange. P=1 kbar (Suzuoki, pers. commun.). Biotite composition is Mg/(Fe+Mg) = 0.89.
59	biotite (Mg-rich) - H_2O	Suzuoki and Epstein (1976)	Ex	-21.3 $(10^6/T^2) - 2.8$	400-850	Direct exchange. 31-100% exchange. P=1 kbar (Suzuoki, pers. commun.). Biotite composition is Mg/(Fe+Mg) = 0.61.
60	biotite (Fe-rich) - H_2O	Suzuoki and Epstein (1976)	Ex	-38.0	650	Direct exchange. 100% exchange. P=1 kbar (Suzuoki, pers. commun.). Biotite composition is Mg/(Fe+Mg) = 0.33.
61	biotite- H_2	Vennemann and O'Neil (1996)	Ex	417.3, 293.0	300, 400	Direct exchange of biotite with H_2 gas. 31-64% exchange. P =2 bars. Extensive reduction of Fe^{3+} and increase in water accompanied isotopic exchange.
	Other Hydrous Phyllosilicates:					
62	chlorite- H_2O	Marumo et al. (1980)	N	-47 to -13	100-250	Based on the analysis of chlorites in drill core samples from the Ohnuma geothermal area, Japan. The authors propose

No.	Mineral pair	Type	Reference	Equation	T (°C)	Comments
63	chlorite- H_2O	N	Taylor (1974)	-45 to -35	300-500	that fractionations are independent of temperature over this temperature range but become systematically more negative with increasing Fe/(Fe+Mg) of chlorite. Graham et al. (1987) question the validity of these conclusions. Based on natural samples.
64	chlorite- H_2O	Ex	Graham et al. (1987)	-28	500-700	Direct exchange. 47-100% exchange. P = 2 kbar. Experiments indicate fractionation factor is independent of temperature over this temperature range.
65	serpentine- H_2O	Mx	Wenner and Taylor (1973)	$1.56(10^6/T^2) - 4.70$	25-400	Based on analysis of natural deweylite, lizardite, antigorite and one preliminary experimental data of Suzuoki and Epstein.
66	serpentine- H_2O	Ex	Suzuoki and Epstein (1976)	-19.8	400	Direct exchange. 98% exchange. P=1 kbar (Suzuoki, pers. commun.).
67	serpentine- H_2O	Ex	Sakai and Tsutsumi (1978)	$2.75(10^7/T^3) - 7.69(10^4/T) + 40.8$	100-500	Direct exchange. 18-93% exchange P = 2 kbar. Starting material is a commercial chrysotile asbestos fiber.
68	serpentine- H_2O	Ex	Grinenko et al (1987) Mineev and Grinenko (1996)	-40 to +35 (100°C) -23 to +6 (200°C)	100, 200	Direct exchange of natural lizardite and water. 11-20% exchange. P = 1-2500 bar. Similar to the semi-empirical calibration of Wenner and Taylor (1973). Attribute the large discrepancy with the results of Sakai and Tsutsumi (1978) to differences in experimental pressures.
69	kaolinite- H_2O	Ex	Suzuoki and Epstein (1976)	-23.7	400	Direct exchange. 90% exchange. P=1 kbar (Suzuoki, pers. commun.). Temperature is beyond the thermal stability limit of kaolinite.
70	kaolinite- H_2O	N	Lambert and Epstein (1980)	$-4.53(10^6/T^2) + 19.36$	<230	Based on analysis of samples from Valles Caldera, New Mexico, geothermal system.
71	kaolinite- $H_2O_{(v)}$	Ex	Liu and Epstein (1984)	-21.20 to -9.6	250-352	Direct exchange with high kaolinite-water H ratio (80). 36-100% exchange. Combined with $H_2O_{(l)}$-$H_2O_{(v)}$ to obtain kaolinite- $H_2O_{(liq)}$ fractionation.
72	kaolinite- H_2	Mx	Gilg and Sheppard (1996)	$-2.2(10^6/T^2) - 7.7$	0-330	Based on the experimental data of Sheppard et al. (1969) and Liu and Epstein (1984) at 300°C and various natural sample data from geothermal systems and weathering profiles (Marumo et al. 1980; Rye et al. 1992; Lawrence and Taylor 1972).
73	kaolinite- H_2	Ex	Vennemann and O'Neil (1996)	$-162.495(10^6/T^2) + 1241.164(10^3/T) - 1219.110$	150-275	Direct exchange of kaolinite with H_2 gas. 32-94% exchange. P=0.3-2 bars.
74	smectite- H_2O	N	Yeh (1980)	$-19.6(10^3/T) + 25$	25-120	Based on the analysis of smectites in deep Gulf Coast wells and assuming Δ=41 at 25°C. See also Savin and Lee (1988) and Sheppard and Gilg (1996) for critique of these and other smectite- H_2O data.
75	illite/smectite- H_2O	N	Capuano (1992)	$-45.3(10^3/T) + 94.7$	0-150	Based on analysis of coexisting clay and water from Gulf Coast geopressured fields.
Sulfates:						
76	alunite- H_2O	Ex	Stoffregen et al. (1994)	-6 to -19	250-450	Cation exchange of natroalunite with 0.7m K_2SO_4. 8-95% exchange.
77	jarosite- H_2O	Ex	Rye and Stoffregen (1995)	-50±12	150-250	Cation exchange of natroanunite with H_2SO_4-K_2SO_4. 40-100% exchange.

Melts-Glass:

78	$H_2O_{(v)}$-silicate glass	Dobson et al. (1989)	Ex	+35 to +51	530-850	Natural rhyolitic obsidian and synthetic albite-orthoclase glasses. P=1.4-2.8 bar.
79	$H_2O_{(liq)}$-rhyolite glass	Friedman et al. (1993)	N	+33	Ambient	Based on analysis of hydrated rhyolitic lava from Idaho and Katmai.
80	$H_2O_{(v)}$-silicate melt	Kuroda et al. (1982)	Ex	-54 to +33	800-1300	All at 20 kbar. D enriched in K-feldspar and albite melts and depleted in quartz and anorthite melts.
81	$H_2O_{(v)}$-rhyolite melt	Taylor and Westrich (1985)	Ex	+23.6	950	Synthesis and reversed isotope exchange experiments. D depleted in the melt.
82	$H_2O_{(v)}$-albite melt	Richet et al. (1986)	Ex	+8 to +25	870-1250	Used natural albite at 2 kbar. Contradict Kuroda et al. (1982).
83	$H_2O_{(v)}$-basalt melt	Shilobreyeva et al. (1992)	Ex	+18 to +25	1250	P=3 kbar. Deuterium depleted in the melt.
84	$H_2O_{(v)}$-basaltic andesite melt	Pineau et al. (1998)	Ex	+20 to +32	1250	Fractionation decreased with pressure from 0.5 to 3 kbar, due to increasing solubility of H_2O.

Notes: 1. Ex = experimental; N = natural sample; M = mixed experimental/natural sample calibration. 2. Equations given with temperature in Kelvin unless otherwise indicated.

2   Rates and Mechanisms of Isotopic Exchange

David R. Cole

Chemical and Analytical Sciences Division
Oak Ridge National Laboratory
Oak Ridge, Tennessee 37831

Sumit Chakraborty

Institut für Geologie, Mineralogie and Geophysik
Ruhr-Universität–Bochum
Bochum, Germany

INTRODUCTION

Background

The proliferation of stable isotope laboratories in recent years has led to a marked increase in the routine use of stable isotope analyses in geochemical studies. This trend is likely to continue because of the development of micro-analytical techniques (e.g. laser ablation, ion microprobe, femtomole-carrier gas methods, etc.) for stable isotope analysis and the potential of these techniques to reveal the detailed sequence of thermal and fluid histories preserved in the rock record (see McKibben et al. 1998). Equilibrium isotope fractionation factors and rates of isotopic exchange form the cornerstones for the interpretation of stable isotope data from natural systems. Indeed, determination of the distribution of the stable isotopes of C-O-H-S in gases, fluids, and minerals has become a standard and extremely powerful approach in elucidating the temperatures, material fluxes, rock and fluid origins, and time scales associated with ancient and active fluid-rock interaction processes in the Earth's crust (see Valley et al. 1986). Although the thermodynamic principles underlying isotopic fractionation behavior were developed more than fifty years ago by Urey (1947), the calibration of fractionation factors has proven to be difficult. During the last four decades, a great deal of progress has been made in determining the equilibrium fractionations of $^{18}O/^{16}O$, D/H, $^{13}C/^{12}C$, and $^{34}S/^{32}S$ among fluid and mineral components at high temperatures ($\geq 400°C$), both from experimental exchange studies, theoretical calculations (using vibrational spectra), semi-empirical calculations (using bond-type considerations), and empirical calibrations based on natural assemblages (see references in O'Neil 1986; Kyser 1987; Criss 1991, 1999; Chacko et al., this volume). Of these, the experimental method is the most direct because it involves the least number of *a priori* assumptions.

Despite the utility of the equilibrium approach in quantifying the isotopic behavior in select water-rock systems, there is mounting evidence that isotopic heterogeneity and disequilibrium may be more widespread than previously appreciated. Even in the absence of sophisticated micro-analytical methods, the evidence for isotopic disequilibrium has been documented. At hydrothermal conditions (~100-300°C) there are several chemical systems where isotopic disequilibrium has been observed—sulfate-sulfide sulfur (Ohmoto and Lasaga 1982), CO_2-CH_4 carbon (Giggenbach 1982; Horita 2001), and whole rock-fluid oxygen (Cole 1994, McConnell et al. 1997). Sluggish reaction rates leading to isotopic disequilibrium in a number of diagenetic systems have also been documented—e.g. detrital feldspars in carbonate hosts (Lawrence and Kastner 1975), partial recrystallization of detrital clays in seawater (Yeh and Savin 1976), and preservation of oxygen isotope compositions in silt-sized quartz in sediments of eolian

1529-6466/00/0043-0002$15.00

and fluvial origin (Clayton et al. 1978). Graham et al. (1996) measured gradients of ~25‰ in $\delta^{18}O$ between quartz overgrowths and detrital grains by ion microprobe. Other examples include the oxygen isotope disequilibrium between dissolved sulfate and seawater (e.g. see Lloyd 1968) and between atmospheric O_2 and seawater (Dole effect; Dole et al. 1954).

This behavior is not necessarily limited to the low temperature systems referred to above. The preferential resetting of isotopic compositions of silicates, carbonates and oxides, each with their own unique blocking temperature, can lead to discordant isotopic fractionations during cooling of high temperature systems (e.g. Giletti 1986, Gregory et al. 1989, Eiler et al. 1992, Farquhar et al. 1993). There is no question that our ability to measure the isotopic signals in smaller and smaller domains with techniques such as laser ablation or the ion microprobe has led to the realization that many grains are heterogeneous on scales of <10 to 100s of microns. Microanalytical methods have been used to document oxygen isotope inhomogeneities in several important crustal or mantle phases—e.g. garnet (Chamberlain and Conrad 1991, 1993; Kohn et al. 1993, Young and Rumble 1993, Crowe et al. 2001), quartz (Elsenheimer and Valley 1993, Hervig et al 1995, Graham et al. 1996, Valley and Graham 1996), feldspar (Elsenheimer and Valley 1993, Riciputi et al. 1998, Mora et al 1999), diopside (Edwards and Valley 1998, Deines and Haggerty 2000), zircon (Bindeman and Valley 2000, Wilde et al. 2001) magnetite (Valley and Graham 1991, 1993; Eiler et al. 1995), calcite (Arita and Wada 1990, Graham et al. 1998), graphite (Wada 1988, Kitchen and Valley 1995), and sulfides (McKibben and Riciputi 1998). The magnitude of isotopic variations observed in these studies can range from less than 1 per mil to over 25 per mil within a single grain, or between like grains within the same thin section. Wherever zoning or irregular distributions are observed, the traditional assumption of isotopic equilibrium may not be valid.

Objectives

This chapter presents an updated, expanded overview of the kinetics of stable isotope exchange reactions originally covered by Cole and Ohmoto (1986). The focus of this chapter is the rates of isotopic exchange reactions, and the mechanisms that control the exchange process. In the previous installment, the emphasis was on the kinetics of high-temperature/high-pressure behavior. High temperature behavior will certainly be the emphasized in this chapter primarily because the bulk of the experimental effort has been in this arena. In addition, we will examine the behavior of systems that traditionally are thought to be fast geologically speaking, but in fact are quite sluggish—namely, isotopic exchange among select groups of gaseous species and/or aqueous species. It is of historic note to point out that some of the earliest applications of stable isotopes were in studies where either oxygen or carbon isotopes were used to track chemical reactions and their rates at room temperatures in systems containing carbonate species (e.g. Mills 1940, Mills and Urey 1940).

Where possible, our examination will be conducted from the viewpoint of elementary kinetic theory, but our experience shows that for many geochemical reactions, detailed breakdown of the exchange kinetics into elementary steps is generally not possible. In the case of diffusion of the light elements in the solid state, details regarding the formation and migration of defects that control the transport processes can not always be quantified, although considerable advances in this area have been made in recent years, and will likely experience more progress in the future. As in our previous review, our intent is to describe the experimental and computational methods used to obtain rates of isotopic exchange for a variety of possible exchange mechanisms. Some redundancy with the previous review is unavoidable in order to put new results in their proper

NOMENCLATURE

A_o	pre-exponential factor from Arrhenius relation for non-diffusion processes
A_j	reaction affinity
A_s	total surface area of solid (m^2)
A'	surface area normalized to density (m^2/cm^3)
$[AX]$, $[BX]$	concentration (or activity if $\gamma=1$) of species with light (common) isotope in phase A and B, respectively
$[AX^*]$, $[BX^*]$	concentration (or activity if $\gamma=1$) of species with heavy (rare) isotope in phase A and B, respectively
C_{tot}	total concentration of water in glass/melt
C^α	concentration of species i in α
d	grain diameter
D_o	pre-exponential factor from Arrhenius relation for diffusion processes
D^*_o	self-diffusion of oxygen recast in terms of defect concentrations
D^α_{tot}	diffusion coefficient relating flux to overall concentration
D^α_i	diffusion coefficient of i in α
D_{gb}	grain boundary diffusion
D_{H2O}	diffusion coefficient of water
Dv	volume (lattice) diffusion
E	isotope effect due to difference in diffusivities of isotopic species (e.g., ^{18}O versus ^{17}O)
E_a	activation energy (kJ/mole)
f	correlation factor
f_i	fugacity of i species
F	fraction of isotopic exchange
h	Planck constant
h	ratio of surface exchange rate to diffusivity
H_f	enthalpy of defect formation
H_m	enthalpy of defect migration
$\Delta H^\#$	enthalpy change during formation of activated complex
$[H_2O]$, $[OH]$, $[O]$	concentrations of individual species in glass/melt
J^α	flux of element/isotope
k	rate constant
k_B	Boltzmann constant
k_+, k_r	forward and reverse rate constants, respectively
k_i	stoichiometric coefficient
k_{sp}	kinetic rate constant of speciation reaction
K_{eq}	equilibrium constant for speciation reaction
K'	equilibrium isotope concentration ratio
L	membrane thickness
M	mass of the solid
M_s	molecular weight of a solid
O^m_i	concentration of oxygen interstitials with m charge
P^α_i	permeability constant of α in i
P_1, P_2, P_o	partial pressures at different points of a permeating species
P_{O2}	partial pressure of oxygen

P	total pressure (generally in MPa)
Q_j	reaction quotient
r	grain radius
r_c; r_o	radius of cation; of oxygen
r*	critical radius
R	proportionality constant between concentration of a species and internal production/consumption of the same
R_{AX}	activity ratio of relevant isotopic species (e.g. [AX*]/[AX])
$\underline{R}$	gas constant (units must be compatible with R or D, and E_a)
R^2	regression fit term
R	overall rate of isotopic exchange
S^{α}	solubility of diffusing species in medium
S	amount of a species produced/consumed in an internal reaction
S_j	van't Hoff slope of the jth reaction
S_i	van't Hoff slope for isotopic reaction
S_c	van't Hoff slope for chemical reaction
$\Delta S^{\#}$	entropy change during formation of activated complex
t	time (usually in sec)
T	temperature (^{o}C or K)
T_c	critical temperature
T_g	glass transition temperature
T_{iso}	isokinetic temperature
V_s	volume of solution
V_o	molar volume of solid
V^n_o	concentration of oxygen vacancies with n charge
$\Delta V^{\#}$	volume change during formation of activated complex
W	mass of water
$(W/S)_m$	mass ratio of water to solid
$(W/S)_o$	mole ratio of oxygen in water to oxygen in solid
x	distance
X^o_w	mole fraction of total water on a single oxygen basis
X_i	mole fraction of i species
Z	ionic porosity
α_{AX-BX}	fractionation factor between two species (AX and BX)
α	element or isotope involved in diffusion or permeation
β	fluid to solid volume ratio
δ	grain boundary width
δ	standard delta notation for isotopic composition
γ_i	activity coefficient of i phase
γ_m	surface area per mole of mineral
$\gamma^{\#}$	activity coefficient of activated complex
κ	ionic conductivity
λ	jump distance
Ω	activity product
Φ	anion porosity
ρ_w; ρ_{soln}	density of water; solution (gm/cm^3)
σ	mineral-fluid interfacial energy (J/m^2)
μ_i	mobility factor for defects
ν	jump frequency

Gaseous or Aqueous System

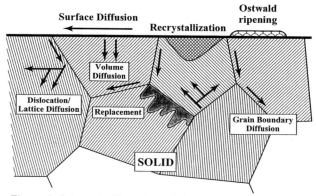

Figure 1. Schematic illustration of the various mass transfer and diffusion processes that can control the rates of isotopic partitioning between a fluid (or gas) and a solid. Many of these are described in this review along with rate processes influential in controlling rates in gaseous and aqueous systems. Modified from Manning (1974).

perspective with earlier studies. Along these lines, we have elected to include the enormous amount of new rate data in with the results presented previously in Cole and Ohmoto (1986). The reasons for this are three-fold: (a) this new volume will supersede the previous volume which may fall out of use, (b) to correct the omissions and mistakes encountered in the earlier compilation, and (c) to put the rate data in a historic perspective for the reader so that if one wishes to revisit a body of work conducted in the past, the tabulation will help point the way. Suffice it to say that many of the earlier efforts still contain very valuable results despite the lack of sophisticated technologies. At this juncture it is important to point out that the tabulations are expansive, but by no means inclusive. For certain chemical systems, the amount of rate data available is truly astounding. Cases in point are the zeolites and oxides—for which data on water and oxygen diffusivities, respectively, are voluminous and deserving of chapters of their own. In a number of cases, we have refit data or recalculated rates using different, but more appropriate rate models. Note that the approach taken in this chapter is to treat the topic of rates in the context of an *Isotopic Exchange Continuum*—i.e. aqueous or gases systems → surface catalyzed or reaction controlled systems → solid state processes (Fig. 1). Abbreviations and symbols used in many of the equations are provided in the Nomenclature section. For the initiated reader, there are a number of excellent texts that present general background, and in many cases, far more detail on various aspects of material covered in this review (e.g. Pilling and Seakins 1995, Schmazlried 1995, Albarede 1995, Lasaga 1998, Criss 1999, Glicksman 2000). Other chapters in this volume (e.g. Valley 2001, Baumgartner and Valley 2001) provide detailed examples of the use of rates in constraining isotopic exchange behavior.

GENERAL KINETIC CONCEPTS OF ISOTOPIC EXCHANGE

Concentration-dependent isotopic rate equation

In the generalized isotopic exchange reaction

$$AX + BX^* \leftrightarrow AX^* + BX \tag{1}$$

it is possible to observe the net extent of reaction, that is, the forward rate (k_f) or the reverse rate (k_r), by use of isotopic tracers (* stands for the heavy isotope) even though the chemical system may be at equilibrium. In a chemical reaction, isotopic exchange occurs where the atoms of a given element (X in the equation above) interchange between two or more chemical forms of the element, or isotopic constituents of molecules such as H_2O and D_2O. An example would be the interchange of oxygen atoms between CO_2 and H_2O. A continuous exchange is presumed to occur when the two are mixed, but it is not observable unless one of the reactants is isotopically "tagged." McKay (1938) was the first to derive a rate law for the simplest case of one exchangeable atom per molecule. Since then, others have published similar derivations, or more complicated variations involving more than two exchanging species (e.g. Norris 1950, Harris 1951, Myers and Prestwood 1951, Bernstein 1952, Luehr et al. 1956, Alberty and Miller 1957, Muzykantov 1980, Friedlander et al. 1981, Kaiser 1991, Chu and Ohmoto 1991).

Although details of the derivation do not require repeating, it is worthwhile showing the essential building blocks of the rate law. Let the concentrations for Equation (1) be designated as follows

$$[AX] + [AX*] = a$$

$$[BX] + [BX*] = b$$

$$[AX*] = x \quad [AX] = a - x$$

$$[BX*] = y \quad [BX] = b - y$$

Then, let R be the rate of exchange, that is, the rate of exchange of all atoms of X whether like or different isotopes. Assume that there is no isotope effect, that is, R is independent of the various isotope masses. Now, regardless of mechanism of the exchange reaction, the rate at which X* in BX* can exchange with X in AX (forward reaction) is proportional to the fraction of AX and AX* that are AX. The situation is similar for the reverse exchange reaction. The net rate is then given by

$$dx/dt = -dy/dt = R\,[y/b][(a-x)/a] - R[(b-y)/b][x/a] \tag{2}$$

or

$$dx/dt = R[(ay - bx)/ab] \tag{3}$$

Since

$$y - y_\infty = x_\infty - x \quad \text{and} \quad x_\infty/y_\infty = a/b$$

then

$$dx/dt = (R/ab)\,[(a-b)(x_\infty - x)] \tag{4}$$

Integrating Equation (4) we get

$$\ln\,[x_\infty /(\,x_\infty - x)] = R/ab(a + b)t = kt \tag{5}$$

where k is the rate constant since R, a and b are constant during an experiment and it is assumed that no AX* was present originally. R is commonly in units of moles per liter (or kilogram) per second. Similarly

$$\ln\,[(y_\infty - y_0)/(\,y_\infty - y)] = kt \tag{6}$$

The result is a first-order rate expression, even though no assumptions have been made about the order of the exchange reaction with respect to the chemical constituents.

R may be any function of a and b. Because the expressions (a − x)/a and (b − y)/b are close to unity, they are commonly omitted from Equation (2), however, Equations (5) and (6) remain the same. Alternatively, Equation (5) may be written in the more familiar form as

$$-\ln (1 - F) = R [(a + b)/ab] \, t \qquad (7)$$

where F is the fraction (progress variable) of isotopic exchange (left-hand term in Eqn. 6) that has occurred. Note that $F = 0$ at the beginning of the exchange process and $F = 1$ at equilibrium (very long times). It is this rate law (Eqn. 7) that Northrop and Clayton (1966) took advantage of when they developed their method of partial isotope exchange for determination of isotopic fractionation factors.

It is important to point out that R is the overall rate of the exchange reaction and not, *senso stricto*, a rate constant (e.g. k_f), because it is a function of one or more factors that govern the rate (e.g. concentrations of [AX] and [BX]). In order to determine the rate of isotopic exchange, the isotopic composition of each phase, AX and BX, are measured as a function of time. By knowing the equilibrium isotope fractionation factors, F values can be calculated and plotted in the form, ln (1 − F) against time, as shown for the system $CO_2(g)$-H_2O-$CaCl_2(aq)$ in Figure 2 modified from Fortier (1994). The slope of a straight line fit in this space yields $-R(a + b)/ab$ which should go through the origin. Deviations from the straight line for short times may be due to an induction process such as either sluggish or rapid reaction while the experiment reaches run temperature. Curvature of the fit at intermediate to long times may be due to the influence of another mechanism (e.g. surface reaction versus volume diffusion), reaction having a different rate order, or that the mechanism is unchanged, but that the experiment is approaching equilibrium with decreasing ΔG_r (Lasaga 1998).

Activity-dependent isotopic rate equation

A useful alternative approach to dealing with the exchange reaction (Eqn. 1) is to derive the rate expression in terms of isotopic values rather than concentrations. Criss (1999) correctly points out that most studies have used extensive variables such as moles which can lead to an unrealistic dependence of exchange rates on the system size. Such problems can be eliminated by the use of intensive variables. Note, however, that the use of concentration units in writing rate laws distinguishes kinetics from thermodynamics. Whereas the thermodynamic concentration or activity determines the equilibrium between thermodynamic components, the spatial concentration (e.g. moles/cm^3) of the colliding molecules determine the molecular collision rates, and hence rates of reaction (Lasaga 1981a). Analogous models of ideality versus non-ideality in kinetics involves ideal and non-ideal collisions defined by collisional cross-sections. Taking the example from Criss (1999) and Criss et al. (1987), we can rewrite the differential rate expression for Equation (1) as

$$d[AX^*]/dt = k_f [AX][BX^*] - k_r[AX^*][BX] \qquad (8)$$

assuming a first-order dependency with respect to (AX); k_f and k_r are the forward and reverse rate constants, respectively. From the definition of the thermodynamic activities of the relevant isotopic species we can write the ratio of the activities as

$$R_{AX} = [AX^*]/[AX] \qquad (9)$$

and

$$R_{BX} = [BX^*]/[BX] \qquad (10)$$

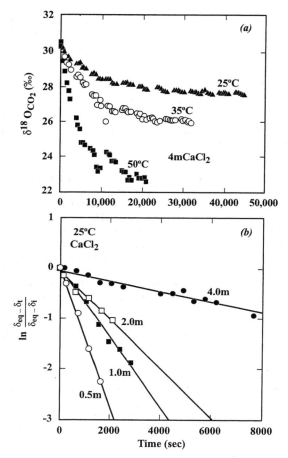

Figure 2. (a) Plot of $\delta^{18}O$ of CO_2 (in per mil relative to VSMOW) reacted with 4 molal (m) $CaCl_2$ versus time in seconds. Symbols: triangles = 25°C, spheres = 35°C, squares = 50°C. Note the greater shift in isotopic values with increasing temperature. (b) Plot of the natural logarithm of the isotopic shifts of CO_2 relative to the equilibrium value versus time (sec) for $CaCl_2$ solutions at 25°C. Symbols: open spheres = 0.5 m, solid squares = 1.0 m, open squares = 2.0 m, solids spheres = 4.0 m. Lines are least-square regression fits to the data. Both figures are modified from Fortier (1994).

where R_{AX} and R_{BX} represent the isotopic ratios. If these ratios are introduced into the activity product of Equation (1), the equilibrium condition is given as

$$\alpha_{AX-BX} = [AX^*][BX]/[AX][BX^*] = R_{AX}/R_{BX} \qquad (11)$$

Equation (8) also reduces to this latter condition when $d[AX^*]/dt$ is zero, conforming to the required condition that the net reaction rate equals zero when equilibrium is attained. The exchange law (Eqn. 8) can now be expressed in terms of isotopic ratios by differentiating Equation (9) giving

$$dR_{AX} = d[AX^*]/[AX] - [AX^*]d[AX]/[AX]^2 \qquad (12)$$

For situations where $[AX^*]$ is a trace isotope that has a very small concentration, the second term on the right-side of Equation (12) is negligible, so that

$$dR_{AX} = d[AX^*]/[AX] \qquad (13)$$

A generalized differential rate equation can now be derived by combining the expressions given in Equations (9), (10) and (13) into Equation (8)

$$dR_{AX}/dt = -k[BX](R_{AX} - \alpha_{AX\text{-}BX}R_{BX}) \tag{14}$$

This expression describes the rate of change of R_{AX} as directly proportional to the difference between the instantaneous isotopic ratio of phase AX and the value that would be in equilibrium with phase BX at that time. The proportionality constant between these quantities, $k[BX]$, represents the product of the rate constant k and the activity BX of the substance with which the phase of interest is exchanging. Criss (1999) points out that when condensed phases are involved, k commonly contains other terms such as surface area or the surface area-to-volume ratio (see also Cole et al. 1983, Cole and Ohmoto 1986).

For cases where pure solids and liquids have unit activities (i.e. $[AX] = 1$ and $[BX] = 1$), Equation (14) simplifies to the rate law of Criss et al. (1987)

$$dR_{AX}/dt = -k\,(R_{AX} - \alpha_{AX\text{-}BX}R_{BX}) \tag{15}$$

Integration of this equation for a case where one of the reservoirs is infinite (e.g. BX) yields the following expression

$$(R_{AX} - R^{eq}_{AX})/(R^{i}_{AX} - R^{eq}_{AX}) = e^{-kt} \tag{16}$$

where k is constant, R^{i}_{AX} and R^{eq}_{AX} refer to the values at $t = 0$ (initial) and $t = \infty$ (equilibrium), respectively. This equation can be recast in terms of standard δ notation $(= [((R_{AX}/R_{std}) - 1)10^3])$, giving

$$(\delta_{AX} - \delta^{eq}_{AX})/(\delta^{i}_{AX} - \delta^{eq}_{AX}) = e^{-kt} \tag{17}$$

The left-hand side of this equation is one of the definitions used to describe $(1 - F)$. Thus

$$\ln(1 - F) = -kt \tag{18}$$

The assumption of the "constancy" of one the species (in this case BX) is analogous to the *method of isolation* described by Lasaga (1998) useful in the manipulation of phenomenological rate laws.

For the case of isotopic exchange in a closed system, the increase of a heavy isotope by one phase must occur at the expense of the heavy isotope content of another, but the total amount of heavy isotope in such a system is constant (Criss 1999). The mass balance requirement for a binary system is given as

$$X_{AX}dR_{AX} = -X_{BX}dR_{BX} \tag{19}$$

where X_{AX} and X_{BX} are the mole fractions of the element of interest for phases AX and BX, respectively. According to Criss et al. (1987), the simultaneous solution to differential Equations (15) and (19) yields

$$(R_{AX} - R^{eq}_{AX})/(R^{i}_{AX} - R^{eq}_{AX}) = e^{-kt(\alpha_{AX\text{-}BX}X_{AX}+X_{BX})X_{BX}} \tag{20}$$

or

$$(1 - F) = e^{-kt(\alpha_{AX\text{-}BX}X_{AX}+X_{BX})X_{BX}} \tag{21}$$

where k is assumed to be constant. For conditions where α is essentially 1.0, and the sum of $X_{AX} + X_{BX}$ is 1.0 for a binary system, Equation (21) can be approximated by

$$(1 - F) = e^{-kt/X_{BX}} \tag{22}$$

Criss (1999) points out that this equation reduces to Equation (15) for an infinite reservoir because $X_{BX} \to 1.0$.

A similar analysis of the rate of reaction (Eqn. 1) can be conducted assuming a second–order dependency rather than first-order as outlined above. The differential equations can be expressed as

$$-d[AX]/dt = -d[BX^*]/dt = d[AX^*]/dt = d[BX]/dt = k \, [AX][BX^*] \qquad (23)$$

The integrated form of this type of differential, cast in terms of isotope activities and F, analogous to the derivation of Equation (18) yields the following expression

$$F/(1 - F) = kt \qquad (24)$$

This type of formalism was used by Graham (1981) and Stoffregen et al. (1994) to assess the rates of isotopic exchange between hydrous silicates and H_2O and alunite and H_2O, respectively. As described above, for condensed phases, k can contain hidden terms for surface area, or surface area-to-volume ratio. The rate constant k should conform to straight line fits to data plotted as $F/(1 - F)$ versus time or sometimes $t^{1/2}$.

Temperature-dependence of rate constants: activation energy and entropy

For most reactions, including those involving isotopic exchange, the rate increases with temperature following the classic equation proposed by Arrhenius given as

$$k = A_o e^{-E_a/\underline{R}T} \qquad (25)$$

or

$$\ln k = \ln A_o - E_a/\underline{R}T \qquad (26)$$

where A_o is the pre-exponential factor (D_o is commonly used in the case of diffusion), E_a is the activation energy, $\underline{R}$ is the gas constant, and temperature is in Kelvin. In the case of Equation (14), the α can have a temperature-dependency (although some examples exist where there is little or no apparent dependency as in the case of hydrogen isotope fractionation in certain hydrous silicate-H_2O systems), as might the chemical activity of BX in Equation (14). Additionally, detailed molecular theory commonly yields a temperature-dependent pre-exponential factor (Lasaga 1981a). Furthermore, the activation energies of overall reactions, E_a, may themselves depend on temperature. This would lead to some form of curvature or inflection in plots of ln k versus 1/T which typically produce straight line fits where the slope determines the E_a and the intercept gives A_o. An example Arrhenius diagram is presented in Figure 3 for representative gaseous and aqueous systems described in Table 1 (see Appendix).

For many geochemical reactions, the activation energy represents the overall reaction and is composed of several activation energies from the elementary reactions. For isotopic exchange reactions in aqueous solution this might involve the formation and exchange of one or more intermediate complexes (Ohmoto and Lasaga 1982). The size of E_a, besides providing the very important temperature-dependence, can also commonly be used to gain insights into the reaction mechanism. For example, isotopic exchange rates in solutions can have rather low activation energies—on the order of 20 kJ/mole, whereas E_a for isotopic exchange reactions involving mineral transformations or surface reactions range from roughly 40 to 80 kJ/mole. The E_a values for solid state diffusion can extend even higher to several hundred kJ/mole. Of course, catalytic effects such as the presence of reactive surfaces or defects can reduce the activation energies (i.e. activation barrier) appreciably. This is particularly true for isotopic exchange among gases or between gases and solid (crystalline or amorphous) surfaces.

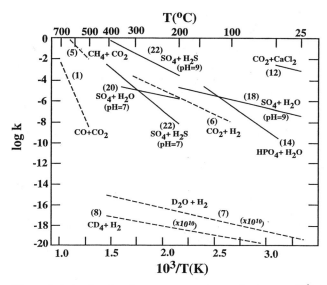

Figure 3. Arrhenius plot of rate constants, given as log k versus $10^3/T$ (K) for select examples of isotopic reactions in gaseous systems (dashed lines) and aqueous systems (solid lines). Numbers refer to rate data for individual reactions summarized in Table 1 (see Appendix). Consult Table 1 for the proper rate units used for these examples because they are not all the same. Also note that rate constants for reactions (7) and (8) have been scaled by 10^{10} for convenience of plotting.

It is instructive to make a closer examination at what E_a and A_o mean in the context of *Transition State Theory* (TST). The rate constant for isotopic reaction can be cast in the following form

$$k = (k_B T / h)[\gamma_{AX}\gamma_{BX} / \gamma^{\#}]e^{\Delta S^{\#}/\underline{R}}e^{-\Delta H^{\#}/\underline{R}T}e^{-(P-1)\Delta V^{\#}/\underline{R}T} \qquad (27)$$

where $\Delta S^{\#}$, $\Delta H^{\#}$ and $\Delta V^{\#}$ are the changes in entropy, enthalpy and volume involving the formation of the activated complex, k_B is the Boltzmann constant and h the Planck constant. For solids and liquids in their standard states with unit activity, the second term in the above equation involving the activity coefficients (γ) drops out. The activation energy is related to the last two terms in the following manner

$$E_a = \underline{R}T + \Delta H^{\#} + (P - 1)\Delta V^{\#} \qquad (28)$$

The term $\underline{R}T$ is usually less than about 4 kJ/mole and the last term in (28) is typically negligible because $\Delta V^{\#}$ is small, so E_a and $\Delta H^{\#}$ are approximately equal (Lasaga 1981b). [However, we will show later that this is not always the case, e.g. select oxygen isotope diffusivities and reaction rates can exhibit an appreciable pressure-dependence.] The pre-exponential term from Equation (25) is composed of the first three terms in Equation (27). Assuming unit activities for the reacting species, A_o can be approximated by

$$\ln A_o = \ln (k_B T/h) + \Delta S^{\#}/\underline{R} \qquad (29)$$

Two important points should be noted here. First, knowledge of A_o from the fit of isotopic rate data to the Arrhenius relation leads to an estimation of $\Delta S^{\#}$. As we will see later, correlation of $\Delta S^{\#}$ against E_a ($\approx \Delta H^{\#}$) serves as a potential predictive tool for estimation of rates as well as insight into exchange mechanisms. Additionally, the magnitude and sign of $\Delta S^{\#}$ has been widely used as an indicator of the configuration of the activated complex. In most cases the values of $\Delta S^{\#}$ derived from A_o are negative because $A_o < k_B T/h$. As a general guide, if the reactant molecules are separated by long bonds in the complex, then the decrease in entropy will not be large and A_o will be high. Conversely, if bond lengths in the activated complex are short then $\Delta S^{\#}$ will be a larger negative value and A_o will be low (Lasaga 1981b). It is important to note that $\Delta S^{\#}$ can be more rigorously estimated from partition functions and hence the vibrational data on isotopic molecules.

Compensation law and the isokinetic temperature

Perhaps the second most widely used empirical observation stemming from the Arrhenius relation is the *compensation law*, which relates the pre-exponential and the activation energy regardless of the reaction mechanism, e.g. diffusion, chemical reaction, etc. (Winchell 1969, Winchell and Norman 1969, Hart 1981). Equations (28) and (29) give qualitative explanation for compensation. In general, $\Delta H^{\#}$ and $\Delta S^{\#}$ tend to change in the same way with temperature or as the chemical system changes, i.e. both increase or both decrease (Lasaga 1981b). Decreases in enthalpy, $\Delta H^{\#}$, are associated with tighter binding of the activated complex (more exothermic). But this tighter binding is also responsible for a decrease in entropy, $\Delta S^{\#}$. When these changes are linearly dependent, then $\Delta S^{\#} = n\Delta H^{\#}$. Using Equations (28) and (29), we get

$$\ln A_o = \ln [k_B T/h] - nT + E_a(n/\underline{R}) \tag{30}$$

or

$$\ln A_o = a + bE_a \tag{31}$$

where T is constant, and a and b are regression coefficients. Typically, experimental data are plotted as E_a versus either $\ln A_o$ (D_o) or $\log A_o$ (see Fig. 4). The distribution of data along any given trend can commonly provide qualitative insight into subtleties related to the compositional and/or structural dependency of rates of exchange. In principle, the results of a set of experiments at one temperature can be used to deduce the activation

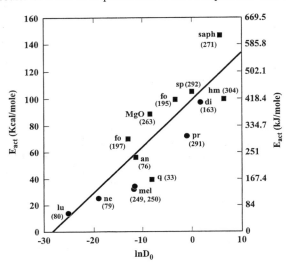

Figure 4. Plot of activation parameters, E_a versus $\ln D_o$, for oxygen diffusion in anhydrous silicates. The correlation of data on this plot conforms to the compensation law described in the text. The numbers refer to data summarized in Table 2 (see Appendix). Modified from Muehlenbachs and Connolly (1991).

energy and therefore the variation of rate with temperature, without conducting any further experiments at several different temperatures. However, there is considerable error associated with this approach. This can be accomplished by solving Equations (26) and (31) for the two activation parameters giving

$$E_a = (\ln k - a)/(b - 1/\underline{R}T) \tag{32}$$

and

$$\ln A_o = \ln k + (\ln k - a)/(b\underline{R}T - 1) \tag{33}$$

Experimentally determined ln k for a given temperature can be substituted into these two equations if the chemical system is expected to conform to both the Arrhenius equation and the compensation law (Lasaga 1998).

One interesting consequence of the compensation law in Equation (31) is the prediction that all rates (chemical, diffusional, etc.) become equal at some universal temperature called the *isokinetic temperature*. From Equation (31)

$$A_o = e^a \, e^{bE_a} \tag{34}$$

Substituting this expression into the Arrhenius relation (Eqn. 25), we get

$$k = e^a \, e^{bE_a} \, e^{-E_a/RT} \tag{35}$$

and it follows that at the isokinetic temperature,

$$T_{iso} = 1/\underline{R}b \tag{36}$$

All the rate constants have the same value,

$$k_{iso} = e^a \tag{37}$$

This parameter can be useful when the estimates for T_{iso} fall within reasonable geological limits, which is not always the case. For example, Muehlenbachs and Connolly (1991) presented an activation parameter plot (Fig. 4; E_a versus ln D_o) that gives

$$\ln D_o = 6.846 \times 10^{-5} \, E_a - 28.1 \tag{38}$$

where E_a is in kJ/mole and D is in cm^2/sec. Using the compensation law in Equation (36) leads to an isokinetic temperature of 1484°C, where the oxygen diffusion will have the same diffusion coefficient, i.e. $D \approx 10^{-12}$cm^2/sec. Based on the spread of data given in Figure 4, the isokinetic temperature is no better than ± a few hundred degrees and ±~0.5 log units in D.

ISOTOPIC EXCHANGE IN GASEOUS AND AQUEOUS SYSTEMS

General background

The mechanisms and rates of isotopic exchange in homogeneous systems (i.e. gases or aqueous solutions) are still poorly understood for a number of reactions of geological relevance such as carbon isotope exchange between CO_2 and CH_4. Although most isotopic reactions taking place between aqueous species in solution or among coexisting gases are rapid geologically speaking, some important reactions are slow enough to warrant further consideration. Examples include carbon isotope exchange between CO_2 and CH_4, sulfur and oxygen isotope exchange among aqueous species in the S-O-H system, oxygen isotope exchange between dissolved phosphate and H_2O, and hydrogen

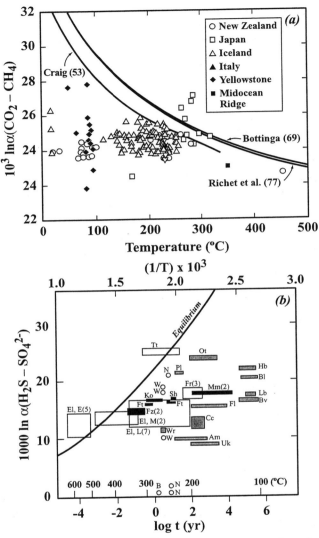

Figure 5. (a) Plot of measured $\delta^{13}C$ differences between geothermal CO_2 and CH_4 gases (corrected for the algebraic difference between Δ and 10^3 lnα) versus measured bore hole or surface temperature. Calculated fractionation curves by Craig (1953), Bottinga (1969) and Richet et al. (1977) are also shown. Note that for most gases the measured fractionation is smaller than that expected for equilibrium at the temperature of the system. Consult Horita (2001) for the sources of the data. (b) Similar plot showing the sulfate-sulfide isotopic fractionation factors for natural hydrothermal systems plotted as a function of their known temperature. Shown on the horizontal axis beside temperature are the times in years to attain 90% approach to equilibrium for solutions containing $\Sigma S = 0.01$ m and pH = 4-7. Consult Ohmoto and Lasaga (1982) for the sources of the data and the equation for the equilibrium curve.

isotope exchange between H_2 and $H_2O_{(v)}$ or $CH_{4(g)}$. The slow isotope exchange rates observed for various isotope pairs in these systems will have a profound effect on their applicability as useful geothermometers. A case in point is the disagreement between carbon isotope temperatures estimated from the pair CO_2-CH_4 and measured temperatures for a variety of natural systems, as shown in Figure 5a. One of the commonly held explanations is that CO_2 and CH_4 were originally in isotopic equilibrium at depth, but during ascent to shallow depths, the original carbon isotopic ratios were "frozen in" due a significant decrease in the isotopic exchange rate (Horita 2001). Alternatively, these species may have never been in equilibrium at depth. A similar kind of "disequilibrium" behavior was noted by Ohmoto and Lasaga (1982) for isotopic exchange between sulfate and sulfide in hydrothermal solutions (Fig. 5b).

The exchange of isotopes between reactants or between reactants and the medium may be categorized into three basic types of reactions: (1) in the absence of an over-all chemical reaction (isotopic exchange in a chemically equilibrated system), (2) in the presence of a reaction proceeding to completion (irreversible), and (3) in the presence of a reversible reaction. The latter two refer to systems that are not at chemical or isotopic equilibrium at the onset of reaction. Superimposed onto these is an added complexity that the reactions affecting isotopic compositions of solution or gaseous species may be catalyzed by mineral surfaces. In fact, a number of the isotopic exchange reactions common to "homogeneous systems" will only undergo exchange in the presence of a mineral (or metal) catalyst. Variables of interest besides the catalytic surface that can influence exchange include temperature, partial pressure (fugacity), ionic strength, species concentration, pH, and time. The intent of this section is to document, by way of selected examples, how changes in the magnitudes of these variables can (a) determine the rate of isotopic exchange and (b) lead to the interpretation of the exchange mechanism(s). Table 1 (see Appendix) provides a summary of isotopic rate data for a select number of gaseous and aqueous systems.

Activity-dependent considerations

As noted above, isotopic exchange of a trace isotope (*) can occur among gaseous species or between a gas and a condensed phase (solid, liquid). Criss (1999) points out that for such systems, the activities of the gaseous species will equal the fugacities of the relevant isotopomers, f^* and f, whereas the solid or liquid will have unit activities for the pure phase. Considering a case where a gaseous phase (AX) is reacting with a condensed phase (BX), the relevant activities of the species are

$$AX = f_{AX} \quad BX = 1 \quad AX^* = f^*_{AX} \quad BX^* = R_{BX} \tag{39}$$

The isotopic ratio R_{AX} is either equivalent or directly proportional to the fugacity ratio of the gas species (f^* and f). Furthermore, the fugacity ratio is equal to the partial pressure of the gas species (P^* and P) due to the cancellation of the fugacity coefficients because of the ideality of isotopic mixtures (Criss 1999). Therefore, the isotopic ratio for AX, R_{AX} of the gas is given by

$$R_{AX} = AX^*/AX = f^*_{AX}/f_{AX} = P^*_{AX}/P_{AX} \tag{40}$$

and where R_{BX} was defined previously by Equation (9). The rate law that describes the rate of isotopic change of the gas will follow Equation (15), where, for illustrative purposes, we see that

$$dR_{AX}/dt = -k \left[(P^*_{AX}/P_{AX}) - \alpha_{AX\text{-}BX}R_{BX} \right] \tag{41}$$

and for the case where BX is an infinite reservoir, Equation (18) is applicable. Historically, it is worth noting that isotopic fractionation factors between vapor and liquid have been determined by measuring the vapor pressure differences for vapor equilibrated with isotopically pure end-member liquids (e.g. D_2O versus H_2O; $H_2{}^{18}O$ versus $H_2{}^{16}O$; e.g. Van Hook 1972).

Criss (1999) concludes that isotopic exchange involving gaseous molecules can be described by the fugacities of the relevant species in every case. That is, the isotopic ratio for R_{BX} can be defined in a similar manner as R_{AX} (Eqn. 40) where

$$R_{BX} = BX^*/BX = f^*{}_{BX}/f_{BX} = P^*{}_{BX}/P_{BX} \qquad (42)$$

He indicates that the rate law given by Equation (14) most likely describes the behavior of gaseous systems. As noted above, isotopic exchange rates for gaseous systems such as CO_2 and CH_4 , or H_2 and CH_4 are extremely sluggish. As we will see in the next section, this is because, unless these species participate in a chemical reaction involving formation of intermediate complexes, dissociate, or are catalyzed by a solid surface, exchange will occur through some form of free-molecule collisional mechanism which can be very slow.

Rates of isotopic exchange between gases

As Chacko et al. (Chapter 1, this volume) point out, of all of the possible isotopic reactions, those involving the gas species are the most straightforward to treat with current theory—i.e. statistical mechanical calculations of equilibrium isotope partition theory. However, of the hundreds of reactions that have been observed to proceed in the gas phase, relatively few, if any, can be described in terms of a single chemical transformation (Benson 1960). The majority proceeds through a more or less complex chemical mechanism that involves the formation and destruction of reaction intermediates and free radicals. By extension, we can presume that isotopic exchange between gaseous species can occur through either: (a) simple, non-catalyzed molecular collisions, or (b) homogeneous or heterogeneous catalyzed reactions, (e.g. oxidation-reduction reactions). Examples of these are presented below.

Non-catalyzed gaseous systems. There are some systems whose isotopic exchange rates are not enhanced by the presence of catalysts. Deuterium-hydrogen exchange between H_2 and either $H_2O_{(v)}$ or CH_4 may exemplify such systems (Lecluse and Robert 1994). Horita (1988a,b) has demonstrated that fast equilibration times (1 to 2 hrs) for hydrogen isotope exchange between H_2 and liquid H_2O (or brines) require a Pt catalyst to allow for this method to be an effective means for measuring the isotopic compositions of solutions. In the absence of noble metal catalysts, however, such is not the case when H_2O coexists as a vapor phase with either H_2 or CH_4 like that encountered in vapor-dominated geothermal systems or solar nebular processes. Lecluse and Robert (1994) have experimentally measured the exchange rates as a function of temperature (25 to 434°C) and an assortment of catalysts (activated charcoal, montmorillonite, iron metal and quartz) for the following reactions

$$D_2O(v) + H_2 \rightarrow HDO(v) + HD \qquad (43)$$

$$CD_4 + H_2 \rightarrow CH_3D + HD \qquad (44)$$

The differential equation for the appearance of HD in Equation (43) is given as

$$d[HD] / dt = k_f [P_{D_2O}]^n [P_{H_2}]^m \qquad (45)$$

where k_f is the rate constant, P stands for partial pressure and t is time. For simplicity, only one set of exchangeable molecules was assumed so that n and m both equal 1. Consideration of the following mass balance relationships involving initial (i) and final (f) measured values

$$[P_{D_2O}]_i = [P_{D_2O}]_f + [P_{HDO}]_f \tag{46}$$

$$[P_{D_2O}]_i = [P_{D_2O}]_f + [P_{HD}]_f \tag{47}$$

$$[P_{H_2}]_i = [P_{H_2}]_f + [P_{HD}]_f \tag{48}$$

can lead to a recasting of Equation (45) in the form

$$d[P_{HD}] / dt = k_f([P_{D_2O}]_i - [P_{HD}])([P_{H_2}]_i - [P_{HD}]) \tag{49}$$

For the experimental conditions employed by Lecluse and Roberts (1994), $[P_{H_2}]_t \gg [P_{HD}]$ such that $[P_{HD}]^2$ can be neglected, and $[P_{H_2}]_i = [P_{H_2}]_f$ due to only minor conversion of H_2 to HD; thus the *method of isolation* can be used. The integrated version of Equation (49) will then become

$$[P_{HD}] / [P_{H_2}] = \exp\{-kt([P_{D_2O}]_i + [P_{H_2}]_i)\}A + X_{D_2O} \tag{50}$$

where

$$A = [P_{HD}]_i / [P_{H_2}]_i - X_{D_2O} \tag{51}$$

$$X_{D_2O} = [P_{D_2O}]_i / ([P_{D_2O}]_i + [P_{H_2}]_i) \tag{52}$$

Equation (50) can be rewritten in terms of (D/H) ratios as

$$2(D / H)_f = [2(D / H)_i - X_{D_2O}]\exp\{-ktP\} + X_{D_2O} \tag{53}$$

A similar expression can be derived for the H_2-CD_4 exchange process. In their experiments, Lecluse and Robert (1994) measured the $(D/H)_f$ in H_2 with conventional mass spectrometry whereas the X_{D_2O} (or X_{CD_4}), the molar fraction of D_2O (CD_4) and P, the total pressure were both measured manometrically. They demonstrated convincingly that neither reaction rate (Eqns. 43 and 44) was greatly influenced by the presence of a catalyst (Figs. 6a,b). Furthermore, the rate of H_2-$D_2O_{(v)}$ was observed to be faster than H_2-CD_4 by nearly two orders of magnitude. Interestingly, the activation energy is higher for water-hydrogen ($E_a \cong 43$ kJ/mole) compared to methane-hydrogen ($E_a \cong 36.4$ kJ/mole). This would be incorrect if isotopic exchange occurred in a single step process for both reactions because the chemical bond for C-D (in CD_4) is more energetic than for O-D (in D_2O). Therefore, it is likely that the deuterium transfer is controlled by several elementary steps which their experimental approach could not delineate. The Arrhenius parameters for reactions given in Equations (43) and (44) are summarized in Table 1 (see Appendix). Ultimately, they used these results to model the hydrogen isotopic composition of water in the solar nebula from −100 to 100°C.

Catalyzed gaseous systems. The reaction between CO and CO_2 at elevated temperatures (~860-920°C) represents a classic example of carbon isotope exchange enhanced by catalysis with surfaces such as quartz, gold and silver (Brander and Urey 1945, Hayakawa 1953). In addition to rate enhancement on catalytic metal or oxide surfaces, Brander and Urey (1945) also observed an increase in rate in the presence of H_2 or $H_2O_{(v)}$. They tested a variety of mechanisms to explain their experiments which involved reaction of ^{13}C-enriched CO_2 with normal CO as a function of temperature, gas

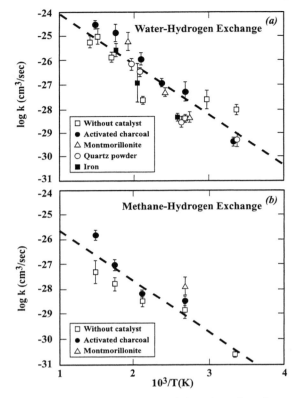

Figure 6. (a) Logarithm of rate constants (k) plotted against $10^3/T$ for hydrogen isotope exchange between water and hydrogen for different types of catalysts. Slope of the linear-regressed line corresponds to an E_a values of 43.8 kJ/mole. (b) Similar plot showing rate constants for the hydrogen isotope exchange between methane and hydrogen. The best-fit line to these data yields an E_a value of 36.4 kJ/mole. Note the lack of dependence on the catalytic surface. Figures modified from Lecluse and Robert (1994).

pressure and composition (gas ratio), and catalyst for reactions of the sort described here:

$$^{13}CO_2 + {}^{12}CO \leftrightarrow {}^{13}CO + {}^{12}CO_2 \tag{54}$$

$$2\,{}^{13}CO_2 \leftrightarrow 2\,{}^{13}CO + O_2 \tag{55}$$

$$^{13}CO + SiO_2 \leftrightarrow {}^{13}CO_2 + SiO \tag{56}$$

$$^{13}CO_2 + H^+ \leftrightarrow {}^{13}CO + OH^- \tag{57}$$

In the case of Equation (54), the rate of disappearance of $^{13}CO_2$ is obtained from the difference in the rates of the forward and reverse reactions, by the equation

$$-d({}^{13}CO_2)/dt = k_f\,[{}^{13}CO_2][{}^{12}CO] - k_r[{}^{12}CO_2][{}^{13}CO] \tag{58}$$

This equation can be used to derive a first-order rate expression for the overall rate, R, as given by

$$R = -\ln(1-F)\,[P(CO_2)\,P(CO)]/\,[P(CO) + P(CO_2)] \tag{59}$$

where P refers to partial pressure and F is the degree of equilibrium based on the isotopic ratios, R, at $t = 0$, $t = f$ (measured) and infinity (∞)

$$F = [R^f CO_2 - R^\infty CO_2]\,/\,[R^i CO_2 - R^\infty CO_2] \tag{60}$$

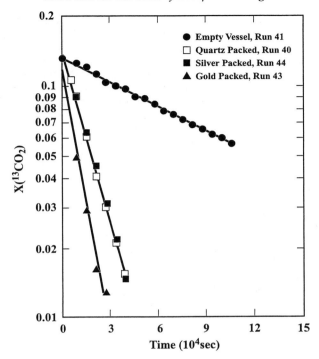

Figure 7. Change in the mole fraction of ^{13}C in unreacted CO_2 plotted against time in (10^4 sec) for the systems $CO-CO_2$. These gases were reacted with different types of potential catalytic surfaces such as SiO_2, Au, Ag, etc. to assess their role in promoting isotopic exchange. The results depicted here were obtained from experiments conducted at 869°C. Data obtained from Brander and Urey (1945).

Hayakawa (1953) found that Equation (59) could be used to adequately determine the rate constant, k, when properly cast in the form of

$$R = k_f [(PCO) \cdot (PCO_2)] / (1 + b\,PCO) \qquad (61)$$

where b is the slope of plots of $1/R$ versus PCO_2 or $1/PCO$. This expression was used to describe results of the anhydrous system only slightly catalyzed by SiO_2 surfaces. Figure 7 demonstrates how the rate of $^{13}CO_2$ changes as a function of time and catalyst type. Note that the results obtained for SiO_2 and Au surfaces can not be distinguished. For the case where carbon isotope exchange was promoted by the presence of H^+ and OH^- sorbed onto quartz surfaces, Brander and Urey (1945) derived a differential rate expression that took into account the total concentration of these ions on the surface (C_1)

$$-d(R_{CO_2}) / dt = \{k_f C_1 [PCO_2 + PCO] / K_{57}[PCO_2 + PCO]\}(R_{CO_2} - R^{\infty}_{CO_2}) \qquad (62)$$

where K_{57} is the equilibrium constant for the chemical reaction represented in Equation (57). Although the carbon isotope exchange in the system $CO-CO_2$ is very fast in geological terms, the reactions illustrate the importance of considering the role of catalytic surfaces on exchange, and furthermore, the possible influence of other volatiles on the rate—in this case dissociated water vapor.

Chemical reaction and isotopic exchange: CO_2-CH_4. Despite the importance of understanding the carbon isotope systematics of the CO_2-CH_4 pair, the reaction rates are not well constrained. This is because of two main reasons: (a) isotopic exchange between these gases is extremely slow in the absence of catalysts, and (b) suitable, geologically relevant catalysts are less effective compared to more conventional, industry-accepted catalysts such as silica gel and certain transition metals (e.g. Ni, Fe) (Horita 2001). Two types of exchange reactions can be envisioned based on the current status of experimental data: (a) carbon isotope exchange (or lack of) for CO_2-CH_4 reacted at elevated temperatures in the presence of non-active catalysts (e.g. Sackett and Chung 1979, Harting and Maass 1980), and (b) reaction in the presence of active catalysts or reaction intermediates that promote both chemical and isotopic reaction, chiefly CO_2 reduction to CH_4 (e.g. Kharaka et al. 1983, Giggenbach 1997, Horita and Berndt 1999, Horita 2001).

Consider the following isotopic exchange reaction

$$^{13}CH_4 + {}^{12}CO_2 \leftrightarrow {}^{13}CO_2 + {}^{12}CH_4 \tag{63}$$

The work of Sackett and Chung (1979) demonstrated that at 500°C for up to 10.5 days, no observable carbon isotope exchange occurred between CO_2 and CH_4 reacted in the presence of water and montmorillonite. In similar experiments, Harting and Maass (1980) observed very limited carbon isotope exchange between normal isotopic CO_2 and 66% $^{13}CH_4$ at temperatures between 500 and 680°C up to 16 hours, also in the presence of montmorillonite and water (probably as water vapor?). The second order rate constants (k_2) reported by Harting and Maass (1980) were derived from the relationships

$$k_2 = X / (X_{^{13}CH_4})(X_{^{12}CO_2})t \tag{64}$$

and

$$k_2 = \delta^{13}C_{CO_2} X_{^{13}CO_2} / 10^3 (X_{^{13}CH_4})(X_{^{12}CO_2})t \tag{65}$$

where X represents the atom fraction of isotope exchanged, $\delta^{13}C_{CO_2}$ is the measured change in isotopic composition of CO_2, $X_{^{13}CH_4}$ and $X_{^{12}CO_2}$ are the atom fractions of these species in the gas mixture, and t the reaction time. Because $X_{^{13}CH_4}$ and $X_{^{12}CO_2}$ can be taken as constants (again, *method of isolation*), Giggenbach (1982) recast Equation (65) as a pseudo-first order equation describing the decrease in $X_{^{13}CH_4}$ in terms of the measured increase in $X_{^{12}CO_2}$ yielding

$$-dX_{^{13}CH_4} / dt = dX_{^{12}CO_2} / dt = k_1 X_{^{13}CH_4} \tag{66}$$

He related his pseudo-first order rate constant to the rate law derived by Harting and Maass (1980) through the following expression

$$k_1 = k_2 [X_{^{12}CO_2}] = k_2 (\%^{12}C_{CO_2})(X_{CO_2}) \tag{67}$$

such that Equation (65) now becomes

$$k_1 = \delta^{13}C_{CO_2} X_{^{13}CO_2} / 10^3 (X_{^{13}CH_4})t \tag{68}$$

In terms of the fraction of isotopic exchange, F, this expression can be cast as

$$k_1 \approx F / [X_{^{13}CH_4}]t \tag{69}$$

Giggenbach (1982) refit the data of Harting and Maass (1980) and reported the temperature-dependence of k_1 as

$$\log k_1 = 11.6 - 10,190/T \tag{70}$$

which differed from the second-order temperature-dependency given as

$$\log k_2 = 9.26 - 10,190/T \tag{71}$$

with T in Kelvin. Giggenbach (1982) correctly points out that Equation (70) is really only valid for carbon isotope exchange between CO_2 and CH_4 where catalytic activity is low or essentially non-existent. In this case, the water and montmorillonite had little or no influence on exchange rate. Figure 8 shows the curve defined by Equation (70), along with a number of other rate curves for chemical processes common to the C-O-H system.

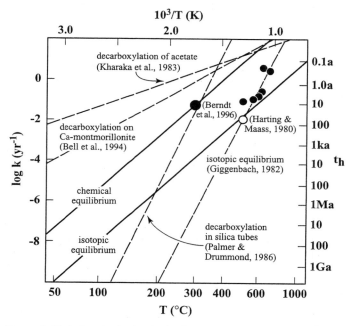

Figure 8. Variations in the log k, the first order rate constant (year^{-1}), for attainment of chemical and isotopic equilibration between CH_4 and CO_2 governed by the oxidation-reduction data derived from Berndt et al. (1996), Harting and Maass (1980), and Giggenbach (1982). Rates of decarboxylation of acetic acid in stainless steel from Kharaka et al. (1983), in silica tubes (Palmer and Drummond 1986), and in the presence of Ca-montmorillonite (Bell et al. 1994). Figure modified from Giggenbach (1997).

Now consider isotopic exchange between CO_2 and CH_4 participating in chemical reactions of the sort

$$CO_2 + 4 H_2 \leftrightarrow CH_4 + 2 H_2O \tag{72}$$

Recent studies indicate that chemical and isotopic exchange may proceed through a number of different intermediate aqueous compounds such acetate or formate, and commonly require the participation of some type of solid catalyst (e.g. Berndt et al. 1996,

Horita and Berndt 1999). Giggenbach (1997) proposed an interesting empirical approach for estimating rates of isotopic exchange for systems driven by chemical reactions like the redox reaction shown in Equation (72). The underlying assumption is that the same activated complexes and rate-determining steps controlling the chemistry influence the isotopic exchange. That is not to say that the rates of approach to equilibrium have to be the same for the chemical and isotopic reactions. Additional key assumptions include: (a) all species are dissolved into a single fluid phase, (b) the kinetics are controlled by a single rate-determining step, and (c) the conversion of CO_2 to CH_4 as shown in Equation (72) is described by a pseudo-first order reaction according to

$$-dX_{CO_2} / dt = k_f = dX_{CH_4} / dt = k_c X_{CO_2} \qquad (73)$$

where X_i are the mole fractions of the species i, t is time, r_f, the rate constant of the forward reaction, and k_c, the pseudo-first order rate constant for the chemical reaction. The key to this approach is the fact that kinetics and thermodynamics can be linked through the following expression (from Boudart 1976)

$$R_j = k_f(1 - e^{-A_j/\sigma \underline{R}T}) \qquad (74)$$

where the forward rate constant, k_f, is related to the net rate, R_j through the use of the reaction affinity, A_j. Assuming the reaction proceeds along a single pathway, σ will be unity. The total affinity is defined as

$$A_j = -\underline{R}T \ln (Q_j/K_j) \qquad (75)$$

where K_j is the equilibrium constant for Equation (72) and Q_j is the reaction quotient at time t, and their ratio is a measure of reaction progress which approaches unity asymptotically near equilibrium (approaching infinite time). With this definition of A_j Equation (74) reduces to

$$R_j = k_f (1 - Q_j/K_j) \qquad (76)$$

The premise behind Giggenbach's method is that the relative rates of approach to chemical and isotopic equilibrium may be provided by the relative rates of response of a system in full chemical and isotopic equilibrium to an incremental change in temperature. This is obtained by differentiating Equation (76) with respect to dT according to

$$dR_j/dT = -k_f (dQ_j/dT)/K_j \qquad (77)$$

The relative difference in rates can be determined by assuming the temperature-dependence of Q_j follows van't Hoff behavior, $\log Q_j = S_j/T +$ constant, and with Q_j close to K_j. The value for the van't Hoff slope for the isotopic exchange reaction, S_i, can be obtained from rearranging the equation for equilibrium fractionation (taken from Bottinga 1968), where

$$\ln\alpha_{CO_2-CH_4} = 25.961 / T - 0.0208 \qquad (78)$$

$$\log\alpha_{CO_2-CH_4} = 11.3 / T - 0.00903 \qquad (79)$$

The van't Hoff slope (isotopic) of 11.3 can then be compared to the van't Hoff slope of the chemical reaction, $S_c = 4625$, taken from the combination of experimental results of Berndt et al. (1996), Huff and Satterfield (1984) and Zimmerman and Bukur (1990), and yields a S_c/S_i value of 409. This result implies that at any temperature, independent of R_f, the isotopic rate may be over 2.5 orders of magnitude slower than the chemical rate. The notion here is that the ratio of chemical to isotopic rates of equilibration for single phase systems not involving a solid may be estimated by comparing the van't Hoff slopes, with $S_j = -\Delta H_j/\underline{R}T$, the reaction enthalpies. For Equation (76), the isotopic rate can be represented by

$$\log k_i = 3.92 - 4440/T \tag{80}$$

assuming that the activation energies are the same for both chemical and isotopic reactions; in this case, roughly 85 kJ/mole. The curve for Equation (80) is shown in Figure 8 along with the curves for the non-catalyzed isotopic carbon isotope exchange reaction, Equation (63) and other curves for various types of decarboxylation reactions that produce CH_4 as a reaction product. This new empirical approach described by Giggenbach (1997) can provide valuable insights into the "near-equilibrium" rates of isotopic exchange reactions for homogeneous systems where rates may be prohibitively slow for experimental investigation at diagenetic and hydrothermal conditions.

Rates of isotopic exchange in aqueous solutions

The rate expressions derived previously (e.g. Eqns. 7, 17 or 22) for isotopic exchange demonstrate the dependence of the rate on concentration or activity of the equilibrating species. The rate law is established by monitoring the time-dependence of the isotopic composition of the species participating in the reaction, and by determining how this dependence is affected by altering the constraints on the reacting system. Depending on the starting conditions, the isotopic composition of one or more species should be measured as a function of time at various temperatures, pressures, and initial concentrations (activities). Isotopic exchange among aqueous species (e.g. SO_4^{2-} and H_2S) should be monitored as a function of not only time and the aforementioned variables, but also pH and ionic strength. The pH-dependency is related to the formation of aqueous species, which can participate, as intermediates, in the isotopic exchange process. Changes in speciation will result from variations in pH, temperature, pressure, and solution composition.

One of the earliest applications of stable isotopes was in isotopic labeling of reactants in order to monitor the progress of chemical reaction in solution and measure the hydration numbers. For example, the rate constant for hydration of CO_2 in acidic solutions was first studied using ^{18}O labels in the pioneering work of Mills and Urey (1940). Since that time, numerous studies have been conducted using ^{18}O (or other stable isotopes such as ^{34}S) tracer techniques to quantify rates of oxygen exchange between aqua $[M(OH_2)^{z+}]$ or oxo-ions $[XO_n]^{z-}$ and H_2O, where M can be atoms such as Al, Cr, Co and Rh, whereas X may represent atoms such as C, S, Se, N, and P. Using isotope tracer methods, quantitative information can be obtained on (a) the number of positions occupied by water, hydroxide and oxide ions around the central atom M or X, (b) the kinetic properties of the aqua-cations and oxo-anions in aqueous solution, and (c) the mechanism(s) by which these exchange processes occur. The advantages of the isotope method are the high accuracy and precision of the mass spectral isotope ratio measurement and the requirement for only low levels of ^{18}O isotope, allowing highly dilute samples (mM) to be handled (Richens 1997). Its disadvantage is that it can only be applied to relatively inert systems, i.e. the rate of oxygen exchange must ideally be several orders of magnitude slower than the time scale for isolation and conversion to the gaseous oxygen or CO_2 form. NMR (^{17}O) is another powerful method used for studying both static and dynamic aspects of metal ion hydration. The determination of the physical chemical properties of these types of reactions has the added benefit of providing potentially useful rates of isotopic exchange that may be applicable to the study of natural systems, particularly for those reactions that are kinetically sluggish. Further details regarding the kinetics of formation and dissociation and isotope exchange of cluster ions in solution and at surfaces are reviewed by Casey (2001). Examples of rate data and Arrhenius parameters for isotopic exchange between aqueous species are summarized in Table 1 (see Appendix).

Oxygen isotope exchange between aqueous species and H_2O. The oxygen exchange between solvent water and simple aqua-metal ions or oxo-anions is closely related since substitution occurs at either the M-O or the X-O bonds, respectively (Gamsjäger and Murmann 1983). General exchange reactions for these can be written as

$$nH_2^{18}O + [M(OH_2)_n]^{z+} \leftrightarrow nH_2O + [M(^{18}OH_2)_n]^{z+} \tag{81}$$

$$nH_2^{18}O + [XO_n]^{z-} \leftrightarrow nH_2O + [X^{18}O_n]^{z-} \tag{82}$$

The isotopic oxygen exchange rates between these ions and H_2O can be represented by a first-order process provided only tracer quantities of ^{18}O labeled reactants are introduced into the chemically equilibrated solutions. Applied to Equations (81) and (82) this rate law results in Equations (83) and (84) as follows

$$-\ln(1-F) = Rt\{(n[M(OH_2)_n^{z+}] + [H_2O])/(n[M(OH_2)_n^{z+} \cdot [H_2O])\} \tag{83}$$

$$-\ln(1-F) = Rt\{(n[XO_n^{z-}] + [H_2O])/(n[XO_n^{z-}] \cdot [H_2O])\} \tag{84}$$

where R is the overall rate of oxygen transfer between H_2O and aqueous ions, t is time, and F is the fraction of exchange completed. In this instance, F can be expressed as

$$F = (X_i - X_t)/(X_i - X_\infty) \tag{85}$$

where X_i, X_t and X_∞ are the mole fractions of ^{18}O in $[M(OH_2)]^{z+}$, $[XO_n]^{z-}$, or the solvent H_2O at times t = 0, t = time of sampling and t = ∞ (at exchange equilibrium). Since usually $n[M(OH_2)_n^{z+}] << [H_2O]$ or $n[XO_n^{z-}] << [H_2O]$, Equations (83) and (84) will reduce, respectively, to

$$-\ln(1-F) = Rt/n[M(OH_2)]_n^{z+} \tag{86}$$

and

$$-\ln(1-F) = Rt/n[XO_n^{z-}] \tag{87}$$

This first-order simplification occurs because in each elementary step, all but one of the concentrations is time independent and the labeled molecules enter each reaction with a molecularity of unity in elementary reaction steps. This is common for many isotopic exchange reactions occurring among solution species, but not all as will be shown below.

In the case of aqua metal ions $[M(OH_2)^{z+}]$, some 20 orders of magnitude cover the present range of water exchange rates from the most labile, Cs^+ (residence time for primary shell water molecules, 25°C ~ 10^{-10} sec), to the most inert, Ir^{3+} (residence time of water molecule exchange, 25°C ~ 317 yr). It is recognized that these factors encompass at the one extreme the size of the metal ion and the magnitude of the cationic charge (electrostatic), thus somewhat correlating with the extent of hydration, and at the other extreme the presence of ligand field effects (transition metal cations). Cs^+ is the largest singly positive metal cation and as such possesses the lowest charge density. On the other hand, Ir^{3+} is a third row transition metal ion with the maximum ligand field stabilization energy (LFSE) for its octahedral field of coordinated waters as a result of its low-spin t_{2g}^6 configuration. Hence water exchange has to overcome an extremely large activation barrier (Richens 1997). In fact, for the first row transition metal divalent ions water exchange rate constants correlate with crystal field activation energy as proportional to the crystal field stabilization energy (CFSE) or LFSE associated with the particular electronic configuration.

The ^{18}O method has proven highly successful for determination of oxygen exchange rates between H_2O and oxo-anions such as $Mo_2^VO_4^{2+}$, V^VO^{2+}, $Pu^{VI}O_2^{2+}$ and $U^{VI}O_2^{2+}$. A wide variation in rates between H_2O and different species of the same metal has been determined. It is likely that the origin of these variations lies in the existence of a range of different exchange mechanisms depending in the species involved (Richens 1997). For example, the half-lives for water exchange on terminal oxo species determined by ^{18}O labeling varies from $\sim 10^{-2}$ sec for $U^VO_2^+$ to 10^{10} sec for $U^{VI}O_2^{2+}$ at 25°C. In the case of vanadium, $V^{IV}O^{2+}$ has a half-life of exchange of 10^5 sec (25°C) whereas $V^VO_2^+$ has a half-life of water exchange of only 0.15 sec (0°C) (Gamsjäger and Murmann 1983). A linear correlation between $\Delta S^{\#}$ and $\Delta H^{\#}$ (isokinetic plot) for some water-oxygen exchange paths of oxo-anions (V, Fe, Mo, Te and W as $[MeO_4^{z-}]$) has been observed. In the case of these metals there is a correlation between increasing $\Delta S^{\#}$ and increasing $\Delta H^{\#}$ that seems to coincide with a subtle increase in the metal-oxygen bond length (e.g V-O > W-O). Interestingly, Casey and Westrich (1992) describe rates of water exchange from the solvent into the first hydration sphere of corresponding cations dissolved from end-member orthosilicates (e.g. Ca_2SiO_4, Mn_2SiO_4, Zn_2SiO_4, Co_2SiO_4, Ni_2SiO_4, etc) that seem to increase with an increase in the metal-oxygen bond distance (Ca-O > Mn-O > Fe-O ≥ Mg-O > Ni-O > Be-O). Rates of ^{18}O exchange between H_2O and simple oxo-anions and oxo-cations are summarized in Table 1 (see Appendix).

The oxo-anion PO_4^{3-} is a key constituent in the Earth's biosphere. Recently, oxygen isotope fractionations (both equilibrium and kinetic) have been measured experimentally between orthophosphate ion $H_2PO_4^-$ and H_2O from 50 to 135°C by Lecuyer et al. (1999). They controlled the experimental conditions such that $H_2PO_4^-$ was the dominant species in solution (pH 5 at ambient conditions). They measured both the oxygen isotope composition of the H_2O and the dissolved phosphate (by precipitation as silver phosphate), and fit the time series data to a simple first order expression similar to Equation (87), where

$$F = 1 - e^{-kt} = (\delta^{18}O_i - \delta^{18}O_t)/(\delta^{18}O_i - \delta^{18}O_{eq}) \qquad (88)$$

where is F is the fraction of oxygen exchange between phosphate and H_2O, and i, t and eq refer to the initial, measured and equilibrium isotopic compositions of phosphate, respectively. Lecuyer et al. (1999) found that this first-order equation adequately described their data for all temperatures (Fig. 9a). The slope of the temperature-dependency of k values yields an activation energy of 133.6±4.6 kJ/mole which is quite high for aqueous reactions (Table 1, Appendix). Model calculations using these rates indicate equilibration times between 10^3 and 10^4 years at 20°C. However, the rates determined from their study are only applicable to natural systems at slightly acidic conditions (pH ≈ 5). At other pH values, the rate of exchange will differ as illustrated in Figure 9b which shows the oxygen isotope exchange rate plotted against pH for the orthophosphate-H_2O system at 100°C and $[PO_4^{3-}]_{tot} = 0.4 - 0.8$ M (Bunton et al. 1961). The overall rate can be defined by the following equation

$$R = k_A[H_3PO_4][H^+]e^{\beta I} + k_N[H_3PO_4] + k_M[H_2PO_4^-] \qquad (89)$$

and by the bold curve in Figure 9b. β is a concentration variable and I refers to ionic strength. Clearly, rates decrease at pH values below ~3 and >7 for this temperature-concentration condition. These results emphasize the obvious that when applying rate data, the experimental data must be closely matched with the natural conditions.

In a similar kind of study, Chiba and Sakai (1985) reacted Na_2SO_4 with ^{18}O-

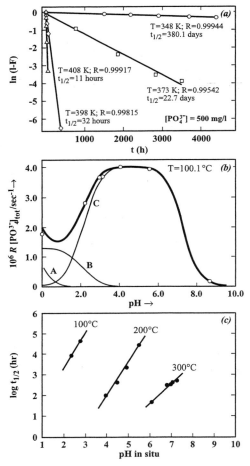

Figure 9. (a) Progress versus time of the oxygen isotope exchange reaction between phosphate and water ($[PO_4^{3-}]$ = 500 mg/l). Slopes from the straight-line fits to these data yield the rate constants for temperatures of 75°, 100°, 125° and 135°C. The half-life of the reaction, $t_{1/2}$ for each temperature was obtained at the time that gives ln $(1 - F)$ = ln 0.5. (b) pH rate profile of the oxygen isotope exchange between orthophosphate and water at 100.1°C. Experimental data shown as open circles. Contributions of the k_A (A), k_N (B) and k_M(C) terms of Equation (90) to total exchange are plotted. (c) Relationship between the experimentally determined log $t_{1/2}$ (hr) and pH calculated at elevated temperatures for oxygen isotope exchange in the system dissolved sulfate-water. The sources of results presented in (a), (b) and (c) are Lecuyer et al. (1999), Bunton et al. (1961) and Chiba and Sakai (1985), respectively.

enriched H_2O and monitored the isotopic changes of both as a function of time at 100 200, and 300°C and different pH values. Not too surprisingly, they found the rate of oxygen exchange to be strongly dependent on temperature and pH. As with the phosphate system, dissolved sulfate could exist in multiple forms: $H_2SO_4^0$, HSO_4^-, SO_4^{2-} and $NaSO_4^-$. Oxygen isotope exchange would proceed through interactions (collisions) among these four species and H_2O and each pathway should have a specific rate constant that may be different in magnitude from each other. An overall exchange rate like that shown for phosphate in Equation (89) can be defined for sulfate where

$$R_{r4} = 4(H_2O)^m[k_1(H_2SO_4)^n + k_2(HSO_4^-)^n + k_3(SO_4^{2-})^n + k_4(NaSO_4^-)^n] \quad (90)$$

which is the reverse rate (fully speciated) for the exchange reaction

$$S^{18}O_3^{16}O + H_2^{18}O = S^{18}O_4 + H_2^{16}O \quad (91)$$

In their study, the half-life of the isotope exchange reaction ($F = 0.5$) was given as

$$\ln 0.5 = -R_{r4}[4(\Sigma SO_4) + (\Sigma H_2O)/4(\Sigma SO_4)(\Sigma H_2O)]t_{1/2} \quad (92)$$

The half-lives of oxygen isotope exchange are plotted in Figure 9c as a function of calculated pH (from mass action, mass balance and proton balance equations among species, H_2O, $H_2SO_4^0$, HSO_4^-, SO_4^{2-}, $NaSO_4^-$, H^+, OH^- and Na^+) for 100, 200, and 300°C. This figure clearly demonstrates the importance of both temperature and pH in controlling the rate of exchange. Complex rate expressions were developed that mechanistically relate the apparent rate constants with H^+ and sulfate speciation. From these, Chiba and Sakai (1985) deduced that the isotopic exchange between dissolved sulfate and water proceeds through collision of HSO_4^0 and H_2O at low pH (<5.5 for 200°C and $<\sim3$ at 100°C), and between HSO_4^- and H_2O at intermediate pH (>6 at 300°C). Their results were in general accord with those of previous studies reported by Hoering and Kennedy (1957) and Lloyd (1968). Table 1 (see Appendix) summarizes rate data and Arrhenius parameters for this system.

Sulfur isotope exchange between sulfate and sulfide. Ohmoto and Lasaga (1982) investigated the kinetics of sulfur isotope exchange between aqueous sulfates and sulfides at hydrothermal conditions (~200 to 400°C). They concluded that the rates of chemical reactions between sulfates and sulfides are essentially identical to the sulfur isotope exchange rates because both the chemical and isotopic reactions involve simultaneous oxidation of sulfide-sulfur atoms and reduction of sulfate-sulfur. They expressed the rate of reaction as a second order rate law

$$R = k [\Sigma SO_4^{2-}] \cdot [\Sigma S^{2-}] \quad (93)$$

where R is the overall rate, k is the rate constant and $[\Sigma SO_4^{2-}]$ and $[\Sigma S^{2-}]$ are in molal concentrations. Based on the available partial exchange data they estimated the rates using the following rate law (similar in form to others described above)

$$\ln [(\alpha_{eq} - \alpha_t)/(\alpha_{eq} - \alpha_o)] = -kt ([\Sigma SO_4^{2-}] + [\Sigma S^{2-}]) \quad (94)$$

where t is time and α_o, α_t and α_{eq} are, respectively, the fractionation factors at $t = 0$ (initial), at the end of the experiment, and at equilibrium ($t = \infty$). Their rate constants are strongly dependent of temperature and pH, but not in a simple manner as outlined by Igumnov (1976). In Na-bearing hydrothermal solutions, rates decrease by about 1 order of magnitude with an increase in pH by 1 unit for pH's < 3, remain essentially constant in the pH range of ~4 to ~7, and decrease again at pH > 7. The activation energies for the reaction also vary as a function of pH: 77±4 kJ/mole at pH = 2, 124±4 kJ/mole for pH = 4 to 7, and between 167 and 197 kJ/mole at pH ≈ 9. They proposed a model involving thiosulfate molecules as reaction intermediates where the intramolecular exchange of sulfur atoms in thiosulfates (i.e. $H_2^{34}S$-SO_3 → H_2S-$^{34}SO_3$) is the rate-limiting (slow) step (see also Lasaga 1998). A more recent assessment of the rates of sulfur isotope exchange involving two chemical compounds (H_2S and thiosulfate) and three exchangeable atoms (H_2S, sulfane, sulfonate) has been described by Chu and Ohmoto (1991). They derived a rate law that was applied to a set of experimental data obtained by Uyama et al. (1985) on sulfur isotope exchange reactions between H_2S and thiosulfate, where the $\delta^{34}S$ values of H_2S, and sulfane and sulfonate sulfur from thiosulfate were measured during reaction for various runs at 120°C and near neutral pH. Chu and Ohmoto (1991) estimated the following rates of exchange:

$$\log k_{H_2S-SH(thiosulfate)} = -1.4 \text{ moles / l / sec; } \log k_{H_2S-SO_3H(thiosulfate)} = -2.6 \text{ moles / l / sec;}$$

and $\log k_{SH-SO_3H(thiosulfate)} = -8$ moles / l / sec.

This kind of behavior where multiple species are involved can also occur with other types of molecules such carboxylic acids (like acetate) where ^{13}C and ^{12}C can interchange between methyl and carboxyl positions as well as with inorganic carbon species in solution (see Kharaka et al. 1983).

REACTION-CONTROLLED MINERAL-FLUID ISOTOPE EXCHANGE

Chemical reaction versus diffusion

A number of microscopic processes have been recognized that influence isotopic (as well as chemical) exchange in mineral-fluid systems. The more important of these include recrystallization, replacement, solution/precipitation, surface diffusion, grain boundary diffusion, and volume (lattice) diffusion. Many of these have been documented in both experimental run products and natural materials through the use of an assortment of physical and/or chemical imaging techniques such as SEM, AFM, CL, HRTEM, LEED, XPS, EMPA, BSE and SIMS. The aforementioned processes can be grouped into two major categories: chemical reaction and diffusion. When minerals and fluids are at or very near chemical equilibrium, isotopic exchange can occur commonly through a diffusional mechanism (Cole et al 1983, Giletti 1985, 1986; Eiler et al. 1992, 1993). In contrast, isotopic exchange in mineral-fluid systems that are initially far from chemical equilibrium are controlled largely by chemical reactions such as recrystallization (Cole et al. 1983, Cole and Ohmoto 1986, Stoffregen 1996). Isotopic exchange controlled by chemical reactions is typically several orders of magnitude faster than rates influenced by diffusion. Despite the significant differences observed in equilibration times between these two major pathways, both are dependent to varying degrees on a number of common factors that include temperature, pressure (or water fugacity), solution-to-solid ratio, grain size (diffusion), surface area (chemical reaction), solution composition, mineral composition and structure, and defects. The discrepancy in the rates of chemical reaction and diffusion means that the fast reaction rates could passivate the surfaces or lead to the formation of reaction rims such that further exchange would require diffusion of matter across this reaction zone into the unreacted crystal—a coupled process.

A large number of studies have now been reported on the measurement of diffusion coefficients of isotopic species (e.g. $^{18}O/^{16}O$) in silicates, oxides and carbonates at elevated temperatures and pressures (see Tables 2, 3 and 4 in this study, or Brady 1995). Those of most relevance, geologically, involved the experimental interaction of a mineral (melt, glass) and either pure water, an alkali chloride solution that is in approximate chemical equilibrium with the solid (e.g. albite and NaCl), or a dry gas such as O_2 or

CO_2. Enough data now exist that empirical correlation methods are providing us with the means to predict oxygen diffusivities for mineral-fluid systems where experimental data are lacking (e.g. Fortier and Giletti 1989, Muehlenbachs and Connolly 1991, Zheng and Fu 1998). Conversely, we know far less about rates of exchange for systems where the mineral undergoes some form of chemical reaction. To date, our knowledge of the rates and mechanisms of isotopic exchange accompanying chemical reactions comes largely from partial isotope exchange experiments designed specifically to measure equilibrium fractionation factors, not rates (e.g. O'Neil and Taylor 1967, O'Neil et al. 1969). Many of these used some form of mineralizing electrolyte solution to enhance the reaction rates (e.g. O'Neil and Taylor 1967). Other partial exchange experiments on rock-fluid systems (Cole et al. 1987, 1992), carbonates-$H_2O\pm NaCl\pm CO_2$ (Anderson and Chai 1974, Chai 1975, Beck et al. 1992, Cole 1992, 2000; Burch et al., in review), quartz-water (Matsuhisa et al. 1978, Matthews and Beckinsale 1979, Matthews et al. 1983a), layer silicates-H_2O (Cole 2000) and alunite-water (Stoffregen et al. 1994) have intentionally targeted the measurement of rates of isotopic exchange controlled by some form of chemical reaction. Collectively, these studies have provided insight into the mechanisms of exchange and permitted limits to be placed on the rates of oxygen isotope exchange accompanying chemical reaction as a function of temperature, pressure, solution composition, surface area, fluid-to-solid ratio and time. The discussion in this section will focus on the important concepts and/or parameters pertinent to the nature of isotopic exchange controlled by chemical reaction. The intent here is not to review the data summarized in Table 2 (see Appendix), but to use examples from the compilation to illustrate important points about the fundamental behavior of isotopic exchange.

Rate models

Isotopic exchange between mineral and fluids that are far from chemical equilibrium can result in noticeable changes in the nature of the solid. The occurrence of secondary product phases or measurable grain coarsening based on petrographic, SEM, HRTEM and/or XRD data obtained from many partial exchange experiments indicate that changes in surface area should be accounted for in the development of rate laws. Regardless of whether the reactants simply recrystallize or become replaced by a new phase, it is clear that the intimate coupling between chemical and isotopic exchange through surface reactions involves dissolution and precipitation leading to crystal growth. With regard to the overall process, because the enthalpies of common mineral-fluid reactions are several orders of magnitude larger than for isotopic exchange, the reaction rates should depend on the rates of attainment of chemical equilibrium. Beck et al. (1992) demonstrated this point with novel experiments that monitored the exchange of Ca and Sr isotopes between a fluid and calcite that experienced dissolution and precipitation.

Simplified surface-area based rate model. A simple isotope exchange rate model derived by Northrop and Clayton (1966) was modified by Cole et al. (1983) to account for the surface area of the solid in experimental mineral-fluid systems. As a first approximation, this model assumes that the rate-limiting step involves the addition and removal of atoms (O, H, C) from the surface of the solid. The overall rate of reaction, R, can be expressed in the following pseudo-first order equation with the inclusion of a factor, A_s, representing the total surface area (m^2) of the mineral

$$R = [-\ln (1 - F) \ W \ S]/(W + S) \ A_s \ t \tag{95}$$

or

$$R = [-(1 - F) \ X_w \gamma_m]/t \tag{96}$$

where W and S represent the total moles of O (or H, C) in the fluid and solid, respectively, X_w is the mole fraction of H_2O in the system, γ_m the surface area per mole of mineral, t is the run time in seconds, and R is in units of moles $O/m^2/sec$. F, the fraction of isotopic exchange, as derived by Northrop and Clayton (1966) can be expressed as

$$F = (\alpha^i - \alpha^f)/(\alpha^i - \alpha^{eq}) \tag{97}$$

where the α's are isotopic fractionation factors between mineral and fluid at time t (α^t), t = 0 (α^i) and t = ∞ (α^{eq}). Experimentally, isotopic data are obtained as a function of time at constant temperature, pressure, W and S. The value for R can be determined from the slope of least squares regression fits to data plotted as -ln (1 – F) versus t. Deviations from straight line fits on such plots may signify multiple mechanisms, a second-order rate-dependence rather than the pseudo-first order behavior described by Equation (95), or simply that the rate is decreasing as the system approaches equilibrium (or steady state) as rapid grain growth gives way to pronounced slowing of grain ripening. Equation (95) implies a linear dependence of the exchange rate on surface area. This relationship is probably most applicable for cases where grain growth is moderately slow, typical of low temperature mineral-fluid interactions involving silicates and oxides, or for short duration results where chemical reactions dominate the exchange process (akin to the initial rate method). Because surface areas can change during the course of reaction, BET surface area measurements should also be made as a function of time. Realistically, the amount of material available for such measurements is generally too small, so some kind of grain-size based geometric model is used (e.g. sphere, plate, cylinder). This issue will be discussed separately in a later section. Alternatively, if surface areas are not known, then for a spherical grain, Equation (95) may be recast as

$$R = [-\ln (1 - F) (W/S)_o (X_s)(a)(\rho)]/\{3[1+(W/S)](t)\} \tag{98}$$

where $(W/S)_o$ is the solution to mineral oxygen mole ratio, a is grain radius (cm), ρ is the density of the solid (g/cm^3), and X_s is the number of moles of O per gram of mineral. The units for R are moles $O/cm^2/sec$, which may be converted to moles $O/m^2/sec$ by multiplying by 10^4. Examples of the temperature-dependency of the oxygen isotope exchange rates for the feldspar-fluid and layer silicate-fluid systems estimated from Equation (95) are shown in Figures 10a and 10b, respectively. Note that these rates have been recalculated in order to accommodate changes in surface area with varying F and are not necessarily in perfect accord with those presented previously by Cole et al. (1983) or Cole and Ohmoto (1986). Additionally, for those experiments conducted at high temperatures (>500°C), the F values probably include contributions from chemical reaction and diffusion, i.e. a coupled process. Holding all variables constant except F, we see that an increase in F from 0.5 to 0.9 leads to an increase in rate by ~0.5 log units, whereas an increase in F from 0.1 to 0.5 leads to an increase in rate by nearly one order of magnitude. Because most of the experiments modeled with Equation (98) exhibit F values in excess of 0.5, we conclude that the rates are probably over estimated by no more than about 0.5 log units. Activation energies for these and similar chemical-controlled mineral-fluid isotope exchange systems range from about 50 to 125 kJ/mole. The significance of these values and other Arrhenius parameters will be discussed at length in a later section.

Dissolution-precipitation models . Dubinina and Lakshtanov (1997) developed a kinetic model that describes isotopic fractionation between a mineral and fluid involved in one of three types of dissolution-precipitation processes (Fig. 11). Type I (mineral synthesis) considers successive dissolution of an unstable phase, A, of uniform isotopic composition and precipitation (crystallization) of phase B. Type II (Ostwald ripening) involves the partial dissolution of phase B which has a non-uniform isotopic composition

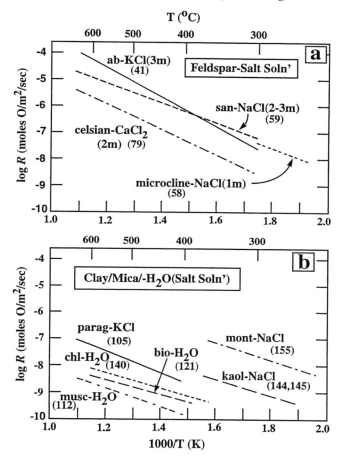

Figure 10. Arrhenius plots of oxygen isotope rate constants for select feldspar-salt solution (a) and clay/mica-fluid systems based on the simplified pseudo-first order surface area rate model (Eqn. 96). Concentrations of electrolytes are given in molality (m). The data used to generate these curves can be found in Table 2 (see Appendix) by referring to the numbers in ( ).

resulting from Type I interaction at the point when the last crystals of A disappear. The isotopic implications of this process were also modeled by Chai (1975) and Stoffregen (1996). Type III process consists of the repeated dissolution and precipitation of phase B from the Type II case. It is typically assumed in these calculations that (a) the system is closed to fluid, (b) the stoichiometry of the solid remains constant, (c) the isotopic fractionation factor between the freshly formed portion of the mineral and the fluid is constant (not necessarily equilibrium), and (d) isotopic exchange before dissolution and after precipitation is negligible. As a basis for their modeling, Dubinina and Lakshtanov (1997) considered the silica gel-quartz-water system.

For the Type I process, they derived a simple mass balance expression that relates the fraction of exchange, F (as defined above in Eqn. 97) for any mass of oxygen (m) in phase B as

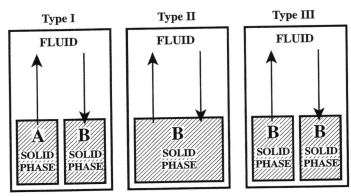

Figure 11. Schematic representation of the processes Types I-III described by Dubininia and Lakshtanov (1997) in their models of isotopic exchange promoted by dissolution-precipitation reactions.

$$F = [1 - \exp(-m/M)] \{1 + (w/M)\} \qquad (99)$$

where w and M are the total masses of oxygen in the fluid and solid phase, respectively, and m is the mass of oxygen in the portion of solid transferred through the fluid. This equation describes the change in F between the entire solid and fluid. The behavior of isotopic composition of the fluid, the increment of B precipitated, bulk B, and bulk solid (A+B) during this process strongly depends on the initial compositions of A and fluid, m, and the ratio (M/w). As Equation (99) indicates, the fraction of exchange depends on both the mineral-to-fluid ratio (M/w) and m, as shown in Figure 12a. It is possible for the F value to exceed a value of 1.0 (equilibrium) before the mineral synthesis ceases. Note that the dependency of F on M/w is generally linear for small values of solid to fluid ratio, but non-linear at larger values of M/w, as mass is transferred from phase A to B.

In Dubinina and Lakshtanov's (1997) Type II model, dissolution-precipitation proceeds by the Ostwald ripening after mineral synthesis. Upon completion of mineral synthesis (Type I), the phase B consists of an aggregate of crystals of different grain sizes. The recrystallization of these mineral grains will take place by Ostwald ripening, i.e. growth of large grains at the expense of dissolution of smaller grains. The result of this process is that the total number of grains in the system decreases and the average grain size increases. In order to calculate oxygen isotope compositions of material transferred and F during this recrystallization, it is necessary to know the crystal size distribution and average grain radius, preferably as a function of time. The link between grain radius and solubility is given by the Gibbs-Kelvin equation which describes the one crystal size (r*, the critical radius) in true equilibrium with the solution as

$$r^* = 2\sigma V_o/[\underline{R}T \ln(\Omega/K_{eq})] \qquad (100)$$

where Ω is the activity product of phase B, K_{eq} is the equilibrium constant for a grain of B, σ is the mineral-fluid interfacial energy (J/m^2), and V_o is the molar volume (m^3/mole) (Chai 1974). An estimation is made of the growth rate from solution which requires knowledge of the critical radius for a given grain size distribution, the degree of supersaturation of the solution with respect to phase B, and the molecular diffusion (in this case SiO_2) in solution. Because the critical radius cannot be measured directly, the average radius is generally used to specify the growth rate. From the growth rate, the amount of mass transferred back to B from solution can be calculated for each grain size

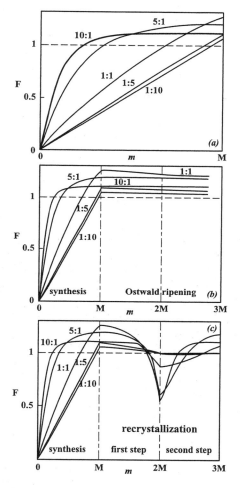

Figure 12. (a) Variation in the fraction of exchange (F) as the mass of oxygen (m) in synthesized solid changes from 0 to M at different initial masses of fluid and starting material. The numbers by the curves indicate the oxygen mole ratio between solid (M) and fluid (m). F values can exceed 1 under conditions of local equilibrium during synthesis. (b) Variation in the fraction of exchange as transferred mass of oxygen in the solid through the fluid changes upon synthesis (0 < m < M) and subsequent Ostwald ripening of the synthesized sample (m > M) at different initial mass ratios of oxygen between starting solid and water. (c) The behavior of the fraction of isotopic exchange as a function of the mass of oxygen (m) in the solid transferred through the water upon synthesis (0 < m < M) and subsequent single (M < m < 2M) and replicate (2M < m < 3M) recrystallization of synthesized sample of varying mass ratios of oxygen in the solid and water. Based on models presented by Dubininia and Lakshtanov (1997).

distribution, and ultimately, the F value may be estimated. The behavior of F for this type of process is shown in Figure 12b for different values of M/w. The rate of dissolution-crystallization during synthesis depends primarily on the degree of supersaturation, i.e. the difference in solubility of phase A and B. In Ostwald ripening, the rate depends on the difference in solubility of variously sized grains of same phase B. For systems where the starting phase does not go through a synthesis step, the trajectories of F versus m have very similar shapes as those given in Figure 12a, but typically with less steep initial slopes (Chai 1975, Stoffregen 1996). A final point is that the pseudo-first order model given by Equation (96) is adequate for interpretation of experimental data, but does not provide a physically meaningful description of exchange controlled by dissolution-precipitation leading to Ostwald ripening (Stoffregen 1996). However, the Type II process requires data on interfacial energies and chemical reaction rates that are lacking for mineral-fluid systems at hydrothermal conditions except quartz-H_2O.

Type III dissolution-precipitation process involves the repeated transfer of mass M of mineral B in a series of steps resulting in a non-uniform isotopic composition. Unlike

Ostwald ripening, this process considers dissolution of all crystals of phase B to proceed simultaneously, irrespective of their size. The outer grain zones produced during the final stages of synthesis are the first to dissolve. The grain core, produced in early stages of mineral formation, dissolves at the end of the process. F in this case is a function of m and M/w only, as in Type I behavior. As shown in Figure 12c from Dubinina and Lakshtanov (1997) the F values drift near 1.0 at each step of mass M dissolution-precipitation. This implies that the same values of F can be obtained more than once during repeated redeposition of a mineral. The amplitude of the variation in F depends on the initial M/w ratio.

Documentation of the isotopic heterogeneities attributable to the kind of Type I-III processes outlined above requires the use of microanalytical methods. The study by Reeder et al. (1997) has described the isotopic fractionations for different growth zones in calcites precipitated experimentally from solution. Within the precision of the ion microprobe, they found no systematic isotopic differences among different crystallographic growth surfaces. They concluded that surface-site preferences during growth are not significantly sensitive to the slight mass and vibrational differences among the light isotopes. Their results differed from other reports of measurable isotopic variation for crystal surfaces from natural samples based on different analytical methods and much larger grains (Dickson 1991).

Surface area, grain size and fluid-to-solid ratio

There are a number of methods that can be used to address the issue of surface area. Obviously, the most accurate method is to measure the surface areas of the run products with N_2 or Kr Brunauer-Emmett-Teller (BET) surface area analysis. However, this is not possible because sample size is prohibitively small in most cases involving isotopic exchange experiments. In their study of oxygen isotope exchange between alunite and water, Stoffregen et al. (1994) assumed as a first approximation that total surface area, A, is a linear function of F, the fraction of exchange where $A = A_o (1 - F)$; A_o is the initial surface area. This relationship has not been confirmed experimentally. In a study of oxygen isotope exchange between granite and fluid, Cole et al. (1992) demonstrated that the rates of exchange (Eqn. 95) were in excellent agreement for like systems that differed only in their starting surface areas (0.13 to 5 m^2/gm) and fluid-to-solid mass ratios (0.2 to 6). In the absence of measured surface areas, the most common approach utilizes the mean grain size and geometry of the grains, such as sphere, plate or cylinder. The calculated surface areas are typically less than the measured surface areas, with the magnitude of the difference dependent on mineral type, surface roughness and grain size distribution. For example, Leamnson et al. (1969) presented results for silica that clearly showed the calculated surface areas were underestimated by 3 to 4 times compared to the BET values. Similarly, White and Peterson (1990) presented comparisons between measured and geometric-calculated surface areas that showed a systematic under-estimation of true surface area by a factor of 5 or more when using the geometric approach. However, they documented a linear, empirical relationship between BET measured surface areas corrected for mineral density (gm/cm^3) and mean particle size (cm) for a variety of mineral types, e.g. carbonates, oxides, and clays. Some mineral groups (e.g. carbonates) appeared to have distinctive linear trajectories.

Cole (2000) described surface area estimates of run products from an experimental study of carbonate and layer silicate-H_2O interaction based on a similar approach as described by White and Peterson (1990). Data on BET surface areas and mean grain diameters for a variety of carbonates and layer silicates were normalized to density (A' in m^2/cm^3) for each phase and regressed against the mean grain diameters ($\bar{d}$ in cm). For carbonates, the linear regression equation is

$$\log A' \ (m^2/cm^3) = -1.7647 - 0.6105 \ (\log \bar{d}) \qquad (101a)$$

with an R^2 of 0.996 (n=4). For layer silicates, the linear regression equation is

$$\log A' \ (m^2/cm^3) = -2.095 - 1.0404 \ (\log \bar{d}) \qquad (101b)$$

with an R^2 of 0.998 (n = 5). The mean grain diameters measured for each run product are used to estimate the specific surface areas, which are converted to total surface areas by dividing by mineral density, then multiplying by sample mass. This empirical approach does not imply any unique geometry for the grains, but simply compares the mean diameters with a set of "standard" minerals of the same mineral group whose BET surface areas have been accurately determined. Note also that these estimates are for physical, not reactive, surface areas.

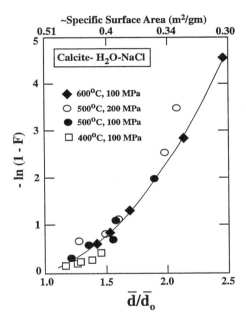

Figure 13. Relationship between the degree of equilibration [$-\ln (1 - F)$] and the ratio of average measured grain diameter (d) to the average diameter of the starting calcite (d_o) for the system calcite-H_2O-NaCl. Five different fluid compositions (0, 0.1, 0.3, 1, 4 m NaCl) are represented on this plot for each temperature. Increase in grain diameter is accompanied by a decrease in the specific surface area of the solid, shown calibrated at the top of this figure (based on Eqn. 101a). Data from Cole (1992).

To date, we know of no experimental study where isotopic rates were monitored as a function of changing surface areas measured directly with BET methods. However, a reasonable correlation was observed between increasing degree of oxygen isotope exchange ln (1 − F) and increasing grain diameter (Fig. 13) in the system calcite-H_2O±NaCl for 400 to 600°C (Cole 1992). Grain sizes were also measured by Chai (1975; calcite-H_2O-NaCl) and Matthews et al. (1983a; quartz-water, albite-water, and anorthite-water). These two studies document a positive correlation between the extent of isotope reaction and grain ripening. Matthews et al (1983a) attempted to estimate the surface areas of run products from grain size measurements for isotope reactions between quartz-water and albite-water. They observed an increase in calculated surface areas with time for fractions of exchange up to about 0.1, for greater values of F the calculated surface areas either remained constant or decreased.

It is important to note that even if surface area is known, other factors may influence the kinetics such as dislocation density or the presence of disturbed surface layers

(Petrovich 1981). The lack of a direct proportionality between surface area and reaction rate observed in some cases requires a distinction between the "physical" surface area and that surface area which participates directly in chemical reactions (Petrovich 1981, Helgeson et al. 1984). The term "reactive" surface area is used to denote the surface containing chemically-reactive sites. These sites can consist of features such as etch pits or crystallographically-controlled extended defects such as edge and screw dislocations (Brantley et al. 1986). Dislocation density data for natural minerals are quite limited, but generally range from 10^3 to 10^{12} dislocations cm^{-2} (White and Peterson 1990). High resolution TEM (HRTEM) imaging of dislocations, which is the principle measurement method, is difficult below a dislocation density of about 10^6 cm^{-2} (Veblen 1992). But common methods of studying dislocation densities are etching and decorating the dislocations—very common in metallurgy and used, for example, for olivines by Kohlstedt et al. (1975). Physical surface areas run between 1 and 2 orders of magnitude greater than reactive surface areas regardless of the material. A limited number of chemical rate studies have shown that a 3 to 4 order of magnitude increase in the dislocation density translates to only about a factor of 2 increase in rate (e.g. Blum et al. 1990).

In most partial exchange experiments, it is common to hold the fluid-to-solid ratio constant while varying temperature, pressure, solution composition and/or grain size. If the first-order rate model is valid, then the amounts of fluid and solid may be varied without affecting the final calculated rate constant (Cole and Ohmoto 1986). Experiments varying the fluid-to-solid ratio by as much as 100 (by mass) have been conducted [i.e. basalt-seawater (Cole at al. 1987) and granite-fluid (Cole et al. 1992)], and indicate that this may be the case. However, because the data from these experiments are limited, these results can only be considered preliminary in nature. Other than Chai's (1975) study of calcite and our work, we know of no other isotopic exchange rate experiments where the fluid-to-solid ratio was varied systematically that would permit a more rigorous test of the pseudo-first order, or any other rate model.

Influence of solution composition

It has long been appreciated that dissolved ions exert a major influence on both the fractionation factors (e.g. Truesdell 1974, O'Neil and Truesdell 1991, Horita et al. 1993a,b; 1995) and rates of stable isotope exchange between liquid water and other phases (e.g. minerals, H_2O, vapor, CO_2). In general, electrolytes (e.g. NaCl, KCl, NH_4Cl) and/or acids have been used to enhance the reaction rates in studies where the main intent was to establish the equilibrium isotope fractionation factors. Recent attention has been focused on the role of solution composition on the nature of fluid - solid dihedral angles (e.g. Holness and Graham 1991). The overall concept is that addition of a salt may increase both the solubility and rate of dissolution, thus leading to faster isotope exchange rates measurable in reasonable experimental times. A wide variety of solution types have been used including, but not limited to: NH_4Cl or NaCl (calcite—O'Neil et al. 1969, Anderson and Chai 1974, Chai 1975, Cole 1992, dolomite—Northrop and Clayton 1966), NaF (quartz—Clayton et al. 1972, Ligang et al. 1989), NaCl or KCl (feldspars—O'Neil and Taylor 1967; muscovite—O'Neil and Taylor 1969), NaCl and/or H_2SO_4 (barite—Kusakabe and Robinson 1977; anhydrite—Chiba et al. 1981) and HCl (anhydrite—Chiba et al. 1981). In the majority of these cases it can be shown, either microscopically or with SEM, that recrystallization or grain replacement has taken place, accompanied by isotopic exchange.

Very few isotope exchange studies have been conducted that purposely varied the concentration of salt or acid. Rates estimated from exchange data for select mineral-fluid systems reported in the literature generally increase with increasing salinity or acidity

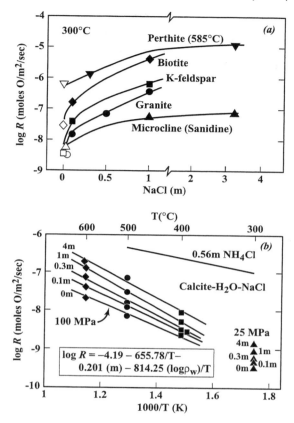

Figure 14. (a) Relationship between the oxygen isotope exchange rate (R) for granite, K-feldspar, biotite, perthite, and microcline (sanidine) and molality of the reacting NaCl fluid. Data from O'Neil and Taylor (1967) have been used to estimate the rates for perthite and sanidine (3m). With the exception of the 585°C perthite experiments, all data are for experiments conducted at 300°C. Taken from Cole et al. (1992). (b) Arrhenius plot of experimentally determined oxygen isotope rate constants for the calcite-H_2O±NaCl system at 100 MPa. The solid lines are least-squares fits for 400-600°C data. Data at 300°C and 25 MPa are shown for comparison. The line for calcite-H_2O-NH_4Cl was derived by Cole et al. (1983) from data given by O'Neil et al. (1969).

(Cole and Ohmoto 1986). In Figure 14a, we show a plot of log R for oxygen versus the molal NaCl (aq) concentration for granite, K-feldspar, biotite, perthite and microcline from Cole et al. (1992). Although the data are limited, we observe that the log R values increase significantly with an increase in NaCl concentration. For example, in the granite-fluid system we observe an increase in log R from about -8.25 at NaCl = 0 m to -6.75 at NaCl = 1 m at 300°C. These data demonstrate that even dilute salt concentrations on the order of 0.1 m NaCl (KCl) have a profound influence on rates of oxygen isotope exchange. In these examples, however, the Na^+ and K^+ participated in the reaction of silicates, resulting in a complex series of alteration reactions involving formation of new phases. A better approach is to use a single phase whose stoichiometry does not change during the course of experimental interaction with solutions of varying salt content. Chai (1975) explored the relationship between crystal growth rates, and salt type and concentration, as well as the accompanying isotopic exchange in the system calcite-(H_2O ±NaCl±$MgCl_2$±$CaCl_2$). He showed that the rate of oxygen isotope exchange between calcite and 2 M NaCl at 585°C 200 MPa was approximately 5 times faster than the rate between calcite and pure water. Cole (1992) observed a pronounced increase in the fraction of oxygen isotope exchange (F) with increasing temperature, pressure, salinity, and degree of grain rounding and regrowth for the system calcite-H_2O-NaCl from 300 to 600°C. There can be no doubt that the greater degree of recrystallization observed in these experiments is related to enhanced solubility of the solid in the presence of NaCl (aq). The rate of change in F is initially steep, but tends to flatten out with increasing

NaCl at ≥ 1 m. Rate constants assuming a pseudo first-order dependence (Fig. 14b) range from a maximum value of -6.75 at 600°C, 100 MPa, 4m NaCl to a minimum rate of -9.46 at 300°C, 0 m NaCl. The temperature-salinity-dependence exhibited by the calcite data, and data from the granite-fluid system can be adequately represented by a four parameter expression of the form:

$$\log R = a - b(T^{-1}) + c\,(m) - d\,(\log \rho_w)\,T^{-1} \tag{102}$$

where T is in Kelvin, m is NaCl molality and ρ_w is the density of fluid, in kg/m³. Density data were taken from: (1) Haar et al. (1984) for pure water; (2) Pitzer et al. (1984) for NaCl-H$_2$O up to 300°C; and (3) Potter and Brown (1977) for the NaCl-H$_2$O system above 300°C. The fits (R^2) of the calcite and granite data to this expression are generally better than 0.99 (Cole et al. 1992, Cole 1992).

Additionally, in order to accurately calculate F values from exchange data we need to account for the salt effect on the oxygen and hydrogen equilibrium isotope fractionation between minerals and fluids, based on results reported by Horita et al. (1993a,b; 1995) and Chacko et al. (this volume). These results indicate the mineral - salt fractionation may be 0.6 to 1 per mil smaller than the mineral-pure water fractionation at 300°C for a 5 m NaCl solution (Horita et al. 1995). This type of data will play an important role in the ultimate accuracy of the rate constants calculated from partial exchange experiments involving minerals and salt solutions.

Pressure effect

The results from hydrothermal diffusion studies (e.g. Goldsmith 1987, Farver and Yund 1990, Farver and Yund 1991a,b; Graham and Elphick 1991) serve to illustrate the possible rate-dependence on pressure or (f_{H_2O}), and perhaps more importantly, the care required in preparing and characterizing starting materials. In the case of isotopic exchange accompanying chemical reactions, there has been no systematic study of the pressure effect. Preliminary data, however, have been obtained by Cole (1992) who investigated the pressure effect on rates of oxygen isotope exchange in the system calcite-H$_2$O-NaCl, Burch et al (in review, AJS) who observed pressure effects in the system calcite-H$_2$O-CO$_2$, and Matthews et al. (1983a) who reported fractional oxygen isotope exchange data as a function of pressure for albite-water and quartz-water. Figure 15a summarizes the pressure effect in the calcite-fluid system at 500°C for three pressures: 55, 100 and 200 MPa. There is a measurable increase in the oxygen isotope exchange rate with increasing pressure (as well as substantial grain growth). The slopes for these trends typically decrease with increasing NaCl concentration. Additionally, we observe a curvature in the trends for the 1 and 4 m solutions between 55 and 100 MPa. Because oxygen diffusion into the solid at 500°C is negligible for the duration of these experiments, and calcite undergoes only minor compressibility with increasing pressure, the effects observed may be a consequence of changes in the behavior of ions or ion pairs in solution, e.g. Ca^{2+}, CaCl$^-$. From the work of Ellis and McFadden (1972) and our work (Cole et al., in preparation) in a system calcite-dolomite-H$_2$O-CO$_2$, we know that pressure can have a strong effect on solubility, particularly at higher temperatures (>100°C).

From these preliminary data, we can estimate the volume of activation, $\Delta V^{\#}$, which is commonly used as a mechanistic criterion in kinetic studies of homogeneous organic and inorganic reactions (e.g. Palmer and Kelm 1981, Merbach 1987, van Eldick 1987). The $\Delta V^{\#}$ is the difference between the partial molar volumes of the transition state and the reactants, and is related to the pressure derivative of log R by the following equation:

$$(\partial \log R / \partial P)_T = -\Delta V^{\#} / 2.3\,\underline{R}T \tag{103}$$

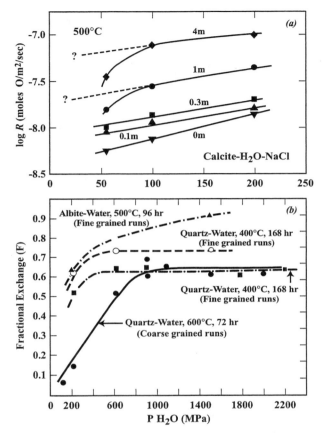

Figure 15. (a) Summary of the pressure effect on the rate of oxygen isotope exchange due to recrystallization at 500°C for the calcite-$H_2O \pm NaCl$ system. In addition to the pronounced increase in rate with increasing salinity, there is a more subtle increase in log R with increasing pressure. Note that the slopes of these lines become flatter with increasing salinity. The slopes are used to estimate the volumes of activation for exchange. (b) The effects of pressure P_{H_2O} on the fractional extent of isotopic exchange in quartz- and albite-water experiments (Matthews et al. 1983a). Run conditions are given adjacent to each curve. It is presumed that the steeper slopes correspond, in large part, to reaction controlled by recrystallization, whereas the flatter slopes correspond to behavior controlled by diffusion of oxygen-bearing species.

If we assume that the $\Delta V^{\#}$ is independent of pressure, that is, we ignore the highly pronounced curvature observed at 1 and 4 m, 55 MPa (Fig. 15a), then the slope of straight-line fits to these data will give us the minimal, absolute $\Delta V^{\#}$. We estimate $\Delta V^{\#}$ values of -40, -18, -26, -23, and -9 cm³/mole for 0, 0.1, 0.3, 1 and 4 m NaCl, respectively. To our knowledge, these are the first data of this type determined from isotope exchange experiments involving the chemical reaction mechanism. The fact that the $\Delta V^{\#}$ values are negative and clearly decrease with increasing NaCl concentration strongly suggests that solvation of ions in solution during the activation step is the main process being influenced by pressure. We suspect that during calcite-fluid interaction, some type of Ca^{2+}-hydrated species forms on the mineral surface. This species may become detached from the surface, solvating more completely with H_2O or form ion pairs with either Cl^- or some carbonate ligand, then become re-attached as $CaCO_3$ during grain growth. Calcite surface reactions involving restructuring and hydration of surface atoms are suggested from studies utilizing X-ray photoelectron spectroscopy (XPS) and low-energy electron diffraction (LEED) (Stipp and Hochella 1991).

In the systems quartz-water and albite-water, Matthews et al. (1983a) observed an increase in the fraction of exchange (F) with increasing pressure up to about 600 MPa, then a near constant F with increasing pressure (Fig. 15b) They concluded that the increase of F with pressure was the result of a recrystallization process, whereas the plateau effect was due to the competing effects of recrystallization (rate enhancement) and volume diffusion (rate inhibition). The dissolution-precipitation mechanism (Ostwald ripening) was more effective for grain radii less than a few microns, and in the coarse grained runs only those portions of crystals with high free energy surfaces experienced any reaction.

Activation parameters and empirical correlation with mineral chemistry

Activation parameters. Oxygen isotope exchange rates have been calculated from available partial exchange accompanying obvious recrystallization for a number of mineral groups—e.g. framework silicates, layer silicates, chain silicates, etc.—and these data are presented in Table 2 (see Appendix). In general, it is difficult to directly compare rates estimated from one mineral group to another because of differences in experimental conditions (e.g. temperature and/or pressure range, solution composition, etc.). However, within a given mineral group there is a systematic relationship between rate and mineral chemistry (Cole and Ohmoto 1986). In the carbonate group, for example, Ca and (Ca + Mg) phases exchange oxygen much slower than either Ba or Sr carbonate (Cole 2000). Si, Na and K-dominated silicates generally exchange oxygen slower than Ca, Mg or Fe-rich silicates. This relationship between exchange rate and mineral chemistry is not entirely unexpected, because we know that equilibrium isotope fractionation factors among isostructural minerals, e.g. $[Me^{2+}]CO_3$; $[Me^{2+}]SO_4$, $[Me^{2+}]SiO_3$, vary according to the specific cation-oxygen bonds (O'Neil 1986) or, in the case of carbonates, cation radius (r_c) and mass (Chacko et al., this volume). In fact, Smyth (1989) has demonstrated a good correlation between oxygen isotope fractionation factors of silicates and oxides with mean oxygen site potentials.

Insight into the mechanistic link between mineral chemistry and isotopic exchange can be realized by comparing the respective activation parameters (Walther and Wood 1986, Cole and Ohmoto 1986). In the past, most workers have estimated the activation energies (E_a) and the pre-exponential factors $(A_o$ or $D_o)$, and plotted these against one another (see Fig. 4). Data falling on a common trend generally indicate a common reaction mechanism. An alternative approach, based on transition state theory (Kreevoy and Truhlar 1986), is to consider the relationship between the rate of reaction and the enthalpy $(\Delta H^{\#})$ and entropy $(\Delta S^{\#})$ of activation, as in the Eyring - Polanyi relationship:

$$\ln R = \ln(k_B T/h) + (\Delta S^{\#}/\underline{R}) - (\Delta H^{\#}/\underline{R}T) \qquad (104)$$

where k_B and h are the Boltzmann and Planck constants, respectively, $\underline{R}$ is the universal gas constant and T is in K. The empirical conclusion is that a linear relationship between $\Delta H^{\#}$ and $\Delta S^{\#}$ may exist if changes within a reaction series do not result in changes in mechanism, or in the nature of the transition state. Note that the first two terms on the right-side of Equation (104) comprise the A_o and that $\Delta H^{\#}$ is only approximately equal to

E_a (see also Eqn. 28).

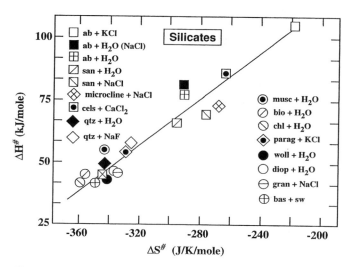

Figure 16. Summary plot of activation parameters (enthalpy of activation, $\Delta H^{\#}$; entropy of activation, $\Delta S^{\#}$) for oxygen isotope exchange in silicates reacted with either pure water or salt solutions. The Eyring-Polanyi relationship (Eqn. 102) was used with data described by Cole et al. (1983), Cole and Ohmoto (1986), Cole et al. (1987, 1992), and in Table 2 (see Appendix). Note that data falling on a linear trend generally indicate that a common reaction mechanism predominates. ab = albite, san = sanidine, cels = celsian, qtz = quartz, musc = muscovite, bio = biotite, chl = chlorite, parag = paragonite, woll = wollastonite, diop = diopside, gran = granite, bas = basalt.

We have refit silicate oxygen isotope rate data to Equation (104) in order to retrieve values of $\Delta H^{\#}$ and $\Delta S^{\#}$. The majority of these data are shown in Figure 16, and exhibit a reasonably good correlation with values ranging from about 42 to 105 kJ/mole and $\Delta S^{\#}$ values ranging from about -375 to -210 J/K/mole. The range in activation enthalpies is consistent with that encountered for chemical reactions involving fluids and mineral surfaces (Lasaga 1990). Given our earlier discussion regarding the relationships between rates, surface areas, salinity and pressure, we need to look at the data in Figure 16 in the context of the possible steps of crystal growth, i.e. those involving solvation of ions, attachment to surfaces, desolvation, etc. Lower $\Delta H^{\#}$ values are typically associated with "tighter" binding of solution and/or surface complexes and this restrictive motion also produces a decrease in $\Delta S^{\#}$—i.e. the loss in degrees of freedom (e.g. vibrational, rotational) (Lasaga 1981b). In general, the large negative $\Delta S^{\#}$ is indicative of surface-complex formation—hence—the loss in degrees of freedom. Furthermore, the negative values are favored by shorter bond lengths—smaller ions typically with higher charge, such as Ca^{2+}, Mg^{2+}, Fe^{2+}, and Al^{3+}. The trends observed in Figure 16 for silicates fit a general pattern predicted by these considerations—namely high charge—smaller cation-bearing phases exhibit low $\Delta H^{\#}$ and more negative $\Delta S^{\#}$ (e.g. diopside, wollastonite, zoisite, kaolinite), whereas smaller charge, large cation-bearing phases such as feldspars are at the opposite end of the trend. Similar relationships between the activation parameters and cation type have been observed for the carbonates (not plotted).

Crystal chemical controls on isotopic exchange. Cole (2000) reported on the relationship between rates of oxygen isotope exchange and mineral chemistry for carbonate-H_2O and layer silicate-H_2O systems reacted at hydrothermal conditions (300°C, 10 MPa and 350°C, 25 MPa, respectively). The rates for the carbonate-H_2O systems increase in order from calcite to strontianite to witherite. This order clearly reflects the influence of the change in cation chemistry, i.e. Ba > Sr > Ca, which qualitatively corresponds to an increase in cation-oxygen distance. A similar pattern is observed for the layer silicate-H_2O systems, where chlorite > biotite > muscovite. The link between cation chemistry and rate is more complicated in this case, but in general, the order follows a concentration trend where Mg-Fe > K-Mg > K, with an associated increase in Si and Al, and decrease in hydroxyl. The isotopic-chemical relations suggest that oxygen isotope exchange behavior monitored experimentally in this study is the net result of bond-breaking and dissolution of the mineral, complex ion formation in solution and growth of the mineral, whose structure is controlled, in large part, by the lattice energy. The greater the lattice energy, the greater the energy required to break up the crystal into constituent ions, and the more sluggish the rate of reaction. Site energies, computed for oxygen by Smyth (1989) and for cations by Smyth and Bish (1988) for silicates and carbonates were used to calculate, U' (in Kcal/mole), the total electrostatic attractive lattice energies (neglecting the nearest-neighbor repulsive forces, van der Waals attractive forces, the distortion energies of electron distributions, and dynamic effects of lattice vibrations—all of which constitute less than 15% of the total), normalized per number of cations, as described in the Born-Mayer relation (Cole 2000).

The oxygen isotope exchange rate (ln R) for each phase has been plotted against its normalized electrostatic lattice energy in Figures 17a and 17b for carbonates and layer silicates, respectively. Data on Figure 17a demonstrates a good linear relationship between ln R and U', where rates increase (witherite > strontianite > calcite) with decreasing electrostatic lattice energy. This is despite the fact that witherite and strontianite (orthorhombic) are in a different structural group from calcite (trigonal). Qualitatively, the decreasing lattice energy reflects an increase in the r_c + r_o distances, where r_{Ba} + r_O = 2.80 Å, r_{Sr} + r_O = 2.65 Å and r_{Ca} + r_O = 2.36 Å. Similarly, we observe a good correlation between an increase in the isotopic exchange rate, ln R for chlorite > biotite > muscovite, with decreasing electrostatic lattice energy (Fig. 17b). This order is consistent with the model and reflects, in part, a greater proportion of Si-O and Al-O bonds in muscovite compared to either chlorite or biotite. The order is also consistent with the order in the rate of weathering of micas, where Fe-rich, silica-poor varieties react much faster than their more alkali and silica-rich counter parts (Brady and Walther 1989). Note that structurally, biotite and muscovite belong to the mica group, whereas chlorite belongs to the chlorite series. Despite this difference, the correlation between rates and U' is very good.

The correlations presented in Figure 17 are generally only valid when temperature, pressure, and initial fluid chemistry are the same and only minor structural differences exist among phases being compared, i.e. isostructural or close to it. Additionally, it is important that the stoichiometry of the phase remains unchanged during the reaction. The experiments for both carbonate-H_2O and layer silicate-H_2O have met these requirements. The good correlations observed in this study suggest the possibility of using this approach to predict the rates of oxygen isotope exchange for phases for which data are lacking. To illustrate this point, we have calculated the normalized electrostatic lattice energies for a number of carbonates and layer silicates from site potential data given by Smyth and Bish (1988; cations) and Smyth (1989; oxygen), and projected these to the least squares regression lines given in Figure 17. The order of rates predicted for carbonate-H_2O oxygen isotope exchange (Fig. 17a), from fastest to slowest is: witherite

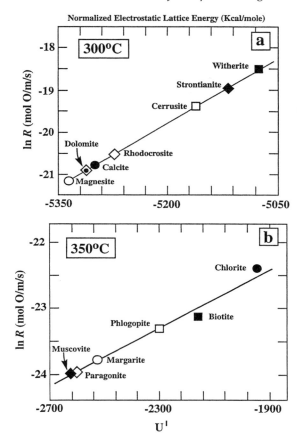

Figure 17. Oxygen isotope exchange rates for metal carbonates (a) and layer silicates (b) recrystallized through reaction with pure water at 300° and 350°C, respectively, plotted against the normalized electrostatic lattice energy (U′, Kcal/mole). U′ represents the sum of electrostatic site potentials for each cation and anion normalized by dividing by the number of cations per formula unit in each phase. The solid symbols represent experimentally determined rates (Cole 2000). The good correlations exhibited in these plots suggest that an empirical method may be used to estimate the rates of other phases common to a mineral group. Lattice energies have been calculated for phases for which rate data are lacking. These have been projected to the lines regressed through the experimental data and are shown as open symbols.

(Ba) > strontianite (Sr) > cerrusite (Pb) > rhodocrosite (Mn) > calcite (Ca) ≥ dolomite (Ca,Mg) > magnesite (Mg). This order follows approximately the magnitude in the $[r_c + r_o]$ distances, from greatest to smallest. The order for the layer silicates, from fastest to slowest, is chlorite > biotite > phlogopite > margarite > paragonite ≥ muscovite (Fig. 17b). Qualitatively, this order follows a pattern of increasing Si, Al, Na, K and Ca and decreasing Mg and Fe. The order of oxygen isotope rates for these layer silicates parallels the order for decreased dissolution rates described by Brady and Walther (1989) that is used to explain the inverse mineral reaction series sequence of surface weathering. By establishing an unambiguous relationship between rate, lattice energy, and ultimately temperature, we can begin to develop empirical equations useful in predicting rates of isotopic exchange for minerals for which experimental data are lacking.

DIFFUSION-CONTROLLED MINERAL-FLUID ISOTOPE EXCHANGE

The rate of isotopic exchange under many geological conditions is dependent on the type and rate of diffusion of the isotopic-bearing species in the system (e.g. CO_2, H_2O). The study of diffusion of isotopic-bearing species is a subset of the much broader topic of diffusion-controlled transport of elements in mineral-fluid, mineral-gas, mineral-mineral, mineral-melt and melt-volatile systems. Indeed, it is now recognized that intercrystalline diffusion is fast enough to play a major role in controlling metamorphic and hydrothermal reactions. Diffusion can lead to profound chemical and isotopic heterogeneities whose distributions can be used to decipher the thermal histories of individual mineral assemblages (Valley, this volume). Of interest is whether or not the mineral grains have maintained their mutual chemical (and isotopic) equilibrium by diffusion and when, during cooling of the rock, the mineral compositions became frozen in (Dodson 1973). This issue of chemical and isotopic compositions continuing to readjust during cooling below a peak thermal event—i.e. closure temperature (and pressure)—complicates the interpretation of the thermal history of any rock system, but, if properly constrained, can also provide more quantitative insight into the pressure-temperature-time paths. However, to do this requires appropriate diffusion data and an understanding of the mechanisms controlling diffusion. The purpose of this section is to (a) describe the fundamental concepts important for understanding isotopic exchange controlled by diffusion in crystalline phases, (b) outline the methods used to determine diffusivities, and (c) document in some detail the factors that control diffusion.

Let us consider the definition of diffusion in general terms as the random jump of particles of matter (atoms, ions, molecules) from one site to another in the medium. This definition clearly presumes the existence of sites—i.e. locations within the crystal (or melt), characterized by a certain geometry and local structure—where an atom of a specific isotope may prefer to reside (Chakraborty 1995). Preference in this case refers to locations of low potential energy where the atom or ion spends more time occupying or vibrating around these sites relative to the time of jump from one site to another, which is a fundamental difference between diffusion in solids from that in an aqueous solution or a gas, with silicate melts/glass providing a case of transition between the two kinds of behavior. A number of mechanisms have been proposed for the elementary atomic jump (Fig. 18, from Bocquet et al. 1996). Most involve one or more kinds of point defects, without which diffusion is impossible in a crystalline solid. For crystalline materials, the generally accepted mechanisms involve the motion of one of two types of point defects, vacancies caused by the absence of an atom or ion at a lattice site (monovacancies, divacancies, and higher order vacancy clusters) or interstitials where an additional atom, ion or molecule occupies a normally unoccupied site (single interstitials, dumbbell interstitials, interstitialcies, and crowdions) (Peterson 1975). These lattice defects which control lattice (or volume) diffusion, are generated either thermally (intrinsic; e.g. Schottky, Frenkel; see Lasaga 1981c) or by the presence of impurities (extrinsic). Identification of the individual atomistic mechanisms is, however, not straight forward. Isotope mass effect studies are one approach and these have been conducted primarily on oxides that compared diffusivities of ^{18}O with ^{17}O. A discussion of this technique will be described below in more detail.

In addition to isotopic exchange occurring through a volume diffusion mechanism,

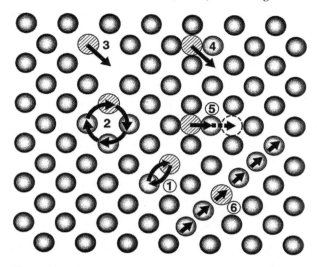

Figure 18. Mechanisms of diffusion, modified from Bocquet et al. (1996): (1) direct exchange, (2) cyclic exchange, (3) vacancy, (4) interstitial, (5) interstitialcy, (6) crowdion.

faster short-circuit pathways may also play a role depending on the nature of the solid (Lasaga 1981c). Macroscopic defects having dimensions exceeding 10^{-3} cm can be present as cracks, fissures and empty pore space. On the microscopic scale, dislocations or other line defects may facilitate diffusion and can consist of displacements of lattice planes along a defect boundary line. Planar defects such as stacking faults, twin boundaries and low and high angle grain boundaries may also be present. As we will see later, grain boundary diffusion has been the focus of a number of recent studies where isotopically–enriched water was reacted with fine-grained polycrystalline aggregates (e.g. Faver and Yund 1995). Non-planar defects have also been identified and include pipes (Yund et al. 1981) and tubes (Hacker and Christie 1991). Sitzman et al. (2000) imaged fast pathways in magnetite by TEM which correlated with $\delta^{18}O$ variations determined by the ion microprobe. Finally, surface diffusion can occur when free surfaces are exposed to fluids (liquids or gases), and under these conditions mineral-fluid exchange may also involve dissolution and precipitation (see e.g Robertson 1969, Steele and Kilner 1982, Grabke 1994, Martin and Duprez 1996, Doornkamp et al. 1998). Numerous surface diffusion studies have been conducted on dry systems where a gas was reacted with a solid, usually an oxide, in order to assess quantitatively the catalytic properties of the solid surface. Discussion of this topic is beyond the scope of this review and will not be considered further. The reader is encouraged to consult the complete texts devoted to diffusion theory and practice, including Jost (1960), Manning (1968), Crank (1975), Allnatt and Lidiard (1993), Lasaga (1998) and Glicksman (2000).

Fick's laws

To measure the motion of atoms (or any other particle) one defines the physical quantity flux (or more rigorously, flux density), J, which is a vector characterized by a magnitude and direction. A flux, J_i, gives the quantity of a diffusing atom (i), which passes per unit time through unit area of a plane perpendicular to the direction of diffusion. Representing the concentration, or amount per unit volume, as C_i, Fick's first law may be written as

$$J_i = -D\nabla C_i \tag{105}$$

where ∇ is the mathematical shorthand for gradient, $\partial C_i/\partial x$ in one dimension (e.g. tabular crystal) or $\partial C_i/\partial r$ for radial diffusion (e.g. spherical grain with radius r). This law is similar in form to Fourier's law stating that the flow of heat is proportional to the temperature gradient. The factor D, known as the diffusivity or diffusion coefficient, was introduced as a proportionality factor. The negative sign is a matter of convention that ensures that for positive diffusion coefficients, flux is positive in the direction of decreasing concentration, i.e. net motion is down concentration. Because the gradient involved is a "local gradient" (in space and time), Fick's first law is by no means a "steady state" or a "time-independent" law (in the sense that it is valid only at steady state), as some text books would like to imply (Chakraborty 1995). It should be noted that D has the dimensions of (area/time), where D is usually given as cm^2/sec or m^2/sec in the S.I. system. In an isotropic mineral, the diffusion coefficient is independent of direction and there is only one value for D. For an anisotropic mineral the diffusion coefficient may vary many orders of magnitude in different directions.

Fick's first law relates flux to concentration gradient at a given place and time, but it does not explicitly describe how the concentration at a point evolves with time. To obtain a complete description of the process as a function of both position and time, it is necessary to combine Equation (105) with the equation that describes the conservation of atoms of type i

$$\partial C_i/\partial t + \nabla \cdot J_i = 0 \tag{106}$$

Elimination of J_i between Equations (105) and (106) gives Fick's second law for one dimension

$$\partial C_i/\partial t = D \nabla^2 C_i \tag{107}$$

and

$$\partial C_i/\partial t = D [\nabla^2 + (2/r) \nabla] \tag{108}$$

for radial diffusion in a sphere. The physical interpretation of these equations is quite straightforward. The terms $\nabla^2 C_i$ or $[\nabla^2 + (2/r) \nabla]$ represent the curvature of the diffusion profile. Therefore, whenever the concentration profile (i.e. C_i versus x) is concave downwards in a given region, the concentration decreases with time. Solutions of this partial differential equation giving C_i as a function of time and position can be obtained by various means once the boundary conditions have been specified. In simple cases (e.g. where D is a constant and the conditions are highly symmetric) it is possible to obtain explicit analytic solutions of Equation (107); see e.g. Carslaw and Jaeger (1959). Experiments can often be designed to satisfy these boundary conditions, but in many other practical situations numerical methods must be used which accommodate different grain geometries (e.g. Crank 1975).

In the context of motion and flux, it is clear that the flux should be defined with respect to a reference frame (Chakraborty 1995). In crystalline silicates, because diffusion of oxygen and silicon can be much slower than that of other cations (with possible exception of Al), this can be achieved quite easily by using the fixed silicate lattice as a reference frame in which ions jump from site to site. This is the so-called lattice fixed frame, which commonly coincides with the laboratory frame. Note that the motion of a dilute isotope of oxygen (e.g. ^{18}O) can still be treated in this frame. Fick's first law can readily be modified to take into account the variability of reference frames.

It can be further extended to describe simultaneous diffusion of multiple species in a multicomponent solid. However, Fick's law may not effectively handle individual atomic jumps at short spatial scales or on very short time scales (as determined by some spectroscopic methods or computer simulations), nor can it address cases where diffusion occurs in a medium whose structure is changing, like the glass transition region. These may lead to what is termed non-Fickian behavior (e.g. Crank 1975).

Diffusion coefficients

Solutions to Fick's equations for a variety of boundary conditions and grain geometries, i.e. infinite plate, cylinder, sphere, etc., have been derived by numerous workers (e.g. Jost 1960, Crank 1975). Many of these solutions demonstrate how the degree of equilibration, F (see e.g. Eqn. 97), is related to the diffusion coefficient, as well as grain size, grain geometry and solution to solid volume ratio. Diffusion rates are sensitive to a number of factors which can be broadly divided into (a) environmental and (b) crystal chemical (Freer 1980). The former includes temperature, pressure, water fugacity, and oxygen activity. The latter includes chemical purity, defect concentration, crystal orientation for anisotropic phases, and porosity of either the single crystal or polycrystalline aggregate.

A number of different types of diffusion coefficients (D) are encountered in the literature and the type of diffusion must be known if the determined D is to be a meaningful value. In some commonly used equations to solve for D, it is assumed that D is a function of C_i, which does not change appreciably during the experiment (i.e. the composition of the solid is essentially the same after reaction). This is accomplished either (a) by using a measurement technique (commonly involving radioactive tracers) that can detect very small changes in C_i, or (b) by using diffusional exchange of stable isotopes of the same element that leave the element concentration unchanged (Brady 1995). Approach (a) yields a tracer diffusion coefficient, D^*, which, according to some authors, refers to transport of atoms present only in dilute quantities. Approach (b) yields a self-diffusion coefficient for the isotopically doped element that is also specific to the bulk chemical composition. Some confusion exists regarding which of these two terms describe transport of an isotope of the host's own species through itself, where the isotope is present at a trace concentration. Consequently, the two have been used interchangeably to some extent in the literature. Both approaches generally ignore the opposite or exchange flux that must occur in dominantly ionic phases such as silicate minerals, glasses and melts. Experiments that use isotopically labeled species interacted with a mineral or mineral aggregate (grain boundary diffusion) are by far the most prevalent with regard to the determination of diffusivities (Table 2, Appendix), and will be the focus of our remaining discussion.

The other major category of D values are termed interdiffusion or chemical diffusion coefficients. Experimentally, C_i does change significantly and D is a function of C_i. Diffusion of one or more chemical species is dependent on the opposing diffusion of another species in order to maintain a constant matrix volume and/or electrical neutrality. The diffusion in olivine of Mg^{2+} in one direction and the complementary diffusion of Fe^{2+} in the opposite direction represent one example. Rarely does this type of experiment employ the use of isotopically labeled species. However, in some cases isotopically-enriched H_2O (T, D and/or ^{18}O) has been used where the composition of the solid (melt) became significantly modified by incorporation of water into the structure.

With some loss of accuracy, therefore, "tracer" and "self" diffusion coefficients may be used interchangeably to refer to diffusion of a species in the *absence* of a driving force, e.g. a chemical potential gradient in a non-ideal solution. Similarly, "chemical" and "inter-diffusion" coefficients may be used interchangeably to refer to diffusion of the same species in the *presence* of a driving force, e.g. a chemical potential gradient in a non-ideal solution (i.e. concentration gradients in most real, multicomponent systems). The important point to note is: the use of one or the other kind of diffusion coefficient to describe a process does not depend on the species concerned (element or specific isotope of an element), but rather on the nature of the boundary conditions of the specific diffusion problem, e.g. whether there is a concentration gradient present or not. This is in contradiction to the fairly common practice of using "tracer" or "self" diffusion coefficients wherever modeling transport of any isotope is concerned.

Secondly, under specific circumstances, the "tracer" or "self" diffusion coefficients may be numerically very similar to the "chemical" or "inter"- diffusion coefficients. This is in particular the case when the diffusing species is very dilute and the diffusing medium is a nearly ideal solution with respect to mixing of the diffusing units. In these cases, all four diffusion coefficients can be used interchangeably for practical purposes, although their physical significance remains different in spite of the numerical similarity. Relationships between tracer and chemical diffusion coefficients in multicomponent amorphous silicates have been derived by a number of authors over the years; we refer the reader to Liang et al. (1997) who suggest their own model and then compare various models with experimental data to decide on the most appropriate one for silicate melts, and to Lasaga (1998) who elucidates the theoretical relationships between the different kinds of models. For crystalline silicate systems, the model of Lasaga (1979a) has been shown to be quite successful and for aqueous solutions, the model given by Lasaga (1979b) has been demonstrated to be almost exact (Applin and Lasaga 1984).

A related question that arises automatically is: do isotopic gradients homogenize more rapidly than the corresponding elemental concentration gradients? Some experimental studies on silicate melts demonstrated faster homogenization of *isotopic ratios* relative to the corresponding elemental gradients (e.g. Lesher 1994, van der Laan et al. 1994), and this has led to a variety of discussions in the literature. In answering this question, it is useful to remember that the diffusive flux of a species depends on the diffusion coefficient as well as the concentration gradient, $J = -D(\partial C/\partial x)$ [see below for a more detailed form of the flux equation in glasses/melts]. Thus, a smaller concentration gradient, e.g. that of one particular isotope of an element, may be homogenized faster than the overall gradient of the element concerned. This observation, however, does not imply that diffusion of that particular isotope is governed by a different diffusion coefficient (e.g. tracer diffusion coefficient)—rather, it is a simple consequence of the abundance of the specific isotope and the magnitude of the relevant concentration gradients (overall elemental as well as that of the specific isotope). Quantitative analyses to clarify this point for silicate melts may be found in Richter et al. (1999) or Ozawa and Nagasawa (2001); Lesher (1994) presents and discusses experimental data to address this issue. A corollary of this discussion is that both—chemical, as well as tracer diffusion coefficients—are of interest for modeling the evolution of stable isotopes in condensed systems. Incidentally, a consequence of the experimental (Richter et al. 1999) and numerical simulation (Tsuchiyama et al. 1994) work is that the isotopic fractionation during diffusion in melts is much lower (depending on a mass exponent of 0.1 or less, as opposed to ~0.5 in the gaseous state), such that detectable mass fractionation is unlikely to result from diffusion in molten systems.

Determination of D

As noted above, in the case of isotopic exchange, we generally measure the self-diffusion coefficient which is defined as the rate of movement of a given isotopic species through a host with a bulk composition that is composed, in large part, of that species (e.g. ^{18}O in silicates, ^{13}C in carbonates). A wide variety of experimental configurations have been used to determine D values, and these can be divided into essentially three groups: bulk techniques, single-crystal techniques, and mineral aggregates. The experimental boundary conditions are commonly dictated, in large part, by the choice of the analytical method available to characterize the change in isotopic composition, either as a function of time, or preferably, space. Bulk exchange methods typically use isotope ratio gas source mass spectrometry (IRMS), as well as NMR, FTIR, liquid scintillation, and even less sophisticated approaches such as weight gain (gravimetry). The single-crystal and mineral aggregate studies rely heavily on secondary ion mass spectrometry (SIMS or the ion microprobe) and nuclear reaction analysis (NRA), and to a lesser extent Rutherford Backscattering (RBS), FTIR and Mössbauer spectroscopy. Grain boundary diffusion studies typically involve the use of mineral aggregates, which are analyzed with the ion microprobe. Interestingly, under certain circumstances single crystal studies can also be used to address fast diffusion pathways such as dislocations, as well as document multiple diffusion mechanisms within single crystals (e.g. Moore et al. 1998).

It is important to note that most of the diffusion data summarized in Table 2 (see Appendix) were obtained in order to quantify either the transport rate of the element of interest (e.g. O, C) or the solid state properties of the crystalline phase, chiefly, the nature of defects. Because most of these studies used isotopically labeled compounds, we assume that the rates of isotopic exchange can be adequately represented by these diffusivities. Therefore, the utility of diffusion data in modeling natural systems depends on selection of the appropriate D and its quality. What constitutes a successful (ideal) diffusion experiment?

1. Precise and accurate measurement of concentration as function of distance (depth) and time,
2. measurement of D at various temperatures and pressures, water, oxygen and hydrogen fugacity,
3. control of the P-T-X conditions so that the solid remains unreacted with either the gas or fluid phase (buffering if necessary; use SEM and/or TEM to characterize the solid),
4. use of different chemical compositions within a particular mineral series (e.g olivine, feldspar) to assess role of cation chemistry,
5. the solid phase is chemically and isotopically homogeneous and the defect state is well-characterized (e.g. phase has been properly annealed prior to reaction), and for a synthesized phase, its formation conditions are well constrained,
6. for bulk exchange experiments, detailed knowledge of the grain size and shape before and after the run are a must,
7. selective use of doped impurities (at ~10 to 10^3 ppm level) can help constrain the defect migration rates that influence diffusion,
8. use of multiple isotopes (e.g. ^{18}O and ^{17}O) to quantify the correlation factor.

Bulk exchange methods. Prior to about 1980 the majority of the diffusion coefficients reported in the literature were obtained from a bulk exchange technique. In this approach, fine-grained crystalline powders of silicate, oxide, carbonate, etc. are reacted with a volatile phase such as H_2O, O_2, CO_2, or CO_2+CO. In numerous examples, at ambient pressures (0.1 MPa), copious amounts of gas or water vapor ($\pm H_2$) are passed over the mineral powder such that the isotopic composition of the volatile does not change appreciably during the course of the experiment. Therefore, only the isotopic change in the mineral need be determined. Jost (1960) described a method for calculating the diffusion coefficients from partial exchange experiments using a simple equation for the boundary conditions of an *infinite reservoir* of isotopically constant gas (H_2O). For a spherical grain geometry, this expression can be cast in terms of isotopic compositions as

$$\frac{\delta^{18}O_f - \delta^{18}O_{eq}}{\delta^{18}O_i - \delta^{18}O_{eq}} = \frac{6}{\pi^2} \sum_{n=1}^{n=\infty} \frac{1}{n^2} (\exp[-n^2 t / \tau]) \qquad (109a)$$

$$\ln\left\{\frac{\pi^2}{6}\left(\frac{\delta^{18}O_f - \delta^{18}O_{eq}}{\delta^{18}O_i - \delta^{18}O_{eq}}\right)\right\} \cong -t / \tau \qquad (109b)$$

where D is the diffusion coefficient (cm^2/sec), t is time in sec, $\tau = (r^2/\pi^2 D)$, r is the grain radius in cm, and the $\delta^{18}O$ compositions are designated by i (the isotopic composition of the starting material), f (isotopic composition of the exchanged solid), and eq (the equilibrium composition of the solid). Diffusion coefficients can be calculated from the slope of a plot of degree of exchange (left hand side of Equation 109b) versus time for several partial exchange experiments (see e.g. Muehlenbachs and Kushiro 1974). Conventional gas source isotope ratio mass spectrometry is used to determine the isotopic compositions. The bulk method was used by Muehlenbachs and Kushiro (1974) and Connolly and Muehlenbachs (1988) to determine diffusivities in silicates reacted with isotopically normal CO_2 at elevated temperatures. Some studies have used this same approach, but instead reacted small volumes of a volatile with a large quantity of solid and measured only the isotopic changes in the gas phase. Alternatively, ^{18}O-enriched O_2, CO_2 or H_2O have been reacted in small to modest quantities with solids so that the condition of an infinite ^{18}O reservoir is still satisfied. In many of these cases, the solid is commonly pre-annealed (equilibrated) at the desired run temperature with a gas of normal $^{18}O/^{16}O$ ratio prior to the introduction of the isotopically enriched gas.

A significant number of diffusion studies that utilized the bulk isotope exchange approach were conducted in systems where diffusion occurred from a well-mixed gas or fluid reservoir of limited volume. Analytical expressions have been derived by Crank (1975) that relate the extent of exchange with diffusivity as a function of grain geometry, grain radius (sphere or cylinder) or grain width (plate), and the volume ratio of volatile (solution) to solid corrected for equilibrium isotope partitioning (see Cole et al. 1983, Lasaga 1998). This approach is valid, however, only for experimental conditions where the solution to solid mass ratio does not exceed approximately 10 for oxygen isotope exchange, and about one for hydrogen isotope exchange. For small values of $(Dt/r^2)^{1/2}$ in a sphere ($<<0.1$), the expression for F was given by Carmen and Haul (1954), and reexamined by Crank (1975, Eqn. 6.33) as

$$F = \frac{M_t}{M_\infty} = (1+\beta)\left[1 - \frac{\gamma_1}{\gamma_1 + \gamma_2}\text{eerfc}\left\{\frac{3\gamma_1}{\beta}\left(\frac{Dt}{r^2}\right)^{\frac{1}{2}}\right\} - \frac{\gamma_2}{\gamma_1 + \gamma_2}\text{eerfc}\left\{\frac{3\gamma_1}{\beta}\left(\frac{Dt}{r^2}\right)^{\frac{1}{2}}\right\}\right] \qquad (110)$$

where

$$\gamma_1 = \frac{1}{2}\left\{(1+\frac{4}{3}\beta)^{\frac{1}{2}} + 1\right\} \quad \text{and} \quad \gamma_2 = \gamma_1 - 1$$

and M_t/M_∞ is the ratio of the total amount of ^{18}O exchanged at time t and after infinite duration (∞) and is referred to as the final fractional uptake, β is the ratio of the volumes of solution (gas) and solid corrected for the equilibrium isotope partitioning (K') between the sphere and solution ($\beta = 3V_s/(4\pi r^3 K')$, r is the grain radius, and eerfc(Z) = exp(Z^2) $\times$ erfc(Z). For known values of r, t, and β, D is solved iteratively until convergence on the measured value of F is obtained. For exchange with aqueous solutions, the value for K' can be estimated from

$$K' = \alpha \frac{X_o / V_o}{(1/18)\rho_{soln}} \tag{111}$$

where $K' = {}^{18}O_{solid} / {}^{18}O_{soln}$, α is the isotopic fractionation factor, X_o the moles of O per mole of mineral, V_o is the molar volume, ρ_{soln} is the density of the solution (or gas), and 1/18 arises from the molecular weight of water. Even though α is a number very close to unity, K' will not be. The β value may be obtained from the more familiar water/solid mass ratio $(W/R)_m$ using

$$\beta = \frac{(W/R)_m / \rho_{soln}}{(V_o / M_s) / K'} \tag{112}$$

where M_s is the molecular weight of the solid. Similar expressions can also be used for hydrogen diffusion. By way of example, Ando and Oishi (1974) and Haneda et al. (1984) used Equation (110) to determine diffusivities of ${}^{18}O$ in Mg-spinel and Y-bearing garnet, respectively (see Table 2, Appendix). Similar forms of Equation (110) have been derived for other grain geometries, most notably plate and cylinder for different magnitudes of β and t (see Crank 1975, Eqns. 5.36 and 5.37 for a cylinder, and Eqn. 4.43 for a plate). The appropriate expression must be selected based on the boundary conditions dictated by the experimental design (e.g. short versus long duration runs, high W/R versus low W/R).

For experiments of longer duration (weeks, months, and even hundreds to thousands of years with proper care), Crank (1975) derives a number of expressions that relate the now familiar F value to diffusion of a solute (H_2O, O_2) from a well-mixed solution (gas) of limited volume into a sphere (or cylinder, plate). The total amount of solute in the sphere after time t is expressed as a fraction (M_t/M_∞ or F) of the corresponding quantity after infinite time (equilibrium) by the following equation (Eqn 6.3 from Crank 1975)

$$F = 1 - \sum_{n=1}^{\infty} \frac{6\beta(\beta+1)e^{-Dq_n^2 t / r^2}}{9 + 9\beta + q_n^2 \beta^2} \tag{113a}$$

where the q_n^2's are the non-zero roots of $\tan q_n = 3q_n / (3 + \beta q_n^2)$. D can be solved iteratively from the known values of r, t and F. The iteration is done in concert with the calculation of the q_n values which uses a Newton-Raphson method (see Press et al. 1986). Each root is approximately π away from the next, therefore the procedure is to add π to the last root and initiate a new Newton-Raphson iteration. Similar expressions have been presented for other grain geometries. A less accurate, but certainly quicker approach is to solve for D graphically utilizing plots provided by Crank (1975) for select boundary conditions. For example, he provided graphs of F plotted against either $(Dt/r^2)^{1/2}$ for a sphere and cylinder, or $(Dt/l^2)^{1/2}$ for a plate for the case of diffusion from a well-mixed solution of limited volume into a solid at modest to long times. The contours in these plots represent the percentage of total solute finally taken up by the solid, where % contour = $[1/(1+\beta)] \times 100$. One can use approximations for Equation (113a) provided by Criss et al. (1987)

$$1 - F = \exp\left[-6Z / (\sqrt{\pi X_w}\right] \quad \text{for } Z < 0.1 \tag{113b}$$

or

$$1 - F = \exp-\left[\pi^2 Z^2 + 0.4977 / (X_w)^{2/3}\right] \quad \text{for } Z > 0.2 \tag{113c}$$

where $Z = \sqrt{Dt}/r^2$ and X_w is oxygen mole fraction for water in the system. Alternatively, the process can be simulated with either finite difference or finite element methods and a best fit found using a non-linear least squares approach.

The bulk exchange approach has certain obvious advantages and disadvantages in terms of the precision, accuracy and ease by which D can be determined. The degree of analytical sophistication is considerably less than that needed for single crystal research, which typically requires microbeam technologies. In fact, by using ^{18}O-enriched gas (or other volatiles such as H_2O vapor), some studies have used gravimetric methods (i.e. microbalance) to measure the very small weight gains resulting from uptake of the heavy isotope. The use of bulk powders precludes the need to obtain or grow high quality single crystals which can be labor intensive, and for some phases, very difficult to accomplish. Some phases are not well suited for study by microbeam methods. For example, diffusion of water in zeolites must be addressed by other means such as NMR. Xu and Stebbins (1998) used ^{17}O spin echo NMR experiments to study the oxygen isotope exchange kinetics between framework oxygen and oxygen in channels waters in stilbite. Wide-line ^{1}H NMR spectra were used by Moroz et al. (2001) to estimate the H_2O diffusion jump frequencies in a variety of fibrous zeolites (e.g. natrolite, mesolite). Further, the bulk exchange methods do not provide an accurate and direct measure of transport rates as a function of position. The bulk method is not particularly accurate for very small D values which are reflected in small measured values of F. The approximation of regular geometric shapes with an average grain radius (width) to represent the solid are main contributors to error in the estimation of D values. Clearly, the effect of anisotropy (i.e. different diffusivities in different crystallographic directions) can not be readily tested. This is particularly problematic for diffusion in hydrous silicates such as micas, chlorite and amphiboles where a high degree of anisotropy is anticipated based on crystal structure arguments. For the layer silicates, a geometric plate model is used to estimate rates parallel to the c-crystallographic direction, whereas the cylinder model is used for transport rates perpendicular to c (e.g. see Giletti and Anderson 1975, Graham 1981). These are, at best, mathematical contrivances that must be tested with direct, *in situ* microbeam methods. Furthermore, for small values of F it may be equivocal as to whether diffusion was the controlling mechanism for exchange, and not some type of shallow surface reaction process? This is not to say that D values determined by bulk methods should be ignored. On the contrary, for phases which are isotropic to oxygen diffusion (but not crystallographically) such as feldspars, the D's measured with bulk techniques are remarkably similar to those determined by more sophisticated microbeam analyses, such as with ion probe (e.g. compare the results from Yund and Anderson 1974 with those of Giletti et al. 1978).

Single-crystal exchange methods. Many problems associated with the bulk exchange method can be overcome if the isotope is analyzed *in situ*. The two most commonly used methods include nuclear reaction analysis (NRA) commonly in concert with Rutherford Backscattering (RBS) and secondary ion mass spectrometry (SIMS or ion microprobe). These have both been described in detail previously so only a brief description is provided here (see e.g. NRA: Robin et al. 1973, Ryerson et al. 1989, SIMS: Giletti et al. 1978, Freer and Dennis 1982, Kilner 1986, Valley et al. 1998). In both methods, single-crystal fragments (commonly prepared with specific crystal orientations polished or cleaved) are annealed in a solution, gas phase, or thin-film coating whose oxygen isotope is highly enriched compared to the solid (10-99% ^{18}O). It is common to thermally anneal the crystal prior to exchange to insure that point defect chemistry does not change while diffusion of the isotopic species is taking place (Ryerson et al. 1989). After exchange, both methods provide an ^{18}O concentration profile with depth beneath the oriented crystal surface. The spot size for SIMS is typically on the order of a few hundred sq. microns, whereas for NRA, spots can be several orders of magnitude greater in area. The depth over which accurate profiling data may be obtained ranges from 25 nm to over 10 μm, appropriate for measurement of diffusion coefficients in the range of 10^{-10} to 10^{-20} cm^2/sec. Within one experiment, these microbeam methods facilitate the interpretation of (a) surface exchange kinetics, (b) dissolution-precipitation phenomena, (c) dislocation- and damage-enhanced diffusion, (d) diffusional anisotropy, and (e) non-Fickian type transport (Freer and Dennis 1982). A third non-isotopic method, FTIR, has also been used, particularly in studies of hydrogen diffusion and water speciation (e.g. Kohlstedt and Mackwell 1998).

<u>F</u>ourier <u>T</u>ransform <u>I</u>nfrared <u>S</u>pectroscopy (FTIR) has been used in a number of studies to determine the mobility of water or hydrogen-bearing species in both crystalline and amorphous solids. IR spectra are obtained from polished slices of sample typically 1-mm thick at a resolution of ~1 cm^{-1}. In many systems a special sample holder is used for diffusion profile measurements so that the sample can be translated from the outside of the FTIR sample chamber, which is maintained at roughing pump vacuum during measurement of IR spectra, to limit contamination due to atmospheric moisture. To obtain a diffusion profile, the sample is translated incrementally past a fixed aperture by adjusting a micrometer attached to the sample stage, but external to the vacuum. After each adjustment of the micrometer (usually at a spacing of about 50 to 100 μm), an IR spectrum is taken. Hydrogen concentrations are determined by subtracting a reference spectrum taken either near the middle of the sample or from a dry reference sample from spectrum at each point across the sample, then fitting a linear background and integrating across remaining O-H bands in the 3-micron region of the IR spectrum. The hydrogen concentration of the reference spectrum is determined by integrating the spectrum across the O-H bands after correcting for background absorbance. Diffusivities of the mobile water can be determined from plots of the hydroxyl concentration versus distance across the sample. For examples of more detailed discussion of this method refer to studies described by Kronenberg et al. (1986), Mackwell and Kohlstedt (1990) and Kohlstedt and Mackwell (1998).

<u>N</u>uclear <u>R</u>eaction <u>A</u>nalysis (NRA) involves the use of a particle accelerator to direct a beam of monoenergetic protons at the sample surface. The protons penetrate the sample and interact with the ^{18}O nuclei in the near-surface region to produce ^{15}N plus detectable α particles, or ^{17}F and neutrons. These particles have a specific energy at the site of generation but they lose energy at a known rate as they travel outward through the sample lattice. Thus the energy of the particle at the sample surface (where it is detected) is a function of depth at which it was produced. The particle count rate at a particular energy depends upon the concentration of ^{18}O at the depth appropriate to that energy, so that the overall energy spectrum of particles can be readily translated into a concentration versus depth profile for ^{18}O in the sample (e.g. Jaoul et al. 1991, Watson and Cherniak 1997). The main advantage to this method is its ability to resolve concentrations over very small depths on the order of a few hundred nanometers. The main disadvantages of this approach are the lack of accurate data for the atomic interaction parameters and its sensitivity to ^{18}O determination only, thus not allowing determination of complementary isotopes (in this case ^{17}O) or other chemical information relevant to the transport process, such as generation of chemical inhomogeneities due to near surface reaction.

<u>S</u>econdary <u>I</u>on <u>M</u>ass <u>S</u>pectrometry (SIMS) is an extremely sensitive technique for the measurement of depth profiles in a wide variety of materials, particularly conductors, for numerous isotopic atoms (e.g. ^{18}O, ^{13}C, D, ^{34}S). Atoms from a small area of the target surface are sputtered off by a primary ion beam usually composed of Cs$^+$, O$^-$ or Ar$^+$. This erodes a crater in the sample and a fraction of the ejected atoms is ionized, collected by

the application of a suitable acceleration potential, and then passed through a mass analyzer. A depth profile is obtained by continuously recording the secondary ion intensities as the primary ion beam erodes a crater into the sample. The depth scale is normally calibrated after analysis using either a profilometer or by an interferometric method. The main advantages of SIMS include: (a) direct determination of isotopic concentration versus depth profiles, (b) simultaneous determination of multi-element profiles to ensure chemical homogeneity throughout the diffusion zone, (c) depth profiles from 25 nm upwards, (d) imaging capability to ensure lateral homogeneity of diffusion throughout planes oriented perpendicular to the diffusion direction, and (e) small spot size (Freer and Dennis 1982). Surface charging and sample topography (rough versus smooth surfaces) are but a few of the major potential disadvantages to this method. Uniform surfaces can be obtained from appropriate cleavage planes or by polishing. However, mechanical polishing can produce a cold-worked layer (damage zone) that may lead to unusual-shaped profiles or enhance rates as deep as one micron (Ryerson et al. 1989). Chemical etching or ion milling can be used to mitigate this problem. Near surface reaction layers, if they form on the scale of a few tens of nanometers, cannot be seen using this technique, because these are commonly sputtered away during the time the beam stabilizes.

A shortcoming of both NRA and SIMS is that the spatial resolution is not high in all three directions. While depth resolutions on the order of a few hundred nanometers or less are routine, the profiles are measured over surface areas of a few hundred square micrometers or even millimeters. Thus, the concentration profile can be an average from various localized reaction or fast diffusion (e.g. pipe diffusion). The consequences of dual-diffusion mechanisms on the geometry of isotopic profiles is discussed in more detail below. There is no available method to by pass these problems for isotopic diffusion, although for elemental diffusion, Analytical Transmission Electron Microscopy (ATEM) offers an alternative (Meissner et al. 1997).

Experiments are typically designed so that resulting concentration profiles can be modeled by the analytical "constant surface concentration" solution to the diffusion equation for a semi-infinite medium

$$\frac{C_t - C_i}{C_s - C_i} = \mathrm{erfc}\left\{ \frac{x}{2\sqrt{Dt}} \right\} \tag{114}$$

where C_t, C_s and C_i are, respectively, the concentrations of the isotope of interest as a depth x from the surface, at the surface of the crystal and the natural isotopic background value of the solid, t is the experiment duration, and D is the diffusion coefficient. The oxygen (or C, D) isotopic composition of the fluid does not change during the experiment, which typically makes C_s constant. This expression works equally well for the assessment of both kinds of profiling methods-SIMS and NRA. Profile data on oxygen diffusion into calcite reacted with a 50:50 mixture of H_2O and CO_2 at 700°C and 80MPa are given in Figure 19a. By taking the inverse error function of the concentration ratio from Equation (114), a straight line fit to the data by least-squares regression analysis will have a slope proportional to $(2\sqrt{Dt})$ as shown in Figure 19b. An alternative approach to solving for D is to use a non-linear least-squares technique (e.g. Levensberg-Marquardt), which minimizes the measure of merit, X^2, fit numerically by adjusting the value of specific parameters, such as C_s and D (Labotka et al. 2000). However, cases have been encountered where the projection of the ion probe profile data back to the surface of the crystal does not match that expected for a surface equilibration scenario. Both of these situations are depicted schematically in Figure 20. The behavior in question arises from the following type of expression (from Kilner 1986)

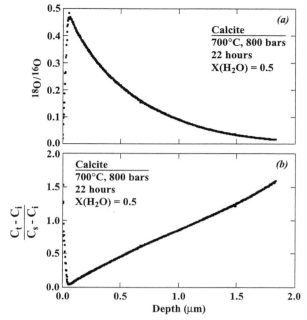

Figure 19. (a) Representative example of a diffusion profile of oxygen isotopes measured in calcite reacted at 700°C and total pressure of 800 bars with a fluid containing $I_{CO_2} = 0.5$ for 22 hours (Labotka et al., 1999). Individual isotope ratio measurements were taken as the Cs^+ ion beam sputtered from the surface into the interior of the crystal. The aberrant data at the beginning of the profile represents isotope ratios collected while the ion beam was sputtering through the gold coating and established a steady state of implantation into the crystal. (b) Corresponding to this profile is a plot in which the isotopic data are reduced using the solution to the diffusion equation for transport normal to the surface of a semi-infinite volume with a planar surface (e.g. Crank 1975). The slope of the inverse error function of the concentration ratio versus depth plot is proportional to $(4Dt)^{-0.5}$ and can be used to extract a diffusion coefficient from the profile. C_i, C_s and C_t are the ^{18}O concentrations at the crystal surface, crystal interior (initial value) and at depth x, respectively.

$$\frac{C_t - C_i}{C_s - C_i} = \mathrm{erfc}\left\{\frac{x}{2\sqrt{Dt}}\right\} - \exp(hx + h^2Dt)\,\mathrm{erfc}\left\{\frac{x}{2\sqrt{Dt}} + h\sqrt{Dt}\right\} \quad (115)$$

where h is the ratio of the surface exchange rate (units of velocity) to diffusivity, K/D. From Figure 20 one can see that for large values of h (solid curve) the kinetics of surface exchange are fast enough for diffusion to be the limiting process and the Equation (115) reduces that given for (114). However, for small values of h (dashed curve), the full version of Equation (115) can be used in a least-squares analysis to obtain both D and K. For $h < \sim 10$, the system is frozen in, whereas for $h > 100$, the system is like that defined in Equation (114). Equation (115) is used for cases where h is in between these limits (Dohmen et al., in press).

Double-diffusion. Another profile-related behavior that has been observed in single crystal diffusion studies is known as "tailing" or double-diffusion. This is observed in

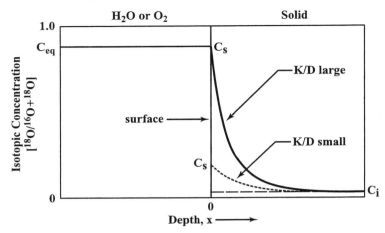

Figure 20. Schematic oxygen isotope concentrations in both solid and water (gas), for solid-fluid exchange experiment. The diagram shows the effect of fast (large K/D) and slow (small K/D) surface exchange on the isotope profile in the solid. K is the surface exchange rate constant and D is the diffusion coefficient. Modified from Kilner (1986).

profiles that exhibit more [18]O (also ^{13}C or D etc.) at depth than can be explained by a single volume diffusion mechanism. This is most obvious in plots of isotopic concentration versus depth where the data exhibit two distinct slopes, one steep near the crystal surface and a more shallow-sloping trajectory deeper into the grain. Of course, this L-shaped trend may be subtle such that the break in slope, or inflection point is less apparent. The inverse error-function plot of all the data typically shows pronounced concave-downward curvature, particularly for the deeper portion of the profile (e.g. Yurimoto et al. 1989). The origin of this curvature is difficult to determine because, in a perfect crystal, there should be no mechanisms other than lattice diffusion controlling exchange. The alternative is that the crystal is imperfect and either contains a large number of dislocations, or a small number of grain boundaries (Kilner 1986). Separate error-function analysis can be performed on each segment to retrieve the diffusion coefficients, or a linear combination of error functions can be considered (Moore et al. 1998). Figure 21 shows examples of these approaches for oxygen diffusion (via NRA) into rutile reacted with ^{18}O-enriched H_2O at 750°C for about 339 hours. Moore et al. (1998) carried out an analysis of variance using the F-test statistic where a "one error-function" fit was compared to that of a "two error-function" fit, the latter of which is given as

$$C_t = a_1 C_{s,1}\left[1 - erf\left(\frac{x}{2\sqrt{D_1 t}}\right)\right] + a_2 C_{s,2}\left[1 - erf\left(\frac{x}{2\sqrt{D_2 t}}\right)\right] \qquad (116)$$

where a_1 and a_2 are the coefficients of each error-function term and represent the interdependence of the two functions. In the case of the rutile study, Moore et al. (1998) determined that both a_1 and a_2 were, within error, equal to 1, suggesting that the two error functions were statistically independent. As they point out, this does not mean that the mechanisms responsible for the functions are physically independent, only that the interaction between mechanisms is small enough to not be measurable. Their time series experiments suggest that both the near-surface and tail regions of the profile were the result of lattice diffusion. They concluded the profiles represented O diffusion that

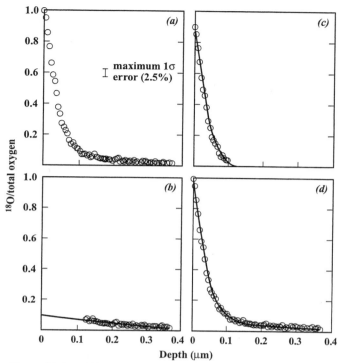

Figure 21. (a) Diffusion profile determined by NRA of oxygen in rutile reacted at 700°C with ^{18}O-eniched water at reduced conditions (Ni-NiO). Notice the "tailing" at depths greater than about 0.1 micron. Maximum error in data is indicated. (b) Error-function fit (solid line) to the tail portion of the profile. (c) Error-function fit to the residual profile. Symbols are residual-concentration data calculated by subtracting the error-function fit in b from data in a, (d) Fit of data is by the linear combination of the error functions shown in b and c. Taken from the work of Moore et al. (1998).

occurred via two simultaneously acting, statistically independent mechanisms. However, it is not uncommon for the D values calculated from the flatter profile to be more on the order expected for grain boundary or dislocation-assisted diffusion and much greater than those calculated from the steep, near surface profiles. Further details regarding numerical recipes for short-circuit pathways can be obtained from the work of Le Claire and Rabinovitch (1981, 1984).

Grain-boundary diffusion. Fast transport processes in many rocks involve fluid flow in transgranular fractures and along open grain boundaries, or through zones of chemical alteration. In the absence of fractures, and for transport between fractures, grain boundary diffusion may be the dominant process controlling isotopic exchange. Over the last decade or so a substantial amount of experimental work has been conducted to determine grain boundary diffusivities of oxygen and major elements in geological materials, and how fluid distribution influences transport. The results of many of these studies are summarized in Table 3 (see Appendix). The classical technique of determining grain boundary diffusion is to deposit a thin layer of the diffusing species on the surface of a synthetic bicrystal with the interface perpendicular to the surface. During annealing the deposited material diffuses into the bicrystal producing V-shaped iso-concentration

(a)

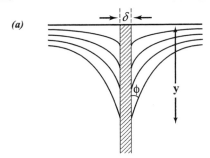

(b)

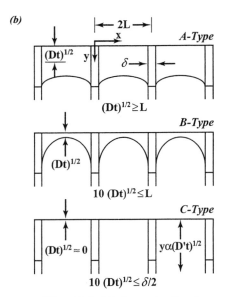

Figure 22. Schematic drawing of grain-boundary iso-concentration contours about a vertical planar boundary (a). (b) is a schematic representation of three types of kinetic scenarios, A-, B- and C-Type illustrating the position of the iso-concentration contours depending on extent of transport along and adjacent to the grain boundary (vertical lines). Modified from Farver and Yund (1991).

contours (Fig. 22a). Mathematically, the grain boundary is represented as an isotropic slab of width, δ, within which diffusion is much faster than in the crystal lattice. The total amount of diffusing species is measured as a function of depth from the surface with SIMS (e.g. Peterson 1975). This method can be extended to the treatment of grain aggregates with multiple grain boundaries normal to the free (exposed) surface provided the ratio of the grain size to the lattice diffusion distance ($2L = \sqrt{Dt}$) is taken into consideration (e.g. see Farver and Yund 1991a, Joesten 1991). Three different diffusion scenarios can be envisioned, designated Type A, B and C kinetics by Harrison (1961), shown schematically in Figure 22b. Type C involves negligible lattice diffusion, which is ideal for the determination of D_{gb}, but is seldom realized because either exchange is too rapid at elevated temperatures, or too slow at low temperatures leading to immature profiles. In Type A, the lattice diffusion distance is much larger than grain size, and the diffusion fields overlap. In general experimental conditions are selected which lead to Type B kinetics such that the lattice diffusion distance is much less than the grain size ($2L \gg \sqrt{Dt}$) and diffusion along each grain boundary is not influenced by diffusion along neighboring grain boundaries.

A mathematical solution for Type B kinetics was derived by Whipple (1954) and modified by Le Claire (1963) which is applicable to a polycrystalline material with a constant surface concentration, given as

$$D_{gb}\delta = \left(\frac{\partial \ln \overline{C}}{\partial y^{6/5}}\right)^{-5/3} \left(\frac{4D_v}{t}\right)^{1/2} (0.661)$$ (117)

where D_{gb} is the diffusion coefficient in the sub-boundary, D_v is the volume diffusion coefficient, δ is the effective boundary thickness, y is the penetration depth of the isotopic tracer, $\overline{C}$ is the average concentration in a layer parallel to the sample surface at depth y, and t is the time of the anneal. A plot of ln $\overline{C}$ versus $y^{6/5}$ yields a straight line in the region where isotopic exchange is dominated by grain-boundary diffusion (typically the flatter portion of the L-shaped profile). The slope of the line, along with the appropriate value for D_v and the duration of the experiment, provides a direct determination of $D_{gb}\delta$ (as m^3/sec or cm^3/sec). Experimental data are commonly reported as $D_{gb}\delta$ and δ is not independently evaluated. The value of δ, and how it varies with the diffusing species or for different solids, is an important parameter and should be determined if possible and compared to the average physical grain-boundary width in the sample. High-resolution TEM observations of sintered ceramics and oxides, and hot-pressed mineral aggregates indicate that the physical-grain boundary widths are typically $\leq$ 5nm (Farver and Yund 1995). Equation (117) has been applied successfully to systems with realistic grain boundary geometries despite the random orientations of grain boundaries which can lead to complex diffusion pathways. Grain-boundary oxygen isotope exchange experiments have been conducted on a number of geologic aggregate materials (see Fig. 23) including calcite (Farver and Yund 1998), quartz (Farver and Yund 1991, 1992), forsterite (Yurimoto et al. 1992), feldspar (Nagy and Giletti 1986, Farver and Yund 1995), and melilite (Yurimoto et al. 1989). Values for $D_{gb}\delta$ typically range from about 10^{-25} m^3/s at low temperatures (300-400°C) to as high as 10^{-15} m^3/sec at high temperatures (800-1200°C). Results can vary considerably depending on whether the aggregate was pre-annealed and under what conditions, i.e. T, P, reacting species, etc. (e.g. Farver and Yund 1991a). Rates for short-circuit pathways in single crystals (e.g. dislocations) are also described in the literature, primarily for oxides. These studies utilize the "tailing" concept described previously for retrieving diffusivities, and the values are typically reported in units of cm^2/sec (see Table 3, Appendix).

One final note. When applying diffusion rates to transport through a rock it is useful to consider the bulk or effective diffusivities (Farver and Yund 1995, Wang 1993). In a polycrystalline material, the bulk diffusivity is equal to the sum of the diffusivities of a given species through all available diffusional pathways. For fluid-absent or fluid unsaturated cases, there are two diffusional pathways available: volume lattice diffusion through the mineral grains and diffusion along grain boundaries. The bulk diffusivity through each available path is equal to the product of the diffusion rate, the concentration of the diffusing species, and fractional cross-sectional area normal to the diffusion direction for each path. For oxygen transport in a silicate aggregate, the concentration will be similar in the two paths and the relationship for bulk diffusion, D_{bulk} is given as

$$D_{bulk} = D_{gb}\delta/a + D_v(1 - \delta/a)$$ (118)

where a is the grain size and δ/a is an approximation of the fractional cross-sectional area occupied by the grain boundaries (Farver and Yund 1995).

Correlation factors. The diffusion coefficient of an atom is different from that of a labeled species of the same element by a factor called the correlation factor, f, which lies for most practical cases between 0.1 and 1 (Chakraborty 1995). Thus, $D_A = f{\cdot}D^*_A$ where D^*_A is the diffusivity of the labeled species of the atom A. Such a relation exists

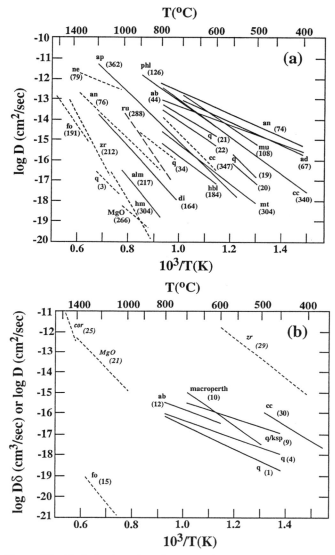

Figure 23. (a) Arrhenius plot of oxygen isotope volume diffusion coefficients for an assortment of silicates, oxides and carbonates. Numbers along side each line refer to data summarized in Table 2 (see Appendix). Data from phases reacted under wet conditions (typically 100 MPa) are depicted with solid lines, whereas dry systems are represented by dashed lines. Note that the rates for the dry systems are generally slower and have higher activation energies (steeper slopes). (b) Arrhenius plot of oxygen isotope grain boundary diffusivities plotted either in units of cm^3/sec or cm^2/sec. δ is the width of the grain boundary. Variability in pre-annealing conditions, pressure, oxidation state, presence or absence of a fluid, etc. can lead to profound differences in diffusivity for any given phase. It is important to select rate data obtained at conditions compatible to those assumed for the natural system being modeled.

irrespective of the concentration of A. In terms of microscopic mechanisms of diffusion by jumping from one site to another, the factor f is related to the correlation between one jump and the next, and is a measure of the randomness of the jump process. The measurement of f can frequently identify the mechanism of self-diffusion or provide considerable detail concerning the diffusion process, and has been used extensively in the study of alloys and oxides (Peterson 1975). For example, f values estimated for diffusion in a face centered cubic lattice have been related to diffusion mechanism where 1.0 implies interstitials, 0.781 vacancies, and 0.475 divacancy-controlled diffusion. A measurement of the mass effect (isotope effect) on diffusion is the most powerful method available for determining f in a broad spectrum of materials. Using oxygen as an example, one approach consists of determining independently the diffusivities of ^{18}O and ^{17}O from ion microprobe profiles described above. The isotope effect E may be written as

$$E \equiv \left[(D_{^{17}O} / D_{^{18}O}) - 1\right] / \left[(m_{^{18}O} / m_{^{17}O})^{1/2}\right] - 1 = f_{^{17}O} \tag{119a}$$

for diffusion mechanisms in which only one atom undergoes a jump displacement where $m_{^{18}O}$ and $m_{^{17}O}$ are atomic masses of ^{18}O and ^{17}O, respectively (e.g. Manning 1968, Peterson 1975). Niu and Millot (1999) estimated an f value of 0.72 for magnetite reacted with ^{18}O-enriched O_2 at 800°C. Another approach for estimating f described by Park and Olander (1991) for ZrO_2 involves taking the ratio of self-diffusivity determined from conductivity measurements to tracer diffusivity determined by single crystal ^{18}O profiles. In this case the f values ranged from 0.61 to 0.69 for temperatures between 1290 and 1550°C, respectively, and are indicative of a vacancy mechanism in cubic crystals. The link between self diffusion (D*) and ionic conductivity (κ) involving f can be described by

$$\frac{\kappa}{D^*} = \frac{nq^2}{k_B T f} \tag{119b}$$

where n is the number of charge carriers, $q \equiv$ valence x elementary charge, and k_B is the Boltzmann constant. The quantity $D^*/(k_B T \kappa/nq^2)$ is sometimes referred to as the Haven ratio (see Chakraborty 1995 for more details regarding this relationship). Methods available for the estimation of correlation factors have not been widely used in diffusion studies involving isotopic exchange reported in the geochemical literature. Perhaps the best example is the study of oxygen diffusion in sanidine by Freer et al. (1997) who examined ^{18}O and ^{17}O diffusivities under hydrothermal conditions from 550-850°C and 100 MPa. They estimated E values in the range of 0.83-1.03, suggesting an interstitial mechanism for oxygen diffusion.

Factors influencing diffusion

Temperature-dependence. Almost all diffusion data obtained on crystalline silicates, carbonates, and oxides exhibit Arrhenian temperature-dependence, where a plot of log D versus 1/T gives a straight line over a significant temperature range, i.e. $D = D_o \exp[-E_a/RT]$. Both the activation energy, E_a (in units of kJ/mole or Kcal/mole) and the pre-exponential factor, D_o (in units of cm^2/sec or m^2/sec) are independent of temperature, but do depend on the identity of the diffusing species, composition and structure of the matrix crystal. From a simple theoretical point of view, D_o is related to the jump frequency (v), jump distance (λ), defect structure, and geometry of the crystal (Manning 1968). E_a is related to the energy of formation of defects (H_f) and the energy required to cause an atom to jump from one site to another (H_m). According to Manning (1968), the tracer diffusion coefficient for a cubic crystal can be described by

$$D^* = [1/6 \, \lambda^2 \, v_o \, f] \exp[-(H_f + H_m)/k_B T] \tag{120}$$

where f, the correlation factor introduced previously, accounts for the fact that the jump probability commonly depends on preceding jumps (non-random walk diffusion).

The energies required for the formation and migration of defects (i.e. vacancies and interstitial atoms) actually can vary as a function of temperature, and hence the slope of the Arrhenius plot can change as different diffusion regimes are encountered. At the highest temperatures, the defect concentration controlling the rate of volume diffusion will be dominated by thermally generated defects (Shewmon 1989). Under these conditions, an *intrinsic diffusion regime* is said to operate, and E_a has its maximum value (steepest slope). As temperature is lowered, the proportional contribution from impurity ions and other defects (e.g. dislocations and grain boundaries) increase until they eventually exceed the intrinsically generated defects. When this occurs, the *extrinsic diffusion regime* is reached, and the change is marked by an inflection in the Arrhenius plot, leading to lower E_a values (shallower slopes).

(1) Experimentally, intrinsic diffusion may be distinguished from extrinsic diffusion because in the intrinsic regime, diffusion is a function of P and T only and is independent of any chemical activities. In the extrinsic regime, diffusion is a function of some chemical potentials in addition to P and T.

(2) At least for Fe-bearing silicates, the binary classification into intrinsic and extrinsic is inadequate, and a third regime is required called "Transition Metal Extrinsic Diffusion" (TaMeD), and most of the observed diffusion occurs in this regime. Thus, diffusion mechanisms may be different in Fe-bearing versus Fe-free systems—an important consideration in planning experiments and extrapolating to nature. This has been demonstrated for olivine by Chakraborty et al. (1995) and discussed in more general terms in Chakraborty (1997).

(3) There is no recorded case of true intrinsic diffusion in any silicate so far. The classic example of Buening and Buseck (1973) is not intrinsic at high T, as has been claimed, because they show D dependent on f_{O_2} both above and below the kink in the Arrhenius plot, which violates condition (1). There are, however, many examples of intrinsic-extrinsic transition among the oxides.

Examples of the temperature-dependence of diffusion in select silicate, oxide and carbonate systems are shown in Figure 23a for volume diffusion and Figure 23b for grain boundary diffusion. A comparison of diffusivities and activation data compiled in Table 2 (see Appendix) leads to the following general observations regarding diffusion behavior in crystalline-fluid (gas) systems.

(1) Anhydrous systems (e.g. O_2 or CO_2–mineral) are characterized by high E_a values, and hence, slower diffusivities compared to mineral-H_2O systems. Dependence of increasing diffusivity with increasing f_{H_2O} has been observed for a number of mineral-H_2O systems, but are not always confirmed by work conducted independently by different investigators.

(2) The best examples of a transition from extrinsic to intrinsic behavior are observed in data from oxides (e.g. Al_2O_3, MgO, FeO) reacted with a gas phase, O_2 or CO_2.

(3) Diffusional anisotropy for select phases has been demonstrated using microbeam profiling along different crystallographic orientations.

(4) Diffusion in select phases, notably oxides, can exhibit a complex dependence on the oxygen activity (P_{O_2} or f_{O_2}) of the reacting gas phase.

(5) Differences among diffusivities measured for identical systems (i.e. T, P, X, t,

analytical method) may sometimes be attributed to differences in the impurity content and annealing history of the phase.

(6) Hydrogen diffusion rates are faster than oxygen in hydrous silicates reacted with water by several orders of magnitude, but not as fast as the bulk-exchange method might suggest based on new results obtained from SIMS (De et al. 2000).

The activation energies can range from as low as a few tens of kJ/mole, as in H_2O transport through zeolite channels, to several hundred kJ/mole for volume diffusion of O in oxides reacted with a gas phase under dry conditions. Data from hydrothermal experiments are characterized by modest E_a values that range from approximately 40 to 100 kJ/mole, and relatively high diffusivities of oxygen compared to the anhydrous systems. The dry data may represent intrinsic diffusion, and the hydrothermal data extrinsic diffusion (Graham and Elphick 1994). The E_a values determined from studies of hydrogen diffusion in hydrous phases such as epidote and biotite are similar in magnitude to those measured for oxygen (i.e. 60 to 80 kJ/mole), but the rates are one to two orders of magnitude less than hydrogen diffusivities estimated from the bulk exchange experiments (De et al. 2000). This large difference may be attributed to a number of factors that include: (a) the correct selection of the diffusion model for bulk exchange experiments on the basis of grain geometry, (b) errors in grain size estimation, (c) exchange controlled by mixed recrystallization-diffusion processes in the bulk powders, and (d) retrograde exchange during quenching of fine powders at the higher temperatures. Great care must be exercised when selecting the proper diffusion coefficient needed to model isotopic behavior in a natural system. The D values or activation parameters used to calculate them, summarized in Tables 2 and 3 (see Appendix), can vary considerably for the same mineral-fluid system because of differences in the starting material (e.g. pre-annealing, natural versus synthetic crystal, orientation, impurities, surface damage due to grinding), chemical environment [e.g. oxidation state, fluid or gas chemistry], pressure range, and temperature range.

Pressure-dependence. The pressure-dependence of diffusion rates is usually described by an equation similar in form to the Arrhenius equation where

$$D_{(P,T)} = D_{(0.1MPa,T)}exp(-P\Delta V^{\#}/\underline{R}T) \tag{121}$$

where $\Delta V^{\#}$ is the activation volume, a parameter that provides a measure of pressure-dependence (Lazarus and Nachtrieb 1963). The pressure- and temperature-dependencies can be conveniently combined into a single equation for practical purposes. $\Delta V^{\#}$ is not a volume of any species, but is equal to the sum of the activation volumes of the change in volume of the crystal upon formation of the defect, ΔV_f, and the lattice dilation attending the elementary diffusion jump, ΔV_m. Note that if diffusion is extrinsic, as appears to be commonly the case, then $\Delta V^{\#} \approx \Delta V_m$. A negative $\Delta V^{\#}$ implies that the activated complex is denser than the ground state, whereas a positive value indicates diffusion rates decrease with increasing pressure. A slope of data plotted as D versus P yields the magnitude and sign of $\Delta V^{\#}$.

Under normal circumstances it is expected that the diffusion rate will decrease with increasing pressure. This is typically the case with the diffusion of cations in crystalline phases such as garnet and olivine (Freer 1993). For anhydrous gas-mineral isotope exchange studies, pressures are generally near 0.1MPa and are not varied much from this level. However, in a number of limited cases, pressures have been varied considerably. In the system calcite-CO_2 at 800°C, Labotka et al. (1999) observed a decrease of 2.5 orders of magnitude in log D (10^{-14} to $10^{-16.6}$) for carbon with increasing pressure from 0.1 MPa to 200 MPa, but nearly constant D's for oxygen (Fig. 24). Similarly, for the system CO_2-

quartz reacted at 900°C, Sharp and Giletti (1991) measured nearly constant oxygen log D values (cm²/sec) for pressures between 10 and 345 MPa, $10^{-14.63}$ and $10^{-14.65}$, respectively, but did observe a slight increase in log D to $10^{-14.09}$ at 720 MPa. There is some conjecture that a small amount of water might have been present to account for this unexpected increase.

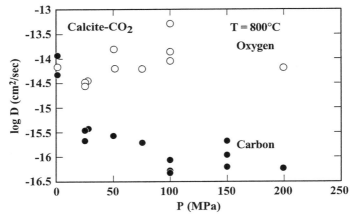

Figure 24. Logarithm of D plotted against pressure of CO_2 for oxygen (open circles) and carbon (solid circles) in calcite reacted with $^{13}C^{18}O_2$ at 800°C. Note that the diffusivity for carbon exhibits a significant decrease with increasing $P(CO_2)$, whereas the oxygen shows no obvious trend because of considerable scatter in the data. Decreasing D with increasing P is expected for dry systems and yields a positive activation volume. Results from Labotka et al. (1999).

Dependence of diffusion on H_2O. The most profound changes in D with pressure have been observed in hydrothermal experiments involving exchange of a mineral with H_2O or H_2O-CO_2 mixtures. The presence of water obviously has a major effect of oxygen and hydrogen diffusion but details of the mechanism are not clear and the role of hydrogen is still controversial (e.g. see Giletti 1994, Graham and Elphick 1994). The importance of water pressure in the physics of mineral behavior has been of great interest for many years since the discovery of the "water-weakening" behavior in quartz (Griggs 1967, 1974; Blacic 1981). Quartz can be more easily deformed under hydrothermal conditions at high pressure relative to dry systems. According to Kronenberg (1994) hydrogen species at the surfaces and interior of quartz serve to interrupt its strong Si-O-Si network (hydrolyze bonds) and assist processes of inelastic deformation that depend on breaking bonds, ranging from crack growth and frictional gliding to solution transfer and dislocation creep. An increase in oxygen and hydrogen diffusivities is also a consequence of this weakening process. The common link has been the proposal that Si-O bonds would be easier to break in the presence of species that would link to Si or O, thereby helping to weaken the Si-O bond.

Numerous species, OH^-, H^+, H_2O, etc., have been suggested as potential candidates in the diffusion process involving water. Zhang et al. (1991a) described an expression for the "apparent" diffusivity of total oxygen in a solid (crystal, glass) or melt ($D^*_{\Sigma O}$) as a function of water speciation given here as

$$D^*_{\Sigma O} \approx D_{O,dry} + D_{H_2O}[H_2O] + D_{OH}[OH] \tag{122}$$

where $[H_2O]$ and $[OH]$ represent mole fractions on a single oxygen basis. $D^*_{\Sigma O}$ will be where [H_2O] and [OH] represent mole fractions on a single oxygen basis. $D^*_{\Sigma O}$ will be directly dependent on which species is the most dominant. A more detailed description of the Zhang et al. (1991a) approach is given in the section of Diffusion in Melts and Glass.

A number of studies have been conducted that have attempted to address the relation among oxygen diffusivity, pressure and water, most notably involving transport in quartz and feldspars. We can not do justice to all of the complexities of this subject. The interested reader is encouraged to consult the following references in order to gain a better perspective on the behavior of water in silicates, contradictions among different experimental results and proposals on the dominant species involved in isotopic exchange (Freer and Dennis 1982, Giletti and Yund 1984, Dennis 1984, Elphick and Graham 1988, Farver and Yund 1990, 1991; Graham and Elphick 1991, 1994; Giletti 1994, McConnell 1995, Doremus 1998, 1999; Kronenberg et al. 1996, Zhang et al. 1991a, Watson and Cherniak 1997, Moore et al. 1998). We will attempt to summarize some of the key observations gleaned from these studies.

(1) The diffusion of oxygen in silicates (also many oxides and carbonates) under wet conditions are many orders of magnitude higher than dry diffusion in the presence of O_2 or CO_2, but many orders of magnitude lower than diffusion of hydrogen (Fig. 25). In fact, hydrogen diffusion in wet systems may be 10 or more orders of magnitude higher than dry oxygen diffusion in the same phase. One notable except is the diffusion of oxygen in rutile which is greater in the dry system (Moore et al. 1998).

(2) A significant number of studies have observed a correlation between increasing P_{H_2O} or f_{H_2O} and oxygen diffusion which is either linear or curvilinear (quartz: Giletti and Yund 1984, Farver and Yund 1991; feldspar: Yund and Anderson 1978, Farver and Yund 1990; amphibole: Farver and Giletti 1985; apatite: Farver and Giletti 1989). See examples for quartz and adularia in Figure 26.

(3) A limited number of similar studies have observed no dependence of D on P_{H_2O} or f_{H_2O} (quartz: Dennis 1984; feldspar: Freer and Dennis 1982).

(4) Comparisons of oxygen log D against estimated $\log(a_{H^+})$ for quartz-H_2O (Farver and Yund 1991) and adularia-H_2O (Farver and Yund 1990) indicate that the there is no correlation with hydrogen ion activity. Additionally, these studies found no correlation between D and either f_{O_2} or f_{H_2}.

(5) Conversely, studies by Elphick and Graham (1988) and Graham and Elphick (1991) on quartz suggest that oxygen diffusion is enhanced by high a_{H^+} (so-called proton transients) because of their ability to move rapidly through the crystal lattice, but inhibited by high f_{H_2} in an analogous manner to Si-Al interdiffusion in feldspars.

(6) Spectrosopic studies indicate that for quartz, hydrogen interstitials make up the majority of the hydrogen defects (e.g. Kronenberg 1994), whereas molecular H_2O (Kronenberg et al. 1996) is the dominant hydrogen-bearing species in K-feldspar.

(7) Counter arguments have been put forth suggesting molecular H_2O plays a major role in diffusion of oxygen in quartz (McConnell 1995, Doremus 1998, 1999), and that hydronium ion H_3O^+ may be important in diffusion in feldspar (Doremus 1996).

(8) Molecular-based simulations suggest that water in albite can be accommodated in the structure as both OH and H_2O with solution energies of 70.4 kJ/mole and 86.6 kJ/mole, respectively (Wright et al. 1996).

Despite the various disagreements among different studies highlighted above, experimental data on oxygen diffusion in select silicates using the ^{18}O tracers in the form

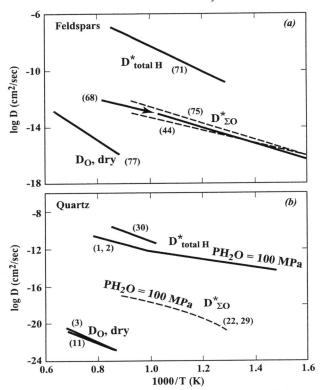

Figure 25. Comparison of diffusion coefficients of total hydrogen, oxygen under hydrothermal conditions, and oxygen under dry conditions. (a) Diffusion coefficients in feldspars. Numbers in ( ) correspond to data summarized in Table 2 (see Appendix). Line 77, O diffusion in anorthite exchanged with ^{18}O-enriched O$_2$ atmosphere; lines 44, 68, and 75 represent data from ablite, adularia and anorthite exchanged with ^{18}O-enriched water at P(H$_2$O) = 100MPa; line 71, total hydrogen diffusion in adularia at unspecified pressure. (b) Diffusion coefficients in quartz. Lines 3 and 11, oxygen diffusion during exchange with ^{18}O-enriched O$_2$ atmosphere; lines 22 and 29, oxygen diffusion during exchange with ^{18}O-enriched water at P(H$_2$O) = 100MPa; lines 1 and 2, H-D interdiffusion with ^{18}O-enriched water at P(H$_2$O) = 2.5 MPa; line 30, total hydrogen diffusion at P(H$_2$O) = 1Gpa. Plot modified from Zhang et al. (1991a).

of H$_2$O are consistent with the idea that ^{18}O transport is dominated by diffusion of H$_2$O molecules even at lower water contents (ppm or less) (Zhang et al. 1991a). This explains why oxygen transport depends on the presence of water and generally depends on water fugacity linearly (e.g. Farver and Yund 1990, 1991). According to their model, if H$_2$O is the dominant diffusing species, $D^*_{\Sigma O}$ is expected to be proportional to f_{H_2O} (or to $[H_2O]^2$) at low water content. This seems to hold for quartz and alkali feldspar. If, on the other hand, OH is the diffusing species, $D^*_{\Sigma O}$ is linearly related to $(f_{H_2O})^{1/2}$ The possible proportionality between $D^*_{\Sigma O}$ and $(f_{H_2O})^{1/2}$ in amphibole and apatite is consistent with the hypothesis that OH is the diffusing species in these phases. This to be expected because OH is the major structural constituent and relatively weakly bound to six or higher coordinated divalent cations. There are examples where the diffusivity of oxygen is

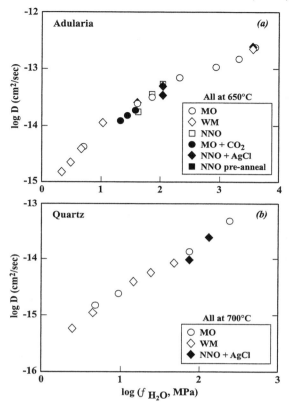

Figure 26. (a) Oxygen diffusion coefficients as function of log f_{H_2O} for adularia reacted with ^{18}O-enriched water at 650°C. The symbols distinguish different buffer assemblages: open circles = Mn_2O_3 + Mn_3O_4, open diamonds = wustite + magnetite, open squares = Ni + NiO, solid circles = Mn_2O_3 + Mn_3O_4 with CO_2, solid diamonds = Ni + NiO and Ag + AgCl, and solid squares = Ni + NiO pre-annealed. (b) Oxygen diffusion coefficients as a function of f_{H_2O} for quartz reacted with ^{18}O-enriched water at 700°C. The symbols represent different buffer assemblages: open circles = Mn_2O_3 + Mn_3O_4, open diamonds = wustite + magnetite, solid diamonds = Ni + NiO and Ag + AgCl. Results are taken from Farver and Yund (1990) [adularia] Farver and Yund (1991) [quartz].

enhanced by presence of water, but does not increase with increasing P_{H_2O}. Watson and Cherniak (1997) observed no dependence of D on P_{H_2O} in the range 7-1000 MPa for zircon exchanged under wet conditions at 925°C.

The results from many studies suggest that the oxygen-bearing transport species is molecular water, which has to diffuse faster than the observed diffusion rate for oxygen. Once the [18]O from a water molecule exchanges with the structure, the water molecule must return to the surface or exchange its oxygen with another water molecule bearing an [18]O. On average each diffusing water molecule has to travel twice as far as the length of the diffusion profile; hence its diffusion coefficient would be four time faster than the measured rate for oxygen. If the rate limiting step is the exchange of oxygen with the silicate structure (e.g. quartz, feldspar), then the rate of water diffusion could be even greater than four times the measured oxygen diffusion rate (Farver and Yund 1990, 1991).

A final note regarding the potential role of [H+] is in order. Graham and Elphick (1994) summarized their view that very rapid transport of [H+] in quartz, and perhaps feldspar, may provide a mechanism for bringing a (+) charge to an oxygen ion so that the Si-O bond might be more easily broken. One can speculate that the rapid transport of this released oxygen might then depend on its continued transport as a neutral water molecule? The idea here is that both [H+] and H_2O play a vital role in controlling oxygen isotope exchange via diffusion between minerals and water (Giletti 1994). It is a fact that H_2O readily dissociates at the temperatures and pressures of these experiments ensuring that increased water pressure translates to increased [H⁺]. Clearly, experimental studies conducted in concert with molecular dynamic simulations and *ab initio* calculations are needed to help resolve this issue.

Oxidation state and defects. In compounds containing atoms with multiple valence, which include many silicates, oxides and even carbonates, the oxygen fugacity can be an important factor influencing the defect structure, and hence, diffusion. The extrinsic versus intrinsic behavior described above is intimately linked to the nature of point defects in solids. A key concept is the nature of majority versus minority defects. The most dominant defects responsible for the charge balance conditions (majority defects) may be unrelated to the oxygen sublattice (e.g. cation vacancies balanced by interstitial cations), in which case it is the minority defects, related to the oxygen sublattice, that influence oxygen diffusion. Diffusion of a cation may be extrinsic while that of oxygen may be intrinsic at the same P-T condition.

Most commonly, defects are not observed as a consequence of their static properties, but because of their mobility. In the case of oxides, oxygen diffusion has been used as a means to quantify the mobilities of minority point defects in the oxygen sublattice just as isotopically-labeled cations have been used to constrain mobilty of majority point defects in the cation sublattice (Monty 1986). In fact, diffusion studies are one of the only ways to identify and characterize these point defects at thermodynamic equilibrium (indirect measurements such as creep give quantities proportional to the slower moving species, hence oxygen in oxides). The nature of defects in the oxygen sublattice for a select number of oxides has been described by assuming the coexistence of oxygen vacancies V_o and oxygen interstitials O_i (e.g. Millot and Niu 1997). The argument used to propose these oxygen defects is based on the analysis of the variation of D^*_O with log P_{O_2}, which can show a complex behavior where D^*_O decreases with increasing P_{O_2} for lower oxygen pressures, but then increases for higher oxygen pressures (Fig. 27). These changes have been attributed to a gradual transition of predominantly oxygen defects from oxygen vacancies V_O^n to oxygen interstitials $O_j^{m'}$, on account of ideal mass action laws for their respective formation.

The relation between the self-diffusion coefficient D^*_O and the concentration of the various defects can be given as a linear combination of the defect concentrations as

$$D^*_O = \sum \mu_{V_O^n} \left[V_O^n \right] + \sum_m \mu_{O_i^{m'}} \left[O_i^{m'} \right] \tag{123}$$

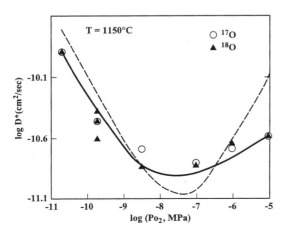

Figure 27. Self-diffusion of ^{17}O and ^{18}O in Fe_3O_4 exchanged with variable partial pressures of oxygen at 1150°C (dry). The dashed line represents the theoretical curve with a consideration of oxygen vacancy defects and interstitials ($V_o + O_i$). The solid curve represents the theoretical curve with consideration of both oxygen vacancies and stable cluster defects involving oxygen and iron [$V_o + (V_oV_{Fe})$]. Modified from Millot and Niu (1997). At 1150°C, Ni-NiO and the QFM (Quartz-Fayalite-Magnetite) buffers lie at P_{O_2} values of roughly 10^{-8} and 10^{-9}, respectively.

where μ_i is a mobility factor for defect i, oxygen vacancies with the effective charge n are represented by

$$[V_O^n] = k_1(T)P_{O_2}^{-1/n} \tag{124}$$

and the formation of an interstitial oxygen which carries the charge m′ is given as

$$[O_j^{m'}] = k_2(T)P_{O_2}^{1/m} \tag{125}$$

Because the activity of electrons is equal to unity due to the disorder of electrons no matter how large the deviation from stoichiometry, the P_{O_2}-dependence of the minority defects in the oxide is generally not affected either by the properties of majority defects or by their electrical charge (Millot and Niu 1997). If oxygen vacancies and interstitial oxygen defects contribute to diffusion in an oxide, such as Fe_3O_4, one could expect experimental results to be described by a relation that combines Equations (123-125) in the form

$$D_O^* = aP_{O_2}^{1/2} + bP_{O_2}^{-1/2} \tag{126}$$

where a and b are parameters that include ideal mass action law and mobilities of defects. This relation would produce a symmetric shaped V on a plot of log D versus log P_{O_2}. Inspection of Figure 27, which shows oxygen data for magnetite, indicates that the lower P_{O_2} portion of the curve adheres to the second term in Equation (126), but not the first. In this case, the data are better represented by an equation of the form

$$D_O^* = aP_{O_2}^{1/6} + bP_{O_2}^{-1/2} \tag{127}$$

According to Niu and Millot (1999), at low P_{O_2}, where the majority defects are iron interstitials, the prevailing oxygen defects are free oxygen vacancies. At high P_{O_2}, where the majority defects are cation vacancies, the observed slope of 1/6 supports the presence of clusters of iron vacancies with oxygen vacancies formed as a result of Coulombic attractive forces. The main two points here are (a) oxygen diffusivities in select oxides (but not all) may follow a complex pattern as a function of P_{O_2} and (b) information of this type is necessary in order to assess the relation between rates and defect structures.

General observations regarding defects. Simply stated, without point defects, one would not have any diffusion in a crystalline solid, and by extension, any process that depends on diffusion would not be possible without defects. An exhaustive discussion of this topic in the context of isotopic exchange controlled by diffusion is beyond the scope of this chapter. Rather, we present here a very brief overview of general behavior linking oxygen and hydrogen diffusion with defects in select silicates, oxides and carbonates. More in-depth coverage may be found in Flynn (1972), Wollenberger (1996), Lasaga (1981c, 1998), Allnatt and Lidiard (1993), Schmalzried (1995), and Glicksman (2000).

As described above, it is now well established for most phases, that the presence of water leads to an enhancement of the diffusion rate of oxygen and hydrogen. One notable exception is the diffusivity of oxygen in rutile (e.g. Dennis and Freer 1993, Moore et al. 1998). Although by no means unanimous, it appears as though water moves through silicate lattices (e.g. feldspar, quartz, pyroxenes) in what has been described as interstitial "water" defects, whereas hydrogen is mobile through hydrogen (or proton) interstitials. Atomistic simulation studies on alkali feldspars seem to support this view (e.g. Wright et al. 1996, Derdau et al. 1998). Oxygen interstitials are thought to play a major role in transport in many, but not all, dry silicate systems (e.g. olivine, Ryerson et al. 1989; sanidine, Derdau et al. 1998). In the case of oxides such as Fe_3O_4 (Millot and Niu 1997) and TiO_2 (Moore et al, 1998), oxygen diffusion under "wet" conditions involves $[Me^{n+}]$ interstitials with a subordinate contribution from oxygen vacancies. Conversely, for dry diffusion in TiO_2 and numerous other oxide phases (including super-conducting oxides), oxygen transport seems to be controlled by the migration of oxygen vacancies. Oxygen diffusion in Y-bearing Al garnet appears to be controlled by oxygen vacancies in a dry system above ~1100°C, but interstitialcies at lower T's affected by trace impurities, in this case Fe (Haneda et al. 1984). In fact, diffusion of oxygen in dry gas-oxide systems reacted at high temperatures is strongly influenced by the presence of trace impurities (see e.g. Atkinson 1994). For calcite reacted with CO_2 containing a minor amount of water, carbon diffusivity seems to be controlled by Frenkel pairs, whereas oxygen transport involves H_2O or OH mobile species (water interstitials?) (Kronenberg et al. 1984). Labotka et al. (2000) reported that for the dry CO_2-calcite system (T's = 600 to 800°C at 100 MPa), oxygen diffusivity is consistent with vacancy migration in the intrinsic region, whereas carbon seems to diffuse as a carbonate anion.

Anisotropy. The use of SIMS and other microbeam methods has allowed the determination of diffusivities for different crystallographic orientations. Diffusion coefficients for anisotropic phases are not simple constants, but rather second-order tensors. In general, the direction of transport will not coincide with the concentration gradient. Pronounced anisotropies for oxygen diffusion have been reported for a number of silicates. Anisotropies of a factor of ~1000 have been measured in micas (more rapid parallel to the layers) (Fortier and Giletti 1989) and apatite (more rapid parallel to the c axis) (Farver and Giletti 1989). For quartz, the oxygen diffusional anisotropy is approximately a factor of 100 (more rapid parallel to the c axis) (Giletti and Yund 1984, Dennis 1984), and in hornblende, the factor is about 10 (more rapid parallel to the c axis) (Farver and Giletti 1985). In the case of diopside, diffusion coefficients are 100 times greater parallel to the c axis than perpendicular to c, over the temperature range of 1000 to 1200°C (Farver 1989). The structures of pyroxene and amphibole minerals are based upon single- and double-chains of SiO_4 tetrahedra that extend parallel to the c crystallographic axis. These chain structures provide a faster diffusional transport path parallel to the c direction. Again, the availability of suitable point defects is a key consideration, because it is not just the channel structure that facilitates diffusion, but also the ease with which vacancies can be formed. The ease refers to the energetic cost—as calculated, for example, by Lasaga (1981c) and the observed jump frequencies—as assessed, for example, by Niemeier et al. (1996). In the case of alkali feldspars, Giletti et al. (1978) used SIMS to measure D_{oxygen} normal to (001) cleavage in Amelia albite and normal to (111), (130) and (010) crystal faces of Cretan albite samples. Atomistic simulations, however, seem to indicate that transport along the a axis may be facilitated more readily compared to either the b or c directions (Derdau et al. 1998). For anorthite, transport parallel to the (010) direction is nearly identical to a direction normal to (001), despite the fact that the activation energies differ by a factor of 1.5 (see Elphick et al. 1988 versus Ryerson and McKeegan 1994).

Empirical methods

We have already described diffusion as a thermally activated process that is commonly represented by the Arrhenius relations (see Eqn. 25). A method of estimating diffusion coefficients takes advantage of the linear interdependence of the activation energy, E_a and the pre-exponential factor, D_o (or A_o). This *compensation law* is an empirical observation that indicates that diffusion rates of different species tend to converge at a particular temperature, generally for materials with similar structure. In addition, the validity of diffusion compensation rests on the assumption that the diffusion mechanism is the same for all the minerals and species being considered (Fortier and

Giletti 1989). An example of this approach was presented previously (Fig. 4) from data provided by Muehlenbachs and Connolly (1991).

An alternative approach was suggested by Dowty (1980), who proposed that the amount of open space in a mineral (closeness of packing of ions) should be an important factor in controlling diffusion through the structure. He demonstrated that there was a qualitative relationship between anion porosity (fraction of space in the unit cell not occupied by anions) and diffusivity, as measured by the relative ease of reequilibration of oxygen isotopes in natural systems. More recently, a number of studies have taken this idea and quantitatively related either ionic or anionic porosity with diffusivity or one of the activation parameters, E_a or D_o, most notably Fortier and Giletti (1989), Muehlenbachs and Connolly (1991), Gautason and Muehlenbachs (1993), and Zheng and Fu (1998). Total ionic porosity defined by Fortier and Giletti (1989) is given as

$$Z = [1 - (V_{anion} + V_{cation})/V_{unit\ cell}] \times 100 \tag{128a}$$

where V refers to volume in units of $\mathring{A}^3$. Unit cell data can be obtained directly from Smyth and Bish (1988). Unit cell volumes can be scaled to the appropriate temperature through the use of thermal expansion data assuming volumes of the cations and anions remain constant where

$$V_{unit\ cell,T} = V_i + (\partial V_T/\partial T)T \tag{128b}$$

V_i is the intercept (assumed to be constant) and $(\partial V_T/\partial T)$ is the slope of the linear relation between unit cell volume and temperature for $400° \leq T \leq 800°C$. Others (Muehlenbachs and Connolly 1991, Gautason and Muehlenbachs 1993, Zheng and Fu 1998) have elected to ignore the cation volumes and estimate anion porosity (Φ) using the relation

$$\Phi = [1 - (V_{anion})/V_{unit\ cell}] \times 100 \tag{129}$$

Obviously, differences between values calculated for Z and Φ reflect this consideration of whether or not to use cation volumes. Disagreements among the calculated values of Φ can also arise from differences in the radii selected for anions. The standard reference for these values is Shannon and Prewitt (1969). The oxygen radius is usually set at between 1.36 and $1.38\mathring{A}$ (Muehlenbachs and Connolly 1991, Gautason and Muehlenbachs 1993). A comparison of Φ values from these studies with those provided by Zheng and Fu (1998) indicate a substantial difference. This is because Zheng and Fu (1998) took into account the coordination numbers (CN) of oxygen. In particular, they used the crystal radii of O^{2-} which can range from 1.21 (CN = II) to 1.28 (CN = VIII), as given by Shannon (1976).

Fortier and Giletti (1989) compared oxygen diffusion data from silicates reacted with H_2O at 100 MPa with ionic porosities corrected for thermal expansion at 500 and 800°C. The straight-line fits to data at each temperature were reasonably good, from which they proposed an overall empirical equation that describes diffusion in select silicates at 100 MPa given as

$$\log D\ (cm^2/sec) = -2.0 - (3.4 \times 10^{-4})(T) + [(-0.13 + (6.4 \times 10^2)T)\ Z] \tag{130}$$

Muehlenbachs and Connolly (1991) and Gautason and Muhlenbachs (1993) demonstrated that oxygen diffusion in dry systems exhibits reasonably good correlations between anion porosity and the activation parameters, E_a and $\ln D_o$, as shown in Figure 28. These studies indicate that as anion porosity increases the values for both E_a and D_o decrease as expected from the compensation relation. Guatason and Muehlenbachs

(1993) and Zheng and Fu (1998) proposed a slightly different equation that relates D with Φ and T given in the general form as

$$\log D = A + B\Phi - [(L + M\Phi)10^3/\underline{R}T] \tag{131}$$

For anhydrous systems Gautason and Muehlenbachs (1993) obtained the following values for the coefficients in Equation (131): A = 36.5±7.7; B = -1.01±0.17; L = 1116±60; M = -18.6±1.3.

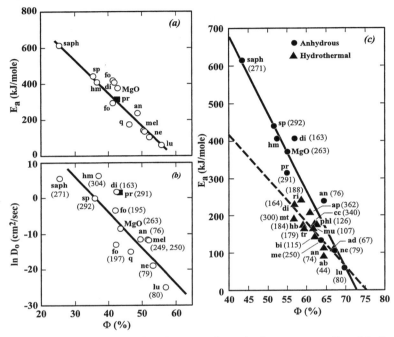

Figure 28. Empirical correlation plots comparing activation parameters, E_a and $\ln D_o$, with anion porosity. Numbers in ( ) refer to data summarized in Table 2 (see Appendix). (a) Relation between activation energy and anion porosity (Φ) for oxygen isotope exchange between anhydrous phases and ^{18}O-enriched gas. (b) Relation between the pre-exponential factor, D_o, and anion porosity, again for anhydrous systems. Plots modified after Muehlenbachs and Connolly (1991). (c) Comparison of activation energies for oxygen diffusion between dry (solid circles) and wet (solid triangles) systems plotted against anion porosity. Plot modified from Zheng and Fu (1998).

Somewhat different values were given by Zheng and Fu (1998) for dry conditions: A = 24.7, B = -0.48, L = 657.3, M = -9.02. Zheng and Fu (1998) also presented values useful for estimating wet systems (Fig. 28): A = 11.14, B = -0.28, L = 388.7, and M = -5.16. While empirical expressions such as these can be used to estimate D values (or activation parameters), it must be recognized that errors associated with such calculations may be as large as several orders of magnitude in D.

A key point here is that the correlations based on either ionic or anion porosity, however well they work, are strictly empirical and cannot have much of a microscopic basis. This is because of the fact that the porosities of nominally anhydrous minerals do not change at all due to the presence of trace amounts of water, but oxygen diffusion rates do, typically by orders of magnitude. One can speculate that this behavior is linked, indirectly, to how easy it is to produce point defects (e.g. vacancies or interstitials) related to oxygen. A consideration of correlations observed between diffusivities and both anion porosity and assorted activation parameters (compensation law), might lead one to conclude that oxygen diffusion in all the widely different silicates takes place via a similar kind of mechanism. Assuming that this is true, and knowing that it is an interstitial mechanism controlling oxygen diffusion in olivine (Ryerson et al. 1989), then one might infer that interstitials control oxygen diffusion in all silicates. Clearly this will be a fruitful area for future investigation.

DIFFUSION-CONTROLLED MELT(GLASS)-VOLATILE ISOTOPE EXCHANGE

Fractionation of stable isotopes in silicate melts and glasses (= condensed amorphous systems) have been used to understand the nature of a variety of geological (e.g. fractionation of S-isotopes during volcanic activity; Baker and Rutherford 1996), cosmochemical (e.g. fractionation of oxygen isotopes during chondrule formation; Yu et al. 1995), environmental (fractionation of hydrogen and oxygen during hydration of obsidian and man-made glass; Anovitz et al. 1999) and technological processes (e.g. in the studies of chemical vapor deposited oxide films, on Si-based chips used for integrated circuits, Cox et al. 1993). Although mass-dependent fractionation affects all multiple isotope-bearing elements and fractionation of elements as heavy as Fe is detectable using modern mass spectrometric techniques (e.g. Richter et al. 1999, Humayun and Clayton 1995), most applications to date have focussed on the isotopes of C, O, H and S. These are, therefore, the elements that are dealt with in this review. Of these, diffusion of H-related species and in particular water, S-related species and C and carbonates have been reviewed in a recent *Reviews in Mineralogy* (Vol. 30) by Watson (1994); oxygen diffusion has been dealt with in *Reviews in Mineralogy*, Volume 32, by Chakraborty (1995). These works also cover the experimental techniques used to measure diffusion of these elements in melts/glasses in great detail. Consequently, to avoid repetition, the following strategy has been adopted in this review. We provide updated tables of data, which include newer results as well as studies already reported in the above works. We discuss only the newer advances in understanding and some aspects specific to isotopic diffusion in amorphous systems. We do not discuss multicomponent coupling during chemical diffusion, because these are too system specific to be covered in general terms.

The mode of transport-diffusion versus percolation

The generalized flux equation for diffusion in amorphous condensed medium may be written in the form of Fick's first law

$$J^\alpha = -D_{tot}^\alpha \frac{\partial C^\alpha}{\partial x} \tag{132}$$

where J^α is the flux of element/isotope α and D_{tot}^α is a diffusion coefficient that relates the flux to the overall concentration (C^α) gradient of α. Although this description is purely macroscopic and empirical, usage has established the implication that the diffusion process referred to above is the outcome of random motion, such as may be expected in a thermally activated process where a particle has to cross an energy barrier. The height of such a barrier is described by the activation energy or enthalpy, which is measured through the temperature-dependence of D_{tot}^α. An alternate expression for flux may be obtained for the case where a gas diffuses across a membrane of thickness L

$$J^\alpha = -P^\alpha \left(\frac{P_1 - P_2}{LP_o} \right) \tag{133}$$

In this case, the constant of proportionality, P^α, is termed the permeability constant. Permeation and diffusion, although related, are not identical processes. Assuming diffusion rate to be independent of concentration, the constants for the two processes may be related to each other by $P^\alpha = D^\alpha_{tot} S^\alpha$, where S^α is the solubility of the diffusing species in the medium, e.g. O_2 in a silicate melt/glass. Permeation is the overall steady-state diffusional flow of a solute across a membrane driven by an external pressure gradient, whereas diffusion, even if it is in response to an externally imposed gradient, is more intimately connected to the structure and energetics of different points in the interior of the medium. Analogous to Lamkin et al. (1992), we will use the term "permeation" to describe processes where diffusive flow occurs without significant interaction with the lattice (i.e. a more physical process) while we reserve the term "diffusion" for processes where reorganization of chemical bonds is a significant component. This distinction is necessary because, as we shall see, the (apparent) activation energies for the former process are much lower and consequently the process is much faster, particularly at lower temperatures. In other words, this process may dominate mass transport at least at lower temperatures so that conclusions based on extrapolation of high temperature "diffusion" data may be invalid (see the section of oxygen diffusion below for examples). From a more fundamental standpoint too, as we have noted, the processes are distinct so that their study reveal different aspects of structure and energetics of amorphous materials.

Glasses versus melts: the glass transition

The glass transition is another concept around which considerable misconception and confusion has nucleated. The nature of the glass transition has been adequately discussed in the recent mineralogical literature, for example, in *Reviews in Mineralogy*, Volume 32 (edited by Stebbins et al. 1995). For our current purpose it is sufficient to note that the glass transition temperature for any given composition is not a unique temperature but rather a function of various dynamic factors such as frequency of observation and cooling rate. The frequently quoted glass transition temperature, at which a given composition attains an arbitrarily chosen viscosity, e.g. 10^{12} Pa, is usually the glass transition temperature corresponding to low frequency observations at moderate, laboratory cooling rates corresponding to relaxation times of ca. 100 sec. Clearly, it is more meaningful to speak of a *glass transition region* rather than a specific temperature. Indeed, the concept is quite analogous to that of the familiar closure temperature—the closure in this case relating to structural, rather than some form of chemical, change. For future reference, a supercooled liquid refers to the state above the glass transition but below the melting point for a particular composition.

Glass forming liquids have been classified as "strong" or "fragile" (Angell 1985) depending on their behavior across the glass transition region. The viscosity of strong liquids show a (near) Arrhenian behavior across the glass transition and these are characterized by relatively small peaks in heat capacity and thermal expansivity at the glass transition. In other words, strong glasses and liquids belong to a structural and energetic continuum with smooth transition in properties across the glass transition. Fragile liquids, on the other hand, are characterized by marked changes across the glass transition and sudden jumps in certain properties. As a rule of thumb, polymerized silicate melts such as SiO_2 and $NaAlSi_3O_8$ (albite) are strong; depolymerized melts such as diopside and anorthite are relatively fragile. We will return to these distinctions in our discussion of diffusion behavior of light stable isotopes later, where we illustrate two extremes in the range of diffusion behavior that may be observed in silicate systems in Figure 29.

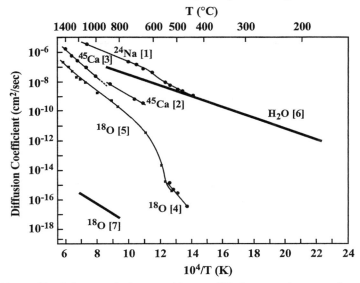

Figure 29. Influence of glass transition on diffusion rates in strong (heavy lines) versus fragile liquids (light lines). A soda-lime-silica glass ($16Na_2O\cdot12CaO\cdot72SiO_2$) is used as an example of a fragile liquid here. Oxygen diffusion data are taken from [5] Terai and Oishi (1977) and [4] Kingery and Lecron (1960). Other data are from [1] Johnson et al., [2] Johnson, and [3] Wakabayashi, as cited in Terai and Oishi (1977). All diffusivities show pronounced change at the glass transition, the Z-shaped form for oxygen diffusion is quite characteristic for fragile liquids. In contrast, a strong liquid is exemplified by SiO_2, which shows a linear Arrhenian behavior for diffusion rate of molecular H_2O, as shown by [6] from Doremus (1995) and network oxygen by [7] from Kalen et al. (1991). These diffusion rates, which differ by many orders of magnitude, show an Arrhenian behavior and even relatively similar activation energies.

The diffusion equation in glasses/melts

Glasses and melts differ fundamentally from crystalline substances in that for any given element (or isotope) a multitude of possible "sites" are available, many of which may be energetically sufficiently distinct from one another. Thus, the overall diffusive flux, as given by the Equation (133), may be decomposed into a sum of diffusive fluxes of all the individual species of which α is a part, i.e

$$J^\alpha = -\sum_i D_i^\alpha \frac{k_i \partial C_i^\alpha}{\partial x} \tag{134}$$

where D_i^α is the diffusion coefficient and C_i^α is the concentration of species i, respectively, and any coupling between the diffusion of different species have been ignored (e.g. Zhang et al. 1991a). k_i is a stoichiometric coefficient in the above equation. Any diffusion process in which new atoms/ions enter/leave such an amorphous condensed medium must always be accompanied, in principle, by speciation reactions. These homogeneous reactions (re)distribute the available atoms/ions among the various available "sites" so as to attain the minimum energy configuration overall. In most general terms, the C_i^α may be described as some function, $g(C^\alpha)$, of the total amount of α in the system at a given P and T. Thus, we have a constitutive equation of the form $C_i^\alpha = g(C^\alpha)$ (e.g. Wasserburg 1988). In general, two kinds of functions may be postulated for $g(C^\alpha)$, depending on the kinetics of the speciation reaction:

Case 1: When the kinetics of the speciation reaction is fast compared to the kinetics of diffusive transport of the fastest species i of α, the equilibrium constant of the speciation reaction, K_{eq}, may be used to relate the concentrations of the different species to one another and to the overall concentration of α.

Case 2: When the kinetics of the speciation reaction are comparable or slower than the diffusion kinetics, then a kinetic rate equation of the speciation reaction, and the related kinetic rate constant, k_{sp}, needs to be used to relate the concentrations of the different species to one another and to the overall concentration of α.

As a specific example, one may consider the diffusion of water into silicate glasses and melts, which has been described in a of recent studies (Watson 1994, Zhang 1999, Zhang and Behrens 2000, Doremus 1995). Assuming the total water content in a glass/melt speciates into two forms—molecular water, H_2O, and hydroxyl groups, (OH)—the diffusion flux equation may be written as (e.g. Zhang et al. 1991b)

$$J_{totH_2O} = -\left(D_{H_2O} \frac{\partial [H_2O]}{\partial x} + \frac{1}{2} D_{OH} \frac{\partial [OH]}{\partial x} \right) \tag{135}$$

The concentrations of the individual species denoted by $[H_2O]$ and $[OH]$ in Equation(135) may be given either by the equilibrium constant, K_{eq}, (Case 1) of the speciation reaction $H_2O + O = 2\,OH$

$$K_{eq} = \frac{[OH]^2}{[H_2O][O_{dry}]} \tag{136}$$

or by some kinetic rate equation (Case 2) such as

$$[OH] = k_{sp} f\left(H_2O_{tot,time} \right) \tag{137}$$

where the concentration of a species such as $[OH]$ is some function of the total water content and time, related via a kinetic rate constant, k_{sp}. Clearly, (137) is a more complicated case and often (136) can be used, even with some further simplifications (see below for examples of both). However, (137) can be used to understand time-dependent diffusion behavior and non-linear compositional dependencies.

To complete the analysis, one applies the continuity relation to the flux equation of the form given by Equation (134) or Equation (135), to obtain an equation where the factor time appears explicitly, i.e. the equivalent of Fick's second law of diffusion. For example, specifically for the case of water treated above (Zhang et al. 1991b):

$$\frac{\partial C_{tot}}{\partial x} = \frac{\partial}{\partial x}\left[D_{H_2O} \frac{\partial [H_2O]}{\partial x} + D_{OH} \frac{\partial [OH]/2}{\partial x} \right] \tag{138}$$

Note that in this case the speciation reaction does not appear explicitly in Equation (138)—any production or consumption of a given species by homogeneous reaction is constrained by conservation laws and does not affect the overall concentration change, i.e. the source/sink terms cancel out. The pre-condition for using this approach, however, is that all forms in which a given element/isotope may occur must be known and accounted for. The advantage, of course, is that gradients in overall concentrations may be described in terms of speciation, even when such speciation cannot be directly measured. The various speciation parameters may then be allowed to float as fit parameters in an analysis. Such analysis often reduces/eradicates many irregularities in diffusion data, as shown by Zhang et al. (1991b) for the case of water diffusion in rhyolite glass.

An alternate approach to treat the same problem is to follow the flux of any one given species and treat the problem as a reaction-diffusion one, e.g.

$$\frac{\partial [H_2O]}{\partial t} = \frac{\partial}{\partial x}\left[D_{H_2O} \frac{\partial [H_2O]}{\partial x} \right] - \frac{\partial S}{\partial t} \tag{139}$$

where S is a function of concentration of molecular water that describes the loss/gain of molecular water due to homogeneous reactions, i.e. interconversion to/from [OH] (e.g. Doremus 1995). This can be an equilibrium (e.g. Eqn. 136) or a kinetic (e.g. Eqn. 137) function. Note that if an equilibrium function is used the time-dependence still remains (i.e. the second term in Eqn. 139 is not zero), because S depends on the instantaneous concentration of molecular water, which of course changes with time by diffusion. In general terms, then, one has

$$\frac{\partial C_i^{\alpha}}{\partial t} = \frac{\partial}{\partial x}\left[D_i^{\alpha} \frac{\partial C_i^{\alpha}}{\partial x} \right] + \frac{\partial S}{\partial t} \tag{140}$$

where the first term is due to diffusion flux and the second term is a source/sink term which is positive if C_i^{α} is produced internally.

The above examples may easily be extended to various species involving C and/or S. It should be noted, as pointed out by Zhang et al. (1991a), that even in the absence of H, simple O diffusion in "dry" silicate melts is also subject, at least in principle, to the same kind of variability because oxygen in a silicate melt or glass can "speciate" into bridging, non-bridging, non-tetrahedrally coordinated (i.e. bonded to network modifiers only) and molecular (i.e. not bonded to the silicate structure) oxygens. In the presence of volatile species such as H_2O, it has been concluded by Palin et al. (1996) that most of the oxygen isotopic exchange takes place through the mediation of these volatile species rather than through a direct exchange of network-bonded oxygen, at least for polymerized silicate melts/glasses, e.g. rhyolites. Similarly, in the presence of transition metal cations, speciation resulting from homogeneous redox reactions control diffusive transport of oxygen (e.g. Beerkens and de Waal 1990). In combination, it is unlikely that in any terrestrial geological problem the speciation effects can be totally ignored *a priori*.

Before we address the behavior of individual systems, it is worth noting some general consequences of the above analysis where diffusion occurs concurrently with speciation reactions. First, as pointed out by Wasserburg (1988), the diffusion equation (e.g. Eqn. 134) can easily take a non-linear form depending on the nature of $g(C^{\alpha})$. Second, since the speciation depends on the total amount of available α, the rate of in-diffusion (e.g. hydration, oxidation) may be substantially different from the rate of out-diffusion (e.g. dehydration, reduction) when the total amount of α only is monitored, as is often the case. When a particular species, C_i^{α}, is monitored, there is a simple relationship between apparent or observed diffusion rates and true diffusion coefficients, as long as the internal production/consumption is described by a linear function of composition, e.g. $S = R. C_i^{\alpha}$ and the diffusion coefficient is independent of composition: $D_{effective} = D_{True}/(R+1)$ (e.g. Crank 1975). Third, because the kinetics of the speciation reactions may be non-Arrhenian, the resulting diffusion rate of overall α may be non-Arrhenian, and this needs to be distinguished from an intrinsically non-Arrhenian, i.e. fragile structural behavior of a melt. Finally, the nature of the speciation reaction itself may change as a function of temperature or content of α, leading to further non-linearities in diffusion behavior. In particular, the dependence of the speciation reaction on the content of α may lead to time-dependent (i.e. as more and more α enters the system, the speciation changes) diffusion behavior. We will see examples of all of these kinds of behavior in specific examples below.

Diffusion caused by gradients other than chemical: electrical, stress and thermal

Although diffusion in response to chemical potential gradients is the most common form to be encountered, atomic migration can occur in response to other driving forces as well. Three of these, which have been shown to have an influence on geological processes, are electrical, stress and thermal gradients. Of these, effects of electrical gradients on stable isotope distribution have not been explicitly addressed although studies of electrical gradients have become relevant in issues such as monitoring the activities of volcanoes. Stable isotope transport in the presence of thermal as well as electrical gradients may become an interesting field of study to elucidate the interaction between magma and wall rock at the borders of near surface magma chambers. There is a relatively large literature on the role of stress gradients in the migration of species such as O_2 and in particular, H_2O, in glasses. The basic idea is that incorporation of species such as H_2O significantly alters the molar volume of a glass, leading to "swelling". Depending on the geometry of the diffusion problem, this may lead to generation of stresses. In the most extreme cases, for example, continued incorporation of water at a given temperature can lead to the water rich part being above the glass transition temperature while the water poor part is still glassy, leading to an extremely non-linear rheological behavior. The combination of volume change and stress alters diffusion rates, although decoupling the effects of simple compositionally induced volume change (i.e. a compositional dependence of diffusion rates) and true influences of stress may sometimes be difficult. This behavior is somewhat analogous to diffusion behavior in some polymers and can lead to such effects as non-Fickian diffusion and time-dependent diffusion rates, at least when the overall, effective diffusion rates are observed (e.g. Tomozawa and Morinelli 1984, Tomozawa and Davis 1999). However, the recent analysis by Doremus (1995) discussed above suggests that when the speciation reactions are suitably accounted for, the residual, purely stress related effects are relatively secondary. Nonetheless, generation of a stressed state in glasses by diffusing in large ions or by hydration is an important industrial process, for example to modify the optical properties of glasses. Similarly, internal friction peaks in glasses have been shown to be related to diffusive motion of ions, including (non-bridging) oxygen, due to stress (e.g. Sakai et al. 1995).

Temperature gradients can cause thermal diffusion (Soret effect), which has been measured by Kyser et al. (1998) above the solvus in silicate liquids that are immiscible at lower temperatures. Additional isothermal oxygen diffusion experiments were performed below the solvus in the immiscible liquids and the results from the two kinds of experiments were compared. Although the magnitude and direction of oxygen isotope fractionation was found to be different from that expected, the authors conclude that this process is unlikely to play a significant role in natural processes such as mantle metasomatism.

Water diffusion in SiO_2 glass: a prototype for a wide range of behavior

Because of the simplicity in chemistry and stability in the amorphous form over a large temperature interval, SiO_2 glass has been studied as a model system for a long time. The fact that SiO_2 is one of the strongest liquids known, i.e. its viscosity is Arrhenian and other properties show small, if any, changes across the glass transition makes it ideally suitable for tracking the various changes related to speciation that can take place in a given amorphous medium with temperature. More recently, water diffusion in amorphous silica has attracted renewed attention because thin films of amorphous SiO_2 may form on

Si based chips and integrated circuits, and understanding water diffusion through such films is crucial for ensuring the longevity of these electronic devices. Oxidation of silicon nitride and carbide based glass ceramics as well as various silicides is also another important Materials Science concern. As a result, data on diffusion of "water" in silica glass are available over a wide range of temperature (160-1200°C) and as a function of a number of variables. Some of the dependencies have been cross checked and improved by recent careful measurements (Davis 1994) and Doremus (1995) provides an integrated model to explain a variety of observations.

At temperatures below about 500°C, equilibrium speciation is not attained on the time scale of the diffusion experiments and a kinetic formulation of the speciation has to be used. At the very lowest temperatures, the [OH] groups are essentially immobile and are able to react with only the nearest neighbors to give a first order reaction. At somewhat higher temperatures (e.g. 450°C), the [OH] groups are still relatively immobile but able to move far enough to sample other nearby [OH] groups, at least for diffusion experiments of long duration, leading to a bi-molecular reaction, i.e. the kinetics of the speciation reaction follows a different law now. Two consequences of these "kinetic regimes" are that (a) the concentration of [OH] at the surface of a glass block where diffusion is taking place changes gradually with time instead of attaining the equilibrium value instantaneously and (b) *apparent* diffusion rates of [OH] are time-dependent. At higher temperatures still, e.g. above 700°C, the speciation reaction reaches equilibrium on the time scale of diffusion and equilibrium relations such as Equation(136) may be used (for example, Zhang et al. 1991b rhyolitic melts, a strong liquid). Note that the mobility of [OH] over diffusion profile length scales may still be largely ignored, as also found by Zhang et al. in their study of rhyolites. With further rise of temperature, the [OH] groups become mobile themselves and their transport needs to be accounted for explicitly. The mobility of [OH] groups even on vibrational time scales have been demonstrated by Keppler and Bagdassarov (1993) in rhyolitic melts at 1300°C. Clearly, the change in diffusion behavior observed as a function of temperature is a consequence of various thermodynamic (e.g. speciation) and kinetic processes with different temperature-dependencies (enthalpy changes or activation energies) competing with each other—in any given temperature interval, some of these dominate over the others with transitional regimes in between. For example, diffusion of [OH] units, presumably involving rupture of relatively strong Si-OH bonds, is a high activation energy process that is significant only at the highest temperatures. Clearly, the speciation reactions can be different depending on the amount of available water, e.g. the ambient partial pressure of H_2O. The significance of properly accounting for speciation has been elegantly demonstrated by Zhang et al. (1991b) in a geologically relevant system, although they did not access as wide a range of behavior as discussed here. Specific aspects of water diffusion in rhyolitic systems are discussed further below.

The main message from this analysis is that with proper consideration of speciation and various processes that affect speciation, diffusion of species such as "water" may be described by simple, Arrhenian diffusion coefficients over a wide range of conditions at least in strong (roughly equivalent to polymerized, in mineralogical terms) liquids and glasses (Fig. 30). Lack of consideration of speciation can lead to apparent highly complex (e.g. non-Fickian diffusion) and irregular behavior (e.g. time-dependent diffusion, strong composition-dependence), which is particularly unsuitable for extrapolations that are so essential in dealing with geological problems. This is not to deny that effects such as strain relaxation exist during diffusion of water—it merely appears that the primary influence on diffusion behavior is that due to speciation. The overall range of behavior that can be expected is well illustrated by the SiO_2 system because of the reasons outlined above.

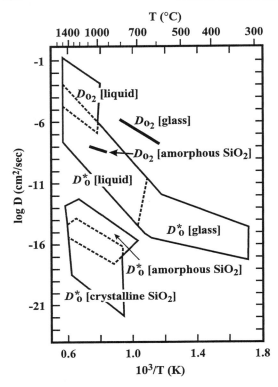

Figure 30. Broad overview of oxygen diffusion data in crystalline and amorphous silicon dioxide as well as silicate glasses and liquids over a wide temperature range, redrawn from Lamkin et al. (1992). Note how permeation rates, denoted by heavy lines and D_{O_2}, are faster than diffusion rates involving interaction with network oxygen, D^*_O, in any given type of medium.

Diffusion of water and other hydrogen-bearing species in natural molten systems

There have been a number of studies of diffusion of "water" in natural molten systems, beginning with the classic work of Shaw (1973), which have been reviewed in Watson (1994). The state of affairs at the time of that review was that once speciation of water in silicate melts was taken into account (Zhang et al. 1991b) along the lines already referred to above, the strong composition-dependence of diffusion rates disappeared. The transport of water could be reasonably described by a relatively compositionally insensitive diffusion rate of the molecular water species. Subsequently, considerable progress has been made in the field through the introduction of techniques to observe the speciation of water in melts *in situ* (Nowak and Behrens 1995) and a new surge of experimental work (Nowak and Behrens 1996, Behrens and Nowak 1997, Zhang and Behrens 2000). These studies have substantially broadened the database (400-1200°C, 0.1–810 MPa and 0.1-7.7 wt % total water content) as well as opened new questions and controversies, which have been discussed in Zhang (1999) and Zhang and Behrens (2000). It remains true that most of the work and discussion refers to polymerized melts such as rhyolites and granitic compositions. The main observations that emerge from these studies may be summarized as follows: (1) diffusion rate of molecular water, $D_{H_2O_m}$ is not a strong function of anhydrous composition for polymerized melts. (2) $D_{H_2O_m}$ does depend on the total water content, H_2O_t, above about 2% H_2O_t in contradiction to the assumptions of Zhang et al. (1991b). This dependence increases with pressure. Thus, at high pressures and water contents the Zhang et al. (1991b) model is inadequate to describe "water" diffusion in rhyolitic melts. (iii) Diffusion of "water" appears to involve motion of molecular H_2O only but not just by discrete jumps, interaction with network oxygen is also involved in addition to the "trapping" effect through the formation of OH complexes. (iv) The activation energy associated with $D_{H_2O_m}$ decreases as the total water content increases and it is also dependent on pressure. Finally, Zhang and Behrens (2000) provide the following expressions to calculate the diffusivity of molecular H_2O, $D_{H_2O_m}$, and total "water", $D_{H_2O_m}$, at all conditions covered experimentally to date in rhyolitic and other highly polymerized silicate systems:

$$D_{H_2O_m} = \exp[(14.08 - 13,128/T - 2.796P/T) + (-27.21 + 36,892/T + 57.23P/T) X_w^o]$$

(141)

where $D_{H_2O_m}$ is in $\mu m^2/sec$, T is in Kelvin, P is in MPa and X_w^o is the mole fraction of H_2O_t on a single oxygen basis.

$$D_{H_2O_m} = X_w^o \exp(m)\left\{1 + \exp\left[56 + m + X_w^o\left(-34.1 + \frac{44,620}{T} + \frac{57.3P}{T}\right) - \sqrt{X_w^o}\left(0.091 + \frac{4.77 \times 10^6}{T^2}\right)\right]\right\}$$

(142)

where $m = -20.79 - 5030/T - 1.4P/T$.

In addition to these diffusion studies, various recent spectroscopic studies (e.g. Eckert et al. 1987, Richet and Polian 1998, Zarubin 1999) reveal that weak hydrogen bonding with strong dynamics even at lower temperatures is a characteristic of silicate melts. Although such dynamics are unlikely to cause translational diffusive motion directly (e.g. Helmich and Rauch 1993), they can modify the nature of speciation reactions and thereby influence diffusion (e.g. see Behrens and Nowak 1997, Doremus 1995). Computer simulations suggest (Kanzaki 1997) that the major bottleneck responsible for the activation energy of ca. 100 kJ/mole for diffusion of H_2O in polymerized silicate glasses/melts is related to the passage of such a molecule through a 5- or 6-membered ring of (SiO) tetrahedral units. The activation energy for the diffusion of total "water" of course is an apparent value that incorporates the enthalpic changes involved in the speciation reactions in some form.

In contrast to all of these studies related to more polymerized melts, the status of water diffusion in basaltic systems remains unchanged since the review of Watson (1994). The only available study is that of Zhang and Stolper (1991) which indicates that the model of Zhang et al. (1991b) is applicable to basaltic melts with low water contents (0.2 wt%) as well, but the diffusivity of water in these melts is about two orders of magnitude faster than that in the more polymerized melts at 1300-1500°C. The Arrhenius equation describing this diffusion behavior is

$$D_{water} = 3.8_{-3.4}^{+35} \times 10^{-2} (cm^2/sec) \exp(-126,000/RT)$$

(143)

where the gas constant, R is 8.314 J mole^{-1} K^{-1}, T is in Kelvin and the activation energy (in this case 126,000) is in J/mole. Among other hydrogen bearing species, macroscopic diffusion of protons by themselves are considered unlikely (Behrens and Nowak 1997) and Stanton (1990) found some indication of hydrogen transport as hydronium ions, H_3O^+, in the presence of an electrical field which needs to be confirmed through detailed work. Chekhmir et al. (1985) found that the diffusion of "dry" H_2 in albite glass between 750-800°C may be described by $D_{H_2} = 3.7 \times 10^{-3}$ cm^2/sec exp (-113,000/RT). In molten albite doped with Mn, the diffusion rates inferred by Chekhmir et al. (1985) for H_2 are on the order of 10^{-4} cm^2/sec, which is extraordinarily high. However, in view of the possible effects of redox reactions on speciation, it is unclear exactly which species the diffusion rate determined in these experiments represents.

Diffusion of C and S related species

No new results are available to complement the discussion of Watson (1994) on diffusion of CO_2 in silicate melts, which is found to be unusually insensitive to composition (ranging from granitic to basaltic) and is well described by (Blank 1993)

$$D_{CO_2} = (6.2^{+4.3}_{-2.6}) \times 10^{-3} (cm^2/sec) \exp(-144,600 \pm 4100 / \underline{R}T) \tag{144}$$

Nishina et al. (1993) report H_2 and O_2 diffusion rates in $(Li,K)CO_3$ melts at 650°C of 8×10^{-6} cm^2/sec, while CO_2 was found to diffuse about half as fast.

New results are available for diffusion of sulfur in silicate melts (Watson 1994, Baker and Rutherford 1996, Winther et al. 1998). While speciation of sulfur can be substantially more complex than that of water because of the number of possible species, it appears that nature has worked to make things simpler in this case. First, the speciation does not depend strongly on sulfur content itself—rather, it is a function of oxygen fugacity. Second, the speciation reactions appear to be too slow in most cases to interfere with the diffusion process. And finally, it appears that under any given set of conditions (melt composition, oxygen fugacity, temperature etc.), one diffusing species seems to dominate. Watson (1994) reports the following results from his preliminary studies: in an ultramafic lunar melt composition (44% SiO_2) under reducing conditions (IW-0.7), D_{sulfur} was found to be 6.6×10^{-8} cm^2/sec at 1300°C. This rate probably corresponds to diffusion of sulfur as sulfide. In a dry reduced dacite composition, $D_{sulfur} = 1.4 \exp(-263,000/\underline{R}T)$; in dry reduced andesite, $D_{sulfur} = 1.0 \times 10^{-2} \exp(-191,000/\underline{R}T)$, both in cm^2/sec. Diffusion rates appear to be faster and activation energies lower in more depolymerized melts. Addition of water increases the diffusion rates, linearly in proportion to the added water in weight percent: D_{sulfur} (5.5 % water, andesite) $= 3.2 \times 10^{-4}$ (cm^2/sec) $\exp(-115,000/\underline{R}T)$. A large decrease in activation energy is also observed when water is present.

Baker and Rutherford (1996) studied diffusion of sulfur in rhyolite melts at 800-1100°C, water contents of 0-7.3 wt % and f_{O_2} ranging from QFM to air. They dissolved anhydrite or pyrrhotite in the melts and measured the resulting concentration profiles—so their results are for chemical diffusion and some amount of diffusive coupling with fluxes of Fe and Ca are inevitably embedded in their results. For the dry rhyolite melt, they obtain, $D_{sulfur} = 5 \times 10^{-2}$(cm^2/sec) $\exp(-221,000 \pm 80,000/\underline{R}T)$. Diffusion rates in water bearing melts are 1-2 orders of magnitude faster, depending on temperature, as found by Watson (1994) as well. They found that over the range of f_{O_2} studied, diffusion rates are relatively insensitive to f_{O_2}. They conclude that sulfide ion must be the dominant diffusing species and homogeneous reactions involving sulfur must be slow compared to diffusion rates. Sulfur diffusion rates measured by them are slower than diffusion rates of other common magmatic volatiles, so that sulfur may be fractionated from other volatiles during magmatic processes.

Finally, Winther et al. (1998) studied diffusion of sulfur in dry albitic melt at 1300-1500°C and 1 GPa. They found that in this highly polymerized melt, sulfate ions are unusually stable, even at low f_{O_2}. However, these are relatively immobile, and rapid speciation reaction takes place in this case, to form some S_2^- and S_3^- ions. The sulfate ions are still less mobile than sulfides but are nevertheless the dominant sulfur transport species in albite melt. The resulting transport rates, given by $D_{sulfur} = 1.47 \times 10^5$(cm^2/sec) $\exp(-458,100/\underline{R}T)$ are significantly slower than the other rates reported above, where diffusion is presumably controlled by motion of sulfides. As expected, these diffusivities are closest to the rates for dry rhyolite and obsidian obtained by Baker and Rutherford (1996) and Watson (1994), respectively.

Diffusion of oxygen: new results

There is a large body of diffusion and permeation data for oxygen in silica and silicate melts and glasses, which have been summarized and discussed before (Frischat 1975, Bansal and Doremus 1986, Mazurin et al. 1983, Dunn 1986, Chakraborty 1995 and Lamkin et al. 1992). Therefore, we will not review all of the available data here—instead, we present the data in Table 4 (see Appendix) and a summary of available data in Figures 30-33 adapted from Lamkin et al. (1992). In the following, we discuss general trends from these figures and some newer results in more detail, in particular when they provide information that deviates from these trends. In dealing with oxygen transport in melts, it is necessary to differentiate between permeation, tracer-diffusion and chemical diffusion. Further, within the category of tracer diffusion itself there are diffusion rates of various individual species as discussed above, as well as overall, macroscopically observed "effective" diffusion rates. And all of these can take place either in the glassy state (below macroscopic T_g) or in the molten state.

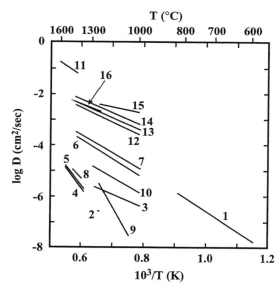

Figure 31. Summary of molecular oxygen self diffusion, including permeation, data in silicate glasses and liquids, redrawn from Lamkin et al. (1992). The references are: [1, 3, 4, 5]: Koros and King (1962); [2]: Goldman and Gupta (1983); [6, 7, 12, 13, 14, 15, 16]: Sasabe and Goto (1974); [8]: Frank et al. (1967), [9] Schrieber et al. (1986); [10]: Green and Kitano (1959); [11]: Semkow and Haskin (1985). Since most of these data are for permeation behavior they are not summarized in Table 4 (see Appendix) which focuses on diffusivities.

In silica glass and in thin amorphous silicate films, the process of permeation is important and is faster, particularly at lower temperatures (Fig. 30). Low activation energies, on the order of 100 kJ/mole, indicate that breaking of strong Si-O bonds is not a significant part of the process. The process tends to become less significant as the glass/melt composition becomes more complex involving various network modifying cations. At the other extreme, when isotopically tagged oxygen is used to monitor exchange with network oxygen and diffusion experiments are carried out such that there is no ready access to gaseous oxygen, high activation energies on the order of 450 kJ/mole are observed in SiO_2 glass (e.g. Mikkelsen 1984), which is close to the Si-O bond energy of 440 kJ/mole. The most common situation lies between these two extremes, where diffusion of oxygen takes place in the presence of gaseous oxygen, and is a coupled process of diffusion of molecular oxygen (often termed "interstitial oxygen") and its interaction with network oxygen, analogous to the transport of H_2O discussed above in detail (e.g. see Doremus 1996). Activation energies on the order of 140 kJ/mole in silica glass, according to the most recent study of Kalen et al. (1991), suggest that motion of the interstitial or molecular oxygen is mainly responsible for transport. Doremus (1996) argues that almost an exact analogy can be made between transport of water by the molecular species with internal "trapping" reactions to form OH and transport of oxygen by a molecular species accompanied by exchange with strongly bonded network oxygen. The linear dependence of transport rates on P_{O_2} has been used as an indication that the interstitial species is molecular, rather than atomic, oxygen. A P_{O_2}-dependence can result either because permeation dominates (Eqn. 133) or because P_{O_2} influences the equilibrium as well as kinetics of the homogeneous reaction between molecular, physically dissolved oxygen and oxygen which is chemically bonded to the melt structure. The nature of the dependence can (i.e. the power law exponent), however, provide insights into the cause.

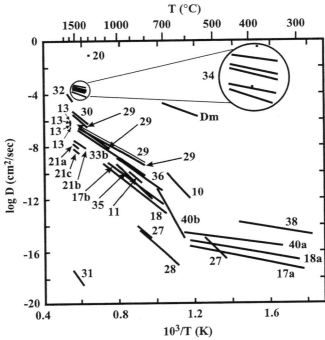

Figure 32. Oxygen self diffusion in silicate glasses and liquids involving interaction with network oxygen, redrawn from Lamkin et al. (1992). Note the similarity in diffusion rates within a given temperature range for a wide variety of compositions, and the overall Z-shaped feature characterizing the glass transition. The pronounced Z-shape occurs because of over-representation of fragile liquids, studied in the glass industry, in the data set. Sources of the data for each line or curve are in Table 4 (see Appendix), except for [5], which is from DiMarcello (1966).

Clearly, as the glass composition becomes more complex with the addition of various network formers, a variety of bonded oxygen species become available (see above) and accordingly activation energies show a wider spread as shown in Figure 32. In particular, note from Figure 29 and Figure 32 that more fragile glasses show a strong non-linear change in diffusion rates across the glass transition region while the stronger melts appear to show a more continuous Arrhenian behavior. The non-linearity clearly precludes the use of relationships such as the compensation law over large ranges of temperature and composition. Different lines of evidence support the postulated motion of "interstitial oxygen" with homogeneous reaction in silicate melts. For example, *in situ*, high temperature NMR spectroscopic studies have now clearly demonstrated that polymeric units involving oxygen bonded to other species do not exist on the time scales of oxygen diffusion in these melts (e.g. see Stebbins 1995 for a summary). Yet, Young et al. (1994) find that the measured profiles for oxygen isotope tracers are different, in the same CoO-SiO_2 melts, depending on whether the tracer was introduced via CoO or SiO_2 to the melt! Such observations are easily reconciled if one considers that the oxygens more weakly bonded to Co are more likely to participate in the homogeneous exchange reactions, resulting in longer tracer isotope profiles for the same experimental conditions. This, however, has an interesting implication for geological systems—even if actual polymeric species or structural units do not survive the melting event, their "memory" may be recorded in the distribution of tracer isotopes. Clearly, further detailed experimental studies are required to explore this possibility.

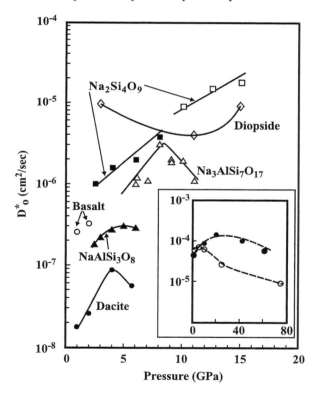

Figure 33. Pressure dependence of oxygen diffusion rates in silicate melts, modified after Tinker and Lesher (2001). Data for the sodium-silicate melts are taken from Poe et al. (1997), dacite from Tinker and Lesher (2001), basalt from Lesher et al. (1996) and diopside from Reid et al. (2001). Inset shows results from computer simulations (Angell et al. 1982 — $NaAlSi_2O_6$, solid; Nevins and Spera 1998 — $CaAl_2Si_2O_8$, open) at much higher temperatures of 4000 to 5000°C which show qualitatively similar behavior. Note how the maxima in pressure dependence shifts to lower pressures for more polymerized, Al-bearing melts whereas in diopside there is actually a minima in the data set; see text for details.

Composition-dependence of oxygen diffusion has been carefully characterized by Liang et al. 1996 who show that across a given compositional join, oxygen diffusivity increases as one goes to less viscous, more depolymerized (in the sense of higher NBO/T ratios) melts. Oxygen diffusion in basaltic to andesitic melts appear to have activation energies on the order of 170-260 kJ/mole (Table 4, Appendix), i.e. between the permeation and "pure" network diffusion limits. For more silicic liquids such as dacites, activation energies are higher (293-380 kJ/mole; Tinker and Lesher 2001), i.e. the values are more dominated by the "network interaction" component, which is no surprise. Note however that the Si-O bond energy itself changes (lowered, generally speaking) as more network modifiers are introduced into pure SiO_2, as evidenced, for example, by the accompanying drop in viscosity. Activation energies decrease with increasing pressure between 1-4 GPa for the dacitic composition studied by Tinker and Lesher (2001), but this may be valid over a limited pressure range, because pressure-dependence of oxygen diffusion rates are complex, as discussed below. One of the strongest lines of evidence for the involvement of network or bonded oxygen in the diffusion processes comes from the fact that the tracer diffusion rates of Si and O in most melts are very close to each other, with similar dependencies on pressure, temperature and composition (e.g. Shimizu and Kushiro 1991, Lesher et al. 1996). Usually, oxygen diffusion rates are slightly faster than Si diffusion, by about a factor of two. However, as one progresses to more polymerized melts, e.g. the dacites studied by Tinker and Lesher (2001), a tendency to decouple is apparent—oxygen and silicon diffusion rates show the same temperature- and pressure-dependencies, but oxygen diffusion rates are faster by about a factor of 5. This is similar to the observations of Shimizu and Kushiro (1984) on jadeites, another relatively polymerized, strong melt.

Finally, Sabioni et al. (1998) carried out a detailed study of oxygen diffusion in a Mg (14.6 wt % MgO) cordierite composition across the glass transition, as a function of different variables. They find a P_{O_2}-dependence of oxygen diffusion and the non-linear, non-Arrhenian behavior across the glass transition as expected. In the glassy region, they obtain an activation energy of 113 kJ/mole, indicative of a permeation dominated process. However, they obtain an unexpectedly high activation energy in the molten regime of 480 kJ/mole, which does not correlate with activation energy for viscous flow or any other process, in contrast to any other similar recent study. While this may be an interesting effect in this particular composition, it is also possible that they measured diffusion rates only along the steep limb of the Z-shaped curve that obtains on an Arrhenius plot for fragile melts (e.g. Figs. 30, 31, 33). Hence, even though they had an excellent fit over a narrow temperature interval (6 points between 826-880°C, with a correlation coefficient of 0.988), the result may be an artifact (in the sense that it is not suitable for extrapolation and does not provide meaningful physical insight into the diffusion process). This underscores the necessity to appreciate the distinction between strong vs. fragile behavior and the nature of the glass transition (Z-shaped transition rather than a kink) in the planning of diffusion experiments and the extrapolation of diffusion data.

The exact nature of the glass transition in polymerized silicate liquids is, however, far from well understood. Mode coupling theories have been successful in understanding many aspects of liquid state behavior and attempts have been made to extend these to strong liquids such as silicate melts (e.g. Rössler and Sokolov 1996). Some general features expected from this theory are exhibited by silicate melts, although a quantitative description was not possible. However, if the theory applies to silicate melts, then a second transition and change of transport mechanism is to be expected at a critical temperature, T_c, well above the glass transition temperature, T_g. It is interesting to speculate if the second bend in the flattening out part of the Z-shaped Arrhenius plots for relatively weak melts (Fig. 29, Fig. 32), far above T_g, correspond to T_c.

We have not discussed chemical diffusion rates in any detail here, because these depend on the nature of the concentration gradient and can in principle be derived from tracer diffusion coefficients, with some additional data. However, it is worth mentioning here that the actual extent of diffusion or mixing in a multicomponent system such as silicate melts often depends not only on a single diffusion coefficient for oxygen, but on the coupling of oxygen transport with other species. These so called cross coefficients are often non-negligible, as has been shown, for example, by Schmalzried et al. (1981) and more recently by Chakraborty et al. (1995) and Liang et al. (1996).

Pressure-dependence of diffusion rates

Pressure-dependence of diffusion rates of oxygen in silicate melts has been a topic of active investigation ever since the pioneering studies by researchers such as Kushiro, Shimizu and Angell showed that diffusivities: (a) may be related to melt viscosity and provide a means to measure viscosities under high pressure, (b) may increase (e.g. in jadeite) or decrease (e.g. in diopside) with pressure, and (c) provide interesting insight into the microscopic mechanisms of diffusion in silicate melts involving coordination changes of Si and O. These studies have received a more recent boost through the direct identification of high coordinated Si and O using spectroscopic techniques (see Stebbins 1995 for a review), advances in high pressure experimental technology (which allow experiments at pressures up to 15 GPa, 2527°C—see Poe and Rubie 2000) and improved computer simulations of diffusion processes (which use potentials derived from *ab initio* calculations in longer runs on larger systems; see Tsuneyuki and Matsui 1995, Poole et al. 1995). The diffusion behavior of oxygen under pressure has been reviewed by Chakraborty (1995) and Poe and Rubie (2000) and we will cover some salient aspects and newer results here. We do not provide any tables for the pressure-dependence of diffusion rates because these dependences show complex non-linearity.

The main feature that emerged from the early studies noted above is that diffusion rates of oxygen may increase in some melts (e.g. jadeite) and decrease in others (e.g. diopside) with increasing pressure. The difference in behavior may be related to degree of polymerization, strength and/or presence/absence of Al. It has been shown that formation of highly coordinated intermediates (not activated complexes in the sense of transition state theory, see Chakraborty 1995), e.g. 5-coordinated Si (or Al) or 3-coordinated O are responsible for the enhancement of diffusion. Spectroscopic studies, molecular dynamic simulations and actual diffusion measurements by Poe and coworkers (1997) have subsequently documented that Al plays a key role in the enhancement of diffusion with pressure, because Al forms highly coordinated species more easily than Si. The degree of polymerization is of course another key factor, because it determines the propensity of Si and O to form highly coordinated species. The combined effect of polymerization, strength, and Al content is that diffusion rates go through extrema as a function of pressure (Fig. 33) and a maximum is encountered at lowest pressures for polymerized, Al bearing melts such as albite (10 GPa) and dacite (5 GPa) (Tinker and Lesher 2001). The occurrence of the maximum can be explained (e.g see Poe and Rubie 2000 for a summary) as being due to "saturation" of a melt with respect to a given highly coordinated species. Thus, for example, the diffusive maxima in $Na_3AlSi_7O_{17}$ liquid at 8 GPa is explained as being due to "saturation" of the melt in Al^V (i.e. the concentration of Al^V drops above this pressure), as the concentration of Si^V continues to increase. This explains the continued increase in diffusivity in the Al-free, $Na_2Si_4O_9$ liquid over this pressure range. Computer simulations provide further insight into the process (Diefenbacher et al. 1998)—the increase in diffusivity in the Al-free melt is related to two different mechanisms—the first involving formation of Si^V through coordination with non-bridging oxygens in the melt (i.e. increased polymerization); the second involving formation of Si^V in conjunction with an already bridging oxygen, i.e. formation of O^{III}. A change of control from one to the other mechanism generates a weak maxima in diffusivity as a function of pressure at around 15 GPa (where experiments, carried out at lower temperatures, have not observed this maxima—maxima tend to shift to lower pressures with higher temperatures). Finally, a totally different form of non-linear behavior is observed in the depolymerized diopside melt, where diffusion rates initially decrease with increasing pressure. Here, one observes a *minima* as a function of pressure, at about 11 GPa at 2000°C (Reid et al. 2001). The initial decrease in diffusivity with pressure in this more depolymerized, "basalt-like" liquid may be the simple normal response to compression, i.e. migration pathways become constricted. The subsequent increase in diffusivity at very high pressures may occur as highly coordinated species begin to appear even in this formally depolymerized melt, but such inferences remain speculations at this stage and need to be confirmed through more transport, spectroscopic and computer simulation data.

While the correlation between the presence of highly coordinated species and diffusion of oxygen in silicate melts is well established now through simulations and experiment, the exact mechanism of this enhancement is also being clarified. For example, for the simpler SiO_2 system (which is special in that there are no non-bridging oxygen atoms) the computer simulations of Tsuneyuki and Matsui (1995), based on a potential obtained from first principles calculations, helps to address this issue. Increased coordination occurs as a transient, diffusionless transformation process because melt structures respond locally to compression. This contributes to diffusion because the oxygen atom ends up with different partners before and after the formation of the intermediate high coordinated complex. The interesting feature of this mechanism is that it involves a continuous change in Si-O coordination and does not involve any isolated atoms at any point, i.e. it bypasses the energetically expensive process of stripping an oxygen from a fourfold coordinated silicon. A similar mechanism had been suggested earlier by Kubicki and Lasaga (1993) for silicate melts, although their interatomic potentials and run times were not as carefully constrained.

Such an understanding of the diffusion process helps us to understand different aspects of transport in silicate melts. For example, while oxygen diffusion in silicate melts has been found to be well correlated with viscosity in general through the Eyring equation (Glasstone et al. 1941), there has been some ambiguity about the "jump distance"—jump distances slightly larger than the radius of the oxygen anion, and closer to the size of a SiO_4 unit, are commonly thought to provide a better agreement (e.g. Dunn 1986, Tinker and Lesher 2001) with independent measurements. The work of Tsuneyuki and Matsui (1995) indicates how this may be so and that indeed, in this case, one cannot really speak of a "jump" distance. This also explains why the Stokes-Einstein relation has not been found to be as suitable for relating diffusion and viscosity in these melts (Kirkaldy and Young 1987). Similarly, although not studied specifically for oxygen, this mode of migration involving larger units also explains why the isotopic fractionation during diffusion may be much smaller than that expected from a $m_e^{1/2}$-dependence, where m_e is the mass of an isotope of element e. And finally, it explains how the diffusion time scale might coincide with the time scale of making and breaking Si-O bonds as observed through NMR spectroscopy. This might reconcile the paradox of using larger units in the Eyring equation when no "polymers" are seen in NMR spectroscopy, if such a mechanism is also found for silicate (as opposed to pure silica) melts.

SUMMARY AND RECOMMENDATIONS

(1) Central to the interpretation of the degree of isotopic equilibration and factors influencing exchange in natural systems is a quantitative understanding of the rates and mechanisms of isotopic exchange among gases, fluids and solids.

(2) General concepts were presented that outline the kinetic rate laws based on both concentrations and isotopic activities. However, these general rate laws must be modified and properly tailored to the geochemical conditions encountered experimentally or in natural settings. There is a need for the identification of the reactive intermediates, which play such a crucial role in the formulation of rate laws and hence the extrapolation of laboratory data to geological conditions, through the use of *in situ* spectroscopic and computational tools.

(3) Isotopic rates of exchange among free gas species are generally quite sluggish except at high temperatures, and even at high P and T, a catalytic surface is commonly required before exchange can proceed. However, not all isotopic exchange reactions are enhanced by a catalyst, nor are they required. Oxidation-reduction reactions among gas species can lead to appreciable rates of exchange. At present there is a poor understanding of the role of geologic catalysts in controlling isotopic exchange among the common crustal gases, e.g. CO_2, CH_4, H_2O, H_2.

(4) Most isotopic reactions between gas and aqueous species or among different aqueous species are geologically fast, e.g. $^{18}O/^{16}O$ between CO_2 and H_2O, D/H between H_2 and H_2O, certain aqua- and oxo-anions and H_2O. However, a few select systems are quite sluggish and exhibit a marked dependence on temperature, pressure, pH and ionic strength. For example, slow rates of oxygen isotope exchange have been measured between H_2O and either aqueous phosphate or sulfate. However, there is still a need for more detailed studies of these types of systems as a function of T, P, pH, ionic strength, etc., that require a full understanding of the aqueous speciation.

(5) At low to modest temperatures (up to 400°C or so), geologically reasonable rates of isotopic exchange are limited by reactions involving Ostwald ripening, recrystallization or transformation to a new phase. Rough estimates of rates controlled by these mechanisms can be made using a pseudo-first order model that accounts for the surface area of the solid. However, physically more meaningful models can be formulated provided detailed data on dissolution/precipitation rates and interfacial free energies are available, which is generally not the case for most phases, particularly at elevated temperatures and pressures. Clearly an area worth further investigation is in the quantitative coupling of chemical reaction rates with isotopic exchange. Focus areas needing attention include physical versus reactive surface areas, role of surface dislocations, steps and other defects. Both observational (e.g. AFM, STM, etc.) and computational tools can help improve our understanding of the role of surfaces in the reaction kinetics.

(6) By far the best-studied rate behavior of the light elements involves diffusion in crystals, glasses and melts. For crystalline materials, diffusion-controlled exchange is controlled by temperature, pressure, fluid (gas)/solid ratio, impurities, presence or absence of water, oxidation state, and crystal anisotropy. Considerable variations in diffusivities can arise from experiments conducted at the same P, T and X because of small differences in the concentration of impurities and whether the solid was pre-annealed prior to experimentation. Diffusion-controlled isotopic exchange in crystalline solids is typically several orders of magnitude slower than exchange accompanying chemical reaction. The majority of studies have focused on oxygen diffusion, whereas very little is known about the behavior of hydrogen. An

exploration of hydrogen diffusion in single crystal hydrous silicates is still at a very early stage. With regard to oxygen, one fruitful area of investigation involves examination of both ^{18}O and ^{17}O profiles, a comparison of which can lead to estimation of correlation factors that provide valuable insight into the nature of point defects.

(7) The Eyring relation describes the link between diffusivity and viscosity in amorphous systems. For which compositions is the Eyring relation valid or invalid, and to what extent (e.g. factor of two, four, an order of magnitude)? Why does the Eyring equation work at all, i.e. why a simple jump of oxygen atoms should describe such complex kinetics? Investigation into these questions may provide the quantitative link between time-scale data obtained from NMR spectroscopy and macroscopic rate/diffusion measurements.

(8) The measurement of diffusion rates as functions of various activities, in particular oxygen fugacity, is extremely important to clarifying the diffusion mechanisms and aid in extrapolation. The work of Ryerson et al. (1989) is the only geochemical example of such type of work. Questions of concern include: under what conditions does the f_{O_2}-dependence occur, and what does it tell us about the main mechanism of O_2 transport (e.g. permeation versus diffusion, network versus molecular oxygen diffusion etc.).

(9) Almost no data exist for oxygen diffusion in wet molten systems. These data are needed for comparison with the dry data.

(10) Data on S, H and D diffusion are also quite limited. Such data need to be obtained as a function of various parameters, particularly pressure because this is a key in degassing in volcanic systems.

ACKNOWLEDGMENTS

Research support to Chakraborty has been provided by the German Science Foundation (DFG). Support for Cole has been provided by the Geosciences Research program of the Division of Chemical Sciences, Geosciences and Biosciences, Office of Basic Energy Research, U.S. Department of Energy under contract DE-AC05-00OR22725 to Oak Ridge National Laboratory, managed and operated by University of Tennessee-Battelle, LLC. The authors gratefully acknowledge the helpful reviews provided by John Valley, Ilya Bindeman, John Farver, Juske Horita, Bill Casey, Chip Lesher and Dave Tinker. S. De (University of Alberta) provided valuable input on the behavior of hydrogen transport in hydrous silicates. The typing talents of Regina Violet and Marlene Leamon and the drafting skill of Rosemary Adams are greatly appreciated.

REFERENCES

Albarede F (1995) Introduction to Geochemical Modeling. Cambridge University Press, Cambridge, UK, 543 p

Alberty RA and Miller WG (1957) Integrated rate equations for isotopic exchange in simple reversible reactions. J Chem Phys 26:1231-1237

Allnatt AR, Lidiard AB (1993) Atomic Transport in Solids. Cambridge University Press, p 572

Amami B, Labidi M, Dolin C, Sabioni A, Millot F, Dhalenne G, Jolles E, Kusinski J, Orewczyk J, Jasienska S, Monty CJA (1994) Oxygen self-diffusion in pure and Ca doped FeO. Metallurgy Foundary Eng 20:247-257

Amami B, Addou M, Millot F, Sabioni A, Monty C (1999) Self-diffusion in α-Fe_2O_3 natural single crystals. Ionics 5:358-370

Anderson TF (1969) Self-diffusion of carbon and oxygen in calcite by isotope exchange with carbon dioxide. J Geophys Res 74:3918-3932

Anderson TF (1972) Self-diffusion of carbon and oxygen in dolomite. J Geophys Res 77:857-861

Anderson TF, Chai BTH (1974) Oxygen isotopic exchange between calcite and water under hydrothermal conditions. *In* Hofmann AW, Giletti BJ, Yoder HS, Jr., Yund RA (eds) Geochemical Transport and Kinetics. Carnegie Inst Wash Publ 634:219-227

Anderson TF, Kaspar RB (1975) Oxygen self-diffusion in albite under hydrothermal conditions. EOS, Trans Am Geophys Union 56:459

Ando K, Oishi, Y (1974) Self-diffusion coefficients of oxygen ion in single crystals of MgO·ηAl$_2$O$_3$ spinels. J Chem Phys 61:625-629

Ando K, Oishi Y, Hidaka Y (1976) Self-diffusion of oxygen in single crystal thorium oxide. J Phys Chem 65:2751-2755

Ando K, Kurokawa H, Oishi Y, Takei H (1981) Self-diffusion coefficient of oxygen in single- crystal forsterite. Commun Am Ceram Soc, p C-30

Ando K, Kurokawa Y, Oishi Y (1983) Oxygen self-diffusion in Fe-doped MgO single crystals. J Chem Phys 78:6890-6892

Angell C, Scamehorn C, Phifer C, Khadiyala R, Cheeseman P (1982) Pressure enhancement of ion mobilities in liquid silicates from computer simulation studies to 800 kilobars. Science 218:885-887

Angell C (1985) Strong and fragile liquids. *In* Ngai K, Wright G (eds) Relaxation in complex systems. National Technical Information Service, Springfield, Virginia, p 3-11

Anovitz L, Elam JM, Riciputi LR, Cole DR (1999) The failure of obsidian hydration dating: Sources, implications, and new directions. J Archaeol Sci 26:735-752

Applin KR, Lasaga AC (1984) The determination of SO$_4^{2-}$, NaSO$_4^-$, and MgSO$_4^0$ tracer diffusion coefficients and their application to diagenetic flux calculations. Geochim Cosmochim Acta 48: 2151-2162

Arita Y, Wada H (1990) Stable isotope evidence for migration of metamorphic fluids along grain boundaries of marbles. Geochem J 24:173-186

Arita M, Hosoya M, Kobayashi M, Someno M (1979) Depth profile measurement by secondary ion mass spectrometry for determining the tracer diffusivity of oxygen in rutile. J Am Ceram Soc 62:443-446

Bagshaw AN, Hyde, BG (1976) Oxygen tracer diffusion in the magneli phases Ti$_n$O$_{2n-1}$. J Phys Chem Solids 37:835-838

Baker LL, Rutherford MJ (1996) Sulfur diffusion in rhyolite melts. Contrib Mineral Petrol 123:335-344

Bansal N, Doremus R (1986) Handbook of Glass Properties. Academic Press, Orlando, Florida

Barrer RM, Fender BEF (1961) The diffusion and sorption of water in zeolites—II. Intrinsic and self-diffusion. J Phys Chem Solids 21:12-24

Beck JW, Berndt ME, Seyfried WE (1992) Application of isotopic doping techniques to evaluate the reaction kinetics and fluid/mineral distribution coefficients: An experimental study of calcite at elevated temperatures and pressures. Chem Geol 97:125-144

Beerkens RGC, Waal HD (1990) Mechanisms of oxygen diffusion in glassmelts containing variable-valence ions. J Am Ceram Soc 73:1857-1861

Behrens H, Nowak M (1997) The mechanisms of water diffusion in polymerized silicate melts. Contrib Mineral Petrol 126:377-385

Bell, JS, Palmer DA, Barnes HL, Drummond SE (1994) Thermal decomposition of acetate: III Catalysis by mineral surfaces. Geochim Cosmochim Acta 58:4155-4177

Benson SW (1960) The Foundations of Chemical Kinetics. McGraw-Hill, New York

Berndt ME, Allen DE, Seyfried WE (1996) Reduction of CO$_2$ during serpentinization of olivine at 300°C and 500 bars. Geology 24:351-354

Bernstein RB (1952) Enrichment of isotopes by difference in rates for irreversible isotopic reactions. J Phys Chem 56:893-896

Bindeman IN, Valley JW (2000) Formation of low-δ^{18}O rhyolites after caldera collapse at Yellowstone, Wyoming, USA. Geology 28:719-722

Blacic JD (1981) Water diffusion in quartz at high pressure: Tectonic implications. Geophys Res Lett 8:721-723

Blank J (1993) An experimental study of the behavior of carbon dioxide in rhyolitic melt. PhD Dissertation, California Institute of Technology, Pasadena

Blum AE, Yund RA, Lasaga AC (1990) The effect of dislocation density on the dissolution of albite. Geochim Cosmochirm Acta 54:283-297

Bocquet JL, Brebec G, Limoge Y (1996) Diffusion in metals and alloys. *In* Cahn RW, Haasen P (eds) Physical Metallurgy, p 535-668

Bottinga Y (1969) Calculated fractionation factors for carbon and hydrogen exchange in the system calcite-carbon dioxide-graphite-methane-hydrogen-water vapor. Geochim Cosmochim Acta 33:49-64

Boudart M (1976) Consistency between kinetics and thermodynamics. J Phys Chem 80:2869-2870

Brady JB (1995) Diffusion data for silicate minerals, glasses, and liquids. *In* Ahrens TJ (ed) Mineral Phys and Crystallography. A Handbook of Physical Constants. Am Geophys Union, Ref Shelf 2:269-290

Brady PV, Walther JV (1989) Controls on silicate dissolution rates in neutral and basic pH solution at 25°C. Geochim Cosmochim Acta 53:2823-280

Brandner JD, Urey MC (1945) Kinetics of the isotopic exchange reaction between carbon monoxide and carbon dioxide. J Chem Phys 13:351-362

Brantley SL, Crane SR, Crerar DA, Hellmann R, Stallard R (1986) Dissolution at dislocation etch pits in quartz. Geochim Cosmochim Acta 50:2349-2361

Brossman U, Wurschum R, Sodervall U, Schaefer H (1999) Oxygen diffusion in ultrafine grained monoclinic ZrO_2. J Appl Phys 85:7646-7654

Buening DK, Buseck PR (1973) Fe-Mg lattice diffusion in olivine. J Geophys Res 78:6852-6862

Bunton CA, Llewellyn DR, Vernon CA, Welch VA (1961) The reactions of organic phosphates. Part IV. Oxygen exchange between water and orthophosphoric acid. J Chem Soc Part II:1636-1640

Burch TE, Cole DR, Bolton E, Rye D, Lasaga AC (in review) The effect of fluid composition on the rates and mechanisms of isotope exchange in the system calcite-CO_2-H_2O to 700°C, 1 kbar. Am J Sci

Canil D, Muehlenbachs K (1990) Oxygen diffusion in a Fe-rich basalt melt. Geochim Cosmochim Acta 54:2947-2951

Carlslaw HS, Jaeger JC (1959) Conduction of Heat in Solids. Clarendon Press, Oxford, UK

Carmen PC, Haul RAW (1954) Measurement of diffusion coefficients. Proc Royal Soc London 222A: 109-118

Casey WH (2001) Clusters in solution and at interfaces: Kinetics of formation and dissociation, and isotope exchange. In Banfield JF, Navrotsky A (eds) Nanoparticles in the Environment and Technology. Rev Mineral Geochem (in press)

Casey WH, Westrich HR (1992) Control of dissolution rates of orthosilicate minerals by divalent metal-oxygen bonds. Nature 355:157-159

Castle JE, Surman PL (1967) The self-diffusion of oxygen in magnetite. Techniques for sampling and isotopic analysis of micro-quantities of water. J Phys Chem 71:4255-4259

Castle JE, Surman PL (1969) The self-diffusion of oxygen in magnetite. The effect of anion vacancy concentration and cation distribution. J Phys Chem 73:632-634

Cawley JD, Boyce RSA (1988) A solution to the diffusion equation for double oxidation in dry oxygen including lazy exchange between network and interstitial oxygen. Phil Mag 58A:586-601

Chacko T, Mayeda TK, Clayton RN, Goldsmith JR (1991) Oxygen and carbon isotope fractionations between CO_2 and calcite. Geochim Cosmochim Acta 55:2867-2882

Chakraborty S (1995) Diffusion in silicate melts. In Stebbins J, McMillan P, Dingwell D (eds) Structure, Dynamics and Properties of Silicate Melts. Rev Mineral 32:411-504

Chakraborty S (1997) Rates and mechanisms of Fe-Mg interdiffusion in olivine at 980°-1300°C. J Geophys Res 102:12317-12331

Chakraborty S, Dingwell D, Rubie D (1995) Multicomponent diffusion in ternary silicate melts in the system K_2O-Al_2O_3-SiO_2: I. Experimental measurements. Geochim Cosmochim Acta 59:255-264

Chakraborty S, Farver JR, Yund RA, Rubie DC (1994) Mg tracer diffusion in synthetic forsterite and San Carlos olivine as a function of P, T and fO_2. Phys Chem Minerals 21:489-500

Chai BHT (1974) Mass transfer of calcite during hydrothermal recrystallization. In Hofmann AW, Giletti BJ, Yoder HS, Jr., Yund RA (eds) Geochemical Transport and Kinetics. Carnegie Inst Wash Pub 634:205-227

Chai BHT (1975) The kinetics and mass transfer of calcite during hydrothermal recrystallization process. PhD Dissertation, Yale University, New Haven, Connecticut, p 203

Chamberlain DP, Conrad ME (1991) Oxygen-isotope zoning in garnet. Science 254:403-406

Chamberlain DP, Conrad ME (1993) Oxygen-isotope zoning in garnet: A record of volatile transport. Geochim Cosmochim Acta 57:2613-2629

Chekhmir AS, Epel'baum MB (1991) Diffusion in magmatic melts: New study. Adv Phys Geochem 9: 99-119

Chen WK, Jackson RA (1969) Oxygen self-diffusion in undoped and doped cobaltous oxide. J Phys Chem Solids 30:1309-1314

Chiba H, Sakai H (1985) Oxygen isotope exchange rate between dissolved sulfate and water at hydrothermal temperatures. Geochim Cosmochim Acta 49:993-1000

Chiba H, Kusakabe M, Hirano S-I, Matsuo S, Somiya S (1981) Oxygen isotope fractionation factors between anhydrite and water from 100 to 550°C. Earth Planet Sci Lett 53:55-62

Choudhury A, Palmer DW, Amsel G, Curien H, Baruch P (1965) Study of oxygen diffusion in quartz by using the nuclear reaction O^{18} (p, α) N^{15}. Solid State Comm 3:119-122

Clayton RN, Jackson ML, Sridhar K (1978) Resistance of quartz silt to isotopic exchange under burial and intense weathering conditions. Geochim Cosmochim Acta 42:1517-1522

Clayton RN, O'Neil JR, Mayeda TK (1972) Oxygen isotope exchange between quartz and water. J Geophys Res 77:3057-3067

Cole DR (1992) Influence of solution composition and pressure on the rates of oxygen isotope exchange in the system calcite-H_2O-NaCl at elevated temperatures. Chem Geol 102:199-216

Cole DR (1994) Evidence for oxygen isotope disequilibrium in selected geothermal and hydrothermal ore deposit systems. Chem Geol 111:283-296

Cole DR (2000) Isotopic exchange in mineral-fluid systems. IV. The crystal chemical controls on oxygen isotope exchange rates in carbonate-H_2O and layer silicate-H_2O systems. Geochim Cosmochim Acta 64:921-931

Cole DR, Fortier SM (1996) Oxygen isotope exchange in the systems carbonate-water and sheet silicate-water: A link between exchange rates and mineral chemistry. Geol Soc Am Ann Mtg, Prog with Abstr 28:A-160

Cole DR, Ohmoto H (1986) Kinetics of isotopic exchange at elevated temperatures. *In* Valley JW, Taylor HP, O'Neil JR Jr (eds) Stable Isotopes in High Temperature Geological Processes. Rev Mineral 16: 41-90

Cole DR, Mottl MJ, Ohmoto H (1987) Isotopic exchange in mineral-fluid systems. II. Oxygen and hydrogen isotopic investigation of the experimental basalt-seawater system. Geochim Cosmochim Acta 51:1523-1538

Cole DR, Ohmoto H, Jacobs GK (1992) Isotopic exchange in mineral-fluid systems. III. Rates and mechanisms of oxygen isotope exchange in the system granite-$H_2O\pm NaC\pm KCl$ at hydrothermal conditions. Geochim Cosmochim Acta 56:445-466

Cole DR, Ohmoto H, Lasaga AC (1983) Isotopic exchange in mineral-fluid systems. I. Theoretical evaluation of oxygen isotopic exchange accompanying surface reactions and diffusion. Geochim Cosmochim Acta 47:1681-1693

Cole DR, Simonson MJ, Corti H, Drummond SE (in prep) Solubility of calcite and dolomite in H_2O-CO_2 fluids to 300°C and 130 MPa (to be submitted to Geochim Cosmochim Acta)

Condit RH, Weed HC, Piwinskii AJ (1985) A technique for observing diffusion along grain boundary regions in synthetic forsterite. *In* Schock RN (ed) Point Defects in Minerals, Geophys Monogr 31, Mineral Phys 1:97-105

Connolly C, Muehlenbachs K (1988) Contrasting oxygen diffusion in nepheline, diopside and other silicates and their relevance to isotopic systematics in meteorites. Geochim Cosmochim Acta 52: 1585-1591

Coughlan RAN (1990) Studies in diffusional transport: Grain boundary transport of oxygen in feldspars, strontium and REE's in garnet, and thermal histories of granitic intrusions in south-central Maine using oxygen isotopes. PhD dissertation, Brown University, Providence, Rhode Island

Cox JN, Hwang K, Kwok KW, Downing RG, Lamaze GP (1993) The spatial resolution of water diffusion and trapping in silicate-glass thin films by micro-FTIR and neutron depth profiling. Proc Electrochem Soc 93:336-342

Craig H (1953) The geochemistry of stable carbon isotopes. Geochim Cosmochim Acta 3:53-92

Crank J (1975) The Mathematics of Diffusion. 2nd edition, Oxford University Press, Oxford, UK

Criss RE (1991) Temperature-dependence of isotopic fractionation factors. Geochim Soc Spec Pub 3:11-16

Criss RE (1999) Principles of Stable Isotope Distribution. Oxford University Press, Oxford, UK

Criss RE, Gregory RT, Taylor HP (1987) Kinetic theory of oxygen isotope exchange between minerals and water. Geochim Cosmochim Acta 51:1099-1108

Crowe DE, Riciputi LR, Bezenek S, Ignatiev A (2001) Oxygen isotope and trace element zoning in hydrothermal garnets: Windows into large-scale fluid flow. Geology 29:479-482

Davis KM (1994) The diffusion of water into silica glass at low temperature. PhD dissertation, Rensselaer Polytechnic Institute

De SK, Cole DR, Riciputi LR, Chacko T, Horita J (2000) Experimental determination of hydrogen diffusion rates in hydrous minerals using the ion microprobe. Tenth VM Goldschmidt Conf Abstr 5:340

de Berg KCD, Lauder I (1977) The diffusion of oxygen from water vapor-oxygen mixtures into lead silicate glass above the transformation temperature. Phys Chem Glasses 19:83-88

de Berg KCD, Lauder I (1978) Oxygen tracer diffusion in lead silicate glass above the transformation temperature. Phys Chem Glasses 19:78-82

de Berg KCD, Lauder I (1980) Oxygen tracer diffusion in a potassium silicate glass above the transformation temperature. Phys Chem Glasses 21:106-109

Deines P, Haggerty SE (2000) Small-scale oxygen isotope variations and petrochemistry of ultradeep (300 km) and transition zone xenoliths. Geochim Cosmochim Acta 64:117-131

Dennis PF (1984) Oxygen self-diffusion in quartz under hydrothermal conditions. J Geophys Res 89: 4047-4057

Dennis PF, Freer R (1993) Oxygen self-diffusion in rutile under hydrothermal conditions. J Mater Sci 28:4804-4810

Derdau D, Freer R, Wright K (1998) Oxygen diffusion in anhydrous sanidine feldspar. Contrib Mineral
    Petrol 133:199-204
Derry DJ, Lees DG, Calbert JM (1981) A study of oxygen self-diffusion in the c-direction of rutile using a
    nuclear technique. J Phys Chem Solids 42:57-64
Dickson JAD (1991) Disequilibrium carbon and oxygen isotope variations in natural calcite. Nature
    353:842-844
Diefenbacher J, McMillan PF, Wolf GH (1998) Molecular dynamics simulations of $Na_2Si_4O_9$ liquid at high
    pressure. J Phys Chem B 102:3003-3008
DiMarcello FV (1966) Oxygen diffusion in a sodium silicate glass. Am Ceram Soc Bull 45:420
Dodson MH (1973) Closure temperature in cooling geochronological and petrological systems. Contrib
    Mineral Petrol 40:259-274
Dohmen R, Chakraborty S, Palme H, Rammensee W (in press) The role of element solubility on the
    kinetics of element partitioning: in situ observations and a thermodynamic model. J Geophys Res
Dole M, Lane GA, Rudd DP, Zaukelies DA (1954) Isotopic composition of atmospheric oxygen and
    nitrogen. Geochim Cosmochim Acta 6:65
Doornkamp C, Clement M, Ponec V (1999) The isotopic exchange reaction of oxygen on metal oxides.
    J Catalysis 182:390-399
Doremus RH (1983) Diffusion-controlled reaction of water with glass. J Noncrystal Solids 55:143-147
Doremus RH (1995) Diffusion of water in silica glass. J Mater Res 10:2379-2389
Doremus RH (1996) Diffusion of oxygen in silica glass. J Electrochem Soc 143:1992-1995
Doremus RH (1998) Diffusion of water and oxygen in quartz: Reaction-diffusion model. Earth Planet Sci
    Lett 163:43-51
Doremus, RH (1999) Diffusion of water in crystalline and glassy oxides: Diffusion-reaction model. J Mater
    Res 14:3754-3758
Dowty E (1980) Crystal-chemical factors affecting the mobility of ions in minerals. Am Mineral 65:
    174-182
Dubinina EO, Lakshtanov LZ (1997) A kinetic model of isotopic exchange in dissolution-precipitation
    processes. Geochim Cosmochim Acta 61:2265-2273
Dunn T (1982) Oxygen diffusion in three silicate melts along the join diopside-anorthite. Geochim
    Cosmochim Acta 46:2293-2299
Dunn T (1986) Diffusion in silicate melts: An introduction and literature review. In Scarfe C, (ed) Silicate
    Melts Short Course Handbook 12:57-92. Mineral Assoc Canada, Ottawa, Ontario
Dyer A, Molyneux A (1968) The mobility of water in zeolites—I. Self-diffusion of water in analcite.
    J Inorg Nucl Chem 30:829-837
Dyer A, Faghihian H (1998a) Diffusion in heteroionic zeolites: Part 1. Diffusion of water in heteroionic
    natrolites. Microporous Mesoporous Materials 21:27-38
Dyer A, Faghihian H (1998b) Diffusion in heteroionic zeolites: Part 2. Diffusion of water in heteroionic
    stilbites. Microporous Mesoporous Materials 21:39-44
Eckert H, Yesinowski J, Stolper E, StantonT, Holloway J (1987) The state of water in rhyolitic glasses: A
    deuterium NMR study. J Non-Crystalline Solids 93:93-114
Edwards KJ, Valley JW (1998) Oxygen isotope diffusion and zoning in diopside: Empirical constraints on
    water fugacity during cooling. Geochim Cosmochim Acta 62:2265-2277
Eiler JM, Baumgartner LP, Valley JW (1992) Intercrystalline stable isotope diffusion: A fast grain
    boundary model. Contrib Mineral Petrol 112:543-557
Eiler JM, Valley JW, Baumgartner LP (1993) A new look at stable isotope geothermometry. Geochim
    Cosmochim Acta 57:2571-2583
Eiler JM, Valley JW, Graham CM, Baumgartner LP (1995) Ion microrpobe evidence for the mechanisms
    of stable isotope retrogression in high-grade metamorphic rocks. Contrib Mineral Petrol 118:365-378
Elsenheimer D, Valley JW (1993) Submillimeter scale zonation of $\delta^{18}O$ in quartz and feldspar, Isle of
    Skye, Scotland. Geochim Cosmochim Acta 57:3669-3676
Ellis AJ, McFadden IM (1972) Partial molal volumes of ions in hydrothermal solutions. Geochim
    Cosmochim Acta 36:413-426
Elphick SC, Graham CM (1988) The effect of hydrogen on oxygen diffusion in quartz: Evidence for fast
    proton transients? Nature 335:243-245
Elphick SC, Graham CM (1990) Hydrothermal oxygen diffusion in diopside at 1 kb, 900-1200°C, a
    comparison with oxygen diffusion in forsterite, and constraints on oxygen isotope disequilibrium in
    peridotite nodules. Terr Abstracts 7:72
Elphick SC, Dennis PF, Graham CM (1986) An experimental study of the diffusion of oxygen in quartz
    and albite using an overgrowth technique. Contrib Mineral Petrol 92:322-330

Elphick SC, Graham CM, Dennis PF (1988) An ion microprobe study of anhydrous oxygen diffusion in anorthite: A comparison with hydrothermal data and some geological implications. Contrib Mineral Petrol 100:490-495

Faiia AM, Feng X (2000) Kinetics and mechanism of oxygen isotope exchange between analcime and water vapor and assessment of isotopic preservation of analcime in geological formations. Geochim Cosmochim Acta 64:3181-3188

Farquhar J, Chacko T, Frost BR (1993) Strategies for high-temperature oxygen isotope thermometry: A worked example from the Laramie Anorthosite Complex, Wyoming USA Earth Planet Sci Lett 117:479-492

Farver JR (1989) Oxygen self-diffusion in diopside with applications to cooling rate determinations. Earth Planet Sci Lett 92:322-330

Farver JR (1994) Oxygen self-diffusion in calcite: Dependence on temperature and water fugacity. Earth Planet Sci Lett 121:575-587

Farver JR, Giletti BJ (1985) Oxygen diffusion in amphiboles. Geochim Cosmochim Acta 49:1403-1411

Farver JR, Gilleti BJ (1989) Oxygen and strontium kinetics in apatite and potential applications to thermal history determinations. Geochim Cosmochim Acta 53:1621-1631

Farver JR, Yund RA (1990) The effect of hydrogen, oxygen, and water fugacity on oxygen diffusion in alkali feldspar. Geochim Cosmochim Acta 54:2953-2964

Farver JR, Yund RA (1991a) Measurement of oxygen grain boundary diffusion in natural, fine-grained, quartz aggregates. Geochim Cosmochim Acta 55:1597-1607

Farver JR, Yund RA (1991b) Oxygen diffusion in quartz: Dependence on temperature and water fugacity. Chem Geol 90:55-70

Farver JR, Yund RA (1992) Oxygen diffusion on a fine-grained quartz-aggregate with wetted and nonwetted microstructures. J Geophys Res 97:14017-14029

Farver JR, Yund RA (1995a) Interphase boundary diffusion of oxygen and potassium in K-feldspar/quartz aggregates. Geochim Cosmochim Acta 59:3697-3705

Farver JR, Yund RA (1995b) Grain boundary diffusion of oxygen, potassium and calcium in natural and hot-pressed feldspar aggregates. Contrib Mineral Petrol 118:340-355

Farver JR, Yund RA (1998) Oxygen grain boundary diffusion in natural and hot-pressed calcite aggregates. Earth Planet Sci Lett 161:189-200

Farver JR, Yund RA (1999) Oxygen bulk diffusion measurements and TEM characterization of a natural ultramylonite: Implications for fluid transport in mica-bearing rocks. J Metamorphic Geol 17:669-683

Feng X, Savin SM (1993) Oxygen isotope studies of zeolites—stilbite, analcime, heulandite, and clinoptilolite: II. Kinetics and mechanisms of isotopic exchange between zeolites and water vapor. Geochim Cosmochim Acta 57:4219-4238

Flynn CP (1972) Point Defects and Diffusion. Clarendon Press, Oxford, UK

Fortier SM (1994) An on-line experimental/analytical method for measuring the kinetics of oxygen isotope exchange between CO_2 and saline/hypersaline salt solutions at low (25-50°C) temperatures. Chem Geol (Isotope Geosci Section) 116:155-162

Fortier SM, Giletti BJ (1989) An empirical model for predicting diffusion coefficients in silicate minerals. Science 245:1481-1484

Fortier SM, Giletti BJ (1991) Volume self-diffusion of oxygen in biotite, muscovite, and phlogopite micas. Geochim Cosmochim Acta 55:1319-1330

Freer R (1980) Bibliography: Self-diffusion and impurity diffusion in oxides. J Mater Sci 15:803-824

Freer R (1981) Diffusion in silicate minerals and glasses: A data digest and guide to the literature. Contrib Mineral Petrol 76:440-454

Freer R (1993) Diffusion in silicate minerals. Defect & Diffusion Forum 101-102:1-17

Freer R, Dennis PF (1982) Oxygen diffusion studies. I. A preliminary ion microprobe investigation of oxygen diffusion in some rock-forming minerals. Mineral Mag 45:179-192

Freer R, Wright K, Kroll H, Gottlicher J (1997) Oxygen diffusion in sanidine feldspar and a critical appraisal of oxygen isotope-mass-effect measurements in non-cubic materials. Phil Mag A 75:485-503

Friedlander G, Kennedy JW, Macias ES, Miller JM (1981) Nuclear and Radiochemistry. John Wiley and Sons, New York

Frischat G (1975) Ionic Diffusion in Oxide Glasses. Trans Tech Publications, Aedermannsdort, Switzerland

Gamsjäger H, Murmann RK (1983) Oxygen-18 exchange studies of aqua- and oxo-ions. *In* Sykes AG (ed) Advances Inorganic and Bioinorganic Mechanisms 2:317-380. Academic Press, New York

Gautason B, Muehlenbachs K (1993) Oxygen diffusion in perovskite: Implications for electrical conductivity in the lower mantle. Science 260:518-520

Gaye H, Riboud P (1976) Diffusion coefficient of oxygen in $CaO-SiO_2$-iron oxides liquid systems. Circ Inf Tech, Cent Doc Sider 33:109-13

Gerard O, Jaoul O (1989) Oxygen diffusion in San Carlos olivine. J Geophys Res 94:4119-4128

Giggenbach WF (1982) Carbon-13 exchange between CO_2 and CH_4 under geothermal conditions. Geochim Cosmochim Acta 46:159-165

Giggenbach WF (1997) Relative importance of thermodynamic and kinetic processes in governing the chemical and isotopic composition of carbon gases in high-heat flow sedimentary basins. Geochim Cosmochim Acta 61:3763-3785

Giletti BJ (1985) The nature of oxygen isotope transport within minerals in the presence of hydrothermal water and role of diffusion. Chem Geol 53:197-206

Giletti BJ (1986) Diffusion effects on oxygen isotope temperatures of slowly cooled igneous and metamorphic rocks. Earth Planet Sci Lett 77:218-228

Giletti BJ (1994) Isotopic equilibrium/disequilibrium and diffusion kinetics in feldspars. In Parsons I (ed) Feldspars and Their Reactions. NATO ASI Series. Kluwer Academic Publishers, Dordrecht, The Netherlands, p 351-382

Gilleti BJ, Anderson TF (1975) Studies in diffusion—II. Oxygen in phlogopite mica. Earth Planet Sci Lett 28:225-233

Giletti BJ, Hess KC (1988) Oxygen diffusion in magnetite. Earth Planet Sci Lett 89:115-122

Giletti BJ, Tullis J (1977) Studies in diffusion. IV. Pressure dependence of Ar diffusion in phlogopite mica. Earth Planet Sci Lett 35:180-183

Giletti BJ, Yund RA (1984) Oxygen diffusion in quartz. J Geophys Res 89:4039-4046

Giletti BJ, Hickey JH, Tullis TE (1979) Oxygen diffusion in olivine under hydrous conditions. EOS, Trans Am Geophys Union 60:370

Giletti BJ, Semet MP, Yund RA (1978) Studies in diffusion -III. Oxygen in feldspars: An ion microprobe determination. Geochim Cosmochim Acta 42:45-57

Glasstone G, Laidler KJ, Eyring H (1941) The Theory of Rate Processes. McGraw-Hill, New York

Glicksman ME (2000) Diffusion in Solids: Field Theory, Solid-State Principles, and Applications. John Wiley & Sons, New York

Goldman D, and Gupta P (1983) Diffusion-controlled redox reaction in a glass melt. J Am Ceram Soc 66:188-190

Goldsmith JR (1987) Al/Si interdiffusion in albite: Effect of pressure and the role of hydrogen. Contrib Mineral Petrol 95:311-321.

Goto KS, Kurahashi T, Sasabe M (1977) Oxygen pressure-dependence of tracer diffusivities of Ca and Fe in liquid $CaO-SiO_2-Fe_xO$ systems. Metallurgical Trans B 8B:523-528

Grabke HJ (1994) Kinetics of gas-solid interactions. Mater SciForum 154:69-86

Graham CM (1981) Experimental hydrogen isotope studies. Diffusion of hydrogen in hydrous minerals, and stable isotope exchange between minerals in metamorphic rocks. Contrib Mineral Petrol 76:216-228

Graham CM, Elphick SC (1991) Some experimental constraints on the role of hydrogen in oxygen and hydrogen diffusion and Al-Si interdiffusion in silicates. In Ganguly J (ed) Diffusion, Atomic Ordering, and Mass Transport—Selected Topics in Geochemistry. Adv Phys Geochem 8:248-285

Graham CM, Elphick SC (1994) Hydrogen in feldspars and related silicates. In Parsons I (ed) Feldspars and Their Reactions. NATO ASI Series. Kluwer Academic Publishers, Dordrecht, The Netherlands, p 383-413

Graham CM, Sheppard SMF (1980) Experimental hydrogen isotope studies, II. Fractionations in the systems epidote-NaCl-H_2O, epidote-$CaCl_2$-H_2O and epidote-seawater, and the hydrogen isotope composition of natural epidotes. Earth Planet Sci Lett 49:237-251

Graham CM, Harmon RS, Sheppard SMF (1984) Experimental hydrogen isotope studies: Hydrogen isotope exchange between amphibole and water. Am Mineral 69:128-138

Graham CM, Sheppard SMF, Heaton THE (1980) Experimental hydrogen isotope studies—I. Studies of hydrogen isotope fractionation in the systems epidote-H_2O, zoisite-H_2O and $AlO(OH)$-H_2O. Geochim Cosmochim Acta 44:353-364

Graham CM, Valley JW, Winter BL (1996) Ion microprobe analysis of $^{18}O/^{16}O$ in authigenic and detrital quartz in the St. Peter sandstone, Michigan Basin and Wisconsin Arch, USA: Contrasting diagenetic histories. Geochim Cosmochim Acta 60:5101-5116

Graham CM, Viglino JA, Harmon RS (1987) Experimental study of hydrogen-isotope exchange between aluminous chlorite and water and of hydrogen diffusion in chlorite. Am Mineral 72:566-579

Graham CM, Valley JW, Eiler JM, Wada H (1998) Time scales and mechanisms of fluid infiltration in a marble. An ion microprobe study. Contrib Mineral Petrol 132:371-389

Greene CH and Kitano I (1959) Rate of solution of oxygen bubbles in commercial glasses. Glastechn Ber 32:44-48.

Gregory RT, Criss RE, Taylor HP (1989) Oxygen isotope exchange kinetics of mineral pairs in closed and open systems: Applications to problems of hydrothermal alteration of igneous rocks and Precambrian iron formations. Chem Geol 75:1-42

Griggs D (1967) Hydrolytic weakening of quartz and other silicates. Geophys J 14:19-31

Griggs D (1974) A model of hydrolytic weakening in quartz. J. Geophys Res 79:1653-1661

Gruenwald TB, Gordon G (1971) Oxygen diffusion in single crystals of titanium dioxide. J Inorg Chem 33:1151-1155

Haar L, Gallagher JS, Kell GS (1984) NBS/NRC Steam Tables: Thermodynamic and transport properties and computer programs for vapor and liquid states of water in SI units. Hemisphere Publ, Washington, DC, 320 p

Hacker BR, Christie JM (1991) Observational evidence for a possible new diffusion path. Science 251: 67-70

Hagel WC and MacKenzie JD (1964) Electrical conduction and oxygen diffusion in aluminoborate and aluminosilicate glasses. Phys Chem Glasses 5:113-19.

Hallwig D, Schachtner R, Sockel HG (1982) Diffusion of magnesium, silicon and oxygen in Mg_2SiO_4 and formation of the compound in the solid state. *In* Dyrek K, Habor J, Nowotry J (eds) Reactivity in Solids. Proc Int'l Symp (9th), p 166-169

Haneda H, Miyazawa Y, Shirasaki S (1984) Oxygen diffusion in single crystal yttrium aluminum garnet. J Crystal Growth 68:581-588

Haneda H, Yamamura H, Watanabe A, Shirasaki S (1987) Relationship between oxygen self-diffusion and Debye temperature in the polycrystalline magnesium aluminum ferrite series, $MgAl_{2-x}Fe_xO_4$. J Solid State Chem 68:273-284

Haneda H, Watanabe A, Kitami Y, Shirasaki S (1993) Oxygen self-diffusion in single and polycrystalline ytterbium iron garnet. Defect and Diffusion Forum 95-98:1065-1070

Harris GM (1951) Kinetics of isotopic exchange reactions. Trans Faraday Soc 47:716-721

Harrison LG (1961) Influence of dislocations on diffusion kinetics in solids with a particular reference to alkali halides. Trans Faraday Soc 57:1191-1199

Hart SR (1981) Diffusion compensation in natural silicates. Geochim Cosmochim Acta 45:279-291

Harting P, Maass I (1980) Neue Ergebrisse zum Kohlenstoff-Isotopenaustausch im System CH_4-CO_2. *In* Mitteilungen zur 2. Arbeitstagung "Isotope in der Natur" Nov. 1979. 2b:13-24. Leipzig

Hashimoto H, Hama M, Shirasaki S (1972) Preferential diffusion of oxygen along grain boundaries in polycrystalline MgO. J Appl Phys 43:4828-4829

Haul RAW, Dumbgen G (1962) Investigation of oxygen diffusion in TiO_2, quartz and quartz glass by isotope exchange. Z Electrochem 66:636-641

Haul RAW, Stein LH (1955) Diffusion in calcite crystals on the basis of isotopic exchange with carbon dioxide. Trans Faraday Soc 51:1280-1290

Hayakawa T (1953) Kinetics of the isotopic exchange reaction between carbon monoxide and carbon dioxide. Bull Chem Soc Japan 26:165-172

Hayashi T, Muehlenbachs K (1986) Rapid oxygen diffusion in melilite and its relevance to meteorites. Geochim Cosmochim Acta 50:585-591

Heggie MI, Jones R, Latham CD, Maynard SCP, Tole P (1992) Molecular diffusion of oxygen and water in crystalline and amorphous silica. Philosophical Magazine B 65:463-471

Helgeson HC, Murphy WM, Aagaard P (1984) Thermodynamic and kinetic constraints on reaction rates among mineral and aqueous solutions. II. Rate constants, effective surface area, and hydrolysis of feldspar. Geochim Cosmochim Acta 48:2405-2432

Helmic M, Rauch F (1993) On the mechanism of diffusion of water in silica glass. Glastech Berichte 66:192-200

Hercule S, Ingrin J (1999) Hydrogen in diopside: Diffusion, kinetics of extraction-incorporation, and solubility. Am Mineral 84:1577-1587

Heuer AH, Lagerlof KPD (1999) Oxygen self-diffusion in corundum (α-Al_2O_3): A conundrum. Phil Mag Lett 79:619-627

Hervig RL, Williams LB, Kirkland IK, Longstaffe FJ (1995) Oxygen isotope microanalysis of diagenetic quartz: Possible low temperature occlusion of pores. Geochim Cosmochim Acta 59:2537-2543

Hoering TC, Kennedy JW (1957) The exchange of oxygen between sulfuric acid and water. J Am Chem Soc. 79:56-60

Holness MB, Graham CM (1991) Equilibrium dihedral angles in the system H_2O-CO_2-NaCl calcite, and implications for fluid flow during metamorphism. Contrib Mineral Petrol 108:368-383

Horita J (1988a) Procedure for the hydrogen isotope analysis of water from concentrated brines. Chem Geol (Isotope Geosci Section) 72:85-88

Horita J (1988b) Hydrogen isotope analysis of natural waters using an H_2-water equilibration method: A special implication to brines. Chem Geol (Isotope Geosci Section) 72:89-94

Horita J (2001) Carbon isotope exchange in the system CO_2-CH_4 at elevated temperatures. Geochim Cosmochim Acta 65:1907-1919

Horita J, Berndt ME (1999) Abiogenic methane formation and isotopic fractionation under hydrothermal conditions. Science 285:1055-1057

Horita J, Wesolowski DJ, Cole DR (1993a) The activity-composition relationship of oxygen and hydrogen isotopes in aqueous salt solutions: I. Vapor-liquid water equilibration of single salt solutions from 50 to 100°C. Geochim Cosmochim Acta 57:2797-2817

Horita J, Cole DR, Wesolowski DJ (1993b) The activity-composition relationship of oxygen and hydrogen isotopes in aqueous salt solutions: II. Vapor-liquid water equilibration of mixed salt solutions from 50 to 100°C. Geochim Cosmochim Acta 57:4703-4711

Horita J, Cole DR, Wesolowski DJ (1995) The activity-composition relations of oxygen and hydrogen isotopes in aqueous salt solutions. III. Vapor-liquid water equilibration of NaCl solutions to 350°C. Geochim Cosmochim Acta 59:1139-1151

Houlier B, Jaoul O, Abel F, Liebermann RC (1988) Oxygen and silicon self-diffusion in natural olivine at $T = 1300°C$. Phys Earth PlanetInteriors 50:240-250

Huff GA, Satterfield CN (1984) Intrinsic kinetics of the Fischer-Tropsch synthesis on a reduced fused-magnetite catalyst. Ind Eng in Chem Prod Res Develop 23:851-954

Humayun M, Clayton R (1995) Potassium isotope cosmochemistry—genetic-implications of volatile element depletion. Geochim Cosmochim Acta 59:2131-2148

Igumnov SA (1976) Sulfur isotope exchange between sulfide and sulfate in hydrothermal solutions. Geokhimiya 4:497-503

Ikuma Y, Shimada E, Sakano S, Oishi M, Yokoyama M (1999) Oxygen self-diffusion in cylindrical single-crystal mullite. J Electrochem Soc 146:4672-4675

Jaoul O, Houlier B, Abel F (1983) Study of ^{18}O diffusion in magnesium orthosilicate by nuclear microanalysis. J Geophys Res 88:613-624

Jaoul O, Sautter V, Abel F (1991) Nuclear microanalysis: A powerful tool for measuring low atomic diffusivity with mineralogical applications. In Ganguly J (ed) Diffusion, Atomic Ordering, and Mass Transport—Selected Topics in Geochemistry. Adv Phys Geochem 8:248-285

Jaoul O, Froidevaux C, Durham WB, Michaut M (1980) Oxygen self-diffusion in forsterite: Implications for the high temperature creep mechanism. Earth Planet Sci Lett 47:613-624

Jibao G, Yaqian Q (1997) Hydrogen isotope fractionation and hydrogen diffusion in the tourmaline-water system. Geochim Cosmochim Acta 21:4679-4688

Joesten R (1991) Grain-boundary diffusion kinetics in silicates and oxide minerals. In Ganguly J (ed) Diffusion, Atomic Ordering, and Mass Transport. Adv Phys Chem 8:345-395. Springer-Verlag, New York

Johnson JR, Bristow RH, Blau HH (1951) Diffusion of ions in some simple glasses. J Am Ceram Soc 34:165-172

Jost W (1960) Diffusion in Solids, Liquids, Gases. Third Printing. Academic Press, New York

Kaiser CJ (1991) Analysis of isotope-transfer kinetics during sulfate reduction by dextrose under hydrothermal conditions. Chem Geol 87:247-263

Kalen JD, Boyce RS, Cawley JD (1991) Oxygen tracer diffusion in vitreous silica. J Am Ceram Soc 74:203-209

Kanzaki M (1997) Activation energies of H_2O and H_2 diffusion in silica glass: Semi-empirical molecular orbital study. Mineralogical J 19:13-19

Kats A, Haven Y, Stevels JM (1962) Hydroxyl groups in β-quartz. Phys Chem Glasses 3:69-75

Keller H, Schwerdtfeger K, Petri H, Holzle R, Hennesen K (1982) Tracer diffusivity of O^{18} in CaO-SiO$_2$ melts at 1600°C. Metallurgical Trans B 13B:237-240

Keneshea FJ, Douglas DL (1971) The diffusion of oxygen in zirconia as a function of oxygen pressure. Oxidation of Metals 3:1-14

Keppler H, Bagdassarov N (1993) High temperature FTIR spectra of H_2O in rhyolite melt to 1300°C. Am Mineral 78:1324-1327

Kharaka YK, Carothers WW, Rosenbauer RJ (1983) Thermal decarboxylation of acetic acid: Implications for origin of natural gas. Geochim Cosmochim Acta 47:397-402

Kilner JA (1986) New techniques for studying mass transport in oxides. In Freer R, Dennis PF (eds) Mater SciForum 7:205-222

Kilner JA, Steele BCH (1984) Oxygen self-diffusion studies using negative-ion secondary ion mass spectrometry (SIMS). Solid State Ionics 12:89-97

King TB, Koros PJ (1959) Diffusion in liquid silicates. In Kingery WD (ed) Kinetics of High Temperature Processes. Wiley, New York, p 80-85

Kingery WD, Lecron JA (1960) Oxygen mobility in two silicate glasses. Phys Chem Glasses 1:87-89

Kirkaldy JS, Young DJ (1987) Diffusion in the Condensed State. The Institute of Metals

Kitchen NE, Valley JW (1995) Carbon isotope thermometry in marbles of the Adirondack Mountains, New York. J Metamorphic Geol 13:577-594

Kohlstedt DL, Mackwell SJ (1998) Diffusion of hydrogen and intrinsic point defects in olivine. Z Physik Chemie 207:147-162

Kohlstedt DL, Goetz C, Durham WB, VanderSande JB (1975) A new technique for decorating dislocations in olivine. Contrib Mineral Petrol 53:13-24

Kohn MJ, Valley JW, Elsenheimer D, Spicuzza MJ (1993) O isotope zoning in garnet and staurolite: Evidence for closed-system mineral growth during regional metamorphism. Am Mineral 78:988-1001

Koros PJ, King TB (1962) The self-diffusion of oxygen in a lime silica-alumina slag. Trans Metall Soc, AIME 224:229-306

Kreevoy MM, Truhlar DG (1986) Transition state theory. *In* Bermasconi CF (ed) Investigation of Rates and Mechanisms of Reactions. Part 1, Techniques of Chemistry, Vol. VI:13-96. John Wiley and Sons, New York

Kronenberg AK (1994) Hydrogen speciation and chemical weakening of quartz. *In* Heaney PJ, Prewitt CT, Gibbs GV (eds) Silica—Physical Behavior, Geochemistry and Materials Applications, Rev Mineral 29:123-176

Kronenberg AK (1996) Hydrogen speciation of α-quartz. *In* Massoud Z, Poindexter EH, Helms CR (eds), The Physics and Chemistry of SiO_2 and the Si-SiO_2 Interface—3, 96:163-171

Kronenberg AK, Tullis J (1984) Flow strengths of quartz aggregates: Grain size and pressure effects due to hydrolytic weakening. J Geophys Res 89:4281-4297

Kronenberg AK, Yund, RA, Giletti BJ (1984) Carbon and oxygen diffusion in calcite: Effects on Mn content and P_{H_2O}. Phys Chem Minerals 11:101-112

Kronenberg AK, Yund RA, Rossman GR (1996) Stationary and mobile hydrogen defects in potassium feldspar. Geochim Cosmochim Acta 60:4705-4094

Kronenberg AK, Yund RA, Rossman GR (1998) Comment on "Stationary and mobile hydrogen defects in potassium feldspar". Geochim Cosmochim Acta 62:377-378

Kronenberg AK, Kirby SH, Aines RD, Rossman GR (1986) Solubility and diffusional uptake of hydrogen in quartz at high water pressures: Implications for hydrolytic weakening. J Geophys Res 91: 12,723-12,744

Kubicki J, Lasaga A (1993) Molecular dynamics simulation of interdiffusion in $MgSiO_3$-Mg_2SiO_4 melts. Phys Chem Minerals 20:255-262

Kusakabe M, Robinson BW (1977) Oxygen and sulfur isotope equilibria in the $BaSO_4$-HSO_4^- system from 110 to 350°C and applications. Geochim Cosmochim Acta 41:1033-1040

Kyser TK (1987) Equilibrium fractionation factors for stable isotopes. *In* Kyser TK (ed) Stable Isotope Geochemistry of Low Temperature Fluids. Mineral Assoc Canada Short Course 13:1-84

Kyser TK, Lesher CE, Walker D (1998) The effects of liquid immiscibility and thermal diffusion on oxygen isotopes in silicate liquids. Contrib Mineral Petrology 133:373-381

Laan, SVD, Zhang Y, Kennedy A, Wyllie P (1994) Comparison of element and isotope diffusion of K and Ca in multicomponent silicate liquids. Earth Planet Sci Lett 123:155-166

Labotka TC, Cole D R, Riciputi L R (1999) Influence of pressure on the diffusivity of C and O in calcite. Abstr, 9th Annual V M Goldschmidt Conf, Boston, Massachusetts

Labotka TC, Cole DR, Riciputi LR (2000) Diffusion of C and O in calcite at 100 MPa. Am Mineral 85:488-494

Lagerlof KPD, Mitchell TE, Heuer AH (1989) Lattice diffusion kinetics in undoped and impurity-doped sapphire (α-Al_2O_3): A dislocation loop annealing study. J Am Ceram Soc 72:2159-2179

Lamkin, MA, Riley, FL, Fordham RJ (1992) Oxygen mobility in silicon dioxide and silicate glasses: A review. J Eur Ceram Soc 10:347-367

Lasaga AC (1979a) Multicomponent exchange and diffusion in silicates. Geochim Cosmochim Acta 43:455-469

Lasaga AC (1979b) The treatment of multicomponent diffusion and ion pairs in diagenetic fluxes. Am J Sci 279:324-346

Lasaga AC (1981a) Rate laws of chemical reactions. *In* Lasaga AC, Kirkpatrick RJ (eds) Kinetics of Geochemical Processes. Rev Mineral 8:1-68

Lasaga AC (1981b) Transition state theory. *In* Lasaga AC, Kirkpatrick RJ (eds) Kinetics of Geochemical Processes. Rev Mineral 8:135-170

Lasaga AC (1981c) The atomistic basis of kinetics: Defects in minerals. *In* Lasaga AC, Kirkpatrick RJ (eds) Kinetics of Geochemical Processes. Rev Mineral 8:261-320

Lasaga AC (1990) Atomic treatment of mineral-water surface reactions. *In* Hochella M, White AF (eds) Mineral-Water Interface Geochemistry. Rev Mineral 23:17-86

Lasaga AC (1998) Kinetic Theory in the Earth Sciences, Princeton University Press, Princeton

Lawrence JR, Kastner M (1975) ^{18}O to^{16}O feldspars in carbonate rocks. Geochim Cosmochim Acta 39: 97-102

Lazarus D, Nachtrieb NH (1963) Effect of high pressure on diffusion. *In* Paul W, Warschauer DM (eds) Solids Under Pressure. McGraw-Hill, New York, p 43-69

Leamnson RN, Thomas J, Ehrlinger HP (1969) A study of the surface areas of particulate microcrystalline silica and silica sand. Illinois State Geol Surv Cir 44:12

LeClaire AD (1963) The analysis of grain boundary diffusion measurements. Brit J Appl Phys 14:351-356

LeClaire AD, Rabinovitch A (1981) A mathematical analysis of diffusion in dislocations: I. Application to concentration 'tails.' J Phys C: Solid Phys 14:3863-3879

LeClaire AD, Rabinovitch A (1984) The mathematical analysis of diffusion in dislocations. *In* Murch FE, Noweick AS (eds) Diffusion in Crystalline Solids, p 257-318

Lecluse C, Robert F (1994) Hydrogen isotope exchange reaction rates: Origin of water in the inner solar system. Geochim Cosmochim Acta 58:2927-2939

Lecuyer C, Grandjean P, Sheppard MF (1999) Oxygen isotope exchange between dissolved phosphate and water at temperatures ≤135°C: Inorganic versus biological fractionations. Geochim Cosmochim Acta 63:855-862

Le Gall M, Huntz AM, Lesage B (1996) Self-diffusion in α-Al_2O_3 III. Oxygen diffusion in single crystals doped with Y_2O_3. Phil Mag A 73:919-934

Lesher C (1994) Kinetics of Sr and Nd exchange in silicate liquids: Theory, experiments and applications to uphill diffusion, isotopic equilibration, and irreversible mixing of magmas. J Geophys Res B 99:9585-9605

Lesher C, Hervig RL, Tinker D (1996) Self diffusion of network formers (silicon and oxygen) in naturally occurring basaltic liquid. Geochim Cosmochim Acta 60:405-413

Liang Y, Richter FM, Chamberlain L (1997) Diffusion in silicate melts: III. Empirical models for multicomponent diffusion. Geochim Cosmochim Acta 61:5295-5312

Liang Y, Richter FM, Watson E (1996) Diffusion in silicate melts: II. Multicomponent diffusion in CaO-Al_2O_3-SiO_2 at 1500°C in 1 GPA. Geochim Cosmochim Acta 60:5021-5035

Liang Y, Richter FM, Davis AM, Watson EB (1996) Diffusion in silicate melts: I. Self diffusion in CaO-Al_2O_3-SiO_2 at 1500°C in 1 GPa. Geochim Cosmochim Acta 60:4353-4367

Liberatore M, Wuensch BJ (1995) Oxygen self-diffusion in MgO grain boundaries. Mater Res Soc Symp Proc 357:389-393

Ligang Z, Jingxiu L, Huanbo Z, Zhensheng C (1989) Oxygen isotope fractionation in the quartz-water system. Econ Geol 84:1643-1650

Liu K-K, Epstein S (1984) The hydrogen isotope fractionation between kaolinite and water. Chem Geol (Isotope Geosci Sect) 2:335-350

Lloyd RM (1968) Oxygen isotope behavior in the sulfate-water system. J Geophys Res 73:6099-6110

Luehr CP, Challenger GE, Masters BJ (1956) Isotopic exchange in non-stable systems. J Am Chem Soc 78:1314-1317

Mackwell SJ, Kohlstedt DL (1990) Diffusion of hydrogen in olivine: Implications for water in the mantle. J Geophys Res 95:5079-5088

Mackwell SJ, Paterson MS (1985) Water-related diffusion and deformation effects in quartz at pressures of 1500 and 300 MPa. *In* Schock RN (ed) Point Defects in Minerals. Am Geophy Union, Geophys Monograph 31, Mineral Phys 1:141-150

Manning JR (1968) Diffusion Kinetics for Atoms in Crystals. Van Nostrand, Princeton, New Jersey, 257 p

Manning JR (1974) Diffusion kinetics and mechanisms in simple crystals. *In* Hofmann AW, Giletti BJ, Yoder HS, Jr., Yund RA (eds) Geochemical Transport and Kinetics. Carnegie Inst Wash Pub 634:3-14

Manning JR, Sirman JD, Kilner JA (1997) Oxygen self-diffusion and surface exchange studies of oxide electrolytes having the fluorite structure. Solid State Ionics 93:125-132

Marin JF, Contamin P (1969) Uranium and oxygen self-diffusion in UO_2. J Nucl Mater 30:16-25

Martin D, Duprez D (1996) Mobility of surface species on oxides. 1. Isotopic exchange of $^{18}O_2$ with ^{16}O of SiO_2, Al_2O_3, ZrO_2, MgO, CeO_2, and CeO_2-Al_2O_3. Activation by noble metals. Correlation with oxide basicity. J Phys Chem 100:9429-9438

Matsuhisa Y, Goldsmith JR, Clayton RN (1978) Mechanisms of hydrothermal crystallization of quartz at 250°C and 15 Kbar. Geochim Cosmochim Acta 42:173-182

Matsuhisa Y, Goldsmith JR, Clayton RN (1979) Oxygen isotopic fractionation in the system quartz-albite-anorthite-water. Geochim Cosmochim Acta 43:1131-1140

Matthews A, Beckinsale RD (1979) Oxygen isotope equilibration systematics between quartz and water. Am Mineral 64:232-240

Matthews A, Goldsmith JR, Clayton RN (1983a) On the mechanisms and kinetics of oxygen isotopic exchange in quartz and feldspars at elevated temperatures and pressures. Bull Geol Soc Am 94:396-412

Matthews A, Goldsmith JR, Clayton RN (1983b) Oxygen isotope fractionations involving pyroxenes: The calibration of mineral pair geothermometers. Geochim Cosmochim Acta 49:631-644

Matthews A, Goldsmith JR, Clayton RN (1983c) Oxygen isotope fractionation between zoisite and water. Geochim Cosmochim Acta 47:645-654

Matthews A, Palin JM, Epstein S, Stolper EM (1994) Experimental study of $^{18}O/^{16}O$ partitioning between crystalline albite, albitic glass, and CO_2 gas. Geochim Cosmochim Acta 58:5255-5266

May HB, Lauder I, Wollast R (1974) Oxygen diffusion coefficients in alkali silicates. J Am Ceram Soc 57:197-200

Mazurin O, Streltsina M, Shvaiko-Shvaikovskaya T (1983) Handbook of Glass Data. Elsevier, Amsterdam

McConnell JDC (1995) The role of water in oxygen isotope exchange in quartz. Earth Planet Sci Lett 136:97-107

McKay HAC (1938) Kinetics of exchange reactions. Nature 142:997-998

McKibben MA, Shanks, III, WC, Ridley WI (eds) (1998) Applications of Microanalytical Techniques to Understanding Mineralizing Processes. Rev Econ Geol, Vol 7

Meissner E, Sharp TG, Chakraborty S (1998) Quantitative measurement of short compositional profiles using analytical transmission electron microscopy. Am Mineral 83:546-552

Merbach H (1987) Kinetics of solvent exchange reactions at high pressure. *In* van Eldick R, Honas J (eds) High Pressure Chemistry and Biochemistry. Reidel, Dordrecht, The Netherlands, p 311-331

Merigoux H (1968) Etude de la mobilite d l'oxygen dans les feldspaths alcalins. Bull Soc fr Minéral Crystallogr 91:51-64

Mikkelsen J (1984) Self-diffusivity of network oxygen in vitreous SiO_2. Appl Phys Lett 45:1187-1189

Millot F, Niu Y (1997) Discussion of O^{18} in Fe_3O_4: An experimental approach to study the behavior of minority defects in oxides. J Phys Chem Solids 58:63-72

Millot F, Picard C (1988) Oxygen self-diffusion in non-stoichiometric rutile TiO_{2-x} at high temperature. Solid State Ionics 28-30:1344-1348

Millot F, Lorin JC, Klossa B, Niu Y, Tarento JR (1997) Oxygen self-diffusion in Fe_3O_4: An experimental example of interactions between defects. Ber Bunsenges Phys Chem 101:1351-1354

Mills GA (1940) Oxygen exchange between water and inorganic oxyanions. J Am Chem Soc 62:2833-2838

Mills GA, Urey HC (1940) The kinetics of isotopic exchange between carbon dioxide, bicarbonate ion, carbonate ion and water. J Am Chem Soc 62:1019-1026

Misener DJ (1974) Cationic diffusion in olivine to 1400°C and 35 kbar. *In* Hofmann AW, Giletti BJ, Yoder HS, Jr., Yund RA (eds) Geochemical Kinetics and Transport. Carnegie Inst Wash Publ 634:117-129

Monty C (1986) Diffusion in oxides. *In* Chadwhich AV, Terenzi M (eds) Defects in Solids: Modern Techniques. NATO ASI Series, Series B, Phys 147:377-394

Moore DK, Cherniak DJ, Watson EB (1998) Oxygen diffusion in rutile from 750 to 1000°C and 0.1 to 1000 MPa. Am Mineral 83:700-711

Mora CI, Riciputi LR, Cole DR (1999) Short-lived oxygen diffusion during hot, deep-seated meteoric alteration of anorthosite. Science 286:2323-2325

Mori K, Suzuki K (1969) Diffusion in iron oxide melts. Trans Iron Steel Inst Japan 9:405-412

Morishita Y, Giletti B, Farver (1996) Volume self-diffusion of oxygen in titanite. Geochem J 30:71-79

Moroz NM, Kholopov EV, Belitsky IA, Fursenko BA (2001) Pressure-enhanced molecular self-diffusion in microporous solids. Microporous Mesopourous Materials 42:113-119

Motzfeld K (1964) On the rates of oxidation of silicon and silicon carbide in oxygen, and correlation with permeability of silica glass. Acta Chemica Scandinavica 18:1596-1606

Moulson AJ, Roberts JP (1960) Water in silica glass. Trans Br Ceram Soc 59:388-399

Muehlenbachs K, Connolly C (1991) Oxygen diffusion in leucite: Structural controls. *In* Taylor HP, O'Neil JR, Kaplan IR (eds) Stable Isotope Geochemistry: A Tribute to Samuel Epstein. The Geochemical Society, Spec Publ 3:27-34

Muehlenbachs K, Kushiro I (1974) Oxygen isotopic exchange and equilibrium of silicates with CO_2 or O_2. Carnegie Inst Wash Yearb 73:232-236

Muehlenbachs K, Schaeffer HA (1977) Oxygen diffusion in vitreous silica —ulitization of natural isotopic abundances. Can Mineral 15:179-184

Murch GE, Bradhurst DH, de Bruin HJ (1975) Oxygen self-diffusion in non-stoichiometric uranium dioxide. Phil Mag 32:1141-1150

Murch GE (1980) Atomic diffusion theory in highly defective solids. Diffusion Defect Monogr Ser No 6. Trans Tech Publications, Rockport, Massachusetts

Murch GE (1983) Oxygen diffusion in uranium oxide:An overview. Diffusion Defect Data 32:9-19

Murch GE, Catlow CR (1987) Oxygen diffusion in UO_2, ThO_2, and PuO_2: A review. J Chem Soc, Faraday Trans 83:1157-1169

Muzykantov VS (1980) Kinetic equations of isotope redistribution in an elementary reaction. React Kinet Catal Lett 14:113-118

Myers OK, Prestwood RJ (1951) Isotopic exchange reactions. *In* Wahl AC (ed) Radiochemistry Applied to Chemistry, p 6-43

Nagy, KL and Giletti, BJ (1986) Grain boundary diffusion of oxygen in a macro perthitic feldspar. Geochim Cosmochim Acta 50:1151-1158

Nemec L (1969) Diffusion-controlled dissolving water vapour bubbles in molten glass. Glass Technology 10:176-181

Nemec L (1980) The behavior of bubbles in glass melts: Part 1. Bubble size controlled by diffusion. Glass Technology 21:134-138

Nemec L, Klouzek J (1995) Determination of diffusion coefficients of gases in glass melts using the method of absorbed gas volume. Cerams-Silikaty 39:1-40

Nevins D, Spera F (1998) Molecular dynamics simulations of molten $CaAl_2Si_2O_8$: Dependence of structure and properties on pressure. Am Mineral 83:1220-1230

Niemeier D, Chakraborty S, Becker KD (1996) A high-temperature Mössbauer study of the directional geometry of diffusion in fayalite, Fe_2SiO_4. Phys Chem Minerals 23:284

Nishina T, Masuda Y, Uchida I (1993) Gas solubility and diffusivity of H_2, CO_2, and O_2 in molten alkali carbonates. Proc Electrochem Soc 93:424-435

Niu Y, Millot F (1999) Oxygen self-diffusion and interactions between defects in Fe_3O_4. Acta Metallurgica Sinica 12:137-142

Norris TH (1950) The kinetics of isotopic exchange reactions. J Phys Colloid Chem 54:777-783

Northrop DA, Clayton RN (1966) Oxygen isotope fractionations in systems containing dolomite. J Geol 74:174-196

Norton FJ (1961) Permeation of gaseous oxygen through vitreous silica. Nature 191:701

Nowak M, Behrens H (1995) The speciation of water in haplogranitic glasses and melts determined by *in situ* near infrared spectroscopy. Geochim Cosmochim Acta 59:3445-3450

Nowak M, Behrens H (1996) An experimental investigation on diffusion of water in haplogranitic melts. Contrib Mineral Petrol 126:365-376

Ohmoto H, Lasaga AC (1982) Kinetics of reactions between aqueous sulfates and sulfides in hydrothermal systems. Geochim Cosmochim Acta 46:1727-1746

Oishi Y, Ando K (1984) Oxygen diffusion in MgO and Al_2O_3. Adv Ceram Structural Prop 10:379-395

Oishi Y, Kingery WD (1960) Self-diffusion of oxygen in single crystal and polycrystalline aluminum oxide. J Chem Phys 33:480-486

Oishi Y, Terai R, Ueda H (1975) Oxygen diffusion in liquid silicates and relation to their viscosity. *In* Cooper AR, Heuer AH (eds) Mass Transport Phenomena in Cerams 9:297-310. Plenum Press, New York

Oishi Y, Ueda H, and Terai R (1974) Self-diffusion coefficient of oxygen and its relation to viscosity of molten silicates. Int'l Congr Glass, 10th ed. Ceram Soc Japan 8:30-5

Oishi Y, Ando, K, Kurokawa H, Hiro Y (1983) Oxygen self-diffusion in MgO single crystals. Comm Am Ceram Soc 66:C60-C62

O'Neil JR (1986) Theoretical and experimental aspects of isotopic fractionation. *In* Valley JW, Taylor HP, O'Neil JR (eds) Stable Isotopes in High Temperature Geological Processes. Rev Mineral 16:1-40

O'Neil JR, Kharaka YK (1976) Hydrogen and oxygen isotope exchange between clay minerals and waters. Geochim Cosmochim Acta 40:241-246

O'Neil JR, Taylor HR, Jr (1967) The oxygen isotope and cation exchange chemistry of feldspars. Am Mineral 52:1414-1437

O'Neil JR, Taylor HP, Jr (1969) Oxygen isotope equilibrium between muscovite and water. J Geophys Res 74:6012-6022

O'Neil JR, Truesdell, AH (1991) Oxygen isotope fractionation studies of solute-water interactions. *In* Taylor HP, O'Neil JR, Kaplan IR (eds) Stable Isotope Geochemistry: A Tribute to Samuel Epstein. Geochem Soc Spec Publ 3:17-25

O'Neil JR, Clayton RN, Mayeda TK (1969) Oxygen isotope fractionation in divalent metal carbonates. J Chem Phys 51:5547-5558

Ozawa K, Nagasawa H (2001) Chemical and isotopic fractionations by evaporation and their cosmochemical applications. Geochim Cosmochim Acta 65:2171-2199

Pacaud L, Ingrin J, Jaoul O (1999) High-temperature diffusion of oxygen in synthetic diopside measured by nuclear reaction analysis. Mineral Mag 53:673-686

Paek S-H (1974) Diffusion of hydrogen and deuterium in rutile. PhD Dissertation, University of Utah, Provo, Utah

Paladino AE, Maguire EA, Rubin LG (1964) Oxygen ion diffusion in single-crystal and polycrystalline yttrium irn garnet. J Am Ceram Soc 47:280-282

Palin JM, Epstein S, Stolper EM (1996) Oxygen isotope partitioning between rhyolitic glass/melt and CO_2: An experimental study at 550-950°C and 1 bar. Geochim Cosmochim Acta 60:1963-1973

Palmer DA, Drummond SE (1986) Thermal decarboxylation of acetate. Part I. The kinetics and mechanism of reaction in aqueous solution. Geochim Cosmochim Acta 50:813-823

Palmer DA, Kelm H (1981) Activation volumes of the reactions of transition metal compounds in solution. Coord Chem Rev 36:89-153

Palmer DW (1965) Oxygen diffusion in quartz studied by proton bombardment. Nucl Inst Methods 38: 187-191

Park K, Olander DR (1991) Oxygen diffusion in single-crystal tetragonal zirconia. J Electrochm Soc 138:1154-1159

Peterson NL (1975) Isotope effects in diffusion. *In* Nowick AS, Nurton JJ (eds) Diffusion in Solids, Recent Developments, p 115-170

Petrovich R (1981) Kinetics of dissolution of mechanically comminuted rock-forming oxides and silicates-II. Deformation and dissolution of oxides and silicates in the laboratory and at the Earth's surface. Geochim Cosmochim Acta 45:1675-1686

Pilling MJ, Seakins PW (1995) Reaction Kinetics. Oxford University Press, Oxford, UK

Pitzer KS, Peiper JC, Busey RH (1984) Thermodynamic properties of aqueous sodium chloride solutions. J Phys Chem Ref Data 13:1-102

Poe B, McMillan P, Rubie D, Chakraborty S, Yarger J, Diefenbacher J (1997) Silicon and oxygen self-diffusivities in silicate liquids measured to 15 gigapascals and 2800 Kelvin. Science 276:1245-1248

Poe B, Rubie D (2000) Transport processes of silicate melts at high pressure. *In* Aoki H, Syono Y, Hemley R (eds) Physics Meets Mineralogy—Condensed Matter Physics in Geosciences. Cambridge University Press, Cambridge, UK, p 340-353

Poole P, McMillan P, Wolf G (1995) Computer simulations of silicate melts. *In* Stebbins J, McMillan P, Dingwell D (eds) Structure, Dynamics and Properties of Silicate Melts. Rev Mineral 32:563-616

Potter, RW, Brown DL (1977) The volumetric properties of aqueous sodium chloride solutions from 0° to 500°C at pressures up to 2000 bars based on regression of available data in the literature. U S Geol Surv Bull 1421C:C1-C36

Poulton DJ, Baldwin HW (1967) Oxygen exchange between carbonate and bicarbonate ions and water. I. Exchange in the absence of added catalysts. J Chem 45:1045-1050

Press WH, Flannery BP, Teukolsky SA, Vetterling WT (1986) Numerical Recipes: The Art of Scientific Computing. Cambridge University Press, Cambridge, UK

Prot D, Monty C (1996) Self-diffusion in α-Al_2O_3 II. Oxygen diffusion in "undoped" single crystals. Phil Mag A 73:899-917

Prot D, LeGall M, Lesage B, Huntz AM, Monty C (1996) Self-diffusion in α-Al_2O_3. IV. Oxygen grain-boundary self-diffusion in undoped and yttria-doped alumina polycrystals. Phil Mag 73:935-949

Ramirez R, Gonzalez R, Colera I, Chen Y (1997) Diffusion of deuterons and protons in α-Al_2O_3 crystals enhanced by an electric field. Mater Sci Forum 239-241:395-398

Rawal BS, Cooper AR (1979) Oxygen self-diffusion in a potassium strontium silicate glass using proton activation analysis. J Mater Sci 14:1425-1432

Reed DJ, Wuensch BJ (1980) Ion-probe measurement of oxygen self-diffusion in single-crystal Al_2O_3. J Am Ceram Soc 63:88-92

Reed DJ, Wuensch BJ, Bowen, HK (1978) Research in Materials. Annual Report MIT

Reeder RJ, Valley JW, Graham CM, Eiler JM (1997) Ion microprobe study of oxygen isotopic compositions of structurally nonequivalent growth surfaces on synthetic calcite. Geochim Cosmochim Acta 61:5057-5063

Reddy KPR, Cooper AR (1976) Diffusion of oxygen in sapphire. Am Ceram Soc Bull 55:402

Reddy KPR, Cooper AR (1982) Oxygen diffusion in sapphire. J Am Ceram Soc 65:634-638

Reddy KPR, Cooper AR (1983) Oxygen diffusion in MgO and α-Fe_2O_3. J Am Ceram Soc 66:664-666

Reddy KPR, Oh SM, Major LD Jr, Cooper AR (1980) Oxygen diffusion in forsterite. J Geophys Res 85:322-326

Reid J, Poe B, Rubie D, Zotov N, Wiedenbeck M (2001) The self-diffusion of silicon and oxygen in diopside ($CaMgSi_2O_6$) liquid up to 15 GPa. Chem Geol 174:77-86

Revesz AG, Schaeffer HA (1982) The mechanism of oxygen diffusion in vitreous SiO_2. J Electrochem Soc 129:357-361

Richens DT (1997) The Chemistry of Aqua Ions. John Wiley and Sons, New York

Richet P, Polian A (1998) Water as a dense ice-like component in silicate glasses. Science 281:396-398

Richet P, Bottinga Y, Javoy M (1977) A review of hydrogen, carbon, nitrogen, oxygen, sulfur, and chlorine stable isotope fractionation among gaseous molecules. Ann Rev Earth Planet Sci 5:65-110

Richter FM, Liang Y, Davis AM (1999) Isotope fractionation by diffusion in molten oxides. Geochim Cosmochim Acta 63:2853-2861

Riciputi LR, Cole DR, Larson P, Mora CI (1998) A SIMS investigation of chemical and oxygen isotope exchange associated with syntaxial feldspar replacement in the Rico Dome, CO. Geol Soc Am Annual Mtg Abstr with Progr 30:A272

Robin R, Cooper AR, Heuer AH (1973) Application of a nondestructive single-spectrum proton activation energy to study oxygen diffusion in zinc oxide. J Appl Phys 44:3770-3777

Robertson WM (1969) Surface diffusion of oxides. J Nucl Mater 30:36-39

Rosenbaum JM (1994) Stable isotope fractionation between carbon dioxide and calcite at 900°C. Geochim Cosmochim Acta 58:3747-3753

Rössler E, Sokolov A (1996) The dynamics of strong and fragile glass formers. Chem Geol 128:143-153

Routbort JL, Tomlins GW (1995) Atomic transport of oxygen in nonstoichiometric oxides. Radiation Effects and Defects in Solids 137:233-238

Rovetta MR, Holloway JR, Blacic JD (1986) Solubility of hydroxyl in natural quartz annealed in water at 900°C and 1.5 GPa. Geophys Res Lett 13:145-148

Rovetta MR (1989) Experimental and spectroscopic constraints on the solubility of hydroxyl in quartz. Phys Earth Planet Interiors 55:326-334

Rovetta MR, Blacic JD, Hervig RL, Holloway JR (1989) An experimental study of hydroxyl in quartz using infrared spectroscopy and ion microprobe techniques. J Geophys Res 94:5840-5850

Rubie DC, Ross II CR, Carroll MR, Elphick SC (1993) Oxygen self-diffusion in $Na_2Si_4O_9$ liquid up to 10 GPa and estimation of high-pressure melt viscosities. Am Mineral 78:574-582

Ruthven DM, Derrah RI (1975) Diffusion of monatomic and diatomic gases in 4 Å and 5 Å zeolites. J Chem Soc Faraday Trans 75:2031-2044

Ruthven DM, Xu Z (1993) Diffusion of oxygen and nitrogen in 5Å zeolite crystals and commercial 5 Å pellets. Chem Eng Sci 48:3307-3312

Ryerson FJ, Durham WB, Cherniak DJ, Lanford WA (1989) Oxygen diffusion in olivine: Effect of oxygen fugacity and implications for creep. J Geophys Res 94:4105-4118

Ryerson FJ, McKeegan KD (1994) Determination of oxygen self-diffusion in akermanite, anorthite, diopside, and spinel: Implications for oxygen isotopic anomalies and the thermal histories of Ca-Al-rich inclusions. Geochim Cosmochim Acta 58:3713-3734

Sabioni ACS, Freire FL, Barros Leite CV (1993) Study of oxygen self-diffusion in oxides by ion beam techniques: Comparison between nuclear reaction analysis and SIMS. Nucl Inst Meth Phys Res B73:85-89

Sabioni ACS, Huntz AM, Philibert J, Lesage B (1992) Relation between the oxidation growth rate of chromia scales and self-diffusion in Cr_2O_3. J Mater Sci 27:4782-4790

Sabioni ACS, Zanotto E, Millot F, Tuller H (1998) Oxygen self-diffusion in a cordierite glass. J Non-Crystalline Solids 242:177-182

Sackett WM, Chung HM (1979) Experimental confirmation of the lack of carbon isotope exchange between methane and carbon oxide at high temperatures. Geochim Cosmochim Acta 43:273-276

Sakaguchi I, Haneda H, Tanaka J, Yanagitani T (1996) Effect of composition on the oxygen tracer diffusion in transparent yttrium aluminum garnet (YAG) Cerams. J Am Ceram Soc 79:1627-1632

Sakai H, Tsutsumi M (1978) D/H fractionation factors between serpentine and water at 100°C to 500°C and 2000 bar water pressure, and the D/H ratios of natural serpentines. Earth Planet Sci Lett 40: 231-242

Sakai T, Takizawa K Eguchi T, Horie J (1995) High-temperature internal friction peak and oxygen diffusion in Li_2O-$2SiO_2$ glass. J Mater Sci Lett 14:1126-1128

Sasabe M, Goto KS (1974) Permeability, diffusivity and solubility of oxygen gas in liquid slag. Metall Trans 5:2225-2261

Schachtner R, Sockel HG (1977) Study of diffusion in quartz by activation analysis. In Wood J, Lindquist IO, Helgeson C, Vannerburg NG (eds) Reactivity of Solids. Proc International Symp (8th), p 605-609

Schaeffer HA (1980) Saerstoff- und Siliciumdiffusion in silicatischen Gläsern. Technische Fakultät, Friedrich-Alexander-Universität, Erlangen, 138 p

Schaeffer HA (1980) Oxygen and silicon diffusion-controlled processes in vitreous silica. J Non-Crystalline Solids 38-39:545-550

Schaeffer HA (1984) Diffusion-controlled processes in glass forming melts. J Non-Crystalline Solids 67:19-33

Schaeffer HA, Muehlenbachs K (1978) Correlations between oxygen transport phenomena in non-crystalline silica. J Mater Sci 13:1146-1148

Schaeffer HA, Oel HJ (1969) Oxygen-18 diffusion in lead glass. Glastechn Ber 42:493-498

Schmalzried H (1995) Chemical Kinetics of Solids. VCH Pub, Weinheim, Germany

Schmalzried H, Takada Y, Langanke B (1981) Tracer diffusion coefficient of Pb-, Si- and O- ions in molten leadsilicates. Z Physik Chemie Neue Folge 128:205-212

Schreiber HD, Kozak SJ, Fritchman AL, Goldman DS, Schaeffer HA (1986) Redox kinetics and oxygen diffusion in a borosilicate melt. Phys Chem Glasses 27:152-177

Seibert A (1975) Salzschmelzen von thiocyanaten und thiosulfaten. II. Untersuchungen uber den platzwechsel der schwefelatome in natrium thiosulfat. Z Phys Chem, NF 97:11-22

Semkow KW and Haskin LA (1985) Concentrations and behavior of oxygen and oxide ion in melts of composition CaO-MgO-xSiO2. Geochim Cosmochim Acta 49:1897-1908

Shaffer EW, Sang S-L J , Cooper AR, Heuer AH (1974) Diffusion of tritiated water in β-quartz. *In* Hofmann AW, Giletti BJ, Yoder HS, Jr., Yund RA (eds) Geochemical Kinetics and Transport. Carnegie Inst Wash Publ 634:131-138

Shannon RD (1976) Revised effective ionic radii and systematic studies of interatomic distances in halides and chalcogenides. Acta Crystallogr A32:751-767

Shannon RD, Prewitt CT (1969) Effective ionic radii in oxides and fluorides. Acta Crystallogr B25: 925-946

Sharp ZD (1991) Determination of oxygen diffusion rates in magnetite from natural isotopic variations. Geology 19:653-656

Sharp ZD, Jenkin GRT (1994) An empirical estimate of the diffusion rate of oxygen in diopside. J Metamorphic Geol 12:89-97

Sharp ZD, Giletti BJ, Yoder HS (1991) Oxygen diffusion rates in quartz exchanged with CO_2. Earth Planet Sci Lett 107:339-348

Shaw HR (1973) Diffusion of H_2O in granitic liquids: Part 1. Experimental data: Part 2; mass transfer in magma chambers. *In* Hofmann, AW, Giletti BJ, Yoder HS, Yund RA (eds) Geochemical Transport and Kinetics, Carnegie Inst Wash Publ 634:139-170

Shewmon PG (1989) Diffusion in Solids. McGraw-Hill, New York

Shimizu N, Kushiro I (1984) Diffusivity of oxygen in jadeite and diopside melts at high pressures. Geochim Cosmochima Acta 48:1295-1303

Shimizu N, Kushiro I (1984) Self-diffusion of silicon and oxygen in silicate melts: An experimental study. Springer Ser Chem Phys 36:469-70

Shimizu N, Kushiro I (1991) The mobility of Mg, Ca and Si in diopside-jadeite liquids at high pressures. *In* Perchuk LL, Kushiro I (eds) Physical Chemistry of Magmas. Adv Physical Geochem 9:192-212

Shiraishi Y, Nagahama H, Ohta H (1983) Self-diffusion of oxygen in CaO-SiO₂ melt. Canadian Metallurgical Quart22:37-43

Shirasaki S, Hama H (1973) Oxygen diffusion characteristics of loosely sintered polycrystalline MgO. Chem Phys Lett 20:361-365

Sitzman SD, Banfield JF, Valley JW (2000) Microstructural characterization of metamorphic magnetite crystals with implications for oxygen isotope distribution. Am Mineral 85:14-21

Smyth JR (1989) Electrostatic characterization of oxygen sites in minerals. Geochim Cosmochim Acta 53:1101-1110

Smyth JR, Bish DL (1988) Crystal Structure and Cation Sites in Rock-Forming Minerals. Allen and Unwin, Win-chester, Massachusetts

Someno M, Kobayashi M (1981) SIMS measurements of tracer diffusivity of oxygen in titanium dioxides and sulfur in a calcium sulfide. *In* Secondary Ion Mass Spectrometry. Springer Series in Chem Phys 9:222-224

Sonder E, Martinelli JR, Zuhr RA, Weeks RA (1987) The use of ion beam analysis for measuring ion transport in oxides. Cryst Latt Def Amorph Mater 15:277-282

Staschewski D (1969) Kinetik des mit Bicarbonat katalysierten Sauerstoff-Austausches zwischen Kohlendioxid und Wasser. Chemie-Ing-Techn 41:1111-1118

Stanton TR (1990) High pressure isotopic studies of the water diffusion mechanism in silicate melts and glasses. PhD Dissertation, Arizona State University, Tempe, Arizona

Stebbins J (1995) Dynamics and structure of silicate and oxide melts: Nuclear magnetic resonance studies. *In* Stebbins J, McMillan P, Dingwell D (eds) Structure Dynamics and Properties of Silicate Melts, Rev Mineral 32

Stebbins J, McMillan P, Dingwell D (eds) (1995) Structure Dynamics and Properties of Silicate Melts, Rev Mineral 32:191-246

Steele BCH, Kilner JA (1982) Some characteristics of oxygen transport and surface exchange reactions in selected oxides. Mater Sci Monogr 15:308-324

Stipp S, Hochella MF (1991) Structure and bonding environments at the calcite surface as observed with X-ray photoelectron spectroscopy (XPS) and low energy electron diffraction. Geochim Cosmochim Acta 55:1723-1736

Stoffregen RE (1996) Numerical simulation of mineral-water isotope exchange via Ostwald ripening. Am J Sci 296:908-931

Stoffregen RE, Rye TO, Wasserman MD (1994) Experimental studies of alunite: II. Rates of alunite-water alkali and isotope exchange. Geochim Cosmochim Acta 58:917-929

Sucov EW (1963) Diffusion of oxygen in vitreous silica. J Am Ceram Soc 46:14-20

Suzuoki T, Epstein S (1976) Hydrogen isotope fractionation between OH-bearing minerals and water. Geochim Cosmochim Acta 40:1229-1240

Suzuki S, Nakashima S (1999) *In situ* IR measurements of OH species in quartz at high temperatures. Phys Chem Minerals 26:217-225

Tang XP, Geyer U, Busch R, Johnson WL, Wu Y (1999) Diffusion mechanism in metallic supercooled liquids and glasses. Letters to Nature 402:160-162

Terai R, Oishi Y (1977) Self-diffusion of oxygen in soda-lime silicate glass. Glastechn Bericht 50:68-73

Thode HG, Cragg CB, Hulston JR, Rees CE (1971) Sulfur isotope exchange between sulfur dioxide and hydrogen sulfide. Geochim Cosmochim Acta 35:35-45

Tinker D, Lesher C (2001) Self diffusion of Si and O in dacitic liquid at high pressures. Am Mineral 86: 1-13

Tiselius A (1934) Die diffusion von Wasser in einem Zeolithkristall. Z Physik Chem A 169:425-458

Tokuda T, Ito T, Yamaguchi T (1971) Self diffusion in a glassformer melt, oxygen transport in boron trioxide. Z Naturforschung 26a:2058-2060

Tomozawa M, Molinelli J (1984) Non-fickian diffusion of water in glass. Rivista della Staz Sper Vetro 5:33-37

Tomozawa M, Davis KM (1999) Time-dependent diffusion coefficient of water into silica glass at low temperatures. Mater Sci Eng A272:114-119

Truesdell AH (1974) Oxygen isotope activities and concentration in aqueous salt solutions at elevated temperatures: Consequences for isotope geothermometry. Earth Planet Sci Lett 23:387-396

Tsuchiyama A, Kawamura K, Nakao T, Uyeda C (1994) Isotopic effects on diffusion in MgO melt simulated by the molecular dynamics (MD) method and implications for isotopic mass fractionation in magmatic systems. Geochim Cosmochim Acta 58:3013-3021

Tsuneyuki S, Matsui Y (1995) Molecular dynamic study of pressure enhancement of ion mobilities in liquid silica. Phys Rev Lett 74:3197-3200

Ueda H, Oishi Y (1970) Self-diffusion coefficient of oxygen in molten $CaO-Al_2O_3-SiO_2$-system glasses. Ashai Garasu Kogyo Gijutsu Shoreikai Kenkyu Kokuku 16:201-220

Urey HC (1947) The thermodynamic properties of isotopic substances. J Chem Soc (London), p 562-581

Uyama F, Chiba H, Kusakabe M, Sakai H (1985) Sulfur isotope exchange reactions in the aqueous system: Thiosulfate-sulfide-sulfate at hydrothermal temperature. Geochem J 19:301-315

Valley JW, Graham CM (1991) Ion microprobe analysis of oxygen isotope ratios in granulite facies magnetites: Diffusive exchange as a guide to cooling history. Contrib Mineral Petrol 109:38-52

Valley JW, Graham CM (1993) Cryptic grain-scale heterogeneity of oxygen isotope ratios in metamorphic magnetite. Science 259:1729-1733

Valley JW, Graham CM (1996) Ion microprobe analysis of oxygen isotope ratios in quartz from Skye Granite: Healed micro-cracks, fluid flow, and hydrothermal exchange. Contrib Mineral Petrol 124: 225-234

Valley, JW, Taylor Jr. HP, O'Neil JR (eds) (1986) Stable Isotopes in High-Temperature Geological Processes. Rev Mineral 16

Valley JW, Graham CM, Harte B, Eiler JM, Kinny PD (1998) Chapter 4-Ion microprobe analysis of oxygen, carbon, and hydrogen isotopic ratios. *In* McKibben MA, Shanks WC, Ridley WI (eds) Applications of Microanalytical Techniques to Understanding Mineralizing Processes. Rev Econ Geol 7:73-98

Van Eldick R (1987) High pressure studies of inorganic reactions. *In* Van Eldick R, Jonas J (eds) High Pressure Chemistry and Biochemistry. Reidel, Dordrecht, The Netherlands, p 333-356

Van Eldick R, Palmer DA (1982) Effects of pressure on the kinetics of the dehydration of carbonic acid and the hydrolysis of CO_2 in aqueous solution. J Solution Chem 11:339-346

Van Hook WA (1972) Vapor pressure isotope effect in aqueous systems. III. The vapor pressure of HOD (-60 to 200°C). J Phys Chem 76:3040-3043

Veblen DR (1992) Electron microscopy applied to nonstoichiometry, polysomatism, and replacement reactions in minerals. *In* Buseck PR (ed) Minerals and Reactions at the Atomistic Scale: Transmission Electron Microscopy. Rev Mineral 27:181-230

Vennemann TW, O'Neil JR, Deloule E, Chaussidon M (1996) Mechanism of hydrogen isotope exchange between hydrous minerals and molecular hydrogen: Ion microprobe study of D/H exchange and calcuations of hydrogen self-diffusion rates. Abstr, Sixth VM Goldschmidt Conf, Heidleberg, p 648

Wada H (1988) Microscopic isotopic zoning in calcite and graphite crystals in marble. Nature 331:61-63

Walther JV, Wood BJ (1986) Mineral-fluid reaction kinetics. *In* Walther JV, Woods BJ (eds) Fluid-Rock Interaction During Metamorphism. Springer-Verlag, Inc. 8:194-211

Wang HB (1993) A double medium model for diffusion in fluid-bearing rock. Contrib Mineral Petrol 114:357-364

Wang L, Zhang Y, Essene EJ (1996) Diffusion of the hydrous component in pyrope. Am Mineral 81: 706-718

Wasserberg GJ (1988) Diffusion of water in silicate melts. J Geol 96:363-367

Watson EB (1994) Diffusion in volatile bearing magmas. *In* Carroll M, Holloway J (eds) Volatiles in Magmas. Rev Mineral 30:371-412

Watson EB, Cherniak DJ (1997) Oxygen diffusion in zircon. Earth Planet Sci Lett 148:527-544

Wendlandt RW (1991) Oxygen diffusion in basalt and andesite melts: Experimental results and discussion of chemical versus tracer diffusion. Contrib Mineral Petrol 108:463-471

West AR (1984) Solid State Chemistry and Its Applications. John Wiley and Sons, New York

Whipple RTP (1954) Concentration contours in grain boundary diffusion. Phil Mag 45:1225-1236

White AF, Peterson MI (1990) Role of reactive-surface area characterization in geochemical kinetic models. *In* Melchior DC, Bassett RL (eds) Chemical Modeling of Aqueous Systems. II. Am Chem Soc Symp 416:461-475

Wilde SA, Valley JW, Peck WA, Graham CM (2001) Evidence from detrital zircons for the existence of continental crust and oceans on Earth 4.4 Gyr ago. Nature 409:175- 178

Williams EL (1965) Diffusion of oxygen in fused silica. J Am Ceram Soc 48:190-194

Winchell P (1969) The compensation law for diffusion in silicates. High Temp Sci 1:200-215

Winchell P, Norman JH (1969) A study of the diffusion of radioactive nuclides in molten silicates at high temperature. Proc 3rd Int'l Symp High Temp Technology, p 479-492

Windhager HJ, Borchardt G (1975) Tracer diffusion und Fehlordnung in dem Orthosilikat CO_2SiO_4. Ber Bunsenges Phys Chem 79:1115-1119

Winther KT, Watson EB, Korenowski GM (1998) Magmatic sulfur compounds and sulfur diffusion in albite melt at 1 GPa and 1300-1500°C. Am Mineral 83:1141-1151

Wollenberger HJ (1996) Point defects. *In* Cahn RW, Haasen P (eds) Physical Metallurgy 18:1621-1721

Woods SC, Mackwell S, Dyar D (2000) Hydrogen in diopside: Diffusion profiles. Am Mineral 85:480-487

Wright K, Freer R, Catlow CRA (1996) Water-related defects and oxygen diffusion in albite: A computer simulation study. Contrib Mineral Petrol 125:161-166

Wyart J, Sabatier G (1958) Mobilite des ions silicium et aluminum dans les cristaux de feldspath. Bull Soc fr Minéral Cristallogr 81:223-226

Xu A, Stebbins JF (1998) Oxygen site exchange kinetics observed with solid state NMR in a natural zeolite. Geochim Cosmochim Acta 62:1803-1809

Yang MH, Flynn CP (1994) Intrinsic diffusion properties of an oxide: MgO. Phys Rev Lett 73:1809-1812

Yaqian Q, Jibao G (1993) Study of hydrogen isotope equilibrium and kinetic fractionation in the ilvaite-water system. Geochim Cosmochim Acta 57:3073-30082

Yeh HW, Savin SM (1976) The extent of oxygen isotope exchange between clay minerals and seawater. Geochim Cosmochim Acta 40:743-748

Yinnon H (1979) A proton-activation study of oxygen in multicomponent glass-forming systems. PhD Dissertation, Case Western Reserve University, Cleveland, Ohio

Yinnon H, Cooper ARJ (1980) Oxygen diffusion in multicomponent glass forming silicates. Phys Chem Glasses 21:201-211

Yoo HI, Wuensch BJ, Petuskey WT (1985) Secondary ion mass spectrometric analysis of oxygen self-diffusion in single-crystal MgO. Adv Ceram Structural Prop 10:394-405

Young ED, Rumble D (1993) The origin of correlated variations in *in situ* $^{18}O/^{16}O$ and elemental concentrations in metamorphic garnet from southeastern Vermont, USA. Geochim Cosmochim Acta 57:2585-2597

Young TF, Keiffer J, Borchardt G (1994a) Tracer diffusion of oxygen in $CoO-SiO_2$ melts. J Phys Condens Matt 6:9825-9834

Yu Y, Hewins RH, Clayton RN, Mayeda TK (1995) Experimental study of high temperature oxygen isotope exchange during chondrule formation. Geochim Cosmochim Acta 59:2095-2104

Yund RA, Anderson TF (1974) Oxygen isotope exchange between potassium feldspar and KCl solutions. *In* Hofmann AW, Giletti BJ, Yoder HS, Jr., Yund RA (eds)Geochemical Transport and Kinetics. Carnegie Inst Wash Publ 634:99-105

Yund RA, Anderson TF (1978) The effect of fluid pressure on oxygen isotope exchange between feldspar and water. Geochim Cosmochim Acta 42:235-239

Yund RA, Smith BM, Tullis J (1981) Dislocation-assisted diffusion of oxygen in albite. Phys Chem Minerals 7:185-189

Yurimoto H, Morioka M Nagasawa H (1989) Diffusion in single crystals of melilite: I. Oxygen. Geochim Cosmochim Acta 53:2387-2394

Yurimoto H, Morioka M Nagasawa H (1992) Oxygen self-diffusion along high diffusivity paths in forsterite. Geochem J 26:181-188

Zarubin D (1999) Infrared spectra of hydrogen bonded hydroxyl groups in silicate glasses. A re-interpretation. Phys Chem Glasses 40:184-192

Zhang Y (1999) H_2O in rhyolitic glasses and melts: Measurement, speciation, solubility, and diffusion. Rev Geophys 37:493-516

Zhang Y, Behrens H (2000) H₂O diffusion in rhyolitic melts and glasses. Chem Geol 169:243-262

Zhang Y, Stolper EM (1991) Water diffusion in basaltic melt. Nature 351:306-309

Zhang Y, Stolper EM, Wasserburg GJ (1991a) Diffusion of a multi-species component and its role in oxygen and water transport in silicates. Earth Planet Sci Lett 103:228-240

Zhang Y, Stolper EM, Wasserburg GJ (1991b) Diffusion of water in rhyolitic glasses. Geochim Cosmochim Acta 55:441-456

Zheng Y-F, Fu B (1998) Estimation of oxygen diffusivity from anion porosity in minerals. Geochem J 32:71-89

Zimmerman WH, Bukur DB (1990) Reaction kinetics over iron catalysts used for Fischer-Tropsch synthesis. Can J Chem Eng 68:292-301

On the following pages ....

APPENDIX — TABLES 1-4

Table 1. Summary of rate data for selected gaseous and aqueous systems.

Isotopic Exchange Reaction (a)	Temperature Range (°C)	Pressure Range (MPa)	Catalyst or pH (b)	Rate Units	E_a (c)	A_o (d)	References (e)
I. Gaseous Systems							
1 $^{13}CO_2 + {}^{12}CO \equiv {}^{13}CO + {}^{12}CO_2$	868-920	0.01	SiO_2	min^{-1}	453 ± 71	7.5E21	Brandner & Urey (1945)
2 $^{13}CO_2 + {}^{12}CO \equiv {}^{13}CO + {}^{12}CO_2$	864-885	0.01	Gold	min^{-1}	323 ± 71	5.5#16	Brandner & Urey (1945)
3 $^{13}CO_2 + {}^{12}CO \equiv {}^{13}CO + {}^{12}CO_2$	710-900	vacuum	(SiO_2)	min^{-1}	303.9	4.56E11	Hayakawa (1953)
4 $^{13}CH_4 + {}^{12}CO_2 \equiv {}^{13}CO_2 + {}^{12}CH_4$	500-685	(?)	Clay; H_2O	at.$\%^{-1}$ year^{-1}	197.5	2.63E9	Harting & Maas (1980)
5 $^{13}CH_4 + {}^{12}CO_2 \equiv {}^{13}CO_2 + {}^{12}CH_4$	500-685	(?)	Clay; H_2O	year^{-1}	195	1.445E11	*Giggenbach (1982)
6 $^{13}CO_2 + H_2 \equiv {}^{12}CH_4 + H_2O$	100-300	(50)	Fe-Mg Silicate	year^{-1}	85	8.317E3	Giggenbach (1997)
7 $D_2O_{(v)} + H_2 \rightarrow HDO_{(v)} + HD$	25-434	0.002-0.013	C, SiO_2 Clay, Fe	cc/s^{-1}	42.97	2.0E-22	Lecluse & Robert (1994)
8 $CD_4 + H_2 \rightarrow CD_3H + HD$	25-400	0.0015-0.05	C, Clay	cc/s^{-1}	36.4	6.8E-25	Lecluse & Robert (1994)
II. Aqueous Species							
9 $C^{18}O_2 + H_2{}^{16}O \equiv H_2C^{18}O^{16}O_2$	0-38	0.1	(?)	s^{-1}	71.7	1.14E11	Mills & Urey (1940)
10 $C^{18}O_2 + H_2{}^{16}O \equiv H_2C^{18}O^{16}O_2$	25	0.1	(?)	s^{-1}	log k = -0.084(I)-1.41 (I = ionic strength)		Poulton & Baldwin (1967)
11 $C^{16}O_2 + H_2{}^{18}O \equiv H_2C^{18}O^{16}O_2$	5-90	l/v	(?)	hr^{-1}	56.1	1.29E12	Staschewski (1969)
12 $C^{18}O_2 + H_2{}^{16}O \equiv C^{16}O_2 + H_2{}^{18}O_2$	25-50	0.1	(7.0)	s^{-1}	lnk = -6.39 – 180m/T (m = molality $CaCl_2$)		Fortier (1994)
13 $H_2P^{18}O_4 + H_2{}^{16}O \equiv H_2P^{16}O^{18}O_4 + H_2{}^{18}O$	100.1	l/v	4.6	s^{-1}	k = 4.03E-6 (1 – 1.37)		Bunton et al. (1961)
14 $H_2P^{18}O_4 + H_2{}^{16}O \equiv H_2P^{16}O_4 + H_2{}^{18}O$	50-135	l/v	5	s^{-1}	133.6 ± 5	2.24E12	Lecuyer et al. (1999)
15 $KHS^{16}O_4 + H_2{}^{18}O \equiv KHS^{18}O_4 + H_2{}^{16}O$	100	l/v	(7)	hr^{-1}	k = 1.2-2.4E-5		Mills (1940)
16 $HS^{16}O_4 + H_2{}^{18}O \equiv HS^{18}O_4 + H_2{}^{16}O$	230	l/v	(3.8)	hr^{-1}	k = 7.04E-4		Lloyd (1968)
17 $H_2S^{16}O_4 + H_2{}^{18}O \equiv H_2S^{18}O_4 + H_2{}^{16}O$	448	v	(3.8)	hr^{-1}	k = 2.39E-2		Lloyd (1968)
18 $S^{16}O_4^{2-} + H_2{}^{18}O \equiv S^{18}O_4^{2-} + H_2{}^{16}O$	25-198	l/v	(7.0)	hr^{-1}	40.2	0.45	Lloyd (1968)

II. Aqueous Species (continued)

	Isotopic Exchange Reaction (a)	Temperature Range (°C)	Pressure Range (MPa)	Catalyst or pH (b)	Rate Units	E_a (c)	A_o (d)	References (e)
19	$HS^{16}O_4^- + H_2^{18}O \equiv HS^{18}O_4^- + H_2^{16}O$	448	v	(7.0)	hr^{-1}	$k = 7.17E\text{-}3$		Lloyd (1968)
20	$S^{16}O_4^{2-} + H_2^{18}O \equiv S^{18}O_4^{2-} + H_2^{16}O$	198-325	l/v	(9.0)	hr^{-1}	47.7	0.36	Lloyd (1968)
21	$S^{18}O_3^{16}O^{2-} + H_2^{18}O \equiv S^{18}O_4^{2-} + H_2^{16}O$	100-200	100	2.3-5.5	kg/mole/hr	87.6	1.53E12	Chiba & Sakai (1985)
		300	100	6.1-7.3	kg/mole/hr	$k_2 = 0.0426$		Chiba & Sakai (1985)
22	$^{32}SO_4^{2-} + H_2^{34}S \equiv {}^{34}SO_4^{2-} + H_2^{32}S$	100-320	l/v	2.0-2.4	kg/mole/hr	77±4	1.59E8	Ohmoto & Lasaga (1982)
		228-320	l/v-100	4-7	kg/mole/hr	123.9±4	1.32E10	Ohmoto & Lasaga (1982)
		200-405	l/v(?)	8.5-9.1	kg/mole/hr	183±11	1.59E12	Ohmoto & Lasaga (1982)
23	$CH_3C^{16}O^{16}O^- + H_2^{18}O \equiv CH_3C^{16}O^{18}O^- + H_2^{16}O$	101	0.1	4.5	s^{-1}	$k_o = 3.98E\text{-}7 \; (I = 1M)$		Gamsjager & Murmann (1983)
24	$As^{16}O_4^{3-} + H_2^{18}O \equiv As^{18}O_4^{3-} + H_2^{16}O$	30	0.1	>10	s^{-1}	$k_o = 1.5E\text{-}6 \; (I = 0.55M)$		op. cit.
25	$Se^{16}O_3^{2-} + H_2^{18}O \equiv Se^{18}O_3^{2-} + H_2^{16}O$	0	0.1	8.7-12.5	s^{-1}	$k_o = 1.5E\text{-}4 \; (I = 0.16; 0.54 M)$		op. cit.
26	$Mo^{16}O_4^{2-} + H_2^{18}O \equiv Mo^{18}O_4^{2-} + H_2^{16}O$	25	0.1	11.5-13	s^{-1}	$k_o = 0.33$		op. cit.
27	$Mn^{16}O_4^- + H_2^{18}O \equiv Mn^{18}O_4^- + H_2^{16}O$	25	0.1	1-8	s^{-1}	$k_o = 1.9E\text{-}5 \; (I = 0.28M)$		op. cit.
28	$Cr^{16}O_4^{2-} + H_2^{18}O \equiv Cr^{18}O_4^{2-} + H_2^{16}O$	25	0.1	7-12	s^{-1}	$k_o = 3.2E\text{-}7 \; (I = 3.2E\text{-}7 \; (I = 0.2; 1M)$		op. cit.
29	$U^{16}O_2^{2+} + H_2^{18}O \equiv U^{18}O_2^{2+} + H_2^{16}O$	25	0.1	1-2	s^{-1}	$k_o = 9.9E\text{-}9 \; (I = 3.8M)$		op. cit.

(a) General isotopic reaction refers to the dominant isotopic species involved in the interaction. No mechanism of exchange is implied with these equations. Only simple examples of oxo-anions or oxo-cations are presented.

(b) Catalysts: including possible surfaces of reaction chamber which may have acted as catalysts during reaction. Material in ( ) are suspected as catalysts. pH of solution sometimes buffered; generally not. pH values in ( ) are estimated based on description of experimental solutions.

(c) E_a: Activation Energy in Arrhenius equation in kJ/mole, I is ionic strength; T in Kelvins.

(d) A_o: Pre-exponential factor; for units, refer to Rate Units column.

(e) * in front of reference refers to use of raw data to recalculate the rate constant Arrhenius relation.

Cole & Chakraborty

Table 2. Summary of rate data from

No.	Solid (a)	Temp Range (°C)	Pressure Range (Mpa) (b)	Sample Type (c)	Reacting Species (d)	Isotope (e)	
I. Framework Silicates							
A. Silica Group							
1	α Quartz	400-620	2.5	SNX	H_2O	H-D	
2	β Quartz	700-900	2.5	SNX	H_2O	H-D	
3	β Quartz	1010-1220	<0.1	PNX	O_2	^{18}O	
4	β Quartz	667	82	SNX	H_2O	^{18}O	
5	β Quartz	667	82	SNX	H_2O	^{18}O	
6	α-β Quartz	350-550	100	PNX	0.7NaF	^{18}O	
7	β Quartz	720-850	0.06(2.19E-5OH/SiO_2)	SSX	THO	^3H	
8	β Quartz	721-850	0.06(3.27E-5OH/SiO_2)	SSX	THO	^3H	
9	β Quartz	722-850	0.06(4.37E-5OH/SiO_2)	SSX	THO	^3H	
10	β Quartz	723-850	0.06(5.46E-5OH/SiO_2)	SSX	THO	^3H	
11	β Quartz	870-1180	2.1	SNX	O_2	^{18}O	
12	βTridymite	1070-1280	2.1	SNX	O_2	^{18}O	
13	α Quartz	250	1,500	PNX	H_2O	^{18}O	
14	α-β Quartz	400-700	1,500	PNX	H_2O	^{18}O	
15	β Quartz	820-1000	1,500	SSX	H_2O	-	
16	β Quartz	900	1,500	SNX	H_2O	-	
17	β Quartz	900	300	SNX	H_2O	-	
18	α Quartz	600;750	100	SNX	H_2O	^{18}O	
19	α Quartz	500-550	100	SNX	H_2O	^{18}O	
20	α Quartz	500-550	100	SNX	H_2O	^{18}O	
21	β Quartz	600-800	100	SNX	H_2O	^{18}O	
22	β Quartz	600-800	100	SNX	H_2O	^{18}O	
23	β Quartz	600-800	100	SNX	H_2O	^{18}O	
24	β Quartz	600-800	100	SSX	H_2O	^{18}O	
25	α-β Quartz	500-800	100	SNX	H_2O	^{18}O	
26	β Quartz	500-800	100	SNX	H_2O	^{18}O	
27	β Quartz	700	25-350	SN/SX	H_2O	^{18}O	
28	β Quartz	700-850	100 (NNO)	SNX	H_2O	^{18}O	
29	β Quartz	700-850	100 (NNO)	SNX	H_2O	^{18}O	
30	β Quartz	700-900	890-1550	SNX	H_2O	-	
31	β Quartz	900	1500(MBA)	ANX	H_2O	-	
32	β Quartz	700	100 (NNO)	SNX	H_2O	^{18}O	
33	β Quartz	800	100 (1:1 H_2/Ar)	SNX	H_2/Ar	^{18}O	
34	β Quartz	745-900	10	SNX	CO_2	^{18}O	

select mineral-fluid or mineral-gas systems.

Mechan. (f)	Ea kJ/mole (g)	A_o (or D_o) (h)	Transport Direction (i)	Method (j)	Reference (k)
TD	79.5	5.00E-05	//c	IR	Kats et al. (1962)
TD	175.7	5	//c	IR	Kats et al. (1962)
SD	230	3.70E-09	bulk	BE-MS	Haul and Dumgen (1962)
SD	$D = 4.1E-12$		//c	$^{18}O(p,\alpha)^{15}N$	Choudhury et al. (1965)
SD	$D = 8.4E-14$		nc	$^{18}O(p,\alpha)^{15}N$	Choudhury et al. (1965)
CR	62.8 ± 8.4	3.56E-04	isotropic	BE-MS	*Clayton et al. (1972)
TD	93.7	2.80E-07	//c	SS	Shaffer et al. (1974)
TD	100	6.40E-07	//c	SS	Shaffer et al. (1974)
TD	104.2	9.10E-07	//c	SS	Shaffer et al. (1974)
TD	108.4	8.50E-07	//c	SS	Shaffer et al. (1974)
SD	195	1.10E-10	n&//c	$^{18}O(p,n)^{18}F$	Schachtner and Sockel (1977)
SD	195	1.10E-09	n&//c	$^{18}O(p,n)^{18}F$	Schachtner and Sockel (1977)
CR	$R = 1.45 \pm 0.44E-7$		sphere	BE-MS	*Matsuhisa et al. (1978)
CR(?)	46 ± 4	3.46E-05	sphere	BE-MS	*Matsuhisa et al. (1979)
SD	469	4.30E+13	??	Optical(deformation)	Blacic (1981)
SD	$D = 1E-8$ to $1E-7$		//c	SS-IR	Mackwell and Paterson (1985)
SD	$D = 1E-15$ to $1E-14$		//c	SS-IR	Mackwell and Paterson (1985)
SD	133.9	1.10E-11	//c	SIMS	Freer and Dennis (1982)
SD	284 ± 92	190	//c	SIMS	Giletti and Yund (1984)
SD	238 ± 12	8.00E-02	n(10/1)	SIMS	Giletti and Yund (1984)
SD	142 ± 4	4.00E-07	//c	SIMS	Giletti and Yund (1984)
SD	155 ± 8	9.00E-07	n(10/1)	SIMS	Giletti and Yund (1984)
SD	234 ± 8	1.00E-04	nc	SIMS	Giletti and Yund (1984)
SD	205 ± 12	2.00E-06	nc	SIMS	Giletti and Yund (1984)
SD	171 ± 8	1.00E-05	//c	SIMS	Giletti and Yund (1984)
SD	184 ± 4	2.00E-05	n(10/1)	SIMS	Giletti and Yund (1984)
SD	$logD = 1.13(logfH_2O)-16.08$		//c	SIMS	*Giletti and Yund(1984)
SD	138 ± 19	2.09E-11	//c	SIMS	Dennis (1984)
SD	204 ± 2	3.86E-10	n(10/0)	SIMS	Dennis (1984)
SD	200 ± 20	1.40E-01	//c	SS-IR	Kronenberg et al. (1986)
SD	$D = 1E-6$ to E-7		//c	SS-IR	Rovetta et al. (1986)
SD	$D = 1.7E-14$		//c	SIMS	Elphick and Graham (1988)
SD	$D = 2.5E-14$		//c	SIMS	Elphick and Graham (1988)
SD	159 ± 13	2.10E-08	//c	SIMS	Sharp et al. (1991)

35	β Quartz	900	345	SNX	CO_2	^{18}O
36	β Quartz	900	727	SNX	CO_2	^{18}O
37	β Quartz	835	10	SNX	$XCO_2 = 0.85$	^{18}O
38	α Quartz	450-590	100	SNX	H_2O	^{18}O
39	β Quartz	600	150-350(MBA)	SNX	H_2O	^{18}O
40	β Quartz	700	5-350(MBA)	SNX	H_2O	^{18}O
B. Feldspar Group						
41	Albite	420-650	100	PN/SX	3mKCl	^{18}O
42	Albite	440-805	25-60	PSX	H_2O	^{18}O
43	Albite	600-800	200	PNX	$H_2O.NaCl$	^{18}O
44	Albite	350-805	100	SNX	H_2O	^{18}O
45	Albite	807	200	SNX	H_2O	^{18}O
46	Albite	807	200	SNX	H_2O	^{18}O
47	Albite	807	200	SNX	H_2O	^{18}O
48	Albite	400-700	700-1,500	PNX	H_2O	^{18}O
49	Albite	450-750	200	PNX	H_2O	^{18}O
50	Albite	450-750	200	PNX	H_2O	^{18}O
51	Albite	600	100	SNX	H_2O	^{18}O
52	Albite	600	50-800	SNX	H_2O	^{18}O
53	Albite	600	1,500	PNX	H_2O	^{18}O
54	Albite	500	200-1,500	PNX	H_2O	^{18}O
55	Albite	650	100-1,500(MBA)	SNX	H_2O	^{18}O
56	Albite	750-950	0.1	PNX	CO_2	^{18}O
57	Microcline	400-700	200	PNX	2M KCl	^{18}O
58	Microcline	250-290	4-7.5	PNX	1M NaCl	^{18}O
59	Sanidine	350-800	100	PSX	2-3M NaCl	^{18}O
60	Sanidine	550-850	100	SNX	H_2O	$^{18}O,^{17}O$
61	Sanidine	869-1053	0.1	SNX	O_2	^{18}O
62	Adularia	520-800	32.5-60	PSX	H_2O	^{18}O
63	Adularia	400-700	200	PNX	2M KCl	^{18}O
64	Adularia	712-843	0.1	PNX	CO_2	^{18}O
65	Adularia	1107	0.1	PNX	Air	^{18}O
66	Adularia	650	25-400	PNX	H_2O	^{18}O
67	Adularia	350-700	100	SNX	H_2O	^{18}O
68	Adularia	650	5-11	SNX	H_2O	^{18}O
69	Adularia	650	25-1,500(MBA)	SNX	H_2O	^{18}O
70	Adularia	500-900	0.1	SNX	Air	H
71	Oligoclase	550;800	100	SNX	H_2O	^{18}O
72	Labradorite	550;700	100	SNX	H_2O	^{18}O
73	Anorthite	1280-1480	0.1	PSX	CO_2,O_2	^{18}O

SD	D = 2.24E-15		//c	SIMS	Sharp et al. (1991)
SD	D = 8.13E-15		//c	SIMS	Sharp et al. (1991)
SD	D = 1.32E-15		//c	SIMS	Sharp et al. (1991)
SD	243 ± 17	2.90E-01	//c	SIMS	Farver & Yund (1991b)
SD	logD = 0.565(logfH$_2$O)-20.04		//c	SIMS	*Farver & Yund (1991b)
SD	logD = 0.783(logfH$_2$O)-19.3		//c	SIMS	*Farver & Yund (1991b)
CR	107.1	1.51E+02	cation ex.	BE-MS	*O'Neil & Taylor (1967)
SD	154.8	4.50E-05	sphere	BE-MS	Merigoux (1968)
SD	154.8 ± 8	2.50E-05	sphere	BE-MS	Anderson & Kasper (1975)
SD	87.5	1.69E-09	n(001)	SIMS	*Giletti et al. (1978)
SD	D = 2.11E-13		n(/11)	SIMS	Giletti et al. (1978)
SD	D = 1.74E-13		n(130)	SIMS	Giletti et al. (1978)
SD	D = 2.42E-13		n(010)	SIMS	Giletti et al. (1978)
CR(?)	20.2	3.1E-2	sphere	BE-MS	Matsuhisa et al. (1979)
SD	139.8 ± 3	9.80E-06	undeformed	BE-MS	Yund et al. (1981)
SD	129.3 ± 5	7.60E-06	deformed	BE-MS	Yund et al.(1981)
SD	D = 4.2E-15		n(001)	SIMS	Freer & Dennis (1982)
SD	D = 3.89E-15		n(001)	SIMS	*Freer & Dennis (1982)
SD(?)	D = 4.65E-13		sphere	BE-MS	Matthews et al. (1983a)
SD(?)	logD = 0.305(logfH$_2$O)-14.95		sphere	BE-MS	*Matthews et al. (1983a)
SD	logD = 0.732(logfH$_2$O)-15.145		n(001)	SIMS	*Farver & Yund (1990)
SD	90	2.00E-16	sphere	BE-MS	Matthews et al. (1994)
SD	123.9 ± 4	2.80E-06	sphere	BE-MS	Yund & Anderson (1974)
CR	76.6	4.99E-01	sphere/BET	BE-MS	Cole et al. (1992)
CR	73.22	0.3	cation ex.	BE-MS	*O'Neil & Taylor (1967)
SD	109.7 ± 4	3.95E-08	n(001)	SIMS	Freer et al. (1997)
SD	245 ± 15	8.40E-07	n(001)	SIMS	Derdau et al. (1998)
SD	133.9	9.00E-07	sphere	BE-MS	Merigoux (1968)
SD	123.9 ± 5	5.3E-7	sphere	BE-MS	Yund & Anderson (1974)
SD	D = 1E-16		sphere	BE-MS	Yund & Anderson (1974)
SD	D = 1E-15		sphere	BE-MS	Yund & Anderson (1974)
SD	logD = 0.604(logfH$_2$O)-15.16		sphere	BE-MS	*Yund & Anderson (1978)
SD	107.1	4.51E-08	n(001)	SIMS	Giletti et al. (1978)
SD	logD = 1.18(logfH$_2$O)-15.20		n(001)	SIMS	*Farver & Yund (1990)
SD	logD = 0.516(logfH$_2$O)-14.48		n(001)	SIMS	*Farver & Yund (1990)
TD	172 ± 15	6.20E-04	plate	IR	Kronenberg et al. (1996)
SD	129.1	3.49E-07	n(001)	SIMS	*Giletti et al. (1978)
SD	130.6	9.82E-07	n(001)	SIMS	*Giletti et al. (1978)
SD	344	1.59E-05	sphere	BE-MS	Muehlenbachs & Kushiro (1974)

74	Anorthite	350-800	100	SNX	H_2O	^{18}O	
75	Anorthite	600	700	PNX	H_2O	^{18}O	
76	Anorthite	850-1298	0.1	SNX	O_2	^{18}O	
77	Anorthite	1008-1295	0.1(QFM)	SNX	$CO-CO_2$	^{18}O	
78	Celsian	420-800	100	PSX	2M $CaCl_2$	^{18}O	
C. Nepheline Group							
79	Nepheline	1000-1300	0.1	PNX	CO_2	^{18}O	
80	Leucite	1000-1300	0.1	PNX	CO_2	^{18}O	
D. Zeolite Group							
81	Chabazite	32-64	<0.1	PNX	H_2O	D	
82	Gmelinite	32-65	<0.1	PNX	H_2O	D	
83	Heulandite	35-65	<0.1	PNX	H_2O	D	
84	Heulandite	20-75	<0.1	PNX	H_2O	-	
85	Heulandite	21-75	<0.1	PNX	H_2O	-	
86	Analcime	65-105	<0.1	PSX	THO	^3H	
87	Analcime	66-105	<0.1	PSX	0.2M NaCl	^3H	
88	Analcime	75	<0.1	PSX	H_2O	^{18}O	
89	Analcime	400;450	<0.1	PNX	H_2O	^{18}O	
90	Analcime	400;451	<0.1	PNX	H_2O	^{18}O	
91	Analcime	400	<0.1	PNX	H_2O	^{18}O	
92	Analcime	400	<0.1	PNX	H_2O	^{18}O	
93	Stilbite	140-220	<0.1	PNX	H_2O	^{18}O	
94	Stilbite	140-220	<0.1	PNX	H_2O	^{18}O	
95	Na-Stilbite	25-65	0.1	PNX	THO	^3H	
96	Ca-Stilbite	25-65	0.1	PNX	THO	^3H	
97	Zeolite 4A	-73to177	<0.01-0.79	PSX	O_2	-	
98	Zeolite 5A	-73to177	<0.01-0.79	PSX	O_2	-	
99	Zeolite 5A	-99to30	<0.01-0.79	PSX	O_2	-	
100	K-Natrolite	25-65	0.1	PNX	THO	^3H	
101	Natrolite	25	0.1-800	PNX	H_2O	-	
102	Mesolite	25	0.1-800	PNX	H_2O	-	
103	Scolecite	25	0.1-800	PNX	H_2O	-	
II. Sheet Silicates							
A. Mica Group							
104	Paragonite	350-600	100-150	PSX	2-3M KCl	^{18}O	
105	Muscovite	450-750	100(?)	PNX	H_2O	H	
106	Muscovite	450-750	100(?)	PNX	H_2O	H	
107	Muscovite	512-700	100	PNX	H_2O	^{18}O	
108	Muscovite	600;700	100	PNX	H_2O	^{18}O	
109	Muscovite	300	0.04-0.2	PNX	H_2	H	

SD	109.6	1.39E-07	n(001)	SIMS	Giletti et al. (1978)
SD(?)	D = 2.443E-12		sphere	BE-MS	Matthews et al. (1983a)
SD	236 ± 8	1.00E-05	n(001)	SIMS	Elphick et al. (1988)
SD	162 ± 36	8.40E-09	//(010)	SIMS	Ryerson & McKeegan (1994)
CR	91.6	0.59	cation ex.	BE-MS	*O'Neil & Taylor (1967)
SD	104.6 ± 10	5.90E-09	sphere	BE-MS	Connolly & Muehlenbachs (1988)
SD	58.6 ± 12.5	1.30E-11	sphere	BE-MS	Muehlenbachs & Connolly (1991)
SD	36.4 ± 1.2	1.20E-01	rhomb	D-H exch./Gravimetry	Barrer & Fender (1961)
SD	33.9 ± 1.2	2.00E-02	rhomb	D-H exch./Gravimetry	Barrer & Fender (1961)
SD	46 ± 1.2	7.60E-01	plate	D-H exch./Gravimetry	Barrer & Fender (1961)
SD	21.4	1.75E-03	n(201)	Light Refraction	*Tiselius (1935)
SD	38.4	0.151	n(001)	Light Refraction	*Tiselius (1935)
SD	60.1	1.13E-03	sphere	Liquid Scintillation	*Dyer & Molyneux (1968)
SD	54.2	1.64E-04	sphere	Liquid Scintillation	*Dyer & Molyneux (1968)
SD	D = 2.64E-12		sphere	BE-MS	Dyer & Molyneux (1968)
SD(CR)	83.9 ± 8	1.55E-03	sphere	BE-MS	Feng & Savin (1993)
CR(SD)	28.9 ± 21	2.18E+08	sphere	BE-MS	Feng & Savin (1993)
SD(CR)	D = 2.8 ± 0.3E-9		sphere	BE-MS	Faiia & Feng (2000)
CR(SD)	kSv = 0.011cm^3/mol/s		sphere	BE-MS	Faiia & Feng (2000)
SD(CR)	66.5 ± 4	4.50E-05	sphere	BE-MS	Feng & Savin (1993)
CR(SD)	149.7 ± 13	1.30E+14	sphere	BE-MS	Feng & Savin (1993)
SD	19	1.63E-07	sphere	Liquid Scintillation	*Dyer & Faghihian (1998b)
SD	11.8	1.98E-08	sphere	Liquid Scintillation	*Dyer & Faghihian (1998b)
SD	18.95	6.60E-06	inf. media	Gravimetry	Ruthven & Derrah (1975)
SD	4.18	2.60E-09	inf. media	Gravimetry	Ruthven & Derrah (1975)
SD	15.3	0.3087	sphere	ZLC-Therm.Conduct.	*Ruthven & Xu (1993)
SD	25.4	1.41E-06	sphere	Liquid Scintillation	*Dyer & Faghihian (1998a)
SD	logD = 0.623(logfH$_2$O)-9.28		?	NMR	*Moroz et al. (2001)
SD	logD = 0.344(logfH$_2$O)-10.28		?	NMR	*Moroz et al. (2001)
SD	logD = 0.177(logfH$_2$O)-11.01		?	NMR	*Moroz et al. (2001)
CR	61.9	2.07E-04	cation ex	BE-MS	*O'Neil & Taylor (1969)
SD	119.7	1.00E-07	//c (plate)	BE-MS	**Suzuoki & Epstein (1976)
SD	121.3	1.05E+05	nc (cyl)	BE-MS	**Suzuoki & Epstein (1976)
SD	163.2 ± 21	7.74E-05	nc (cyl)	BE-MS	Fortier & Giletti (1991)
SD	212	3.74E-06	//c	SIMS	*Fortier & Giletti (1991)
SD	D = 6.1E-15		nc (cyl)	BE-MS	Vennemann et al. (1996)

110	Muscovite	400	0.2	PNX	H_2	H	
111	Muscovite	350	25	PNX	H_2O	^{18}O	
112	Muscovite	350-550	25,100	PNX	H_2O	^{18}O	
113	Biotite	450-800	100(?)	PNX	H_2O	H	
114	Biotite	450-800	100(?)	PNX	H_2O	H	
115	Biotite	500-800	100	PNX	H_2O	^{18}O	
116	Biotite	600;700	100	PNX	H_2O	^{18}O	
117	Biotite	250-300	35	PNX	0.1mNaCl	^{18}O	
118	Biotite	300	0.2	PNX	H_2	H	
119	Biotite	400	0.2	PNX	H_2	H	
120	Biotite	350	25	PNX	H_2O	^{18}O	
121	Biotite	350-500	25,100	PNX	H_2O	^{18}O	
122	Phlogopite	500-800	200	PNX	H_2O	^{18}O	
123	Phlogopite	500-800	200	PNX	H_2O	^{18}O	
124	Phlogopite	575-650	100(?)	PNX	H_2O	H	
125	Phlogopite	575-650	100(?)	PNX	H_2O	H	
126	Phlogopite	600-900	100	PNX	H_2O	^{18}O	
127	Phlogopite	800;900	100	PNX	H_2O	^{18}O	
128	Phlogopite	700	20-200	PNX	H_2O	^{18}O	
129	Lepidolite	650	100(?)	PNX	H_2O	H	
130	Lepidolite	650	100(?)	PNX	H_2O	H	
	B. Serpentine & Chlorite Groups						
131	Serpentine	400	100(?)	PNX	H_2O	H	
132	Serpentine	400	100(?)	PNX	H_2O	H	
133	Serpentine	100-300	200	PNX	H_2O	H	
134	Serpentine	100-300	200	PNX	H_2O	H	
135	Chlorite	500-700	200-500	PNX	H_2O	H	
136	Chlorite	500-700	200-500	PNX	H_2O	H	
137	Chlorite	500	155	PNX	D_2O	D	
138	Chlorite	500	250	PNX	D_2O	D	
139	Chlorite	350	25	PNX	H_2O	^{18}O	
140	Chlorite	350-550	25,100	PNX	H_2O	^{18}O	
	C. Clay Group						
141	Kaolinite	400	100(?)	PNX	H_2O	H	
142	Kaolinite	400	100(?)	PNX	H_2O	H	
143	Kaolinite	100;200	100(?)	PNX	0.04N NaCl	H	
144	Kaolinite	100;200	100(?)	PNX	0.04N NaCl	H	
145	Kaolinite	100;200	100(?)	PNX	0.04N NaCl	^{18}O	
146	Kaolinite	100;200	100(?)	PNX	0.04N NaCl	^{18}O	
147	Kaolinite	350	100(?)	PNX	0.04N NaCl	^{18}O	

SD	D = 2.95E-14		nc (cyl)	BE-MS	Vennemann et al. (1996)
CR	R = 3.81E-11		plate/BET	BE-MS	Cole (2000)
CR	63.8	1.50E-05	plate/BET	BE-MS	Cole (1996, unpublished)
SD	116.3	3.40E-07	//c (plate)	BE-MS	*Suzuoki & Epstein (1976)
SD	122.6	7.60E-04	nc (cyl)	BE-MS	*Suzuoki & Epstein (1976)
SD	142.3 ± 8	9.10E-06	nc (cyl)	BE-MS	Fortier & Giletti (1991)
SD	149.1	2.22E-08	//c	SIMS	*Fortier & Giletti (1991)
CR (to chl)	46.03	2.94E-03	plate	BE-MS	Cole et al. (1992)
SD	D = 4.0E-14		nc (cyl)	BE-MS	Vennemann et al. (1996)
SD	D = 5.0E-13		nc (cyl)	BE-MS	Vennemann et al. (1996)
CR	R = 8.92E-11		plate/BET	BE-MS	Cole (2000)
CR	53	6.53E-06	plate/BET	BE-MS	Cole (1996, unpublished)
SD	121.3 ± 8	1.03E-09	//c (plate)	BE-MS	Giletti & Anderson (1975)
SD	150.6 ± 8	1.20E-05	nc (cyl)	BE-MS	Giletti & Anderson (1975)
SD	128.6	7.85E-07	//c (plate)	BE-MS	*Suzuoki & Epstein (1976)
SD	155.2	0.02226	nc (cyl)	BE-MS	*Suzuoki & Epstein (1976)
SD	175.7 ± 13	1.40E-04	nc (cyl)	BE-MS	Fortier & Giletti (1991)
SD	249.1	1.02E-05	//c	SIMS	*Fortier & Giletti (1991)
SD	D = 5.248E-14		nc (cyl)	BE-MS	Fortier & Giletti (1991)
SD	D = 6.37E-12		//c (plate)	BE-MS	*Suzuoki & Epstein (1976)
SD	D = 8.05E-15		nc (cyl)	BE-MS	*Suzuoki & Epstein (1976)
SD	D = 3.41E-11		//prism	BE-MS	*Suzuoki & Epstein (1976)
SD	D = 3.39E-12		n prism	BE-MS	*Suzuoki & Epstein (1976)
SD	48.8	1.25E-06	//prism	BE-MS	*Sakai & Tsutsumi (1978)
SD	47.2	7.61E-08	n prism	BE-MS	*Sakai & Tsutsumi (1978)
SD	166.9	4.79E-08	//c (plate)	BE-MS	Graham et al. (1987)
SD	171.7	6.17E-06	nc (cyl)	BE-MS	Graham et al. (1987)
SD	D = 8.05E-16		//c (plate)	SIMS	Cole & Riciputi (unpublished)
SD	D = 9.84E-16		//c (plate)	SIMS	Cole & Riciputi (unpublished)
CR	R = 1.87E-10		plate/BET	BE-MS	Cole (2000)
CR	49.1	6.51E-06	plate/BET	BE-MS	Cole (1996, unpublished)
SD	D = 5.03E-15		//c (plate)	BE-MS	*Suzuoki & Epstein (1976)
SD	D = 4.59E-12		nc (cyl)	BE-MS	*Suzuoki & Epstein (1976)
SD	74.8	2.13E-08	//c (plate)	BE-MS	*O'Neil & Kharaka (1976)
SD	60.9	1.39E-08	nc (cyl)	BE-MS	*O'Neil & Kharaka (1976)
SD(?)	36.2	9.18E-15	//c (plate)	BE-MS	*O'Neil & Kharaka (1976)
SD(?)	32.8	7.76E-14	nc (cyl)	BE-MS	*O'Neil & Kharaka (1976)
CR	R = 2.66E-9		plate	BE-MS	*O'Neil & Kharaka (1976)

148	Kaolinite	350	100(?)	PNX	0.04N NaCl	H
149	Kaolinite	200-352	0.032-0.126	PNX	H_2O	H
150	Kaolinite	200-352	0.032-0.126	PNX	H_2O	H
151	Kaolinite	150-275	0.03-0.08	PNX	H_2	H
152	Illite	100-350	100(?)	PNX	0.04N NaCl	H
153	Illite	100-350	100(?)	PNX	0.04N NaCl	H
154	Illite	100-350	100(?)	PNX	0.04N NaCl	^{18}O
155	Illite	100-350	100(?)	PNX	0.04N NaCl	^{18}O
156	Montmor	100;200	100(?)	PNX	0.04N NaCl	H
157	Montmor	100;200	100(?)	PNX	0.04N NaCl	H
158	Montmor	350	100(?)	PNX	0.04N NaCl	H

III. Chain Silicates

    A. Pyroxene Group

159	Enstatite	1280	0.1	PSX	O_2	^{18}O
160	Diopside	1280	0.1	PSX	O_2	^{18}O
161	Diopside	1280	0.1	PSX	CO_2	^{18}O
162	Diopside	600-800	1300-1500	PSX	H_2O	^{18}O
163	Diopside	1150-1350	0.1	PSX	CO_2	^{18}O
164	Diopside	700-1200	100	SNX	H_2O	^{18}O
165	Diopside	1000-1200	100	SNX	H_2O	^{18}O
166	Diopside	900-1200	100	SNX	H_2O	^{18}O
167	Diopside	1104-1251	0.1(NNO)	SNX	CO_2	^{18}O
168	Diopside	800	700-800(?)	PNX	$CaCO_3$	^{18}O
169	Diopside	1200-1370	1E-3 to 1E-14	SSX	$H_2O(v)$	^{18}O
170	Diopside	600-900	0.01	SNX	D_2	D
171	Diopside	600-900	0.01	SNX	D_2	D
172	Diopside	700-850	0.1($logfO_2$ = -14.1)	NSX	$H_2O(v)$	-
173	Diopside	700-850	0.1($logfO_2$ = -14.1)	NSX	$H_2O(v)$	-
174	Diopside	750	0.1($logfO_2$ = -14.1)	NSX	$H_2O(v)$	-
175	Diopside	800	0.1($logfO_2$ = -14.1)	NSX	$H_2O(v)$	-
176	Wollastonite	500-800	900-1500	PNX	H_2O	^{18}O

    B. Amphibole Group

177	Tremolite	650-850	200-400	PNX	H_2O	H
178	Tremolite	350-800	200-400	PNX	H_2O	H
179	Tremolite	650-800	100	SNX	H_2O	^{18}O
180	Actinolite	400-700	100(?)	PNX	H_2O	H
181	Actinolite	400-700	100(?)	PNX	H_2O	H
182	Hornblende	350-550	100(?)	PNX	H_2O	H
183	Hornblende	350-550	100(?)	PNX	H_2O	H
184	Hornblende	650-800	100	SNX	H_2O	^{18}O

CR	R = 4.67E-8		plate	BE-MS	*O'Neil & Kharaka (1976)
SD	83.7	8.70E-09	//c (plate)	BE-MS	*Lui & Epstein (1984)
SD	72.4	5.50E-08	nc (cyl)	BE-MS	*Lui & Epstein (1984)
SD	80	9.60E-06	nc (cyl)	BE-MS	Vennemann et al. (1996)
SD	34.72	6.60E-13	//c (plate)	BE-MS	*O'Neil & Kharaka (1976)
SD	36.4	1.80E-11	nc (cyl)	BE-MS	*O'Neil & Kharaka (1976)
SD(?)	13.4	5.01E-17	//c (plate)	BE-MS	*O'Neil & Kharaka (1976)
SD(?)	16.8	3.39E-15	nc (cyl)	BE-MS	*O'Neil & Kharaka (1976)
SD	49.8	6.40E-10	//c (plate)	BE-MS	*O'Neil & Kharaka (1976)
SD	52.7	4.45E-08	nc (cyl)	BE-MS	*O'Neil & Kharaka (1976)
CR	R = 1.021E-8		plate	BE-MS	*O'Neil & Kharaka (1976)

SD	D = 6.0E-12		sphere	BE-MS	Muehlenbachs & Kushiro (1974)
SD	D = 1.8E-12		sphere	BE-MS	Muehlenbachs & Kushiro (1974)
SD	D = 2.9E-12		sphere	BE-MS	Muehlenbachs & Kushiro (1974)
CR(?)	71.5	2.34E-04	sphere	BE-MS	*Matthews et al. (1983b)
SD	404.6	6.3	sphere	BE-MS	Connolly & Muehlenbachs (1988)
SD	225.9 ± 21	1.50E-06	//c	SIMS	Farver (1989)
SD	225.9	2.80E-08	nc	SIMS	Farver (1989)
SD	351 ± 21	9.00E-03	//c	SIMS	Elphick & Graham (1990)
SD	457 ± 26	4.3	//c	SIMS	Ryerson & McKeegan (1994)
SD	D = 5.6E-20		sphere	BE-MS(empirical)	Sharp & Jenkin (1994)
SD	310 ± 30	6.31E-06	//b	$^{18}O(p,\alpha)^{15}N$	Pacaud et al. (1999)
SD	149 ± 16	3.98	//(001;100)	FTIR	Hercule & Ingrin (1999)
SD	143 ± 33	1.00E-01	//(010)	FTIR	Hercule & Ingrin (1999)
SD	181 ± 38	79.43	//(100)	FTIR	Woods et al. (2000)
SD	153 ± 32	3.981	//(001)	FTIR	Woods et al. (2000)
SD	D = 1.5 (± 0.5)E-8		//(010)	FTIR	Woods et al. (2000)
SD	D = 2 (± 1)E-7		//(010)	FTIR	Woods et al. (2000)
CR(?)	66.1	2.81E-04	sphere	BE-MS	*Matthews et al. (1983b)

SD	71.1	8.88E-09	//c prism	BE-MS	Graham et al. (1984)
SD	72.4	1.77E-07	//c prism	BE-MS	Graham et al. (1984)
SD	163.2 ± 21	2.00E-08	//c	SIMS	Farver & Giletti (1985)
SD	102.5	2.39E-08	//c prism	BE-MS	*Suzuoki & Epstein (1976)
SD	98.75	6.30E-06	nc prism	BE-MS	Graham (1981)
SD	79.5	1.58E-07	//c prism	BE-MS	Graham et al. (1984)
SD	84.1	2.39E-08	nc prism	BE-MS	Graham et al. (1984)
SD	171.6 ± 25	1.00E-07	//c prism	SIMS	Farver & Giletti (1985)

185	Hornblende	800	100	SNX	H_2O	^{18}O	
186	Hornblende	800	100	SNX	H_2O	^{18}O	
187	Hornblende	800	20-200	SNX	H_2O	^{18}O	
188	Richterite	650-800	100	SNX	H_2O	^{18}O	

IV. Ortho & Ring Silicates

A. Olivine Group

189	Forsterite	1280	0.1	PSX	O_2	^{18}O	
190	Forsterite	1000	400	SSX	H_2O	^{18}O	
191	Forsterite	1275-1560	0.1	SNX	O_2	^{18}O	
192	Forsterite	1275-1628	0.1	SNX	O_2	^{18}O	
193	Forsterite	1425-1628	0.1	SNX	O_2	^{18}O	
194	Forsterite	1150-1600	$0.1(\log PO_2 = -9.2;-4.4)$	SNX	$Ar/H_2O(v)$	^{18}O	
195	Forsterite	1472-1734	0.02	SSX	O_2	^{18}O	
196	Forsterite	1000-1500	$PO_2 = 0.02$	SSX	O_2	^{18}O	
197	Forsterite	1300-1600	$0.1(\log PO_2 = -11$ to $-4)$	SSX	$H_2/H_2O(v)$	^{18}O	
198	Forsterite	1325	0.1	SSX	O_2	^{18}O	
199	Forsterite	1300	$0.1(\log PO_2 = -8.3$ to $-7)$	SNX	$H_2/H_2O(v)$	^{18}O	
200	Forsterite	1300	$0.1(\log PO_2 = -8.3)$	SNX	$H_2/H_2O(v)$	^{18}O	
201	Fo(0.90)	1300	$0.1(\log PO_2 = -9.5)$	SNX	$H_2/H_2O(v)$	^{18}O	
202	Fo(0.90)	1090-1500	$0.1(\log PO_2 = -12$ to $-7)$	SNX	$H_2/H_2O(v)$	^{18}O	
203	Fo(0.92)	1198-1401	0.1(NNO;IW)	SNX	CO/CO_2	^{18}O	
204	Fo(0.91)	800-1000	300(IW)	SNX	H_2O	-	
205	Fo(0.91)	800-1000	300(IW)	SNX	H_2O	-	
206	Fo(0.91)	800-1000	300(IW)	SNX	H_2O	-	
207	Forsterite	1100	0.1(?)	SSX	CO_2	^{18}O	
208	Forsterite	1200	0.1(?)	SSX	CO_2	^{18}O	
209	Fo(0.91)	800-1000	300(IW;NNO)	SNX	H_2O	-	
210	Fo(0.91)	800-1000	300(IW;NNO)	SNX	H_2O	-	
211	Fo(0.91)	800-1000	300(IW;NNO)	SNX	H_2O	-	

B. Zircon & Sphene

212	Zircon	1100-1500	0.1	SNX	$Si^{18}O_2$	^{18}O	
213	Zircon	1100-1500	7-70	SNX	H_2O/SiO_2	^{18}O	
214	Titanite	700-900	100	SNX	H_2O	^{18}O	

C. Garnet Group

215	Hessonite	850	200	SNX	H_2O	^{18}O	
216	Hessonite	1050	800	SNX	H_2O	^{18}O	
217	Almandine	800-1000	100	SNX	H_2O	^{18}O	

SD	D = 4.05E-17		//a	SIMS	Farver & Giletti (1985)
SD	D = 2.1E-17		//b	SIMS	Farver & Giletti (1985)
SD	logD = 0.434(logfH$_2$O)-16.08		//c prism	SIMS	Farver & Giletti (1985)
SD	238.5 ± 8	3.00E-04	//c prism	SIMS	Farver & Giletti (1985)

SD	D = 1.1E-14		sphere	BE-MS	Muehlenbachs & Kushiro (1974)
SD	D = 2.0E-16		n(110)	SIMS	Giletti et al. (1979)
SD	324.3	9.30E-05	//(100)	^{18}O(p,α)^{15}N	*Reddy et al. (1980)
SD	364.4	1.86E-03	//(010)	^{18}O(p,α)^{15}N	*Reddy et al. (1980)
SD	379.9	1.38E-02	//(001)	^{18}O(p,α)^{15}N	*Reddy et al. (1980)
SD	328.4	1.46E-04	isotropic?	SIMS; ^{18}O(p,α)^{15}N	*Jaoul et al. (1980)
SD	416	2.85E-02	sphere	BE-MS	Ando et al. (1981)
SD	377.8	5.90E-04	unknown	SIMS	*Hallwig et al. (1982)
SD	292.9	2.30E-06	isotropic?	^{18}O(p,α)^{15}N	Jaoul et al. (1983)
SD	D = 1.8E-11		slab(?)	^{18}O(p,n)^{17}F	Condit et al. (1985)
SD	D = 2.56E-15		//(100)	^{18}O(p,α)^{15}N	Houlier et al. (1988)
SD	D = 6.26E-15		//(010)	^{18}O(p,α)^{15}N	Houlier et al. (1988)
SD	D = 4.31E-14		//(001)	^{18}O(p,α)^{15}N	Houlier et al. (1988)
SD	318 ± 17	-2.874 + 0.34 x log(PO_2)	//(001)	^{18}O(p,α)^{15}N	Gerard & Jaoul (1989)
SD	266 ± 11	-5.585 + 0.21 x log(PO_2)	//(100;010)	^{18}O(p,α)^{15}N	Ryerson et al. (1989)
SD	130 ± 30	0.6 ± 0.3	//(100)	FTIR	Mackwell & Kohlstedt (1990)
SD	130 ± 30	0.05 ± 0.04	//(010)	FTIR	Mackwell & Kohlstedt (1990)
SD	130 ± 30	5.0 (± 4.0)E-3	//(001)	FTIR	Mackwell & Kohlstedt (1990)
SD	D = 2.76 (± 0.9)E-17		//c	SIMS	*Yurimoto et al. (1992)
SD	D = 8.15 (± 1.5)E-17		//c	SIMS	*Yurimoto et al.(1992)
SD	145 ± 30	1.30	//(100)	FTIR	*Kohlstedt & Mackwell (1998)
SD	110 ± 50	1.68E-03	//(010)	FTIR	*Kohlstedt & Mackwell (1998)
SD	110 ± 50	1.72E+00	//(001)	FTIR	*Kohlstedt & Mackwell (1998)

SD	448.3	1.33	//;nc	^{18}O(p,α)^{15}N	Watson & Cherniak (1997)
SD	210.2	5.50E-08	//;nc	^{18}O(p,α)^{15}N	Watson & Cherniak (1997)
SD	254 ± 28	1.00E-04	//a,b,c	SIMS	Morishita et al. (1996)

SD	D = 4.8(± 0.3)E-17			SIMS	Freer & Dennis (1982)
SD	D = 2.5(± 0.1)E-16			SIMS	Freer & Dennis (1982)
SD	301 ± 46	5.99E-05	isotropic	SIMS	Coughlan (1990); via Brady (1995

218	Pyrope	700-950	0.1(Air:N_2)	SNX	H_2O	
	Py0.67Al0.18Gr0.14					
219	Pyrope	700-950	0.1(Air:N_2)	SNX	H_2O	
	Py0.69Al0.16Gr0.14					
220	$Y_3Fe_5O_{12}$	1100-1473	0.1(?)	S/PSX	O_2	
221	$Y_3Fe_5O_{12}$	800-1100	0.1(?)	S/PSX	O_2	[18]O
222	YAG	1060-1550	0.1(?)	PSX	O_2	[18]O
223	YAG	1190-1490	0.1(?)	PSX(O_2)	O_2	[18]O
224	YAG	1290	0.1(?)	PSX(N_2)	O_2	[18]O
225	YAG	1410	0.1(?)	PSX(N_2)	O_2	[18]O
226	YAG	1140-1430	0.1(?)	PSX(Al)$_\lambda$	O_2	[18]O
227	YAG	1100-1385	0.0173	PSX	O_2	[18]O
228	YAG	1100-1385	0.0173	PSX	O_2	[18]O
229	$Yb_3Fe_5O_{12}$	825-1100	0.005	S/PNX	O_2	[18]O
230	$Yb_3Fe_5O_{12}$	1100-1400	0.005	S/PNX	O_2	[18]O
D. Aluminum Silicate Group						
231	Mullite	1100-1300	0.0066	SSX	O_2	[18]O
E. Epidote-Zoisite Group						
232	Epidote	200-350	200-400	PNX	H_2O	H
233	Epidote	450-650	400	PNX	H_2O	H
234	Epidote	200-350	200-400	PNX	H_2O	H
235	Epidote	450-650	400	PNX	H_2O	H
236	Epidote	200-650	200-400	PNX	1-4M NaCl	H
237	Epidote	200-650	200-400	PNX	1-4M NaCl	H
238	Epidote	200-600	200	SNX	D_2O	D
239	Zoisite	350-650	200-400	PNX	H_2O	H
240	Zoisite	350-650	200-400	PNX	H_2O	H
241	Zoisite	200-650	200-400	PNX	H_2O	H
242	Zoisite	200-650	200-400	PNX	H_2O	H
243	Zoisite	600-700	1340	PNX	H_2O	[18]O
244	Zoisite	600-700	1340	PNX	H_2O	[18]O
245	Zoisite	400-600	1340	PNX	H_2O	[18]O
246	Zoisite	400-600	1340	PNX	H_2O	[18]O
247	Ilvaite	350-650	9-20	PNX	H_2O	H
248	Ilvaite	350-650	9-20	PNX	H_2O	H
F. Melilite Group						
249	Ak0.5Gh0.5	700-1300	0.1	PSX	CO_2	[18]O
250	Ak0.75Gh0.25	700-1300	0.1	PSX	CO_2	[18]O
251	Akermanite	800-1303	0.1(NNO:IW)	SSX	CO/CO_2	[18]O

TD	254 ± 12	1.77E+04		FTIR	Wang et al. (1996)
TD	241 ± 32	1.36E+04		FTIR	Wang et al. (1996)
SD	273	0.4		BS-MS	Paladino et al. (1964)
SD	321.7	0.0912		SIMS	*Kilner et al. (1984)
SD	325	5.24E-03		BE-MS	Haneda et al. (1984)
SD	315	2.34E-04		BE-MS	Haneda et al. (1984)
SD	D = 5.86E-18			BE-MS	*Haneda et al. (1984)
SD	D = 3.08E-17			BE-MS	*Haneda et al. (1984)
SD	297	8.13E-03		BE-MS	Haneda et al. (1984)
SD	304.1	4.30E-04	isotropic	SIMS	Sakaguchi et al. (1996)
SD	265.8	2.60E-06	isotropic	SIMS	Sakaguchi et al. (1996)
SD	67	2.39E-10		SIMS; BE-MS	Haneda et al. (1993)
SD	511	8.74E+06		SIMS; BE-MS	Haneda et al. (1993)
SD	397 ± 45	1.32(± 0.4)E-2	inf. cyl	BE-MS	Ikuma et al. (1999)
SD	128	1.23	nc (prism)	BE-MS	Graham (1981)
SD	57.7	3.31E-06	nc (prism)	BE-MS	Graham (1981)
SD	128.5	9.332	//c (prism)	BE-MS	Graham (1981)
SD	52.3	9.77E-06	//c (prism)	BE-MS	Graham (1981)
CR(?)	92.9	3.47	plate	BE-MS	*Graham & Sheppard (1980)
CR(?)	92.9	19.1	cylinder	BE-MS	*Graham & Sheppard (1980)
SD	81.38	1.55E-09	//b	SIMS	De et al.(2000)
SD	102.5	4.47E-05	nc (prism)	BE-MS	Graham (1981)
SD	100	1.62E-04	//c (prism)	BE-MS	Graham (1981)
CR(?)	52.3	3.23E-05	nc (prism)	BE-MS	*Graham et al. (1980)
CR(?)	52.3	1.74E-04	//c (prism)	BE-MS	*Graham et al. (1980)
CR(?)	61.5	4.40E-05	nc (prism)	BE-MS	*Matthews et al. (1983c)
CR(?)	61.5	4.00E-06	//c (prism)	BE-MS	*Matthews et al. (1983c)
CR(?)	59.4	2.60E-03	nc (prism)	BE-MS	*Matthews et al. (1983c)
CR(?)	59.4	2.20E-04	//c (prism)	BE-MS	*Matthews et al. (1983c)
SD	118.5	4.27E-04	nc (prism)	BE-MS	Yaqian & Jibao (1993)
SD	115.5	9.68E-04	//c (prism)	BE-MS	Yaqian & Jibao (1993)
SD	140.2	8.60E-06	sphere	BE-MS	Hayashi & Muehlenbachs (1986)
SD	133.5	7.20E-06	sphere	BE-MS	Hayashi & Muehlenbachs (1986)
SD	278 ± 33	4.70E-07	nc	SIMS	Ryerson & McKeegan (1994)

252	Akermanite	1000-1300	0.0064	SSX	CO_2	^{18}O
253	Akermanite	1000-1300	0.0064	SSX	CO_2	^{18}O
254	Gehlenite	1000-1300	0.0064	SSX	CO_2	^{18}O
G. Tourmaline						
255	Schorlite	450-800	15-25	PNX	H_2O	H
256	Schorlite	450-800	15-25	PNX	H_2O	H
V. Oxides						
257	MgO	1100-1400	0.1	PSX	O_2	^{18}O
258	MgO	1100-1400	0.1	PSX	O_2	^{18}O
259	MgO	1020-1450	0.1	PSX	O_2	^{18}O
260	MgO	1020-1280	0.1	PSX	O_2	^{18}O
261	MgO	1280-1450	0.1	PSX	O_2	^{18}O
262	MgO	1500-1750	0.1(?)	SSX	O_2	^{18}O
263	MgO	1307-1547	0.1	SSX	O_2	^{18}O
264	MgO	1000-1650	0.1(?)	SSX(^{18}O)	$Mg^{16}O$	^{18}O
265	MgO	1300-1500	0.1(^{16}O in air)	SSX(^{18}O)	O_2	^{18}O
266	MgO	845-974	0.1	SSX	O_2	^{18}O
267	Corundum	1580-1840	0.1(?)	SSX	O_2	^{18}O
268	Corundum	1585-1840	vacuum	SSX	^{18}O film	^{18}O
269	Corundum	1650-1780	0.02	PSX	O_2	^{18}O
270	Corundum	1400-1650	0.02	PSX	O_2	^{18}O
271	Sapphire	1477-1677	0.1(?)	SSX	O_2	^{18}O
272	Corundum	1220-1470	?	SSX	O_2	^{18}O
273	Corundum	1000-1100	vapor (?)	SSX	D_2O	D
274	Corundum	1500-1720	0.2-0.5	<u>SSX</u>	O_2	^{18}O
275	Y-Al_2O_3	1110-1630	0.2-0.5	<u>SSX</u>	O_2	^{18}O
276	Rutile	550-700	0.1	SSX	H_2	H
277	Rutile	302-550	0.1	SSX	H_2	H
278	Rutile	806	0.1(?)	SSX	O_2	^{18}O
279	Rutile	806	0.1(?)	SSX	O_2	^{18}O
280	Rutile	1000	0.1(?)	PSX	O_2	^{18}O
281	Rutile	966	0.1(?)	PSX	O_2	^{18}O
282	Rutile	877-1177	0.006	SSX	O_2	^{18}O
283	Rutile	900-1200	$PO_2 = 0.0049$-0.02	SSX	O_2	^{18}O
284	Rutile	600	100	SNX	H_2O	^{18}O
285	Rutile	1050	100	SNX	H_2O	^{18}O
286	Rutile	600-1100	100	SNX	H_2O	^{18}O
287	Rutile	600-1100	100	SSX	H_2O	^{18}O
288	Rutile	750-1000	100 (Ti reduced by H_2O)	<u>SSX</u>	H_2O	^{18}O
289	Rutile	750-1000	0.1-100 (non-reduced) (or reduced, in air)	<u>SSX</u>	H_2O or dry	^{18}O

SD	215 ± 51	9.41E-06	//c	SIMS	Yurimoto et al. (1989)
SD	300 ± 37	6.96E-02	//a	SIMS	Yurimoto et al. (1989)
SD	186 ± 16	4.36E-08	//c	SIMS	Yurimoto et al. (1989)
SD	122.9	2.29E-06	//c (plate)	BE-MS	Jibao & Yaqian (1997)
SD	127.9	1.10E-06	nc (cyl)	BE-MS	Jibao & Yaqian (1997)
SD	252	4.50E-05	sphere	BE-MS(80-115µm)	Hashimoto et al. (1972)
SD	252	1.30E-05	sphere	BE-MS(170-200µm)	Hashimoto et al. (1972)
SD	233 ± 21	2.40E-07	sphere	BE-MS(well sintered)	Shirasaki & Hama (1973)
SD	252 ± 25	1.60E-07	sphere	BE-MS(poorly sintered)	Shirasaki & Hama (1973)
SD	426 ± 42	0.99	sphere	BE-MS(poorly sintered)	Shirasaki & Hama (1973)
SD	536	5.75E+06		SIMS	Oishi et al. (1983)
SD	370 ± 20	1.90E-04		$^{18}O(p,\alpha)^{15}N$	Reddy & Cooper (1983)
SD	312.6 ± 13	1.80E-06		SIMS	Yoo et al. (1984)
SD	385.9	2.90E-04		SIMS	Liberatore & Wuensch (1995)
SD	180.8	1.14E-11		$^{18}O(p,\alpha)^{15}N$	*Sonder et al. (1987)
SD	788 ± 29	1.56E-06		SIMS	Reed et al. (1978)
SD	786.6 ± 29	6.40E+05	n(1/02)	SIMS	Reed & Wuensch (1978)
SD	635 ± 105	1.90E-03		BE-MS	Oishi & Kingery (1960)
SD	241	1.90E-08		BE-MS	Oishi & Kingery (1960)
SD	615 ± 42	266	n(10/2)	$^{18}O(p,\alpha)^{15}N$	Reddy & Cooper (1982)
SD	588.6 ± 19	6.8		TEM(disloc. loops)	Lagerlof et al. (1989)
SD	437.2	1.04E+10	//c	FTIR	Ramirez et al. (1997)
SD	636	206	//c	SIMS	Prot & Monty (1996)
SD	590	67	//c	SIMS	LeGall et al. (1996)
SD	123.4	0.38	//c	Conductivity	Paek (1974)
SD	56.9	1.80E-03	nc	Conductivity	Paek (1974)
SD	D = 3.2E-16		//c	$^{18}O(p,\alpha)^{15}N$	Gruenwald & Gordon (1971)
SD	D = 1.7E-15		nc	$^{18}O(p,\alpha)^{15}N$	Gruenwald & Gordon (1971)
SD	D = 2.15E-16 (75-150µm)		sphere	BE-MS	Bagshaw & Hyde (1976)
SD	D = 6.78E-17 (150-300µm)		sphere	BE-MS	Bagshaw & Hyde (1976)
SD	251	3.40E-03	//c	SIMS	Arita et al. (1979)
SD	282.6	2.40E-02	//c	$^{18}O(p,n)^{18}F$	Derry et al. (1981)
SD	D = 1.7(± 0.1)E-16		//c	SIMS	Freer & Dennis (1982)
SD	D = 3.2(± 0.2)E-15		//c	SIMS	Freer & Dennis (1982)
SD	168.8	1.14E-07	//c	SIMS	Dennis & Freer (1993)
SD	175.2	2.41E-08	//c	SIMS	Dennis & Freer (1993)
SD	330 ± 15	5.90E-01	//c	$^{18}O(p,\alpha)^{15}N$	Moore et al. (1998)
SD	258 ± 22	4.70E-03	//c	$^{18}O(p,\alpha)^{15}N$	Moore et al. (1998)

290	TiO_{2-x}	1100	$\log(PO_2) = -9.5$ to -15.8	SSX	$H_2/H_2O(v)$	^{18}O
291	$CaTiO_2$	900-1300	0.1	PNX	CO_2	^{18}O
292	$MgAl_2O_4$	1432-1739	0.024-0.027	PSX	O_2	^{18}O
293	$MgAl_{2.4}O_{4.6}$	1501-1716	0.024-0.027	PSX	O_2	^{18}O
294	$MgAl_2O_4$	1415-1660	$PO_2 = 0.0053$	PSX	O_2	^{18}O
295	$MgAl_2O_5$	1135-1310	$PO_2 = 0.0053$	PSX	O_2	^{18}O
296	$MgFe_2O_4$	1135-1415	$PO_2 = 0.0053$	PSX	O_2	^{18}O
297	$MgFe_2O_5$	985-1095	$PO_2 = 0.0053$	PSX	O_2	^{18}O
298	Magnetite	302-550	<0.1	PSX	$H_2O(v)$	^{18}O
299	Magnetite	302-550	$PH_2/PH_2O = 1E-5$ to 3.8	PSX	$H_2/H_2O(v)$	^{18}O
300	Magnetite	500-800	100	SNX	H_2O	^{18}O
301	Magnetite	500-655	$f(H_2O)low$	SNX	$CaCO_3$	^{18}O
302	Magnetite	1150	$PO_2 = 1E-10$ to $1E-4$	S/PSX	$H_2/H_2O(v)$	^{18}O
303	Magnetite	800	$PO_2 = 1E-17$ to $1E-9$	S/PSX	$H_2/H_2O(v)$	^{18}O
304	Hematite	852-1077	0.1	S/NSX	O_2	^{18}O
305	Hematite	890-1227	$a(O_2) = 4.5E-4$ to $6.5E-1$	SNX	O_2	^{18}O
306	Wustite	700-1150	$PH_2O/PH_2 = 0.6$	SSX	$H_2/H_2O(v)$	^{18}O
307	Wustite	700-1150	$PH_2O/PH_2 = 0.86$	SSX	$H_2/H_2O(v)$	^{18}O
308	Wustite	700-1150	$PH_2O/PH_2 = 2.4$	SSX	$H_2/H_2O(v)$	^{18}O
309	Wustite	700-1150	$PH_2O/PH_2 = 4.0$	SSX	$H_2/H_2O(v)$	^{18}O
310	Wustite	700	$PO_2 = 1.22E-21$	SSX	$H_2/H_2O(v)$	^{18}O
311	Wustite	700	$PO_2 = 2.82E-21$	SSX	$H_2/H_2O(v)$	^{18}O
312	Wustite	700	$PO_2 = 5.15E-22$	SSX	$H_2/H_2O(v)$	^{18}O
313	Cr_2O_3	1100	$P(O_2) = 1E-11$	SSX	$H_2O/N_2/H_2$	^{18}O
314	Cr_2O_3	1300	$P(O_2) = 1E-11$	SSX	$H_2O/N_2/H_2$	^{18}O
315	ZnO	940-1140	?	SSX	O_2	^{18}O
316	ZnO	1040	?	PSX	O_2	^{18}O
317	CoO	1175-1560	$P(O_2) = 0.021$	PSX	O_2	^{18}O
318	ZrO_2	600-1000	$P(O_2) = 0.092$	PSX	O_2	^{18}O
319	$ZrO_2(Fe,Al)$	1300-1600	<0.1	SSX	$He/H_2O(v)$	^{18}O
320	$ZrO_2(Fe,Al)$	1310-1435	<0.1	SSX	$H_2/H_2O(v)$	^{18}O
321	ZrO_2	1315-1550	<0.1	SSX	$H_2/H_2O(v)$	^{18}O
322	ZrO_2	450-950	$P(O_2) = 0.001$	ASX	O_2	^{18}O
323	$Y-ZrO_2$	450-1100	$P(O_2) = 0.001$	SSX	O_2	^{18}O
324	UO_2	778-1247	<0.1(H_2)	P/SSX	$U^{18}O_2$	^{18}O
325	UO_{2-x} $\times(0.005-0.021)$	1275-1650	$P(O_2) = 1E-7$ to $3E-4$	P/SSX	Ar/O_2	^{18}O
326	$UO_{2.006}$	400-900	vacuum	SSX	$U^{18}O_{2-x}$	^{16}O
327	$UO_{2.02}$	400-800	vacuum	SSX	$U^{18}O_{2-x}$	^{16}O

SD	$logD = 0.1235(logPO_2)-17.02$	nc	Gravimetry	Millot & Picard (1988)
SD	313 ± 10	5	sphere	BE-MS	Gautaso & Muehlenbachs (1993)
SD	439 ± 67	0.89(± 1.9)	sphere	BE-MS	Ando & Oishi (1974)
SD	443 ± 25	2.2(± 0.68)	sphere	BE-MS	Ando & Oishi (1974)
SD	467	1.23E-04	sphere	BE-MS	Haneda et al. (1987)
SD	289	2.21E-10	sphere	BE-MS	Haneda et al. (1987)
SD	328	1.52E-05	sphere	BE-MS	Haneda et al.(1987)
SD	159	1.20E-11	sphere	BE-MS	Haneda et al.(1987)
SD	71.1 ± 7	3.2(± 1.6)E-14	sphere	BE-MS	Castle & Surman (1967)
SD	$logD = -3.715(1000/T) - 0.27log(PH_2/PH_2O) - 12.74$		sphere	BE-MS	Castle & Surman (1969)
SD	188.3	3.50E-06		SIMS	Giletti & Hess (1988)
SD	211 ± 20	4.30E-09	sphere	Empirical-natural data	Sharp (1991)
SD	$logD = -14.82 - 0.5log(PO_2)$		n(111)	SIMS	Millot & Niu (1997)
SD	$logD = -21.39 - 0.5log(PO_2)$		n(111)	SIMS	Niu & Millot (1999): Millot et al. (1997)
SD	405 ± 25	6.3(+58/-5.7)E-2	///(0001); ///(10/2)	$^{18}O(p,\alpha)^{15}N$	Reddy & Cooper (1983)
SD	$logD = 8.431 - 0.26log(aO_2) - 28.312(1000/T)$			SIMS	Amami et al. (1999)
SD	286.6 ± 2.9	2.00E-01		SIMS	Amami et al. (1994)
SD	250.9 ± 2.9	8.70E-02		SIMS	Amami et al. (1994)
SD	225.8 ± 2.9	8.90E-03		SIMS	Amami et al. (1994)
SD	218.1 ± 2.9	7.00E-03		SIMS	Amami et al. (1994)
SD	$D = 3.2E-15$			$^{18}O(p,\alpha)^{15}N$	Sabioni et al. (1993)
SD	$D = 2.1E-15$			$^{18}O(p,\alpha)^{15}N$	Sabioni et al. (1993)
SD	$D = 3.2 (\pm 0.4)2E-15$			SIMS; $^{18}O(p,u)^{15}N$	*Sabioni et al. (1993)
SD	$D = 3.8E-18$		n(01/2)	SIMS	Sabioni et al. (1992)
SD	$D = 6.59E-17$		n(01/2)	SIMS	Sabioni et al. (1992)
SD	124 ± 7	1.20E-10		$^{18}O(p,\alpha)^{15}N$	Robin et al. (1973)
SD	$D = 4.8E-15$		//a	SIMS	Routbort & Tomlins (1995)
SD	397 ± 21	50	plate	BE-MS	Chen & Jackson (1969)
SD	189.4	2.34E-02		BE-MS	Kenesha & Douglass (1971)
SD	55.6 ± 5	5.4(± 1.8)E-7		Gravimetry	Park & Olander (1991)
SD	58.6 ± 3	7.3(± 1.3)E-7		Gravimetry	Park & Olander (1991)
SD	56.1 ± 5	3.7(± 1.2)E-7		Gravimetry	Park & Olander (1991)
SD	220.9	2.50E-03		SIMS	Brossmann et al. (1999)
SD	106.2	0.00708		SIMS	Manning et al. (1997)
SD	248.1	0.26		$^{18}O(p,\alpha)^{15}N$; BE-MS	Marin & Contamin (1969)
SD	$logD = 2.114 + 5logx - 18.75(1000/T)$			$^{18}O(p,\alpha)^{15}N$; BE-MS	Marin & Contamin (1969)
SD	87.8	5.00E-05		$^{18}O(p,n)^{17}F$; MS	Murch et al. (1975)
SD	87.8	1.00E-04		$^{18}O(p,n)^{17}F$; MS	Murch et al. (1975)

328	$UO_{2.1}$	500-800	vacuum	<u>SSX</u>	$U^{18}O_{2-x}$	^{16}O	
329	ThO_2	1200-1646	$PO_2 = 0.024\text{-}0.027$	P/SSX	O_2	^{18}O	
330	ThO_2	845-1200	$PO_2 = 0.024\text{-}0.027$	P/SSX	O_2	^{18}O	

VI. Carbonates

331	Calcite	606-848	0.066-0.095	<u>PNX</u>	CO_2	^{13}C	
332	Calcite	250-550	<0.1	PNX	CO_2	^{14}C	
333	Calcite	550-750	<0.1	<u>PNX</u>	CO_2	^{13}C	
334	Calcite	650-850	<0.1	<u>PNX</u>	CO_2	^{18}O	
335	Calcite	197-493	100	PSX	H_2O/NH_4Cl	^{18}O	
336	Calcite	305-700	200	PNX	H_2O	^{18}O	
337	Calcite	500-800	$PCO_2 = 0.1\text{-}0.5;PH_2O = 0.002\text{-}2.4$	SNX	CO_2/H_2O	^{13}C	
338	Calcite	500-800	$PCO_2 = 0.1\text{-}0.5;PH_2O = 0.002\text{-}2.4$	SNX	CO_2/H_2O	^{13}C	
339	Calcite	300-600	25-100	PNX	0-4m NaCl	^{18}O	
340	Calcite	400-800	100	SNX	H_2O	^{18}O	
341	Calcite	700	$logf(H_2O) = 0.6\text{-}2.38$	SNX	H_2O	^{18}O	
342	Calcite	900	1250	PNX	CO_2	^{18}O	
343	Calcite	900	1250	PNX	CO_2	^{13}C	
344	Calcite	900	1250	PSX	CO_2	^{18}O	
345	Calcite	900	1250	PSX	CO_2	^{13}C	
346	Calcite	600-800	100	<u>SNX</u>	CO_2	^{13}C	
347	Calcite	600-800	100	<u>SNX</u>	CO_2	^{18}O	
348	Calcite	300	25	PSX	H_2O	^{18}O	
349	Strontianite	60-200	100	PSX	0.5m NH_4Cl	^{18}O	
350	Strontianite	300	10	PSX	H_2O	^{18}O	
351	Witherite(Ba)	103-205	100	PSX	0.5m NH_4Cl	^{18}O	
352	Witherite(Ba)	300	10	PSX	H_2O	^{18}O	
353	Dolomite	255-660	100	PNX	0.4m NH_4Cl	^{18}O	
354	Dolomite	645-785	12-93.5	PNX	CO_2	^{18}O	
355	Dolomite	645-785	12-93.5	PNX	CO_2	^{13}C	

VII. Sulfates & Phosphates

356	Barite	110-350	Psat	PSX	1m NaCl	^{18}O	
357	Barite	110-350	Psat	PSX	1m NaCl + 1m H_2SO_4	^{18}O	
358	Anhydrite	197-550	1-100	PSX	1m NaCl	^{18}O	
359	Anhydrite	140-350	0.4-100	PSX	1m HCl	^{18}O	
360	Anhydrite	146-350	0.4-100	PSX	1m H_2SO_4	^{18}O	
361	Alunite	100-300	0.1-100	PS/NX	H_2O	H	
362	Fluorapatite	550-1200	100	SNX	H_2O	^{18}O	
363	Fluorapatite	800-1100	100	SNX	H_2O	^{18}O	

SD	89.9	2.70E-04		$^{18}O(p,n)^{17}F$: MS	Murch et al. (1975)
SD	208.8 ± 6	5.73(± 0.18)E-2		BE-MS	Ando et al. (1976)
SD	73.6 ± 9	1.0(± 0.4)E-6		BE-MS	Ando et al. (1976)
SD	242.7	4.50E-04	plate	BE-MS	Haul & Stein (1955)
SD	7.11E+01	4.60E-16	plate	BE-MS	Anderson (1969)
SD	368.2	1.30E+03	plate	BE-MS	Anderson (1969)
SD	421.9	2.05E+06	plate	BE-MS	Anderson (1969)
CR	27.2	2.60E-05	rhomb	BE-MS	*O'Neil et al. (1969)
CR	43.5	4.72E-04	rhomb	BE-MS	*Anderson & Chai (1974)
SD	364 ± 8	9(+12/-5)E2	//c	SIMS	Kronenberg et al. (1984)
SD	364 ± 8	5(+6/-3)E2	nc	SIMS	Kronenberg et al. (1984)
CR	$\log R = -4.185-6.55.88/T+$ $0.201m-814.25(\log \rho_w)/T$	rhomb/BET	BE-MS	Cole (1992)
SD	173 ± 6	7.00E-05	isotropic	SIMS	Farver (1994)
SD	$\log D = 0.835\log f(H_2O)-19.03$	isotropic	SIMS	Farver (1994)
SD	$D = 9E-13$	plate	BE-MS	Rosenbaum (1994)
SD	$D = 4E-13$	plate	BE-MS	Rosenbaum (1994)
SD	$D = 2E-11$	plate	BE-MS	Rosenbaum (1994)
SD	$D = 6E-12$	plate	BE-MS	Rosenbaum (1994)
SD	166 ± 16	7.77E-09	//c	SIMS	Labotka et al. (2000)
SD	242 ± 39	7.50E-03	//c	SIMS	Labotka et al. (2000)
CR	$R = 9.74(+5.36/-3.47)E-10$	rhomb/BET	BE-MS	Cole (2000)
CR	26.8	1.00E-03	rhomb	BE-MS	*O'Neil et al. (1969)
CR	$R = 5.89(+5.01/-2.72)E-9$	rhomb/BET	BE-MS	Cole (2000)
CR	42.7	4.30E-02	rhomb	BE-MS	*O'Neil et al. (1969)
CR	$R = 9.14(+5.66/-3.49)E-9$	rhomb/BET	BE-MS	Cole (2000)
CR	45.6	1.10E-03	rhomb	BE-MS	*Northrop & Clayton (1966)
SD	485.96	2.81E+11	plate	BE-MS	*Anderson (1972)
SD	468.6	3.73E+10	plate	BE-MS	*Anderson (1972)
CR	56.1	4.75E-02	sphere	BE-MS	*Kusakabe & Robinson (1977)
CR	38.49	3.05E-03	sphere	BE-MS	*Kusakabe & Robinson (1977)
CR	81.17	3.89E-02	sphere	BE-MS	*Chiba et al. (1981)
CR	29.29	2.40E-05	sphere	BE-MS	*Chiba et al. (1981)
CR	40.17	4.80E-04	sphere	BE-MS	*Chiba et al. (1981)
SD	26.4 ± 7.1	5.01E-14	sphere/cyl	BE-MS	Stoffregen et al. (1994)
SD	205 ± 12.5	9.00E-05	//c	SIMS	Farver & Giletti (1989)
SD	125.5 ± 33	3.00E-11	nc	SIMS	Farver & Giletti (1989)

VIII. Whole Rocks							
364	Basalt	300-500	60-100	PNX	Seawater	^{18}O	
365	Granite	250-300	Psat-35	PNX	0.1m NaCl	^{18}O	

(a) Mineral names or compositions are given as described by each source, for details regarding specific compositional data refer to the original reference; typical mineral abbreviations used in some figs.: quartz(q), albite (ab), anorthite (an), sanidine (san), adularia (ad), nepheline (ne), leucite (lu), paragonite (parag), muscovite (mu), biotite (bi), phlogopite (ph), chlorite (chl), diopside (di), hornblende (hb), forsterite (fo), zircon (zr), almandine (al), mullite (ml), epidote (ep), montmorillonite (mont), kaolinite (kaol), akermanite (ak), gehlenite (ge), corundum (co), rutile (ru), sapphire (saph), perovskite (pv), magnetite (mt), hematite (hm), calcite (cc), dolomite (dol)

(b) Pressures given as total water, O_2, CO_2, H_2 or mixed gas pressure, or as partial pressures of select species; in many cases pressures are not reported so estimates have been made, MBA refers to multiple buffer assemblage, NNO refers to nickel-nickel oxide, QFM refers to quartz-fayalite-magnetite, IW refers to iron-wustite buffer, Psat = saturated water pressure at T.

(c) NX = natural crystals, SX = synthetic crystals, prefix S = single crystal, prefix P = powdered material, A = aggregates, SN/SX = combined data from single natural and synthetic crystal experiments, PN/SX refers to experimental data sets including both natural powders and single crystals; underlining means the phase has been annealed prior to experimentation, (^{18}O) refers to phases that have been preequilibrated with ^{18}O-rich gas or fluid

(d) H_2O = pure water, $H_2O(v)$ = water vapor, THO = tritiated water, D_2O = deuterated water, H_2/H_2O = mixed hydrogen gas/water systems, O_2 = pure oxygen gas, CO_2 = pure carbon dioxide gas, H_2 = hydrogen gas, D_2 = deutrium gas; assorted aqueous electrolytes such NaCl, NaF, $CaCl_2$, H_2SO_4, etc. are given in either molal (m) or molar (M) concentrations, XCO_2 = mole fraction of CO_2 in CO_2-H_2O system.

(e) Isotopic species use as tracer for rate given as ^{18}O, H, D, ^3H (tritium), ^{13}C; no isotopic species is listed when transport of the entire reacting species is being monitored.

(f) SD = self-diffusion, TD = tracer diffusion, CR = chemical reaction (e.g. recrystallization, cation exchange, etc.), CR(SD) and SD(R) refer to a coupled process where rate data are given for first mentioned process, (?) refers to case where there is uncertainty as to the controlling mechanism.

CR	48.2	1.58E-05	sphere/BET	BE-MS	Cole et al. (1987)
CR	52.2	1.95E-04	sphere/BET	BE-MS	Cole et al. (1992)

(g) Activation energy in kJ/mole (can be converted to Kcal/mole by dividing by 4.1842), consult original references for the source of errors about these values, shown as (±). In some cases equations were fit for D as a function of $P(O_2)$ or $f(H_2O)$.

(h) Pre-exponential factor in the Arrhenius equation (for diffusion studies D_o is commonly used), where diffusion $D = A_o(\text{or } D_o)\exp(-E_a/RT)$ typically in units of either cm^2/s or m^2/s. We have attempted to recast all E_a values in cm^2/s units; values for rates controlled by chemical reaction (CR) are given in moles O/m/s where $R = A_o\exp(-E_a/RT)$.

(i) Crystallographic orientation of diffusion can be determined from microbeam methods (e.g. SIMS, NRA), here n = normal to and ‖ = parallel to the direction indicated such as the *c* direction or a particular (hkl) plane—e.g. (010), axis of rotoinversion is shown in italic (e.g. *1*11), isotropic is used for phases where rates are uniform regardless of direction. In cases where orientation is not known, the grain geometry selected in the calculation for diffusion coefficient may imply a particular orientation. For micas and clays, a "plate" model implies transport ‖ to *c*, where as a "cylinder" model implies an orientation normal to c. In the case of prismatic crystals such as epidote, hornblende, etc. the plate model implies transport ‖ to the prism long direction, whereas the cylinder model implies transport normal to the prism long dimension. For rates calculated for chemical reaction, these geometries denote the shapes used to estimate the surface areas of the grains based either on sieve size or SEM observation, when orientation is not known or assumed we have left the space blank. BET refers to the Brunaner-Emmett-Teller gas adsorption-surface area method.

(j) SIMS = secondary ion mass spectrometry (ion microprobe), IR = infrared absorption spectrometry, SS-IR = serial section IR, FTIR = Fourier Transform infrared adsorption spectrometry, NRA = nuclear reaction analysis -$(^{18}O(p,\alpha)^{15}N)$. BE-MS = bulk isotopic exchange with conventional mass spectometry, NMR = nuclear magnetic resonance (typical for zeolites), ZLC = zero length column – thermal conductivity cells (for zeolite beds).

(k) * in front of reference refers to either the use of raw data to calculate the rate constant, *R*, from partial exchange experiments where chemical reaction occurred, or the D values were refit as a function of either T, $P(O_2)$, $f(H_2O)$, etc. ** in front of reference refers to calculation by Graham (1981).

Table 3. Summary of select rate data for short-circuit diffusion (grain-

No.	Solid	Temperature Range (°C)	Pressure Range (MPa) (a)	Sample Type (b)	Reacting Species (c)
1	Novaculite	450 - 800	100	ANX (1.2 μm)	H_2O
2	Novaculite	550 - 800	100	ANX (4.9 μm)	H_2O
3	Novaculite	500 - 800	100 (pre-anneal w/CO_2)	ANX (1.2 μm)	H_2O
4	Novaculite	450 - 800	100(pre-anneal w/H_2O)	ANX (1.2 μm)	H_2O
5	Novaculite	450 - 800	100(pre-anneal w/NaCl)	ANX (1.2 μm)	H_2O
6	Novaculite	600	100(pre-anneal w/NaCl)	ANX (1.2 μm)	CO_2
7	Novaculite	600	100 (as is texture)	ANX (1.2 μm)	6 M NaCl
8	Novaculite	600	100 (as is texture)	ANX (1.2 μm)	CO_2:H_2O
9	Quartz/Kspar	450 - 700	100	ANX	H_2O
10	Macroperthite	500 - 700	100	SNX	H_2O
11	Tanco Albite	600 - 800	100	ANX	H_2O
12	Amelia Albite	600 - 800	100	ANX	H_2O
13	Albite (from gl)	600 - 800	100	ASX	H_2O
14	Orthoclase (gl)	450 - 800	100	ASX	H_2O
15	Forsterite	1100	0.064	SSX	CO_2
16	Forsterite	1100	0.064	SSX (// c)	CO_2
17	Forsterite	1200	0.064	SSX (// c)	CO_2
18	Forsterite	1200	0.064	SSX (// c)	CO_2
19	Gehlenite	1000 - 1300	0.064	SSX (// c)	CO_2
20	Akermanite	1000 - 1300	0.064	SSX (// c)	CO_2
21	MgO	1050 - 1438	0.1 (?)	ASX (75 μm)	O_2
22	MgO	1050 - 1438	0.1 (?)	ASX (40 μm)	O_2
23	MgO	1300 - 1500	0.1	SSX (18O)	$^{16}O_2$ (air)
24	Corundum	1500 - 1720	0.2 - 0.5	SSX	O_2
25	Corundum	1460 - 1720	0.2 - 0.5	ASX (100-200nm)	O_2
26	Y-Al_2O_3	1460 - 1720	0.2 - 0.5	ASX (100-200nm)	O_2
27	Y-Al_2O_3	1110 - 1630	0.2 - 0.5	SSX	O_2
28	Hematite	890 - 1227	a(O_2) = 4.5E-4 to 6.5E-1	SNX	O_2
29	ZrO_2	450 - 950	P(O_2) = 0.001	ASX	O_2
30	Calcite	300 - 500	100	ANX	H_2O
31	Ultramylonite	250 - 550	100 (// foliation)	ANX	H_2O
32	Ultramylonite	250 - 550	100 (n foliation)	ANX	H_2O

boundary; dislocations) determined by the ion microprobe.

Mechansim (d)	E_a kJ/mole (e)	D_o (or log D_o) (cm^3/sec) (f)	References (g)
GB	113 ± 4	2.60E-11	Farver & Yund (1991a)
GB	109 ± 12.5	3.40E-11	Farver & Yund (1991a)
GB	102.3	4.32E-12	*Farver & Yund (1992)
GB	82.6	1.04E-12	*Farver & Yund (1992)
GB	85.98	6.03E-12	*Farver & Yund (1992)
GB	Dδ = (1.25 ± 0.33)E-17		Farver & Yund (1992)
GB	Dδ = (3.91 ± 0.72)E-18		Farver & Yund (1992)
GB	Dδ = 7.78E-19 to 2.34E-17		Farver & Yund (1992)
GB	75 ± 13	3.60E-12	Farver & Yund (1995a)
GB	180 ± 29	4.7(+240/-4.6)E-6	Nagy & Giletti (1986)
GB	83 ± 20	3.20E-12	Farver & Yund (1995b)
GB	73 ± 14	1.10E-12	Farver & Yund (1995b)
GB	68 ± 12	1.10E-12	Farver & Yund (1995b)
GB	78 ± 8	2.60E-12	Farver & Yund (1995b)
GB	Dδ = (2.77 ± 1.77)E-22		*Yurimoto et al. (1992)
Dislocation	D = (3.43 ± 2.03)E-17 cm^2/s		*Yurimoto et al. (1992)
GB	Dδ = (1.2 ± 0.57)E-21		*Yurimoto et al. (1992)
Dislocation	D = (1.69 ± 0.9)E-16 cm^2/s		*Yurimoto et al. (1992)
Dislocation	285 ± 20	0.46 cm^2/s	Yurimoto et al. (1989)
Dislocation	262 ± 23	54 cm/s	Yurimoto et al. (1989)
GB	255.9	4.5E-5 cm^2/s	Hashimoto et al. (1972)
GB	255.9	1.3E-5 cm^2/s	Hashimoto et al. (1972)
GB	376.3	0.31 cm^2/s	Liberatore & Wuensch (1995)
Dislocation	896	3.1E14 cm^2/s	Prot & Monty (1996)
GB	921	1.6E16 cm^2/s	Prot et al. (1996)
GB	800	7.0E10 cm^2/s	Prot et al. (1996)
Dislocation	980	1.0E17 cm^2/s	Le Gall et al. (1996)
Dislocation	log D = 25.51 - 0.4log(aO_2) - 47.59(1000/T)		Amami et al. (1999)
GB	188.1	0.33 ± 0.15 cm^2/s	Brossmann & Wurschum (1999)
GB	127 ± 17	3.80E-08	Farver & Yund (1998)
"Bulk"	30 ± 6	2.0E-7cm^2/s	Farver & Yund (1999)
"Bulk"	30 ± 14	5.6E-8cm^2/s	Farver & Yund (1999)

Table 4. Summary of diffusion data from select glass and

No.	Glass/Melt Composition	Temperature (°C)	Pressure (MPa)	fO_2 (MPa)/fH_2O (MPa)	Sample Type /Type of Experiment	Source of Diffusant
1	Fused silica, SiO_2	850-1250	0.1	0.025 - 0.4	Glass bulb, bulk uptake	Enriched O_2 gas
2	Fused silica, SiO_2	950-1080	0.1		Membrane, bulk uptake	Enriched O_2 gas
3	Fused silica, SiO_2	900-1250	0.1		Rate uptake	Enriched O_2 gas
4	Fused silica, SiO_2	925-1225	0.1		Tracer loss	Enriched O_2 gas
5	Fused silica, SiO_2	1150-1430	0.1		Rate of exchange	Enriched O_2 gas
6	Fused silica, SiO_2	815-1018	0.1		Profile measurement	Enriched O_2 gas
7	Fused silica, SiO_2	1200-1400	0.1		Profile measurement	Enriched SiO_2 glass diffusion couple , thin films
8	Fused silica, SiO_2	900-1200	0.1		Profile measurement	Enriched SiO_2 glass diffusion couple , thin films
9	Fused silica, SiO_2	800-1200	0.1		Profile measurement	Enriched O_2 gas
10	Lead Silicate Glass (24.96 SiO_2-67.74 PbO-7.47K_2O by wt.)	578-678	0.1	0.007-0065	Gas exchange	Enriched O_2 gas
11	Potassium Silicate Glass (62.2SiO_2-37.2K_2O)	820-902	0.1	0.007-0.055	Gas exchange	Enriched O_2 gas
		820-873		$fH_2O=$ 0.0013-0.007		O_2 Enriched water vapor
		673-820		$fH_2O=$ 0.0013-0.007		O_2 enriched water vapor
12	Borosilicate glass SRL-131	1050-1250	0.1		Redox kinetics, chemical diffusion	O_2 Enriched water vapor
13	CaO-SiO_2 melt	1600	0.1		Capillary technique	[18]O Enriched diffusion couple

melt systems at elevated temperatures and pressures.

Isotope Exchanged	Analytical Method	E_a Energy (kJ/mole)	Pre-exponential (cm²/sec)	Activation Volume (cm³/mole)	Reference	Year
^{18}O	Mass spectrometry	121	2.00E-09		Williams	1965
^{18}O	Mass spectrometry	113	2.80E-04		Norton	1961
^{18}O	Mass spectrometry	234	4.30E-06		Haul and Dumgen	1962
^{18}O	Mass spectrometry	298	1.50E-02		Sucov	1963
^{18}O	Mass spectrometry	82	4.40E-11		Muehlenbachs and Schaeffer,	1977
^{18}O	Nuclear reaction analyses	110	2.10E-10		Yinnon	1979
^{18}O	SIMS	454	2.60E+00		Mikkelson	1984
^{18}O	SIMS	280	2.80E-05		Cawley and Boyce	1988
^{18}O	Nuclear reaction analyses + SIMS	143	5.54E-11		Kalen, Boyce and Cawley	1991
^{18}O	Mass spectrometry	300	3.20E+06		de Berg and Lauder	1978
^{18}O	Mass spectrometry	250	1.70E+01		de Berg and Lauder	1980
		238	5.40E+00			
		126	2.30E-05			
^{18}O	Mass spectrometry	90	6.30E+08		Schreiber et al.	1986
^{18}O	Nuclear reaction analysis				Keller et al.	1982

14	Iron Oxide Melt	1430-1550	0.1	CO-CO_2 gas mixes	Gas exchange	Fe-O exchange, Fe^{2+}/Fe^{3+} measured
15	$16Na_2O$-$12CaO$-$72SiO_2$ Glass	460-525	0.1		Gas exchange	Enriched gas
16	$20Al_2O_3$-$40CaO$-$40SiO_2$ Glass	765-845	0.1		Gas exchange	Enriched gas
17a	Commercial Soda-lime glass	563-893 (below Tg)	0.1		Gas exchange	Enriched gas
17b		973-1356 (above Tg)				
18a	$20K_2O$-$20SrO$-$60SiO_2$	568-853 (below Tg)	0.1		Gas exchange	Enriched gas
18b		953-1394 (above Tg)				
19	Basalt	1320-1500	0.1	O_2 or CO_2 gas	Gas exchange	Enriched gas
20	Jadeite	1400	500 - 2000		Diffusion couple	Doped couple
	Diopside					
21a	Diopside	1396-1474	0.1		Degassing	Enriched glass
21b	$Di_{58}An_{42}$					
21c	$Di_{40}An_{60}$					
22	$40CaO$-$20Al_2O_3$-$40SiO_2$	1320-1540		0.02-0.04	Gas-liquid exchange	Enriched gas
	$16Na_2O$-$12CaO$-$72SiO_2$	800-1450 (liquid), measured down through supercooled liquid, where it is non-linear	0.1			
	$20Na_2O$-$80SiO_2$	1061-1395				
23	Basalt	1160-1360	0.1	pure CO_2 to IW+0.05	Redox kinetics, Chemical diffusion	
	Andesite					
24	Basalt	1320-1600	1000-2000		Doped diffusion couple	Enriched glass /melt
25	Dacite	1355-1662	1000-5700		Doped diffusion couple	Enriched glass/melt

	Wet chemistry	45 (12% Fe^{3+}) 69.5 (33% Fe^{3+})	6.6E-3 (12%),1.61E-2 (33%)		Mori and Suzuki	1969
^{18}O	Mass spectrometry	278	2.00E+03		Kingery and Lecron	1960
^{18}O	Mass spectrometry	289	4.00E+00		Kingery and Lecron	1960
^{18}O	Proton activation analysis	57.8	9.00E-13		Yinnon and Cooper	1980
		208	6.30E-02			
^{18}O	Proton activation analysis	50.6	9.30E-13		Yinnon and Cooper	1980
		219	8.50E-02			
^{18}O	Mass spectrometry	251	1.40E+01		Canil and Muehlenbachs	1990
^{18}O	SIMS			-6.3	Shimizu and Kushiro,	1984
				8.48		
^{18}O	Mass spectrometry	264	1.64E+01		Dunn	1982
		196	1.35E-01			
		185	1.29E-02			
^{18}O	Mass spectrometry	227	1.80E+01		Oishi, Terai and Ueda	1975
		161.5	4.50E-02			
		186	7.90E-02			
	Weight gain or loss	215	3.64E+01		Wendlandt	1991
		251	5.25E+01			
^{18}O	SIMS	170	3.20E-09	-6.7	Lesher, Hervig, Tinker	1996
^{18}O	SIMS	293 (1 GPa)	1.26E-05	-14.5 (1460°C)	Tinker and Lesher	2001
		264 (2 GPa)	6.30E-07	-9.8 (1561 °C)		
		155 (4 GPa)	1.26E-12	-8.8 (1662 °C)		

26	Cordierite	650-880	0.1	$N_2/H_2/H_2O$ or Ar/O_2	Gas-liquid/ glass exchange	Enriched gas
27	SiO_2-Al_2O_3-CaO (42.3-12.4-45.3)	765-845	0.1			Enriched gas
28	SiO_2-Al_2O_3-CaO (41.7-13.3-45.0)	625-830	0.1			Enriched gas
29	SiO_2-Al_2O_3-CaO (42.3-12.4-45.3)	1320-1540	0.1			
30	SiO_2-Al_2O_3-CaO (42.3-12.4-45.3)	1320-1540	0.1			Enriched gas
31	SiO_2-Al_2O_3-CaO (42.3-12.4-45.3)	1370-1520	0.1			Enriched solid
32	SiO_2-CaO (44-56)	1550-1650	0.1			Capillary
33a	SiO_2-CaO-Na_2O (72-12-16)	(a) 460-525	0.1			
33b		(b) 800-1470				
34	SiO_2-CaO-Fe_2O_3	1350-1550	0.1			
35	SiO_2-K_2O (82.5-17.5)	750-1000	0.1			Enriched gas
36	SiO_2-K_2O (73.6-26.4)	700-1000	0.1			Enriched gas
37	SiO_2-K_2O (72.4-27.6)	820-902	0.1			Enriched gas
38	SiO_2-K_2O-PbO (53.4-8.7-33.0)	275-425 (Glass)	0.1			Enriched gas
39	SiO_2-K_2O-PbO (53.4-8.7-33.0)	578-678	0.1			Enriched gas
40a	SiO_2-K_2O-SrO (71.1-15.1-13.8)	327-600 (<Tg)	0.1			Enriched gas
40b		600-727 (> Tg)				

^{18}O	SIMS	113 (<Tg)	3.00E-10		Sabioni et al	1998
		481 (>Tg)	8.00E+07			
		255	7.49E-03		Kingery and Lecron	1960
		245	2.79E-03		Hagel and MacKenzie	1964
		257	9.30E+01		Ueda and Oishi	1970
		227	1.80E+01		Oishi, Ueda and Terai	1974
		410	4.00E-06		King and Koros	1959
	Nuclear reaction analysis	408	1.50E+07		Shiraishi, Nagahama and Ohta	1983
		278	2.00E+03		Kingery and Lecron	1960
		186	7.90E-02		Terai and Oishi	1977
		48.5-106 (depending on composition)	D(1450)= 1E-3 to 1E-4, (depending on composition)		Gaye and Riboud	1976
		266	4.00E+01		May, Lauder and Wollast	1974
		199	2.40E-01		May, Lauder and Wollast	1974
		250	1.70E+01		DeBerg and Lauder	1980
		50	1.00E-10		Schaeffer and Oel	1969
		300	3.20E+06		deBerg and Lauder	1978
		42	1.00E-12		Rawal and Cooper	1979
		498	7.60E+14			

3   Fractionation of Carbon and Hydrogen Isotopes in Biosynthetic Processes[*]

John M. Hayes

National Ocean Sciences Accelerator Mass Spectrometry Facility
and *Department of Geology and Geophysics*
Woods Hole Oceanographic Institution
Woods Hole, Massachusetts 02543

This review is concerned with the isotopic relationships between organic compounds produced by a single organism, specifically their enrichments or depletions in ^{13}C relative to total-biomass carbon. These relationships are biogeochemically significant because

(1) An understanding of biosynthetically controlled, between-compound isotopic contrasts is required in order to judge whether plausibly related carbon skeletons found in a natural mixture might come from a single source or instead require multiple sources.

(2) An understanding of compound-to-biomass differences must underlie the interpretation of isotopic differences between individual compounds and total organic matter in a natural mixture.

My approach is pedagogic. The coverage is meant to be thorough, but the emphases and presentation have been chosen for readers approaching this subject as students rather than as research specialists. In common with the geochemists in my classes, many readers of this paper may not be very familiar with biochemistry and microbiology. I have not tried to explain every concept from those subjects and I have not inserted references for points that appear in standard texts in biochemistry or microbiology. Among such books, I particularly recommend the biochemistry text by Garrett and Grisham (1999) and the microbiology text by Madigan et al. (2000). The biochemistry text edited by Zubay (1998) is also particularly elegant and detailed. White (1999) has written a superb but condensed text on the physiology and biochemistry of prokaryotes.

ISOTOPE FRACTIONATION

A schematic overview of the relevant processes is shown in Figure 1. Plants and other autotrophs fix CO_2. Animals and other heterotrophs utilize organic compounds. If the assimilated carbon is a small molecule (like CO_2, CH_4, or acetate), significant isotopic fractionation is likely to accompany the fixation or assimilation of C. Such fractionations establish the isotopic relationship between an organism and its carbon source. Those associated with photosynthesis encode information about chemical and physical conditions in the environment of fixation. Logically, therefore, they are treated here in the chapter dealing with the biogeochemistry of marine basins (Freeman, this volume).

The initial products formed by autotrophs are small carbohydrates ("the photosynthate"). In heterotrophs, if a carbon skeleton is not destined for incorporation in biomass with little or no alteration, it is commonly broken down to yield two- and three-carbon "metabolic intermediates." Depending on whether we are talking about an autotroph or a heterotroph, either the small carbohydrates or the metabolic products then serve as inputs to downstream processes. These involve oxidation, reduction, bond

---

[*] Dedicated to my father and mother on his 90th birthday and their 65th wedding anniversary, 26 March 2001. Of all the privileges and good fortunes that I have enjoyed, none equals that of being his son and their child.

1529-6466/00/0043-0003$05.00

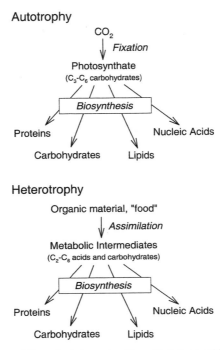

Figure 1. The roles of biosynthesis in autotrophic and in heterotrophic organisms.

formation, and bond cleavage. Some of these processes produce energy and others consume it. They are balanced against each other, and the organism will remain viable as long as net yields of energy are adequate for maintenance. Growth—the production of additional biomass—will be possible if photosynthate or metabolites remain after adequate quantities of energy have been produced. This will depend not only on the magnitude of the supply but also on the range of energy-producing processes accessible (for example, whether the availability of O_2 permits aerobic respiration). Remaining supplies, and the mixture of further energy-producing and energy-consuming processes, will then be managed so that a portion of the available carbon flows through the network of reactions required to produce the mixture of nucleic acids, proteins, carbohydrates, and lipids that together form biomass. It is the isotopic budget for that reaction network that concerns us here.

Isotopes in reaction networks

Isotope effects and mass balance. An arbitrary sequence of reactions is indicated schematically in Figure 2. Analysis of this system will provide a useful introduction to studies of real organisms. Reactant **A** could, for example, represent a particular carbon position in a two- or three-carbon product of metabolism or photosynthesis. If there were no isotope effect associated with Reaction (1) ($\varepsilon_1 = 0$), then the isotopic composition of the carbon being transmitted by Reaction (1) would be equal to that of reactant **A**. This follows from a general relationship which can be written as

$$\delta_{P'} = \delta_{R'} - \varepsilon \qquad (1)$$

where $\delta_{P'}$ and $\delta_{R'}$ are the instantaneous isotopic compositions of a product and reactant

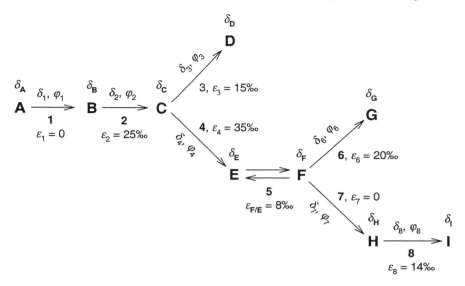

Figure 2. A network of chemical reactions. Letters indicate carbon positions within reactants and products. Isotopic compositions of these positions are indicated by δs with alphabetical subscripts. Reactions are designated by numbers and the δs, φs, and εs with numerical subscripts indicate respectively isotopic compositions of the carbon being transmitted by a reaction, the flux of carbon being transmitted (moles/time), and the isotope effect associated with the reaction. The latter value is expressed in ‰ and ε is defined in the text accompanying Equation (1).

and ε (expressed in ‰) is the isotope effect associated with the reaction linking R and P.[1] Here we have $\delta_1 = \delta_A - \varepsilon_1$ and, from $\varepsilon_1 = 0$, $\delta_1 = \delta_A$. To provide a quantitative basis for downstream δ values, we will define $\delta_A = 0$ ‰.

The example depicted in Figure 2 indicates that **B** is an intermediate between **A** and **C**. There is no division of the carbon flow, but there is a significant isotope effect associated with Reaction (2), which transforms **B** to yield **C**. Two requirements then seem to collide. From Equation (1), we know that $\delta_2 = \delta_B - \varepsilon_2$. From conservation of mass we have $\delta_2 = \delta_1$. Since $\varepsilon_2 = 25$‰, this requires $\delta_B = +25$‰ even though δ_1 (the δ value of the carbon flowing to **B**) = 0‰. At steady state, this *must* be true. Initially, it *cannot* be. The resolution of these requirements is shown in Figure 3, which depicts isotopic compositions as a function of time. Initially $\delta_1 = 0$ and $\delta_2 = -25$‰. As a result of this imbalance, δ_B rises until $\delta_2 = \delta_1$. The time constant for an adjustment of this kind is given by $\tau = m/\varphi$, where m represents the quantity of the intermediate (**B** in this case) present at steady state and φ is the flux (e.g. moles/time) of material through the intermediate pool. Whenever standing stocks are small relative to throughputs, steady-state conditions must be dominant in biosynthetic-reaction networks.

The flow of carbon divides at **C**, which lies at branch point in the network. From Equation (1) and the isotope effects specified for Reactions (3) and (4), we have $\delta_3 = \delta_C - 15$‰ and $\delta_4 = \delta_C - 35$‰. Mass balance requires also that the total amounts of ^{12}C and ^{13}C flowing to and from **C** are equal. In mathematical terms, the mass balance for total

---

[1] In order to focus on the principles, we will adopt the approximation represented by Equation (1). The precise relationship, which should be used in order to avoid errors in practical work, is given by $\varepsilon \equiv 10^3(\alpha - 1)$, where $\alpha \equiv [(\delta_R + 1000)/(\delta_P + 1000)]$.

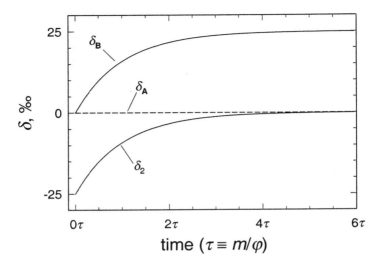

Figure 3. Isotopic compositions as a function of time. Time is expressed in units of τ, the time constant for the system. As noted, $\tau = m/\varphi$, where m is the quantity of intermediate present at steady state. Values of δ refer to reactant **A** (broken line), product **B** (upper solid line), and the carbon transmitted by Reaction (2) (lower solid line).

carbon is specified by

$$\varphi_2 = \varphi_3 + \varphi_4 \tag{2}$$

The mass balance for ^{13}C is specified by

$$\varphi_2 \delta_2 = \varphi_3 \delta_3 + \varphi_4 \delta_4 \tag{3}$$

At steady state, the flux terms can be replaced by a coefficient that describes the division of the carbon flow in terms of a branching ratio, $f_3 \equiv \varphi_3/\varphi_2$, where f_3 is the fraction of **C** that flows through Reaction (3) to yield **D**.

$$\delta_2 = f_3 \delta_3 + (1 - f_3)\delta_4 \tag{4}$$

The resulting relationship, in which the isotopic composition of the carbon flowing to **D** and **E** depends on f_3, is summarized graphically in Figure 4. As shown, at steady state, **C** can be either enriched or depleted in ^{13}C relative to **B**.

Isotopic shifts imposed on **E** propagate downstream. As a result, the isotopic compositions of **E**, **F**, **G**, **H**, and **I** are all affected by f_3. Any investigator seeking to understand the isotopic composition of **I**, and perhaps focusing his or her attention on Reactions (1), (2), (4), (5), (7), and (8) (the pathway linking **A** and **I**), would have to remember also that the *yield* of **D**, a byproduct in this reaction network, could have an important effect on δ_I.

The appearance of graphs like Figure 4 is controlled largely by the relative magnitudes of the isotope effects. The lines representing the isotopic compositions of the branching flows will be separated vertically by the difference between the isotope effects. If f expresses the fraction of the depleted product, the lines will have a positive slope. The slope will be negative if f expresses the fraction of the enriched product (compare Figs. 4 and 5).

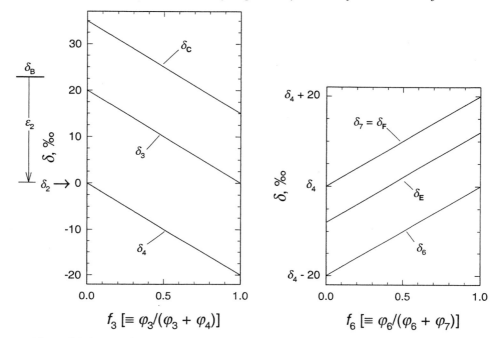

Figure 4 (left). Isotopic compositions of carbon flows, and of reactant **C**, at the **C** branch point. As shown by the graph, values of δ vary with f_3, a parameter equivalent to the fractional yield of **D**. The zero point of the δ scale is arbitrarily set by the isotopic composition of the input carbon flow, δ_1. The diagram at the left of the graph indicates δ_2, the isotopic composition of the carbon flowing to the branch point, and its relationship to δ_B.

Figure 5 (right). Isotopic compositions of carbon flows, and of components of the **E-F** equilibrium, at the **F** branch point. Values of δ vary with f_6, the fractional yield of **G**, and are expressed relative to δ_4, the isotopic composition of the carbon flowing to the **E** $\Leftrightarrow$ **F** equilibrium. Values of δ shown on this graph cannot be expressed relative to δ_1 unless f_3 has been specified (see Fig. 4, which shows that δ_4 is a function of f_3).

Substances **E** and **F** are in equilibrium. As a result, their isotopic compositions will vary together, the difference being controlled by the isotope effect associated with the equilibrium. In the example depicted by Figure 2, **F** will always be enriched in ^{13}C by 8‰ relative to **E**. The branch point at **F** is similar to that at **C**. In this case, the absence of an isotope effect at Reaction (7) means that the carbon flowing to **H** will have an isotopic composition equal to that of **F**, but the presence of a branching pathway with an unequal isotope effect ($\varepsilon_6 \neq \varepsilon_7$) means that significant fractionation can still occur. The resulting isotopic relationships are summarized graphically in Figure 5 and show that **G** provides a second example of a byproduct whose yield can influence the isotopic composition of **I**.

A substantial isotope effect is associated with Reaction (8), but the absence of a branching pathway means that the isotopic composition of **I** is effectively controlled by δ_7 ($\delta_I = \delta_7$). The only effect of $\varepsilon_8 > 0$ is to require that, at steady state, **H** is enriched relative to **I**.

Carbon positions vs. whole molecules. Important words appear in the caption to Figure 2: "Letters indicate carbon positions within reactants and products." For example, **C** could represent the C-2 (carbonyl) position in pyruvate. **D** and **E** would then represent the fates of that position in products derived from the pyruvate. Reactions occur between molecules but isotopic selectivity is expressed as chemical bonds are made or broken at particular carbon positions. Isotope effects pertain to those specific processes and control fractionations only at the reaction site, not throughout the molecule. To calculate changes in the isotopic compositions of whole molecules, we must first calculate the change at the reaction site, then make allowance for the rest of the molecule. In doing so, we inevitably find that the site-specific isotopic shift is diluted by admixture of the carbon that was just along for the ride. It follows that the isotopic differences that can be observed *between* molecules "must be related to, indeed, must be the attenuated and superficial manifestations of, isotopic differences *within* molecules" (Monson and Hayes 1982a).

Useful lessons. Generalities can be drawn from the foregoing discussion. First, isotopic compositions of intermediates (e.g. **B**) can differ substantially from those of final products (e.g. **I**). Second, the division of carbon flows at branch points can strongly affect isotopic compositions downstream. As a result, it is practically impossible to predict the isotopic compositions of final, biosynthetic products on the basis of observed isotopic compositions of intermediates. In plain words, if you want to know about the isotopic composition of lipids, it is very risky to rely entirely on analyses of acetate, even though most lipids are produced from acetate.

On the other hand, if you want to know, for example, *why* lipids are depleted in ^{13}C relative to other biosynthetic products, it will be necessary to examine evidence capable of revealing the structure and characteristics of the related network of reactions. Two complementary approaches have been developed thus far. Most straightforwardly, DeNiro and Epstein (1977) devised experiments for the determination of isotope effects at key points in the reaction network that links *n*-alkyl lipids to metabolites derived from carbohydrates and from some amino acids. In contrast, Abelson and Hoering (1961) pioneered the examination of intramolecular patterns of isotopic order. They studied the biosynthesis of amino acids, analyzing only the end products. However, they determined not only the δ values of the individual molecules but also (to the extent possible) the distributions of ^{13}C *within* the molecules. A similar approach was later used by Monson and Hayes (1982a) to study isotopic fractionations in lipid biosynthesis.

These reaction-based and product-based lines of investigation are complementary because neither can be perfect or complete and because inferences drawn from their results must be consistent. Even if it were possible to determine the isotope effect at each carbon position for every single step in a reaction sequence, and even if it were possible to recognize in advance every significant feature of the network, it would never be possible to be sure that studying the reactions in isolation was a reliable guide to their characteristics *in vivo*. On the other hand, even if it were possible to determine the isotopic composition at every single carbon position within all the related products of a network, it would never be possible to assign each observed variation reliably to causes that might lie many steps upstream in the sequence of reactions. More positively, results of intramolecular isotopic analyses could show that reaction-based investigations must have overlooked a key step. And, if multiple reaction steps have significantly influenced isotopic compositions in end products, quantitative interpretation of intramolecular patterns of isotopic order requires information developed from reaction-based studies. Examples will be provided in subsequent sections of this review.

Further general factors affecting isotopic compositions

Compartmentalization. A cell is not a stirred reaction vessel. In eukaryotic algae, for example, CO_2 is fixed and initial carbohydrates are produced in the chloroplast (also called plastid). Fatty acids are produced in the chloroplast and then exported for use elsewhere in the cell (Ohlrogge and Jaworski 1997) and all of the C_{20} and C_{40} isoprenoid carbon skeletons required by the photosynthetic apparatus are plastidic products (Kleinig 1989). In contrast, sterols, which are derived from a C_{30} isoprenoid carbon skeleton, are produced in the cytosol (Lichtenthaler 1999). In higher plants, the carbon feedstocks required to support biosynthesis outside the chloroplast are exported mainly as the C_3 carbohydrate derivative, dihydroxyacetone phosphate (Schleucher et al. 1998). As a result of these factors, plastidic fatty acids and cytosolic sterols, both derived from acetate, can have different isotopic starting points. Moreover, even when starting points and downstream processes are closely similar, the separation of pathways between compartments can mean that divisions of carbon flows at branch points differ significantly and, therefore, that final isotopic compositions differ sharply.

Smaller and larger organisms can present correspondingly simpler and more complex isotopic relationships. Prokaryotic cells, those of Archaea and Bacteria, are much smaller and lack the internal boundaries that stabilize chemical compartmentalization in Eukarya. As a result, compounds with similar structures and deriving from the same biosynthetic pathway *usually* have similar isotopic compositions (but see "Timing," below). In contrast, for higher organisms with differentiated cells, it is entirely possible that two compounds with identical carbon skeletons and biosynthetic origins (same pathway of synthesis, same location within the cell) but from different cell types (leaf epidermis *vs.* palisade cell) would have very different isotopic compositions even though both were products of the same organism.

Timing and reversibility. The concept of steady state was stressed in the discussion of Figures 2-5. But even bacterial populations have phases of growth (lag, exponential, stationary, and death) and, within a single cell, the production of one or more enzymes need not be uniform over the life of a cell. As a result, isotope effects at one or more important branch points may change. If an enzyme with a small isotope effect substitutes for one with a large effect, the isotopic compositions of downstream products will change. A specific example of this has been identified in methanotrophic bacteria (Summons et al. 1994) and is discussed in a concluding section of this review.

Reversibility of carbon flows must also play a role in controlling isotopic compositions of biosynthetic products. If carbon flows only *to* a particular product, it will be necessary only to consider isotope effects and branching ratios on the synthetic pathway. On the other hand, the isotopic compositions of compounds that are constantly being degraded as well as produced—compounds which "turn over" within the organism—will be shaped in addition by the isotopic characteristics of the degradative processes. Moreover, the *degree* of reversibility will be important. Molecular catalysts for which needs change over the lifetime of an organism—enzymes—turn over rapidly. Structural components such as lignin, cellulose, and some proteins (built from the same amino acids also found in enzymes) turn over much more slowly. There is evidence that the flow of carbon to the fatty acids in bacterial membranes is essentially unidirectional (Cronan and Vagelos 1972) and that some fatty acids involved in the regulation of membrane fluidity in higher plants turn over quite rapidly (Monson and Hayes 1982b), perhaps in response to diurnally cycling temperatures.

Isotopic compositions of compound classes relative to biomass

The most basic sorting of isotopes occurs in the distribution of carbon among nucleic acids, proteins, carbohydrates, and lipids, the major classes of compounds present within most cells. The most important pathways of carbon flow are indicated schematically in Figure 6. In autotrophs, these lead from carbohydrates (the direct products of carbon fixation) to proteins, nucleic acids, and lipids. Biomass—the organic material that

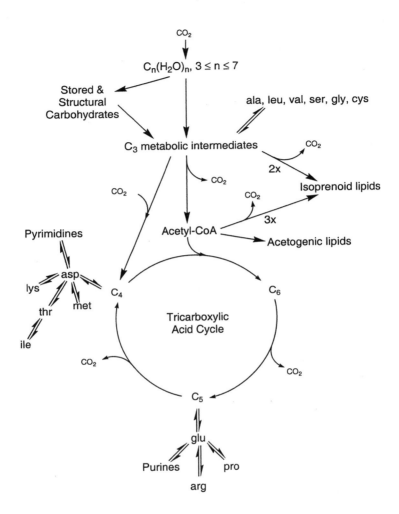

Figure 6. An overview of metabolic relationships. The arrows indicate dominant directions of carbon flow in, for example, a growing algal unicell. Most of the pathways indicated are, however, found in nearly all living organisms and are reversible to some degree. The three-letter abbreviations designate amino acids and are decoded in Table 1. As indicated, isoprenoid lipids can be made from either C_2 (acetyl-CoA) or C_3 precursors (see discussions of Table 4 and Fig. 29). The reversibility of the pathways related to the amino acids is indicated explicitly because those compounds turn over most rapidly and in practically all organisms.

comprises a living organism—is a mixture of these products. Two kinds of isotopic relationships are often discussed, those between various classes (e.g. "lipids are depleted in ^{13}C relative to proteins") and those between classes and biomass (e.g. "abundances of ^{13}C in nucleic acids and in biomass are essentially equal"). The first of these comparisons expresses a relationship like that between δ_3 and δ_4 in Figure 4. As carbon flows either to lipids or to proteins, we can expect the isotopic difference between them to remain roughly constant even though the δ values themselves will change. That change occurs

Table 1. Abbreviations for amino acids[a]

Amino acid	Single-letter code	Three-letter code	Amino acid	Single-letter code	Three-letter code
Alanine	A	ala	Isoleucine	I	ile
Arginine	R	arg	Leucine	L	leu
Aspartic acid	D	asp	Lysine	K	lys
Asparagine	N	asn	Methionine	M	met
asp *and/or* asn		asx	Phenylalanine	F	phe
Cysteine	C	cys	Proline	P	pro
Glutamic acid	E	glu	Serine	S	ser
Glutamine	Q	gln	Threonine	T	thr
glu *and/or* gln		glx	Tryptophan	W	trp
Glycine	G	gly	Tyrosine	Y	tyr
Histidine	H	his	Valine	V	val

[a] The more compact and less memorable single-letter codes are widely used in molecular biology and in proteomics. Most biogeochemical papers are still using the three-letter codes and this review follows that precedent.

relative to the available carbon supply (δ_2 in Fig. 4). Such changes must affect class-to-biomass comparisons. If more carbon flows to lipids, the cell will become more lipid-rich and its isotopic composition will approach that of the lipids. The depletion of ^{13}C in the lipids, expressed relative to biomass rather than relative to proteins, will decrease. This phenomenon can be considered quantitatively.

The isotopic composition of biomass carbon is given by Equation (5).

$$\delta_{Biomass} = X_{CNA}\delta_{NA} + X_{CProt}\delta_{Prot} + X_{CSacc}\delta_{Sacc} + X_{CLip}\delta_{Lip} \qquad (5)$$

where the subscripts NA, Prot, Sacc, and Lip respectively refer to nucleic acids and related materials such as nucleoside cofactors, proteins and amino acids, mono- and polysaccharides, and lipids of all kinds. The X_C terms, which sum to 1.0, refer to mole fractions of carbon and the δ terms refer to mass-weighted average isotopic compositions for all of the compounds within the indicated classes. Referring specifically to marine phytoplankton and considering a wide range of available observations, Laws (1991) concluded that $X_{CProt}/X_{CNA} = 8.6$ and discussed further regularities: at maximal rates of growth, X_{CProt} approaches 0.54; if the rate of growth is limited by availability of nutrients, X_{CProt} declines to values as low as 0.15 (with parallel declines in X_{CNA}); at low light levels, X_{CSacc} increases relative to X_{CLip}.

Regularities prevail also in isotopic relationships. Isotopic compositions of compound classes reported by Blair et al. (1985), Coffin et al. (1990), Benner et al. (1987), and Sternberg et al. (1986) converge to yield estimates of $\delta_{NA} \approx \delta_{Prot}$, $\delta_{Prot} - \delta_{Sacc} \approx$ -1‰, and $\delta_{Lip} - \delta_{Sacc} \approx$ -6‰. Figure 7, which expresses the depletion of ^{13}C in lipids relative to biomass, has been prepared using these relationships and Equation (5). *Note*: the indicated depletions have been calculated, not measured. The intent is to illustrate and

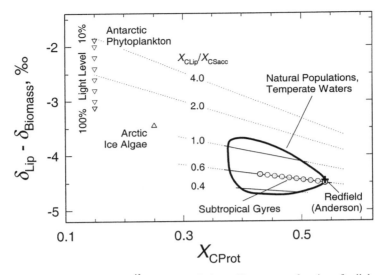

Figure 7. The depletion of ^{13}C in lipids relative to biomass as a function of cellular composition, where X_{CProt}, X_{CLip}, and X_{CSacch} are the mole fractions of carbon in proteins, lipids, and carbohydrates, respectively (see Eqn. 5 and related discussion). The indicated relationships are based on isotopic mass-balance requirements and on concepts outlined by Laws (1991). The cross marked "Redfield (Anderson)" indicates the position of cells with C/N/P = 106/16/1 but with lower (and much more realistic) abundances of H and O than those specified by the conventional Redfield formula (Anderson 1995).

to consider factors controlling δ_{Lip} - $\delta_{Biomass}$. If the saccharide-to-lipid fractionation in a particular organism were appreciably different from -6‰, the vertical axis would expand or contract. The results show both that the depletion of lipids in ^{13}C relative to biomass *can* vary widely and that, in nature, a consistent depletion can be understood quite readily. The dotted and solid lines represent relationships between the isotopic depletion and X_{CProt} at varying values of the lipid/carbohydrate ratio (X_{CLip}/X_{CSacc}). They show that the isotopic depletion becomes more marked as lipids become less abundant (i.e. as X_{CProt} increases) and that the relative abundance of carbohydrates can also be an important controlling factor.

The biomass of phytoplankton successfully competing in marine environments generally falls within the area enclosed by the gray envelope (Laws 1991). Compositions can vary more widely in laboratory cultures and in some extreme environments. The range which has been observed is roughly indicated by the extent of the dotted lines and by the points and ranges indicated for Arctic and Antarctic phytoplankton (Laws 1991 and references therein). For the particular case of phytoplankton in subtropical gyres, Laws (1991) estimates values of X_{CProt} and X_{CLip}/X_{CSacc} in the range indicated by the open circles. The maximum-protein end of this range lies close to the composition derived by Anderson (1995) as an updated and improved estimate of the Redfield composition ($C_{106}H_{175}O_{42}N_{16}P$ instead of $C_{106}H_{263}O_{110}N_{16}P$).

Isotopic compositions of carbohydrates

Mechanisms of production. Organic carbon is depleted in ^{13}C relative to inorganic carbon largely because of isotopic fractionations associated with the fixation of CO_2. Moreover, some of the hydrogen incorporated in this process becomes the first non-exchangeable H that shapes the hydrogen isotopic composition of organic materials.

Not all carbon fixation is photosynthetic. Table 2 provides an overview that has been organized in terms of the substrates, enzymatic catalysts, and first stable products. It does not include all processes in which inorganic carbon is incorporated in organic molecules, but it is meant to include all processes that contribute to net carbon fixation in organisms that can grow autotrophically. It thus includes pathways that occur not only in plants but also in Bacteria and Archaea that build biomass from inorganic carbon while deriving energy from chemical reactions.

For the moment, we will focus on photoautotrophs that utilize rubisco. Rubisco is the official name of an enzyme for which the systematic name—ribulose-1,5-bisphosphate carboxylase oxygenase—is inconveniently long. As indicated by the first activity specified in the systematic name, this enzyme catalyzes the carboxylation of ribulose-1,5-bisphosphate, "RuBP," a five-carbon molecule. A six-carbon product is formed as a transient intermediate, but the first stable products are two molecules of "PGA," 3-phosphoglyceric acid, $C_3H_7O_7P$. The carbon number of this compound gives the process its shorthand name, "C_3 photosynthesis." At physiological pH, the acidic functional groups on the reactants and products are ionized as shown below.

$$
\begin{array}{llll}
\text{CH}_2\text{OPO}_3^= & & \text{CH}_2\text{OPO}_3^= & \\
\text{C=O} & & \text{C-C-OH} & \\
\text{CHOH} & + \text{CO}_2 \xrightarrow{\text{Rubisco}} & \text{C=O} & \longrightarrow \\
\text{CHOH} & & \text{CHOH} & \\
\text{CH}_2\text{OPO}_3^= & & \text{CH}_2\text{OPO}_3^= &
\end{array}
\qquad (6)
$$

Ribulose-1,5-bisphosphate   3-phosphoglyceric acid

This reaction fixes carbon but there is no net change in oxidation number. The CO_2 is reduced to carboxyl but one of the carbon atoms in the RuBP is oxidized to yield the second carboxyl group. In subsequent steps, each mole of PGA reacts with a mole of NADPH in order to produce two moles of 3-phosphoglyceraldehyde, a product in which average oxidation number of carbon is 0. NADPH is the reduced form of nicotinamide adenine dinucleotide phosphate (see any biochemistry text for structures and further details). In biosynthetic processes, it functions as a hydride donor or reductant. A typical reaction is shown below. Note that NADPH + H^+ is equivalent to $NADP^+ + H_2$.

$$
\text{NADPH} + \text{H}^+ + \text{H}_3\text{C-C-C} \longrightarrow \text{H}_3\text{C-C-C} + \text{NADP}^+ \qquad (7)
$$

The isotopic composition of the H^- transferred from NADPH to biosynthetic substrates must be one of the most important factors controlling the hydrogen-isotopic composition of organic matter.

For contrast and completeness, NAD^+ and NADH should be introduced here. NAD^+ is the oxidized form of nicotinamide adenine dinucleotide. It is structurally identical to $NADP^+$ except that it lacks a phosphate group at a key point. Order and control are brought to biochemical oxidations and reductions by this seemingly trivial distinction. NAD^+ is generally an oxidant. NADPH is generally a reductant. Each is present within a cell at only microscopic concentrations. Specialized mechanisms exist for the oxidation

Table 2. Enzymatic isotope effects and overall fractionations associated with fixation of inorganic carbon during autotrophic growth.

Pathway, enzyme[a]	Reactant and substrate	Product	ε^b, ‰	Notes
C_3			10-22	1
Rubisco, form I, green plants and algae	CO_2 + Ribulose-1,5-bisphosphate	3-Phosphoglyceric acid (two moles)	30	2
Rubsico, form II, bacteria and cyanobacteria	CO_2 + Ribulose-1,5-bisphosphate	3-phosphoglyceric acid (two moles)	22	3
Phosphoenolpyruvate (PEP) carboxyase	HCO_3^- + PEP	Oxaloacetate	2	4
PEP carboxykinase	CO_2 + PEP	Oxaloacetate		5
C_4 and CAM			2-15	6
Phosphoenolpyruvate (PEP) carboxyase	HCO_3^- + PEP	Oxaloacetate	2	4
Rubisco, form I, green plants and algae	CO_2 + Ribulose-1,5-bisphosphate	3-Phosphoglyceric acid (two moles)	30	1
Acetyl-CoA			15-36	7
Formate dehydrogenase	CO_2 + 2 [H]	HCO_2^-		
Carbon monoxide dehydrogenase	CO_2 + 2 [H] + H_4Pt-CH_3 + CoASH	$CH_3COSCoA$ + H_2O + $H_4Folate$	52	8
Pyruvate synthase	CO_2 + Acetyl-CoA + 2 [H]	Pyruvate + CoASH		9
PEP carboxylase	HCO_3^- + PEP	Oxaloacetate	2	3
PEP carboxykinase	CO_2 + PEP	Oxaloacetate		10
Reductive or reverse TCA Cycle			4-13	11
α-Ketoglutarate synthase	CO_2 + Succinyl-CoA	α-Ketoglutarate + CoASH		
Isocitrate dehydrogenase	CO_2 + α-Ketoglutarate + 2 [H]	Isocitrate		
Pyruvate synthase	CO_2 + Acetyl-CoA + 2 [H]	Pyruvate + CoASH		
PEP carboxylase	HCO_3^- + PEP	Oxaloacetate		
3-Hydroxypropionate Cycle			0	12
Acetyl-CoA carboxylase	HCO_3^- + Acetyl-CoA	Malonyl-CoA		
Propionyl-CoA carboxylase	HCO_3^- + Propionyl-CoA	Methylmalonyl-CoA		
unknown		$\rightarrow C_3$		13
unknown		$\rightarrow C_4$		13

[a] For the C_3 and C_4 pathways, the enzymes are listed in the order in which they process carbon. For the bacterial pathways, the enzymes are listed in order of their quantitative contributions to net carbon fixation.

[b] $\equiv 1000[(^{12}k/^{13}k) - 1]$. Values of ε reported for pathways rather than for specific, enzymatic reactions reflect overall fractionation of ^{13}C vs. CO_2 and must represent a weighted average of the related enzymatic isotope effects.

Table 2 Notes. 1. The tabulated range is an estimate chosen to include both tropical phytoplankton with $\delta \approx$ -18‰ and land plants with $\delta \approx$ -30‰. 2. Guy et al. 1993; also reported as 29‰, Roeske and O'Leary 1984. 3. Guy et al. 1993; also reported as 18‰; Roeske and O'Leary 1985. 4. O'Leary et al. 1981. PEP carboxylase maintains supplies of C_4 metabolic intermediates (= *anaplerotic* fixation) during growth of C_3 plants and many bacteria. In C_4 plants it fulfills the same function *and* catalyzes the initial C-fixing reaction. 5. According to Descolas-Gros and Fontugne (1990), PEP carboxykinase is the main anaplerotic producer of C_4 intermediates in diatoms and chrysophytes. 6. The tabulated range is an estimate chosen to include both C_4 plants with $\delta \approx$ -10‰ and CAM plants with $\delta \approx$ -23‰. 7. Preuß et al. (1989); Belyaev et al. (1983). 8. Gelwicks et al. (1989), isotope effect observed at both methyl and carboxyl carbons of acetate. 9. Pyruvate synthase is distinct from pyruvate dehydrogenase, for which the isotope effect at C-3 is *ca.* 9‰ (Melzer and Schmidt 1987). Both enzymes use thiamine pyrophosphate as a cofactor. 10. Preuß et al. (1989) identify PEP carboxykinase as important in completion of the acetyl-CoA pathway in *Acetobacterium woodii*. 11. Range reported by Preuß et al. (1989). 12. van der Meer et al. (2001b). 13. Although the immediate C-fixing processes yield malonyl- and methylmalonyl-CoA, the net product of the 3-hydroxypropionate cycle is glyoxylate, a C_2 molecule. The processes by which glyoxylate is assimilated and converted to biomass are not known (Strauss and Fuchs 1993).

of NADH and for the reduction of $NADP^+$. If they are not activated and do not continuously regenerate NAD^+ or NADPH, the amount of oxidation or reduction that can occur is very strictly limited. The nicotinamide adenine dinucleotides are not the *only* biochemical reductants and oxidants, but they are by far the most common.

Rubisco's second activity—oxygenase—can be troublesome. If concentrations of O_2 are high, as they are likely to be in a brightly illuminated plant, the reaction between RuBP and O_2 can compete with carboxylation (Zelitch 1975). This process of *photorespiration* has a doubly negative result: CO_2 is not fixed and a mole of RuBP is destroyed. The problem is particularly severe for plants that grow in environments in which they must minimize amounts of water lost by evapotranspiration. If they close their stomata, they also tend to retain O_2 and to impede access to atmospheric CO_2. The ratio of CO_2 to O_2 declines and photorespiration is exacerbated. Two strategies, distinctly different but closely related, have evolved to overcome this problem.

In C_4 plants (Hatch 1977), the light (O_2-producing) and dark (carbohydrate-producing) reactions of photosynthesis are separated spatially. Moreover, a sort of molecular turbocharger is used to raise the pressure of CO_2 at rubisco's reaction site. Carbon is initially fixed in mesophyll cells, relatively close to the surface of the leaf. The reaction utilized is shown below.

$$HCO_3^- + \text{Phosphoenolpyruvate} \xrightarrow{\text{Phosphoenolpyruvate Carboxylase}} \text{Oxaloacetate} + HOPO_3^= \quad (8)$$

As shown, rubisco is avoided in favor of phosphoenolpyruvate carboxylase, an enzyme that has no competing, oxygenase activity. The first stable product is oxaloacetate, $C_4H_2O_5^{2-}$. The carbon number of this product gives this pathway its name. In a second step, NADPH is used to reduce the keto group on the oxaloacetate to CHOH (see Reaction 7). The malate thus produced diffuses to cells that sheath the bundle of fluid-transporting canals that make up a vein within the leaf. In these "bundle-sheath cells," the malate is reoxidized to oxaloacetate, thus reclaiming the reducing power that was expended in the mesophyll, and decarboxylated to yield CO_2 and pyruvate. The pyruvate returns to the mesophyll and is reactivated by isomerization and phosphorylation to form the phosphoenolpyruvate required in Reaction (8). Because the bundle-sheath cells are relatively impermeable to CO_2, concentrations of CO_2 within them rise to levels as high as 1 mM, nearly 100× higher than the concentrations in equilibrium with atmospheric concentrations of CO_2. Rubisco now catalyzes the production of phosphoglyceric acid and the NADPH derived from the oxidation of the malate is used to reduce the PGA to 3-phosphoglyceraldehyde.

CAM plants use the same chemistry but package it differently. Specifically, they lack the "Kranz anatomy" that is the defining characteristic of the C_4 plants. Kranz is the German word for wreath and refers to the appearance—in a cross-sectioned leaf—of the cells which sheath the vascular bundles in C_4 plants. CAM stands for Crassulacean Acid Metabolism. There is no such thing as crassulacean acid. The name instead refers to the initial discovery of this pathway of carbon fixation, in which oxaloacetic, malic, and pyruvic *acids* play key roles, in plants from the family Crassulaceae. CAM plants open their stomata, take in CO_2, and produce malate at night. Temperatures and, consequently, water losses are lower. During the day, the stomata are closed and the malate is processed as in the bundle-sheath cells of C_4 plants. Diffusive losses of CO_2 are, however, greater than those in C_4 plants.

Very recently, it has been demonstrated that some unicellular algae utilize the C_4 pathway of carbon fixation (Reinfelder 2000). Accordingly, it is necessary to distinguish clearly between C_4 *plants*, which have the Kranz anatomy, and the C_4 *pathway*, which is apparently quite widely distributed.

The Calvin Cycle. In all organisms that use rubisco to fix CO_2, whether initially (C_3 plants and most aerobic, chemoautotrophic bacteria) or after release of CO_2 from oxaloacetate (plants using the C_4 pathway), the 3-phosphoglyceraldehyde resulting from the reduction of the PGA is processed by the enzymes of the Calvin Cycle (synonyms = Calvin-Benson Cycle, Calvin-Bassham Cycle, reductive pentose phosphate cycle or pathway; Melvin Calvin received the Nobel Prize in chemistry in 1961 for his work on the pathway of carbon in photosynthesis—Andrew Benson and James Bassham were prominent among his coworkers). The reactions within the Calvin cycle are schematically summarized in Figure 8. As shown, RuBP is regenerated, an obvious requirement if carbon fixation is to be sustained. The reactions are complicated (Calvin and Bassham 1962) but the overall plan is simple. The net result of numerous rearrangements (none of which involve oxidation or reduction of carbon) can be summarized by the equation:

$$3\,C_5 + 3\,CO_2 \rightarrow 6\,C_3 \rightarrow 3\,C_5 + C_3 \tag{9}$$

Where C_5 represents RuBP and C_3 is a three-carbon carbohydrate, either glyceraldehyde or dihydroxy acetone. The C_3 can either be exported from the chloroplast directly, used for the synthesis of lipids and proteins within the chloroplast, or used to produce starch (a polymer of glucose, a C_6 carbohydrate) that can be retained in the chloroplast (in cells without chloroplasts, starch can occur in the cytosol).

Isotopic fractionations, C_3 pathway. The reaction network is shown below:

$$\delta_a \underset{\varphi_3,\,\delta_3}{\overset{\varphi_1,\,\delta_1}{\rightleftharpoons}} \overset{\delta_i}{C_i} \underset{\varepsilon_f}{\overset{\varphi_2,\,\delta_2}{\longrightarrow}} \overset{\delta_f}{C_f} \tag{10}$$

where C_a, C_i, and C_f represent respectively the ambient CO_2 (an infinite, stirred reservoir), the CO_2 inside the cell (at the active site of rubisco), and the fixed carbon (the carboxyl groups in PGA that derive from CO_2). The isotope effects ε_t and ε_f are

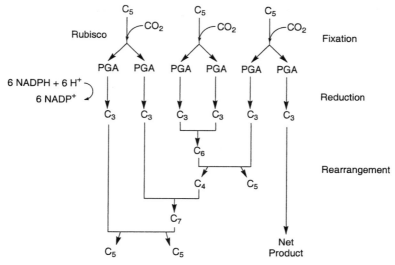

Figure 8. A schematic view of the flow of carbon in the Calvin Cycle. Subscripted Cs represent carbohydrates for which the net oxidation number for carbon is zero. PGA represents 3-phosphoglyceric acid. Six moles of NADPH + H$^+$ are required to provide the 12 electrons required to reduce three moles of carbon initially in the form of CO_2 to the level of carbohydrate. Energy provided by the hydrolysis of ATP (not shown) is required to drive the overall process.

respectively those associated with mass transport of CO_2 to and from the reaction site and with the fixation of CO_2. At steady state, mass balance must prevail at the C_i branch point. By analogy with Equation (4), we can write

$$\delta_1 = f_2\delta_2 + (1 - f_2)\delta_3 \tag{11}$$

where $f_2 \equiv \varphi_2/(\varphi_2 + \varphi_3)$. Substituting $\delta_1 = \delta_a - \varepsilon_t$, $\delta_2 = \delta_i - \varepsilon_f$, and $\delta_3 = \delta_i - \varepsilon_t$ yields an equation relating δ_a, f_2, the isotope effects, and δ_i. Substitution of $\delta_i = \delta_f + \varepsilon_f$ allows elimination of δ_i and provides the result

$$\varepsilon_p = \delta_a - \delta_f = \varepsilon_f - f_2(\varepsilon_f - \varepsilon_t) \tag{12}$$

where ε_p is the overall fractionation between C_a and C_f.

The form of Equation (12) differs from that of standard presentations. In considerations of emergent plants, f_2 is usually cast in terms of p_i and p_a, the internal and ambient partial pressures of CO_2 (Farquhar et al. 1989). For aquatic plants, various substitutions allow f_2 to be expressed in terms of c_e, the concentration of dissolved CO_2 in the water in which the plant is growing (Laws et al. 1995; Popp et al. 1998b). In all cases, however, the result is equivalent to that called for by Equation (12). As $f_2 \to 0$ (as p_i approaches p_a or as concentrations of dissolved CO_2 become so high that only the enzymatic reaction limits the rate of fixation), $\varepsilon_p \to \varepsilon_f$. Alternatively, if $f_2 \to 1$, indicating that nearly every molecule of CO_2 that enters the plant is fixed, then $\varepsilon_p \to \varepsilon_t$. These extremes can also be summarized phenomenologically: if the rate of fixation is limited only by the enzymatic reaction, then the resulting free exchange of CO_2 between the cell and its surroundings allows maximal expression of the enzymatic isotope effect. On the other hand, if every molecule of CO_2 that enters the plant is fixed, then the only isotope effect visible will be that of mass transport.

In the context established by Equation (10), ε_f is the net isotope effect associated with the fixation of inorganic carbon. As noted in Table 2, for C_3 plants this includes not only the rubisco-catalyzed carboxylation of RuBP but also the fixation of bicarbonate by PEP carboxylase. The latter process has a much smaller isotope effect *and* it utilizes a carbon source that is enriched in ^{13}C relative to dissolved CO_2. These factors are indicated schematically in Figure 9, which shows that carbon fixed by PEP carboxylase will be enriched in ^{13}C relative to dissolved CO_2. As a result of such phenomena, ε_f is smaller than the ε listed in Table 2 for rubisco itself. The difference depends on the amount of C fixed by PEP carboxylase. For land plants, Farquhar et al. (1989) estimate ε_f = 27‰. For unicellular, eukaryotic phytoplankton, Popp et al. (1998b) find ε_f = 25‰. The difference indicates either that fixation of C by PEP carboxylase is more important in algae or that the rubisco in those organisms has a slightly lower isotope effect than the enzyme present in higher plants. Values of ε_t are 4.4 ‰ for emergent plants in which the rate-determining transport of CO_2 occurs in air and 0.8‰ for plants that grow under water (0.7‰, O'Leary 1984; 0.87‰, Jähne et al. 1987). Rubisco exists in multiple forms with differing isotope effects (Robinson and Cavanaugh 1995). Values of $\varepsilon_{rubisco}$ are 29‰ for higher plants (Roeske and O'Leary 1984), and apparently somewhat lower for most phytoplankton and bacteria (Roeske and O'Leary 1985; Guy et al. 1993; Robinson and Cavanaugh 1995; Popp et al. 1998b).

Isotopic fractionations, C_4 pathway. The reaction network for a C_4 plant is:

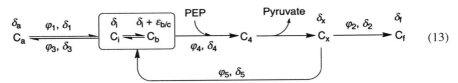

$$\text{(13)}$$

Terms are defined in parallel with those in Reaction (10), with the addition of C_b, C_4, and C_x which indicate respectively bicarbonate in equilibrium with CO_2 in mesophyll cells, C added to phosphoenolpyruvate (PEP) to produce malate and oxaloacetate, and CO_2 released by decarboxylation of oxaloacetate. Leakage of CO_2 from the site of decarboxylation is represented by φ_5. Taking this into account is a key factor if we are to understand variations in fractionation associated with the C_4 pathway. Two branch points appear in this network. For carbon in mesophyll cells we require

$$\varphi_1\delta_1 + \varphi_5\delta_5 = \varphi_4\delta_4 + \varphi_3\delta_3 \tag{14}$$

and, for carbon released by decarboxylation of oxaloacetate:

$$\varphi_4\delta_4 = \varphi_5\delta_5 + \varphi_2\delta_2 \tag{15}$$

Defining a leakage parameter, $L \equiv \varphi_5/\varphi_4$, we obtain

$$\delta_4 = L(\delta_5 - \delta_2) + \delta_2 \tag{16}$$

Relating the isotopic compositions of the specified fluxes of carbon to those of the various carbon pools, we can write $\delta_1 = \delta_a - \varepsilon_{ta}$, $\delta_3 = \delta_i - \varepsilon_{ta}$, $\delta_4 = \delta_i + \varepsilon_{b/d} - \varepsilon_c$, $\delta_2 = \delta_s - \varepsilon_f$, and $\delta_5 = \delta_x - \varepsilon_{tw}$, where ε_{ta} and ε_{tw} are respectively the isotope effects associated with mass transport of CO_2 in air and in water (as specified above, 4.4 and 0.8‰, respectively), $\varepsilon_{b/d}$ is the equilibrium isotope effect relating bicarbonate and dissolved CO_2 (Mook et al. 1974), and ε_c is the isotope effect associated with fixation of bicarbonate by phosphoenolpyruvate carboxylase (O'Leary et al. 1981; here we use a compromise value of 2.2‰ implied by O'Leary 1981).

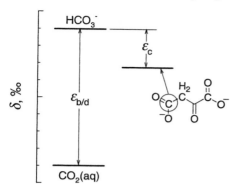

Figure 9. Diagram summarizing the isotopic relationships between dissolved CO_2, bicarbonate, and the carbon added to phosphoenolpyruvate in order to produce C-4 in oxaloacetate.

Using the relationships specified, our task is to rearrange and combine Equations (14) to (16) in order to produce an expression for ε_{P4} ($\approx \delta_a - \delta_f = \delta_a - \delta_2$), the net isotopic fractionation associated with the C_4 pathway. Using $\delta_5 - \delta_2 = \varepsilon_f - \varepsilon_{tw}$, Equation (16) can be rewritten as $\delta_4 = L(\varepsilon_f - \varepsilon_{tw}) + \delta_2$. Substituting for δ_4 yields an expression incorporating ε_c:

$$\delta_i + \varepsilon_{b/d} - \varepsilon_c = L(\varepsilon_f - \varepsilon_{tw}) + \delta_2 \tag{17}$$

Terms relating to the mass balance at C_i must be incorporated and δ_i, which can't be measured, must be eliminated. To accomplish these steps, we substitute $\delta_i = \delta_3 + \varepsilon_{ta}$, replacing δ_3 by an expression obtained by rearranging Equation (14). The result is

$$\varphi_1\delta_1 + \varphi_5\delta_5 - \varphi_4\delta_4 + \varphi_3\varepsilon_{ta} + \varphi_3(\varepsilon_{ta} + \varepsilon_{b/d} - \varepsilon_c) = \varphi_3 L(\varepsilon_f - \varepsilon_{tw}) + \varphi_3\delta_2 \tag{18}$$

Division by φ_1 yields important terms with the coefficient φ_3/φ_1, which is equal to $1 - f_2$. A final substitution for δ_5 based on rearrangement of Equation (15) yields

$$\delta_a - \delta_2 = \varepsilon_{P4} = \varepsilon_{ta} + [\varepsilon_c - \varepsilon_{b/d} + L(\varepsilon_f - \varepsilon_{tw}) - \varepsilon_{ta}](1 - f_2) \tag{19}$$

This equation relates ε_{P4} to the minimum number of controlling variables, specifically the isotope effects and the branching ratios (L, f_2) within the reaction network. Earlier, Farquhar (1983) presented a physiologically based derivation of an expression for ε_{P4}. Equation (19) is precisely equivalent except for the coefficient for L. The earlier derivation adopted $\varepsilon_{tw} \approx 0$ and thus yielded $L\varepsilon_f$ rather than $L(\varepsilon_f - \varepsilon_{tw})$. Expressed in terms of the internal and ambient partial pressures of CO_2, $1 - f_2 = p_i/p_a$.

Two important aspects of the carbon isotopic fractionation imposed by the C_4 pathway are summarized graphically in Figures 9 and 10. The first schematically indicates the carbon-isotopic relationships between dissolved CO_2, bicarbonate, and the carbon that is added to phosphoenolpyruvate in the reaction catalyzed by phosphoenolpyruvate carboxylase. It shows that, because the kinetic isotope effect associated with PEP carboxylase is smaller than the equilibrium isotope effect between dissolved CO_2 and bicarbonate, the fixed carbon is enriched in ^{13}C relative to that in the dissolved CO_2. As result, the CO_2 subsequently made available to rubisco in CAM and C_4 plants is enriched in ^{13}C relative to atmospheric CO_2. If that carbon were fixed with perfect efficiency, the biomass of the plant would be enriched in ^{13}C relative to CO_2 from the atmosphere—ε_{P4} would be negative, indicating an inverse fractionation.

Values of δ_i and δ_x (see Reaction 13) are, however, influenced strongly by the branching ratios at the C_i and C_x branch points. The first pertains to exchange of CO_2 between the atmosphere and the plant. As long as the partial pressure of CO_2 in the

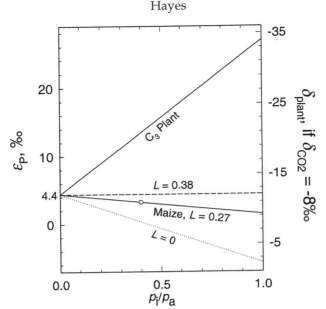

Figure 10. Overall isotopic fractionation (essentially the difference
between CO_2 and biomass) as a function of p_i/p_a, the ratio of partial
pressures of CO_2 inside and outside C-fixing cells. The scale on the
right-hand axis is inverted and shows $\delta_{biomass}$ for plants using CO_2 with
$\delta = -8‰$. The solid lines represent C_3 plants and maize, a representative
C_4 plant, with the point indicating its typical value of p_i/p_a (Marino and
McElroy 1991). The horizontal dashed line shows that ε_p is
independent of p_i/p_a when L, the leakiness coefficient for bundle-sheath
cells in a C_4 plant, is 0.38 (see Eqns. 13-19 and accompanying discussion). The dotted line represent a (hypothetical) plant with gas-tight
bundle-sheath cells and indicates that the biomass of such a plant could
be enriched in ^{13}C relative to CO_2 for $p_i/p_a > 0.45$.

interior air spaces of the leaf is not zero, CO_2 will be diffusing out of the plant as well as
into it. The second pertains to leakage of CO_2 away from the site at which oxaloacetate is
decarboxylated. Depending on whether that leakage is minimal or extreme, the isotope
effect associated with rubisco will be expressed to a lesser or greater degree at the C_x
branch point. Net fractionations are plotted as a function of L and p_i/p_a in Figure 10. For
all cases, $p_i/p_a = 0$ corresponds to a situation in which carbon fixation is limited entirely
by the transport of CO_2 into the plant. Accordingly, $p_i/p_a = 0$ requires $\varepsilon_{P4} = \varepsilon_{ta} = 4.4‰$,
independent of L. If *no* CO_2 leaked away from the site of decarboxylation of oxaloacetate
but, at the same time, the plant was very freely exchanging CO_2 with the atmosphere—an
implausible combination of circumstances corresponding to $p_i/p_a \approx 1$ and $L =$
0—maximal inverse fractionation would result, with the fixed carbon being enriched in
^{13}C relative to atmospheric CO_2 by 5.7‰ (for $\varepsilon_{b/d} = 7.9‰$ at 25°C). In contrast to these
extremes, maize, a representative C_4 plant, typically has $L = 0.27$ and $p_i/p_a = 0.4$ (Marino
and McElroy 1991). Consistent with these values, measurements of the fractionation
between atmospheric CO_2 and cellulose from corn kernels consistently yielded
$\varepsilon_{P4} = 3.28‰$ (Marino and McElroy 1991). Given the weak dependence of ε_{P4} on p_i/p_a (a
hypothetical C_4 plant with $L = 0.38$ would yield ε_{P4} independent of p_i/p_a), Marino and
McElroy (1991) proposed that C_4-plant debris could provide a record of the isotopic
composition of atmospheric CO_2. For CAM plants, which lack the specialized, bundle-

sheath cells found in C_4 plants, values of L are significantly higher and, as a result, isotopic depletions are significantly greater.

Hydrogen isotopes in carbohydrates. Hydrogen is introduced by the reduction of PGA, with NADPH serving as the hydride donor (cf. Reaction 7). As shown in Figure 8, the fixed carbon and its accompanying, "non-exchangeable" hydrogen (the H directly bonded to C) are then shuffled among C_3, C_4, C_5, C_6, and C_7 sugars. The net effects are regeneration of ribulose bisphosphate and repackaging of fixed C and H into trioses (glyceraldehyde, dihydroxyacetone) and hexoses (fructose, glucose). These products can be used in biosynthesis, exported to the cytosol, or stored within the chloroplast as starch. The inventory of C-bound H remains constant within the Calvin Cycle. Isotopic fractionations can, therefore, only affect the distribution of D among positions, not the overall δ value. The hydrogen can, however, be partly exchanged with H_2O in the course of the numerous rearrangements within the Calvin Cycle. Yakir and DeNiro (1990) have reviewed earlier work and have experimentally isolated the production of photosynthate well enough to estimate $\varepsilon_{P/w} = -171‰$, where P designates photosynthate and w the water used by the plant. That isotopic relationship between water and photosynthate is indicated schematically at the left of Figure 11. The sketch includes some broken lines and a question mark to indicate that the fractionation might represent the net of two or more processes. These could include the transfer of H from H_2O to $NADP^+$, possibly favoring H that was strongly depleted in D (Luo et al. 1991); the transfer of H from NADPH to PGA; and subsequent exchanges of C-bound H with H_2O, a process which is known to favor partitioning of D into the carbohydrate ($1.00 \leq K \leq 1.18$; Cleland et al. 1977).

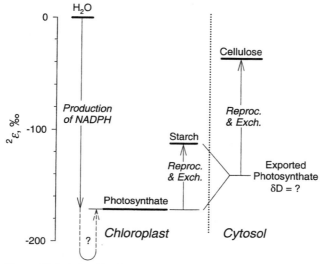

Figure 11. Relationships between the hydrogen isotopic compositions of initial photosynthate and related products and of water available within plant cells. The uncertainty indicated on the pathway between water and photosynthate reflects the possibility of offsetting fractionations in the production of NADPH and the subsequent transfer of H^- to photosynthate.

There are ample opportunities for further hydrogen-isotopic fractionations. These have been well studied for the specific case of the non-exchangeable H in cellulose. As demonstrated by the examples cited in Table 3, they attenuate the fractionation imposed

Table 3. Hydrogen isotopic fractionations measured in aquatic plants.

	$\varepsilon_{C/w}$ [a]	$\varepsilon_{L/w}$ [b]	n [c]	Reference
Cyanobacteria		-198 ± 17	4	Estep & Hoering 1980
Macrophytes		-152 ± 27	16	Estep & Hoering 1980
Red algae	-45 ± 31	-139 ± 25	13	Sternberg et al. 1986
Brown algae	-42 ± 42	-178 ± 11	21	Sternberg et al. 1986
Green algae	-141 ± 56	-132 ± 18	6	Sternberg et al. 1986
Macrophytes	1 ± 40	-125 ± 20	21	Sternberg 1988

[a] $\equiv 1000[(\delta_{Cellulose} + 1000)/(\delta_{water} + 1000) - 1]$ (non-exchangeable H only)
[b] $\equiv 1000[(\delta_{Lipids} + 1000)/(\delta_{water} + 1000) - 1]$
[c] number of specimens; indicated uncertainties are standard deviations of populations
[d] Tabulated entry refers to saponifiable lipids. Fractionation for saponifiable lipids in the same samples = $-232 \pm 32‰$.

during the initial production of photosynthate. None of the values of $\varepsilon_{C/w}$ is as large as 171‰. Yakir (1992) has reviewed evidence showing that this apparent recovery of D is associated with the "heterotrophic processing" of photosynthate. This occurs when carbohydrates are degraded to yield carbon, energy, and reducing power for the biosynthesis of lipids and proteins. The oxidative pentose phosphate pathway is particularly important in this regard. As shown in Figure 12, it involves the partial oxidation of C_6 sugars and serves as a source for NADPH outside the chloroplast (in green algae, it also operates within the chloroplast; Falkowski and Raven 1997, p. 241). The C_5 products, pentoses, are repackaged to form new hexoses which can be condensed to form cellulose. Up to 50% of the C-bound H exchanges with water during this repackaging (Yakir and DeNiro 1990). To examine the effects of the reprocessing of carbohydrates in the cytoplasm, Luo and Sternberg (1991) compared the hydrogen-isotopic composition of plastidic starch with that of cellulose. As noted in Figure 11, the cellulose—built from exchange-affected cytosolic glucose—was consistently enriched in D, the average difference for five CAM plants and six C_3 plants being about 80‰.

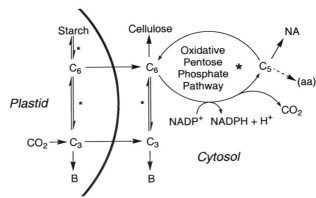

Figure 12. The flow of H from initial photosynthate (C_3 within the chloroplast) to stored carbohydrates and to cytosolic NADPH. Asterisks mark processes in which exchange of C-bound H with water is likely. B, biomass; NA, nucleic acids; (aa), minor contribution to amino acids.

Intramolecular carbon-isotopic order in carbohydrates. The carbon-isotopic fractionations described thus far have pertained to fixed carbon entering the Calvin Cycle. The extensive transfers of carbon within that cycle might be expected to act as an isotopic mixer, so that the distribution of ^{13}C within carbohydrates from photosynthate would be uniform. To examine this hypothesis, Rossman et al. (1991) used chemical degradations and microbial fermentations to produce CO_2 from each of the six carbon positions within glucose. One sample of glucose was from corn starch (maize, a C_4 plant) and the other was from sucrose from sugar beets (C_3). Both samples represented cytoplasmic, storage forms of carbohydrate and both yielded closely similar results. Relative to the average for the molecule, positions C-3 and C-4 were enriched in ^{13}C by 2‰ and 5‰, respectively, and position C-6 was depleted by 5‰. The precision of the analyses was such that the apparent imbalance (more enrichment than depletion) is not significant, but the consistency between plants and between the chemical and fermentative degradations is impressive. The evidence for isotopic inhomogeniety is substantial.

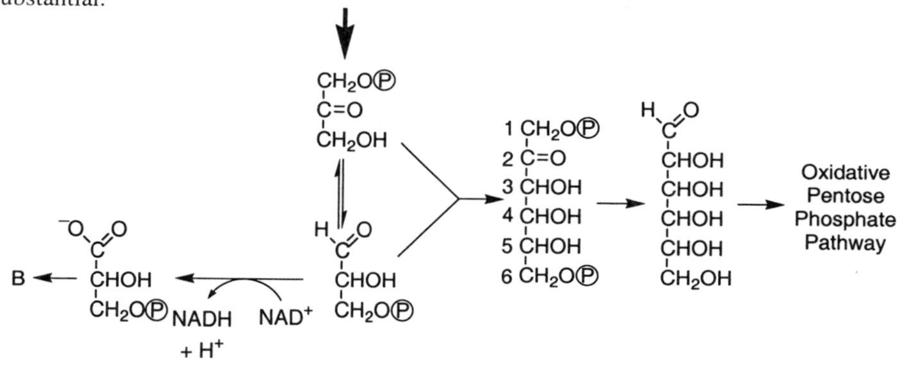

Figure 13. Flow of carbon from dihydroxyacetone phosphate to phosphoglyceraldehyde (and, with oxidation, to 3-phosphoglyceric acid and biomass) and to C_6 sugars. As shown, fructose-1,6-bisphosphate can be formed by combining dihydroxyacetone phosphate and phosphoglyceraldehyde. The further sugar shown has the structure of glucose (stereochemistry not shown). Carbon positions in such aldohexoses are numbered 1-6, starting with the aldehydic carbon. The processes indicated are possibly related to the intramolecular distribution of ^{13}C in glucose observed by Rossmann et al. (1991).

The relevant reaction network is shown in Figure 13. The input to the system is dihydroxyacetone phosphate, exported from the chloroplast. In the cytosol, its equilibration with glyceraldehyde-3-phosphate is catalyzed by triose-phosphate isomerase. As the first step in respiratory metabolism or in biosyntheses requiring C_2 units, a portion of the glyceraldehyde-3-phosphate is dehydrogenated to yield PGA. Rossmann et al. (1991) suggest that an isotope effect associated with this reaction could lead to enrichment of ^{13}C at the aldehyde carbon in glyceraldehyde-3-phosphate and that, by action of the triose phosphate isomerase, the enrichment could propagate to the free-alcohol carbon in dihydroxyacetone phosphate. Since these carbons flow to positions C-4 and C-3, respectively, in glucose, the mechanism provides a good explanation for the isotopic enrichment at those positions, even accounting for the greater enrichment at C-4. The depletion of ^{13}C at C-6 is unexplained. Rossmann et al. (1991) speculate that it arises because C-6 is sheltered from all of the bond-making and -breaking in the oxidative pentose phosphate pathway, which provides an independent means of affecting isotopic compositions within glucose. This might be the beginning of a correct explanation, but C-5 is also relatively sheltered and seems not to be isotopically unusual.

Isotopic compositions of amino acids

Amino acids are unique in that the first report of their isotopic compositions also provided intramolecular analyses (n.b. *intra*molecular = position-specific). In fact, the paper by Abelson and Hoering (1961) introduced the concept of intramolecular isotopic order and thus established a new basis for understanding the isotopic compositions of organic compounds. The breakthrough resulted in part from the development of ion-exchange-chromatographic techniques for the separation of amino acids. This high-resolution, liquid chromatographic technique provided quantities that were large enough for isotopic analyses using the techniques available in 1960. In conventional analyses, amino acids were detected by mixing the column effluent stream with a solution of ninhydrin. The resulting reaction, essentially quantitative (Moore and Stein 1951), produces a strongly colored product that can be detected spectrophotometrically. In their preparative separations, Abelson and Hoering spiked initial hydrolysates with traces of radiocarbon-labeled amino acids and followed them through the separations. Peaks were collected and aliquots were treated with ninhydrin:

$$2 \quad \text{(structure)} + R\text{-}C\text{-}CO_2^- \rightarrow \text{(structure)} + CO_2 + R\text{-}C\overset{O}{\underset{H}{}} + 3\,H_2O \quad (20)$$

As shown in the equation above, CO_2 is produced from the carboxyl group of the amino acid. Because the reaction is essentially quantitative, the isotopic composition of the CO_2 specifically reflects the abundance of ^{13}C in the carboxyl group.

[13]C in individual amino acids. More recent results from the Geophysical Laboratory—the site of Abelson and Hoering's (1961) seminal investigations—provide the best starting point for a consideration of the carbon-isotopic fractionations associated with the biosyntheses of amino acids. These later results, summarized graphically in Figure 14, come from a by-then-highly-experienced team of isotopic and amino-acid analysts (Macko et al. 1987). They relate to the simplest possible autotrophic system, a cyanobacterium grown under optimal conditions with continuous illumination and harvested in late log phase. Under these conditions, complications arising from the reversible carbon flows generally characteristic of amino acid biosyntheses must be minimized.

The intramolecular isotopic distributions reported by Abelson and Hoering (1961), though less precise and based on less carefully cultured organisms, provide further information. The results are summarized graphically in Figure 15. Positive vertical bars reflect enrichment of ^{13}C in the carboxyl group relative to the molecule as a whole. Even assigning an uncertainty of ±3‰ to each result, a few observations appear robust: (1) carboxyl groups of amino acids from autotrophs are generally enriched in ^{13}C relative to the rest of the molecule, with particularly strong enrichments in asx and glx, in the aromatic amino acids (phe and tyr), and in arg; and (2) intramolecular contrasts in amino acids produced by organisms growing heterotrophically are generally smaller, with the notable exception of carboxyl groups in leu, which are strongly *depleted* in ^{13}C relative to the rest of the molecule.

Additional intramolecular analyses of aspartate have been reported by Melzer and O'Leary (1987 1991). These authors examined aspartate produced by C_3 higher plants and showed that C-4, the carboxyl group not involved in peptide bonding, is commonly

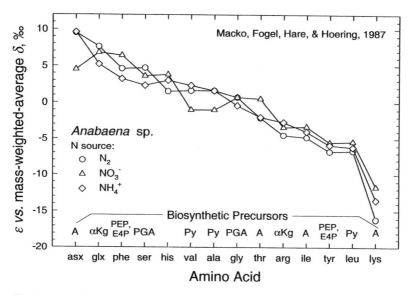

Figure 14. Carbon-isotopic compositions of individual amino acids synthesized by the cyanobacteria *Anabaena* sp. Strain IF (N_2-fixing and NO_3^--utilizing cultures) and Strain CA (NH_4^+-utilizing culture). Abundances of ^{13}C are expressed in terms of ε, the fractionation relative to the mass-weighted isotopic composition of all of the amino acids analyzed from each culture (Macko et al. 1987). The abbreviations specified in Table 1 are used to designate the amino acids. Abbreviations above the horizontal axis specify biosynthetic families: A, aspartic acid; αKg, α-ketoglutarate; PEP+E4P, phosphoenolpyruvate + erythrose-4-phosphate; PGA, phosphoglyceric acid; and Py, pyruvate. Histidine (his) is synthesized by a unique pathway. Lysine (lys) is a member of the aspartic-acid family except in *Euglena* and fungi, where it is a member of the α-ketoglutarate family.

enriched in ^{13}C by 10-12‰ relative to the rest of the molecule. The enrichment at C-4 in aspartate from C_4 plants is about half as large.

The amino acids are commonly divided into families based on their biosynthetic origins. These are indicated above the horizontal axis in Figure 14. The abbreviations and gaps are explained in the caption and the pathways are described below. The zero point of the ε scale in Figure 14 is set by the mass-weighted average isotopic composition of all of the amino acids analyzed. That point is probably about 1‰ above the biomass-average δ (see Eqn. 5 and related discussion). With allowances for respiratory losses of fixed carbon, the isotopic composition of photosynthate—the δ value of C_3 units being supplied to biosynthetic processes—is probably another 0.5‰ lower, near -1.5‰ on the ε scale defined in Figure 14. By this measure, thr is practically unfractionated relative to photosynthate and val, ala, and gly are only slightly enriched in ^{13}C. The left-to-right ordering of the amino acids was arranged to yield declining abundances of ^{13}C. Very clearly, it has *not* separated the amino acids into their biosynthetic families. Representatives of the aspartic acid family, for example, are spread from one end of the axis to the other. Do such depletions represent fractionations caused by successive isotope effects or must other explanations be sought?

The tricarboxylic acid cycle (TCA cycle) provides a logical starting point. Relationships between carbon skeletons in that cycle are shown in Figure 16. The TCA

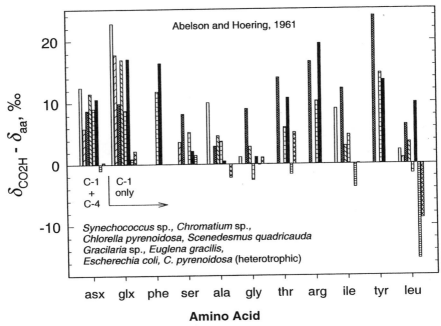

Figure 15. Enrichments (and rare depletions) of ^{13}C in carboxyl groups of amino acids as measured by Abelson and Hoering (1961). Enrichments and depletions are expressed relative to the overall average for each amino acid, *not* relative to the non-carboxyl carbon. Within each cluster, the bars refer in left-to-right sequence to the organisms listed (*Synechococcus* sp. is the present name of the organism identified as *Anacystis nidulans* in the original publications), with the last two representing heterotrophic cultures. As noted, the analytical procedure utilized produced CO_2 from both carboxyl groups in asp but only from the C-1 carboxyl group in glu.

cycle is normally thought of as the central carbon-processing facility of respiratory metabolism. As shown, acetyl groups are transferred from coenzyme A to oxaloacetate in order to produce citrate. The citrate is oxidatively decarboxylated to yield α-ketoglutarate which is in turn oxidatively decarboxylated to yield succinate. The succinate is further oxidized to yield oxaloacetate, thus regenerating the reactant needed for the next turn of the cycle. In effect, acetate is burned to yield CO_2. The oxygen required comes from H_2O. The redox budget is balanced by removal of H_2 (in the form of NADH + H[+], not shown in Fig. 16). A separate system (the "electron-transport chain") uses an inorganic electron acceptor (O_2, NO_3^-, SO_4^{2-}, among others) to oxidize the H_2 and conserves the energy produced by the oxidation of the acetate. Quite apart from its role in oxidative metabolism, the TCA cycle is important to amino-acid biosynthesis because, by producing oxaloacetate and α-ketoglutarate, it provides the carbon skeletons for aspartic and glutamic acids (notably the amino acids most strongly enriched in ^{13}C; see Fig. 14).

The key step in amino-acid biosynthesis is production of the carbon skeleton rather than the amino acid itself. Given the carbon skeleton, the process is completed by transamination. An example is shown in Reaction (21) below. In this case, the availability of phenylpyruvic acid allows production of phenyl-alanine. As shown, amino groups are generally provided by glutamic acid. A special system exists for the production of glutamate from α-ketoglutarate + NH_4^+. With this basic information we can now consider specific biosyntheses.

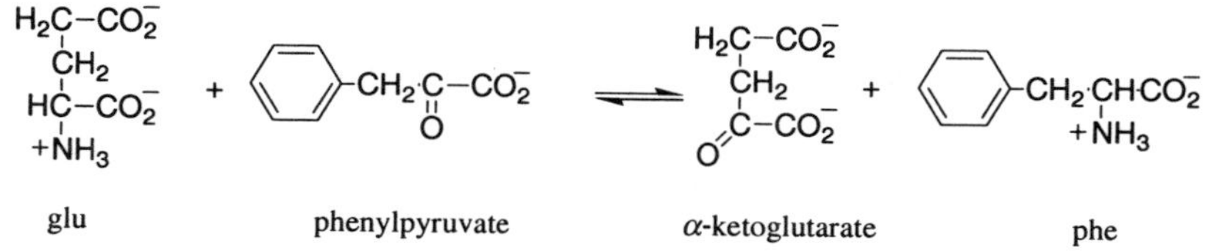

$$\begin{array}{c}
\text{H}_2\text{C}-\text{CO}_2^- \\
\text{CH}_2 \\
\text{HC}-\text{CO}_2^- \\
+\text{NH}_3
\end{array}
+ \;\;\boxed{}-\text{CH}_2\cdot\text{C}-\text{CO}_2^- \;\;\rightleftharpoons\;\;
\begin{array}{c}
\text{H}_2\text{C}-\text{CO}_2^- \\
\text{CH}_2 \\
\text{C}-\text{CO}_2^- \\
\end{array}
+ \;\;\boxed{}-\text{CH}_2\cdot\text{CHCO}_2^- \;\;\;\;(21)$$

glu   phenylpyruvate   α-ketoglutarate   phe

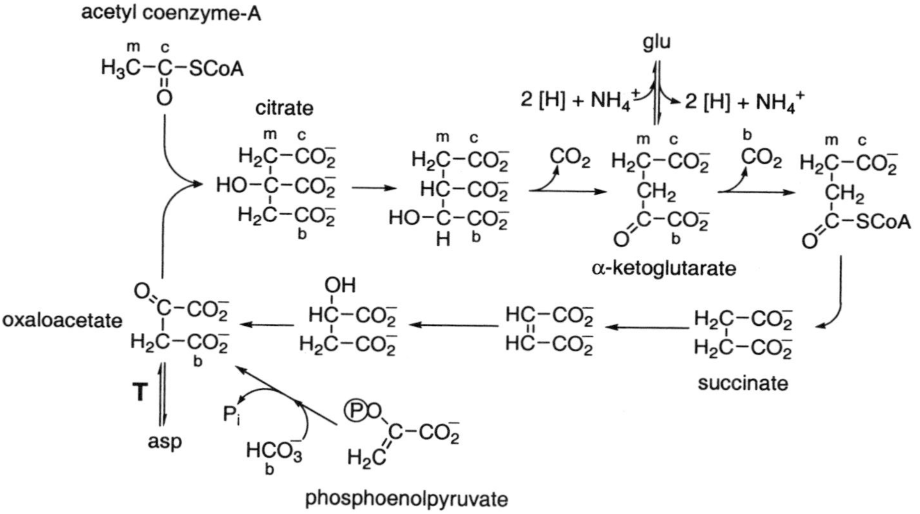

Figure 16. A schematic view of the flow of carbon within the tricarboxylic-acid cycle. The labels m, c, and b designate respectively carbon from the methyl and carboxyl positions of acetyl-CoA and from bicarbonate used to produce oxaloacetate from phosphoenolpyruvate. Inputs and outputs of water, of redox cofactors (NAD^+, *etc.*), and of coenzymes (CoASH) have been omitted in order to focus on the carbon skeletons. The amination of α-ketoglutarate in order to produce glu is the principal means of importing N for use in the amino-acid pool. The process involves multiple steps, including a reduction (indicated by addition of [H]) and is represented here only schematically. The boldface T indicates transamination, an example of which is shown in equation 21. The circled P represents a phosphate group, PO_3^{2-}. P_i represents inorganic phosphate, HPO_4^{2-}.

The aspartic-acid family can be considered first. The relationship of aspartic acid to oxaloacetate (OAA) and thus to the TCA cycle creates a problem, particularly for growing photoautotrophs. Aspartic acid is a very common amino acid. Relatively large quantities are required for the synthesis of proteins. If all the OAA is used to make aspartate, the TCA cycle will be shut down because citrate cannot be synthesized. An alternate source of OAA is therefore required. As shown in Figure 16, this is provided by the carboxylation of phosphoenolpyruvate (PEP), which can be formed directly from the products of C_3 photosynthesis. The resulting isotopic relationships are indicated schematically in Figure 17, which considers a representative C_3 plant with $\delta_a = -8‰$ and $\delta_f = -25‰$. The internal CO_2 will have $\delta_i = 2‰$ (from $\delta_i = \delta_f + \varepsilon_f$, see Eqn. 11 and accompanying discus-sion) and the bicarbonate available to PEP carboxylase will be further enriched as a result of the isotope effect associated with the CO_2-HCO_3^- equilibrium. The carbon at C-4 in the OAA is depleted in ^{13}C by only 2‰ relative to that source and is thus enriched by 32‰ relative to the other three carbon atoms in the molecule, assuming that their isotopic composition

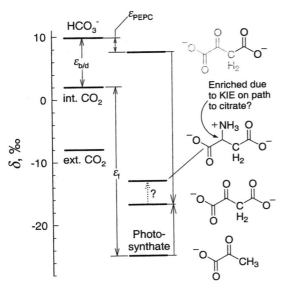

Figure 17. Diagram summarizing isotopic relationships between inorganic carbon pools and carbon in photosynthate and C_4 carbon skeletons. $\varepsilon_{b/d}$ is the equilibrium isotopic fractionation between dissolved CO_2 and bicarbonate (Mook et al. 1974). ε_f is the isotope effect associated with carbon fixation. The estimate of δ for internal CO_2 is based on $\delta_{photosynthate} = -25‰$ and $\varepsilon_f = 27‰$. ε_{pepc} is the isotope effect associated with phosphenolpyruvate carboxylase (O'Leary et al. 1982). The carbon pools represented in the right-hand column are, from top to bottom, the carbon added by the carboxylation of phosphoenolpyruvate, the total carbon in aspartic acid and in oxaloacetate, and the total carbon in pyruvate.

does not differ from that of photosynthate. The average δ for the four-carbon molecule is thus pulled up by 8‰. This is already in good agreement with the enrichments in asp relative to photosynthate (roughly represented in Fig. 14 by gly or thr). Further enrichment is possible if there is an isotope effect at C-2 in OAA during the formation of citrate. At steady state, such an effect would lead to a pile-up of ^{13}C at C-2 in the OAA pool. The 10-12‰ enrichment at C-4 in asp from higher plants measured by Melzer and O'Leary (1987 1991) was interpreted as indicating partial formation of OAA by PEP carboxylase with the remainder coming from the TCA cycle. Macko et al. (1987) did not measure intramolecular patterns in their cyanobacterial products, but the enrichment found by Abelson and Hoering (1961, see Fig. 15) was about 13‰. For the particular case of asp, the latter analyses represent both the C-1 and C-4 carboxyl groups. If the enrichment were localized at C-4, it might have been as great as 26‰ and thus approached the relationship indicated in Figure 17.

The pathways leading to other members of the aspartic acid family are shown in Figure 18. Isotopic compositions of three of these, thr, ile, and lys, are shown in Figure 14. Threonine (thr) retains all of the C present in asx. Why is it not similarly enriched? In fact, thr, ile, and lys are all significantly depleted in ^{13}C relative to asx. The first branch point occurs at asp itself, which can be used either to produce other amino acids or to synthesize proteins. A large isotope effect at C-4 in the production of aspartyl-β-phosphate could wholly or partly neutralize enrichment of ^{13}C at that position. But an isotope effect associated with the attachment of C-1 to aspartyl-transfer RNA via an ester linkage, the first step in protein synthesis, could lead to enrichment of ^{13}C at C-1. The second branch point occurs at β-aspartyl semialdehyde, which can flow either to homoserine or, with addition of pyruvate, to a seven-carbon intermediate that is rapidly cyclized to form 2,3-dihydropicolinate. Fractionation at this point could explain the depletion of ^{13}C in thr relative to asp, but would have the effect of sending a ^{13}C-enriched stream toward lys, which is in fact even more strongly depleted. The 2,3-dihydropicolinate is later opened to yield diaminopimelate. The reactions leading from β-aspartyl semialdehyde to diaminopimelate add carbon that dilutes the enrichment at C-4 in asp and involve very significant changes in bonding at four carbon atoms. These processes

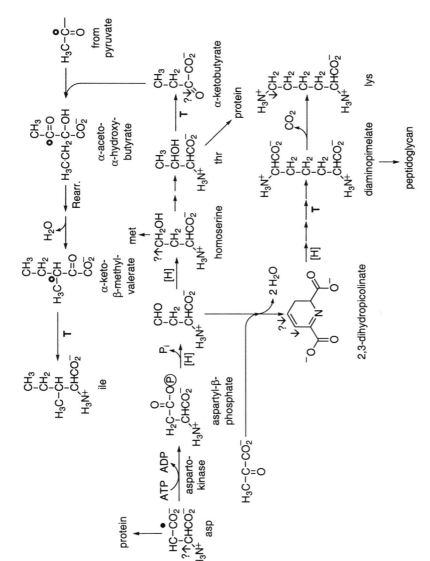

Figure 18. Biosynthetic pathways for amino acids in the aspartate family. Circled Ps represent phosphate groups, PO_3^{2-}. P_i represents inorganic phosphate, HPO_4^{2-}. [H] indicates reduction. T indicates transamination. Branch points and likely sites of isotopic fractionation are discussed in the text. In this and other reaction schemes, filled and open circles mark positions enriched or depleted in ^{13}C as a result of the source of the carbon flowing to that position. Upward and downward arrows mark positions likely to be enriched or depleted in ^{13}C relative to precursor positions as a result of fractionations induced by isotope effects.

and a further branch point at which diaminopimelate can be used to produce peptidoglycan (of which it is an important constituent in Gram-negative bacteria and in cyanobacteria) must play a role in explaining the depletion of ^{13}C in lys. A third branch point occurs at homoserine. Both paths involve largely irreversible reactions at the OH group (phosphorylation, acylation). Significant isotopic fractionations are not likely.

Isoleucine (ile) is the last member of the aspartic acid family represented in Figure 14. Two processes are potentially responsible for its depletion in ^{13}C relative to thr. The first lies at the thr branch point (Fig. 18), where an isotope effect on the transamination leading to α-ketobutyrate could lead to depletion of ^{13}C at C-2 in that product. The second is the addition of an acetyl group to produce α-aceto–α-hydroxybutryate. The carbonyl position in the acetyl group is likely to be depleted in ^{13}C (Melzer and Schmidt 1987).

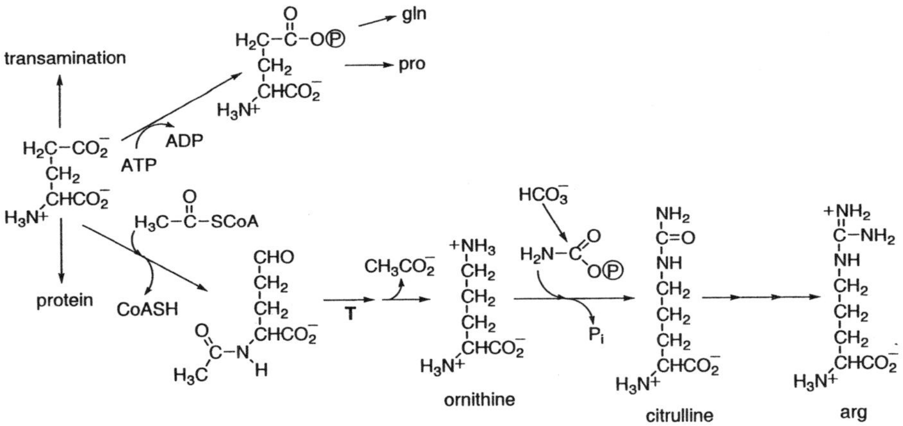

Figure 19. Biosynthetic pathways for amino acids in that glutamate family. Abbreviations as in Figure 18. The reactant formed from bicarbonate and used in the production of citrulline is carbamoyl phosphate.

Two members of the α-ketoglutarate family of amino acids are represented in Figure 14. Carbon in glu + gln (glx) is only slightly less strongly enriched in ^{13}C than asx. The intramolecular analyses which, in the case of glx, refer only to C-1, indicate enrichment of nearly 20‰ at that position. Tracking this carbon backward through the TCA cycle leads to C-4 of OAA. Comparison to the carboxyl-group enrichment measured for the sum of positions C-1 and C-4 in asp (Fig. 15) suggests that most of the excess ^{13}C in OAA must have been at C-4. As indicated in Figure 19, at least four pathways lead away from newly synthesized glu. Any kinetic isotope effect associated with transamination (in which glu serves as the NH₂ source for the synthesis of other amino acids) would enrich C-2. Any isotope effect associated with production of glutamate-5-phosphate has the potential to enrich C-5. While both of these mechanisms might contribute to the enrichment observed in glu, the first step leading toward ornithine and ultimately to arginine involves a reaction at the amino group and is not likely to cause any carbon-isotopic fractionation. Moreover, from that point, no carbon is removed from the set of atoms flowing to arg and only one, for which HCO_3^- serves as the source, is added. The depletion of ^{13}C in arg is thus unexplained.

Valine (val), alanine (ala), and leucine (leu) are members of the pyruvate family.

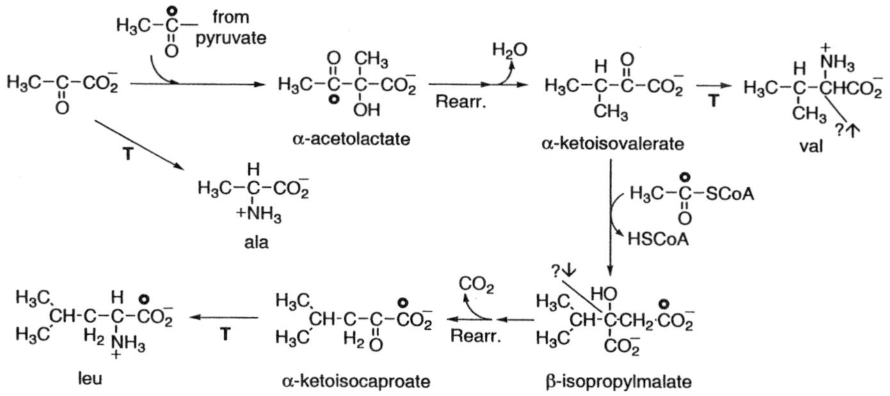

Figure 20. Biosynthetic pathways for amino acids in the pyruvate family. Abbreviations as in Figure 18.

Their biosynthetic pathways are summarized in Figure 20. Pyruvate is closely related to the C_3 products of photosynthesis. At least one major pathway leading from it, the production of acetyl groups by pyruvate dehydrogenase, has a significant carbon kinetic isotope effect (Melzer and Schmidt 1987). The slight enrichment observed in ala relative to the estimated isotopic composition of photosynthate fits into this picture nicely. As shown (Fig. 20), val is produced by addition of an acetyl group to the ala carbon skeleton. In spite of the expectation that this acetyl group would be depleted in ^{13}C relative to the pyruvate, val is slightly enriched in ^{13}C relative to ala. The branch point at α-ketoisovalerate provides an opportunity for the required fractionation. A carbon kinetic isotope effect at C-2 in that intermediate would send a ^{13}C-depleted stream to leu and a relatively enriched stream to val. The average content of ^{13}C in leu would be further decreased by the addition of a ^{13}C-depleted acetyl group. Leucine is the only amino acid which derives its carboxyl group from the carboxyl position in acetate. The uniquely large depletion of ^{13}C in the carboxyl group of leu from heterotrophic organisms (Fig. 15) provided the first evidence for the large carbon kinetic isotope effect associated with pyruvate dehydrogenase, now held responsible for the depletion of ^{13}C in *n*-alkyl lipids (DeNiro and Epstein 1977; Monson and Hayes 1982a). Notably, the carboxyl group of leu from photoautotrophs is *not* commonly depleted in ^{13}C (Fig. 14). It follows that the overall depletion of ^{13}C in leu must be due also to depletions at other sites within the molecule.

Serine (ser) and glycine (gly) are members of the phosphoglycerate family, for which biosynthetic pathways are summarized in Figure 21. A close isotopic relationship to photosynthate would be expected, but ser is enriched by roughly 5‰ (Fig. 14; even larger enrichments have been reported by Abelson and Hoering 1961, and by Winters 1971). Particularly because there is no evidence for strong enrichment of ^{13}C in the carboxyl group (Fig. 15), attention is due to the reaction by which gly is produced from ser. In it, C-3 of serine is transferred to tetrahydrofolate, from which it flows to provide C_1 units in a wide variety of biosynthetic reactions. A significant isotope effect associated with the transfer would be consistent with the enrichment of ^{13}C in ser relative to gly and with the relationships of these products to photosynthate.

The aromatic amino acids, phe and tyr remain. Their carbon derives from phosphoenolpyruvate and erythrose-4-phosphate. As required by the large, structural

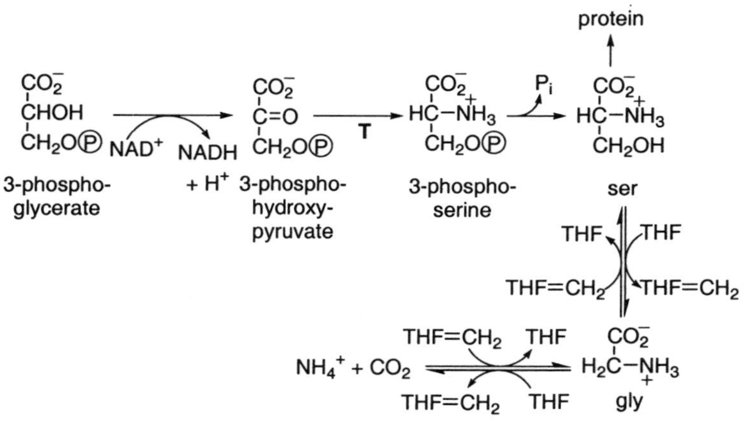

Figure 21. Biosynthetic pathways for amino acids in the 3-phosphoglycerate family. THF represents tetrahydrofolate, a cofactor which can accept and donate methylene groups. Other abbreviations as in Figure 18.

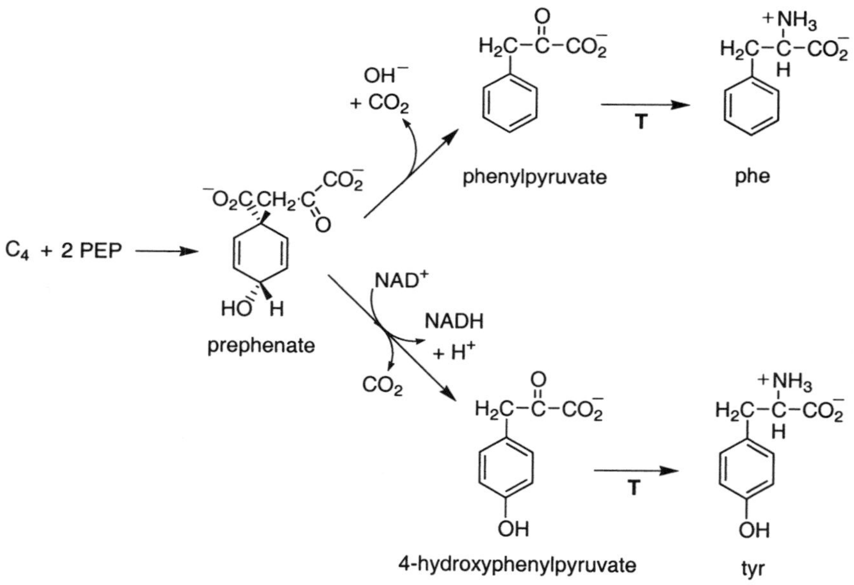

Figure 22. Terminal stages of the biosyntheses of the aromatic amino acids phe and tyr. As noted in the text, it is unlikely that isotope effects associated with this branch point alone can account for the isotopic difference observed between phe and tyr.

differences between those materials and the aromatic amino acids, there are many reactions between the biosynthetic feedstocks and the ultimate products. All but one of these, however, affect both phe and tyr which, as shown in Figure 22, share an immediate precursor. It is not possible to explain the intermolecular difference of more than 10‰ based on fractionations localized at the OH-bearing carbon in prephenate. The products each contain nine carbon atoms so that the required isotopic difference at that position would be 90‰. Pools of prephenate separated in time or space are indicated.

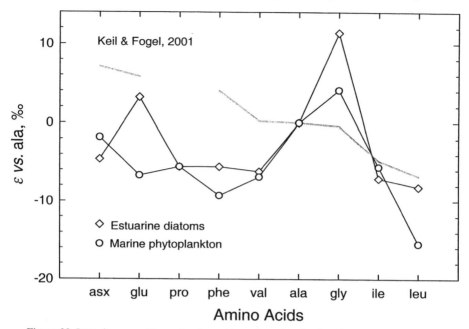

Figure 23. Isotopic compositions of amino acids, expressed as fractionations relative to alanine, from natural populations of estuarine diatoms and marine phytoplankton analyzed by Keil and Fogel (2001). For comparison, the heavy gray line depicts the average of the analyses reported in Figure 14. The range of ε values is identical to that in Figure 14. The horizontal scale covers only the amino acids analyzed by Keil and Fogel (2001).

Amino acids* not *from cultured cyanobacteria. Strikingly different isotopic distributions are found when amino acids from multicellular plants or from natural populations of algal unicells are analyzed. In general, the trend to greater downstream depletion discernible in Figure 14 is less marked. The generally flatter isotopic distribution is, however, interrupted by some notable enrichments and depletions. These points, along with some notable inconsistency and variability, are exemplified by the distributions shown in Figures 23 and 24. Amino acids in the sequence from asx to ala are commonly depleted and those in the sequence from ala to lys are commonly enriched relative to the trend defined by *Anabaena*. One feature of this flattening is removal of the puzzling isotopic contrast between phe and tyr. Notable enrichments of ^{13}C appear frequently in gly. On the other hand, the depletion of ^{13}C in leu noted originally in the *Anabaena* data appears to be a robust feature.

The turnover of amino acids in microbial cultures increases in stationary phase. If it is, at least, more important during slow growth and under natural stresses, it follows that isotopically depleted products from the ends of the biosynthetic pathways will be re-circulated more frequently than in the continuously illuminated, optimal cultures of *Anabaena*. Greater recycling will reduce the removal of carbon skeletons from the tricarboxylic acid cycle and result in the importation of less ^{13}C-enriched material from that source. Together, these processes should lead to the accumulation of ^{13}C-enriched residues at the ends of the biosynthetic pathways and to the delivery of ^{13}C-depleted feedstocks to the first steps in the pathways.

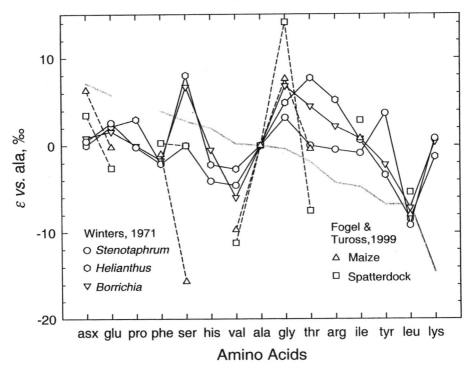

Figure 24. Isotopic compositions of amino acids, expressed as fractions relative to alanine, from the vascular plants *Stenotaphrum* (St. Augustine grass), *Helianthus* (sunflower), and *Borrichia* (sea daisy) reported by Winters (1971) and from maize and spatterdock reported by Fogel and Tuross (1999). For comparison, the heavy gray line depicts the average of the analyses reported in Figure 14.

The enrichment of ^{13}C in gly is probably linked to C_1 metabolism. As shown in Figure 21, methylene groups for use in other biosynthetic processes can be produced by the ser $\rightarrow$ gly and gly $\rightarrow CO_2 + NH_4^+$ pathways. Conversely, if THF=CH_2 is in over-supply, it can be consumed *via* reversal of those pathways. As a result, isotopic relationships between gly, ser, and the other amino acids should respond to production or consumption of THF=CH_2. Enrichment of ^{13}C in gly relative to ser seems most likely to reflect synthesis of gly from CO_2, NH_4^+, and THF=CH_2.

Isotopic compositions of nucleic acids

The nucleic acids are polynucleotides. Each nucleotide contains one phosphate group, a C_5 sugar (ribose or deoxyribose), and an aromatic, heterocyclic "base," either a one-ring pyrimidine (four or five carbon atoms + N, O, and H) or a two-ring purine (five carbon atoms + N, O, and H). The carbon isotopic compositions of nucleic acids, therefore, are expected to represent an average of the ribose and the bases. In the four-carbon pyrimidines uracil and cytidine, three carbons come from C-2, 3, and 4 from asp and the fourth comes from carbamoyl phosphate (see structure in Fig. 19). The five-carbon pyrimidine thymidine adds one C from a tetrahydrofolate carrier (cf. discussion of the biosynthesis of glycine). The inclusion of C-4 from asp and the insertion of C from carbamoyl phosphate, for which HCO_3^- serves as the carbon source, both suggest that the pyrimidines, with the possible exception of thymidine, would be enriched in ^{13}C relative to photosynthate. The purines contain both carbons from gly, two C_1 units from THF, and a C from CO_2. If one dared to make a prediction about such a salad, it would be that it was depleted in ^{13}C relative to photosynthate.

Observations indicate that the enrichments in the pyrimidines (likely, but of unknown magnitude) and the depletions in the purines (hypothetical, but apparently needed to explain the results) balance at least roughly. Pioneering analyses of the carbon-isotopic compositions of bacterial nucleic acids were reported by Blair et al. (1985), who found nucleic acids enriched in ^{13}C relative to biomass by 0.6‰. More recently, this work has been very nicely extended by Richard Coffin and his coworkers, who report that bacterial nucleic acids are enriched relative to biomass by about 0.3‰ (Coffin et al. 1990). The isotopic compositions closely follow those of heterotrophic carbon sources, and this can be exploited to provide new information about the roles of bacteria in aquatic food webs (Coffin et al. 1994, 1997).

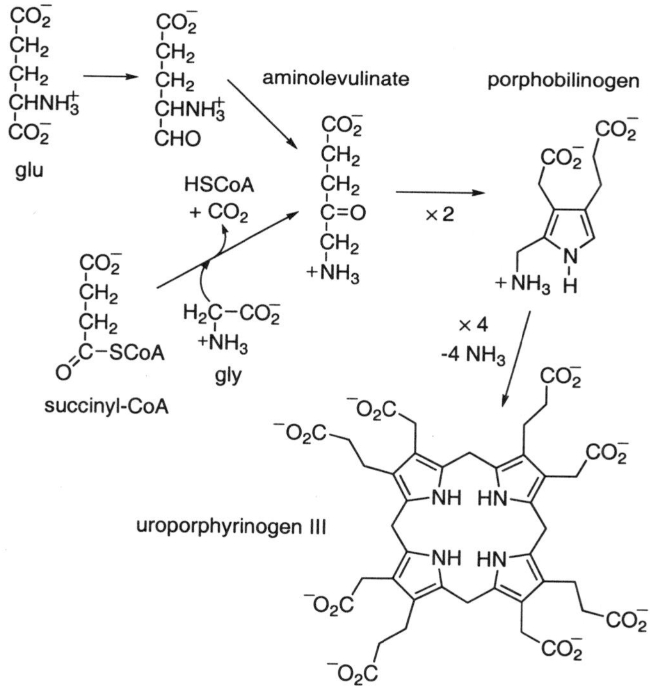

Figure 25. Diagram summarizing the relationships between carbon atoms in a tetrapyrrole ring system and in various precursors, either glu or succinate + the methylene carbon in gly.

Isotopic compositions of tetrapyrroles

The heteroaromatic ring system in chlorophyll is a prominent tetrapyrrole and is of particular interest since related products are often preserved in sediments as porphyrins. As shown in Figure 25, its immediate precursor is aminolevulinic acid (C_5) which is in turn synthesized from glycine and succinyl-CoA with loss of CO_2 ($C_2 + C_4 \rightarrow C_5 + C_1$) or from glutamate ($C_5$). The carbon flowing to tetrapyrroles is thus closely related to that in amino acids and in intermediates from the TCA cycle and is expected to have an isotopic composition close to that of biomass. In specific analyses of chlorophyllides (i.e. the non-phytol carbon from chlorophyll) from the photosynthetic bacteria *Rhodopseudomonas capsulata* and *Chromatium vinosum*, Takigiku (1987) found enrichments of 0.0 and 0.7 relative to biomass. Chloroplasts from beech tree leaves used to test procedures in the same investigation also yielded chlorophyllides enriched in ^{13}C relative to plastid biomass by 0.7‰. Madigan et al. (1989) found chlorophyllides from *Chromatium tepidum* enriched in ^{13}C by 0.5‰ relative to biomass. Considering diverse analyses that provided indirect evidence about the isotopic compositions of chlorophyllides relative to biomass, Laws et al. (1995) and Bidigare et al. (1999) concluded that 0.5‰ was a good estimate of the enrichment of chlorophyllides relative to biomass in marine plankton. In a single investigation of chlorophyllide from a cyanobacterium, *Synechocystis* sp., Sakata et al. (1997) found an enrichment of 2.7‰ relative to biomass and attributed the difference to shifted carbon flows that increased the abundance of proteins relative to lipids and other cellular components in cyanobacteria.

Isotopic compositions of lipids

Lipids have either linear or isoprenoidal carbon skeletons. All of the linear skeletons derive from the same biosynthetic pathway but there are two ways to produce isoprenoids, one of which has been discovered so recently that it is not mentioned in standard textbooks. The linear, or *n*-alkyl, lipids are often termed "acetogenic" because of their relationship to acetate, which is both their immediate biosynthetic precursor and the main product of their metabolic degradation. Structural variations within the acetogenic lipids are exemplified in Figure 26. Products containing no hydrolyzable linkages are "simple lipids," those containing one or more ether or ester bonds are "complex lipids." As shown, structural variations that affect the carbon skeleton are restricted to methyl branching and rare, cycloalkyl substituents.

The structural range of isoprenoid carbon skeletons is far greater. All are based on the isoprene, but there are two means of variation: (1) the connections between isoprene units, whether head-to-tail, head-to-head, tail-to-tail, or irregular (involving isoprene positions C-2 or C-3); and (2) cyclization. Only the first of these is considered in Figure 27. Assuming that no carbon atoms have been added to or trimmed from the basic carbon skeleton, the carbon number for any isoprenoid is some multiple of five. A C_{10} isoprenoid is a monoterpene, a C_{15} is a sesquiterpene, a C_{20} is a diterpene ... etc. In eukaryotic plants, including phytoplankton, the C_{20} and C_{40} isoprenoids are formed in the chloroplast and the C_{30} products are of cytosolic origin. The head-to-head linkage is generally restricted to the Archaea.

Turning to cyclization, isolated rings are common in C_{40} isoprenoid structures (carotenoids and archaeal tetraethers) but are not key points of classification. Structures with more than two fused rings are almost exclusively restricted to isoprenoids based on squalene (gibberelins, based on phytol, are a rare exception). Two major classes —tetracyclic and pentacyclic—are shown in Figure 28. Sterols are tetracyclic structures that are important constituents of all eukaryotic membranes. Although they have been widely reported as present in heterotrophic bacteria and cyanobacteria (see, for example, references cited by Kohl et al. 1983), proven instances of their *biosynthesis* are limited to aerobic, methanotrophic bacteria (Bird et al. 1971; Summons et al. 1994) and to *Nannocystis exedens* (Kohl et al. 1983). The emphasis on biosynthesis, demonstrated by transmission of an isotopic label from a precursor to steroidal products stems from (1) the demonstration by Levin and Bloch (1964) that sterols isolated from cyanobacterial cultures were not biosynthetic products but instead contaminants apparently derived from the media and (2) the observation that some bacteria which incorporate sterols in their membranes nevertheless require an exogenous source for the sterolic carbon skeleton (Razin 1978). Pentacyclic isoprenoid alcohols, often with highly polar, non-lipid

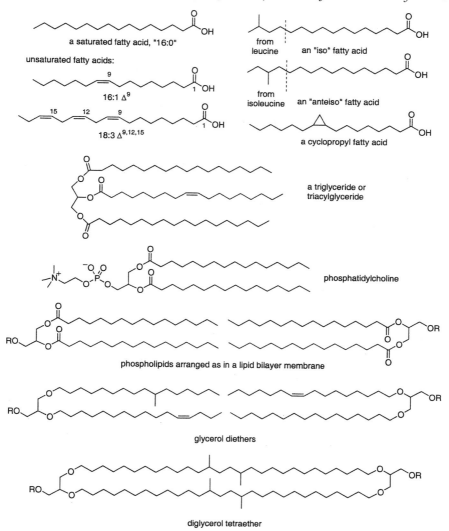

Figure 26. Representative *n*-alkyl, or "acetogenic" lipids. Each linear carbon skeleton is built up from C$_2$ units supplied in the form of acetyl-CoA. The saturated carbon skeletons represented here by carboxylic acids also occur commonly as alcohols. As shown, lipid bilayer membranes are comprised of diglycerides with polar "head groups" represented by R in the shorthand structures shown in the lower portion of the figure. An example of a polar head group is shown in the structure of phosphatidylcholine. Although ester linkages are most common in complex lipids, ether linkages are being recognized with increasing frequency among bacterial products (Hayes 2000).

substituents that raise their carbon number to 35 or more, are produced by many aerobic bacteria and apparently serve as sterol surrogates in their membranes (Rohmer et al. 1984). The dominant family of products is based on the hopane carbon skeleton (Fig. 28). Two additional groups of pentacyclic triterpenoids are comprised entirely of six-membered rings. Those with an OH group on the fifth ring are commonly products of protozoans. Tetrahymanol (Fig. 28) is a prominent example (Raederstorff and Rohmer

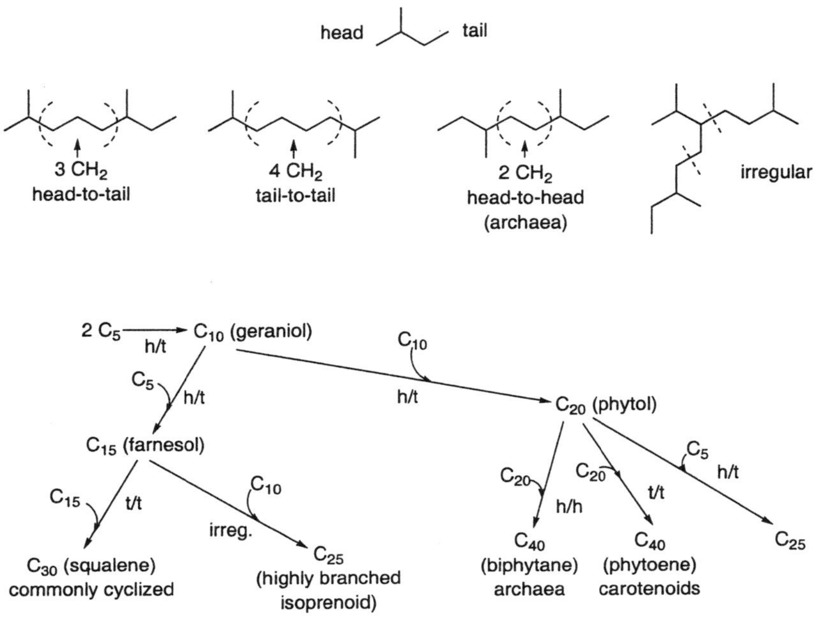

Figure 27. Molecular structures showing the that junctions between isoprene units can be recognized by counting the number of CH_2 groups between methyl branches. The branching diagram indicates biosynthetic relationships between various isoprenoid carbon skeletons. Names of typical products appear in parentheses (*n. b.*, geraniol and phytol are merely examples and are far from the only C_{10} and C_{20} isoprenoids). The irregular-junction case is an example based on Rowland et al. (1995).

1988). A thus-far-unique bacterial occurrence of this compound has also been reported in *Rhodopseudomonas palustris* (Kleeman et al. 1990). Pentacyclic triterpenoids with an OH group on the first ring are commonly products of higher plants.

Processes affecting the carbon-isotopic compositions of n-alkyl lipids. *n*-Alkyl carbon skeletons are essentially acetate polymers derived from acetyl-coenzyme A. In plants and in heterotrophs, that building block is formed by the oxidative decarboxylation of pyruvate, which is produced by the degradation of carbohydrates (including the immediate, C_3 products of photosynthesis). The reaction is catalyzed by pyruvate dehydrogenase and is associated with a significant, carbon kinetic isotope effect (Melzer and Schmidt 1987). Because pyruvate has multiple fates, isotopic fractionation is expected and, in fact, the general depletion of ^{13}C in *n*-alkyl lipids has been attributed to this step (DeNiro and Epstein 1977; Monson and Hayes 1982a). Additional factors must be important and may help to explain variations in lipid isotopic compositions (Blair et al. 1985). These factors include variations in the branching ratios at pyruvate; alternate sources of acetyl-CoA, which is also produced by the degradation of some amino acids (ile, leu, thr, and trp); and downstream isotope effects. Acetyl-CoA itself has multiple fates. If the reaction pathways competing for it have different isotope effects, the isotopic compositions of the *n*-alkyl lipids will vary from that of the acetyl-CoA. The fates include oxidation to CO_2 *via* the TCA cycle; biosynthesis of some amino acids (leu, lys); the production of *n*-alkyl-lipid carbon skeletons; the production of mevalonic acid, a precursor of isoprene; and, in many aquatic unicells, hydrolysis to yield acetate which leaks from the cell before being utilized (Roberts et al. 1955; Blair et al. 1985).

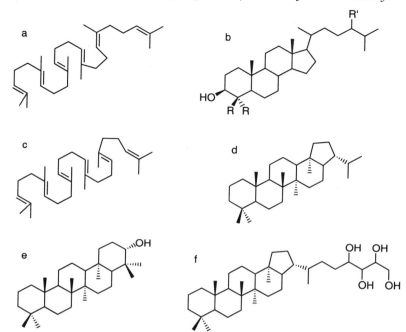

Figure 28. Cyclic isoprenoid lipids and their relationships to squalene. (a) Squalene folded so as to demonstrate its relationship to a sterol carbon skeleton (b). In sterols that occur commonly, some of the methyl groups resulting from the cyclization of squalene move from one carbon to another and others are lost. The substituents marked R can be either methyl or H and the substituent marked R' can be H, methyl, ethyl, or propyl. In specific natural products, double bonds often occur within the ring system. (c) Squalene folded so as to demonstrate its relationship to the pentacyclic ring systems in structures d, e, and f. (d) The hopane carbon skeleton. (e) Tetrahymanol. (f) Bacteriohopane tetrol, an extended hopanoid that includes five carbons deriving from the C$_5$ sugar ribose. (g) β-Amyrin, a typical pentacyclic isoprenoid produced by higher plants.

The site of fatty-acid synthesis varies very significantly. If a cell has chloroplasts—if it is an algal unicell or a carbon-fixing cell within a differentiated photoautotroph—all fatty acids with 18 or fewer carbon atoms are produced within it (Cavalier-Smith 2000). If a cell lacks chloroplasts the fatty acids are produced in the cytosol using (in eukaryotic cells) acetate exported from the mitochondria. The consequences of these points can be discussed more completely and systematically in parallel with a consideration of isoprenoid lipids.

Processes affecting the carbon-isotopic compositions of isoprenoid lipids. The isoprene carbon skeleton is indicated schematically in Figure 27. The corresponding biosynthetic reactant—equivalent in its role to acetyl-CoA—is isopentenyl pyrophosphate. As shown in Figure 29, this compound can be made by two different and fully independent pathways. The mevalonic-acid pathway was until recently thought to be the only route to isoprenoids. The deoxyxylulose-phosphate, or methylerythritol-phosphate, pathway was first discovered in Bacteria by Rohmer and coworkers (Flesch and Rohmer

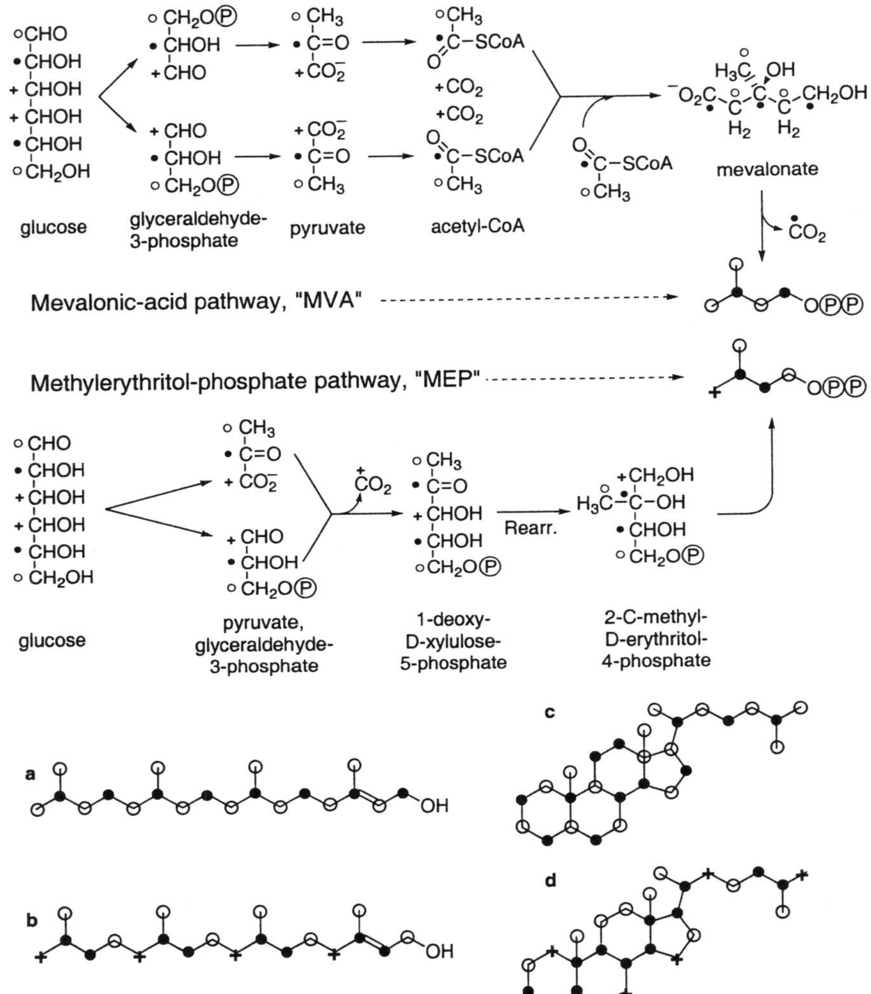

Figure 29. Relationships between the carbon positions in isopentenyl pyrophosphate and their sources. In the mevalonic-acid pathway, all five carbon positions in isopentenyl pyrophosphate derive from acetate and, in turn from the C-1 + C-6 and C-2 + C5 positions of glucose. In the methylerythritol-phosphate pathway, one carbon derives from the C-3 + C-4 position in glucose. Moreover, the mapping of positions from precursors into products of the two pathways differs sharply, as indicated by structures of acyclic and steroidal carbon skeletons based on the MVA (a, c) and MEP pathways (b, d).

1988; Rohmer et al. 1993). Subsequent investigations (reviewed by Lichtenthaler 1999) have shown that this pathway is widely distributed in prokaryotes, in the chloroplasts of eukaryotic algae and higher plants, and in the cytosol of members of the Chlorophyta (including the Trebouxiophyceae, Chlorophyceae, and Ulvophyceae; see summary in Table 4 and note very recent clarification by Schwender et al. 2001). In contrast to acetyl-CoA, isopentenyl pyrophosphate is a dedicated product flowing only to the biosynthesis

Table 4. Pathways used for the biosynthesis of isoprenoid lipids.

Organism	Pathway		Reference[a]
Prokaryotes			Lange et al. 2000
Bacteria			Boucher & Doolittle 2000
Aquificales, Thermotogales	MEP		
Photosynthetic bacteria			
Chloroflexus	MVA		Rieder et al. 1998
Chlorobium	MEP		Boucher & Doolittle 2000
Gram positive eubacteria			
Commonly	MEP		
Streptococcus, Staphylococcus	MVA		Boucher & Doolittle 2000
Streptomyces	MEP & MVA		Seto et al. 1996
Spirochaetes			Boucher & Doolittle 2000
Borrelia burgdorferi	MVA		
Treponema pallidum	MEP		
Proteobacteria			
Commonly	MEP		
Myxococcus, Nannocystis	MVA		Kohl et al. 1983
Cyanobacteria	MEP		Disch et al. 1998
Archaea	MVA		Lange et al. 2000
Eukaryotes			
Non-plastid-bearing	MVA		Lange et al. 2000
Plastid-bearing	*Plastid*	*Cytosol*	
Chlorophyta[b]	MEP	MEP	Schwender et al. 2001
Streptophyta[c]	MEP	MVA	Lichtenthaler et al. 1997
Euglenoids	MVA	MVA	Lichtenthaler 1999

[a] Where no reference is cited, either the assignment is based on generalizations introduced and supported in three major reviews—Lange et al. (2000), Boucher & Doolittle (2000), and Lichtenthaler (1999)—or it is based on work cited in those reviews, the first two of which are based on genetic analyses rather than labeling studies.

[b] Chlorophytes include the Trebouxiophyceae, Chlorophyceae, and Ulvophyceae.

[c] Streptophytes include higher plants and eukaryotic algae other than those named above.

of a single class of lipids. The isotopic compositions of the polyisoprenoid lipids will therefore be controlled by the isotopic composition of the isopentenyl pyrophosphate.

Isotope effects potentially associated with the syntheses of mevalonic acid or of methylerythritol have not been directly investigated. Nor have intramolecular patterns of isotopic order been measured in polyisoprenoids. As a result, stepwise or process-related isotopic fractionations associated with the biosynthesis of polyisoprenoid lipids can only be estimated from observed isotopic compositions of whole molecules.

Carbon isotopic compositions of lipids from heterotrophic bacteria. Four separate investigations of carbon isotopic fractionation in aerobic growth of *Escherichia coli* provide consistent and complementary results that provide a foundation for understanding the isotopic compositions of lipids. DeNiro and Epstein (1977) grew separate cultures of *E. coli* using glucose, pyruvate, and acetate as carbon sources. "Lipids" produced by the bacteria were depleted in ^{13}C by 6 to 8‰ relative to glucose or pyruvate but nearly unfractionated relative to acetate. Accordingly, they concluded that the reaction that produced acetate from pyruvate must be responsible for the depletion of ^{13}C in lipids. To examine that process specifically, they made site-specific analyses of all of the carbon positions in the reactants and products, substituting readily available yeast pyruvate carboxylase for *E. coli* pyruvate dehydrogenase. They found normal kinetic isotope effects at all carbon positions in pyruvate: $\varepsilon_{C-1} = 7.8$, $\varepsilon_{C-2} = 14.7$, and $\varepsilon_{C-3} = 1.0$‰. C-2 in pyruvate becomes the carboxyl carbon in acetyl-CoA. C-3 becomes the methyl carbon. Because of the strongly differing isotope effects, depletion of ^{13}C at the carboxyl position should be 15 times greater than that at the methyl position. The chain of chemically indistinguishable CH_2 groups within the fatty acid should contain two isotopically distinct subsets of carbon atoms, with those derived from the carboxyl carbon being depleted in ^{13}C relative to those derived from the methyl carbon. The depletion of ^{13}C in the molecule overall would be the average of the small depletion at the even-numbered positions and the large depletion at the odd-numbered positions.

Monson and Hayes (1980, 1982a) grew *E. coli* on glucose, isolated the individual fatty acids, and used chemical techniques to obtain CO_2 from specific positions within the molecules. Key reactions produced CO_2 quantitatively from carboxyl groups (Vogler and Hayes 1978; 1979) and oxidized double bonds quantitatively to produce carboxyl groups at the terminal positions of cleavage products (Monson and Hayes 1982a). The double bonds in unsaturated fatty acids produced by *E. coli* do not result from the action of a desaturase enzyme. Instead, even when O_2 is available, they are produced by an anaerobic mechanism that prevents complete hydrogenation of the alkyl chain as it is lengthened during biosynthesis. No mechanism exists for the isotopic fractionation of the doubly bonded carbons—for example, at C-9 and C-10 in *n*-hexadec-9-enoic acid—relative to the other odd- and even-numbered carbon positions in the molecule. Like DeNiro and Epstein (1977), Monson and Hayes (1982a) found that the crude lipids were depleted in ^{13}C by 7‰ relative to the carbon supplied by glucose. The individual fatty acids were, however, depleted by only 3‰ (the difference being explained by the presence of more strongly depleted neutral components in the lipid fraction). Production of CO_2 from positions within the carbon chain showed that the odd-numbered carbon positions, derived from the carboxyl group of acetyl-CoA, were depleted relative to the supplied glucose by 6±1‰ and that the even-numbered carbon positions, derived from the methyl position of acetyl-CoA, were enriched by 0.5±1.4‰. These results decisively confirmed the existence of intramolecular isotopic order in *n*-alkyl lipids. The pattern, moreover, was consistent with the outline provided by DeNiro and Epstein (1977) in that fractionation was localized at the carboxyl carbon of acetyl-CoA.

To interpret their results quantitatively in terms of isotope effects and reaction pathways, Monson and Hayes (1982a) adopted DeNiro and Epstein's (1977) most basic finding and considered the branch point shown in Figure 30. The existence of well-supported estimates of the carbon flows (Roberts et al. 1955) allowed construction of a fractionation plot based on the assumption that the isotopic composition of C-2 in the pyruvate was equal to that of the glucose supplied to the culture. The resulting fractionation factor, 23‰, is an estimate of the *difference between* the isotope effect at C-2 for pyruvate dehydrogenase *in vivo* and the weighted-average isotope effect at C-2 on the pathways leading to ala, val, and OAA. If those were greater than zero, then ε_{C-2} for pyruvate dehydrogenase must be greater than 23‰. The interpretation also assumed that no isotopic fractionation occurred on the pathway leading from acetyl-CoA to fatty acids and that the steady-state isotopic composition of acetyl-CoA was unaffected by any isotope effects associated with the pathways leading from that reactant to leucine, to the TCA cycle, or to acetate leaking from the cell (a point supported in part by the observation

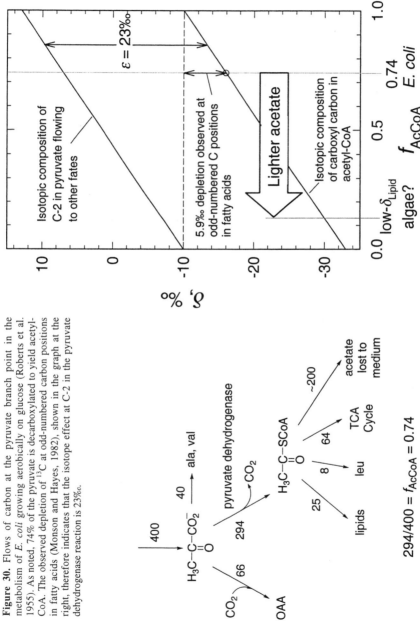

Figure 30. Flows of carbon at the pyruvate branch point in the metabolism of *E. coli* growing aerobically on glucose (Roberts et al. 1955). As noted, 74% of the pyruvate is decarboxylated to yield acetyl-CoA. The observed depletion of ^{13}C at odd-numbered carbon positions in fatty acids (Monson and Hayes, 1982), shown in the graph at the right, therefore indicates that the isotope effect at C-2 in the pyruvate dehydrogenase reaction is 23‰.

that even-numbered carbon positions in the fatty acids were unfractionated). The 23‰ effect was significantly larger than the 15‰ effect found by DeNiro and Epstein (1977), but it pertained to a different enzyme—pyruvate dehydrogenase in *E. coli* rather than pyruvate decarboxylase from yeast—and to *in-vivo* rather than *in-vitro* conditions.

A third approach was taken by Blair et al. (1985) who grew *E. coli* on glucose while obtaining a complete mass balance. They confirmed the finding of Monson and Hayes (1982a) that the fatty acids are depleted in ^{13}C by 3‰ relative to the glucose carbon supply. In their experiments, however, the excreted acetate was found to be enriched in ^{13}C relative to the carbon source by 12‰. Intramolecular analyses showed that virtually all of the enrichment was at the carboxyl carbon, which was enriched by 24‰. They also isolated and analyzed citrate (after quickly killing the cells at mid-log phase) and found that it was unfractionated relative to the carbon source. The latter result can be accommodated by proposing that the depletion of ^{13}C in the acetyl-CoA called for by the earlier results was neutralized because the citric acid contained not only the depleted acetate but also carbon from ^{13}C-enriched OAA, the enrichment resulting from utilization of the pyruvate represented by the upper line in the graph shown in Figure 30. The strikingly enriched extracellular acetate, however, remains unexplained. The authors point out that acetate would not only be leaking from the cells but also being reabsorbed by them and that the loss rate indicated in Figure 30 would be the net of those opposing flows. An isotope effect associated with the assimilation of acetate may, therefore, play a role.

In the final investigation, Melzer and Schmidt (1987) simply made a good measurement of the isotope effects associated with pyruvate dehydrogenase from *E. coli*. For ε_{C-2} they reported 21‰, in excellent agreement with the estimate provided by Monson and Hayes (1982a). In sum, these four investigations provide a roughly quantitative view of the processes responsible for the depletion of ^{13}C in *n*-alkyl lipids but also show how many factors might have importances that are not yet understood.

Isotopic fractionations in lipid biosynthesis during aerobic and anaerobic growth of *Shewanella putrefaciens* on lactate have recently been compared by Teece et al. (1999). Under aerobic conditions, fatty acids were depleted in ^{13}C relative to biomass by 2 to 3‰, roughly duplicating the characteristics of *E. coli* discussed above. In contrast, fatty acids produced under anaerobic conditions were depleted in ^{13}C by 5 to 10‰ relative to biomass. Earlier, Scott and Nealson (1994) had concluded that, under anaerobic conditions, *S. putrefaciens* metabolized lactate to yield acetyl-CoA and formate, with most of the acetate being excreted and a portion of the formate being assimilated *via* the serine pathway. If the latter pathway were fully functional, it would provide a second source of acetyl-CoA. The carboxyl carbon in that product is derived from HCO_3^- *via* PEP carboxylase, but Teece et al. (1999) do not report the isotopic composition of the CO_2 in their cultures. Together, the metabolic and isotopic investigations provide an excellent demonstration of anaerobic-bacterial complexity and its related hazards. Good work has led to good data but not to firm conclusions. The wide range of depletions (5-10‰) might indicate varying contributions from the two sources of acetyl-CoA (i.e. lactate and the serine pathway). The relatively large depletion must be kept in mind when considering the isotopic compositions of fatty acids in natural systems.

Carbon isotopic compositions of lipids from cyanobacteria. Although prokaryotic and thus uncompartmentalized, cyanobacteria differ from the organisms just considered in that they are photosynthetic, and thus obtain from light much of the energy required to drive biosynthesis. Moreover, they produce isoprenoidal as well as *n*-alkyl lipids. Sakata et al. (1997) recently reported results of the first investigation of lipid-biosynthetic fractionations in a cultured cyanobacterium, specifically *Synechocystis*, which uses the

MEP pathway for synthesis of isoprenoids (Lichtenthaler et al. 1997; Disch et al. 1998).

Extractable *n*-alkyl lipids were depleted in ^{13}C relative to total biomass by 9.1‰, a fractionation three times greater than that in *E. coli*. If localized at the odd-carbon positions, the depletion of 18‰ would require $f_{AcCoA} = 0.22$ (see Fig. 30). Factors suggested as responsible for the 3.4-fold decrease in f_{AcCoA} were (1) much lower needs for acetyl-CoA in energy production *via* the TCA cycle and (2) very low concentrations of lipids in *Synechocystis* (2% of biomass C). These are good points, but it will not be surprising if further investigations show that the large depletion is due to multiple factors.

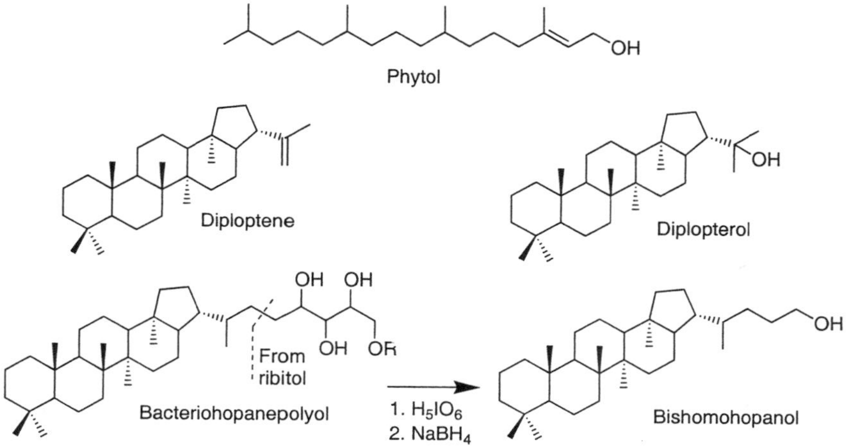

Figure 31. Polyisoprenoids produced by *Synechocystis* sp. and analyzed isotopically by Sakata et al. (1997). The bacteriohopanepolyol is too polar to be isolated in good yield. As is common in such cases, the procedure introduced by Rohmer et al. (1984) was therefore applied, with periodic acid being used to cleave the bonds between OH-bearing carbon atoms. Treatment with sodium borohydride then produces a C_{32} alcohol that can be recovered in good yield for further studies.

Three subsets of polyisoprenoids, shown in Figure 31, were analyzed. Phytol, which accounted for 90% of the total polyisoprenoids and 1% of biomass C, was depleted relative to biomass by 6.8‰. Diplopterol and diploptene, comprising 0.04% of biomass C, were similarly depleted (6.9‰, 6.5‰). Bishomohopanol (0.2% of biomass C), a degradation product of bacteriohopanepolyol, was depleted by 8.5‰. The latter result is striking because, as noted in Figure 31, bishomohopanol contains two carbons that are derived from carbohydrate (Rohmer 1993) and which are expected to be enriched in ^{13}C relative to lipid carbon. It suggests that the triterpenoid portion of the bishomohopanol is significantly more strongly depleted (~9‰) than the other isoprenoids. Noting that phytol, diplopterol, and diploptene were all resident mainly in membranes and that the abundance of bacteriohopanepolyol was more strongly correlated with cellular volume (Jürgens et al. 1992), Sakata et al. (1997) attributed the difference to changing branching ratios over the life of the cell, with greater depletion being favored in the later (volume-correlated) products.

The results provide some information about the isotopic characteristics of the MEP pathway of isoprenoid biosynthesis. As shown in Figure 29, there are two processes by which MEP might be depleted in ^{13}C relative to photosynthate. The first is in the decarboxylation of pyruvate to yield the C_2 unit which is condensed with glyceraldehyde to yield deoxyxylulose. The second—if, as seems likely for a C_5 carbohydrate, the deoxyxylulose has alternate fates—is in the rearrangement of the linear carbon skeleton to yield methylerythritol. If only the first of these steps were effective, only the tertiary C would be depleted in each isoprene unit and, to account for the observed overall depletion of 6.8‰, the depletion at that position would be 34‰. Since this exceeds the isotope effect found or estimated for any enzyme-catalyzed decarboxylation of pyruvate, it appears very likely that further fractionation, affecting at least one if not two additional carbon positions, is associated with the second step.

Table 5. Observed depletions of ^{13}C in lipids relative to biomass, prokaryotes.

Organism[a]	C source	Metabolism	$\varepsilon_{Biomass/Lipid}$, ‰ n-Alkyl	Isopren.[b]	Reference
E. coli	glucose	het. O_2	3		Monson & Hayes 1982
E. coli	glucose	het. O_2	3		Blair et al. 1985
S. putrefaciens	lactate	het. O_2	2-3		Teece et al. 1999
S. putrefaciens	lactate	het. NO_3^-	5-10		Teece et al. 1999
Synechocystis sp.	CO_2	C_3	9	6-8 E	Sakata et al. 1997
C. tepidum	CO_2	C_3		3.5 E	Madigan et al. 1989
C. tepidum	acetate	photohet.		5 E	Madigan et al. 1989
C. vinosum	CO_2	C_3		4.1 E	Takigiku 1987
Rps. capsulata	CO_2	C_3		4.9 E	Takigiku 1987
C. limicola	CO_2	rev. TCA	-16 - -11[c]	-2 - -3 E	van der Meer et al. 1998
T. roseopersicina	CO_2	C_3 + rev. TCA	-4 - 2[d]	4-5 E	van der Meer et al. 1998
C. aurantiacus	CO_2	3-OH prop.	0.7	4 V	van der Meer et al. 2001b
N. europea	CO_2	C_3 NH_3/O_2	7-13	9.3 E	Sakata et al. 2001
T. ruber	CO_2	unknown	-2		Jahnke et al. 2001
M. capsulatus	CH_4	RuMP	2-4	6-11[e] E	Summons et al. 1994
M. capsulatus	CH_4	RuMP	2-4	6-10 E	Jahnke et al. 1999
M. trichosporium	CH_4	serine	12	2 E	Jahnke et al. 1999
CEL1923	CH_4	RuMP	5	6.4 E	Jahnke et al. 1999
M. barkeri	$N(CH_3)_3$	M'gen		18 V	Summons et al. 1998
M. burtonii	$N(CH_3)_3$	M'gen		29 V	Summons et al. 1998
M. thermoauto-trophicum	CO_2	M'gen		13 V	Takigiku 1987
M. sedula	CO_2	3-OH prop.		-2	van der Meer et al. 2001a

[a] For genera see text.   [b] E = MEP, V = MVA.   [c] 17:1 unique at -5‰.   [d] 17:0 unique at 4‰.

[e] Tabulated depletions pertain only to end products formed early in life of a culture, when only the membrane-bound form of methane monooxygenase (MMO) is present. Squalene is enriched relative to end products. Fractionations decrease when soluble MMO is also present.

A survey of carbon isotopic compositions of lipids from other organisms. The lipid fractionations discussed in the preceding paragraphs provide the first entries in Table 5, which also summarizes results from many further investigations and provides a basis for discussion.

Lipids from several photosynthetic bacteria have been studied (photosynthetic bacteria are anaerobic phototrophs that use electron donors other than water and which are distinct from cyanobacteria). The Chromatiaceae (*Chromatium, Thiocapsa*) can grow photoheterotrophically (using light energy but an organic carbon source) as well as autotrophically. As autotrophs they use rubisco and the Calvin Cycle to fix carbon and to produce photosynthate (*T. roseopersicina* is reported to use in addition the reversed TCA cycle; Ivanovsky 1985). The corresponding fractionations of isoprenoids summarized in Table 5 must refer to biosyntheses *via* the MEP pathway, which is assumed to be universal in Bacteria. There is no obvious reason that *C. tepidum, C. vinosum, Rhodopseudomonas capsulata,* and *T. roseopersicina* should fractionate less strongly than *Synechocystis* (~4‰ vs. 6-8‰) and the similar fractionation obtained when *C. tepidum* uses acetate as the carbon source is remarkable, given the relatively complex network of reactions required to produce biosynthetic intermediates under such conditions (typically the glyoxalate cycle; Madigan et al. 2000, p. 629). Van der Meer et al. (1998) note that the inverse fractionation (= enrichment) of ^{13}C in *n*-alkyl carbon skeletons produced by *T. roseopersicina* may be due to use of acetate produced by the reversed TCA cycle (see discussion of *C. limicola,* below).

Chemoautotrophs are microbes which obtain energy by catalyzing an exergonic chemical reaction (commonly at interfaces between aerobic and anaerobic environments) and which produce biomass by fixing inorganic carbon. *Nitrosomonas europea* oxidizes ammonia and uses rubisco and the Calvin Cycle to fix carbon (the most common sulfide-oxidizing organisms also use the Calvin Cycle). Its isotopic characteristics could therefore, be expected to be similar to those of other aerobic, prokaryotic autotrophs such as *Synechocystis*. As shown in Table 5, this expectation is fulfilled.

Nearly all of the remaining organisms listed in Table 5 do not produce C_3–C_6 carbohydrates as immediate products of carbon fixation nor do they commonly assimilate them (*M. capsulatus* is the exception). As a result, pathways of carbon flow differ strongly from those in the organisms already discussed. *Chlorobium limicola,* for example, is a member of the Chlorobiaceae, which use the reversed TCA Cycle for fixation of carbon. Van der Meer et al. (1998) point out that the flow of carbon in such organisms proceeds from CO_2 to acetate and *then* to C_3 and larger compounds. Since each further step brings a chance for more fractionation, it should not be surprising that *n*-alkyl carbon skeletons are *enriched* in ^{13}C relative to biomass, although the magnitude of the enrichment in comparison to that observed in isoprenoids is both remarkable and unexplained.

Chloroflexus aurantiacus fixes carbon using the 3-hydroxypropionate pathway (Strauss and Fuchs 1993). Remarkably, it is known also to produce its isoprenoids from acetate *via* the MVA pathway (Rieder et al. 1998). Van der Meer et al. (2001b) point out that, if the same pool of acetate is used to produce the *n*-alkyl lipids, the isotopic compositions at the methyl and carboxyl positions of the acetate groups used in biosynthesis can be calculated (e.g. there are 12 methyl and 8 carboxyl carbons in the C_{20} isoprenoid and 8 of each in the C_{16} fatty acid). The fractionations reported in Table 5 then indicate that the carboxyl carbon is enriched in ^{13}C relative to the methyl carbon by 40‰! The pathways of carbon flow in *C. aurantiacus* are very incompletely known and the branch points potentially responsible for such a large fractionation cannot be identified.

Thermocrinus ruber is a hyperthermophilic, hydrogen-oxidizing chemoautotroph. The pathway by which it assimilates inorganic carbon is unknown. Its biomass is depleted relative to CO_2 by only 3‰. The *n*-alkyl lipids are in turn slightly enriched relative to biomass so that their isotopic composition is very close to that of the CO_2. In this case, fractionations may be minimized both by the high temperature, which reduces the magnitude of equilibrium isotope effects, and by the structure of the metabolic reaction network, which apparently leads to similar isotopic compositions for the precursors of both amino acids and lipids.

The remaining entries in Table 5 are involved in the production and consumption of methane. *Methylococcus capsulatus, Methylosinus trichosporium,* and the isolate designated as CEL1923 are methylotrophic bacteria. Such organisms assimilate methane (and, in many cases, other "C_1 compounds," such as methanol, formaldehyde, formic acid, methyl amines, and methyl sulfides) using either a pathway that yields carbohydrates (the ribulose monophosphate or RuMP pathway) or one which yields acetyl-CoA after the C_1 unit is initially added to glycine to yield serine (thus called the serine pathway). The methyl carbon in that acetyl-CoA derives from the C_1 substrate but the carboxyl carbon derives from CO_2. Variations are possible within the serine pathway. 3-Phosphoglyceric acid, an oxidation product of serine, can be withdrawn from the pathway much as C_4 and C_5 acids are withdrawn from the TCA cycle (Jahnke et al. 1999). The acetyl-CoA product can reenter the cycle and lead to the production of succinate for use in biosyntheses (White 1995, p. 265).

The RuMP- and serine-pathway methylotrophs yield sharply contrasting lipid-biosynthetic fractionations. In *M. capsulatus,* the *n*-alkyl carbon skeletons are depleted by 2 to 4‰ and the isoprenoids are depleted by 6 to 11‰. This resembles patterns seen in other prokaryotes in which biosynthesis starts from carbohydrates. The *n*-alkyl carbon skeletons are depleted as in *E. coli* and the isoprenoids are depleted roughly as in *Synechocystis* and *N. europea.* But *M. capsulatus* does not have a complete TCA cycle. Processes unique to the RuMP-cycle methylotrophs must be controlling the isotopic compositions of the lipid precursors in *M. capsulatus.* The serine pathway used by *M. trichosporium* leads straightforwardly to acetyl-CoA but not to building blocks for isoprene units. Why, then, are the *n*-alkyl carbon skeletons from serine-cycle methanotrophs so much more strongly depleted than the isoprenoids? If phosphoglyceric acid is withdrawn from the pathway and condensed with a C_2 unit to obtain deoxyxylulose, where is that C_2 unit coming from? It cannot be the same acetyl-CoA that is yielding the strongly depleted *n*-alkyl carbon skeletons. Very interestingly, the strong depletion in the *n*-alkyl carbon skeletons resembles that in anaerobic cultures of *Shewanella,* in which serine-pathway metabolism has been invoked on the basis of enzymological studies (Scott and Nealson 1994).

The initial study of *M. capsulatus* (Summons et al. 1994) provided the best example of temporal variations in isotopic fractionations. Two variants of methane mono-oxygenase, one soluble, the other membrane-bound, mediate the assimilation of methane. The membrane-bound enzyme is dominant at low cell densities (i.e. during the exponential phase of growth) and has an isotope effect that is large compared to that of the soluble enzyme, which becomes important later. As a result, all of the carbon initially available for production of biomass is depleted in ^{13}C relative to that assimilated later. The mixture of lipids produced also changes as the culture ages. Specifically, 3-methyl bacteriohopanepolyol and 4,4-dimethylsterols increase in abundance relative to bacteriohopanepolyol and 4-methyl sterols. As a result of these changes, the late-synthesized methyl hopanoids and dimethyl sterols are enriched in ^{13}C relative to the nonmethylated hopanoids and monomethyl sterols even though all derive from the same precursor (namely, squalene). The effect could be observed only by harvesting cultures at varying cell densities and making repeated analyses, and that very laborious task was undertaken only after initial observations (based, of course, on large amounts of material from fully grown cultures) had yielded "impossible" results (e.g. monomethyl sterols strongly depleted in ^{13}C relative to their dimethyl homologs). These elegant investigations also showed the squalene itself was commonly enriched in ^{13}C by 5 to 7‰

relative to its products, indicating that an isotope effect significant even for a 30-carbon molecule must be associated with the cyclization reaction.

Individual lipids have been isotopically analyzed in only three Archaea, all of them methanogens (see entries for *Methanosarcina barkeri*, *Methanococcoides burtonii*, and *Methanobacterium thermoautotrophicum* in Table 5). All reflect strong depletion of ^{13}C in lipids relative to biomass. Methanogens are commonly described as fixing carbon by use of the acetyl-CoA pathway. This can provide the feedstock required for synthesis of lipids, but C_3 and C_4 carbon skeletons are required for the synthesis of amino acids (which, in the form of proteins, account for most of the biomass). In methanogens, these are produced by additional CO_2-fixing steps (White 1995, p. 261). If the isotope effects associated with those reactions are much smaller than those associated with the production of acetyl-CoA, the isotopic contrast between the lipids and the biomass can be accounted for.

Table 6. Observed depletions of ^{13}C in lipids relative to biomass, eukaryotic algae.

Organism	$\varepsilon_{Biomass/Lipid}$, ‰		Reference
	n-*Alkyl*[a]	*Isoprenoids*[b]	
Scenedesmus communis	7-9, 12	5.3 Ep, 7 Ec	Schouten et al. 1998
Tetraedron miniimum	3-6, 6	0-5 Ep, -2 Ec	Schouten et al. 1998
Chlamydomonas monoica	7.5	4.9 Ep, 5.1 Ec	Schouten et al. 1998
Dunaliella sp.	3.5	0.5 Ep, -2.7 Ec	Schouten et al. 1998
Rhizosolenia setigera	5-7	2.9 Ep, 6-7 Vc	Schouten et al. 1998
Chaetoceros socialis	0-2	-1 Ep	Schouten et al. 1998
Thalassiosira weissflogii	1-2	1 Vc	Schouten et al. 1998
Gymnodinium simplex	1-3	4.5 Vc	Schouten et al. 1998
Isochrysis galbana	6-8, 3.1	2.8 Ep, 7 Vc	Schouten et al. 1998
Chrysochromulina polylepis	6	2.7 Ep	Schouten et al. 1998
Tetraselmis sp.	8	4.2 Ep, 0.4 Ec	Schouten et al. 1998
Rhodomonas sp.	5-7	2.2 Ep, 1 Vc	Schouten et al. 1998
Phaeodactylum tricornutum	9, 7	3.4 Ep, 7.2 Vc	Bidigare et al. 1997
Emiliania huxleyi	6, 4.1	4.2 Ep, 7.3 Vc	Bidigare et al. 1997
Emiliania huxleyi	4.2[c]		Popp et al. 1998a
Emiliania huxleyi	2.5, 4.0, 5.4[d]	2.0 Ep, 8.3 Vc	Riebesell et al. 2000

[a] In multiple entries, the first number or range pertains to C_{14}-C_{18} carbon skeletons presumably of plastidic origin and the second pertains to extended chains that may be of cytosolic origin.
[b] E, MEP pathway; V, MVA pathway; p, plastid; c, cytosol.
[c] C_{37} alkadienone only.
[d] In sequence: C_{14} and C_{16} fatty acids, C_{18} and C_{36} fatty acids, C_{22} fatty acid + C_{37} and C_{38} alkenones.

The entries in Table 6 pertain to eukaryotic algae. *S. communis*, *T. minimum*, *C. monoica*, and *Dunaliella* sp. are green algae and are expected to synthesize isoprenoids *via* the MEP pathway in the cytosol as well as in the plastids. The remaining species are expected to utilize the MEP pathway within the chloroplast and the MVA pathway in the cytosol. By far the most thorough and precise investigation was that of biosynthesis in *E. huxleyii* undertaken by Riebesell et al. (2000). Their analyses resolved four lipid groups: phytol and the C_{14} and C_{16} fatty acids, all depleted relative to biomass by 2 to 2.5‰ in spite of their contrasting origins from acetyl-CoA and methyerythritol; the unsaturated C_{18} fatty acids and a C_{36} alkadiene (apparently produced from two C_{18} acid skeletons), all depleted by 4‰; the C_{37} and C_{38} alkenones, depleted by 5.2‰; and a dominant sterol depleted by 8.3‰. Comfort can be taken from the fact that the other entry that comes closest to duplicating this pattern refers to additional analyses of *E. huxleyi*.

Depletions of ^{13}C in the C_{37} alkenones produced by *E. huxleyi* are of particular interest and illustrate a key point. The abundances of the alkenones relative to other products vary substantially depending on conditions of growth. For example, Riebesell et al. (2000) found that the alkenones comprised 1.0% of biomass C at low concentrations of dissolved CO_2 and up to 5.9% of biomass C at higher concentrations of dissolved CO_2. In spite of this variation in abundance, the depletion of ^{13}C relative to biomass in the C_{37} alkadienone consistently averaged 5.4‰ (s. d. = 0.3‰; $n = 10$) and was not correlated with [$CO_2(aq)$]. If, on the other hand, the isotopic composition of the same compound was expressed relative to that of the fully saturated C_{16} fatty acid it varied systematically from a depletion of 2.5‰ at low concentrations of dissolved CO_2 to a depletion of 4.0‰ at high concentrations, apparently reflecting a progressive redistribution of carbon within the lipid-biosynthetic reaction network. Although they found a significantly different fractionation (4.2‰ vs. 5.4‰, light and nutrient limitations varied between the two treatments), Popp et al. (1998a) found that the isotopic composition of the C_{37} alkadienone relative to biomass did not vary systematically with rate of growth although the isotopic composition of the biomass relative to the source CO_2 varied strongly.

Patterns of isotopic depletion among compound classes for other species frequently differ. It appears common, but far from universal, for MVA-pathway sterols to be depleted relative to biomass by 5-8‰. Phytol is nearly always less strongly depleted (2 to 5.5‰). Both of these relationships are consistent with field data summarized by Popp et al. (1999). If extended ($\geq C_{18}$) *n*-alkyl carbon skeletons are abundant, they are often isotopically depleted relative to the C_{14}-C_{17} fatty acids. Differences of more than about 1‰ cannot reasonably be attributed to the addition of a single, isotopically exotic acetyl unit (i.e. during the extension of a C_{16} chain to C_{18}). Such contrasts, therefore, probably indicate the use of distinct pools of acetate. There is a problem, however, in asserting that the lighter *n*-alkyl carbon skeletons must derive from cytosolic acetate, since the MVA-derived isoprenoids must also come from that pool and they are generally lighter still. To explore these relationships more securely, some immediate objectives can be defined. We need determine how much of the variability evident among the eukaryotic algae is experimental (analytical noise, stressed cultures under unnatural conditions, etc.) and how much is biological (response to subtle factors that may well vary in natural environments). We need to obtain a better view of the comparative isotopic characteristics of the MEP and MVA pathways. More analyses of sterols from green algae and *any* analyses of products of *Euglena* (in which the MVA pathway is found in both the plastid and the cytosol) would be helpful.

EPILOGUE

It often seems that isotopic fractionations provide *too much* information about *too many* processes, combining it all in a package that is unmanageably intricate. In response, investigators keep increasing the complexity of the available data by providing more and more detailed analyses. The proliferation of compound-specific isotopic analyses is a prime example of this phenomenon. Does it increase the information-carrying capacity of the isotopic channel or is it another case of the triumph of entropy? To obtain the preferred result, we will have to understand biosynthetic fractionations like those reviewed here.

ACKNOWLEDGMENTS

Support for this work and for the author's molecular-isotopic studies in general has come from the Programs in Exobiology and in Astrobiology at the National Aeronautics and Space Administration. It is a privilege also to acknowledge advice, comments, manuscripts, and reviews provided by Bob Bidigare, Marilyn Fogel, Ed Laws, Alex Sessions, Roger Summons, and Marcel van der Meer.

REFERENCES

Abelson PH, Hoering TC (1961) Carbon isotope fractionation in formation of amino acids by photosynthetic organisms. Proc Nat'l Acad Sci 47:623-632

Anderson LA (1995) On the hydrogen and oxygen content of marine phytoplankton. Deep-Sea Research 42:1675-1680

Belyaev SS, Wolkin R, Kenealy WR, DeNiro MJ, Epstein S, Zeikus JG (1983) Methanogenic bacteria from the Bondyuzhskoe oil field: general characterization and analysis of stable-carbon isotopic fractionation. Appl Environ Microbiol 45:691-697

Benner R, Fogel ML, Sprague EK, Hodson RE (1987) Depletion of ^{13}C in lignin and its implications for stable carbon isotope studies. Nature 329:708-710

Bidigare RR, Hanson KL, Buesseler KO, Wakeham SG, Freeman KH, Pancost RD, Millero FJ, Steinberg P, Popp BN, Latasa M, Landry MR, Laws EA (1999) Iron-stimulated changes in ^{13}C fractionation and export by equatorial Pacific phytoplankton: toward a paleogrowth rate proxy. Paleoceanogr 14: 589-595

Bidigare RR, Popp BN, Kenig F, Hanson K, Laws EA, Wakeham SG (1997) Variations in the stable carbon isotopic composition of algal biomarkers. Abstracts, 18th Int'l Meeting on Organic Geochemistry, Maastricht, The Netherlands, p 119-120

Bird CW, Lynch JM, Pirt FJ, Reid WW, Brooks CJW, Middleditich BS (1971) Steroids and squalene in *Methylococcus capsulatus* grown on methane. Nature 230:473-474

Blair N, Leu A, Muñoz E, Olsen J, Des Marais D (1985) Carbon isotopic fractionation in heterotrophic microbial metabolism. Appl Environ Microbiol 50:996-1001

Boucher Y, Doolittle WF (2000) The role of lateral gene transfer in the evolution of isoprenoid biosynthesis pathways. Mol Microbiol 37:703-716

Calvin M, Bassham JA (1962) The Photosynthesis of Carbon Compounds. W A Benjamin, New York

Cavalier-Smith T (2000) Membrane heredity and early chloroplast evolution. Trends Plant Sci 5:174-182

Cleland WW, O'Leary MH, Northrop DB (eds) (1977) Isotope Effects on Enzyme-Catalyzed Reactions. Appendix A: A note on the use of fractionation factors versus isotope effects on rate constants. University Park Press, Baltimore, Maryland

Coffin RB, Velinsky DJ, Devereux R, Price WA, Cifuentes LA (1990) Stable carbon isotope analysis of nucleic acids to trace sources of dissolved substrates used by estuarine bacteria. Appl Environ Microbiol 56:2012-2020

Coffin RB, Cifuentes LA, Elderidge PM (1994) The use of stable carbon isotopes to study microbial processes in estuaries. *In* K Lajtha, RH Michener (eds) Stable Isotopes in Ecology and Environmental Science, p 222-240. Blackwell Scientific Publications, Oxford, UK

Coffin RB, Cifuentes LA, Pritchard PH (1997) Assimilation of oil-derived carbon and remedial nitrogen applications by intertidal food chains on a contaminated beach in Prince William Sound, Alaska. Mar Environ Res 44:27-39

Cronan JE, Vagelos PR (1972) Metabolism and function of the membrane phospholipids of *Escherichia coli*. Biochimica et Biophysica Acta 265:25-60

DeNiro MJ, Epstein S (1977) Mechanism of carbon isotope fractionation associated with lipid synthesis. Science 197:261-263

Descolas-Gros C, Fontugne M (1990) Stable carbon isotope fractionation by marine phytoplankton during photosynthesis. Plant Cell Environment 13:207-218

Disch A, Schwender J, Müller C, Lichtenthaler HK, Rohmer M (1998) Distribution of the mevalonate and glyceraldehyde phosphate/pyruvate pathways for isoprenoid biosynthesis in unicellular algae and the cyanobacterium *Synechocystis* PCC 6714. Biochem J 333:381-388

Estep MF, Hoering TC (1980) Biogeochemistry of the stable hydrogen isotopes. Geochim Cosmochim Acta 44:1197-1206

Falkowski PG, Raven JA (1997) Aquatic Photosynthesis. Blackwell Science

Farquhar GD (1983) On the nature of carbon isotope discrimination in C$_4$ species. Aust J Plant Physiol 10:205-226

Farquhar GD, Ehleringer JR, Hubick KT (1989) Carbon isotope discrimination and photosynthesis. Ann Rev Plant Physiol Plant Mol Biol 40:503-537

Flesch G, Rohmer M (1988) Prokaryotic hopanoids: the biosynthesis of the bacteriohopane skeleton. Eur J Biochem 175:405-411

Fogel ML, Tuross N (1999) Transformation of plant biochemicals to geological macromolecules during early diagenesis. Oecologia 120:336-346

Garrett RH, Grisham CM (1999) Biochemistry, 2nd edition. Harcourt College Publishers, Philadelphia, Pennsylvania

Gelwicks JT, Risatti JB, Hayes JM (1989) Carbon isotope effects associated with autotrophic acetogenesis. Org Geochem 14:441-446

Guy RD, Fogel ML, Berry JA (1993) Photosynthetic fractionation of the stable isotopes of oxygen and carbon. Plant Physiol 101:37-47

Hatch MD (1977) C_4 pathway of photosynthesis: mechanism and physiological function. Trends Biochemical Sci 2:199-202

Hare PE, Fogel ML, Stafford Jr. TW, Mitchell AD, Hoering TC (1991) The isotopic composition of carbon and nitrogen in individual amino acids isolated from modern and fossil proteins. J Archaeol Sci 18:277-292

Hayes JH (2000) Lipids as a common interest of microorganisms and geochemists. Proc Nat'l Acad Sci 97:14033-14034

Ivanovsky RN (1985) Carbon metabolism in phototrophic bacteria under different conditions of growth. In IS Kulaev, EA Dawes, DW Tempest (eds) Environmental Regulation of Microbial Metabolism. FEMS Symp 23:263-272. Academic Press, New York

Jähne B, Heinz G, Dietrich W (1987) Measurement of the diffusion coefficients of sparingly soluble gases in water. J Geophys Res 92:10,767-10,10,776

Jahnke LL, Summons RE, Hope JM, Des Marais DJ (1999) Carbon isotopic fractionation in lipids from methanotrophic bacteria II: The effects of physiology and environmental parameters on the biosynthesis and isotopic signatures of biomarkers. Geochim Cosmochim Acta 63:79-93

Jahnke LL, Summons RE, Hope JM, Eder W, Huber R, Stetter KO, Hinrichs K-U, Hayes JH, Des Marais DJ, Cady S (2001) Composition of hydrothermal vent microbial communities as revealed by analyses of signature lipids, stable carbon isotopes and *Aquificales* cultures. Submitted to Appl Env Microbiol

Jürgens UJ, Simonin P, Rohmer M (1992) Localization and distribution of hopanoids in membrane systems of the cyanobacterium *Synechocystis* PCC6714. FEMS Microbiol Lett 92:285-288

Keil RG, Fogel ML (2001) Reworking of amino acid in marine sediments: Stable carbon isotopic composition of amino acids in sediments along the Washington coast. Limnol Oceanogr 46:14-23

Kleemann G, Poralla K, Englert G, Kjøsen H, Liaaen-Jensen S, Neunlist S, Rohmer M (1990) Tetrahymanol from the phototrophic bacterium *Rhodopseudomonas palustris*: first report of a gammacerane triterpene from a prokaryote. J Gen Microbiol 136:2551-2553

Kleinig H (1989) The role of plastids in isoprenoid biosynthesis. Ann Rev Plant Physiol Plant Mol Biol 40:39-59

Kohl W, Gloe A, Reichenbach H (1983) Steroids from the Myxobacterium *Nannocystis exedens*. J Gen Microbiol 129:1629-1635

Lange BM, Rujan T, Martin W, Croteau R (2000) Isoprenoid biosynthesis: The evolution of two ancient and distinct pathways across genomes. Proc Nat'l Acad Sci 97:13172-13177

Laws EA (1991) Photosynthetic quotients, new production and net community production in the open ocean. Deep-Sea Research 38:143-167

Laws EA, Popp BN, Bidigare RR, Kennicutt MC, Macko SA (1995) Dependence of phytoplankton carbon isotopic composition on growth rate and $[CO_2]_{aq}$: Theoretical considerations and experimental results. Geochim Cosmochim Acta 59:1131-1138

Levin EY, Bloch K (1964) Abscence of sterols in blue-green algae. Nature 202:90-91

Lichtenthaler HK, Schwender J, Disch A, Rohmer M (1997) Biosynthesis of isoprenoids in higher plant chloroplasts proceeds via a mevalonate-independent pathway. FEBS Letters 400:271-274

Lichtenthaler HK (1999) The 1-deoxy-D-xylulose-5-phosphate pathway of isoprenoid biosynthesis in plants. Ann Rev Plant Physiol Plant Mol Biol 50:47-65

Luo Y, Sternberg L (1991) Deuterium heterogeneity in starch and cellulose nitrate of CAM and C_3 plants. Phytochemistry 30:1095-1098

Luo Y-H, Sternberg L, Suda S, Kumazawa S, Mitsui A (1991) Extremely low D/H ratios of photoproduced hydrogen by cyanobacteria. Plant Cell Physiol 32:897-900

Macko SA, Fogel ML, Hare PE, Hoering TC (1987) Isotopic fractionation of nitrogen and carbon in the synthesis of amino acids by microorganisms. Chem Geol 65:79-92

Madigan MT, Takigiku R, Lee RG, Gest H, Hayes JH (1989) Carbon isotope fractionation by thermophilic phototrophic sulfur bacteria: Evidence for autotrophic growth in natural populations. Appl Env Microbiol 55:639-644

Madigan MT, Martinko JM, Parker J (2000) Brock Biology of Microorganisms, 9th edition. Prentice Hall, New Jersey

Marino BD, McElroy MB (1991) Isotopic composition of atmospheric CO_2 inferred from carbon in C_4 plant cellulose. Nature 349:127-131

Melzer E, Schmidt H-L (1987) Carbon isotope effects on the pyruvate dehydrogenase reaction and their importance for relative carbon-13 depletion in lipids. J Biol Chem 262:8159-8164

Melzer E, O'Leary MH (1987) Anapleurotic CO_2 fixation by phosphoenolpyruvate carboxylase in C_3 plants. Plant Physiol 84:58-60

Melzer E, O'Leary MH (1991) Aspartic-acid synthesis in C_3 plants. Planta 185:368-371

Monson KD, Hayes JM (1980) Biosynthetic control of the natural abundance of carbon 13 at specific positions within fatty acids in *Escherichia coli*. J Biol Chem 255:11,435-11,441

Monson KD, Hayes JM (1982a) Carbon isotopic fractionation in the biosynthesis of bacterial fatty acids. Ozonolysis of unsaturated fatty acids as a means of determining the intramolecular distribution of carbon isotopes. Geochim Cosmochim Acta 46:139-149

Monson KD, Hayes JM (1982b) Biosynthetic control of the natural abundance of carbon-13 at specific position within fatty acids in *Saccharomyces cerevisiae*. Isotope fractionations in lipid synthesis as evidence for peroxisomal regulation. J Biol Chem 257: 5568-5575

Mook WG, Bommerson JC, Staverman MH (1974) Carbon isotope fractionation between dissolved bicarbonate and gaseous carbon dioxide. Earth Planet Sci Lett 22:169-175

Moore S, Stein WH (1951) Chromatography of amino acids on sulfonated polystyrene resins. J Biol Chem 192:663-681

Ohlrogge JB, Jaworski JG (1997) Regulation of fatty acid synthesis. Annu Rev Plant Physiol Plant Mol Biol 48:109-136

O'Leary MH (1981) Carbon isotope fractionation in plants. Phytochemistry 20:553-567

O'Leary MH, Rife JE, Slater JD (1981) Kinetic and isotope effect studies of maize phosphoenolpyruvate carboxylase. Biochem 20:7308-7314

O'Leary MH (1984) Measurement of the isotope fractionation associated with diffusion of carbon dioxide in aqueous solution. J Phys Chem 88:823-825

Popp BN, Kenig F, Wakeham SG, Laws EA, Bidigare RR (1998a) Does growth rate affect ketone unsaturation and intracellular carbon isotopic variability in *Emiliania huxleyi*? Paleoceanogr 13:35-41

Popp BN, Laws EA, Bidigare RR, Dore JE, Hanson KL, Wakeham SG (1998b) Effect of phytoplankton cell geometry on carbon isotopic fractionation. Geochim Cosmochim Acta 62:69-77

Popp BN, Trull T, Kenig F, Wakeham SG, Rust TM, Tilbrook B, Griffiths FB, Wright SW, Marchant HJ, Bidigare RR, Laws EA (1999) Controls on the carbon isotopic composition of Southern Ocean phytoplankton. Global Biogeochem Cycles 13:827-843

Preuß A, Schauder R, Fuchs G, Stichler W (1989) Carbon isotope fractionation by autotrophic bacteria with three different CO_2 fixation pathways. Z Naturforsch 44:397-402

Raederstorff D, Rohmer M (1988) Polyterpenoids as cholesterol and tetrahymanol surrogates in the ciliate *Tetrahymena pyriformis*. Biochimica Biophysica Acta 960:190-199

Razin S (1978) The mycoplasmas. Microbiol Rev 42:414-470

Reinfelder JR, Kraepiel AML, Morel FMM (2000) Unicellular C_4 photosynthesis in a marine diatom. Nature 407:996-999

Riebesell U, Revill AT, Holdsworth DG, Volkman JK (2000) The effects of varying CO_2 concentration on lipid composition and carbon isotope fractionation in *Emiliania huxleyi*. Geochim Cosmochim Acta 64:4179-4192

Rieder C, Strauß G, Fuchs G, Arigoni D, Bacher A, Eisenreich W (1998) Biosynthesis of the diterpene verrucosan-2β-ol in the phototrophic eubacterium *Chlorflexus aurantiacus*. J Biol Chem 273:18,099-18,108

Roberts RB, Abelson PH, Cowie DB, Bolton ET, Britten RJ (1955) Studies of Biosynthesis in *Escherichia coli*. Carnegie Institution of Washington Publ 607, Washington, DC

Robinson JJ, Cavanaugh CM (1995) Expression of form I and form II Rubisco in chemoautotrophic symbioses: Implications for the interpretation of stable carbon isotope values. Limnol Oceanogr 40:1496-1502

Roeske CA, O'Leary MH (1984) Carbon isotope effects on the enzyme-catalyzed carboxylation of ribulose bisphosphate. Biochemistry 23:6275-6284

Roeske CA, O'Leary MH (1985) Carbon isotope effect on carboxylation of ribulose bisphosphate catalyzed by ribulosebisphosphate carboxylase from *Rhodospirillum rubrum*. Biochem 24:1603-1607

Rohmer M, Bouvier-Nave P, Ourisson G (1984) Distribution of hopanoid triterpenes in prokaryotes. J Gen Microbiol 130:1137-1150

Rohmer M (1993) The biosynthesis of triterpenoids of the hopane series in the Eubacteria: A mine of new enzyme reactions. Pure Appl Chem 65:1293-1298

Rohmer M, Knani M, Simonin P, Sutter B, Sahm H (1993) Isoprenoid biosynthesis in bacteria: a novel pathway for the early steps leading to isopentenyl diphosphate. Biochem J 295:517-524

Rossmann A, Butzenlechner M, Schmidt H-L (1991) Evidence for a nonstatistical carbon isotope distribution in natural glucose. Plant Physiol 96:609-614

Rowland SJ, Belt ST, Cooke DA, Hird SJ, Neeley S, Robert J-M (1995) Structural characterisation of saturated through heptaunsaturated C_{25} highly branched isoprenoids. *In* JO Grimalt, C Dorronsoro (eds) Organic Geochemistry: Developments and Applications to Energy, Climate, Environment and Human History. A.I.G.O.A., The Basque Country, Spain, p 581-582

Sakata S, Hayes JH, McTaggart AR, Evans RA, Leckrone KJ, Togasaki RK (1997) Carbon isotopic fractionation associated with lipid biosynthesis by a cyanobacterium: Relevance for interpretation of biomarker records. Geochim Cosmochim Acta 61:5379-5389

Sakata S, Hayes JH, Rohmer M, Hooper AB, Seeman M (2001) Molecular and carbon isotopic compositions of lipids isolated from an ammonia-oxidizing chemoautotroph. Submitted to Proc Nat Acad Sci USA

Schleucher J, Vanderveer PJ, Sharkey TD (1998) Export of carbon from chloroplasts at night. Plant Physiol 118:1439-1445

Schouten S, Klein Breteler WCM, Blokker P, Schogt N, Rijpstra WIC, Grice K, Baas M, Sinninghe Damasté JS (1998) Biosynthetic effects on the stable carbon isotopic compositions of algal lipids: Implications for deciphering the carbon isotopic biomarker record. Geochim Cosmochim Acta 62:1397-1406

Schwender J, Gemünden C, Lichtenthaler HK (2001) Chlorophyta exclusively use the 1-deoxyxylulose 5-phosphate/2-Cmethylerythritol 4-phosphate pathway for the biosynthesis of isoprenoids. Planta 212:416-423

Scott JH, Nealson KH (1994) A biochemical study of the intermediary carbon metabolism of *Shewanella putrefaciens.* J Bacteriol 176:3408-3411

Seto H, Watanabe H, Furihata K (1996) Simultaneous operation of the mevalonate and non-mevalonate pathways in the biosynthesis of isopentenyl diphosphate in *Streptomyces aeriouvifer.* Tetrahedron Lett 37:7979-7982

Sternberg LdSL, DeNiro MJ, Ajie HO (1986) Isotopic relationships between saponifiable lipids and cellulose nitrate prepared from red, brown and green algae. Planta 169:320-324

Sternberg LdSL (1988) D/H ratios of environmental water recorded by D/H ratios of plant lipids. Nature 333:59-61

Strauss G, Fuchs G (1993) Enzymes of a novel autotrophic CO_2 fixation pathway in the phototrophic bacterium *Chloroflexus aurantiacus,* the 3-hydroxypropionate cycle. Eur J Biochem 275:633-643

Summons RE, Jahnke LL, Roksandic Z (1994) Carbon isotopic fractionation in lipids from methanotrophic bacteria: Relevance for interpretation of the geochemical record of biiomarkers. Geochim Cosmochim Acta 13:2853-2863

Summons RE, Franzmann PD, Nichols PD (1998) Carbon isotopic fractionation associated with methylotrophic methanogenesis. Org Geochem 28:465-475

Takigiku R (1987) Isotopic and molecular indicators of origins of organic compounds in sediments. PhD Dissertation, Indiana University, Bloomington, Indiana, 248 p

Teece MA, Folge ML, Dollhopf ME, Nealson KH (1999) Isotopic fractionation associated with biosynthesis of fatty acids by a marine bacterium under oxic and anoxic conditions. Org Geochem 30:1571-1579

van der Meer MTJ, Schouten S, Sinninghe Damasté JS (1998) The effect of the reversed tricarboxylic acid cycle on the ^{13}C contents of bacterial lipids. Org Geochem 28:527-533

van der Meer MTJ, Schouten S, Rijpstra WIC, Fuchs G, Sinninghe Damasté JS (2001a) Stable carbon isotope fractionations of the hyperthermophilic crenarchaeon *Metallosphaera sedula.* FEMS Microbiol Lett (in press)

van der Meer MTJ, Schouten S, van Dongen W, Rijpstra WIC, Fuchs G, Sinninghe Damasté JS, de Leeuw JW, Ward DM (2001b) Biosynthetic controls on the ^{13}C-contents of organic components in *Chloroflexus aurantiacus.* J Biol Chem (in press)

Vogler EA, Hayes JM (1978) The synthesis of carboxylic acids with carboxyl carbons of precisely known stable isotopic composition. International J Appl Radiation Isotopes 29:297-300

Vogler EA, Hayes JM (1979) Carbon isotopic fractionation in the Schmidt decarboxylation: evidence for two pathways to products. J Org Chem 44:3682-3686

White DW (1999) The Physiology and Biochemistry of Prokaryotes, 2nd edition. Oxford University Press, New York

Winters JK (1971) Variations in the natural abundances of ^{13}C in proteins and amino acids. PhD Dissertation, University of Texas, Austin, Texas, 76 p.

Yakir D (1992) Variatons in the natural abundance of oxygen-18 and deuterium in plant carbohydrates. Plant, Cell and Environment 15:1005-1020

Yakir D, DeNiro MJ (1990) Oxygen and hydrogen isotope fractionation during cellulose metabolism in *Lemna gibba* L. Plant Physiol 93:325-332

Zelitch I (1975) Improving the efficiency of photosynthesis. Science 188:626-633

Zubay G (1998) Biochemistry, 4th edition. WCB/McGraw-Hill, New York

4   Stable Isotope Variations in Extraterrestrial Materials

Kevin D. McKeegan

Department of Earth & Space Sciences
University of California, Los Angeles
Los Angeles, California 90095

Laurie A. Leshin

Department of Geological Sciences
Arizona State University
Tempe, Arizona 85287

INTRODUCTION

The materials of the planets and the small bodies of the solar system contain a rich record of stable isotope variations in the light elements. As in terrestrial isotope geochemistry, this record reflects physical and chemical processes involving isotopic mixing among different reservoirs as well as fractionations arising in chemical reactions. The processes that influence the isotopic records of extraterrestrial materials range widely in environmental conditions from very high-energy events such as formation of refractory inclusions and chondrules by evaporation, condensation and melting in the solar nebula to lower temperature fluid-rock interactions in asteroids and planets. In addition, however, stable isotope cosmochemistry must consider issues that are beyond the scope of isotope geochemistry. For example, in the terrestrial sphere one may assume the existence of an isotopic reservoir that was originally homogenized during planet formation and the actual isotope compositions of the bulk earth do not need to be known in order to study the differences in stable isotope compositions that have been generated subsequently by various geochemical processes.

This assumption of homogenization cannot be made for extraterrestrial samples, and in fact stable isotopes in meteorites preserve some of the most dramatic evidence for the incomplete nature of the mixing of distinct presolar materials during formation of the solar system. Such 'isotopic anomalies' are present in the isotopic distributions of H, C, N, and O on all spatial scales—from microscopic zoning in certain meteoritic minerals to the bulk compositions of asteroids and planets. Thus, some of the isotopic heterogeneities of 'primitive' solar system materials represent vestiges of primordial differences that could not be fully erased during the processing of presolar materials. In other cases, isotopic heterogeneities reflect the preservation of unique clues to processes occurring during formation of the solar system and planetary accretion, including early 'geologic activity' on planetesimals and even pre-biotic organic chemistry. Deciphering whether an isotopic signature is primordial or has been modified or even generated by solar system processes is often possible (e.g. by comparison to astronomical observations), but is sometimes fraught with ambiguity.

The most extreme example of isotopic anomalies is provided by the laboratory analyses of individual preserved presolar dust grains extracted from primitive meteorites (Anders and Zinner 1993). These micron-size or smaller grains of SiC, graphite, and (less commonly analyzed) refractory oxides formed in the outflows of evolved stars and their isotopic compositions of C, N, O and other major and even trace elements quantitatively reflect the unique nucleosynthetic environment of that particular star, which may differ from average solar system compositions by one or more orders of magnitude. That such

1529-6466/00/0043-0004$10.00

materials could have survived the journey from star, through the interstellar medium, the solar accretion disk and residence in the asteroidal parent bodies of meteorites, and finally to terrestrial laboratories is one of the most remarkable discoveries in all of twentieth century astronomy. The isotopic compositions of these particles have great importance for constraining models of stellar evolution and nuclear astrophysics, but because these materials are so rare and are distinctive by virtue of their lack of interaction with average solar materials, their study gives little indication of the nature of the most important processes that transformed primordial materials into the stuff that built the planets. Hence the subject of isotopic variations in presolar grains is beyond the scope of this chapter, and the reader is directed to Zinner (1998), Ott (1998), and Hoppe and Zinner (2000) for excellent reviews of this rapidly evolving subject.

This chapter concentrates on the isotopic records of H and O, with some additional discussion of C and N, in materials that formed within the solar system (see Pillinger 1984 for a review of the earliest history of this subject). As alluded to above, even though these materials formed 'locally,' their isotopic records may preserve, with varying degrees of fidelity, memories of primordial heterogeneities that were established in presolar matter from various astrophysical sources. With the possible exception of oxygen, one type of isotopic memory that we do not discuss are those associated with demonstrably nucleosynthetic processes (e.g. in neutron-rich isotopes of Ca, Ti, Cr, etc.) (see e.g. Lee 1979; Begemann 1980; Ireland 1990) or with 'anomalies' due to decay of short-lived radioisotopes (see Wasserburg and Papanastassiou 1982; Podosek and Nichols 1997). We also choose to focus on laboratory analyses as these measurement methods are most similar to what isotope geochemists are familiar with, although the fine-grained nature and limited supply of some types of extraterrestrial samples has stimulated new instrumental developments. Some of these methods, e.g. stable isotope analysis by ion microprobe, have been increasingly finding application in terrestrial geochemistry (e.g. Riciputi and Paterson 1994; Hervig et al. 1995; Riciputi et al. 1998; Valley et al. 1998). We begin by considering the general behavior of these isotope systems from a cosmochemical point of view.

Isotope cosmochemistry of the light stable isotopes

The volatile elements H, C, N, and O are incompletely condensed in meteorites and planetary bodies, and thus their abundances are fractionated throughout the solar system (see discussion in Anders and Grevesse 1989). It is therefore not unreasonable to expect that the isotopic compositions of these elements may also be variable on a solar-system-wide scale. Large isotopic variations are known on planetary scales for H and N, and more subtle, but nevertheless very significant variations occur in C and O as well. These variations reflect heterogeneities inherited from presolar materials or established during planetary accretion, as well as those developed by specific planetary processes (e.g. atmospheric escape). In any discussion of extraterrestrial stable isotopes it is important, and sobering, to note that the average (or solar) isotopic composition is not necessarily well determined for any of these elements, where by 'well determined' we mean something close to the levels of precision that isotope geochemists tend to consider. Here we briefly summarize the main astrophysical sources of each isotope system, current estimates of their solar values, and the ranges encountered in both remote and *in situ* observations of various solar system objects.

Hydrogen. The stable isotopes of hydrogen, ^1H and ^2H (deuterium, hereafter denoted D), are truly primordial, having been essentially completely produced in the big bang (Burbidge et al. 1957; Epstein et al. 1976). They are the only isotopes considered here that lack the potential for specific stellar input into presolar materials. That is not to imply that there do not exist significant presolar isotope variations in D/H. On the contrary, because of the very low temperatures in molecular clouds and the large differences in vibrational frequency levels between deuterated and protonated molecules, huge mass fractionations occur in condensed phases (e.g. organic ices) in these environments. As we shall see, a memory of these presolar D/H enrichments is preserved in organic phases of chondritic meteorites, in interplanetary dust particles (IDPs), in cometary volatiles and, possibly, even in water from chondrites and on Earth (Robert et al. 2000).

It has long been known that the terrestrial D/H ratio of ~1.5×10^{-4} (or 150 ppm; the D/H value of the Standard Mean Ocean Water reference is 1.5576×10^{-4}) is highly elevated relative to cosmic values (see review by Geiss and Reeves 1981). The original solar system D/H value cannot be determined by direct observations of the Sun because D was completely burned by the convecting proto-Sun as it contracted toward an interior density where stable H fusion was achieved (the main sequence). Estimates of D/H in protosolar matter are instead based on $^3He/^4He$ abundances in the solar wind and astronomical observations of galactic hydrogen (Geiss and Reeves 1981). Within relatively large uncertainties (~25%), the average protosolar D/H ratio is calculated to have been ~20 ppm (Geiss and Gloeckler 1998), leading to the conclusion that Earth's oceans are enriched by a factor of about 8 in D/H relative to cosmic values.

With the exception of the gaseous giants Jupiter and Saturn, variable enrichments of D/H relative to the protosolar reservoir is a general characteristic of the planets, although the mechanisms responsible for this fractionation are not known definitively and may be different in each case (see Yung and Dissly 1992 for a review). The terrestrial planets exhibit the highest D/H values: spacecraft and ground-based observations of the atmospheres of Venus and Mars show D/H ratios of ~2×10^{-2} and ~8×10^{-4}, respectively, probably reflecting the significant loss of the original inventory of hydrogen by Jeans escape or similar mass-fractionating processes (Yung and Dissly 1992 and references therein). These values are much higher, by factors of 100 and 5, respectively, than the terrestrial D/H, which is thought not to have been measurably affected by atmospheric loss of H over geologic time (Robert et al. 2000). The D/H values on Mars may be considered to be known in much greater detail than remote observations permit owing to laboratory analyses of the martian (SNC) meteorites, which are igneous rocks that are highly likely to have been derived from Mars (McSween 1994). These analyses offer insights into planetary process that may have affected D/H ratios in different martian reservoirs (atmosphere, crust, mantle) and have important implications for understanding the sources and evolutionary history of water on the planet (e.g. Leshin et al. 1996a; Leshin 2000; Sugiura and Hoshino 2000).

Spectroscopic observations of Jupiter and Saturn made with the Infrared Space Observatory are consistent with other estimates of protosolar D/H (see Robert et al. 2000 and references therein), as is the D/H value determined by the Galileo atmospheric entry probe mass spectrometer (26±7 ppm; Mahaffy et al. 1998). This result is expected as the jovian planets are thought to have formed by quantitative capture of gas from the solar nebula. On the other hand, the outer solar system planets Uranus and Neptune appear to be significantly enriched in D/H by factors of ~3 compared to the protosolar value (Feuchtgruber et al. 1999). This enrichment is interpreted to reflect the mixing of material from the more D-rich icy cores of these planets, which constitutes a significant fraction of their mass (as opposed to the jovian case where the planetary mass is dominated by the gaseous envelope).

Much larger D/H values are observed in comets. The hydrogen isotopic composition of water has been determined with about 10 to 30% precision in 3 comets: P/Halley

(Eberhardt et al. 1995), Hyakutake (Bockelée-Morvan et al. 1998), and Hale-Bopp (Meier et al. 1998). All three have D/H $\sim 3 \times 10^{-4}$ or $\delta D \approx 1000$ ‰. Measurements of DCN/HCN in Hale-Bopp (Meier et al. 1998) suggest that organic species might be more highly enriched in deuterium by as much as an order of magnitude, but it is not yet known how representative this single result might be. The D/H value for methane in the atmosphere of Titan, Saturn's largest satellite, is close to the terrestrial value (de Bergh et al. 1988), suggesting that a cometary component to Titan's atmosphere is present but is of limited abundance (Owen 2000).

Carbon. Carbon has two stable isotopes, ^{12}C and ^{13}C, which have distinct nucleosynthetic origins. The former, abundant, isotope is produced during He-burning by the triple-alpha reaction ($3 \, ^4He \rightarrow \, ^{12}C + \gamma$) whereas ^{13}C requires seed nuclei and hence was not initially present in the first generation of stars. As further generations of stars process interstellar materials in their nuclear furnaces, the relative abundances of such secondary isotopes increase (e.g. Briley et al. 1997; Busso et al. 1999). This chemical evolution of the galaxy resulted in a $^{12}C/^{13}C$ ratio of ~ 90 in the molecular cloud fragment from which the solar system formed.

Little precise information is available regarding the carbon isotopic composition in major planetary reservoirs in the solar system (with the exception of Mars, discussed below), or in the Sun. Determination of isotopic abundances from photospheric observations is difficult; a high resolution study of CO molecular lines yields $^{12}C/^{13}C = 84 \pm 5$ (Harris et al. 1987). Lacking a precise and meaningful reference value for average solar system, by tradition carbon isotope ratios in extraterrestrial materials are defined relative to the Pee Dee Belemnite with $^{13}C/^{12}C \equiv 0.011237$ (Craig 1957).

Nitrogen. The abundances of the two stable isotopes of nitrogen, ^{14}N and ^{15}N, are modified by hydrogen burning in various types of stars during their life-cycles. Both isotopes can also be produced by He-burning, but ^{15}N is additionally produced by explosive H-burning in type II supernovae. Thus, with various distinct stellar sources, it is perhaps not surprising that N exhibits very large isotope heterogeneities, reaching more than 3 orders of magnitude in the $^{15}N/^{14}N$ ratio of circumstellar SiC grains wherein N is a major impurity (Zinner 1998). In addition to nucleosynthetic effects (of which the SiC grains are an extreme example), the $^{15}N/^{14}N$ relative abundances can be affected by mass fractionation at low-T in interstellar clouds and by planetary atmospheric processes in much the same manner as hydrogen isotopes. As pointed out by Kerridge (1995), the result of these various effects is that nitrogen isotopic records among different solar system reservoirs are quite variable and it is not possible to precisely define a representative solar system mean initial value. Consequently, the $\delta^{15}N$ scale is referenced to the terrestrial atmosphere, $^{15}N/^{14}N = 3.68 \times 10^{-3}$.

The $^{15}N/^{14}N$ of the Sun would serve as an appropriate reference point from which to interpret isotopic variations observed in meteorites and planetary bodies. As with C, direct spectroscopic or spacecraft observations suffer from lack of precision, e.g. the Solar and Heliospheric Observatory (SOHO) mission determined $^{15}N/^{14}N$ ratios compatible with the terrestrial atmosphere within errors of 300 to 500 ‰ (Kallenbach et al. 1998). High precision measurements can best be made in terrestrial laboratories on returned samples, and indeed analyses of the noble gas contents of soils returned by the Apollo missions demonstrated that the lunar regolith provides an effective capture medium for solar material. However, in the case of nitrogen, the interpretation of the record of surface-correlated material in lunar soils is complex. The $^{15}N/^{14}N$ ratios of different lunar samples are highly variable, and it has been argued that the variations in $\delta^{15}N$ exhibit a secular trend with soil 'antiquity' implying that the nitrogen isotopic composition of the solar wind has systematically become heavier over the past several aeons (Becker and Epstein 1982; Kerridge 1993). However, the Sun is not the only source of N on the lunar surface, and mixing of isotopically distinct reservoirs may also explain the observed $\delta^{15}N$ variability (Wieler et al. 1999). An important advance in this long-standing problem has recently been made by Hashizume and co-workers who utilized an ion microprobe to perform direct depth-profiling analyses of $\delta^{15}N$ in individual lunar minerals that were exposed on the Moon's surface at epochs differing by $\sim 10^9$ years (Hashizume et al. 2000). These analyses provide convincing evidence for a solar-wind component of nitrogen that is highly depleted in ^{15}N ($\delta^{15}N < -240$ ‰) in a sample irradiated between 1 and 2 Ga, suggesting that the solar $^{15}N/^{14}N$ ratio has not changed markedly over time but that the terrestrial nitrogen isotopic composition is significantly different than the average solar system value.

Oxygen. Among the elements considered in this chapter, oxygen is unique by virtue of its having 3 stable isotopes, ^{16}O, ^{17}O, and ^{18}O. This typically allows the possibility to distinguish between processes that partition isotopes from a common isotopic reservoir according to the usual thermodynamically-driven mass-dependent fractionation and processes involving the mixing of isotopically distinct reservoirs that had never been previously equilibrated. (This is possible with the other elements having only one stable isotope pair only when the degree of isotopic heterogeneity greatly exceeds the plausible range of possible mass-dependent isotopic fractionation given other circumstantial evidence regarding temperature, etc.) On the three-isotope plot of $\delta^{17}O$ vs. $\delta^{18}O$, the former set of processes, which include physical changes of state and most chemical reactions, result in a series of isotopic compositions that are related by a straight line of slope ~0.52 for the normally observed small degrees of isotopic fractionation of a few percent or less (Fig. 1a). In contrast, partial isotopic exchange between two unequilibrated reservoirs results in compositions lying on the tie-line between the two reservoirs which can, in principle, result in any slope on the three-isotope diagram (Fig. 1a) depending on the initial compositions. Slight deviations from a strictly linear relation between two such reservoirs can occur due to the superposition of slope ~1/2 mass-dependent fractionation effects on the overall mixing line.

The oxygen isotopic compositions of nearly all (see next paragraph) terrestrial samples fall along a single correlation line of slope ~0.52, as expected, because of the highly effective initial homogenization of the terrestrial reservoir. This empirically defined relationship is called the terrestrial fractionation line (TF in subsequent figures) and by definition it passes through the reference composition for oxygen, which is Standard Mean Ocean Water (SMOW) with a composition $^{18}O/^{16}O = 0.0020052$ (Baertschi 1976) and $^{17}O/^{16}O = 0.0003829$ (Fahey et al. 1987). It is useful to characterize deviations from this line by the quantity $\Delta^{17}O \equiv \delta^{17}O - 0.52 \times \delta^{18}O$ (Clayton et al. 1991), which is a constant for any suite of samples that are related by mass-dependent fractionation from a common reservoir (Fig. 1b). Samples with $\Delta^{17}O < 0$ are therefore depleted in $^{17}O/^{16}O$ compared to terrestrial samples having the same $^{18}O/^{16}O$ values, which (as will be seen below in the discussion of isotope records of nebular components of chondrites) is best interpreted as reflecting an excess of ^{16}O in these materials compared to terrestrial values. Note that the addition or depletion of ^{16}O only follows a trajectory of slope of 1.0 on the three-isotope diagram (Fig. 1a).

The first observations of any samples with $\Delta^{17}O \neq 0$ were for anhydrous mineral separates from carbonaceous chondrites measured by Clayton and colleagues (Clayton et al. 1973). This remarkable discovery was considered prima facie evidence for the preservation of a distinctive nucleosynthetic component, rich in ^{16}O, in these meteoritic phases (Clayton et al. 1973; Clayton et al. 1977) and it stimulated the discovery of many

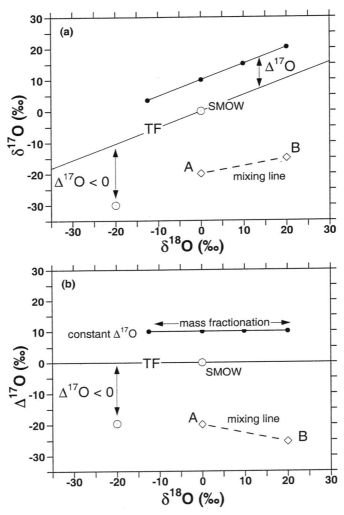

Figure 1. (a) Schematic representation of the traditional 3-isotope plot for oxygen with $\delta^{18}O$ on the abscissa and $\delta^{17}O$ on the ordinate. On such a plot, mixtures of 2 isotopic reservoirs (e.g. A and B) lie on a straight line connecting the end-member compositions. Most chemical reactions partition isotopes in a mass-dependent manner which, for small degrees of fractionation, results in a dispersion along a line of slope = 0.52. Terrestrial rocks and fluids lie on such a fractionation line (TF), passing through the origin defined by standard mean ocean water (SMOW). The deviation of a point above or below the TF line is expressed by $\Delta^{17}O$, which indicates the difference between its $\delta^{17}O$ value and the $\delta^{17}O$ of a point on the TF line at the same value of $\delta^{18}O$. Components which are fractionated from a homogeneous reservoir only by normal, mass-dependent processes lie on a line of slope = 0.52 that is parallel to the TF line, and hence such compositions are characterized by a constant $\Delta^{17}O$ value. (b) Alternate representation of the same oxygen 3-isotope data with $\delta^{18}O$ remaining on the abscissa and $\Delta^{17}O$ now plotted on the ordinate. Compositions that are related by mass-dependent fractionation to a common reservoir plot on horizontal lines with the degree of mass fractionation indicated by the difference in their $\delta^{18}O$ values. Mixing lines between unequilibrated compositions can still exhibit any possible slope in this representation.

isotope anomalies of nucleogenetic origin in primitive refractory phases of chondrites (e.g. Lee 1979). A decade later, however, it was demonstrated by laboratory experiment that it is possible to produce non-mass dependent isotopic fractionations in certain gas-phase chemical reactions, leading to changes in $\Delta^{17}O$ of the products from initially isotopically equilibrated (i.e. constant $\Delta^{17}O$) reactants (Thiemens and Heidenreich 1983). Chemically-produced non-mass dependent isotopic signatures have now been recognized in natural terrestrial atmospheric samples (Thiemens et al. 1995) and in certain minerals formed by interactions with isotopically anomalous atmospheric species (Bao et al. 2000), and have also been posited to have affected minerals formed near the martian surface (e.g. Farquhar and Thiemens 2000). Thus, an unambiguous assignment of departures from mass-dependent isotopic fractionation of oxygen isotopes as due to nuclear, as opposed to chemical, effects is not possible, and arguments regarding the origin of the premier isotopic anomaly discovered by Clayton must rely on interpretations of circumstantial, often complex evidence. A quarter century after its discovery, the interpretation of the origin of oxygen isotopic anomalies in solar system materials remains as arguably the most important outstanding problem in cosmochemistry.

An excellent summary of the cosmochemically significant properties of oxygen is given by Clayton (1993). Of prime importance is the fact that oxygen in the solar nebula is more abundant than carbon, allowing it to exist simultaneously as a significant gas-phase species (primarily CO and H_2O) and as the major constituent of rocks. In addition, the nucleosynthetic sources that are responsible for producing the primary isotope, ^{16}O, are distinct from those synthesizing the secondary, rare isotopes, ^{17}O and ^{18}O, which have abundances of only about 400 and 2000 ppm, respectively. The abundance of these secondary isotopes ought to increase with the chemical evolution of the galaxy, so that if there exists a difference in mean age between interstellar dust and gas, it could be expected that the older component (i.e. the dust) would, on average, be ^{16}O-rich compared to the younger component (Clayton 1988). The potential for distinct isotopic signatures of oxygen in dust compared to gas, regardless of whether they be inherited from presolar materials or generated by solar system gas-phase chemistry, combined with oxygen's unique chemical abundance properties, make the distribution oxygen isotope anomalies a powerful tracer of high-temperature gas-dust interactions in the solar nebula (Clayton 1993).

The interpretation of oxygen isotope records in differentiated planetary bodies (Earth, Mars, asteroidal sources of achondritic meteorites), as well as in distinct nebular components of chondrites, would be more straightforward if an average solar system composition could be defined. However, as with the other stable isotopes, the solar composition cannot be directly determined with anything approaching a precision useful for planetary science purposes. An estimate of the solar oxygen isotope composition has been constructed by Wiens et al. (1999), but it is based on model-dependent interpretations of meteoritic data. The *Genesis* Discovery mission, to be launched in summer of 2001, is designed to capture and return (in 2004) a solar wind sample to terrestrial laboratories for analysis of its oxygen isotopic composition (see discussion in Wiens et al. 1999).

To date, there exists very little quantitative information concerning oxygen isotope compositions in major solar system reservoirs that is obtained by remote (spectroscopic) observation or spacecraft measurements. A measurement of water ice from comet P/Halley, made by the *Giotto* mission, yields $\delta^{18}O = 12\pm75$ ‰ (Balsiger et al. 1995; Eberhardt et al. 1995) but no measurement of $\delta^{17}O$ is available. Precise data are obtained for the Moon, of course, from returned Apollo samples, and the oxygen isotope composition of Mars and the largest asteroid, Vesta, may be inferred from laboratory mass spectrometric measurements of meteorites that are thought to derive from these bodies. These data are discussed in detail below in the section on isotope records in planetary materials.

Analysis methods

Measurements of stable isotope abundances have traditionally been accomplished by means of gas-source mass spectrometry. In the 'conventional' method, gases are evolved by either thermal or chemical means from 'bulk samples' to produce on the order of micromoles of purified analyte for high-precision isotope-ratioing mass spectrometry. For C and N, mass spectrometry is sometimes done in the 'static mode' to improve sensitivity to the sub-nanomole regime (Pillinger 1984). Accuracy and precision are both very high by these gas-source methods, and for most types of sufficiently large samples are in the few ‰ range for δD, better than ~0.1 ‰ for $\delta^{13}C$ and $\delta^{15}N$, and <0.2 ‰ for $\delta^{17}O$ and $\delta^{18}O$. For oxygen isotopes, mineral separates (often < 100 % purity) may be analyzed on suitably coarse-grained materials (e.g. Clayton et al. 1977). Differential dissolution or other chemical means are sometimes utilized for studying specific components of meteorites (eg., H, C, N in soluble organic matter, or C and O in carbonates). Stepped combustion or pyrolysis may be employed for H, C, and N to discriminate among different carrier phases.

In the last decade, lasers have been used to assist in the chemical extraction of stable isotopes from individual mineral samples (see chapter by Valley, this volume). Most of the laser analyses of extraterrestrial materials concern oxygen isotope measurements (e.g. Young et al. 1998a; Young et al. 1998b), where there are significant advantages over conventional methods because of the reduction in system blank. The resulting increase in sensitivity allows sub-μmole quantities of O_2 to be analyzed which can be obtained from individual mineral grains in favorable circumstances. The IR laser method is used for physically isolated samples and can improve precision and accuracy of analyses of some difficult-to-react minerals, such as olivine, to sub-ε levels (e.g. <0.05 ‰ in $\Delta^{17}O$; Franchi et al. 1999). For *in situ* work, a UV laser is utilized because it does not suffer from the partial reaction aureoles that limit IR laser *in situ* work. Typical spot sizes analyzed are ~100 μm in lateral dimension and ~50 μm deep. Precision and accuracy are also high (~±0.3‰) for *in situ* O isotope work (Young et al. 1998a).

Because of its high sensitivity and spatial resolution, the ion microprobe has become a key instrument for the analysis of stable isotopes in a wide variety of extraterrestrial materials. Analyses can be made on very small samples, e.g. individual interplanetary dust grains (McKeegan et al. 1985), and other samples can be examined in thin-section with a spatial resolution of ~10 μm. The ion microprobe is used for measurements of all isotopes considered in this review, although N is only measured in C-rich materials (because of the high yield of the CN^- molecular ion). Analytical methods for ion microprobe isotopic measurements have been reviewed by Zinner (1989), Ireland (1995), and Valley (1998); developments (relevant to this chapter) since that time involve primarily analyses of D/H in hydrous phases (Deloule et al. 1991; Deloule et al. 1998), of oxygen isotope compositions measured *in situ* in insulating minerals (Lorin et al. 1990; Yurimoto et al. 1994; Leshin et al. 1998a; Engrand et al. 1999b; Hiyagon and Hashimoto 1999; Simon et al. 2000) and of large D/H anomalies with approximately micron spatial resolution by ion microscope imaging techniques (Messenger and Walker 1997; Aléon et al. 2001). Precision and accuracy obtained by ion microprobe is considerably poorer than that achievable by gas mass spectrometric methods (on much larger samples) and is somewhat dependent on sample characteristics. When good standards are available, and analyses are not counting statistic limited, precision and accuracy of one to several permil is typically realized for H, C, N, or O isotopic analyses.

STABLE ISOTOPE RECORDS IN PRIMITIVE MATERIALS

The record of variability in primordial isotopic reservoirs, i.e. those inherited from the presolar molecular cloud, and of the isotopic compositions and interactions between important components of the solar nebula is preserved only in so-called 'primitive' materials. These materials are also referred to as having a 'chondritic' composition, a term which originally denoted chondrule-bearing, but which has evolved to encompass any chemically-undifferentiated material having approximately solar concentrations of all but the most volatile elements (Anders and Grevesse 1989). Three types of primitive, chondritic extraterrestrial materials are distinguished on the basis of size, which determines modes of collection and is possibly indicative of distinctive origins. Chondritic interplanetary dust particles (IDPs) are decelerated in the upper atmosphere without melting owing to their small sizes (< ~30 μm) and are collected by high-flying aircraft (e.g. Brownlee 1981). In the next size range (~50 μm to ~200 μm) are cosmic dust particles, which constitute the dominant fraction of the total extraterrestrial material presently accreted by the Earth (Love and Brownlee 1993). Depending on their velocity, these dust particles are either melted during atmospheric entry becoming cosmic spherules or suffer only modest levels of heating insufficient to cause total melting. The latter are termed micrometeorites, and are efficiently collected from Antarctic ice (Maurette et al. 1994). While the cosmic spherules have generally undergone either chemical differentiation and/or sufficient mass loss so that most are no longer chondritic, many Antarctic micrometeorites (AMMs) are totally unmelted and retain chondritic compositions. Macroscopic meteorites are also found efficiently in Antarctica (e.g, Cassidy et al. 1992) and other desert environments (e.g. Folco et al. 2000) and, occasionally, they are observed to fall (e.g. Brown et al. 2000). A small percentage (~15%) of observed meteorite falls are samples of chemically-differentiated objects, but most are chondrites and are thought to derive from remnant, relatively unprocessed planetesimals in the asteroid belt (e.g, Wetherill and Chapman 1988).

The chondrites may be conveniently thought of as cosmic sediments, formed by aggregation and compaction of a hodge-podge of rocky and (sometimes) icy materials existing near the midplane of the solar nebula (most likely at several AU). Most chondrites consist primarily of chondrules, whence their name, and fine-grained dust. The chondrules are typically rounded, mm-sized solidified melt droplets consisting of ferromagnesian silicate minerals and glass, often with minor amounts of FeNi-metal and sulfides. The astrophysical sites and processes responsible for chondrule formation are enigmatic, but the most widely-accepted models involve some form of flash-heating mechanism (Rubin 2000). Chondritic meteorites are generally not strong rocks and are lithified only by a 'matrix' of chondrule fragments and fine-grained dust grains (which includes a tiny fraction of dust grains from other stellar systems). In addition to chondrules, an important component of some chondrites are inclusions containing refractory oxide and silicate minerals that also formed as free-floating objects within the solar nebula. These Ca-Al-rich inclusions (CAIs) are most abundant (~5%) and the largest (up to several cm) in the CV3 carbonaceous chondrites (Grossman 1980), but a wide spectrum of size ranges and mineralogical types exist among the various chondrite classes (e.g. MacPherson et al. 1988). CAIs occur only rarely in meteorites outside the carbonaceous chondrite class (e.g. Bischoff and Keil 1984; Fagan et al. 2000).

Some chondrites have been thermally metamorphosed while resident on an asteroidal parent body, and others (especially the volatile-rich carbonaceous chondrites) have undergone variable degrees of interaction with low-temperature aqueous fluids [see

Brearley and Jones (1998) for a comprehensive review of chondrite chemical and mineralogical properties and classification]. While thermal and aqueous alteration generally have not appreciably altered bulk chemical compositions of chondrites, both types of processes are imprinted on the mineral chemistry and stable isotope compositions of chondritic components. Thus, even among chondrites, there exists varying degrees of 'primitiveness' which is reflected in their isotope compositions. Those meteorites which exhibit the best preserved nebular records are considered to be the most primitive; these include the unequilibrated ordinary chondrites and enstatite chondrites of petrologic type 3, as well as various classes of carbonaceous chondrites (e.g. CR, CH, CB, reduced CV). Based primarily on mineralogical and petrographic observations, it is thought that particles of the anhydrous subset of IDPs are in some sense even more 'primitive' than any solar nebula components of macroscopic meteorites (i.e. excluding circumstellar dust grains), and may even comprise true presolar materials (e.g. Bradley 1994). Stable isotopic evidence (see below) has been found to support a view of the extremely primitive nature of certain IDPs that, in turn, bolsters the hypothesis that many of these types of (anhydrous, low density) particles are derived from comets.

Hydrogen

Hydrogen isotopic compositions of various components of primitive extraterrestrial materials preserve evidence of extreme isotopic heterogeneity among presolar carriers, and track the processes associated with the partial equilibration of these materials in the formation of the solar system. The hydrogen isotopic compositions of these materials have been reviewed recently by Robert et al. (2000) and are summarized graphically in Figure 2.

Bulk D/H analyses of chondrites are useful for interpreting the relative contributions of these materials to the hydrogen budgets of the Earth and other terrestrial planets. Such analyses show a relatively restricted range of D/H values (Robert et al. 1979; Kolodny et al. 1980; McNaughton et al. 1981; Robert and Epstein 1982; Kerridge 1985; Robert et al. 1987), with 2/3 of the samples falling between δD of -165 and +90 ‰, and a mean δD value of -100 ‰ (Robert et al. 2000), essentially identical the composition estimated for the bulk Earth (Lécuyer et al. 1998). Put another way, many chondrites have D/H values very similar to Earth, and do not show evidence for significant internal isotopic variation.

The relatively homogenous hydrogen isotopic compositions of bulk chondrites belie the complexity of the distribution of deuterium in some extraterrestrial materials. Analyses of separated components from chondrites (e.g. matrix, organic matter) reveal δD variations from ~ -500 to +6000 ‰. Organic matter in meteorites consistently shows D-enrichment at levels that are extreme compared to terrestrial values, but still well below observed compositions of molecular clouds (Fig. 2). As discussed earlier, the high D/H values in molecular clouds are interpreted to reflect very low T ion-molecule reactions. The D/H values of the organic matter in meteorites most likely reflects the partial isotopic equilibration of precursor organic species with hydrogen during solar system formation (which occurs at higher temperatures). The organic matter in meteorites can be broadly divided into a soluble component, which includes amino and carboxylic acids and other hydrocarbons, and an insoluble component with a complex structure, often referred to as "kerogen" (e.g. Kerridge 1999 and references therein). D/H analyses of amino acids separated from the soluble portion of the organic matter from the Murchison meteorite (the best studied sample in this respect) show δD values of ~+1700 ‰, and hydrocarbons and carboxylic acids show somewhat lower δD values (Pizzarello et al. 1991; Cronin and Chang 1993; Cronin et al. 1993; Pizzarello et al. 1994). The bulk kerogen from carbonaceous chondrites has δD values of ~+1400 to

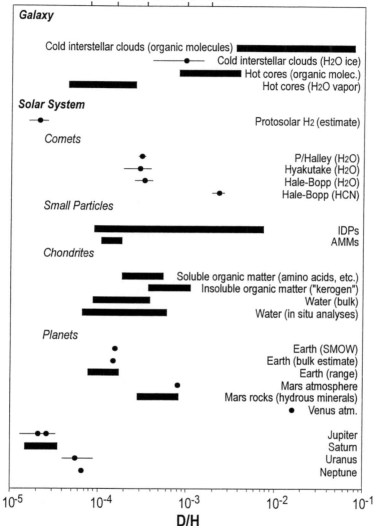

Figure 2. Overview of the hydrogen isotopic compositions of different objects, materials, and reservoirs in the solar system and beyond. The data are discussed in different portions of this review paper, and most are summarized in tabular form in the recent review of Robert et al. (2000). Points with error bars represent single values with a quoted uncertainty. Solid bars reflect the range of multiple measurements (~10 or more of analyses in most cases) . The two astronomical settings are cold, dense interstellar clouds (T ~ 10 K) and so-called "hot cores," such as found in Orion, where temperatures are ~70K and above. Behind the data for solar system objects lies much complexity which cannot be represented on a single summary figure. For example, even though the range of D/H in water from bulk chondrites is relatively large, most of the samples fall in a very narrow range, as described in the text.

3000 ‰, with CR chondrite Renazzo showing the highest D-enrichment, while kerogen from LL3 ordinary chondrites studied shows even higher δD values of ~+4000 to +6000 ‰ (Robert and Epstein 1982; Yang and Epstein 1983; Kerridge et al. 1987).

Water in components of chondrites also shows large isotopic variations, whether analyzed from physically separated water-bearing minerals such as phyllosilicate, or *in situ* using an ion microprobe. The δD values range from ~-400 to +3700 ‰ (Robert et al. 1979; McNaughton et al. 1981; Deloule and Robert 1995; Deloule et al. 1998). The most striking feature of these data is that almost the entire range is observed on the scale of single chondrules from LL3 chondrites when traverses are performed with a 10 μm ion probe beam (Deloule and Robert 1995; Deloule et al. 1998). The isotopic variations are too large to be caused by isotopic fractionation at variable nebular temperatures associated with chondrite formation, and thus imply mixing of water with variable histories in the nebula and on meteorite parent bodies. Specifically, Robert et al. (2000) have recently proposed that the range in D/H values results from variable sampling of water as it isotopically evolves from a high-δD "presolar water" component with time and space in the nebula through progressive isotopic exchange with protosolar H_2 (Deloule et al. 1998; Robert et al. 2000). The data upon which this model is based are somewhat limited, however, and confirmation of the large isotopic effects in meteoritic water is worth pursuing.

Ion microprobe analyses of Antarctic micrometeorites (AMMs) show a relatively restricted range in δD values from ~-300 to +200 ‰ (Engrand et al. 1999a). These data support the idea that AMMs are related to aqueously altered carbonaceous chondrites, which also show a similar restricted range in D/H. Interplanetary dust particles (IDPs) collected in the stratosphere, on the other hand, show the largest range in D/H of any solar system materials analyzed, from ~-500 to >+15000 ‰ (McKeegan et al. 1985; Messenger and Walker 1997; Messenger 2000; Aléon et al. 2001). Many particles have elevated D/H ratios, but these are most dramatically observed in so-called "cluster IDPs" which consist of loose agglomerations of anhydrous, fine-grained, highly porous, carbon-rich objects (Messenger 2000). The carriers of the D-enrichments in most IDPs are not well constrained, as particles are usually analyzed in "bulk" with the ion microprobe. However, observation of the most extreme D-enrichments in very C-rich IDPs is suggestive of one or more organic carrier phases, although associated large isotopic anomalies in C are lacking (McKeegan et al. 1985; Messenger 2000; Aléon et al. 2001). Cluster IDPs that are D-rich also tend to show moderate enrichments in ^{15}N, but the correlation between δD and $\delta^{15}N$ exhibits a large amount of scatter (see Messenger and Walker 1997) possibly due to multiple carrier phases of the presolar materials or differing degrees of isotopic exchange with solar system matter.

For some particles it has been possible to utilize the ion microscope (McKeegan et al. 1985, 1987; Messenger and Walker 1997; Messenger 2000) or scanning ion imaging (Aléon et al. 2001) to examine the distribution of hydrogen isotopes within a single IDP at the ~1 μm spatial scale. Surveys by these ion imaging methods indicate that D-rich "hot spots" (Fig. 3) occur with great frequency among cluster IDPs. The D/H values inferred for some of these microscopic phases approach those observed astronomically in molecular clouds, suggesting that these IDPs may be carriers of essentially unaltered presolar material (e.g. Messenger 2000). This is consistent with the notion that at least some IDPs are samples of relatively unprocessed comets (e.g. Bradley 1994). However, most IDPs have D/H values more consistent with meteoritic (and terrestrial) compositions than cometary values, suggesting that IDPs sample both asteroids and comets. Correlations of D/H heterogeneities with major element distributions, obtained by high-resolution scanning ion imaging, provide further quantitative evidence for

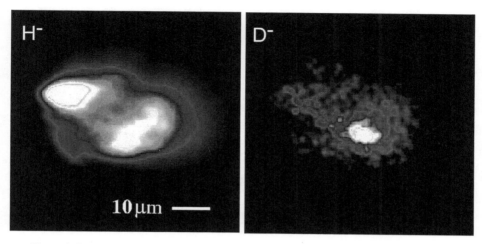

Figure 3. Secondary ion images of hydrogen isotope distributions within a cluster IDP that has a 'bulk' D/H value of ~11,000 ‰ (Messenger 2000). The images are obtained with a Cameca ims 3F ion microscope with a spatial resolution (given by the ion optics) of 1 to 2 μm. The deuterium distribution is highly heterogeneous, much more so than the H, with most of the D in this particle being concentrated in a small region that is comparable to the image resolution. A minimum estimate of the D/H in this 'hot spot' is ~4 times the average value of the particle, or close to 50 times the terrestrial value. [Figure adapted from Messenger (2000); used with author's permission.]

multiple hydrogen isotopic components within IDPs, including water with D/H similar to that of carbonaceous chondrites and two types of D-rich organic matter (Aléon et al. 2001). These investigations support the notion that IDPs sample a variety of primitive materials that may have formed over a range of heliocentric distances in the solar nebula.

Oxygen

The premier isotope anomaly. In 1973, R. Clayton and co-workers discovered correlated variations in $\delta^{18}O$ and $\delta^{17}O$ in anhydrous minerals from the Allende carbonaceous meteorite that indicated mixing of isotopically "normal" oxygen with an "exotic" component enriched in ^{16}O (Clayton et al. 1973). By disaggregating the meteorites and analyzing individual components, it was quickly learned (Clayton et al. 1977) that the anomalous oxygen was most prevalent in the CAIs of carbonaceous chondrites, many of which are highly enriched in ^{16}O by values approaching 5% relative to terrestrial minerals (Fig. 4). Subsequent analyses, performed by Clayton, Mayeda and colleagues, demonstrated a systematic trend in the oxygen isotopic compositions of the nebular constituents of the various chondrite groups (see Clayton 1993 and references therein). As can be seen in Fig. 4, the chondrules and CAIs of carbonaceous chondrites are variably enriched in ^{16}O, sometimes (for CAIs) to a large degree, compared to either terrestrial minerals or other classes of primitive or evolved meteorites. On the other hand, individual chondrules from the ordinary chondrites plot above the TF line ($\Delta^{17}O > 0$) indicating that they have small relative depletions in ^{16}O compared to terrestrial minerals. Although the data are more limited, the oxygen isotopic compositions of chondrules separated from the highly reduced enstatite chondrites lie within error of the TF line ($\Delta^{17}O = 0$) and thus are intermediate in this sense between their CC and OC counterparts.

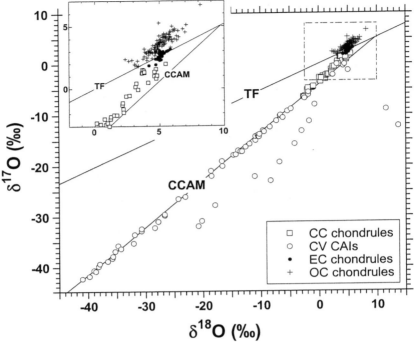

Figure 4. Oxygen isotopic compositions of the nebular components of chondrites plotted on the traditional 3-isotope diagram. Shown are high precision bulk and mineral separate analyses (data sources: Bridges et al. 1998; Clayton and Mayeda 1977, 1983, 1984; Clayton et al. 1991; Clayton et al. 1977; Davis et al. 1991; Gooding et al. 1983; Lee et al. 1980; Li et al. 2000; Mayeda et al. 1986; Rubin et al. 1990; Weisberg et al. 1991) of individual CAIs and chondrules from the three main chondrite classes: ordinary (OC), enstatite (EC), and carbonaceous (CC). The ^{16}O-mixing line discovered by R. Clayton and colleagues (Clayton et al. 1973), primarily from analyses of anhydrous minerals from CAIs from the Allende (CV) meteorite, is indicated as the carbonaceous chondrites anhydrous minerals line, or 'CCAM.' Although it is now recognized that many anhydrous minerals in carbonaceous chondrites do not plot exactly on this line, it is nevertheless a convenient reference line and is thus retained in subsequent plots. In fact, the majority of CAIs from CV chondrites do plot on this line, but data from six so-called FUN CAIs fall significantly to the right of CCAM.

The preponderance of the data plotting approximately along what is essentially a single linear trend (Fig. 4) strongly suggests that, to first order, oxygen isotope compositions of primitive solar system materials are determined by variable degrees of mixing and isotopic exchange between two dominant reservoirs (Clayton et al. 1977). Three independent lines of evidence have been utilized to argue that these major reservoirs consist of an ^{16}O-rich dust source and the solar nebula gas which is characterized by an oxygen isotopic composition near or above the TF line. The first is a simple appeal to mass balance—although we will discuss heterogeneities in detail below, in the first approximation most of the samples of planetary bodies that we have measurements for (Earth, Moon, and probably Mars and Vesta) plot near the TF line, as do most of the ordinary chondrites, and even (roughly speaking) the bulk carbonaceous chondrites. The relative abundances of oxygen and the major rock-forming elements dictates that most (~80-85%) of oxygen in the solar nebula was sequestered in gas-phase molecules, and thus it would have to be the minor dusty component that was anomalous. As discussed above, this circumstantial argument would be even more compelling if the solar composition were known with better precision (Wiens et al. 1999).

A second line of evidence for a gaseous composition that is somewhat ^{16}O-depleted relative to the Earth comes from the analysis of secondary minerals in both carbonaceous and ordinary chondrites that are considered, on very strong petrographic grounds, to have formed by interaction with aqueous fluids, presumably on an asteroidal 'meteorite parent body'. Several classes (CI, CM) of carbonaceous chondrites have been extensively mineralogically altered by chemical reaction with relatively large amounts of aqueous fluids at low temperatures, resulting in the formation of hydrous minerals (phyllosilicates), carbonate, and magnetite which exhibit elevated Δ^{17}O values relative to the precursor anhydrous minerals in the same meteorites (Clayton and Mayeda 1984; Rowe et al. 1994). A quantitative estimate of the isotopic composition of the solar nebula gas from which these fluids condensed can be derived based on a model of the coupled evolution of the fluid and the mineral phases of CM (Clayton and Mayeda 1984, 1999) and CI (Rowe et al. 1994; Leshin et al. 1996b) chondrites during the partial approach to equilibrium at low temperatures in an asteroidal setting. The gas itself is assumed to have evolved from an initial composition due to partial isotopic exchange with ^{16}O-enriched dust at high temperatures in the nebula. If the ^{16}O-rich component is modeled as having a starting composition of δ^{18}O $\sim \delta^{17}$O ~ -40 ‰, then the mass balance calculations yield compositions in the range δ^{18}O $\sim +25$ to $+30$ ‰ and δ^{17}O $\sim +20$ to $+25$ ‰ (Δ^{17}O $\sim +7$ to $+9$ ‰) for the initial solar nebula gas. This Δ^{17}O value is remarkably consistent with an estimate of fluid composition (Δ^{17}O $\sim +7$ ‰) derived from ion microprobe analyses of small magnetite nodules from the highly unequilibrated ordinary chondrite Semarkona (Choi et al. 1998). Magnetite is a useful probe of volatile reservoirs because it results from the oxidation of Fe-Ni metal, which likely occurs by reaction with aqueous fluids in an asteroidal parent body, and hence all the oxygen in magnetite is derived from the fluid (with little temperature dependence of the fractionation factors in the temperature range of interest for aqueous alteration of chondrites, ~50 to 300°C). The magnetite data of Choi et al. (1998) represent the highest Δ^{17}O values measured in solar system materials (~+5 ‰); this comes about essentially because of the small degree of fluid-rock interaction in Semarkona which leads to the preservation in the magnetite of a vestige of the high Δ^{17}O values of the solar nebula gas from which the fluid is derived.

A third line of evidence regarding the isotopic compositions of the initial dust and of the dominant gaseous reservoir in the solar nebula was actually historically the first to be developed, but perhaps ironically it is the most complicated to understand and fundamental aspects of this model remain enigmatic to this day. One of the more astounding features of Clayton's discovery is that most of the extreme range of O isotopic variations among nebular components (Fig. 4) can be spanned by the internal isotopic distributions among the various minerals of an individual CAI. This remarkable result was first observed for the large (~cm-sized) CAIs from the CV carbonaceous chondrites (e.g. Allende) which are sufficiently coarse-grained that mineral separates of several CAIs could be obtained for conventional fluorination analysis (Clayton et al. 1977). Spinel and Ti-bearing clinopyroxene (termed 'fassaite') were always found to be highly ^{16}O-enriched, down to values of δ^{18}O ~ -40‰, δ^{17}O ~ -41 ‰, whereas melilite was invariably much closer to the TF line in composition (Fig. 5a). Such extreme O isotopic heterogeneity (up to ~20 ‰ in Δ^{17}O) in objects that formed by crystallization from a melt at high temperatures (> 1400 °C) can only be understood as reflecting open-system behavior, most plausibly post-crystallization isotopic exchange with an external reservoir which has been presumed to be nebular gas (Clayton 1993). Because spinel is an early crystallizing phase (Stolper 1982), and because the same extreme composition of spinel

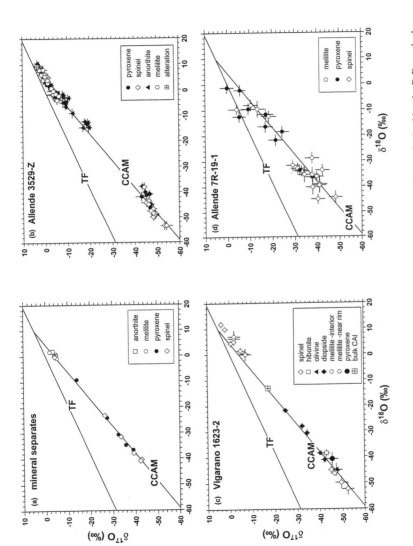

Figure 5. Oxygen isotopic compositions of minerals in CAIs from CV chondrites. (a) Data obtained by the BrF$_5$ method on ~mg quantities of mineral separates from coarse-grained type B inclusions from Allende (Clayton et al. 1977; Mayeda et al. 1986). To facilitate comparison with the ion microprobe, only data for relatively pure mineral separates are plotted; error bars are smaller than the symbol sizes. (b) Ion microprobe *in situ* data for minerals of Allende 3529-Z, a coarse-grained type B CAI; error bars are 1σ (McKeegan et al. 1996 and unpublished data). (c) Ion microprobe measurements of mineral phases in the interior and rim phases of inclusion 1623-3, a coarse-grained type A CAI from the reduced CV chondrite Vigarano (McKeegan et al. 1998a). (d) Ion microprobe measurements of individual mineral phases in Allende 7R-19-1 (Yurimoto et al. 1998).

could be found whether the spinel was enclosed within ^{16}O-rich pyroxene or ^{16}O-poor melilite (Clayton et al. 1977), it can be concluded that the initial isotopic composition of the melt had to have been ^{16}O-enriched and thus the second reservoir (i.e. solar nebula gas) was ^{16}O-poor.

Two further results are obtained by conventional mineral separate analyses that have had a large significance for estimating the composition of the initial dust reservoir prior to isotopic exchange with the gas and for confirming the general nature of the dust-gas fractionation that characterized the early solar nebula. The first is the oxygen isotopic composition of a spinel mineral separate obtained by acid-dissolution of the CM2 Murchison meteorite (Clayton and Mayeda 1984). Although it is now recognized as an average composition, slightly contaminated by a minor population of ^{16}O-poor spinel (McKeegan 1987; Simon et al. 1994; Simon et al. 2000), the close agreement of this value ($\delta^{18}O = -40$ ‰, $\delta^{17}O = -42$ ‰) with the most ^{16}O-enriched spinel compositions determined from mineral separates of individual Allende CAIs (Fig. 5a), suggested this as an appropriate composition for the end-member of the mixing line corresponding to refractory solar system dust. Additional evidence for the existence of an important reservoir near this -40 ‰ value is provided by a model of the internal distributions of O isotopes within the rare CAIs that do not fall on the Allende mixing line (Fig. 4). These so-called FUN (Fractionated with Unknown Nuclear effects) inclusions are isotopically (but not chemically or mineralogically) unique in that they have strongly mass-fractionated compositions of their major elements (e.g. Mg and Si) accompanied by correlated and unusually large apparent nuclear (i.e. nucleosynthetic) isotopic anomalies in several minor and trace elements (Lee and Papanastassiou 1974; Wasserburg et al. 1977; Lee et al. 1978). The melilite in EK-1-4-1, one of the 'classical' FUN inclusions for which mineral separations were possible, has essentially the same oxygen isotopic composition as 'normal' Allende CAIs (Clayton and Mayeda 1977) which is also the case for the rim phases of a unique Allende hibonite inclusion called HAL (Lee et al. 1980). These data can be understood as resulting from an overprint of the same post-solidus isotopic exchange process that affected the non-FUN type B CAIs in Allende and other CV chondrites, except in the FUN case operating on a starting composition that is mass fractionated from the -40 ‰ end-member (Fig. 6a). The degree of mass fractionation inferred for oxygen correlates in a semi-quantitative fashion with that measured for Mg and Si in several of the classical FUN CAIs (Clayton and Mayeda 1977).

The hypothetical mass fractionation trajectory on the O three-isotope plot for the FUN CAIs was recently confirmed by ion microprobe analyses (Davis et al. 2000) of primary minerals in a forsterite-rich inclusion from CV Vigarano (Fig. 6b). This CAI must have been crystallizing while still suffering large amounts of mass loss and kinetic isotope fractionation accompanying evaporation from a molten state (Rayleigh distillation). Isotopic exchange of the melilite followed after crystallization of the CAI. An extrapolation of the mass fractionation trend intercepts the CCAM line at a composition $\delta^{18}O \sim \delta^{17}O \sim$ -45 to -50 ‰. Although somewhat more ^{16}O-enriched compositions ($\Delta^{17}O \sim$ -25 ‰ or more) can be found among oxide minerals, such as hibonite and corundum (Fahey et al. 1987; Virag et al. 1991), that are still more refractory than the silicate-dominated CAIs typical of CV chondrites, this composition of -45 to -50 ‰ seems to characterize the extreme values found in a sufficiently large number of the refractory inclusions found in CV chondrites (including the FUN CAIs) that it must constitute an important solar system reservoir. Thus, there is abundant evidence that confirms the basic systematics of the two reservoir mixing model with analyses of individual phases by microprobe techniques providing slightly modified estimates of the end-member compositions: ^{16}O-enriched dust with $\Delta^{17}O \sim$ -25 ‰ and

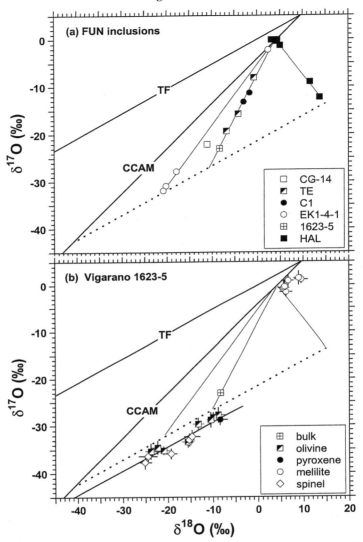

Figure 6. (a) Oxygen isotopic compositions of FUN inclusions from the CV chondrites Allende and Vigarano obtained by gas-source mass spectrometry (Clayton et al. 1984; Clayton and Mayeda 1977; Clayton et al. 1977; Davis et al. 1991; Lee et al. 1980). Isotopic mixing arrays are seen for 4 FUN CAIs for which mineral separates or different aliquots were analyzed (for CG-14 and Vigarano 1623-5 only bulk data exist). The dashed line indicates a hypothetical mass fractionation line from the composition of the spinel fraction of CM Murchison (Clayton and Mayeda 1984) through an estimate of the ^{16}O-rich terminus of each individual mixing line. (b) Oxygen isotope compositions of individual minerals in FUN CAI Vigarano 1623-5, obtained by ion microprobe (Davis et al. 2000). Olivine-pyroxene-spinel define a mass fractionation trend close to the postulated initial FUN line, whereas melilite exchanged oxygen isotopes to a more ^{16}O-poor composition compatible with melilite in other FUN CAIs and 'normal' CAIs.

^{16}O-depleted nebular gas with $\Delta^{17}O \sim +7$ ‰.

The isotopic compositions of chondrules also exhibit variable degrees of ^{16}O-enrichments (Clayton 1993 and references therein), however falling in a more restricted range between the end-members reflecting their greater degree of processing and equilibration with solar compositions. The greatest degree of inter-chondrule and intra-chondrule isotopic heterogeneity documented thus far is exhibited by Al-rich chondrules from ordinary chondrites (Russell et al. 2000). These objects, while not representing simple mixtures of CAIs and precursor materials of typical ferromagnesian chondrules, nevertheless provide evidence for partial isotopic exchange between refractory precursors with moderate ^{16}O-enrichments and ^{16}O-depleted nebular gas in a manner similar to that proposed for CAIs. Additionally, it is recognized that the interpretation of the isotopic mixing patterns in chondrules such as these is complicated by an overprinting due to multiple events, some involving aqueous processing (Russell et al. 2000). The complexities in interpreting nebular records caused by secondary effects, which probably originate on the parent body, has been the focus of high precision laser fluorination analyses (Ash et al. 1998, 1999; Bridges et al. 1998). As of this writing, a consensus on the relative importance of precursor heterogeneity vs. isotopic exchange during chondrule formation, as well as the confounding effects of secondary alteration in determining the preserved O isotopic records in chondrules from both CC and OC has not yet emerged, and much recent *in situ* work utilizing new developments in both laser and ion microprobe analyses has not yet appeared in the reviewed literature (e.g. Leshin et al. 1998b, 2000; Young and Ash 2001) although some has (Saxton et al. 1998b; Sears et al. 1998; Jones et al. 2000). Given this situation, we defer a detailed discussion on oxygen isotope compositions of chondrules, and in the remainder of this review concentrate on new constraints obtained by spatially resolved analyses since the last comprehensive review by Clayton (1993). As is often the case, observations made on the microscopic scale provide a new perspective for understanding the large scale issues relating to the origin and distribution of the oxygen isotopic anomaly within the solar system.

The nature of dust-gas isotopic exchange: a microscopic view. The first phases to be investigated by high spatial resolution methods were coarse-grained type B Allende CAIs similar those that Clayton and colleagues had previously demonstrated preserve a record of extraordinary isotopic heterogeneity. The ion microprobe and UV-laser fluorination results broadly confirm the inferences from mineral separate data, but also provide new constraints that point to the complexity of the isotope exchange process(es). The pattern of mineral-controlled isotopic heterogeneity is maintained in microscale analyses of CV CAIs (Figs. 5b,c,d). The same basic pattern is observed in a coarse-grained type A CAI from the reduced CV chondrite Vigarano, which argues against isotopic exchange in melilite from Allende CAIs being the result of late-stage metasomatic alteration of the Allende chondrite (because Vigarano shows very little mineralogical alteration compared to Allende). In contrast to this is the observation of a correlation of degree of aqueous alteration among whole CO chondrites with the oxygen isotopic composition of CAI melilite from those meteorites (Wasson et al. 2000).

The isotopic contrast in type B CAIs between near end-member ^{16}O-enriched spinel and melilite near the TF line is maintained over extremely fine spatial scales, in some cases <10 μm. This is true whether the grains are located in the center of a large (2 cm-diameter) inclusion (Fig. 7a) or are at the periphery where the CAI is rimmed by a Wark-Lovering mineral sequence (Fig. 7b). Significantly, no gradients are observed in melilite $\delta^{18}O$ and $\delta^{17}O$ as would be expected under any scenario in which gas-solid diffusion was the primary mechanism for isotopic exchange in melilite (Ryerson and McKeegan 1994). Zoning is observed in anorthite (Fig. 5b), but its geometry is generally inconsistent with

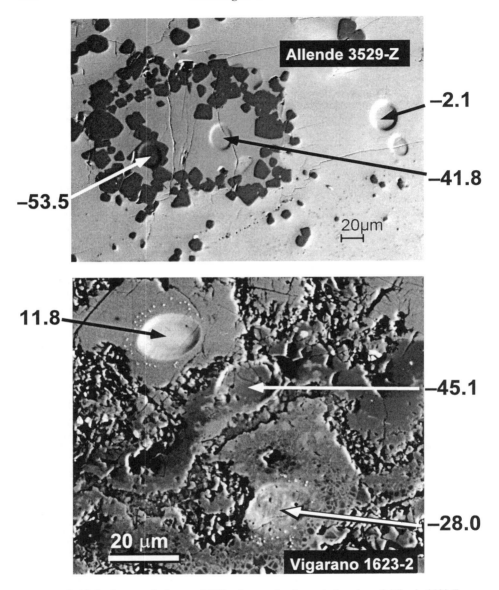

Figure 7. (a) Backscattered electron (BSE) micrograph of a central region of Allende 3529-Z, a ~2 cm-diameter type B CAI, showing ion microprobe sputtered holes in a large melilite crystal (light-gray) that encompasses a cluster of spinel grains which, in turn, surround a pocket of late-crystallizing fassaite (darker gray). Analysis spots are labeled by their $\delta^{18}O$ values; typical 1σ uncertainties are ~2 ‰ or better. The scale bar is 20 μm. (b) BSE image of a portion of the rim sequence surrounding Vigarano 1623-2, a Type A CAI, showing the locations of three ion probe craters where oxygen isotope compositions were determined (McKeegan et al. 1998a). Shown are $\delta^{18}O$ values (typical errors are ~1 ‰) indicating large isotopic heterogeneity (along a slope ~1 mixing line) between different mineral phases in the Wark-Lovering rim sequence. The inclusion interior is toward the top of the image; the light-gray material at the top is melilite. Data from the UCLA ion microprobe laboratory.

diffusive exchange (McKeegan et al. 1996). A sharp zoning profile (essentially a step-function) within a single melilite crystal, and the first instance of ^{16}O-enriched melilite in a CV CAI, was discovered by Yurimoto and colleagues (Yurimoto et al. 1998) who invoked a re-melting episode to explain the isotopic exchange pattern. Certainly many type B CAIs have experienced multiple partial melting events (Connolly and Burnett 1999; Davis et al. 1991), as well as impulsive heating that produced the coarse-grained Wark-Lovering rims. The difficulty with appealing to partial melting to explain the isotopic patterns in most type-B CV CAIs, however, is that pyroxene should melt prior to melilite upon reheating of the CAI (Stolper and Paque 1986), but *in situ* data found so far indicates that pyroxene nearly always preserves its ^{16}O-rich composition (e.g. data in Fig. 5).

A hint is provided by high-precision laser fluorination analyses of an Allende CAI that demonstrates a distinct difference in the oxygen isotopic composition of melilite which is inferred on chemical and petrographic evidence to have been partially altered by interaction with alkali-bearing aqueous fluids (Young and Russell 1998). Melilite crystals lacking these indications of alteration lie on a mixing line (terminating at ^{16}O-rich spinel) that has a slope exactly equal to unity, whereas the altered melilite in the CAI trends along the CCAM line with slope = 0.94 (Fig. 8a). An extension of this pure ^{16}O-mixing line (Fig. 8b) passes through the data for the (anhydrous) components of ordinary chondrites, suggesting the presence of a "primitive oxygen reservoir" in the solar nebula (Young and Russell 1998). Recent ion microprobe analyses of very small CAIs from CH (Sahijpal et al. 1999; McKeegan et al. 2001) and (to a lesser extent) enstatite (Fagan et al. 2001) chondrites corroborate the existence of a slope = 1.0 mixing line coincident (within error) with the Young and Russell (Y&R) line (see Fig. 10 below). In contrast to the large CV CAIs that individually define the CCAM line, individual CH CAIs are essentially homogeneous in their O-isotopic compositions, and it is the ensemble of CAIs that indicates isotopic mixing of gaseous reservoirs along a pure ^{16}O trajectory (McKeegan et al. 2001). As the CH CAIs are free of mineralogical alteration, these data lend support to the hypothesis (Young and Russell 1998) that isotopic exchange facilitated by alteration is somehow responsible for lowering the slope of mixing lines, even those exhibited by high-temperature nebular phases, such as CV CAIs. This view differs substantially from the long-held interpretation of the CCAM line (e.g. Clayton 1993), and this controversy will likely continue to fuel a vigorous level of research activity directed toward understanding the significance of O-isotopic heterogeneities amongst and within the high-temperature nebular components of chondrites.

Oxygen isotopic exchange that occurs during aqueous alteration reactions at low temperatures, presumably in an asteroidal setting, has long been recognized as a significant process overprinting the nebular signatures of bulk carbonaceous chondrites (e.g. Clayton and Mayeda 1984). As discussed above, oxygen isotopic abundances in secondary minerals of carbonaceous chondrites can be used to estimate the fluid compositions, and with some knowledge of the isotopic compositions of primary phases, models can be constructed (e.g. Clayton and Mayeda 1984; Rowe et al. 1994; Leshin et al. 1996b) to constrain the environmental conditions of the alteration process(es). Recently, Young and co-workers have developed a physical model that follows the thermal, mineralogical, and isotopic evolution of the components of various classes of carbonaceous chondrites as resulting from variable degrees of fluid-rock interaction on asteroidal parent bodies (Young et al. 1999). A detailed discussion of the mechanisms and implications of these competing classes of models is beyond the scope of this review, and the reader is directed to the recent primary literature (e.g. Clayton and Mayeda 1999; Young et al. 1999) for further information on this still-developing application of stable isotopes to early asteroid geology.

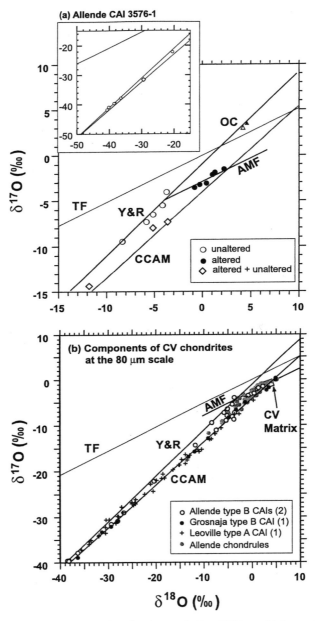

Figure 8. Oxygen three-isotope plots showing results from UV laser ablation analysis of CV meteorite components: (a) Mineralogically altered and unaltered portions of a CAI from the Allende meteorite. The unaltered points were used by Young and Russell (1998) to suggest that the primordial oxygen reservoir(s) for the early solar system define a line with a slope of 1.00 and an intercept of -1.0 (the Y&R line). (b) CAIs and chondrules from the CV meteorites Allende, Grosnaja, and Leoville. O isotopic data for these

primitive meteorites occupy a region bounded by the Y&R line, the mass fractionation line described by Young et al. (1999) (termed the Allende mass fractionation line, AMF), and the CCAM line. Also shown is the field for CV matrix (Clayton and Mayeda 1999). Young and Russell (1998) suggested that the CCAM line is a secondary and not a primary (i.e. nebular) mixing line in that it results from the inclusion of both altered and unaltered components in mineral separate analyses. As indicated in (a), alteration shifts data off the Y&R line along a mass-fractionation trajectory. Analytical uncertainties are smaller than the symbol sizes; laser ablation data are from Young and Russell (1998), Young et al. (1999, 2000, 2001).

Distribution of the oxygen isotope anomaly in the inner solar system and implications for the provenance of CAIs. The oxygen isotopic heterogeneity which is the single most pervasive feature characterizing unequilibrated chondritic materials on microscopic size scales is also reflected in the bulk compositions (Fig. 9) of chondrites, differentiated meteorites, and even of entire planets (Clayton 1993 and references therein). Although the fundamental reason for this heterogeneous distribution of the oxygen isotopes on planetary size scales is not understood, it is presumably related in some sense to distinct accretion regions in the solar nebula, and it provides a practical aid for classification purposes and deciphering possible genetic relationships among chondrites.

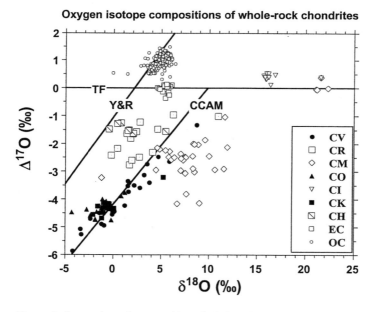

Figure 9. Oxygen isotopic composition of whole-rock chondrites according to group. Plotted are $\Delta^{17}O$ vs. $\delta^{18}O$ obtained by the BrF_5 method; most data are obtained in the laboratory of R. Clayton (data sources: Bischoff et al. 1993; Clayton and Mayeda 1984, 1999; Clayton et al. 1976, 1977, 1984, 1991, 1995; Rowe et al. 1994; Rubin et al. 1983; Weisberg et al. 1993) with error bars generally similar to the symbol sizes. Each meteorite group occupies a distinctive field on this diagram enabling oxygen isotope compositions to be used as a classification tool, although there is a significant degree of overlap between certain groups (e.g. CV-CO-CK, and CR-CH) possibly indicating genetic affinities. On this plot, mass-dependent fractionation lines are horizontal. The carbonaceous chondrite anhydrous minerals (CCAM: Clayton et al. 1977) and the Y&R (Young and Russell 1998) mixing lines are indicated.

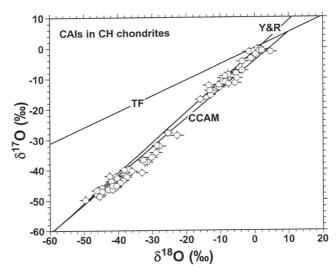

Figure 10. Oxygen isotopic compositions of CAIs in CH chondrites (data sources: Sahijpal et al. 1999; McKeegan et al. 2001). Isotopic heterogeneity within an individual inclusion is limited and the ensemble of CH CAIs are distributed along a line of slope = 1.0 on the three-isotope plot.

The ion microprobe allows measurement of samples that are too small and/or rare to be studied by fluorination techniques. We and our colleagues have used an ion microprobe to examine the oxygen isotopic records of previously unstudied types of refractory calcium-aluminum-rich inclusions (CAIs) from ordinary (McKeegan et al. 1998b) and enstatite chondrites (Fagan et al. 2001; Guan et al. 2000), CH chondrites (McKeegan et al. 2001), as well as micrometeorites (Engrand et al. 1999b) and IDPs (McKeegan 1987). With the notable exception of a group of CH CAIs (Fig. 10) (Sahijpal et al. 1999; McKeegan et al. 2001), the data (from unaltered samples) show consistently large ^{16}O-enrichments with $\delta^{18}O \approx \delta^{17}O \approx -40$ to -50 ‰ in all CAI minerals across all chondrite groups (Fig. 11a). Unlike the larger coarse-grained CAIs from CV chondrites (Fig. 5), isotopic heterogeneity within these small (or even microscopic) CAIs is relatively limited suggesting that these materials had simpler thermal histories and that the large, better studied, CV CAIs may actually be exceptional. The O isotopic compositions of the most refractory components of AMMs and IDPs (Fig. 11b) are also consistent with these highly ^{16}O-enriched values (McKeegan 1987; Greshake et al. 1996; Engrand et al. 1999b). In addition, Hiyagon and Hashimoto (1999) have discovered a class of (relatively common) forsterite-bearing inclusions (sometimes called 'amoeboid olivine aggregates' or AOAs because of their irregular shapes) that are isotopically the same as more refractory CAIs.

The relative consistency of the O-isotopic compositions of most CAIs from several different classes of chondrites stands in contrast to the isotopic behavior of other components of chondrites which fall into reasonably distinct and well-defined groups (see Fig. 4 and Fig. 9). For example, while all chondrules have relatively ^{16}O-poor isotopic compositions compared to CAIs, these compositions vary significantly between the major classes of chondrites, with ordinary chondrite chondrules being the most ^{16}O-poor, followed by enstatite chondrite chondrules with $\Delta^{17}O \approx 0$, and then carbonaceous chondrite chondrules with $\Delta^{17}O < 0$ (Fig. 4). The observation that rare CAIs in the

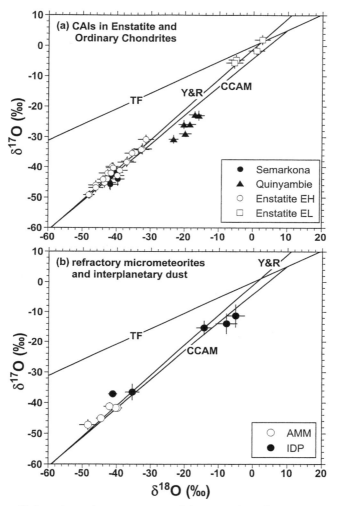

Figure 11. Ion microprobe measurements of the oxygen isotopic compositions of refractory materials not from carbonaceous chondrites. (a) Data from small and rare CAIs in enstatite (EL3, EH3) and the unequilibrated ordinary chondrites Semarkona (LL3.0) and Quinyambie (LL3.4). Most CAIs show high degrees of ^{16}O-enrichment independent of mineral phase (e.g. melilite is also ^{16}O-rich). The Quinyambie data are affected by alteration and brecciation on the meteorite parent body, and possibly by matrix effects in the ion microprobe. The relatively ^{16}O-poor composition of the only CAI analyzed from an EL chondrite appears indigenous (i.e. not affected by terrestrial weathering), but it is not known how representative this single result may be. (Data sources: McKeegan et al. 1998b; Guan et al. 2000; Fagan et al. 2001.) (b) Data from <10 µm-sized melilite and spinel grains in Antarctic micrometeorites (AMMs) and from fragments of ~10 to 20 µm-diameter interplanetary dust particles (IDPs) bearing refractory minerals (spinel, hibonite, corundum, melilite, perovskite). Dust particles generally share a remarkable similarity of ^{16}O-rich isotopic compositions with most CAIs; analyses of two IDPs with less ^{16}O-rich compositions may possibly be 'contaminated' by isotopically 'normal' chondritic material. (Data sources: McKeegan 1987; Engrand et al. 1999b).

ordinary (McKeegan et al. 1998b) and enstatite (Guan et al. 2000; Fagan et al. 2001) chondrites have oxygen isotopic compositions essentially identical to carbonaceous chondrite CAIs implies that these materials are genetically related. Additionally, since these O-isotopic compositions are distinct from those of all other materials in those chondrites, it seems likely that these CAIs (and AOAs) did not form in the same isotopic reservoirs as other matter (e.g. chondrules, matrix) accreted by the ordinary and enstatite chondrite parent bodies. Thus, the oxygen isotope data suggest that CAIs from all chondrite groups, probably including micrometeorites and IDPs, sampled a restricted isotopic reservoir in the solar nebula that was outside the formation region of the other chondritic materials. This is most easily understood if nearly all CAIs formed in a single, isotopically unique nebular locale and were subsequently transported to the chondrite accretion regions with differing efficiencies such that carbonaceous chondrites acquired a far greater abundance of CAIs than did ordinary or enstatite chondrites (McKeegan et al. 1998b). The X-wind model of Shu and co-workers (Shu et al. 1996) links the petrogenesis of CAIs to localized high-energy environments near the protosun (~0.06 AU) and provides a natural transport mechanism, in the form of bipolar outflows (X-winds) driven by the magnetic interactions between the forming star and its accretion disk, for scattering these materials out to large radial distances in the solar nebula where they may later be incorporated into the parent asteroids of chondrites. The hypothesis that CAIs and, possibly, other components of chondrites were formed in the violent environment of the X-wind is highly controversial, in part because of the many collateral implications (e.g. production of short-lived radioactivity; Lee et al. 1998) for models of the early evolution of the solar system. Further detailed work exploring for correlations of radioactive and stable isotope signatures in various components of primitive meteorites (CAIs, chondrules, etc.) is required to better elucidate the nebular sites where these materials originated and the processes that they witnessed.

The origin of the oxygen isotope anomaly. This is a key question, the answer to which has remained elusive despite 25 years of intensive study. The new *in situ* data detailing the microscopic distribution of 'anomalous' oxygen in CAIs and chondrules (not discussed here) and the O isotopic compositions of other types of primitive materials too small for study by conventional fluorination techniques (e.g. microscopic CH CAIs, AOAs, IDPs, etc.) have provided additional indirect arguments, but no firm conclusions, regarding the ultimate origin of the ^{16}O enrichments in these objects. For example, the observations of slope = 1.0 mixing lines in unaltered portions of CAIs (Young and Russell 1998) and in CH CAIs (McKeegan et al. 2001) that lack other evidence of isotopic anomalies of nucleosynthetic origin, as well as the seeming ubiquity of ^{16}O enrichments in forsterite from AOAs to the same levels as in many CAIs (Hiyagon and Hashimoto 1999), have all been cited as evidence favoring a local, chemical origin (Thiemens 1996, 1999) of ^{16}O enrichments. However, a chemical mechanism that can directly produce and preserve ^{16}O enrichments in condensed refractory dust is not known, and it even remains unclear what kind of environment in the solar nebula should plausibly be considered as a potential site for these types of processes.

The alternative hypothesis, that of a nucleosynthetic component carried into the early solar system by ^{16}O-rich presolar dust (see discussion in Clayton 1993) remains plausible; after all, as mentioned in the Introduction, trace amounts of presolar dust is known to exist in the matrices of chondrites. However, all the high spatial (ion microprobe) measurements of the many types of primitive refractory materials discussed here have failed to find any very extreme oxygen isotopic anomalies such as might be expected if pure ^{16}O dust (e.g. from a supernova) were incorporated into CAIs or AOAs along with local condensates. Even among the ~100 true interstellar oxide grains that have been examined by Nittler (1997, 1998), only a single grain with an extremely ^{16}O-rich composition was found (most grains instead exhibiting a wide range of isotopic effects in both ^{17}O and ^{18}O). Thus, it seems that if interstellar dust was the source of the ^{16}O-enrichments in solar system materials, then it most likely is characterized by oxygen isotopic compositions that are not too different from the maximum anomalies observed in the most refractory CAIs. A model for how such an isotopic memory, perhaps carried into the solar system by only moderately ^{16}O-enriched silicate dust, could have been preferentially implanted in refractory minerals has been developed by Scott and Krot (2001). The model calls for formation of CAI precursors as condensates from a gas that is enriched in ^{16}O by evaporation of presolar dust in a region of the nebula that is depleted, relative to cosmic abundances, in (^{16}O-poor) nebular gas. In the sense that such a nebular region must necessarily be spatially restricted, this model is compatible with inferences regarding the provenance of CAIs discussed above.

STABLE ISOTOPE RECORDS IN EVOLVED MATERIALS

The achondritic meteorites are generally igneous rocks and, thus, their stable isotope records can be utilized to investigate petrogenesis in the same manner as is done for analogous terrestrial samples (Clayton 1986). These meteorites are derived from differentiated asteroids and planetary bodies. The oxygen isotope records of achondritic meteorites from the asteroid belt have been utilized to establish genetic relationships and to investigate thermal histories (e.g, isotopic equilibration temperatures). This extensive body of data has been summarized and reviewed from various contexts in the recent literature (Clayton 1993; Clayton and Mayeda 1996; Mittlefehldt et al. 1998), and the reader is directed to these sources. Here we discuss the H and O isotopic records from only those samples that are considered to represent planetary bodies.

Moon

Analyses of lunar samples returned via spacecraft and of lunar meteorites have a bearing on the origin of the largest satellite in the inner solar system. Oxygen isotopic compositions show a narrow range in $\delta^{18}O$ values from +4.2 to +6.4 ‰ and suggest a lunar bulk isotopic composition very similar to Earth (e.g. Clayton and Mayeda 1975; Clayton and Mayeda 1996; Taylor and Epstein 1970). Especially significant are the $\Delta^{17}O$ values that lie within analytical uncertainty (~0.1 ‰) of the terrestrial mass fraction line (Clayton and Mayeda 1975; Robert et al. 1992; Clayton and Mayeda 1996). Some "Giant Impact" models for lunar formation (e.g. Benz et al. 1986), in which an approximately Mars-sized body impacts the early Earth and forms the Moon from the residue of the event, predict that a large fraction of the Moon represents impactor material. The oxygen isotopic compositions of lunar materials suggest that if this hypothesis is correct, the impactor had $\Delta^{17}O$ value very similar to the Earth. An impactor with a Mars-like $\Delta^{17}O$ value (see below), for example, should have been detected with the available data. New analyses, utilizing the most modern and precise laser fluorination techniques applied to a wide-variety of lunar materials, require even more stringent constraints on the similarity the $\Delta^{17}O$ values of lunar and terrestrial materials (Weichert et al. 2000). These data show that 33 lunar samples have very uniform $\Delta^{17}O$ values of -0.01±0.03 ‰. Implications of these data for Giant Impact models of the Moon's formation have yet to be discussed in detail in the reviewed literature, and remain a future important direction in lunar research.

The Moon is greatly depleted in volatile elements including C and H. Isotopic analyses of H in lunar materials typically reflect the isotopic composition of the solar wind (essentially D/H = 0) modified to varying degrees by terrestrial contamination (e.g. Epstein and Taylor 1971). Very rare indigenous C components have $\delta^{13}C \approx$ -19 to -26 ‰, and are also easily compromised by terrestrial contamination (e.g. DesMarais 1978).

Mars

Martian (SNC) meteorites, remote sensing, and spacecraft exploration have provided insights into the origin, evolution and interaction of several light-element reservoirs on Mars including the lithosphere, hydrosphere and atmosphere. Beyond Earth, Mars is certainly the most exhaustively investigated body from a stable isotopic perspective.

Atmosphere. Owen et al. (1988) were the first to report detection of D on Mars. They derived a value of D/H of 6±3 times the terrestrial value (SMOW) by observing from Mauna Kea the intensity of a fundamental HDO absorption at 3.7 μm relative to H_2O lines in the martian atmospheric spectrum. The precision of this result was improved by Bjoraker et al. (1989), who made measurements from the Kuiper Airborne Observatory. They report a D/H of 5.2±0.2 times terrestrial, corresponding to a δD value of +4200±200 ‰, but this value has only been published in abstract form and its high precision is somewhat questionable. It is appropriate to note that the range of δD values commonly found on Earth is approximately -500 to +100 ‰. Thus, the value of D/H of water in the martian atmosphere is significantly higher than the D/H of any hydrogen found on the Earth. It is generally assumed that Mars originally contained water of similar D/H to the Earth and most meteorites (see "primitive materials" section for a review) and that the present atmosphere D/H value is a result of preferential loss of H relative to heavier D from the martian atmosphere throughout the planet's history. The details of the timing of water loss and the evolution of the deuterium enrichment, however, remain highly unconstrained.

Carbon dioxide is the predominant constituent of the current martian atmosphere composing ~95% of the ~7 mbar of gas present in the atmosphere today. Additionally, CO_2 is thought to have played an important role in martian climate in the past, possibly contributing to an early warmer climate due to the greenhouse effect (Fanale et al. 1992). The Viking entry mass spectrometer measured the values of $^{13}C/^{12}C$ and $^{18}O/^{16}O$ in martian atmospheric CO_2. Within the uncertainty of ±5 %, the values of both of these ratios were found to be indistinguishable from terrestrial (Nier and McElroy 1977), confirming the results of earlier, less precise measurements as summarized by Owen (1992). More recently, Krasnopolsky et al. (1996) measured C and O isotope ratios in CO_2 consistent with these observations using the 4 m telescope at Kitt Peak. The lack of enrichment of the oxygen isotopes is an interesting observation, since currently one oxygen atom could escape from Mars with each two hydrogen atoms. Considering the extremely fractionated hydrogen isotopes, the oxygen in the atmospheric water must be buffered by an additional reservoir of oxygen to explain the lack of heavy oxygen isotope enrichment.

Studies of martian meteorites can also provide insights in to the composition of the atmospheric reservoir. These are discussed below.

Stable isotopic composition of martian meteorites. At this writing there are 18 known members of the martian meteorite clan. All are igneous rocks that have been affected to differing degrees by secondary processing on Mars, including impact events and interactions with fluids in the martian crust. Evidence for the martian origin of these rocks is rooted in isotopic compositions and relative abundances of gases in impact-produced glasses in EETA 79001 and other martian meteorites (e.g. Bogard and Johnson 1983; Becker and Pepin 1984; Marti et al. 1995) which closely match the composition of the current martian atmosphere. McSween and Treiman (1998) provide the most recent overview of the mineralogy and petrology of these samples, and review in detail the evidence for martian origin.

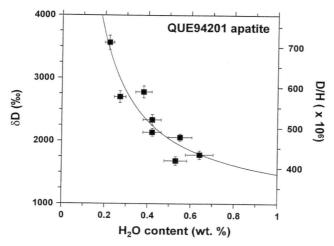

Figure 12. Laser fluorination analyses of whole-rock martian (SNC) meteorites after Franchi et al. (1999). The data demonstrate that the meteorites sample a common parent body with a $\Delta^{17}O$ value of +0.32 ‰.

Oxygen isotopes provide the most compelling evidence that the mineralogically diverse suite of martian meteorites represent samples from a single parent body. Clayton and Mayeda (1983; 1986; 1996), Romanek et al. (1998), and Franchi et al. (1999) report high-precision whole-rock 3-isotope oxygen analysis of martian meteorites. The most recent and precise data, laser fluorination analysis of powdered whole rocks from Franchi et al. (1999), show relatively restricted $\delta^{18}O$ values from +4.2 to +5.2 ‰, consistent with the igneous origin of the samples and their variable primary mineralogy. The other above works report slightly offset (by up to ~0.3 ‰) but similar ranges in $\delta^{18}O$ values, with interlaboratory differences probably attributable to slight sample heterogeneities and differences in reference gas calibration methods (Franchi et al. 1999). $\Delta^{17}O$ values reported by Franchi et al. (1999) are very tightly clustered, with all 13 samples analyzed falling in a range of 0.30 to 0.34 ‰ (Fig. 12). Thus, all the martian meteorites fall on a common mass fractionation trend (the Mars silicate fractionation line or MSFL) with $\Delta^{17}O = 0.32\pm0.02$ ‰, strongly supporting the hypothesis that these meteorites sample a common parent body. Again, these $\Delta^{17}O$ values are consistent with previous whole-rock data (Clayton and Mayeda 1983, 1986, 1996; Romanek et al. 1998), which are somewhat less precise. Given the ancient age of martian meteorite ALH84001 (4.5 Ga), and the large range in ages of the martian meteorites as a clan (from 180 Ma to 4.5 Ga, Shih et al. 1982; Jagoutz 1991; Jagoutz et al. 1994; Nyquist et al. 1995) the whole-rock oxygen isotopic data suggest that the lithosphere of Mars was isotopically homogenized very early in martian history.

Recent, more detailed investigations of oxygen isotopic compositions of martian meteorite components have focussed on secondary minerals and water contained in the samples. Studies that report analyses of all three oxygen isotopes consistently show that secondary minerals and water in the martian meteorites do not lie on the MSFL. Karlsson et al. (1992) observed elevated $\Delta^{17}O$ values, up to +0.9 ‰ (or ~0.6 ‰ above the MSFL) in water extracted by stepped heating from Nakhla, Lafayette and Chassigny, with the most pronounced effects observed in Nakhla. The water extracted from the shergottites generally records $\Delta^{17}O$ values between than the MSF and the TF line. Farquhar and colleagues (2000; 1998) found $\Delta^{17}O$ values ranging from +0.7 to +1.3 ‰ in carbonates and sulfates from ALH84001, Nakhla and Lafayette, also out of isotopic equilibrium with martian silicates. Secondary "iddingsite" in Nakhla, a mixture of secondary Fe-rich oxides, hydroxides and clays, also shows elevated $\Delta^{17}O$ of ~+1.4 ‰ (Romanek et al. 1998). Taken together, these analyses indicate that oxygen in the martian hydrosphere and lithosphere are not fully equilibrated. The most plausible hypothesis for the origin of the oxygen isotopic signature in secondary materials is that atmospheric mass independent isotopic effects, similar to those observed in a wide variety of terrestrial atmospheric species (see Thiemens 1999 for a recent review), have been transferred to fluids in the martian hydrosphere that are sampled by the secondary minerals. This atmosphere-hydrosphere-lithosphere interaction is also supported by hydrogen and carbon isotopic studies discussed below.

Oxygen isotopic compositions can also be used to gain insight into the environmental history of secondary materials in martian samples, similar to their use in terrestrial materials (although somewhat complicated by the observation that they are not strictly in isotopic equilibrium with their host rocks). In general the oxygen isotopic compositions of secondary carbonates and silicates have elevated $\delta^{18}O$ values relative to their host rocks, and the waters contained in the rocks (Carr et al. 1985; Clayton and Mayeda 1988; Wright et al. 1988; Romanek et al. 1994; Romanek et al. 1998; Farquhar and Thiemens 2000), suggestive of relatively low temperatures of formation. In the one case where detailed *in situ* isotopic studies have been performed, however, the situation has proven to be quite complex. Carbonates in ALH 84001 show $\delta^{18}O$ variation of ~20 to 25 ‰, from values of ~ 0 to +25 ‰ (Valley et al. 1997; Leshin et al. 1998a; Saxton et al. 1998a; Eiler et al. 2001). These isotopic values correlate with mineral chemistry with the earliest-forming, Ca-rich carbonates having low the lowest $\delta^{18}O$ values, and later-forming Mg-rich carbonates having the highest $\delta^{18}O$ values. These very large isotopic effects are observed over distances of ~100 μm using ion microprobe techniques, and are explained as resulting from large temperature variations during carbonate formation, or fluid compositional variations either due to evaporation of a water-rich fluid or closed system evolution (Raleigh distillation) during carbonate formation from a CO_2-rich fluid. Comparable effects in secondary minerals from other martian meteorites have not been observed, although studies are much more limited.

Hydrogen isotopic systematics of bulk martian meteorites are reported by Leshin et al. (1996). Stepwise extraction of water from the samples generally shows terrestrial contamination at low temperatures (below 300-400°C), and what is interpreted to be martian water, with elevated δD values, at higher temperatures. The high-temperature component ranges in δD from +50 ‰ for Chassigny to +2100 for Shergotty. All samples except Chassigny show a clear extraterrestrial signature in their bulk D/H values, and the Chassigny water arguably represents terrestrial contamination (Leshin et al. 1996a). However, interpreting these values is complicated by the presence, probably to differing degrees, of addition of or isotopic exchange with terrestrial water, and by the varied geological history of these samples (e.g. differing degrees of interaction with groundwaters on Mars).

The D/H values of magmatic, hydrous minerals apatite, Ti-rich amphibole, and biotite have been analyzed in five martian meteorites by ion microprobe (Leshin 2000; Rubin et al. 2000; Watson et al. 1994). The elevated and variable D/H values of water discovered in the minerals (δD ~ +800 to +4300) are interpreted qualitatively as representing a mixture of magmatic water in the minerals with a D-enriched component derived from the martian atmosphere (with a D/H value ~5 times terrestrial), through isotopic exchange with D-enriched groundwaters introduced after the phases crystallized.

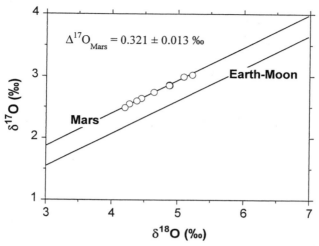

Figure 13. D/H and water contents of apatite grains from martian meteorite QUE94201 after Leshin (2000). The data are interpreted to represent a mixture of two end members, and most plausibly represent addition (or exchange) of water with an atmospheric D/H signature (δD ~ +4000 ‰) to minerals which initially uniformly contained water with δD of ~+900±250 ‰, or approximately twice the D/H value commonly assumed for magmatic water on Mars. The curve shows the mixing model from which the initial D/H of the minerals was calculated.

The presence of a "meteoric" (atmosphere-derived) D-rich component is also indicated by ion microprobe analyses of carbonates in ALH84001, which have δD values up to ~+2000 ‰ (Sugiura and Hoshino 2000). The mixture between magmatic water and high δD water is shown most quantitatively in apatite grains from shergottite QUE 94201 (Leshin 2000). These data (Fig. 13) show 2-component mixing, with the high-δD component plausibly derived from the atmosphere (δD ~ +4000 ‰), and the low-δD component representing martian magmatic water. These data show that martian magmatic water has a D/H ratio of ~twice the terrestrial value, and is consistent with a two-stage history of martian volatiles where an initial D-enrichment of originally "chondritic" water is achieved via early hydrodynamic escape, and the subsequent enrichment to the current atmospheric value is achieved through Jeans (thermal) escape, which continues today (Leshin 2000).

There have been many reports of the abundances and $^{13}C/^{12}C$ values of CO_2 extracted from martian meteorites and their probable implications for the evolution of CO_2 on Mars (Carr et al. 1985; Clayton and Mayeda 1988; Wright et al. 1986; Wright et al. 1988, 1990, 1992; Jull et al. 1995, 1997; Romanek et al. 1994; Leshin et al. 1996a). These studies suggest the presence of three main carbon-bearing components in martian meteorites, distinguished by their temperature of release (or release by reaction with orthophosphoric acid) and their $\delta^{13}C_{PDB}$. The first component, released at temperatures below ~500°C, is interpreted as low-temperature carbonaceous material mostly from contamination by terrestrial organic matter but possibly mixed with extraterrestrial organics. This component has a $\delta^{13}C$ of -20 to -30 ‰ (Carr et al. 1985; Wright et al. 1986; Wright et al. 1992; Grady et al. 1994; Jull et al. 1998, 2000; Leshin et al. 1996a). The second carbon-bearing component, released between 400 and 700°C in heating experiments, or by reaction of samples with orthophosphoric acid, originates from breakdown of carbonate. This component is associated with ^{13}C-enriched CO_2 in several of the martian meteorites: in Nakhla it has δ^{13}C up to ~+35 ‰ (Carr et al. 1985; Wright et al. 1992; Jull et al. 1995; Leshin et al. 1996a); in orthopyroxenite ALH84001 it has δ^{13}C up to +46 ‰ (Grady et al. 1994; Romanek et al. 1994; Jull et al. 1995; Leshin et al. 1996a; Jull et al. 1997; Valley et al. 1997). The ^{13}C-enriched nature of this component provides strong evidence of a preterrestrial origin (since terrestrial carbonates generally have δ^{13}C near or slightly below zero). The third component consists of carbon released at temperatures above 700°C, generally with δ^{13}C values of -20 to -30 ‰, and has been interpreted as a "magmatic" carbon component representative of the isotopic composition of CO_2 in the martian interior (Carr et al. 1985; Wright et al. 1986; Wright et al. 1990 1992). If the high-temperature fraction discussed above really does represent magmatic carbon, the δ^{13}C value of -20 to -30 ‰, which could represent the martian "primitive" value given the probably lack of plate tectonics on Mars, is distinct from the proposed value for terrestrial carbon bulk δ^{13}C (-5 ‰). However, it remains difficult to rule out the concern that this value represents terrestrial organic contamination that lingers even to high temperature in the bulk heating experiments.

Taken together the stable isotopic data on all martian meteorite components produce a relatively coherent picture of martian volatile cycles. The solid planet was isotopically homogenized very early in martian history, as evidenced by the homogeneity of the Δ^{17}O values of the igneous silicates in samples which range in age from 4.5 to 0.18 Ga. Atmospheric C, H and O isotopic signatures are preserved in many phases in martian meteorites, thus the crustal and atmospheric volatile reservoirs are clearly in isotopic communication. Martian atmospheric plus crustal volatile reservoir is clearly not in isotopic equilibrium with the planetary interior based on the δD, δ^{13}C, and Δ^{17}O measurements discussed above, which preserve a degree of isotopic heterogeneity unmatched in any suite of terrestrial igneous rocks. The lack of equilibration of the crustal and interior volatile reservoirs argues against a large degree of crustal recycling on Mars.

UNRESOLVED ISSUES AND FUTURE DIRECTIONS

We have attempted a review of a small portion of a rapidly developing field, one which is poised to achieve breakthroughs on several problems of fundamental importance for understanding the formation and evolution of our planetary system. After a decade of rapid progress, laboratory techniques for the precise measurement of stable isotope abundances at an appropriate scale of spatial resolution or sensitivity for specific compounds in many extraterrestrial materials are operational in a couple of dozen laboratories worldwide, and new developments are underway in anticipation of the first returned samples from known extraterrestrial bodies in more than a generation. First order discoveries await with respect to solar isotopic abundances (Genesis mission), the correlated isotopic and mineralogic compositions of known cometary dust particles (Stardust mission), and, hopefully, sample returns from asteroid(s) and from Mars. Other important advances can be anticipated in the near term through application of the maturing techniques of laser and ion microprobe stable isotope analysis, among others, to samples already in hand. We close by noting some of the more significant outstanding problems which we expect will occupy the creative energies of that cadre of isotope geochemists who turn their instruments on the strange and wonderful materials which come to us from the heavens.

- **Origin and distribution of the oxygen isotopic anomaly:** It is presently unclear whether the premier isotopic anomaly in meteorites is ultimately due to early solar system chemistry or to stellar nucleosynthesis. The resolution of this issue has first-order astrophysical implications for the origin of the solar system. Because oxygen is the dominant rock-forming element, and the oxygen isotopic anomaly is even manifest on a planetary scale within at least the inner solar system, if it turns out that this effect is ultimately due to symmetry dependent non-equilibrium chemistry, this would represent a major process during solar system formation about which we know precious little.

- **Origin of solar system water:** Assessment of the degree of the isotopic variability of water on various solar system bodies is essential to understanding the origin of water in our solar system. Remote isotopic analysis of water (and other species) from more comets, as well as the analysis of returned, pristine cometary samples is required. Further analysis of meteorite samples with modern analytical techniques is also crucial. Specifically, H isotopic variability of water over micron scales in primitive meteorites must be confirmed by further analysis, because it is this observation that is driving the direction of models of water evolution in the solar system.

- **The origin of the Moon and planets:** Stringent constraints on the composition of the hypothesized Giant Impactor from new, high precision laser fluorination 3-oxygen isotopic analyses of lunar materials have yet to be fully explored. Similar isotopic analyses of samples returned from Mars would quickly confirm (or not) the hypothesis that the martian meteorites are indeed samples of our planetary neighbor. Returned samples from Venus and Mercury would permit complete examination of O isotopic heterogeneity in the inner solar system, and would allow hypotheses about the mixing of materials in the inner solar system during planetary formation to be tested. In addition such data could hold important clues to the origin of the O isotope anomaly.

- **Mars astrobiology:** Understanding the distribution of water on Mars is key to assessing Mars' biological potential. Further H isotopic studies are needed to address the origin of isotopic variations observed in several martian meteorites. Carbon reservoirs on Mars also need to be explored, as C isotope data are currently limited predominantly to stepped heating analyses of bulk samples. Isolation and isotopic analysis of C carriers in martian meteorites is an important step towards understanding the origin of the isotopic variation observed in bulk samples. Stable isotopic analysis will obviously play a key role in study of any Mars samples returned by spacecraft. Such analyses will ultimately be needed to unravel decisively the origin, distribution, and history of martian volatiles.

ACKNOWLEDGMENTS

Reviews by François Robert, John Eiler, John Valley, and Mark Thiemens improved the manuscript and are gratefully acknowledged. Helpful discussions of all things isotopic with colleagues too numerous to mention are also valued and appreciated. NASA grants for support of the work discussed herein are acknowledged by both authors.

REFERENCES

Aléon J, Engrand C, Robert F, Chaussidon M (2001) Clues to the origin of interplanetary dust particles from the isotopic study of their hydrogen-bearing phases. Geochim Cosmochim Acta (in press)

Anders E, Grevesse N (1989) Abundances of the elements: Meteoritic and solar. Geochim Cosmochim Acta 53:197-214

Anders E, Zinner E (1993) Interstellar grains in primitive meteorites: diamond, silicon carbide, and graphite. Meteoritics 28:490-515

Ash RD, Rumble D, III, Alexander CMOD, MacPherson GJ (1998) Oxygen isotopes in isolated chondrules from the Tieschitz ordinary chondrite: Initial compositions and differential parent body alteration. Lunar Planet Sci XXIX, #1854 (CDROM)

Ash RD, Young ED, Rumble D, III, Alexander CMOD, MacPherson GJ (1999) Oxygen isotope systematics in Allende chondrules. Lunar Planet Sci XXX, #1836 (CDROM)

Baertschi P (1976) Absolute ^{18}O content of standard mean ocean water. Earth Planet Sci Lett 31:341-344

Balsiger H, Altwegg K, Geiss J (1995) D/H and ^{18}O/^{16}O ratio in the hydronium ion and in the neutral water from *in situ* ion measurements in comet Halley. J Geophys Res 100:5827-5834

Bao HM, Thiemens MH, Farquhar J, Campbell DA, Lee CCW, Heine K, Loope DB (2000) Anomalous O-17 compositions in massive sulphate deposits on the Earth. Nature 406:176-178

Becker RH, Epstein S (1982) Carbon, hydrogen and nitrogen isotopes in solvent-extractable organic-matter from carbonaceous chondrites. Geochim Cosmochim Acta 46:97-103

Becker RH, Pepin RO (1984) The case for a martian origin of the shergottites: Nitrogen and noble gases in EETA79001. Earth Planet Sci Lett 69:225-242

Begemann F (1980) Isotopic anomalies in meteorites. Rep Prog Phys 43:1309-1356

Benz W, Slattery WL, Cameron AGW (1986) The origin of the Moon and the single impact hypothesis. Icarus 66:515-535

Bischoff A, Keil K (1984) Al-rich objects in ordinary chondrites: Related origin of carbonaceous and ordinary chondrites and their constituents. Geochim Cosmochim Acta 48:693-709

Bjoraker GL, Mumma MJ, Larson HP (1989) Isotopic abundance ratios for hydrogen and oxygen in the martian atmosphere. Bull Am Astron Soc 21:991

Bockelée-Morvan D, Gautier D, Lis DC, Young K, Keene J, Phillips T, Owen T, Crovisier J, Goldsmith PF, Bergin EA, Despois D, Wootten A (1998) Deuterated water in comet C 1996 B2 (Hyakutake) and its implications for the origin of comets. Icarus 133:147-162

Bogard DD, Johnson P (1983) Martian gases in an Antarctic meteorite. Science 221:651-654

Bottazzi P, Ottolini L, Vannucci R (1992) SIMS analyses of rare earth elements in natural minerals and glasses: an investigation of structural matrix effects on ion yields. Scanning 14:160-168

Bradley JP (1994) Chemically anomalous, preaccretionally irradiated grains in interplanetary dust from comets. Science 265:925-929

Brearley A, Jones RH (1998) Chondritic meteorites. *In* JJ Papike (ed) Planetary Materials. Rev Mineral 36:3.01-03.370

Bridges JC, Franchi IA, Hutchison R, Sexton AS, Pillinger CT (1998) Correlated mineralogy, chemical compositions, oxygen isotopic compositions and size of chondrules. Earth Planet Sci Lett 155:183-196

Briley MM, Smith VV, King J, Lambert DL (1997) Isotopic carbon abundances in M71. Astron J 113:306-310

Brown PG, Hildebrand AR, Zolensky ME, Grady M, Clayton RN, Mayeda TK, Tagliaferri E, Spalding R, MacRae ND, Hffman EL, Mittlefehldt DW, Wacker JF, Bird JA, Campbell MD, Carpenter R, Gingerich H, Glatiotis M, Greiner E, Mazur MJ, McCausland PJ, Plotkin H, Mazur TR (2000) The fall, recovery, orbit,and composition of the Tagish Lake meteorite: A new type of carbonaceous chondrite. Science 290:320-325

Brownlee DE (1981) Interplanetary dust—Its physical nature and entry into the atmosphere of terrestrial planets. I EC Ponnamperuma (ed) Comets and the Origin of Life. D.Reidel Publishing Company, Dordrecht, Holland, p 63-70

Burbidge EM, Burbidge GR, Fowler WA, Hoyle F (1957) Synthesis of the elements in stars. Rev Modern Physics 29:547-650

Busso M, Gallino R, Wasserburg GJ (1999) Nucleosynthesis in Asymptotic Giant Branch stars: relevance for galactic enrichment and solar system formation. Ann Rev Astron Astrophys 37:239-309

Carr RH, Grady MM, Wright IP, Pillinger CT (1985) Martian atmospheric carbon dioxide and weathering products in SNC meteorites. Nature 314:248-250

Cassidy W, Harvey R, Schutt J, Delisle G, Yanai K (1992) The meteorite collection sites of Antarctica. Meteoritics 27:490-525

Choi BG, McKeegan KD, Krot AN, Wasson JT (1998) Extreme oxygen-isotope compositions in magnetite from unequilibrated ordinary chondrites. Nature 392:577-579

Clayton DD (1988) Isotopic anomalies: Chemical memory of galactic evolution. Astrophys J 334:191-195

Clayton RN (1986) High temperature isotopic effects in the early solar system. Rev Mineral 16:129-140

Clayton RN (1993) Oxygen isotopes in meteorites. Ann Rev Earth Planet Sci 21:115-149

Clayton RN, Mayeda TK (1975) Genetic relations between the Moon and meteorites. Proc 6th Lunar Sci Conf, p 1761-1769

Clayton RN, Mayeda TK (1977) Correlated oxygen and magnesium isotope anomalies in Allende inclusions, I: Oxygen. Geophys Res Lett 4:295-298

Clayton RN, Mayeda TK (1983) Oxygen isotopes in eucrites, shergottites, nakhlites, and chassignites. Earth Planet Sci Lett 62:1-6

Clayton RN, Mayeda TK (1984) The oxygen isotope record in Murchison and other carbonaceous chondrites. Earth Planet Sci Lett 67:151-161

Clayton RN, Mayeda TK (1986) Oxygen isotopes in Shergotty. Geochim Cosmochim Acta 50:979-982
Clayton RN, Mayeda TK (1988) Isotopic composition of carbonate in EETA 79001 and its relation to parent body volatiles. Geochim Cosmochim Acta 52:925-927
Clayton RN, Mayeda TK (1996) Oxygen isotope studies of achondrites. Geochim Cosmochim Acta 60:1999-2017
Clayton RN, Mayeda TK (1999) Oxygen isotope studies of carbonaceous chondrites. Geochim Cosmochim Acta 63:2089-2104
Clayton RN, Grossman L, Mayeda TK (1973) A component of primitive nuclear composition in carbonaceous meteorites. Science 182:485-488
Clayton RN, Onuma N, Grossman L, Mayeda TK (1977) Distribution of the pre-solar component in Allende and other carbonaceous chondrites. Earth Planet Sci Lett 34:209-224
Clayton RN, Mayeda TK, Goswami JN, Olsen EJ (1991) Oxygen isotope studies of ordinary chondrites. Geochim Cosmochim Acta 55:2317-2337
Clayton RN, MacPherson GJ, Hutcheon ID, Davis AM, Grossman L, Mayeda TK, Molini-Velsko C, Allen JM, El Goresy A (1984) Two forsterite-bearing FUN inclusions in the Allende meteorite. Geochim Cosmochim Acta 48:535-548
Connolly HC, Burnett DS (1999) A study of the minor element concentrations of spinels from two type B calcium-aluminum-rich inclusions: An investigation into potential formation conditions of calcium-aluminum-rich inclusions. Meteoritics Planet Sci 34:829-848
Craig H (1957) Isotopic standards for carbon and oxygen and correction factors for mass spectrometric analyses of carbon dioxide. Geochim Cosmochim Acta 12:133-149
Cronin JR, Chang S (1993) Organic matter in meteorites: Molecular and isotopic analyses of the Murchison meteorite. *In* JM Greenberg, CX Mendoza-Gómez, V Pirronello (eds) The Chemistry of Life's Origins. Kluwer Academic Publishers, Dordrecht, p 209-258
Cronin JR, Pizzarello S, Epstein S, Krishnamurthy RV (1993) Molecular and Isotopic Analyses of the Hydroxy Acids, Dicarboxylic Acids, and Hydroxydicarboxylic Acids of the Murchison Meteorite. Geochim Cosmochim Acta 57:4745-4752
Davis AM, McKeegan KD, MacPherson GJ (2000) Oxygen-isotopic compositions of individual minerals from the FUN inclusion Vigarano 1623-5. Meteoritics Planet Sci 35:A47-A47
Davis AM, MacPherson GJ, Clayton RN, Mayeda TK, Sylvester PJ, Grossman L, Hinton RW, Laughlin JR (1991) Melt solidification and late-stage evaporation in the evolution of a FUN inclusion from the Vigarano C3V chondrite. Geochim Cosmochim Acta 55:621-637
de Bergh C, Lutz BL, Owen T, Chauville J (1988) Monodeuterated methane in the outer solar system. III. Its abundance on Titan. Astrophys J 329:951-955
Deloule E, Robert F (1995) Interstellar water in meteorites? Geochim Cosmochim Acta 59:4695-4706
Deloule E, France-Lanord C, Albarede F (1991) D/H analysis of minerals by ion probe. Geochim Cosmochim Acta Spec Publ 3:53-62
Deloule E, Robert F, Doukhan JC (1998) Interstellar hydroxyl in meteoritic chondrules: Implications for the origin of water in the inner solar system. Geochim Cosmochim Acta 62:3367-3378
DesMarais DJ (1978) Carbon, nitrogen and sulfur in Apollo 15, 16 and 17 rocks. Proc 9th Lunar Sci Conf, p 2451-2467
Eberhardt P, Reber M, Krankowsky D, Hodges RR (1995) The D/H and $^{18}O/^{16}O$ ratios in water from comet P/Halley. Astron Astrophys 302:301-316
Eiler JM, Valley JW, Graham CM, Fournelle J (2001) Two populations of carbonate in ALH84001: Geochemical evidence for discrimination and genesis. Geochim Cosmochim Acta (in press)
Engrand C, DeLoule E, Robert F, Maurette M, Kurat G (1999a) Extraterrestrial water in micrometeorites and cosmic spherules from Antarctica: An ion microprobe study. Meteoritics Planet Sci 34:773-786
Engrand C, McKeegan KD, Leshin LA (1999b) Oxygen isotopic compositions of individual minerals in Antarctic micrometeorites: Further links to carbonaceous chondrites. Geochim Cosmochim Acta 63:2623-2636
Epstein RI, Lattimer JM, Schramm DN (1976) The origin of deuterium. Science 263:198-202
Epstein S, Taylor HP, Jr (1971) $^{18}O/^{16}O$, $^{30}Si/^{28}Si$, D/H, and $^{13}C/^{12}C$ ratios in lunar rocks and soils. Proc 2nd Lunar Sci Conf, p 1421-1441
Fagan TJ, Krot AN, Keil K (2000) Calcium-aluminum-rich inclusions in enstatite chondrites(I): Mineralogy and textures. Meteoritics Planet Sci 35:771-781
Fagan TJ, McKeegan KD, Krot AN, Keil K (2001) Calcium-, aluminum-rich inclusions in enstatite chondrites (2): Oxygen isotopes. Meteoritics Planet Sci 36:223-230
Fahey AJ, Goswami JN, McKeegan KD, Zinner EK (1987) ^{16}O excesses in Murchison and Murray hibonites: A case against a late supernova injection origin of isotopic anomalies in O, Mg, Ca, and Ti. Astrophys J Lett 323:L91-L95

Fanale FP, Postawko SE, Pollack JB, Carr MH, Pepin RO (1992) Mars: Epochal climate change and volatile history. *In* HH Kieffer, BM Jakosky, CW Snyder, MS Matthews (eds) Mars. University of Arizona Press, Tucson, p 1135-1179

Farquhar J, Thiemens MH (2000) Oxygen cycle of the martian atmosphere-regolith system: $\Delta^{17}O$ of secondary phases in Nakhla and Lafayette. J Geophys Res-Planets 105:11991-11997

Farquhar J, Thiemens MH, Jackson T (1998) Atmosphere-surface interactions on Mars: $\Delta^{17}O$ measurements of carbonate from ALH 84001. Science 280:1580-1582

Feuchtgruber H, Lellouch E, Bezard B, Encrenaz T, de Graauw T, Davis GR (1999) Detection of HD in the atmospheres of Uranus and Neptune: a new determination of the D/H ratio. Astron Astrophys 341:L17-L21

Folco L, Franchi IA, D'Orazio M, Rocchi S, Schultz L (2000) A new martian meteorite from the Sahara: The shergottite Dar al Gani 489. Meteoritics Planet Sci 35:827-839

Franchi IA, Wright IP, Sexton AS, Pillinger CT (1999) The oxygen-isotopic composition of Earth and Mars. Meteoritics Planet Sci 34:657-661

Geiss J, Reeves H (1981) Deuterium in the solar system. Astron Astrophys 93:189-199

Geiss J, Gloeckler G (1998) Abundances of deuterium and helium-3 in the protosolar cloud. Space Sci Rev 84:239-250

Gooding JL, Mayeda TK, Clayton RN, Fukuoka T (1983) Oxygen isotopic heterogeneities, their petrological correlations, and implications for melt origins of chondrules in unequilibrated ordinary chondrites. Earth Planet Sci Lett 65:209-224

Grady MM, Wright IP, Douglas C, Pillinger CT (1994) Carbon and nitrogen in ALH 84001. Meteoritics 29:469

Greshake A, Hoppe P, Bischoff A (1996) Mineralogy, chemistry, and oxygen isotopes of refractory inclusions from stratospheric dust particles and micrometeorites. Meteoritics Planet Sci 31:739-748

Grossman L (1980) Refractory inclusios in the Allende meteorite. Ann Rev Earth Planet Sci 8:559-608

Guan Y, McKeegan KD, MacPherson GJ (2000) Oxygen isotopes in calcium-aluminum-rich inclusions from enstatite chondrites: New evidence for a common CAI source in solar nebula. Earth Planet Sci Lett 181:271-277

Harris MJ, Lambert DL, Goldman A (1987) The $^{12}C/^{13}C$ and $^{16}O/^{18}O$ ratios in the solar photosphere. Monthly Notices Royal Astron Soc 224:237-255

Hashizume K, Chaussidon M, Marty B, Robert F (2000) Solar wind record on the moon: Deciphering presolar from planetary nitrogen. Science 290:1142-1145

Hervig RL, Williams LB, Kirkland IK, Longstaffe FJ (1995) Oxygen isotope microanlayses of diagenetic quartz: Possible low temperature occlusion of pores. Geochim Cosmochim Acta 59:2537-2543

Hiyagon H, Hashimoto A (1999) ^{16}O excesses in olivine inclusions in Yamato-86009 and Murchison chondrites and their relation to CAIs. Science 283:828-831

Hoppe P, Zinner E (2000) Presolar dust grains from meteorites and their stellar sources. J Geophys Res-Space Phys 105:10371-10385

Ireland TR (1990) Presolar isotopic and chemical signatures in hibonite-bearing refractory inclusions from the Murchison carbonaceous chondrite. Geochim Cosmochim Acta 54:3219-3237

Ireland TR (1995) Ion microprobe mass spectrometry: Techiques and applications in cosmochemistry, geochemistry, and geochronology. *In* Advances in Analytical Geochemistry 2:1-118. JAI Press Inc

Jagoutz E (1991) Chronology of SNC meteorites. Space Sci Rev 56:13-22

Jagoutz E, Sorowka A, Vogel JD, Wanke H (1994) ALH 84001: Alien or progenitor of the SNC family? Meteoritics 29:478-479

Jones RH, Saxton JM, Lyon IC, Turner G (2000) Oxygen isotopes in chondrule olivine and isolated olivine grains from the CO3 chondrite Allan Hills A77307. Meteoritics Planet Sci 35:849-857

Jull AJT, Eastoe CJ, Cloudt S (1997) Isotopic composition of carbonates in the SNC meteorites, Allan Hills 84001 and Zagami. J Geophys Res–Planets 102:1663-1669

Jull AJT, Beck JW, Burr GS (2000) Isotopic evidence for extraterrestrial organic material in the martian meteorite, Nakhla. Geochim Cosmochim Acta 64:3763-3772

Jull AJT, Eastoe CJ, Xue S, Herzog GF (1995) Isotopic composition of carbonates in the SNC meteorites, Allan Hills 84001 and Nakhla. Meteoritics 30:311-318

Jull AJT, Courtney C, Jeffrey DA, Beck JW (1998) Isotopic evidence for a terrestrial source of organic compounds found in Martian meteorites Allan Hills 84001 and Elephant Moraine 79001. Science 279:366-369

Kallenbach R, Geiss J, Ipavich FM, Gloeckler G, Bochsler P, Gliem F, Hefti S, Hilchenbach M, Hovestadt D (1998) Isotopic composition of solar wind nitrogen: First *in situ* determination with the CELIAS/MTOF spectrometer on board SOHO. Astrophys J 507:L185-L188

Kerridge JF (1985) Carbon, hydrogen and nitrogen in carbonaceous chondrites: Abundances and isotopic compositions in bulk samples. Geochim Cosmochim Acta 49:1707-1714

Kerridge JF (1993) Long term compositional variation in solar corpuscular radiation: Evidence from nitrogen isotopes in the lunar regolith. Rev Geophys 31:423-437

Kerridge JF (1995) Nitrogen and its isotopes in the early solar system. *In* Volatiles in the Earth and Solar System. American Institute of Physics, p 167-173

Kerridge JF (1999) Formation and processing of organics in the early solar system. Space Sci Rev 90: 275-288

Kerridge JF, Chang S, Shipp R (1987) Isotopic characterization of kerogen-like material in the Murchison carbonaceous chondrite. Geochim Cosmochim Acta 51:2527-2540

Kolodny Y, Kerridge JK, Kaplan IR (1980) Deuterium in carbonaceous chondrites. Earth Planet Sci Lett 46:149-158

Krasnopolsky VA, Mumma MJ, Bjoraker GL, Jennings DE (1996) Oxygen and carbon isotope ratios in Martian carbon dioxide: Measurements and implications for atmospheric evolution. Icarus 124: 553-568

Lécuyer C, Gillet P, Robert F (1998) The hydrogen isotope composition of seawater and the global water cycle. Chem Geol 145:249-261

Lee T (1979) New isotopic clues to solar system formation. Rev Geophys Space Phys 27:1591-1611

Lee T, Papanastassiou DA (1974) Mg isotopic anomalies in the Allende meteorite and correlation with O and Sr effects. Geophys Res Lett 1:225-228

Lee T, Russell WA, Wasserburg GJ (1978) Calcium isotopic anomalies and the lack of aluminum-26 in an unusual Allende inclusion. Astrophys J 228:L93-L98

Lee T, Mayeda TK, Clayton RN (1980) Oxygen isotopic anomalies in Allende inclusion HAL. Geophys Res Lett 7:493-496

Lee T, Shu FH, Shang H, Glassgold AE, Rehm KE (1998) Protostellar cosmic rays and extinct radioactivities in meteorites. Astrophys J 506:898-912

Leshin LA (2000) Insights into martian water reservoirs from analyses of martian meteorite QUE94201. Geophys Res Lett 27:2017-2020

Leshin LA, Rubin AE, McKeegan KD (1996b) Oxygen isotopic compositions of olivine and pyroxene from CI chondrites. Lunar Planet Sci XXVII:745-746

Leshin LA, Epstein S, Stolper EM (1996a) Hydrogen Isotope Geochemistry of SNC Meteorites. Geochim Cosmochim Acta 60:2635-2650

Leshin LA, McKeegan KD, Benedix GK (2000) Oxygen isotope geochemistry of olivine from carbonaceous chondrites. Lunar Planet Sci XXXI, #1918 (CDROM)

Leshin LA, McKeegan KD, Carpenter PK, Harvey RE (1998a) Oxygen isotopic constraints on the genesis of carbonates from martian meteorite ALH84001. Geochim Cosmochim Acta 62:3-13

Leshin LA, McKeegan KD, Engrand C, Zanda B, Bourot-Denise M, Hewins RH (1998b) Oxygen isotopic studies of isolated and chondrule olivine from Renazzo and Allende. Meteoritics Planet Sci 33: A93-A94

Li CL, Bridges JC, Hutchison R, Franchi IA, Sexton AS, Ouyang ZY, Pillinger CT (2000) Bo Xian (LL3.9): Oxygen-isotopic and mineralogical characterisation of separated chondrules. Meteoritics Planet Sci 35:561-568

Lorin JC, Slodzian G, Dennebouy R, Chaintreau M (1990) SIMS measurement of oxygen isotope-ratios in meteorites and primitive solar system matter. *In* A Benninghoven, CA Evans, KD McKeegan, HA Storms, HW Werner (eds) Secondary Ion Mass Spectrometry, SIMS VII:377-380. Wiley, Monterey

Love SG, Brownlee DE (1993) A direct measurement of the terrestrial mass accretion rate of cosmic dust. Science 262:550-553

MacPherson GJ, Wark DA, Armstrong JT (1988) Primitive material surviving in chondrites: refractory inclusions. *In* JF Kerridge, M. Matthews (eds) Meteorites and the Early Solar System. Universityof Arizona Press, Tucson, p 746-807

Mahaffy PR, Donahue TM, Atreya SK, Owen TC, Niemann HB (1998) Galileo probe measurements of D/H and He-3/He-4 in Jupiter's atmosphere. Space Sci Rev 84:251-263

Marti K, Kim JS, Thakur AN, McCoy TJ, Keil K (1995) Signatures of the martian atmosphere in glass of the Zagami meteorite. Science 267:1981-1984

Maurette M, Immel G, Hammer C, Harvey R, Kurat G, Taylor S (1994) Collection and curation of IDPs from the Greenland and Antarctic ice sheets. *In* ME Zolensky, TL Wilson, FJM Rietmeijer, GJ Flynn (eds) Analysis of Interplanetary Dust. Am Inst Phys Conf Proc 310:277-289.

Mayeda TK, Clayton RN, Nagasawa H (1986) Oxygen isotope variations within Allende refractory inclusions. Lunar Planet Sci Conf XVII:526-527

McKeegan KD (1987) Oxygen isotopic abundances in refractory stratospheric dust particles: proof of extraterrestrial origin. Science 237:1468-1471

McKeegan KD, Walker RM, Zinner E (1985) Ion microprobe isotopic measurements of individual interplanetary dust particles. Geochim Cosmochim Acta 49:1971-1987

McKeegan KD, Leshin LA, MacPherson GJ (1998a) Oxygen isotopic stratigraphy in a Vigarano type-A calcium-aluminum-rich inclusion. Meteoritics Planet Sci 33:A102-A103

McKeegan KD, Leshin LA, Russell SS, MacPherson GJ (1998b) Oxygen isotopic abundances in calcium-aluminum-rich inclusions from ordinary chondrites: Implications for nebular heterogeneity. Science 280:414-418

McKeegan KD, Leshin LA, Russell SS, MacPherson GJ (1996) *In situ* measurement of O isotopic anomalies in a type B Allende CAI. Meteoritics Planet Sci 31:A86-A87

McKeegan KD, Swan P, Walker RM, Wopenka B, Zinner E (1987) Hydrogen isotopic variations in interplanetary dust particles. Lunar Planet Sci XVIII:627-628

McKeegan KD, Sahijpal S, Krot AN, Weber D, Ulyanov AA (2001) Oxygen isotopic compositions of Ca, Al-rich inclusions from CH chondrites: preservation of the primary oxygen isotopic reservoirs of the solar system. Earth Planet Sci Lett (in press)

McNaughton NJ, Borthwick J, Fallick AE, Pillinger CT (1981) Deuterium/hydrogen ratio in unequilibrated ordinary chondrites. Nature 294:639-641

McSween HY, Jr, Treiman AH (1998) Martian meteorites. Rev Mineral 36:6.01-6.53

McSween HYJ (1994) What we have learned about Mars from SNC meteorites. Meteoritics 29:757-779

Meier R, Owen TC, Matthews HE, Jewitts DC, Bockelée-Morvan D, Biver N, Crovisier J, Gautier D (1998) A determination of the HDO/H_2O ratio in comet C/1995 O1 (Hale-Bopp). Science 279:842-844

Messenger S (2000) Identification of molecular-cloud material in interplanetary dust particles. Nature 404:968-971

Messenger S, Walker RM (1997) Evidence for molecular cloud material in meteorites and interplanetary dust. *In* TJ Bernatowicz, E Zinner (eds) Astrophysical Implications of the Laboratory Study of Presolar Materials. American Institute of Physics, Woodbury, CT, p 545-563

Mittlefehldt DW, McCoy TJ, Goodrich CA, Kraher A (1998) Non-chondritic meteorites from the asteroid belt. Rev Mineral 36:3.01-4.195

Nier AO, McElroy MB (1977) Composition and structure of Mars' upper atmosphere: Results from the neutral mass spectrometers on Viking 1 and 2. J Geophys Res 82:4341-4349

Nittler LR, Alexander CMO, Wang J, Gao X (1998) Meteoritic oxide grain from supernova found. Nature 393:222

Nittler LR, Alexander CMOD, Gao X, Walker RM, Zinner E (1997) Stellar sapphires: the properties and origins of presolar Al_2O_3 in meteorites. Astrophys J 483:475-495

Nyquist LE, Bansal BM, Wiesmann H, Shih C-Y (1995) "Martians" young and old: Zagami and ALH84001. Lunar Planet Sci XXVI:1065-1066

Ott U (1998) On laboratory studies of grains from outside the solar system. Proc Indian Acad Science–Earth Science 107:379-390

Owen T (1992) The composition and early history of the atmosphere of Mars. *In* HH Kieffer, BM Jakosky, CW Snyder, MS Matthews (eds) Mars. University of Arizona Press, Tucson, p 818-834

Owen T, Maillard JP, de Bergh C, Lutz BL (1988) Deuterium on Mars: The abundance of HDO and the value of D/H. Science 240:1767-1770

Owen TC (2000) On the origin of Titan's atmosphere. Planet Space Sci 48:747-752

Pillinger CT (1984) Light element stable isotopes in meteorites—from grams to picograms. Geochim Cosmochim Acta 48:2739-2766

Pizzarello S, Krishnamurthy RV, Epstein S, Cronin JR (1991) Isotopic analysis of amino acids from the Murchison meteorite. Geochim Cosmochim Acta 55:905-910

Pizzarello S, Feng X, Epstein S, Cronin JR (1994) Isotopic analyses of nitrogenous compounds from the Murchison meteorite—ammonia, amines, amino acids, and polar hydrocarbons. Geochim Cosmochim Acta 58:5579-5587

Podosek FA, Nichols RHJ (1997) Short-lived radionuclides in the solar nebula. *In* TJ Bernatowicz, EK Zinner (eds) Astrophysical Implications of the Laboratory Study of Presolar Materials. American Inst. Physics, Woodbury, CT, p. 617-647

Riciputi LR, Paterson BA (1994) High spatial-resolution measurement of O-isotope ratios in silicates and carbonates by ion microprobe. Am Mineral 79:1227-1230

Riciputi LR, Paterson BA, Ripperdan RL (1998) Measurement of light stable isotope ratios by SIMS: Matrix effects for oxygen, carbon, and sulfur isotopes in minerals. Int'l J Mass Spectrom 178:81-112

Robert F, Epstein S (1982) The concentration and isotopic composition of hydrogen, carbon, and nitrogen in carbonaceous meteorites. Geochim Cosmochim Acta 46:81-95

Robert F, Merlivat L, Javoy M (1979) Deuterium concentration in the early Solar System: hydrogen and oxygen isotope study. Nature 282:785-789

Robert F, Rejou-Michel A, Javoy M (1992) Oxygen isotopic homogeneity of the Earth: New evidence. Earth Plan Sci Lett 108:1-9

Robert F, Gautier D, Dubrulle B (2000) The solar system D/H ratio: observations and theories. Space Sci Rev 92:201-224

Robert F, Javoy M, Halbout J, Dimon B, Merlivat L (1987) Hydrogen isotope abundances in the solar system; Part II, Meteorites with terrestrial-like D/H ratio. Geochim Cosmochim Acta 51:1807-1822

Romanek CS, Perry EC, Treiman AH, Socki RA, Jones JH, Gibson EK (1998) Oxygen isotopic record of silicate alteration in the Shergotty-Nakhla-Chassigny meteorite Lafayette. Meteoritics Planet Sci 33:775-784

Romanek CS, Grady MM, Wright IP, Mittlefehldt DW, Socki RA, Pillinger CT, Gibson EKJ (1994) Record of fluid-rock interactions on Mars from the meteorite ALH84001. Nature 372:655-657

Rowe MW, Clayton RN, Mayeda TK (1994) Oxygen isotopes in separated components of CI and CM meteorites. Geochim Cosmochim Acta 58:5341-5347

Rubin AE (2000) Petrologic, geochemical and experimental constraints on models of chondrule formation. Earth-Science Rev 50:3-27

Rubin AE, Wasson JT, Clayton RN, Mayeda TK (1990) Oxygen isotopes in chondrules and coarse-grained chondrule rims from the Allende meteorite. Earth Planet Sci Lett 96:247-255

Rubin AE, Warren PH, Greenwood JP, Verish RS, Leshin LA, Hervig RL, Clayton RN, Mayeda TK (2000) Los Angeles: The most differentiated basaltic martian meteorite. Geology 28:1011-1014

Russell SS, MacPherson GJ, Leshin LA, McKeegan KD (2000) [16]O enrichments in aluminum-rich chondrules from ordinary chondrites. Earth Planet Sci Lett 184:57-74

Ryerson FJ, McKeegan KD (1994) Determination of oxygen self-diffusion in åkermanite, anorthite, diopside, and spinel: implications for oxygen isotopic anomalies and the thermal histories of Ca-Al-rich inclusions. Geochim Cosmochim Acta 58:3713-3734

Sahijpal S, McKeegan KD, Krot AN, Weber D, Ulyanov AA (1999) Oxygen isotopic compositions of Ca-Al-rich inclusions from the CH chondrites, Acfer182 and Pat 91546. Meteoritics Planet Sci 34:A101

Saxton JM, Lyon IC, Turner G (1998) Oxygen isotopes in forsterite grains from Julesburg and Allende: Oxygen-16-rich material in an ordinary chondrite. Meteoritics Planet Sci 33:1017-1027

Saxton JM, Lyon IC, Turner G (1998) Correlated chemical and isotopic zoning in carbonates in the Martian meteorite ALH84001. Earth Planet Sci Lett 160:811-822

Scott ERD, Krot AN (2001) Oxygen isotopic compositions and origins of Ca-Al-rich inclusions and chondrules. Meteoritics Planet Sci (in press)

Sears DWG, Lyon I, Saxton J, Turner G (1998) The oxygen isotopic properties of olivines in the Semarkona ordinary chondrite. Meteoritics Planet Sci 33:1029-1032

Shih C-Y, Nyquist LE, Bogard DD, McKay GA, Wooden JL, Bansal BM, Weismann H (1982) Chronology and petrogenesis of young achondrites Shergotty, Zagami and ALHA 77005: Late magmatism on a geologically active planet. Geochim Cosmochim Acta 46:2323-2344

Shu FH, Shang H, Lee T (1996) Toward an astrophysical theory of chondrites. Science 271:1545-1552

Simon SB, McKeegan KD, Ebel DS, Grossman L (2000) Complexly zoned Cr-Al spinel found *in situ* in the Allende meteorite. Meteoritics Planet Sci 35:215-228

Simon SB, Grossman L, Podosek FA, Zinner EK, Prombo CA (1994) Petrography, composition, and origin of large, chromian spinels from the Murchison meteorite. Geochim Cosmochim Acta 58:1313-1334

Stolper E (1982) Crystallization sequences of Ca-Al-rich inclusions from Allende : and experimental study. Geochim Cosmochim Acta 46:2159-2180

Stolper E, Paque JM (1986) Crystallization sequences of Ca-Al-rich inclusions from Allende: The effects of cooling rate and maximum termperature. Geochim Cosmochim Acta 50:1785-1806

Sugiura N, Hoshino H (2000) Hydrogen-isotopic compositions in Allan Hills 84001 and the evolution of the martian atmosphere. Meteoritics Planet Sci 35:373-380

Taylor HP, Jr., Epstein S (1970) 18-O/16-O ratios of Apollo 11 lunar rocks and soils. Proc Apollo 11 Lunar Sci Conf, p 1613-1626

Thiemens M, Jackson T, Zipf EC, Erdman PW, Van Egmond C (1995) Carbon dioxide and oxygen isotope anomalies in the mesosphere and stratosphere. Science 270:969-972

Thiemens MH (1996) Mass-independent isotopic effects in chondrites: the role of chemical processes. *In* RH Hewins, RH Jones, ERD Scott (eds) Chondrules and the protoplanetary disk. Cambridge University Press, New York, p 107-118

Thiemens MH (1999) Mass-independent isotope effects in planetary atmospheres and the early solar system. Science 283:341-345

Thiemens MH, Heidenreich JE (1983) The mass independent fractionation of oxygen: A novel isotope effect and its possible cosmochemical implications. Science 219:1073-1075

Valley JW, Graham CM, Harte B, Eiler JM, Kinny PD (1998) Ion microprobe analysis of oxygen, carbon, and hydrogen isotope ratios. *In* MA McKibben, WC Shanks, WI Ridley (eds) Applications of Microanalytical Techniques to Understanding Mineralizing Processes 7:73-98

Valley JW, Eiler JM, Graham CM, Gibson EK, Romanek CS, Stolper EM (1997) Low-temperature carbonate concretions in the martian meteorite ALH84001: Evidence from stable isotopes and mineralogy. Science 275:1633-1638

Virag A, Zinner E, Amari S, Anders E (1991) An ion microprobe study of corundum in the Murchison meteorite: Implications for ^{26}Al and ^{16}O in the early solar system. Geochim Cosmochim Acta 55:2045-2062

Wasserburg GJ, Papanastassiou DA (1982) Some short-lived nuclides in the early solar system—a connection with the placental ISM. In CA Barnes, DD Clayton, DN Schramm (eds) Essays in Nuclear Astrophysics. Cambridge University Press, Cambridge, p 77-140

Wasserburg GJ, Lee T, Papanastassiou DA (1977) Correlated oxygen and magnesium isotopic anomalies in Allende inclusions: II. Magnesium. Geophys Res Lett 4:299-302

Wasson JT, Yurimoto H, Russell SS (2000) The Abundance of ^{16}O-rich melilite in CO chondrites; The possible role of aqueous alteration. Lunar Planet Sci Conf XXXI, #2075 (CDROM)

Watson LL, Hutcheon ID, Epstein S, Stolper EM (1994) Water on Mars—Clues from deuterium/hydrogen and water contents of hydrous phases in SNC meteorites. Science 265:86-90

Weichert U, Halliday AN, Lee D-C, Snyder GA, Taylor LA, Rumble D (2000) Oxygen isotope homogeneity of the Moon. Lunar Planet Sci XXXI, #1699 (CD-ROM)

Weisberg MK, Prinz M, Kojima H, Yanai K, Clayton RN, Mayeda TK (1991) The Carlisle Lakes-type chondrites—A new grouplet with high delta-O-17 and evidence for nebular oxidation. Geochim Cosmochim Acta 55:2657-2669

Wetherill GW, Chapman CR (1988) Asteroids and meteorites. In JF Kerridge, MS Matthews (eds) Meteorites and the Early Solar System. University of Arizona Press, Tucson, AZ, p 35-70

Wieler R, Humbert F, Marty B (1999) Evidence for a predominantly non-solar origin of nitrogen in the lunar regolith revealed by single grain analyses. Earth Planet Sci Lett 167:47-60

Wiens RC, Huss GR, Burnett DS (1999) The solar oxygen-isotopic composition: Predictions and implications for solar nebula processes. Meteoritics Planet Sci 34:99-107

Wright IP, Carr RH, Pillinger CT (1986) Carbon abundance and isotopic studies of Shergotty and other shergottite meteorites. Geochim Cosmochim Acta 50:983-991

Wright IP, Grady MM, Pillinger CT (1988) Carbon, oxygen and nitrogen isotopic compositions of possible martian weathering products in EETA 79001. Geochim Cosmochim Acta 52:917-924

Wright IP, Grady MM, Pillinger CT (1990) The evolution of atmospheric CO_2 on Mars: The perspective from carbon isotope measurements. J Geophys Res 95:14789-14794

Wright IP, Grady MM, Pillinger CT (1992) Chassigny and the nakhlites—carbon-bearing components and their relationship to martian environmental conditions. Geochim Cosmochim Acta 56:817-826

Yang J, Epstein S (1983) Interstellar organic matter in meteorites. Geochim Cosmochim Acta 47:2199-2216

Young ED, Russell SS (1998) Oxygen reservoirs in the early solar nebula inferred from an Allende CAI. Science 282:452-455

Young ED, Ash RD (2001) An oxygen isotopic link between chondrules, matrix, and CAIs in the Allende meteorite. Earth Planet Sci Lett (in press)

Young ED, Coutts DW, Kapitan D (1998a) UV laser ablation and irm-GCMS microanalysis of O-18/O-16 and O-17/O-16 with application to a calcium-aluminium-rich inclusion from the Allende meteorite. Geochim Cosmochim Acta 62:3161-3168

Young ED, Fogel ML, Rumble D, Hoering TC (1998b) Isotope-ratio-monitoring of O_2 for microanalysis of $^{18}O/^{16}O$ and $^{17}O/^{16}O$ in geological materials. Geochim Cosmochim Acta 62:3087-3094

Young ED, Ash RD, England P, Rumble D (1999) Fluid flow in chondritic parent bodies: Deciphering the compositions of planetesimals. Science 286:1331-1335

Yung YL, Dissly RW (1992) Deuterium in the Solar System. In JA Kaye (ed) Isotope Effects in Gas-Phase Chemistry. Am Chem Soc Symp Ser 502:370-389.,

Yurimoto H, Ito M, Nagasawa H (1998) Oxygen isotope exchange between refractory inclusion in Allende and solar nebula gas. Science 282:1874-1877

Yurimoto H, Nagasawa H, Mori Y, Matsubaya O (1994) Micro-distribution of oxygen isotopes in a refractory inclusion from the Allende meteorite. Earth Planet Sci Lett 128:47-53

Zinner E (1989) Isotopic measurements with the ion microprobe. In WCI Shanks, RE Criss (eds) New Frontiers in Stable Isotope Research: Laser Probes, Ion Probes, and Small-Sample Analysis, U S Geol Surv Bull 1890:145-162

Zinner E (1998) Stellar nucleosynthesis and the isotopic composition of presolar grains from primitive meteorites. Ann Rev Earth Planet Sci 26:147-188

5   Oxygen Isotope Variations of Basaltic Lavas and Upper Mantle Rocks

John M. Eiler

Division of Geological and Planetary Sciences
California Institute of Technology
Pasadena, California 91125

INTRODUCTION

This chapter summarizes the oxygen isotope geochemistry of terrestrial basalts and their mantle sources, including the conceptual framework for interpreting such data and the phenomenology of known variations. In particular, the first section outlines the motivations for and first-order results of oxygen isotope studies of terrestrial and lunar basalts over the last 30 years; the second section reviews oxygen isotopic fractionations among phases relevant for studying basalts and mantle rocks; the third summarizes variations in $\delta^{18}O$ of various crustal rocks that may contribute to the petrogenesis of basalts either as subducted source components or lithospheric contaminants; and the final and longest section describes observed oxygen isotope variations of major classes of terrestrial basalts and related mantle nodules with an emphasis on data generated within the last six years using laser-based fluorination techniques. In the interests of brevity, I do not describe in detail methods for oxygen isotope analysis or changes in $\delta^{18}O$ of volcanic rocks caused by sub-solidus alteration; however, these issues are important practical considerations for anyone studying oxygen isotope compositions of basalts and interested readers are directed to the following references: analytical methods: Sharp (1990), Mattey and Macpherson (1993), and Valley et al. (1995); basalt alteration: Muehlenbachs (1986), Alt (1993), and Staudigel et al. (1995).

GUIDING PRINCIPLES

Geochemical studies of basaltic rocks are concerned with (among other issues) identifying subducted crustal materials in their mantle sources and characterizing extent and mechanisms of their interactions with the crustal rocks they intrude. Both aims require geochemical tools capable of discriminating rocks that are now or once were part of the crust from those that have always resided in the mantle. Many crustal materials have oxygen isotope ratios that differ strongly from those characteristic of the mantle because of isotopic fractionations associated with low-temperature weathering and water-rock interaction. Therefore, oxygen is among the isotopic systems (others include H, Li, B, C, N, Cl and S) that are likely to provide measures of the role of crustal rocks in basalt petrogenesis, either as subducted source components or contaminants from the current lithosphere. This line of reasoning was perhaps put best by Taylor and Sheppard (1986), who wrote: *"We here assert the 'central dogma' of the oxygen isotope geochemistry of igneous rocks, as follows: All relatively ^{18}O-rich or ^{18}O-depleted silicate melts on Earth ... must have in part been derived from, or have exchanged with, a precursor material that once upon a time resided on or near the Earth's surface."* In practical terms, 'near' the earth's surface should be defined as the portion of the lithosphere that contains hydrostatically pressured fluids; i.e. within ~10 km of the surface, where rocks undergo aqueous alteration at high water-rock ratios. The question of one's reference frame for defining ^{18}O-enriched and ^{18}O-depleted compositions will be considered at various points throughout this chapter.

1529-6466/00/0043-0005$05.00

concentration and therefore mixing proportions of isotopically distinct components contributing to the chemistry of magmas or mantle rocks are more easily defined for oxygen than for most other isotopic systems. However, sensitivity is correspondingly poor—in general more than 1 wt % of components that are or once were from the crust must be sampled in order for a basaltic rock to differ in $\delta^{18}O$ from a value in equilibrium with upper mantle peridotites. In this sense, oxygen isotope geochemistry differs from and potentially complements a large number of geochemical tracers involving highly incompatible elements (i.e. those elements that partition into silicate melts during melting or crystallization—such as Sr, Nd, Hf, or Pb). The principles and mathematics of mixing relationships in the combined oxygen—strontium isotope system are discussed by Taylor (1980), and Taylor and Sheppard (1986); these principles are easily extrapolated to mixing relationships involving oxygen and other elements and their mathematics will not be reviewed again here except in reference to the interpretation of particular data sets.

Finally, although the motivations for oxygen isotope studies of basaltic rocks are similar to those inspiring studies of the isotope geochemistry of several other elements, oxygen isotope variations of mantle and crustal rocks are known with greater breadth and detail than are isotopic variations of C, S and, particularly, H, Li, B, and Cl because it has been technically possible to efficiently and precisely measure the $^{18}O/^{16}O$ ratio of most major rock types for nearly 40 years; therefore new observations can be placed in a relatively rich context.

EARLY STUDIES OF OXYGEN ISOTOPE VARIATIONS
IN MAFIC IGNEOUS ROCKS

The potential strength of oxygen isotope geochemistry to illuminate the role of crustal rocks in the petrogenesis of basalts was recognized more than 30 years ago (Taylor 1968; Anderson et al. 1971) and since this time has been explored extensively by 'conventional' (resistance-heated) fluorination analyses of whole rock samples. Most of these data are not reviewed in detail here both because they were recently reviewed by Harmon and Hoefs (1995), who supply an excellent data base of these measurements on request, and because geochemical records based on such measurements generally differ significantly from those produced by the more recent laser-based analyses that are emphasized in this chapter. However, it is useful to briefly review the first-order description of oxygen isotope variations of terrestrial basalts, their mantle sources, and their lunar analogues provided by these studies:

- Most terrestrial basalts and lunar rocks are similar to one another in $^{18}O/^{16}O$ ratio and all such rocks define a common mass-dependent fractionation line in a plot of $^{17}O/^{16}O$ vs. $^{18}O/^{16}O$ (Clayton et al. 1972; Robert et al. 1992; Harmon and Hoefs 1995; Clayton and Mayeda 1996). These observations are generally interpreted to mean that: (1) The silicate mantles in the Earth–Moon system are made of a single, originally well-mixed oxygen reservoir; and (2) no large (1 ‰) mass-dependent fractionations preferentially influenced oxygen in the lunar mantle or the terrestrial mantle during or after the formation of the Earth–Moon system (i.e. if such mass-dependent fractionations occurred they influenced the lunar and terrestrial mantles approximately equally).
- Fresh, Cenozoic oceanic basalts typically span a narrow range in $\delta^{18}O$ (~5 to 7 ‰ for whole rocks), restricting the abundance of subducted crust in their mantle sources and crustal contaminants added to them during differentiation to amounts less than ~10 wt %. Furthermore, these data suggest that isotopic fractionations during melting, metasomatism, and metamorphism in the upper mantle are small

(1 ‰) on the scales sampled by erupted basalts (Harmon and Hoefs 1995).

- The weight of evidence suggests that unaltered oceanic basalts have not varied in $\delta^{18}O$ by more than ~1 to 2 ‰ since the mid-Archean, requiring that the integrated effects of plate tectonic evolution have not produced large shifts in the $\delta^{18}O$ of the upper mantle (e.g. Kyser 1986; Muehlenbachs 1986).

Collectively, these conclusions can be interpreted as meaning that oxygen isotope compositions of primitive basaltic rocks and their mantle sources usually vary within narrow limits and that compositions outside this range, if observed, are most likely caused by sub-solidus alteration or contamination of evolving magmas in the lithosphere. However, several studies suggested that these generalizations are wrong or incomplete, at least with respect to subtle (~1 to 2 per mil) variations in $\delta^{18}O$. For instance, Woodhead et al. (1987) suggested that oceanic arc basalts are systematically ^{18}O-enriched relative to MORBs due to addition to their sources of components from recently subducted lithosphere. Similarly, ~0.5 to 1.0 per mil ^{18}O-depletions in some ocean-island basalts relative to MORBs were interpreted as a property of their mantle sources, although the cause could not be confidently inferred (Kyser et al. 1982; Harmon and Hoefs 1995). These suggestions did not strongly impact other areas of mantle geochemistry because of the lack of clear and consistent relationships between $\delta^{18}O$ and other geochemical properties that are recognized as having significance for whole-earth evolution (e.g. radiogenic isotopes). However, this field has changed in the last six years because a technical innovation—laser-based fluorination (Sharp 1990; Mattey and Macpherson 1993; Valley et al. 1995)—has produced improved records of oxygen isotope variations of mafic and ultramafic rocks. This chapter will particularly emphasize recent observations made using these techniques. It is important to note at the outset that clear records of small oxygen isotope variations in mafic and ultramafic rocks are a relatively new thing; therefore, although detailed interpretations of these records naturally have been put forward, it is likely (perhaps inevitable) that we have yet to understand some of the processes that drive subtle isotopic variations in mantle rocks and basaltic magmas. The following section reviews current knowledge of high-temperature oxygen isotope fractionations among relevant phases and the processes by which they might express themselves.

ISOTOPIC FRACTIONATIONS RELEVANT TO STUDY OF MANTLE ROCKS AND BASALTIC LAVAS

Oxygen isotope fractionations among silicate melts, their phenocrysts and exsolved gases, and minerals in their residues are small at magmatic temperatures (typically 1 to 2 ‰ for major phases) and can be neglected in studies examining large variations in $\delta^{18}O$, such as those exhibited by some suites of crustally contaminated igneous rocks (e.g. Davidson and Harmon 1989). However, these fractionations are similar in magnitude to the range in $\delta^{18}O$ exhibited by relatively primitive, mantle-derived magmas and some fractionations involving minor phases (e.g. oxides, CO_2) can be quite large even at high temperature. Therefore equilibrium isotopic partitioning must be considered when interpreting subtle variations in $\delta^{18}O$ typical of basaltic lavas and mantle rocks.

Experimental, theoretical and empirical constraints on fractionation factors

Figures 1 and 2 summarize select data relevant to discussion of oxygen isotope systematics of mantle rocks and high-temperature silicate liquids and volatiles, shown as plots of the difference in $\delta^{18}O$ between a given phase and olivine as a function of temperature. I have chosen olivine as a reference because it is a common early-liquidus

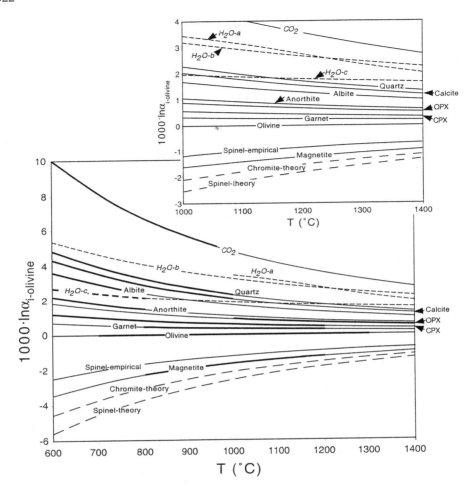

Figure 1. Oxygen isotope fractionations among minerals and vapors at mantle temperatures. Thick curve segments mark ranges of direct experimental constraints; thin curve segments are extrapolations or empirical estimates. Long-dash curves are theoretical estimates based on the model of Zheng (1991). Several estimates of the olivine-water fractionation are shown (see text for details); these are emphasized by plotting them as short-dash curves. 'CPX' stands for clinopyroxene; 'OPX' for orthopyroxene. Natural feldspars should have reduced partition function ratios intermediate between albite and anorthite (in mafic rocks generally near the anorthite extreme). Similarly, natural mantle pyroxenes should be intermediate between CPX and OPX, co-existing natural garnets and clinopyroxenes should be more similar to one another than experiments suggest, and various spinels and Fe-Ti oxides should span a range between magnetite and spinel depending upon their Fe, Cr, Mg, and Al contents (although constraints on this effect are poor).

Fractionations plotted in this figure were calculated based on data from the following references: Experiments from Clayton et al. (1989), Chiba et al. (1989), Rosenbaum and Mattey (1995), Rosenbaum (1994), Rosenbaum et al. (1994a), Matthews et al. (1983), and Matthews et al. (1994); theoretical calculations from Richet et al. (1977), Zheng (1991), and Rosenbaum et al. (1994b); and natural data for mantle peridotites from Chazot et al. (1997). Details of data selection and combination are discussed further in the text.

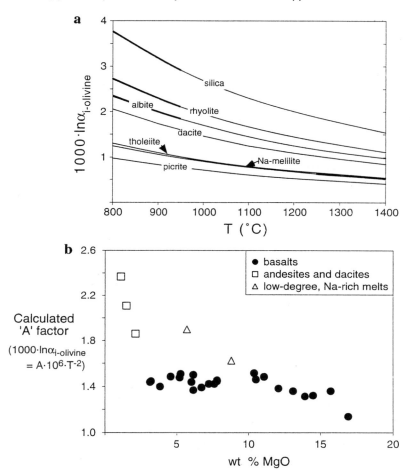

Figure 2. (a) Oxygen isotope fractionations between olivine and various silicate melts at mantle temperatures. Thick curve segments mark ranges of direct experimental constraints; thin curve segments are extrapolations, empirical estimates, or theoretical estimates based on the assumption that fractionation factors for silicate melts equal the weighted sum of those for their normative mineral constituents (taken from Fig. 1). Experimental data in addition to that referenced in Figure 1 come from: Palin et al. (1996), Appora et al. (2000), and Stolper and Epstein (1991). Details of data selection and combination are discussed further in the text. (b) Calculated temperature coefficient, 'A,' for the relationship: $1000 \cdot \ln\alpha_{\text{i-olivine}} = A \cdot 10^6 \cdot T^{-2}$, based on the assumption that fractionation factors for silicate melts equal the weighted sum of those for their normative mineral constituents and using measured compositions of basaltic and more evolved melts from peridotite partial melting experiments (Baker and Stolper 1994; Baker et al. 1995; Hirschmann et al. 1998), basalt crystallization experiments (Tormey et al. 1987; Juster and Grove 1989), and estimated primary Hawaiian picrite (Hauri 1996).

● melts of broadly basaltic composition; □ andesites and dacites produced by high degrees of basalt crystallization; △ Na- and Si-rich, low-degree peridotite melts. Note: all basalts between ~10 and ~3 wt % MgO have similar calculated melt-olivine fractionations.

phase in the crystallization of basaltic melts and the major component of most mantle rocks; furthermore its solid solution is relatively simple, involving only one major exchange component ($FeMg^{-1}$) that is not thought to impact high-temperature reduced partition function ratios of solids (Kieffer 1982). The plotted curves are calculated as $1000 \cdot \ln\alpha_{i\text{-olivine}}$, where $\alpha_{i\text{-olivine}} = (^{18}O/^{16}O)_i/(^{18}O/^{16}O)_{olivine}$; these numbers are closely similar to $\delta^{18}O_i - \delta^{18}O_{olivine}$ unless this difference is large (10 ‰) or if $\delta^{18}O_{olivine}$ differs greatly from 0 (O'Neil 1986a). In most cases I have had to use informed prejudices to chose among and combine various experimental, theoretical and empirical constraints. The general principle I have used is to adopt the experimental mineral-mineral fractionations of Clayton et al. (1989) and Chiba et al. (1989) wherever possible and to estimate fractionations involving phases not considered in those studies by combining their data with other experiments (preferred) or theoretical or empirical estimates (in last resort). For example, the CO_2-olivine fractionation shown in Figure 1 is based on the albite-olivine fractionation reported by Chiba et al. (itself actually a combination of observed albite-calcite and olivine-calcite fractionations) and the CO_2-albite fractionation experimentally determined at 1-bar pressure by Matthews et al. (1994). References for this and other fractionations are given in the captions to Figures 1 and 2. This method of combining experimental data sets increases overall uncertainty because of the need to add errors from each experiment in quadrature; I make no effort to present a detailed error analysis here but those using such data in their own work are encouraged to conduct such an analysis on a case-by-case basis. This method of combination is likely to introduce little systematic bias when it involves solid phases examined using similar methods; the potential for error is greater when experiments involving volatile phases at different pressures are combined or extrapolated. In both Figures 1 and 2 thick black curve segments mark the range of direct experimental constraint for each fractionation and thin black curve segments are extrapolations or empirical estimates. Long-dashed curves show estimates based only on theory or (in the case of melt-olivine fractionations) empirical algorithms discussed below. Two sets of estimated fractionations need particular comment:

(1) Several estimates for the olivine-H_2O fractionation are shown; these are emphasized by plotting them as short-dashed curves. One estimate ('a') is taken from Rosenbaum et al. (1994) and is based on combination of theoretical fractionations; a second (curve 'b') combines measured CO_2-albite and albite-olivine fractionations (Chiba et al. 1989; Matthews et al. 1994) with the CO_2-H_2O fractionation calculated by (Richet et al. 1977); the third ('c') combines the albite-water fractionation measured by Matsuhisa et al. (1979) with the albite-olivine fractionation from Chiba et al. (1989). All three of these approaches have their strengths and weaknesses; I will consider the full range of values in the following discussion.

(2) Estimates of the fractionations between olivine and either spinel or chromite are also problematic; long-dashed curves are based on combining the measured olivine-magnetite fractionation (Chiba et al. 1989) with Zheng's (1991) calculations of fractionations among various metal oxides. However, the curve for the resulting olivine-spinel fractionation is in poor agreement with the sparse data available from studies of mantle xenoliths (Chazot et al. 1997), suggesting a systematic error in one or the other of the constraints combined to make this estimate. Until this discrepancy is resolved, an empirical estimate based on data from Chazot et al. (1997) arguably offers the best constraint. The fine black curve for olivine-spinel fractionations is based on the assumption that olivine-spinel fractionations measured by Chazot et al. in mantle peridotites (~1 per mil) are in isotope exchange equilibrium at the ~1100°C temperatures suggested by other co-existing oxygen isotope thermometers in those same rocks and scaling of that fractionation with T^{-2}.

Oxygen isotope fractionations involving silicate melts or glasses (e.g. $\alpha_{mineral/melt}$ and $\alpha_{vapor/melt}$) are poorly known—in fact, I am aware of only two experimentally 'bracketed' determinations involving natural melt compositions (Mulenbachs and Kushiro 1974; Palin et al. 1996) and only three 'bracketed' determinations involving synthetic melt compositions (Stolper and Epstein 1991; Matthews et al. 1994; Appora et al. 2000); bracketing is the practice of approaching the equilibrium fractionation between two phases from both directions and is generally regarded as a pre-requisite for confident experimental determination (O'Neil 1986b). These data are plotted as a function of temperature in Figure 2; as in Figure 1, all fractionations are relative to olivine and thick segments of those curves mark the range of experimental constraints. Comparison of experimental data in Figures 1 and 2 shows that reduced partition function ratios for silicate glasses and melts are similar to those one would expect for mixtures of minerals in proportion to their normative abundances in those glasses and melts. For example, albite glass is indistinguishable from crystalline albite and rhyolite glass is intermediate to crystalline quartz and albite. However, an important exception to this generalization is the estimated olivine-silica glass fractionation, which is ~0.3 ‰ larger than the estimated olivine-quartz fractionation in Figure 1 at a given temperature (Stolper and Epstein 1991). This difference is comparable in magnitude to fractionations between chemically dissimilar minerals at magmatic temperatures (e.g. forsterite-diopside). Nevertheless, there is a general similarity between isotopic fractionations involving experimental melts and those involving minerals with similar cation chemistry. This similarity has been previously noted (Matthews et al. 1994; Palin et al. 1996) and used as the basis for models of fractionation involving melt compositions that have not yet been examined experimentally (e.g. Matthews et al. 1998). These models are generally indistinguishable from the assumption that silicate melts have reduced partition functions equal to the weighted sum of those for their normative mineral constituents. This assumption involves a more familiar translation of melt chemistry than calculations involved in some of the alternative methods (e.g. the 'Garlick' index; Matthews et al. 1998), and so I will adopt it throughout this chapter wherever calculated melt-mineral fractionations are discussed. Thin curves in Figure 2 show melt-olivine fractionations calculated by combining this assumption with the data for mineral-olivine fractionations in Figure 1. The fractionation calculated for 'picrite' is similar to Matthews et al.'s (1998) model estimate for komatiite; the fractionation calculated for 'basalt' is representative of those calculated for a wide range of basaltic and basaltic-andesite compositions. Note also that the model calculation for basalt is quite similar to the experimental determination for Na-melilite—a synthetic basalt analogue. Figure 2b shows the temperature-dependent coefficients for the melt-olivine fractionation of a wide range of natural and experimental melt compositions, calculated using our assumption that melts have fractionation factors similar to the weighted sum of their normative constituents. All values are plotted vs. MgO; a reasonably close estimate of fractionation factors for other basaltic melt compositions can be made by comparison with these results.

There is a large data base of oxygen isotope fractionations measured between phenocrysts and co-existing glass or groundmass in volcanic rocks. It is possible (perhaps likely) that these data are compromised by incorporation of xenocrysts in erupted lavas and sub-solidus alteration of groundmass or glass. Nevertheless, they make up a relatively large data set covering a diverse range in lava chemistry and tectonic setting and therefore typical fractionations observed in these data should offer useful tests of the reasonableness of the experimental, theoretical and empirical estimates described above and summarized in Figures 1 and 2. Figure 3 presents recent data on plagioclase-olivine fractionation in Central American arc lavas, as determined

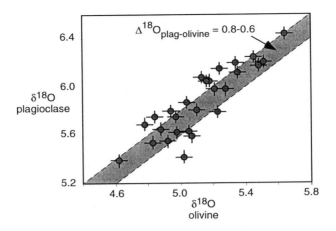

Figure 3. Comparison of oxygen isotope compositions of co-existing olivine and plagioclase in 27 samples of basalt from the Central American volcanic arc. The experimentally-determined range in this fractionation of approximately 0.8 to 0.6 ‰ (assuming An_{60-80} feldspar, a T of 1200-1300°C, and data from Chiba et al. 1989) is shown as a gray field with dashed black border. All measurements are unpublished data from the Caltech lab; similar results from smaller suites of samples are presented in Eiler et al. (1995 and 2000a). With few exceptions, data are consistent with a single fractionation near the experimentally determined range.

by laser fluorination in the Caltech laboratories. To the best of my knowledge, this is the largest data set of its type on closely related rocks. The results agree well with the experimental constraints assuming that the plagioclase has an anorthite content of ~60 to 80 mole % (within the range typical of phenocrysts in magnesian arc lavas), suggesting that the data base used to construct Figure 1 provides a reasonably good guide to isotopic partitioning in natural magmatic systems, at least in the case of feldspar-olivine fractionationss. There is considerably less laser-based data constraining fractionations between olivine and basaltic glass; most of those data that do exist suggest a melt-olivine fractionation of ~0.4 to 0.5 ‰ (Eiler et al. 2000a), consistent with experiments and model calculations illustrated in Figure 2. However, rigorous experimental constraints on the partitioning of oxygen isotopes between high-temperature minerals or volatiles and melts lag far behind the use of subtle oxygen isotope variations to understand the petrogenesis of such materials. Therefore, laboratory studies of partitioning in these systems are among of the most lasting contributions one could make to this field in the near future.

Oxygen-isotope systematics of crystallization, partial melting and degassing

Oxygen isotope fractionations among melts, minerals and magmatic volatiles (Figs. 1 and 2) are large enough to generate observable variations in the $\delta^{18}O$ of erupted lavas due to crystallization, melting and/or magmatic degassing. These processes have been previously discussed (Matsuhisa et al. 1973; Muehlenbachs and Byerly 1982; Kalamarides 1986; Taylor and Sheppard 1986), but with an emphasis on oxygen isotope changes associated with high extents of differentiation and expressed in highly evolved melts. This section presents models for analogous effects over ranges of relatively primitive compositions based on data presented in Figures 1 and 2. Most previous models of processes of this sort have assumed that isotopic fractionations among co-existing phases (e.g. melt vs. phenocrysts) always have the same value and then used those fractionations in the equations for equilibrium distribution among co-existing phases or for Rayleigh distillation for certain idealized cases of crystallizing assemblages or melting reactions. What I do here instead is to make one assumption about the nature of melt-mineral fractionations—melts behave as the weighted sums of their normative constituents—and apply that principle in combination with experimental constraints in Figure 1 to calculate the expected isotopic distribution among all co-existing phases in observed assemblages from melting and crystallization experiments. This approach is more work (i.e. one must calculate normative abundances for the melt to estimate its fractionation behavior, calculate mineral-mineral and mineral-melt fractionations at the appropriate temperature, and use the measured modal abundances to calculate the $\delta^{18}O$ of each phase for an assumed bulk system $\delta^{18}O$), and it cannot directly describe the effects of Rayleigh distillation because melting and crystallization experiments generally attempt to produce equilibrium assemblages. However, it has the advantage of correctly describing effects of natural, multiply-saturated melting and crystallization processes and of taking into account temperature- and composition-dependence of fractionations. There are a variety of thermodynamic and empirical models of phase proportions and chemistries in mineral-melt systems (e.g. the 'Melts' algorithm of Ghiorso et al. 1994) that could also be used for this purpose; these are not developed here in the interests of simplicity.

Crystallization

These calculations are based on experimental studies of the equilibrium crystallization sequence of tholeiitic basalt at 1 bar pressure and between 1205 and 1040°C. Two sets of experiments are included: one describing high extents of crystallization of Galapagos basalt (Juster and Grove 1989), dominated by plagioclase and pyroxene crystallization and leading to a pronounced Fe-enrichment trend and very SiO_2-rich final liquids, and a second describing lesser extents of crystallization of MORB from the Kane fracture zone (Tormey et al. 1987), dominated by olivine and plagioclase fractionation and producing weaker Fe-enrichment and negligible changes in SiO_2. The chosen experiments illustrate the range of behaviors expected of both 'Fenner' (Fe-enrichment) and 'Bowen' (calc-alkaline) differentiation trends. Calculated melt $\delta^{18}O$ values for the first of these are plotted vs. measured melt MgO contents as filled triangles in Figure 4; calculated melt $\delta^{18}O$ values for the second set of experiments are shown as filled squares. It is important to note that melt-mineral fractionations are not expected to be constant over the course of basalt crystallization; this is illustrated by also plotting the $\delta^{18}O$ of olivine in equilibrium with each calculated melt as circles. Olivine is a major component of the crystallizing assemblage in Kane fracture zone basalt (filled circles) but is present in traces or absent in Galapagos basalt (unfilled circles). Therefore olivine is a 'fictive' component in some experiments but is shown for consistency with other figures in this chapter; calculated variations in olivine $\delta^{18}O$ values are representative of those for other phenocrysts. In all cases I assume that the entire experimental system (melt plus minerals) has a bulk $\delta^{18}O$ of 5.5 ‰. The results of this calculation are that equilibrium crystallization of common basalt compositions will not change the $\delta^{18}O$ values of their residual melts by more than ~0.1 ‰ between MgO contents of 8 and 3 wt %. This is principally due to the fact that such basalts are multiply saturated in both higher-$\delta^{18}O$ and lower-^{18}O phases (plagioclase and olivine or pyroxene), such that the bulk crystallizing assemblage differs little in $\delta^{18}O$ from the melt. For this reason, Rayleigh distillation (i.e. continuous removal of crystals from residual melt) will not appreciably magnify changes in melt $\delta^{18}O$ shown in Figure 4

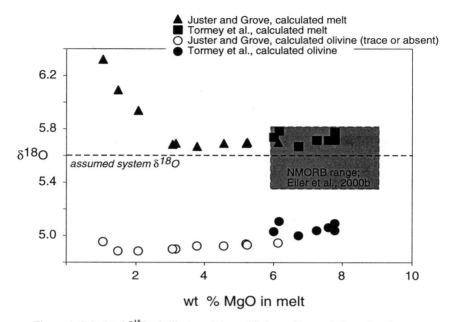

Figure 4. Calculated $\delta^{18}O$ of silicate melt in equilibrium with co-existing minerals over the course of equilibrium crystallization, plotted vs. measured MgO contents of those melts. Calculated $\delta^{18}O$ of co-existing olivine is shown for comparison (other minerals follow patterns similar to olivine and are excluded for simplicity). All calculations are based on measured phase chemistries and proportions in experiments of Tormey et al. (1987) and Juster and Grove (1989), which examine the 1-bar phase equilibria of Galapagos and Kane fracture zone basalts. Calculations assume mineral and melt oxygen isotope fractionation behavior illustrated in Figures 1 and 2; see text for details. Note that the $\delta^{18}O$ of melt is <u>nearly</u> constant over a large range in MgO due to 'buffering' of $\delta^{18}O$ during multiply saturated crystallization; only when residual melt becomes andesitic to dacitic in composition does crystallization generate large changes in melt $\delta^{18}O$. The $\delta^{18}O$ values of precipitated minerals are relatively insensitive to equilibrium crystallization. The gray field with dashed black border shows the range in $\delta^{18}O$ and MgO contents of NMORB glasses from Eiler et al. (2000b).

except under extreme circumstances. This situation changes for MgO contents less than 2 to 3 wt %, when SiO_2 contents of the melt become high and ^{18}O partitions into the melt relative to plagioclase. At this point, melt $\delta^{18}O$ can rise sharply (presumably until it begins crystallizing large proportions of quartz). Importantly, the melt-mineral fractionations also change at low MgO contents, such that phenocrysts in equilibrium these high-$\delta^{18}O$, evolved melts will not be significantly different in $\delta^{18}O$ from those minerals in equilibrium with more primitive melts of the same source. However, mixing of the most evolved melts with less evolved melts would produce nearly linear mixing curves in Figure 4 and therefore could produce melts with measurable ^{18}O enrichments at MgO contents of ~3 to 5 wt %. Models of crystallization of magmas with far higher MgO contents than those shown in Figure 4 can be reasonably approximated by Rayleigh distillation of olivine using the picrite-olivine fractionation of ~0.4 ‰. Even highly magnesian melts generally will not crystallize more than 20 to 30 wt % olivine before becoming multiply saturated, and thus the effect of olivine-only fractional crystallization will generally be ^{18}O-enrichments of residual melt by amounts 0.1 ‰.

For comparison, the gray field in Figure 4 shows the range in $\delta^{18}O$ and MgO for NMORB glasses recently examined by Eiler et al. (2000b); $\delta^{18}O$ is uncorrelated with MgO in this suite and varies by more than can be accounted for by low-pressure crystallization alone. I am aware of no high-precision data sets of significant size that clearly document changes in $\delta^{18}O$ with increasing extent of differentiation among basalts and basaltic andesites, perhaps because such studies have generally focused on the samples with 6 to 9 wt % MgO in which these effects are expected to be small (~0.1 ‰). However, several studies document increases in $\delta^{18}O$ with increasing extent of differentiation among components of layered basic intrusions (Kalamarides 1986) and evolved lavas (Matsuhisa et al. 1973; Muehlenbachs and Byerly 1982) that are generally consistent with the size and direction of changes in melt $\delta^{18}O$ illustrated in Figure 4.

Partial melting

This section describes the results of a model calculation that follows the same methods and assumptions described above (i.e. mineral proportions and melt chemistries are taken from experiments and mineral-melt fractionations are based on assumptions described with reference to Fig. 2) but applies them to experiments examining partial melting of fertile spinel peridotite at 10 kbar and temperatures of 1390 to 1250. The experiments considered used the 'diamond aggregate' method (Baker and Stolper 1994; Baker et al. 1995; Hirschmann et al. 1998) and extend to very low melt fractions (1.4 %) where melts are enriched in Na and Si compared to typical basalts. The results of these experiments are controversial because of the unusual compositions of the lowest-temperature melts, but they are useful for this purpose because they provide an extreme case of oxygen isotope variations that might be produced by peridotite partial melting. The empirical model of olivine-spinel fractionations in Figure 1 (thin black curve) was used to calculate the effects of changing oxide abundances in solid residues. As with the preceding calculations, co-existing minerals and melts are assumed to be in isotopic equilibrium with one another at all times. This is a reasonable assumption for a model of this type, although some studies of mantle peridotites have suggested that fluid-rock and/or melt-rock interaction may take place under local disequilibrium conditions (Zhang et al. 2000).

The results of this calculation are shown in Figure 5 as a plot of the calculated $\delta^{18}O$ of melt and olivine in the residual peridotite as a function of the measured Na_2O content of the melt—a useful and widely used indicator of the extent of melting of mantle rocks (e.g. Langmuir et al. 1992). The range in measured $\delta^{18}O$ and fractionation-corrected Na_2O in NMORBs ('$Na_{8.0}$' of Langmuir et al. 1992—Na_2O content corrected to 8.0 wt % MgO) is shown for comparison (Eiler et al. 2000b); $\delta^{18}O$ and Na_2O are uncorrelated in this suite. Our model calculations predict that the $\delta^{18}O$ of partial melts of upper-mantle rocks could vary by up to 0.3 ‰ due to melting processes alone, increasing approxi-mately linearly with increasing Na_2O as melt fraction decreases from ~30 to ~1 %; the $\delta^{18}O$ values of residual minerals are predicted to be relatively constant (<0.1 ‰ range). This effect is principally due to increasing abundance of 'albite' component in the melt with decreasing melt fraction. However, most of the calculated range in $\delta^{18}O$ is expressed only in melts with unusual compositions; for Na_2O contents of 2 to 4 wt % (the range typical of oceanic tholeiites), a range in $\delta^{18}O$ of only 0.1 ‰ is predicted. Because the predicted relationship between $\delta^{18}O$ and Na_2O is nearly linear, mixing of partial melts (for instance, due to 'dynamic melting' processes thought to be important for MORB melting; Langmuir 1992) cannot magnify (or decrease) this effect. The Na_2O contents of NMORBs examined by Eiler et al. (2000b) span too small a range for this effect to account for their variations in $\delta^{18}O$.

Note that at very low degrees of melting (<1 %) melts of metasomatized or mafic

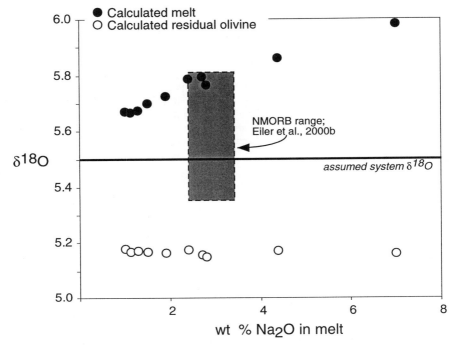

Figure 5. Calculated $\delta^{18}O$ of silicate melt in equilibrium with co-existing minerals during 'batch' melting of peridotite, plotted vs. measured Na_2O contents of those melts. Calculated $\delta^{18}O$ of co-existing olivine is shown for comparison (other minerals follow patterns similar to olivine and are excluded for simplicity). All calculations are based on measured phase chemistries and proportions in experiments of Baker and Stolper (1994), Baker et al. (1995) and Hirschmann et al. (1998), which examine the phase equilibria of partially molten synthetic spinel peridotite. Calculations assume mineral and melt oxygen isotope fractionation behavior illustrated in Figures 1 and 2; see text for details. Note that the $\delta^{18}O$ of partial melt increases linearly with increasing Na_2O due to increasing 'albite' component of partial melt with decreasing melt fraction; however, common tholeiitic basalts having 2-4 wt % Na_2O are expected to vary little in $\delta^{18}O$ due to this effect. The $\delta^{18}O$ values of residual minerals are insensitive to the extent of batch melting. The gray field with dashed black border shows the range in $\delta^{18}O$ and Na_2O contents of NMORB glasses from Eiler et al. (2000b).

mantle rocks may be strongly enriched Na, Si and perhaps other components (CO_2, H_2O, SO_2) and therefore even more extreme in their oxygen isotope compositions than the Na-rich end in Figure 5. Carbonatitic, rhyolitic and dacitic liquids in equilibrium with average mantle olivine should have $\delta^{18}O$ values of ~7 to 8 ‰—approximately 1 to 2 ‰ higher than basaltic high-degree melts of those same sources (Fig. 2). However, these exotic melts will also have unusual isotopic fractionations with respect to their precipitated phenocrysts; e.g. the first olivine to precipitate from a chemically exotic melt should be, to first order, equal in $\delta^{18}O$ to olivine in the mantle source of that magma. Therefore, isotopic fractionations of this type should not be expressed in the oxygen isotope variability of phenocrysts except under special circumstances. There are currently no strong observational constraints on the variation in $\delta^{18}O$ of partial melts of mafic or ultramafic rocks with variations extent of melting of a source that can be reasonably inferred to be constant in $\delta^{18}O$ (studies of arc-related rocks suggest such a relationship, but these effects are inconsistent with those discussed here and have a different interpretation discussed below). Eiler et al. 1996a, report oxygen isotope compositions of olivines and pyroxenes from extremely depleted harzburgites from the Voykar ophiolite, Russia which are interpreted to be residues of high degrees of melting in a ridge environment. These rocks are indistinguishable in $\delta^{18}O$ from average mantle peridotites (Mattey et al. 1994), consistent with the calculations illustrated in Figure 5.

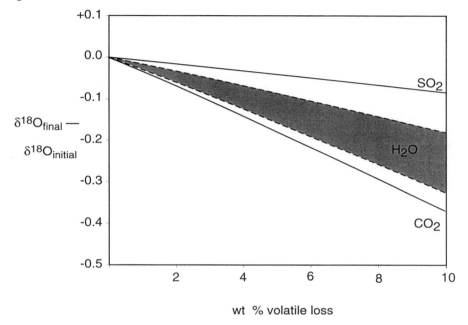

wt % volatile loss

Figure 6. Calculated change in $\delta^{18}O$ of basaltic melt as a function of volatile loss (in wt %). This calculation assumes that the conditions for Rayleigh distillation apply, olivine-volatile fractionations shown in Figure 1, a CO_2-SO_2 fractionation given by Richet et al. (1977), an olivine-melt fractionation of 0.5 per mil, and a temperature of 1200°C. A range of water-melt fractionations were used (short-dash curves in Fig. 1), leading to a range of predicted isotopic effects of de-watering. Note that the $\delta^{18}O$ of the initial magma likely depends on initial volatile content (i.e. CO_2-rich melts might be higher in $\delta^{18}O$ than CO_2 poor melts equilibrated with the same residual mantle minerals); this effect will tend to make degassed but initially volatile-rich melts more similar to melts that were always volatile poor than suggested by this calculation.

Devolatilization

Oxygen isotope changes during degassing of basalts are generally neglected because the volatile contents of submarine glasses are so low that even complete degassing should be incapable of producing large shifts in $\delta^{18}O$. However, these changes could be non-trivial in special cases (e.g. dehydration of water-rich melts in subduction zone environments; Sisson and Layne 1993) and therefore they are a potential source of oxygen isotope variability in erupted silicate melts. This section presents a relatively simple, idealized model in which the $\delta^{18}O$ of residual melt is calculated as a function of the wt % of various volatiles lost assuming a constant melt-volatile fractionation for each species. Figure 6 presents the expected change in $\delta^{18}O$ of a basaltic melt during degassing of H_2O, CO_2 and SO_2. Losses of reduced gasses such as H_2, H_2S and CH_4 obviously have no direct effect on the $\delta^{18}O$ of residual melt, although it is possible that they could have indirect effects such as driving fractional crystallization of oxides which in turn change the $\delta^{18}O$ of residual melt. All curves assume Rayleigh distillation and melt-vapor fractionation factors cited in the caption ('batch' devolatilization will have similar effects over the plotted range in extents of degassing; Valley 1986); a range of melt-water fractionations were assumed to capture the range of possible effects implied by diverse experimental and theoretical constraints (Figs. 1 and 2). Degassing of H_2O, CO_2 and SO_2 over the range of abundances for those species in most submarine basalts (0 to 1 wt % H_2O; 0 to 0.1 wt % CO_2; 0 to 0.3 wt % SO_2) should produce variations in $\delta^{18}O$ that are small relative to analytical precision and the natural variability of basaltic rocks (< 0.1 ‰). The same should be true of SO_2 degassing from even strongly sulfate-enriched melts. However, the expected shift in $\delta^{18}O$ due to degassing of CO_2-rich melts and, under some assumptions of melt-water fractionation, H_2O-rich melts can be many times analytical precision. Arc-related lavas in particular are suspected in some cases of being descended from primary melts with many wt % of dissolved H_2O (Sisson and Layne 1993) and therefore there is reason to suspect that devolatilization alone could promote the generation of ^{18}O-depleted convergent-margin lavas (~0.2 to 0.3 per mil lower than basalt in equilibrium with their mantle sources). The initial CO_2 abundances required for similar extents of magma ^{18}O depletion (>1 to 2 wt %) are higher than those believed likely based on current constraints on the volatile contents of tholeiitic basalts (Holloway and Blank 1994); however, the primary CO_2 contents of mantle melts are not well known and in some cases might be very high (Wyllie 1996); therefore this effect could be important in some settings.

ISOTOPIC SIGNATURES OF CRUSTAL CONTAMINANTS AND SUBDUCTED LITHOSPHERE

Although issues of high-temperature isotopic fractionation discussed in the preceding sections are important for interpreting variations in $\delta^{18}O$ of basaltic lavas and mantle rocks (and may become increasingly important as our understanding of these fractionations improves), the principle motivation for studying oxygen isotope variations in such samples is to detect and characterize subducted crustal components in the mantle and/or crustal contaminants that have influenced magmatic differentiation. This section describes the known isotopic variability of crustal rocks and sediments and their inferred subducted equivalents; Figures 7a and 7b summarize these data. This section can only briefly summarize the great deal of relevant data; interested readers are encouraged to read review articles cited below for further information.

Oceanic crust and sediments

It is generally believed that most of the crustal material injected into the asthenosphere and deeper mantle comes from the sediments and altered basaltic and gabbroic components of the subducted oceanic lithosphere (Hofmann 1997), although delamination of deep crustal plutonic rocks may also provide a major source of recycled crust (Kay and Kay 1993; Ducea and Saleeby 1998). The oxygen isotope composition of old oceanic crust and underlying lithospheric mantle is complex and portions of it are only known by studying ophiolites, which may not be representative of the ocean floor generally. Nevertheless, the following generalizations appear to be robust: Phanerozoic oceanic carbonate sediments have $\delta^{18}O$ values of 25 to 32 ‰, siliceous oozes 35 to 42 ‰, and pelagic clays 15 to 25 ‰ (Kolodny and Epstein 1976; Arthur et al. 1983); weathered and hydrothermally altered upper oceanic crust generally has $\delta^{18}O$ values of 7 to 15 ‰ and hydrothermally altered lower oceanic crustal gabbos 0 to 6 ‰ (Gregory and

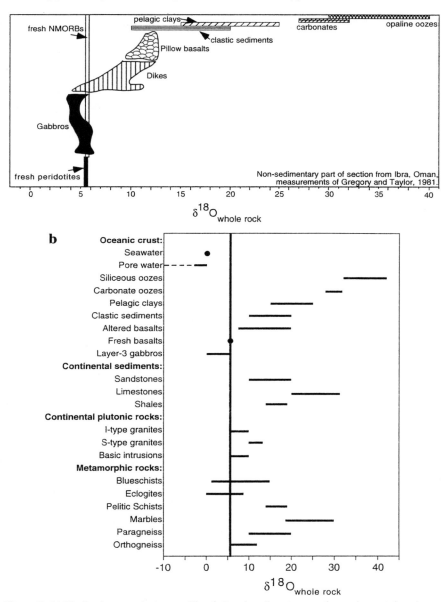

Figure 7. (a) Idealized oxygen isotope profile of altered, sediment-covered oceanic crust; based on Gregory and Taylor's (1981) study of the Ibra section of the Oman ophiolite and data for marine sediments taken from data sources cited in the text. The vertical black bar marks the range in $\delta^{18}O$ typical of mantle peridotites; the vertical white bar marks $\delta^{18}O$ typical of fresh oceanic basalts. (b) Values of $\delta^{18}O$ typical of various major rock types in the Earth's crust. Data sources are listed in the text. These ranges emphasize typical, representative values and purposefully exclude extreme examples of many rock types. The isotopic composition of most mantle peridotites (Fig. 8) and all NMORBs (Fig. 9) spans a range equal to the thickness of the vertical black line.

Taylor 1981; Alt et al. 1986; Muehlenbachs 1986; Stakes 1991; Staudigel et al. 1995; see Fig. 7a). It is also worth noting that pore-waters in the oceanic crust are initially low in $\delta^{18}O$ (0 near the sediment-water interface to -3 ‰ at depths of several hundred meters; Perry et al. 1976), although it would be reasonable to assume that these waters are quantitatively expelled and/or isotopically equilibrated with co-existing rocks before the oceanic lithosphere is deeply subducted. Because lower crustal gabbros (and pore fluids) on one hand and upper crustal basalts and sediments on the other deviate in $\delta^{18}O$ from typical peridotites ($\delta^{18}O = 5.5\pm0.2$; detailed below) in opposite directions, the oxygen isotope system has the potential to discriminate between subducted components derived from different parts of the oceanic crust. Of course, it is also possible that a mixture of ocean crust components could balance to have no contrast in $\delta^{18}O$ with the mantle. However, this case would require fortuitous mixtures of these materials and is not to be generally expected.

Subduction-zone metamorphic rocks

Subduction of oceanic lithosphere produces profound changes in its chemistry, principally because volatiles and their dissolved solutes are driven off during metamorphic reactions, and in some cases due to partial melting of the 'slab'. Several lines of evidence indicate that these processes have little impact on the bulk $\delta^{18}O$ of thick sections (10's of meters or thicker) of subducted oceanic crust, although they can homogenize meter-scale and smaller oxygen-isotope contrasts between adjacent rocks. The best evidence comes from studies of blueschist and eclogite facies metamorphic terrains (e.g. Bebout and Barton 1989; Putlitz et al. 2000), which document that deeply subducted oceanic crust retains both ^{18}O-enriched and ^{18}O-depleted components, each with $\delta^{18}O$ values and in relative proportions broadly similar to those in unsubducted oceanic crust. Additional constraints that lead to the same conclusion come from studies of eclogite xenoliths and mantle xenocrysts believed to be related to recycled crust (Garlick et al. 1971; Neal et al. 1990; Nadeau et al. 1993; Valley et al. 1998), mafic rocks associated with alpine peridotites (Pearson et al. 1991), and eclogite-type inclusions in mantle diamonds (Lowry et al. 1999), although this last suite of materials is not known to include ^{18}O-depleted compositions. A subset of these data are shown in Figure 8 along with data for mantle peridotites, which are discussed below.

It was recently discovered that ultra-high-pressure metamorphic complexes in northeast China contain mafic metamorphic rocks—plausibly deeply subducted basalts—with some of the lowest $\delta^{18}O$ values ever observed in terrestrial rocks (down to -10 ‰; Yui et al. 1995). These values can be interpreted as a result of high-temperature hydrothermal alteration by low-$\delta^{18}O$, perhaps high-latitude, meteoric waters before subduction. In this case, these $\delta^{18}O$ values are an historical accident unrelated to the extreme tectonic processes the rocks subsequently underwent and are likely not representative of recycled crustal materials.

Upper continental crust

The oxygen isotope geochemistry of the upper continental crust is dominated in most cases by the relative proportions of carbonate and siliciclastic sediments ($\delta^{18}O = 20$ to 30 ‰ and 10 to 20 ‰, respectively; Kolodny and Epstein 1976; Arthur et al. 1983) and granitoids ($\delta^{18}O = 7$ to 14, with higher values systematically observed in those generated by melting pre-existing sediments; Taylor 1968; Taylor and Sheppard 1986), or their metamorphosed equivalents. I am not aware that anyone has estimated the globally-averaged $\delta^{18}O$ of the upper continental crust (beyond broad approximations). Such an estimate would require a careful inventory of the relative proportions of the relevant rock types and is beyond the scope of this work. However, the most important

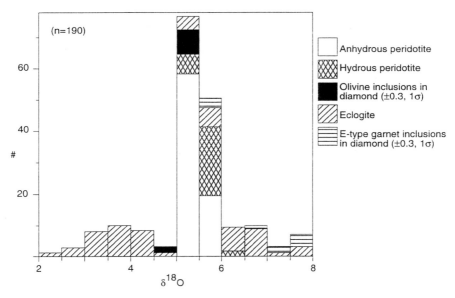

Figure 8. Oxygen isotope compositions of representative mafic and ultramafic mantle xenoliths. All data are reported on a whole-rock basis where possible and as named minerals when necessary; however, fractionations among co-existing minerals are small (0.5 ‰) in most cases and introduce no significant bias at the scale plotted. Note that all ultramafic rock types span a small range in $\delta^{18}O$ centered around ~5.5 ‰; eclogites and 'E-type' inclusions in diamonds span much greater ranges, presumably reflecting a relationship to subducted oceanic crust. Data sources: Garlick and MacGregor (1971), Ongley et al. (1987), Neal et al. (1990), Kempton et al. (1988), Fourcade et al. (1994), Mattey et al. (1994), Chazot et al. (1997), and Lowry et al. (1999).

point to consider for studies of $\delta^{18}O$ values in basalts that have interacted with continental crust is the makeup of the local crustal section: Cratonic margins and mobile belts rich in clastic and (particularly) chemical sediments will tend to be high in $\delta^{18}O$, whereas cratonic cores of plutonic rocks and mobile belts rich in juvenile igneous rocks will tend to be closer to mantle $\delta^{18}O$ values. An example of this difference can be seen in the difference in oxygen isotope composition between crustally-derived granitoids in the Grenville province (high $\delta^{18}O$; metasediment-rich mobile belt) on one hand and the Superior province (low $\delta^{18}O$, orthogneiss- rich craton) on the other (Peck et al. 2000).

Lower continental crust

The oxygen isotope geochemistry of the lower continental crust depends principally on the relative proportions of juvenile plutonic rocks (i.e. those produced from basaltic underplating) on one hand and tectonically emplaced upper-continental crustal rocks on the other. The first of these are best represented by mafic xenoliths from depths of 30 to 50 km (granulites and eclogites) whereas the latter are arguably best represented by regional granulite facies terrains, most of which sample the middle of doubly-thickened orogenic crust (Harley 1989). The oxygen isotope geochemistry of deep-crustal orthogneisses and paragneisses is reviewed by Valley (1986); tectonically emplaced sediments, granitoids and their metamorphosed equivalents are, to first order, similar to their upper-crustal progenitors described above. Clemens-Knott (1992) reports studies of the oxygen isotope geochemistry of the interface between juvenile plutonic lower crust and pre-existing metamorphic lower crust in the Sierra Nevada and Ivrea zone. The most primitive members of juvenile plutonic rocks in these settings have $\delta^{18}O$ values close to those typical of upper mantle rocks and oceanic basalts (i.e. $\delta^{18}O$ = 6 to 7 ‰). However, parts of the plutonic complex in the Ivrea zone are subject to extensive contamination by stoped blocks and partial melts of the pre-existing crustal rocks they intrude and can take on $\delta^{18}O$ values of 8 to 10 ‰ due to this pollution. Kempton and Harmon (1992) provide the most extensive available study of the oxygen isotope geochemistry of deep-crustal xenoliths, showing that plutonic complexes intruded into the base of the continental crust generally undergo extensive interaction with pre-existing metasediments and felsic igneous rocks, resulting in large (several per mil) ^{18}O-enrichments of mafic granulites compared to upper mantle rocks and oceanic basalts.

OXYGEN ISOTOPE COMPOSITIONS OF MANTLE PERIDOTITES

The oxygen isotope geochemistry of mantle peridotites was until recently a controversial subject (Kyser et al. 1981; Gregory and Taylor 1986). Many conventional fluorination measurements indicated that pyroxenes in most upper-mantle rocks span a relatively narrow range in $\delta^{18}O$ (~1 ‰) whereas olivines in those same rocks span a much larger range (~3 ‰) that is uncorrelated with the $\delta^{18}O$ of coexisting pyroxenes (Kyser et al. 1981). This result was interpreted as either a consequence of complex temperature dependence to the olivine-pyroxene fractionation (Kyser et al. 1981) or of differences between olivine and pyroxene in the rate of isotopic exchange with metasomatizing fluids and/or melts (Gregory and Taylor 1986). In either case, the implication was that the oxygen isotope compositions of upper mantle rocks are usually variable and internally complex. Other studies suggested that olivine, garnets and pyroxenes are generally in mutual oxygen isotope equilibrium, but still vary by approximately 2 to 3 per mil from sample to sample (Harmon et al. 1986).

Subsequent study of peridotites (including many of the same samples investigated by Kyser et al. 1981) using laser fluorination techniques contradicts much of these earlier data, instead showing that most mantle peridotites span a relatively narrow range in $\delta^{18}O_{whole-rock}$ (5.5±0.2 ‰) and almost universally consist of minerals that are in mutual oxygen isotope equilibrium predicted by experimental constraints (Mattey et al. 1994; Ionov et al. 1994; Chiba et al. 1989). The scope of Mattey et al.'s (1994) survey of mantle peridotites and the internal consistency of their results suggests that the complexity apparent in earlier studies is an artifact of low analytical yields in conven- tional fluorination of refractory minerals and that the upper mantle principally consists of peridotites of nearly fixed and internally equilibrated oxygen isotope composition. This conclusion is re-affirmed by a subsequent comparison of hydrated (i.e. amphibole- bearing) and anhydrous mantle peridotites (Chazot et al. 1997), which found them to be indistinguishable in $\delta^{18}O$ from one another and from typical mantle peridotites previously analyzed by Mattey et al. (1994). These data are summarized in Figure 8 along with the compositions of mantle eclogites and 'E-type' inclusions in diamonds. A dissenting view that re-interprets the Kyser et al. (1981) results was presented by Rosenbaum et al. (1994a). Finally, it should be noted that most of the mantle peridotites that have been measured for $\delta^{18}O$ come from the continental lithospheric mantle and thus may be unrepresentative of the mantle as a whole.

More recently there have been several indications that the oxygen isotope geochemistry of mantle xenoliths from certain exotic settings can be more variable than indicated by Mattey et al.'s (1994) survey and subsequent work of Chazot et al. (1997). Zhang et al. (2000) present oxygen isotope measurements on variably metasomatized and re-crystallized ultramafic xenoliths from the Kaapvaal craton, South Africa, which document inter-mineralic disequilibrium and intra-crystalline zonation that presumably reflect metasomatic fluid-rock interaction. The direction of these changes is toward ^{18}O-depletions in the most metasomatized portions of the rocks. Similarly, Deines and Haggerty (2000) report complex intra-crystalline zonation in the oxygen isotope compositions of ultramafic xenoliths thought to come from unusually great depths. Finally, Valley et al. (1998) interpret variability in δ^{18}O among some zircon metacrysts in kimberlites as a consequence of recycling of subducted material into mantle melts. Collectively, these results suggest that there are discoveries to be made in the oxygen isotope geochemistry of metasomatized ultramafic xenoliths; nevertheless, it is important to keep in mind that such samples are rare.

Additional insight about the oxygen isotope geochemistry of the earth's mantle comes from lunar rocks, which sample a system that plausibly began its evolution with an oxygen isotope composition like that of the earth but has differentiated without weathering, hydrothermal alteration or subduction (although contamination of lunar magmas by the regolith could potentially influence their oxygen isotope compositions). Only one study has examined the δ^{18}O of lunar rocks using laser-based techniques (Wiechert et al. 2000), but much of the conventional data generated in response to the Apollo program was of unusually high quality and was not affected by the post-eruptive weathering that compromises whole-rock analyses of terrestrial volcanic rocks (Onuma et al. 1970; Clayton et al. 1971; Clayton et al. 1972; Clayton and Mayeda 1996). Results of both laser-based and conventional studies indicate that lunar anorthosites and minerals and glass in most lunar basalts span a narrow range in δ^{18}O that is close to high-temperature equilibrium with the average value for terrestrial MORBs (detailed below); this result supports the view that the earth's upper mantle is within ~0.1 to 0.2 ‰ of its primordial ^{18}O/^{16}O ratio. There is, however, one remarkable exception: Wiechert et al. (2000) found several high-Ti lunar basalts that are ^{18}O-depleted by several tenths of per mil relative to most lunar and terrestrial basalts. This observation was attributed to re-melting of Fe-Ti-oxide cumulates in the lunar upper mantle, which could be low in δ^{18}O because of high-temperature isotopic fractionations between Fe-Ti oxides and common silicates and silicate melts (Fig. 1). It should also be noted that the analyzed glasses are far richer in FeO and TiO_2 and poorer in SiO_2 and Al_2O_3 than the terrestrial and lunar basalts to which they are compared; therefore their ^{18}O-depleted compositions could also reflect their unusual fractionations relative to minerals in their residues (Fig. 2) rather than their extraction from a low-δ^{18}O source. This second hypothesis will be difficult to evaluate without experimental constraints on fractionation factors involving melts having such exotic major element compositions.

MID-OCEAN RIDGE BASALTS

Mid-ocean-ridge basalts (MORBs) are the most abundant and widespread terrestrial magmas and span relatively small ranges in composition for many geochemical indices (e.g. Sr isotope ratios; Hofmann 1997); therefore they are a natural reference frame for defining the geochemical properties of the upper mantle. Early studies of the oxygen isotope geochemistry of MORBs established that they span a limited range in δ^{18}O (generally ~5 to 7 per mil; □ in Figure 9; see figure caption for references) and, as mentioned above, have an average value that is near those for mafic lunar rocks. For these reasons, and in light of the complexity of oxygen isotope variations believed at that time to be characteristic of terrestrial peridotites, MORBs were widely used to define the δ^{18}O value of the earth's mantle. This interpretation of the oxygen isotope geochemistry of MORBs was re-enforced by the study of Ito et al. (1987)—until recently the most extensive and systematic survey of the oxygen isotope composition of any one major class of terrestrial basalts. Ito et al. (1987) found that fresh MORB glasses vary in δ^{18}O between 5.3 and 6.2 ‰, but that most fall within the range 5.5 to

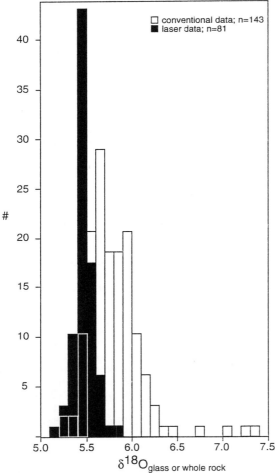

Figure 9. Oxygen isotope compositions of nominally fresh MORB glasses and whole-rocks. Unfilled boxes are data collected using conventional (resistance heated) fluorination methods between 1966 and 1993; filled boxes are data collected only on glass using laser-based methods. Where these two data types overlap, conventional fluorination data are shown as white-outlined boxes. Data sources: Taylor (1968), Muehlenbachs and Clayton (1972), Pineau et al. (1976), Kyser et al. (1982), Muehlenbachs and Byerly (1982), Ito et al. (1987), Barrat et al. (1993), Harmon and Hoefs (1995) and references therein, Eiler et al. (2000b), and Eiler and Kitchen (unpublished data).

5.9 ‰. No systematic correlations were found between $\delta^{18}O$ and other geochemical indices thought significant for mantle processes and no statistically significant differences were found between major geographic and compositional groups (e.g. E-MORB vs. N-MORB; Pacific vs. Atlantic). A simple interpretation of these results is that the upper mantle varies little in $\delta^{18}O$ and that any variations that do exist are unrelated to other geochemical properties. Nevertheless, Ito et al. emphasized that MORBs vary in $\delta^{18}O$ by more than analytical precision and thus it is possible that isotopic 'signals' exist that simply were not successfully resolved by their study.

Eiler et al. (2000b) examined oxygen isotope variations of nominally fresh mid-ocean-ridge glasses using laser-based methods. The suite they examined included 28 approximately normal ('N-') MORBs distributed evenly among the Pacific, Atlantic and Indian ocean basins; these data are shown as black boxes in Figure 9. This work was recently followed by a study of 51 samples from a portion of the mid-Atlantic ridge, also analyzed in the Caltech lab; these data are unpublished but included as black boxes in Figure 9 for the purposes of completeness. Eiler et al.'s (2000b) study found that NMORBs vary in $\delta^{18}O$ by less than suggested by previous studies (5.4 to 5.8 ‰—this range extends to values of 5.2 to 5.3 ‰ when three exotic samples from their more recent work on the mid-Atlantic ridge are included), that they have a slightly lower average value than suggested by previous work (5.5 vs. ~5.7 ‰) and, most importantly, that their $\delta^{18}O$ values correlate with abundances and abundance ratios of various

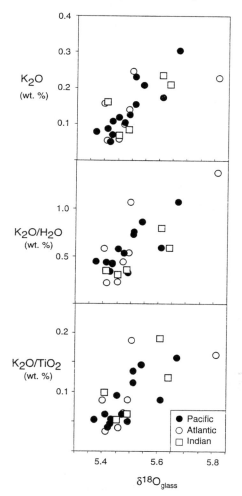

Figure 10. Comparison of $\delta^{18}O$ values with various minor-element abundances and abundance ratios in NMORBs; data from Eiler et al. (2000b). Lavas from all three major ocean basins (Pacific, Atlantic and Indian) define common trends of increasing $\delta^{18}O$ with increasing concentrations of K (and other incompatible elements) and increasing ratios of K abundance to other incompatible element abundances. Model calculations detailed in Eiler et al. (2000b) indicate that these trends are consistent with mixing between low-$\delta^{18}O$ depleted peridotite and high-$\delta^{18}O$, subducted upper oceanic crust. These are the same data shown as gray fields in Figures 4 and 5; their range in $\delta^{18}O$ is too great to be accounted for by known fractionations during partial melting and crystallizaton-differentiation given their relatively small ranges in major element chemistry.

incompatible minor and trace elements (including but not restricted to K_2O; K_2O/H_2O and K_2O/TiO_2; Fig. 10). The data in Figure 10 are inconsistent with models in which oxygen isotope variations reflect only fractional crystallization or variations in degree of partial melting of an isotopically homogenous source (these are the data shown as gray fields in Figs. 4 and 5) but agree with models in which the upper mantle consists of ~98 % depleted peridotite with a $\delta^{18}O$ of 5.4 ‰ and variable amounts (~2 %, on average) of recycled upper oceanic crust and sediments with a bulk $\delta^{18}O$ of ~10 ‰ (see Eiler et al. 2000b for details of this model). Similarities among the three major ocean basins in Figure 10 suggest that if recycled crustal components are responsible for differences in $\delta^{18}O$ among fresh NMORBs, then they are present in a roughly equal range of abundances in the mantle beneath all three major ocean basins. This interpretation reinforces previous conclusions based on of the coupled Pb–Os isotope systematics of NMORBs (Schiano et al. 1997) but runs contrary to common inferences that the sources of NMORBs are a depleted residue of prior melting with little or no back-mixing of recycled crust (see review of Hofmann 1997). This interpretation also suggests that enrichments in ^{18}O are different in character from at least some causes of radiogenic-isotope 'enrichment,' which have strong regional variations. For example, Indian ocean MORB's are systematically higher in $^{87}Sr/^{86}Sr$ than Pacific and Atlantic MORBs but conform to the same trends as other MORBs in Figure 10. Finally, these data, whatever their ultimate explanation, require that one make subjective decisions when defining the $\delta^{18}O$ of the upper mantle (e.g. should one use the average MORB value, the minimum MORB value, or the $\delta^{18}O$ of MORBs at some specific concentration of minor elements?). The differences between these choices are subtle but analytically resolvable and could influence interpretation of oxygen isotope variations in other tectonic settings where 'signals' of exotic components might be superimposed on a template of upper-mantle variability. Collectively, these results suggest one should treat oxygen isotopes as one does other geochemical properties that vary systematically in MORBs (e.g. Nd and Pb isotopes), namely by allowing that the upper mantle in any given region can differ from the global average by ~0.1 to 0.2 per mil.

OCEANIC INTRAPLATE BASALTS

Ocean-island basalts (OIB's) provide perhaps the most diverse record of chemical and isotopic variability in the mantle (Hofmann 1997), reflecting, in part, the presence in their mantle sources of constituents that are absent or present in smaller abundances in the upper mantle sources of MORBs. Plausible causes of geochemical diversity in the sources of OIBs include recycled oceanic crust and sediments (Hofmann and White 1982), delaminated continental lithosphere (McKenzie and O'Nions 1983), and relatively primitive or metasomatically enriched rocks that have always resided in the non-lithospheric mantle (DePaolo and Wasserburg 1976). As outlined above, oxygen isotopes can potentially trace subducted material in the mantle sources of basalts and therefore can help discriminate among these sources of geochemical variations among OIBs. Furthermore, there is growing evidence that major-element variations of OIBs constrain the mineralogy and extents and conditions of melting of the sources of ocean-island volcanoes (Elliott et al. 1991; Hauri 1996); oxygen provides an isotopic tracer that is also a major element and therefore could be particularly useful in testing and advancing such studies. Other classes of oceanic intraplate basalts—e.g. oceanic plateaus; off-axis and isolated seamounts—have not been examined in any detail for their oxygen isotope geochemistry and will not be discussed here (although they are an attractive target for future studies).

Many studies have characterized the oxygen isotope ratios of ocean island basalts using conventional whole-rock analyses. The most comprehensive of these studies is the review paper of Harmon and Hoefs (1995), which concludes that, after filtering the data base to remove differentiated and altered rocks, approximately half of all OIBs have $\delta^{18}O$ values outside the range of fresh MORBs and that the average value for OIBs is ^{18}O depleted compared to MORBs. These authors suggest that low $\delta^{18}O$ values of OIBs are particularly associated with 'HIMU'-type radiogenic isotope signatures (the most distinctive property of which is high $^{206}Pb/^{204}Pb$; Zindler and Hart 1986) and are caused by high concentrations of recycled lower oceanic crust in their mantle sources. They also note that the two volcanic centers that are the sources of most known low-$\delta^{18}O$ OIB's (Hawaii and Iceland) are also locations that erupt basalts with unusually high $^3He/^4He$ ratios and thus low $\delta^{18}O$ values might also be a property of un-degassed lower mantle. Finally, subducted sediment is called on as the cause of high $\delta^{18}O$ values in some 'enriched' OIBs, principally based on the results of Woodhead et al. (1993) who found an increase in $\delta^{18}O$ with decreasing $^{143}Nd/^{144}Nd$ among glasses from the Pitcairn seamounts. However, oxygen isotope variations among OIB whole-rocks and glasses are ambiguous both because there is substantial overlap between $\delta^{18}O$ values of OIBs having very different radiogenic isotope compositions and because (with the exception of the Pitcairn seamounts) $\delta^{18}O$ generally does not correlate well with other geochemical properties within suites of related lavas.

Oxygen isotope ratios of ocean-island basalts have been re-examined using laser-based techniques in detailed studies of Hawaiian and Icelandic lavas (both aimed at establishing the origin of ^{18}O-depleted compositions in some OIBs) (Eiler et al. 1996a,b), the Canary islands (Thirlwall et al. 1997) and Tristan da Cunha/Gough (Harris et al. 2000), and one survey of approximately 60 globally-distributed OIBs (Eiler et al. 1996a); these data are summarized in the following sections:

Survey of ocean islands

This discussion is organized by sub-dividing samples into groups that easily fit end-member categories previously defined by radiogenic isotope systematics (Zindler and Hart 1986); 'low-$^3He/^4He$' lavas discussed in Eiler et al. (1996a) are excluded from this discussion and are instead mentioned below in the context of work on Hawaiian and Icelandic lavas. Figure 11 summarizes the data discussed in the following paragraphs.

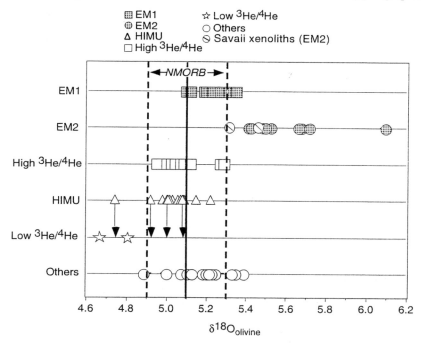

Figure 11. Oxygen isotope compositions of olivines in ocean island basalts, organized by geochemical sub-type as defined by radiogenic isotope compositions (Zindler 1986; see text for details). The vertical band between heavy dashed lines marks the range in $\delta^{18}O$ of olivine in equilibrium with NMORBs (assuming a melt-olivine fractionation of 0.5 ‰); the lower-$\delta^{18}O$ half of this range is typical of incompatible-element poor MORBs and the upper-$\delta^{18}O$ half is typical of incompatible-element rich MORBs (see Fig. 10). The 'low-$^3He/^4He$' category refers to lavas with $^3He/^4He$ ratios lower than 8 Ra; this is a characteristic often associated with ^{18}O-depleted compositions (regardless of other radiogenic-isotope properties) and is discussed in the text in reference to Hawaiian and Icelandic lavas. Arrows indicate samples that arguably belong to more than one group. These data are taken from Eiler et al. (1996a).

EM1. 'EM1'-type ocean-island lavas are most simply characterized by strongly sub-chondritic ('enriched') $^{143}Nd/^{144}Nd$ ratios but $^{87}Sr/^{86}Sr$ ratios within the range of or only slightly higher than estimates for bulk Earth (Zindler and Hart 1986). These characteristics have been attributed to subducted pelagic sediments in their mantle sources (perhaps with a component of subducted oceanic basalt; Zindler and Hart 1986; Weaver 1991; Woodhead et al. 1993), although Sun and McDonough (1989) argued that the trace-element abundances of EM1 lavas are more consistent with intra-mantle metasomatic or magmatic differentiation. Woodhead et al. (1993) present oxygen isotope data for submarine glasses from the Pitcairn seamounts (perhaps the most well characterized, end member 'EM1' hotspot), which document correlations between $\delta^{18}O$ and various radiogenic isotope ratios and suggest that their mantle sources contain up to 10 % recycled sediment. However, the most extreme radiogenic isotope compositions in the Pitcairn seamounts are found in relatively evolved lavas, raising the question of whether they are ^{18}O-rich because of the composition of their sources or because of their extent of magmatic evolution. Subsequent examination of olivine and plagioclase phenocrysts in basalts from nearby Pitcairn island, where the most extreme radiogenic isotope compositions are found in relatively magnesian basalts, found no evidence for $\delta^{18}O$ values outside the range for MORBs (Eiler et al. 1995). These data demonstrate that the enriched radiogenic isotope signature of EM1 can exist in the mantle sources of OIBs without any deviations in oxygen isotope ratio from typical upper mantle values, and thus that this end member is likely produced by little ($< \sim 1$ %) or no recycled crust. This conclusion is consistent with oxygen isotope compositions of less extreme EM1-type ocean island lavas from Kerguelan and Heard island (Eiler et al. 1996a).

EM2. 'EM2' type ocean island basalts are characterized by $^{87}Sr/^{86}Sr$ ratios that are greater than (more 'enriched' than) that estimated for bulk earth but a $^{143}Nd/^{144}Nd$ ratio that is near or slightly above (more 'depleted' than) chondritic (Zindler and Hart 1986). The compositions of EM2-type OIB has been widely ascribed to the presence of subducted sediments in their mantle sources (White and Hofmann 1982; Zindler and Hart 1986; Hofmann 1997); they arguably present the least controversial case for recycled crust in the sources of OIBs. Nine near-end-member 'EM2' lavas from the Society islands and Samoa (as well as two related Samoan xenoliths) were included in Eiler et al.'s (1996a) survey. Olivine in these lavas is consistently ^{18}O-enriched relative to typical terrestrial basalts, averaging 5.6±0.2 ‰ and having almost no overlap in $\delta^{18}O$ with MORBs and most other OIB's (Fig. 11). The only other oceanic basalts to have such consistently high $\delta^{18}O$ values are OIB from Koolau, Hawaii, and certain arc-related basalts having distinctive chemistries suggesting high degrees of melting; both of these are detailed below. The association of ^{18}O-enrichment with high $^{87}Sr/^{86}Sr$ is further emphasized in Figure 12, which compares the $\delta^{18}O$ values of olivines to the $^{87}Sr/^{86}Sr$ ratios measured on host whole rocks for representative OIBs from Eiler et al.'s (1996a) survey; curves illustrating the expected trends for mixing recycled sediment with depleted mantle are shown for comparison (this calculation is a minor modification of that detailed in Eiler et al. 1996, changed to reflect new constraints on the $\delta^{18}O$ of MORBs; Fig. 9).

HIMU. The 'HIMU' radiogenic isotope signature is most distinctive for its high $^{206}Pb/^{204}Pb$ ratio, which is frequently explained as a consequence of sampling large mass fractions (10's of %) of ~ 1 to 2 Ga recycled oceanic crust (e.g. Hauri and Hart 1993). Sun and McDonough (1989) instead suggest that distinctive trace-element concentrations of HIMU lavas are better explained by processes of magmatic and/or metasomatic differentiation within the mantle; a similar proposal was put forward by

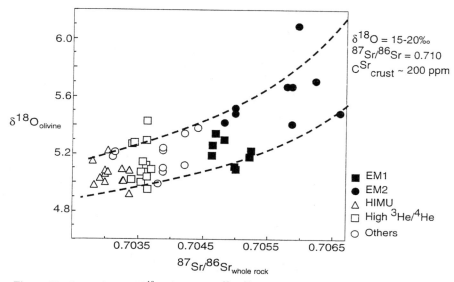

Figure 12. Comparison of $\delta^{18}O$ values with $^{87}Sr/^{86}Sr$ ratios in representative ocean-island basalts, organized by sub-type as defined by radiogenic isotope systematics (see Fig. 11; data from Eiler et al. 1996a). Low-$^{3}He/^{4}He$ lavas are a special case discussed in Eiler et al. (1996a) and in the text of this chapter and are excluded here. The heavy dashed curves mark the range of mixing hyperbolae between depleted mantle (based on data for NMORBs; Figs. 9 and 10) and high-$\delta^{18}O$ siliciclastic sediments.

Halliday et al. (1990) based on ^{207}Pb-^{206}Pb-^{204}Pb systematics. Eiler et al. (1996a) measured the $\delta^{18}O$ of olivine phenocrysts from 14 near-end-member HIMU basalts ($^{206}Pb/^{204}Pb$ from 20.0 to 21.8) and found them to be, with one exception, the most homogenous independently defined group of OIB (10 span a range of only 0.11 ‰). They are also the only group to have an average $\delta^{18}O$ that is statistically indistinguishable from that of the 'depleted' (low $\delta^{18}O$, incompatible-element-poor) end of the range for NMORBs; Fig. 10). At face value, this result is difficult to reconcile with models in which the mantle sources of HIMU lavas contain large mass fractions of recycled seafloor basalt: One might argue that the ^{18}O-rich and ^{18}O-poor parts of the oceanic crust simply balance to have no contrast in $\delta^{18}O$ with upper mantle peridotites, but this explanation requires an implausibly delicate balance of the abundances of two materials that differ strongly in $\delta^{18}O$ from each other and from the upper mantle. This point is emphasized further in Figure 13, which plots $\delta^{18}O$ vs. $^{187}Os/^{186}Os$ for the subset of OIBs from Eiler et al.'s (1996a) survey that were also characterized for Os isotope composition and compares the data to calculated mixing trends between depleted upper mantle (based on data for NMORBs; Fig. 10) and either upper or lower oceanic crust (this calculation assumes a model for Os abundances and isotope ratios in the end members given by Hauri and Hart 1993). In this figure, 'EM2' (and certain 'enriched' lavas from Hawaii also shown for comparison) are consistent with derivation from sources rich in recycled upper oceanic crust, but 'HIMU' and 'EM1' lavas are not derived from sources rich in recycled crust, barring fortuitous mixtures of upper and lower crust as mentioned above. On this basis alone, one would conclude that HIMU and EM1 isotopic signatures are a consequence of intra-mantle metasomatic enrichments or trace components of subducted materials.

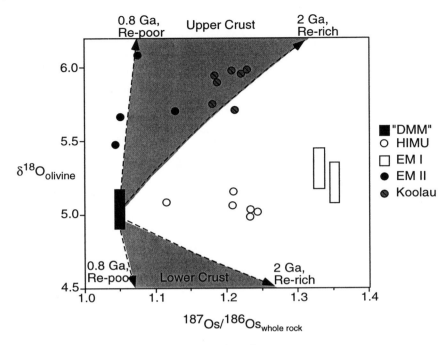

Figure 13. Comparison of $\delta^{18}O$ values with $^{187}Os/^{186}Os$ ratios in representative ocean-island basalts, organized by sub-type as defined by radiogenic isotope systematics (see Fig. 11; data from Eiler et al. 1996a, Hauri and Hart 1993, Reisberg et al. 1993, Lassiter and Hauri 1997, and Hauri, unpublished data). Fields for 'EM1' lavas are based on comparisons of $\delta^{18}O$ values and $^{187}Os/^{186}Os$ ratios from closely related samples. Samples from Koolau, Hawaii, are shown for comparison. Dashed curves bounding gray fields mark the range of mixing hyperbola between depleted mantle (based on data for NMORBs; Figs. 9 and 10) and either high-$\delta^{18}O$ oceanic crust (i.e. altered basalt) or low-$\delta^{18}O$ oceanic crust (i.e. hydrothermally altered layer 3 gabbros). Re/Os ratios and Os isotope compositions are taken from models of crustal recycling presented in Hauri and Hart (1993).

High $^3He/^4He$

One of the more longstanding conjectures about the oxygen isotope geochemistry of the mantle is that lower-mantle reservoirs rich in primitive noble gases are ^{18}O depleted compared to the upper mantle. This idea was originally based on the observations that low-$\delta^{18}O$ basalts are common in Hawaii and Iceland (Kyser et al. 1982; Kyser 1986; Harmon and Hoefs 1995)—locations that erupt basalts with unusually high $^3He/^4He$ ratios, which are in turn interpreted as an indication that their mantle sources contain relatively un-degassed materials, perhaps from the lower mantle (Kurz 1991). Oxygen isotope compositions of olivine phenocrysts have been measured for 21 lavas from Loihi, Kauai, and the Juan Fernandez islands having $^3He/^4He$ ratios of 15 Ra or greater (Fig. 11; see also Eiler et al. 1996b)—a range that is unambiguously distinct from MORB and most other ocean-island basalts. Olivines from these lavas vary little in $\delta^{18}O$ and have an average value that is indistinguishable from those in most basalts from other tectonic settings or from olivine in upper-mantle xenoliths. Although these results may simply reflect decoupling of volatiles from their long-term mantle sources (for instance, by intra-mantle metasomatism), they certainly provide no positive evidence that mantle reservoirs having high $^3He/^4He$ ratios differ in $\delta^{18}O$ from typical upper mantle rocks.

Thirlwall et al. (1999) and McPherson et al. (2000) recently defended the hypothesis of [18]O depletion in high [3]He/[4]He reservoirs based on studies of basalts from the Reykjanes ridge and Manus basin, respectively. The evidence in these studies is suggestive but not easily reconciled with the observations that lavas with strongly elevated [3]He/[4]He ratios are 'normal' in δ^{18}O (Fig. 11). Nevertheless, the question of vertical stratification of δ^{18}O in the earth's mantle is an important one and will doubtless continue to be the focus of new work and diverse interpretations.

Detailed studies

The preceding paragraphs describe general patterns in the oxygen isotope geochemistry of oceanic-island lavas, but conclusions based on broad overviews of this type can be superficial without also considering detailed studies of individual volcanic centers. In the following paragraphs I summarize the results of recent work on the three oceanic-island volcanic centers that have been studied in detail: Hawaii, Iceland, and the Canary islands. All three of these sets of studies share a focus on exploration of ^{18}O-depleted compositions in oceanic islands (but differ in various other important details).

Hawaii. It has been recognized for nearly 20 years that Hawaiian lavas are ^{18}O-depleted compared to MORBs and most terrestrial and lunar basalts generally (Kyser et al. 1982). Given the importance of the Hawaiian islands to studies of ocean-island volcanism, this anomaly has been the subject of several detailed studies (e.g. Garcia et al. 1989), but until recently remained enigmatic because of the lack of obvious relationships between δ^{18}O and other geologic and geochemical properties. A series of laser-fluorination studies of phenocrysts and glasses (Eiler et al. 1996b; Eiler et al. 1996c; Wang and Eiler 2000) revealed several such relationships, although their significance remains debated; the most important of these are: (1) Among sub-aerial, shield building lavas, low-δ^{18}O compositions are restricted to 'Kea trend' volcanoes (Kilaeua, Mauna Kea, Kohala, and Haleakala; Fig. 14); (2) Koolau and Lanai (Fig. 14) are characterized by high δ^{18}O values, comparable to those in 'EM2' lavas (Figs. 11 to 13 above; Koolau lavas are included in Fig. 13); and (3) δ^{18}O correlates with various radiogenic isotope ratios, although these correlations are dominated by differences between volcanoes rather than trends exhibited by lavas from any one volcano; see Eiler et al. (1996b) for details of these correlations. Eiler et al. (1996b) proposed that ^{18}O-enriched compositions in Koolau and Lanai lavas reflect a component of recycled upper-crustal rocks and/or sediments in their mantle sources, in agreement with previous suggestions that such components can explain 'enriched' trace-element and radiogenic-isotope compositions of these lavas (Roden et al. 1994). This interpretation has been followed by more recent papers adding Os isotope constraints (Hauri 1996; Lassiter and Hauri 1996), but argued against on the basis of some of the same data (Bennett et al. 1996).

The origin of ^{18}O-depleted compositions has been more hotly debated and remains more ambiguous: Eiler et al. (1996b) noted similarities between radiogenic isotope compositions of low-δ^{18}O, Kea trend lavas on one hand and modern MORBs on the other (e.g. decreasing δ^{18}O is associated with decreasing [3]He/[4]He) and concluded that ^{18}O-depleted compositions are a consequence of contamination by layer-3 gabbros in the Pacific plate. Lassiter and Hauri (1998) instead suggested that ^{18}O-depleted compositions sample recycled hydrothermally altered, ultramafic portions of oceanic lithosphere en-trained in the Hawaiian 'plume.' However, both of these interpretations are inconsistent with the most recent data from Wang and Eiler (2000) which show that the ^{18}O-depleted character of Mauna Kea lavas goes away sharply at the submarine-to-subaerial transition traversed by the Hawaiian Scientific Drilling Project ('HSDP'; Fig. 15) without the accompanying change in radiogenic isotope compositions one would

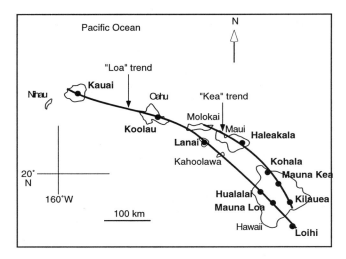

Figure 14. Map of the Hawaiian Islands. Volcanoes whose oxygen isotope geochemistry has been examined are marked with filled circles and labeled in bold. The 'Loa' and 'Kea' trends of volcanoes are indicated with solid curves; lavas from these two trends are distinct from one another in several geochemical properties (i.e. radiogenic isotope ratios) and also differ systematically in oxygen isotope composition: 'Loa' trend volcanoes are generally similar in $\delta^{18}O$ to MORBs and other ocean island lavas and include two edifices (Koolau and Lanai) with high $\delta^{18}O$ lavas; 'Kea' trend volcanoes have abundant low-$\delta^{18}O$ lavas.

predict from either the Eiler et al. or Lassiter and Hauri hypotheses. Wang and Eiler (2000) suggest that ^{18}O-depleted compositions in Kea-trend Hawaiian volcanoes are instead a product of contamination by hydrothermally altered rocks in rift zones in the volcanic edifice (analogous to processes inferred from study of Puu Oo lavas in Kilauea's east rift zone; Garcia et al. 1997). This hypothesis explains the lack of ^{18}O-depleted lavas in the 'Loa trend' volcanoes because of the low supply of meteoric water to their rift zones (such that hydrothermally altered crustal rocks are less abundant) and

~~~~~~~~~~~~~~~~~~~~~~~~~~

Figure 15 (opposite page). (a) Stratigraphic profile of the core recovered from the 2nd phase of the Hawaiian Scientific Drilling Project (HSDP). Relative abundances of rock types are indicated in the left panel using symbols described in the key; sulfur concentrations in glass (a measure of extent of degassing) are shown in the right panel; major boundaries in volcanic evolution and degassing history are indicated by heavy dashed lines. This figure is taken from Stolper, Depaolo and Thomas' (2000) description of the HSDP core. (b) Profile of oxygen isotope compositions of olivines from the HSDP core. The vertical field defined by the dashed line marks the range in $\delta^{18}O$ typical of upper-mantle basalts (based on data for MORBs and an assumed melt-olivine fractionation of 0.5 per mil). Values of $\delta^{18}O$ for Mauna Loa lavas (unfilled circles) are typical of those for lavas from other Loa trend volcanoes (excepting Koolau and Lanai, which are higher in $\delta^{18}O$). Low values of $\delta^{18}O$ at the top of the Mauna Kea section of the core are typical of subaerial lavas from other Kea trend volcanoes, including recent lavas from the Puu Oo east rift zone (Garcia et al. 1998). The 'subaerial to submarine' transition marks the boundary between Mauna Kea lavas erupted above sea level and underlying Mauna Kea lavas erupted underwater (identified based on vessicularity and other petrographic criteria). Below this boundary, Mauna Kea lavas are indistinguishable in $\delta^{18}O$ from most Loa-trend lavas and typical terrestrial basalts generally. No such discontinuity exists at this boundary in other source-related geochemical properties (i.e. radiogenic isotope ratios).

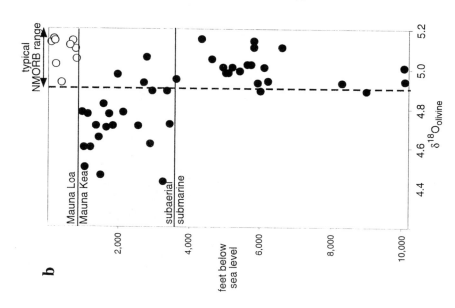

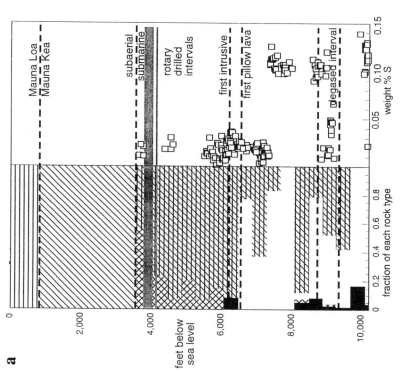

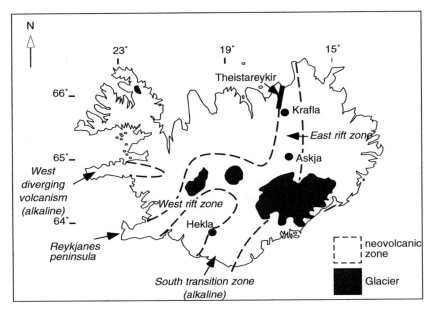

Figure 16. Map of Iceland, emphasizing major volcanic features. The dashed curves outline neovolcanic provinces that have been active since the last glacial maximum. Central volcanic complexes (characterized by abundant evolved magmas) are marked as filled circles and labeled. Theistareykir is a segment of the northern neovolcanic province; data for post-glacial lavas from here are shown in Figure 19. Large glaciers are marked in gray.

the absence of low-δ^{18}O values in submarine lavas because of the presumed isolation of their vents from low-pressure meteoric-hydrothermal systems that promote generation of ^{18}O-depleted rocks. These samples are still being characterized for other geochemical properties and thus detailed tests of this or other explanations of the data in Figure 15 have yet to be performed.

Iceland. Iceland is a section of the mid-Atlantic ridge that is anomalous for its unusually thick crust and the 'enriched' compositions of its lavas compared to adjacent sections of the mid ocean ridge; these features are widely attributed to a focused mantle upwelling or 'plume' in the sub-Icelandic mantle (e.g. Schilling et al. 1982; see Fig. 16). Iceland is arguably similar in importance to Hawaii in the general study of ocean-island volcanism because of its excellent exposure, close relationship to a seafloor spreading center, and the unusually high ^{3}He/^{4}He ratios of some of its lavas (Condomines et al. 1983; Poreda et al. 1986).

Iceland sits at high latitudes and, therefore, receives strongly ^{18}O-depleted precipi-tation. In addition, it is undergoing high rates of extension and volcanism, promoting deep-seated meteoric-hydrothermal systems in the crust (Bjornsson 1985). As a result, Iceland has a greater contrast in δ^{18}O between the crust and mantle than any other well-studied oceanic volcanic center (Gautason and Muehlenbachs 1998). Because the oxygen isotope 'signals' of crustal contamination are so strong there, conventional studies of whole-rock specimens are adequate for producing clear records of relationships between δ^{18}O and other geochemical properties among highly evolved and contaminated rocks (andesites, dacites and rhyolites). There have been many such studies conducted on a variety of scales (i.e. single eruptive centers vs. surveys of all of Iceland); a particularly

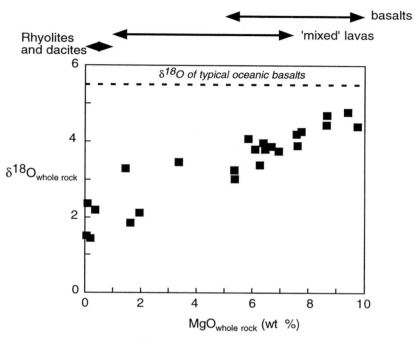

Figure 17. Comparison of $\delta^{18}O$ values with MgO concentrations in whole-rock samples from Krafla, a central volcano complex in Iceland's northern neovolcanic zone; data from Nicholson et al. (1991). MgO-poor, low-$\delta^{18}O$ rhyolites and dacites are interpreted as melts of hydrothermally altered crustal rocks; mixing between these melts and basalts generates the trend of decreasing $\delta^{18}O$ with decreasing MgO. It is not clear from studies of such highly evolved suites whether subtle differences in $\delta^{18}O$ between the most magnesian lavas (MgO of between 8 and 9 wt %) and typical terrestrial basalts are caused by subtle crustal contamination or are instead a property of the sub-Icelandic mantle.

detailed and revealing example is the work of Nicholson et al. (1991), who documented the relationship between indices of increasing differentiation (e.g. decreasing MgO) and increasing extent of contamination by hydrothermally altered crustal rocks or their melts (as monitored by decreasing $\delta^{18}O$); these data are reproduced in Figure 17 and a cartoon illustrating a common model for such contamination processes is shown in Figure 18. Other studies establish that low $\delta^{18}O$ Icelandic lavas are characterized by low $^{3}He/^{4}He$ ratios (Condomines et al. 1983)—re-enforcing similar results from Hawaii, described above, and the finding that low $\delta^{18}O$ values in other OIB volcanic centers are also associated with low $^{3}He/^{4}He$ (Eiler et al. 1996a). Low $^{3}He/^{4}He$ ratios are characteristic of even relatively young crustal rocks and thus this signature is consistent with pollution of basalts by ^{18}O-depleted crust.

The oxygen isotope geochemistry of relatively primitive Icelandic lavas is more controversial. There is abundant evidence that basalts and basaltic andesites from this setting are often (although not ubiquitously) ^{18}O-depleted compared to MORBs (Hemond et al. 1993), but there is no consensus as to whether this is because most such lavas have undergone small amounts of crustal contamination known to produce profound ^{18}O-depletions in more evolved lavas or instead because the mantle beneath Iceland contains one or more ^{18}O-depleted components. It is this author's prejudice that

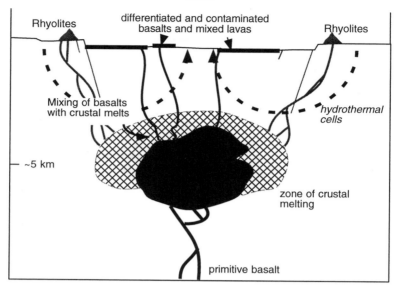

Figure 18. Schematic illustration of models for the contamination of Icelandic basalt in crustal hydrothermal / magmatic systems. Mafic magma is shown in black; the hatched area marks the region of the crust adjacent magma chambers where hydrothermally altered rocks are partially melted. Rhyolites (presumed erupted crustal melts) are shown in gray. Mafic magmas are presumed to mix with siliceous melts while passing through the zone of partially melted crustal rocks. Dashed arrows schematically indicate flow of water through the crust. Adapted from models described by Condomines et al. (1983), Nicholson et al. (1991), Sigmarsson et al. (1992), Hemond et al. (1993), and Jonasson (1994).

the second hypothesis demands a high burden of proof in a system with such extreme ^{18}O depletion in hydrothermally altered rocks of the pre-existing crust, but both explanations are plausible and worth exploring. Gee et al. (1998) found evidence that both 'depleted' and 'enriched' primitive basalts in the Reykjanes peninsula (Fig. 16) have δ^{18}O values within the range of MORBs, suggesting that the compositional heterogeneity in the sub-Icelandic mantle is not associated with variations in δ^{18}O, but that lavas with intermediate compositions are frequently ^{18}O-depleted, suggesting that contamination by ^{18}O-depleted rocks accompanies crustal mixing processes. Similarly, Eiler et al. (2000c) examined oxygen isotope variations in recent lavas from Theistareykir in Iceland's northern neovolcanic zone (Fig. 16) and found subtle, ubiquitous ^{18}O-depletions relative to MORBs that were interpreted as a consequence of contamination during crustal mixing processes. This study revealed usually simple correlations between δ^{18}O and other chemical indices; Figure 19 illustrates a representative view of these data and various models of crustal contamination that plausibly explain them, calling on assimilation and mixing processes previously suggested to explain the oxygen isotope geochemistry of evolved lavas from Krafla and similar central volcanic complexes (Figs. 17 and 18). Note that the conclusions of this study contrast with interpretations of the major and minor element chemistry of Theistareykir lavas as a consequence of dynamic melting of the Iceland plume (Slater et al. 1998) and are likely to be debated. Finally, Thirlwall et al. (1999) recently reported ^{18}O-depleted compositions in submarine lavas from the Reykjanes ridge and suggested that they provided evidence for an ^{18}O-depleted component of the mantle beneath Iceland,

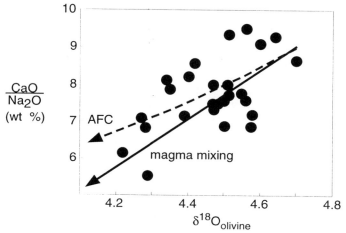

Figure 19. Comparison of $\delta^{18}O$ values of olivines with CaO/Na_2O ratios of whole-rocks for recent Theistareykir basalts, northern Iceland (data of Eiler et al. 2000c). The positive correlation is consistent with subtle decreases in $\delta^{18}O$ accompanying increases in degree of magmatic differentiation. The observed trend can be modeled as either magma mixing (solid curve, calculated assuming a low-$\delta^{18}O$ end member similar to Krafla 'mixed' lavas; Fig. 17) or as an Assimilation-Fractional-Crystallization process (dashed curve, calculated assuming an assimilant similar in composition to typical Icelandic upper-crustal altered volcanics). Details of both model curves are provided in Eiler et al. (2000c).

perhaps related to entrainment in that mantle of outer core material. This result has yet to be fully explored but will doubtless provide an interesting new insight on the problem of ^{18}O-depleted oceanic magmas once this is done.

Canary Islands. The Canary Islands do not have the prominent role in studies of oceanic intraplate volcanism shared by Hawaii and Iceland but have been the focus of a detailed oxygen isotope study in an effort to understand the origin and evolution of their lavas (Thirlwall et al. 1997). This work revealed relationships among isotopic compositions of Pb, Nd and O consistent with primary Canary Island magmas being relatively low in $\delta^{18}O$ (equal to or less than values of NMORBs) and 'HIMU'-like in their radiogenic isotope signatures, and that magmas were extensively contaminated by high-$\delta^{18}O$ low-$^{143}Nd/^{144}Nd$, low-$^{206}Pb/^{204}Pb$ crustal rocks during their differentiation. Thirlwall et al. favor an interpretation that the primary magmas are significantly ^{18}O-depleted compared to most upper mantle rocks and, thus, might sample recycled lower oceanic crust; however, they noted that low-$\delta^{18}O$ lavas are significantly evolved and might have taken on their ^{18}O-depleted character during contamination by hydrothermally altered rocks in the current lithosphere (i.e. a different contaminant from that suspected to be responsible for the properties of unusually high-$\delta^{18}O$ lavas).

To summarize, recent research using high-precision laser-based methods suggests that there is no systematic difference in $\delta^{18}O$ between the mantle sources of most ocean-island lavas on one hand and MORBs on the other (Fig. 11), but that many ocean-island centers erupt lavas higher and/or lower than the MORB range (Figs. 12 to 19), plausibly reflecting mantle processes in some cases and crustal processes in others. The diversity of ocean island lavas does not support simple generalizations; nevertheless, the weight of evidence suggests that ^{18}O-enriched compositions in magnesian lavas are most often explained by high-$\delta^{18}O$, enriched mantle components (although ^{18}O enrichments of comparable magnitude can result from high extents of crystallization-differentiation and crustal contamination; Harris et al. 2000). Anomalously low $\delta^{18}O$ values are debated, although the two most extensively studied examples (Hawaii and Iceland) reveal evidence implicating crustal contamination processes. These results suggest that a key to producing ^{18}O-depleted ocean-island lavas may be contamination of magmas in extensional rift zones that receive sufficient meteoric water to have vigorous hydrothermal systems. This conclusion mirrors that of Taylor and Sheppard 1986, regarding the origin of much more extreme ^{18}O depletions characteristic of some continental silicic volcanic centers (e.g. Yellowstone; see also Bindeman and Valley 2000) but not others (e.g. central Nevada and San Juan volcanic fields). Nevertheless, it is clear that ^{18}O-depleted rocks are subducted and occasionally found as high-pressure mantle xenoliths. Therefore we should expect to eventually find evidence for their oxygen isotope signatures in oceanic basalts derived from sources in which these rocks are entrained.

CONTINENTAL INTRAPLATE BASALTS

Continental intra-plate lavas are good candidates for oxygen isotope studies because of the likelihood that they have been significantly contaminated by ^{18}O-rich continental crustal rocks (i.e. the 'signals' in the oxygen isotope compositions of such lavas should be strong and well-correlated with important processes in their petrogenesis). This expectation is born out by studies of continental flood basalts based on conventional analyses of whole-rock specimens (e.g. Carlson 1984; Peng et al. 1994; Harmon and Hoefs 1995, and references therein). However, these remain the only major class of terrestrial basalts that have not yet been explored systematically using methods of high-precision laser-fluorination. A recent study of continental flood basalts from Yemen using laser-based measurements (Baker et al. 2000) illustrates the potential for oxygen isotopes to illuminate the role of even small amounts of crustal contamination in the geochemistry of continental intraplate basalts. Baker et al. compare the $\delta^{18}O$ values (and other properties) of green clinopyroxenes that crystallized early in the differentiation history of flood basalts from Yemen to later-crystallizing dark-brown clinopyroxenes in the same rocks. The former were found to have $\delta^{18}O$ values similar to those of most primitive basaltic lavas whereas the latter are relatively ^{18}O-enriched, consistent with progressive contamination by continental crust over the course of crystallization-differentiation. This is an excellent example of the potential power of mineral-specific oxygen isotope analysis to reveal the details of contamination and/or magma mixing processes and will doubtless serve as a model for future studies of similarly complex lavas.

ARC-RELATED LAVAS

There is unambiguous evidence that the sources of subduction-zone-related volcanic rocks incorporate to greater or lesser degrees the rocks, sediments, and fluids of the down-going plate (Gill 1981). Therefore, the contrast in $\delta^{18}O$ between the upper mantle on one hand and various components of old oceanic crust (and their metamorphosed equivalents) on the other could provide a tool for studying the extent and mechanisms of this incorporation. However, convergent-margin igneous rocks are frequently evolved when compared to lavas from other tectonic settings, their evolution often involves complex processes of crystallization and magma mixing that may confound attempts to link phenocryst $\delta^{18}O$ to bulk lava properties, and the crust through which they erupt is generally thicker than the oceanic crust and thus offers greater opportunity for contamination than is the case for MORBs and OIBs (Gill 1981).

Therefore arc-related lavas present special opportunities but also special challenges for oxygen isotope studies.

Some of the first comprehensive studies of oxygen isotope variations in island-arc lavas were those of Matsuhisa (Matsuhisa et al. 1973; Matsuhisa 1979), who established that calc-alkaline basalts and basaltic andesites from Japan have $\delta^{18}O$ values that are broadly within the range of mafic lavas from other tectonic settings and, thus, that Japanese island arc basalts likely originate by partial melting of peridotites in the mantle 'wedge' rather than by melting ^{18}O-enriched sediments and altered basaltic rocks in the top of the down-going slab. This is the prevalent view of arc-related magmatism today, but was an important contribution to the debate over the origin of calc-alkaline andesites at the time it was made. These studies also provide one of the best early records of the relationship between $\delta^{18}O$ and major-element chemistry among differentiated lavas. The first-order conclusions of the work of Matsuhisa et al. were corroborated by a subsequent study comparing the $\delta^{18}O$ of submarine glasses from the Marianas arc and MORBs (Ito and Stern 1985). However, contemporaneous studies (Pineau et al. 1976; Woodhead et al. 1987) suggested that the oceanic arc basalts and andesites are subtly, but systematically, ^{18}O enriched compared to MORBs due to subducted components in their mantle sources—a conclusion also reached by Harmon and Hoefs (1995) in their review of conventional fluorination data generated up to 1995.

Early studies of variations in whole-rock $\delta^{18}O$ provided a somewhat clearer view of the oxygen isotope geochemistry of continental arcs—e.g. the Andes (Harmon et al. 1981)—and oceanic arcs in which the overriding plate is thick and rich in sediments—e.g. the Antilles (Davidson and Harmon 1989). Studies of these systems revealed large ranges in $\delta^{18}O_{whole\ rock}$ (from values comparable to most terrestrial basalts up to +14 ‰ in several systems) that were correlated with increasing extents of magmatic differentiation and almost certainly reflect crustal assimilation and/or mixing with crustal melts. Recent laser-based studies of mafic lavas from the Andes (Feeley and Sharp 1995), Antilles (Thirlwall et al. 1996) and the Kermadec-Hikurangi arc (Macpherson et al. 1998) corroborate these interpretations (e.g. Fig. 20), suggesting that the oxygen isotope geochemistry of arc-related lavas erupted through thick sequences of sediments and/or continental crust are unlikely to provide information about variations in the $\delta^{18}O$ of their mantle sources, but can provide important insights on the evolution and crustal contamination of convergent-margin magmas.

However, use of laser based methods of oxygen isotope analysis to study mafic lavas in arcs has also produced significant differences with the results of previous work. Most simply, the total range in $\delta^{18}O$ of phenocrysts in oceanic arc basalts is found to be one-tenth the range observed in whole-rock samples of similar rock types (Fig. 21). Some of this difference must almost certainly be due to the greater diversity of samples in the data base of whole-rock analyses; however, much of this difference is likely due to the fact that analyses of early-liquidus phenocrysts more effectively 'see through' effects of sub-solidus alteration and crystallization-differentiation. Arguably, the most interesting results from recent studies of oxygen isotope variations in convergent margins have come from those focused on mafic lavas from oceanic arcs with thin crust and from back-arc environments. The key result of this work is that values of $\delta^{18}O$ appear to increase with decreasing abundance of moderately incompatible elements and, therefore, plausibly with increasing extent of melting of their mantle sources. Figure 22a illustrates this result by comparing Na_2O contents and $\delta^{18}O$ values of glasses from the Lau basin (modified from Macpherson and Mattey 1998) and Figure 22b presents a similar comparison of average $\delta^{18}O$ values and fractionation-corrected TiO_2 contents of several suites of lavas from the Marianas, South Sandwich arc and Vanuatu (modified

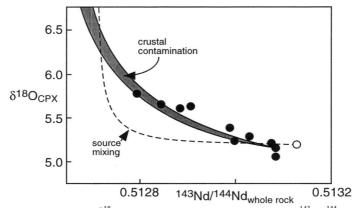

Figure 20. Comparison of $\delta^{18}O$ values of clinopyroxenes ('CPX') with $^{143}Nd/^{144}Nd$ ratios of whole rocks for arc-related lavas from the Kermadec-Hikurangi margin; adapted from MacPherson et al. (1998). ● = arc lavas; ○ = inferred composition of clinopyroxene in equilibrium with local depleted mantle. Solid curves are calculated trends for models of magma contamination by continental crust; the dashed curve is the calculated mixing trend between local depleted mantle and subducted upper oceanic crust components. The approximately linear increases in $\delta^{18}O$ with decreasing $^{143}Nd/^{144}Nd$ is characteristic of crustal contamination processes.

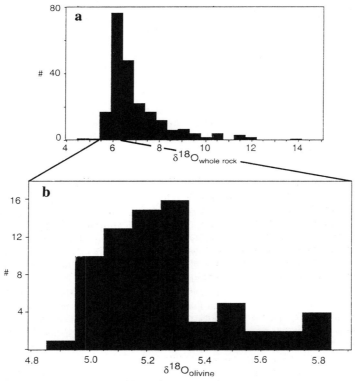

Figure 21. Oxygen isotope compositions of oceanic arc lavas, as measured by (a) conventional fluorination analyses of whole rocks (data compiled by Harmon and Hoefs 1995) and (b) laser fluorination of olivines from mafic lavas (data from Eiler et al. 2000a).

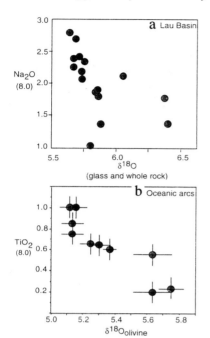

Figure 22. (a) Comparison of $\delta^{18}O$ of glass or whole rocks from the Lau basin with the fraction-corrected Na$_2$O contents of those same rocks; data from Pineau et al. (1976), Vallier et al. (1991), Lichtenstein and Loock (1993) and Macpherson and Mattey (1998). (b) Comparison of the average $\delta^{18}O$ of olivine phenocrysts in suites of island-arc tholeiites and calc-alkaline basalts with fractionation-corrected TiO$_2$ contents of those same suites; data from Eiler et al. (2000a). Inverse correlations in both (a) and (b) are consistent with increased integrated extents of melting of relatively high-$\delta^{18}O$ mantle sources; a model of 'fluxed' mantle melting describing such trends is presented in Eiler et al. (2000a).

from Eiler et al. 2000a). These trends have several possible explanations, but clearly differ from predictions of the effects of peridotite melting alone (Fig. 5) and can be explained by models in which mantle sources of arc lavas initially have $\delta^{18}O$ values within the range for MORBs and are 'fluxed' (i.e. driven to melting by addition of components that lower the peridotite solidus) by high-$\delta^{18}O$, slab derived fluids (Eiler et al. 2000a; see Fig. 23). This model has features in common with previous explanations of the relationship between water content and elemental abundances in Mariana back-arc lavas (Stolper and Newman 1994) and, like that model, can be used to infer the absolute abundances of minor elements in volatile-rich phases extracted from the downgoing slab.

Comparisons between igneous rocks from different tectonic settings are often compromised by confounding variables (e.g. systematic differences in extent and conditions of melting and/or differentiation); nevertheless, Figure 24 shows that MORBs and back-arc basin basalts may define a continuous spectrum of pollution of the upper mantle by subducted components. This figure plots the $\delta^{18}O$ values vs. $^{87}Sr/^{86}Sr$ ratios for NMORB glasses from the Pacific and Atlantic oceans and back-arc-basin basalts from the Mariana trough; the results suggest that a subtle but systematic increase in $\delta^{18}O$ of the back-arc mantle with respect to NMORB sources accompanies the subtle enrichment of the back arc mantle in radiogenic Sr from the down going slab. This trend resembles the correlation between $\delta^{18}O$ and $^{87}Sr/^{86}Sr$ defined by the Eiler et al. 1996a, survey of ocean island basalts (Fig. 12), although it is also important to note that there are exceptions to these trends that presumably reflect sources of geochemical enrichment other than stirring of high-$\delta^{18}O$ recycled crust into the upper mantle; e.g. Indian ocean MORBs are systematically higher in $^{87}Sr/^{86}Sr$ than Pacific and Atlantic MORBs—the so-called 'Dupal' anomaly—but are indistinguishable from other MORBs in average $\delta^{18}O$ or relationships between $\delta^{18}O$ and minor-element abundances (Fig. 10).

A recent study of mantle xenoliths from a convergent-margin setting found evidence for metasomatism of the sub-arc mantle by ^{18}O-enriched phases, consistent with the inferences outlined above based on studies of erupted lavas. Eiler et al. (1998) used ion microprobe methods to determine the $\delta^{18}O$ values of jadeitic glasses in metasomatic veins in xenoliths from Lihir island, Papua New Guinea, and found them to have $\delta^{18}O$ values up to 12 ‰—strikingly ^{18}O-enriched compared to values of most

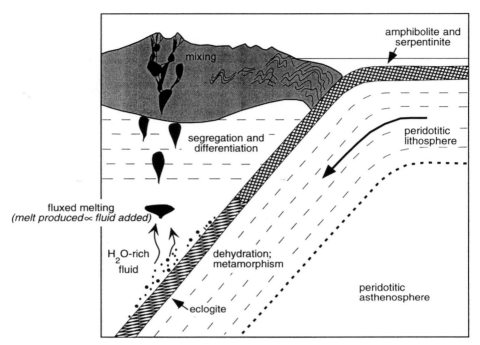

Figure 23. Schematic illustration of subduction-zone processes that can lead to trends such as those illustrated in Figure 22. Subduction injects hydrated mafic and ultramafic rocks (and accompanying marine sediments) into the upper mantle. Dehydration metamorphism transforms these rocks into eclogitic assemblages with lower water contents, releasing water-rich fluid. This fluid ascends into the mantle 'wedge' (between the over-riding and down-going plates), where it fluxes peridotite melting. The degree of melting is inferred to be proportional to the fluid/rock ratio in the mantle wedge; thus, decreasing concentrations of some incompatible elements (TiO_2, Na_2O) are associated with oxygen isotope signatures of the subducted slab (typically increasing $\delta^{18}O$).

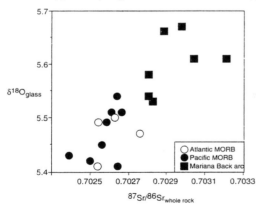

Figure 24. Comparison of $\delta^{18}O$ values for NMORB and back-arc-basin glasses with $^{87}Sr/^{86}Sr$ ratios measured on the same samples; data from Eiler et al. (2000a,b). Indian MORB have been excluded in order to avoid confounding effects of the Dupal anomaly, which is marked by large differences in Sr isotope ratios between Indian MORBs and other MORBs and appears to have no signature in $\delta^{18}O$ (Fig. 10).

erupted basalts and consistent with the derivation of the metasomatic melts from a source rich in subducted, upper-crustal oxygen.

Collectively, the recent work on oxygen isotope variations in arc-related rocks suggests that $\delta^{18}O$ values of primitive lavas and mantle nodules preserve subtle but important records of the transfer of subducted components through the mantle wedge and into the volcanic arc, whereas the overprinting influences of crustal contamination dominate oxygen isotope records in more evolved lavas. Naturally, one should expect that these competing processes will interfere with one another in some cases and, thus, future studies will need to exercise caution to make sure the influences of one process (mantle enrichment or crustal contamination) are not 'mapped' onto interpretations of the other.

SUMMARY AND FUTURE DIRECTIONS

View of the crust/mantle system through stable isotope geochemistry

This chapter began by stating the axioms that lead us to believe oxygen isotope variations of mantle rocks and basalts can provide unusual insights on problems of crustal recycling and lithospheric contamination. So what has this approach shown so far? It is never safe to over-generalize such a broad field, particularly given that it underwent a first-order change in data quality and quantity over the last several years. Nevertheless, the observations presented in the preceding sections suggest the following themes:

- Oxygen isotope variations are closely related to variations in radiogenic-isotope ratios and abundances of trace and major elements in many basaltic systems. Interpretations of these relationships vary from place to place and from time to time as debates evolve; however, the observations themselves are better than would have been imagined five or more years ago.

- There are two 'styles' of geochemical enrichment in the earth's mantle that are commonly associated with lavas having $\delta^{18}O$ values higher than those of common peridotites: (1) High concentrations of incompatible minor elements (particularly K) in MORBs; and (2) radiogenic Sr in ocean-island basalts and perhaps back-arc-basin basalts. These phenomena have broadly similar interpretations: Addition of upper oceanic crust and sediment to the mantle raises the abundance of alkali's and other incompatible elements and increases $\delta^{18}O$ in that mantle. Simple mixing relationships in the oxygen-isotope system allow relatively confident estimates of the amounts of recycled material in each case (e.g. ~0 to 5 %, average ~2 %, in the sources of NMORBs, Eiler et al. 2000b; ~2 to 8 %, average ~5 %, in the sources of 'EM2' ocean-island lavas, Eiler et al. 1996a).

- Perhaps equally importantly, there are other 'styles' of mantle enrichment that are unrelated to oxygen isotope variations, requiring that those enrichments are produced by high-temperature differentiation (metasomatism, melting and crystallization) or by incorporation of only traces (< ~1 %) of subducted crustal material in those regions of the mantle. These include the 'EM1' and (with some debate; Thirlwall et al. 1997) 'HIMU' types of enriched OIB sources, as well as the 'Dupal' anomaly that distinguishes Indian ocean MORBs from other MORBs. It has been common to explain radiogenic-isotope properties of all of these types of enrichment as a consequence of crustal recycling and, therefore, negative evidence from oxygen isotopes creates a dilemma, particularly given the apparent sensitivity of oxygen isotopes to recycled crust in the sources of other types of 'enriched' basalts (e.g. 'EM2' lavas in Figs. 11 to 13).

- Oxygen isotope variations in magnesian arc lavas from convergent margins erupted through thin crust appear dominated by a processes related to the extent of melting of their mantle sources. The precise balance of adiabatic upwelling vs. 'fluxing' of the mantle wedge in producing primary arc magmas is one of the great outstanding issues in igneous petrology (e.g. Plank and Langmuir 1988; Stolper and Newman 1994) and it appears that oxygen isotopes have something to say about this problem. The existing data can be explained by models of 'fluxed' melting of a static mantle wedge by high-δ^{18}O fluids (Eiler et al. 2000a), but alternatives have been suggested (e.g. MacPherson and Mattey 1998, call on the effects of oxygen isotope fractionations between melts and residues) and further tests and refinements of this interpretation are needed.

• There are many environments in which oxygen isotope variations are dominated by crustal contamination. Clear examples include convergent-margin volcanic centers erupted through thick and/or sediment-rich crust, continental intra-plate magmas, central volcano complexes on Iceland, and evolved, ^{18}O-enriched lavas from the Canary islands. Controversial cases that have also been argued to reflect crustal contamination are subtly ^{18}O-depleted Hawaiian and Icelandic basalts.

• Our understanding of the oxygen isotope geochemistry of actual mantle rocks is in a state of rapid change. Several studies leave little doubt that most mantle peridotites span a small range in δ^{18}O and are internally equilibrated. However, recent work documents exceptions to this generalization, including the identification of extremely ^{18}O-enriched metasomatic phases (Eiler et al. 1998) and complex isotopic zonation and ^{18}O-depletion in re-crystallized and cryptically metasomatized peridotites (Deines and Haggerty 2000; Zhang et al. 2000). It is challenging to concretely link small-scale diversity and complexity of metasomatized xenoliths to compositions of related basaltic lavas, which naturally average out small-scale properties of their sources; however, the connection between these observations and the oxygen isotope geochemistry of basalts is clearly a fruitful area for future research.

Outstanding problems and future directions

The oxygen isotope geochemistry of mantle-derived melts and rocks presents much opportunity for new work simply because the field has undergone a recent revolution in analytical approach, but many easily imagined applications have not yet been done. I conclude this chapter with a list of particularly outstanding research opportunities in this field (based, of course, on the author's prejudices and limitations!); this list could clearly be expanded by adding questions that are raised by recent work described above but not yet explored in detail:

Experiments. The experimental basis for interpreting subtle oxygen isotope variations among silicate melts and their phenocrysts is extremely poor, principally because of the lack of data on fractionation factors involving relevant silicate liquids. The directions and magnitudes likely for such fractionations are explored in the first half of this chapter, but without direct experimental constraints it is almost inevitable that misinterpretations of natural oxygen isotope variations have been or will be made. Therefore, new experimental work on this subject is arguably the most important lasting contribution that could be made to this field at this point.

Subtle secular variations in upper mantle? There is strong evidence that the isotopic composition of oxygen in the upper mantle has not varied greatly (>1 ‰) over Earth history (e.g. Kyser 1986), but it is not known whether average mantle δ^{18}O has varied by one or more tenths of per mil. If the δ^{18}O of the upper mantle has changed through time, even subtly, it would be a unique constraint on the integrated history of subduction and, combined with other information, the size of the mantle reservoir into which subducted crust is stirred. This question is naturally made difficult by the sub-solidus changes in δ^{18}O undergone by old volcanic rocks and likely diversity in the δ^{18}O of magmas at almost any point in earth history (i.e. 'false positive' results will be easy to find), but the importance of the problem almost demands that attempts to answer it be made. Insights of this type could provide an important complement to recent constraints on the variation with time of $\delta^{18}O$ values of continental granites (Peck et al. 2001).

Continental flood basalts. A first-pass has been made at applying laser-based methods to study the oxygen isotope systematics of most major classes of terrestrial basalts. However, continental flood basalts are an exception and the one detailed study that has been made (Baker et al. 2000) suggests that much will be learned by such work.

Oxygen isotope variations of the Moon and achondrite parent bodies. There is a long history of taking lessons from the study of terrestrial igneous rocks and applying them to samples from the Moon and the parent bodies of non-chondritic meteorites. Previous work on the oxygen isotope geochemistry of these samples has emphasized mass-independent differences in composition (Clayton and Mayeda 1996); however, a recent study of lunar basalts (Wiechert et al. 2000) indicates that subtle mass-dependant variations in oxygen isotope ratio may have been generated during the evolution of the moon's mantle and it is plausible that similar subtle but coherent 'signals' exist in the oxygen isotope variations of the parent bodies of the SNC ('Martian') and Howardite-Eucrite-Diogenite meteorite clans. These suites provide potentially enlightening points of comparison with the earth's mass-dependent oxygen isotope systematics but are virtually unexplored from this perspective.

ACKNOWLEDGMENTS

This chapter benefited from thorough reviews by I. Bindeman, R. Harmon and J. Valley. Thanks also are due to R. Kessel, I. Appora and Z. Wang for their comments on an early draft. Work reported in this paper was supported by NSF grant EAR 9805191.

REFERENCES

Alt JC (1993) Low-temperature alteration of basalts from the Hawaiian Arch, leg 136. Proc Ocean Drilling Program, Scientific Results 136:133-145
Alt J C, Muehlenbachs K, Honnorez J (1986) An oxygen isotopic profile through the upper kilometer of the oceanic crust, DSDP hole 504B. Earth Planet Sci Letters 80:217-229
Anderson AT, Clayton RN, Mayeda TK (1971) Oxygen isotope thermometry of mafic igneous rocks. J Geol 79:714-729
Appora I, Eiler JM, Stolper EM (2000) Experimental determination of oxygen-isotope fractionations between CO_2 vapor and Na-melilite melt. Goldschmidt Conference (Oxford, UK) Abstracts 5:149
Arthur MA, Anderson TF, Kaplan IR (1983) Stable isotopes in sedimentary geology. SEPM Short Course. 10:432 pp.
Baker JA, MacPherson CG, Menzies MA, Thirlwall MF, Al-Kadasi M, Mattey DP (2000) Resolving crustal and mantle contributions to continental flood volcanism, Yemen: constraints from mineral oxygen isotope data. J Petrol 41:1805-1820
Baker MB, Stolper EM (1994) Determining the composition of high-pressure mantle melts using diamond aggregates. Geochim Cosmochim Acta 58:2811-2827
Baker MB, Hirschmann MM, Ghiorso MS, Stolper EM (1995) Compositions of near-solidus peridotite melts from experiments and thermodynamic calculations. Nature 375:308-311
Barrat JA, Jahn BM, Fourcade S, Jaron JL (1993) Magma genesis in an ongoing rifting zone—the Tadjoura gulf (Afar area). Geochim Cosmochim Acta 57:2291-2302
Bebout GE, Barton MD (1989) Fluid flow and metasomatism in a subduction zone hydrothermal system: Catalina Schist terrane, California. Geology 17:976-980
Bennett VC, Esat TM, Norman MD (1996) Two mantle-plume components in Hawaiian picrites inferred from correlated Os-Pb isotopes. Nature 381:221-223
Bindeman I, Valley JW (2000) Formation of low-$\delta^{18}O$ rhyolites after caldera collapse at Yellowstone. Geology 28:719-722
Bjornsson A (1985) Dynamics of crustal rifting in NE Iceland. J Geophys Res 90:10,151-10,162
Carlson RW (1984) Isotopic constraints on Columbia River flood basalt genesis and the nature of the subcontinental mantle. Geochim Cosmochim Acta 48:2357-2372

Chazot G, Lowry D, Menzies M, Mattey D (1997) Oxygen isotopic composition of hydrous and anhydrous mantle peridotites. Geochim Cosmochim Acta 61:161-169

Chiba H, Chacko T, Clayton RN, Goldsmith JR (1989) Oxygen isotope fractionations involving diopside, forsterite, magnetite and calcite; application to geothermometry. Geochim Cosmochim Acta 53:2985-2995

Clayton RN, Onuma N, Mayeda TK (1971) Oxygen isotope fractionation in Apollo 12 rocks and soils. Proc 2nd Lunar Sci Conf, p 1417-1420

Clayton RN, Hurd JM, Mayeda TK (1972) Oxygen isotope abundances in Apollo 14 and 15 rocks and minerals. Lunar Sci Inst Contrib 88:141-143

Clayton RN, Goldsmith JR, Mayeda TK (1989) Oxygen isotopic fractionations in quartz, albite, anorthite and calcite. Geochim Cosmochim Acta 53:725-733

Clayton RN, Mayeda TK (1996) Oxygen isotope studies of achondrites. Geochim Cosmochim Acta 60:1999-2017

Clemens-Knott D (1992) Geologic and isotopic investigations of the early Cretaceous Sierra Nevada batholith, Tulane Co., CA, and the Ivrea zone, NW Italian alps: Examples of interaction between mantle-derived magma and continental crust. PhD Thesis, Division of Geological and Planetary Sciences, California Institute of Technology, Pasadena, 349 p

Condomines M, Gronvold K, Hooker P, Muehlenbachs K, O'Nions R, Oskarsson N, Oxburgh R (1983) Helium, Oxygen, strontium, and neodymium isotopic relationships in Icelandic volcanics. Earth Planet Sci Letters 66:125-136

Davidson JP, Harmon RS (1989) Oxygen isotope constraints on the petrogenesis of volcanic arc magmas from Martinique, Lesser Antilles. Earth Planet Sci Letters 95:255-270

Deines P, Haggerty SE (2000) Small-scale oxygen isotope variations and petrochemistry of ultradeep (>300 km) and transition zone xenoliths. Geochim Cosmochim Acta 64:117-131

DePaolo DJ , Wasserburg GJ (1976) Inferences about magma sources and mantle structure from variations in ^{143}Nd/^{144}Nd. Geophys Res Letters 3:743-746

Ducea M, Saleeby J (1998) A case for delamination of the deep batholithic crust beneath the Sierra Nevada, California. International Geology Rev 40:78-93

Eiler JM, Farley KA, Valley JW, Stolper EM, Hauri E, Craig H (1995) Oxygen isotope evidence against bulk recycled sediment in the source of Pitcairn island lavas. Nature 377:138-141

Eiler JM, Farley KA, Valley JW, Hauri E, Craig H, Hart S, Stolper EM (1996a) Oxygen isotope variations in ocean island basalt phenocrysts. Geochim Cosmochim Acta 61:2281-2293

Eiler JM, Farley KA, Valley JW, Hofmann A, Stolper EM (1996b) Oxygen isotope constraints on the sources of Hawaiian volcanism. Earth Planet Sci Letters 144:453-468

Eiler JM, Valley JW, Stolper EM (1996c) Oxygen isotope ratios in olivine from the Hawaii scientific drilling project. J Geophys Res 101:11,807-11,814

Eiler JM, McInnes B, Valley JW, Graham CM, Stolper EM (1998) Oxygen isotope evidence for slab-derived fluids in the sub-arc mantle. Nature 393:777-781

Eiler JM, Crawford A, Elliott T, Farley KA, Valley JW, Stolper EM (2000a) Oxygen isotope geochemistry of oceanic arc lavas. J Petrol 41:229-256

Eiler JM, Schiano P, Kitchen N, Stolper EM (2000b) Oxygen isotope evidence for recycled crust in the sources of mid ocean ridge basalts. Nature 403:530-534

Eiler JM, Gronvold K, Kitchen K (2000c) Oxygen isotope evidence for the origin of geochemical variations in lavas from Theistareykir volcano in Iceland's northern neovolcanic zone. Earth Planet Sci Letters 184:269-286

Elliott TR, Hawkesworth CJ, Gronvold K (1991) Dynamic melting of the Iceland plume. Nature 351:201-206

Feeley TC, Sharp ZD (1995) O-18/O-16 isotope geochemistry of silicic lava flows erupted from volcano llague, Andean central volcanic zone. Earth Planet Sci Letters 133:239-254

Fourcade S, Maury RC, Defant MJ, McDermott F (1994) Mantle metasomatic enrichment versus arc crust contamination in the Philippines: Oxygen isotope study of Batan ultramafic nodules and northern Luzon arc lavas. Chem Geol 114:199-215

Garcia MO, Muenow DW, Aggrey KE, O'Neil JR (1989) Major element, volatile and stable isotope geochemistry of Hawaiian submarine tholeiitic glasses. J Geophys Res 94:10,525-10,538

Garcia MO, Ito E, Eiler JM, Pietruszka AJ (1997) Crustal contamination of Kilauea volcano magmas revealed by oxygen isotope analysis of glass and olivine from Puu Oo eruption lavas. J Petrol 39:803-817

Garlick GD, MacGregor ID, Vogel DE (1971) Oxygen isotope ratios in eclogites from kimberlites. Science 172:1025-1027

Gautason B, Muehlenbachs K (1998) Oxygen isotope fluxes associated with high-temperature processes in the rift zones of Iceland. Chem Geol 145:275-286

Gee MAM, Thirlwall MF, Taylor RN, Lowry D, Murton BJ (1998) Crustal processes; major controls on Reykjanes Peninsula lava chemistry, SW Iceland. J Petrol 39:819-839

Ghiorso MS, Hirschmann MM, Sack RO (1994) New software models thermodynamics of magmatic systems. EOS Trans Am Geophys Union 75:571-576

Gill JB (1981) Orogenic Andesites and Plate Tectonics. Springer-Verlag, New York, 390 p

Gregory RT, Taylor HP (1981) An oxygen isotope profile in a section of Cretaceous oceanic crust, Samail ophiolite, Oman: evidence for $\delta^{18}O$ buffering of the oceans by deep (>5 km) seawater-hydrothermal circulation at mid-ocean ridges. J Geophys Res 86:2737-2755

Gregory R, Taylor HP (1986) Possible non-equilibrium oxygen isotope effects in mantle nodules, an alternative to the Kyser-O'Neil-Carmichael $^{18}O/^{16}O$ geothermometer. Contrib Mineral Petrol 93:114-119

Halliday AN, Davidson JP, Holden P, DeWolf C, Lee D-C, Fitton JG (1990) Trace element fractionation in plumes and the origin of HIMU mantle beneath the Cameroon line. Nature 347:523-528

Harley SL (1989) The origins of granulites—a metamorphic perspective. Geological Mag 126:215-247

Harmon RS, Hoefs J (1995) Oxygen isotope heterogeneity of the mantle deduced from global ^{18}O systematics of basalts from different geotectonic settings. Contrib Mineral Petrol 120:95-114

Harmon RS, Thorpe RS, Francis PW (1981) Petrogenesis of Andean andesites from combined O-Sr isotope relationships. Nature 290:396-399

Harmon RS, Kempton PD, Stosch HG, Kovalenko VI, Eonov D (1986) $^{18}O/^{16}O$ ratios in anhydrous spinel lherzolite xenoliths from the Shavaryn-Tsaram volcano, Mongolia. Earth Planet Sci Letters 81:193-202

Harris C, Smith HS, le Roex AP (2000) Oxygen isotope composition of phenocrysts from Tristan da Cunha and Gough Island lavas: variation with fractional crystallization and evidence for assimilation. Contrib Mineral Petrol 138:164-175

Hauri EH (1996) Major-element variability in the Hawaiian plume. Nature 382:415-419

Hauri EH, Hart SR (1993) Re-Os isotope systematics of EMII and HIMU ocean island basalts from the south Pacific Ocean. Earth Planet Sci Letters 114:353-371

Hemond C, Arndt NT, Lichtenstein U, Hofmann AW, Oskarsson N, Steinthorsson S (1993) The heterogeneous Iceland plume: Nd-Sr-O isotopes and trace element constraints. J Geophys Res 98(B9):15,833-15,850

Hirschmann MM, Baker MB, Stolper EM (1998) The effect of alkalis on the silica content of mantle-derived melts. Geochim Cosmochim Acta 62:883-902

Hofmann AW (1997) Mantle geochemistry: the message from oceanic volcanism. Nature 385:219-229

Hofmann AW, White WM (1982) Mantle plumes from ancient oceanic crust. Earth Planet Sci Letters 57:421-436

Holloway JR, Blank JG (1994) Application of experimental results to C-O-H species in natural melts. Rev Mineral 30:187-230

Ionov DA, Harmon RS, Francelanord C, Greenwood PB, Ashchepkov IV (1994) Oxygen-isotope composition of garnet and spinel peridotites in the continental mantle—evidence from the Vitim xenolith suite, southern Siberia. Geochim Cosmochim Acta 58:1463-1470

Ito E, Stern RJ (1985) Oxygen and strontium isotopic investigations of subduction zone volcanism, the case of the Volcano arc and the Marianas island arc. Earth Planet Sci Letters 76:312-320

Ito E, White EM, Gopel C (1987) The O, Sr, Nd and Pb isotope geochemistry of MORB. Chem Geol 62:157-176

Jonasson K (1994) Rhyolite volcanism in the Krafla central volcano, north-east Iceland. Bull Volcanology 56:516-528

Juster TC, Grove TL (1989) Experimental constraints on the generation of FeTi basalts, andesites and rhyodacites at the Galapagos spreading center, 85°W and 95°W. J Geophys Res 94:9251-9274

Kalamarides RI (1986) High temperature oxygen isotope fractionation among the phases of the Kiglapait intrusion, Labrador, Canada. Chem Geol 58:303-310

Kay RW, Kay SM (1993) Delamination and delamination magmatism. Tectonophysics 219:177-189

Kieffer SW (1982) Thermodynamics and lattice vibrations of minerals: 5. Applications to phase equilibria, isotopic fractionation, and high-pressure thermodynamic properties. Rev Geophys Space Physics 20:827-849

Kempton PD, Hawkesworth CJ, Fowler M (1991) Geochemistry and isotopic composition of gabbros from layer 3 of the Indian ocean crust, hole 735B. Proc Ocean Drilling Program, Scientific Results 118:127-142

Kempton PD, Harmon RS (1992) Oxygen isotope evidence for large-scale hybridization of the lower crust during magmatic underplating. Geochim Cosmochim Acta 56:971-986

Kolodny Y, Epstein S (1976) Stable isotope geochemistry of deep sea cherts. Geochim Cosmochim Acta 40:1195-1209

Kurz MD (1991) Noble gas isotopes in oceanic basalts: controversial constraints on mantle models. *In* Short Course Handbook on Applications of Radiogenic Isotope Systems to Problems in Geology. L Heamar, JN Ludden (eds) Mineral Assoc Canada, p 259-286

Kyser TK (1986) Stable isotope variations in the mantle. Rev Mineral 16:141-164

Kyser TK, O'Neil JR, Carmichael ISE (1981) Oxygen isotope thermometry of basic lavas and mantle nodules. Contrib Mineral Petrol 77:11-23

Kyser TK, O'Neil JR, Carmichael ISE (1982) Genetic relations among basic lavas and ultramafic nodules: Evidence from oxygen isotope compositions. Contrib Mineral Petrol 81:88-102

Langmuir CH, Klein EM, Plank T (1992) Petrologic systematics of mid-ocean ridge basalts: constraints on melt generation beneath ocean ridges. *In* Mantle Flow and Melt Generation at Mid-Ocean Ridges. JP Morgan, DK Blackman, JM Sinton (eds) Washington, DC, American Geophysical Union, p 183-280

Lassiter JC, Hauri EH (1996) Os-isotope constraints on Hawaiian plume composition and melt/lithosphere interaction: results from Mauna Kea and Koolau volcanoes. EOS Trans Am Geophys Union 77:S287

Lassiter JC, Hauri EH (1998) Osmium-isotope variations in Hawaiian lavas: evidence for recycled oceanic lithosphere in the Hawaiian plume. Earth Planet Sci Letters 164:483-496

Lowry D, Mattey DP, Harris JW (1999) Oxygen isotope composition of syngenetic inclusions in diamond from the Finsch mine, RSA. Geochim Cosmochim Acta 63:1825-1836

Macpherson CG, Mattey DP (1998) Oxygen isotope variations in Lau Basin lavas. Chem Geol 144:177-194

Macpherson CG, Gamble JA, Mattey DP (1998) Oxygen isotope geochemistry of lavas from an oceanic to continental arc transition, Kermadec-Hikurangi margin, SW Pacific. Earth Planet Sci 160:609-621

Macpherson CG, Hilton DR, Mattey DP, Sinton JM (2000) Evidence for an O-18-depleted mantle plume from contrasting O-18/O-16 ratios of back-arc lavas from the Manus basin and Mariana trough. Earth Planet Sci Letters 176:171-183

Magaritz M, Whitford DJ, James DE (1978) Oxygen isotopes and the origin of high $^{87}Sr/^{86}Sr$ andesites. Earth Planet Sci Letters 40:220-230

Matsuhisa Y (1979) Oxygen isotopic compositions of volcanic rocks from the east Japan island arcs and their bearing on petrogenesis. J Volcanol Geothermal Res 5:271-296

Matsuhisa Y, Matsubaya O, Sakai H (1973) Oxygen isotope variations in magmatic differentiation processes of the volcanic rocks in Japan. Contrib Mineral Petrol 39:277-288

Matsuhisa Y, Goldsmith JR, Clayton RN (1979) Oxygen isotopic fractionation in the system quartz-albite-anorthite-water. Geochim Cosmochim Acta 43:1131-1140

Mattey D, Macpherson C (1993) High-precision oxygen isotope microanalysis of ferromagnesian minerals by laser-fluorination. Chem Geol 105:305-318

Mattey D, Lowry D, Macpherson C (1994) Oxygen isotope composition of mantle peridotite. Earth Planet Sci Letters 128:231-241

Matthews A, Goldsmith JR, Clayton RN (1983) Oxygen isotope fractionations involving pyroxenes: The calibration of mineral pair geothermometers. Geochim Cosmochim Acta 47:631-644

Matthews A, Palin JM, Epstein S, Stolper EM (1994) Experimental study of $^{18}O/^{16}O$ partitioning between crystalline albite, albitic glass and CO_2 gas. Geochim Cosmochim Acta 58:5255-5266

Matthews A, Stolper EM, Eiler JM, Epstein S (1998) Oxygen isotope fractionation among melts, minerals and rocks. Mineral Mag 62A:971-972 (abstr)

McKenzie D, O'Nions RK (1983) Mantle reservoirs and ocean island basalts. Nature 301:229-231

Muehlenbachs K (1986) Alteration of the oceanic crust and the ^{18}O history of seawater. Rev Mineral 16:425-444

Muehlenbachs K, Clayton RN (1972) Oxygen isotope studies of fresh and weathered submarine basalts. Canadian J Earth Sci 9:72-184

Muehlenbachs K, Byerly G (1982) ^{18}O enrichment of silicic magmas caused by crystal fractionation at the Galapagos spreading center. Contrib Mineral Petrol 79:76-79

Nadeau S, Philppot P, Pineau F (1993) Fluid inclusion and mineral isotopic compositions (H-C-O) in eclogitic rocks as tracers of local fluid migration during high-pressure metamorphism. Earth Planet Sci Letters 114:431-448

Neal CR, Taylor LA, Davidson JP, Holden P, Halliday AN, Nixon PH, Paces JB, Clayton RN, Mayeda TK (1990) Eclogites with oceanic crustal and mantle signature from the Bellsbank kimberlite, South Africa, part 2: Sr, Nd, and O isotope geochemistry. Earth Planet Sci Letters 99:362-379

Nicholson H, Condomines M, Fitton JG, Fallick AE, Gronvold K, Rogers G (1991) Geochemical and isotopic evidence for crustal assimilation beneath Krafla, Iceland. J Petrol 32:1005-1020

O'Neil JR (1986a) Appendix: Terminology and standards. Rev Mineral 16:561-570

O'Neil JR (1986b) Theoretical and experimental aspects of isotopic fractionation. Rev Mineral 16:1-40

Ongley JS, Basu AR, Kyser TK (1986) Oxygen isotopic study of coexisting garnets, clinopyroxene and phlogopites from Roberts Victor eclogite xenoliths. EOS (Trans Am Geophys Union) 67:394

Onuma N, Clayton RN, Mayeda TK (1970) Oxygen isotope fractionation between minerals and an estimate of the temperature of formation. Science 167:536-538

Palin J.M, Epstein S, Stolper EM (1996) Oxygen isotope partitioning between rhyolitic glass/melt and CO_2: An experimental study at 550-950°C and 1 bar. Geochim Cosmochim Acta 60:1963-1973

Pearson DG, Davies GR, Nixon PH, Greenwood PB, Mattey DP (1991) Oxygen isotope evidence for the origin of pyroxenites in the Beni Bousera peridotite massif north Morocco—derivation from subducted oceanic lithosphere. Earth Planet Sci Letters 102:289-301

Peck WH, Valley JW, Wilde SA, Graham CM (2001) Oxygen isotope ratios and rare earth elements in 3.3 to 4.4 Ga zircons: Ion microprobe evidence for high $\delta^{18}O$ continental crust in the Early Archean. Geochim Cosmochim Acta (in press)

Peck W.H., King EM, Valley JV (2000) Oxygen isotope perspective on Precambrian crustal growth and maturation. Geology 28:363-366

Peng ZX, Mahooney J, Hooper P, Harris C, Beane J (1994) A role for lower continental crust in flood basalt genesis? Isotopic and incompatible element study of the lower six formations of the western Deccan traps. Geochim Cosmochim Acta 58:267-288

Perry EA, Gieskes JM, Lawrence JR (1976) Mg, Ca and $^{18}O/^{16}O$ exchange in the sediment-pore water system, Hole 149, DSDP. Geochim Cosmochim Acta 40:413-423

Pineau F, Javoy M, Hawkins JW, Craig H (1976) Oxygen isotopic variations in marginal basin and ocean-ridge basalts. Earth Planet Sci Letters 28:299-307

Plank T, Langmuir CH (1988) An evaluation of the global variations in the major element chemistry of arc basalts. Earth Planet Sci Letters 90:349-370

Poreda R, Schilling J-G, Craig H (1986) Helium and hydrogen isotopes in ocean-ridge basalts north and south of Iceland. Earth Planet Sci Letters 78:1-17

Putlitz B, Matthews A, Valley JW (2000) Oxygen and hydrogen isotope study of high-pressure metagabbros and metabasalts (Cyclades, Greece): Implications for the subduction of oceanic crust. Contrib Mineral Petrol 138:114-126

Reisberg L, Zindler A, Marcantonio F, White W, Wyman D, Weaver B (1993) Os isotope systematics in ocean island basalts. Earth Planet Sci Letters 120:149-167

Richet P, Bottinga Y, Javoy M (1977) A review of hydrogen, carbon, nitrogen, oxygen, sulfur, and chlorine stable isotope fractionation among gaseous molecules. Ann Rev Earth Planet Sci 5:65-110

Robert F, Rejoumichel A, Javoy M (1992) Oxygen isotopic homogeneity of the earth—new evidence. Earth Planet Sci Letters 108:1-9

Roden MF, Trull T, Hart SR, Frey FA (1994) New He, Nd, Pb, and Sr isotopic constraints on the constitution of the Hawaiian plume: Results from Koolau volcano, Oahu, Hawaii, USA. Geochim Cosmochim Acta 58:1431-1440

Rosenbaum JM (1994) Stable isotope fractionation between coexisting carbon dioxide and calcite at 900°C. Geochim Cosmochim Acta 58:3747-3753

Rosenbaum JM, Walker D, Kyser TK (1994a) Oxygen isotope fractionation in the mantle. Geochim Cosmochim Acta 58:4767-4777

Rosenbaum JM, Kyser KT, Walker D (1994b) High temperature oxygen isotope fractionation in the enstatite-olivine-BaCO$_3$ system. Geochim Cosmochim Acta 58:2653-2660

Rosenbaum JM, Mattey D (1995) Equilibrium garnet-calcite oxygen isotope fractionation. Geochim Cosmochim Acta 59:2839-2842

Schiano P, Birck J, Allegre CJ (1997) Osmium-strontium-neodymium-lead isotopic covariations in mid-ocean ridge basalt glasses and the heterogeneity of the upper mantle. Earth Planet Sci Letters 150:363-379

Schilling J-G, Meyer PS, Kingsley RH (1982) Evolution of the Iceland hotspot. Nature 296:313-320

Sharp ZD (1990) A laser-based microanalytical method for the in situ determination of oxygen isotope ratios of silicates and oxides. Geochim Cosmochim Acta 54:1353-1357

Sigmarsson O, Condomines M, Fourcade S (1992) A detailed Th, Sr and O isotope study of Hekla: differentiation processes in an Icelandic volcano. Contrib Mineral Petrol 112:20-34

Sisson TW, Layne GD (1993) H_2O in basalt and basaltic andesite glass inclusions from four subduction-related volcanoes. Earth Planet Sci Letters 117:619-635

Slater L, Jull MG, McKenzie D, Gronvold K (1998) Deglaciation effects on melting beneath Iceland: Results from the Northern Volcanic Zone. Earth Planet Sci Letters 164:151-164

Stakes DS (1991) Oxygen and hydrogen isotope compositions of oceanic plutonic rocks; high-temperature deformation and metamorphism of oceanic layer 3. Geochemical Society, Spec Pub No. 3. HP Taylor, JR O'Neil, IR Kaplan (eds) p 77-90

Staudigel H, Davies GR, Hart SR, Marchant KM, Smith BM (1995) Large-scale isotopic Sr, Nd and O isotopic anatomy of altered oceanic crust- DSDP/ODP sites 417/418. Earth Planet Sci Letters 130:169-185

Stolper EM, Epstein S (1991) An experimental study of oxygen isotope partitioning between silica glass and CO_2 vapor. Geochemical Society, Spec Pub No. 3. HP Taylor, JR O'Neil, IR Kaplan (eds) p 35-51

Stolper EM, Newman S (1994) The role of water in the petrogenesis of Mariana trough magmas. Earth Planet Sci Letters 121:293-325

Sun SS, McDonough WF (1989) Chemical and isotopic systematics of oceanic basalts: implications for mantle composition and processes. *In* Magmatism in the Ocean Basins. AD Saunders, MJ Norry (eds) Geol Soc Spec Pub 42:313-345

Taylor HP (1968) The oxygen isotope geochemistry of igneous rocks. Contrib Mineral Petrol 19:1-71

Taylor HP (1980) The effects of assimilation of country rocks by magmas on $^{18}O/^{16}O$ and $^{87}Sr/^{86}Sr$ systematics in igneous rocks. Earth Planet Sci Letters 47:243-254

Taylor HP, Sheppard SMF (1986) Igneous rocks: I. Processes of isotopic fractionation and isotope systematics. Rev Mineral 16:227-272

Thirlwall MF, Graham AM, Arculus RJ, Harmon RS, Macpherson CG (1996) Resolution of the effects of crustal assimilation, sediment subduction, and fluid transport in island arc magmas: Pb-Sr-Nd-O isotope geochemistry of Grenada, Lesser Antilles. Geochim Cosmochim Acta 60:4785-4810

Thirlwall MF, Jenkins C, Vroon PZ, Mattey DP (1997) Crustal interaction during construction of ocean islands: Pb-Sr-Nd-O isotope geochemistry of the shield basalts of Gran Canaria. Chem Geol 135:233-262

Thirlwall MF, Taylor RN, Mattey DP, Macpherson CG, Gee MAM, Murton BJ (1999) Oxygen-isotope systematics of Reykjanes ridge mid-ocean-ridge basalt: constraints on the origin and composition of the Icelandic plume mantle. 9th Ann Goldschmidt Conf, Harvard Univ, p 296-297 (abstr)

Tormey DR, Grove TL, Bryan WB (1987) Experimental petrology of normal MORB near the Kane fracture zone: 22°-25° N, mid-Atlantic ridge. Contrib Mineral Petrol 96:121-139

Valley JW (1986) Stable isotope geochemistry of metamorphic rocks. Rev Mineral 16:445-490

Valley JW, Kitchen N, Kohn MJ, Niendorff CR, Spicuzza MJ (1995) Strategies for high precision oxygen isotope analysis by laser fluorination. Geochim Cosmochim Acta 59:5223-5231

Valley JW, Kinny PD, Schulze DJ, Spicuzza MJ (1998) Zircon megacrysts from kimberlite: oxygen isotope variability among mantle melts. Contrib Mineral Petrol 133:1-11

Wang Z, Eiler JM (2000) Oxygen isotope compositions of lavas recovered from the second phase of HSDP drilling. Fall 2000 meeting Am Geophys Union, San Francisco (abstr)

Weaver BL (1991) The origin of ocean island basalt end-member compositions: Trace element and isotopic constraints. Earth Planet Sci Letters 104:381-397

White WM, Hofmann AW (1982) Sr and Nd isotope geochemistry of oceanic basalts and mantle evolution. Nature 296:821-825

Wiechert UH, Halliday AN, Lee DC, Snyder GA, Taylor LA, Rumble D (2000) Oxygen- and tungsten-isotopic constraints on the early development of the moon. Meteoritics Planetary Sci 35:A169

Woodhead JD, Greenwood P, Harmon RS, Stoffers P (1993) Oxygen isotope evidence for recycled crust in the source of EM-type ocean island basalts. Nature 362:809-813

Woodhead JD, Harmon RS, Fraser DG (1987) O, S, Sr, and Pb isotope variations in volcanic rocks from the northern Mariana islands; implications for crustal recycling in intra-oceanic arcs. Earth Planet Sci Letters 83:39-52

Wyllie P (1996) Carbonate and carbonate-rich liquids in the Earth's interior, and critical fluids in diamond inclusions. Proc 3rd NIRIM International Symposium on Advanced Materials, National Institute for Research in Inorganic Materials

Yui TF, Rumble D, Lo CH (1995) Unusually low $\delta^{18}O$ ultra-high pressure metamorphic rocks from the Sulu Terrain, eastern China. Geochim Cosmochim Acta 59:2859-2864

Zhang HF, Mattey DP, Grassineau N, Lowry D, Brownless M, Gurney JJ, Menzies MA (2000) Recent fluid processes in the Kaapvaal craton, South Africa: Coupled oxygen isotope and trace element disequilibrium in polymict peridotites. Earth Planet Sci Letters 176:57-72

Zheng Y (1991) Calculation of oxygen isotope fractionation in metal oxides. Geochim Cosmochim Acta 55:2299-2307

Zindler A, Hart S (1986) Chemical geodynamics. Ann Rev Earth Planet Sci 14:493-571

6   Stable Isotope Thermometry at High Temperatures

John W. Valley

Department of Geology & Geophysics
University of Wisconsin
1215 West Dayton Street
Madison, Wisconsin 53706

INTRODUCTION

Determination of accurate temperatures for geological events has been the grail of stable isotope geochemistry since the seminal 1947 paper by Urey. In theory, this should be simple. Calibrated mineral pairs are common, analysis is rapid, and there is no significant pressure correction. However, in spite of widespread application, the promise of reliable thermometry has been elusive. Stable isotope temperatures in metamorphic and igneous rocks are often controversial; is the fractionation between two phases a thermometer, speedometer (rate dependant), hygrometer ($P_{(H_2O)}$-dependant), or chimera? A number of factors have contributed to this uncertainty including: incomplete and sometimes conflicting calibration of isotope fractionation factors; limited understanding of the kinetics of mineral exchange; and the lack of microanalytical techniques. This situation has improved dramatically over the past decade, permitting reliable and detailed thermal histories to be inferred.

In a landmark paper, Bottinga and Javoy (1975) empirically calibrated the oxygen isotope fractionation factors among ten rock-forming minerals at T > 500°C. Their compilation shows that 84% of analyzed samples contain at least three minerals meeting a concordance test for equilibrium. They concluded "the great majority of igneous and metamorphic rocks .... has conserved a state of oxygen isotope exchange equilibrium." If correct, this bodes well for thermometry. However, this conclusion was contested by Deines (1977) who critically examined the existing data and concluded that "less than half of the rocks analyzed to date would yield concordant ^{18}O-derived temperatures." While seeming irreconcilable, both conclusions are based on fact. All rocks contain some degree of disequilibrium, but many also preserve a thermometric record. It is the purpose of this review to discuss the new strategies for reliable isotope thermometry and to explore disequilibrium processes so they may be recognized and properly interpreted.

The organization of this paper will be to discuss the assumptions of thermometry, processes of isotope exchange, and approaches for recovering temperature estimates of identifiable geological events. This is followed by applications of oxygen and carbon isotope thermometry to metamorphic and igneous rocks. While many careful studies have been published since 1947, emphasis will be placed on studies that have benefited from technical, experimental, and theoretical advances of the past decade.

REQUIREMENTS OF STABLE ISOTOPE THERMOMETRY

In theory, temperature can be estimated if minerals of interest record equilibrium conditions for a specific geologic event. Analysis of any two coexisting phases defines the isotope fractionation and uniquely determines the temperature of equilibration.

In practice, rocks follow a temperature-time path, equilibration is difficult to prove, and a number of conditions must be met for stable isotope thermometry to be accurate. The assumption that these conditions have been met is implicit in all thermometry, but

1529-6466/00/0043-0006$05.00

may not be fully evaluated. Similar requirements apply to petrologic geothermometry which is based on the partitioning of cations (see Essene 1989, Spear and Florence 1992).

There are three basic conditions for successful stable isotope thermometry.

(1) The sample domain was equilibrated in a specific event and compositions are "frozen-in". Thus, at the scale of measurement, minerals do not contain growth zonation or retrograde alteration.

(2) Analysis of isotope ratio is accurate at the appropriate scale.

(3) Isotope fractionation is sufficiently sensitive to temperature and is accurately calibrated, including the effects of solid solution, mixed fluids, and pressure.

Thus, the best stable isotope thermometers will have large fractionations, involve relatively common minerals, and occur in rocks where equilibration can be evaluated. The ideal case is a rock where all minerals crystallized at a specific temperature (diagenetic, metamorphic or igneous), and then cooled rapidly in a closed system. In other situations, knowledge of the exchange kinetics of the minerals in question can be important in the choice of thermometer.

If the first condition for thermometry (above) is not met, then stable isotope data contain information on interesting and more complex aspects of the thermal and fluid history. Rocks commonly preserve partial records of the varying conditions they have enjoyed: P-T-time-fluid-deformation. In well defined situations, multiple events can be identified or characterized in the evolution of a terrane, including: conditions during polymetamorphism; development of multiple fabrics; temperatures at specific reactions along the prograde or retrograde P-T path; sources of fluid infiltration during prograde, peak, or retrograde metamorphism; open or closed system recrystallization; crystallization of anatectic melts; assimilation during fractional crystallization of magmas; water fugacity during cooling; prograde heating rate; or cooling rate during uplift. The ability to study seemingly "invisible" retrograde history or to "see through" the peak of metamorphism is evolving rapidly. Such work will be discussed here as it applies to thermometry, but a full review is beyond the scope of this paper.

ANALYSIS OF STABLE ISOTOPE RATIOS

The first discoveries of stable isotope geochemistry were made possible by development of the Nier-type (gas-source, double-collecting) mass-spectrometer (Nier 1947), which improved analytical precision of light stable isotopes from ~2 ‰ to 0.05 ‰ (Murphey and Nier 1941, McCrea 1950). Modern gas-source mass-spectrometers are capable of measuring $\delta^{18}O$ to within a standard error of 0.003 ‰ (3 per meg, Luz et al. 1999). The analytical accuracy associated with thermometry is limited by sample preparation, not mass-spectrometry.

Table 1 summarizes the capabilities of laser and ion microprobe analysis in comparison to the time-honored conventional techniques. Figure 1 shows the advantages and trade-offs involved in the newer techniques: spatial resolution vs. accuracy and precision vs. cost of analysis. These factors will be discussed further under 'Microanalysis.' Continuous flow mass-spectrometry (Merritt and Hayes 1994); IR-spectroscopy (Kerstel et al. 1999, Esler et al. 2000); large radius, multi-collector ion probes (McKeegan and Leshin, this volume); automation; and shorter wavelength UV lasers (Young et al. 1998, Farquhar and Rumble 1998, Fiebig et al. 1999, Jones et al. 1999) also hold promise for enhanced analytical capabilities and better thermometry.

Table 1. Characteristics of various techniques for light stable isotope analysis (Valley 1998).

Technique	Isotopes, Minerals	Sample Prep.	Typical Sample Size	Spatial Resolution		Precision (1 sd)	References
CONVENTIONAL TECHNIQUES							
Ni reaction vessels	O silicates, oxides	Powder or chips	10-20 mg	-		0.2‰[A]	1
Phosphoric acid	O,C carbonates	Powder or chips	> 10µg	-		0.03-0.1‰[B]	2
Combustion	C graphite, diamonds	Powder or chips	> 1µg	-		0.03‰	3
Combustion	S	Powder or chips	> 1 mg	-		0.1‰	4
Fusion	H solids	Powder or chips	50-100 mg	-		1-2‰	5
LASER PROBE / MASS-SPECTROMETER							
Laser probe, IR	O silicates, oxides	Powder or chips	≥ 0.5 mg	500	µm	0.07‰	6,7
Laser probe, IR	O silicates, oxides	In situ	≥ 0.5 mg	500	µm (5mm)[C]	0.3-0.5‰	6
Laser probe, IR	S	In situ, chips	≥0.1 mg	200	µm	0.2‰	8
Laser probe, IR	O,C carbonates	In situ	≥0.5 mg	500	µm	[D]	9
Laser probe, IR	O phosphates	Powder or chips	≥0.5 mg	500	µm	0.1‰	10
Laser probe, UV	O silicates, oxides	In situ	~ 0.5 mg	~500	µm	0.1‰	11
Laser probe, UV	O phosphates	In situ		300x400	µm	0.4‰	12
ION MICROPROBE / SIMS							
Ion probe	O oxides (conductors)	In situ	0.4 ng	8	µm	0.5-1‰	13
Ion probe	O silicates or conductors	In situ	5 ng	20	µm	0.5-1‰	14
Ion probe	O carbonates	In situ	5 ng	20	µm	0.5-1‰	15
Ion probe	S	In situ	1-5 ng	10-20	µm	0.25-1‰	16
Ion probe	H	In situ	5-10 ng	20-30	µm	10‰	17
Ion probe	C	In situ	1-5 ng	10-30	µm	0.5-1‰	18

Footnotes; [A]some labs analyze 2 mg and attain precision ≤0.1 per mil; [B]precision for C is typically better than for O; [C]in situ pits are surrounded by 5mm haloes that prevent high analysis density, and causes fractionation with IR lasers. UV lasers may prevent such edge effects for improved spatial resolution and precision (Weichert and Hoefs, 1995); [D]precision and accuracy may be sample and procedure dependent.

REFERENCES: 1. Clayton and Mayeda, 1963; 2. McCrea, 1950; Sharma and Clayton, 1965; Rosenbaum and Sheppard, 1986; 3. Craig, 1953; DesMarais and Moore, 1984; 4. SO₂: Rafter, 1957; Robinson and Kusakabe, 1975; SF₆: Rees, 1978; 5. Bigeleisen et al., 1952; Godfrey, 1962; 6. Sharp, 1990; Elsenheimer and Valley, 1992; 7. Valley et al., 1995; 8. Crowe et al., 1990; 9. Smalley et al., 1989; Sharp and Cerling, 1996; 10. Kohn et al., 1996; 11. Weichert and Hoefs 1995; Farquhar and Rumble 1998; Fiebig et al., 1999; 12. Jones et al., 1999; 13. Valley and Graham, 1991a; 14. Hervig et al., 1992; Eiler et al., 1997a; 15. Valley et al., 1998; 16. Deloule et al., 1986; Eldridge et al., 1988; Riciputi, 1996; 17. Deloule et al., 1991a,b; 18. Zinner et al., 1989; Harte and Otter, 1992; Eiler et al., 1997b.

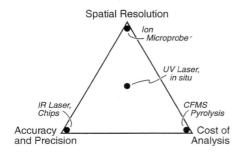

Spatial Resolution

Ion Microprobe

UV Laser, in situ

IR Laser, Chips

CFMS Pyrolysis

Accuracy and Precision

Cost of Analysis

Figure 1. Advantages and trade-offs of new analytical techniques for stable isotope analysis (see Table 1). At present, the best accuracy and precision is achieved for $\delta^{18}O$ by IR-laser fluorination of chips or powdered samples; the fastest and least expensive analyses are made by automated pyrolysis systems with continuous flow mass-spectrometers (CFMS); and the smallest samples and best *in situ* spatial resolution is attained by ion microprobe. The capabilities of *in situ* UV-laser fluorination are intermediate.

Microanalytical techniques are increasingly useful for stable isotope thermometry. Analytical precision for $\delta^{18}O$ of 1 ‰ can be obtained from 0.3-0.4 ng samples ($\cong 75$ µm³) by single collecting, small radius ion microprobe at high mass-resolution (Valley and Graham 1991, Valley et al. 1998a). Double collecting, large radius instruments offer enhanced counting efficiency and precision a factor of 2 better for similar spot sizes (McKeegan and Leshin, this volume). This is a dramatic improvement over the mg-size samples required before 1991. The question arises; how much smaller will be possible? The ultimate limit comes from counting statistics for a Gaussian distribution; 10^8 atoms of ^{18}O must be counted to obtain one standard deviation equal to 0.1 ‰. At natural abundances, this oxygen will be mixed with 5×10^{10} atoms of ^{16}O, the equivalent of approximately 3×10^{-3} ng of mineral. Perfect efficiency will never be attained and spots smaller than 1 μm^3 are unlikely, though sub-micron linear resolution is possible today by depth profiling. This spatial resolution is competitive with that of elemental analysis by electron microprobe, and brings the length scales of oxygen diffusion for igneous and metamorphic processes within the reach of stable isotope analysis. It also opens the door for investigation of a wide range of thin overgrowths and small particles of interest to sedimentary, environmental, or biological studies.

CALIBRATION OF ISOTOPE FRACTIONATION

The theory of stable isotope exchange was described by Urey (1947) and Bigeleisen and Mayer (1947), and has been reviewed and elaborated many times (Javoy 1977, Clayton 1981, O'Neil 1986, Hoefs 1997, Criss 1999; Chacko et al., this volume; Cole and Chakraborty, this volume). Only the basic equations need be given here.

For any two minerals in isotopic equilibrium, an exchange reaction relates the isotopic end-member compositions. This is illustrated by the exchange of oxygen isotopes between quartz and rutile (Reaction 1). The example is simplified by writing the mineral formulae with one oxygen, but the approach is general for SiO_2-TiO_2, or any mineral pair containing an element of interest.

$$Ti_{0.5}{}^{18}O + Si_{0.5}{}^{16}O = Ti_{0.5}{}^{16}O + Si_{0.5}{}^{18}O \tag{1}$$

The equilibrium constant for Reaction (1) is also the fractionation factor:

$$K_{eq} = \left[\frac{\left({}^{18}O/{}^{16}O\right) \text{quartz}}{\left({}^{18}O/{}^{16}O\right) \text{rutile}} \right] = \alpha {}^{18}O\,(Qt - Rt) \tag{2}$$

$$1000\left(\frac{-\Delta G^{\circ}}{RT}\right) = 1000 \ln K_{eq} = 1000 \ln\alpha\,(Qt - Rt) \cong \Delta {}^{18}O\,(Qt - Rt) \tag{3}$$

Thus, fractionation is mostly dependent on temperature. Since isotopes differ only in their nucleus, the volume change of Reaction (1) is very small. From the relation $\delta G = -S\delta T + V\delta P$, it follows that there is generally little or no measurable effect of pressure on fractionation (Clayton et al. 1975, Polyakov and Kharlashima 1994; Chacko et al., this volume).

The principle means of calibrating mineral thermometers are experiment, theoretical calculation, the increment method, and empirical comparison with other thermometers. It is common practice to assume linearity of 1000 ln α vs. T^{-2} according to the relation:

$$1000 \ln \alpha(a\text{-}b) = A_{a\text{-}b} \times 10^6 / T^2 \tag{4}$$

Most mineral systems can be approximated by Equation (4). However, for fluids it has been shown theoretically (Stern et al. 1968) and experimentally (see Friedman and O'Neil 1977) that cross-overs and other complex behavior can occur and thus higher order polynomials are sometimes fit to fractionation data. When available, high pressure, solid-solid exchange experiments are most generally reliable (see Clayton and Kieffer 1991, Matthews 1994, Rosenbaum and Mattey 1995). Figure 2 shows experimentally determined fractionations for calcite vs. some common rock-forming minerals (Clayton and Kieffer 1991). Experimental calibrations involving aqueous fluids are not fully understood and may depend on additional factors including solutes, fluid speciation, mixed data for vapor and supercritical fluid, recrystallization, and pressure (Clayton et al. 1989, Matthews 1994). Increment method calculations are a useful first approximation,

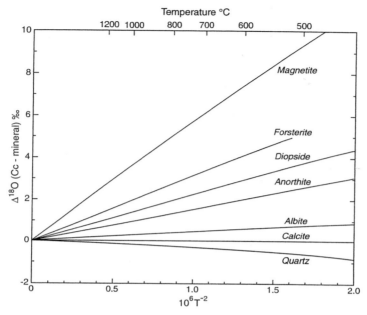

Figure 2. Experimentally determined oxygen isotope fractionations between calcite and quartz, albite, anorthite, diopside, forsterite and magnetite (Clayton and Kieffer 1991).

but should be independently evaluated. This subject is reviewed by Chacko, Cole and Horita (this volume) and only empirical calibrations, which depend on independent thermometry (e.g. petrologic estimates of T), will be discussed in detail below.

A useful Excel spreadsheet compiles published calibrations (uncritically) for easy calculation (J.D. Martin, unpublished) as does a web site (http://www.ggl.ulaval.ca/cgi-bin/isotope/generisotope.cgi) by Beaudoin and Therrien. The critical evaluation of various calibrations of equilibrium fractionation is essential to accurate thermometry. This topic is discussed in detail by Chacko, Cole and Horita (this volume).

KINETICS OF MINERAL EXCHANGE

Apparent temperature

The measurement of isotope fractionation between two equilibrated phases yields one estimate of temperature, and n phases yield (n - 1) temperatures. In the real world where equilibrium may be imperfectly attained, recorded, or preserved, analysis of n minerals provides at least (n - 1) estimates of apparent temperature (T_A) and often many more. These values of T_A may represent the true temperature of a meaningful geologic event or they may be spurious. The choices of both the appropriate strategy of analysis and the correct interpretation will depend on the history of the rock.

Post-crystallization exchange can take many forms that will complicate thermometry. Figure 3 shows four cartoons of the isotope distribution predicted for various processes of hydrothermal alteration of quartz by a low $\delta^{18}O$ fluid. If diffusion is radial and inward from the grain boundary then a concentric zonation of $\delta^{18}O$ is predicted with a smooth error function shaped gradient (Fig. 3A and Fig. 4). If the rim of the crystal

Figure 3. Schematic representation of the oxygen isotope zonation developed in a mineral by exchange with low $\delta^{18}O$ fluids through different mechanisms: (A) volume diffusion inward from the grain boundary; (B) dissolution and reprecipitation; (C) exchange along a set of microfractures; and (D) exchange along multiple sets of microfractures. Distinguishing among these processes requires microanalysis (Elsenheimer and Valley 1992).

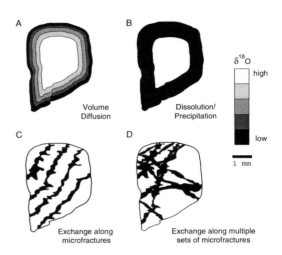

is recrystallized to form an overgrowth then the concentric pattern should have a steep gradient against the core of the crystal, i.e. a step function partially smoothed by any diffusional exchange (Fig. 3B). If micro fractures cause recrystallization within the crystal, then more complex patterns of heterogeneity are possible combining recrystallization and diffusion (Figs. 3C and 3D). Some amount of diffusion is inescapable in all samples and the magnitude of diffusive exchange must always be considered, although it may be insignificant at the scale of observation in some cases.

Diffusion

Diffusion is a significant and predictable process of intercrystalline (and intracrystalline) isotope exchange. Discussion will concentrate on self-diffusion involving exchange of isotopes of major elements such as oxygen, which is distinct from chemical diffusion or tracer diffusion (Lasaga 1998). The Arrhenius equation

$$D = D_o e^{-E/RT} \tag{5}$$

relates D (diffusion coefficient, cm^2/s or m^2/s), D_o (pre-exponential factor, cm^2/s or m^2/s), E (activation energy, sometimes represented as Q, J/mol), R (gas constant, $8.3143J/°K·mol$), and T (temperature, K). The use of different units, especially in earlier literature, can be confusing.

Diffusion rates are highly variable. Volume diffusion through the crystal lattice is often the rate-limiting step in inter-mineral exchange. The extent of exchange by volume diffusion depends on the phases present, the nature of the grain boundary, diffusion coefficients, activation energies, crystal size (or diffusion distance), and thermal history. Large differences in D exist from mineral to mineral (see review by Cole and Chakraborty, this volume). Diffusion rates of oxygen correlate with ionic porosity for many anhydrous minerals (Fortier and Giletti 1989, Zheng and Fu 1998) such that less dense tectosilicates like feldspar and quartz have faster diffusion than more densely packed ino- or orthosilicates like pyroxene or olivine. Thus, in a dry rock, diffusion is retarded by increasing pressure which tends to reduce ionic porosity, but the effect of increased $P(H_2O)$ or $f(H_2O)$ is to speed diffusion. Volume diffusion can also be strongly dependent on mineral chemistry. Diffusion rates may also be highly anisotropic, and experimental calibrations of anisotropy are needed for many minerals.

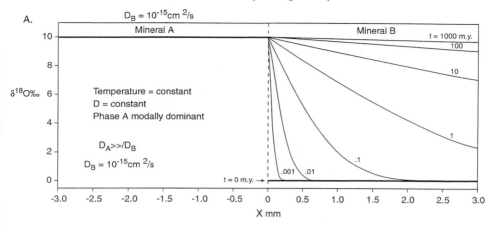

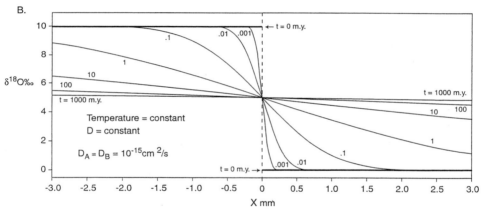

Figure 4. One-dimension diffusion profiles across the boundary of two touching minerals at constant temperature and variable times. Calculations assume an initial step in $\delta^{18}O$ of 10 ‰ and a final equilibrium fractionation of $\Delta(A\text{-}B) = 0$ ‰. (A) Pinned boundary condition where Mineral A is modally dominant and has fast oxygen diffusion such that its $\delta^{18}O$ does not vary during exchange. (B) a diffusion couple where the abundance and diffusion properties are the same in both minerals, and $\Delta(A\text{-}B) = 0$ ‰.

In contrast to volume diffusion through a crystal lattice, diffusion through a fluid, along a grain boundary, or on a surface is more rapid (Fig. 5). Thus, grain boundaries will be significant pathways for exchange, especially if they are hydrated, discordant, or filled with hydrous alteration or fluid. Even apparently anhydrous grain boundaries between unlike minerals in granulite facies gneisses are zones of high ionic porosity and promote intercrystalline exchange (Eiler et al. 1995a). Diffusion rates along dry coherent crystal boundaries may be significantly slower, close to volume diffusion. Thus some exsolution lamellae will create fast pathways for diffusive exchange (Nagy and Giletti 1986, Farquhar and Chacko 1994) while others are not measurably different (Schwarcz 1966).

Effect of deformation

Isotope exchange is enhanced by deformation (Kerrich et al. 1977, Farquhar et al.

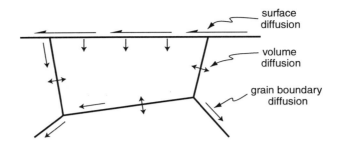

Figure 5. Pathways for diffusion. The relative diffusion rates decrease: surface > grain boundary > volume diffusion. In many rocks, diffusion and exchange is so much faster along grain boundaries than through the crystal lattice that grain boundaries can be modeled as equilibrated at temperatures where intermineral exchange is important.

1996). Commonly deformation is accompanied by recrystallization and minerals will readily exchange with fluids or other exchangeable phases, but volume diffusion rates will be enhanced by any process that facilitates breaking of oxygen-cation bonds. Mylonites are often shifted in $\delta^{18}O$ relative to their protolith (Cartwright et al. 1993). Farver and Yund (1999) experimentally studied bulk oxygen exchange in a natural sample of mylonite. Combined TEM and ion microprobe analysis showed faster oxygen diffusion parallel to the foliation of this very fine grained (~5 µm), micaceous (15-20% biotite) sample, consistent with fast pathways of exchange via a thin interconnected aqueous film along grain boundaries. These results also indicate that oxygen transport is several orders of magnitude faster through micaceous lithologies than in experiments of mono-mineralic quartzites or marbles (Farver and Yund 1991a, 1992, 1995a,b; 1998).

The fine grain size and differing mechanisms of exchange complicate thermometry in highly deformed rocks. O'Hara et al. (1997) correlate oxygen exchange with quartz and feldspar textures. Heterogeneous strain is associated with partial exchange while homogeneous strain correlates to isotopic homogenization on a grain scale. Reasonable, self-consistent temperatures have been estimated in favorable cases for deformation (350-400°C, O'Hara et al. 1997) and for retrograde fluid infiltration of granulites (600°C, Hoernes et al. 1995). Morrison and Anderson (1998) found that quartz-epidote fractionations increase approaching (50 to 12 m) the detachment fault of the Whipple mountains metamorphic core complex due to cooling by circulating surface-derived fluids.

Effect of water fugacity

Many minerals have been shown to experience oxygen diffusion that is several orders of magnitude faster in hydrothermal vs. anhydrous experiments (see Cole and Chakraborty, this volume). The exact causes of this relation are controversial: do protons diffuse rapidly through the crystal hydrating bonds, is H_2O the diffusing species, is the effect of $f(H_2O)$ inherited via charge-balanced substitutions from the time of crystallization or does diffusion rate (D) adjust rapidly to changing fluid environments, and does D scale linearly with $f(H_2O)$ (Giletti and Yund 1984, Elphick and Graham 1988, Farver and Yund 1991b, Zhang et al. 1991, McConnell 1995, Doremus 1998, Zheng and Fu 1998). Grain boundary diffusion rates are also enhanced by the presence of aqueous fluid (Farver and Yund 1992).

The scale of isotope equilibration and the magnitude of diffusive retrogression is critically dependent on the fluid conditions of a rock. Dry, granulite facies metamorphism has been invoked to explain very slow diffusion and the preservation of high oxygen isotope temperatures in some rocks (Sharp et al. 1988, Farquhar et al. 1996) while variable $f(H_2O)$ during cooling has explained faster diffusion and resetting of $\delta^{18}O$ in diopsides from granulite facies marbles (Edwards and Valley 1998), and of amphibolites and pelites (Kohn 1999). The rapid decrease in water activity that is common during closed-system cooling after the peak of metamorphism will cause chemical quenching in rocks that are water saturated during prograde metamorphic dehydration and anhydrous during cooling.

DIFFUSION MODELS

Diffusion distance

Figure 4A shows the $\delta^{18}O$ profile developed after isothermal diffusion for varying lengths of time based on the relation:

$$C = C_o \, \text{erfc}\left(\frac{X}{2\sqrt{Dt}}\right) \tag{6}$$

where C_o, the composition of mineral A, is held constant; C, the composition at distance X in mineral B, is initially zero; D is the diffusion constant, t is time, and erfc is the inverse error function (Crank 1975, Eqn. 2.45). This equation and (3.13) assume one-dimensional diffusion in a semi-infinite medium. Equations (6.18) to (6.20) in Crank (1975) model spherical geometry. For oxygen isotopes, $\delta^{18}O$ can be substituted for composition or isotope ratio. In Figure 4, D is set at 10^{-15} cm^2/s in mineral B (a mid-range for many igneous and metamorphic minerals at mid-crustal temperatures) and in 4A the profile is "pinned" by assuming $D_A \gg D_B$ so the composition remains constant at the grain boundary. It is assumed that $\Delta(A-B) = 0$ and that the mass of A is infinite, i.e. large enough to maintain a constant composition after exchange. These last two conditions would also be met if the grain boundary transport is fast, or if phase A is a fluid that is abundant in comparison to the exchanged volume of B. Calculations with Equation (6) are aided by the useful approximation:

$$\text{erf}(a) = \sqrt{1 - e^{-\left(4a^2/\pi\right)}} \tag{7}$$

where erfc $a = 1 - \text{erf } a$ (Lasaga 1998). From Figure 4A, it is seen that an initial step profile decays into a migrating diffusion profile.

It is often useful to refer to "diffusion distance" in order to approximate the magnitude of diffusive exchange for given Dt, but Figure 4A shows that there is no unique distance unless the percentage of exchange is defined. The simplest diffusion distance is defined for 50% exchange, i.e. the distance X to the point in mineral B where 50% of the initial difference has exchanged ($C/C_o = 0.5$, $\delta^{18}O = 5$ ‰); then Equation (6) reduces to:

$$X = \sqrt{Dt} \tag{8}$$

Figure 4B shows the diffusion profiles that will develop in semi-infinite sheets from the same initial step of 10‰ if $D_A = D_B = 10^{-15}$ cm^2/s and $\Delta(A-B) = 0$.

Dodson's closure temperature

The concept of closure or blocking temperature (T_C) below which exchange effectively ceases has been applied to both stable and radiogenic isotopes, and to cations. While many processes can contribute to exchange, the most successful treatments of T_C have been in systems that are dominated by volume diffusion into a crystal from its grain boundary.

The Dodson equation (Eqn. 9) defines T_C strictly in terms of diffusion. Use of T_C (Eqn. 9) implicitly assumes that other processes are not important. The strong dependence of diffusion coefficient on temperature (Eqn. 5) causes the transition from rapid to slow diffusion during cooling to be over a relatively small temperature interval. Dodson (1973, 1979) presented an equation for the closure temperature, T_C of a single mineral:

$$T_c = \left\{ \frac{E/R}{\ln\left(\frac{-A\,R\,T_c^{2}\left(D_o/a^2\right)}{E\left(\delta T/\delta t\right)} \right)} \right\} \tag{9}$$

where A = diffusional anisotropy parameter, a = radius of mineral grain, and $\delta T/\delta t$ = linear cooling rate (see Appendix 1 in Giletti 1986). Thus closure temperatures are higher for lower values of diffusion coefficient (D in Eqn. 5), larger grain size, or faster cooling. Figure 6 shows values of T_C for oxygen in magnetite with variable radii from 1 to 0.01 mm and cooling rates from 1 to 1000°C/m.y. (Giletti and Hess 1988).

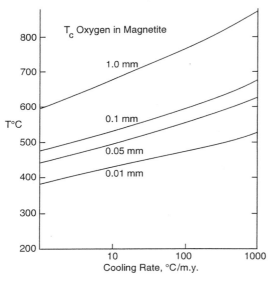

Figure 6. Calculated values of closure temperature for oxygen diffusion in magnetite of 1 to 0.01 mm in radius at cooling rates of 1 to 1000°C/m.y. (Giletti and Hess 1988).

Closure temperature is a useful concept for comparing the relative diffusivities of different minerals, or of the same mineral in different situations, but its applicability is limited by certain assumptions (Dodson 1973, Giletti 1986, Eiler et al. 1992, 1993). Dodson (1973) defined T_C for a radiogenic system as the temperature at the time corresponding to a mineral's apparent age, but its meaning may be somewhat different for oxygen isotopes. Equation (9) contains terms for the diffusion characteristics of only one mineral. It is implicit that this mineral is surrounded by an infinite, well-mixed reservoir of the element of interest. For many radiogenic systems, the loss of a trace element (e.g. Ar, Pb) is complete once it reaches a grain boundary and this condition is met. However, for major elements, the grain boundary in not a large reservoir and a mineral can only change in composition if there is another phase with which to exchange. Thus, the properties of other minerals or fluids in a rock influence the actual closure of diffusive exchange. Furthermore, the transition to closure is not a single temperature; diffusion slows, but does not stop at any geologic temperature (Eqn. 5). Values of T_C are calculated for oxygen and carbon diffusion in selected minerals of thermometric interest under representative conditions in Table 3 (below).

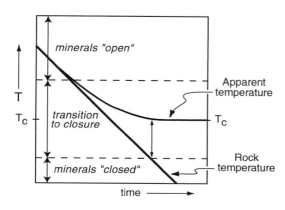

Figure 7. Plot of temperature vs. time during slow cooling. At high temperatures, minerals are "open" to exchange and the actual rock temperature (T_R) is equal to the apparent temperature (T_A) that would be estimated for the measured isotope fractionation between two minerals at that time. At intermediate temperatures, during the transition to closure: diffusion becomes slower; exchange is no longer able to keep up with increasing equilibrium fractionation; and T_A lags behind and is higher than T_R. At the end of the transition, minerals are closed, and T_A undergoes no further change and equals T_C. Note that the rock passes through T_C at an earlier time than T_A records T_C.

Figure 7 shows the apparent temperatures recorded by a mineral pair vs. the actual cooling path of a rock. As temperature approaches T_C, diffusion slows, exchange no longer keeps pace with declining temperature, and T_A diverges from actual rock T to asymptotically approach T_C. It is significant that closure occurs as a transition and not at a distinct temperature; once diffusion slows to the point where it retards exchange, the actual T of the rock becomes increasingly lower than apparent temperature or T_C. The time at which the rock finally records an apparent temperature equal to T_C is much later (and at lower T) than the time that the rock cooled through this temperature. Furthermore, diffusive exchange continues well below T_C near grain boundaries and will create isotopic profiles. Thus, T_C is an approximation for the transition to closure. These considerations are significant for interpretation of radiogenic as well as stable isotopes; the significance for stable isotope exchange will be discussed below.

The Giletti model

The first systematic treatment of stable isotope exchange by diffusion was applied to slowly cooled igneous and metamorphic rocks. Giletti (1986) used the Dodson closure temperature and hydrothermal experimental data for diffusion rates to calculate when minerals would cease exchanging with the rock during cooling. The computer program, COOL, can be used to make these calculations (Jenkin et al. 1991a,b). This is schematically represented in Figure 8 for a granite comprised of quartz, feldspar and hornblende. The relative values for equilibrium $\delta^{18}O$ decrease, Qt > Fsp > Hb, and the diffusion coefficients decrease Fsp > Qt > Hb. Thus hornblende has the lowest value of $\delta^{18}O$ and the slowest diffusion rate. The vertical dashed lines separating Periods I and II, and II and III are T_C for hornblende and quartz respectively at defined crystal size and cooling rate. At the initial temperature of 750°C, $\delta^{18}O$ values are equilibrated among minerals, representing magmatic crystallization or the thermal peak of metamorphism, and fractionations are small. During cooling through Period I, all minerals are open to diffusive exchange and reequilibration is continuous such that fractionations all increase

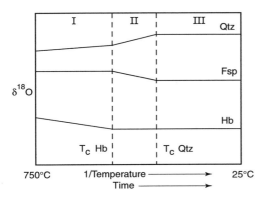

Figure 8. Schematic representation of $\delta^{18}O$ of quartz, feldspar and hornblende in an idealized granite undergoing diffusive exchange while cooling as described by the model of Giletti (1986). At high temperatures, above $T_C(Hb)$, all three minerals are open to exchange and maintain equilibrium (Period I). At lower temperatures (Period II) hornblende is "closed" to oxygen diffusion and only quartz and feldspar continue to equilibrate. In this period, $\Delta(Fs-Hb)$ decreases although no feldspar-hornblende exchange occurs. Finally, below $T_c(Qt)$, no further exchange can take place (Period III) and final fractionations are preserved (from Eiler et al. 1992).

with $1/T$. During Period II, hornblende is closed and does not change in $\delta^{18}O$, but quartz and feldspar continue to exchange with one another. During Period III, only feldspar is still open to exchange and since exchange requires at least two phases, $\delta^{18}O(Fsp)$ does not change.

Figure 8 illustrates important predictions of the Giletti model:

(1) None of the measured fractionations or T_A's record the formation temperature of the rock and are useful for thermometry, in the classical sense. This results because all minerals formed above their T_C.

(2) Only the fractionation between the two minerals with fastest diffusion in the rock (Qt-Fsp) has geologic significance. It records the closure temperature of the second fastest mineral, quartz, [i.e. $T_A(Qt\text{-}Fsp) = T_C(Qt)$].

(3) The final $\Delta(Fsp\text{-}Hb)$ is established at the end of Period II after hornblende has closed. This fractionation does not correspond to a temperature of equilibration or closure and, in the case presented by Figure 8; the apparent temperature is higher than any temperature experienced by the rock.

Other more general conclusions derive from the Giletti model.

(1) Reliable thermometry is possible if at least one mineral has formed below its closure temperature and other conditions are met.

(2) In minerals reset by diffusion, the apparent temperatures will be a function of grain size and mode in the rock, water activity that affects D, and thermal history.

(3) Cooling rates can be estimated in well-defined situations.

An important limitation of the Giletti model results from its reliance on the Dodson equation (Eqn. 9). This equation only considers one mineral in a rock and for major elements such as oxygen, a significant violation of mass balance can result.

The Fast Grain Boundary diffusion model (FGB)

The FGB model calculates exchange by diffusion of all minerals in a rock during cooling or heating (Eiler et al. 1992, 1993). This model considers any number of minerals and grain sizes in a rock, the diffusion characteristics of each mineral, the isotope fractionations, and various forms of $\delta T/\delta t$. It can be calculated with the computer program FGB (Eiler et al. 1994).

The main simplifying assumption of the FGB model is that all grain boundaries maintain isotopic equilibrium while the interior of each grain exchanges with its edge according to the diffusional properties of that grain. This is a good approximation for oxygen exchange in many rocks because the grain boundary region is oxygen-rich and diffusion is many orders of magnitude faster than volume diffusion. In practice for most rocks, this criterion is only important over distances of about three grain diameters and must be maintained only at temperatures above T_C for the fastest diffusing minerals involved (samples from within diffusive exchange distance of rock contacts may be more complex). Eiler et al. (1995a) tested the assumption of fast grain boundary diffusion in an anhydrous granulite facies rock with coarse mineral banding and found that grain boundary exchange of oxygen was very rapid over distances of several centimeters. Thus, even in a worst-case scenario, the FGB assumption was upheld. The assumption should also apply in greenschist or amphibolite facies rocks where grain size is smaller and water activities higher, or in igneous rocks where temperature is higher.

The FGB model has significant advantages over the Giletti model which is limited by reliance on Dodson's closure temperature. FGB calculations can evaluate the importance of variable mineral mode, intracrystalline diffusion profiles, variable grain size, fluid infiltration, thermometry, and cooling rate for the interaction of all minerals in a rock (Eiler et al. 1992, 1993, 1995a,b). FGB calculations can also be used to predict which mineral systems and thermal conditions are most appropriate for accurate thermometry (Eiler et al. 1993, Ghent and Valley 1998, Putlitz et al. 2001), empirical estimation of isotope fractionations (Kohn and Valley 1998, King et al. 2001) or diffusion coefficients (Edwards and Valley 1998, Peck 2000), or cooling rate.

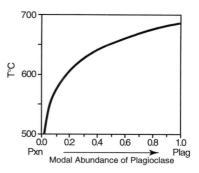

Figure 9. Plot of apparent temperature recorded between a mineral with slow oxygen diffusion (pyroxene) and a mineral with fast diffusion (plagioclase) vs. the modal percentage of plagioclase in a bi-mineral rock. Cooling is at 5°C/m.y. after peak equilibration at 750°C. In feldspar dominated rocks, T_A may preserve peak temperature, but for a rock dominated by pyroxene, T_A can be hundreds of degrees below peak T approaching 250°C, T_C(plag). The mode effect is modeled by the Fast Grain Boundary diffusion model and predicts that refractory accessory mineral (RAM) thermometers are most reliable (from Eiler et al. 1993).

The mode effect

A surprising and significant prediction of FGB calculations is that apparent temperature (T_A) is strongly dependent on mineral proportions. This is shown in Figure 9 where mineral mode is plotted against the T_A predicted by FGB for a simplified, two-mineral system: pyroxene-plagioclase, cooling (5°C/m.y.) after equilibration at 750°C. Plagioclase, the mineral with faster oxygen diffusion is plotted on the right and good agreement is predicted between T_C(Px) and T_A for a rock of this composition (anorthosite). This agreement arises because the rock is dominated by the faster mineral which acts as the well-mixed, infinite reservoir envisioned by Dodson (1973). Thus Δ(Plg-Pxn) thermometry is predicted to work in anorthosites if initial T is less than T_C(Pxn). However, a significant difference emerges as the proportion increases of pyroxene, the mineral with slower oxygen diffusion. For a pyroxene-rich rock, it is predicted that T_A will approach T_C(Plag). This difference arises because T_C (Eqn. 9) considers the diffusional characteristics of only one mineral. Thus, the Giletti approach ignores diffusion over short distances near the boundary of the slower mineral and incorrectly assumes that no further exchange is possible once the mineral with faster diffusion cools below T_C of the second slowest mineral. In many rocks, the exchangeable, grain boundary regions of refractory minerals are a small reservoir and T_C (Eqn. 9) $\cong T_C$ (FGB). However, due to grain boundary diffusion, a significantly larger difference is predicted between T_C and T_A for the mineral proportions represented on the left side of Figure 9. In the case of a pyroxenite with traces of feldspar, the difference can be over 200°C. In this case, significant exchange will continue below T_C(Pxn) because the larger percentage of pyroxene rims is sufficient in mass to shift the $\delta^{18}O$ of the smaller amounts of plagioclase. This generalization leads to important predictions for thermometry as described for refractory accessory minerals.

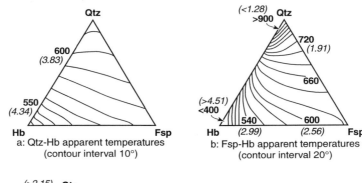

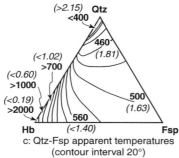

Figure 10. Contour plots of apparent temperatures and fractionations (‰ in italics) as a function of modal percentages in an idealized three-mineral rock composed of quartz, feldspar, and hornblende, predicted by the Fast Grain Boundary diffusion model to be recorded by: (a) quartz and hornblende, (b) feldspar and hornblende, and (c) quartz and feldspar. These figures show the relation of discordance and the mode of each mineral in a rock (from Eiler et al. 1993).

Mode effects can be complex for rocks composed of more than two minerals. This is shown in Figure 10, which predicts the apparent temperatures, recorded by Qtz-Hb, Fsp-Hb, and Qtz-Fsp pairs as a function of modal proportions in a three mineral rock that underwent slow cooling from 750°C. This figure illustrates some common problems of stable isotope thermometry in plutonic or high grade metamorphic rocks. Diffusion is relatively fast in each of these three minerals (i.e. T_C < 750°C) and the abundance of each mineral determines T_A. Peak temperatures will not be recorded, but erroneous apparent temperatures come fortuitously close to 750°C in a number of instances. While cooling rate can be estimated in a granite or other quartz-feldspar-hornblende rock, other minerals should be sought for reliable thermometry in this temperature range. Thus, T_C for each mineral and the mode effects for the rock can be used to predict which mineral systems will be accurate thermometers.

STRATEGIES FOR SUCCESSFUL THERMOMETRY

The theoretical models of inter-mineral and intra-mineral exchange provide a foundation for the interpretation of fractionation data for thermometry. However, thermometry presents the inverse problem. In modeling, one assumes a temperature vs. time path and calculates the mineral compositions (and zonation) of the resulting rock. In thermometry, one is presented with measured compositions and must deduce the thermal history.

Isotope exchange trajectories

One approach to thermometry employs the whole rock $\delta^{18}O$, assuming no open system modification, as well as mineral data. The evolution of isotope ratio can be calculated for each mineral in the rock if inter-mineral equilibrium exchange is maintained during cooling at $T > T_C$ as shown in Figure 11. Farquhar et al. (1993) present equations for thermometry based on understanding the changing $\delta^{18}O$ for each mineral during cooling. They call this trend an Isotope Exchange Trajectory (IET) (Fig. 11). Calculation of IET's requires knowledge of the whole rock isotope ratio, and the modal abundance and fractionation factor for each mineral. For minerals in a single sample, this approach derives in part from that of Gregory and Criss (1986) and Gregory et al. (1987) who apply δ_i vs. δ_j diagrams to evaluate open and closed systems. Model temperatures calculated by the IET method will either be T_C for the highest T_C mineral or the peak T if one or more minerals formed at a temperature below its T_C. The precision of this approach is sensitive to the fractionation factor of the mineral with the highest T_C. Because all minerals must be accounted for, the accuracy of this approach depends on the complete absence of open system alteration, a requirement not made by other techniques.

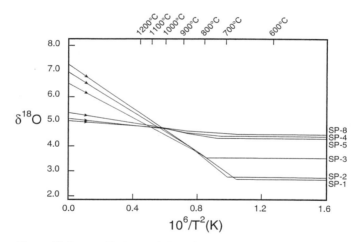

Figure 11. Isotope Exchange Trajectories (IET) for magnetite from six rocks from the Sybille Fe-Ti-oxide quarry, Laramie anorthosite complex. The average temperature of intersection, 1077±25°C, is higher than T_C for any mineral in these rocks, and it is interpreted as the solidus temperature of the magma (from Farquhar et al. 1993).

The IET approach can also be applied to multiple closely spaced hand samples if it is assumed that they were equilibrated with each other at the peak temperature (Farquhar et al. 1993, 1996). If each sample has different modal proportions, then the IET for any given mineral will be different in each lithology due to mass-balance effects. If the IET's for a single mineral are plotted together, they will intersect at the closure temperature for the multi-rock system, $T_C(sys)$. $T_C(sys)$ can be higher than T_C for any individual mineral. If more than two rocks are plotted, the redundant information serves as a check for concordance (Fig. 11). Krylov and Mineev (1994) analyzed closely spaced samples of granulite and derived a temperature of 720°C, well above the mineral temperatures that were generally lower. Loucks (1996) presents an alternative graphical and arithmetic approach to multiple closely spaced hand samples such as banded gneisses and discusses applications for cation as well as isotope thermometry.

RAM thermometers

Diffusion modeling predicts that accurate thermometry can be obtained from Refractory Accessory Minerals (RAM, i.e. minerals resistant to the effects of temperature that are modally subordinate) if they occur in a rock that is modally dominated by a readily exchangeable mineral (i.e. the right side of Fig. 9). RAM thermometers can be applied without additional diffusion modeling, especially in nearly monomineralic rocks such as quartzite, marble, or anorthosite. The choice of accessory mineral is constrained to phases with T_C above the peak temperature of interest and may include minerals that are relatively common (garnet, pyroxene, olivine), have simple mineral chemistry (aluminosilicates, corundum, magnetite, graphite), and/or which can be dated (zircon, titanite, rutile, baddelyite, monazite). In the right matrix, many gems are appropriate including ruby/sapphire, tourmaline, diamond, or emerald. In lower temperature sedimentary or hydrothermal systems, minerals such as calcite, quartz and clays can serve as RAM thermometers if a fluid with known isotope ratio is the dominant isotope reservoir.

The basis of RAM thermometry is that the accessory mineral preserves the isotope ratio from crystallization because of slow diffusion, while the dominant mineral preserves its isotope ratio by mass-balance because there are no other sufficiently abundant, exchangeable phases. This assumes that the accessory mineral does not exchange by some faster process such as recrystallization. A number of tests can be applied for recrystallization such as imaging by optics, CL, BSE, IR, X-ray mapping, or chemical etching. Applications of these tests are discussed in later sections. Accuracy is improved if the refractory mineral is not abundant as this minimizes the exchangeable reservoir near its grain boundaries. This approach assumes that the rock has been a closed system with respect to the element of interest, and there is no growth zonation, cryptic alteration, or recrystallization of the RAM.

There are a large number of potential RAM thermometers and some guidelines and tests evaluate which are best in a certain situation. Mineral pairs with a large fractionation will generally have a larger temperature coefficient and the effects of analytical uncertainty are minimized. This consideration favors rocks where relatively high $\delta^{18}O$ minerals such as calcite, quartz, or feldspar are the exchangeable phase. Growth zonation of isotope ratio can be evaluated by microanalysis (graphite, Wada 1988; garnet, Kohn et al. 1993, 1994; Chamberlain and Conrad 1993); by comparison of large vs. small crystals (Valley and Graham 1991, Sharp 1991, Kitchen and Valley 1995, Edwards and Valley 1998); or by analysis of rocks with different mineral proportions (kyanite-quartz, Putlitz et al. 2001). Partial or complete recrystallization of the RAM after the event of interest can be evaluated by microanalysis (magnetite, Valley and Graham 1993, Eiler et al. 1995a,b, Sitzman et al. 2000; quartz, Graham et al. 1996, Valley and Graham 1996; calcite, Graham et al. 1998); or by geochronology (zircon, Valley et al. 1994, Peck et al. 2001; titanite, King et al. 2001).

One variant of RAM thermometry applies to rocks with more than one exchangeable mineral, but a single refractory mineral, such as sphene in a quartzo-feldspathic gneiss or granitoids, or garnet in granulite facies gneiss (Hoernes et al. 1995). An average fractionation factor is calculated for the exchangeable minerals based on mode and this exchangeable reservoir is compared to the refractory mineral. This approach has the advantage of being applicable to many more common rocks than the strict RAM thermometer described above, but the use of whole rock data has the disadvantage that temperature sensitivity is reduced by averaging of high and low $\delta^{18}O$ minerals, and open system retrograde fluid effects are more commonly problematic due to analysis of all components of the rock.

Microscopic versus macroscopic models

The development of accurate techniques for microanalysis of stable isotope ratios permits studies of intra-crystalline zonation and applications of isotope exchange models at the microscopic scale. For isotopically complex samples, these results can differ significantly from macroscopic models that depend on averaged, bulk-mineral data. However, if minerals are homogeneous or have exchanged by a well-understood process, then the added effort of detailed microanalysis may not be necessary. One goal of future studies will be to determine which samples, processes, and geologic environments require microanalysis and which can reliably be interpreted with more rapid analysis at the macroscopic scale.

There are a number of processes that create fast pathways of exchange and effectively short-circuit volume diffusion into a crystal. Thus, the "real world" may be influenced by crystal defects and dislocations, mineral inclusions, exsolution lamellae, kink bands, microcracks, and other cryptic features (Fig. 12C). Diffusion is always active on a scale that can be modeled (Fig. 12B) and thus a world-view where all minerals are perfectly equilibrated and homogeneous (Fig. 12A) is generally a figment of imagination. In thermometry, these factors all potentially contribute to the compositions that are measured. Major advances have been made in determining when the macroscopic "model world" accurately predicts the microscopic "real world" situation. However, more work may be necessary to accurately deconvolute complex cases and tests should always be applied to evaluate thermometry.

TESTS OF THERMOMETRY

Concordance

The concordance test is commonly applied to evaluate equilibrium and the accuracy of stable isotope thermometry (Bottinga and Javoy 1975, Deines 1977). For three

Imaginary World   Model World   Real World?

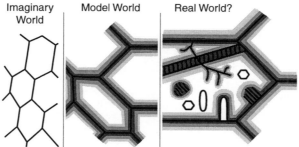

Figure 12. Cartoons of minerals showing gradients of isotope ratio due to exchange in shades of gray. (A) the "imaginary world" where all minerals are homogeneous and equilibrated, and there are no gradients. (B) the idealized "model world" where exchange takes place radially inward from grain boundaries by diffusion. Actual gradients would be smooth. (C) the "real world" will show effects of diffusion and may also have fast pathways of exchange due to recrystallization, micro-cracks, dislocations, inclusions or deformation.

minerals in equilibrium, a plot of Δ_{i-j} vs. Δ_{j-k} will be a smooth curve. However, this is a necessary but not sufficient criterion. The calibrations of Δ_{i-j} and Δ_{j-k} vs. T should be independently known (Clayton 1981). Furthermore, a correlation can be observed even though individual samples may plot substantially off the equilibrium line because processes of resetting may be systematic (e.g. Giletti 1986, Eiler et al. 1992, 1993; Cole and Chakraborty, this volume).

In practice, concordance should be judged based on units of $\delta^{18}O$ (‰) rather than temperature (Deines 1977). For example, an uncertainty in Δ_{i-j} of ±0.2 ‰ would yield a large temperature uncertainty for high temperature minerals with small fractionations like sanidine and quartz, but ±0.2 ‰ would seem very accurate for a system that is quite sensitive to temperature variation such as Δ(quartz-magnetite).

δ-δ and δ-Δ diagrams

A number of papers have explored the use of δ_i vs. δ_j diagrams for determining the extent of isotope disequilibrium and open system fluid exchange (Gregory and Criss 1986, Gregory and Taylor 1986a,b; Criss et al. 1987, Gregory et al. 1989). Figure 13 plots $\delta^{18}O$(feldspar) vs. $\delta^{18}O$(pyroxene) for hydrothermally altered gabbros (Criss et al. 1987). Under equilibrium magmatic conditions, these minerals would have small high temperature fractionations falling along an array with slope = 1. The steeper slopes shown here result from open system hydrothermal alteration which has affected feldspar more extensively than pyroxene.

Several papers discuss the uses and pitfalls of δ_i vs. Δ_{i-j} diagrams (Shelton and Rye 1982, Ohmoto 1986, Gregory and Criss 1986, Gregory et al. 1989). Zheng (1992) compares δ-δ and δ-Δ diagrams for calcite- graphite thermometry.

Imaging

Imaging often reveals processes that reset stable isotope thermometers or that short-circuit pathways of exchange such that diffusion models do not apply. Optical examination may show evidence of: inclusions including exsolution lamellae, crack healing cements, late mineral growth on grain boundaries,

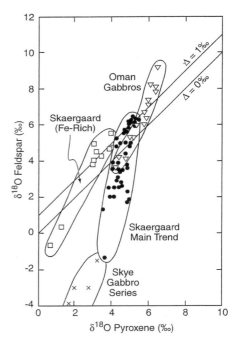

Figure 13. A plot of $\delta^{18}O$(fdsp) vs. $\delta^{18}O$(pyroxene) for various gabbros. If equilibrated at magmatic temperatures, these minerals would form a linear array with small fractionation and a slope of 1. The steeper slope and highly variable fractionations shown here indicate that the rocks have been altered by infiltration of hydrothermal fluids and that feldspar has exchanged more completely than pyroxene. (from Criss et al. 1987).

deformation lamellae, or kink bands (Fig. 12C). If not visible optically, these textures and other more cryptic features such as dislocations, subtle chemical zoning, or submicroscopic porosity may be detected by use of: X-ray diffraction, back-scattered electron imaging, secondary electron microscopy (Waldron et al. 1994), transmission electron microscopy (Farver and Yund 1999, Sitzman et al. 2000), hot or cold cathode cathodoluminescence (Valley and Graham 1996, Graham et al. 1998), chemical etching (Eiler et al. 1995a, King et al. 1998), ion milling (Valley and Graham 1993, Sitzman et al. 2000), infra-red spectroscopy (Johnson et al. 2000), or elemental mapping by electron or ion microprobe (Walker 1990). In addition to identifying processes of alteration, X-ray imaging by electron microprobe can reveal cation zonation which affects isotope fractionation or correlates to isotope zonation (Kohn et al. 1993). The inherent assumptions of thermometry imply that samples be "well-behaved," i.e. relatively devoid of these features. When observed, such features can aid interpretation of stable isotope data to elucidate otherwise unrecognized events. The importance of carefully looking at one's samples cannot be over emphasized.

The outcrop test

The simplest and most direct test of thermometry is the outcrop test. This is a necessary but not sufficient criterion for accurate thermometry. If temperature estimates from a small area of constant T do not agree, the lack of self-consistency proves that there is a problem in interpretation. Conversely, if temperatures are self-consistent, especially for a suite of samples that spans a range of $\delta^{18}O$ values, this is commonly interpreted as strong support that a thermometer is working. However, precision does not prove accuracy, and other interpretations should still be evaluated.

Table 2. Outcrop tests for self-consistency in calcite-graphite (Cc-Gr) thermometry from Adirondack marbles (Kitchen and Valley 1995).

Locality	Facies	$\Delta^{13}C(Cc\text{-}Gr)$	n	$T(°C) \pm 1\ std\ dev$ [a]
Train Wreck	amphibolite	4.2± 0.2‰	5	652± 13
Fish Creek	amphibolite	4.0± 0.4‰	29	660± 29
Valentine Mine	amph/gran	3.8± 0.3‰	9	677± 21
Bloomingdale	granulite	3.2± 0.1‰	8	722± 9
Willsboro	granulite	3.1± 0.1‰	3	738± 11

[a] Temperature calculated from Dunn and Valley (1992).

Kitchen and Valley (1995) report $\Delta^{13}C$(calcite-graphite) temperatures for 144 amphibolite and granulite facies marbles from the Adirondacks, including outcrop tests at five localities (Table 2). Each outcrop was smaller than 100m in dimension. For a regional metamorphic terrane such as the Adirondacks, peak temperature is expected to be constant on this scale. Four of the Adirondack outcrops yielded reproducible temperatures within ±10-20°C, however, samples adjacent to the Fish Creek alaskite body (leucogranite) were more variable, ±29°C. This contact was investigated in depth because of controversy whether the metamorphosed alaskite was originally deposited as rhyolite or intruded as granite (McLelland et al. 1992). To explore the cause of the ±0.36 ‰ variability in $\Delta^{13}C$(Cc-Gr) at Fish Creek, calcite was microsampled from the 5-cm^3 domains that were analyzed for Cc-Gr thermometry and heterogeneity was found to range from 0.0 to 0.30 ‰. Subsequent examination of thin sections by cathodoluminescence (CL) revealed that the heterogeneity in $\delta^{13}C$ is caused by variable amounts of late-stage calcite veining that is common in this area. Late calcite veining is commonly seen by CL in granulite facies orthogneisses (Morrison and Valley 1991) and marbles (Graham et al. 1998), and may degrade the precision and accuracy of thermometry elsewhere.

Microanalysis

Analysis at the appropriate scale for thermometry often requires very small samples, either as mineral separates, chips, or *in situ*. Table 1 reviews techniques for stable isotope microanalysis and Figure 1 shows the analytical trade-offs for different systems of microanalysis as they are generally applied in 2001: ion microprobe analysis; IR-wavelength laser fluorination of chips or powder; UV-wavelength laser analysis *in situ*; and continuous flow mass-spectrometry (CFMS) which can be coupled to pyrolysis or laser systems.

Laser heating, fluorination, or vaporization is increasingly used for isotope analysis. A variety of wavelengths have been employed from mid-IR (λ = 10.6 μm) to UV (0.266-0.157 μm, Kelley and Fallick 1990, Sharp 1990, Crowe et al. 1990, Farquhar and Rumble 1998, Eiler unpbd., see Rumble and Sharp 1998, Shanks et al. 1998). *In situ* analysis of oxygen or sulfur by IR laser yields a rind of partial reaction and fractionation encircling the laser pit. These edge effects can be corrected for sulfur isotopes (Crowe et al. 1990), but lead to degraded precision for oxygen (Elsenheimer and Valley, 1992). Analysis of mineral separates is quantitative and removes fractionation due to edge effects. The CO_2 laser (10.6 μm) yields the best accuracy and precision of any technique for $\delta^{18}O$ in silicate and oxide minerals when applied to 1-2 mg of mineral separate. Significant other advantages include ability to analyze refractory minerals, speed, and reliability of analysis, and better standardization (Sharp 1992, Valley et al. 1995).

Mineral separates for laser analysis can be prepared in a number of ways so as to micro-sample specific features in a single crystal or to concentrate a specific region from many similar crystals. Thin diamond saw blades can easily cut a checkerboard of chips from specially prepared thick microscope sections permitting correlation of $\delta^{18}O$ to optical or electron microprobe imaged features (Elsenheimer and Valley 1993). Over 200 analyses per square centimeter are possible by this approach (Kohn et al. 1993). Carbide or diamond tipped drills can be precisely controlled by computer. In soft minerals, elongate features as thin as 10-20 μm can be sampled if length is sufficient for sample needs (Patterson et al. 1993). Soft minerals can also be shaved at the 10-20 μm-scale by microtome (Wada 1988). Minerals with good cleavage like mica or graphite can be delaminated at scales less than 100 μm with a razor blade (Wada 1988, Kitchen and Valley 1995). Air abrasion permits analysis of cores or thin overgrowths on equant minerals such as zircon (Bindeman and Valley 2000) or pyroxene (Edwards and Valley 1998). For minerals that dissolve congruently without residue, partial etching in acid can remove unwanted features without compromising accuracy. HF etching removes overgrowths and cements on quartz (Forrester and Taylor 1977, Graham et al. 1998) and zones of radiation damage in zircon (King et al. 1998).

In situ analysis by UV laser is more precise than *in situ* analysis by IR because there is less heating (see Rumble and Sharp 1998, Farquhar and Rumble 1998, Young et al. 1998, Fiebig et al. 1999, Jones et al. 1999). The spot diameter of a UV laser and the sample size of a continuous flow mass-spectrometer can rival that presently used for oxygen isotope analysis by ion microprobe, however, the spatial resolution by laser is limited by the quantity of oxygen necessary for accurate purification during conversion of mineral to gas. At present, the analytical precision and the spot size attainable by *in situ* lasers are intermediate between the capabilities of the ion microprobe (best spatial resolution) and the CO_2 laser (best precision and accuracy, Fig. 1). For some projects this offers the best analytical compromise.

The best spatial resolution for stable isotope analysis is obtained by ion microprobe, but at a trade-off against precision (see Valley et al. 1998a, McKibben and Riciputi 1998,

McKeegan and Leshin, this volume). Typically, spot size is 10-30 μm in diameter and 1-5 μm deep. Over short distances normal to a flat surface, sub-micron spatial resolution is possible by depth profiling (Valley and Graham 1991). The recent development of multiple collectors for ion microprobes has lead to a major reduction in analysis time and promises enhanced accuracy and precision.

Ion microprobe analysis has an additional advantage over *in situ* analysis by UV laser. Even in the absence of any edge effect, *in situ* laser analysis of silicates requires that the entire sample be immersed in fluorinated gas and that only the area of laser illumination reacts to liberate oxygen. This situation may be met for a few rocks, but most rocks contain minor amounts of highly reactive material along grain boundaries and many rocks contain one or more minerals that are reactive at such low temperatures that pre-reaction of unlased material occurs leading to unavoidable sample contamination. Elsenheimer and Valley (1992) coated thin wafers with gold prior to laser analysis, but high blanks were still a problem for many rocks. Thus, background blanks should be carefully evaluated and reported for *in situ* laser data. In contrast, the zone of ion microprobe sputtering is well defined, and with proper precautions, no significant signal is derived from outside the pit.

Correlations to mode or crystal size

For minerals that have not equilibrated, isotope ratio may correlate to the size of an individual crystal. Growth zonation is to be expected in metamorphic minerals such as garnet due to temperature change during growth (Kohn et al. 1993) or in magmatic phenocrysts such as zircon due to evolving magma chemistry (Bindeman and Valley 2000). Retrograde exchange by diffusion causes concentric zonation affecting smaller crystals more than large crystals due to the difference in surface area/ volume. Thus, $\delta^{18}O$ has been shown to correlate with crystal size for metamorphic magnetites (Valley and Graham 1991, Sharp 1991), diopside (Edwards and Valley 1998, but see Sharp and Jenkin 1994), graphite (Kitchen and Valley 1995), and detrital zircons in granulite facies quartzite (Valley et al. 1994, Peck et al. 2001). Thus a correlation of isotope ratio to crystal size indicates a departure from the requirements of accurate thermometry and should be avoided for that purpose.

The Fast Grain Boundary diffusion model predicts that retrograde exchange will create correlations of isotope ratio to the modal proportions of a rock (Figs. 9 and 10; Eiler et al. 1993). These predictions have been verified for a range of amphibolite to granulite facies rocks (Hoffbauer et al. 1994, Eiler et al. 1995a,b; Farquhar et al. 1996, Edwards and Valley 1998, Ghent and Valley 1998, Putlitz et al. 2001). Mode effects also indicate pertur-bations that should be avoided for thermometry, but which may be useful for other studies. Knowledge of such correlations can distinguish polymetamorphic overprints, or estimate the rate of diffusive exchange or cooling. In turn, understanding retrograde kinetics may permit retrograde effects to be avoided and more accurate thermometry to result.

OXYGEN ISOTOPE THERMOMETRY

RAM thermometers

Several oxygen isotope Refractory Accessory Mineral (RAM) thermometers have been applied, including aluminosilicate, magnetite, garnet, or rutile in quartzite; and magnetite, titanite or diopside in marble. Graphite in marble forms a commonly applied RAM carbon isotope thermometer. It is important to keep in mind that all thermometers have an optimum temperature range for applicability. The peak temperature should be below the closure temperature of the RAM. However, for some minerals, at temperatures too far below T_C, the possibility of growth zoning becomes greater.

Refractory minerals are defined based on the relative diffusion rates of the RAM vs. its matrix. Thus Δ(Plag-Mt) or Δ(Plag-Rt) may be excellent RAM thermometers in amphibolite or eclogite facies anorthosites or metabasalts, but fail in the granulite facies. Likewise, Δ(Qt-Mt) and Δ(Cc-Mt) should be used with caution because of the smaller, possibly reversed, contrast in diffusivity. Diffusion data and values of T_C are given in Table 3 as a quick guide for choosing an appropriate thermometer. Cole and Chakraborty (this volume) provide a more detailed presentation of stable isotope diffusion data and exchange kinetics. However, such theoretical predictions do not, at present, consider the full complexity of the real world. There is no substitute for careful measurements.

Table 3. Diffusion characteristics (D_o, E) and closure temperature (T_o; Eqn. 9) for minerals useful in stable isotope thermometry. Cooling rate is varied from 1°C/Ma to 106°C/Ma for a mineral radius of 0.1 mm.*

Mineral	Orientation	P(H2O)	Reference	D_o cm²/s	E KJ/mol	A	T_c °C * 1C/Ma	T_c °C * 10²C/Ma	T_c °C * 10⁴C/Ma	T_c °C * 10⁶C/Ma
OXYGEN										
anorthite	~isotropic	1 Kb	Giletti et al. 1978	1.39E-07	110	55	151	219	312	445
anorthite		dry	Elphick et al.	9.00E-06	234	55	515	624	766	961
anorthite	//(010)	dry	Ryerson+McKeegan 1994	8.40E-09	162	55	402	520	685	931
apatite	//C	1 Kb	Farver and Giletti 1989	9.00E-05	205	55	379	464	573	719
biotite	//C	1 Kb	Fortier and Giletti 1991	9.10E-06	142	8.7	237	312	413	553
calcite	~isotropic	1 Kb	Farver 1994	7.00E-05	173	55	283	356	451	578
calcite	~isotropic	dry	Labotka et al. 2000	7.50E-03	242	55	417	497	597	727
corundum	//C		Cawley 1984	1.51E+01	527	55	989	1111	1258	1439
diopside	//C	dry	Ryerson+McKeegan 1994	4.30E+00	457	55	851	963	1099	1267
diopside	//C	1 Kb	Farver 1989	1.50E-06	226	55	526	642	797	1011
garnet	isotropic	wet	Coghlan 1990	6.00E-05	301	55	685	810	971	1185
hornblende	//C	1 Kb	Farver and Giletti 1985	1.00E-07	172	8.7	426	545	711	954
kyanite		1 Kb	est. *	8.60E-06	363	8.7	997	1180	1422	1757
magnetite	isotropic	1 Kb	Giletti and Hess 1988	3.50E-06	188	55	380	474	597	767
Mg-spinel	isotropic		Ryerson+McKeegan 1994	2.20E-03	404	55	898	1036	1210	1435
muscovite	//C	1 Kb	Fortier and Giletti 1991	7.70E-05	163	8.7	277	353	453	588
phlogopite	//C	1 Kb	Fortier and Giletti 1991	1.40E-04	176	8.7	308	387	491	630
quartz	//C	1 Kb	Dennis 1984a	2.00E-07	138	8.7	282	375	503	691
quartz		dry	Dennis 1984b	3.00E-07	222	8.7	594	735	928	1207
quartz	//C	dry	Sharp et al. 1991	2.10E-08	159	8.7	411	534	709	973
richterite	//C	1 Kb	Farver and Giletti 1985	3.00E-04	239	55	495	596	727	903
rutile	//C	1 Kb	Moore et al. 1998	5.90E-01	330	8.7	611	707	826	976
sillimanite		1 Kb	est. *	6.20E-07	255	8.7	696	850	1058	1355
sphene	isotropic	1 Kb	Morishita et al. 1996	1.00E-04	254	55	528	632	765	942
tremolite	//C	1 Kb	Farver and Giletti 1985	2.00E-08	163	8.7	429	556	736	1008
zircon	~isotropic	70-700 b	Watson+ Cherniak 1997	5.50E-08	210	55	546	679	861	1124
zircon	~isotropic	dry	Watson+ Cherniak 1997	1.33E+00	448	55	856	971	1112	1287
zircon		empirical	Peck et al. 2001				>700	>700	>700	>700
CARBON										
calcite	~isotropic	0.02-24 b	Kronenberg et al. 1984	7.00E+02	364	55	544	618	706	813
calcite	~isotropic	dry	Labotka et al. 2000	7.77E-09	166	55	420	541	711	964
graphite	~isotropic		Thrower+ Mayer 1978	9.10E-01	651	55	1367	1534	1738	1993

*Footnotes: Note that cooling rate scales inversely to radius squared in Eqn. (9). Thus, Tc for 10⁶C/Ma and 0.1mm is equivalent to Tc for 10⁴C/Ma and 1mm, 10²C/Ma and 1cm, etc.

See Cole and Chakraborty (2001) for a discussion of alternate calibrations of D_o and E.

Do and A estimated by the technique of Fortier and Giletti 1989 (Ghent and Valley 1998).

Aluminosilicate-quartz

Assemblages of quartz plus kyanite, sillimanite or andalusite have yielded exceptionally precise and apparently accurate temperatures when applied as a RAM thermometer. Diffusion of oxygen in aluminosilicates is very slow, yielding $T_C > 800°C$ for moderate to coarse grain sizes, while diffusion is comparatively fast in quartz (Table

3). The effects of solid solution are generally nil and the temperature coefficient relative to quartz is moderate (Zheng 1993b, Sharp 1995, Tennie et al. 1998).

Putlitz et al. (2001) analyzed coarse kyanite from deformed quartz veins in pelites from the kyanite (+sillimanite) zone on the island of Naxos, Greece (Table 4). Six samples from an outcrop test within 100m at Stavros yielded highly precise fractionations (±0.06 ‰) and temperatures in excellent agreement with published estimates for coexisting kyanite + sillimanite. Temperatures above $T_C(Qt)$ are preserved because there is not another low T_c mineral in the veins. The host pelitic rocks at Stavros also contain quartz and kyanite; however, Δ(Qt-Ky) is larger in pelites (3.03 vs. 2.62 ‰) due to diffusive exchange between quartz, feldspars, and micas yielding reset temperatures that are 65°C lower. Fibrolitic sillimanite was also analyzed from three samples at Stavros, yielding about the same average temperature, but more variable results (634-726°C), possibly reflecting exchange due to the fine grain size. Samples of Qt-Ky from other localities accurately reproduce the petrologic temperatures. These accurate and precise results demonstrate the potential of RAM thermometry.

Table 4. Aluminosilicate- quartz thermometry in quartzites, pelites, and quartz veins.

Ref.	Location	*Average* Δ(Qt-AS)	n	*Qt-AS* Ave T*	*Independent* T estimate	
1.	**Naxos, Greece, deformed quartz veins, coarse kyanite**					
	Stavros	2.62±0.06	6	659±11	660	
	Komiaki	2.48	1	685	670	
	Sifones	2.64	1	656	630	
	Moni	2.65	1	649	630	
	Appollon	2.76	1	635	630	
	Appollon Village	2.77	1	634	620	
2.	**Mica Creek, BC, pelites and quartz nodules**					
	Qt-Ky, nodule	2.55	2	665±10	645	Ky-Si zone
	Qt-Ky, pelite	3.0±0.24	4	596±35	637±28	St-Ky zone
	Qt-Ky, pelite	2.9±0.25	3	605±38	605±26	Ky-Si zone
	Qt-Si, pelite	2.7±0.25	3	648±43	702±23	Si zone

References: **1**. Putlitz et al. (2001); **2**. Ghent and Valley (1998).
* T calculated from Sharp (1995).

In an experimental study, Tennie et al. (1998) challenge the interpretations of Sharp (1995) and Ghent and Valley (1998) for aluminosilicate-quartz thermometry. They propose that the refractory nature of kyanite prevents metamorphic equilibration and makes kyanite generally unfit for thermometry. They estimated $A_{Qt-Ky} = 3.00$ (Eqn. 4) based on 11 of 17 piston cylinder experiments for calcite-kyanite exchange, and analysis using externally heated nickel reaction vessels. This calibration yields higher temperatures than the A value empirically estimated by Sharp (A = 2.25, 1995). In many instances, temperatures based on A = 3.00 are in significant disagreement with independent estimates. For instance, at Stavros (Naxos, Table 4), the average temperature is raised from 659 to 797°C. At 797°C, there should be widespread melting in these water-rich metasediments, but no evidence of *in situ* melting is observed on Naxos except at much higher grade. Tennie et al. ascribe such discrepancies to slow diffusion in kyanite (see Table 3) which could prevent exchange and cause kyanite to preserve $\delta^{18}O$ from the lower temperatures of first crystallization rather than the peak of metamorphism. However, this explanation would yield Qt-Ky temperatures in quartzite that are too low rather than too high as is observed, and it does not explain the excellent agreement of results from six different hand samples in the outcrop test at Stavros (Table 4).

Discrepancies between empirical and experimental calibrations exist for several systems, including quartz-alumino-silicate. In this case, one should ask if an empirically derived thermometer falsely seems to record accurate temperature because the rocks used for calibration were retrogressed by an equal amount as those being studied, or if all rocks really preserve equilibrium compositions. This distinction is important because a non-equilibrated system cannot be relied upon and unequilibrated apparent temperatures may actually be controlled by other variables such as fluid composition, time, or deformation. Conversely, if careful thermometry yields self-consistent results in agreement with other systems, as concluded by Sharp (1995), Ghent and Valley (1998), Vannay et al. (1999), and Putlitz et al. (2001), then it may be that the experiments are in error.

Values of $\Delta(Qt-Al_2SiO_5)$ can yield more than temperature information. Larson and Sharp (2000) analyzed coexisting quartz + sillimanite + kyanite + andalusite to show that "triple-point" assemblages from New Mexico did not form in equilibrium. Texturally equilibrated quartz + andalusite + sillimanite from the Front Range, Colo, yield estimates of pressure as well as temperature (Cavosie et al. 2000). Quartz + kyanite and quartz + sillimanite pairs from British Columbia yield precise temperatures from quartz nodules, but lower reset temperatures from assemblages in poly-mineralic pelites (Table 4). Fast Grain Boundary diffusion calculations based on the difference between RAM and pelite temperatures suggest that water activity was low during slow retrograde cooling (Ghent and Valley 1998).

Other applications of quartz-aluminosilicate fractionations include migmatitic, amphibolite or granulite facies gneisses (van Haren et al. 1996, Kohn et al. 1997, Vannay et al. 1999, Moecher and Sharp 1999), and eclogites (Sharp et al. 1992, 1993; Rumble and Yui 1998, Zheng et al. 1998, 1999). Moecher and Sharp (1999) compared the results of aluminosilicate-quartz and aluminosilicate-garnet thermometry in pelites and report variable retrograde resetting as predicted by diffusion modeling.

Magnetite-quartz

The magnetite-quartz pair is the most commonly applied oxygen isotope thermometer. It is very promising as a RAM thermometer below $T_C(Mt)$ or when used in low to moderate grade metamorphic, rapidly cooled, or very dry rocks. It is frequently discussed in reviews of isotope thermometry (O'Neil and Clayton 1964, Chiba et al. 1989, Gregory et al. 1989, Zheng and Simon 1991). The fractionation is large yielding good temperature sensitivity and the effects of solid solution and crystal chemistry are relatively small (Ti, Fe^{3+} in magnetite, Bottinga and Javoy 1975, Zheng and Simon 1991; SiO_2 polymorphs, Kawabe 1978, Zheng 1993b; spinel structure, Zheng 1995).

Rumble (1978) measured Mt-Qt fractionations from eight amphibolite facies quartzites on the summit of Black Mountain, New Hampshire. These closely-spaced samples were subjected to the same P-T-time conditions. All but one sample contains at least 80% quartz and smaller amounts of magnetite. Coexisting minerals indicate a peak metamorphic temperature of 530°C: kyanite + staurolite + chloritoid + chlorite + muscovite + quartz + magnetite ± ilmenite ± garnet. The fractionations are self-consistent (9.46±0.26‰) yielding nearly parallel tie lines (Fig. 14). The oxygen isotope temperatures range from 530-561°C (542±11, Table 5). This study is an outcrop test and demonstrates the accuracy and precision obtainable from Mt-Qt when applied to appropriate rocks for a RAM thermometer.

The magnetite/hematite-quartz thermometer has been extensively applied to banded

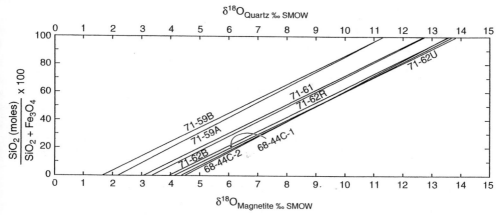

Figure 14. Plot of $\delta^{18}O$(magnetite) vs. $\delta^{18}O$(quartz) for amphibolite facies quartzites from Black Mtn., New Hampshire. The nearly parallel tie-lines indicate that fractionations are self-consistent for these eight samples from one outcrop. The average oxygen isotope temperature of 542°C is in excellent agreement with that from petrology (530°C). (from Rumble (1978).

Table 5. Comparison of $\Delta^{18}O$(Qt-Mt) temperature estimates from quartzite, granitic gneiss, and banded iron formation. Amphibolite facies Black Mtn. quartzite yields accurate and precise RAM temperatures. Granulite facies temperatures are often low relative to petrologic estimates.

Reference, Location	*Average* $\Delta(Qt\text{-}Mt)$	n	*Ave* T^a °C	*Comments*
1. Shuksan, N. Cascades	13.8±1.50	6	402±37	BSF, lawsonite, pumpellyite
2. Black Mtn., N.H., Qtzt,	9.46±0.26	8	542±11	AF, 530°C, 500 × 100 m, RAM
3. Isua, SW Greenland	11.62±0.42	12	463±13	AF, BIF, 1 × 2 km
4. Minas Gerais, Brazil	10.79±0.39	16	491±14	GSF, BIF, s_1 only[b]
	8.14±0.77	38	606±40	AF, BIF, s_1 only[b]
	6.32±0.41	9	724±33	GF, BIF, s_1 only[b]
	8.89±2.47	16	568±88	GF, BIF, $s_1+s_2+s_3$[c]
5. Ruby Range, Mont., Kelly	8.42±0.19	3	591±10	GF, 745±50°C, BIF, 1 km²
Carter Creek	9.67±0.43	3	533±18	GF, 675±45°C, BIF, 9 km²
6. Adirondack Mts, unsheared	6.93±0.58	11	680±38	GF, 675-700°C, GG
sheared	8.8±1.22	11	603±71	GF, 675-700°C, GG
7. Wind River Range, Wyoming	9.24±0.93	16	552±44	GF, 750±50°C, BIF

References: 1. Brown and O'Neil (1982), 2. Rumble (1978); 3. Perry et al. (1978); 4. Muller et al. (1986b); 5. Dahl (1979); 6. Cartwright et al. (1993); 7. Sharp et al. (1988); Sharp and Essene (1991).

Footnotes: [a] calculated using the calibrations of Chiba et al. (1989); [b] data only for samples of magnetite/hematite showing primary schistosity s_1; [c] data for samples of magnetite/hematite with variable schistosity s_1, s_2, or s_3

Abbreviations: AF = amphibolite facies; BIF = banded iron formation; BSF = blueschist facies; GF = granulite facies; GG = granitic gneiss or charnockite; GSF = greenschist facies; RAM = refractory accessory mineral thermometer; Qtzt = quartzite.

iron formations in order to determine conditions of deposition, diagenesis and metamorphism: Animikie Basin, U.S. (Clayton and Epstein 1958, James and Clayton 1962); Duluth gabbro contact aureole, U.S. (James and Clayton 1962, Perry and Bonnichsen 1966); Hamersley Basin, Western Australia (Becker and Clayton 1976); Isua, SW Greenland (Perry et al. 1978); Krigoy Rog, Ukraine (Perry and Ahmad 1981); Mesabi

Range, U.S. (James and Clayton 1962, Perry et al. 1973); Minas Gerais, Brazil (Hoefs et al. 1982, Muller et al. 1986a,b); Ruby Range, Montana (Dahl (1979); Urucum area, Brazil (Hoefs et al. 1987); and the Wind River Range, Wyoming (Sharp et al. 1988, Sharp and Essene 1991). These data include samples from different metamorphic grades: unmetamorphosed to sub-greenschist (Animikie, Hamersley, Mesabi, Urucum); greenschist facies (Krivoy Rog, Minas Gerais); amphibolite facies (Isua, Minas Gerais); granulite facies (Minas Gerais, Ruby Range, Wind River); and contact metamorphism (Duluth gabbro aureole). Reexamination of these samples, using more recent criteria for thermometry, can be expected to yield improved accuracy. The temperatures found by these studies should be recalculated with newer experimental data (Chiba et al. 1989, Clayton et al. 1989). Furthermore, it is possible that isotope heterogeneity exists at the scale of the large samples analyzed, and that equilibrated domains can be identified and analyzed with microanalysis.

Magnetite-quartz thermometry has also been applied in igneous rocks (Hildreth et al. 1984, Bindeman and Valley 2001) and high grade gneisses (Fourcade and Javoy 1973, Shieh and Schwarcz 1974, Shieh et al. 1976, Li et al. 1991, Cartwright et al. 1993, Hoffbauer et al. 1994, Farquhar et al. 1996), blueschists (Brown and O'Neil 1982), and greenschist or amphibolite facies pelites (Schwarcz et al. 1970, Hoernes and Friedrichsen 1974, Kerrich et al. 1977, Goldman and Albee 1977).

There are several cautionary notes for thermometry involving Fe-Ti oxides and quartz. Minerals may differentially exchange with circulating fluids as clearly demonstrated by Gregory et al. (1989) with $\delta^{18}O(Qt)$ vs. $\delta^{18}O(Mt)$ plots. For Hamersley and Mesabi, it is proposed that sedimentary Fe-oxides exchanged with fluids during recrystallization to magnetite, while quartz remained unaffected. The calibration of $1000\ln\alpha^{18}O(Qt-Mt)$ vs. T is uncertain at low temperatures. Values of $\Delta(Mt-hematite)$ and $\Delta(ilmenite-Mt)$ are small, but not insignificant (Bottinga and Javoy 1975, Zheng 1991, Zheng and Simon 1991). In Ti- or Fe^{+3}-rich oxides, exsolution effects are minimized for lower grade rocks, but may be significant in granulites (Bohlen and Essene 1977, Farquhar and Chacko 1994). Fine grain size has sometimes precluded complete mineral separation and various projection schemes have been applied to banded iron formation (see, Yapp 1990). The moderate diffusion rates of oxygen in both quartz and magnetite, and measured fractionations indicate that temperatures will be reset above T_C, for granulite and upper amphibolite facies rocks (Chiba et al. 1989, Valley and Graham 1991, Sharp 1991, Sharp and Essene 1991, Eiler et al. 1993, 1995a,b). In addition to diffusive processes that operate in all samples, both magnetite and quartz can exchange $\delta^{18}O$ by recrystallization or crack-healing (Valley and Graham 1993, 1996; Eiler et al. 1995a, Sitzman et al. 2000). Likewise, different generations of quartz are recognized from textures in Archean Onvervacht cherts and small differences in $\delta^{18}O$ have been measured (Knauth and Lowe 1978).

In spite of the many potential pitfalls, Mt-Qt is an accurate and reliable thermometer when carefully applied to appropriate rocks in an appropriate temperature regime. Samples should be selected with regard to diffusive exchange, discussed above. Temperature estimates are consistent with petrology for blueschist, greenschist and lower amphibolite facies samples in Table 5. In spite of the moderate T_C for both quartz and magnetite (Table 3), some granulite facies gneisses preserve peak metamorphic temperatures (Muller et al. 1986b, Cartwright et al. 1993) and others yield temperatures that are reset, but higher than T_C (Sharp et al. 1988). However, most granulite and upper amphibolite facies samples yield temperatures that are too low in comparison to petrologic thermometers, though reset results can be precise (Table 5). The variable retention of peak temperature is predicted if water fugacity is low (promoting slow diffusion) in some rocks *during cooling*, but high in others (Sharp et al. 1988, Cartwright et al. 1993, Edwards and Valley 1998). Improved results have been obtained through careful attention to rock and mineral fabric in regionally metamorphosed terranes (Muller et al. 1986a, Sharp et al. 1988, Cartwright et al. 1993). Crystal dislocations in magnetite that can facilitate retrograde oxygen exchange are easily seen in polished thin sections that have been etched in HCl (Valley and Graham 1993, Eiler et al. 1995b, Sitzman et al. 2000). Healed microfractures in quartz, if present, are often imaged by cathodoluminescence using a SEM or electron microprobe, though sensitivity varies greatly with instrument and operating conditions (Valley and Graham 1996).

Rutile-quartz

The Ru-Qt thermometer has the same theoretical advantages as Mt-Qt. The temperature coefficient of fractionation is large and solid-solution is minor. The greatest limitation of this system may be the rate of oxygen diffusion (Table 3).

Quartz-rutile pairs have been measured most commonly from eclogites (Vogel and Garlick 1970, Desmons and O'Neil 1978, Matthews et al. 1979, Agrinier et al. 1985, Agrinier 1991, Sharp 1995, Rumble and Yui 1998, Zheng et al. 1998, 1999). A few analyses are reported for blueschists and pelites (Matthews and Schliestedt 1984, Sharp 1995), and nelsonite (Addy and Garlick 1974). Vogel and Garlick (1970) report very high precision: Δ(Qt-Ru)= 6.46±0.05‰ for 5 unrelated type B eclogites (609±4°C, Matthews 1994) and 7.30‰ for one type C (556°C). However, other studies are less sucessful. Sharp (1995) reports analyses from five granulite facies samples from terranes where the reported temperatures average 755°C and Δ(Qt-Ru) yields average temperatures of 567±150°C. The only granulite that yields above 700°C is a rapidly quenched xenolith from La Joya Honda maar. Plots of Δ(Qt-Ru) vs. Δ(Qt-garnet) and Δ(Qt-Ru) vs. Δ(Qt-kyanite) for the data referenced above show considerably larger scatter than Δ(Qt-garnet) vs. Δ(Qt-kyanite) for the same rocks, suggesting that in spite of the smaller temperature sensitivity, these other pairs may be more reliable. The effect of diffusion cannot be predicted for most of these samples without information on modes and grain size that is not published. Furthermore, rutile has the unusual property that hydrous experiments yield slower diffusion coefficients than dry experiments in the same lab (Moore et al. 1998). Since quartz is a minor phase in many of these mafic rocks, it is likely that variable resetting to lower temperatures has occurred by exchange with micas or chlorite. In a few samples with high Qt-Ru temperatures, it is possible that retrograde fluids were present.

Rutile is common in certain quartz veins, quartzites, and some massif-type anorthosites. These assemblages should be sought as a test of the RAM thermometers. Within the temperature range dictated by diffusion in rutile (Table 3) it is predicted to be highly accurate and precise.

CARBON ISOTOPE THERMOMETRY

Calcite-graphite

Cc-Gr is the most commonly applied RAM thermometer (Table 6). The percentage of graphite is generally less than 1% in marble, other carbon-bearing phases such as scapolite or dolomite are usually minor in abundance, and the diffusion rate of carbon is very slow. Furthermore, Δ(Cc-Scap) and Δ(Cc-Dol) are small at high temperatures, minimizing the effect of neglecting small amounts of these minerals. With routine care, this is an easily applied and accurate thermometer for high-grade marbles.

A number of empirical, experimental or theoretical calibrations of $\Delta^{13}C$(Cc-Gr) vs. T

Table 6. Comparison of $\Delta^{13}C(Cc-Gr)$ temperature estimates (°C) from granulite and amphibolite facies terranes.

Ref.	Location	Average $\Delta(Cc-Gr)$	n	Cc-Gr K+V	Ave T's[b] D+V	Independent estimate T °C[c]
1.	Anabar Shield, Russia	3.83	4	691	677	<850-950 gf
2.	Bohemian Massif	4.81	9	590	613	700-780, 530-620
3.	Central Adirondack Mtns.	3.44[a]	38	744	707	675-775 gf
	near anorthosite	3.02	17	813	743	polymetamorphic[d]
4.	Central Adirondack Mtns	3.49	10	737	703	675-775 gf
5.	NW Adirondacks	4.08	89	661	659	<675 af
6.	Cucamonga terrane, California	3.54	10	746	705	710-820 gf
7.	Franklin marble, amph. facies	3.60	3	722	695	af
	gran. facies	3.09	3	801	737	gf
8.	Gour Oumelalen, Algeria	3.45	10	743	706	≥710 gf
9.	Hida Belt, Japan	3.46	5	742	706	polymetamorphic[d]
10.	Ivrea Zone, Italy	4.9-2.3	14	580-970	600-800	
11.	Kamioka area, Japan	3.54	40	729	699	
12.	Kerala Belt, S. India	3.33	10	760	716	650-750 gf
13.	Kurobegawa area, Japan	5.77	9	513	560	St+Ky af
14.	Lutzow Holm Bay, E Antarctica	2.85	2	845	759	700-800 gf
15.	Madurai, southern India	2.67	18	881	776	Sap+Qt gf
16.	Panamint Mtns, marbles	6.40	17	473	530	400-650 af
	schists	8.78	18	364	441	400-650 af
17.	Sanbagawa terrain, Japan	8.37	15	379	455	500 Gt-zone af
18.	Skallen marble, E. Antarctica	2.85	5	844	758	760-830 gf
19.	Southern Grenville Province, ON	6.88	31	447	510	500-650 af
20.	Sri Lanka, SW Group marbles	4.49	14	618	632	600-750 gf
21.	Sri Lanka, granulite facies	3.79	19	696	680	gf
	altered calcite	2.93	6	829	751	disequilibrium
22.	Tudor Gabbro aureole, <1km	4.31	17	636	644	polymetamorphic
	>1km	7.32	8	424	492	490 lower af
23.	Madurai, Tamil Nadu	4.06	3	662	660	gf

References: 1. Galimov et al. (1990); 2. Schrauder et al. (1993); 3. Valley and O'Neil (1981), Kitchen and Valley (1995); 4. Weis et al. (1981); 5. Kitchen and Valley (1995), Gerdes and Valley (1994); 6. Morrison and Barth (1993); 7. Crawford and Valley (1990); 8. Pineau et al. (1976); 9. Arita and Wada (1990); 10. Baker (1988); 11. Wada (1977); 12. Satish-Kumar et al. (1997); 13. Wada (1977); 14. Satish-Kumar et al. (1998); 15. Satish-Kumar (2000); 16. Bergfeld et al. (1996); 17. Wada et al. (1984); 18. Satish-Kumar and Wada (2000); 19. Rathmell et al. (1999); 20. Elsenheimer (1988); 21. Hoffbauer and Spiering (1994); 22. Dunn and Valley (1992); 23. Pandey et al. (2000).

Footnotes: [a] does not include zoned, polymetamorphic graphites; [b] calculated using the calibrations of Kitchen and Valley (1995) and Dunn and Valley (1992); [c] some terranes contain gradients in metamorphic temperature and these differences are averaged; [d] graphites from plutonic contacts are zoned in $\delta^{13}C$.

Abbreviations: af = amphibolite facies, gf = granulite facies, Gt = garnet, Ky = kyanite, Qt = quartz, Sap = sapphirine, S t= staurolite

have been proposed (Fig. 15). Kitchen and Valley (1995) compared 38 measured fractionations with petrologic thermometry for granulite facies metamorphism of the central Adirondack Highlands, avoiding complications due to contact metamorphism, to derive:

$$\Delta^{13}C\ (Cc-Gr) = 3.56 \times 10^6 / T^2\ (K) \tag{10}$$

At high temperatures, >600°C, independent petrologic thermometry is in excellent agreement with this calibration (Table 6) and with that of Chacko et al. (1991, CMCG). The theoretical calculations of Polyakov and Kharlishina (1994, 1995) yield a slightly smaller partition function ratio than used by Chacko et al. (1991) and suggest a small

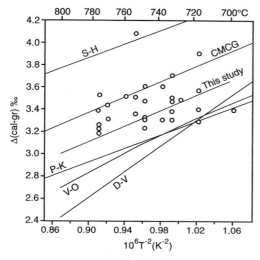

Figure 15. Temperature dependence of $\Delta^{13}C(Cc\text{-}Gr)$ as determined by various workers: This Study = Kitchen and Valley (1995); CMCG = Chacko et al. (1991); S-H = Scheele and Hoefs (1992); P-K = Polyakov and Kharlishina (1995); D-V = Dunn and Valley (1992); V-O = Valley and O'Neil (1981). Data points are for granulite facies marbles from the Adirondacks that were sampled over 2 km away from contacts with massif-type anorthosite (from Kitchen and Valley 1995).

pressure correction. The effect of these changes is to reduce slightly the CMCG calibration (R Clayton, pers. comm. 9-94). The experimental calibration of Scheele and Hoefs (1992) leads to temperature estimates that are unreasonably high and which would promote wholesale melting in several granulite facies terrains (Morrison and Barth 1993, Kitchen and Valley 1995, Satish-Kumar and Wada 2000, Satish-Kumar 2000).

The empirical calibration of Dunn and Valley (1992) is fit to data including lower temperature amphibolite facies marbles:

$$\Delta^{13}C \ (Cc\text{-}Gr) = 5.81 \times 10^6 / T^2 - 2.61 \ (K) \tag{11}$$

This is a significantly steeper slope that implies curvature at higher temperature in order to pass through the origin (Fig. 15). Agreement is excellent with the other calibrations at 600-700°C, but temperatures from (11) are increasingly higher than those from (10) below 650°C. Several studies at lower temperatures show that petrologic thermometry agrees better with the results of (11) (Wada and Suzuki 1983, Wada et al. 1984, Bergfield et al. 1996). Two interpretations are possible: there is a significant change in slope that is not predicted by theory, or lower temperature samples used for empirical calibration are not fully equilibrated (see, Kitchen and Valley 1995). Isotopic disequilibrium is well known below 500-600°C, so this possibility cannot be discounted. Regardless of whether the Dunn and Valley calibration is a true thermometer based on equilibrium or a pseudo-thermometer that is dependent on reaction kinetics of graphitization, it is the calibration that has yielded the most reasonable temperature estimates below 600°C (Table 6).

Graphite crystallinity and morphology

Graphite can form by a variety of processes. Commonly, graphite in marbles is derived from sedimentary organic matter that has undergone progressive maturation during metamorphism. This process begins with diagenesis and may not be complete until above 500-600°C (see Dunn and Valley 1992). Changes due to maturation include: the coalescence and coarsening of single crystals (flakes) of graphite; decrease in the elemental ratio, $(N+H+S)/C$; sharpening and shifts of X-ray diffraction peaks; shifts in Raman spectra (Luque et al. 1998), and increased vitrinite reflectance. High resolution transmission electron microscopy reveals that in the final stages of graphitization (400 to

>500°C), crystallinity increases at the atomic-scale from moderately well-organized rims on grains with immature cores to well-organized crystalline flake graphite (Buseck and Huang 1985).

Graphitization in marble is also accompanied by carbon isotope exchange. Carbon diffusion in graphite is very slow (Thrower and Mayer 1978) and $T_C \gg 800°C$ (Table 3). High T_C is further indicated by isotopically zoned flakes of graphite that were not homogenized by granulite facies metamorphism (Kitchen and Valley 1995, Farquhar et al. 1999, Satish-Kumar et al. 2000). Thus, carbon isotope exchange between graphite and carbonate is sluggish, rate-limited by graphite, and can only occur upon crystallization or recrystallization of graphite. In contrast, carbon diffusion in calcite is relatively rapid. Accordingly, Cc-Gr thermometry should only be attempted with visible flakes of graphite and the presence of growth-zoning within flakes should be evaluated.

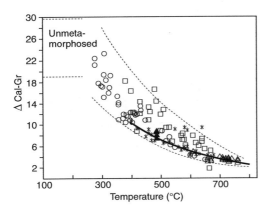

Figure 16. Measured values of Δ(Cc-Gr) versus metamorphic temperature independently determined by petrologic thermometers. Data are from: Valley and O'Neil (1981), open triangles; Wada and Suzuki (1983), asterisks; Kreulen and van Beek (1983), squares; Morikiyo (1984), circles; and Dunn and Valley (1992), solid triangles and heavy curved line. The unmetamorphosed calcite-organic matter fractionation is from Eichmann and Schidlowski (1975). Samples show a successive approach to equilibrium at higher temperatures rate-limited by slow exchange in graphite (from Dunn and Valley 1992).

A number of studies have measured Cc-Gr fractionations that appear too large in relation to Equation (10), resulting in temperatures that are lower than expected (Hoefs and Frey 1976, Kreulen and van Beek 1983, Wada and Suzuki 1983, Morikiyo 1984, Wada et al. 1984, Schrauder et al. 1993, Bergfield et al. 1996). These studies support the hypothesis that either the calibration has a steeper slope at low temperatures than is shown in Figure 15 or that the samples are not fully equilibrated. The data in Figures 16 and 17 show that exchange begins at 300°C, but that below 500-600°C, large values of $\Delta^{13}C$(Cc-Gr) are seen that are highly variable indicating incomplete exchange. Equilibrium is not assured below 650°C. Dunn and Valley (1992) found that samples with flakes of graphite (vs. less mature carbonaceous material) plot along the bottom of this envelope of fractionations (Fig. 16) and they suggested that the lowest values were equilibrated. Accordingly, the heavy line in Figure 16 is their calibration (Eqn. 11). For these reasons, it is unlikely that values of $\Delta^{13}C$(Cc-Gr) > 8 ‰ represent equilibrium and values between 5 and 8 ‰ should be interpreted with extra caution.

Once coarsening and recrystallization stops, graphite flakes will preserve isotope composition due to slow diffusion. Intracrystalline zoning of $\delta^{13}C$ has been detected in some crystals by most studies that have evaluated it. This has been demonstrated by delamination of single flakes at the 20-100 μm-scale (Wada 1988, Arita and Wada 1990, Kitchen and Valley 1995, Satish-Kumar 2000), by ion microprobe (Farquar et al. 1999) or, more simply, by analysis of large vs. small flakes from the same rock (Kitchen and Valley 1995).

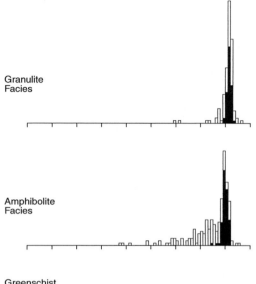

Granulite
Facies

Amphibolite
Facies

Greenschist
Facies

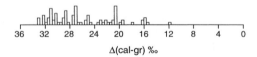

Unmetamorphosed

Δ(cal-gr) ‰

Figure 17. Carbon isotope fractionations as a function of metamorphic grade. (A, left) Compilation of measured values of Δ(Cc-Gr) shows scatter and disequilibrium at greenschist facies and lower temperature conditions. Many amphibolite facies and most granulite facies samples show a tight clustering of values consistent with isotope equilibration above 600°C. Values in black are from the Adirondacks (from Kitchen and Valley 1995). (B, below) Values of $\delta^{13}C$ for a Liassic black shale formation (Hoefs and Frey 1976) showing successive approach to equilibrium at maximum T = 500-600°C (from Sharp et al. 1995).

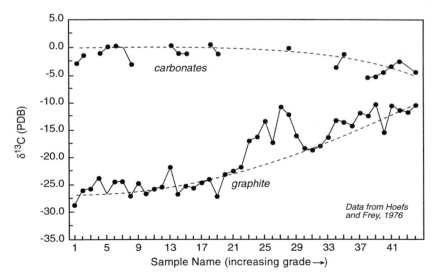

Graphite inclusions wholly within silicate minerals have also been found to have slightly lower $\delta^{13}C$ than graphite in the calcite matrix suggesting that they were armored against prograde exchange (Wada and Suzuki 1983). This effect will not affect thermometry if care is taken to only select graphite that is disaggregated by acid dissolution of carbonate. Conversely, the analysis of selected inclusions from different minerals, or from different positions within a zoned mineral, may provide temperature history of prograde mineral growth.

Skeletal graphite or flakes with etched or pitted surfaces are frequently seen, suggesting retrograde recrystallization. Skeletal graphites with larger $\Delta^{13}C(Cc-Gr)$ than metallic flakes support this interpretation (Weis 1980, Arita and Wada 1990). Van der Pluijm and Carlson (1989) determined temperatures were less than 450°C during mylonitization along the Bancroft shear zone in the Grenville Province by reequilibration of $\Delta^{13}C(Cc-Gr)$ in 650-700°C marbles. Likewise, graphite spheres with radial texture have been described in nearby marble and Cc-Gr thermometry suggests lower temperature growth (Jaszczak and Robinson 1998; S. Dunn, unpublished 1999). Samples of charnockite from southern India have been shown to contain three generations of graphite varying by up to 10‰ in $\delta^{13}C$ by ion microprobe (Farquhar et al. 1999).

Carbonate/graphite ratio

Accurate thermometry is most likely if the ratio of carbon-in-carbonate to carbon-in-graphite is high. The mode effects on diffusive resetting, described above, are a criterion for RAM thermometry. Also, retrograde precipitation of carbonate is common and may be more significant in carbonate-poor lithologies. Considering the concentration of carbon in each mineral, 100% in graphite vs. 12 wt % in calcite, mode ratios on the order of 100/1 for carbonate/graphite are prudent. This is seldom a concern in marbles where 0.1 to 1.0 vol % graphite is typical, but care should be considered for graphite-rich or carbonate-poor lithologies.

Elsenheimer (1988, 1992) analyzed 11 Cc-Gr pairs from calcite-poor calcsilicates (<3% calcite) in the same region as the 14 marbles reported in Table 6 (#20). The average temperature was the same in each case. However, the marbles yielded a reasonable range from 510 to 740°C while the calcsilicates yielded many unreasonable temperatures from 400 to 1130°C. Likewise, Bergfield et al. (1996) found that Cc-Gr temperatures were lower and more variable in schists than marbles from the Panamint Range. In these two studies, the calcsilicates and schists are graphite-rich and calcite-poor, violating the criteria for a RAM thermometer, and only the temperatures from marble were judged to represent the peak of metamorphism.

Contact and polymetamorphism

There have been a number of applications of Cc-Gr thermometry to contact metamorphism. Wada (1978) studied Cc-Gr pairs in skarns from the Kamioka Mining District and found that graphite in the skarn was similar in $\delta^{13}C$ to that of the country rocks. However skarn calcite was 5-10 ‰ lower than the equilibrium value with graphite and gradients of up to 10 ‰/2cm were measured in $\delta^{13}C(Cc)$. Wada and Suzuki (1983) found a good correlation of $\Delta^{13}C(Cc-Gr)$ and temperature from 400 to 680°C, however, as discussed for (11), it is not certain if this correlation is controlled by equilibrium or kinetics.

Arita and Wada (1990) studied the effects of contact metamorphism overprinted on granulite facies marbles of the Hida belt and found that the cores of graphite and calcite retained granulite facies fractionations, but that profound gradients in composition and disequilibrium existed from core to rim within single crystals.

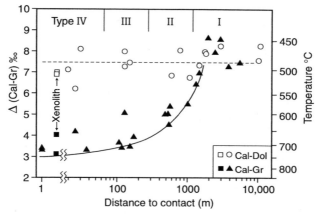

Figure 18. The results of calcite-graphite isotope thermometry (filled symbols) and calcite-dolomite solvus thermometry (open symbols) plotted against distance to the Tudor Gabbro contact, Ontario. The solvus thermometry is reset by regional metamorphism at 490°C while the isotope thermometry preserves high temperatures due to earlier contact metamorphism. The solid curved line is the predicted thermal gradient for contact metamorphism (from Dunn and Valley 1992).

Two studies have demonstrated the preservation of high contact metamorphic temperatures adjacent to mafic plutons in terranes that subsequently experienced lower temperatures during regional metamorphism. Cc-Gr temperatures in marble smoothly increase from 500 to 700°C approaching the Tudor gabbro, Ontario. The increase in temperature correlates with a change from fine grained gray marble to coarser white marble and with mineral isograds (Dunn and Valley 1992, 1996). However, the temperatures recorded by calcite-dolomite solvus thermometry are uniform at 490°C (Fig. 18). Thus, the Cc-Gr thermometer has seen through the regional metamorphic overprint that reequilibrated the Mg content of calcite and dolomite. Kitchen and Valley (1995) delaminated single flakes of graphite in granulite facies marbles from the central Adirondacks and found gradients in $\delta^{13}C$ in crystals from within 10 km (ca. 0-1 km in 3-D) of exposed massif-type anorthosite. The cores of these grains suggest relict temperatures as high as 890°C while the rims record the subsequent regional metamorphism at 700-750°C.

Fluid flow

Calcite-graphite thermometry requires that carbon-bearing fluids have not infiltrated with contrasting $\delta^{13}C$. Conversely, $\Delta^{13}C$(Cc-Gr) can be used to infer fluid flow. Several studies have shown that calcite and/or graphite is zoned in $\delta^{13}C$ (Wada 1988, Arita and Wada 1990, Kitchen and Valley 1995, Satish-Kumar et al. 1998, Farquhar et al. 1999). Most commonly, this results from open system exchange with grain boundary fluids that have also caused a significantly larger zonation in $\delta^{18}O$ (Wada 1988, Graham et al. 1998, Satish-Kumar et al. 1998), indicating that fluids were H_2O rich. Likewise, if temperature is independently known, then discordant or variable Cc-Gr T's may reflect fluid infiltration. Conversely, systematic and self-consistent Cc-Gr thermometry in rocks of variable $\delta^{13}C$ demonstrates that significant amounts of carbon-rich fluid (CO_2 or CH_4) have not pervasively infiltrated large areas of granulite facies terrains (Adirondacks, Anabar Shield, Cucamonga Terrane in San Gabriel Mtns, E. Antarctica, Ivrea Zone, or southern India, see Table 6).

Other minerals

Some studies have included limited analysis of graphite pairs with other carbon-bearing minerals, including dolomite (Wada and Oana 1975, Wada 1977, Wada and Suzuki 1983, Kitchen and Valley 1995, Brady et al. 1998), siderite (Perry and Ahmad 1977), and cordierite (Vry et al. 1988, Fitzsimons and Mattey 1994). Moecher et al. (1992, 1994) estimate that $\Delta^{13}C$(Cc-scapolite) $\cong$ 0.1 ‰ in granulites, suggesting that small amounts of scapolite may not adversely affect Cc-Gr thermometry and that scapolite-graphite thermometry is possible in selected rocks.

Biogenic versus abiogenic graphite

The controversy over biogenic vs. abiogenic formation of graphite has a long history and implications for interpreting the earliest record of life on Earth and possibly Mars (Schidlowski 2001), subduction of sediments into the mantle (Kyser 1986, Schulze et al. 1997), and controversy over genesis of fossil fuels (Gold 1987, Valley et al. 1988). While low values of $\delta^{13}C$ are generally accepted as evidence for biogenic origin, this evidence can be obscured by metamorphic exchange in carbonate-bearing rocks. For instance, intermediate values of $\delta^{13}C$(Gr) = -9.3 to -10.7 in 3.8 Ga amphibolite facies meta-sediments from Isua, Greenland are interpreted as abiogenic and in equilibrium with coexisting siderite ($\Delta^{13}C$(Sid-Gr) = 5.8-6.0 ‰; Perry and Ahmad 1977). Subsequent studies of graphite in rocks that do not contain carbonate from Isua show a range of values to –28 ‰ that are interpreted as biogenic (Schidlowski et al. 1979, Hayes et al. 1983). These low values have been verified in related rocks by ion microprobe, strengthening the interpretation of earliest life (Mojzsis et al. 1996, Schidlowski 2001).

The initial sedimentary value of $\delta^{13}C$ of carbonaceous matter can be estimated from mass-balance even in marbles that have undergone inter-mineral metamorphic exchange at high temperature. Kitchen and Valley (1995) found a strong inverse correlation of graphite/calcite ratio and $\delta^{13}C$(Gr). They calculated that even though all graphite in their study now has $\delta^{13}C$ > -10 ‰, this value results from metamorphic exchange with calcite. In all rocks, the protolith carbonaceous matter is estimated at $\delta^{13}C$ < -25, indicative of a biogenic origin. Likewise, Morrison and Barth (1993) and Bergfield et al. (1996) show that mass-balance and exchange with carbonate controls $\delta^{13}C$(Gr) in marbles.

Adirondack Mountains—A case study

Graphite-bearing calcite marbles are common throughout the upper amphibolite facies NW Adirondack Lowlands and locally in the granulite facies central Adirondack Highlands. Figure 19A shows the results of Cc-Gr thermometry for 89 samples from the NW and 55 from the central Adirondacks (Kitchen and Valley (1995). All graphites were flakes freed by dissolution of carbonate. Five outcrop tests were made and reproducibility of temperatures is ±10-20°C except where calcites are heterogeneous (Table 2).

The 55 granulite facies samples include 17 that are from near the plutonic contacts of massif-type anorthosite bodies. These samples contain zoned graphite crystals with higher $\delta^{13}C$ cores that record the pre-regional metamorphism igneous contact temperatures.

The data for the NW Adirondacks show that the central zone of the region (B in Fig. 19B) experienced systematically lower temperatures, averaging 640°C and that temperatures increase systematically towards transitions to granulite facies conditions to the SE across the Carthage-Colton Line and to the NW across the St. Lawrence River. A temperature of 675°C is recorded for the orthopyroxene isograd to the SE. The two 675°C isotherms parallel the strike of dominant lithologies in the area and are normal to the NW vergence, in good agreement with recent petrologic thermometry (Liogys and

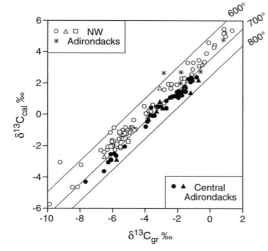

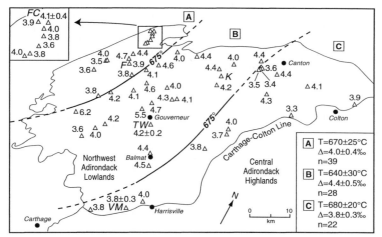

Figure 19. Carbon isotope thermometry for Adirondack marbles. (A, left) Plot of $\delta^{13}C$(calcite) vs. $\delta^{13}C$(graphite). Open symbols are for upper the amphibolite facies NW Adirondacks and closed symbols are for the granulite facies Adirondack Highlands. (B, below) Values of Δ(Cc-Gr) in per mil and resulting isopleths for 675°C parallel the strike of the terrane and, show lower metamorphic temperatures centered on Gouverneur and increasing towards granulite facies rocks to the SE and to NW. Note outcrop tests at Fish Creek (FC), the Train Wreck (TW) and the Valentine Mine (VM) (from Kitchen and Valley 1995).

Jenkins 2000). No zones of higher temperature were found at exposed igneous contacts, though the locally low values of $\Delta^{13}C$(Cc-Gr) 5 km W of Canton may be an unrecognized contact aureole. The Cc-Gr RAM thermometer provides the best estimates for metamorphic temperature in this terrane.

SULFUR ISOTOPE THERMOMETRY

Several common sulfide mineral pairs have fractionations that are calibrated and appropriate for thermometry, especially in sulfide deposits. However thermometry has often been disappointing, and sulfur isotope studies have tended to concentrate on elucidating fluid conditions, sources of sulfur, and ore mineral paragenesis (Ohmoto 1986). Recent microanalysis by laser probe and ion microprobe documents one of the main problems. Sulfides forming at low to intermediate temperatures show extreme zonation of $\delta^{34}S$, over 50 ‰ in less than 100 µm in some diagenetic pyrites (McKibben and Riciputi 1998) and 4-5 ‰/cm in a hydrothermal black smoker vent (Shanks et al. 1998).

Crowe (1994) analyzed coexisting pyrite-chalcopyrite pairs from the stockwork feeder zones of several volcanogenic massive sulfide deposits. Individual 200-μm diameter grains were analyzed by laser. Sulfides that were in grain to grain contact were found to have exchanged during greenschist to upper amphibolite facies metamorphism and during cooling, while sulfides that were isolated by quartz matrix preserved original hydrothermal $\delta^{34}S$ values. Temperatures of 170-424°C were calculated for three different hydrothermal systems. Thus, reliable thermometry will require assessment of sub-mm scale isotope zonation and exchange kinetics in sulfide minerals.

SKARNS

Skarns and hydrothermal veins associated with shallow igneous activity offer excellent opportunities for stable isotope thermometry because minerals commonly co-precipitate directly from a fluid and cool quickly. Minerals that are not zoned in chemical composition are most promising. Tests can be applied for mineral zoning or precipitation from evolving fluids.

Bowman (1998) reviewed stable isotope studies of skarns (Table 7). In a number of skarns, temperatures are reported that are generally consistent with the results of phase equilibria and fluid inclusion thermometry. However, minerals that have not co-precipitated do not typically yield self-consistent temperature estimates with the different techniques. Stable isotope concordance is also a good test. For instance, quartz, calcite and magnetite appear texturally to be co-precipitated and were analyzed from ten rocks in the Hanover Zn-Pb skarn (Table 7, #8, Turner and Bowman 1993). The Qt-Mt temperatures are in good agreement with independent thermometry, but the Cc-Qt and Cc-Mt temperatures are too low and too high, respectively (Table 7) because of late-stage depletion of $^{18}O/^{16}O$ in calcite (Bowman 1998).

Stable isotope zonation at the cm-scale is well documented in skarns, but few samples have been tested at mm to μm-scale (Jamtveit and Hervig 1994, Bezenek et al. 1995, Clechenko and Valley 2000). A careful study would include petrography and imaging, and other tests of temperature reliability. Microanalysis should be employed when other tests fail. It offers a powerful new tool for resolving the thermal and fluid evolution of skarns.

ONE MINERAL THERMOMETERS

Isotope fractionation can occur among crystallographically distinct sites within single minerals creating a potential one mineral thermometer. However, techniques for extraction and analysis of the isotope ratio of a specific site vary greatly in difficulty and reliability, and more experimental calibrations of fractionation are needed. Specific studies for oxygen isotopes include: clays (Hamza and Epstein 1980, Bechtel and Hoernes 1990, Sheppard and Gilg 1996); phyllosilicates (Savin and Lee 1988); analcime

NOTES for Table 7 (next page)

[1]Mineral abbreviations: Am = amphibole, Anh = anhydrite, Bt = biotite, Cal = calcite, Ccp = chalcopyrite, Chl = chlorite, Ep = epidote, Grt = garnet, Gn =galena, Hem = hematite, Mag = magnetite, Ms = muscovite, Pl = plagioclase, Px = pyroxene,. Py = pyrite, Qtz = quartz, Sp = sphalerite.

[2]Number of pairs analyzed in parentheses.

[3]Pressure (MPa) used in study for calculating phase equlibria and for correcting fluid-inclusion microthermometry data..

*References: 1. Taylor & O'Neil 1977; 2. Shelton & Rye 1982; Shelton 1983; 3. Bowman et al. 1985a; 4. Jamtveit et al. 1992a,b; 5. Bowman et al. 1985b; 6. Brown et al. 1985; 7. Layne & Spooner 1991; Layne et al. 1991; 8. Turner & Bowman 1993; 9. Kemp 1985; 10. Shimizu & Iiyama 1982.

Table 7. Comparison of temperature estimates (°C) from stable isotopes, phase equilbria, and fluid inclusions in skarns (from Bowman 1998).

			Stage	Isotope temperature			Skarn *(reference)	Type	Phase equil.	Fluid inclusion			P_{Corr} (Mpa)
				Ave.	Max.	Min.				Ave.	Max.	Min.	
*1.	Osgood Mtns, Nevada	W	I	381	640	330	Qtz-Grt (11)	575					150
		Cu, Au	II	523	540	480	Qtz-Am (6)	475					150
				636	1000	505	Qtz-Cal (7)						
2.	Mines Gaspe, Quebec	Cu	II	412	552	366	Py-Ccp (6)	500	340	350	425	200	30
			III	688	801	550	Anh-Py, Ccp (20)						
3.	Cantung, NWT	W	I	478	500	460	Qtz-Px (6)	575			410	360	100
			II	505	525	485	Qtz-Bt (2)	475			400	350	100
4.	Oslo Rift, Norway	W, Zn / Cu	II	360	360	360	Qtz-Grt (1)	380		350	400	300	50
5.	Elkhorn, Montana	-	I	497	660	420	Pl-Px (11)	560	490				100
6.	Pine Creek, California	W	I	494	560	400	Qtz-Px (8)	590					150
7.	JC, Yukon	Sn	III	460	504	416	Qtz-Ep (2)	550			500	410	75
			IV	446	446	446	Qtz-Bt (1)	450			560	430	75
			V	441	441	441	Cal-Ms (1)	450					
8.	Hanover, New Mexico	Zn, Pb	I	391	405	377	Grt-Mag (2)	425		355	400	305	40
				432	506	337	Cal-Ep,Chl (13)						
				625	1128	353	Qtz-Cal,Ep (9)						
			III	358	483	291	Qtz-Mag (12)	380		315	375	255	40
				172	377	-3	Qtz-Cal (10)	0					
				401	533	336	Cal-Mag (10)						
9.	Alta, Utah	Cu	I	538	595	500	Px-Mag (5)	575			>580	380	150
			II		Reversal		Qtz-Cal (5)			395	445	345	150
10.	Nakatatsu, Japan	Zn, Pb	II	329	364	278	Sp-Gn (4)	390					30

(Karlsson and Clayton 1990, Cheng et al. 2000); micas and chlorite (Hamza and Epstein 1980); scapolite (Moecher et al. 1984); alunite (Ustinov and Grinenko 1985, O'Neil and Pickthorn 1988, Rye et al. 1992); sahaite and spurrite (Ustinov and Grinenko 1985); and apatites (Shemesh et al. 1983, 1988, Ustinov and Grinenko 1985, Zheng 1996). Such studies hold great promise, especially in sedimentary rocks where co-precipitated minerals are rare.

ACKNOWLEDGMENTS

I thank James R. O'Neil for introducing me to the rigor and the fun of stable isotope geochemistry, and for many years of friendship and inspiration. Much of the research reported here was supported by NSF and DOE. Figures were drafted by Mary Diman. John Eiler, James Farquhar, Yaron Katzir, William Peck, and Zach Sharp are thanked for helpful reviews.

REFERENCES

Addy SK, Garlick GD (1974) Oxygen isotope fractionation between rutile and water. Contrib Mineral Petrol 45:119-121
Agrinier P (1991) The natural calibration of $^{18}O/^{16}O$ geothermometers: Application to the quartz-rutile mineral pair. Chem Geol 91:49-64
Agrinier P, Javoy M, Smith DC, Pineau F (1985) Carbon and oxygen isotopes in eclogites, amphibolites, veins and marbles from Western Gneiss Region, Norway. Isotope Geosci 52:146-162
Anderson AT (1967) The dimensions of oxygen isotope equilibrium attainment during prograde metamorphism. J Geol 75:323-332
Arita Y, Wada H (1990) Stable isotopic evidence for migration of metamorphic fluids along grain boundaries of marbles. Geochem J 24:173-186
Baker AJ (1988) Stable isotope evidence for limited fluid infiltration of deep crustal rocks from the Ivrea Zone, Italy. Geology 16:492-495
Bechtel Z, Hoernes S (1990) Oxygen isotope fractionation between oxygen of different sites in illite minerals: A potential single-mineral thermometer. Contrib Mineral Petrol 104:463-470
Becker RH, Clayton RN (1976) Oxygen isotope study of a Precambrian banded iron-formation, Hamersley Range, Western Australia. Geochim Cosmochim Acta 40:1153-1165
Bergfeld D, Nabelek PI, Labotka TC (1996) Carbon isotope exchange during polymetamorphism in the Panamint Mountains, California, USA. J Metamor Geol 14:199-212
Bestmann M, Kunze K, Matthews A (2000) The evolution of calcite marble shear zone complex on Thassos island, Greece; Microstructural and textural fabrics and their kinematic significance. J Struc Geol 22:1789-1807
Bezenek SR, Crowe DE, Riciputi LR (1995) Evidence of protracted growth history of skarn garnet using SIMS oxygen isotope, trace element and rare earth element data. Geol Soc Am Abstr Prog 27:67
Bigeleisen J, Mayer MG (1947) Calculation of equilibrium constants for isotopic exchange reactions. J Chem Phys 15:261-267
Bigeleisen J, Perlman ML, Prosser HC (1952) Conversion of hydrogenic materials for isotopic analysis. Analyt Chem 24:1356
Bindeman IN, Valley JW (2000) Formation of low-$\delta^{18}O$ rhyolites after caldera collapse at Yellowstone, Wyoming, USA. Geology 28:719-722
Bindeman IN, Valley JW (2001) Oxygen isotope study of phenocrysts in zoned tuffs and lavas from Timber Mountain/Oasis Valley caldera complex: Generation of large volumes of low-$\delta^{18}O$. Contrib Mineral Petrol (in review)
Bohlen SR, Essene EJ (1977) Feldspar and oxide thermometry of granulites in the Adirondack Highlands. Contrib Mineral Petrol 62:153-169
Bottinga Y, Javoy M (1973) Comments on oxygen isotope geothermometry. Earth Planet Sci Letters 20:250-265
Bottinga Y, Javoy M (1975) Oxygen Isotope Partitioning Among the Minerals in Igneous and Metamorphic Rocks. Rev Geophys Space Phys 13:401-418
Bottinga Y, Javoy M (1987) Comments on stable isotope geothermometry: The system quartz-water. Earth Planet Sci Letters 84:406-414
Bowman JR (1998) Stable-Isotope Systematics of Skarns. In Lentz DR (ed) Mineral Assoc Canada Short Course Series: Mineralized Intrusion-Related Skarn Systems, p 99-145

Bowman JR, Covert JJ, Clark AH, Mathieson GA (1985a) The Can Tung scheelite skarn orebody, Tungsten, Northwest Territories, Canada: Oxygen, hydrogen and carbon isotope studies. Econ Geol 80:1872-1895

Bowman JR, O'Neil JR, Essene EJ (1985b) Contact skarn formation at Elkhorn Montana. II: Origin and evolution of C-O-H skarn fluids. Am J Sci 285:621-660

Brady JB, Cheney JT, Larson Rhodes A, Vasquez A, Green C, Duvall M, Kogut A, Kaufman L, Kovaric D (1998) Isotope geochemistry of Proterozoic talc occurrences in Archean marbles of the Ruby Mountains, southwest Montana, U.S.A. Geol Materials Res 1:1 [http://www.minsocam.org]

Brown EH, O'Neil JR (1982) Oxygen Isotope Geothermometry and Stability of Lawsonite and Pumpellyite in the Shuksan Suite, North Cascades, Washington. Contrib Mineral Petrol 80:240-244

Brown PE, Bowman JR, Kelly WC (1985) Petrologic and stable isotope constraints on the source and evaluation of skarn-forming fluids at Pine Creek, California. Econ Geol 80:72-95

Buseck PR, Huang BJ (1985) Conversion of carbonaceous material to graphite during metamorphism. Geochim Cosmochim Acta 49:2003-2016

Cartwright I, Valley JW, Hazelwood A (1993) Resetting of oxybarometers and oxygen isotope ratios in granulite facies orthogneisses during cooling and shearing, Adirondack Mountains, New York. Contrib Mineral Petrol 113:208-225

Cavosie AJ, Sharp ZD, Selverstone J (2000) Application of a stable isotope geobarometer: Co-existing aluminosilicates in isotopic equilibrium from the Northern Colorado front range. Geol Soc Am Annual Meeting, p A-115

Cawley JD (1984) Oxygen Diffusion in Alpha Alumina. PhD Dissertation, Case Western Reserve University, Cleveland, Ohio

Chacko T, Mayeda TK, Clayton RN, Goldsmith JR (1991) Oxygen and carbon isotope fractionations between CO_2 and calcite. Geochim Cosmochim Acta 55:2867-2882

Chamberlain CP, Conrad ME (1991) The relative permeabilities of quartzites and schists during active metamorphism at mid-crustal levels. Geophys Res Letters 18:959-962

Chamberlain CP, Conrad ME (1992) Oxygen-isotope zoning in garnet: A record of volatile transport. Geochim Cosmochim Acta 57:2613-1619

Chamberlain CP, Ferry JM, Rumble D (1990) The effect of net-transfer reactions on the isotopic composition of minerals. Contrib Mineral Petrol 105:322-336

Cheng X, Zhao P, Stebbins JF (2000) Solid state NMR study of oxygen site exchange and Al-O-Al site concentration in analcime. Am Mineral 85:1030-1037

Chiba H, Chacko T, Clayton RN, Goldsmith JR (1989) Oxygen isotope fractionations involving diopside, forsterite, magnetite, and calcite: Application to geothermometry. Geochim Cosmochim Acta 53:2985-2995

Clayton RN (1981) Isotopic thermometry. *In* Newton RC, Navrotsky A, Wood BJ (ed) Advances in Physical Geochemistry. Springer-Verlag, Berlin, p 85-109

Clayton RN, Epstein S (1958) The relationship between O^{18}/O^{16} ratios in coexisting quartz, carbonate, and iron oxides from various geological deposits. J Geol 66:352-373

Clayton RN, Mayeda TK (1963) The use of bromine pentafluoride in the extraction of oxygen from oxides and silicates for isotopic analysis. Geochim Cosmochim Acta 27:43-52

Clayton RN, Kieffer SW (1991) Oxygen isotopic thermometer calibrations. *In* Taylor HP, O'Neil JR, Kaplan IR (ed) Stable Isotope Geochemistry: A Tribute to Samuel Epstein. Geochem Soc Spec Publ 3:3-10

Clayton RN, Goldsmith JR, Mayeda TK (1989) Oxygen isotope fractionation in quartz, albite, anorthite and calcite. Geochim Cosmochim Acta 53:725-733

Clayton RN, Goldsmith J, Karel KJ, Mayeda T, Newton RC (1975) Limits on the effect of pressure on isotopic fractionation. Geochim Cosmochim Acta 39:1197-1201

Clechenko CC, Valley JW (2000) Oscillatory zoned skarn garnet adjacent to massif anorthosite, Willsboro wollastonite mine, N.E. Adirondack Mts, N.Y. Geol Soc Am Abstr Prog, p A-295

Coghlan RAN (1990) Studies in diffusional transport: Grain boundary transport of O in feldspars, diffusion of O, strontium, and the REEs in garnet and thermal histories of granitic intrusions in south-central Maine using O isotopes. PhD Dissertation, Brown University, Providence, Rhode Island

Craig H (1953) The geochemistry of the stable carbon isotopes. Geochim Cosmochim Acta 3:53-92

Crank J (1975) The Mathematics of Diffusion. Clarendon Press, Oxford, UK

Crawford WA, Valley JW (1990) Origin of graphite in the Pickering gneiss and the Franklin marble, Honey Brook Upland, Pennsylvania Piedmont. Geol Soc Am Bull 102:807-811

Criss RE (1999) Principles of Stable Isotope Distribution. Oxford Press, Oxford, UK

Criss RE, Gregory RT, Taylor HP (1987) Kinetic theory of oxygen isotopic exchange between minerals and water. Geochim Cosmochim Acta 51:1099-1108

Crowe DE (1994) Preservation of original hydrothermal $\delta^{34}S$ values in greenschist to upper amphibolite volcanogenic massive sulfide deposits. Geology 22:873-876

Crowe DE, Valley JW, Baker KL (1990) Microanalysis of sulfur isotope zonation by laser microprobe. Geochim Cosmochim Acta 54:2075-2092

Dahl PS (1979) Comparative geothermometry based on major-element and oxygen isotope distributions in Precambrian metamorphic rocks from southwestern Montana. Am Mineral 64:1280-1293

Deines P (1977) On the oxygen isotope distribution among mineral triplets in igneous and metamorphic rocks. Geochim Cosmochim Acta 41:1709-1730

Deloule E, Allegre CJ, Doe B (1986) Lead and sulfur isotope microstratigraphy in galena crystals from Mississippi Valley-type deposits. Econ Geol 81:1307-1321

Deloule E, Albarede F, Sheppard SMF (1991a) Hydrogen isotope heterogeneities in the mantle from ion probe analysis of amphiboles from ultramafic rocks. Earth Planet Sci Letters 105:543-553

Deloule E, France-Lanord C, Albarede F (1991b) D/H analysis of minerals by ion probe. In Taylor HP, O'Neil JR, Kaplan IR (eds) Stable Isotope Geochemistry. Geochem Soc Spec Publ 3:53-62

Dennis PF (1984a) Oxygen self-diffusion in quartz under hydrothermal conditions. J Geophys Res 89:4047-4057

Dennis PF (1986) Oxygen self diffusion in quartz. Prog Exp Petrol, NERC Publ D 25:260-265

Des Marais DJ, Moore JG (1984) Carbon and its isotopes in mid-oceanic basaltic glasses. Earth Planet Sci Letters 69:43-57

Desmons J, O'Neil JR (1978) oxygen and hydrogen isotope compositions of eclogites and associated rocks from the eastern Sesia Zone (Western Alps, Italy). Contrib Mineral Petrol 67:79-85

Dodson MH (1973) Closure temperature in cooling geochronological and petrologic systems. Contrib Mineral Petrol 40:259-274

Dodson MH (1979) Theory of cooling ages. In Jager E, Hunziker JC (eds) Lectures in Isotope Geology. Springer-Verlag, p 194-202

Doremus RH (1998) Diffusion of water and oxygen in quartz: Reaction-diffusion model. Earth Planet Sci Letters 163:43-51

Dunn SR, Valley JW (1992) Calcite-graphite isotope thermometry: A test for polymetamorphism in marble, Tudor gabbro aureole, Ontario, Canada. J Metamor Geol 10:487-501

Dunn SR, Valley JW (1996) Polymetamorphic fluid-rock interaction of the Tudor Gabbro and adjacent marble, Ontario. Am J Sci 296:244-295

Edwards KJ, Valley JW (1998) Oxygen isotope diffusion and zoning in diopside: The importance of water fugacity during cooling. Geochim Cosmochim Acta 62:2265-2277

Eichmann R, Schidlowski M (1975) Isotopic fractionation between coexisting organic carbon-carbonate pairs in Pre-Cambrian sediments. Geochim Cosmochim Acta 39:585-595

Eiler JM, Valley JW (1994) Preservation of premetamorphic oxygen isotope ratios in granitic orthogneiss from the Adirondack Mountains, New York, USA. Geochim Cosmochim Acta 58:5525-5535

Eiler JM, Baumgartner LP, Valley JW (1992) Intercrystalline stable isotope diffusion: A fast grain boundary model. Contrib Mineral Petrol 112:543-557

Eiler JM, Baumgartner LP, Valley JW (1993) A new look at stable isotope thermometry. Geochim Cosmochim Acta 57:2571-2583

Eiler JM, Baumgartner LP, Valley JW (1994) Fast grain boundary: A Fortran-77 program for calculating the effects of retrograde interdiffusion of stable isotopes. Computers Geosciences 20:1415-1434

Eiler JM, Graham CW, Valley JW (1997a) SIMS analysis of oxygen isotopes: Matrix effects in complex minerals and glasses. Chem Geol 138:221-244

Eiler JM, Valley JW, Graham CM (1997b) Standardization of SIMS analysis of O and C isotope ratios in carbonate from ALH84001. 28th Lunar Planet Sci Conf, p 327-328

Eiler JM, Valley JW, Graham CM, Baumgartner LP (1995a) The oxygen isotope anatomy of a slowly cooled metamorphic rock. Am Mineral 80:757-764

Eiler JM, Valley JW, Graham CM, Baumgartner LP (1995b) Ion microprobe evidence for the mechanisms of stable isotope retrogression in high-grade metamorphic rocks. Contrib Mineral Petrol 18:365-378

Eldridge CS, Compston W, Williams IS, Both RA, Walshe JL, Ohmoto H (1988) Sulfur isotope variability in sediment-hosted massive sulfide deposits as determined using the ion microprobe SHRIMP: I. An example from the Rammelsberg orebody. Econ Geol 83:443-449

Elphick SC, Graham CM (1988) The effect of hydrogen on oxygen diffusion in quartz: Evidence for fast proton transients? Nature 335:243-245

Elphick SC, Graham CM, Dennis PF (1988) An ion probe study of anhydrous oxygen diffusion in anorthite: A comparison with hydrothermal data and some geological implications. Contrib Mineral Petrol 100:490-495

Elsenheimer DW (1988) Petrologic and Stable Isotopic Characteristics of Graphite and Other Carbon-bearing Minerals in Sri Lankan Granulites. MS Thesis, University of Wisconsin, Madison

Elsenheimer DW (1992) Development and Application of Laser Microprobe Techniques for Oxygen Isotope Analysis of Silicates and, Fluid/Rock Interaction During and After Granulite-Facies Metamorphism, Highland Southwestern Complex, Sri Lanka. PhD Dissertation, University of Wisconsin, Madison

Elsenheimer DW, Valley JW (1992) *In situ* oxygen isotope analysis of feldspar and quartz by Nd: YAG laser microprobe. Chem Geol Isotope Geosci Section 101:21-42

Elsenheimer DW, Valley JW (1993) Submillimeter scale zonation of $\delta^{18}O$ in quartz and feldspar, Isle of Skye, Scotland. Geochim Cosmochim Acta 57:3669-3676

Esler MB, Griffith DWT, Wilson SR, Steele LP (2000) Precision trace gas analysis by FT-IR spectroscopy. 2. The $^{13}C/^{12}C$ isotope ratio of CO_2. Analyt Chem 72:216-221

Essene EJ (1989) The current status of thermobarometry in metamorphic rocks. *In* Daly JS, Cliff RA, Yardley BWD (eds) Evolution of Metamorpic Belts, p 1-44

Farquhar J, Chacko T (1994) Exsolution-enhanced oxygen exchange: Implications for oxygen isotope closure temperatures in minerals. Geol 22:751-754

Farquhar J, Rumble D (1998) Comparison of oxygen isotope data obtained by laser fluorination of olivine with KrF excimer laser and CO_2 laser. Geochim Cosmochim Acta 62:3141-3149

Farquhar J, Chacko T, Frost BR (1993) Strategies for high-temperature oxygen isotope thermometry: A worked example from the Laramie Anorthosite Complex, Wyoming, USA. Earth Planet Sci Letters 117:407-422

Farquhar J, Chacko T, Ellis DJ (1996) Preservation of oxygen isotope compositions in granulites from Northwestern Canada and Enderby Land, Antarctica: Implications for high-temperature isotopic thermometry. Contrib Mineral Petrol 125:213-224

Farquhar J, Hauri E, Wang J (1999) New insights into carbon fluid chemistry and graphite precipitation: SIMS analysis of granulite facies graphite from Ponmudi, South India. Earth Planet Sci Letters 171:607-621

Farver JR (1989) Oxygen self-diffusion in diopside with application to cooling rate determinations. Earth Planet Sci Letters 92:386-396

Farver JR (1994) Oxygen self-diffusion in calcite: Dependence on temperature and water fugacity. Earth Planet Sci Letters 121:575-587

Farver JR, Giletti BJ (1985) Oxygen diffusion in amphiboles. Geochim Cosmochim Acta 49:1403-1411

Farver JR, Giletti BJ (1989) Oxygen and strontium diffusion kinetics in apatite and potential applications to thermal history determinations. Geochim Cosmochim Acta 53:1621-1631

Farver JR, Yund RA (1990) The effect of hydrogen, oxygen, and water fugacity on oxygen diffusion in alkali feldspar. Geochim Cosmochim Acta 54:2953-2964

Farver JR, Yund RA (1991) Oxygen diffusion in quartz: Dependence on temperature and water fugacity. Chem Geol 90:55-70

Farver JR, Yund RA (1991) Measurement of oxygen grain boundary diffusion in natural, fine-grained, quartz aggregates. Geochim Cosmochim Acta 55:1597-1607

Farver JR, Yund RA (1992) Oxygen diffusion in a fine-grained quartz aggregate with wetted and nonwetted microstructures. J Geophys Res 97:14,017-14,029

Farver JR, Yund RA (1995) Grain boundary diffusion of oxygen, potassium and calcium in natural and hot-pressed feldspar aggregates. Contrib Mineral Petrol 118:340-355

Farver JR, Yund RA (1995) Interphase boundary diffusion of oxygen and potassium in K-feldspar/quartz aggregates. Geochim Cosmochim Acta 59:3697-3705

Farver JR, Yund RA (1998) Oxygen grain boundary diffusion in natural and hot-pressed calcite aggregates. Earth Planet Sci Letters 161:189-200

Farver JR, Yund RA (1999) Oxygen bulk diffusion measurements and TEM characterization of a natural ultramylonite: Implications for fluid transport in mica-bearing rocks. J Metamor Geol 17:669-683

Fiebig J, Wiechert U, Rumble D, Hoefs J (1999) High precision, insitu oxygen isotope analysis of quartz using an ArF laser. Geochim Cosmochim Acta 63:687-702

Fitzsimons ICW, Mattey DP (1994) Carbon isotope constraints on volatile mixing and melt transport in granulite-facies migmatites. Earth Planet Sci Letters 134:319-328

Forester RW, Taylor Jr. HP (1976) ^{18}O depleted igneous rocks from the Tertiary complex of the Isle of Mull, Scotland. Earth Planet Sci Letters 32:11-17

Forester RW, Taylor HP (1977) $^{18}O/^{16}O$, D/H, and $^{13}C/^{12}C$ studies of the Tertiary igneous complex of Skye, Scotland. Am J Sci 277:136-177

Fortier SM, Giletti BJ (1989) An empirical model for predicting diffusion coefficients in silicate minerals. Science 245:1481-1484

Fortier SM, Giletti BJ (1991) Volume self-diffusion of oxygen in biotite, muscovite, and phlogopite micas. Geochim Cosmochim Acta 55:1319-1330

Fortier SM, Luttge A, Satir M, Metz P (1994) Oxygen isotope fractionation between fluorphlogopite and calcite: An experimental investigation of temperature dependence and F/OH⁻ effects. Eur J Mineral 6:53-65

Fourcade S, Javoy M (1973) Rapports $^{18}O/^{16}O$ dans les roches du vieux socle catazonal d'In Ouzzal (Sahara Algerien). Contrib Mineral Petrol 42:235-244

Freer R, Dennis PF (1982) Oxygen diffusion studies. I. A preliminary ion microprobe investigation of oxygen diffusion in some rock-forming minerals. Mineral Mag 45:179-192

Friedman I, O'Neil JR (1977) Compilation of Stable Isotope Fractionation Factors of Geochemical Interest. U S Geol Surv Prof Paper 440-KK

Galimov EM, Rozen OM, Belomestnykh AV, Zlobin VL, Kromtsov IN (1990) The nature of graphite in Anabar-Shield metamorphic rocks. Geokhimiya 3:373-384

Gerdes ML, Valley JW (1994) Fluid flow and mass transport at the Valentine wollastonite deposit, Adirondack Mountains, N.Y. J Metamor Geol 12:589-608

Ghent ED, Valley JW (1998) Oxygen isotope study of quartz-Al₂SiO₅ pairs from the Mica Creek area, British Columbia: Implications for the recovery of peak metamorphic temperatures. J Metamor Geol 16:223-230

Giletti BJ (1986) Diffusion effects on oxygen isotope temperature of slowly cooled igneous and metamorphic rocks. Earth Planet Sci Letters 77:218-228

Giletti BJ, Yund RA (1984) Oxygen diffusion in quartz. J Geophys Res 89:4039-4046

Giletti BJ, Hess KC (1988) Oxygen diffusion in magnetite. Earth Plant Sci Letters 89:115-122

Giletti BJ, Semet MP, Yund RA (1978) Studies in diffusion—III. Oxygen in feldspars: An ion microprobe determination. Geochim Cosmochim Acta 42:45-57

Godfrey JD (1962) The deuterium content of hydrous minerals from the East-Central Sierra Nevada and Yosemite National Park. Geochim Cosmochim Acta 26:1215-1245

Gold T (1987) Power from the Earth. Dent & Sons Ltd, London, Melbourne, 208 p

Goldman DS, Albee AL (1977) Correlation of Mg/Fe partitioning between garnet and biotite with $^{18}O/^{18}O$ partitioning between quartz and magnetite. Am J Sci 277:750-767

Graham CM, Valley JW, Winter BL (1996) Ion microprobe analysis of $^{18}O/^{16}O$ in authigenic and detrital quartz in St. Peter sandstone, Michigan Basin and Wisconsin Arch, USA: Contrasting diagenetic histories. Geochim Cosmochim Acta 24:5101-5116

Graham CM, Valley JW, Eiler JM, Wada H (1998) Timescales and mechanisms of fluid infiltration in a marble: An ion microprobe study. Contrib Mineral Petrol 132:371-389

Gregory RT, Criss RE (1986) Isotopic exchange in open and closed systems. Rev Mineral 16:91-127

Gregory RT, Taylor HP (1986a) Possible non-equilibrium oxygen isotope effects in mantle nodules, an alternative to the Kyser-O'Neil-Carmichael $^{18}O/^{16}O$ geothermometer. Contrib Mineral Petrol 93:114-119

Gregory RT, Taylor HP (1986b) Non-equilibrium, metasomatic $^{18}O/^{16}O$ effects in upper mantle mineral assemblages. Contrib Mineral Petrol 93:124-135

Gregory RT, Criss RE, Taylor HP (1989) Oxygen isotope exchange kinetics of mineral pairs in closed and open systems: Applications to problems of hydrothermal alteration of igneous rocks and Precambrian iron formations. Chem Geol 75:1-42

Hamza MS, Epstein S (1980) Oxygen isotopic fractionation between oxygen of different sites in hydroxyl-bearing silicate minerals. Geochim Cosmochim Acta 44:173-182

Harte B, Otter ML (1992) Carbon isotope measurements on diamonds. Chem Geol 101:177-183

Hayes JM, Kaplan IR, Wedeking KW (1983) Precambrian organic geochemistry: Preservation of the record. In Schopf JW (ed) Earth's Earliest Biosphere: Its Origin and Evolution. Princeton University Press, Princeton, NJ, p 93-134

Hervig RL, Williams P, Thomas RM, Schauer SN, Steele IM (1992) Microanalysis of oxygen isotopes in insulators by secondary ion mass spectrometry. Int'l J Mass Spect Ion Proc 120:45-63

Hildreth W, Christiansen RL, O'Neil JR (1984) Catastrophic isotopic modification of rhyolitic magma at times of caldera subsidence, Yellowstone plateau volcanic field. J Geophys Res 89:8339-8369

Hoefs J (1997) Stable Isotope Geochemistry, 4th edn. Springer-Verlag, Berlin, 201 p

Hoefs J, Frey M (1976) The isotopic composition of carbonaceous matter in a metamorphic profile from the Swiss Alps. Geochim Cosmochim Acta 40:945-951

Hoefs J, Muller G, Schuster AK (1982) Polymetamorphic relations in iron ores from the iron quadrangle, brazil: The correlation of oxygen isotope variations with deformation history. Contrib Mineral Petrol 79:241-251

Hoefs J, Mueller G, Schuster KA, Walde D (1987) The Fe-Mn ore deposits of Urucum, Brazil; an oxygen isotope study. Chem Geol, Isotope Geosci Section 65:311-319

Hoernes S, Friedrickson H (1978) Oxygen and hydrogen isotope study of the polymetamorphic area of the Northern Otztal-Stubal Alps. Contrib Mineral Petrol 67:305-315

Hoernes S, Lichtenstein U, van Reenen DD, Mokgatlha K (1995) Whole-rock/mineral O-isotope fractionations as a tool to model fluid-rock interaction in deep seated shear zones of the Southern Marginal Zone of the Limpopo Belt, South Africa. S Afr J Geol 98:488-497

Hoffbauer R, Hoernes S, Fiorentini E (1994) Oxygen isotope thermometry based on a refined increment method and its applications to granulite-grade rocks from Sri Lanka. *In* Raith M, Hoernes S (eds) Tectonic, Metamorphic and Isotopic Evolution of Deep Crustal Rocks, With Special Emphasis on Sri Lanka, p 199-220

Hoffbauer R, Spiering B (1994) Petrologic phase equilibria and stable isotope fractionations of carbonate-silicate parageneses from granulite-grade rocks of Sri Lanka. Precambrian Res 66:325-349

James HL, Clayton RN (1962) Oxygen isotope fractionation in metamorphosed iron formations of the Lake Superior region and in other iron-rich rocks. *In* Petrologic Studies: Buddington Volume. Geol Soc Am, p 217-239

Jamtveit B, Hervig RL (1994) Constraints on transport and kinetics in hydrothermal systems from zoned garnet crystals. Science 263:505-508

Jamtveit B, Bucher-Nurminen K, Stijfhoorn DE (1992a) Contact metamorphism of layered shale carbonate sequences in the Oslo Rift: I. Buffering infiltration and mechanisms of mass transport. J Petrol 33:377-422

Jamtveit B, Grorud HF, Bucher-Nurminen K (1992b) Contact metamorphism of layered shale—carbonate sequences in the Oslo Rift: II. Migration of isotopic and reaction fronts around cooling plutons. Earth Planet Sci Letts 114:131-148

Jaszczak J, Robinson G (1998) Spherical graphite from Gooderham, Ontario. 17th Int'l Mineral Assoc Meeting, Toronto, Aug 1998 (abstr)

Javoy M (1977) Stable isotopes and geothermometry. J Geol Soc London 133:609-636

Jenkin GRT, Linklater C, Fallick AE (1991) Modeling of mineral $\delta^{18}O$ values in an igneous aureole: Closed-system model predicts apparent open-system $\delta^{18}O$ values. Geology 19:1185-1188

Jenkin GRT, Fallick AE, Farrow CM, Bowes GE (1991) Cool: A Fortran-77 computer program for modeling stable isotopes in cooling closed systems. Computers and Geosciences 17:391-412

Johnson EA, Rossman GR, Valley JW (2000) Correlation between OH content and oxygen isotope diffusion rate in diopsides from the Adirondack Mountains, New York. Geol Soc Am Abstr Progr 32:A-114

Jones AM, Iacumin P, Young ED (1999) High resolution $\delta^{18}O$ analysis of tooth enamel phosphate by isotope ratio monitoring gas chromatography and UV laser fluorination. Chem Geol 153:241-248

Karlsson HR, Clayton RN (1990) Oxygen and hydrogen isotope geochemistry of zeolites. Geochim Cosmochim Acta 54:1369-1386

Kawabe I (1978) Calculation of oxygen isotope fractionation in quartz-water system with special reference to the low temperature fractionation. Geochim Cosmochim Acta 42:613-621

Kelley SP, Fallick AE (1990) High precision spatially resolved analysis of $\delta^{34}S$ in sulphides using a laser extraction technique. Geochim Cosmochim Acta 54:883-888

Kemp WM (1985) A Stable Isotope and Fluid Inclusion Study of the Contact Al(Fe)-Ca-Mg-Si Skarns in the Alta Stock Aureole, Alta, Utah. MS Thesis, University of Utah, Provo

Kerrich R, Beckinsale RD, Durham JJ (1977) The transition between deformation regimes dominated by intercrystalline diffusion and intracrystalline creep evaluated by oxygen isotope thermometry. Tectonophysics 38:241-257

Kerstel ERT, van Trigt R, Dam N, Reuss J, Meijer HAJ (1999) Simultaneous determination of the $^2H/^1H$, $^{17}O/^{16}O$, and $^{18}O/^{16}O$ isotope abundance ratios in water by means of laser spectrometry. Analyt Chem 71:5297-5303

King EM, Valley JW, Davis DW, Edwards G (1998) Oxygen isotope ratios of Archean plutonic zircons from granite-greenstone belts of the Superior Province: Indicator of magmatic source. Precambrian Res 92:47-64

King EM, Valley JW, Davis DW, Kowallis BJ (2001) Empirical determination of oxygen isotope fractionation factors for titanite with respect to zircon and quartz. Geochim Cosmochim Acta, in press

Kitchen NE, Valley JW (1995) Carbon isotope thermometry in marbles of the Adirondack Mountains, New York. J Metamor Geol 13:577-594

Knauth LP, Lowe DR (1978) Oxygen isotope geochemistry of cherts from the Onverwacht group (3.4 billion years), Transvaal, South Africa, with implications for secular variations in the isotopic composition of cherts. Earth Planet Sci Letters 41:209-222

Kohn MJ (1999) Why most "dry" rocks should cool "wet." Am Mineral 84:570-580

Kohn MJ, Valley JW (1994) Oxygen isotope constraints on metamorphic fluid flow, Townshend Dam, Vermont, USA. Geochim Cosmochim Acta 58:5551-5566

Kohn MJ, Valley JW (1998) Obtaining equilibrium oxygen isotope fractionations from rocks: Theory and examples. Contrib Mineral Petrol 132:209-224

Kohn MJ, Schoeninger MJ, Valley JW (1996) Herbivore tooth oxygen isotope compositions: Effects of diet and physiology. Geochim Cosmochim Acta 60:3889-3896

Kohn MJ, Spear FS, Valley JW (1997) Dehydration-melting and fluid recycling during metamorphism: Rangeley Formation, New Hampshire, USA. J Petrol 9:1255-1277

Kohn MJ, Valley JW, Elsenheimer D, Spicuzza M (1993) Oxygen isotope zoning in garnet and staurolite: Evidence for closed system mineral growth during regional metamorphism. Am Mineral 78:988-1001

Kreulen R, van Beek PCJM (1983) The calcite-graphite isotope thermometer; data on graphite bearing marbles from Naxos, Greece. Geochim Cosmochim Acta 47:1527-1530

Kronenberg AK, Yund RA, Giletti BJ (1984) Carbon and oxygen diffusion in calcite: Effects of Mn content and PH_2O. Phys Chem Minerals 11:101-112

Krylov DP, Mineev, S.D. (1994) The concept of model-temperature in oxygen isotope geochemistry: An example of a single outcrop from the Rayner Complex (Enderby Land, East Antarctica). Geochim Cosmochim Acta 58:4465-4473

Kyser TK (1986) Stable isotope variations in the mantle. Rev Mineral 16:141-164

Labotka TC, Cole DR, Riciputi LR (2000) Diffusion of C and O in calcite at 100 MPa. Am Mineral 85:488-494

Larson TE, Sharp ZD (2000) Isotopic disequilibrium in the classic triple point localities of New Mexico. Geol Soc Am Annual Meeting, p A-297

Lasaga AC (1998) Kinetic Theory in the Earth Sciences. Princeton University Press, Princeton, New Jersey, 811 p

Layne GD, Longstaffe FJ, Spooner ETC (1991) The JC tin skarn deposit, southern Yukon Territory: II. A carbon, oxygen, hydrogen, and sulfur stable isotope study. Econ Geol 86:48-65

Layne GD, Spooner ETC (1991) The JC tin skarn deposit, southern Yukon Territory: I. Geology, paragenesis, and fluid inclusion microthermometry. Econ Geol 86:29-47

Li H, Schwarcz HP, Shaw DM (1991) Deep crustal oxygen isotope variations: The Wawa-Kapuskasing crustal transect, Ontario. Contrib Mineral Petrol 107:448-458

Liogys VA, Jenkins DM (2000) Hornblende geothermometry of amphibolite layers of the Popple Hill gneiss, north-west Adirondack Lowlands, New York, USA. J Metamor Geol 18:513-530

Loucks RR (1996) Restoration of the elemental and stable-isotopic compositions of diffusionally altered minerals in slowly cooled rocks. Contrib Mineral Petrol 124:346-358

Luque FJ, Pasteris JD, Wopenka B, Rodas M, Barrenechea JF (1998) Natural fluid-deposited graphite: Mineralogical characteristics and mechanisms of formation. Am J Sci 298:471-498

Luz B, Barkan E, Bender ML, Thiemens MH, Boering KA (1999) Triple-isotope composition of atmospheric oxygen as a tracer of biosphere productivity. Nature 400:547-550

Manning JR (1974) Diffusion kinetics and mechanisms in simple crystals. In Hofmann AW et al. (eds) Geochemical Transport and Kinetics. Carnegie Inst Washington, Washington, DC, p 3-13

Matthews A (1994) Oxygen isotope geothermometers for metamorphic rocks. J Metamor Geol 12:211-219

Matthews A, Schliestedt M (1984) Evolution of the blueschist and greenschist facies rocks of Sifnos, Cyclades, Greece. Contrib Mineral Petrol 88:150-163

Matthews A, Beckinsale RD, Durham JJ (1979) Oxygen isotope fractionation between rutile and water and geothermometry of metamorphic eclogites. Mineral Mag 43:406-413

McConnell JDC (1995) The role of water in oxygen isotope exchange in quartz. Earth Planet Sci Letters 136:97-107

McCrea JM (1950) On the isotopic chemistry of carbonates and a paleotemperature scale. J Chem Phys 18:849-857

McKibben MA, Riciputi LR (1998) Sulfur Isotopes by Ion Microprobe. In McKibben MA, Shanks III WC, Ridley WI (eds) Applications of Microanalytical Techniques to Understanding Mineralizing Processes. Soc Econ Geol Rev Econ Geol 7:121-140

McLelland J, Chiarenzelli J, Perham A (1992) Age, field, and petrological relationships of the Hyde School gneiss, Adirondack Lowlands, New York: Criteria for an intrusive igneous origin. J Geol 100:69-90

Merritt DH, Hayes JM (1994) Factors controlling precision and accuracy in isotope-ratio-monitoring mass spectrometry. Analyt Chem 66:2336-2347

Moecher DP, Sharp ZD (1999) Comparison of conventional and garnet-aluminosilicate-quartz O isotope thermometry: Insights for mineral equilibrium in metamorphic rocks. Am Mineral 84:1287-1303

Moecher DP, Essene EJ, Valley JW (1992) Stable isotopic and petrological constraints on scapolitization of the Whitestone meta-anorthosite, Grenville, Province, Ontario. J Metamor Geol 10:745-762

Moecher DP, Valley JW, Essene EJ (1994) Extraction and carbon isotope analysis of CO_2 from scapolite in deep crustal granulites and xenoliths. Geochim Cosmochim Acta 58:959-967

Mojzsis SJ, Arrhenius G, McKeegan KD, Harrison TM, Nutman AP, Friend RL (1996) Evidence for life on Earth before 3800 million years ago. Nature 384:55-59

Moore DK, Cherniak DJ, Watson EB (1998) Oxygen diffusion in rutile from 750 to 1000 C and 0.1 to 1000 MPa. Am Mineral 83:700-711

Morikiyo T (1984) Carbon isotopic study on coexisting calcite and graphite in the Ryoke metamorphic rocks, northern Kiso district, central Japan. Contrib Mineral Petrol 87:251-259

Morishita Y, Giletti BJ, Farver JR (1996) Volume self-diffusion of oxygen in titanite. Geochem J 30:71-79

Morrison J, Valley JW (1991) Retrograde fluids in granulites: Stable isotope evidence of fluid migration. J Geol 99:559-570

Morrison J, Barth AP (1993) Empirical tests of carbon isotope thermometry in granulites from southern California. J Metamor Geol 11:789-800

Morrison J, Anderson JL (1998) Footwall refrigeration along a detachment fault: Implications for the thermal evolution of core complexes. Science 279:63-66

Muller G, Schuster A, Hoefs J (1986a) The metamorphic grade of banded iron-formations: Oxygen isotope and petrological constraints. Fortschr Mineral 64:163-185

Muller G, Hans-Joachim L, Hoefs J (1986b) Sauerstoff- und Kohlenstoff-Isotopenuntersuchungen an Mineralen aus gebanderten Eisenerzen und metamorphen Gesteinen nordostlich des Eisernen Vierecks in Brasilien. Geologische Jahrb D 79:21-40

Murphey BF, Nier AO (1941) Variations in the relative abundance of the carbon isotopes. Phys Rev 59:771-772

Nagy KL, Giletti BJ (1986) Grain boundary diffusion of oxygen in a macroperthitic feldspar. Geochim Cosmochim Acta 50:1151-1158

Nier AO (1947) A mass spectrometer for isotope and gas analysis. Rev Sci Instr 18:398

O'Hara KD, Sharp ZD, Moecher DP, Jenkin GRT (1997) The effect of deformation on oxygen isotope exchange in quartz and feldspar and the significance of isotopic temperatures in mylonites. J Geol 103:193-204

O'Neil JR (1986) Theoretical and experimental aspects of isotopic fractionation. Rev Mineral 16:1-41

O'Neil JR, Clayton RN (1964) Oxygen isotope geothermometry. *In* Craig H, Miller SL, Wasserburg GJ (eds) Isotope and Cosmic Chemistry, p 148-157

O'Neil JR, Pickthorn WJ (1988) Single mineral oxygen isotope thermometry. Chem Geol 71:369

Ohmoto H (1986) Stable isotope geochemistry of ore deposits. Rev Mineral 16:491-556

Pandey UK, Chabria T, Krishnamurthy P, Viswanathan R, Kumar B (2000) Carbon isotope and X-ray diffraction studies on calcite-graphite system in calc-granulites around Usilampatti Area, Madurai, Tamil Nadu. J Geol Soc India 55:37-46

Patterson WP, Smith GR, Lohmann KC (1993) Continental Paleothermometry and Seasonality Using the Isotopic Composition of Aragonitic Otoliths of Freshwater Fishes. *In* Swart PK, Lohmann KC, McKenzie J, Savin S (eds) Climate Change in Continental Isotopic Records. Geophysical Monogr, p 191-202

Peck WH (2000) Oxygen isotope studies of Grenville Metamorphism and Magmatism. PhD Dissertation, University of Wisconsin, Madison

Peck WH, Valley JW, Graham CM (2001) Slow oxygen diffusion in igneous zircons from metamorphic rocks (in review)

Perry EC, Bonnichsen B (1966) Quartz and magnetite oxygen-18–oxygen-16 fractionation in metamorphosed Biwabik Iron Formation. Science 153:528-529

Perry EC, Ahmad SN (1977) Carbon isotope composition of graphite and carbonate minerals from 3.8-AE metamorphosed sediments, Isukasia, Greenland. Earth Planet Sci Letters 36:280-284

Perry EC, Ahmad SN (1981) Oxygen and carbon isotope geochemistry of the Krivoy Rog iron formation, Ukranian SSR. Lithos 14:83-92

Perry EC, Tan FC, Morey GB (1973) Geology and Stable Isotope Geochemistry of the Biwabik Iron Formation, Northern Minnesota. Econ Geol 68:1110-1123

Perry EC, Ahmad SN, Swulius TM (1978) The oxygen isotope composition of 3,800 M.Y. old metamorphosed chert and iron formation from Isukasia, West Greenland. J Geol 86:223-239

Pineau F, Latouche L, Javoy M (1976) L'origine du graphite et les fractionnements isotopiques du carbone dans les marbres metamorphiques des Gour Oumelalen (Ahaggar, Algerie), des Adirondacks (New Jersey, U.S.A.), et du Damara (Namibie, Sud-Ouest africain). Bull Soc Geol France 7 t. XVIII:1713-1723

Polyakov VB, Kharlashina NN (1994) Effect of pressure on equilibrium isotopic fractionation. Geochim Cosmochim Acta 58:4739-4750

Polyakov VB, Kharlashina NN (1995) The use of heat capacity data to calculate isotope fractionation between graphite, diamond, and carbon dioxide: A new approach. Geochim Cosmochim Acta 59:2561-2572

Putlitz B, Valley JW, Matthews A (2001) Oxygen isotope thermometry of quartz-Al_2SiO_5 veins in high grade metamorphic rocks on Naxos. Contrib Mineral Petrol (in press)

Rafter TA (1957) Sulphur isotopic variations in nature: Part I. The preparation of sulphur dioxide for mass spectrometer examination. New Zealand J Sci Techn B38:849

Rathmell MA, Streepey M, M., Essene EJ, van der Pluijm BA (1999) Comparison of garnet-biotite, calcite-graphite, and calcite-dolomite thermometry in the Grenville Orogen; Ontario, Canada. Contrib Mineral Petrol 134:217-231

Rees CE (1978) Sulphur isotope measurements using SO_2 and SF_6. Geochim Cosmochim Acta 42:383-389

Riciputi LR (1996) A comparison of extreme energy filtering and high mass resolution techniques for measurement of $^{34}S/^{32}S$ ratios by ion microprobe. Rapid Comm Mass Spectrom 10:282-286

Robinson BW, Kusakabe M (1975) Quantitative preparation of sulphur dioxide for $^{34}S/^{32}S$ analyses from sulphides by combustion with cuprous oxide. Analyt Chem 47:1179

Rosenbaum J, Sheppard SMF (1986) An isotopic study of siderites, dolomites and ankerites at high temperatures. Geochim Cosmochim Acta 50:1147-1150

Rosenbaum JM, Mattey D (1995) Equilibrium garnet-calcite oxygen isotope fractionation. Geochim Cosmochim Acta 59:2839-2842

Rumble D (1978) Mineralogy, petrology, and oxygen isotopic geochemistry of the Clough Formation, Black Mountain, Western New Hampshire, U.S.A. J Petrol 19:317-340

Rumble D, Sharp ZD (1998) Laser microanalysis of silicates for $^{18}O/^{17}O/^{16}O$ and of carbonates for $^{18}O/^{16}O$ and $^{13}C/^{12}C$ ratios. In McKibben MA, Shanks III WC, Ridley WI (eds) Soc Econ Geol Rev Econ Geol 7:99-119

Rumble D, Yui T-F (1998) The Qinglongshan oxygen and hydrogen isotope anomaly near Donghai in Jiangsu Province, China. Geochim Cosmochim Acta 62:3307-3321

Rye RO, Bethke PM, Wasserman MD (1992) The stable isotope geochemistry of acid sulfate alteration. Econ Geol 87:225-262

Ryerson FJ, McKeegan KD (1994) Determination of oxygen self-diffusion in akermanite, anorthite, diopside, and spinel: Implications for oxygen isotopic anomalies and the thermal histories of Ca-Al-rich inclusions. Geochim Cosmochim Acta 58:3713-3734

Santosh M, Wada H (1993) A carbon isotope study of graphites from the Kerala Khondalite Belt, Southern India: Evidence for CO_2 infiltration in granulites. J Geol 101:643-651

Satish-Kumar M (2000) Ultrahigh-temperature metamorphism in Madurai granulites, Southern India: Evidence from carbon isotope thermometry. J Geol 108:479-486

Satish-Kumar M, Wada H (2000) Carbon isotopic equilibrium between calcite and graphite in Skallen Marbles, East Antarctica: Evidence for the preservation of peak metamorphic temperatures. Chem Geol 166:173-182

Satish-Kumar M, Santosh M, Wada H (1997) Carbon isotope thermometry in marbles of Ambasamudram, Kerala Khondalite Belt, southern India. J Geol Soc India 49:523-532

Satish-Kumar M, Yoshida M, Wada H, Niitsuma N, Santosh M (1998) Fluid flow along microfractures in calcite from a marble from East Antarctica: Evidence from gigantic oxygen isotopic zonation. Geology 26:251-254

Savin SM, Lee M (1988) Isotopic studies of phyllosilicates. Rev Mineral 19:189-223

Scheele N, Hoefs J (1992) Carbon isotope fractionation between calcite, graphite and CO_2: An experimental study. Contrib Mineral Petrol 112:35-45

Schidlowski M (2001) Carbon isotopes as biogeochemical recorders of life over 3.8 Ga of Earth history: Evolution of a concept. Precambrian Res 106:117-134

Schidlowski M, Appel PWU, Eichmann R, Junge CE (1979) Carbon isotope geochemistry of the 3.7 x 10^9 yr. old Isua sediments, West Greenland: Implications for the Archaean carbon and oxygen cycles. Geochim Cosmochim Acta 43:189-199

Schrauder M, Beran A, Hoernes S, Richter W (1993) Constraints on the Origin and the Genesis of Graphite-Bearing Rocks from the Variegated Sequence of the Bohemian Massif (Austria). Mineral Petrol 49:175-188

Schulze DJ, Valley JW, Viljoen KS, Stiefenhofer J, Spicuzza M (1997) Carbon isotope composition of graphite in mantle eclogites. J Geol 105:379-386

Schwarcz HP (1966) Oxygen isotope fractionation between host and exsolved phases in perthite. Geol Soc Am Bull 77:879-882

Schwarcz HP, Clayton RN, Mayeda T (1970) Oxygen isotopic studies of calcareous and pelitic metamorphic rocks, New England. Geol Soc Am Bull 81:2299-2316

Shanks WC, Crowe DE, Johnson C (1998) Sulfur isotope analyses using the laser microprobe. In McKibben RA, Shanks WC, Ridley WI (eds) Soc Econ Geol Rev Econ Geol 7:141-153

Sharma T, Clayton RN (1965) Measurement of O^{18}/O^{16} ratios of total oxygen of carbonates. Geochim Cosmochim Acta 29:1347-1353

Sharp ZD (1990) A laser-based microanalytical method for the in situ determination of oxygen isotope ratios of silicates and oxides. Geochim Cosmochim Acta 54:1353-1357

Sharp ZD (1991) Determination of oxygen diffusion rates in magnetite from natural isotopic variations. Geology 19:653-656

Sharp ZD (1992) *In situ* laser microprobe techniques for stable isotope analysis. Chem Geol 101:3-19

Sharp ZD (1995) Oxygen isotope geochemistry of the Al_2SiO_5 polymorphs. Am J Sci 295:1058-1076

Sharp ZD, Essene EJ (1991) Metamorphic conditions of an Archean core complex in the northern Wind River Range, Wyoming. J Petrol 32:241-273

Sharp ZD, Jenkin GRT (1994) An empirical estimate of the diffusion rate of oxygen in diopside. J Metamor Geol 12:89-97

Sharp ZD, Cerling TE (1996) A laser GC-IRMS techniques for *in situ* stable isotope analyses of carbonates and phosphates. Geochim Cosmochim Acta 60:2909-2916

Sharp ZD, Giletti, BJ, Yoder HS (1991) Oxygen diffusion rates in quartz exchanged with CO_2. Earth Planet Sci Letters 107:339-348.

Sharp ZD, Essene EJ, Smyth JR (1993) Ultra-high temperatures from oxygen isotope thermometry of a coesite-sanidine grospydite. Contrib Mineral Petrol 112:358-370

Sharp ZD, Essene EJ, Hunziker JC (1993) Stable isotope geochemistry and phase equilibria of coesite-bearing whiteschists, Dora Maira Massif, western Alps. Contrib Mineral Petrol 114:1-12

Sharp ZD, Frey M, Levi KJT (1995) Stable isotope variations (H, C, O) in a prograde metamorphic Triassic red bed formation, Central Swiss Alps. Schweiz mineral petrogr Mitt 75:147-161

Sharp ZD, O'Neil JR, Essene EJ (1988) Oxygen isotope variations in granulite-grade iron formations: Constraints on oxygen diffusion and retrograde isotopic exchange. Contrib Mineral Petrol 98:490-501

Shelton KL (1983) Composition and origin of ore-forming fluids in a carbonate-hosted porphyry copper and skarn deposit: A fluid inclusion and stable isotope study of Mines Gaspe, Quebec. Econ Geol 78:387-421

Shelton KL, Rye DM (1982) Sulfur isotopic compositions of ores from Mines Gaspe, Quebec: An example of sulfate-sulfide isotopic disequilibria in ore-forming fluids with applications to other porphyry-type deposits. Econ Geol 77:1688-1709

Shemesh A, Kolodny Y, Luz B (1983) Oxygen isotope variations in phosphate of biogenic apatites, II: Phosphorite rocks. Earth Planet Sci Letters 64:405-416

Shemesh A, Kolodny Y, Luz B (1988) Isotope geochemistry of oxygen and carbon in phosphate and carbonate of phosphorite francolite. Geochim Cosmochim Acta 52:2565-2572

Sheppard SMF, Gilg HA (1996) Stable isotope geochemistry of clay minerals. Clay Minerals 31:1-24

Shieh Y-S, Schwarcz HP (1974) Oxygen isotope studies of granite and migmatite, Grenville province of Ontario, Canada. Geochim Cosmochim Acta 38:21-45

Shieh Y-S, Schwarcz HP, Shaw DM (1976) An Oxygen Isotope Study of the Loon Lake Pluton and the Apsley Gneiss, Ontario. Contrib Mineral Petrol 54:1-16

Shimizu M, Iiyama JT (1982) Zinc-lead skarn deposits of the Nakatatsu mine, central Japan. Econ Geol 77:1000-1012

Sitzman SD, Banfield JF, Valley JW (2000) Microstructural characterization of metamorphic magnetite crystals with implications for oxygen isotope distribution. Am Mineral 85:14-21

Smalley PC, Stijfhoorn DE, Raheim JH, Dickson JAD (1989) The laser microprobe and its application to the study of C and O isotopes in calcite and aragonite. Sedimentary Geol 65:1-11

Spear FS, Florence FP (1992) Thermobarometry in granulites: Pitfalls and new approaches. Precambrian Res 55:209-241

Stern MJ, Spindel W, Monse EU (1968) Temperature dependence of isotope effects. J Chem Phys 48:2908-2919

Stoffregen RE, Rye RO, Wasserman MD (1994) Experimental studies of alunite: I, ^{18}O-^{16}O and D-H fractionation factors between alunite and water at 250-450°C. Geochim Cosmochim Acta 58:903-916

Taylor BE, O'Neil JR (1977) Stable isotope studies of metasomatic Ca-Fe-Al-Si skarns and associated metamorphic and igneous rocks, Osgood Mountains, Nevada. Contrib Mineral Petrol 63:1-49

Tennie A, Hoffbauer R, Hoernes S (1998) The oxygen isotope fractionation behavior of kyanite in experiment and in nature. Contrib Mineral Petrol 133:346-355

Thrower PA, Mayer RM (1978) Point defects and self-diffusion in graphite. Physica Status Solidi 47:11-37

Turner DR, Bowman JR (1993) Origin and evolution of skarn fluids, Empire zinc skarns, Central Mining District, New Mexico, USA. Appl Geochem 8:9-36

Urey HC (1947) The thermodynamic properties of isotopic substances. J Chem Soc (1947):562-581

Ustinov VI, Grinenko VA (1985) Intrastructural isotope distribution during mineral formation. Geochem Int'l 22:143-149

Valley JW, O'Neil JR (1980) $^{13}C/^{12}C$ exchange between calcite and graphite: A possible thermometer in Grenville marbles. Geochim Cosmochim Acta 45:411-419

Valley JW, Graham CM (1991) Ion microprobe analysis of oxygen isotope ratios in granulite facies magnetites: Diffusive exchange as a guide to cooling history. Contrib Mineral Petrol 109:38-52

Valley JW, Graham CM (1993) Cryptic grain-scale heterogeneity of oxygen isotope ratios in metamorpic magnetite. Science 259:1729-1733

Valley JW, Graham CM (1996) Ion microprobe analysis of oxygen isotope ratios in quartz from Skye granite: Healed micro-cracks, fluid flow, and hydrothermal exchange. Contrib Mineral Petrol 124:225-234

Valley JW, Chiarenzelli J, McLelland JM (1994) Oxygen isotope geochemistry of zircon. Earth Planet Sci Letters 126:187-206

Valley JW, Kinny PD, Schulze MJ (1998b) Zircon megacrysts from kimberlite: Oxygen isotope heterogeneity in mantle melts. Contrib Mineral Petrol 133:1-11

Valley JW, Kitchen N, Kohn MJ, Niendorf CR, Spicuzza MJ (1995) UWG-2, a garnet standard for oxygen isotope ratio: Strategies for high precision and accuracy with laser heating. Geochim Cosmochim Acta 59:5223-5231

Valley JW, Graham CM, Harte B, Eiler JM, Kinny PD (1998a) Ion Microprobe Analysis of Oxygen, Carbon, and Hydrogen Isotope Ratios. In McKibben MA, Shanks III WC, Ridley WI (eds) Applications of Microanalytical Techniques to Understanding Mineralizing Processes. Soc Econ Geol Rev Econ Geol 7:73-98

Valley JW, Komor SC, Baker K, Jeffrey AWA, Kaplan IR, Raheim A (1988) Calcite crack cements in granite from the Siljan Ring, Sweden: Stable isotopic results. In Boden A, Eriksson KG (eds) Deep Drilling in Crystalline Bedrock. Springer-Verlag, New York, p 156-179

van der Pluijm BA, Carlson KA (1989) Extension in the central metasedimentary belt of the Ontario Grenville: Timing and tectonic significance. Geology 17:161-164

van Haren JLM, Ague JJ, Rye DM (1996) Oxygen isotope record of fluid infiltration and mass transfer during regional metamorphism of pelitic schist, Connecticut, USA. Geochim Cosmochim Acta 60:3487-3504

Vannay J-C, Sharp ZD, Grasemann B (1999) Himalayan inverted metamorphism constrained by oxygen isotope thermometry. Contrib Mineral Petrol 137:90-101

Vogel DE, Garlick GD (1970) Oxygen-isotope ratios in metamorphic eclogites. Contrib Mineral Petrol 28:183-191

Vry J, Brown PE, Valley JW, Morrison J (1988) Constraints on granulite genesis from carbon isotope compositions of cordierite and graphite. Nature 332:66-68

Wada H (1977) Isotopic studies of graphite in metamorphosed carbonate rocks of central Japan. Geochem J 11:183-197

Wada H (1978) Carbon isotopic study on graphite and carbonate in the Kamioka Mining District, Gifu Prefecture, Central Japan, in relation to the role of graphite in the pyrometasomatic ore deposition. Mineral Deposita 13:201-220

Wada H (1988) Microscale isotopic zoning in calcite and graphite crystals in marble. Nature 331:61-63

Wada H, Oana S (1975) Carbon and oxygen isotope studies of graphite bearing carbonates in the Kasuga area, Gifu Prefecture, central Japan. Geochem J 9:149-160

Wada H, Suzuki K (1983) Carbon isotope thermometry calibrated by dolomite-calcite solvus temperatures. Geochim Cosmochim Acta 47:697-706

Wada H, Enami M, Yanagi T (1984) Isotopic studies of marbles in the Sanbagawa metamorphic terrain, central Shikoku, Japan. Geochem J 18:61-73

Waldron K, Lee MR, Parsons I (1994) The microstructures of perthitic alkali feldspars revealed by hydrofluoric acid etching. Contrib Mineral Petrol 116:360-364

Walker FDL (1990) Ion microprobe study of intragrain micropermeability in alkali feldspars. Contrib Mineral Petrol 106:124-128

Watson EB, Cherniak DJ (1997) Oxygen diffusion in zircon. Earth Planet Sci Letters 148:527-544

Weis PL (1980) Graphite skeleton crystals—A newly recognized morphology of crystalline carbon in metasedimentary rocks. Geology 8:296-297

Weis PL, Friedman I, Gleason JP (1981) The origin of epigenetic graphite: Evidence from isotopes. Geochim Cosmochim Acta 45:2325-2332

Wiechert U, Hoefs J (1995) An excimer laser-based micro analytical preparation technique for in situ oxygen isotope analysis of silicate and oxide minerals. Geochim Cosmochim Acta 59:4093-4101

Wright K, Freer R, Catlow CRA (1995) Oxygen diffusion in grossular and some geological implications. Am Mineral 80:1020-1025

Yapp CJ (1990) Oxygen isotopes in iron (III) oxides 2. Possible constraints on the depositional environment of a Precambrian quartz-hematite banded iron formation. Chem Geol 85:337-344

Young ED (1993) On the $^{18}O/^{16}O$ record of reaction progress in open and closed metamorphic systems. Earth Planet Sci Letters 117:147-167

Young ED, Fogel ML, Rumble D, Hoering TC (1998) Isotope ratio monitoring of O_2 for microanalysis of $^{18}O/^{16}O$ and $^{17}O/^{16}O$ in geological materials. Geochim Cosmochim Acta 62:3087-3094

Zhang Y, Stolper EM, Wasserburg GJ (1991) Diffusion of a multi-species component and its role in oxygen and water transport in silicates. Earth Planet Sci Letters 103:228-240

Zheng Y-F (1991) Calculation of oxygen isotope fractionation in metal oxides. Geochim Cosmochim Acta 55:2299-2307

Zheng Y-F (1992) Discussion on the use of δ–Δ diagram in interpreting stable isotope data. Eur J Mineral 4:635-643

Zheng Y-F (1993) Calculation of oxygen isotope fractionation in hydroxyl-bearing silicates. Earth Planet Sci Letters 120:247-263

Zheng Y-F (1993b) Oxygen isotope fractionation in SiO_2 and Al_2SiO_5 polymorphs: Effect of crystal structure. Eur J Mineral 5:651-658

Zheng Y-F (1993c) Calculation of oxygen isotope fractionation in anhydrous silicate minerals. Earth Planet Sci Letters 120:247-263

Zheng Y-F (1995) Oxygen isotope fractionation in magnetites: Structural effect and oxygen inheritance. Chem Geol 121:309-316

Zheng Y-F (1996) Oxygen isotope fractionations involving apatites: Application to paleotemperature determination. Chem Geol 127:177-187

Zheng Y-F, Fu B (1998) Estimation of oxygen diffusivity from anion porosity in minerals. Geochem J 32:71-89

Zheng Y-F, Simon K (1991) Oxygen isotope fractionation in hematite and magnetite: A theoretical calculation and application to geothermometry of metamorphic iron-formations. Eur J Mineral 3:877-886

Zheng Y-F, Fu Bin YL, Xiao Y, Shuguang L (1998) Oxygen and hydrogen isotope geochemistry of ultrahigh-pressure eclogites from the Dabie Mountains and the Sulu terrane. Earth Planet Sci Letters 155:113-129

Zheng Y-F, Fu B, Yilin X, Yiliang L, Bing G (1999) Hydrogen and oxygen isotope evidence for fluid-rock interactions in the stages of pre- and post-UHP metamorphism in the Dabie Mountains. Lithos 46:677-693

Zinner E, Ming T, Anders E (1989) Interstellar SiC in the Murchison and Murray meteorites: Isotopic composition of Ne, Xe, Si, C, and N. Geochim Cosmochim Acta 53:3273-3290

7

Stable Isotope Transport and
Contact Metamorphic Fluid Flow

Lukas P. Baumgartner

Institute of Geosciences
Johannes Gutenberg University of Mainz
Johann-Joachim-Becher-Weg 21
55099 Mainz – Germany

John W. Valley

Department of Geology and Geophysics
University of Wisconsin
Madison, Wisconsin 53706

INTRODUCTION

Stable isotopes are a powerful tool for deciphering the fluid histories of metamorphic terranes. The nature of fluid flow, fluid sources, and fluid fluxes can be delineated in well-constrained studies. Observed isotopic gradients in metamorphic rocks and minerals can thus shed light on many processes involved in mass-transport including diffusion, recrystallization, fluid infiltration, volatilization, metasomatism, and heat flow. Modeling of fluid flow and mineral exchange kinetics offers greatly enhanced understanding of metamorphic processes that can be tested and refined by application of new micro-analytical techniques. This review will concentrate on the principles of stable isotope fluid-rock interaction with an emphasis on fluid-rock interaction and fluid flow in contact metamorphism. Earlier reviews discuss some aspects of regional metamorphism and hydrothermal systems (Valley 1986; Kerrich 1987; Nabelek 1991; Young 1995; Ferry and Gerdes 1998; Bowman 1998).

Isotopic studies are especially useful for defining the scale of fluid migration. The intensity of interaction between fluids and the minerals in rocks can be assessed. During metamorphism, the scale of isotopic exchange can vary from less than a micrometer to over 10 kilometers. Many fluid-driven processes are characterized by the degree to which fluid flow is concentrated into zones of high permeability. Thus, the definition of two end-member situations is useful. The flow of a *pervasive fluid* is distributed throughout the pores in a rock. Pervasive flow can be along grain boundaries or fine-scale crack networks and the effect is to homogenize the chemical potential of all components, including stable isotopes, at a macroscopic scale. In contrast, the flow of a *channeled fluid* is along vein systems, shear zones or other channelways such as rock contacts or more permeable lithologic units. Channeled flow leads to local chemical heterogeneity, allowing some rocks to remain unaffected while others are extensively infiltrated and modified isotopically. If flow is channeled, open and closed systems can occur in close proximity and one-dimensional flow models are not sufficient (e.g. Gerdes et al. 1995b; Baumgartner et al. 1996; Cartwright and Buick 1996; Bolton et al. 1999). Accurate fluid budgets require knowledge of the degree of channelization, fluid pathways, and the fluid flux.

The stable isotope composition of a metamorphic rock is controlled by six factors: (1) the composition of the pre-metamorphic protolith; (2) the effects of volatilization; (3) the temperature of exchange; (4) exchange kinetics; (5) fluid composition, and (6) fluid flux. These factors are best evaluated in studies of contact metamorphism because of the

Abbreviations and Symbols

α_{f-r}	isotope fractionation factor between fluid and rock
α_L	longitudinal dispersivity
β_i^O	stoichiometry of oxygen in phase i
$C_i^{^{18}O}$	concentration of ^{18}O in phase I
$\bar{c}$	normalized concentration
D	diffusion coefficient
D_L	longitudinal dispersion coefficient
D_T	transverse dispersion coefficient
$\Delta x, \Delta y, \Delta z$	small increments in x, y, z direction
δ_f^f, δ_r^f	final isotope value of fluid (f) or rock (r)
δ_f^i, δ_r^i	initial isotope value of fluid (f) or rock (r)
$\delta^{13}C_f^f, \delta^{13}C_r^f$	final ^{13}C value of fluid (f) or rock (r)
$\delta^{13}C_f^i, \delta^{13}C_r^i$	initial ^{13}C value of fluid (f) or rock (r)
$\delta^{18}O_f^f, \delta^{18}O_r^f$	final ^{18}O value of fluid (f) or rock (r)
$\delta^{18}O_f^i, \delta^{18}O_r^i$	initial ^{18}O value of fluid (f) or rock (r)
F	mole fraction of element of interest remaining in the rock
F_1, F_2	flux in x-direction through surface 1 or 2 (see Fig. 6, below)
$\Delta F^{^{18}O}$	change in flux of ^{18}O
Φ	porosity of rock
h	fresh water equivalent head
g	gravitational acceleration constant
Γ	tortuosity of pore space
$J_{dif}^{^{18}O}$	diffusive flux of ^{18}O
$J_{disp}^{^{18}O}$	dispersive flux of ^{18}O
$J_{inf}^{^{18}O}$	infiltrative flux of ^{18}O
K_D^j	molar equilibrium constant between solid and fluid for isotope j
L	characteristic length scale of system
μ_r	relative fluid viscosity
N_{Pe}	Peclet number
P	pressure
Q	fluid production rate per unit volume
r_k	reaction rate of the mineral/fluid isotope reaction k
R_f^r, R_r^f	final isotope ratio of fluid (f) or rock (r)
R_f^i, R_r^i	initial isotope ratio of fluid (f) or rock (r)
ρ_f	fluid density
ρ_r	relative fluid density
ρ_o	fluid density at reference pressure
T	temperature
t	time, or dimensionless time
V	volume
$\bar{V}$	molar volume
v_f	Darcy velocity of the fluid phase
v_p	average linear pore fluid velocity
x, y, z	coordinates
X_i	mole fraction of species I
$\bar{x}$	transposed coordinate

good geologic control and the common occurrence of fluids with distinct initial isotopic compositions in these systems. One example is the large initial difference in $\delta^{18}O$ between intrusive igneous rocks and carbonates which facilitates the study of fluid sources, CO_2/H_2O ratios, direction of flow, fluid flux, and temperature. The effect of volatilization on stable isotope composition of rocks is assessed first. In the second section of this chapter, the equations for modeling fluid flow are derived and their use reviewed. A short review of stable isotope systematics in contact aureoles is presented in the final part of the chapter. Special attention is given to the Alta aureole, where detailed measurements and theory have been applied by several groups working in fluid/rock interaction. This focus was chosen because the insight gained from contact metamorphism is necessary to attack the more difficult problems related to regional metamorphism (see Valley 1986).

"CLOSED SYSTEM" METAMORPHIC VOLATILIZATION

Prograde metamorphism of sediments (and to a lesser degree igneous and metamorphic rocks) causes the liberation of "volatile" components by the reaction of lower temperature, volatile rich minerals. If no externally derived fluids infiltrate the rock, volatilization is often referred to as "closed system" even though it is clear that evolved fluids have left the rock. Dehydration is most common, but decarbonation also occurs in carbonate-bearing lithologies (Ferry and Burt 1982) and desulfidation can locally be important (see also Cartwright and Oliver 2000).

Volatilization reactions typically have large, positive volume changes. Hence the volume of the produced fluid and the residual solids is greater than that of the initial solids and pore fluids, creating a fluid overpressure sufficient for fluid expulsion (Hanson 1992; Hanson et al. 1995a,b; Ferry 1995; Connolly 1997; Connolly and Podladchikov 2000). Small, transient fluid over-pressures increase permeability allowing buoyant fluids to rise in the crust (Walther and Orville 1982; Walther and Wood 1984; Connolly 1997). In shallow environments, if permeability is sufficiently high to permit convection (see Etheridge et al. 1983; Brace 1984; Gerdes et al. 1998; Bolton et al. 1999; Cook et al. 1997), it is also possible that hydrothermal fluids will migrate laterally or downward. Most evolved fluids are expelled from their rock of origin. The general absence of voids in metamorphosed rocks shows that the amount of retained fluid is almost nothing. In most rocks, fluid inclusions are the sole remnants of metamorphic fluids.

The liberation of metamorphic fluids can have a profound effect on the stable isotope composition of the residual rock. Thus, isotopic ratios provide information about the nature and amounts of volatilization that have occurred. The effects of volatilization can be modeled as one of two end-member equilibrium processes: (1) *batch volatilization,* where all fluid is evolved before any is permitted to escape and (2) *Rayleigh volatilization,* where each volatile molecule is immediately isolated from its rock of origin due to steady and perhaps slow expulsion. Most natural processes at high temperatures fall between these extremes and the models provide useful limits. At low temperature and in environments with rapidly changing conditions, isotopic disequilibrium can be important (Lasaga and Rye 1993).

BATCH VOLATILIZATION

If a rock volatilizes in the absence of infiltration and all evolved fluid equilibrates with the rock before being expelled, then the isotopic ratio of the rock will increase or decrease depending on whether the fluid preferentially partitions the light or the heavy isotope. This process is termed "batch" volatilization (Rumble 1982; Valley 1986). Depletion of the rock in the heavy isotope is most common. Assuming equilibrium, the magnitude of this effect will vary directly with the amount of volatilization in accord with mass balance:

$$\delta_r^f = \delta_r^i - (1-F)1000 \ln\alpha_{f-r} \qquad (1)$$

where F is the mole fraction of the element of interest that remains in the rock after volatilization; α_{f-r} is the fractionation factor (fluid-rock); and δ_r^i, δ_r^f are the initial and final isotopic values of the rock in standard permil notation.

The amount of ^{18}O depletion relative to ^{16}O caused by batch volatilization of a siliceous dolomite ($\delta_r^i = 22‰$, $\alpha_{f-r} = 1.0060$) is shown by the straight lines in Figure 1. The application of this calculation requires a number of assumptions that will be discussed in more detail later. Most importantly, the large volume increase that accompanies volatilization requires that fluids escape more or less continuously under normal conditions and thus true batch volatilization is unlikely. Nevertheless, the simplicity of Equation (1) makes it very useful. In most instances of volatilization involving the isotopes of oxygen, and for many cases involving C, H and S, the difference between the results of a calculation using Equation (1) and those of a more complex Rayleigh model are minimal compared to other uncertainties involved. In any case, Equation (1) yields a minimum estimate of the isotopic effect due to equilibrium volatilization.

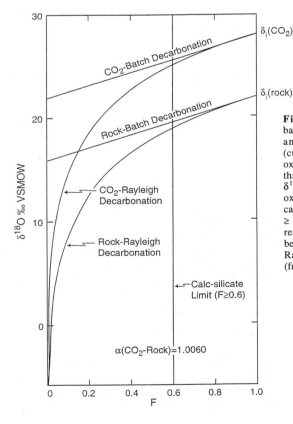

Figure 1. Lowering of $\delta^{18}O$ by batch decarbonation (straight line) and Rayleigh decarbonation (curves). F is the mole fraction of oxygen remaining in the rock. Note that for Rayleigh decarbonation, $\delta^{18}O$ tends toward -1000‰ if all oxygen is volatilized, but that a calc-silicate limit exists such that F ≥ 0.6 for most metamorphic reactions. There is little difference between the results of the batch and Rayleigh models above F = 0.6 (from Valley 1986).

Rayleigh volatilization

The process of Rayleigh volatilization or distillation involves the continuous exchange and removal of infinitely small aliquots of fluid, each before the volatilization of the next (see Rumble 1982; Valley 1986). Rayleigh volatilization may closely approximate the isotopic effects of dehydration and decarbonation (Shieh and Taylor 1969; Rumble 1982; Valley and O'Neil 1984; Valley 1986).

The following equations are equivalent and quantify the isotopic change due to Rayleigh volatilization.

$$\frac{R_r^f}{R_r^i} = F^{(\alpha_{f-r}-1)} \tag{2}$$

$$\delta_r^f - \delta_r^i = 1000 \left(F^{(\alpha_{f-r}-1)} - 1 \right) \tag{3}$$

R_r^f and R_r^i are the final and initial isotopic ratios (i.e. $^{18}O/^{16}O$, etc.) of the rock system and other terms are as defined for Equation (1).

The importance of Rayleigh volatilization is seen from the curves in Figure 1, which solve Equation (3) for $\alpha_{f-r} = 1.0060$ and $\delta_r^i = 22\%o$. This calculation is made to model the effect of decarbonation on $\delta^{18}O$ in siliceous dolomites, but in general Figure 1 may be applied to any volatilization process involving O, C, H or S (changing the scale on the ordinate axis allows for different values of δ^i and α). In the case of decarbonation, Figure 1 shows that when F is close to 1.0 the majority of oxygen is still in the rock, only small amounts of CO_2 have been liberated, and $\delta^{18}O$ of the rock is slightly decreased as the escaping CO_2 preferentially partitions ^{18}O over ^{16}O. For values of $F \geq 0.6$ the amount of $^{18}O/^{16}O$ depletion by a Rayleigh process is very similar to that of batch volatilization.

In processes where volatilization nears completion, the Rayleigh model becomes important and substantial deviations from the batch model occur. For instance, in Figure 1 at $F = 0.05$, 95% of the oxygen has been volatilized as CO_2 and $\delta^{18}O_{rock} \approx 4$, which is over 12 permil lower than the corresponding "batch" value. This magnification of the heavy isotope depletion arises because each successive molecule of volatilizing CO_2 partitions with an increasingly ^{18}O-depleted rock, compounding the effect.

Because of the stoichiometry of most volatilization reactions, values of F below 0.6 are hypothetical for oxygen and are very unlikely in nature (see for example Reactions 4 and 5). This is called the calc-silicate limit (Valley 1986). However, for C, H and S, any value of F from 1.0 to 0.0 may be expected and thus Rayleigh-controlled processes are particularly important for these elements.

A rigorous application of Equations (2) and (3) can be complex. At a given moment, α_{f-r} is the stoichiometrically weighted average fractionation between fluid and rock (Rumble 1982). To gain an exact value of α_{f-r} may require knowledge of fractionation factors relative to a mixed-volatile fluid. Further complication arises because α_{f-r} is not constant during reaction—it varies with temperature, the modal abundance of minerals, and the fluid composition. Pressure thus indirectly affects α_{f-r} by controlling temperature of reaction and fluid composition; larger fractionations are expected in low-pressure environments where most metamorphic reactions (with positive P-T slopes) occur at a lower temperature. Furthermore, the exact pre-metamorphic composition of a rock is difficult to know and can only be approximated in a regionally metamorphosed terrane. The situation is often better controlled in a contact aureole, where the non-metamorphic protolith is available.

Equilibrium values are often accepted as good approximations for α during regional metamorphism, but growth zonation and heterogeneity can be preserved (Kohn et al. 1993; Kohn and Valley 1994). This documents at least partial disequilibrium between the minerals and the fluid. At lower temperatures, minerals are generally not equilibrated during hydrothermal alteration and the approach to equilibrium is uncertain during contact metamorphism.

Dehydration

Dehydration is the best-known and most common example of metamorphic volatilization. The magnitude of the isotopic effect is controlled by the amounts of water expelled; high-grade rocks contain less water than lower grade equivalents due to progressive dehydration (Spear 1993). In contrast to shales, which are the most water-rich protolith commonly available for metamorphic dehydration (up to 5 wt %), igneous rocks typically contain only 0.5 to 0.8 wt % H_2O (Wedepohl 1969).

The effect of dehydration reactions on the value of $\delta^{18}O$ in a rock will always be small, less than 1 permil. The magnitude of this change can be calculated with a few simplifying assumptions, all conservatively made so as to maximize the isotope effect. If it is assumed that 5 wt % H_2O is driven off, then Equation (1) for batch volatilization can be used with $F \approx 0.9$ (H_2O is 89 wt % oxygen while rocks are approximately 50%). The temperature effect on $\alpha_{f,r}$ is important due to the crossover in the sign of fractionation. At low temperature (T < 400-500°C) H_2O is isotopically lighter than an average rock and dehydration will cause enrichment in $^{18}O/^{16}O$ in the remaining rock. At T > 500°C, oxygen in H_2O is heavier than in most minerals and dehydration causes $^{18}O/^{16}O$ depletion. Thus, reactions may tend to cancel each other depending on the details of the prograde reaction path. Furthermore, even if the entire 5 wt % H_2O is evolved at a low T (300°C) and a large fractionation is chosen ($\alpha_{f,r} = 0.994$), then Equation (1) yields only $\delta_r^f - \delta_r^i = $ +0.6 permil for oxygen. Given the isotopic heterogeneity of metasedimentary rocks, it is unlikely that such small effects can be recognized or are significant.

The effect of dehydration on δD may be much larger because, in contrast to $\delta^{18}O$, values of F can range down to 0.0. The details of this process will depend greatly on the value of α, but as an example consider $\alpha_{H_2O-muscovite}^{D/H} = 1.018$ at 500°C (Suzuoki and Epstein 1976). If a rock initially contains 3.0 wt % water and liberates 2.7% then F = 0.1 and from Equation (1) (batch) $\delta_r^f - \delta_r^i = $ -16.2 permil (for the rock). In contrast, as gradual fluid escape is more likely, Rayleigh dehydration (Eqn. 3) yields $\delta_r^f - \delta_r^i = $ -40.6 permil, a substantially larger depletion.

Mixed volatile reactions

Carbon dioxide is frequently evolved with H_2O during metamorphism of siliceous carbonates and marls. Large volumes of rock can be volatilized (up to 40%) with the potential for significant isotopic shifts.

In an extreme example of simple decarbonation, the appropriate modal proportions of quartz react with either dolomite or calcite to form a rock that is 100% diopside or wollastonite by reactions such as:

$$CaMg(CO_3)_2 + 2\ SiO_2 = CaMgSi_2O_6 + 2\ CO_2 \tag{4}$$
dolomite   quartz   diopside

$$CaCO_3 + SiO_2 = CaSiO_3 + CO_2 \tag{5}$$
calcite   quartz   wollastonite

For complete decarbonation, these reactions have volume increases (for rock plus fluid) of approximately +40% at metamorphic P-T conditions showing that fluids must escape during reaction. However, ΔV of reaction for the rock only is $\sim-35\%$, potentially creating significant permeability. In spite of the large amounts of reaction indicated by these volume changes, there is a "calc-silicate limit" of $F \geq 0.6$ for oxygen (Eqns. 1-3; Fig. 1). This limit exists because the dominant oxygen reservoir of a silicate rock will always be the silicate minerals that remain as the residual metamorphosed rock, even if decarbonation is complete.

Many complex reactions or sequences of reactions can be considered in an effort to accurately calculate the isotopic effects of mixed volatile reactions in a metamorphic rock. Sometimes the results are not significantly different from those calculated for Reactions (4) or (5). Volatilization effects always depend on the amounts of fluid evolved by reaction.

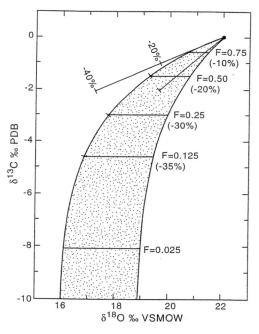

Figure 2. Plot of the coupled $\delta^{13}C$ vs. $\delta^{18}O$ trends that result from batch (straight lines) and Rayleigh (curves) volatilization assuming normal calc-silicate decarbonation (see Fig. 3). Average fractionation factors are appropriate for metamorphic temperatures; $\alpha^{13}C(CO_2\text{-rock}) = 1.0022$ and $\alpha^{18}O(CO_2\text{-rock}) = 1.006$ and 1.012 (larger $\delta^{18}O$ shifts correspond to $\alpha = 1.012$). Values of F(carbon) and the approximate volume reduction of the rock are also shown. The resulting $\delta^{18}O$ is little affected by choice of batch versus Rayleigh model, though large differences in estimated $\delta^{13}C$ result (from Valley 1986).

COUPLED O-C DEPLETIONS

The magnitude of $^{18}O/^{16}O$ and $^{13}C/^{12}C$ depletions is directly linked and can be calculated if reaction stoichiometry is known. Thus, when Equations (1) or (3) are applied to a specific reaction sequence, a coupled O-C trend results, as seen in Figure 2, where $\alpha^{13C}_{CO_2-rock} = 1.0022$, $\alpha^{18O}_{CO_2-rock} = 1.0120$ to 1.0060, $\delta^{13}C^i_r = 0.0$, and $\delta^{18}O^i_r = 22.0$. The two straight lines model batch volatilization (Eqn. 1 and $\alpha^{18O}_{f-r} = 1.0060, 1.0120$). The curved lines model Rayleigh volatilization using Equation (3), the same values of α^{18O}_{f-r}, and a normal calc-silicate decarbonation trend for F (see discussion for Fig. 3). For $\delta^{18}O$, only small differences (≤ 1.3 permil) are seen between the Rayleigh and batch calculations, but for $\delta^{13}C$, the limit of the Rayleigh calculation is $-1000\%o$ as F $\to$ 0.

Note that this value is an artifact of the approximation introduced by replacing the isotope ratios in Equation (2) by the delta-values in Equation (3). The ratio R_r^f / R_r^i trends towards 0 (zero). F_{carbon} and $\Delta V_{reaction}$ for solids are shown on the Rayleigh curves in Figure 2 predicting that large depletions in $^{13}C/^{12}C$ only occur as decarbonation nears completion and as nearly all carbon is converted to CO_2. Thus, the relation of F_{oxygen} vs. F_{carbon} controls the shape and magnitude of coupled depletion under Rayleigh conditions.

Figure 3 illustrates different relations of F_{oxygen} to F_{carbon}, which depend on stoichiometry, and kinetics of reaction. The normal calc-silicate decarbonation trend applies if all minerals in the rock are fully equilibrated during a reaction such as 4 or 5, in which case all carbon in the rock is liberated as CO_2 ($F_{carbon} \rightarrow 0$), while only 40% of the oxygen is released ($F_{oxygen} \rightarrow 0.6$). However, if the rock has (for example) 50% excess silicates that are not involved in the reaction, but which still equilibrate isotopically, then $F_{oxygen} \rightarrow 0.8$ as $F_{carbon} \rightarrow 0.0$ along the 50% inert oxygen trend and the amount of $^{18}O/^{16}O$ depletion will be smaller. Likewise, if 50% of a rock's carbon does not participate due to stoichiometric excess, a 50% inert carbon trend will be followed and the depletion of $^{13}C/^{12}C$ will also be less. In practice any trend is possible from ~100% excess oxygen to ~100% excess carbon, but any deviation from normal decarbonation will diminish the total isotopic change in either O or C and the "normal" trend yields the largest isotope variations and is thus quantitatively most important.

Depletions in $^{18}O/^{16}O$ that are larger than those that result from "normal" calc-silicate decarbonation trends may be accomplished by reactions that exceed the calc-silicate limit of $F_{oxygen} = 0.6$ (Fig. 3). Some rare decarbonation reactions achieve this in the absence of silicates, such as the breakdown of calcite to lime:

$$CaCO_3 = CaO + CO_2 \tag{6}$$

which follows a silicate-absent decarbonation trend (Fig. 3). Still larger $^{18}O/^{16}O$ depletions would be possible if carbonates volatilize without maintaining isotopic equilibrium with coexisting silicates (Lattanzi et al. 1980; Lasaga and Rye 1993).

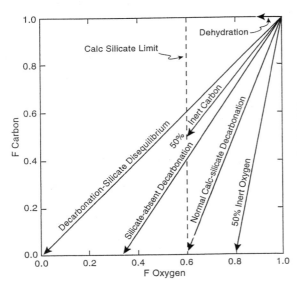

Figure 3. Values of F(carbon) vs. F(oxygen) along various reaction paths. F is the mole fraction of O or C remaining in the rock after reaction. Most marbles, if equilibrated, will follow a F-F path intermediate between 50% inert oxygen and 50% inert carbon (from Valley 1986).

Coupled O-C isotope depletions are seen in many metamorphic systems involving carbonate rocks. Table 1 and Figures 4 and 5 summarize results for 28 studies of marbles, mostly in contact metamorphic settings. These results will be discussed in more detail later, but one conclusion is general and deserves emphasis. In each of these localities, the O-C trend has a negative slope in Figures 4 and 5, qualitatively similar to the effects of devolatilization. However, in each area, the magnitude of the depletions is too large to be explained by the closed-system devolatilization processes discussed here. Significant fluid infiltration and exchange involving low $\delta^{18}O$, low $\delta^{13}C$ fluids is indicated by the stable isotope data (see Valley 1986).

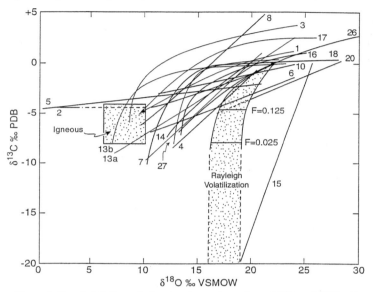

Figure 4. Coupled O-C trends showing decreasing values of $\delta^{18}O$ and $\delta^{13}C$ with increasing metamorphic grade from localities in Table 1. Trends mimic one another, starting at values for normal marine limestone ($\delta^{18}O$ = 20 to 26‰, $\delta^{13}C$ = -2 to 4‰) and decreasing towards igneous values. The stippled field, labeled Rayleigh Volatilization, models the effect of volatilization as in Figure 2. Most samples are from contact metamorphic wall rocks or skarns. These trends indicate large amounts of fluid flow and infiltration by magmatic and/or meteoric fluids (Valley 1986).

OPEN SYSTEM FLUID-ROCK INTERACTION:
CONTINUUM MECHANICS MODELS OF STABLE ISOTOPE
TRANSPORT IN HIGH TEMPERATURE CRUSTAL SYSTEMS

Continuum mechanics models for stable isotope fluid-rock exchange have been developed over the last three decades to accurately describe fluid flow in the Earth's crust. The geometry of crustal rocks, especially in areas of active metamorphism and deformation requires that fluid flow in the Earth is modeled in three dimensions. Evaluation of such aspects as the importance of fluid flow as driving force for metamorphism (e.g. Fyfe et al. 1978; Ferry 1980, 1986, 2000; Baumgartner and Ferry 1991; Ague 1997), and the influence of metamorphic fluid fluxes on climate (Kerrick and Caldeira 1999) depend on an accurate appraisal of the amount and the timing of fluid percolating through the crust. Significant progress has been made applying continuum

Table 1. Studies demonstrating coupled O-C depletion trends in metamorphosed carbonates. Locality numbers are the same as in Figure 4.

		Width of aureole or traverse	Pressure/ Depth	Maximum temperature °C	$X(CO_2)$	$\delta^{18}O$ range ‰	$\delta^{13}C$ range ‰
1.	Trenton limestone Mount Royal, Quebec	100m				14	6
2.	Marysville, Montana	1-3 m	1 kbar	525	0.95	18	6
3.	Pine Creek, California W-skarn	roof pendant	<2 kbar	600	<0.25	22	14
4.	Osgood Mts., Nevada	<1 km	<2 kbar	>550	<0.15	9	9
5.	Skye, Scotland					20	7
6.	Elkhorn, Montana	marble 240m skarn 21m	1 kbar	525	<0.25	15	7
7.	Notch Peak Stock, Utah	3 km	1.5 kbar	600	0.75 to 0.001	13	13
8.	Weolag W-Mo Deposit, Korea	2.5 km	1.0-2.4 km	>400		15	11
9.	Tauern Area, Austria	3m	>4 kbar	450-600		8	2
10.	Birch Creek, California	750 m		540		13	4
11.	McArthur R. Pb-Zn Deposits, Australia			350		7	2.5
12.	Providencia Pb-Zn Deposits Mexico			365		10	10
13.	Gaspé Cu Deposits, Quebec			350		17	14
14.	CanTung W-skarn, N.W. Territories			500		9	9
15.	Mottled Zone, Israel		<25 bar	1300		13	25
16.	Bergell Aureole, Italy	<20 cm	2 kbar	400	0.1-0.25	15.3	8
17.	Alta Stock, Utah	1.5 km on S	1.5 kbar	575	>0.54 to <0.04	20	10
18.	Costabonne, Pyrenees	meters	2 kbar	>500	0.05	16	12
19.	Empire Zn skarns, New Mexico	0.5 km	<0.4 kbar	700	<0.2	19	8
20.	Ubehebe Peak, Death Valley, California	2 km	1.5 kbar	665	<0.05	11	5
21.	Utsubo granite, Japan	<3 cm	1.2 kbar	620	<0.6	5	5
22.	Bufa del Diente, Mexico	10 cm	0.75 kbar	500-600	>0.5 to 0.02	10	4
23.	Isle of Skye, Scotland	1 km	0.5 kbar	>600	<0.2	22	5
24.	Isle of Skye, Scotland	80 m	0.5 kbar	710	<0.61	16	5
25.	Adamello (SW) Italy	meters	2.8 kbar	630	<0.16	7	4
26.	Adamello (SE) Italy	1 km				20	10
27.	Adamello (SE) Italy	1 m			0.01		
28.	Ritter Range Pendant, California	2 km	1.5 kbar	600	<0.28	5	7

References:

(1) Deines & Gold (1969): (2) Lattanzi et al. (1980), Lasaga & Rye (1993); (3) Brown et al. (1985); (4) Taylor & O'Neil (1977); (5) Forester & Taylor (1977); (6) Bowman et al. (1985a); (7) Hover et al. (1983), Nabelek et al. (1984, 1988, 1992), Labotka et al. (1988); (8) So et al. (1983); (9) Schoell et al. (1975); (10) Shieh & Taylor (1969); (11) Rye & Williams (1981); (12) Rye (1966); (13) Shelton (1983); (14) Bowman et al. (1985b); (15) Kolodny & Gross (1974); (16) Taylor & Bucher-Nurminen (1986); (17) Moore & Kerrick (1976), Bowman et al. (1994), Cook & Bowman (1994, 2000), Cooke et al. (1997), Bowman & Valley (1999); (18) Guy et al. (1988); (19) Turner & Bowman (1993); (20) Roselle et al. (1999); (21) Wada (1988), Arita & Wada (1990), Graham et al. (1998), Wada et al. (1998); (22) Heinrich et al. (1995), Romer & Heinrich (1998); (23) Holness & Fallick (1997); (24) Ferry & Rumble (1997); (25) Abart (1995); (26) Gerdes et al. (1999); (27) Gerdes et al. (1995a); (28) Ferry et al. (1998).

Table 1, continued

	Pluton	Rock types, Comments
1.	essexite, nepheline-syenite	limestone, marble, calcite from intrusion
2.	granodiorite	marble, hydrothermal assemblages, largest isotope shift at diopside isograd
3.	quartz-monzonite	marble, calc-silicate, skarn in pendent, gradients in $\delta^{18}O$ up to 10 permil/10 cm across skarn; depletion attributed to mixing with magmatic O+C
4.	granodiorite	marble, calc-silicate hornfels, skarn, 3 stages of skarn formation
5.	granite	marble, skarn, some depletion due to meteoric water
6.	quartz-diorite	dolostone, marble, skarn
7.	granite to qt-monzonite	calcareous argillite and marble; woll. up to 1.5km from contact; fluids highly channeled; Δ(Cc-silicate) disequilibrium at >cm-scale.
8.	granite	limestone, calc-silicate, skarn
9.	---	traverse across a 3m thick marble layer
10.	granite	marble, skarn
11.	---	dolomite, values change in relation to distance to the Emu Fault; range in values due to variable T = 350-150°C and constant fluid composition
12.		limestone, late-stage hydrothermal calcites variable T=365-200°C
13.		13a = drill hole GMS4, limestone, marble, vein calcite, skarn
		13b = drill hole GMS2, limestone, marble, variable T=350-150°C
14.	monzo-granite	limestone, marble, calc-silicate, skarn, variable skarn T=400-270°C
15.	none	natural combustion of bituminous marl
16.	tonalite	metasomatic zone around veins in marble, sharp isotopic gradients, 5-14 permil/cm at infiltration fronts
17.	granodiorite	fluids channeled by Alta-Grizzly thrust; flux and $\delta^{18}O$ variable at cm-scale; $\Delta^{18}O$(Cc-Forst) disequilibrium in periclase zone aquifers, heat and fluid flow models indicate zones of up and down T flow with down T flow dominating the inner aureole
18.	Hercynian granite	W-skarn in dolomitic marble, magmatic/metamorphic fluid stage followed by meteoric stage; $\delta^{13}C$ to −16 at Lacourt corresponds to reaction of graphite; $\delta^{18}O$ lower in W-skarn than in barren skarn
19.	granodiorite	fluids channeled in limestone beneath shale; max. T = 400°C in skarn; low $\delta^{18}O$ indicates meteoric skarn fluids
20.	quartz-monzonite	limited heterogeneous magmatic fluid flow vertically away from pluton; isotope gradients are on sides of flow system; higher $X(CO_2)$ inferred from isotopes (0.3) than petrology
21.	granite	microsampling of grain boundaries in calcite; granite intrusive into granulite facies marble; $\delta^{18}O$ correlates to cathodoluminescence; isotope disequilibrium at <100 μm-scale; $\delta^{18}O$ gradients up to 10 ‰/200μm
22.	syenite	magmatic brine infiltration along meta-chert layers in limestone
23.	granite	magmatic Si-F-bearing fluid infiltration into siliceous dolostone and dolomitic marble heterogeneously along fractures and grain boundaries; $\Delta^{18}O$(Cc-Dol) not equilibrated
24.	granite	magmatic fluids channeled ~vertically along bedding and pre-metamorphic dike; flux <500 mol/cm^2; brucite after periclase correlates with low $\delta^{18}O$
25.	granodiorite	garnet-vesuvianite veins in marble; graphite is oxidized by fluids at $\delta^{18}O$ gradients of 4‰/3 cm, $\Delta^{18}O$(Cc-Gt) disequilibrium
26.	tonalite	localized magmatic fluid infiltration into dominantly closed system silica-poor dolostones
27.	granodiorite	layer parallel flow of magmatic fluid into marble; 10‰/40 cm gradient of $\delta^{18}O$; flow parallel to gradient
28.	granodiorite	magmatic fluids channeled along thin siliceous carbonate layers in metavolcanics; prograde flow structurally controlled, layer parallel and upwards; flux = 245-1615 mol/cm^2; retrograde Trem, and Cc+Qt+Wol

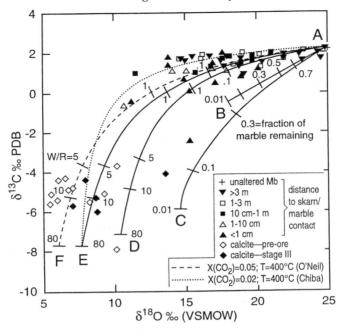

Figure 5. Plot of $\delta^{18}O$ and $\delta^{13}C$ values for the Tierra Blanca Limestone and hydrothermal calcites. Symbols represent distance from the skarn/marble contact. Curves A-B and A-C show depletion resulting from Batch and Rayleigh decarbonation, respectively. Curves A-D, A-E represent progressive depletion in limestone resulting from isotope exchange with meteoric water in equilibrium with the Hanover-Fierro pluton at 315°C and water-rock ratios of 0.01-0.001. Curve A-E (dotted line) is calculated at 400°C using the fractionation factor of Chiba et al. (1989), assuming $X(CO_2) = 0.02$. Curve A-F (dashed line) defines progressive depletion resulting from water/rock interaction at 400°C using the fractionation factor of O'Neil et al. (1969) (from Turner and Bowman 1993).

mechanics models to the interpretation of stable isotope alteration patterns in metamorphic terrains.

Fluid-rock interaction is modeled by considering the stable isotope mass balance between fluid and rock, following the flow direction (Bear 1972). The principle prediction of these models is that stable isotope fronts, similar to chromatographic fronts, should develop when fluids infiltrate rocks that are not in equilibrium with the infiltrating fluid composition. The stable isotope ratios (e.g. $^{18}O/^{16}O$) increase or decrease abruptly at the front, depending on the initial ratio in the rock and infiltrating fluid. The rocks have a stable isotope ratio that is close to equilibrium with the infiltrating fluid from the front towards the source or inlet side of the flow system. This part of the flow system is called fluid dominated. The rocks down stream of the front have an unchanged (or nearly so) stable isotope composition. This part of the system is called rock dominated. The exact shape of the predicted isotope concentration ratio, as well as the speed of the front depends on the mechanism of stable isotope transport, the isotope exchange kinetics, and the composition of the fluid and the rock. Stable isotopes can diffuse due to the chemical potential gradient of the isotope. Diffusion of isotopes takes place through the individual mineral grains (volume diffusion) and on grain boundaries (grain boundary diffusion).

Grain boundary diffusion (Joesten 1991) is typically several orders of magnitude faster than volume diffusion (Joesten 1991; Cole and Chakraborty, this volume; Valley, this volume). Nevertheless, volume diffusion becomes important at high temperatures since it increases strongly with temperature, and since the proportion of the grain boundary volume (porosity) is much smaller than the proportion of the mineral grains. In addition, grain boundary diffusion follows increasingly tortuous paths as the grain size increases (Joesten 1991). Large grain sizes are typical for high temperature rocks. This increases the path length.

Transport of stable isotopes in a moving fluid phase is called advection. Here infiltrating fluids move the isotopic species of interest. Fluid flow is restricted to connected pore spaces. The amount of connected pore space and the manner of connection determines the permeability of a rock. Mixing of stable isotope ratios by a flowing fluid on grain boundary intersections, micro crack intersections, and fracture intersections results in dispersion. Dispersion is similar to diffusion (at least mathe-matically), since this is a mixing process.

The basic transport equations for stable isotopes will be derived here, in order to facilitate a basic understanding of the mathematical formalism describing the transport of stable isotopes in metamorphic and igneous environments, and the resulting fluid-rock interaction. There are several excellent treatments of most aspects in the literature (e.g. Baumgartner and Rumble 1988; Lassey and Blattner 1988; Bowman and Willet 1991; Bowman et al. 1994; Lichtner 1996; Lasaga 1998). Stable isotopes in a high temperature fluid-rock system behave like reactive tracers, a term often used in hydrology. In fact, much of the following material was developed in hydrology and is summarized in excellent textbooks and papers (e.g. Ogata 1964; Bear 1972; Cameron and Klute 1977; de Marsily 1986; Domenico and Schwartz 1990; Appelo and Postma 1999). In the present paper, the focus will be on the mass balance equations that describe fluid-rock exchange. The first step will be to derive these equations. The mass transport equations suffice to describe one-dimensional fluid flow, or flow in homogeneous media. In addition to the mass balance equations, one needs to solve a set of differential equations that govern fluid flow, temperature distribution, etc, as soon as transient or hetero-geneous, multi-dimensional systems are modeled. The necessary differential equations are only mentioned, not derived. The reader is referred to the hydrologic literature (e.g. Bear 1972; Appelo and Postma 1999), and some detailed papers dealing with hydrothermal flow in contact aureoles (e.g. Furlong et al. 1991; Gerdes et al. 1995a,b, 2000; Cook et al. 1997). The petrologic and geochemical communities have used the formulations developed by hydrologists (e.g. Bear 1972; de Marsily 1986; Domenico and Schwartz 1990; Appelo and Postma 1999), despite the fact that the environment of groundwater hydrology is in many respects different from that encountered in the deep crust. Pore space and permeability are bound to be transient under higher temperatures and pressures of the crust (e.g. Connolly 1997; Connolly and Podladchikov 2000; Balashov and Yardley 1998). Permeability and porosity are not well understood in high temperature regimes (Nur and Walder 1990; Oelkers 1996; Connolly 1997; Connolly and Podladchikov 2000). The authors suspect that it is here where significant advances will occur in the near future. Nevertheless, stable isotope patterns in one dimension can be fully investigated through the mass balance equations describing stable isotope transport and fluid-rock exchange. Hence, we choose to focus on this aspect of stable isotope transport.

The mass balance equation for stable isotope transport

The theory is developed for a simple transport system that only involves oxygen isotopes. This will help to minimize the clutter of superscripts and subscripts in the equations. Nevertheless, the equations are perfectly general. Transport and fluid-rock exchange equations for other stable isotopes can be obtained by replacing the oxygen isotope specific properties in the equations with those of the isotope under consideration. Care will be taken to identify possible problems with other isotopes, as the theory is developed. Potential pitfalls for two and three-dimensional mass transport are also discussed along the way.

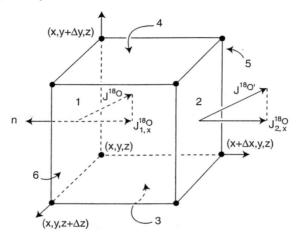

Figure 6. The microscopic mass balance for fluid flow is schematically illustrated with a cube. The change in mass of an isotope is equal to the sum of all fluxes entering (+) or leaving (–) the cube.

We start by considering the mass changes of oxygen isotopes in a small volume. A mathematical statement is developed that will relate the change of the concentration of isotopes in the small volume to the change of the mass of stable isotopes entering and exiting the surface of the small volume. To mathematically evaluate this balance, Green's (or Gauss') theorem is used. It mathematically relates the surface integral of an arbitrarily shaped body to the volume integral of this body. This is a crucial step in understanding the individual terms of the partial differential equation used for stable isotope fluid-rock interaction. Hence we will follow the approach outlined by Feynman et al. (1963). Instead of a general shape, a small cube is used. Mass balance requires that the change in mass (concentration) within a cube is equal to the sum of all fluxes entering (+) and leaving (–) the cube through its surfaces (Fig. 6). The flux of stable isotopes by any kind of process outwards through surface 1 is given by:

$$F_1 = -J_{1,x}^{^{18}O} \Delta y \Delta z \tag{7}$$

Here $J_{1,x}^{^{18}O}$ is the moles of ^{18}O per unit area flowing in the x-direction through surface 1. The flux exiting the cube through surface 2 is related to the flux at surface 1 by:

$$J_2^{^{18}O} = J_{1,x}^{^{18}O} + \frac{\partial}{\partial x} J_{1,x}^{^{18}O} \Delta x \tag{8}$$

Hence the flow exiting the small cube through the second surface can be approximated by:

$$F_2^{18}O = \left[J_{1,x}^{18}O + \frac{\partial}{\partial x} J_{1,x}^{18}O \right] \Delta x \Delta y \Delta z \tag{9}$$

The sum of the two fluxes of isotopes results in the change of stable isotope flux, $\Delta F_x^{18}O$, through the cube in the x-direction. Hence we have:

$$\Delta F_x^{18}O = F_2^{18}O + F_1^{18}O = \frac{\partial}{\partial x} J_1^{18}O \Delta x \Delta y \Delta z \tag{10}$$

Similar equations can be obtained by adding the fluxes through the corresponding surfaces in the other two axis directions. The overall stable isotope flow out of the (infinitesimal) cube is then given by:

$$\Delta F^{18}O = \left[\frac{\partial}{\partial x} J^{18}O + \frac{\partial}{\partial y} J^{18}O + \frac{\partial}{\partial z} J^{18}O \right] \Delta x \Delta y \Delta z = \nabla J^{18}O dV \tag{11}$$

The change of mass (or moles) of oxygen isotope within the small cube with volume dV can alternatively be calculated from the change of mass, dm, of the system. Mass gain or loss of ^{18}O can be calculated by multiplying the concentration of ^{18}O by the volume of the system, dV. Hence we obtain for the basic mass balance equation:

$$\frac{\partial C_{sys}^{18}O}{\partial t} dV = -\nabla J^{18}O dV \tag{12}$$

Note that the concentration gain of an isotope in the system is equal to the negative of the change in flux of this isotope. The dV terms can be dropped to yield the basic mass balance equation for stable isotope transport (or any other element):

$$\frac{\partial C_{sys}^{18}O}{\partial t} = -\nabla J^{18}O \tag{13}$$

The mass balance Equation (13) is written in terms of mass of fluid or rock per unit time and volume. Hence the concentration units should be in mass (or moles) per unit volume. The fluxes are to be written in mass per unit time and area. Note that the system concentration is the concentration of ^{18}O in both the rock and the fluid phase, which occupies the pore space.

The art of understanding water/rock interaction in metamorphic systems is to write appropriate mathematical terms for the mass transport mechanism(s) operating in the geologic system of interest and to relate the fluid and rock isotope concentrations through appropriate equilibrium or rate law equations. The resulting partial differential equation must be solved. Some analytical solutions are available for geometrically simple systems (e.g. Ogata 1964; Golubev and Garibyants 1971; Cameron and Klute 1983; Van Genuchten and Parker 1984; Baumgartner and Rumble 1988; Blattner and Lassey 1989; Bowman and Willet 1991; Appelo and Postma 1999). The solution of more complex geometries, complicated geologic boundary conditions, and transient variations in temperature, pressure, and reaction kinetics along the fluid flow paths typically require numerical solutions. Mineral net transfer reactions also occur simultaneously with stable isotope fluid/rock exchange, which requires numerical solutions to the differential equation. In fact, many geochemists find it more convenient to solve Equation (13) by numerical tools, even if analytical solutions are available, because most analytical solutions are expressed as infinite series or numerical integrals.

MATHEMATICAL FORMULATION OF THE TRANSPORT MECHANISMS

The transport of stable isotopes is assumed to be the result of diffusion or infiltration. In stagnant fluids, or when fluids are absent, transport is dominated by diffusion. Diffusion will occur along grain boundaries or through crystals (volume diffusion). Grain boundary diffusion is typically assumed to be the dominant mechanism at medium temperatures if a fluid phase is present (e.g. Eiler et al. 1992; Valley 2001). A fluid phase is typically present during prograde metamorphism due to volatilization reactions, but it is absent or only intermittently present during prolonged high temperature metamorphism, or retrograde metamorphism (Yardley and Valley 1997). Volume diffusion of stable isotopes becomes increasingly important with higher temperature, or in the absence of a fluid phase on the grain boundary. This topic is not considered further in this communication since it is covered by Valley (2001). Alternatively, stable isotopes are swept along in the fluid phase during infiltration. Advection of stable isotopes is always accompanied by dispersion. The mathematical treatment of each transport mechanism is reviewed and the final transport equation derived.

Infiltration. Transport of stable isotopes or other solutes by movement of the bulk fluid is referred to as advection. The rate at which an isotope is advected to a location is given by the fluid velocity, v_f, and the concentration C of the isotope in the fluid.

$$J_{inf}^{18}{}^{O} = v_f C_f^{18}{}^{O} \tag{14}$$

Typically, the velocity is interpreted to be the Darcy velocity, which is defined to be a vector describing the direction and average of the linear velocity of the fluid in the pore, v_p, multiplied by the porosity. The equation assumes that the solid, e.g. the minerals, are not moving in the system and hence the concentration of isotopes in the fluid phase is used in the equation (indicated by the subscript f) for calculating the flux. If other phases (solids, melts or a second fluid) are moving in the chosen reference system, additional terms for the fluxes have to be added. Such fluxes could occur due to deformation of the mineral framework in response to tectonic movements and pore collapse (Person and Baumgartner 1995; Oliver 1996). The fluid velocity is measured in units of meters per second (or similar length per time units). It is usual to measure concentration in terms of mass per volume [kg/m^3] or moles per volume [moles/m^3]. The range of values for the velocity used by workers for hydrothermal high temperature systems is given in Table 2.

There has been much discussion (though little written) about the ability of Equation (14) to describe focused flow or episodic flow. However, this equation only describes fluxes of fluids moving through a surface—without considering the geometry of the pore space or fractures. One problem is the physical meaning of 'velocity' in cases, where flow is pulsating or in large-scale fractures. Here it definitely does not represent the Darcy velocity. This equation does not define or depend on the rate of fluid-rock exchange. In fact, rocks might not record the actual amount of fluid flow if fluid-rock exchange rates are slow (e.g. Cole and Chakraborty, this volume).

Diffusion. The diffusive flux of stable isotopes along grain boundaries can be written according to Fick's law:

$$J_{dif}^{18}{}^{O} = -\Phi\Gamma D\nabla C_f^{18}{}^{O} \tag{15}$$

The diffusive flux along grain boundaries is obtained by scaling the diffusion in a free aqueous fluid by the porosity (Φ) and the tortuosity (Γ) of the rock in the diffusion direction. Values for the diffusion coefficient (D) in the free aqueous phase have been estimated by several workers. They are summarized in Table 2. In principle, it is possible to calculate the diffusion coefficient in an aqueous fluid phase, if speciation and

Table 2. Published estimates of transport parameters in metamorphic flow systems.

Transport parameter	Published range	Source
Porosity Φ	10^{-6} to 10^{-3}	2, 8, 10, 13, 20
Permeability K $\left[m^2\right]$	10^{-23} to 10^{-12}	1, 2, 3, 5, 6, 8, 13, 14, 19, 20
Fluid Velocity v_f $[m/s]$	10^{-14} to 10^{-9}	1, 2, 4, 5, 8, 9, 13
Molecular Diffusion in Fluid Phase D $\left[m^2/s\right]$	10^{-17} to 10^{-7}	7, 12, 15, 16, 18, 20
Longitudinal Dispersivity α_l $[m]$	10^{-2} to 10^2	11, 17, 20
Transverse Dispersivity α_t $[m]$	10^{-3} to 10^1	11, 17, 20

References:
(1) Baumgartner and Ferry (1991); (2) Bickle and Baker (1990a); (3) Brace (1980); (4) Cartwright (1994); (5) Cartwright and Weaver (1993); (6) Clauser (1992); (7) Farver and Yund (1992); (8) Ferry and Dipple (1991); (9) Ferry and Dipple (1992); (10) Ganor et al. (1989); (11) Gelhar et al. (1992); (12) Ildefonse and Gabis (1976); (13) Leger and Ferry (1993); (14) Manning et al. (1993); (15) Nagy and Gilleti (1986); (16) Walton (1960); (17) Gerdes et al. (1995); (18) Joesten, 1991; (19) Manning and Ingebritsen, 1999; (20) Oelkers (1996) and references therein.

hydration of the complexes that incorporate the stable isotope of interest are known (e.g. Oelkers 1996). Diffusion along grain boundaries follows a tortuous path. Hence, the path actually traveled by the isotopes is longer than the shortest distance between two points. The tortuosity factor is used to scale this effect. It varies in most rocks between 0.1 and 1.0 (Dullien 1979; Brady 1983; Oelkers 1996). The rock fabric, for example schistosity, can have an important role by rendering the rock anisotropic with respect to diffusion. Diffusion will always occur in the rock matrix as well as along grain boundaries.

Dispersion. Dispersion is a composite effect that is due to several different physical processes occurring in a rock through which fluid is flowing. Fluid flowing through a rock, in fractures or along grain boundaries, follows tortuous paths of different lengths. When individual batches of fluid arrive at intersections of pores or fractures these fluids can mix. If the individual paths have resulted in different isotopic compositions, mixing will occur. This mixing is called dispersion. On the scale of a grain boundary cross section (or diameter), there are boundary layer effects. The fluid velocity in a pore is slowest close to the fluid-grain interface, its actual values depend on the pore surface roughness. Diffusion between individual fluid flow lines results in exchange of stable isotopes between fluids flowing at different speeds. Hereby a longitudinal, D_L, and a transverse dispersivity, D_T, will occur (see Oelkers 1996; Appelo and Postma 1999). The dispersion of stable isotope concentrations will depend on the structure of the rock. Not much is known about the magnitude of dispersion in high temperature metamorphic systems. Detailed studies with column experiments have shown that dispersion coefficients depend on the fluid flow rate. Figure 7 reveals that there are two drastically different dispersion realms. When the pore fluid velocity is small, diffusion dominates the dispersion of tracer species in a column. Species diffusing through a rock column stuffed with solid obstacles (in the experiments they were spheres!) will follow a tortuous path. Thus they are slowed slightly. This is reflected by the fact that the dispersion value is slightly smaller than the diffusion coefficient. Dispersion increases linearly with increasing pore velocity in cases where the product of the fluid pore velocity and the

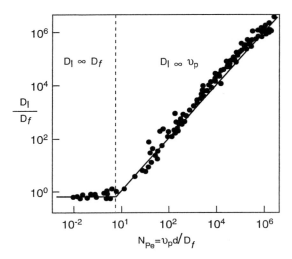

Figure 7. Longitudinal dispersion (D_L) divided by the diffusion coefficient (D_f) for tracers measured in column experiments as a function of the particle scale Peclet number (N_{Pe}). It is defined as the product of the average pore fluid velocity, v_p, and the grain diameter, d, divided by the free fluid diffusion coefficient, D_f. The magnitude of the dispersion is independent of the pore fluid velocity (v_p) for very small Peclet numbers (or fluid velocities). Note that the effective diffusion coefficient in a porous media is smaller than the diffusion coefficient in a free fluid phase due to the tortuosity. The dispersion increases linearly with increasing flow velocity (increasing Peclet number). Modified from Appelo and Postma (1999).

grain diameter is much larger than that of the diffusion coefficient. This dimensionless ratio is referred to as the (grain scale) Peclet number.

The mathematical formulation of material transport by dispersion in transport direction (longitudinal dispersion) is:

$$J_{disp}^{^{18}O} = -\Phi D_L \nabla C_f^{^{18}O} \tag{16}$$

$$D_L = \Gamma D + \alpha_L v_p \tag{17}$$

The longitudinal dispersivity (α_L) is a material property of the rock. It is an interesting aspect of the dispersivity that its numerical value depends on the size of the system studied and sampled (Fig. 8). As a rule of thumb, one can assume that dispersivity is approximately one tenth of the system size (Fig. 8; Appelo and Postma 1999). This diagram is based on data reported and summarized in Gelhar et al. (1985) for near surface flow systems. The macro-dispersion in large aquifers is due to the heterogeneity of the natural rock sequences. Aquifer rocks are composed of lenses and layers of different rock types, like carbonates, sandstones, and gneisses, each one with a different permeability. Grain size variations, localization of fractures into zones, and many other factors increase this heterogeneity. Hence, in such macroscopic systems one should view dispersion to be the result of heterogeneity on the scale of the model (e.g. Gelhar et al. 1985; Appelo and Postma 1999; Gerdes et al. 1995b). Care should be taken to use the appropriate value for the dispersion coefficient. It should reflect the heterogeneity scale, which is not modeled

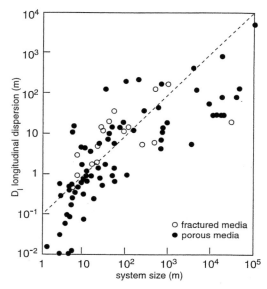

Figure 8. The longitudinal dispersion for tracers from column experiments and groundwater tracer studies is proportional to the length of the flow system studied. A typical value for the dispersion is roughly one tenth of the length of the system. While not much is known about dispersion in high temperature flow systems, it is nevertheless likely that a similar correlation exists. Hence dispersion will be important for a system of any size where mass transport occurs by fluid infiltration. Data from Gelhar et al. (1985).

in detail (Appelo and Postma 1999). Another aspect adding to dispersion that is not well studied is the heterogeneous reactivity of rock units with respect to stable isotope transport. It will be apparent from the following sections that rocks will retard most isotopic concentration fronts. The retardation depends on the isotope studied and the composition of the rock through which the fluid traveled. If fluids that have experienced different isotope retardation are mixed further down stream this will result in dispersion. Thus the dispersion coefficient will be different for each isotope! The magnitude of the dispersion coefficient reflects the isotope under investigation, the rock composition, the shape of the pore space (flow path), and the fluid composition. Less is known about the transverse dispersivity even in hydrologic, near surface systems. In hydrology, it is assumed to be approximately one tenth of the longitudinal dispersivity (Appelo and Postma 1999). The fact that the dispersion coefficient is about 10% of the system size under consideration (Appelo and Postma 1999; and see above) implies that dispersion cannot be neglected in any flow system.

The transport equation

The general continuum mechanics equation of stable isotope transport can now be assembled. All individual mass transport fluxes need to be summed in the mass balance Equation (13):

$$\frac{\partial C_{sys}^{^{18}O}}{\partial t} = -\nabla J^{^{18}O} = -\nabla \left[J_{dif}^{^{18}O} + J_{disp}^{^{18}O} + J_{inf}^{^{18}O} \right] \tag{18}$$

Equations (14)-(17) then can be inserted into Equation (18). Assuming that dispersion reduces to the diffusion coefficient in the limiting case of no infiltration (Eqn. 17), we obtain:

$$\frac{\partial C_{sys}^{^{18}O}}{\partial t} = -\nabla \left[-\Phi D_L \nabla c_f^{^{18}O} + v_f C_f^{^{18}O} \right] \tag{19}$$

In one-dimensional systems it is often assumed that the porosity and the fluid velocity are constant in the transport direction. Equation (19) reduces with a constant dispersion/diffusion coefficient to:

$$\frac{\partial C_{sys}^{^{18}O}}{\partial t} = \Phi D_L \frac{\partial^2 C_f^{^{18}O}}{\partial x^2} - v_f \nabla C_f^{^{18}O} \tag{20}$$

For a large velocity, the dispersion-diffusion coefficient is no longer constant, and is dominated by the dispersivity (Eqn. 17), while for slow fluid velocity the equation reduces to the time-dependant diffusion equation.

FLUID- ROCK INTERACTION:
THE TIME DEPENDENCE OF THE SYSTEM'S ISOTOPIC COMPOSITION

The rock consists of several mineral species and pore space filled with fluid. Hence the system isotope concentration is the sum of the individual isotope concentrations of each phase, multiplied by the abundance of the phase. For example, the system concentration of ^{18}O, $C_{sys}^{^{18}O}$, in a rock containing quartz, calcite and a fluid in the pore space composed of H_2O and CO_2 is calculated from:

$$C_{sys}^{^{18}O} = \frac{1}{V_{sys}}\left[\frac{3V_{cc}}{\bar{V}_{cc}}C_{cc}^{^{18}O} + \frac{2V_{qtz}}{\bar{V}_{qtz}}C_{qtz}^{^{18}O} + \frac{1V_{H_2O}}{\bar{V}_{H_2O}}C_{H_2O}^{^{18}O} + \frac{2V_{CO_2}}{\bar{V}_{CO_2}}C_{CO_2}^{^{18}O}\right] \tag{21}$$

Here we have chosen a mole per volume scale, or a molar concentration scale. The concentration of ^{18}O in mineral i, $C_i^{^{18}O}$, is with respect to the pure phase i. For this geologically simple system, the time differential of Equation (13) can be rewritten as:

$$\frac{\partial C_{sys}^{^{18}O}}{\partial t} = \frac{1}{V_{sys}}\frac{\partial}{\partial t}\left[\frac{3V_{cc}}{\bar{V}_{cc}}C_{cc}^{^{18}O} + \frac{2V_{qtz}}{\bar{V}_{qtz}}C_{qtz}^{^{18}O} + \frac{1V_{H_2O}}{\bar{V}_{H_2O}}C_{H_2O}^{^{18}O} + \frac{2V_{CO_2}}{\bar{V}_{CO_2}}C_{CO_2}^{^{18}O}\right] =$$
$$3\frac{\partial(1-\Phi)X_{cc}C_{cc}^{^{18}O}}{\partial t} + 2\frac{\partial(1-\Phi)X_{qtz}C_{qtz}^{^{18}O}}{\partial t} + \frac{\partial\Phi X_{H_2O}C_{H_2O}^{^{18}O}}{\partial t} + 2\frac{\partial\Phi X_{CO_2}C_{CO_2}^{^{18}O}}{\partial t} \tag{22}$$

Here the mole fraction for the mineral phase, X_i, refers to the molar abundance fraction of the mineral i compared to the sum of oxygen moles in all the minerals. Similarly, the mole fraction X_j of the fluid species j is the molar fraction of oxygen in species j, compared to the fluid phase. Further on, we introduce the porosity, Φ. It is here that the chemical model of fluid-rock interaction enters the governing equations. It is obvious from Equation (22) that any change of the modal abundance through mineral reactions, as well as changes in porosity and fluid composition will lead to a change in the system composition, unless the system is closed. In the latter case a change in one phase will be compensated exactly by adjustments in other phases.

The isotope concentration of each mineral and the fluid phase—and even the individual aqueous species containing the isotope of interest—are linked to each other through thermodynamic relations describing equilibrium or kinetic isotope exchange. The actual mechanism of isotopic exchange between the components in the fluid phase and each mineral must be accurately known to obtain a valid fluid-rock interaction model. Several models describing fluid-rock exchange have been presented in the literature. These include: (a) transport of an inert component (e.g. Bowman et al. 1994; Appelo and Postma 1999); (b) local equilibrium (e.g. Baumgartner and Rumble 1988; Bowman et al. 1994); (c) kinetic stable isotope exchange (Lassey and Blattner 1988; Bowman and

Willett 1991; Cook et al. 1997; Bolton et al. 1999). Models (a) and (b) are end-member cases rarely realized in nature. If the kinetics of stable isotope exchange are extremely slow, or the flow very fast, model case (a), the inert tracer, is a reasonable approximation. On the other hand, if the reaction is fast compared to stable isotope movement, the reaction maintains equilibrium between the fluid phase and the solid. Typically it is assumed, in models of high temperature geological settings, that isotope exchange between fluid species is fast. Hence, homogeneous equilibrium is maintained in the fluid phase. Nevertheless, this assumption should be independently assessed for each case. To solve the partial differential Equation (22), the time-dependent concentrations of each phase have to be linked by thermodynamic relations describing their change as a function of time. Below, several cases are discussed.

In most cases for oxygen, carbon and sulfur, it is permissible to use the δ-notation instead of the concentration, or concentration ratio (e.g. Baumgartner and Rumble 1988). This is the case for relatively small stable isotope variations, since the δ-units are linearly related to the ratio of two stable isotopes over a range of approximately 10 δ-units in permil.

Inert tracer movement. Consider the case of the movement of ^{13}C through the pore space of a quartzite, pelite, or granitic rock, all of which do not contain any carbon (e.g. no calcite, scapolite or graphite). In this case, there will be no exchange of carbon isotopes between the mineral phases and the fluid. Hence the time derivative of the system composition reduces to that of the fluid phase, since the solid does not change its isotopic composition:

$$\frac{\partial C_{sys}^{^{13}C}}{\partial t} = \frac{\partial \Phi C_f^{^{13}C}}{\partial t} \tag{23}$$

The same simple equation is obtained in the case, where the rock composition is changing very slowly compared to that of the fluid. This limiting assumption occurs in many low temperature (hydrology) environments (e.g. Appelo and Postma 1999) and may apply to some metamorphic rocks (e.g. Eiler et al. 1992). This assumption might also be valid where fluid flow is very rapid, a situation often occurring in fractured fluid flow systems. In that case a fluid composition can be imposed on a rock that is in disequilibrium with it. The fluid isotope composition will become virtually independent of the rock isotope composition when isotopes are advected much faster than isotopes can exchange with the rock. This will lead to a constant fluid composition throughout the fluid/rock interaction system. It is here that box models (0-dimensional models) with detailed kinetic chemical exchange models can be used successfully (e.g. Criss et al. 1987). The passing of a truly inert tracer cannot be detected by analysis of the rock since only the fluid phase can change its composition.

Local equilibrium between fluid and rock. A common assumption of stable isotope transport is that of local equilibrium (Baumgartner and Rumble 1988, Bowman et al. 1994; Gerdes et al. 1995 a, b). Equilibrium is assumed for a small volume of the transport system. Within this volume, all phases are in equilibrium at all times (Thompson 1959). A molar equilibrium constant, K_D can be defined to describe the thermodynamic stable isotope equilibrium between two phases or species. The stable isotope concentration of a mineral (s) can be related to that of a fluid species through the equilibrium constant, K_D:

$$C_s^{^{18}O} = K_D C_f^{^{18}O} \tag{24}$$

Here the concentrations of the oxygen isotope in the fluid phase (subscript f) and the solid phase (subscript s) are measured in moles per volume. Hence K_D is not equivalent to the dimensionless fractionation factor (α) used in the isotope literature that relates isotope ratios of two phases to each other (see Eqn. 1). The fractionation factor (α) can be related to the equilibrium constant. For this, molar concentrations of the rare isotope in the solid and fluid phase (e.g. C_s^{18O} and C_f^{18O}) are written in terms of the ratio of the rare and the common isotope (R_s^{18O}, R_f^{18O}):

$$C_s^{18O} = R_s^{18O} \frac{\beta_s^O}{\bar{V}_s} \tag{25}$$

$$C_f^{18O} = R_f^{18O} \frac{\beta_f^O}{\bar{V}_f} \tag{26}$$

Here β is the stoichiometry of oxygen in the fluid (f) and solid (s), and $\bar{V}$ the molar volume of the phase. The above relationships hold for all isotopes, in which the isotope of interest is present in small quantities (Rumble 1982). Dividing one equation by the other and solving for the solid concentration yields:

$$C_s^{18O} = \alpha_{s-f}^{18O} \frac{\beta_s^O \bar{V}_f}{\beta_f^O \bar{V}_s} C_f^{18O} \tag{27}$$

Hence the molar equilibrium constant for ^{18}O is related to the fractionation factor by:

$$K_D^{18O} = \alpha_{s-f}^{18O} \frac{\beta_s^O \bar{V}_f}{\beta_f^O \bar{V}_s} \tag{28}$$

An equation similar to Equation (27) can be written between every phase and an arbitrarily picked reference phase (Rumble 1982; Baumgartner and Rumble 1988). A convenient approach for mass transport calculation is to pick the composition of a fluid species, for example water in the case of oxygen or hydrogen isotope modeling. The concentration change of the system with time can now be expressed by replacing all concentrations in the time derivative by an equation similar to (27). For a monomineralic rock infiltrated by a fluid we can write:

$$\frac{\partial\left[\Phi C_f^{18O} + (1-\Phi)C_s^{18O}\right]}{\partial t} = \frac{\partial\left[\Phi C_f^{18O} + (1-\Phi)K_D C_f^{18O}\right]}{\partial t} = \left[\Phi + (1-\Phi)K_D\right]\frac{\partial C_f^{18O}}{\partial t} \tag{29}$$

Equation (29) has been used in modeling studies that have assumed local equilibrium (Baumgartner and Rumble 1988; Bickle 1992; Bickle and Baker 1990a,b; Bowman et al. 1994; Cartwright 1997; Dipple and Ferry 1992; Ferry and Dipple 1992; Gerdes et al. b 1995 a,b; Nabelek 1991; Norton and Taylor 1979; Roselle et al. 1999). The equation assumes time invariant porosity and an equilibrium constant that is not changing with time, though it may change with distance, as is the case for fluid/rock interaction along a thermal gradient.

Kinetic exchange of stable isotopes between fluid and rock. The exchange of stable isotopes between fluid and rock can be slow enough that the fluid/rock exchange cannot keep up with the transport of stable isotopes into or out of the system. This will result in significant disequilibrium between the isotopic composition of the fluid and the rock. This exchange is typically modeled by a first order rate law (Lasaga 1998; Lassey and Blattner 1988; Bowman and Willett 1991; Gerdes et al. 1995a; Cook et al. 1997). The first order rate law for stable isotope exchange is written as:

$$\frac{\partial C_r^{^{18}O}}{\partial t} = r_k\left(K_D C_f^{^{18}O} - C_s^{^{18}O}\right)$$

(30)

Here r_k is the first order kinetic rate constant for the stable isotope exchange between fluid f and mineral s, and K_D is the molar equilibrium constant defined previously (Eqns. 26 and 28). Hence, the rate of exchange is linearly dependant on the concentration difference between the actual isotope concentration in the rock at a given time and the concentration of the rock that would be in equilibrium with the fluid filling the pore space at this time.

In principle, one can imagine many other possible rate laws. The exchange of stable isotopes between an intact mineral and a surrounding fluid phase will by diffusion. Volume diffusion is very slow (Cole and Ohmoto 1986; Criss et al. 1987; Eiler et al. 1992; Cole and Chakraborty 2001), so that disequilibrium on mineral grain scales should be observed in most fluid-rock systems unless significant recrystallization has followed fluid infiltration or exchange was by a different mechanism. Exchange of stable isotopes will also occur during recrystallization of the mineral containing the element of interest. Recrystallization driven fluid-mineral exchange is typically faster by several orders of magnitude (Cole and Ohmoto 1986; Criss et al. 1987; Cole and Chakraborty, this volume).

The kinetic exchange equation used in stable isotope geochemistry (Eqn. 30) assumes that the stable isotope difference is driving stable isotope exchange. On the other hand, it is has been well established that recrystallization of minerals can be driven by many different mechanisms, including such diverse forces as the mineral surface energy, deformation state of the mineral, or mineral net transfer and exchange reactions (Lasaga 1981, 1998; Fisher and Lasaga 1981; Joesten 1991; Lasaga et al. 1991; Passchier and Trouw 1998). Recrystallization and mineral net transfer reactions involve the breaking up of the mineral into its constituents. It is likely that isotopes are rapidly exchanged between the constituents of the mineral and the fluid phase. In this case, the mineral recrystallization will promote rapid stable isotope exchange. The rate of exchange does not generally depend on the stable isotope disequilibrium, but rather the recrystallization rate. Equation (30) cannot be used to describe these reacting systems. For example, Valley and Graham (1996) used ion microprobe analysis of oxygen isotope ratios to document crack healing as an important process of oxygen exchange in hydrothermally altered quartz from granite on the Isle of Skye and show that simple rate laws are not applicable to such complex processes.

The isotopic rate of exchange will depend on the mechanism and the driving force of the net transfer reaction. Roselle (1997) found in a study of contact metamorphic rocks from the olivine zone of the Ubehebe Peak contact aureole, that calcite and dolomite do not record equilibrium stable isotope compositions (Fig. 9). Values of $\delta^{18}O_{calcite}$ are too low compared to dolomite in the same sample. Roselle (1997) concludes, that a large part of the calcite was newly formed by the olivine net transfer reaction, which consumes dolomite and tremolite. Petrologic data suggest that this reaction occurred during infiltration (Roselle 1997). It is likely that stable isotope exchange equilibration between calcite and fluid (but not dolomite) occurred during the prograde mineral reaction that formed olivine and calcite. In this case, stable isotope exchange kinetics were limited (or at least influenced) by the rate at which the net transfer reaction occurred (Baumgartner et al. 2001). Hence, the advection of H_2O and the transport of heat, both of which drive the net transfer reaction, have to be modeled to obtain an adequate rate law for stable isotope exchange of calcite. On the other hand, many individual dolomite grains in these

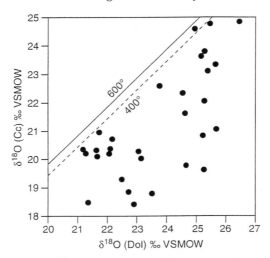

Figure 9. Plot of $\delta^{18}O$ values for dolomites and calcites from individual samples from the olivine zone of the Ubehebe Peak contact aureole, Death Valley (USA). Based on phase petrology, Roselle (1997) estimated peak metamorphic temperatures to be near 620°C in these samples collected within 50 meters of the contact with the quartz-monzonite. Equilibrium fractionation lines are shown for 600°C and 400°C (fractionation factor of Sheppard and Schwarcz 1970). The isotope data demonstrate disequilibrium between dolomite and calcite. The $\delta^{18}O$ values in calcite are consistently lower than those expected from dolomite composition. This plot suggests that calcite, which is a metamorphic reaction product during the formation of olivine and periclase, equilibrated with isotopically light fluids infiltrating the dolomites during prograde reaction, but that dolomite did not completely exchange. Thus, stable isotope exchange kinetics are coupled to the progress of metamorphic reactions. Stable isotope kinetics in reacting high temperature systems are not described by Equation (30).

rocks probably predated fluid infiltration. Thus, for dolomite, isotopic exchange might have been governed by Equation (30). Simplifying the above situation, one might assume that net transfer is relatively fast, and thus calcite crystallized in equilibrium with the fluid. This would justify the use of Equation (30) for this part of the fluid-rock exchange. Lackey (2000) described a similar situation for the Mt. Morrison pendant, Sierra Nevada, to explain wollastonite that formed during infiltration and exchanged with fluid more completely than detrital quartz. Additional research is needed before kinetically limited stable isotope exchange can be quantitatively modeled.

Geologically relevant insights from simple analytical solutions to the stable isotope transport equations

Inert tracer movement: pure infiltration. Non-reactive transport of stable isotopes by pure infiltration serves as an example of the solution approach. The rock will not exchange with the fluid phase, since non-reactive fluid flow is assumed. Isotope geologists working in high temperature, fossil hydrothermal systems cannot sample the fluid phase, though advances in analyzing individual fluid inclusions have been made. Thus, non-reactive stable isotope transport will go completely unnoticed by the geochemist. This limiting case is nevertheless of considerable interest because reactive transport can be compared to this case. It also results in a simple differential equation, which is readily solved by a transformation of the space coordinates.

A realistic geologic model would be the one-dimensional transport of carbon isotopes by CO_2 infiltration through a quartzite with constant porosity. Note that oxygen isotope transport is described by reactive transport equations, since oxygen isotopes will at least partially exchange with the quartzite (see below). The initial carbon isotope composition of the fluid in the pore spaces of the rock is $C_{f,0}^{13C}$. The column is infiltrated by a pure CO_2 fluid with a Darcy velocity v_f. The infiltrating fluid composition is $C_{f,1}^{13C}$. The governing transport equation for this one-dimensional, inert tracer transport problem by pure infiltration is obtained by inserting Equation (23) into Equation (20). The diffusion/dispersion term is dropped (assumption of pure infiltration, but see discussion above) to yield:

$$\frac{\partial \Phi C_f^{13C}}{\partial t} = -v_f \frac{\partial C_f^{13C}}{\partial x} \tag{31}$$

The boundary compositions and initial compositions specified above are stated mathematically by:

$$C(x,0) = C_{f,0}^{13C}$$
$$C(0,t) = C_{f,1}^{13C} \tag{32}$$

To proceed, a new coordinate system is introduced. Its origin is assumed to move with the average pore fluid velocity, V_p, in the x-direction. The new x-axis is denoted by $\bar{x}$. We have:

$$\bar{x} = x - v_p t \ ; \ \frac{\partial \bar{x}}{\partial x} = 1 \ ; \ \frac{\partial \bar{x}}{\partial t} = -v_p \tag{33}$$

The above equations are substituted into the partial differential Equation (31). We obtain for the left hand side:

$$\frac{\partial \Phi C^{13C}}{\partial \bar{x}} \frac{\partial \bar{x}}{\partial t} = -\Phi v_p \frac{\partial C^{13C}}{\partial \bar{x}} \tag{34}$$

The right side is unchanged in its form by the coordinate transformation. Thus, we have for Equation (31):

$$-\Phi v_p \frac{\partial C^{13C}}{\partial \bar{x}} = -v_f \frac{\partial C^{13C}}{\partial \bar{x}} \tag{35}$$

Recall that the pore fluid velocity multiplied by the porosity is equal to the Darcy velocity, v. Hence Equation (35) is correct for all time and space values without further manipulation. This result is not surprising, since the fluid is not moving in this transposed coordinate system. Hence, the initial conditions—here assumed to be a step in isotope composition—are displaced downstream by the fluid velocity, v_p. Solutions to more complicated initial and boundary conditions can be obtained (Dipple and Ferry 1992; Ferry and Dipple 1992; Nabelek 1991).

Local equilibrium exchange. The case of stable isotope transport with very fast fluid-rock exchange kinetics is of more interest. It is approximated by local equilibrium. The geologic example could be the transport of oxygen isotopes in the quartzite column discussed above.

The oxygen isotope composition of the pore fluid in the rock column is initially everywhere $C_{f,0}^{^{18}O}$. The infiltrating fluid composition is denoted with $C_{f,1}^{^{18}O}$. For a model system governed by pure infiltration, Equation (29) reduces to:

$$\frac{\partial C_f^{^{18}O}}{\partial t} = -\frac{v_f}{\left[\Phi + (1-\Phi)K_D\right]}\nabla C_f^{^{18}O} \qquad (36)$$

This equation can be solved again by transforming the solution domain into a moving coordinate system. The velocity of the coordinate system is slowed by the isotope exchange reaction. The following transformation is used:

$$\bar{x} = x - \frac{\Phi v_p}{K_D(1-\Phi)+\Phi}t \; ; \qquad \frac{\partial \bar{x}}{\partial x} = 1 \; ; \qquad \frac{\partial \bar{x}}{\partial t} = -\frac{\Phi v_p}{K_D(1-\Phi)+\Phi} \qquad (37)$$

The above equations are substituted into Equation (36), which describes equilibrium exchange transport. We obtain for the left hand side:

$$\frac{\partial C^{^{18}O}}{\partial \bar{x}}\frac{\partial \bar{x}}{\partial t} = -\frac{\Phi v_p}{K_D(1-\Phi)+\Phi}\frac{\partial C^{^{18}O}}{\partial \bar{x}} \qquad (38)$$

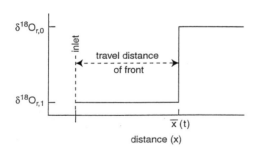

Figure 10. Plot of $\delta^{18}O$(quartz) versus distance for a mono mineralic, one-dimensional rock column undergoing equilibrium exchange. The initial rock composition is given as $\delta^{18}O_{r,0}$. Quartz with an isotope composition of $\delta^{18}O_{r,1}$ is in isotopic equilibrium with the infiltrating fluid composition $\delta^{18}O_{f,1}$. The solution assumes local equilibrium; hence the fluid composition in the column can be calculated with the help of the quartz-fluid fractionation factor. The distance traveled by the oxygen front is given by the equation $x = \Phi v_p / \{K_D(1-\Phi)+\Phi\}t$.

This is equal to the right hand side of Equation (36), written for the one-dimensional case under investigation. Thus, again the initial and boundary conditions move along with the fluid flow system, but the isotopic exchange front moves more slowly. The front is retarded relative to the fluid. The solution is illustrated in Figure 10. Here the concentrations are given for the rock composition (in δ-notation). Fluid is in equilibrium at every space coordinate within the column, so that its composition mirrors that of the rock. Fluid composition is calculated from the fractionation factor between fluid and rock. The advance of a concentration front is slowed by the retardation factor $[K_D(1-\Phi)+\Phi]/\Phi$ (e.g. Appelo and Postma 1999). The larger the retardation factor, the slower an isotopic front advances through the rock.

Care should be taken when comparing the formulas for the retardation factor, since the mathematical form critically depends on the concentration units used. For example, concentration of adsorbed solute in the hydrology literature is measured in weight of adsorbed material per volume or mass of the fluid, rather than the solid!

In many geologic systems, several isotopes can exchange between fluid and the rock. It is possible to obtain information on the relative abundances of the elements in the infiltrating fluid, assuming the exchange of each isotope obeys local equilibrium. Consider a calcite column that is infiltrated by a binary H_2O-CO_2 fluid. In this case both carbon and oxygen isotopes can exchange between fluid and rock. The advance of each isotopic front (oxygen and carbon isotope front) is given by its retardation factor. The relative distance each front traveled after infiltration of fluid over a time span of t_0 is given by:

$$\frac{t_0 v_{^{13}C}^{front}}{t_0 v_{^{18}O}^{front}} = \frac{\dfrac{\Phi v_p}{K_D^{^{13}C}(1-\Phi)+\Phi}}{\dfrac{\Phi v_p}{K_D^{^{18}O}(1-\Phi)+\Phi}} = \frac{K_D^{^{18}O}(1-\Phi)+\Phi}{K_D^{^{13}C}(1-\Phi)+\Phi} \approx \frac{K_D^{^{18}O}}{K_D^{^{13}C}} \tag{39}$$

Here $v_{^{13}C}^{front}$ and $v_{^{18}O}^{front}$ are velocities of the C and O isotope fronts, and $K_D^{^{13}C}$ and $K_D^{^{18}O}$ the molar distribution coefficients (Eqns. 26, 28) for carbon and oxygen isotopes. The right most approximation assumed in Equation (39) is that of a small porosity, Φ. Recall that the molar distribution coefficient, $K_D^{^{18}O}$ (Eqn. 28) depends on the fractionation factors and the relative abundances of the element in the fluid and solids.

$$\frac{v_{^{13}C}}{v_{^{18}O}} \approx \frac{\alpha_{s-f}^{^{18}O}\dfrac{\beta_s^{^{18}O}\bar{V}_f}{\beta_f^{^{18}O}\bar{V}_s}}{\alpha_{s-f}^{^{13}C}\dfrac{\beta_s^{^{13}C}\bar{V}_f}{\beta_f^{^{13}C}\bar{V}_s}} = \frac{\alpha_{s-f}^{^{18}O}\dfrac{\beta_s^{^{18}O}}{\beta_f^{^{18}O}}}{\alpha_{s-f}^{^{13}C}\dfrac{\beta_s^{^{13}C}}{\beta_f^{^{13}C}}} \tag{40}$$

The fractionation factor (α_{s-f}) is very close to 1 (0.99 to 1.01) at high temperatures for most isotopic systems of interest (i.e. Δ_{i-j} 10‰). Hence the movement of the front depends mainly on the relative abundance of the element in the mineral and the fluid phase, rather than the isotopic concentrations. Since the relative mobility of the fronts is independent of fluid velocity and porosity (if it is small), one can use the relative location of stable isotope fronts to determine the concentration ratio of these elements in the fluid phase (e.g. Baumgartner and Rumble 1988; Gerdes et al. 1995a). The relative retardation of the ^{13}C and ^{18}O fronts recorded by the model calcite column infiltrated by a binary H_2O-CO_2 fluid is approximated by:

$$\frac{v_{^{13}C}}{v_{^{18}O}} \approx \frac{\alpha_{s-f}^{^{18}O}\dfrac{3}{(1-X_{CO_2})+2X_{CO_2}}}{\alpha_{s-f}^{^{13}C}\dfrac{1}{X_{CO_2}}} \approx \frac{3X_{CO_2}}{1+X_{CO_2}} \tag{41}$$

Thus the relative speed of the fronts is determined only by the mole fraction of CO_2 in the fluid phase for this simple case. The oxygen and the carbon front will advance at equal speed if $3\,X_{CO_2} = 1 + X_{CO_2}$. This is the case only for X_{CO_2} equal to 0.5 for infiltration of CO_2-H_2O fluids into a pure calcite rock. The molar ratio of oxygen to carbon is the same in the fluid and the rock for $X_{CO_2} = 0.5$. Isotopic data from systems that are infiltrated by a fluid with molar abundance ratios of carbon:oxygen identical to the rock will plot as a straight array between unaltered (e.g. sedimentary) and the infiltrating fluid composition in Figure 4. It is surprising that many systems indeed plot as straight arrays in Figure 4, since this implies large quantities of carbon in the igneous fluids (e.g. Roselle et al. 1999; Baumgartner et al. 2001).

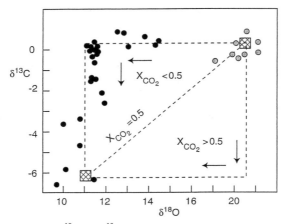

Figure 11. Plot of $\delta^{18}O$ and $\delta^{13}C$ values samples from a 10 m by 10 m outcrop of hydrothermally altered marble surrounding a 0.5 m to 1m gabbro dike in the Valle Paghera, Adamello, Italy (Baumgartner 1986; Baumgartner et al. 1989). The oxygen and carbon isotopic composition of the contact metamorphic calcite marbles outside the hydrothermal alteration zone have sedimentary compositions (open circles). Towards the gabbroic dikes, first the $\delta^{18}O$ of the calcite shifts towards the igneous signature, and finally the $\delta^{13}C$ value also decreases towards igneous values, due to magmatic fluid infiltration. The difference in retardation factors for carbon and oxygen isotopes is responsible for the position of the fronts for carbon and oxygen (Eqns. 42, 41). This is reflected in the δ–δ diagram (compare to Fig. 4). The data array is characteristic for water dominated infiltrative fluids (e.g. Valley 1986; Baumgartner and Rumble 1988; Gerdes et al. 1995a).

It is often difficult to identify the exact location of the beginning of the infiltration column. Also, stable isotope data are typically collected along two-dimensional outcrops. It is unlikely that these outcrops are oriented exactly in the direction in which the fluid infiltrated the rocks (e.g. Gerdes et al. 1995a). In general the outcrops cut the infiltration direction obliquely. This results in sample profiles that are often referred to as "sides of fluid flow systems" (e.g. Gerdes and Valley 1994; Yardley and Lloyd 1995; Gerdes et al. 1995a, Skelton et al. 2000). In these cases it is not possible to directly apply the one-dimensional calculations outlined above to obtain fluid amounts or fluid composition. Nevertheless, it is possible to obtain a qualitative estimate of fluid composition. Figure 11 reports isotope data from a meter scale hydrothermal alteration system, which developed around a small gabbroic dike in the Adamello contact aureole, Northern Italy (Baumgartner 1986; Baumgartner et al. 1989). The oxygen and carbon isotopic composition of the contact metamorphic calcite marbles display largely unchanged sedimentary compositions. Towards the gabbroic dyke, $\delta^{18}O$ of the calcite first shifts towards the igneous signature, then $\delta^{13}C$ value also decreases towards igneous values. The difference in retardation factors for carbon and oxygen isotopes is reflected in this upper L-shaped curve in a $\delta^{18}O$ – $\delta^{13}C$ plot (Fig. 11). This shape is characteristic for water-dominated (i.e. $X_{CO_2} < 0.5$) infiltrative fluids (e.g. Valley 1986; Baumgartner and Rumble 1988; Gerdes et al. 1995a). Similar relations have been demonstrated for $\delta^{18}O$ vs. δD in water-dominated systems (Criss and Taylor 1986).

The above example is easily generalized to more complicated systems involving more than two fluid species and several mineral phases and different stable isotopes. In each case, care should be taken to evaluate the fractionation factor. For some isotopes, like D/H, with large fractionation between the phases one might have to include the fractionation factor in the model calculations.

Chemical dispersion due to heterogeneity of rocks. The concept of stable isotope retardation is important for understanding of chemical retardation. In large and small-scale systems, a component of the dispersion is due to the heterogeneous distribution of reactive material (Dipple 1998). Some fluid flowing along a specific grain boundary or microfracture path will encounter more reactive mineral surface, while fluid along another path might encounter less, due to heterogeneous distribution of minerals. At junctions of the grain boundary or microfracture networks, fluids that travel different paths will be mixed. Since the retardation of a fluid batch is dependant on the amount of reaction, this will result in mixing of different compositions. This is referred to as chemical dispersion. Similarly, on a much larger scale, the reactivity of fluid paths will vary due to heterogeneity of geologic formations, rock layers or fracture systems. Again a chemical component will be added to the large-scale dispersion of solutes (Dipple 1998). The effect of chemical dispersion is strongly dependent on the stable isotope species under consideration.

Diffusion-reaction equation for stable isotopes. The diffusion-reaction equation is obtained by substituting Equation (27) or (30) into Equation (20) and dropping the infiltrative term. The partial differential equation then reads for the case of local equilibrium:

$$\left[\Phi + (1-\Phi)K_D\right]\frac{\partial C_f^{^{18}O}}{\partial t} = \Phi\Gamma D\frac{\partial^2 C_f^{^{18}O}}{\partial x^2} \tag{42}$$

$$\frac{\partial C_f^{^{18}O}}{\partial t} = \frac{\Phi\Gamma D}{\Phi + (1-\Phi)K_D}\frac{\partial^2 C_f^{^{18}O}}{\partial x^2} = D_r^*\frac{\partial^2 C_f^{^{18}O}}{\partial x^2} \tag{43}$$

Here, D_r^* is called the effective diffusion coefficient (e.g. Baumgartner and Rumble 1988; Cartwright and Valley 1991) whose value is modified by both the tortuosity of the diffusion path and by fluid-rock exchange. The advantage of rearranging the partial differential equation with an effective diffusion coefficient is that solutions for the partial differential Equation (43) are formally identical to those of the diffusion equation for a continuous, non-reactive media. For non-reactive diffusion we have:

$$\frac{\partial C_f^{^{18}O}}{\partial t} = \Gamma D\frac{\partial^2 C_f^{^{18}O}}{\partial x^2} = D^*\frac{\partial^2 C_f^{^{18}O}}{\partial x^2} \tag{44}$$

The effective diffusion coefficient, D_r^*, contains the tortuosity of the diffusion path, but not the porosity or any chemical factors of the host rock.

The diffusion of stable isotopes is retarded by the fluid-rock interaction. The retardation factor for diffusion with equilibrium fluid/rock exchange is:

$$\frac{D^*}{D_r^*} = \frac{\Phi + (1-\Phi)K_D}{\Phi} = 1 + \frac{(1-\Phi)}{\Phi}K_D \tag{45}$$

Thus, the effect of reaction on diffusion is similar to the effect it has on infiltration. In principle it is possible to determine the relative abundance of the major isotopes in the fluid compared to the rock, though the separation of curves is much less pronounced (Baumgartner and Rumble 1988). Solutions to the local equilibrium reaction/diffusion Equations (42) and (43) and the non-reactive diffusion Equation (44) have the same form. A wealth of solutions are available in the classic reference of Crank (1975).

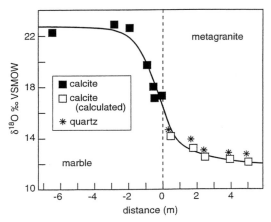

Figure 12. Oxygen isotope data from a metagranite-marble contact ("Steer's Head", Adirondack Mountains, New York, USA) plotted against distance. Oxygen isotope data for calcite was measured in the marbles (filled squares) and calculated (open squares) from measured quartz values (asterisks) using the calibration of Clayton et al. (1989). The oxygen isotope data for calcite is fit by the diffusion equation solution for two touching infinite half plates of different isotopic composition (after Cartwright and Valley 1991).

The solution for two touching infinite half spaces (touching at x = 0) is shown in Figure 12, along with data by Cartwright and Valley (1991) for oxygen isotope ratios determined across three closely spaced granite/marble contacts in the NW Adirondacks (Fig. 12). The data show the same sigmoidal exchange profile of approximately 10‰ across 6 m at each contact. The point of inflection is centered on the contact showing that there was no significant fluid flux normal to this surface and the complex geometry of the contacts in three dimensions argues against a significant flux along the contacts. Cartwright and Valley interpret these data in terms of the polymetamorphic history of the area where shallow contact metamorphism was overprinted by burial and upper amphibolite facies metamorphism. Fluid-filled porosity of 10^{-3} to 10^{-6} would be sufficient for these profiles to form in 10^3 to 10^7 years via diffusion through a static grain boundary fluid, most likely in or shortly after periods of fluid production during contact metamorphism. A slight asymmetry to the profile may be the result of different porosity in marble vs. granite; different rates of diffusive exchange between minerals and the grain boundary fluid; or different rates of mineral recrystallization and exchange.

Infiltration with diffusion/dispersion. In real flow systems, both diffusion and dispersion accompany fluid flow. The general equation of open system fluid-rock interaction is Equation (20). It has an infiltration term, together with a dispersion/diffusion term. Fluid-rock exchange is generally described by a kinetic formulation. Equation (30) is used here, for illustrative purposes. Substituting Equation (30) into Equation (20) one obtains:

$$\Phi \frac{\partial C_f^{^{18}O}}{\partial t} + (1-\Phi)r_k\left(K_D C_f^{^{18}O} - C_s^{^{18}O}\right) = -\Phi D_f \frac{\partial^2 C_f^{^{18}O}}{\partial x^2} + \Phi v_f \frac{\partial C_f^{^{18}O}}{\partial x} \qquad (46)$$

Here, we have written the equation for one mineral and one fluid species only, assuming constant porosity. Summation over all fluid and mineral species is necessary to describe fluid-rock exchange in polymineralic rocks (Bowman et al. 1994). Analytical solutions to this equation have been obtained for several boundary conditions (e.g. Ogata 1964; Golubev and Garibyants 1971; Cameron and Klute 1977; Van Genuchten and Parker 1984; Blattner and Lassey 1989; Bowman and Willet 1991; Appelo and Postma 1999). Most solutions are reported in terms of non-dimensional variables.

Non-dimensional transport variables. The one dimensional transport equation can be reorganized, so that the results are described in a convenient way by a few dimensionless variables. First, dimensionless variables for distance, concentration, and time are introduced (e.g. Bowman et al. 1994; Lasaga 1998). Dimensionless distance is defined by $z = x/L$. In principle, an arbitrary distance (L) can be used here. Nevertheless, it is useful to use a length characteristic of the length scale of the study or the operating process. For small-scale studies, one typically uses the grain diameter (e.g. Bear 1972). One might alternatively choose the maximum distance a reaction front has traveled, or fraction of the model scale, or the whole length of the model. The time (t) is scaled according to $\tau = t\Phi/(\upsilon_f L)$. The concentration is normalized to values between 0 (fully exchanged) and 1 (unexchanged) by introducing the dimensionless concentration,

$$\bar{c} = (\delta_r^{18}{}^{O} - \delta_{r,1}^{18}{}^{O})/(\delta_{r,0}^{18}{}^{O} - \delta_{r,1}^{18}{}^{O}).$$

Here the index (r) refers to the local rock concentration (r,1) to the fully changed (final) composition of the rock (e.g. the rock in equilibrium with the fluid inlet composition), and (r,0) to the initial rock composition.

Peclet-number. The Peclet number (N_{Pe}) is a dimensionless transport variable defined as (Bear 1972; Bowman and Willett 1991):

$$N_{Pe} = \frac{\upsilon_p L}{D} \tag{47}$$

N_{Pe} is the ratio of infiltration to diffusion in the fluid phase. The distance, L, is a characteristic transport length (see above), either taken as the size of the transport system, or the grain boundary diameter (Appelo and Postma 1999). Transport is dominated by infiltration for large Peclet numbers, $N_{Pe} > 1$, while small Peclet numbers characterize diffusive systems (N_{Pe} does not consider diffusion through mineral phases as volume diffusion is much slower than grain boundary diffusion). The solutions of pure infiltration discussed in the previous section are described by very large Peclet numbers. Very large Peclet numbers are unlikely, since the dispersion coefficient itself increases with pore fluid velocity (Fig. 7, Eqn. 17). In addition, the dispersion coefficient increases with system size (Fig. 8), so that dispersion will be important even for large-scale systems. Hence sharp infiltrative reaction fronts, such as are predicted by many models, will typically not occur in nature. Most relatively sharp isotope gradients are likely preexisting features or the sides of flow systems, since the transverse dispersivity is smaller than the longitudinal dispersivity (e.g. Gerdes and Valley 1994; Yardley and Lloyd 1995; Gerdes et al. 1995a; Skelton et al. 2000).

Damköhler I number. A further dimensionless number is the Damköhler I number (Boucher and Alves 1959; Lassey and Blattner 1988; Bowman and Willett 1991; Lasaga 1998). It describes the importance of the kinetic limitation of mass transfer relative to the advection of mass by the fluid flow. It is defined by:

$$N_D^k = \frac{r_k L}{\upsilon_p} \tag{48}$$

Here the rate, r_k, of the kth fluid-mineral exchange reaction is compared to the pore velocity of the fluid, υ_p. A high Damköhler I number implies fast reaction kinetics relative to fluid infiltration rates, and with increasing values local equilibrium is approached.

The general kinetic-dispersion-diffusion-infiltration equation. Equations (47) and (48) can be inserted into Equation (46), along with the definitions of dimensionless distance and time:

$$\frac{\partial \bar{c}_f^{18o}}{\partial \tau} = \frac{1}{N_{Pe}} \frac{\partial^2 \bar{c}_f^{18o}}{\partial z^2} - \frac{\partial \bar{c}_f^{18o}}{\partial z} - (1-\Phi) \sum_{k=1}^{n} N_D^k \left(K_D^k \bar{c}_f^{18o} - \bar{c}_k^{18o} \right)$$

$$= \frac{1}{N_{Pe}} \frac{\partial^2 \bar{c}_f^{18o}}{\partial z^2} - \frac{\partial \bar{c}_f^{18o}}{\partial z} - (1-\Phi) N_D \left(K_D \bar{c}_f^{18o} - \bar{c}_k^{18o} \right)$$

(51)

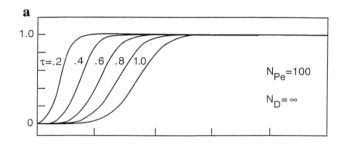

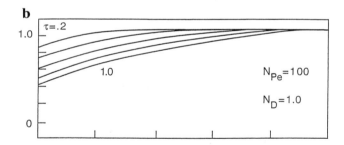

Figure 13. Normalized stable isotope composition of a rock column infiltrated by a reactive fluid. (a) The solution assumes local equilibrium, with a Peclet number (N_{Pe}) of 100 and infinite Damkohler I number. (b) Solution calculated for the case of a Peclet number of 100 and a Damköhler I number (N_D) of 1. Dimensionless parameters are given by: normalized concentration $\bar{c} = (\delta_r^{18o} - \delta_f^{18o}) / (\delta_i^{18o} - \delta_f^{18o})$; distance by $z = x/L$; and dimensionless time by $\tau = t\Phi / (\upsilon_f L)$. After Bowman and Willet (1991)—note that captions for their Figures 1 and 3 are switched.

The summation included for completeness is over all mineral species k. It reduces to the last term for the case of a single fluid and mineral species containing oxygen. Figure 13 shows the solutions of the equation for different Peclet and Damköhler I numbers calculated for a fluid composed of water and a one-dimensional rock column. Bowman et al. (1994) presented many more examples. The curves shown in Figure 13a are solutions for an infinite Damköhler I number at a Peclet number of 100. In this case of rapid reaction, e.g. local equilibrium, the rock and the fluid composition are different by the amount of the fractionation factor. Once either rock or fluid composition is known, it is simple to calculate the other. The effect of the Damköhler I number is illustrated for a value of 1.0 (Fig. 13 b). Here the composition of the fluid phase and the rock composition depend on the fluid velocity. Without solving the differential equation, it is not possible to relate the solid and fluid composition (a function of the Peclet number) in a simple way.

Fluid-rock reactions in rocks with heterogeneous permeability. Fluid flow in geologic systems nearly always requires solutions in three dimensions. The heterogeneity of rocks in space with respect to mineralogy, porosity, permeability, and mineral composition, as well as continuous change in the value of permeability, porosity, mineral mode, temperature, and other properties complicates flow paths and calculation of retardation of isotopes. The isotope signature measured in fossil hydrothermal systems is the integral effect of all these variations. Fluid flow is focused into areas of high permeability (Phillips 1991). Stable isotope exchange fronts move relatively fast in zones of high permeability, while zones of low permeability will be (nearly) devoid of fluid flow, resulting in slow front movement. Also, advection of stable isotopes is often the dominant mass transport mechanism in high permeability zones, like veins, while diffusion might dominate mass transport in the low permeability zones. Thus, for real geologic systems one will have to solve the equations governing fluid flow, along with those of the mass conservation for stable isotopes. In these cases it is generally necessary to solve the equations numerically. Even if analytical solutions are available (as in the case of the double porosity model discussed below) it is usually more convenient to solve the equations numerically with finite difference or finite element methods (e.g. Wang and Anderson 1982).

Double porosity model. The double porosity model (Wang 1993; Skelton et al. 2000) attempts to describe the heterogeneous media as a homogeneous media. If the distribution of heterogeneities is on a small enough scale, and the distribution of the heterogeneities is homogenous on a larger scale, one can define a media, which is composed of two interpenetrated medias. This is achieved by modeling transport along the high permeability zone using the transport equation introduced above, but allow for 'leakage' into the rock surrounding the high permeability zones. The flux into the low permeability zone is set equal to the difference between the equilibrium value of the rock, calculated from the concentration of the fluid in high permeability zone at any given time and location, and the actual (average) rock composition. This implies a steady state diffusional flux into the low permeability zones. For this model one can write for the change in rock composition a first order-like rate expression:

$$\frac{\partial C_r^{^{18}O}}{\partial t} = \bar{r}_r\left(K_D C_f^{^{18}O} - C_r^{^{18}O}\right) \tag{50}$$

Here $\bar{r}_r$ is a phenomenological coefficient characteristic of the geometry of the high and low permeability zones in the rock. Its value depends on the geometry of the high permeability and low permeability zones, as well as the mineral reaction kinetics. It is thus not readily linked to the reaction rate for stable isotope exchange between minerals and a fluid phase for most geometries. The kinetic coefficient of Equation (50) has to be treated as a phenomenological coefficient, which is adjusted to model the field data.

Examples of double porosity models applied to stable isotope transport in regional and contact metamorphism are given by Bowman et al. (1994) (see discussion of Alta contact aureole below), though this is not explicitly stated. Curves calculated for one-dimensional transport using a first order rate law were fitted to observed profiles with a large scatter of isotope data by adjusting the Damköhler I number. The resulting Damköhler I number indeed contains the effect of the double porosity model. The advantage of this kind of model is that the complicated transport system is reduced to one described by the one-dimensional kinetic transport Equation (51), for which solutions are readily obtained for many boundary conditions and geometries (e.g. Ogata 1964; Golubev and Garibyants 1971; Cameron and Klute 1983; Van Genuchten and Parker 1984; Blattner and Lassey 1989; Bowman and Willett 1991; Wang 1993; Appelo and Postma 1999; Lasaga 1998).

Two-dimensional models of heterogeneous permeability. The effect of the geometry of heterogeneous permeability can be modeled quantitatively with numerical models. In a conceptual way, one has to first calculate the fluid velocity distribution for a specific heterogeneous permeability distribution. This is achieved by solving equations developed in hydrology (e.g. Bear 1972; Appelo and Postma 1999). The velocity distribution at any given time can then be used to numerically solve the mass balance equation of stable isotope transport (Eqns. 19 and 20). For details of the derivation of the fluid flow equations and their solutions, the reader is referred to the literature (e.g. Bear 1972; Appelo and Postma 1999; Cook et al. 1997; Lasaga 1998; Gerdes et al. 1998). Here only a short summary is given.

The conservation of mass of fluid can be written with a sink/source term. One formulation of these equations is (e.g. Bear 1972; Gerdes et al. 1998):

$$\nabla\left(\rho_f \upsilon_f\right) = -\frac{\partial\left(\phi\rho_f\right)}{\partial t} + Q \tag{51}$$

Here ρ_f is the fluid density, and Q is the source or sink term for fluid. The flow velocity, υ_f is calculated from Darcy's equation for variable density fluid flow:

$$\upsilon_f = K\mu_r\left(\nabla h + \rho_r\nabla Z\right) \tag{52}$$

$$h = \frac{P}{\rho_o g} + Z \tag{53}$$

h is the fresh water equivalent head (Lusczinski 1961), P the fluid pressure, ρ_r the relative fluid density, ρ_o the reference density, and μ_r the relative fluid viscosity. K is the conductivity tensor, and Z the depth (or elevation). The solution of these equations yields fluid velocities, which can be used to calculate stable isotope transport and fluid rock exchange with Equation (22). A transient model of thermal driven fluid flow will also require the solution of the heat transport equation; e.g. Norton and Taylor 1979; Cathles 1981; Manning and Bird 1991; Hanson 1995a).

Deterministic permeability models. Application of the above principles to high temperature stable isotopes was pioneered by Norton and Taylor (1979) in their models of isotopic alteration of the Skaergaard layered intrusion and its host rocks. They used discreet zones and layers to which they assigned individual permeability values. Cartwright (1997) presented two-dimensional cases in which he modeled individual high permeability networks (fractures). Cook et al. (1997) used multiple, constant permeability zones to model the distribution of lithologies in the Alta stock area (see detailed discussion below). The advantage of this approach is that the calculated stable isotope patterns can be compared directly with measured patterns provided the permeability structure is adequately known. Permeability is also a function of time. Bolton et al. (1999) presented results of simulations in which they couple the chemistry of precipitation and dissolution with a variable permeability field. They demonstrated that no steady flow field can be reached in this case, resulting in migrating alteration plumes.

Stochastic permeability models. The effect of heterogeneous permeability can be investigated through the use of stochastic modeling (e.g. Gerdes et al. 1995b; Baumgartner et al. 1996). The basic idea of these models is that permeability is correlated in space and can be described by a random, but spatially correlated distribution. In these models, the likelihood of finding a high permeability value next to another high permeability value is larger, than finding a low permeability value next to a high one. Often the permeability distributions are assumed to be log normal and characterized by a standard deviation and a correlation length. For a discussion of the details the reader is referred to books on geostatistics (e.g. de Marsily 1986; Isaaks and Srivistava 1989; Gelhar 1993). Stochastic techniques are widely used in hydrology to evaluate possible flow and pollutant movements. Many individual permeability distributions, so called realizations, are generated, all obeying a specific set of permeability statistics and the spatial fluid velocity distribution is calculated. These velocity distributions are then in turn used to generate geochemical alteration patterns. No single realization will reproduce the actual data set—hence the comparison to any specific case is difficult. The intent is to study the possible alteration patterns that develop. Figure 14 shows results of calculations using Equation (36) for stable isotope transport (Gerdes et al. 1995b). The models were developed to describe the stable isotope data of the Alta contact aureole (Cook 1992; Cook and Bowman 1994; Bowman et al. 1994; Gerdes et al. 1995b; Cook et al. 1997).

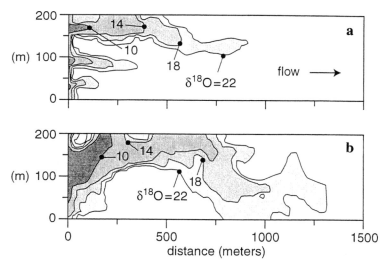

Figure 14. Patterns of $^{18}O/^{16}O$ depletion resulting from fluid infiltration into two different heterogeneous permeability distributions. The two-dimensional solution domain consists entirely of dolomite, is 200 m high and 1500 m long. The temperature was assumed to vary linearly from 600°C at the inlet (left) to 300°C at the outlet. Initial dolomite composition is $\delta^{18}O = 25‰$, and the infiltrating fluid composition was assumed to be $\delta^{18}O = 8‰$. Both diagrams were calculated for a variance of the logarithm of the permeability of 4.0 with a horizontal correlation length (x-axis direction) for the permeability of 300 m. (a) illustrates the solution for a vertical (y-axis) correlation length of 30 m, and (b) the solution for a vertical correlation of the permeability of 100 m. The models were designed by Gerdes et al. (1995b) to simulate the stable isotope data of the Alta contact aureole (Bowman et al. 1994; Cook et al. 1997). The permeability heterogeneity produces strong fingering of isotopic depletion, resulting in complicated alteration patterns.

The actual amount of breaking up of the front, the so called fingering of a front (Ortoleva et al. 1987; Steefel and Lasaga 1990; Skelton et al. 2000), as well as the orientation of the fingers, critically depend on the orientation and size of the high permeability zones (Gerdes et al. 1995b; Marchildon and Dipple 1998) and the temporal variation of the permeability (Bolton et al. 1999).

CONTINUUM MECHANICS AND FLUID-ROCK RATIOS

Fluid-rock interaction has been traditionally quantified using the concept of fluid-rock ratio (Ferry 1980; Ferry 1986). This concept was pioneered by Taylor (1977) for stable isotopes. The conceptual model of fluid-rock ratios is that of a volume of rock, that interacts with a volume of fluid, thereby changing the composition of the rock (and the fluid). A mass balance equation is written for this situation, in which one assumes a complete equilibration of the rock with the whole batch of fluid ("closed system"). This is a 0-dimensional model and is visualized by a beaker experiment, in which a specific amount of rock is enclosed together with an amount of fluid. After extensive shaking and exchange, equilibrium will be reached and the fluid can be extracted. From mass balance, one can calculate the composition of the rock after alteration, δ_r^f. A second model ("one-pass" model) describes a variation of the first. Here the rock is allowed to equilibrate with successive (infinitesimally) small batches of fluid, which are extracted before the next batch enters. While fluid-rock ratios can be easily visualized, their exact physical meaning is nevertheless ambiguous, as will be demonstrated below. In the following section, we present the mathematical models for calculating fluid/rock ratios. Then the accuracy and meaning of traditional fluid-rock ratio calculations are compared to model calculations using continuum mechanics calculations.

Calculation of fluid/rock ratios. The amount of fluid that has infiltrated and exchanged with a rock is often estimated from stable isotope data by mass-balance calculations assuming equilibrium stable isotope exchange. Taylor (1977, p. 523-524) derived the following relation for a 0-dimensional or box model:

$$(f/r)_{cs} = \frac{\delta_r^f - \delta_r^i}{\delta_f^i - \delta_r^f + \Delta_{r-f}} \tag{54}$$

where δ_r^f and δ_f^i are the initial rock and fluid isotope ratios. The final isotope ratio of the rock is denoted δ_r^f (all in standard permil notation). $(f/r)_{cs}$ (closed system) is the atom concentration ratio of the element of interest in the fluid and the rock for the entire closed fluid/rock system (e.g. molar units), and $\Delta_{r-f} = \delta_r^f - \delta_f^f$ In practice, this is applied by measuring δ_r^f of the rock and δ_r^i of its possible protolith. The initial fluid composition, δ_f^i, is estimated by inferring the fluid source. Finally Δ_{r-f} is estimated from the independently estimated equilibration temperature. If some minerals in a rock have not equilibrated or if temperature is known to have changed during the infiltration event, an appropriately weighted value of Δ_{r-f} can be used.

Equation (54) applies in a "beaker system," where all fluid arrives unreacted and simultaneously equilibrates with the rock within a defined volume. In the isotope literature (e.g. Taylor 1977) this is referred to as "closed system" isotope exchange. Alternatively, infiltration has been considered an "open-system" process such that each infinitely small aliquot of fluid equilibrates and then leaves the system. In this case the appropriate relation is

$$(f/r)_{os} = \ln[(f/r)_{cs} + 1] \tag{55}$$

where $(f/r)_{cs}$ is calculated for a "closed-system" from Equation (54) (Taylor 1977; Criss and Taylor 1986).

Applications of f/r calculations to continuum mechanics models. The fluid-rock ratio equations yield molar ratios, f/r, of oxygen or carbon in the fluid and the rock. The relationship between the f/r values and the continuum mechanics calculations are best illustrated by re-arranging Equation (54) in terms of the dimensionless concentration. The equation for the multi-pass fluid-rock ratio (in mole equivalents of the isotope under consideration) can be recast in terms of the dimensionless concentration of the rock,

$$\bar{c} = (\delta_r^{18}{}^O - \delta_{r,1}^{18}{}^O)/(\delta_{r,0}^{18}{}^O - \delta_{r,1}^{18}{}^O).$$

The concentration is normalized to values between 0 (fully changed) and 1 (unchanged). Here the index (r) refers to the local rock concentration, $(r,1)$ to the fully altered composition of the rock (e.g. the rock in equilibrium with the fluid inlet composition) and $(r,0)$ to the initial rock composition. With this definition it is possible to simplify Equation (54) to:

$$(f/r)_{mp} = \frac{1}{\bar{c}_r} - 1 \tag{56}$$

Similarly, the single pass fluid-rock ratio equation can be recast to yield:

$$(f/r)_{sp} = \ln\left(\frac{1}{\bar{c}_r}\right) \tag{57}$$

Figure 15 shows the results of three model calculations of isotopic shifts in rock that result from fluid infiltration along a one-dimensional column of homogeneous rock. The infiltration-reaction curves are given for pure infiltration and equilibrium exchange ($N_D =$) (curve a, Fig. 15), and a Peclet number of 100 (curve b, Fig. 15) assuming local equilibrium (Eqns. 27, 29). They are drawn in terms of the dimensionless variables introduced in the previous sections. Curve c in Figure 15 assumes a Damköhler number of 1. Before proceeding further, some comments about the continuum mechanics model have to be made. The infiltration profiles were calculated for a dimensionless time of $\tau = 1.0$, which determines the flux that entered the column (see previous section). The amount of fluid that entered the column percolated the entire column. The physical or actual fluid/rock ratio is the time integrated fluid flux that infiltrated the column, divided by the rock volume. Thus, the physical fluid-rock ratio depends on the size of the sample chosen for analysis. Nevertheless, the physical fluid-rock ratios are constant throughout the column, once a sample size is chosen. Thus a constant fluid-rock ratio is expected for this simple geometric problem.

In the following section we will apply the fluid-rock ratio Equations (56) and (57) to these infiltration-reaction curves by inserting values from the calculated continuum mechanics profiles into these equations. The idea is, that each calculated value represents a sample collected from a one-dimensional hydrothermal alteration profile.

The f/r values calculated for the local equilibrium cases (curves a and b, lower panel, Fig. 15) are infinite or at least very large at the inlet side (distance = 0) of the column, as the normalized concentration is close to or at zero, since the rocks in the column have completely exchanged with the fluid. In contrast, extremely low or zero fluid-rock ratios are obtained for rocks collected down stream, to the right of the infiltration front, near the outlet. Intermediate fluid-rock ratios are obtained for rock samples placed in the region near the infiltration front. Intermediate fluid-rock ratios are also calculated for the case of kinetic fluid-rock exchange (Fig. 15b). The ratio f/r steadily decreases to smaller values as samples from locations further down the column,

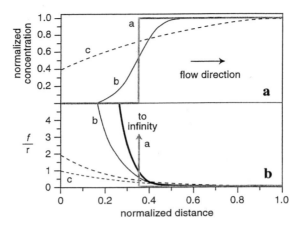

Figure 15. Normalized stable isotope compositions (upper diagram) and fluid-rock ratios calculated using Equations (54) and (55) (lower diagram) along one-dimensional rock columns. The curves given in the upper diagram were calculated for different model systems. a) reactive transport by infiltration only (N_{Pe} very large); b) local equilibrium reactive transport with a Peclet number of $N_{Pe} = 100$; c) kinetically controlled stable isotope exchange using Equation (30), with a Damköhler I number (N_D) of 1.0 and a Peclet number (N_{Pe}) of 100. The curves in the lower diagram reflect calculated fluid-rock ratios for the corresponding theoretical alteration curves. Dashed lines were calculated with Equation (8), while solid lines were calculated with Equation (7). Both Equations yield the same results in the case of pure infiltration, because the continuum mechanics solution predicts either complete exchange (infinite f/r for both cases) or non-exchanged (zero f/r for both cases). Note that the calculated fluid-rock ratios depend on the dimensionless transport parameters (N_{Pe} and N_D) since different fluid-rock ratios are obtained for the three calculated curves (a), (b), (c), which were all calculated for the same fluid flux! The fluid/rock ratio calculated by Equations (7) and (8) do not reflect physical fluid quantities and are best viewed as reaction progress variables for isotope exchange (Baumgartner and Rumble 1988; Bowman et al. 1994). Profiles are calculated for the dimensionless time $\tau = 1.0$. See Figure 13 for definition of dimensionless variables.

towards the outlet, are considered because the rock is at or near its original isotopic composition.

The sample calculations clearly demonstrate the limitations of the fluid-rock ratio concept. The calculated fluid-rock ratio depends on the location in the column, which is in contrast to the actual physical situation, which requires that fluid-rock ratios are constant throughout the column. It was shown previously that the shape, and the speed of a front crucially depend on the Peclet and the Damköhler numbers. Hence fluid-rock ratios calculated using Equations (56) and (57) are also dependant on these dimensionless transport numbers (see also Bowman et al. 1994). A calculated f/r ratio does not necessarily represent either a minimum, or a maximum physical fluid-rock ratio, since the calculated values vary between zero and infinity, while the physical fluid-rock ratio is a finite value. Hence f/r ratios are not an adequate measure of the amounts of fluid that infiltrated a given rock sample, nevertheless, they can be useful. They are a measure of how much isotopic exchange has occurred in a rock. As such, these ratios are better thought of as reaction progress variables for stable isotope exchange. This is clearly seen from Equations (56) and (57), where the ratio is linked to the normalized concentration. In fact, the normalized concentration is a reaction progress variable that varies from zero (no exchange) to one (fully exchanged). Fluid-rock ratio varies between zero (no exchange) and infinity (fully exchanged).

In conclusion, fluid-rock ratios should be used for a rough calculation to demonstrate open or closed system behavior. These ratios can yield qualitative information on stable isotope fluid-rock exchange. Their use should be limited to cases where continuum mechanics approaches to stable isotope transport are not applicable. This is often the case when field relations do not provide evidence of the geometry of the flow system. But one should keep in mind while interpreting the data, that the values do not correspond to actual, physical fluid amounts, but just represent a measurement of exchange (reaction) progress.

CONTACT METAMORPHISM

Studies of contact metamorphism generally have better geologic control than those of regional metamorphism. Most importantly, protolith compositions can be tightly constrained if clearly correlative, non-metamorphic equivalents are available for analysis. Orientation of temperature gradient (i.e. increasing towards pluton) and constant fluid and rock pressures can be assumed. Fluid flow patterns may vary systematically in relation to the igneous contact. All of these factors aid in the interpretation of isotopic results. However, the kinetics of isotope exchange are less certain.

In many contact metamorphic aureoles, values of $\delta^{18}O$ and $\delta^{13}C$ vary systematically. The coupled O-C trend is towards lower $\delta^{18}O$ and $\delta^{13}C$ values at higher metamorphic grade approaching the igneous contact. Thus, δ-values decrease progressively as metamorphism increases. This trend may also correlate with bleaching of the rock, coarsening of calcite and the development of calc-silicate minerals. This may be the result of one or more volatilization reactions, variable temperature, or fluid infiltration.

Table 1 summarizes 28 studies, mostly of contact aureoles, that show coupled O-C trends. In all cases, values of $\delta^{18}O$ and $\delta^{13}C$ decrease as the degree of metamorphism increases at scales ranging from <3 cm to 3 km. Values are plotted for representative trends in Figure 4. Some variation is evident among the unmetamorphosed equivalents in Figure 4 and this variation causes a scattering of values along the coupled O-C trends. In many areas this scatter is significant. However, the main conclusion from this diagram is that most rocks show similar trends. The internal consistency of each trend, with high values for protoliths and low values for samples from near contacts with igneous rocks, rules out a pre-metamorphic origin for the trends themselves. The important question that remains to be answered is: to what extent were these δ-values set by: (1) volatilization, (2) infiltration, or (3) changing P-T-X conditions?

In general, the effects of volatilization, fluid infiltration, and temperature are all in the same direction; calc-silicates and marbles tend toward lower values of $\delta^{18}O$ and $\delta^{13}C$ at higher grades of metamorphism. Thus, these processes mimic one another on a $\delta^{18}O$–$\delta^{13}C$ diagram. Detailed analysis shows that: (1) volatilization is always a factor in heavy isotope depletion, but is not the dominant cause of large shifts; (2) polythermal or disequilibrium effects can be identified and may be important; and (3) most contact aureoles studied (Table 1) have been infiltrated by fluids and O-C-H isotopic ratios frequently enable identification of fluid sources (see reviews by Valley 1986; Bowman 1998).

Volatilization during contact metamorphism

Most reactions in siliceous carbonates are stoichiometrically restricted to some variation of the normal calc-silicate decarbonation (Fig. 3) between "50% excess oxygen" and "50% excess carbon"; the "calc-silicate limit" of $F_{oxygen} \geq 0.6$ is not often surpassed.

Under these conditions the isotopic effects of volatilization are predicted to be similar to those shown by the upper portion of the ladder-like Rayleigh curves shown in Figure 4. Variations in each rock's path are controlled by initial isotopic ratios, values of α_{f-r}, the amount of reaction progress, temperature, and fluid/rock exchange kinetics. Although these calculations show that a component of heavy isotope depletion due to volatilization is present in the O-C trends of Figure 4, other processes must generally be dominant to cause the large depletions that are observed.

Figure 5 shows measured values of $\delta^{18}O$ and $\delta^{13}C$ for the Tierra Blanca limestone and for hydrothermal calcites adjacent to the Hanover-Fierro granodiorite (Turner and Bowman 1993). Additional intrusion related skarn systems are compiled in Table 2 of Bowman (1998). Curves are calculated for closed system devolatilization and for infiltration of magmatic or meteoric fluids (see the following sections on how to calculate the latter). Fluid flow, as evidenced by the distribution of skarn, was controlled by local stratigraphy and structure. While several end-member models approximate the values of $\delta^{18}O$ and $\delta^{13}C$ shown in Figure 5, the magnitude of isotope depletion increases with proximity to the pluton. This change is attributed to progressing upstream in the flow channels towards source located at the intrusive contact.

The flat slopes of some O-C trends, which pass above the field of igneous values, in Figure 4 requires exchange with infiltrating meteoric water (or seawater). This is most clearly seen for the Marysville (#2) and Skye (#5) data (Fig. 4), which extend to $\delta^{18}O < 5$.

The combusted bituminous marls of the Mottled Zone, Israel (#15), show an extreme depletion trend characterized by a steep slope in Figure 4 and very low values of $\delta^{13}C$ (Kolodny and Gross 1974). The depletion in ^{18}O shows a strong correlation with the conversion of Ca-carbonate to Ca-silicate over a wide range of temperatures (200 to 1300°C). The negative correlation between $\delta^{18}O$ and wt % CaO in silicate suggests that volatilization may have been the cause of the $^{18}O/^{16}O$ depletion (Matthews and Kolodny 1978). The low pressure environment required for natural combustion of these rocks makes them an extreme case where each specific metamorphic reaction will have taken place at as low a temperature as possible because the volatilization reactions have positive P-T slopes. Furthermore, exchange is likely with air or groundwater, which are involved as oxidants. Values of $\delta^{13}C$ in Mottled Zone calcites cannot be explained by volatilization. Low values of $\delta^{13}C$, and the unusual slope in Figure 4, reflect high temperature exchange of carbonates with bituminous matter (Kolodny and Gross 1974; Matthews and Kolodny 1978).

Fluid infiltration during contact metamorphism

The magnitude of $^{18}O/^{16}O$ or $^{13}C/^{12}C$ depletion caused by volatilization is not sufficient to cause the large isotopic shifts seen in Figure 4. Hence, these shifts must result largely from exchange with infiltrating fluids. The effect of infiltration is greatest when sufficient fluids are available, when contrasts in isotopic composition are great, when permeability is high, and fluid-rock exchange is not slowed by exchange kinetics. Convecting fluids may be derived from surface waters (meteoric or seawater), meta-morphic reactions, magmatic crystallization, or sedimentary formation waters. Permeability is highest in certain rock types, like uncemented sandstones (Brace 1984; Oelkers 1996) and fracture zones. Furthermore, permeability is higher at low pressure or if enhanced by metamorphic reactions. Fluid flow is driven by topography, fluid production, or the influence of temperature and fluid composition (salinity) on fluid density in near surface hydrology where the pressure is close to the hydrostatic pressure (see the review by Person and Baumgartner 1995). Here, the importance of infiltration is enhanced by fluid convection. Gravity driven fluid flow is limited to near surface environments, where thermal gradients are small (Furlong et al. 1991; Roselle and Bowman 2001). In many cases below the zone of hydrostatic fluid pressure, infiltration is driven by local pressure gradients caused by fluid production during igneous crystallization and metamorphic volatilization, or by deformation and other factors.

Controls of permeability. Differences in permeability and hence degree of infiltration may relate to a number of factors including: (1) the size, number, distribution and orientation of fractures, (2) connected porosity, (3) stress, (4) fluid pressure, and (5) the dissolution and precipitation of minerals (see, Valley 1986; Rutter and Brodie 1985; Baumgartner et al. 1996; Cartwright and Buick 1994; Cartwright 1997; Gerdes et al. 1998; Balashov and Yardley 1998; Bolton et al. 1999; Manning and Ingebritsen 1999). Petrologic and isotopic studies show that permeability decreases with depth in the crust (see Yardley and Valley 1997; Manning and Ingebritsen 1999). In shallow, permeable environments, values of P_{H2O} are approximately equal to the hydrostatic pressure of an overlying column of water, and rocks have sufficient strength to hold fluid conduits open. At greater depth and higher temperature, rock ductility increases and fluids are compressed to the lithostatic pressure of overlying rock ($P_{lithostatic} \approx 3 \times P_{hydrostatic}$). The transition from hydrostatic to lithostatic pressure is thus related to the brittle/ductile transition for rock deformation. These changes are time-dependent and there is no single depth at which these important transitions take place.

The permeability of a rock will evolve in time, since most of the above mentioned properties change with the history of the rock. For example, the permeability of a metamorphic rock can respond transient to fluid production during volatilization itself (Fyfe et al. 1978; Rumble and Spear 1983; Valley 1986; Nur and Walder 1990; Hanson 1992; Connolly 1997; Balashov and Yardley 1998; Connolly and Podladchikov 2000). During volatilization reactions, small fluid over-pressures are created by the positive ΔV(reaction), dilating the pore space or hydro-fracturing the rock. At the same time, the volume of the solid rock is reduced by evolution of fluid and the dissolution of solids. Fluid flow increases due to increased permeability, draining the rock, which in turn leads to collapse of the pore space in the ductile environment. This can result in either hydrofracturing or fluid/porosity waves (Nur and Walder 1990; Hanson 1992; Connolly 1997; Balashov and Yardley 1998; Connolly and Podladchikov 2000). Mineral dissolution and precipitation further changes the porosity and permeability of rocks leading to continuously changing flow conditions (Bolton et al. 1999).

It is also important to note that these factors influence permeability on different scales as described by Brace (1984), permeabilities measured on the scale of a drill hole (30-300 m) can be much higher, due to widely spaced fractures, than those measured on a laboratory sample (5-15 cm). Thus "drill hole" permeabilities may be appropriate for estimating the flux of channeled fluids through the crust, but "laboratory" scale permeability, if properly measured, affords a better estimate of truly pervasive fluid flow.

The infiltration of surface derived fluids into a contact aureole requires that fluid pressures be close to hydrostatic. Thus, if stable isotope ratios indicate exchange with large amounts of meteoric water, P_{H2O} must have been much less than $P_{lithostatic}$ at some time during the metamorphic history. The deepest known instances of meteoric water exchange represent the transition towards lithostatic fluid pressure. All known zones of meteoric water infiltration are shallower than ~15 km and most are less than 6 km (Valley and O'Neil 1982). Likewise, penetration of seawater to depths of 6-7 km in oceanic crust is well documented (Gregory and Taylor 1981). Possibly the deepest known penetration of surface-derived fluids is in veins and in faults of extensional terranes (Fricke et al. 1992; Nesbitt and Muehlenbachs 1995; Morrison and Anderson 1998).

An example: The Alta contact aureole

The Alta contact aureole in the Wasatch Mts, Utah, U.S.A. (Fig. 16) has been studied and modeled by many workers (Smith 1972; Moore and Kerrick 1976; Cook 1992; Cook and Bowman 1994; Bowman et al. 1994; Ferry 1994; Gerdes et al. 1995b; Baumgartner et al. 1996; Cook et al. 1997; Marchildon and Dipple 1998; Dipple 1998; Bowman and Valley 1999; Cook and Bowman 2000). A short review of the pertinent geology, phase petrology, and stable isotope geochemistry will be presented, based on the data of Smith (1972), Moore and Kerrick (1976), Cook (1992), Cook and Bowman (1994), Bowman et al. (1994), Cook et al. (1997), and Cook and Bowman (2000). The focus here is on the stable isotope geochemistry of the aureole. The Alta stock intruded into a sub-horizontal sequence of Paleozoic carbonates and pelites (Fig. 16) at a depth corresponding to a pressure of 100-200 MPa. Intrusion temperature was estimated at 825 to 925°C, with contact temperatures in the host rock of approxi-mately 575 to 625°C. Ambient host rock temperature is estimated at 300°C.

The location of the periclase isograd and/or associated stable isotope depletion patterns have been interpreted by all workers (Moore and Kerrick 1976; Cook 1992; Cook and Bowman 1994; Bowman et al. 1994; Cook et al. 1997; Cook and Bowman 2000; Gerdes et al. 1995b; Baumgartner et al. 1996; Ferry 1994; Dipple and Ferry 1992; Marchildon and Dipple 1998; Dipple 1998) to indicate that the direction of fluid flow in the inner aureole was away from the intrusion (i.e. down-temperature flow) and channeled into the host rocks below the flat-lying Alta-Grizzly thrust (Fig. 16). The most likely causes for this are low permeability layers above the Alta-Grizzly thrust, or a low permeability of the thrust itself (Cook 1992; Cook and Bowman 1994; Bowman et al.

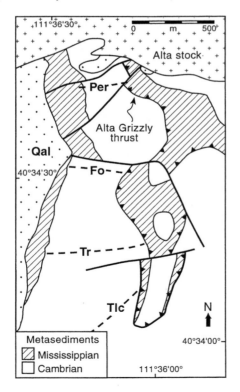

Figure 16. Map of the southern portion of the Alta contact aureole, Utah (USA). The talc (Tlc), tremolite (Tr), forsterite (Fo) and periclase (Per) isograds mark the first appearance of these index minerals in siliceous dolomites. After Cook (1992).

1994; Cook et al. 1997; Cook and Bowman 2000). The fluids carry a mag-matic oxygen and carbon isotope signature. Hence the fluid source is probably the crystallizing magma of the Alta stock.

Stable isotope data. The $\delta^{18}O$ values of carbonates in the outer aureole (23‰ to 28‰, SMOW) are typical for marine carbonates (Fig. 17). Values of both $\delta^{18}O$ and $\delta^{13}C$ decrease towards the intrusive contact in a characteristic coupled trend (#17 in Fig. 4; Table 1). There are dramatic differences in the spatial distribution of ^{18}O depletion and periclase abundance from location to location within the periclase zone. Samples near the contact, below the pre-existing Alta-Grizzly thrust, contain abundant periclase, along with low $\delta^{18}O$ values, while periclase is

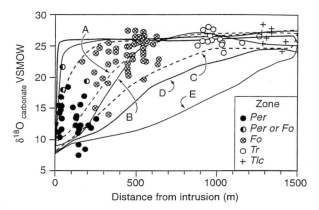

Figure 17. Comparison of the oxygen isotope data of whole rocks (dolomite/calcite) for the southern Alta contact aureole, plotted as a function of distance from the contact (see Fig. 16), with model predictions. Curve (A) denotes the alteration predicted by Bowman et al. (1994), and curve (B) is the best fit model of Cook et al. (1997). The ranges denoted (C), (D), (E) are expected isotope variations from the heterogeneous permeability models of Gerdes et al. (1995b). Compare Figure 14 for two-dimensional results of the model calculations. Model C: 300 m horizontal and 30 m vertical correlation length for the permeability, with a variance of the logarithm of the permeability of 2.0; Model D: 300 m horizontal and 30 m vertical correlation length for the permeability, with a variance of the logarithm of the permeability of 4.0; Model E: 300 m horizontal and 100 m vertical correlation length for the permeability, with a variance of the logarithm of the permeability of 4.0.

very rare above the thrust and $\delta^{18}O$ values of carbonates cluster in the sedimentary range. Much of the isotopic and petrologic data (Fig. 17) has thus been collected from samples below the thrust from a 200 m-thick section of sub-horizontal carbonates sampled in the Mississippian Fitchville formation and lower parts of the Deseret/Gardison formation. Samples span the distance from the periclase/brucite zone near the intrusive contact to the talc zone 1500 m away (Fig. 16).

Near the intrusion, the $\delta^{18}O$ values decrease, while the heterogeneity of the values increases. Values of $\delta^{18}O$ as low as 9.0‰ are found near the contact (Fig. 17), but nearly unaltered values of up to 22‰ are found as well (Cook et al. 1997). A detailed profile taken perpendicular to the sedimentary layers (Fig. 8 from Cook et al. 1997) reveals that layers containing abundant periclase have low $\delta^{18}O$ values. Heterogeneity of stable isotope compositions is on the 10 cm- to 1 m-scale corresponding to the scale of sedimentary layering (Cook et al. 1997). Calcite/forsterite pairs are not isotopically equilibrated at the sub-cm-scale in the Periclase zone (Bowman and Valley 1999)

Modeling. Bowman et al. (1994) interpreted the low and variable $\delta^{18}O$ values near the intrusion as the result of horizontal, down–temperature infiltration of H_2O–rich magmatic fluids ($X_{CO_2} < 0.03$) through rocks with heterogeneous permeabilities. The low X_{CO_2} is indicated both by mineral stabilities and by the concave downwards curve of line #17 in Figure 4. They argue that the observed isotopic depletion patterns cannot be produced by isotopic fractionation effects accompanying the decarbonation reactions occurring in these carbonate rocks. They also demonstrate that up-temperature fluid flow cannot explain the observed stable isotope patterns. Bowman et al. (1994) attempted to describe the average stable isotope composition of the carbonates using a one-dimensional, local equilibrium continuum mechanics model (see Eqn. 29). Assuming fluid flow down the temperature gradient (x = 0, T = 600°C; x = 2.4 km, T = 200°C) away from the intrusion. Infiltration of fluids with a $\delta^{18}O$ of 8‰ with a Peclet number of about 20 yields a fit to the data (Fig. 17, curve A). A fit to the average oxygen isotope composition is obtained by adjusting the Damköhler number, assuming small values for the dispersion coefficient (Peclet number > 20). A very low Damköhler number of 1.2 to 0.02 was required to fit the data in this model. Bowman et al. (1994) argue that exchange kinetics are fast for stable isotopes at these temperatures, since oxygen exchange is dominated by recrystallization accompanying decarbonation reactions. Hence the model with a Peclet number of 20 was preferred by the authors. A time integrated fluid flux of $516 m^3/m^2$ (on average) is required to produce the observed isotopic alteration patterns. An average fluid flux of 0.1 m^3/m^2 per year was calculated using an estimate for the prograde heating in the inner aureole (periclase zone). These models do not attempt to describe the isotopic composition of the individual samples, but rather the average composition is fitted. In this way, much of the information in the data is lost. It is clear from these studies, that multi-dimensional stable isotope models are required to model the scatter of the data. At least a two dimensional model is required to model hetero-geneous permeability of individual sedimentary layers.

In a follow-up publication, Cook et al. (1997) presented a sophisticated numerical study of coupled heat, fluid, and isotope fluid-rock interaction for the whole Alta aureole. They used a radial-symmetry formulation of the transport equations. Stable isotope fluid-rock exchange was modeled using a kinetic formulation (Eqn. 30). They used the rate data summarized in Cole and Ohmoto (1986) for carbonate-water exchange, since the rate constant should reflect surface rate kinetics, rather than volume diffusion in minerals. This is due to the fact that the minerals have undergone significant recrystallization associated with the prograde silicate-carbonate reactions. The use of Equation (30) is inappropriate, since the driving force for recrystallization is not the stable isotope disequilibrium assumed in Equation (30). The rate of exchange is controlled by mineral reaction affinities and the rate equation should reflect this. In addition, multiple primary and retrograde fluid pulses might overprint the prograde stable isotope alteration pattern.

Cook et al. (1997) conducted a sensitivity study with several different heuristic permeability distributions. The thermal profiles and the location of the periclase and tremolite isograds, as well as the aureole scale pattern of $^{18}O/^{16}O$ depletions, served as criteria for the selection of the preferred model. The best-fit model requires several low permeability horizons above the Alta-Grizzly thrust and intrusions with permeabilities less than 10^{-18} m^2 to prevent large vertical fluid fluxes. Also, a high-permeability zone of at least $2x10^{-16}$ m^2 is required below the Alta-Grizzly thrust. The resulting fit to the data is shown in Figure 17 (curve B). An average time integrated fluid flux of roughly 3000 m^3/m^2 was required to produce the observed isotopic depletion zone, over a period of 5000 years. Soon after this time of prograde heating of the innermost aureole, fluid flow decreased during cooling due to volume increase upon hydration of periclase to brucite. This prevents the fluid from advancing the isotopic front farther out into the aureole. Again, the models are unable to reproduce the scatter of the isotopic data quantitatively, since their permeability variations are on to large a scale (100-1000 m). Permeability variations in ten meter or even hand specimen scale are required to obtain a fit to the data.

Stable isotope transport models using stochastic permeability models were applied to the carbonate unit at Alta (Gerdes et al. 1995b; Baumgartner et al. 1996; Marchildon and Dipple 1998). Figure 14 shows results from variable permeability calculations for a 200 m-thick section just below the Alta-Grizzly thrust (Fig. 16). The effects of permeability variance and horizontal and vertical permeability correlation distances were investigated in the studies. The permeability field used had a variance of the logarithm of the permeability of 2 to 4 log units. For comparison, the variance for shallow aquifers and soils ranges between ~0.04 to 3 log units (Gelhar 1993). The spatial correlation in the direction of the lithology was assumed to be 300 m for all the Alta simulations, consistent with Fogg and Lucia's (1990) estimate of 300 to 600 m for a layered carbonate unit. The correlation lengths perpendicular to layering were varied between 30 m and 100 m. This correlation length is about 10 to 100 times larger then that observed by Cook et al. (1997, their Fig. 8). These simulations are capable of describing the observed scatter in $\delta^{18}O$ values (Fig. 17; ranges C and D). Complex $\delta^{18}O$ patterns are produced in the carbonate layer (Fig. 14). The lobate, irregular $\delta^{18}O$ patterns shown in the stochastic models clearly show, as expected, the importance of the heterogeneous permeability on the fluid-rock interaction. A permeability variance of at least 2 log units is required to fit the Alta stable isotope data (Gerdes et al. 1995b) and the phase assemblages (Marchildon and Dipple 1998).

A further interesting aspect was added to stable isotope models of the Alta aureole by Dipple (1998). He realized that the production of carbon dioxide by mineral reactions, and changing modal abundances of minerals near the inlet (the igneous contact) of the model system will change the isotopic composition of the fluid down stream of the mineral reaction fronts. The oxygen isotope signature of fluid can be altered by up to 5‰ due to coupling of the isotopic effect of metamorphic mineral reactions, non-steady state thermal evolution of the flow system, and stable isotope transport. The metamorphic reactions add significant amounts of dispersion to the isotope profiles (see section on dispersion). Dipple (1998) argued that part of the variability in isotopic compositions of rocks observed in hydrothermal systems, especially in the outer parts of an aureole, could be due to chemical dispersion of isotopes.

FUTURE WORK:
MERGING THE RESULTS OF MODELING AND MICROANALYSIS

The assumption of local equilibrium at Alta has been tested by detailed laser fluorination analysis of $\delta^{18}O$ in forsterite at one layered outcrop of the periclase and forsterite zones (Bowman and Valley 1999). Figure 17 is based on analysis of carbonates from the earlier work of Bowman and co-workers (Bowman et al. 1994; Cook et al. 1997). From these data and equilibrium Δ(Cc-Fo), predicted values of $\delta^{18}O$ (Fo) are several per mil lower than those actually measured by laser ablation. The magnitude of disequilibrium is largest in the periclase zone where fluid flow was greatest. Thus, local equilibrium does not apply to silicates in these rocks, the exchange of forsterite with infiltrating fluids has not been as complete as that of carbonates, and as a result the oxygen isotope composition of the forsterite has lagged behind calcite. This indicates that the Damkohler number (N_D) was less than infinite and it suggests that N_D varied through time. It is likely that N_D is highest during periods of mineral reaction or deformation, and low at other times. It is also clear that the carbonates have faster exchange kinetics or are easier recrystallized than the silicates. They are more sensitive recorders of fluid flow.

It remains to be tested if local equilibrium fully describes the exchange of carbonates at Alta. Preliminary examination of polished thin sections from the periclase and forsterite zone marbles by cathodoluminescence (CL) reveals thin (<1 mm) interconnected networks of brightly cathodoluminescent calcite cutting across darker CL calcite. The number of crosscutting features and the intensity of CL are greatest in the highly infiltrated, low $\delta^{18}O$ periclase zone calcites. The simplest interpretation of these luminescence zones is that they represent healed microfractures and that the composition of calcites was altered by infiltrating fluids derived from the Alta Stock upon crystallization. This interpretation predicts that large gradients of $\delta^{18}O$ exist at the sub-mm-scale. Similar features were discovered by ion microprobe in marble from adjacent to the Utsubo granite, Japan by Graham et al. (1998; #21 in Table 1). Values of $\delta^{18}O$ in calcites were found to be significantly altered by infiltrating fluids at Utsubo and gradients of up to 10‰ per 200 μm are documented in calcites. A double medium exchange model (e.g. Wang 1993) would be more appropriate than local equilibrium for this system, but no such sophisticated model of mineral-fluid exchange has yet been coupled to a fluid flow code.

Thus, improved models of the fluid flow history at Alta, and elsewhere, must be formulated based on more detailed stable isotope analyses and knowledge of exchange kinetics. The zonation and heterogeneity of $\delta^{18}O$ in specific minerals contain a rich, and relatively untapped, store of information for understanding the complex history of fluid flow in metamorphic rocks.

Two new approaches to the study of metamorphic fluids are presented by mechanistic fluid flow models and by microanalyses of stable isotope ratios. As more accurate and spatially resolved analyses become available, some simplifying assumptions of flow models will be challenged and refined models may become necessary. The challenge is to merge different scales of observation. On one side it is necessary to model the whole hydrothermal system to adequately describe the driving forces for fluid flow and fluid pathways (e.g. Cook et al. 1997; Gerdes et al. 1998). On the other hand, stable isotope analyses are obtained from smaller and smaller aliquots of rocks or minerals. This requires very small-scale resolution of the numerical models. Large scale models are necessary to simultaneously link mineral reactions and thermal flow with fluid flow models, since accurate boundary conditions for flow systems are typically only known far from the thermal (or mechanical) perturbation driving the fluid flow. Such large-scale coupled models require numerical solutions to the governing equations. The consequence is that even with large computers, the numerical resolution is still in the tens-of-meters range (e.g. Gerdes et al. 1998). On the other hand, analytical procedures allow us to investigate smaller and smaller domains. Bridging the gap between detailed, careful analytical work and large-scale modeling will pose a major challenge in the future!

ACKNOWLEGMENTS

John Bowman, Dave Cole, Elizabeth King, Val Fereirra, Brooks Hanson, Benita Putlitz, Jade Star Lackey, Greg Roselle, and Mike Spicuzza are thanked for reviews that helped improve the manuscript. Mary Diman drafted all figures. This research was supported by the U.S. Department of Energy and National Science Foundation, and the German Science Foundation.

REFERENCES

Abart R (1995) Phase equilibrium and stable isotope constraints on the formation of metasomatic garnet-vesuvianite veins (SW Adamello, N Italy). Contrib Mineral Petrol 122:116-133

Ague JJ (1997) Crustal mass transfer and index mineral growth in Barrow's garnet zone, Northeast Scotland. Geology 25:73-76

Ague JJ, Rye DM (1999) Simple models of CO_2 release from metacarbonates with implications for interpretation of directions and magnitudes of fluid flow in the deep crust. J Petrol 40:1443-1462

Appelo CAJ, Postma D (1999) Geochemistry, Groundwater and Pollution. A A Balkema, Rotterdam, 536 p

Arita Y, Wada H (1990) Stable isotope evidence for migration of metamorphic fluids along grain boundaries of marbles. Geochem J 24:173-186

Balashov VN, Yardley BWD (1998) Modeling metamorphic fluid flow with reaction-compaction-permeability feedbacks. Am J Sci 298:441-470

Baumgartner LP (1986) Petrologie und Geochemie stabiler Isotopen von zonierten metasomatischen Gesteinen der SW-Adamello Kontakt-Aureole, Italien. PhD Dissertation, Universität Basel, Basel, Switzerland, 157 p

Baumgartner LP, Gieré R, Trommsdorff V, Ulmer P (1989) Field Guide for the Southern Adamello. *In* Tromsdorff V et al. (eds) Guide Book for the Excursion to the Central Alps, Bergel and Adamello. Universita degli Studi di Siena, p 91-121

Baumgartner LP, Ferry JM (1991) A model for coupled fluid flow and mixed -volatile mineral reactions with applications to regional metamorphism. Contrib Mineral Petrol 106:273-285

Baumgartner LP, Rumble D III (1988) Transport of stable isotopes—Development of continuum theory for stable isotope transport. Contrib Mineral Petrol 98:273-285

Baumgartner LP, Gerdes ML, Person MA, Roselle GT (1996) Porosity and permeability of carbonate rocks during contact metamorphism. *In* Jamtveit B, Yardley B (eds) Fluid Flow and Transport in Rocks: Mechanisms and Effects. London, Chapman & Hall, p 83-98

Baumgartner LP, Clement RM, Putlitz B, Roselle GT, Wenzel TU (2001) Fluid and melt movement in contact aureoles. Eleventh annual V M Goldschmidt Conf Abstract #3746. LPI Contrib 1088, Lunar and Planetary Institute, Houston, Texas (CD-Rom)

Bear J (1972) Dynamics of Fluids in Porous Media. Elsevier, New York. 764 p

Bickle MJ (1992) Transport mechanism by fluid flow in metamorphic rocks: oxygen and strontium decoupling in the Trois Seigneurs Massif: A consequence of kinetic dispersion. Am J Sci 292: 289-316

Bickle JM, Baker J (1990a) Advective-dispersive transport of isotopic fronts: An example from Naxos, Greece. Earth Planet Sci Lett 97: 78-93

Bickle JM, Baker J (1990b) Migration of reaction and isotopic fronts in infiltration zones: Assessments of fluid flux in metamorphic terrains. Earth Planet Sci Lett 98: 1-13

Blattner P, Lassey R (1989) Stable isotope exchange fronts, Damköhler numbers, and fluid-rock ratios. Chem Geol 78: 381-392

Bolton EW, Lasaga AC, Rye DM (1999) Long-term flow/chemistry feedback in a porous medium with heterogeneous permeability: kinetic control of dissolution and precipitation. Am J Sci 299: 1-68

Boucher DF, Alves GE (1959) Dimensionless numbers for fluid mechanics, heat transfer, mass transfer and chemical reaction: Chem Engineer Progr 55:55-64

Bowman JR (1998) Stable-isotope systematics of skarns. *In* Lentz DR (ed) Mineralized Intrusion-Related Skarn Systems. Mineral Assoc Canada Short Course Series 26:99-146

Bowman JR, Covert, JJ, Clark AH, Mathieson GA (1985b) The Cantung E Zone Scheelite Orebody, Tungsten, NW Territories: O, H and C isotope studies. Econ Geol 80:1872-1985

Bowman JR, O'Neil JR, Essene EJ (1985a) Contact skarn formation at Elkhorn, Montana. II: Origin and evolution of C-O-H skarn fluids. Am J Sci 285:621-660

Bowman JR, Willet SD (1991) Spatial patterns of oxygen isotope exchange during one-dimensional fluid infiltration. Geophys Res Lett 18:971-974

Bowman JR, Willett SD, Cook SJ (1994) Oxygen isotope transport and exchange during fluid flow: One-dimensional models and applications. Am J Sci 294:1-55

Bowman JR, Valley JW (1999) Disequilibrium $^{18}O/^{16}O$ exchange of olivine and humite in marbles of the Alta Stock thermal aureole, Utah. Geol Soc Am Abstracts with Program 31:7, A101

Brace WF (1980) Permeability of crystalline and argillaceous rocks. International J Rock Mech Mining Sci Geomech Abstr 17:241-251

Brace WF (1984) Permeability of crystalline rocks: New *in situ* measurements. J Geophys Res 89: 4327-4330

Brady JB (1983) Intergranular diffusion in metamorphic rocks. Am J Sci 283-A:181-200

Brown PE, Bowman JR, Kelly WC (1985) Petrologic and stable isotope constraints on the source and evolution of skarn-forming fluids at Pine Creek, California. Econ Geol 80:72-95

Cameron DR, Klute A (1977) Convective-dispersive solute transport with a combined equilibrium and kinetic adsorption model. Water Resource Res 13:183-188

Cartwright I (1997) The two dimensional patterns of metamorphic fluid flow at Mary Kathleen, Australia: Fluid focusing, transverse dispersion, and implication for modeling fluid flow. Am Mineral 79: 526-535

Cartwright I, Buick IS (1996) Determining the direction of contact metamorphic fluid flow: an assessment of mineralogical and stable isotope criteria. J Metmorph Geol 14:289-305

Cartwright I, Oliver NHS (2000) Metamorphic fluids and their relationship to the formation of metamorphosed and metamorphogenic ore deposits. In Spry P, Marshal B, Vokes FM (eds) Metamorphosed and Metamorphogenic Ore Deposits. Rev Econ Geol 11:81-96

Cartwright I, Valley JW (1991) Steep oxygen-isotope gradients at marble-metagranite contacts in the Northwest Adirondack Mountains, New York, USA; products of fluid-hosted diffusion. Earth Planet Sci Lett 107:148-163

Cartwright I, Weaver TR (1993) Metamorphic fluid flow at Stephen Cross Quarry, Quebec: Stable isotopic and petrologic data. Contrib Mineral Petrol 113:533-544

Cathles LM (1981) Fluid flow and genesis of hydrothermal ore deposits. In Skinner, BJ (ed) Econ Geol, Seventy-fifth Anniversary Volume, 1905-1980. Soc Econ Geol Publication, p 424-457

Chiba H, Chacko T, Clayton RN, and Goldsmith JR (1989) Oxygen isotope fractionations involving diopside, forsterite, magnetite, and calcite: application to geothermometry. Geochim Cosmochim Acta 53:2985-2995

Clayton RN, Goldsmith JR, Mayeda TK (1989) Oxygen isotope fractionation in quartz, albite, anorthite, and calcite. Geochim Cosmochim Acta 53:725-733

Clauser C (1992) Permeability of crystalline rocks. EoS, Trans Am Geophys Union 73:237-238

Cole DR, Ohmoto H (1986) Kinetics of isotope exchange at elevated temperatures and pressures. Rev Mineral 16:41-90

Connolly JAD (1997) Devolatilization-generated fluid pressure and deformation-propagated fluid flow during prograde regional metamorphism. J Geophys Res B, 102:18149-18173

Connolly JAD, Holness MB, Rubie DC, Rushmer T (1997) Reaction-induced microcracking; an experimental investigation of a mechanism for enhancing anatectic melt extraction. Geology 25: 591-594

Connolly JAD, Podladchikov YY (2000) Temperature-dependent viscoelastic compaction and compartmentalization in sedimentary basins. Tectonophys 324:137-168

Cook SJ (1992) Contact metamorphism surrounding the Alta stock, Little Cottonwood Canyon, Utah. PhD Dissertation, University of Utah, Salt Lake City

Cook SJ, Bowman JR (1994) Contact metamorphism surrounding the Alta stock: Thermal constraints and evidence of advective heat transport from calcite + dolomite geothermometry. Am Mineral 79:513-525

Cook SJ, Bowman JR, Forster CB (1997) Contact metamorphism surrounding the Alta stock: finite element model simulation of heat- and $^{18}O/^{16}O$ mass-transport during prograde metamorphism. Am J Sci 297: 1-55

Cook SJ, Bowman JR (2000) Mineralogical evidence for fluid-rock interaction accompanying prograde contact metamorphism of Siliceous Dolomites: Alta Stock Aureole, Utah, USA. J Petrol 41:739-757

Crank J (1975) The mathematics of diffusion. Clarendon Press, Oxford, UK, 347 p

Criss RE, Taylor HP Jr (1986) Stable isotope geochemistry of metamorphic rocks. Rev Mineral 16:373-424

Criss RE, Gregory RT, Taylor HP Jr (1987) Kinetic theory of oxygen isotope exchange between minerals and water. Geochim Cosmochim Acta 51:1099-1108

Deines P, Gold DP (1969) The change in C and O isotopic composition during contact metamorphism of the Trenton Limestone by the Mount Royal pluton. Geochim Cosmochim Acta 33:421-424

De Marsily G (1986) Quantitative Hydrology, Groundwater Hydrology for Engineers. Academic Press, Orlando, Florida, 440 p

Dipple GM, Ferry JM (1992) Fluid flow and stable isotope alteration in rocks at elevated temperatures with applications to metamorphism. Geochim Cosmochim Acta 56:3539-3550

Dipple GM (1998) Reactive dispersion of stable isotopes by mineral reaction during metamorphism. Geochim Cosmochim Acta 62:3745-3752

Domenico PA, Schwartz FW (1990) Physical and chemical hydrology. John Wiley and Sons, New York, 824 p

Dullien FAL (1979) Porous Media: Fluid Transport and Pore Structure. Academic Press, New York

Eiler JM, Baumgartner LP, Valley JW (1992) Intercrystalline stable isotope diffusion: a fast grain boundary model. Contrib Mineral Petrol 112:543-557

Etheridge MA, Wall VJ, Vernon RH (1983) The role of the fluid phase during regional metamorphism and deformation. J Metmorph Geol 1:205-226

Farver JR, Yund RA (1992) Oxygen diffusion in a fine-grained quartz aggregate with wetted and nonwetted microstructures. J Geophys Res B, Solid Earth Planets 97:14,017-14,029

Ferry JM (1980) A case study of the amount and distribution of heat and fluid during metamorphism. Contrib Mineral Petrol 71:373-385

Ferry JM (1986) Reaction progress; a monitor of fluid-rock interaction during metamorphic and hydrothermal events. *In* Walther JV, Wood BJ (eds) Fluid-rock Interactions During Metamorphism. Adv Phys Geochem 5:60-88

Ferry JM (1994) Role of fluid flow in contact metamorphism of siliceous dolomitic limestones. Am Mineral 79:719-736

Ferry JM (1995) Role of fluid flow in the contact metamorphism of siliceous dolomitic limestones—Reply to Hanson. Am Mineral 80:1226-1228

Ferry JM (2000) Patterns of mineral occurrence in metamorphic rocks. Am Mineral 85:1573-1588

Ferry JM, Burt DM (1982) Characterization of metamorphic fluid composition through mineral equilibria. Rev Mineral 10:207-262

Ferry JM, Dipple GM (1991) Fluid flow, mineral reactions and metasomatism. Geology 19:211-214

Ferry JM, Dipple GM (1992) Models for coupled fluid flow, mineral reaction, and isotopic alteration during contact metamorphism: The Notch Peak Aureole, Utah. Am Mineral 77:571-577

Ferry JM, Rumble D III (1997) Formation and destruction of periclase by fluid flow in two contact aureoles. Contrib Mineral Petrol 128:313-334

Ferry JM, Gerdes ML (1998) Chemically reactive fluid flow during metamorphism. Ann Rev Earth Planet Sci 26:255-87

Ferry JM, Sorensen SS, Rumble D III (1998) Structurally controlled fluid flow during contact metamorphism in the Ritter Range Pendant, California, USA. Contrib Mineral Petrol 130:358-378

Feynman RP, Leighton RB, Sands M (1963) The Feynman Lectures on Physics. Vol II. Addison-Wesley

Fisher GW, Lasaga AC (1981a) Irreversible thermodynamics in petrology. Rev Mineral 8:171-260

Fogg GE, Lucia FJ (1990) Reservoir modeling of restricted platform carbonates; geologic/ geostatistical characterization of interwell-scale reservoir heterogeneity, Dune Field, Crane County, Texas. Report of Investigations—University of Texas, Bureau of Econ Geol #66

Forester RW, Taylor HP Jr (1976) ^{18}O depleted igneous rocks from the Tertiary complex of the Isle of Mull, Scotland. Earth Planet Sci Lett 32:11-17

Fricke HC, Wickham SM, O'Neil JR Jr (1992) Oxygen and hydrogen isotope evidence for meteoric water infiltration during mylonitization and uplift in the Ruby Mountains-East Humboldt Range core complex, Nevada. Contrib Mineral Petrol 111/2:203-221

Furlong KP, Hanson RB, Bowers JR (1991) Modeling thermal regimes. Rev Mineral 26:437-505

Fyfe WS, Price NJ, Thompson AB (1978) Fluids in the Earth's crust; their significance in metamorphic, tectonic and chemical transport processes. Developments in Geochemistry 1, Elsevier Science, Amsterdam, 383 p

Ganor J, Matthews A, Paldor N (1989) Constraints on effective diffusivity during oxygen isotope exchange at a marble schist contact, Sifnos (Cyclades), Greece. Earth Planet Sci Lett 94:208-216

Gelhar LW (1993) Stochastic Subsurface Hydrology. Prentice Hall, New York, 390 p

Gelhar LW, Mantoglou A, Welty C, Rehfeldt KR (1985) A review of field scale physical solute transport processes in saturated and unsaturated porous media. EPRI, Palo Alto, California

Gelhar LW, Welty C, Rehfeldt KR (1992) A critical review of data on field-scale dispersion in aquifers. Water Resource Res 28:1955-1974

Gerdes ML, Baumgartner LP, Person M, Rumble D III (1995a) One- and two-dimensional models of fluid flow and stable isotope exchange at an outcrop in the Adamello contract aureole, Southern Alps, Italy. Am Mineral 80:1004-1019

Gerdes ML, Baumgartner LP, Person M (1995b) Stochastic permeability models of fluid flow during contact metamorphism. Geology 23:945-948

Gerdes ML, Baumgartner LP, Person M (1998) Convective fluid flow through heterogeneous country rocks during contact metamorphism. J Geophys Res 103 B10:23,983-24,003

Gerdes ML, Baumgartner LP, Valley JW (1999) Stable isotopic evidence for limited fluid flow through dolomitic marble in the Adamello contact aureole, Cima Uzza, Italy. J Petrol 40:853-872

Gerdes ML, Valley JW (1994) Fluid flow and mass transport at the Valentine wollastonite deposit, Adirondack Mts, NY. J Metmorph Geol 12:589-608

Golubev VS, Garibyants AA (1971) Heterogeneous processes of geochemical migration. Plenum (translated from Russian), 150 p

Graham CM, Valley JW, Eiler JM, Wada H (1998) Timescales and mechanisms of fluid infiltration in a marble: an ion microprobe study. Contrib Mineral Petrol 132:371-389

Granath-Hover VC, Papike JJ, Labotka TC (1983) The Notch Peak contact metamorphic aureole, Utah: Petrology of the Big Horse limestone member of the Orr Formation. Geol Soc Am Bull 94:889-906

Gregory RT, Taylor HP (1981) An oxygen isotope profile in a section of Cretaceous oceanic crust, Samail ophiolite, Oman: Evidence for $\delta^{18}O$ buffering of the oceans by deep (>5 km) seawater-hydrothermal circulation at mid-ocean ridges. J Geophys Res 86:2737-2755

Guy B, Sheppard SMF, Fouillac AM, Le Guyader R, Toulhoat P, Fonteilles M (1988) Geochemical and isotope (H, C, O, S) studies of barren and tungsten-bearing skarns of the French Pyrenees. *In* Boissonnas J, Omenetto P (eds) Mineral Deposits Within the European Community. Springer-Verlag, Berlin-Heidelberg, p 54-75

Hanson RB (1992) Effects of fluid production on fluid flow during regional and contact metamorphism. J Metmorph Geol 10:87-97

Hanson RB (1995a) The hydrodynamics of contact metamorphism. Geol Soc Am Bull 107:595-611

Hanson RB (1995b) Role of fluid flow in the contact metamorphism of siliceous dolomitic limestones— a discussion. Am Mineral 80:1222-1225

Heinrich W, Hoffbauer R, Hubberten H-W (1995) Contrasting fluid flow patterns at the Bufa del Diente contact metamorphic aureole, north-east Mexico: Evidence from stable isotopes. Contrib Mineral Petrol 119:362-376

Holness MB, Fallick AE (1997) Paleohydrology of the calcsilicate aureole of the Beinn an Dubhaich Granite, Skye, Scotland: A stable isotopic study. J Metmorph Geol 15:71-83

Ildefonse JP, Gabis V (1976) Experimental study of silica diffusion during metasomatic reactions in the presence of water at 550°C and 1,000 bars. Geochim Cosmochim Acta 40:297-303

Isaaks EH, Srivastava RM (1989) An introduction to applied geostatistics. Oxford Univ. Press, New York, NY. 561 p

Joesten RL (1991) Kinetics of coarsening and diffusion-controlled mineral growth. Rev Mineral 26: 507-582

Kerrich R (1987) Stable isotope studies of fluids in the crust. *In* Kyser TK (ed) Stable Isotope Geochemistry of Low Temperature Fluids. Mineral Assoc Canada Short Course Notes 13:258-278

Kerrick DM, Caldeira K (1999) Was the Himalayan orogen a climatically significant coupled source and sink for atmospheric CO_2 during the Cenozoic? Earth Planet Sci Lett 173:195-203

Kerrick DM, Lasaga AC, Raeburn SP (1991) Kinetics of heterogeneous reactions. Rev Mineral 26: 583-672

Kohn MJ, Valley JW, Elsenheimer D, Spicuzza M.J (1993) O isotope zoning in garnet and staurolite: Evidence for closed-system mineral growth during regional metamorphism. Am Mineral 78:988-1001

Kohn MJ, Valley JW (1994) Oxygen isotope constraints on metamorphic fluid flow, Townshend Dam, Vermont, USA. Geochim Cosmochim Acta 58:5551-5566

Kolodny Y, Gross S (1974) Thermal metamorphism by combustion of organic matter: isotopic and petrologic evidence. J Geol 82:489-506

Labotka TC, Nabelek PI, Papike JJ (1988) Fluid infiltration through the Big Horse limestone member in the Notch Peak contact-metamorphic aureole, Utah. Am Mineral 73:1302-1324

Lackey JS (2000) Fluid flow in Sandstones and Marbles During Wollastonite Grade Contact Metamorphism at Laurel Mt, E-Central California, MS thesis, University Wisconsin-Madison, 121 p

Lasaga AC (1981) Rate laws of chemical reactions. Rev Mineral 8:1-68

Lasaga AC (1998) Kinetic Theory in Earth Sciences. Princeton University Press, Princeton, NJ, 811 p

Lasaga AC, Rye DM (1993) Fluid flow and chemical reactions kinetics in metamorphic systems. Am J Sci 293:361-404

Lassey KR, Blattner P (1988) Kinetically controlled oxygen isotope exchange between fluid and rock in one-dimensional advective flow. Geochim Cosmochim Acta 52:2169-2175

Lattanzi P, Rye DM, Rice J.M (1980) Behavior of ^{13}C and ^{18}O in carbonates during contact metamorphism at Marysville, Montana: Implications for isotope systematics in impure dolomitic limestones. Am J Sci 280:890-906

Leger A, Ferry JM (1993) Fluid infiltration and regional metamorphism of the Waits River Formation, north-east Vermont, USA. J Metmorph Geol 11:3-30

Lichtner PC (1996) Continuum formulation of multicomponent-multiphase reactive transport. Revi Mineral 34:1-82

Lusczinski NJ (1961) Head and flow of groundwater of variable density. J Geophys Res 66:4247-4256

Manning CE, Bird DK (1991) Porosity evolution and fluid flow in the basalts of the Skaergaard magma-hydrothermal system, East Greenland. Am J Sci 291:201-257

Manning CE, Ingebritsen SE (1999) Permeability of the continental crust: Implications of geothermal data and metamorphic systems. Rev Geophys 37:127-150

Manning CE, Ingebritsen SE, Bird DK (1993) Missing mineral zones in contact metamorphosed basalts. Am J Sci 293:894-938

Marchildon NM, Dipple GM (1998) Irregular isograds, reaction instabilities, and the evolution of permeability during metamorphism. Geology 26:15-18

Matthews A, Kolodny Y (1978) Oxygen isotope fractionation in decarbonation metamorphism: the mottled zone event. Earth Planet Sci Lett 39:179-192

Moore JN, Kerrick DM (1976) Equilibria in siliceous dolomites of the Alta Aureole, Utah. Am J Sci 276:502-524

Morrison J, Anderson JL (1998) Footwall refrigeration along a detachment fault: implications for the thermal evolution of core complexes. Science 279:63-66

Nabelek PI, Labotka, TC, O'Neil JR, Papike JJ (1984) Contrasting fluid/rock interaction between the Notch Peak granitic intrusion and limestones in western Utah: evidence from stable isotopes and phase assemblages. Contrib Mineral Petrol 86:25-34

Nabelek PI, Hanson GN, Labotka TC, Papike JJ (1988) Effects of fluids on the interaction of granites with limestones: The Notch Peak Stock, Utah. Contrib Mineral Petrol 99:49-61

Nabelek PI (1991) Stable isotope monitors. Rev Mineral 26:395-436

Nabelek PI, Labotka TC, Russ-Nabelek C (1992) Stable isotope evidence for the role of diffusion, infiltration, and local structure on contact metamorphism of calc-silicate rocks at Notch Peak, Utah. J Petrol 33:557-583

Nagy KL, Gilleti BJ (1986) Grain boundary diffusion of oxygen in a micro-perthitic feldspar. Geochim Cosmochim Acta 50:1151-1158

Nesbitt BE, Muehlenbachs K (1995) Geochemical studies of the origins and effects of synorogenic crustal fluids in the southern Omineca Belt of British Columbia, Canada. Geol Soc Am Bull 107:1033-1050

Norton DL, Taylor HP (1979) Quantitative simulation of the hydrothermal system of crystallizing magmas on the basis of transport theory and oxygen isotope data: An analysis of the Skaergaard intrusion. J Petrol 20:421-486

Nur AM, Walder J (1990) Time-dependant hydraulics of the Earth's crust. *In* The Role of Fluids in the Crustal Processes. Geophys Study Comm, Commission on Geosciences, Environment, and Resources, National Res Council, p 113-127

Oelkers EH (1996) Physical and chemical properties of rocks and fluids for chemical mass transport calculations. Rev Mineral 34:131-191

Ogata A (1964) Mathematics of dispersion with linear adsorption isotherm. U S Geol Soc Surv Prof Paper 411-H

Ohmoto H (1986) Stable isotope geochemistry of ore deposits. Rev Mineral 16:491-559

O'Neil JR, Clayton RN, Mayeda TK (1969) Oxygen isotope fractionation of divalent metal carbonates. J Chem Phys 51:5547-5558

Oliver NHS (1996) Review and classification of structural controls on fluid flow during regional metamorphism. J Metmorph Geol 14:477-492

Ortoleva PJ (1987) Modeling geochemical self-organization. *In* Nicolis C (ed) Irreversible Phenomena and Dynamical Systems Analysis in Geosciences. NATO ASI Series C: Math Phys Sci 192:493-510

Passchier CW, Trouw RAJ (1998) Microtectonics. Springer-Verlag, Berlin, 289 p

Person M, Baumgartner LP (1995) (1995) New evidence for long-distance fluid migration within the Earth's crust. Rev Geophys Suppl, p 1083-1091

Phillips OM (1991) Flow and Reactions in Permeable Rocks. Cambridge University Press, Cambridge, UK, 285 p

Romer RL, Heinrich W (1998) Transport of Pb and Sr in leaky aquifers of the Bufa del Diente contact metamorphic aureole, northeast Mexico. Contrib Mineral Petrol 131:155-170

Roselle GT (1997) Integrated petrologic, stable isotopic, and statistical study of fluid-flow in carbonates of the Ubehebe Peak contact aureole, Death Valley National Park, California. PhD Dissertation, University Wisconsin-Madison, 279 p

Roselle GT, Bowman JR (2001) Patterns of fluid circulation surrounding cooling plutons: The influence of surface topography and other parameters. *In* Eleventh Annual V.M. Goldschmidt Conference, LPI Contribution #1088 (CD-ROM)

Roselle GT, Baumgartner LP, Valley JW (1999) Stable isotope evidence of heterogeneous fluid infiltration at the Ubehebe Peak contact aureole, Death Valley National Park, California. Am J Sci 299:93-138

Rumble D III (1982) Stable isotope fractionation during metamorphic volatilization reactions. Rev Mineral 10:327-353

Rumble D III, Spear FS (1983) Oxygen-isotope equilibrium and permeability enhancement during regional metamorphism. J Geol Soc London 140:619-628

Rutter EH, Brodie KH (1985) The permeation of water into hydrating shear zones. *In* Thompson AB, Rubie DC (eds) Adv Phys Geochem 4:242-250

Rye RO (1966) The C, H, and O isotopic composition of the hydrothermal fluids responsible for the lead-zinc deposits at Providencia, Zacatecas, Mexico. Econ Geol 61:1399-1427

Rye DM, Williams N (1981) Studies of the base metal sulfide deposits at McArthur River, Northern Territory, Australia: III. The stable isotope geochemistry of the H.Y.C. Ridge and Cooley Deposits. Econ Geol 76:1-26

Schoell M, Morteani G, Hormann PK (1975) $^{18}O/^{16}O$ and $^{13}C/^{12}C$ ratios of carbonates from gneisses, serpentinites and marbles of the Zillertaler Alpen, Western Tauern area, Austria. N Jahrb Mineral Monat 10:444-459

Shelton KL (1983) Composition and origin of ore-forming fluids in a carbonate-hosted porphyry copper and skarn deposit: A fluid inclusion and stable isotope study of Mines Gaspé, Quebec. Econ Geol 78:387-421

Sheppard SMF, Schwarcz HP (1970) Fractionation of carbon and oxygen isotopes and magnesium between coexisting metamorphic calcite and dolomite. Contrib Mineral Petrol 26:161-198

Sheppard SMF, Schwarcz H.P (1986) Fractionation of carbon and oxygen isotopes and magnesium between metamorphic calcite and dolomites. Contrib Mineral Petrol 26:161-198

Shieh YN, Taylor HP (1969) O and C isotope studies of contact metamorphism of carbonate rocks. J Petrol 10:307-331

Skelton ADL, Valley JW, Graham CM, Bickle MJ, Fallick AE (2000) The correlation of reaction and isotope fronts and the mechanism of metamorphic fluid flow. Contrib Mineral Petrol 138:364-375

Smith RK (1972) The mineralogy and petrology of the contact metamorphic aureole around the Alta stock, Utah. PhD Dissertation, University Iowa, Iowa City

So CS, Rye DM, Shelton KL (1983) C, H, O, and S isotope and fluid inclusion study of the Weolag Tungsten-Molybdenum Deposit, Korea: Fluid histories of metamorphic and ore-forming evens. Econ Geol 78:1551-1573

Spear FS (1993) Metamorphic phase equilibria and pressure-temperature-time paths. Monogr, Mineral Soc Am, Washington, DC, 799 p

Steefel CI, Lasaga AC (1990) Evolution of dissolution patterns; permeability change due to coupled flow and reaction. In Chemical Modeling of Aqueous Systems II. Melchior DC, Bassett RL (eds) Am Chem Soc Symp 416:212-225

Suzuoki T, Epstein S (1976) Hydrogen isotope fractionation between OH-bearing minerals and water. Geochim Cosmochim Acta 40:1229-1240

Taylor HP Jr (1977) Water/rock interaction and the origin of H_2O in granitic batholiths. J Geol Soc London 133:509-558

Taylor BE, O'Neil JR (1977) Stable isotope studies of metasomatic Ca-Fe-Al-Si skarns and associated metamorphic and igneous rocks, Osgood Mts, Nevada. Contrib Mineral Petrol 63:1-49

Taylor BE, Bucher-Nurminen K (1986) Oxygen and carbon isotope and cation geochemistry of metasomatic carbonates and fluids-Bergell aureole, N. Italy. Geochim Cosmochim Acta 50:1267-1279

Thompson JB Jr (1959) Local equilibrium in metasomatic process. In Abelson PH (ed) Res Geochem 1:427-457

Turner DR, Bowman JR (1993) Origin and evolution of skarn fluids, Empire zinc skarns, Central Mining District, New Mexico, U.S.A. Applied Geochemistry 8:9-36

Valley JW (1986) Stable isotope geochemistry of metamorphic rocks. Rev Mineral 16:445-489

Valley JW, Graham CM (1996) Ion microprobe analysis of oxygen isotope ratios in quartz from Skye granite: healed micro-cracks, fluid flow, and hydrothermal exchange. Contrib Mineral Petrol 124:225-234

Valley JW, O'Neil JR (1982) Oxygen isotope evidence for shallow emplacement of Adirondack anorthosite. Nature 300:497-500

Valley JW, O'Neil JR (1984) Fluid heterogeneity during granulite facies metamorphism in the Adirondacks: stable isotope evidence. Contrib Mineral Petrol 85:158-173

Van Genuchten, MTh, Parker JC (1984) Boundary conditions for displacement experiments through short laboratory soil columns. Soil Sci Soc Am J 48:703-708

Wada H (1988) Microscale isotopic zoning in calcite and graphite crystals in marble. Nature 331:61-63

Wada H, Ando T, Suzuki M (1998) The role of the grain boundary at chemical and isotopic fronts in marble during contact metamorphism. Contrib Mineral Petrol 132:309-320

Walther JV, Orville PM (1982) Volatile production and transport in regional metamorphism. Contrib Mineral Petrol 79:252-257

Walther JV, Wood BJ (1984) Rate and mechanism in prograde metamorphism. Contrib Mineral Petrol 88:246-259

Walton M (1960) Molecular diffusion rates in supercritical water vapor estimated from viscosity data. Am J Sci 258:385-401

Wang HF (1993) A double medium model for diffusion in fluid-bearing rock. Contrib Mineral Petrol 114:357-364

Wang HF, Anderson MP (1982) Introduction to Groundwater Modeling: Finite Difference and Finite Element Methods. Freeman and Co, 237 p

Wedepohl KH (1969) Composition and abundance of common igneous rocks. *In* Wedepohl KH (ed) Handbook of Geochemistry. Springer-Verlag, New York, p 227-249

Yardley BWD, Lloyd GE (1995) Why metasomatic fronts are really metasomatic sides. Geology 23:53-56

Yardley BWD, Valley JW (1997) The petrologic case for a dry lower crust. J Geophys Res 102(B6): 12,173-12,185

Young ED (1995) Fluid flow in metamorphic environments. Rev Geophys—Suppl, p 41-52

8   Stable Isotopes in Seafloor Hydrothermal Systems:
Vent fluids, hydrothermal deposits,
hydrothermal alteration, and microbial processes

W. C. Shanks, III

U. S. Geological Survey
973 Denver Federal Center
Denver, Colorado 80225

INTRODUCTION

The recognition of abundant and widespread hydrothermal activity and associated unique life-forms on the ocean floor is one of the great scientific discoveries of the latter half of the twentieth century. Studies of seafloor hydrothermal processes have led to revolutions in understanding fluid convection and the cooling of the ocean crust, the chemical and isotopic mass balance of the oceans, the origin of stratiform and statabound massive-sulfide ore-deposits, the origin of greenstones and serpentinites, and the potential importance of the subseafloor biosphere. Stable isotope geochemistry has been a critical and definitive tool from the very beginning of the modern era of seafloor exploration.

Early suggestions of possible submarine hydrothermal activity date from the late 1950s when a number of investigators were debating the importance of "volcanic emanations" as a factor in the widespread occurrence of manganese nodules and other ferromanganese oxide deposits on the seafloor. Arrhenius and Bonatti (1965), in their classic paper, *Vulcanism and Neptunism in the Oceans,* stated the following:

> *"The origin of authigenic minerals on the ocean floor has been extensively discussed in the past with emphasis on two major processes; precipitation from solutions originating from submarine eruptions, and slow precipitation from sea water of dissolved elements, originating from weathering of continental rocks. It is concluded that in several marine authigenic mineral systems these processes overlap."*

Bostrom and Peterson (1966), in another classic, published evidence for extensive and widespread Fe-rich metalliferous sediments on the seafloor with a distribution strongly correlated with the mid-ocean ridges (Fig. 1). They stated:

> *"On the very crest of the East Pacific Rise, in equatorial latitudes--particularly 12° to 16°S, the sediments are enriched in Fe, Mn, Cu, Cr, Ni, and Pb. The correlation of these areas of enrichment to areas of high heat flow is marked. It is believed that these precipitates are caused by ascending solutions of deep-seated origin, which are probably related to magmatic processes at depth. The Rise is considered to be a zone of exhalation from the mantle of the earth, and these emanations could serve as the original enrichment in certain ore-forming processes."*

At about the same time, the discovery of hot brine pools on the floor of the Red Sea indicated the possibility of direct precipitation of metalliferous sediments from high-temperature hydrothermal fluids on the seafloor (Miller et al. 1966; Bischoff 1969). This, more than any other discovery, resulted in a sweeping revolution in the field of ore genesis and a total reassessment of the origin of massive sulfide deposits, which were previously considered in many cases to be large mesothermal vein deposits that formed by replacement of country rocks deep in the earth. Moreover, in the first application of stable isotopes to seafloor hydrothermal processes, oxygen isotope studies of the Red Sea

1529-6466/00/0043-0008$05.00

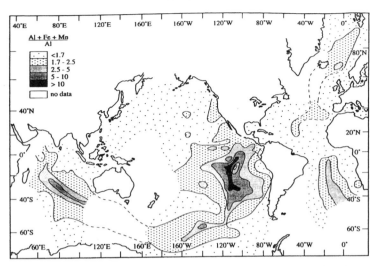

Figure 1. Distribution of metalliferous surficial sediments, enriched in (Al+Fe+Mn)/Al, around divergent plate margin (modified after Bostrom et al. 1969). Used with permission of the American Geophysical Union.

brines indicated $\delta^{18}O$ values close to normal Red Sea waters and eliminated speculation about direct emanations of magmatic water (Miller et al. 1966; Craig 1966). Kaplan et al. (1969) showed that the $\delta^{34}S$ values of sulfide minerals from Atlantis II Deep metalliferous muds have a narrow range that averages ~6±3‰, consistent with hydrothermal processes that require some reduction of seawater-derived sulfate. Taken together, these stable isotopic constraints showed that magmatic water and magmatic sulfur gases (H_2S or SO_2) could be only very minor components in the Red Sea brines.

Following the Red Sea discoveries, several lines of evidence from mid-ocean ridge studies suggested a dominant role for heated seawater circulation through and reaction with ocean crust. First, Muehlenbachs and Clayton (1971, 1972, 1976) used $\delta^{18}O$ studies of greenstones and other metamorphic rocks dredged from the seafloor to demonstrate that subseafloor metamorphism was due to high-temperature seawater reactions with ocean crust. Simultaneously, the discovery that heat flow on the flanks of divergent plate boundaries is low compared to that expected for conductive heat loss implicated convective heat loss due to seawater circulation (Anderson 1972). Laboratory hydrothermal experiments reacting seawater with basalt at 200°C showed, contrary to conventional geochemical expectations, that Mg is quantitatively removed by formation of secondary sheet-silicates, producing acidic solutions capable of transporting ore-forming metals and H_2S (Bischoff and Dickson 1975). Wolery and Sleep (1976) used the heat flow deficit at mid-ocean ridges to calculate that the entire world ocean circulates through the ocean crust approximately every 5-11 Ma. They also concluded that Mg-uptake by the crust during such circulation balances the long-standing problem of excess Mg-supply to the oceans by streams.

These investigations set the stage for the spectacular discoveries of hydrothermal activity on the mid-ocean ridges; discoveries came in rapid succession with the deep submersible *Alvin* dives to warm springs on the Galapagos Rise (Corliss et al. 1979), followed by the French discovery of massive sulfides on the East Pacific Rise at 21°N (Francheteau et al. 1979; Hekinian et al. 1980), followed by direct observation and sampling with *Alvin* of vigorous 350°C "black smoker" vents and massive sulfide "chimneys" at 21°N EPR (Spiess et al. 1980). Edmond et al. (1979a,b; 1982) measured the chemical composition of the evolved seawater hot spring fluids and, using insight from the experimental studies (Bischoff and Dickson 1975), determined end-member compositions by extrapolating to zero Mg. Von Damm et al. (1983, 1985) showed unequivocally that the 21°N vent fluids are a result of low water/rock (mass ratios, 0.3-0.7) reactions with basalt during subsurface convection of heated, evolved seawater. East Pacific Rise Study Group (1981) and Welhan and Craig (1983) published the first oxygen isotope studies of black smoker fluids and showed that the $\delta^{18}O$ values are close to ambient seawater but increased by 1-2‰ due to basalt-seawater reaction at elevated temperatures. Hekinian et al. (1980) demonstrated that chimney sulfides have $\delta^{34}S$ values of 1.9-3.3‰, requiring a significant component of seawater SO_4 ($\delta^{34}S = 21‰$) that is reduced to H_2S during basalt-seawater reactions.

Approximately 35 years of study of seafloor hydrothermal processes using stable isotopes has resulted in a fairly detailed understanding of stable isotope systematics of such systems (Shanks et al. 1995) and an increased realization that stable isotopes provide important constraints on system evolution but must be combined with other geological, geochemical, and biological data to produce reliable interpretations. Active and inactive hydrothermal vent systems and hydrothermal deposits on the seafloor are much more abundant than ever anticipated, especially considering that only a few percent of the area of the ridges and convergent margins have been explored in any detail. Over 150 sites with evidence of significant past or present seafloor hydrothermal activity are now known (Rona and Scott 1993), and Butterfield (2000) tabulates 35 "confirmed presently active" hydrothermal vent sites. Locations of seafloor vent sites discussed in this paper are shown on Figure 2. Several large (millions of tons), high-metal-grade deposits are now known, including the TAG site on the Mid-Atlantic Ridge (MAR), the Middle Valley site on the northernmost Juan de Fuca Ridge (JFR), the 13°N site on the

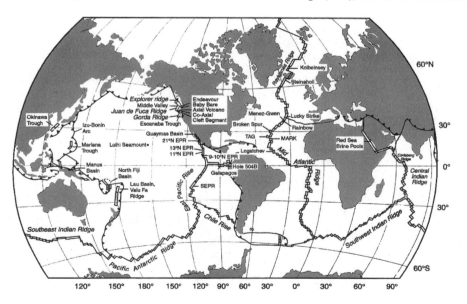

Figure 2. Locations of seafloor hydrothermal vents and deposits discussed in this paper (modified after Butterfield 2000).

East Pacific Rise (EPR), and the Atlantis II Deep, Red Sea (Shanks and Bischoff 1980; Humphris et al. 1995; Fouquet et al. 1996; Zierenberg et al. 1998). Actively venting systems, which can be sampled by deep-diving submersibles and ROVs (remotely operated vehicles) provide information on precipitation rates, physical processes, volatile constituents, and fluid-mineral reaction processes that are unobtainable in studies of ancient hydrothermal deposits on the continents. DSDP/ODP (Deep Sea Drilling Project and its successor, Ocean Drilling Program) drilling and core sampling provide three-dimensional subsurface information on the effects of water-rock interaction in hydrothermal systems but has two distinct disadvantages: (1) drilling has been impossible on bare-rock sites because of technical difficulties, and (2) surface seawater is used as the drilling fluid and uncontaminated downhole water samples cannot be collected except by squeezing pore waters out of sediments. In recent years, seafloor vent studies and ODP drilling have both increasingly focused on the subsurface biosphere; microbial processes in the ocean crust. The effects of microbial processes on stable isotope geochemistry are just beginning to emerge.

The objective of this chapter is to review available stable isotopic data on seafloor hydrothermal systems. However, this goes far beyond a simple literature review because much new, previously unpublished data, collected by the author, is included. In addition, an important goal of this chapter is to interpret the stable isotope systematics of seafloor hydrothermal systems in the context of fluid-rock reactions and geochemical reaction calculations. Boiling and supercritical phase-separation, volcanic eruption and dike-emplacement events, addition of magmatic volatiles, and bacterial fractionation processes will be discussed where applicable. In addition to the commonly measured stable isotopes of C, O, H, and S, stable isotope ratios of B, Li, N, Cl, Cu, and Fe are included where data are available. Much new data has appeared since the last comprehensive overview of stable isotopes in seafloor hydrothermal systems (Shanks et al. 1995). This includes a wealth of information on hydrothermal systems related to volcanic arcs, back-arc spreading centers, seamounts, and serpentinized ultramafics.

METHODS

Study of hydrothermal systems on the ocean floor includes significant advantages and significant limitations relative to active subaerial or ancient inactive systems. The present-day activity of the hydrothermal systems means that fluids can be sampled along with the minerals precipitated at the seafloor. If drilling has been carried out or if there are significant exposures along fault zones, then subsurface rock alteration processes can also be studied. Taken together, at sites that have been extensively investigated, drilling and submersible sampling provide a very powerful package that allows interpretation not only of the reaction products of the hydrothermal systems, but also reaction rates, isotopic and chemical disequilibria, and fluid circulation rates.

However, the vents of interest occur in water depths up to 3600 m, requiring high-pressure submersible capabilities. Vent temperature typically is measured before water sampling using a thermocouple probe deployed in the hydrothermal vent by the mechanical arm of the submersible, but more recently a simultaneous system that attaches to the samplers has been deployed. Water samples are generally taken in specially built 760 cm^3 titanium syringes (Von Damm et al. 1985) equipped with sampling snorkels, flush valves, and a mechanical trigger that opens, fills, and closes the samplers under the control of the submersible pilot. The samplers cool to ambient temperature immediately after sampling and, although they are not strictly gas-tight, they retain most of the volatiles (except at very high volatile content) until samples are withdrawn after the submersible returns to the surface. Specially designed 150 cm^3

evacuated gas-tight samplers are used for studies focused on He and other dissolved gases. Recently, Seewald et al. (2001) have designed new samplers that ameliorate many of the problems of previous sampling technology, especially the need to deploy separate gas-tight samples. The new samplers (1) have a thermocouple probe incorporated so temperature is measured simultaneously with sampling, (2) allow the fill rate to be controlled by the scientists so diffusely flowing vents can be sampled effectively, (3) are all titanium so no potentially contaminating precious metal gaskets contact the fluid, and (4) are gas-tight and designed to maintain pressure at sampling and during shipboard withdrawal of aliquots for chemistry and isotopes.

A disadvantage of underwater sampling in jets of hot water is the inevitable entrainment of cold ambient bottom-water that is oxygenated, sulfate-rich, Mg-rich, and has a pH of about 8. However, as mentioned above, the fact that high temperature vent fluids contain little or no Mg allows a convenient method for reconstructing the end-member composition of the vent fluid by taking multiple samples from a given vent (each will have a different amount of seawater entrainment) and plotting mixing lines to determine the zero-Mg intercept for each element or isotope value. In practice, many isotopic end-members must be determined by the mixing-line method; other isotope ratios, such as S, Cu, and Fe isotopes are not sensitive to mixing. Cu and Fe are essentially absent in ambient bottom waters. Sulfur isotope values of vent fluid H_2S are unaffected by mixing because bottom waters contain SO_4 but no H_2S, and SO_4-H_2S isotopic exchange does not occur in the samplers (this has been verified many times by studies of $\delta^{34}S_{H_2S}$ vs. Mg for a given vent; Shanks, unpublished data). H_2S is quite volatile in acidic vent solutions and samples for $\delta^{34}S_{H_2S}$ must be handled very carefully. In general, H_2S should be taken as early as possible in the shipboard sub-sampling sequence. An excellent sampling procedure includes the use of a nitrogen filled plastic syringe with a plastic gas chromatographic valve, followed by injection of the sample into evacuated, septum-cap glass bottles that contain crystals of Zn or Cd salts. This results in precipitation of all sulfide and poisoning of any potential bacterial activity. It is generally not possible to determine the $\delta^{34}S_{SO_4}$ or $\delta^{18}O_{SO_4}$ of dissolved sulfate in vent fluid because, if present, end-member sulfate is at very low concentrations (<1 mmol/kg) and is swamped by admixed ambient seawater.

Analytical methods for H and O isotopes are well established. Water samples for isotope analyses are stored in tightly capped glass vials. Before analysis, an aliquot of Cu granules are added to each sample to extract H_2S. Water samples are prepared for hydrogen isotopic analyses using the H_2 equilibration method (Coplen et al. 1991) and for oxygen isotope analyses using the CO_2 equilibration technique (Epstein and Mayeda 1953). Values of $\delta^{18}O$ and δD are calibrated relative to the VSMOW standard and have reproducibility of approximately 0.1 and 1.0‰, respectively. δD values of water samples determined by H_2 equilibration methods, which give isotope activity ratios, may be significantly different from those determined by Zn- or U-reduction methods, which give isotope concentration ratios, if the samples contain dissolved salts. The effect of salts on the isotopic activity ratio of seawater may be expected to cause δD values determined by the H_2 equilibration method at 30°C to be approximately 1.1-1.2‰ higher than those determined by a Zn- or U-reduction method (Horita et al. 1993). Results (Table 1) of hydrogen and oxygen isotope values of vent fluids are presented as $\Delta^{18}O_{activity}$ and $\Delta D_{concentration}$ relative to local bottom waters (see Shanks et al. 1995). Oxygen isotope activity values and hydrogen isotope concentration values are chosen because most data in the literature and all of the basic isotope fractionation data from experimental studies are presented on these scales. The Δ approach removes global ocean δD and $\delta^{18}O$ variations and allows focus on processes that affect the isotopes during hydrothermal processes. VSMOW values for δD and $\delta^{18}O$ of bottom waters in various areas are

Table 1. Geochemistry of seafloor hydrothermal vents

Vent	Dive	Date	Depth m	Spreading Rate mm/yr	T C	P bars	Cl mm/kg	SiO2 mm/kg	D18O activity	DD concentration	d34S_H2S CDT	H2S mm/kg
SOUTHERN EAST PACIFIC RISE[a]												
17°24-27'S												
Stanley	3294	10/25/98	2600	153	363	260	412	14.7	0.57	-0.2	6.9	10.9
North Vent	3294	10/25/98	2600	153	376	260	410	9.6	0.77	-0.9	6.2	10.7
Rehu-Marka	3295	10/26/98	2600	153	279	260	737	14.1			8.5	1.1
Nadir	3297	10/28/98	2600	153	343	260	476	14.6	0.54	-2.1	6.1	8.6
Dumbo	3298	10/29/98	2610	153	260	261	1090	20.3	0.85		6.2	4.2
Gumbo	3298	10/29/98	2610	153	294	261	915	21.3	0.99		6.2	6.2
17°35-44'S												
Hobbes	3299	10/30/98	2615	153	349	262	481	18.1	1.00	-0.1	5.9	9.0
Calvin	3299	10/30/98	2615	153	276	262	720	19.4	1.23	0.8	4.8	4.8
Homer	3296	10/27/98	2615	153	347	262	591	17.2	0.35	-0.2	4.7	3.7
Marge	3296	10/27/98	2615	153	350	262	581	16.5	0.42	-0.1	6.0	4.5
Wally	3296	10/27/98	2615	153	314	262	752	18.0	1.10	-0.1	4.9	6.5
Eckels	3301	11/1/98	2640	153	320	264			1.43		5.7	
Minnow	3301	11/1/98	2640	153	350	-264			1.14		4.6	
18°24-26'S												
Simba	3292	10/23/98	2630	153	341	263	654	18.8	0.94	-0.6	6.3	7.8
Sojourn	3292	10/23/98	2630	153	382	263	628	14.9	1.05	0.3		9.7
Mona	3292	10/23/98	2630	153	372	263	616	15.9	1.16		6.3	9.0
Mkr72	3293	10/24/98	2630	153	343	263	668		0.93	1.6	7.1	
Mkr66	3293	10/24/98	2630	153	345	263	665	18.8	1.10	-0.4	8.6	7.6
21°24-27'S												
Tweety	3282	10/12/98	2800	150	343	280	116	15.4	0.74	0.4	7.1	8.9
Tweety	3286	10/16/98	2800	150	343	280	119		0.56	0.0	7.0	
Sylvester	3282	10/12/98	2800	150	319	280	573	19.7	0.90			5.6
Hector	3283	10/13/98	2800	150	295	280	569	20.7	0.65	-0.4	4.3	4.0
Natasha	3283	10/13/98	2800	150	354	280	390	18.9	0.81	0.0	5.8	7.3
Jasmine	3284	10/14/98	2800	150	349	280	371	14.3	0.83	-1.8	5.8	9.6
Boudreaux	3284	10/14/98	2800	150	157	280	394	20.0	0.82		5.7	6.6
Phideaux	3284	10/14/98	2800	150	231	280	545	20.8	0.57			5.0
Scooby	3286	10/16/98	2800	150	276	280	524	20.0	0.70	-0.8		4.8
Taz	3286	10/16/98	2800	150	330	280	635	20.1	0.54		7.2	6.0
Goofy	3291	10/21/98	2800	150	287	280	802	21.1	0.36	0.0	5.2	2.8

Vent	Dive	Date	Depth m	Spreading Rate mm/yr	T C	P bars	Cl mm/kg	SiO2 mm/kg	D18O activity	DD concentration	d34S_H2S CDT	H2S mm/kg
Simon	3288	10/18/98	2800	150	337	280	751	20.2	0.27	-0.6	4.5	3.5
21°33-34'S												
Krasnov	3290	10/20/98	2805	150	370	281	752	18.7	0.70	0.9	6.3	7.4
Krasnov	3306	11/7/98	2805	150		281	755	18.4	0.93	1.6	5.9	7.3
Shorn	3305	11/6/98	2790	150	353	279	511	18.8	0.80	1.4	6.9	9.0
Gromit	3290	10/20/98	2820	150	374	282	269	14.6	0.32		6.8	11.5
Gromit	3305	11/6/98	2820	150	374	282	270	14.0	0.83	1.0	5.8	13.1
Preston	3305	11/6/98	2810	150	383	281	321	12.0	1.00			11.4
Wallace	3302	11/3/98	2810	150	398	281	311	10.0	0.70		4.6	8.1
RoadRunner	3289	10/19/98	2810	150	388	281	432	11.8	0.58	0.4	5.2	9.0
Maryland	3289	10/19/98	2810	150	358	281	453	16.1	0.57	0.1	5.7	10.1
9-10N EAST PACIFIC RISE[a]												
9 17'N - F.1	2365	16-Apr-91	2600	102	388	260	46	6.0	1.46	1.4	6.4	48.8
9 17'N- F.2	2739	15-Mar-94	2600	102	351	260	846	20.0	1.54	-0.1	6.7	8.7
9 33.5'N-D.1	2352	2-Apr-91	2560	102	290	256	846	21.0	0.78	-0.5	3.3	9.9
9 33.5'N-D.2	2367	18-Apr-91	2560	102	308	256	801	20.4	0.77	0.3	3.5	6.4
9 33.5'N-D.4	2738	14-Mar-94	2560	102	264	256	836	17.1	1.75	1.4	4.5	3.2
9 33.5'N-D.5	2744	19-Mar-94	2560	102	264	256	851	19.6	0.87	-0.1	5.1	3.5
9 33.5'N- E.1	2352	2-Apr-91	2560	102	280	256	859	20.0	1.03	-0.3	4.0	9.3
9 33.5'N- E.2	2367	18-Apr-91	2560	102	274	256	841	19.8	0.80	1.1	4.0	6.4
9 33.5'N- E.3	2494	2-Mar-92	2560	102	188	256	858	19.7	0.68	-0.9	4.1	4.9
9 33.5'N- E.4	2738	14-Mar-94	2560	102	260	256	860	20.3	0.97	1.4	3.9	4.3
9 39'N- B.1	2361	11-Apr-91	2550	102	329	255	416	19.3	0.74	0.3	5.3	7.2
9 39'N- B.2	2748	23-Mar-94	2550	102	329	255	365	17.1	0.75	1.5	4.8	9.3
9 39'N- C.1	2361	11-Apr-91	2550	102	345	255	329	19.0	0.52	1.5	3.9	8.7
9 39'N- C.2	2748	23-Mar-94	2550	102	342	255	397	19.4	0.67	1.1	4.8	8.5
9 46.3'N-La.1	2366	17-Apr-91	2540	102	388	254	114	6.0	0.97	0.3	3.8	68.3
9 46.3'N-La.2	2502	10-Mar-92	2540	102	380	254	62	5.9	1.17	0.2	4.4	39.7
9 46.3'N-La.4	2751	26-Mar-94	2540	102	347	254	492	14.7	0.95	0.1	3.5	10.4
9 46.5'N-Aa.1	2360	10-Apr-91	2540	102	390	254	81	7.0	1.59	-0.2	3.8	50.3
9 46.5'N-Aa.2	2366	17-Apr-91	2540	102	396	254	31	3.8	1.91	0.2	3.7	86.2
9 46.5'N-Aa.3	2373	24-Apr-91	2540	102	403	254	43	2.7	2.10	-0.4	3.2	105.0
9 46.5'N-Aa.4	2500	8-Mar-92	2540	102	332	254	286	18.6	0.79	0.0	5.1	28.3
9 46.5'N-Aa.5	2502	10-Mar-92	2540	102	332	254	282	17.6	0.87	-0.6	4.7	30.2
9 46.5'N-Aa.6	2736	12-Mar-94	2540	102	329	254	398	16.5	1.60	1.2	5.0	10.7
9 46.5'N-Aa.7	2751	26-Mar-94	2540	102	330	254	397	17.5	1.12	0.9	4.8	11.2
9 50.3'N-B9.1	2351	1-Apr-91	2540	102	368	254	154	9.9	0.77	-1.8	7.8	23.2

Vent	Dive	Date	Depth m	Spreading Rate mm/yr	T C	P bars	Cl mm/kg	SiO2 mm/kg	$\delta^{18}O$ activity	δD concentration	$\delta^{34}S_{H2S}$ CDT	H2S mm/kg
9 50.3'N- B9.2	2498	6-Mar-92	2540	102	>388	254	76	7.0	0.77	0.3	5.4	15.2
9 50.3'N- B9.4	2734	10-Mar-94	2540	102	359	254	263	12.1	0.70	1.2	5.8	4.1
9 50.3'N- B9.5	2754	29-Mar-94	2540	102	363	254	267	12.6	0.64	1.0	5.7	5.1
9 50.3'N- B9.8	3030	25-Nov-95	2540	102	364	254	498	14.8	0.72		4.3	4.6
TWP.4	3034	28-Nov-95	2540	102	341	254	301	13.8	1.91		5.5	11.3
9 50.3'N- P.1	2357	7-Apr-91	2540	102	369	254	135	8.7	0.88	-1.8	6.4	27.4
9 50.3'N- P.3	2501	9-Mar-92	2540	102	392	254	42	3.9	0.68	0.2	6.0	24.8
9 50.3'N- P.5	2743	18-Mar-94	2540	102	350	254	347	14.2	0.99	2.1	5.8	5.4
9 50.3'N- P.6	2752	27-Mar-94	2540	102	377	254	352	14.3	0.66	0.5	5.3	4.6
9 50.3'N- P.8	3021	16-Nov-95	2540	102	360	254	622	15.8	1.18		2.9	7.0
9 50.3'N- P.10	3033	28-Nov-95	2540	102	367	254	620	16.4	1.18			5.8
9 50.7'N- Ma.1	2492	29-Feb-92	2540	102	321	254	245	17.3	0.28	0.0	5.9	2.8
9 50.7'N- Ma.2	2505	13-Mar-92	2540	102	321	254	285	18.6	0.34	-0.4	6.2	2.0
9 50.7'N- Ma.3	2734	10-Mar-94	2540	102	337	254	554	20.2	0.64	0.8	5.9	1.4
9 50.7'N- Ma.4	2755	30-Mar-94	2540	102	342	254	563	19.6	0.49	1.7	6.3	2.1
9 50.7'N- Ma.6	3019	14-Nov-95	2540	102	361	254	592	17.8	0.50		5.7	4.0
9 50.7'N- Q.1	2368	19-Apr-91	2520	102	371	252	71	7.8	1.12	-0.9	5.5	33.4
9 50.7'N- Q.2	2747	22-Mar-94	2520	102	293	252	490	19.1	0.37	3.0	5.8	3.7
9 50.7'N- Q.3	2755	30-Mar-94	2520	102	297	252	480	18.5	0.61	1.7	3.5	9.9
9 50.7'N- Q.4	3046	18-Nov-95	2520	102	310	252	571	18.8	0.72			----
EAST PACIFIC RISE[b,c,d,e]												
11N-Vent5			2600	98		260	686	20.6	0.41	1.0	4.8	4.4
13N-Chandeliur			2600	91	335	260	898	21.3	0.71	1.6		4.7
13N-Chainette			2600	91		260	739	23.3	0.44	0.7		4.9
EAST PACIFIC RISE[f]												
21N-NGS			2500	61	260	250	576	19.4	1.50	1.2	3.4	6.8
21N-OBS			2500	61	340	250	500	17.6	1.50	1.7	1.4	7.6
21N-SW			2500	61	335	250	525	17.2	1.50	2.5	3.8	8.1
21N-HG			2500	61	331	250	506	16.8	1.80	3.0	2.9	7.9
GUAYMAS BASIN[f]												
Guaymas - East Hill			2000		315	200	599	13.8	0.80	3.6	1.1	4.8
Guaymas - Central Sill			2000		291	200	589	12.5	0.90	3.1	2.3	4.0
Guaymas - South Sill			2000		291	200	601	12.9	1.00	4.1	-2.3	5.8
GORDA RIDGE[g]												

Vent	Dive	Date	Depth m	Spreading Rate mm/yr	T C	P bars	Cl mm/kg	SiO2 mm/kg	$\delta^{18}O$ activity	δD concentration	$\delta^{34}S_{H2S}$ CDT	H2S mm/kg
Escanaba Trough			3300		217	330	668	6.9	0.60	-0.1	7.8	1.1
Escanaba Trough			3300		108	330	660	6.9	0.50			1.1
SOUTHERN JUAN DE FUCA RIDGE[h]												
SJFR-Plume			2200	60	340	220	1087	23.3	0.61	-0.3	5.7	3.5
SJFR-Vent1			2200	60	285	220	896	22.8	0.72	0.5	6.4	3.0
AXIAL SEAMOUNT, JUAN DE FUCA RIDGE[j]												
ASHES-Inferno			1542	60	328	154	624	15.1	1.11	2.5	6.1	7.1
ASHES-Hell			1542	60	301	154	550	15.0	1.28	1.3		6.0
ASHES-Mushroom			1542	60	283	154	520	10.0	1.02	1.7	6.3	9.0
ASHES-Hillock			1542	60	323	154	482	15.0	1.14	2.5	6.5	8.5
ASHES-Crack			1542	60	217	154	258	15.0	1.09	1.5	7.1	10.5
ASHES-Virgin			1542	60	299	154	176	13.5	0.92	2.6	7.3	18.0
ENDEAVOUR SEGMENT, JUAN DE FUCA RIDGE[k]												
Hulk			2210	60	353	221	505	17.0	0.82	1.1	5.0	2.9
Crypto			2210	60	351	221	479	16.6	0.75	1.9	4.5	2.5
TP			2200	60	362	220	448	15.6	0.80	0.7	4.0	4.9
Dante			2200	60	370	220	457	16.2	0.90	1.7	4.3	5.4
Grotto			2200	60	357	220	425	15.7	0.90	1.1	4.4	5.0
LOBO			2200	60	346	220	428	15.9	0.90	2.1	4.4	5.0
Dudley			2205	60		221	349	14.1	0.80	1.2		3.0
S&M			2200	60		220	334	15.0	0.96	1.6	4.6	5.7
Peanut			2200	60	350	220	253	11.0	0.85	1.5	3.8	8.1
North			2200	60	356	220	477	16.8	0.79	0.9	5.7	2.0
MID-ATLANTIC RIDGE[i]												
TAG			3450	24	362	345	659	22.0	1.50	1.6	7.2	5.3
MARK-1			3600	24	350	360	559	18.2	1.90	1.9	4.9	5.9
MARK-2			3600	24	335	360	559	18.3	1.90	1.9	5.0	5.9
LUCKY STRIKE[a]												
Eiffel T	176, 177, 178	07/08,19,20/96	1687	24	323	169	447	15.4	0.91	2.5		1.5
Sintra	179, 180	07/21,26/96	1618	24	222	162	532	13.3	0.75	1.7		2.0
Mkr4	180	07/26/96	1700	24	318	170	441	15.2	1.08	2.1		3.1
2608	183	07/29/96	1719	24	328	172	526	17.5	0.84	3.3		4.6
Jason	183	07/29/96	1644	24	308	164	542	14.8	0.78	2.9		3.0

Vent	Dive	Date	Depth m	Spreading Rate mm/yr	T C	P bars	Cl mm/kg	SiO2 mm/kg	D18O activity	DD concentration	d34S$_{H2S}$ CDT	H2S mm/kg
Crystal	183-2	7/30-31/96	1726	24	281	173	535		0.60	2.1		2.0
WESTERN PACIFIC[m]												
Mariana BAB				30	285		544	14.0			4.2	2.6
Okinawa Trough- Izena Cauldron					320		550	12.5			7.3	13.1
Okinawa Trough- Minami-Ensei					278		527	10.8			3.6	2.4
Manus BAB- DESMOS caldera			1926		120	193	400	18.0			-5.6	9.8

T, SiO2, Cl & H2S data from: [a]Von Damm, personal communication, [b]Bowers et al. (1988), [c]Bluth and Ohmoto (1988), [d]Michard et al. (1984), [e]Merlivat et al. (1987), [f]Campbell et al. (1988), [g]Campbell et al. (1994), [h]Von Damm and Bischoff (1987), [i]Massoth et al. (1989), [j]Butterfield et al. (1990)- (note: SiO2 and H2S data have been estimated from figures in the original paper), [k]Butterfield et al. (1994), [l]Campbell et al. (1988) [m]Ishibashi and Urabe (1995)

tabulated in Shanks et al. (1995).

ZnS or CdS precipitates from vent waters and any sulfur-bearing phases from the solid samples are analyzed for $\delta^{34}S$ using an automated elemental analyzer interfaced to an isotope ratio mass spectrometer. Laser probe $\delta^{34}S$ microanalysis has been used to study zonation in sulfide chimney walls (Crowe and Valley 1992; Shanks et al. 1998). $\delta^{34}S$ error is estimated to be ±0.2‰. $\delta^{18}O$ analysis of silicates, sulfates, and whole rock samples were carried out using the BrF_5 method of Clayton and Mayeda (1963). $\delta^{18}O$ error is estimated to be ±0.2‰.

B, Li, Cl, Cu, and Fe isotope analyses have been carried out by a variety of gas-source and solid-source isotope ratio mass spectrometry techniques and, more recently, by plasma-source magnetic-sector multiple-collector mass spectrometry. The reader is referred to the cited references for specifics.

GEOLOGIC SETTING OF SEAFLOOR HYDROTHERMAL SYSTEMS

Only a small percentage of the divergent and convergent boundaries on the seafloor have been explored to date, but hydrothermal vents are now known from settings that include mid-ocean ridges, island arcs and back-arc basins, seamounts and hotspot-related oceanic islands, and rifted-continental settings (Fig. 2). Hydrothermal vent fluids have been sampled at more than 35 sites (Butterfield 2000), but there are at least 100 additional sites where hydrothermal deposits or vents are indicated but are either presently inactive or not fully explored. For given vent fields, there is a wide spectrum of available information from the earliest exploration stages where hydrothermal activity is known only from water column plume studies or bottom photography to systems that are fully explored and sampled by submersible. Some areas, like 21°N EPR, Guaymas Basin, 9-10°N EPR, the TAG site on the MAR, and the Endeavour Main Field of the JFR have a long history of repeated sampling and observations and a well-established time-series of vent chemistry and isotopic studies. Sites with comprehensive stable isotope studies form a small subset of the known hydrothermal sites on the seafloor. However, it is instructive to briefly describe the full range of geologic settings on the seafloor. Excellent reviews are now available of the geologic setting of MOR hydrothermal systems (Fornari and Embley 1995) and convergent margin systems in the western Pacific (Ishibashi and Urabe 1995).

Slow-spreading ridges, characterized by full spreading rates from 10 to 40 mm/yr, are represented by the Mid-Atlantic Ridge, the southern Gorda Ridge (Escanaba Trough), and the Central and Southwest Indian Ocean Ridges (Fig. 2). Moderate-spreading ridges (50-70 mm/yr) include the Juan de Fuca Ridge, northern Gorda Ridge, Galapagos Rift, the East Pacific Rise at 21°N, and the Southeast Indian Ridge. Fast-spreading ridges (80-120 mm/yr) are represented by the East Pacific Rise south of 13°N and includes a large off-axis sulfide deposit at 13°N and the important vent fields at 9-10°N EPR. Ultra-fast spreading occurs on the southern EPR from 17-22°S, with rates in excess of 150 mm/yr. Recent investigations on the SEPR have shown at least 40 individual high-temperature vent sites, with a broad variety of associated sulfide deposits and large variations in vent chemistry (Von Damm, personal communication).

Unsedimented ridges represent, in some cases, the simplest possible hydrothermal systems with only one rock type (basalt) and only seawater as the reactive fluid. However, even these "simple" systems are quite complicated when the range of tectonic and volcanic processes is considered. Spreading rate seems to be a fundamental variable, and it is closely related to magma effusion rate, but short-term variations at individual ridge segments or vent fields produce large variations in vent-fluid chemistry and stable isotope geochemistry. Investigations to date have shown that extremely high-temperature hydrothermal vents (350-400°C) with strong focused flow can be found on fast, moderate, and slow spreading ridges. Large ($>10^6$ tons), well-developed hydrothermal deposits are known only on slow-spreading (TAG, MAR) and moderate-spreading (Middle Valley and 13°N) ridges, but this may be due to incomplete exploration of other sites. Active seafloor volcanic eruptions have been documented only on moderate- and fast-spreading ridges. Perfit and Chadwick (1998) have shown that spreading rate is inversely proportional to the interval between volcanic eruptions and to the volume of individual flow units.

Earlier studies of mid-ocean ridges suggested that slow-spreading ridges are tectonically dominated and fast-spreading ridges are dominated by volcanic processes. As the observational database has grown, it has become obvious that this model is oversimplified. Perfit and Chadwick (1998) summarize a model that accounts for both magmatic and tectonic phases in the developmental cycles of divergent margins (Fig. 3). In this model, slow, moderate, and fast spreading ridges all undergo cycles that begin with renewed magmatism and progress with time to tectonic-dominated stages, but these cycles occur at different time scales depending on spreading rate. Thus, we have moderate-spreading rate ridges on the northern EPR and the Endeavour Segment of the JFR that have broad, open tectonically-dominated axial valleys, but still have robust, extremely high-temperature hydrothermal systems related to heat from underlying magma chambers. The TAG field on the MAR is quite similar, but in some cases slow-spreading ridges can evolve to purely amagmatic extensional regimes. Continuously replenished axial magma chambers exist only below moderate- and fast-spreading ridges. Other differences include much more frequent eruptions and dike injection events on faster spreading ridges, and a shallow axial magma chamber at about 1.5 km depth that can be continuously imaged using multi-channel seismic reflection by driving along the ridge axis (Detrick et al. 1987). Slow-spreading ridges typically have much broader axial valleys (Fig. 3), much less common and deeper magma chambers (2.5 to >3.5 km), and lower crustal and serpentinized ultramafic rocks emplaced along ridge-parallel structures and transform faults. Thus, the spreading rate and geology of different ridge crests have potentially important influences on the stable isotope geochemistry of vent fluids and

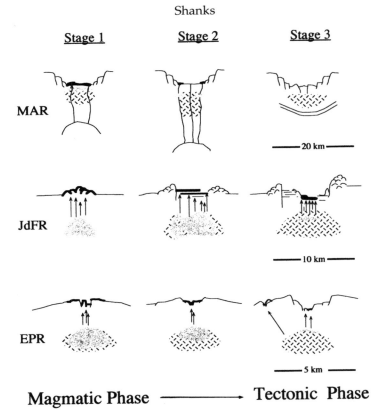

Figure 3. Mid-ocean ridge morphology for slow (MAR), medium (JFR) and fast (EPR) spreading ridges. All ridge types cycle from magmatic to tectonic phases, but there are significant differences related to spreading rate. Faster spreading ridges have continuously-present, shallower axial magma chambers and more frequent eruptive events. Slow-spreading ridges have intermittent, deeper magma chambers and very broad axial valleys that are dominated by tectonism (Perfit and Chadwick 1998; used by permission of the American Geophysical Union).

hydrothermal deposits. There are also a variety of different settings within a typical ridge segment with vent fluid ranging from high-temperature black smokers to diffuse low-temperature seeps and with widely variable hydrothermal deposit types and sizes.

Significant hydrothermal sites are known from a number of on- and off-axis seamounts. These include the Axial Volcano site on the JFR, a large sulfide deposit on a near-axis volcano at 13°N EPR, Loihi seamount in the Hawaiian-Emperor chain, the Lucky Strike hot-spot-related seamount site on the MAR, and a number of other localities. Axial Volcano and Lucky Strike have been studied most thoroughly, and have high-temperature hydrothermal systems. The Ashes vent field on the summit of Axial Volcano was the first to show effects of boiling at the reduced pressures encountered on the seamount relative to a normal ridge crest (Massoth et al. 1989). Many ridge-crest vent fields have been discovered in the last decade that show the effects of phase-separation into low-salinity vapor and more saline fluid (Butterfield 2000).

Sediment-covered mid-oceanic ridges occur where continentally-derived detritus can reach the ridge crest. The two most well-known examples are Middle Valley at the northern end of the JFR and Escanaba Trough at the southern end of the Gorda Ridge (Fig. 2). These deposits are quite similar in setting, but quite different in stage of development. Both have been extensively explored by surface ship and submersible and both were drilled during ODP leg 169. Middle Valley was also drilled during ODP leg 139. Both Escanaba Trough and Middle Valley hydrothermal systems occur near sediment-covered hills that protrude above flat, turbidite-filled axial valleys. The hills that host the deposits are a result of subsurface sill emplacement related to ridge magmatism. The Bent Hill area in Middle Valley hosts at least two thick (>100 m) lenses of massive sulfide with extensively developed underlying sulfide feeder-vein zones and hydrothermal alteration zones in the surrounding sediments. Escanaba is less extensive and was a much weaker hydrothermal system, at least where explored to date. Sulfide deposits at Escanaba are only a few meters thick, but they do contain some remarkable concentrations of exotic elements (for these systems) such as As, Sb, and Bi (Koski et al. 1988). Sulfides from both areas contain substantially more Ba and Pb than bare volcanic systems.

Hydrothermal systems related to rifted continental margins are known from the Red Sea and at Guaymas Basin in the Gulf of California. In both cases continental crust occurs in close-proximity to narrow axial rifts. This, along with unique sedimentary features in these basins, has a strong influence on the hydrothermal systems.

The occurrences of seafloor hydrothermal mineral deposits in arc-backarc settings in the western Pacific are of particular interest at the present time because exploration, submersible sampling, and drilling are less advanced than at MORs and because arc-related intermediate and felsic rocks produce vent fluids with different chemistries and are presumably more enriched in magmatic water and other volatiles (Ishibashi and Urabe 1995). In addition, most of the world's economic accumulations of base-metal (Cu, Pb, Zn) volcanogenic massive sulfide deposits that are mined on the continents were originally formed in arc-related settings (Franklin et al. 1981; Franklin et al. 1998). Important seafloor hydrothermal systems occur related to back-arc spreading centers in Okinawa Trough, Mariana Trough, Manus Basin, North Fiji Basin, and Lau Basin and related to island arc volcanoes of the Izu-Bonin arc.

SEAFLOOR HYDROTHERMAL VENT FLUIDS

Stable isotopic data for vent fluids from selected hydrothermal areas are presented in Table 1. Except as referenced, these data have been generated in USGS studies over the last 18 years and have a high degree of internal consistency. Stable isotope data from the southern East Pacific Rise, some of the more recent data from 9-10°N on the East Pacific Rise, data from the Lucky Strike vent field on the Mid-Atlantic Ridge, and data from pore fluids and hydrothermal minerals from ODP drilling at Middle Valley and Escanaba Trough are presented for the first time here. Chemical data in Table 1 from the southern East Pacific Rise, 9-10°N on the East Pacific Rise, and the Lucky Strike vent field are from Karen Von Damm and associates (Von Damm et al. 1995, 1997, and unpublished data).

Seafloor temperature and pressure data for venting fluids indicate a wide variability relative to the two-phase (vapor-liquid) boundary for seawater (Fig. 4). Chloride concentrations in vent fluids (Table 1) range from 35 to 1090 mmol/kg due to subcritical boiling and/or supercritical phase-separation. For saline fluids such as seawater, phase-separation can occur above the critical point with condensation of brine droplets from a less saline residual "vapor." Obviously, vent fluids decompress on ascent from sub-surface reactions zones, which are inferred to be in cracking fronts just above the ~1-4

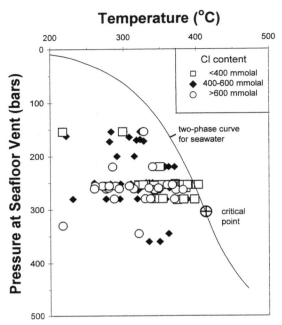

Figure 4. Temperature and seafloor pressure of hydrothermal vents relative to two-phase curve and critical point for seawater (redrawn from Bischoff and Rosenbauer 1985).

km-deep magma chamber that drives convective fluid circulation (Lister 1983; Alt 1995; Perfit and Chadwick 1998; Singh et al. 1999). It is also apparent from the assembled data (Table 1, Fig. 4) that, despite evidence of phase-separation in the widely variable Cl contents of the fluids, measured temperatures of vent fluids are in many cases significantly below the two-phase boundary as defined by the ambient pressure at the seafloor. Conductive cooling of fluids on ascent is an obvious possibility, moving fluids off the two-phase curve to lower temperatures. However, Von Damm and Bischoff (1987), Bischoff and Rosenbauer (1989), Von Damm (1990), and Edmonds and Edmond (1995) have suggested a 3-component mixing model between (1) a "deep-seated" highly-concentrated brine, (2) a low-chloride vapor-phase generated during phase-separation, and (3) normal seawater. These processes have important implications for isotopic exchange reactions with minerals and for effects of phase-separation on ΔD and $\Delta^{18}O$ of vent fluids.

Silica content also varies widely in vent fluids and was originally thought to be an excellent indicator of depth of reaction zones where the fluids last equilibrated with quartz that precipitates due to decompression during ascent. Silica contents are plotted versus temperature in Figure 5, and are overlain on isobars of equilibrium quartz solubility in 0.5 molal NaCl solution (Von Damm et al. 1991). While many of the vent fluids with near-seawater chloride contents do plot reasonably on the 200-600 bar contours (Fig. 5) and may represent subsurface equilibration with quartz precipitates, it is clear that many high-chloride samples have high silica and many low-chloride fluids have low silica contents. Conductive cooling could explain many of the high-silica fluids. Most of the really low-silica fluids are from the 9-10°N EPR site, where active, very shallow phase-separation is taking place and this shows as very low silica in the fluids. In

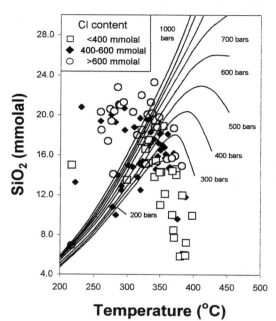

Figure 5. Silica content of end-member hydrothermal vent fluids, superimposed on equilibrium quartz solubility curves for 0.5 wt % NaCl fluid (redrawn after Von Damm et al. 1991).

any case, the most reliable indicator of depth of reaction zone is probably geophysical evidence on the depth to the top of the magma chamber in each particular vent field.

Oxygen and hydrogen isotopes in vent fluids

$\Delta^{18}O$ and ΔD values of vent fluids (Table 1) can best be understood in terms of water-rock interaction within the ocean crust. Field studies, experimental studies, and isotopic exchange computations (Muehlenbachs 1972; Stakes and O'Neil 1982; Bowers and Taylor 1985; Cole et al. 1987; Bowers 1989; Böhlke and Shanks 1994; Shanks et al. 1995) have clearly shown that both $\Delta^{18}O$ and ΔD increase due to water-rock interaction with igneous crust. Oxygen and hydrogen isotope values of end-member vent fluids (Fig. 6) follow a calculated seawater-basalt reaction vector (within ΔD error of $\pm 1.5‰$), due to fluid evolution to decreasing water/rock mass ratios (Shanks et al. 1995).

Bach and Humphris (1999) have recently suggested that ocean ridge hydrothermal vent fluids show a global correlation of both $^{87}Sr/^{86}Sr$ and $\Delta^{18}O$ to spreading rate (or magma supply rate). They concluded that at low spreading rates (or low magma supply rates) there is significantly less seawater-derived Sr in the vent fluids, and vent fluids have higher $\Delta^{18}O$. These effects are attributed to a greater exchange of Sr and O with the crust during hydrothermal circulation, probably due to longer reaction paths at slow-spreading ridges. To further examine possible global isotope systematics, ΔD, $\Delta^{18}O$, and $\delta^{34}S_{H_2S}$ have been plotted against spreading rate (Fig. 7). Linear regression equations and R^2 values indicate that the $\delta^{18}O$ vs. spreading rate correlation suggested by Bach and Humphris (1999) is not statistically significant when the full data set of ridge related vent fluids is utilized. Similarly, values of $\delta^{34}S_{H_2S}$ show no significant correlation with spreading rate. The strongest correlation is for ΔD, but even that, with an R^2 of 0.38, is very poorly constrained. The indicated higher ΔD values for slower spreading ridges are consistent with longer flow paths and more water-rock interaction. However, $\Delta^{18}O$ values should be even more responsive to w/r interaction processes, and $\Delta^{18}O$ shows

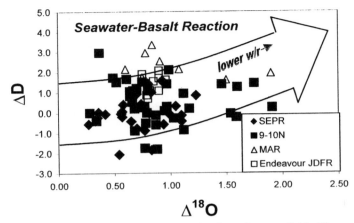

Figure 6. Oxygen and hydrogen isotope values of end-member vent fluids. The seawater-basalt reaction vector indicates calculated fluid evolution to decreasing water/rock mass ratios (Shanks et al. 1995). Most of these data follow the reaction vector within ΔD error of $\pm 1.5\permil$. Exceptions are the Mid-Atlantic Ridge samples, which have quite high ΔD values, and very low-chloride samples from fast spreading ridges, which are influenced by phase separation or magmatic water (see text).

insignificant variation with spreading rate. It seems that, as the database of vent fluids is increased, short-term temporal variations are increasingly important, and short-term variations at a given vent field are more important than global comparisons in terms of stable isotope systematics.

Sulfur isotopes in vent fluids

Sulfur isotope studies of seafloor hydrothermal vent systems have been particularly instructive in terms of understanding ancient hydrothermal systems where coexisting fluids and solids are not available. A generalized schematic (Fig. 8), after Woodruff and Shanks (1988), shows the salient features. Circulating seawater is heated as it approaches the sub-seafloor magma chamber and anhydrite is precipitated at temperatures above about 150°C (Sleep 1991) due to retrograde anhydrite solubility and calcium increase in the fluid due to rock reactions. This removes most of the original 2710 mg/kg of sulfate in the fluid, but a small amount, approximately 100-300 mg/kg remains in solution at equilibrium with the precipitated anhydrite. Most intense water-rock reaction occurs in a high-temperature reaction zone (Seyfried 1987; Seyfried et al. 1991, 1999) in a cracking front near the magma chamber (Singh et al. 1999) where the heated, evolved-seawater fluid becomes reducing, acidic, Mg- and SO_4-depleted, metal-rich, and strongly enriched in H_2S due to a combination of hydrolysis of rock sulfides and sulfate-reduction by reaction with ferrous minerals (Shanks et al. 1981).

As pointed out by Seal et al. (2000), many studies of ancient hydrothermal systems have utilized equilibrium sulfate-sulfide sulfur isotope fractionation models, but these should be applied with great caution. As shown in Figure 9, seafloor hydrothermal vent fluid $\delta^{34}S_{H_2S}$ values do not conform to simple equilibrium fractionation models. Shanks et al. (1981) first showed experimentally that sulfate in seawater-basalt systems is quantitatively reduced at temperatures above 250°C when ferrous minerals like the fayalitic olivine are present. When magnetite is the only ferrous iron-bearing mineral in the system, sulfate-reduction proceeds to sulfate-sulfide equilibrium, but natural basalts contain ferrous iron-bearing olivine, pyroxene, titanomagnetite, and iron-monosulfide solid-solution (mss) (approximately pyrrhotite). It is the anhydrite precipitation step

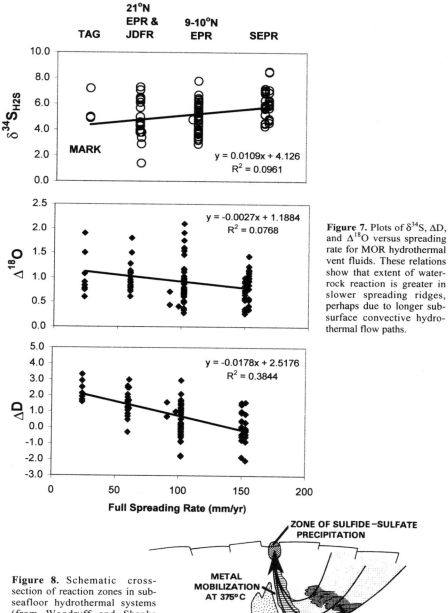

Figure 7. Plots of δ^{34}S, ΔD, and Δ^{18}O versus spreading rate for MOR hydrothermal vent fluids. These relations show that extent of water-rock reaction is greater in slower spreading ridges, perhaps due to longer subsurface convective hydrothermal flow paths.

Figure 8. Schematic cross-section of reaction zones in subseafloor hydrothermal systems (from Woodruff and Shanks 1988). Used by permission of the American Geophysical Union.

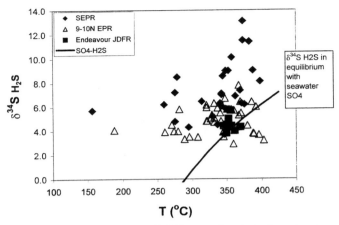

Figure 9. Sulfur isotope values of H$_2$S from seafloor hydrothermal systems relative to equilibrium SO$_4$-H$_2$S fractionation assuming $\delta^{34}S_{SO_4} = 21‰$.

during recharge of the hydrothermal system that controls the $\delta^{34}S_{H_2S}$; most sulfate that enters the high-temperature reaction zone is quantitatively reduced and mixed with H$_2$S hydrolyzed from the rock. Sulfur isotope systematics in seafloor hydrothermal systems cannot be understood without carefully evaluating the chemical reactions taking place during convective fluid circulation. The fundamental seawater-basalt reaction process, as described here, is the same for most seafloor hydrothermal systems, but some systems have additional possible sulfur contributions from bacterial reduction processes, mixing in near-seafloor alteration zones, reaction with sediments that contain isotopically light sulfide, and direct contributions of magmatic SO$_2$.

Carbon isotopes in vent fluids

Dissolved inorganic carbon (DIC) occurs as HCO$_3$ in seawater with a concentration of about 0.5 mmol/kg and a $\delta^{13}C$ value of approximately 0‰; its isotopic composition is largely controlled by equilibrium with atmospheric CO$_2$, which has carbon isotopic composition of $-7‰$. Much of the following discussion of carbon isotopes in vent fluids is taken from Shanks et al. (1995).

At unsedimented, volcanic ridges, such as 21°N EPR the $\delta^{13}C$ value for CO$_2$ is -7.0‰, similar to MORB CO$_2$ (Lilley et al. 1993; Welhan and Craig 1979, 1983). On the southern Juan de Fuca Ridge and the East Pacific Rise at 13°N and 21°N, $\delta^{13}C$ values (Fig. 10) for CH$_4$ are uniform and range from -20.8 to -15.0‰. The ΔCO$_2$-CH$_4$ values for the 21°N fluids range from 8.1 to 10.7, which correspond to apparent temperatures of 620° to 770°C (Welhan and Craig 1983; Richet et al. 1977). On the basis of the high $\delta^{13}C$ values for the CH$_4$ compared to other hydrocarbon-related methane, the carbon isotope thermometry, and the lack of organic matter at 21°N, Welhan and Craig (1983) have proposed an abiogenic origin for the methane through direct derivation from MORB. On the basis of similar CH$_4$/^{3}He ratios for vent fluids between the East Pacific Rise (21°N) and the Mid-Atlantic Ridge (23°N), Jeanbaptiste et al. (1991) also proposed an abiogenic MORB source for the Mid-Atlantic Ridge methane. Most of the CO$_2$ in vent fluids apparently results from magma degassing below the ridge crest, and CH$_4$ also may be from the magmatic gases, or it may be leached from rocks during water/rock interaction. Thus, the carbon isotopic thermometry may indicate very high temperature equilibration, with CO$_2$-CH$_4$ re-equilibration down to a blocking temperature of about 700°C below

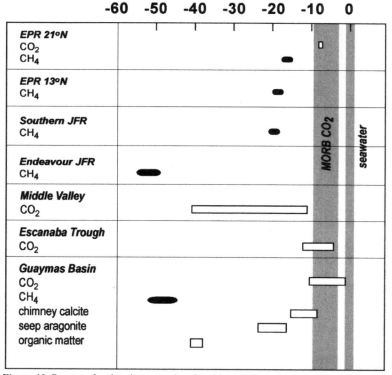

Figure 10. Ranges of carbon isotope values in mid-ocean ridge hydrothermal systems. Data from: Lilley et al. (1993), Welhan and Craig (1979, 1983), Welhan and Lupton (1987), Jeanbaptiste et al. (1991), Charlou and Donval (1993), Taylor (1992), and Peter and Shanks (1992). From shanks et al. (1995); used by permission of the American Geophysical Union.

which isotopic exchange is too slow to permit equilibration. Even at 700°C, equilibration times for CO_2-CH_4 exchange are estimated to be on the order of a month (Giggenbach 1982). Alternatively, the CO_2 and CH_4 may have different sources, degassing and leaching, respectively. In this case, the apparent fractionation has no temperature significance.

At sediment-covered ridges, such as Guaymas Basin (Fig. 11), vent-fluid CH_4 contents range from 2.0 to 6.8 mmol/kg, and CH_4/C2+ ratios are relatively low, ranging from 78 to 690, reflecting the high long-chain hydrocarbon content of the fluids (Lilley et al. 1993; Welhan and Lupton 1987). The $\delta^{13}C$ values for CH_4 range from -50.8 to -43.2‰; the $\delta^{13}C$ values for endmember hydrothermal CO_2 range from -10.5 to -1.096 (Welhan and Lupton 1987). The ΔCO_2-CH_4 values for the fluids range from 34.6 to 41.7, which correspond to unrealistically low temperatures of approximately 150°C (Richet et al. 1977). On the basis of $\delta^{13}C$ values of methane, Welhan and Lupton (1987) proposed thermal degradation (pyrolysis) of organic matter for the origin of Guaymas methane.

Two additional sedimented ridge hydrothermal vent sites, Escanaba Trough and Middle Valley, show quite different $\delta^{13}C$ ranges for CO_2 in vent fluids (Fig. 11).

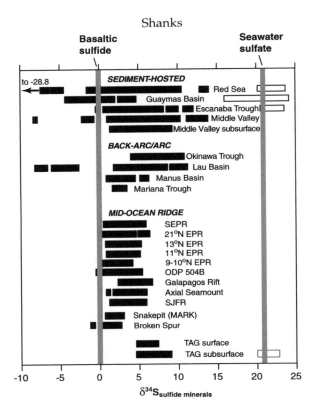

Figure 11. Sulfur isotope values of sulfide minerals from selected seafloor hydrothermal deposits. Modified from Herzig et al. (1998) with additional data from Table 1. See text for reference to individual data sets.

Escanaba Trough has a range from -12.2 to -3.9‰, and $\delta^{13}C$ values increase with CO_2 content, presumably due to increased magmatic contributions (Taylor 1992). The vent fluids at Middle Valley show a $\delta^{13}C$ range from -38.9 to -10.6‰, indicating substantial contributions from pyrolysis of organic carbon in the sediments and dissolution of detrital biogenic carbonates (Taylor 1992).

The $\delta^{13}C$ values of calcite from chimney structures in Guaymas Basin hydrothermal vent fields range from -14.0 to -9.6‰ (Peter and Shanks 1992). This indicates that carbon was derived in approximately equal proportions from the dissolution of marine carbonate minerals and the oxidation of organic matter during fluid flow through the underlying sediments. These data do not preclude a basaltic CO_2 contribution with $\delta^{13}C$ of -7 to -4‰ (Taylor 1986), but it is unlikely that basaltic CO_2 is a major component in the vent fluids or chimney carbonates (Peter and Shanks 1992). Seewald et al. (1994) have shown that CO_2 produced by experimental oxidation and thermocatalytic degradation of Guaymas Basin sedimentary organic matter has $\delta^{13}C$ values similar to the chimney calcites.

The vent fluids from the unsedimented Endeavour segment of the Juan de Fuca Ridge have CH_4 contents (1.8 to 3.4 mmol/kg) and $\delta^{13}C$ values (-55.0 to -48.4‰) that are more similar to the sediment-covered Guaymas Basin than the unsedimented East Pacific Rise. Lilley et al. (1993) attributed this to reaction of circulating hydrothermal fluids with deeply buried (and undiscovered) sediments that were buried in an earlier stage of development of this ridge segment.

STABLE ISOTOPE SYSTEMATICS OF SELECTED
SEAFLOOR HYDROTHERMAL SYSTEMS

Seafloor hydrothermal systems that are reasonably well studied in terms of stable isotope values of fluids, hydrothermal precipitates and hydrothermal alteration are presented as "case studies" in this section. Ranges of sulfur isotope values from given deposits are summarized in Figure 11. It is clear that most of the mid-ocean ridge sulfides (designated $\delta^{34}S_{MeS}$ to indicate metal sulfide minerals), plus those from Manus and Mariana back-arc spreading centers, are quite similar with $\delta^{34}S_{MeS}$ ranges of about 1-5‰. Broken Spur and Snakepit $\delta^{34}S_{MeS}$ values are lower than the others, and TAG values are higher, suggesting more heterogeneity, as might be expected on a slow spreading ridge with intermittent axial magmatism. The $\delta^{34}S_{MeS}$ ranges of sulfide minerals from MOR black smoker chimneys and mounds, which clearly precipitate from black smoker vent fluids, do not show systematic variation with spreading rate. There is generally a larger range of $\delta^{34}S$ values in the sulfide minerals than in the vent fluid H_2S, due mainly to processes that alter $\delta^{34}S$ values in the near-seafloor environment such as seawater mixing and local sulfate reduction and, in larger mounds, recrystallization and modification of sulfide minerals by later hydrothermal fluids. The back-arc spreading systems at Okinawa Trough and Lau Basin have much larger ranges of $\delta^{34}S_{MeS}$, as do all of the sediment-hosted systems. But even in the sediment-hosted systems, most $\delta^{34}S_{MeS}$ values range from about 0 to 15‰, averaging about 5‰. Guaymas Basin uniquely occurs in organic-rich sediments with abundant diagenetic, bacteriogenic sulfide. Remobilization of the isotopically-light bacterial sulfide component, mixed with the typical magmatic and seawater sulfate components, results in sulfides at Guaymas with $\delta^{34}S_{MeS}$ values from -4 to 5‰ (Peter and Shanks 1992).

Bare volcanic ridges: 21°N, 9-10°N, and 17-22°S EPR

Hydrothermal systems on bare volcanic ridge crests, like the 21°N vent field, can be simply represented by water-rock reactions involving seawater and basalt, as indicated by $\Delta^{18}O$ and ΔD relations in vent fluids (Fig. 6). In the 21°N vent field, strongly focused 350°C black smoker fluids typically jet from well-defined 1-30 cm-diameter central conduits at flow rates that often exceed 1 kg/s (Converse et al. 1984). With vertical ascent rates estimated at 1 m/s at the chimney, ascent from the deep reaction zone at the top of the axial magma chamber takes only about 30 minutes, but this assumes upflow in a "pipe" with the same diameter as the chimney conduit. Of course, more diffuse flow from many channelways may occur at depth, but ascent is geologically rapid, probably on the order of days. Hydrothermal deposits at 21°N and many of the bare rock sites consist of individual chimneys or clusters of chimneys less than 10 m-tall built upon mounds of sulfide rubble formed by degradation and collapse of formerly active chimneys. Black smoker chimneys grow by developing an outer carapace of anhydrite followed by concentric inner zones of wurtzite (ZnS) and isocubanite ($CuFe_2S_3$) in the innermost zone (Haymon and Kastner 1981; Hekinian et al. 1981; Styrt et al. 1981; Haymon 1993).

The 21°N vent field exhibits a number of enigmatic features that have been apparent from the earliest geochemical studies of black smoker fluids and sulfide deposits. For example, Kerridge et al. (1983) measured vent fluid $\delta^{34}S_{H_2S}$ values that were generally higher than coexisting chimney $\delta^{34}S_{MeS}$ values. Woodruff and Shanks (1988) found the same effect, even though they carefully sampled Cu-Fe sulfides from the innermost chimney wall immediately adjacent to conduits that carried the black smoker fluids. At equilibrium, most sulfide minerals in chimneys are expected to be ^{34}S-enriched relative to the H_2S from which they precipitate, and at the temperatures of black smokers mineral-H_2S fractionation should be <1‰ (Ohmoto and Goldhaber 1997). How could the vent fluid H_2S have significantly higher sulfur isotope values than the metal sulfides that precipitate from these fluids? Woodruff and Shanks (1988) and Janecky and Shanks (1988) suggested reduction of a small amount of isotopically heavy seawater or anhydrite sulfate in the chimney environment, but Bluth and Ohmoto (1988) argued that $\delta^{34}S_{H_2S}$ values are set in the deep reaction zone and that fluid flow in black smokers is too high to allow local modification of vent fluid chemistry or $\delta^{34}S_{H_2S}$. However, Tivey and McDuff (1990) used a fluid advection model to show that pressure differentials develop where there are irregularities in chimney conduits, and these effects can lead, in some cases, to significant entrainment of seawater through the wall into the chimney conduit, so local, temporal variations in $\delta^{34}S_{H_2S}$ could occur. Janecky and Shanks (1988) demonstrated that reactions in the chimney walls can only increase $\delta^{34}S_{H_2S}$ by about 3 ‰ because of limited availability of reducing agents.

Perhaps the scale of sampling using conventional techniques is too coarse to discern finely zoned $\delta^{34}S$ variations in sulfide minerals in chimney walls. The development of highly focused laser or ion beam techniques to analyze sulfur isotopes in chimney walls to delineate temporal variations in $\delta^{34}S_{MeS}$ values offered a possible answer (Crowe et al. 1990). Laser microprobe analyses of ~150 μm-diameter samples of sulfide chimneys from 21°N show systematic $\delta^{34}S_{MeS}$ variations in the inner 15 mm of isocubanite chimney walls that include the whole range of values expected based on analyses of vent fluid $\delta^{34}S_{H_2S}$ (Shanks et al. 1998). Nonetheless, it is troubling that most seafloor vent fields seem to have $\delta^{34}S_{H_2S} > \delta^{34}S_{MeS}$. An example data set from 9-10°N (Fig. 12) shows that all vent fluids from this extensive area have H_2S that is isotopically heavier than

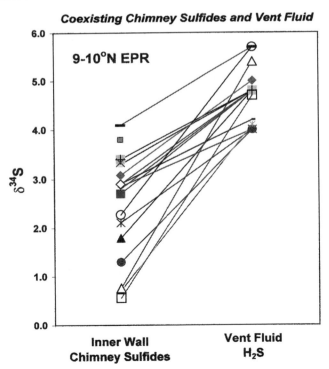

Coexisting Chimney Sulfides and Vent Fluid

9-10°N EPR

Figure 12. Sulfur isotope values of coexisting inner wall chimney sulfides and H_2S from 9-10°N hydrothermal vents. Note that $\delta^{34}S_{H2S}$ values always exceed $\delta^{34}S_{MeS}$ values even though fractionation factors indicate <1‰ difference between Py-Sl-Po-Cp and H_2S at 300 to 400°C (Ohmoto and Goldhaber 1997).

Inner Wall Chimney Sulfides   **Vent Fluid H_2S**

coexisting sulfide minerals. Why are there no vent fluids that are isotopically lighter than coexisting sulfides? These observations suggest that additional careful studies of $\delta^{34}S_{MeS}$ in chimney walls, using recently developed ion microprobe methods (Riciputi 1996) that provide high precision $\delta^{34}S$ analyses (±0.25‰) on ~20 μm-diameter spots keyed to time-series observations of $\delta^{34}S$ in vent fluids might provide additional important information on the isotopic evolution of these systems and processes that control such evolution.

From the first studies of black smoker vents, fluid compositions at 21°N indicated small but significant deviations from normal seawater in chloride concentration. The NGS vent has higher Cl and the OBS, HG, and SW sites have lower chloride than seawater (Table 1). Proposed explanations of these relatively small variations included rock hydration, formation of Cl-bearing minerals, and boiling. Boiling seemed the most logical explanation, but the 350°C, 250 bar vent fluids were not even very close to the boiling point on the seafloor (Fig. 4), and subsurface boiling was not predicted even when adiabatic cooling on ascent was restored. The only way these fluids could have boiled on ascent is if significant conductive cooling had occurred. However, Campbell et al. (1988) demonstrated long term stability of vent fluid chemistry at 21°N (1979 to 1985) and Guaymas Basin (1982 to 1985). Bowers et al. (1988) and Seyfried et al. (1991) suggested fluid compositions were buffered by equilibria with silicate minerals, but this does not explain constant Cl over long time periods, especially if conductive cooling is a factor. How could the vent fluid Cl remain stable over time if it is due to phase-separation processes? Obviously, these processes are closely tied to δD and $\delta^{18}O$ variations in vent fluids. Once again an important role for time-series is indicated, especially in vent fields that show strong variations with time.

The 9-10°N area on the East Pacific Rise has been an ideal area for such studies. It is a fast-spreading ridge segment dominated by volcanic rather than tectonic processes. A schematic diagram of the ridge crest (Fig. 13) indicates a narrow axial cleft, referred to as the axial summit collapse trough (ASCT), that is the locus for ridge crest eruptive activity (Fornari and Embley 1995). Ridge crest eruptions sometimes fill and overflow the ASCT as evidenced by hollow basalt pillars with drain-back selvages and lava flows that can be traced back to the ASCT. Basalt pillars are believed to have formed around escape pathways for seawater trapped under lava flows. Perfit and Chadwick (1998) have documented that off-axis flows are also important and are not necessarily co-magmatic with on-axis lavas.

The 9-10°N ridge crest was mapped in 1989 using deep-towed bottom photography, video imaging, and side-scan sonar (Haymon et al. 1991). This revealed numerous robust vent fields that were targeted in 1991 Alvin submersible investigations. However, Alvin dives beginning on April 1, 1991 revealed that something unusual had occurred on the ridge segment between 9°45'-52'N (Fig. 14). Robust vent fields were gone, replaced with shiny new lava flows, and white bacterial floc was abundant within 60 m of the bottom, in low temperature "snow-blower" vents, and as thick "snow" deposits (up to 20-30 cm-thick) in some areas of the ASCT. Even more remarkable was the discovery of the "Tubeworm Barbecue" (Fig. 14), an area of recently killed vestimentifera tubeworms (no scavenging crabs or fish; no decay of the flesh), fresh lava, and a thin sediment of glass, anhydrite, and chalcopyrite (Haymon et al. 1993). Subsequent dating of the lavas using ^{210}Po in-growth from ^{210}Pb indicated that the eruption took place within a few weeks preceding the April 1991 dive series (Rubin et al. 1994). This fortuitous discovery allowed initiation of a time-series of vent rebirth from zero age that continues to present. This has been an excellent natural laboratory for studying growth of sulfide/sulfate structures and the associated evolution of fluid chemistry and stable isotope systematics. Since 1991, this area has been re-visited with Alvin seven times for the purpose of

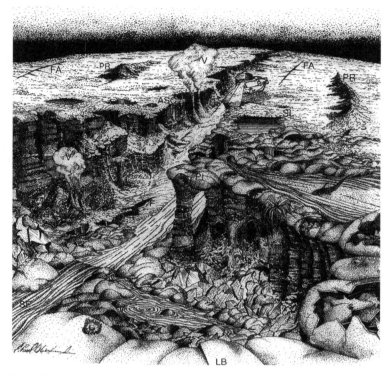

Figure 13. Perspective drawing of the axial summit trough of the East Pacific Rise at 9-10°N. The central cleft or trough is typically 50-150 m-wide and 30-50 m-deep, is the locus of most eruptive events, and is typically floored by sheet flows (SF) that sometimes have eruptive fissures (F). Hydrothermal vents (V) and basalt pillars (PI) are common within the axial trough. Pillars are related to chilling by seawater escape and commonly have drain-back selvages marking lava lake levels. Outside the axial summit trough lavas form pillow ridges (PR) or lobate flow units (LB) that often are hollow and have skylights (SL). From Fornari and Embley (1995); used with permission of the American Geophysical Union.

obtaining geochemical, petrologic, and biological samples.

High-salinity vent fluids were first discovered in 1984 on the southern Juan de Fuca ridge (Von Damm and Bischoff 1987). Studies of dissolved gases strongly suggested addition of a degassed, phase-separated, saline fluid to the vent fluids (Evans et al. 1988), which led Von Damm and Bischoff (1987) to suggest mixing of a very saline, previously phase-separated fluid with ridge-crest hydrothermal fluids of normal seawater salinity. Significant chloride variations were also measured at Axial Seamount (Massoth et al. 1989) and at the Endeavour Main Field (Butterfield et al. 1994b). However, it was the 1991 discovery of nearly pure "vapors" venting at 403°C at the A vent field at 9°46.5′N on the EPR (Table 1) that unequivocally proved phase-separation as the cause of vent chloride variations (Von Damm et al. 1995, 1997). Detailed sampling of both high temperature "smokers" and low temperature diffuse vents in 1991 failed to detect the conjugate brine produced during the vapor-rich venting event. Subsequent studies have shown a time-delay between initial venting of low salinity fluids and later venting of the conjugate brine from the same chimney.

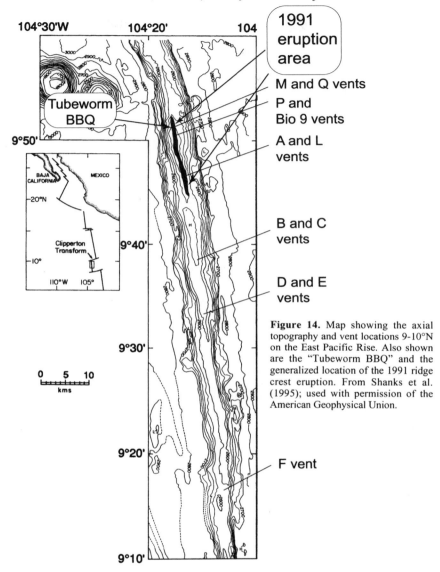

Figure 14. Map showing the axial topography and vent locations 9-10°N on the East Pacific Rise. Also shown are the "Tubeworm BBQ" and the generalized location of the 1991 ridge crest eruption. From Shanks et al. (1995); used with permission of the American Geophysical Union.

The 9-10°N vent field showed dramatic short-term changes in vent chemistry following the 1991 eruption. Figure 15 shows the variations in temperature, chemistry, and stable isotope values for the Aa vent (one of the marked, individual vents in the A vent field) from 1991 to 1994. This vent was sampled on three separate Alvin dives over a two-week period following the 1991 eruption. Temperature increased from 389 to 403°C and chloride and silica dropped to extremely low values (Table 1) indicating shallowing and increased boiling in the hydrothermal system. The silica contents of these vent fluids have been interpreted as indicating depth of quartz equilibration within a few hundred meters of the seafloor (Von Damm et al. 1995). Stable isotope values of these Aa vent fluids also changed significantly during this two week time period;

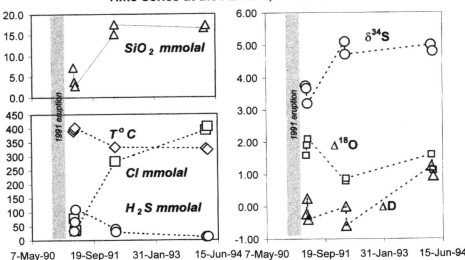

Figure 15. Time-series plots of vent fluid variability at the Aa vent, 9-10°N EPR.

$\delta^{34}S$ decreased, ΔD decreased, and $\Delta^{18}O$ increased. The short term changes in sulfur and oxygen isotope values are consistent with reaction at lower w/r ratios, but the ΔD shift to lower values is probably due to phase separation.

The P and Biomarker 9 vents, located about 40 m apart at 9°50.3′N, also showed a marked and almost identical time-series following the 1991 eruption (Table 1). P vent data from 1991 to 1995 (Fig. 16) show large, gradual increases in chloride (to concentrations greater than normal seawater in 1995), silica, and ΔD, and large decreases in H_2S and $\delta^{34}S$. These changes are generally consistent with deepening and maturing of the hydrothermal system at P vent, but a number of details in this time-series are worth noting. First, there are significant temperature and ΔD increases and Cl and SiO_2 decreases at P vent (and at Biomarker 9 vent) in 1992, which indicate a second eruptive event in this area that caused the hydrothermal system to shallow and intensify. This was confirmed by ^{210}Po dating of fresh lava samples collected in 1992 (Rubin et al. 1994). This 1992 event was only recorded in fluid chemistry of these two vents, so its extent was smaller than the 1991 eruption. The ΔD increase in both P and Bio 9 vents in 1992 might be related in these cases to increased reaction with the rock, which is consistent with the observed increase in $\Delta^{18}O$ and decrease in $\delta^{34}S_{H2S}$ values (Fig. 16; Table 1). However, the generally high $\delta^{34}S_{H2S}$ and H_2S concentrations in 1991 and 1992 are still consistent with a relatively high water-rock system where abundant sulfate ingress and sulfate reduction in contact with hot rock is occurring. Finally, in 1995, a cracking event was detected in this area at a depth of about 1-1.3 km, just above the axial magma chamber melt lens, using a closely spaced array of ocean bottom seismometers (Sohn et al. 1998). Four days after the cracking event at depth, a temperature monitor on the Bio 9 vent recorded a seven-day pulse of temperature increase from 365 to 372°C. Chloride contents increased significantly on both P and Bio 9 vents following the cracking event, and $\delta^{34}S_{H2S}$ decreased sharply in both cases while ΔD and $\Delta^{18}O$ remained relatively constant (Fig. 16, Table 1). These variations suggest that the hydrothermal system continued to deepen as a result of the cracking event, releasing higher salinity fluids (greater than seawater

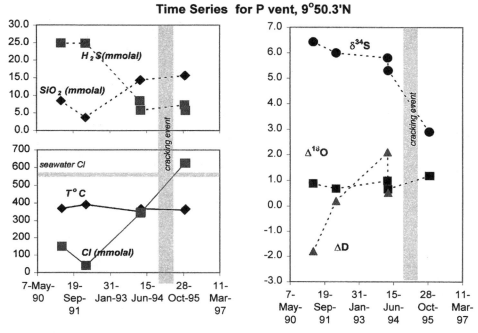

Figure 16. Time-series plots of vent fluid variability at the P vent, 9-10°N EPR.

chloride at P) with lower $\delta^{34}S_{H_2S}$ due to less access to seawater-derived SO_4 deeper in the system.

The very low ΔD values that occur for a short time immediately after volcanic eruptions, especially in the P and Bio 9 vents (Fig. 16, Table 1), could be a result of either phase separation or addition of magmatic water, which has δD values of -40 to -80‰ (Rye and O'Neil 1968; Rye 1993). Horita et al. (1995) and Berndt et al. (1996) have experimentally determined significant hydrogen isotope fractionation in NaCl fluids during subcritical boiling and during supercritical phase-separation. These studies indicate that water vapor at isotopic equilibrium has higher δD than coexisting brine by as much as 7-8‰. Because the 9-10°N low chloride fluids have low ΔD values, it initially appeared that phase separation could not produce the observed values. However, Berndt et al. (1996) have shown that during isobaric heating of hydrothermal fluids, as in the case of a near-surface dike injection event, open-system loss of the vapor phase could lead to brines that become progressively isotopically lighter by a Rayleigh distillation effect. Of course, the δD values of the vapors track the brines, giving low δD, low-chloride fluids. It appears from the field evidence that the vapors are more mobile and tend to segregate from the brines and dominate vent fluids for some period of time following eruption or dike injection events. The conclusion that low δD fluids may be a result of open-system phase-separation is supported in the case of the 9-10°N fluids by Br/Cl systematics (Oosting and Von Damm 1996; Berndt and Seyfried 1997) which show that the fluids may have boiled until the liquid-vapor-solid curve was intersected and NaCl may have formed. The highest Cl values at 9-10°N are about 850 mmol/kg, suggesting that about one-third of the vent fluid has been lost as vapor.

Direct addition of magmatic water to the 9-10°N vent fluids seems unlikely because both the Br/Cl ratios and ΔD values point to open system phase separation. In addition, MORBs tend to be quite dry, with only about 0.2 wt % water (Dixon et al. 1988). Studies of Hawaiian basalt glasses from the submerged portion of the East Rift have shown that seafloor basalts never reach saturation with respect to water at eruption depths of more than a few hundred meters (Dixon et al. 1991). That is, seafloor basaltic lavas do not exsolve water at depths greater than a few hundred meters simply by freezing to form glass. However, complete crystallization of a mid-ocean ridge basalt dike containing only anhydrous phases could release substantial amounts of magmatic water. For example, a dike 20 km long, 1.5 km thick, and 1 m wide could produce 2×10^{11} g of water, assuming complete and anhydrous crystallization. If this magmatic water is mixed to produce a vent fluid with a constant 3 wt % of magmatic water with a δD of -70‰, this would result in the observed 9-10°N vents with ΔD of about -2‰. At a steady rate of mixing, and assuming a total flux of hydrothermal vent fluid of about 7 kg/sec from the Q, G, P, Bio9, and A vents, all of the magmatic water from the crystallizing dike would be exhausted in about three years. Thus, magmatic water contributions to MORB hydrothermal systems cannot be ruled out at this point, but they seem unlikely at 9-10°N.

Several of the vent fluids from the Southern EPR (Table 1) have negative ΔD values, and these values are restricted to vent fields at 17°24-27'S and 21°24-27'S. Considering what we have learned at 9-10°N, it seems likely that these areas of the SEPR have experienced recent eruptive events that produce vent waters with transient negative ΔD. Indeed these vent fields contain some of the lowest Cl vent fluids (especially at 21°24-27'S) and low SiO_2 concentrations (especially at 17°24-27'S), indicating shallow hydrothermal systems and recent phase separation in the shallow subsurface. In general, the SEPR vent fields (Fig. 7) have high $\delta^{34}S$ and low ΔD and $\Delta^{18}O$ values. These stable isotope systematics are all consistent with shallow reaction zones, short flow paths, and relatively high water-rock ratios.

Deep reaction zones: 504B

The deep reaction zone in the ocean crust, like that which underlies MOR hydrothermal vent systems, is best understood as a result of drilling in hole 504B of the DSDP/ODP and by experimental studies of high-temperature seawater-diabase interaction processes. Hole 504B is located in the eastern equatorial Pacific about 200 km south of the Costa Rica Rift in 6 Ma ocean crust (Fig. 2) and it provides an excellent record of past hydrothermal processes in the crust. The hole was first drilled in 1979 during DSDP (Deep Sea Drilling Project) Leg 69 and has been revisited and deepened during DSDP/ODP legs 70, 83, 111, 137, 140, and 148. The hole now penetrates 2111 mbsf (meters below sea floor) through 275 m of pelagic sediments, 572 m of pillows, flows, and volcanic breccias, a 209-m transition zone of pillow basalts and dikes, and 1056 m of sheeted dikes. The upper volcanic section has been altered oxidatively by relative cool seawater, presumably by near seafloor weathering processes over time. In the lithologic transition zone, characterized by the common appearance of dikes injected into pillow basalts, hydrothermal alteration is prevalent and is mostly due to seawater metasomatism along the steep, ridge-related, geothermal gradient that existed during formation of these rocks. In contrast, a mineralized stockwork or sulfide stringer zone containing abundant sulfide and quartz veins in fractured pillows occurs at 910-928 mbsf. This stockwork zone formed near the seafloor as a subjacent "feeder" zone beneath a seafloor vent system. The sheeted dike complex preserves the results of an intensifying hydrothermal system with increasing alteration temperature with depth, but the heterogeneity of the alteration process is impressive, especially in the upper dikes. Current downhole temperatures are well below the 350-400°C temperatures that occurred at the time of ridge-related sub-seafloor alteration. Alteration mineralogy is complex and highly variable and a detailed description is beyond the scope of this paper. The reader is referred to Alt et al. (1996) and references therein for a detailed summary. In general, the transition zone and upper and lower dikes are characterized by abundant secondary chlorite, pyrite, anhydrite, calcite, quartz, epidote, Ca-zeolite, prehnite, albite, titanite, magnetite, actinolite, hornblende, anorthite, and clinopyroxene. Calcite occurs mainly as veins in the volcanic section and transition zone. Epidote, zeolite, and prehnite are largely restricted to the transition zone and upper dikes. Actinolite and secondary hornblende, anorthite, and clinopyroxene occur as hydrothermal alteration minerals mainly in the lower dikes. Anhydrite is present sporadically thoughout the section.

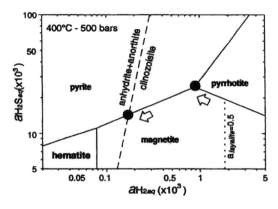

Figure 17. Phase diagram showing mineral reactions that control oxidation-reduction (a_{H_2}) and total sulfur (a_{H_2S}) conditions during seawater-basalt interaction in the deep subseafloor reaction zone. From Seyfried and Ding (1995); used with permission of the American Geophysical Union.

Berndt et al. (1989), Seyfried et al. (1991), and Berndt and Seyfried (1993) have shown that MOR vent fluid compositions are controlled by reactions with calcic-plagioclase solid solution plus epidote, quartz, K-feldspar, and chlorite at approximately 400°C, 400-500 bars. These controlling reactions are similar in many ways to the phase equilibria indicated for the reaction zone in the lower dikes in hole 504B. Some minerals, notably actinolite, are not available in the thermodynamic databases and therefore were not included in the phase equilibrium treatment of Seyfried et al. (1991). Redox conditions are controlled by pyrite-pyrrhotite-magnetite or pyrite-magnetite-anhydrite as shown in Figure 17 (Seyfried and Ding 1995). Phase relations dictate that as long as pyrrhotite (as a proxy for igneous monosulfide solid solution) is present and reactive, a_{H_2S} and a_{H_2} is fixed at the pyrite-pyrrhotite-magnetite join (Fig. 17). However, Berndt and Seyfried (1993) have argued that the high calcium content of seafloor vent fluids requires the presence of relatively high Ca-anorthite observed in deep dikes in hole 504B. As the reaction zone progresses to higher w/r ratio, sulfate-bearing seawater reaction with primary basalts would lead to conversion of pyrrhotite to pyrite and destruction of the fayalite component of olivine. Once this occurs, equilibrium of anorthite and anhydrite with epidote (or clinozoisite) controls a_{H_2S} and a_{H_2} at the anorthite- anhydrite-epidote-pyrite-magnetite intersection (Fig. 17). The ubiquitous presence of magnetite and pyrite in the lower dike section of hole 504B and the sporadic distribution of anhydrite suggest that redox conditions in this reaction zone may fluctuate between the two control points discussed above.

Stable isotope ($\delta^{18}O$, δD, $\delta^{34}S$, $\delta^{11}B$, and $\delta^{13}C$) and $^{87}Sr/^{86}Sr$ data for hole 504B are summarized in Figure 18 from Alt et al. (1996). The important thing to take from this comprehensive data set is that the sheeted dike complex, especially the lower dikes, is representative of the deep reaction zone beneath a mid-ocean ridge hydrothermal vent system. Whole rock $\delta^{18}O$ values decrease with depth in the sheeted dikes to values of

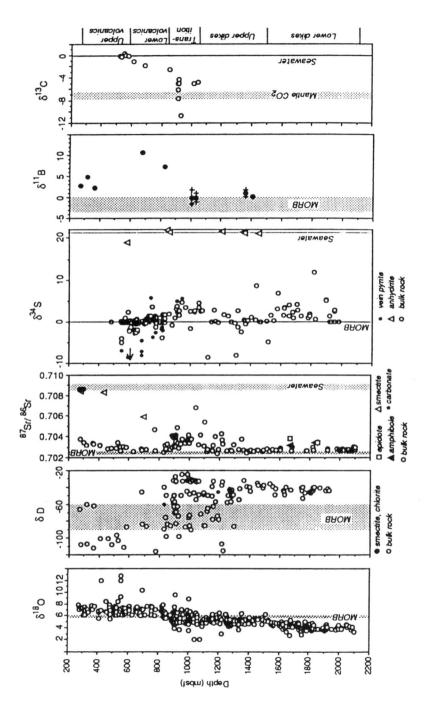

Figure 18. Variations of stable isotopes and $^{87}Sr/^{86}Sr$ in ODP hole 504B, which penetrated nearly 2 km of 6 Ma volcanic and intrusive ocean crust at an off-axis site 200 km south of the Costa Rica rift (see Fig. 2 for location). After Alt et al. (1996); used with permission of the Ocean Drilling Program.

about 4‰ and δD values of hydrous minerals stabilize at about -40‰. These $\delta^{18}O$ and δD systematics are consistent with evolved (~1 to 2‰) seawater alteration at temperatures of ~350 to ≥400°C (Alt et al. 1996) in the lower dikes. Cu, Zn, and S are significantly depleted in the lower dikes and $\delta^{34}S$, $\delta^{11}B$, and $^{87}Sr/^{86}Sr$ values are consistent with seafloor vent systems. Sulfur contents (mostly sulfide) are variably depleted from the original ~1000 mg/kg igneous sulfide contents to values nearly 100x lower in some samples and $\delta^{34}S$ values of sulfides range from -0.5 to 3‰ (higher $\delta^{34}S$ values for some sheeted dike samples shown in Fig. 18 are whole rock $\delta^{34}S$ values and may include a significant sulfate component). Thus, the $\delta^{34}S$ values of sulfides (mostly pyrite) from the lower dikes are significantly higher than igneous values and are significantly lower than sulfides in MOR hydrothermal vents (Alt 1995), which range from ~1-6‰ (Fig. 11). These observations prove that (1) seawater sulfate penetrated and was reduced in this reaction zone which formed at least 1.7 km beneath the seafloor and (2) $\delta^{34}S$ values of seafloor vent-fluid H_2S are increased due to additional sulfate reduction on ascent to the seafloor, probably in the chimney environment or in immediately subjacent alteration/feeder zones as suggested by Janecky and Shanks (1988). Anhydrite in Hole 504B is a common minor mineral in late veins and vugs throughout the transition zone and upper sheeted dike complex; it has $\delta^{34}S$ values close to normal seawater (21‰). In the lower dikes anhydrite is more intimately dispersed in the rocks and samples with $\delta^{34}S$ values as low as 11‰ have been observed (Alt et al. 1996), indicating oxidation of sulfide derived from igneous or secondary sulfide minerals. This indicates moderately oxidizing conditions consistent with Ca-plagioclase-epidote equilibrium as suggested by Seyfried et al. (1991).

Oceanic gabbros: Hess Deep and Southwest Indian Ridge

Ocean gabbros and other intrusive rocks may be important components of the deep reaction zone for MOR hydrothermal fluids. Alt and Teagle (in press) have summarized the stable isotope systematics for oceanic gabbros, which are largely known by ODP drilling in Hess Deep, the Indian Ocean, the Mid-Atlantic Ridge, and from studies of dredged rocks from the seafloor. $\delta^{18}O$ values of gabbroic rocks range from 2.2 to 7.4‰ and $\delta^{34}S$ values range from -16.6 to 6.9‰, but most $\delta^{34}S$ values cluster about normal magmatic values of $0 \pm 2‰$. Agrinier et al. (1995) have shown that most gabbros from Hess Deep have δD values higher than -57‰ and $\delta^{18}O$ values close to magmatic values (~5.6‰). The relatively high δD values suggest some reaction with seawater, even in samples that appear unaltered. Albitized and prehnite-altered rocks have more variable $\delta^{18}O$ values, ranging from 4 to 8‰. Fruh-Green et al. (1996) found that mineral separates from hydrothermally altered gabbros and troctolites from Hess deep have $\delta^{18}O$ values ranging from 1.1 to 6.3 ‰, with chlorite and tremolite having the lowest values due to high temperature reaction with seawater. δD values of chlorite and amphiboles range from -41 to -27‰. In general, these values are consistent with seawater alteration effects at temperatures from about 350-600°C in rock-dominated reaction systems (low integrated w/r ratio).

Shallow alteration zones and seawater entrainment: TAG

Slow-spreading ridges differ fundamentally from fast-spreading ridges (Fig. 3) in terms of volume and periodicity of magma supply, ridge topography and tectonism, and shallow structural emplacement of deep crustal gabbros and ultramafic rocks which may be significantly involved in active hydrothermal systems. The Mid-Atlantic Ridge is the best-studied slow-spreading ridge, and it displays an impressive array of variability in vent fluid and sulfide deposit types. Important high-temperature hydrothermal vents related to seawater-basalt reactions on the MAR occur at the TAG, MARK, Broken Spur, and Lucky Strike sites. In contrast to small bare-rock MOR systems, the TAG site, at

26°8′N on the MAR, is a key example of a modern seafloor hydrothermal site where ongoing shallow seawater entrainment, sub-seafloor hydrothermal mineral precipitation, and hydrothermal rock alteration are critical to deposit formation and vent fluid evolution. It is one of the largest, longest lived, most active, and best studied ridge-related hydrothermal systems. The TAG mound is about 200 m in diameter and about 50 m-high and is comprised of hydrothermal breccias that have formed intermittently over the last 50 Ka (Lalou et al. 1998; Humphris et al. 1995). The TAG mound hosts a 360°C black smoker field, a peripheral 260-300°C white smoker field, and was drilled during ODP Leg 158 (Humphris et al. 1995). Drilling revealed 125 m of heterogenous sulfide, anhydrite, silica, and chloritized- and silicified-basalt breccias in the still-active upflow zone (Fig. 19).

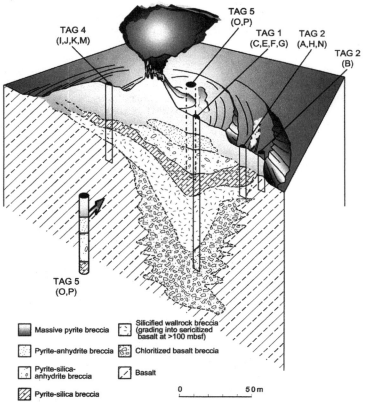

Figure 19. Schematic of subsurface mineralization and alteration associated with hydrothermal upflow beneath the TAG mound, MAR. Cross-section is derived from drilling results of Leg 158 of the Ocean Drilling Project (Humphris et al. 1995).

Stable isotope studies of sulfide and sulfate minerals from TAG have been carried out by Alt and Teagle (1998), Chiba et al. (1998), Friedman (1998), Gemmell and Sharpe (1998), Herzig et al. (1998), Knott et al. (1998), and Teagle et al. (1998a,b). Limited stable isotope data on the black smoker fluids are presented in Table 1. Sulfur isotope studies of minerals in the TAG deposit (Fig. 11) indicate a range from 4.5 to 9‰ for sulfides and 20 to 23‰ for anhydrite. The δ^{34}S of sulfides is somewhat higher than most MOR systems, probably due to reduction of sulfate from the abundant anhydrite in the TAG Mound (Fig. 19). The one analysis of TAG black smoker fluid H_2S shows a $\delta^{34}S_{H_2S}$ value in the middle of the range for precipitated sulfide minerals. However, the black smoker sample analyzed represents the highest temperature, most pristine fluids that vent in the TAG area, and probably represent conditions at considerable depth in the system. Stable isotope studies of white smoker fluids, or fluids from within the mound, if available, would probably reveal strong effects of near-surface seawater entrainment and sulfate-reduction in the near-surface environment. Knott et al. (1998) have presented data that show δ^{34}S increase in sulfide minerals from 5 to 8 ‰ with depth in the TAG Mound. This suggests that sulfate-reduction is most effective near the base of the subjacent alteration zone where sericitized and chloritized basalt breccias are found (Fig. 19). Of course, another interpretation, that cannot be discounted, is simply that the deep reaction zone and fluids derived from it have changed substantially with time and that the sulfides in the deeper portions of the TAG Mound formed at some considerable time in the past when sulfate reduction was more intense in the deep reaction zone, perhaps due to a magmatic or tectonic event.

Alt and Teagle (1998), Chiba et al. (1998), and Teagle et al. (1998a,b) have analyzed ^{87}Sr/^{86}Sr, δ^{18}O, and δ^{34}S in whole rocks, quartz, and anhydrite from the TAG mound. δ^{34}S of anhydrite (Fig 11) ranges from 20 to 23‰, confirming sulfate derivation from admixed seawater in the mound. Precipitation is due mainly to retrograde solubility of anhydrite and $\delta^{34}S_{anhydrite}$ values show only minor effects of sulfate-reduction or sulfide-oxidation processes. Teagle et al. (1998a) using Sr and O isotope systematics have suggested that TAG Mound anhydrite forms by mixing of black smoker fluids with isotopically normal seawater that is conductively heated to 100-180°C before mixing; therefore, anhydrite forms from a heated seawater-vent fluid mixture at temperatures between 230 and 320°C. These predicted temperatures (Fig. 20) are not consistent with homogenization temperatures for fluid inclusions from central mound samples which show consistently higher temperatures in the range from 338-388°C in anhydrite or 318-350°C in quartz (Tivey et al. 1998, Peterson et al. 1998). Alt and Teagle (1998) have shown that quartz separates from the TAG Mound range in δ^{18}O from 7.9 to 16.5, consistent with quartz formation in the central mound at temperatures from 260 to 359°C (mostly >300°C) in equilibrium with hydrothermal fluids with δ^{18}O of 1.7‰ (from Shanks et al. 1995). Sr/O isotope studies of quartz separates suggest precipitation is mainly due to conductive cooling of the black smoker fluids (Teagle et al. 1998b). Thus fluid inclusion temperatures and oxygen isotope temperatures for quartz consistently indicate that conditions in the central mound were in the 300-360°C range when quartz and anhydrite formed (Fig. 20). The ^{87}Sr/^{86}Sr data combined with δ^{18}O data allow identification of mixing versus conductive cooling or heating processes. The Sr and S isotope data prove that anhydrite precipitates during mixing with ambient seawater, but the temperatures of precipitation suggested by Teagle et al. (1998a) are low (Fig. 20), probably because of sulfate oxygen isotope disequilibrium. Rates of SO_4-H_2O oxygen isotope exchange (Chiba and Sakai 1985), extrapolated to 360°C and pH = 5, suggest δ^{18}O equilibration within 50 to 150 hours. The actual time scale of conductive cooling, mixing, and anhydrite precipitation in the TAG mound is not known, but Tivey (1995) has shown that fluids convecting through the TAG mound can be either conductive cooled or heated on time scales of a few 10's of seconds, so SO_4-H_2O oxygen isotope equilibration during rapid precipitation events related to rapid cooling is unlikely.

Stable isotope data for other vent sites on the MAR, including (Fig. 2) Lucky Strike, Menez-Gwen, Rainbow, Broken Spur, and MARK (Snakepit), are incomplete at this time. ΔD and Δ^{18}O data exist for one TAG vent, two MARK (Snakepit) vents, and for

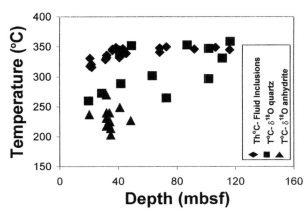

Figure 20. Stable isotope data summary for the TAG mound. Data from Chiba et al. (1998), Petersen et al. (1998), Teagle et al. (1998a,b), and Tivey et al. (1998).

several Lucky Strike vents. Sulfur isotope data exist for sulfide minerals ($\delta^{34}S_{MeS}$) from TAG, Snakepit, and Broken Spur (Fig. 11) and both Snakepit and Broken Spur have low $\delta^{34}S$ values. Butler et al. (1998), Duckworth et al. (1995), and Kase et al. (1990) explain the low $\delta^{34}S_{MeS}$ values as due to a limited supply of seawater sulfate.

Li and B isotopes have been especially helpful in understanding the Mid-Atlantic ridge hydrothermal systems. Palmer and Edmund (1989) suggested that high Cs/Rb ratios in vent fluids at TAG might be the result of high temperature reaction with alkali-enriched weathered basalts. However, Li isotope studies indicated relatively normal Li concentrations and δ^6Li values (Chan et al. 1993), which would be highly unlikely if weathered basalts were involved in the reaction zone (Seyfried et al. 1998). Li and B are both highly soluble elements at 350°C, but both tend to concentrate in alteration phases during low-temperature rock-alteration. For δ^6Li, lower temperature phases such as clay minerals tend to preferentially concentrate ^7Li, giving low δ^6Li values ($^6Li/^7Li$ ratios are used in the δ^6Li calculation, meaning lower values indicate isotopically heavier samples).

Boron isotope studies of MAR vents suggest significant removal of seawater B during low-temperature reactions in the downwelling limb of fluid infiltration (James et al. 1995, Palmer 1996). This differs from EPR vents where $\delta^{11}B$ values are close to seawater values of 39.5‰ and are arrayed along a mixing line with ocean crust basalts that have values of -1.2 to -6.5‰. Because seawater has ~4.6 mg/kg B whereas basalts have <1 mg/kg, $\delta^{11}B$ values of vent fluids are usually dominated by seawater. In the MAR vents (Broken Spur, TAG, and MARK), however, $\delta^{11}B$ values are shifted to lower values due to fractionation during low temperature alteration along the recharge pathway (Spivack and Edmond 1987). Boron is known to partition into alteration phases at low-temperature <150°C and into the hydrothermal fluid at higher temperatures (Seyfried et al. 1984). At Broken Spur, more than 50% of the seawater B has been removed (James et al. 1995). Palmer (1996) has suggested that low-temperature, subsurface serpentinization might be important in the B removal process.

Serpentinization

Recent studies of serpentinization and vent fluids related to serpentinization are producing some of the most exciting new geochemical discoveries on seafloor-hydrothermal systems. Ultramafic-rock-hosted sites have recently been investigated at the

Rainbow and Logatchev vent fields. These are very high in both hydrogen (H_2) and methane (CH_4), but neither the chemical nor isotopic compositions of these vent fluids or associated deposits are available at this time. In addition, very recent (December 2000) submersible investigations at an off-axis Mid-Atlantic Ridge site near 30°N discovered spectacular carbonate- and silica-rich chimneys up to 60 m-tall by ≥10 m-diameter that apparently have grown from serpentine-related low-temperature (~70°C) vents (www.nsf.gov/od/lpa/news/press/00/pr0093.htm). Finally, Conical Seamount, a serpentine mud volcano near the Mariana Arc that has small carbonate and Mg-silicate chimneys related to cold vent fluids, was investigated by Alvin dives in 1987, and drilled on leg 125 of the ODP.

Serpentinized ultramafic rocks in the ocean crust have been drilled in the MARK area, near the Iberian Margin, and in Hess Deep west of the Galapagos Rift. Each of these sites is different and studies of $\delta^{34}S$ of sulfides and other stable isotopes in these rocks provide important information on serpentinization and subsequent retrograde hydrothermal alteration processes. Oxygen and hydrogen isotope studies have been carried out on seafloor serpentinites from Hess Deep and from the MARK site (Agrinier et al. 1995; Fruh-Green et al. 1996). Serpentinized peridotites from Hess Deep have δD from -47 to -83‰ and $\delta^{18}O$ from 3.3 to 4.9‰ (Agrinier et al. 1995). The oxygen isotope data are consistent with high-temperature (~300-400°C) seawater reactions and the highly variable hydrogen isotope data suggest a range of water-rock δD fractionations from about -40 to -80‰. Serpentine mineral separates analyzed by Fruh-Green et al. (1996) give a similar $\delta^{18}O$ range from 3.2 to 5.4‰ and a tighter δD range from -51 to -69‰, consistent with a fairly uniform water rock interaction process. Similarly, Werner and Pilot (1997) found that whole rock $\delta^{18}O$ ranges from 2.9 to 4.6‰ for serpentinites from the MARK area.

Subsurface hydrothermal alteration in serpentinized ultramafic rocks south of the Kane Fracture Zone (MARK) was investigated in Hole 920 of Leg 153 of the Ocean Drilling Program. The location and geologic setting of this site are shown in Figure 21. Serpentinites south of the Kane Fracture Zone occur in an unroofed block of mantle material on the western margin of axial valley of the MAR. The neovolcanic ridge in the central axial valley and an area of gabbro exposures along the Kane Transform intersection with the axial valley are shown for reference (Fig. 21). Approximately 200 m of serpentinite was drilled in hole 920. Most of the original ultramafic rock was harzburgite, with minor amounts of dunite and lherzolite. Small gabbroic dikes and veins intrude the serpentinite in various places. Peridotites are mostly intensely serpentinized to lizardite, chrysotile, magnetite, and brucite. Phase relations and oxygen isotope geothermometry in the serpentinites suggest alteration at about 300-400°C (Agrinier and Cannat 1997).

Alt and Shanks (in prep) have investigated stable isotope systematics and serpentization processes in these rocks. There are a few late carbonate veins that have $\delta^{13}C$ from 1.1-4.5‰ and $\delta^{18}O$ from 7.9-34.1‰. These values indicate calcite formation at temperatures from 10 to 240°C with some methane formation at the higher temperatures. Secondary sulfide and oxide minerals include magnetite, pentlandite ([Fe,Ni]$_9$S$_8$), pyrrhotite, millerite (NiS), and rare chalcopyrite. Sulfur isotope data for these serpentinites (Fig. 22) indicate a range for sulfide minerals from about 3-13‰. Sulfide is enriched in these serpentinites with sulfide S contents ranging from 25 to 9524 mg/kg, but most samples are in the 500 to 1000 mg/kg range. The samples with sulfide sulfur contents >1200 mg/kg contain discrete sulfide veins and therefore the sulfide content analyses may be unrepresentative of the bulk rock sulfide content. For comparison, depleted mantle peridotites have about 100-150 mg/kg sulfide S. Sulfates in these rocks have $\delta^{34}S$ values that are intermediate between the sulfides and seawater sulfate and are

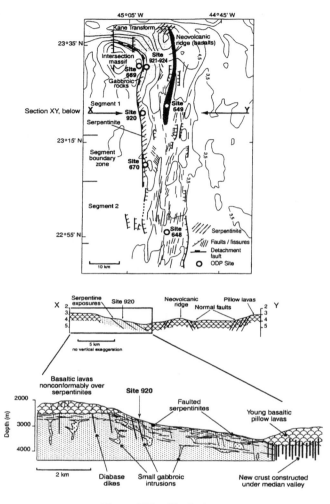

Kane F.Z.   Mark Area

Figure 21. MARK serpentinite location map, showing hole 920 in unroofed serpentinite on the western wall of the MAR. From Lagabrielle et al. (1998); used with permission of the American Geophysical Union.

probably due to late sulfide oxidation.

Alt and Shanks (in prep) have evaluated possible pathways of rock alteration by seawater using incremental reaction modeling. Sulfur isotope systematics are complicated in this system and require a multi-step model because of the high sulfide content of the rocks. A possible conceptual sequence involves serpentinization of harzburgite (90% olivine, 10% enstatite, 250 mg/kg pyrrhotite) by seawater at 400°C (in deeper zones, or zones more proximal to ridge-related gabbro intrusives) followed by cooling to 300°C and additional serpentinization. Results of these computations (Fig. 23) agree reasonably well with the observed δ^{34}S-S$_{sulfide}$ relations. As in the case of MOR systems, most sulfate

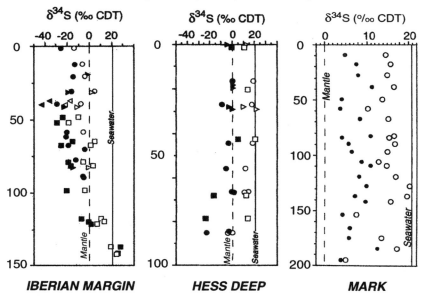

Figure 22. Sulfur isotope data for sulfides (closed symbols) and sulfates (open symbols) form Iberian Margin, Hess Deep, and MARK serpentinites. Data from Alt and Shanks (1998 and in preparation).

is removed from the seawater during heating to 400°C; only 150 mg/kg sulfate enters the high temperature reaction zone, and this is quantitatively reduced to sulfide give high $\delta^{34}S$ values during initial stages of reaction (Fig. 23). Continued reaction at 400°C produces H_2S that gradually approaches $\delta^{34}S$ values of about 4‰ due to increased sulfide contribution from the 0‰ pyrrhotite in the rock, but no sulfides are precipitated at this high temperature. Additional reaction of the fluids and minerals from the 400°C reaction with fresh harzburgite at 300°C produces abundant pyrrhotite in the serpentinite. The 300°C calculations were carried out in two ways to simulate the range of observed $\delta^{34}S$ values. In the first case, reactants simply include pyrrhotite-bearing harzburgite and this produces secondary pyrrhotite with $\delta^{34}S$ values from 4.5 to 2.2‰ and a pyrrhotite content of about 1000 mg/kg. In the secondary case, an additional 250 mg/kg anhydrite with $\delta^{34}S$ of 21‰ is added to the reactants, as might be the case where anhydrite remains from an original prograde heating event. This model (Fig. 23) produces about 3000 mg/kg of secondary pyrrhotite with $\delta^{34}S$ values of 11-12‰. Thus, these reaction models provide a reasonable mechanism for formation of the sulfides and sulfur isotope systematics observed in the MARK serpentinites. An alternative conceptual model, which we do not evaluate here, is the possible participation of gabbroic rocks in the hydrothermal alteration processes. Gabbros have significantly higher original sulfide than peridotites and hydrothermal transport of gabbroic sulfide into the serpentinites may explain some of the extremely high sulfide contents.

Alt and Shanks (1998) have analyzed sulfide and oxide mineralogy and sulfur isotope systematics in serpentinites from Hess Deep in the eastern Pacific and from the Iberian Margin. These data are included in Figure 22 for comparison to the MARK serpentinite data. Most of the Hess Deep data look like typical ocean crust data; $\delta^{34}S$ within a few ‰ of the mantle value of zero. However, at depth in hole 895 there are sulfides with strongly negative $\delta^{34}S$ values. Similarly, in the Iberian Margin drilling most

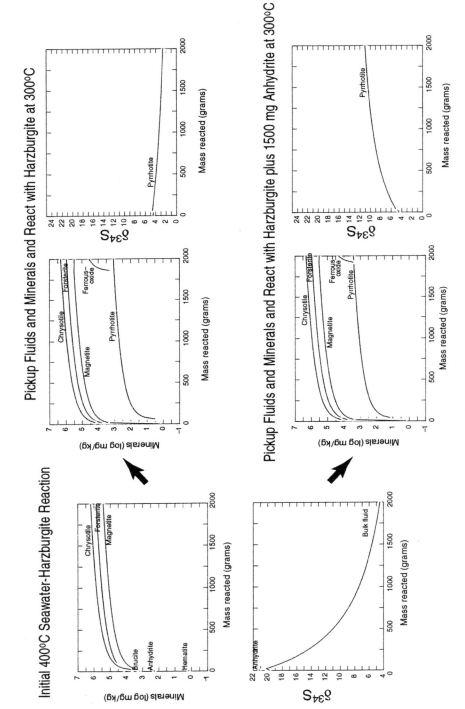

Figure 23. Incremental reaction model of seawater-harzburgite interaction at the MARK serpentinite drilled in ODP hole 920. This model includes high temperature serpentinization at 400°C, followed by additional harzburgite reaction as the fluids cool to 300°C. The model closely duplicates sulfur isotope values and sulfide sulfur contents of MARK serpentinites.

sulfides have $\delta^{34}S$ values between -10 and -30‰. These strongly negative values were quite a surprise in rocks derived from the mantle. Alt and Shanks (1998) interpreted these negative isotope values as due to sulfate reduction by thermophilic or hyperthermophilic micro-organisms that survive chemo-synthetically on the abundant H_2 or CH_4 that is generated during serpentinization. Recently, Canfield et al. (2000) have demonstrated in the laboratory that thermophilic micro-organisms from Guaymas Basin sediments produce sulfur isotope fractionation during sulfate reduction of 13 to 28‰ at 85°C. Fractionations of this sort, coupled with Rayleigh distillation processes could explain the $\delta^{34}S$ data in the Iberian Margin and Hess serpentinites.

Sedimented ridges.

Escanaba Trough on the southern Gorda Ridge and Middle Valley at the northern end of the Juan de Fuca Ridge (Fig. 2) represent the best studied spreading ridges that are sediment-covered because of proximity of continental detrital sources. These two sites are similar in geologic setting, but the hydrothermal systems at Middle Valley have had a much longer history and are much more extensively developed. Active ridge extension and sedment infilling by turbidites were contemporaneous. ODP drilling in legs 139 and 169, augmented by a significant number of Alvin and ROV dives in these areas, has provided excellent opportunities to understand subsurface fluid flow and long term growth, recrystallization, alteration, and surface oxidation of modern sediment-hosted seafloor massive sulfide deposits.

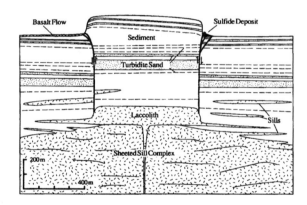

Figure 24. Schematic cross-section of uplifted sediment hill and underlying laccolith at Escanaba Trough, southern Gorda Ridge. From Morton et al. (1994).

The Escanaba Trough hydrothermal systems are centered around one of many topographic highs on the southern Gorda Ridge that result from basaltic sill emplacement into a thick sequence of clastic cover on the spreading ridge. Figure 24 is a schematic diagram that shows laccolith emplacement in the subsurface that has caused uplift of a sediment hill with associated circumferential faults and sulfide deposits (Morton et al. 1994). The sedimentary cover consists of turbidites that are carbonate-free and contain little organic matter or pyrite, except where affected by hydrothermal alteration. Present-day hydrothermal venting is confined to circumferential fault zones around the sediment-covered sills and reaches a maximum observed temperature of 217°C. Fluid chemistries are very similar to MOR vent fluids, but with enrichments of Ba, NO_3, and I due to reaction with sediments (Campbell et al. 1994). Sulfide deposits are small and often contain assemblages of pyrrhotite, anhydrite, and barite, indicating a variable mixing environment at the seafloor. In contrast to MOR deposits, the sulfides contain significant amounts of galena with sediment-derived Pb (LeHuray et al. 1988). Sulfur isotope values of sulfide minerals range from 0 to 12‰.

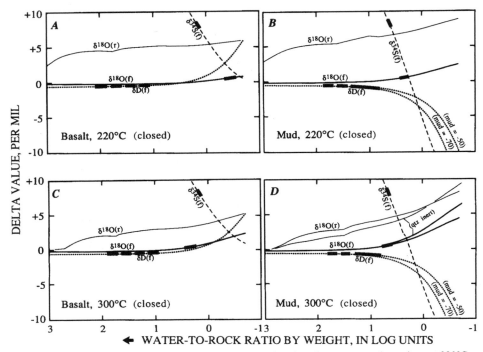

Figure 25. Geochemical reaction models for seawater-basalt and seawater-mud reactions at 220°C and 300°C. From Böhlke and Shanks (1994).

Böhlke and Shanks (1994) have used incremental reaction models to show that the high $\delta^{34}S$ values and large range of values are probably due to sulfate reduction reactions at relatively high w/r ratios with basaltic sills or laccoliths and/or with organic matter in the sediments (Fig. 25). Seawater/mud ratios of about 3-5 (mass basis) and seawater/basalt ratios of about 0.3-2 would produce the observed range of $\delta^{34}S_{MeS}$ values. In comparison, Campbell et al. (1994) argued for relatively high w/r (~5) in the Escanaba hydrothermal system based on concentrations of some soluble species in the vent fluids.

Böhlke and Shanks (1994) also showed that hydrothermal seawater reaction with detrital sediments will produce fluids with strongly negative δD values if reaction proceeds to w/r < 5 (Fig. 25, mud reactions). This occurs because the hydrothermal evolved-seawater fluids are reacting with detrital hydroxyl-bearing clays that formed at low temperatures and have strongly negative δD values (-50 and -70 were used in this model, Fig. 25). The vent fluids at Escanaba Trough have δD values close to 0‰ (Böhlke and Shanks 1994) and vent fluids with strongly negative δD (<-5‰) have not (yet) been observed in sedimented systems (Table 1). However, δD and $\delta^{18}O$ values of pore fluids in hydrothermally altered sediments from Escanaba (Fig. 26) show isotope shifts similar to those predicted by Böhlke and Shanks (1994). Starting from values near local bottom water (δD = -0.8; $\delta^{18}O$ = -0.1) the Escanaba pore waters evolve to δD as low as -10.4 and $\delta^{18}O$ of +3.2, which suggests w/r ratios of less than 1 (Fig. 25).

Figure 27 shows the related shifts in silicates from ODP drilling in the Central Hill

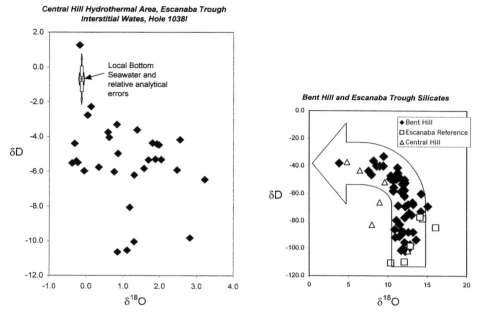

Figure 26 (left). Pore-water stable isotope data from ODP hole 1038I in the Central Hill Vent Field, Escanaba Trough, Gorda Ridge.

Figure 27 (right). Stable isotope data for whole rock samples of altered and unaltered sediments from Escanaba Trough and Middle Valley.

Vent Field (Escanaba; hole 1038), the Escanaba Trough Reference Hole (1037), which was drilled approximately 4.5 km from the nearest known hydrothermal activity, and from the Bent Hill Massive Sulfide system (holes 856 and 1035) in Middle Valley. These data show dramatic shifts in δD and $\delta^{18}O$ from -100 to -40 and from 15 to 5, respectively. Using the equations of Taylor (1974), relatively high w/r mass ratios of about 3-7 are indicated, as expected for the significant change in the δD and $\delta^{18}O$ of the muds. This w/r estimate is higher than indicated in Figure 25 because Böhlke and Shanks (1994) assumed a δD for the starting muds of -50 to -70‰ (assuming detritus from continental weathering reactions), whereas the actual values are about -110‰ (Fig. 27). At Escanaba Trough, the difference in $\delta^{18}O$ between the final altered mud ($\delta^{18}O \sim 8$) and the final evolved pore water ($\delta^{18}O \sim 3$) suggests alteration temperatures of about 300°C (Fig. 25) for isotopic equilibrium with a predicted assemblage of chlorite, quartz, albite, muscovite, and minnesotaite (Böhlke and Shanks 1994). Unfortunately, the isotope compositions of neither vent fluids nor high-temperature pore waters are available at Middle Valley.

Drilling has delineated two major massive sulfide lenses in the Bent Hill area (Fig. 28). The Bent Hill Massive Sulfide (BHMS) deposit (Zierenberg et al. 1998) consists (from top to bottom) of 93 m of massive sulfide, 100 m of sulfide "feeder" or "stringer" veins in highly altered sediments, a stratiform deep copper zone of chalcopyrite and pyrrhotite in a totally altered sandstone, and basaltic lava flows and sills at about 500 mbsf. An additional sulfide deposit, or possibly an extension of the BHMS deposit, occurs 400 m to the south at the ODP mound (Fig. 28). The ODP mound has an active 264°C hydrothermal vent on top (Butterfield et al. 1994a), three separate massive

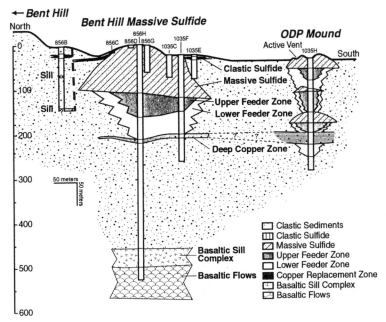

Figure 28. Schematic cross-section of the Bent Hill Massive Sulfide and the ODP
Mound, Middle Valley. Modified from Fouquet et al. (1998), ODP Sci Results
169; used with permission of the Ocean Drilling Project.

sulfide/stringer zone lenses, and a thickened deep copper zone at depth.

Sulfur isotope studies (Fig. 29) show a range of $\delta^{34}S$ for sulfides from about 1.8 to
8.9‰. These values are similar to, but somewhat higher than the range of $\delta^{34}S$ values for
most seafloor hydrothermal deposits. Sulfide values in the deeper portions of BHMS
show a narrower range but average the same, about 5‰. Once again, it is likely that
detrital sediments with ≥0.3 wt % organic carbon are important in stimulating increased
thermochemical sulfate reduction.

Rifted continental settings

Known hydrothermal systems in rifted continental settings include the Red Sea brine
pools and Guaymas Basin, in the central Gulf of California. Both the Red Sea and the
Gulf of California have evolved to the stage where ocean crust can be found in central rift
zones (McKenzie et al. 1970; Larson et al. 1968).

In the Red Sea, the central rift zone is flanked by broad shallow shelves that are
underlain by thin Plio-Pleistocene hemipelagic sedimentary sequences overlying thick
sequences of Upper Miocene evaporite deposits. Most of the evaporite deposits are halite,
anhydrite, and sapropels that formed in deep water, with interlayers of late-stage K and
Mg minerals indicating dessication of the Red Sea basin during certain intervals. The
combination of a narrow central rift zone depression, volcanic heat to drive fluid
convection, and thick flanking halite-bearing evaporites makes the Red Sea unique in the
modern oceans. During subsurface circulation, Red Sea seawater attains high salinity,
about 8 times normal seawater, by leaching evaporites, is heated by rift zone volcanics,
and emerges as negatively buoyant hot fluids that pond in depressions in the central rift.
There are at least 12 pools with supersaline brines in the central Red Sea and there is

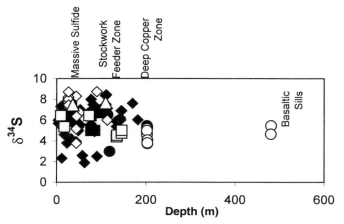

Figure 29. Sulfur isotope data composite for sulfide minerals from Bent Hill Massive Sulfide deposit. Pyrite (diamonds), pyrrhotite (squares), sphalerite/wurtzite (triangles), and chalcopyrite/isocubanite (circles).

some indication that brine pool location may be related to transform faulting (Bäcker and Schoell 1972). By far the most important of these pools is the Atlantis II Deep (Fig. 30), which contains a 200 m-thick layer of about 5 km³ of brine with temperatures that have

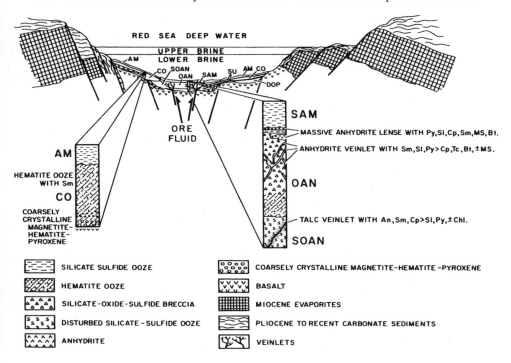

Figure 30. Schematic cross-section of the Atlantis II Deep. From Zierenberg and Shanks (1983); used with permission of the Economic Geology Publishing Company.

ranged from 56°C in 1966 to 67°C in 1992. On the floor of the brine pool are 10-30 m-thick recent deposits of sulfide-, oxide-, and silicate-rich metalliferous sediments covering an area of about 72 km^2.

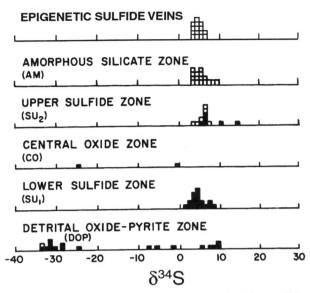

Figure 31. Sulfur isotope distribution in Atlantis II Deep sulfide minerals. Data from Kaplan et al. (1969), Shanks and Bischoff (1980), and Zierenberg and Shanks (1988).

The Atlantis II Deep is unique and important among the known seafloor hydrothermal sites. The Red Sea deeps represent the only active seafloor hydrothermal systems that consist of brines saline enough (up to ~8× seawater) to form negatively buoyant pools in depressions on the seafloor. In the Atlantis II Deep, this provides a stable, extensive environment for deposition of sheetlike layers of metalliferous sediments over a large area compared to black smoker systems (Shanks and Bischoff 1980). The setting of the Atlantis II Deep is summarized in a schematic E-W cross-section (Fig. 30) from Zierenberg and Shanks (1983). The general stratigraphy of the metalliferous muds (after Bäcker and Richter 1973) consists of an upper amorphous silicate zone (AM), upper (SU2) and lower (SU1) sulfide zones separated by the central oxide zone (CO), and at the base of the sequence immediately overlying basalts, the detrital oxide-pyrite zone (DOP). Sulfur isotope data are summarized in Figure 31 (Kaplan et al. 1969; Shanks and Bischoff 1980; Zierenberg and Shanks 1988). In general, the sulfide-rich horizons have $\delta^{34}S$ values that range from about 1 to 10 and average about 5.4. Even though the Atlantis II brine pool is currently sterile, pyrite with very negative, probably bacterial, $\delta^{34}S$ values occurs in the CO and DOP zones, suggesting that the brine pool did not exist or existed intermittently during the time periods when those units were deposited. Shanks and Bischoff (1980) suggested that the hydrothermal sulfide in the Red Sea system formed by thermochemical sulfate-reduction with a non-equilibrium kinetic isotope effect to explain the approximately 15‰ fractionation from seawater sulfate.

In retrospect, the Red Sea system is probably not that much different than mid-ocean ridge systems, with basalt-driven sulfate reduction occurring at 350-400°C in a deep reaction zone. However, it was not until 1971 when Brewer et al. (1971) measured an increase in temperature of the lower brine in the Atlantis II Deep that possible higher subsurface brine temperatures were recognized. Brewer et al. (1971) calculated an input temperature of 110°C. Shanks and Bischoff (1977) used chemical and isotopic geother-mometers to estimate a subsurface temperature of 250°C for brine input and Pottorf and Barnes (1983) used sulfide phase equilibria to suggest temperatures in excess of 334°C. It is now known that inorganic sulfate reduction is rapid at temperatures above 250°C (Shanks et al. 1981), and it is likely that, other than being highly saline, the brine supply into the Atlantis II Deep is quite hot and chemically similar to black smoker fluids (no Mg, low SO_4, low pH, high H_2S and metals). Recent investigators have continued to track temperature increases in the Atlantis II brine pool. Temperatures increased by 0.62°C/yr from 1966-72 and by about 0.30°C/yr from 1972-92 (Blanc and Anschutz 1995). These increases require high temperature input of new brine because brine volume has not changed dramatically. Winckler et al. (2000) have provided rare-gas evidence that argues for boiling of the Red Sea brines on ascent to the brine pool. The boiling point of the ~260g/kg brine is estimated to be 390°C, which agrees with fluid inclusion filling T measurements on epigenetic anhydrite by Ramboz et al. (1988).

The channelways for new brine ingress in the Atlantis II Deep are a series of epigenetic veins that cut the gelatinous metalliferous muds (Zierenberg and Shanks 1983 1988). The veins are lined with and cemented by anhydrite, talc, smectites, and a variety of sulfides (Fig. 30). Sulfur isotope values of vein sulfides (Fig. 31) are similar to those in the sulfide zones, but have a narrower range from about 4 to 6‰, very similar to MOR systems. Considering the new, high temperature estimates and the narrow range of $\delta^{34}S$ values in the epigenetic veins, it is very likely that the generalized sulfur isotope model for mid-ocean ridges (Fig. 7) applies with some modification to the Red Sea brines.

As mentioned above, Craig (1966) showed that the Atlantis II brine with $\delta D = 7.4$ and $\delta^{18}O = 1.2$ is isotopically very similar to Red Sea deep water with $\delta D = 11.5$ and $\delta^{18}O = 1.9$. The Red Sea is a closed basin and evaporation accounts for the higher than normal water isotope values. Craig (1966) suggested that Atlantis II brines originate in the south where Red Sea waters are not as evaporated, but this model requires a very complicated subsurface transport path, with isotope-modifying water-rock interaction precluded. More likely, as documented by Schoell and Faber (1978), the Atlantis II brine is supplied by paleowaters that were less saline during the post-glacial climatic optimum about 5,000 years ago.

The Gulf of California represents a transitional tectonic setting, with a volcanic arc built on continental crust on the oceanward side of the rifted margin, and a geometry that involves long segments of oblique transform faults offset by short rift segments that host basaltic material (Bischoff and Henyey 1974; Saunders et al. 1982). The open geometry of the Gulf of California precludes widespread, thick evaporites; evaporite deposits are limited to coastal sabkas in the northern Gulf. There is also a tremendous supply of clastic material into the northern Gulf by the Colorado River and robust sedimentation of organic-rich diatom and foraminifera pelagic ooze in the central Gulf, where the Guaymas Basin hydrothermal system is located. This results in a unique situation where crustal accretion is by basaltic sill injections at the appropriate level of neutral buoyancy in the sedimentary section (Einsele et al. 1980; Ryan 1994). Sill injection has caused substantial dewatering of adjacent sediments and seawater circulation hydrothermally alters basaltic sills and surrounding sediments (Kastner 1982). The resulting hydrothermal fluids that vent in Guaymas Basin have reacted with both basalts and sediments, resulting in vent fluids rich in organic gases, liquid hydrocarbons, and CO_2.

Chimney structures at Guaymas Basin are mainly calcite with lesser sulfides and form ornate "pagoda" structures consisting of multiple tapered flanges arrayed on a central stalk (Lonsdale and Becker 1985). Hydrothermal vents up to 315°C have been sampled. Cooling and pH increase is ascribed to reaction with sediments upon ascent after vent fluids leave basaltic basement or sills (Campbell et al. 1988). ODP drilling and deeply towed geophysical surveys revealed shallow basaltic sills underlying vent fields that act as cap rocks to focus flow (Curray et al. 1982; Lonsdale and Becker 1985)

Stable isotope studies of Guaymas Basin hydrothermal systems (Peter and Shanks 1992) show a unique range of $\delta^{34}S_{MeS}$ values from -3.4 to 3.8 (Fig. 11), which is due in part to remobilization of a bacteriogenic sulfide from the sediments. Carbonate minerals in the hydrothermal "pagodas" have $\delta^{13}C$ values that range from near -9.6 to -14‰, indicating oxidation of organic matter in the hydrothermal system.

Convergent margins.

Okinawa Trough is an active back-arc basin formed by extension within continental crust behind the Rhukyu island arc. The middle Okinawa Trough contains bimodal volcanism and very high heat flow. Three significant hydrothermal sites have been investigated to date. These include the Jade Field on the NE flank of the Izena Cauldron, where black smoker fluids up to 320°C and Zn-Pb-Ba rich massive sulfides have been sampled; Iheya Ridge a basaltic pillow ridge that hosts the "Clam" hydrothermal vent site with fluids up to 220°C and carbonate-rich chimneys; and the Minami-Ensei knoll in northern Okinawa Trough with fluids up to 278°C venting from anhydrite-rich chimneys.

Mariana Trough is a back-arc spreading center that occurs between a remnant arc and the currently active Mariana arc. On the flanks of axial volcanoes in the central Mariana Trough are several vent fields with measured temperatures up to 287°C and sulfide-sulfate chimneys comprised of sphalerite, galena, and barite, due to metal sources from the underlying andesitic crust. An unusual occurrence in the northeast portion of the Mariana Arc is a serpentinite mud volcano called Conical Seamount. Carbonate (calcite, aragonite) and silicate (Mg silicate) chimneys occur near the mud volcano summit. Associated fluids are cold, sulfate-sulfide-carbonate-silica-rich, and have pHs as high as 12.

Manus Basin is a back-arc basin located north of the New Britain island arc that consists of NE-SW trending ridge segments offset by transform faults. On the Manus Spreading Center in central Manus Basin the Vienna Woods hydrothermal field consists of a 300 m-diameter sulfide mound with an extensive "forest" of active and inactive sphalerite-barite-rich, sulfide-sulfate-oxide chimneys up to 15 m-long atop it. Vent fluids from one of the active chimneys reach temperatures as high as 302°C. In the eastern Manus Basin, two important hydrothermal fields are known. The PACMANUS field includes discontinuous sulfide occurrences over a 3 by 0.8 km area and active venting of unknown nature. Volcanic rocks are andesitic to rhyodacitic.

The DESMOS cauldron vent field has up to 120°C vent fluids that are sulfate-rich and have very low pH (<2.1). Gamo et al. (1997) have suggested that these fluids may be "acid-sulfate" fluids similar to those in some continental epithermal systems due to disproportionation of magmatic SO_2. Disproportionation of SO_2 is a plausible mechanism to explain a measured $\delta^{34}S_{H_2S}$ value of -5.6‰ and a measured δD value of -8.1‰, but these are very unusual fluids and one would like to see more than single determinations of these critical isotope values. Furthermore, the DESMOS vent fluids seem to have nearly normal seawater Mg contents and do not seem to track to zero Mg. Gamo et al. (1997) suggest that these low-pH (~2) fluids may be steam-heated, like those commonly encountered in subaerial acid-sulfate geothermal systems (Fournier 1989), but more data are needed. Most continental acid-sulfate systems result from oxidation of H_2S, not SO_2 disproportionation.

North Fiji Basin is an active, mature marginal basin with an 800 km-long spreading-ridge that produces normal Mid Ocean Ridge basalts (N-MORB). The "White Lady" vent field, named after an almost pure anhydrite chimney that vents low-Cl fluids at 290°C, occurs on the ridge crest in the north-central part of the basin. Immediately north of the White Lady site is a 2-km-wide fossil vent field with chimneys up to 15 m that are typical MOR types; composed of anhydrite, chalcopyrite, and pyrrhotite.

Lau Basin is an active back-arc basin with the Valu Fa Ridge situated between the Lau Ridge remnant arc and the active Tofua Arc. Lau Ridge contains a highly fraction-ated volcanic suite, from Fe-Ti basalts to andesites to dacites approaching rhyolites near the propagating tip at the southernmost end of the ridge. Three vent fields occur on the southern ridge: the White Church vent field with inactive sphalerite-barite chimneys, the Vai Lili field with active black and white smokers and an exposed sulfide stockwork "feeder" zone, and the highly oxidized Mn-rich Hine Hina field. Venting occurs only at Vai Lili, with fluids up to 334°C, pH of about 2, and Cl nearly 1.5 times seawater. Sulfides are Ba-, Pb-, and As-rich relative to MOR smoker deposits. Herzig et al. (1998) have measured $\delta^{34}S$ on sulfide minerals and barite from the White Church, Vai Lili, and Hine Hina fields. Sulfides from White Church range from 2.7-6.8 similar to MOR systems and those from Vai Lili range from 7.1 to 10.9‰. Barite from both of these fields has $\delta^{34}S$ values close to seawater sulfate. In contrast, the Hine Hina field has sulfide minerals that range from -2.8 to -7.7‰, which is very unusual for a volcanic-hosted system. Barites from these same samples also give low $\delta^{34}S$ values of 16.1-16.7 and are quite consistent (5 analyses). Herzig et al. (1993) suggest disproportion of magmatic SO_2 to explain the isotopically light sulfides and sulfates. Disproportionation in a seafloor acid-sulfate environment is consistent with the presence of alunite in the Hine Hina samples. Disproportionation of SO_2 is a plausible explanation for the sulfur isotope systematics, but this mechanism will not be proven until the presence of magmatic SO_2 is demonstrated by analysis of vent fluid gases.

Izu-Bonin arc is an intra-oceanic island arc to the west of the Izu-Ogasawara trench. Calderas on the summits of arc-related submarine volcanoes host a number of hydrothermal deposits and vent sites, and one vent field occurs on the Sumisu rift, an incipient back-arc basin just west of the volcanic front. Volcanic rocks range from basalt to dacite and the most significant active hydrothermal field occurs in the caldera of the dacitic Suiyo seamount. Fluids vent from active chimneys over an area 300 m by 150 m. High Cl vent fluids have temperatures up to 311°C and sulfide chimneys are comprised of an inner zone of chalcopyrite surrounded by sphalerite, barite, and anhydrite.

In addition to submersible investigations, Hole 786B of ODP leg 125 drilled 725 m into volcanic flows, breccias, and basal dikes of the Izu-Bonin forearc basement. Alt et al. (1998) investigated $^{87}Sr/^{86}Sr$, $\delta^{13}C$, δ^8O, and $\delta^{34}S$ systematics of fresh and altered volcanic rocks, dikes, and secondary minerals. Oxygen isotope geothermometry indicates that vein and vug-filling carbonates formed at temperatures that increase downward from 5 to 200°C , assuming precipitation from 0 ‰ seawater. Smectites that occur as alteration veins and selvages in the volcanic section have $\delta^{18}O$ values from 16.6-21.2 and form at temperatures from 45-80°C. Quartz from mineralized and altered dikes have $\delta^{18}O$ values from 11.2-15.9 and formed at temperatures from 180 to 255°C, assuming typical black smoker fluids with 1 to 2 ‰ $\delta^{18}O$ values. These data, and whole rock $\delta^{18}O$ data, all indicate increasing temperature and intensity of alteration downward in the hole, with highest temperatures occurring in the dike section below 800 m depth. An interesting feature of this site is the occurrence of negative $\delta^{34}S$ values (-7.6 to -17.6) for sulfide minerals in the intensely altered dike section and low $\delta^{34}S$ (-3.2 to 12.5 per mil) values for rock SO_4 in the same zone. Alt et al. (1998) suggest that disproportionation of magmatic SO_2 is responsible for these unusual sulfur isotope systematics.

Shallow water sites. Two other arc-related hydrothermal sites, in shallow water, have recently been investigated and are worthy of mention as they will undoubtedly receive significant stable isotope studies in coming years. Lihir Island lies northeast of Manus Basin in the fore-arc region of the New Hanover-New Ireland-Bougainville island arc. Gas-rich hydrothermal (60-96°C) vents occur in shallow water (3-10 m) about 100 m offshore (Pichler et al. 1999). Gases are enriched in CO_2, H_2S, and CH_4 and mineral accumulations of pyrite/marcasite, dolomite, and native S are precipitated in vent orifices. As, Se, Sb, Hg, and Tl are greatly enriched in the iron-sulfide deposits. Carbon and oxygen isotope analyses of the dolomite give $\delta^{13}C_{VPDB}$ of 3.9 and 4.2 and $\delta^{18}O_{VSMOW}$ of 20.8 and 21.8, which indicate formation temperatures of 89-107°C, assuming equilibrium with normal seawater. Methane-CO_2 chemical equilibration temperatures suggest 286-327°C in the subsurface. The other shallow water site of new interest is the Bay of Plenty in the Taupo Volcanic Zone between the north island of New Zealand and White Island. Sarano et al. (1989) and Stoffers et al. (1999) have investigated gas-rich vents in about 200 m of water. Sarano et al. (1989) found conical anhydrite mounds up 8 m high and associated sediments with up to 10 wt % C and strongly enriched As, Sb, Hg, Tl, U, and Mo. Stoffers et al. (1999) observed liquid Hg in vent samples from the same area.

SUMMARY AND CONCLUSIONS

Eruptive events

Some of the most important discoveries in seafloor hydrothermal studies in recent years have come as a result of investigation of sites that have recently experienced eruptive events. The 9-10°N EPR site is a prime example, but event response cruises following eruptions detected by sonar and seismic events have now been carried out at several sites on the Juan de Fuca ridge. In general, these studies provide important information on short term variations in vent fluid temperature and composition, bacterial processes and expulsion of bacteria from the subsurface biosphere, re-colonization of vent sites by vent-specific macrofauna, and re-growth of sulfide chimneys. Stable isotope studies have shown a distinct relationship to volcanic and seismic events at the ridge crests. In particular, vent fluids evolve from low salinity, low ΔD and $\Delta^{18}O$ values and high $\delta^{34}S_{H_2S}$ values immediately following eruptions to higher ΔD and $\Delta^{18}O$ values and lower $\delta^{34}S_{H_2S}$ values as hydrothermal systems "mature" and deepen, typically in a few years following eruptions.

Magmatic components

Magmatic volatiles are certainly important components of seafloor hydrothermal systems. Helium and CO_2 in vent systems are clearly related to degassing of axial magma chambers in MOR systems. H_2S may also be added to hydrothermal systems by direct magmatic degassing, but evidence is ambiguous because H_2S is produced in abundance by leaching of igneous sulfide from basaltic rocks during water-rock interaction. Studies of $\delta^{34}S_{H_2S}$ in black smoker vent fluids has never indicated H_2S with purely magmatic values of 0±0.5 ‰, so, if H_2S is degassed directly from axial magma chambers, it does not make it into vent fluids without incorporating some sulfur from sulfate-reduction reactions.

Controversy arises over magmatic water, especially in MOR basaltic systems. Shanks et al. (1995) discovered vent fluids from the 1991 eruption area at 9°45-52′N on the EPR with ΔD values as low as -2 ‰. These values could result from either a small component of magmatic water with ΔD values of -40 to -80 ‰, or they could be due to an open system isobaric phase-separation process. The latter seems more likely in the 9-10°N case. The only other measurement of vent fluid with negative ΔD values is a single analysis from the DESMOS vent field in Manus Basin. This field is hosted by andesitic to rhyodacitic rocks and the magmas that produce these rocks are significantly more water-rich than MORBs. Felsic rocks are more oxidizing, so SO_2 is expected to be the dominant sulfur gas. Various investigators have suggested SO_2 disproportionation reactions (to H_2S and SO_4) to explain negative $\delta^{34}S$ values of H_2S or sulfide minerals and lower-than-seawater $\delta^{34}S_{SO4}$ values in associated anhydrite and barite. These reactions are plausible and provide an explanation for negative $\delta^{34}S$ values related to volcanic hosted systems, but SO_2 degassing needs to be proven by vent gas sampling. The role of thermophilic chemosynthetic microbes in possibly producing these light values also needs to be assessed.

Phase separation

Phase separation is now an accepted part of the hydrothermal vent process, producing fluids that range from about 6-200‰ of seawater chloride concentrations. Hydrogen isotope fractionation during phase separation has been demonstrated during field and experimental studies. These studies indicate that high temperature (300-400°C) phase-separation produces D-enrichment in the vapor and D-depletion in the conjugate brine. However, the actual ΔD values of vent fluids depend on the details of the phase separation process. Rayleigh distillation during phase separation can lead to significant stable isotope variations in vent fluids. Experimental studies have shown ΔD fractionations of up to 8‰, with vapor isotopically heavier than brine. However, open-system vapor-loss during near-surface eruptive or dike injection events can lead to a subsurface brine with quite negative ΔD values. Mixing of this brine with vapor and/or normal salinity hydrothermal seawater (Von Damm and Bischoff 1987; Bischoff and Rosenbauer 1989; Von Damm 1990; Butterfield et al. 1994b; Edmonds and Edmond 1995) can produce vent fluids with negative ΔD. $\Delta^{18}O$ and $\delta^{34}S$ variations in vent fluids due to phase-separation are small or indiscernible. Basically, a better understanding of hydrothermal fluid flow at extreme temperatures during phase-separation is needed. Seawater convection is the process that cools axial magma chambers and mines heat from the ocean crust. Seawater cannot quench magma without experiencing phase-separation. The detailed evolution of fluid chemistry and stable isotope systematics of phase-separated fluids depends on the mass-balance relations during phase-separation and these will only be evaluated properly when fluid flow modeling, or measurements of fluid flow in the deep crust, can be combined with realistic assessments of the transfer of magmatic heat and the physical-chemical basis of phase separation and attendant effects on composition, density, heat capacity, etc. are considered.

Deep biosphere

Studies of "snowblower" vents following volcanic eruptions, studies of the microbial ecology of black smoker systems, and direct examination of thermophilic organisms in drilled subseafloor hydrothermal systems all indicate that the subsurface biosphere is potentially significant to chemical and biological transformations. Stable isotopes have been and will continue to be an important tool in evaluating these processes. Sulfur isotope studies of sulfides in serpentinites offer the most direct currently available evidence of such processes, showing strongly negative $\delta^{34}S$ values in mantle derived altered ultramafic rocks due to microbial sulfate reduction using H_2 or CH_4 as an energy source (Alt and Shanks 1998 and in prep). This field is just emerging. Application of the traditional stable isotope systems (H, C, N, O, S) and especially the new emerging techniques (Fe, Cu, Mo, Hg, etc.) to studies of microbial processes related to seafloor hydrothermal systems is certain to be an important frontier area.

ACKNOWLEDGMENTS

I thank Craig Johnson, Jeff Alt, Karen Von Damm, Phil Brown, Jim Bischoff, and Dave Cole for excellent reviews of the manuscript that helped refine some of the concepts and improved the presentation. I also thank Bill Seyfried, Meg Tivey, Ian Ridley, Bob Rye, Jeff Alt, and Karen Von Damm for many hours of discussion of seafloor hydrothermal systems. Excellent analytical data were provided by Pam Gemery (USGS, Denver), Dave Winter and Howie Spero (UC Davis), and Tyler Coplen and JK Bohlke (USGS, Reston). I thank Karen Von Damm, Rachel Haymon, Meg Tivey, Scott Brinson, Alison Bray, Laurel Buttermore, and Debbie Colodner for assistance in shipboard sampling. Finally, I acknowledge the captain and crew of R/V Atlantis II and R/V Atlantis and the ALVIN group for excellent and professional operations that make these kinds of study possible.

REFERENCES

Agrinier P, Cannat M (1997) Oxygen-isotope constraints on serpentinization processes in ultramafic rocks from the Mid-Atlantic Ridge (23°N). Proc ODP, Sci Results 153:381-388
Agrinier P, Hekinian R, Bideau D, Javoy M (1995) O and H stable isotope compositions of oceanic crust and upper mantle rocks exposed in the Hess Deep near the Galapagos Triple Junction. Earth Planet Sci Lett 136:183-196
Alt JC (1995) Subseafloor processes in mid-ocean ridge hydrothermal systems *In* SE Humphris, RA Zierenberg, LS Mullineaux, RE Thomson (eds) Seafloor Hydrothermal Systems: Physical, Chemical, Biological, and Geological Interactions. Am Geophys Union Monogr 91:85-114
Alt JC (1995) Sulfur isotopic profile through the oceanic crust: Sulfur mobility and seawater-crustal sulfur exchange during hydrothermal alteration. Geology 23:585-588
Alt JC, Laverne C, Vanko D, Tartarotti P, Teagle DAH, Bach W, Zuleger E, Erzinger J, Honnorez J, Pezard PA, Becker K, Salisbury MH, Wilkens RH (1996) Hydrothermal alteration of a section of upper oceanic crust in the eastern equatorial Pacific: a synthesis of results from Site 504 (DSDP Legs 69, 70, and 83, and ODP Legs 111, 137, 140, and 148). *In* JC Alt, Kinoshita H, Stokking LB, Michael PJ (eds) Proc ODP, Sci Results 148:417-434
Alt JC, Shanks WC III (1998) Sulfur in serpentinized oceanic peridotites: serpentinization processes and microbial sulfate reduction. J Geophys Res 103:9917-9929
Alt JC, Shanks WC III (2001) Uptake of sulfur during serpentinization of abyssal peridotite. Earth Planet Sci Lett (submitted)
Alt JC, Teagle DAH (in press) Hydrothermal alteration and fluid fluxes in ophiolites and oceanic crust. *In* Ophiolites and Oceanic Crust: New Insights from Field Studies and Ocean Drilling Program, Penrose Vol 37, ms. pages
Alt JC, Teagle DAH (1998) Probing the TAG hydrothermal mound and stockwork; oxygen-isotopic profiles from deep ocean drilling. Proc ODP, Sci Results 158:285-295
Alt JC, Teagle DAH, Brewer TS, Shanks WC III, Halliday AN (1998) Alteration and mineralization of an oceanic forearc and the ophiolite-ocean crust analogy. J Geophys Res 103:12,365-12,380
Anderson RN (1972) Petrologic significance of low heat flow on the flanks of slow-spreading mid-ocean ridges. Geol Soc Am Bull 83:2947-2956
Arrhenius G, Bonatti E (1965) Neptunism and vulcanism in the ocean. Prog Oceanogr 3:7-22
Bach W, Humphris SE (1999) Relationship between the Sr and O isotope compositions of hydrothermal fluids and the spreading and magma-supply rates at oceanic spreading centers. Geology 27:1067-1070
Bäcker H, Richter H (1973) Die rezente hydrothermal-sedimentare Lagerstatte Atlantic II-Tief im Roten Meer. Geol Rundsch 62:697-740
Bäcker H, Schoell M (1972) New deeps with brines and metalliferous sediments in the Red Sea. Nature 240:153-158
Berndt ME, Seal RR II, Shanks WC III, Seyfried WE Jr (1996) Hydrogen isotope systematics of phase separation in submarine hydrothermal systems; experimental calibration and theoretical models. Geochim Cosmochim Acta 60:1595-1604

Berndt ME, Seyfried WE Jr, Janecky DR (1989) Plagioclase and epidote buffering of cation ratios in mid-ocean ridge hydrothermal fluids: experimental results in and near the supercritical region. Geochim Cosmochim Acta 53:2283-2300

Berndt ME, Seyfried WE Jr (1993) Calcium and sodium exchange during hydrothermal alteration of calcic plagioclase at 400°C and 400 bars. Geochim Cosmochim Acta 57:4445-4451

Berndt ME, Seyfried WE Jr (1997) Calibration of Br/Cl fractionation during subcritical phase separation of seawater; possible halite at 9 to 10°N East Pacific Rise. Geochim Cosmochim Acta 61:2849-2854

Bischoff JL (1969) Red Sea geothermal brine deposits: their mineralogy, chemistry, and genesis. *In* ET Degens, DA Ross (eds) Hot Brines and Recent Heavy Metal Deposits in the Red Sea. Springer Verlag, New York, p 368-401

Bischoff JL, Dickson FW (1975) Seawater-basalt interaction at 200°C and 500 bars implications for origin of sea-floor heavy metal deposits and regulation of seawater chemistry. Earth Planet Sci Lett 25:385-397

Bischoff JL, Henyey TL (1974) Tectonic elements of the central part of the Gulf of California. Geol Soc Am Bull 85:1893-1904

Bischoff JL, Rosenbauer RJ (1985) An empirical equation of state for hydrothermal seawater (3.2% NaCl). Am J Sci 285:725-763

Bischoff JL, Rosenbauer RJ (1989) Salinity variations in submarine hydrothermal systems by layered double-diffusive convection. J Geol 97:613-623

Blanc G, Anschutz P (1995) New stratification in the hydrothermal brine system of the Atlantis II Deep, Red Sea. Geology 23(6):543-546

Bluth GJ, Ohmoto H (1988) Sulfide-sulfate chimneys on the East Pacific Rise, 11° and 13°N latitudes, Part II: Sulfur isotopes. Can Mineral 26:505-515

Böhlke JK, Shanks WC III (1994) Stable isotope study of hydrothermal vents at Escanaba Trough, NE Pacific: Observed and calculated effects of sediment-seawater interaction *In* JL Morton, RA Zierenberg, CA Reiss (eds) Geologic, Hydrothermal, and Biologic Studies at Escanaba Trough, Gorda Ridge, Offshore Northern California, Gorda Ridge. U S Geol Surv Bull 2022:223-239

Bostrom K, Peterson MNA (1966) Precipitates from hydrothermal exhalations on the East Pacific Rise. Econ Geol 61:1258-1265

Boström K, Peterson MNA, Joensuu O, Fisher DE (1969) Aluminum-poor ferromanganoan sediments on active ocean ridges. J Geophys Res 74:3261-3270

Bowers TS (1989) Stable isotope signatures of water-rock interaction in mid-ocean ridge hydrothermal systems: sulfur, oxygen, and hydrogen. J Geophys Res 94(B5):5775-5786

Bowers TS, Campbell AC, Measures CI, Spivack AJ, Khadem M, Edmond JM (1988) Chemical controls on the composition of vent fluids at 13°N-11°N and 21°N, East Pacific Rise. J Geophys Res 93:4522-4536

Bowers TS, Taylor HP (1985) An integrated chemical and stable isotopic model of the origin of mid-ocean ridge hot spring systems. J Geophys Res 90:12,583-12,606

Butler IB, Fallick AE, Nesbitt RW (1998) Mineralogy, sulphur isotope geochemistry and the development of sulphide structures at the Broken Spur hydrothermal vent site, 29°10'N, Mid-Atlantic Ridge. J Geol Soc London 155:773-785

Butterfield DA (2000) Deep ocean hydrothermal vents. *In* H Sigurdsson, BF Houghton, SR McNutt, H Rymer, J Stix (eds) Encyclopedia of Volcanoes. Academic Press, San Diego, California, p 857-875

Butterfield DA, Massoth GJ, McDuff RE, Lupton JE, Lilley MD (1990) Geochemistry of hydrothermal fluids from Axial Seamount Hydrothermal Emissions Study vent field, Juan de Fuca Ridge; subseafloor boiling and subsequent fluid-rock interaction. J Geophys Res 95:12,895-12,921

Butterfield DA, McDuff RE, Franklin J, Wheat CG (1994a) Geochemistry of hydrothermal vent fluids from Middle Valley, Juan de Fuca Ridge. *In* MJ Mottl, EE Davis, AT Fisher (eds) Proc ODP, Sci Results 139:395-410

Butterfield DA, McDuff RE, Mottl MJ, Lilley MD, Lupton JE, Massoth GJ (1994b) Gradients in the composition of hydrothermal fluids from the Endeavour Segment vent field; phase separation and brine loss. J Geophys Res 99:9561-9583

Campbell AC, Bowers TS, Measures CI, Falkner KK, Khadem M, Edmond JM (1988) A time-series of vent fluid compositions from 21°N East Pacific Rise (1979, 1981, 1985) and the Guaymas Basin, Gulf of California (1982, 1985). J Geophys Res 93:4537-4549

Campbell AC, German CR, Palmer MR, Gamo T, Edmond JM (1994) Chemistry of hydrothermal fluids from Escanaba Trough, Gorda Ridge. *In* JL Morton, RA Zierenberg, CA Reiss (eds) Geologic, Hydrothermal, and Biologic Studies at Escanaba Trough, Gorda Ridge, Offshore Northern California. U S Geol Surv 2022:201-222

Canfield DE, Habicht KS, Thamdrup B (2000) The Archean sulfur cycle and the early history of atmospheric oxygen. Science 288:658-661

Cannat M et al. (eds) (1995) Proc ODP Initial Repts 153, Ocean Drilling Program, College Station, Texas, 798 p

Chan LH, Edmond JM, Thompson G (1993) A lithium isotope study of hot springs and metabasalts from mid-ocean ridge hydrothermal systems. J Geophys Res-Solid Earth 98(B6):9653-9659

Charlou, JL.,Donval JP (1993) Hydrothermal methane venting between 12°N and 26°N along the Mid-Atlantic Ridge. J Geophys Res 98:9625-9642

Chiba H, Sakai H (1985) Oxygen isotope exchange rate between dissolved sulfate and water at hydrothermal temperatures. Geochim Cosmochim Acta 49:993-1000

Chiba H, Uchiyama N, Teagle DAH (1998) Stable isotope study of anhydrite and sulfide minerals at the TAG hydrothermal mound, Mid-Atlantic Ridge, 26°N. Proc ODP, Sci Results 158:85-90

Clayton RN, Mayeda TK (1963) The use of bromine pentafluoride in the extraction of oxygen in oxides and silicates for isotopic analysis. Geochim Cosmochim Acta 27:43-52

Cole DR, Mottl MJ, Ohmoto H (1987) Isotopic exchange in mineral-fluid systems: II. Oxygen and hydrogen isotopic investigation of the experimental basalt-seawater system. Geochim Cosmochim Acta 51:1523-1538

Converse DR, Holland HD, Edmond JM (1984) Flow rates in the axial hot springs of the East Pacific Rise (21°N); implications for the heat budget and the formation of massive sulfide deposits. Earth Planet Sci Lett 69:159-175

Coplen TB, Wildman JD, Chen J (1991) Improvements in the gaseous hydrogen-water equilibration technique for hydrogen isotope ratio analysis. Analyt Chem 63:910-912

Corliss JB, Dymond J, Gordon LI, Edmond JM, Von Herzen RP, Ballard RD, Green K, Williams D, Bainbridge A, Crane K, Van Andel TH (1979) Submarine thermal springs on the Galapagos Rift. Science 203:1073-1083

Craig H (1966) Isotopic composition and origin of the Red Sea and Salton Sea geothermal brines. Science 154:1544-1548

Crowe DE, Valley JW, Baker KL (1990) Micro-analysis of sulfur-isotope ratios and zonation by laser microprobe. Geochim Cosmochim Acta 54:2075-2092

Crowe DE, Valley JW (1992) Laser microprobe study of sulfur isotope variation in a sea-floor hydrothermal spire, Axial Seamount, Juan de Fuca Ridge, eastern Pacific. Chem Geol 101:63-70

Curray JR, Moore DG, Aguayo JE, Aubry MP, Einsele G, Fornari DJ, Gieskes J, Guerrero JC, Kastner M, Kelts K, Lyle M, Matoba Y, Molina-Cruz A, Niemitz J, Rueda J, Saunders AD, Schrader H, Simoneit BRT, Vacquier V (1982) Init Repts DSDP, 1313 p. U S Govt Printing Office, Washington, DC

Detrick RS, Buhl P, Vera E, Mutter J, Orcutt J, Madsen J, Brocher T (1987) Multi-channel seismic imaging of a crustal magma chamber along the East Pacific Rise. Nature 326:35-41

Dixon JE, Stolper E, Delaney JR (1988) Infrared spectroscopic measurements of CO_2 and H_2O in Juan de Fuca Ridge basaltic glasses. Earth Planet Sci Lett 90:87-104

Dixon JE, Clague DA, Stolper EM (1991) Degassing history of water, sulfur, and carbon in submarine lavas from Kilauea Volcano, Hawaii. J Geol 99:371-394

Duckworth RC, Knott R, Fallick AE, Rickard D, Murton BJ, Van Dover C (1995) Mineralogy and sulfur isotope geochemistry of the Broken Spur sulphides, 29°N Mid-Atlantic Ridge. In LM Parson, CL Walker, DR Dixon (eds) Hydrothermal Vents and Processes. Spec Publ Geol Soc London 87:175-190

East-Pacific-Rise-Study-Group (1981) Crustal processes of the mid-ocean ridge. Science 213:31-40

Edmond JM, Measures C, Mangum B, Grant B, Sclater FR, Collier R, Hudson A, Gordon LI, Corliss J (1979a) On the formation of metal-rich deposits at ridge crests. Earth Planet Sci Lett 46:19-30

Edmond JM, Measures C, McDuff RE, Chan L, Collier R, Grant B, Gordon LI, Corliss J (1979b) Ridge crest hydrothermal activity and the balances of the major and minor elements in the ocean the Galapagos data. Earth Planet Sci Lett 46:1-18

Edmond JM, Von Damm KL, McDuff RE, Measures CI (1982) Chemistry of hot springs on the East Pacific Rise and their effluent dispersal. Nature 297:187-191

Edmonds HN, Edmond JM (1995) A three-component mixing model for ridge-crest hydrothermal fluids. Earth Planet Sci Lett 134:53-67

Einsele G, Gieskes JM, Curray J, Moore DG, Aguayo E, Aubry MP, Fornari D, Guerrero J, Kastner M, Kelts K, Lyle M, Matoba Y, Molina-Cruz A, Niemitz J, Rueda J, Saunders A, Schrader H, Simoneit B, Vacquier A (1980) Intrusion of basaltic sills into highly porous sediments, and resulting hydrothermal activity. Nature 283:441-445

Epstein S, Mayeda TK (1953) Variation of [18]O content of waters from natural sources. Geochim Cosmochim Acta 44:213-224.

Evans WC, White LD, Rapp JB (1988) Geochemistry of some gases in hydrothermal fluids from the southern Juan de Fuca Ridge. J Geophys Res 93:15,305-15,313

Fornari DJ, Embley RW (1995) Tectonic and volcanic controls on hydrothermal processes at the mid-ocean ridge; an overview based on near-bottom and submersible studies. In SE Humphris, RA Zierenberg,

LS Mullineaux, RE Thomson (eds) Seafloor hydrothermal systems; physical, chemical, biological, and geological interactions. Am Geophys Union Monogr 91:1-46

Fouquet Y, Knott R, Cambon P, Fallick A, Rickard D, Desbruyeres D (1996) Formation of large sulfide mineral deposits along fast spreading ridges: Example from off-axial deposits at 12°43'N on the East Pacific Rise. Earth Planet Sci Lett 144:147-162

Fouquet Y, Zierenberg RA, Miller DJ, Bahr JM, Baker PA, Bjerkgarden T, Brunner CA, Duckworth RC, Gable R, Gieskes JM, Goodfellow WD, Groeschel-Becker HM, Guerin G, Ishibashi J, Iturrino GJ, James RH, Lackschewitz KS, Marquez LL, Nehlig P, Peter JM, Rigsby CA, Simoneit BRT, Schultheiss PJ, Shanks WC, III, Summit M, Teagle DAH, Urbat M, Zuffa GG, Gieskes J, Iturrino G, Schultheiss P (1998) Proc Ocean Drilling Program, Initial Reports; Sedimented Ridges II, Covering Leg 169 of the Cruises of the Drilling Vessel JOIDES Resolution, Victoria, British Columbia, to San Diego, California, Sites 1035-1038, 21 August-16 October, 1996, p 7-16

Fournier RO (1989) Geochemistry and dynamics of the Yellowstone National Park hydrothermal system. Ann Rev Earth Planet Sci 17:13-53

Francheteau J, Needham HD, Choukroune P, and others (1979) Massive deep-sea sulphide ore deposits discovered on the East Pacific Rise. Nature 77:523-528

Franklin JM, Hannington MD, Jonasson IR, Barrie CT (1998) Arc-related volcanogenic massive sulphide deposits Metallogeny of Volcanic Arcs. British Columbia Geol Survey Short Course Notes 5, Section B, 31 p

Franklin JM, Lydon JW, Sangster DF (1981) Volcanic associated massive sulphide deposits. Econ Geol 75th Anniv Vol, p 485-627

Friedman CT (1998) Analysis of stable sulfur isotopes and trace cobalt on sulfides from the TAG hydrothermal mound. 91pp MSc Thesis, Massachusetts Institute of Technology, Cambridge

Fruh-Green GL, Plas A, Lecuyer C (1996) Petrologic and stable isotope constraints on hydrothermal alteration and serpentinization of the EPR shallow mantle at Hess Deep (Site 895). *In* Mevel C, Gillis KM, Allan JF, Meyer PS (eds) Proc ODP, Sci Results 147:255-291

Gamo T, Okamura K, Charlou J-L, Urabe T, Auzende J-M, Ishibashi J, Shitashima K, Chiba H, Binns RA, Gena K, Henry K, Matsubayashi O, Moss R, Nagaya Y, Naka J, Ruellan E (1997) Acidic and sulfate-rich hydrothermal fluids from the Manus back-arc basin, Papua New Guinea. Geology 25:139-142

Gemmell JB, Sharpe R (1998) Detailed sulfur-isotope investigation of the TAG hydrothermal mound and stockwork zone, 26°N, Mid-Atlantic Ridge. Proc ODP, Sci Results 158:71-84

Giggenbach WF (1982) Carbon-13 exchange between CO_2 and CH_4 under geothermal conditions. Geochim Cosmochim Acta 46:159-166,

Haymon RM (1983) Growth history of hydrothermal black smoker chimneys. Nature 30:695-698

Haymon RM, Kastner M (1981) Hot spring deposits on the East Pacific Rise at 21°N: preliminary description of mineralogy and genesis. Earth Planet Sci Lett 53:363-381

Haymon RM, Fornari DJ, Edwards MH, Carbotte S, Wright D, Macdonald K (1991) Hydrothermal vent distribution along the East Pacific Rise crest (9°09-54'N) and its relationship to magmatic and tectonic processes on the fast-spreading mid-ocean ridges. Earth Planet Sci Lett 104:513-534

Haymon RM, Fornari DJ, Von Damm KL, Lilley MD, Perfit MR, Edmond JM, Shanks WC III, Lutz RA, Grebmeier JM, Carbotte S, Wright D, McLaughlin E, Smith M, Beedle N, Olson E (1993) Volcanic eruption of the mid-ocean ridge along the East Pacific Rise crest at 9°45-52'N; direct submersible observations of seafloor phenomena associated with an eruption event in April 1991. Earth Planet Sci Lett 119:85-101

Hekinian R, Fevrier M, Bischoff JL, Picot P, Shanks WC III (1980) Sulfide deposits from the East Pacific rise near 21°N: A mineralogical study. Science 207:1433-1444

Herzig PM, Hannington MD, Fouquet Y, von Stackelberg U, Petersen S (1993) Gold-rich polymetallic sulfides from the Lau back arc and implications for the geochemistry of gold in sea-floor hydrothermal systems of the Southwest Pacific. Econ Geol 88:2178-2205

Herzig PM, Petersen S, Hannington MD (1998) Geochemistry and sulfur-isotopic composition of the TAG hydrothermal mound, Mid-Atlantic Ridge 26°N. Proc ODP, Sci Results 158:47-70

Horita J, Cole DR, Wesolowski DJ (1995) The activity-composition relationship of oxygen and hydrogen isotopes in aqueous salt solutions; III, Vapor-liquid water equilibration of NaCl solutions to 350°C. Geochim Cosmochim Acta 59:1139-1151

Horita J, Wesolowski DJ, Cole DR (1993) The Activity-Composition relationship of oxygen and hydrogen isotopes in aqueous salt solutions: 1. Vapor-Liquid water equilibration of single salt solutions from 50°C to 100°C. Geochim Cosmochim Acta 57:2797-2817

Humphris SE, Herzig PM, Miller DJ, Alt JC, Becker K, Brown D, Bruegmann G, Chiba H, Fouquet Y, Gemmell JB, Guerin G, Hannington MD, Holm NG, Honnorez JJ, Iturrino GJ, Knott R, Ludwig R, Nakamura K, Petersen S, Reysenbach AL, Rona PA, Smith S, Sturz AA, Tivey MK, Zhao X (1995) The internal structure of an active sea-floor massive sulphide deposit. Nature 377:713-716

Ishibashi J, Urabe T (1995) Hydrothermal activity related to arc-backarc magmatism in the western pacific. *In* B Taylor (ed) Backarc Basins. Plenum, New York, p 451-495

James RH, Elderfield H, Palmer MR (1995) The chemistry of hydrothermal fluids from the Broken Spur site, 29°N Mid-Atlantic ridge. Geochim Cosmochim Acta 59:651-659

Janecky DR, Shanks WC III (1988) Computational modeling of chemical and sulfur isotopic reaction processes in seafloor hydrothermal systems: chimneys, mounds, and subjacent alteration zones. Can Mineral 26:805-825

Jeanbaptiste P, Charlou JL, Stievenard M, Donval JP, Bougault H, Mevel C (1991) Helium and methane measurements in hydrothermal fluids from the Mid-Atlantic Ridge—The Snake Pit site at 23°N, Earth Planet Sci Lett 106:1728

Kaplan IR, Sweeney RE, Nissenbaum A (1969) Sulfur isotope studies on Red Sea geothermal brines and sediments. *In* ET Degens, DA Ross (eds) Hot Brines and Recent Heavy Metal Deposits in the Red Sea: A geochemical and geophysical account. Springer-Verlag, New York, p 474-498

Kase K, Yamamoto M, Shibata T (1990) 13 copper-rich sulfide deposits near 23°N, Mid-Atlantic Ridge: Chemical composition, mineral chemistry, and sulfur isotopes. *In* J Honnerez, WB Bryan, T Juteau (eds) Proc ODP, Sci Results 106/109:163-177

Kastner M (1982) Evidence for two distinct hydrothermal systems in the Guaymas Basin. *In* JR Curray, DG Moore, et al. (eds) Init Repts DSDP 64:1143-1157, U S Gov't Printing Office, Washington, DC

Kerridge J, Haymon RM, Kastner M (1983) Sulfur isotope systematics at the 21°N site, East Pacific Rise. Earth Planet Sci Lett 66:91-100

Knott R, Fouquet Y, Honnorez JJ, Petersen S, Bohn M (1998) Petrology of hydrothermal mineralization; a vertical section through the TAG mound. Proc ODP, Sci Results 158:5-26

Koski RA, Shanks WC III, Bohrson WA, Oscarson RL (1988) The composition of massive sulfide deposits from the sediment-covered floor of Escanaba Trough, Gorda Ridge: Implications for depositional processes. Can Mineral 26:655-673

Lagabrielle Y, Bideau D, Cannat M, Karson JA, Mevel C (1998) Ultramafic-mafic plutonic rock suites exposed on the Mid-Atlantic Ridge (10°N-30°N): Symmetrical-assymetrical distribution and implications for seafloor spreading processes. *In* RW Buck, PT Delaney, JA Karson, Y Lagabrielle (eds) Faulting and Magmatism at Mid-ocean ridges. Geophys Monogr 106:153-176. American Geophysical Union, Washington, DC

Lalou C, Reyss JL Brichet, E, Herzig PM, Humphris SE, Miller DJ, Alt JC, Becker K, Brown D, Bruegmann GE, Chiba H, Fouquet Y, Gemmell JB, Guerin G, Hannington MD, Holm NG, Honnorez JJ, Iturrino GJ, Knott R Ludwig, RJ, Nakamura K-i, Petersen S, Reysenbach A-L, Rona PA, Smith SE, Sturz AA, Tivey MK, Zhao X (1998) Age of sub-bottom sulfide samples at the TAG active mound Proc ODP, Sci Results 158:111-117

Larson RL, Menard HW, Smith SM (1968) Gulf of California: a result of ocean-floor spreading and transform faulting. Science 161:781

LeHuray AP, Church SE, Koski RA, Bouse RM (1988) Pb isotope ratios in sulfides from mid-ocean ridge hydrothermal sites. Geology 16:362-365

Lilley MD, Butterfield DA , Olson EJ, Lupton JE, Macko SA, Mcduff RE (1993)Anomalous CH_4 and NH_4^+ concentrations at an unsedimented mid-ocean-ridge hydrothermal system. Nature 364:45-47

Lister CRB (1983) The basic physics of water penetration into hot rock *In* PA Rona, K Boström, L Laubier, KL Smith Jr (eds) Hydrothermal Processes at Seafloor Spreading Centers. Plenum Press, New York, p 141-168

Lonsdale P, Becker K (1985) Hydrothermal plumes, hot springs, and conductive heat flow in the Southern Trough of Guaymas Basin. Earth Planet Sci Lett 73:211-225

Massoth GJ, Butterfield DA, Lupton JE, McDuff RE, Lilley MD, Jonasson IR (1989) Submarine venting of phase-separated hydrothermal fluids at Axial Volcano, Juan de Fuca Ridge. Nature 340:702-705

McKenzie DP, Davies D, Molnar P (1970) Plate tectonics of the Red Sea and east Africa. Nature 226: 243-248

Michard G, Albarede F, Michard A, Minster JF, Charlou JL, Tan N (1984) Chemistry of solutions from the 13°N East Pacific Rise hydrothermal site. Earth Planet Sci Lett 67:297-307

Miller AL, Densmore CD, Degens ET, Hathaway JC, Manheim FT, McFarlin PF, Pocklington R, Jokela A (1966) Hot brines and recent iron deposits of the Red Sea. Geochim Cosmochim Acta 30:341-359

Morton JL, Zierenberg RA, Reiss CA (1994) Geologic, Hydrothermal, and Biologic Studies at Escanaba Trough, Gorda Ridge, Offshore Northern California. U S Geol Surv Bull 2022, 359 p

Muehlenbachs K, Clayton RN (1972) Oxygen isotope geochemistry of submarine greenstones. Can J Earth Sci 9:471-478

Muehlenbachs K, Clayton RN (1971) Oxygen isotope ratios of submarine diorites and their constituent minerals. Can J Earth Sci 8:1591-1595

Muehlenbachs K, Clayton RN (1976) Oxygen isotope composition of the oceanic crust and its bearing on seawater. J Geophys Res 81:4365-4369

Ohmoto H, Goldhaber M (1997) Sulfur and carbon isotopes. *In* HL Barnes (ed) Geochemistry of Hydrothermal Ore Deposits. John Wiley and Sons, New York, p 517-612

Oosting SE, Von Damm KL (1996) Bromide/chloride fractionation in seafloor hydrothermal fluids from 9-10°N East Pacific Rise. Earth Planet Sci Lett 144:133-145

Palmer MR (1996) Hydration and uplift of the oceanic crust on the Mid-Atlantic Ridge associated with hydrothermal activity; evidence from boron isotopes. Geophys Res Lett 23:3479-3482

Palmer MR, Edmond JM (1989) Cesium and rubidium of submarine hydrothermal fluids: evidence for recycling of alkali elements. Earth Planet Sci Lett 95:8-14

Perfit MR, Chadwick WW Jr (1998) Magmatism at mid-ocean ridges: Constraints from volcanological and geochemical investigations. *In* WR Buck, PT Delaney, JA Karson, Y Lagabrielle (eds) Faulting and Magmatism at Mid-Ocean Ridges. Am Geophys Union Monogr 106:59-116

Peter JM, Shanks WC III (1992) Sulfur, carbon, and oxygen isotope variations in submarine hydrothermal deposits of Guaymas Basin, Gulf of California. Geochim Cosmochim Acta 56:2025-2040

Petersen S, Herzig PM, Hannington MD (1998) Fluid inclusion studies as a guide to the temperature regime within the TAG hydrothermal mound, 26°N, Mid-Atlantic Ridge. Proc ODP, Sci Results 158:163-178

Pichler T, Giggenbach WF, McInnes BIA, Buhl D, Duck B (1999) Fe sulfide formation due to seawater-gas-sediment interaction in a shallow-water hydrothermal system at Lihir Island, Papua New Guinea. Econ Geol 94:281-288

Pottorf RJ, Barnes HL (1983) Mineralogy, geochemistry, and ore genesis of hydrothermal sediments from the Atlantis II Deep, Red Sea. Econ Geol Monogr 5:198-223

Ramboz C, Oudin E, Thisse Y (1988) Geyser-type discharge in Atlantis II Deep, Red Sea: evidence of boiling from fluid inclusions in epigenetic anhydrite. Can Mineral 26:765-786

Richet P, Bottinga Y, Javoy M (1977) A review of hydrogen, carbon, nitrogen, oxygen, sulfur, and chlorine stable isotope fractionation among gaseous molecules. Ann Rev Earth Planet Sci 5:65-110,

Riciputi LR (1996) A comparison of extreme energy filtering and high mass resolution techniques for the measurement of $^{34}S/^{32}S$ ratios by ion microprobe. Rapid Commun Mass Spec 10:282-286

Rona PA, Scott SD (1993) A special issue on sea-floor hydrothemal mineralization: new perspectives. Econ Geol 88:1935-1976

Rubin KH, Macdougall JD, Perfit MR (1994) ^{210}Po-^{210}Pb dating of recent volcanic eruptions on the seafloor. Nature 368:841-844

Ryan MP (1994) Neutral-buoyancy controlled magma transport and storage in mid-ocean ridge magma reservoirs and their sheeted-dike complex; a summary of basic relationships. Magmatic Systems 57:97-138

Rye RO, O'Neil JR (1968) The O^{18} content of water in primary fluid inclusions from Providencia, North-Central Mexico. Econ Geol 63:232-238

Rye RO (1993) The evolution of magmatic fluids in the epithermal environment - the stable isotope perspective. Econ Geol 88:733-753

Sarano F, Murphy RC, Houghton BF, Hedenquist JW (1989) Preliminary observations of submarine geothermal activity in the vicinity of White Island Volcano, Taupo Volcanic Zone, New Zealand. J Royal Soc New Zealand 19:449-459

Saunders AD, Fornari DJ, Morrison MA (1982) The composition and emplacement of basaltic magmas produced during the development of continental-margin basins: the Gulf of California, Mexico. J Geol Soc London 139:335-346

Schoell M, Faber E (1978) New isotopic evidence for the origin of Red Sea brines. Nature 275:436-438

Seal RR II, Alpers CN, Rye RO (2000) Stable isotope systematics of sulfate minerals. *In* CN Alpers, JL Jambor, DK Nordstrom (eds) Sulfate Minerals: Crystallography, Geochemistry and Environmental Significance. Rev Mineral 40:541-602

Seewald JS, Doherty KW, Hammar TR, Liberatore SP (2001) A new gas-tight isobaric sampler for hydrothermal fluids. Deep Sea Research (in press)

Seewald JS, Seyfried WE Jr, Shanks WC III (1994) Variations in the chemical and stable isotope composition of carbon and sulfur species during organic-rich sediment alteration: An experimental and theoretical study of hydrothermal activity in Guaymas Basin, Gulf of California. Geochim Cosmochim Acta 58:5065-5082

Seyfried WE Jr (1987) Experimental and theoretical constraints on hydrothermal alteration processes at mid-ocean ridges. Ann Rev Earth Planet Sci 15:317-335

Seyfried WE Jr, Chen X, Chan L-H (1998) Trace element mobility and lithium exchange during hydrothermal alteration of seafloor weathered basalt; an experimental study at 350°C, 500 bars. Geochim Cosmochim Acta 62:949-960

Seyfried WE Jr, Ding K, Berndt ME (1991) Phase equilibria constraints on the chemistry of hot spring fluids at mid-ocean ridges. Geochim Cosmochim Acta 55:3559-3580

Seyfried WE Jr, Ding K (1995) Phase equilibria in subseafloor hydrothermal systems; a review of the role of redox, temperature, pH and dissolved Cl on the chemistry of hot spring fluids at mid-ocean ridges. *In* SE Humphris, RA Zierenberg, LS Mullineaux, RE Thomson (eds) Seafloor Hydrothermal Systems: Physical, Chemical, Biological, and Geological Interactions. Am Geophys Union, Washington, DC, p 248-272

Seyfried WE Jr, Ding K, Berndt ME, Chen X (1999) Experimental and theoretical controls on the composition of mid-ocean ridge hydrothermal fluids. *In* CT Barrie, MD Hannington (eds) Volcanic-associated massive sulfide deposits; processes and examples in modern and ancient settings. Rev Econ Geol 8:181-200

Seyfried WE Jr, Janecky DR, Mottl MJ (1984) Alteration of the oceanic crust; implications for geochemical cycles of lithium and boron. Geochim Cosmochim Acta 48:557-569

Shanks WC III, Bischoff JL (1977) Ore transport and deposition in the Red Sea geothermal system: a geochemical model. Geochim Cosmochim Acta 41:1507-1519

Shanks WC III, Bischoff JL (1980) Geochemistry, sulfur isotope composition, and accumulation rates of Red Sea geothermal deposits. Econ Geol 74:445-459

Shanks WC III, Bischoff JL, Rosenbauer RJ (1981) Seawater sulfate reduction and sulfur isotope fractionation in basaltic systems: interaction of seawater with fayalite and magnetite at 200-350°C. Geochim Cosmochim Acta 45:1977-1995

Shanks WC III, Crowe DE, Johnson C (1998) Sulfur isotope analyses using the laser microprobe *In* MA McKibben, WC Shanks, III, WI Ridley (eds) Applications of Microanalytical Techniques to Understanding Mineralizing Processes. Soc Econ Geol 7:141-153

Shanks WC III, Bohlke JK, Seal RR (1995) Stable isotopes in mid-ocean ridge hydrothermal systems: Interaction between fluids, minerals and organisms. *In* SE Humphris, RA Zierenberg, LS Mullineaux, RE Thomson (eds) Seafloor Hydrothermal Systems: Physical, Chemical, Biological, and Geological Interactions. Am Geophys Union, Washington, DC, p 194-221

Singh SC, Collier JS, Harding AJ, Kent GM, Orcutt JA (1999) Seismic evidence for a hydrothermal layer above the solid roof of the axial magma chamber at the southern East Pacific Rise. Geology 27:219-222

Sleep NH (1991) Hydrothermal circulation, anhydrite precipitation, and thermal structure at ridge axes. J Geophys Res 96(B2):2375-2387

Sohn RA, Fornari DJ, Von Damm KL, Hildebrand JA, Webb SC (1998) Seismic and hydrothermal evidence for a cracking event on the East Pacific Rise crest at 9°50'N. Nature 396:159-161

Spiess FN, et al (1980) East Pacific Rise:Hot springs and geophysical experiments. Science 207:1421-1432

Spivack AJ, Edmond JM (1987) Boron isotope exchange between seawater and the oceanic crust. Geochim Cosmochim Acta 51:1033-1043

Stakes DS, O'Neil JR (1982) Mineralogy and stable isotope geochemistry of hydrothermally altered oceanic rocks. Earth Planet Sci Lett 57:285-304

Stoffers P, Hannington M, Wright I, Herzig P, de Ronde C, Arpe T, Battershill C, Botz R, Britten K, Browne P, Cheminee JL, Fricke HW, Garbe-Schoenberg D, Hekinian R, Hissman K, Huber R, Robertson J, Schauer J, Schmitt M, Scholten J, Schwarz-Schampera U, Smith I (1999) Elemental mercury at submarine hydrothermal vents in the Bay of Plenty, Taupo volcanic zone, New Zealand. Geology 27:931-934

Styrt MM, Brackmann AJ, Holland HD, Clark BC (1981) The mineralogy and isotopic composition of sulfur in hydrothermal sulfide/sulfate deposits of the East Pacific Rise, 21°N latitude. Earth Planet Sci Lett 53:382-390

Taylor BE (1986) Magmatic volatiles: isotopic variation of C, H, and S. *In* Stable Isotopes in High Temperature Geological Processes. Valley JW, Taylor HP Jr, O'Neil JR (eds) Rev Mineral 16:185-225

Taylor BE (1992) Degassing of CO_2 from Kilauea: Carbon isotope evidence and implications for magmatic contributions to sediment-hosted submarine hydrothermal systems. Rept Geol Surv Japan 279:205-206

Taylor HP Jr (1974) The application of oxygen and hydrogen isotope studies to problems of hydrothermal alteration and ore deposition. Econ Geol 69:843-883

Teagle DAH, Alt JC, Chiba H, Halliday AN (1998a) Dissecting an active hydrothermal deposit; the strontium and oxygen isotopic anatomy of the TAG hydrothermal mound- anhydrite. Proc ODP, Sci Results 158:129-142

Teagle DAH, Alt JC, Humphris SE, Halliday AN (1998b) Dissecting an active hydrothermal deposit; the strontium and oxygen isotopic anatomy of the TAG hydrothermal mound; whole rock and silicate minerals. Proc ODP, Sci Results 158:297-309

Tivey MK (1995) Modeling chimney growth and associated fluid flow at seafloor hydrothermal vent sites. *In* SE Humphris, RA Zierenberg, LS Mullineaux, RE Thomson (eds) Seafloor Hydrothermal Systems:

Physical, Chemical, Biological, and Geological Interactions, p 158-177. American Geophysical Union, Washington, DC

Tivey MK, McDuff RE (1990) Mineral precipitation in the walls of black smoker chimneys: A quantitative model of transport and chemical reaction. J Geophys Res 95:12617-12637

Tivey MK, Mills RA, Teagle DAH (1998) Temperature and salinity of fluid inclusions in anhydrite as indicators of seawater entrainment and heating in the TAG active mound. Proc ODP, Sci Results 158:179-190

Von Damm KL (1990) Seafloor hydrothermal activity; black smoker chemistry and chimneys. Ann Rev Earth Planet Sci 18:173-204

Von Damm KL, Bischoff JL (1987) Chemistry of hydrothermal solutions from the Southern Juan de Fuca Ridge. J Geophys Res 92:11334-11346

Von Damm KL, Bischoff JL, Rosenbauer RJ (1991) Quartz solubility in hydrothermal seawater—an experimental study and equation describing quartz solubility for up to 0.5 M NaCl solutions. Am J Sci 291:977-1007

Von Damm KL, Buttermore LG, Oosting SE, Bray AM, Fornari DJ, Lilley MD, Shanks WC III (1997) Direct observation of the evolution of a seafloor "black smoker" from vapor to brine. Earth Planet Sci Lett 149:101-111

Von Damm KL, Edmond JM, Grant B, Measures CI, Walden B, Weiss RF (1985) Submarine hydrothermal solutions at 21°N, East Pacific Rise. Geochim Cosmochim Acta 49:2197-2220

Von Damm KL, Grant B, Edmond JM (1983) Preliminary report on the chemistry of hydrothermal solutions at 21° North, East Pacific Rise. *In* PA Rona, K Bostroem, L Laubier, KL Smith Jr (eds) Hydrothermal Processes at Seafloor Spreading Centers. Plenum Press, New York, p 369-389

Von Damm KL, Oosting SE, Kozlowski R, Buttermore LG, Colodner DC, Edmonds HN, Edmond JM, Grebmeier JM (1995) Evolution of East Pacific Rise hydrothermal vent fluids following a volcanic eruption. Nature 375:47-50

Welhan JA, Craig H (1983) Methane, hydrogen and helium in hydrothermal fluids at 21°N on the East Pacific Rise. *In* Rona, PA, Bostrom, K, Laubier, L, Smith, KL, Jr (eds) Hydrothermal Processes at Seafloor Spreading Centers. NATO conference series: IV, Marine Sciences. Plenum Press, New York, p 391-409

Welhan, JA, Craig H (1979) Methane and hydrogen in East Pacific Rise hydrothermal fluids. Geophys Res Lett 6:829-831

Welhan, JA, Lupton JE (1987) Light hydrocarbon gases in Guaymas Basin hydrothermal fluids: Thermogenic versus abiogenic origin. Am Assoc Petrol Geol Bull 71:215-223

Werner C-D, Pilot J (1997) 26.Data report: Geochemistry and mineral chemistry of ultramafic rocks from the Kane Area (MARK). *In* Karson JA, Cannat M, Elthon D, Eds, Proc ODP, Sci Results 153:457-470

Wolery TJ, Sleep NH (1976) Hydrothermal circulation and geochemical flux at mid-ocean ridges. J Geol 84:249-275

Winckler G, Kipper R, Aeschbrach-Hertig W, Botz R, Schmidt M, Schuller S, Bayer R (2000) Sub-seafloor boiling of Red Sea brines; new indication from noble gas data. Geochim Cosmochim Acta 64:1567-1575

Woodruff LG, Shanks WC III (1988) Sulfur isotope study of chimney minerals and vent fluids from 21°N, East Pacific Rise:Hydrothermal sulfur sources and disequilibrium sulfate reduction. J Geophys Res 93:4562-4572

Zierenberg RA, Fouquet Y, Miller DJ, Bahr JM, Baker PA, Bjerkgard T, Brunner CA, Duckworth RC, Gable R, Gieskes JM, Goodfellow WD, Groeschel-Becker HM, Guerin G, Ishibashi J, Iturrino GJ, James RH, Lackschewitz KS, Marquez LL, Nehlig P, Peter JM, Rigsby CA, Schultheiss PJ, Shanks WC III, Simoneit BRT, Summit M, Teagle DAH, Urbat M, Zuffa GG (1998) The deep structure of a sea-floor hydrothermal deposit. Nature 392:485-488

Zierenberg RA, Shanks WC III (1983) Mineralogy and geochemistry of epigenetic features in metalliferous sediment, Atlantis II Deep, Red Sea. Econ Geol 78:57-72

Zierenberg RA, Shanks WC III (1988) Isotopic studies of epigenetic features in metalliferous sediment, Atlantis II Deep, Red Sea. Can Mineral 26:737-753

9   # Oxygen- and Hydrogen-Isotopic Ratios of
Water in Precipitation: Beyond Paleothermometry

Richard B. Alley

Environment Institute and *Department of Geosciences*
The Pennsylvania State University
Deike Building
University Park, Pennsylvania 16802

Kurt M. Cuffey

Department of Geography
University of California
McCone Hall
Berkeley, California 947.0

INTRODUCTION

Paleoclimatologists face a dilemma. No sedimentary proxy is a pure recorder of quantitative climate information. Yet climate modelers and policy-makers increasingly seek quantitative comparisons between instrumentally documented, possibly anthropogenic, climate changes and those produced naturally in the past.

In the second edition of his influential book, Bradley (1999, p. 6) discussed the calibration of proxy records to learn past climate changes: "Calibration involves using modern climatic records and proxy materials to understand how, and to what extent, proxy materials are climate-dependent. It is assumed that the modern relationships observed have operated, unchanged, throughout the period of interest (the principle of uniformitarianism)." In other words, one relates the characteristics of sediment to climate at different places for one time, or at a place for short times, and then uses that relation plus characteristics of older sediments to estimate the climatic conditions that produced those sedimentary characteristics. Bradley (1999) then extensively discussed the difficulties in applying this methodology in a complex world with imperfect recorders; nonetheless, the goal of using calibrated proxies for quantitative as well as qualitative paleoclimatic reconstruction is clear.

A prominent recent use of calibrated paleoclimatic data is the assessment of whether the probably-anthropogenic warming of the latter 20th century moved beyond the band of natural variability of the prevailing climate. Bradley (2000) combined recent instrumental records with several longer proxy-based reconstructions of surface temperature including that of Mann et al. (1999). Based on this composite data set, Bradley (2000) argued that "temperatures in the late 20th century were unique in the context of the entire millennium". The proxy records were primarily based on tree-ring data, but included isotopic and major-element geochemistry of corals, and occurrence of melt layers and isotopic ratios of water in ice cores. However, Broecker (2001) questioned the basis for the reconstruction of Mann et al. (1999), arguing that the proxy indicators lack sufficient accuracy to reliably resolve the small climate changes under consideration.

Because of the implications for possible human causation of global warming, this is not just an academic issue. Rather, the ability of paleoclimatic indicators to allow quantitative reconstruction of paleotemperatures and other paleoclimatic variables influences public perceptions and perhaps public policy. It is much easier to believe that an event was human-caused if it is unique over the last millennium than if the recent change duplicates earlier, natural fluctuations.

1529-6466/00/0043-0009$05.00

Our discussions with colleagues indicate that there is a wide range of opinions in the community on these issues. On one end of the spectrum is the desire to use our best understanding of proxies calibrated against modern or recent conditions to produce the quantitative estimates and uncertainties of paleoclimatic conditions desired by modelers and policy-makers. On the other end is the worry that we cannot assess the uncertainties because the uniformitarian hypothesis of time-invariance of the calibration is inherently untestable; instead, we might focus on producing consistent interpretations, multi-parameter reconstructions of likely conditions, or some other less-precise or less-quanti-tative though perhaps still-useful interpretation of the paleoclimatic data.

Reality lies somewhere between these extremes. Some paleoclimatic proxies are so clearly and directly related to paleoclimatic conditions that quantitative estimates of past climate changes and their uncertainties can be made with considerable although not total confidence. In other cases, the tie between proxy and climate is sufficiently complex that direct quantitative interpretation is risky. Such proxies can nonetheless be appropriate for generating hypotheses, for allowing quantitative estimates if used in multi-parameter reconstructions with independent assumptions, and for allowing quantitative hypothesis-testing of models that directly compute proxy characteristics rather than the climate variables related to the proxies. The value of research on paleoclimate proxies thus extends well beyond possible direct quantitative reconstructions.

The range of paleoclimatic proxies is much too broad to allow useful discussion in one short article. We thus choose to focus on one well-understood and prominent system, the isotopic ratios of oxygen or hydrogen in water in precipitation as a proxy for temperature. We summarize some of the evidence for this paleothermometer, and some of the difficulties. Next, we discuss two case studies showing large inaccuracy of this paleothermometer, which argues strongly that combined uncertainties are large enough to cast doubts on reconstructions that are not independently supported. These, and other studies, are motivating an ongoing trend away from paleothermometry based solely on oxygen- and hydrogen-isotopic compositions of water, a trend that is here encouraged. Instead of attempts at direct paleothermometry, oxygen and hydrogen isotopes of water can be used to test Earth-system models focused on the hydrologic cycle, and for paleothermometry if combined with other indicators.

BASIS OF PALEOTHERMOMETRY USING OXYGEN- AND HYDROGEN-ISOTOPIC RATIOS OF WATER

Hydrogen- or oxygen-isotopic ratios of water from precipitation have long been used in estimating past Earth-surface temperatures. A strong correlation is observed between surface temperatures and modern or recent deuterium:hydrogen (D:H) and $^{18}O:^{16}O$ ratios of water from precipitation, on seasonal, interannual, and multiannual scales over the instrumental record (Dansgaard 1964; Rozanski et al. 1993; Fricke and O'Neil 1999). Arguably the most frequently used data sets related to past temperatures are the histories of the ratios $^{18}O:^{16}O$ or D:H of water from the ice of the GRIP and GISP2 cores of central Greenland (Grootes and Stuiver 1997; Johnsen et al. 1997), from the Vostok core of central East Antarctica (Fig. 1) (Petit et al. 1999), and from other ice cores (e.g. Thompson et al. 1998). Typically, isotopic composition of water is reported as $\delta^{18}O$ or δD, respectively the per mil deviation of $^{18}O:^{16}O$ or D:H from an approximation of the oceanic ratio V-SMOW, for Vienna Standard Mean Ocean Water.

Among other sources, water for paleoclimatic reconstructions can be obtained from ice cores, groundwater, and speleothem fluid inclusions. In addition, the $\delta^{18}O$ or δD of water can be inferred from ratios in materials including cellulose and shells that equilibrated with water in the past and have not re-equilibrated (e.g. White 1989).

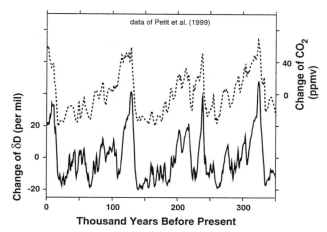

Figure 1. Time variation of carbon dioxide concentration and δD measured on the Vostok ice core (Petit et al. 1999). These data span three complete glacial-interglacial cycles. Carbon dioxide changes must result from global-scale changes in biogeochemical cycling. These are strongly, though not perfectly, related to the water isotope history (covariance of the two is 0.64 for 0-150 kyr b.p. (thousand years before present), and is 0.74 for 150-350 kyr b.p.). A traditional interpretation of the δD curve is that it measures temperature changes, with magnitude approximately 8°C from glacial to interglacial.

Because such materials often can be dated, the history of $\delta^{18}O$ or δD of water becomes a provisional history of surface temperatures.

However, this simple picture is complicated by a large number of factors other than site temperature that can affect $\delta^{18}O$ or δD of "fossil" waters. In aggregate, the effect of these additional factors may be small, or may exceed the effect of temperature. For this reason, much recent work has moved away from paleothermometry based on $\delta^{18}O$ or δD. Instead, $\delta^{18}O$ or δD are increasingly being used as hydrologic tracers. Much of the power of these analyses results from coupling of interpretations of $\delta^{18}O$ or δD of water with temperature reconstructions based on other techniques, which often use multiple parameters and which often exploit simpler systems.

The saturation vapor pressure of $H_2{}^{18}O$ or HDO is lower than for $H_2{}^{16}O$ (e.g. Dansgaard 1964; Gat 1996). The approximately exponential increase of saturation vapor pressure with increasing absolute temperature causes air-mass cooling to be the primary driving force for condensation. Thus, progressive cooling of air masses leads to precipitation of water enriched in ^{18}O or D compared to the vapor remaining in the air mass (Fig. 2). The resulting progressive depletion of ^{18}O or D in the air mass forces water in subsequent precipitation to have lower $\delta^{18}O$ or δD (Fig. 3). The dominant source for water precipitating on land masses is evaporation from warm low- and mid-latitude oceans. Cooling from poleward movement of air away from these sources, or upward movement by convection, frontal activity, or orography, then produces progressive depletion of ^{18}O and D in water in the air mass. Air masses transported to colder geographic regions are cooled systematically and the amount of vapor remaining in the air decreases systematically as the temperature decreases. Hence, colder places have precipitation with lower $\delta^{18}O$ and δD. (A strong linear relation in mean-annual precipitation between $\delta^{18}O$ or δD allows them to be used almost interchangeably,

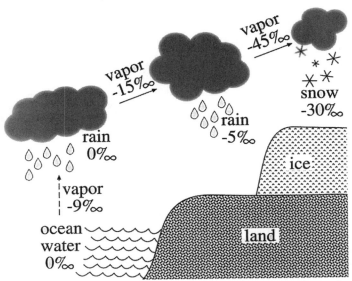

Figure 2. Cartoon of isotopic fractionation during evaporation and precipitation over ocean, and additional precipitation on land. The isotopic changes are illustrative only, and depend on the temperature at the time of condensation, the amount of moisture removed, whether the condensation occurs to water or ice, and other factors. The general trends are accurate.

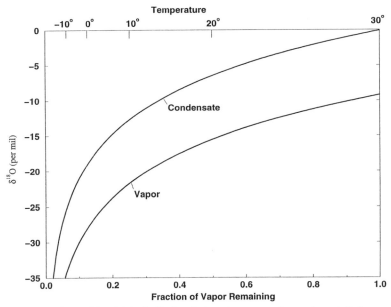

Figure 3. Effect of Rayleigh distillation during air-mass cooling and precipitation on $\delta^{18}O$ of condensate and of vapor remaining in the air mass, calculated following Faure (1977, p. 327) using a temperature-independent fractionation factor.

although as discussed below, subtle deviations from this linear relation provide additional useful information.)

This theoretical basis is supported by the clear relation between $\delta^{18}O$ or δD of precipitation and site temperatures, obtained when considering multi-annual averages, year-to-year variability, and summer-to-winter changes (Dansgaard 1964; Rozanski et al. 1993; Fricke and O'Neil 1999). The correlation between mean-annual $\delta^{18}O$ or δD ratios of precipitation and mean-annual temperatures from different sites is especially strong, with a change per degree Celsius of ~0.6 to 0.7 per mil in $\delta^{18}O$, or of ~5 per mil in δD (Fig. 4). By a substitution of space for time (sometimes referred to as an ergodic substitution), combination of this recent mean-annual correlation with isotopic ratios of older waters yields an admittedly naive first estimate of paleotemperatures (e.g. Dansgaard and Oeschger 1989, p. 296).

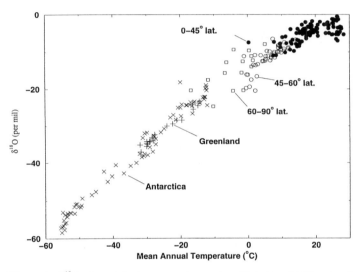

Figure 4a. $\delta^{18}O$ of precipitation versus temperature. Data from IAEA network (Fricke and O'Neil 1999; calculated as the means of the summer and winter data of their Table 1 for all sites with complete data), and from Greenland (× ; Johnsen et al. 1989) and Antarctica (+ ; Dahe et al. 1994). For the IAEA data, □ = poleward of 60° latitude (but with no inland ice-sheet sites), ○ = from 45° to 60°, and ● = equatorward of 45°. About 71% of the Earth's surface area is equatorward of 45°, where dependence of $\delta^{18}O$ on temperature is weak to nonexistent. Only 16% of Earth's surface falls in the 45° to 60° band, with only 13% poleward of 60°. The linear array is clearly dominated by data from the ice sheets.

We note, however, that the very strong cooling and isotopic distillation associated with the relatively small polar regions and especially the ice sheets of Greenland and Antarctica can give a misleading impression of the strength of the relation between temperature and $\delta^{18}O$ or δD over much of the Earth. Figure 4a is an updated plot similar to Dansgaard (1964). Fricke and O'Neil (1999) summarized the data set on precipitation isotopic ratios from the International Atomic Energy Agency (IAEA; www.iaea.org). Figure 4a includes all of the sites with complete data in Fricke and O'Neil (1999), shown for each site as the average of the summer and winter precipitation $\delta^{18}O$ plotted against the average of summer and winter station temperatures. Latitude bands are shown by different symbols as indicated. Additional data from the polar ice sheets are from Johnsen et al. (1989) for Greenland, and Dahe et al. (1994) for Antarctica. More than 70% of the Earth's surface lies equatorward of 45° latitude, so Figure 4a shows that most of the linear relation between $\delta^{18}O$ of precipitation water and temperature is based on data from a small part of the Earth, and especially on results from the ~3% of the Earth covered by the ice sheets of Greenland and Antarctica.

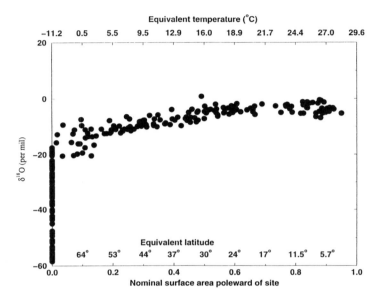

Figure 4b. Area-normalized display of data from Figure 4a. The regression line of temperature versus latitude for the IAEA data of Fricke and O'Neil (1999) was used to assign to each site an equivalent latitude corresponding to its temperature, with points plotting beyond 90° assigned to 90°. Precipitation $\delta^{18}O$ ratios are plotted against the fraction of the Earth's surface poleward of this equivalent latitude. This emphasizes that the systematic variation of the $\delta^{18}O$ of precipitation with temperature is small compared to the variability about that trend for most of the Earth's surface. The equivalent temperature shown along the top is the temperature corresponding to the equivalent latitude of the regression line.

The relative insensitivity of $\delta^{18}O$ of precipitation to temperature is emphasized by replotting the data from Figure 4a in an area-weighted way (Fig. 4b). Mean annual temperature depends both on latitude and altitude. We have made a crude correction for this altitude effect by assigning to each site an equivalent latitude. We formed the regression line between temperature and latitude for the IAEA data, and then assigned to each site in the complete data set the equivalent latitude given by the site temperature and this regression line. The IAEA data set lacks points on the large ice sheets, which are anomalously cold owing to their very high latitude and elevation and to the radiative effects of the permanent snow cover, and so are colder than allowed by the regression; we assign such points an equivalent latitude of 90°. We plot in Figure 4b the mean-annual $\delta^{18}O$ of precipitation at a site against the fraction of the Earth's total surface area poleward of the equivalent latitude corresponding to each site temperature. The abscissa (x axis) in Figure 4b thus is an area-weighted temperature scale. Figure 4b shows that for roughly half of the Earth's surface area there is little dependence of $\delta^{18}O$ of precipitation

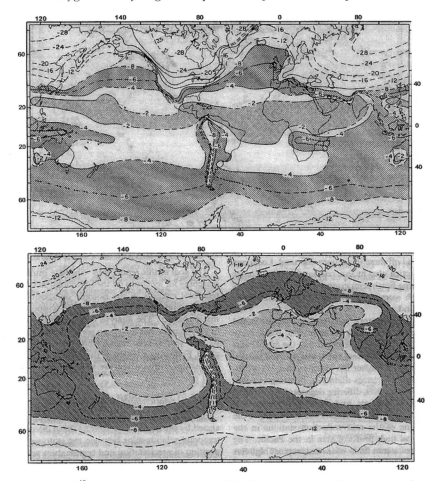

Figure 5. δ^{18}O of precipitation at more than 80 IAEA sites summarized by Lawrence and White (1991), for January (top) and July (bottom). A general tendency is evident for lower values at higher latitudes, and lower values during winter, However, it is clear that much additional information is present in the δ^{18}O data.

on temperature, with most of the dependence restricted to the anomalously cold regions on ice sheets.

Figure 5 shows the spatial distribution of the δ^{18}O of January and July precipitation waters, based on more than 80 IAEA sites as summarized by Lawrence and White (1991). The strong patterns indicate that there must be important information in such isotopic data. Further, these patterns are largely reproduced by GCM models equipped with isotopic tracers (Jouzel et al. 1987), indicating that local physical processes responsible for isotopic distillation are reasonably well understood.

Those sites for which the correlation between monthly mean δD of precipitation water and monthly mean surface temperature is significant are indicated by ● in Figure 6, and the sites with a significant correlation between monthly mean δD of precipitation

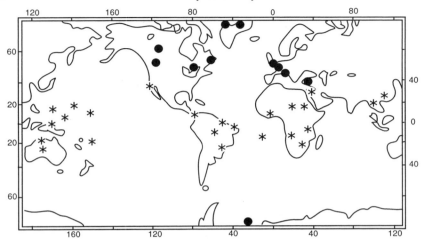

Figure 6. Those sites in the data base of Figure 5 from Lawrence and White (1991) for which there is a significant correlation between monthly means of precipitation δD and temperature (·) (11 of the more than 80 sites considered) and between monthly means of precipitation δD and amount of precipitation (*) 24 of the more than 80 sites considered). Those three sites plotted touching the north or south latitudinal limits actually should plot at higher latitude than included on the map. A general pattern is evident of high-latitude correlation between $\delta^{18}O$ and temperature, and a low-latitude correlation between $\delta^{18}O$ and precipitation amount.

water and amount of monthly precipitation reaching the surface are shown as * in Figure 6. The relatively few sites (11 out of more than 80) with a significant correlation between temperature and $\delta^{18}O$, and their restricted spatial distribution (remembering the gross distortion of the projection used; see Fig. 3), emphasize the shortcomings of using δD or δ^8O of water from precipitation as a paleothermometer. The correlation with precipitation amount is discussed further below.

COMPLICATIONS

Complications to interpretation of water from precipitation in terms of surface temperature arise from processes at the moisture source, during transport and precipitation, and subsequent to precipitation on the surface. The $\delta^{18}O$ or δD of each "packet" of a few drops of precipitation at a site depends on conditions at the moisture source(s) and along the path(s) to the site. The annually (or decadally or other) averaged $\delta^{18}O$ or δD of precipitation at the site is a weighted mean of different packets, and that weighting may change. Local physical processes can alter that weighted mean further, and larger-scale transport processes such as ice flow can disconnect this mean from conditions at a study site. Here, we briefly review each of these in turn, as summarized in Figure 7.

Source effects on $\delta^{18}O$ or δD of precipitation

Most evaporation on Earth occurs from low-latitude oceans. The surface-water $\delta^{18}O$ of the modern open oceans varies spatially by about 1 per mil (about 8 per mil for δD) (Fig. 8), and more in restricted basins (e.g. Craig and Gordon 1965). Surface water has higher $\delta^{18}O$ and δD in regions where evaporation causes greater net removal of vapor, which has lower $\delta^{18}O$ and δD than the water left behind; surface water has lower $\delta^{18}O$ and δD where precipitation or runoff (including melting of icebergs) return more water,

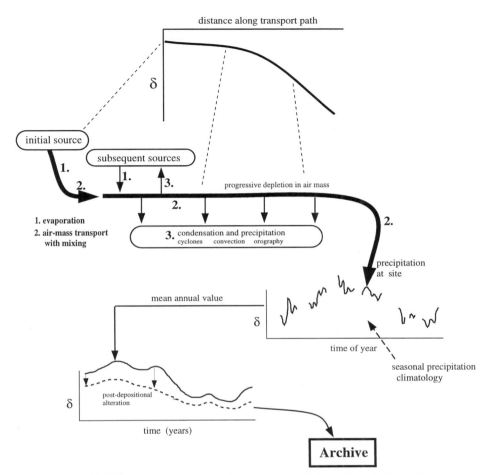

Figure 7. Schematic representation of the processes that contribute to the isotopic composition of water in a paleoclimatic archive.

which has lower $\delta^{18}O$ and δD than the oceans.

Over the most recent ice-age cycle, and probably over all Pleistocene ice-age cycles of the last ~1 million years, the mean $\delta^{18}O$ of water in the oceans has varied about 1 per mil (8 per mil for δD) owing to growth and melting of large ice sheets that sequester water with low $\delta^{18}O$ and δD (Schrag et al. 1996). The glacial-maximum ice sheets grew initially from precipitation on low-elevation land masses that were not especially cold. Cooling associated with the radiative and topographic effects of the ice sheets progressively shifted the precipitation to lower $\delta^{18}O$ and δD, and ice flow caused this precipitation to progressively replace the ice from earlier precipitation with higher $\delta^{18}O$ and δD. However, equilibrium was not reached; had glacial-maximum lasted longer, the mean $\delta^{18}O$ and δD of the water in ice sheets would have become lower while the mean $\delta^{18}O$ and δD of the oceans became higher (Mix and Ruddiman 1984). Hence, there is at least a possibility of larger depletion of the ocean in ^{18}O and D during pre-Pleistocene glacial maxima owing to ice-sheet processes if closer approach to equilibrium was

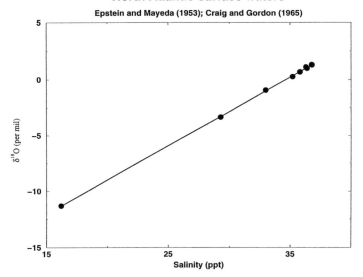

North Atlantic surface waters

Epstein and Mayeda (1953); Craig and Gordon (1965)

Figure 8. Surface-water $\delta^{18}O$ versus salinity for the north Atlantic, from Craig and Gordon (1965) and Epstein and Mayeda (1953). Those samples with the lowest $\delta^{18}O$ are from the coast of Greenland.

achieved, or if larger ice sheets grew than during the Wisconsinan glacial maximum. And, the modern world retains significant ice with low $\delta^{18}O$ and δD, so we are not at the extreme of conditions. In total, then, ice-sheet and evaporative processes are capable of altering open-ocean surface-water $\delta^{18}O$ by a few per mil (and δD about 8-fold more), with a similar effect on precipitation derived from water evaporated from that surface ocean. Water evaporated from restricted marine basins, or from land surfaces, may vary more than this.

Equilibrium isotopic-fractionation factors for ^{18}O and D during evaporation and condensation are temperature-dependent, but the main control on degree of fractionation of precipitation ^{18}O and D from the source composition is the fraction of water extracted from an air mass before condensation of the precipitation of interest (e.g. Dansgaard 1964). This in turn is controlled primarily by the difference between the temperature at which condensation first begins for an air mass, and the temperature at which the precipitation of interest condenses. The first-condensation temperature in turn is tied closely to the ocean surface temperature by the high and more-or-less constant relative humidities over most ocean surfaces (Vimeux et al. 2001). Hence, $\delta^{18}O$ and δD of precipitation depend in part on the temperatures at all contributing moisture sources, and not just on the temperature at the site.

A partial correction can be made for this source-temperature influence on $\delta^{18}O$ and δD of precipitation by using a derived quantity called the deuterium excess, **d**, of that precipitation, although this procedure needs further verification. A plot of precipitation δD versus $\delta^{18}O$ produces a meteoric water line with a slope of about 8 (Craig 1961). Adopting the slope of 8 exactly, the $\delta^{18}O$ and δD of any sample define a line with an intercept called the deuterium excess, **d** (Dansgaard 1964). This varies spatially and temporally in response to both equilibrium and kinetic effects during evaporation, which are in turn functions of temperature and relative humidity in source regions (Merlivat and Jouzel 1979). To the extent that ocean-surface relative humidity and temperature maintain a persistent correlation over time, and to the extent that there is no variability of subsequent disequilibrium fractionations along the transport path, changes in the deuterium excess can be used to estimate changes in the source temperature, and thus their effect on precipitation $\delta^{18}O$ and δD (Johnsen et al. 1989; Cuffey and Vimeux, in press).

Path effects on $\delta^{18}O$ or δD of water in precipitation

The classical assumption used in many isotopic studies is of Rayleigh distillation (Fig. 3): an air mass containing water vapor evaporated from the low-latitude ocean surface moves poleward, with cooling causing condensation and precipitation of water enriched in ^{18}O and D relative to the vapor remaining in the air mass. This is a useful model, but it neglects the complexity of air-mass transport and recharge (e.g. Fisher 1992). The general circulation models of the atmosphere (GCMs) that have been equipped with isotopic tracers are the state-of-the-art for isotopic data interpretation (e.g. Jouzel et al. 1987; Charles et al. 1994; Joussaume and Jouzel 1993; Jouzel et al. 1997; Hoffman et al. 1998).

The complications have been brought into focus by Hendricks et al. (2000), whose zonally averaged model demonstrates that the classic Rayleigh model is an end-member behavior that is unlikely to be fully realized, and that deviations from the assumed behavior will affect $\delta^{18}O$ and δD of water in precipitation. Hendricks et al. (2000) developed a simple, zonally averaged, parameterized model of isotopic ratios of water in precipitation, and noted that much of the variability can be understood in terms of two parameters: the ratio of recycled vapor (re-equilibrated or re-evaporated precipitation) to that originally in the air mass, and the importance of advection versus mixing in the atmospheric transport. The classical model is the zero-recycling, zero-mixing end-member.

If atmospheric transport is purely by eddy diffusion (that is, by turbulent mixing in features smaller than the sub-global time-averaged dominant circulation of the atmosphere), the change in $\delta^{18}O$ or δD of condensate per degree cooling is only half as large as for pure advection assumed in the Rayleigh model, owing to the effects during cooling of mixing of different air parcels containing vapor that is less-depleted and more-depleted in ^{18}O and D (Eriksson 1965). And, in the limit of precipitation being completely balanced by air mass recharge from ocean sources, no progressive depletion of ^{18}O and D occurs in condensate.

Hendricks et al. (2000) found that between about 45°N and S, important re-evaporation of water allows only weak dependence of $\delta^{18}O$ and δD on temperature. This is quite evident in examination of Figure 4, which shows that most of the linear dependence of $\delta^{18}O$ of precipitation on temperature arises from polar and subpolar sites. Poleward of 45°S, re-evaporation is minor, although this is less true in the north. In the south, the data for the dependence of $\delta^{18}O$ of precipitation on surface temperature fall between those slopes expected for purely advective transport (0.95 per mil per degree) and diffusive transport (0.72 per mil per degree) (Hendricks et al. 2000). The geographic simplicity of the high southern latitudes offers the hope of relatively accurate paleothermometry using $\delta^{18}O$ and δD of water; for the rest of the globe, greater uncertainty suggests significant caution, especially at lower latitudes.

Several "path effects" have been given names in the literature. For example, the "amount effect" (Dansgaard 1964) is the observation that locations or times with more precipitation have anomalously lower $\delta^{18}O$ and δD in that precipitation (Fig. 6). This amount effect may have more than one cause. The first precipitate has higher $\delta^{18}O$ and δD than the vapor from which it forms. As precipitation proceeds, the vapor composition evolves towards lower $\delta^{18}O$ and δD, which in turn causes the subsequent precipitate to evolve towards lower $\delta^{18}O$ and δD. If all of the vapor is precipitated, then the average $\delta^{18}O$ and δD of that precipitate must match those of the original vapor. Thus, if only a little precipitation is extracted, it will have relatively high $\delta^{18}O$ and δD; extraction of more condensate from a given air mass without extensive resupply of moisture, perhaps owing to convective activity causing extensive cooling, must extract water with lower and lower $\delta^{18}O$ and δD. Variation of more than 10 per mil in $\delta^{18}O$ has been observed within a single storm (Rindsberger et al. 1990). Convective versus stratiform clouds may exhibit different behavior (Lawrence et al. 1982). Many tropical stations show an annual cycle in $\delta^{18}O$ and δD of precipitation controlled by the amount effect (Rozanski et al. 1992), often with the variation opposite to that expected from the (small) surface-temperature changes (Grootes et al. 1989).

The "altitude effect" and the "continental effect" (Rozanski et al. 1993) refer to the tendencies for $\delta^{18}O$ and δD of precipitation to become progressively lower towards continental interiors and higher elevations. The altitude effect includes important control by temperature. Both may reflect a stochastic tendency for air masses moving away from moisture sources to encounter conditions (frontal/convective) that cause moisture extraction and so isotopic fractionation for $^{18}O{:}^{16}O$ and D:H; it is possible that modeling of $\delta^{18}O$ and δD of precipitation could involve stochastic treatments. As reviewed by Rozanski et al. (1993), the continentality depletion for $\delta^{18}O$ of precipitation reaches 8 per mil over 4500 km into Europe from the Atlantic coast. This continental effect is more pronounced in winter than in summer, perhaps because evapotranspiration returns most summertime precipitation to the atmosphere, whereas runoff is more important in wintertime.

Also of importance are additional in-cloud and sub-cloud processes. A non-equilibrium fractionation occurs during formation of ice condensate in clouds (Jouzel and Merlivat 1984), and this is modeled to be highly sensitive to cloud supersaturations. The temperature at which condensate begins forming as ice rather than liquid is also important for the net depletion of ^{18}O and D of water from the air mass, because the vapor-solid fractionation is different from the vapor-liquid fractionation (Jouzel and Merlivat 1984). This transition temperature may be affected by atmospheric dust content, dust particles being important ice-condensation nuclei. Below the cloud, a small amount of rain falling through dry air may be enriched in ^{18}O and D by evaporation; once the air is nearly saturated, further enrichment does not occur. Because diffusion is slower in ice than in water, evaporative enrichment of ^{18}O and D in falling ice is more limited. Vapor exchange between precipitation and moisture in sub-cloud air thus is more important for $\delta^{18}O$ and δD of liquid precipitation than for frozen precipitation (Jouzel 1986).

Different sources of evaporating water can have different isotopic compositions, and often different path lengths to a site producing different continentality. Changes in moisture source thus can affect $\delta^{18}O$ and δD of precipitation in the absence of a temperature change (e.g. Charles et al. 1994).

Sampling and other site effects

Clearly, many factors in addition to site temperature affect the $\delta^{18}O$ and δD of the water in a "packet" of precipitation delivered to a site. The $\delta^{18}O$ or δD preserved at a site is a weighted average of all the packets of precipitation delivered over some time interval. The relevant interval or the minimum resolvable interval may be determined by human analysts (sample size), but may be determined by physical processes that mix or transport waters from different sites and times.

As we will see in the first case study, below, change in seasonality of precipitation is potentially one of the most important complications in paleothermometry based on $\delta^{18}O$ or δD of water. The difference between summer and winter temperatures, hence $\delta^{18}O$ or δD of precipitation, is often as large as or larger than any plausible climatic change. A shift in the summer:winter ratio of precipitation thus can mimic, or obscure, the effect of any climatic temperature change on $\delta^{18}O$ or δD.

Seasonality effects clearly are not limited to precipitation. In seasonal climates, evapotranspiration during the growing season captures and returns to the air much or most of the precipitation, with little net fractionation (Zimmerman et al. 1967). However, during the dormant season, much of the precipitation enters groundwater or runoff (Rozanski et al. 1993). Hence, lakes, rivers, and groundwater obtain a seasonally biased sample of precipitation, and of any seasonal cycle in the amount or $\delta^{18}O$ or δD of that precipitation. Spatial or temporal changes in the bias of this sample (perhaps caused by changes in vegetation affecting evapotranspiration) will affect interpretations. Even in polar regions, seasonality of sublimation or of wind-drift removal or delivery of snow may affect $\delta^{18}O$ and δD of mean-annual snow accumulation at a site (Fisher et al. 1983).

During and following precipitation, diffusive processes tend to mix signals of different precipitation events (e.g. Whillans and Grootes 1985; Cuffey and Steig 1998; Johnsen et al. 2000; Stute and Schlosser 2000). Diffusive removal of the complexity of change in $\delta^{18}O$ and δD within single storms may make interpretation easier. However, diffusion also can obscure events of interest, especially in older waters.

In glaciers and ice sheets or in groundwater flow systems, water may be moved significant distances and elevations from the site where it fell. This is complicated on ice sheets by changes over time in the ice-sheet thickness (Cuffey 2000), and near ice sheets by isostatic effects of the changing ice load.

This list of potential complications is far from exhaustive. We have on occasion asked students in paleoclimatology classes to list ways that paleothermometry based on $\delta^{18}O$ or δD of water could err, and the long, complex, and stimulating ideas generated by the students have left them uneasy about the use of this technique. The clear result of such an exercise is to place a considerable burden on those using $\delta^{18}O$ and δD of water to infer paleotemperatures, to demonstrate the reliability of their method.

CASE STUDIES

Increasingly, paleotemperatures are being estimated in ways that do not rely on $\delta^{18}O$ or δD of water. As discussed in some examples below, the focus is especially on less-complicated physical systems such as borehole paleothermometry (e.g. Pollack and Huang 2000) and noble-gas concentrations in groundwater (Stute and Schlosser 2000). Focus is also on multiparameter reconstructions, to avoid potentially erroneous assumptions or to test the effects of assumptions against techniques based on different assumptions. Reliable independent information on temperature change allows assessment of the utility of temperatures estimated from $\delta^{18}O$ or δD of water. Significant differences between the temperatures reconstructed from $\delta^{18}O$ and δD of water and from more-robust indicators may show that the water-isotopic ratios contain more-interesting information on changes in the hydrological cycle, as discussed below.

Central Greenland

As discussed above, the spatial relationship between surface temperature and $\delta^{18}O$ or

δD of precipitation is strongest over the large ice sheets, where theory leads us to expect that the relation over time and space will be strongest. Thus, failure as a paleo-thermometer of δ^{18}O and δD of water from ice of the large sheets would cast doubt on the technique everywhere. To date, independent tests of the isotopic thermometer are more rigorous in Greenland than in Antarctica. Work by Salamatin et al. (1998), Delaygue et al. (2000) and Cuffey and Vimeux (in press), among others, indicates that in Antarctica δ^{18}O and δD of water from the ice are excellent paleothermometers over time, but that calibration of δ^{18}O and δD to temperature likely differs at least somewhat from their spatial covariation (Salamatin et al. 1998).

Pending additional work in Antarctica, in this section we present an extended case study of the climatic significance of δ^{18}O and δD of water from ice at one site in central Greenland. For times before instrumental records, thermometric information has been obtained from three different sources, one based on the physical temperatures of the ice sheet, and the other two based on physical processes recorded in isotopic ratios of gases trapped in bubbles in the ice. Each method involves fewer assumptions than thermometry based on the use of δ^{18}O and δD of water from the ice. The results from these independent methods agree quite closely with each other, and provide an interpretation of the δ^{18}O and δD of the water from the ice that is fully consistent with much independent information. We will go into some detail on the techniques involved, as they almost certainly are less familiar to some readers than is the use of δ^{18}O or δD of water as a paleothermometer. The results of application of these independent methods are that the δ^{18}O and δD of water from the ice cores are:

(1) useful climatic indicators with clear, direct relations to important climatic variables;

(2) strongly correlated to surface temperature at time of deposition;

(3) also related to other climatic variables; and

(4) likely to give highly erroneous quantitative results if used to interpret temperature changes without additional guidance from other paleoclimatic indicators or modeling results.

The Greenland Summit site, comprising the GRIP and GISP2 deep ice cores and adjacent shallow cores, pits, automatic-weather-station emplacements, the ongoing Summit observatory, and related sampling arrangements near 72°N, 38°W on the Greenland ice sheet, is especially well studied. A comprehensive review of paleothermometry there is given by Jouzel et al. (1997), with related papers by White et al. (1997), Grootes and Stuiver (1997), Cuffey and Clow (1997), Alley (2000) and additional ones as cited below.

Near-surface snow temperatures in the dry firn zone (the region with little or no melting) of Greenland and Antarctica away from major crevasses are quite close to the mean-annual air temperature (Paterson 1994). Evidence of past melting (e.g. Alley and Anandakrishnan 1995), or of past crevassing (e.g. Voigt et al. 1997), is readily observed in an ice core, so one can tell with confidence that those have not been a problem in central Greenland over at least the last 110,000 years for which the central Greenland ice cores give a coherent climate record (Alley et al. 1995).

Heat transfer in ice sheets is primarily by conduction and advection with the moving ice (see Robin 1983). In central regions of an ice sheet, the ice flow is mainly vertical, and so horizontal advection of heat is not important. (Where both horizontal and vertical flow are important, ice at depth was transported from higher, colder regions and so may be colder than ice above or below for advective reasons even in the absence of climate change. At an ice divide, there is no "upglacier" from which the ice can flow. Modeling studies (Anandakrishnan et al. 1993; Cuffey and Marshall 2000; Marshall and Cuffey 2000) show that the position of the ice divide in central Greenland has been relatively insensitive to climatic changes over at least tens of thousands of years.) The thermal properties of ice and snow are known exceptionally well (e.g. Yen 1981), with little or no uncertainty related to history, structure, or impurity loading of the ice (except perhaps for the lowermost few meters where important dirt loading may occur).

Polar ice-sheet thickness is quite insensitive to climate change, especially on an island such as Greenland for which the continental shelf is relatively narrow and major changes in ice-sheet extent cannot occur (Paterson 1994; Cuffey and Clow 1997). An ice sheet adjusts to remove the snow supplied to it. The gravitational stress causing flow of ice is proportional to the surface slope and the thickness of ice above, and the rapid deformation in deep layers controlling the ice-sheet shape is typically found to increase with approximately the cube of that stress. Integrating this relation from the bed to obtain the velocity, and then integrating through the thickness to obtain the ice flux, gives ice flux proportional to the third power of the surface slope and the fifth power of the thickness. If the ice-sheet half-width is approximately fixed, then the surface slope averaged from the divide to the coast is proportional to the thickness in the center, yielding an eighth-power dependence of ice flux on ice thickness. Hence, very large changes in accumulation rate and thus flux of ice at steady state produce very small changes in ice thickness (Cuffey and Clow 1997). An infinitesimal perturbation of $\varepsilon\%$ in net snow accumulation would cause only an $\varepsilon/8\%$ change in thickness.

This in turn means that the advection term in the heat-flow equation for central regions of the Greenland ice sheet is dominated by the snow accumulation rate with little correction for changing thickness. An increase in snow accumulation initially causes thickening (Whillans 1981; Alley and Whillans 1984), but the flux increases as the eighth power of the thickness and steady state is approached with little thickness change. In steady state, the spreading of the ice sheet causes a downward velocity in the center just equal to the snow accumulation rate.

If accurate dating of an ice core is available, then the thickness of ice between two time lines gives the average accumulation rate for that interval after adjustment for ice-flow thinning, which in turn is closely related to the ice-sheet shape and the accumulation rate (Nye 1963; Schott et al. 1992; Alley et al. 1993; Dahl-Jensen et al. 1993; Cutler et al. 1995; Cuffey and Clow 1997). With considerable confidence and little uncertainty, a well-dated ice core in central Greenland allows calculation of the advective and conductive contributions to heat flow in the ice. The result is dominance by advection over long distances in the upper half of the ice sheet, with conduction important over short distances and in the lower part. (Note that not all ice masses are as "easy" as central Greenland in this regard; large changes in extent of mountain glaciers and even portions of the Antarctic ice sheet, coupled with large changes in bottom sliding not observed in central Greenland owing to very cold basal conditions there (Firestone et al. 1990), complicate interpretations for some other glaciers and ice sheets.)

Temperature measurements in boreholes through ice sheets are now available with millikelvin accuracy (Clow et al. 1997; Cuffey and Clow 1997). The heat from drilling dissipates rapidly, and can be monitored and corrected for by repeat profiling, as can effects of convection in the fluid used to prevent borehole collapse. The temperatures in an ice sheet, as measured in a borehole, in turn contain information on past surface temperatures of the ice sheet. The ice about halfway down through central Greenland has not finished warming from the last ice age, and is observed to be colder than ice above or below (Fig. 9).

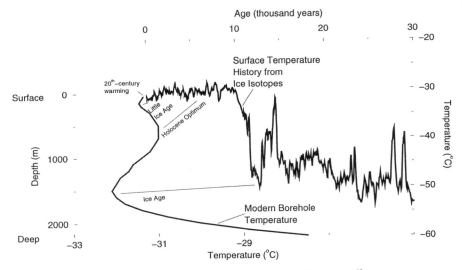

Figure 9. Borehole temperature versus depth (Cuffey and Clow 1997), and $\delta^{18}O$ of ice versus age (Grootes and Stuiver 1997), for GISP2, central Greenland. Approximate correspondence between the imprints of prominent climate events in these records are indicated by tie lines; the diffusive information loss from heat flow is evident in the greater smoothness of the borehole-temperature record. The borehole record is quite smooth, but appears slightly rough because of the plotting package used.

Diffusive information loss limits resolution of borehole-derived paleotemperatures. A borehole temperature profile from central Greenland reveals information about the previous day, the previous year, the previous decade, the Little Ice Age of one to a few centuries ago, the warmth in the middle Holocene a few millennia ago, and the cold of the ice age centered about 20-30 thousand years ago (Alley and Koci 1990; Cuffey et al. 1992, 1994, 1995; Firestone 1995; Johnsen et al. 1995; Cuffey and Clow 1997; Dahl-Jensen et al. 1998).

Inversion of the borehole temperature profile to learn the surface temperatures that produced it can be done in various ways. These include control methods (MacAyeal et al. 1991) or other formal inversion (Clow et al. 1997), and Monte Carlo techniques (Dahl-Jensen et al. 1998). In any of these, a provisional surface-temperature history is adjusted until it does the best possible job of predicting the modern distribution of temperature with depth in the ice sheet. "Best" is usually defined as minimizing some measure of misfit such as the root-mean-square error between observed and calculated temperature profiles in the ice, possibly subject to constraints on timing or size of allowable temperature changes.

Additionally, the history of $\delta^{18}O$ or δD from the ice core can be used as a short cut to interpreting the borehole-temperature record (Paterson and Clarke 1978; Cuffey et al. 1992, 1994, 1995; Johnsen et al. 1995; Cuffey and Clow 1997; Johnsen et al. 1997; also see Beltrami and Taylor 1995). The $\delta^{18}O$ records of ice from GISP2 and GRIP were used for central Greenland. A provisional relation between the $\delta^{18}O$ of ice formed from accumulated snow and the surface temperature translates the $\delta^{18}O$ record of the ice core into a provisional surface-temperature history, which is used to drive a time-dependent heat- and ice-flow model to predict modern temperature versus depth in the ice sheet. The provisional relation between $\delta^{18}O$ of ice and temperature is then adjusted to optimize simulation of the modern temperature profile in the ice sheet. A large $\delta^{18}O$ data set is used to predict an information-rich borehole-temperature profile (Cuffey and Clow 1997; NSIDC 1997), but very few parameters are adjusted in the model. (Often three parameters are adjusted: the slope and intercept of an assumed linear relation between $\delta^{18}O$ of ice and surface temperature, and the geothermal flux.) Such a small number of parameters will not allow a model to find a close relation between $\delta^{18}O$ of ice and surface temperature unless they indeed are related. Thus, successful matching of the modern temperature-depth profile demonstrates a relationship between $\delta^{18}O$ and surface temperature at time of deposition with high confidence, and calibrates that relationship. Were the $\delta^{18}O$ history to prove unable to predict the modern temperature-depth profile accurately, it could indicate that the wrong calibration equation was used (linear versus higher-order, exponential, etc.), that an error existed in data or computation, or that the $^{18}O:^{16}O$ ratios of ice have not been simply related to surface temperature.

Fortunately, in the case of central Greenland, very close matches between predicted and observed temperature-depth profiles were obtained independently for the GISP2 (Cuffey et al. 1995) (Fig. 10) and GRIP (Johnsen et al. 1995) boreholes. Both indicated that: (1) $\delta^{18}O$ of ice did record temperature accurately; (2) a simple linear (Cuffey et al. 1995) or perhaps quadratic (Johnsen et al. 1995) calibration between $\delta^{18}O$ of ice and temperature is adequate; (3) the slope of the relation (per mil per degree) has been less than the slope based on recent spatial dependence of $\delta^{18}O$ of ice on temperature for all times considered; (4) the slope of the relation was steeper over the Little-Ice-Age time scale of the last few centuries than over the ~10,000 years of the Holocene or the older deglacial changes; (5) the long-term slope is just less than half of that inferred from the recent spatial dependence; and (6) deglacial temperature changes were slightly more than twice as large as previously estimated from the measured change in $\delta^{18}O$ of ice and the modern dependence of $\delta^{18}O$ of ice on temperature (e.g. Dansgaard and Oeschger 1989).

The borehole-temperature paleothermometer does not resolve abrupt or short-lived

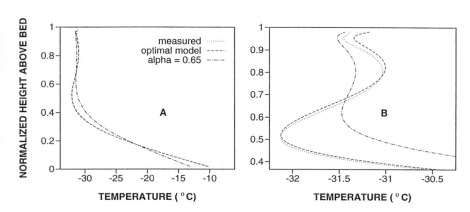

Figure 10. Borehole temperature versus depth as measured in central Greenland, and as simulated by Cuffey et al. (1995). The model was a coupled heat- and ice-flow computation driven by the $\delta^{18}O$ record of the ice converted to temperature assuming a linear relation. The spatial dependence of $\delta^{18}O$ on temperature is approximately 0.65 per mil per degree, which provides a poor fit to the data; the optimal model differs from this by a factor of 2, and provides a much better fit. (A) The profile from the bed (height 0) to the surface (height 1, or ~3050 m); (B) the upper portion the profiles on an expanded temperature scale. In Figure 10A, the optimal and measured curves are indistinguishable; in 10B, the small differences between the optimal and measured curves indicate time-variation of the calibration of $\delta^{18}O$ to temperature.

old events. Fortunately, two new paleothermometers developed by Severinghaus et al. (1998) and Severinghaus and Brook (1999) do resolve abrupt and short-lived events that were sufficiently large. These new techniques have been used to estimate the absolute temperature just before abrupt warmings, and the temperature change in those warmings, and produce results in close agreement to those obtained from the borehole-temperature paleothermometer.

The transformation of snow to ice in central regions of polar ice sheets occurs without melting, in a metamorphic process analogous to sintering/hot-isostatic pressing in materials science. The physical processes are relatively well understood, and semi-empirical models are quite accurate in predicting depth-density and depth-age profiles for conditions that are common on ice sheets today (Herron and Langway 1980; Alley 1987; Barnola et al. 1991). Transformation of snow to ice is faster at higher temperature (owing to dominance of thermally activated processes) and higher accumulation rate (because the load from overlying snow, which provides the main driving force for densification, accumulates more rapidly; surface-tension effects are minor in most cases (Alley 1987)). The old snow during transformation to ice is called firn, and remains permeable to air to a depth of typically 50-100 m before air is isolated from the free atmosphere to form bubbles. Wind or pressure variations affect air in the uppermost few meters of the snow, but cause less mixing than does diffusion at greater depths (Sowers et al. 1989).

The gas through most of the firn is a diffusive column. Gravitational fractionation in this column causes the trapped gas in bubbles to be slightly enriched in heavier elements and isotopes compared to the free atmosphere (Sowers et al. 1989), by an amount that increases with depth in the diffusive zone in the firn. Additionally, if a temperature difference exists between the surface and the bubble-trapping depth, thermal fractionation may separate the gases by weight, with the heavier isotopes and species moving to the colder end (Severinghaus et al. 1998), which may be at the top or the bottom of the firn depending on the sense of the temperature difference.

Exceptionally large, abrupt climate changes have occurred in the past in Greenland and many other places (see review by Alley 2000), with warmings especially prominent. These warmings often followed periods of relatively stable climate that were long enough to allow the depth-density profile of the firn to come to steady-state with the climatic conditions. After an abrupt warming, the firn gases will adjust diffusively in a few years (Severinghaus et al. 1998). However, downward-flowing heat must warm the ice through which it passes, and will not significantly warm the ice at the bubble-trapping depth and remove the temperature gradient across the firn for more than a century. Hence, following an abrupt warming, isotopically anomalous gases will begin to be trapped in a few years, and the anomaly will continue to be trapped for a century or more.

This discovery allows two paleothermometers. First, the number of years between the records of an abrupt warming in the $\delta^{18}O$ of the ice and in the isotopic or elemental ratios of trapped gases, determined by annual-layer counting or other techniques (Alley et al. 1993), shows the age of the firn at the bubble-trapping depth when the abrupt warming occurred. From our knowledge of the transformation of snow to ice, this gas-age/ice-age difference is a well-known function of the snow accumulation rate and temperature that applied before the abrupt warming. In turn, the accumulation rate is available from accurate dating and ice-flow models.

The abrupt warmings of the past in Greenland were sufficiently fast (as shown by many indicators including the rate of change of the gas-phase record of the warming; Severinghaus et al. 1998) compared to the time for adjustment of the depth-density profile (as shown by our physical understanding of the active processes) that the changes in the firn column were minor during the few years when the gas-isotopic composition of air in the firn was adjusting to the abrupt warming. Furthermore, as noted above, many of the abrupt changes in Greenland followed times of relatively stable climate that were longer than the time for snow to change to firn and then to ice, again shown by many indicators, so the depth-density profile was steady before the abrupt change. Thus, it is possible to estimate the absolute temperature just before the warming (Severinghaus et al. 1998).

Different gases show different sensitivities to temperature-gradient versus gravitational-fractionation effects. Following an abrupt warming and any associated changes in windiness, snow accumulation rate, etc., the height of the diffusive gas column in the firn will begin to adjust and affect the gravitational fractionation of gases, even as the temperature gradient affects thermal fractionation of gases. Use of two indicators with differing sensitivities to these effects allows isolation of the temperature effect, thus revealing the size of the warming (Severinghaus et al. 1998). For example, Ar is observed to be half as sensitive to thermal fractionation as N_2 per unit mass difference. Gravitational fractionation is strictly mass-dependent, so a change in the depth of the diffusive column of firn will affect $\delta^{40}Ar$ four times as much as $\delta^{15}N$ of N_2, whereas a change in temperature gradient with no change in the thickness of the diffusive column will affect $\delta^{40}Ar$ only twice as much as $\delta^{15}N$ of N_2. Comparing changes in $\delta^{40}Ar$ and $\delta^{15}N$ of N_2 then allows separation of temperature-gradient from firn-thickness effects (Severinghaus et al. 1998).

Application of these joint techniques has been limited to a few events, but thus far, there is agreement at the 1-2°C level, for both absolute temperatures before abrupt warming and temperature changes during abrupt warming, with those inferred from the borehole-temperature calibration of the record of $\delta^{18}O$ of ice. With abrupt shifts of roughly 10°C, and total deglacial temperature change of more than 20°C at the surface in central Greenland, this indicates accuracies of about 10%, versus the two-fold or more errors from use of the spatial dependence of $\delta^{18}O$ of ice on temperature to interpret the changes in $\delta^{18}O$ of ice in terms of temperature changes over time.

The similarity is striking between the temperature calibration of $\delta^{18}O$ of central-Greenland ice over the short intervals of large, abrupt climate changes and over the long changes of the deglaciation and the orbital shifts in Holocene climate over millennia, at around 0.33 per mil per degree, or even a bit lower over the Holocene. Over somewhat shorter times and smaller changes of the Little Ice Age oscillation, the calibration is somewhat higher, giving 0.53 per mil per degree (Cuffey et al. 1992, 1994) (Fig. 11).

(Most interest has been focused on paleoclimatic issues and the differences between spatial and temporal gradients over long times, but we note that the spatial dependence of $\delta^{18}O$ of water on temperature does not closely match the temporal dependence over shorter times in central Greenland either. Shuman et al. (1995) related $\delta^{18}O$ of water sampled from snow pits covering a few years to instrumental temperature records from combination of automatic-weather-station and satellite-passive-microwave data. A few events per year were identifiable in both records, and correlation of these suggests that $\delta^{18}O$ of accumulated snow increases ~0.51 per mil per degree warming. This number may have been lowered by diffusive smoothing, however, as diffusion affects isotopic ratios in snow but not instrumental records. Over very short times, Grootes and Stuiver (1997) found a strong correlation between $\delta^{18}O$ of near-surface water-vapor samples and of snowfall and hoarfrost during summer months at and near the GISP2 camp, and between these $\delta^{18}O$ data and the air temperature at GISP2. Their dependence of 0.83 per mil per degree (Grootes and Stuiver 1997) is strikingly higher than the recent mean-annual spatial dependence of $\delta^{18}O$ of accumulated ice on temperature. Focus only on major

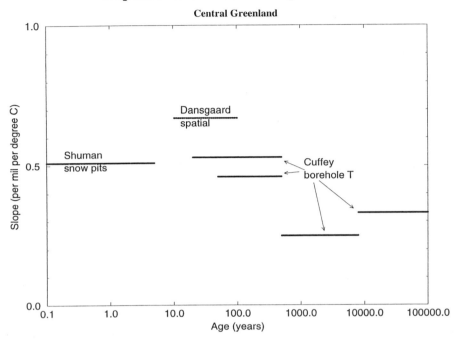

Figure 11. Dependence of $\delta^{18}O$ on temperature for central Greenland, from Cuffey et al. (1992, 1994, 1995) and Shuman et al. (1995). Modified from Alley et al. (1999).

peaks and valleys in the Grootes and Stuiver (1997) data set produces a dependence as high as 1.4 per mil per degree C.)

Much effort has been invested in understanding the relation between $\delta^{18}O$ of ice and paleoclimate in central Greenland (and, of course, for many other sites). A review of contributing factors is given by Cuffey (2000).

Briefly, changes in almost any of the factors discussed above, or in others, might have occurred and caused the time-dependence of $\delta^{18}O$ of ice on temperature to deviate from the modern spatial relation. Prominent hypotheses have especially included changes in seasonality of precipitation in which a smaller fraction of the precipitation falls during the winter during times of lower mean-annual temperature (Fawcett et al. 1997), changes in the source temperature parallel to those at the site (Boyle 1997), changes in the strength of the surface temperature inversion such that ice-age cooling was stronger at the surface than aloft where condensation occurred (Cuffey et al. 1995), and changes in the $\delta^{18}O$ of vapor-source ocean waters (Johnsen et al. 1995). Taken together, these are sufficient, or more, to explain the differences between time and space dependence of $\delta^{18}O$ of ice on temperature.

Several arguments indicate changes in seasonality of precipitation in Greenland (see Alley et al. 1999), with cooling causing a greater reduction in wintertime than in summertime precipitation. Atmospheric modeling provides the strongest evidence of this change (Fawcett et al. 1997; Krinner et al. 1997, Werner et al. 2000). Widespread oceanic cooling coincident with Greenland cooling is well-documented (Boyle 1997), and atmospheric modeling simulates a contribution of this effect to the dependence of $\delta^{18}O$ of precipitation on temperature (Werner et al. 2000). Contribution from changing moisture sources is probably minor (Charles et al. 1994; Werner et al. 2000). A small contribution from changing surface-inversion strength probably occurred (Cuffey and Clow 1997; Werner et al. 2000). Our current understanding of the system is such that the spatial dependence of $\delta^{18}O$ of ice on temperature simply should not be applicable to changes over time (Jouzel et al. 1997; Cuffey 2000).

Southern Africa

A second, shorter case study will further illustrate the weaknesses of traditional paleothermometry based on isotopic ratios of water, at a lower-latitude site. Two aquifers from southern Africa (about 24° and 34°S) have opposite shifts in $\delta^{18}O$ of water between modern and glacial-maximum conditions. The temperature changes, and gradients, implied by direct interpretation of $\delta^{18}O$ of water as a paleothermometer are physically implausible. An independent paleothermometer, based on a physically direct interpretation of the noble-gas concentrations of the water, indicates similar temperature changes for both sites that are in turn consistent with a wide range of other information. The $\delta^{18}O$ changes of the water are plausibly explained based on non-temperature factors.

Groundwater concentrations of the suite of noble gases provide among the most reliable paleothermometers, as reviewed by Stute and Schlosser (2000). The equilibrium solubility of gases in water increases as temperature falls (Henry's Law). Groundwater that recharges into aquifers through water tables a few meters or more deep has the mean-annual ground temperature, which typically is close to the mean-annual air temperature. Recharge is slow enough that dissolved gases can come to equilibrium in the water. A complication is that excess gas may be trapped as bubbles. This excess gas can be recognized in groundwaters, and its effects on noble-gas abundances removed, through measurement of the suite of noble gases (He, Ne, Ar, Kr, Xe), because the ratios of gases in the atmosphere are quite different from their equilibrium solubilities in water.

Some aquifers are confined, with impermeable layers preventing water addition except in one recharge area. Water flows along such aquifers to discharge points or sampling wells. Radiocarbon or other techniques including excess radiogenic ^4He can be used to date the water (Stute and Schlosser 2000). Typical recharge rates are sufficiently rapid compared to diffusion/mixing lengths along aquifers that the ice-age signal is preserved well, although any record of millennial events is lost (Stute and Schlosser 2000).

Noble-gas observations in the Stampriet artesian aquifer in Namibia, southern Africa, show a cooling of between 5 and 6°C at glacial maximum, consistent with results from another aquifer in South Africa (Stute and Talma 1998), and somewhat larger than but broadly consistent with results of atmospheric modeling forced by ice-age boundary conditions (Pollard and Thompson 1997). The $\delta^{18}O$ of the water recharged during the ice age in the South African aquifer is about 1 per mil lower than the recent waters; correcting for the 1 per mil shift in mean oceanic composition indicates a shift of about 2 per mil, or a dependence of $\delta^{18}O$ of water on temperature of about 0.3 to 0.4 per mil per degree. However, the $\delta^{18}O$ of the waters recharged during the ice age in the Namibian case is about 1.5 per mil higher than more recently, and is slightly higher than, or similar to, the $\delta^{18}O$ of recent recharge even after correction for the mean oceanic change in $\delta^{18}O$. Interpreting the change in $\delta^{18}O$ of water in the "normal" manner of paleothermometry would indicate that temperatures in Namibia were warmer during the ice age than recently, whereas conditions were more than 5°C colder there during the ice age (Stute and Talma 1998). A change in moisture source affecting continentality likely explains the odd pattern (Stute and Talma 1998). Regardless of the explanation, the remarkably erroneous results that would be obtained from naïve interpretation of $\delta^{18}O$ as a thermometer are clear.

DISCUSSION

Much (although certainly not all) paleoclimatic research is based on a strict form of the old geologic principle of uniformitarianism: obtain a modern relation (either spatially or over short times at a place) between a climatic variable of interest and a related quantity that is preserved in sedimentary records, measure the age and the sedimentary character at a site, assume that the relation has applied at the site over the times studied, and calculate the paleoclimatic conditions (Bradley 1999; Alley 2001). Most geologists including many paleoclimatologists long ago abandoned this strict interpretation of uniformitarianism for a less-strict actualism: the underlying laws of nature have not changed over time, but the world is too complex to assume that modern quantitative spatial relationships have been time-invariant at a place. Increasingly, tests such as the case studies discussed here are showing the need to further weaken this strict uniformitarian interpretation. When our clear understanding of the climate system includes processes that would cause space and time gradients to differ, it is not appropriate to assume that they are quantitatively the same, for $\delta^{18}O$ or δD of water or for other paleoclimatic indicators.

Obviously, most practitioners in $\delta^{18}O$ or δD analyses of waters are aware of these trends, and indeed are the leaders in developing and applying new techniques and urging caution in use of the old techniques. Such efforts are enjoying significant success. The range of paleoclimatic techniques is increasing rapidly (e.g. Bradley 1999). Robust indicators such as those cited above involve few untestable assumptions, and are more actualistic than uniformitarian. Co-interpretation of multiple indicators subject to independent uncertainties allows testing of assumptions (e.g. Mix et al. 2000).

In this regard, two major recommendations relative to the paleoclimatic interpretation of $\delta^{18}O$ or δD of water are (Cuffey 2000): (1) The conversion of $\delta^{18}O$ or δD of water to temperature should not be based on geographic covariation of modern ratios and temperatures, but on independent historical reconstructions using information such as borehole temperatures; and (2) $\delta^{18}O$ or δD of water rather than the temperatures estimated from them should be preferred for model testing when possible. Fricke and O'Neil (1999) similarly noted that it is better to use $\delta^{18}O$ or δD of water to test predictions of global climate models directly, rather than using isotopic ratios to reconstruct temperatures.

The case studies cited above illustrate a further result, that has been emphasized by workers such as Charles et al. (1994), Jouzel et al. (1997), Delaygue et al. (2000), and others. The $\delta^{18}O$ and δD of water in precipitation are produced by the integrated effect of the hydrologic cycle, not just by paleotemperature. The $\delta^{18}O$ and δD of water thus allow tests of the hydrologic cycles of Earth-system models that include appropriate isotopic parameterizations.

For most of the earth's continental environments, the hydrologic cycle is probably more important to living things than is temperature. For example, the hydrologic cycle is the main determinant of whether cold places are tundra or ice sheets, and whether warm places are rain forests or deserts. $\delta^{18}O$ or δD of water record such truly important variables as amount of precipitation. Given the complexity of the controls on $\delta^{18}O$ and δD, widespread success in simulating them almost certainly demonstrates model skill.

And model skill allows evaluation of such questions as whether past changes in vegetation were forced by changes in water availability or in some other factor such as atmospheric carbon-dioxide concentration. Thus, $\delta^{18}O$ and δD of old water are among the most valuable of paleoclimatic indicators.

Joint interpretation of $\delta^{18}O$ or δD with borehole temperatures or other indicators remains highly useful in paleothermometry (Cuffey and Clow 1997; Delaygue et al. 2000; Cuffey and Vimeux, in press). However, rigorous and meaningful analyses must recognize that $\delta^{18}O$ and δD of water are most directly measuring properties of the hydrologic cycle. Use of $\delta^{18}O$ or δD of water in paleothermometry works only because the hydrologic cycle is sensitive to temperature, and the complications implied by this connection demand that temperature reconstructions be supported by other indicators such as borehole temperatures, noble-gas concentrations in groundwaters, evidence such as presence or absence of permafrost, and multiparameter reconstructions including indicators such as pollen.

Given the great complexity of controls on the $\delta^{18}O$ and δD of precipitation, it is not surprising that $\delta^{18}O$ of groundwater failed as a paleothermometer in the African case study reviewed above. Perhaps more surprising is the Greenland case study, which indicates that $\delta^{18}O$ and δD of accumulated snow have tracked temperature closely over time, though with a very different quantitative scaling than for their spatial covariation. As argued by Cuffey (2000), the covariation of $\delta^{18}O$ or δD of ice and temperature over time is preserved in this case because changes in all of the numerous important controls must have been strongly correlated over time. The effective complexity (or true degrees of freedom) of the Earth system must have been greatly reduced by strong couplings and feedbacks between different system components. In the Greenland case, this is manifest as covariance of Greenland temperature, source-region temperature, and precipitation seasonality.

In the Greenland example, and probably in many cases, $\delta^{18}O$ and δD of water have the practical advantage that they are relatively quick and easy to measure. This permits measurement at very high time resolution over an entire ice core, aquifer, or other archive, which may not be practical with other paleothermometric measurements such as isotopic ratios of ice-core gases. If it can be shown, as was done in Greenland, that $\delta^{18}O$ and δD of water have a quasi-uniform calibration to temperature, then the entire record may be calibrated at just a few representative points with more labor-intensive methods.

In conclusion, we argue that $\delta^{18}O$ and δD of water should not be used for paleothermometry based on modern or recent correlations in the absence of additional supporting information. Fortunately, these other thermometers are increasing steadily in power and diversity. Moreover, recognizing these isotopic ratios as hydrologic indicators opens a tremendous source of new information as a means for testing hydrologic cycles in Earth system models; this direction will be most exciting for future research. By analogy, use of any quantitative paleoclimatic reconstruction based on substitution of space for time or recent time for long time should be approached with caution, with emphasis placed on physically simple and multiparameter reconstructions.

ACKNOWLEDGMENTS

We thank the National Science Foundation for funding; Jeff Severinghaus, Henry Fricke, Melissa Hendricks, and John Valley for reviews; and numerous colleagues for stimulating ideas and discussions.

REFERENCES

Alley RB (1987) Firn densification by grain-boundary sliding: a first model. J Physique 48:249-254

Alley RB (2000) The Younger Dryas cold interval as viewed from central Greenland. Quat Sci Rev 19:213-226

Alley RB (2001) The key to the past? Concepts. Nature 409:289

Alley RB, Whillans IM (1984) Response of the East Antarctic ice sheet to sea-level rise. J Geophys Res 89:6487-6493

Alley RB, Agustsdottir AM, Fawcett PJ (1999) Ice-core evidence of Late-Holocene reduction in North Atlantic ocean heat transport. In Clark PU, Webb RS, Keigwin LD (eds) Mechanisms of Global Climate Change at Millennial Time Scales, American Geophysical Union, Washington DC, p 301-312

Alley RB, Anandakrishnan S (1995) Variations in melt-layer frequency in the GISP2 ice core: implications for Holocene summer temperatures in central Greenland. Ann Glaciol 21:64-70

Alley RB, Koci BR (1990) Recent warming in central Greenland? Ann Glaciol 14:6-8

Alley RB, Meese DA, Shuman CA, Gow AJ, Taylor KC, Grootes PM, White JWC, Ram M, Waddington ED, Mayewski PA, Zielinski GA (1993) Abrupt increase in snow accumulation at the end of the Younger Dryas event. Nature 362:527-529

Alley RB, Gow AJ, Johnsen SJ, Kipfstuhl J, Meese DA, Thorsteinsson Th (1995) Comparison of deep ice cores. Nature 373:393-394

Anandakrishnan S, Alley RB, Waddington ED (1993) Sensitivity of ice-divide position in Greenland to climate change. Geophys Res Letters 21:441-444

Barnola J-M, Pimienta P, Raynaud D, Korotkevich YS (1991) CO_2-climate relationship as deduced from the Vostok ice core—a reexamination based on new measurements and on a reevaluation of the air dating. Tellus B 43:83-90

Beltrami H, Taylor AE (1995) Records of climatic changes in the Canadian Arctic: Towards calibrating oxygen isotope data with geothermal data. Global Planet Change 11:127-138

Boyle EA (1997) Cool tropical temperatures shift the global δ^{18}O-T relationship: An explanation for the ice core δ^{18}O-borehole thermometry conflict? Geophys Res Letters 24:273-276

Bradley RS (2000) 1000 years of climate change. IGBP Global Change Newsletter 44, 5-6

Bradley RS (1999) Paleoclimatology: Reconstructing Climates of the Quaternary, 2nd edn. Academic, San Diego

Broecker, WS (2001) Was the medieval warm period global? Science 291:1497-1499

Charles C, Rind D, Jouzel J, Koster R, Fairbanks R (1994) Glacial-interglacial changes in moisture sources for Greenland: influences on the ice core record of climate. Science 261:508-511

Clow GD, Waddington ED, Gundestrup NS (1997) Reconstruction of past climatic changes in central Greenland from the GISP2 high-precision temperature profile using spectral expansion. EOS Trans Am Geophys Union 78:F6 (abstr)

Craig H (1961) Isotopic variations in meteoric waters. Science 133:1702-1703

Craig H, Gordon LI (1965) Deuterium and oxygen-18 variations in the ocean and the marine atmosphere. In Stable Isotopes in Oceanographic Studies and Paleotemperatures. Spoleto, July 26-27, 1965. Consiglio Nazionale delle Ricerche, Laboratorio di Geologia Nucleare, Pisa, p 1-122

Cuffey KM (2000) Methodology for use of isotopic climate forcings in ice sheet models. Geophys Res Letters 27:3065-3068

Cuffey KM, Clow GD (1997) Temperature, accumulation, and ice sheet elevation in central Greenland through the last deglacial transition. J Geophys Res 102:26383-26396

Cuffey KM, Marshall SJ (2000) Substantial contribution to sea-level rise during the last interglacial from the Greenland ice sheet. Nature 404:591-594

Cuffey KM, Steig EJ (1998) Isotopic diffusion in polar firn: implications for interpretation of seasonal climate parameters in ice-core records, with emphasis on central Greenland. J Glaciol 44:273-284

Cuffey KM, Vimeux F (in press) Validation of deuterium excess thermometry provides new evidence for strong covariation of carbon dioxide and temperature. Nature

Cuffey KM, Alley RB, Grootes PM, Anandakrishnan S (1992) Toward using borehole temperatures to calibrate an isotopic paleothermometer in central Greenland. Global Planet Change 98:265-268

Cuffey KM, Alley RB, Grootes PM, Bolzan JF, Anandakrishnan S (1994) Calibration of the δ^{18}O isotopic paleothermometer for central Greenland, using borehole temperatures. J Glaciol 40:341-349

Cuffey KM, Clow GD, Alley RB, Stuiver M, Waddington ED, Saltus RW (1995) Large Arctic temperature change at the glacial-Holocene transition. Science 270:455-458

Cutler NN, Raymond CF, Waddington ED, Meese DA, Alley RB (1995) The effect of ice-sheet thickness changes on the accumulation history inferred from GISP2 layer thicknesses. Ann Glaciol 21:26-32

Dahe Q, Petit JR, Jouzel J, Stievenard M (1994) Distribution of stable isotopes in surface snow along the route of the 1990 International Trans-Antarctic Expedition. J Glaciol 40:107-118

Dahl-Jensen DJ, Johnsen SJ, Hammer CU, Clausen HB, Jouzel J (1993) Past accumulation rates derived from observed annual layers in the GRIP ice core from Summit, central Greenland. *In* Peltier WR (ed) Ice in the Climate System. Springer-Verlag, Berlin, p 517-532

Dahl-Jensen D, Mosegaard K, Gundestrup N, Clow GD, Johnsen SJ, Hansen AW, Balling N (1998) Past temperatures directly from the Greenland Ice Sheet. Science 282:268-271

Dansgaard W (1964) Stable isotopes in precipitation. Tellus 16:436-468

Dansgaard W, Oeschger H (1989) Past environmental long-term records from the Arctic. *In* H Oeschger, CC Langway Jr (eds) The environmental record in glaciers and ice sheets. Wiley, New York, p 287-317

Delaygue G, Jouzel J, Masson V, Koster RD, Bard E (2000) Validity of the isotopic thermometer in central Antarctica: limited impact of glacial precipitation seasonality and moisture origin. Geophys Res Letters 27:2677-2680

Epstein S, Mayeda TK (1953) Variation of ^{18}O content of waters from natural sources. Geochim Cosmochim Acta 4:213-224

Eriksson E (1965) Deuterium and oxygen-18 in precipitation and other natural waters: some theoretical considerations. Tellus 18:498-512

Faure, G (1977) Principles of Isotope Geology. Wiley, New York

Fawcett PJ, Ágústsdóttir AM, Alley RB, Shuman CA (1997) The Younger Dryas termination and North Atlantic deepwater formation: insights from climate model simulations and Greenland ice core data. Paleoceanography 12:23-38

Firestone J (1995) Resolving the Younger Dryas event through borehole thermometry. J Glaciology 41:39-50

Firestone J, Waddington E, Cunningham J (1990) The potential for basal melting under Summit, Greenland. J Glaciol 36:163-168

Fisher DA (1992) Stable isotope simulations using a regional stable isotope model coupled to a zonally averaged global model. Cold Reg Sci Tech 21:61-77

Fisher DA, Koerner RM, Paterson WSB, Dansgaard W, Gundestrup N, Reeh N (1983) Effect of wind scouring on climatic records from ice-core oxygen-isotope profiles. Nature 301:205-209

Fricke HC, O'Neil JR (1999) The correlation between 18O/16O ratios of meteoric water and surface temperature: its use in investigating terrestrial climate change over geologic time. Earth Planet Sci Letters 170:181-196

Gat JR (1996) Oxygen and hydrogen isotopes in the hydrologic cycle. Ann Rev Earth Planet Sci 24:225-262

Grootes PM, Stuiver M (1997) Oxygen 18/16 variability in Greenland snow and ice with 10^{-3}- to 10^5-year time resolution. J Geophys Res 102:26455-26470

Grootes PM, Stuiver M, Thompson LG, Mosley-Thompson E (1989) Oxygen isotope changes in tropical ice, Quelccaya, Peru. J Geophys Res 94:1187-1194

Hendricks MB, DePaolo DJ, Cohen RC (2000) Space and time variation of δO-18 and δD in precipitation: Can paleotemperature be estimated from ice cores? Global Biogeochem Cycles 14:851-861

Herron MM, Langway CC Jr (1980) Firn densification: an empirical model. J Glaciol 25:373-385

Hoffman G, Werner M, Heimann M (1998) The water isotope module of the ECHAM atmospheric general circulation model—a study on time scales from days to several years. J Geophys Res 103:16,871-16,896

Johnsen SJ, Dansgaard W, White JWC (1989) The origin of Arctic precipitation under present and glacial conditions. Tellus 41B:452-468

Johnsen SJ, Dahl-Jensen D, Dansgaard W, Gundestrup N (1995) Greenland paleotemperatures derived from GRIP bore hole temperature and ice core isotope profiles. Tellus 47B:624-629

Johnsen SJ and 14 others (1997) The δ^{18}O record along the Greenland Ice Core Project deep ice core and the problem of possible Eemian climatic instability. J Geophys Res 102:26,397-26,410

Johnsen SJ, Clausen HB, Cuffey KM, Hoffman G, Schwander J, Creyts T (2000) Diffusion of stable isotopes in polar firn and ice: the isotope effect in firn diffusion. *In* Hondoh T (ed) Physics of Ice Core Records. Hokkaido University Press, Sapporo, p 121-140

Joussaume S, Jouzel J (1993) Paleoclimatic tracers: An investigation using an atmospheric general circulation model under ice age conditions 2. Water isotopes. J Geophys Res 98:2807-2830

Jouzel J (1986) Isotopes in cloud physics: multiphase and multistage condensation processes. *In* Handbook of Environmental Isotope Geochemistry, v 2, The Terrestrial Environment. B. Elsevier, Amsterdam, p 61-105

Jouzel J, Merlivat L (1984) Deuterium and oxygen 18 in precipitation: modeling of the isotopic effects during snow formation. J Geophys Res 89:11,749-11,757

Jouzel J, Russell G, Suozzo R, Koster R, White JWC, Broecker WS (1987) Simulations of the HDO and H$_2^{18}$O atmospheric cycles using the NASA GISS general circulation model: The seasonal cycle for present-day conditions. J Geophys Res 92:14739-14760

Jouzel J, Alley RB, Cuffey KM, Dansgaard W, Grootes P, Hoffmann G, Johnsen SJ, Koster RD, Peel D, Shuman CA, Stievenard M, Stuiver M, White J (1997) Validity of the temperature reconstruction from water isotopes in ice cores. J Geophys Res 102:26471-26487

Krinner G, Genthon C, Jouzel J (1997) GCM analysis of local influences on ice core δ signals. Geophys Res Letters 24:2825-2828

Lawrence JR, White JWC (1991) The elusive climate signal in the isotopic composition of precipitation. In Taylor, HP Jr, O'Neil JR, Kaplan IR (eds) Stable Isotope Geochemistry: A Tribute to Samuel Epstein. Geochemical Society Spec Publ 3:169-185

Lawrence JR, Gedzelman SD, White JWC, Smiley D, Lazov P (1982) Storm trajectories in eastern U.S.: D/H isotopic composition of precipitation. Nature 296:638-640

MacAyeal DR, Firestone J, Waddington ED (1991) Paleothermometry by control methods. J Glaciol 37:326-338

Mann ME, Bradley RS, Hughes MK (1999) Northern hemisphere temperatures during the past millennium: inferences, uncertainties and limitations. Geophys Res Letters 26:759-762

Marshall SJ, Cuffey KM (2000) Peregrinations of the Greenland Ice Sheet divide in the last glacial cycle: implications for central Greenland ice cores. Earth Planet Sci Letters 179:73-90

Merlivat M, Jouzel J (1979) Global climatic interpretation of the deuterium-oxygen-18 relationship for precipitation. J Geophys Research 84:5029-5033

Mix AC, Ruddiman WF (1984) Oxygen-isotope analyses and Pleistocene ice volumes. Quat Res 21:1-20

Mix AC, Bard E, Eglinton G, Keigwin LD, Ravelo AC, Rosenthal Y (2000) Alkenones and multiproxy strategies in paleoceanographic studies. Geochem Geophys Geosystems, Vol. 1, Paper No. 2000GC000056

NSIDC (National Snow and Ice Data Center) (1997) The Greenland Ice Cores CD-ROM. Available from the National Snow and Ice Data Center, University of Colorado at Boulder, and the World Data Center-A for Paleoclimatology, National Geophysical Data Center, Boulder Colorado, http://www.ngdc.noaa.gov/paleo/icecore/greenland/summit/index.html

Nye JF (1963) Correction factor for accumulation measured by the thickness of annual layers in an ice sheet. J Glaciol 4:785-788

Paterson WSB (1994) The Physics of Glaciers. Third edn. Pergamon, Oxford Univ Press, Oxford, UK

Paterson WSB, Clarke GKC (1978) Comparison of theoretical and observed temperature profiles in Devon Island ice cap, Canada. Geophys J Royal Astronomical Soc 55:615-632

Petit JR and 18 others (1999) Climate and atmospheric history of the past 420,000 years from the Vostok ice core, Antarctica. Nature 399:429-436

Pollack, HN, Huang SP (2000) Climate reconstruction from subsurface temperatures. Ann Rev Earth Planet Sci 28:339-365

Pollard D, Thompson SL (1997) Climate and ice-sheet mass balance at the last glacial maximum from the Version 2 Global Climate Model. Quat Sci Rev 16:841-863

Rindsberger M, Jaffe S, Rahamin S, Gat JR (1990) Patterns of the isotopic composition of precipitation in time and source: data from the Israeli storm water collection program. Tellus 42B:263-271

Rozanski K, Araguas-Araguas L, Gonfiantini R (1993) Isotopic patterns in modern global precipitation. In Climate change in continental isotopic records, Geophys Monogr 781-36, American Geophysical Union, Washington DC

Robin G de Q (ed). (1983) The climatic record in polar ice sheets. Cambridge University Press, Cambridge, UK

Salamatin AN, Lipenkov VY, Barkov NI, Jouzel J, Petit JR, Raynaud D (1998) Ice core age dating and paleothermometer calibration based on isotope and temperature profiles from deep boreholes at Vostok Station (East Antarctica). J Geophys Res 103:8963-8977

Schott C, Waddington ED, Raymond CF (1992) Predicted time-scales for GISP2 and GRIP boreholes at Summit, Greenland. J Glaciol 38:162-168

Schrag DP, Hampt G, Murray DW (1996) Pore fluid constraints on the temperature and oxygen isotopic composition of the glacial ocean. Science 272:1930-1932

Severinghaus JP, Brook EJ (1999) Abrupt climate change at the end of the last glacial period inferred from trapped air in polar ice. Science 286:930-934

Severinghaus JP, Sowers T, Brook EJ, Alley RB, Bender ML (1998) Timing of abrupt climate change at the end of the Younger Dryas interval from thermally fractionated gases in polar ice. Nature 391: 141-146

Shuman CA, Alley RB, Anandakrishnan S, White JWC, Grootes PM, Stearns CR (1995) Temperature and accumulation at the Greenland Summit: comparison of high-resolution isotope profiles and satellite passive microwave brightness temperature trends. J Geophys Res 100:9165-9177

Sowers T, Bender ML, Raynaud D (1989) Elemental and isotopic composition of occluded O_2 and N_2 in polar ice. J Geophys Res 94:5137-5150

Stute M, Schlosser P (2000) Atmospheric noble gases. *In* Cook PG, Herczer AL (eds) Environmental Tracers in Subsurface Hydrology, Kluwer, Boston, p 349-377

Stute M, Talma AS (1998) Glacial temperatures and moisture transport regimes reconstructed from noble gases and $\delta^{18}O$, Stampriet Aquifer, Namibia. *In* Isotope Techniques in the Study of Environmental Change, International Atomic Energy Agency, Vienna, IAEA-SM-349/53, p 307-318

Thompson LG, Davis ME, Mosley-Thompson E, Sowers TA, Henderson KA, Zagorodnov VS, Lin PN, Mikhalenko VN, Campen RK, Bolzan JF, Cole-Dai J, Francou B (1998) A 25,000-year tropical climate history from Bolivian ice cores. Science 282:1858-1864

Vimeux F, Masson V, Jouzel J, Petit JR, Steig EJ, Stievenard M, Vaikmae R, White JWC (2001) Holocene hydrological cycle changes in Southern Hemisphere documented in East Antarctic deuterium excess records. Clim Dyn 17:503-513

Voigt DE, Alley RB, Spencer MK, Creyts TT (1997) The flow and shutdown of ice stream C based on ice-core data. EOS Trans Am Geophys Union 78:F244 (abstr)

Werner M, Mikolajewicz U, Heimann M, Hoffmann G (2000) Borehole versus isotope temperatures on Greenland: seasonality does matter. Geophys Res Letters 27:723-726

Whillans IM (1981) Reaction of the accumulation zone portions of glaciers to climate change. J Geophys Res 86:4274-4282

Whillans IM, Grootes PM (1985) Isotopic diffusion in cold snow and firn. J Geophys Res 90:3910-3918

White JWC (1989) Stable hydrogen isotope ratios in plants: A review of current theory and some potential applications. *In* PW Rundel, JR Ehleringer, KA Nagy (eds) Stable Isotopes in Ecological Research. Ecological Studies 68:142-162. Springer-Verlag, New York

White JWC, Barlow LK, Fisher D, Grootes P, Jouzel J, Johnsen SJ, Stuiver M, Clausen H (1997) The climate signal in the stable isotopes of snow from Summit, Greenland: Results of comparisons with modern climate observations. J Geophys Res 102:26,425-26,439

Yen Y-C (1981) Review of thermal properties of snow, ice and sea ice. Cold Regions Research and Engineering Lab Rep 81-10

Zimmermann U, Ehhalt DH, Munnich KO (1967) Soil water movement and evapotranspiration: Changes in the isotopic composition of water. *In* Isotopes in Hydrology. International Atomic Energy Agency, Vienna, p 567-585

10   Isotopic Evolution of the Biogeochemical Carbon Cycle During the Precambrian

David J. Des Marais

Exobiology Branch
Ames Research Center
Moffett Field, California 94035

INTRODUCTION

Carbon is highly important for our biosphere, not just because it forms organic compounds; it also creates atmospheric greenhouse gases, pH buffers in seawater, and redox buffers virtually everywhere. Carbon species can stabilize metamorphic minerals and they can affect plutonism and volcanism. These various C constituents all interact via the biogeochemical C cycle, an array of C reservoirs linked by a network of physical, chemical and biological processes. The overall C cycle actually consists of multiple nested cyclic pathways that differ with respect to some of their reservoirs and processes (Fig. 1). However, all pathways ultimately pass through the hydrosphere and atmosphere, and it is this common course that unites the entire carbon cycle and allows even its most remote constituents to influence our environment and biosphere.

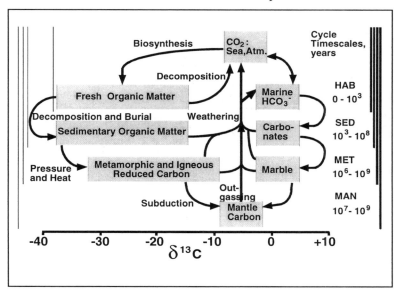

Figure 1. Biogeochemical C cycle, showing principal C reservoirs (boxes) in the mantle, crust, oceans and atmosphere, and showing the processes (arrows) that unite these reservoirs. The range of each of these reservoir boxes along the horizontal axis gives a visual estimate of $\delta^{13}C$ values most typical of each reservoir. The vertical bars at right indicate the timeframes within which C typically completely traverses each of the four C sub-cycles (the HAB, SED, MET and MAN sub-cycles, see text). For example, C can traverse the hydrosphere-atmosphere-biosphere (HAB) sub-cycle typically in the time scale between 0 to 1000 years.

1529-6466/00/0043-0010$05.00

The history of the biogeochemical C cycle has been at least partially recorded in the C isotopic composition ($\delta^{13}C_{PDB}$) of carbonate (δ_{carb}) and reduced C (δ_{org}) in ancient sedimentary and metamorphic rocks. To the extent that sedimentary rocks avoided deep burial and alteration, they have preserved information that indicates the status of the C cycle at the time of their deposition.

THE PRESENT-DAY CARBON CYCLE

The C cycle can be represented as an integrated system of reservoirs and processes (Fig. 1). The reservoirs and processes are shown as boxes and labeled arrows, respectively, and this network delineates the various sub-cycles (see labels "HAB," "SED," "MET" and "MAN," Fig. 1). Representative ranges of δ_{carb} and δ_{org} values for the various reservoirs are depicted by the widths of the boxes (Deines 1980; Weber 1967). The range of time scales typically needed for C to traverse each of these sub-cycles is indicated along the right margin, below their corresponding subcycle labels. Of course, the actual physical boundaries between these sub-cycles are not so sharply delineated in the natural environment; thus the sub-cycles depicted here represent characteristic domains along the continuum of reservoirs and processes that collectively constitute the complete biogeochemical C cycle.

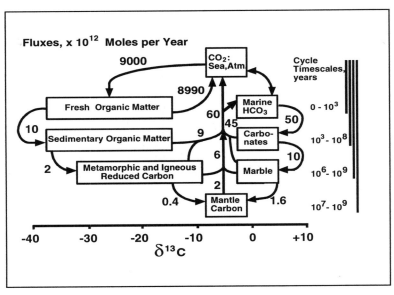

Figure 2. Biogeochemical C cycle (as in Fig. 1), showing principal C reservoirs (boxes) and their isotopic compositions in the mantle, crust, oceans and atmosphere, and the processes (arrows) that unite these reservoirs. Numbers adjacent to the arrows give estimates of present-day fluxes, expressed in the units 10^{12} mol yr^{-1}.

The hydrosphere-atmosphere-biosphere (HAB) sub-cycle

The sub-cycle that includes only the hydrosphere, atmosphere and biosphere (the "HAB" sub-cycle; see Figs. 1 and 2) is characterized by the relatively small sizes of its reservoirs (Table 1), the very high rates of exchange of C between them, and, accordingly, the relatively short time scales required for cycling. The biosphere, which is included in the "fresh organic matter" reservoir (Figs. 1 and 2), dominates the exchange

Table 1. Reservoirs of carbon in the atmosphere, hydrosphere and geosphere.

	Mass, $\times 10^{18}$ moles		
Reservoir	*Reduced C*	*Oxidized C*	*Total C*
Atmosphere	---	0.06[a]	0.06[a]
Biosphere: plants and algae	0.13[b]		0.13
Hydrosphere	---	3.3[c]	3.3
Pelagic sediments	60[d]	1300[d]	1360
Continental margin sediments	>370[e]	>1000[e]	>1370
Sedimentary rocks	750[f]	3500[f]	4250
Crustal metamorphic and igneous rocks	100[g]	?	>100
Mantle			27000[h]

[a] Holland (1984).   [b] Mopper and Degens (1979); Olson et al. (1985).   [c] Holland (1984).
[d] Holser, et al. (1988).   [e] Minimum inventories required for C isotopic mass balance, see text.
[f] Ronov (1980).   [g] Hunt (1972).   [h] Derived from estimates of mantle mass and C concentration, see text.

of C between CO_2 and organic matter. Today, the fixation of CO_2 is achieved principally by photosynthetic organisms, and is shared almost equally by primary production in the marine (4000×10^{12} mol yr^{-1}) and terrestrial (5000×10^{12} mol yr^{-1}) environments (Martin et al. 1987; Olson et al. 1985; Field et al. 1998). Because only about 10×10^{12} mol yr^{-1} of reduced C escapes into sediments (Berner and Canfield 1989), approximately 99.9% of the C fixed by the biosphere is cycled relatively rapidly back to inorganic C in the oceans and atmosphere.

Carbon can traverse the HAB sub-cycle within time scales of minutes to hours, as with the rapid cycling of C within microbial ecosystems (e.g. Des Marais 1995), and, at the thousand-year time scale, the present-day turnover rate of the global ocean (Broecker and Peng 1982).

The sedimentary (SED) sub-cycle

The sedimentary (SED) sub-cycle includes the reservoirs of sedimentary organic matter and carbonates (Table 1; Figs. 1 and 2). The SED sub-cycle strongly influences the HAB sub-cycle. For example, sedimentation limits global productivity by removing nutrients, phosphorus in particular (Holland 1984). The balance between sedimentation of oxidized (carbonate, sulfate, Fe^{3+}) versus reduced (organic C, sulfide, Fe^{2+}) species determines the abundances of O_2 and sulfate in the hydrosphere and atmosphere (Garrels and Perry 1974; Holland 1984). The weathering and transport of igneous and sedimentary rocks deliver nutrients to the oceans, thereby modulating global biological productivity. Organisms can substantially enhance weathering rates (Barker et al. 1997). The rates of weathering and erosion of sediments and rocks, together with the contents of reduced species in those deposits, determine their rates of consumption of O_2.

The rate at which C traverses the SED sub-cycle is determined principally by tectonic controls upon rates of formation and destruction of sedimentary rocks. The globally averaged half-life of sedimentary rocks is about 200 million years (Derry et al. 1992). Organic C is buried principally in terrigenous deltaic-shelf sediments (8.7×10^{12} mol yr^{-1}) and in sediments beneath highly productive open-ocean regions (0.8×10^{12} mol yr^{-1}). Global rates of organic C burial are about 10×10^{12} mol yr^{-1} (Berner and Canfield

1989). Although calcareous plankton presently dominate marine carbonate precipitation, the global net burial rate of carbonate (about 50×10^{12} mol yr^{-1}; Holser et al. 1988) is controlled ultimately by the rate of weathering of Ca^{2+} and Mg^{2+} and by the global rate of carbonate formation by alteration of submarine basalts (Garrels and Perry 1974). The modern rate of CO_2 uptake and carbonate formation during the hydrothermal alteration of submarine basalts has been estimated to be $(2-3) \times 10^{12}$ mol yr^{-1}; Staudigel et al. 1989). Sedimentary C reservoirs are much larger than the C reservoirs of the HAB sub-cycle (Table 1). Therefore, key biogeochemical properties of the oceans and atmosphere, including their nutrient inventories and oxidation states, are ultimately controlled by interactions with the SED sub-cycle. This control is exerted over time scales of typically tens of thousands to millions of years (Figs. 1 and 2).

The metamorphic (MET) sub-cycle

The MET sub-cycle (Figs. 1 and 2) includes the more deeply buried fraction of sedimentary and igneous rocks that experience metamorphism. This includes C in rocks entering subduction zones but that escapes injection into the mantle, either because this C is degassed or because its host rock also escapes subduction (e.g. by lateral accretion into continental crust). The volume of rocks in the MET sub-cycle greatly exceeds those in the SED sub-cycle (Lowe 1992), but the reduced C reservoirs in the MET sub-cycle are smaller (Hunt 1972), due both to losses during metamorphism of sedimentary rocks and to the typically much lower reduced C contents of igneous rocks. For example, hydrocarbon gases and CO_2 can be released by thermal decomposition of sedimentary organic matter (Hunt 1979). Carbonate C is also lost during metamorphism. Carbonates hosted in a range of rock compositions can yield CO_2 at elevated temperatures (Ferry 1991), for example:

Dolomite = Periclase + Calcite + CO_2

Tremolite + 11 Dolomite = 8 Forsterite + 13 Calcite + 9 CO_2 + H_2O

2 Talc + 3 Calcite = Dolomite + Tremolite + H_2O + CO_2

Carbon requires typically tens of millions to billions of years to traverse the MET sub-cycle. These transit times are typically longer than those for the SED sub-cycle (Fig. 2), and reflect the longer residence times of more deeply buried continental rocks.

The mantle-crust (MAN) sub-cycle

The deepest (MAN) sub-cycle includes the mantle C reservoir (Figs. 1 and 2) and the processes of subduction and mantle outgassing. Even though rocks that are subducted into the mantle become metamorphosed in the process, the associated C inventories that also reach the mantle become part of the MAN, not the MET, sub-cycle. The modern global rate of C outgassing from the mantle is approximately 2×10^{12} mol yr^{-1} (Des Marais 1985; Marty and Jambon 1987). The values shown in Figure 2 for subduction of carbonate and reduced C reflect the assumption that rates of C exchange between the mantle and the crust/hydrosphere are balanced (at near-steady state), and that subduction does not discriminate between oxidized and reduced C species. Some tens of millions to billions of years are required for C to traverse the MAN sub-cycle (Fig. 2). Assuming a C content of about 80 μg g^{-1} in the source region of mid-ocean ridge magmas (Pawley et al. 1992, Holloway, pers. comm.), the mantle C inventory is roughly 3×10^{22} mol (Table 1), which greatly exceeds the crustal C inventory. Accordingly, over time scales of tens of millions to billions of years, the processes of mantle-crust exchange would have controlled both the abundance and the overall oxidation state of the much smaller C reservoirs in the HAB and SED sub-cycles. In addition, thermal emanations of other reduced species (principally sulfides, H_2, and Fe^{2+}) consume O_2 and contribute reducing power for biosynthesis by chemoautotrophs. Thus the overall oxidation state of the surface environment and the crustal carbon reservoirs is controlled by the balance between biological productivity and decomposition, sediment cycling, and thermal processes.

Biota that dwell at or below the sea floor along the mid-ocean ridges obtain their energy principally from oxidation-reduction reactions involving reduced hydrothermal emanations (e.g. Jannasch and Wirsen 1979). Accordingly, their primary productivity cannot exceed the flux of reduced species from thermal sources. The principal reduced species include reduced sulfur, H_2 (derived from water-rock reactions), and Fe^{2+}. Today, this total flux, expressed as O_2 equivalents, is in the range (0.2 to 2.1) $\times 10^{12}$ mol yr^{-1} (Elderfield and Schultz 1996).

The heat flow from Earth's interior was substantially greater during the earlier Precambrian (e.g. Lambert 1976). During the past 3.0 billion years, radioactive decay of the elements U, Th and ^{40}K has been the principal source of this heat, therefore their decay over time has caused global heat flow to decline. Heat flow 3.0 billion years ago has been estimated to have been 2.2 times its modern value (Turcotte 1980), therefore the midocean ridge mantle CO_2 flux was substantially larger during the Archean Eon (discussed later).

ISOTOPIC INDICATORS OF CARBON BUDGETS AND PROCESSES

Biological isotopic discrimination

The $\delta^{13}C$ values of freshly-deposited sedimentary C are established by the "exogenic" part of the C cycle, namely those processes and reservoirs associated with erosion and outgassing of C, transport and chemical transformations of C within the hydrosphere and atmosphere, and C sedimentation and burial. As it flows through the exogenic domain, C can be oxidized or reduced through interactions with other oxidation/reduction-sensitive elements, principally O, N, S, Fe, and Mn. Values of δ_{carb} are determined by the $\delta^{13}C$ value of inorganic C dissolved in seawater. δ_{carb} values of carbonates deposited in open marine environments are generally representative of the average global δ_{carb} value. The $\delta^{13}C$ values of inorganic C dissolved in modern seawater can vary regionally by a few permil (Kroopnick 1985), accordingly, coeval marine carbonates can also vary isotopically. Values of δ_{org} in marine settings are determined principally by isotopic discrimination associated with biological fixation of CO_2, which today is determined primarily by photosynthetic populations and then modified by various consumers (Hayes et al. 1989; also, see Hayes, this volume). Although marine δ_{org} values vary regionally (Rau et al. 1989), samples taken from a globally representative array of environments can give a reasonable estimate of global mean marine δ_{org} values. Thus the isotopic difference

$$\varepsilon_{TOC} \sim \delta_{carb} - \delta_{org} \tag{1}$$

where TOC is Total Organic Carbon. The term ε_{TOC} reflects the net effect of the following: (1) isotopic equilibrium between inorganic carbon species, (2) the metabolic pathways of CO_2 fixation and C metabolism by autotrophic biota, and (3) the pathways and mechanisms that transform and remineralize organic C.

Isotopic mass balance and the sedimentary cycle

Values of δ_{carb} and δ_{org} can also be incorporated into an isotopic mass balance to quantify relative fluxes of carbonate and organic C to sediments, as follows:

$$\delta_{in} = f_{carb}\delta_{carb} + f_{org}\delta_{org} \tag{2}$$

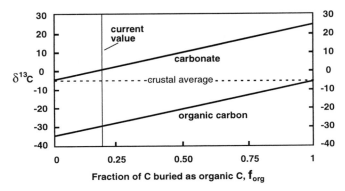

Figure 3. Relationship between isotopic composition (δ_{carb} and δ_{org}) and the fraction of carbon buried as organic matter. The vertical separation between the lines depicts ε_Δ, and thus reflects the combined effects of equilibria between inorganic C species and biological isotope discrimination (see text). A value $\varepsilon_{TOC} = 30$ is depicted here, and represents the long-term average value during the past 800 Ma (Hayes et al. 1999). The vertical line represents the value of $f_{org} = 0.2$, which represents the current value for the global C cycle.

where δ_{in} represents the isotopic composition of C entering the surface environment (including land, oceans and atmosphere), and the equation's right side represents the weighted-average isotopic composition of C buried in sediments; f_{carb} and f_{org} represent the fractions of C buried as carbonates and organic matter ($f_{carb} = 1 - f_{org}$), respectively. Rearranging Equation (2) and substituting for f_{carb} yields:

$$f_{org} = (\delta_{carb} - \delta_{in})/(\delta_{carb} - \delta_{org}) \tag{3}$$

Values of f_{org} reflect biological productivity, organic decomposition, and also those processes that control sedimentation and burial of C. Given the requirement that isotopic mass balance be preserved, variations in f_{org} can be linked to variations in values of δ_{carb} and δ_{org}, as illustrated in Figure 3.

If inorganic C is extensively converted to organic C that is then efficiently buried (that is, f_{org} is high), it follows that other redox-sensitive species must become oxidized in the process. Thus, changes in δ_{carb} and δ_{org} can reflect changes in the production and/or sedimentation of carbonates and organic matter, as well as changes in the oxidation state of the environment.

Preservation of the carbon isotopic record

Thermal alteration or maturation of organic C chemically alters this C and even depletes it from sediments. To the extent that such losses are isotopically selective, the δ_{org} values of the reduced C ("kerogen") that remains can change (Hayes et al. 1983). Kerogen δ_{org} values can increase during thermal alteration, due to the preferential breakage of $^{12}C\text{-}^{12}C$ bonds, relative to $^{13}C\text{-}^{12}C$ bonds. This selectivity leads to loss of ^{12}C-enriched material and retention of a ^{13}C-enriched kerogen residuum. In order to reconstruct the δ_{org} value of the original, unaltered sediment, the relationship must be assessed between a rock's thermal maturity and the degree to which its δ_{org} value has increased.

Fortunately, thermal maturity of organic C can be quantified by measuring its elemental H/C value. Thermal alteration eliminates organic hydrogen more readily than

C, lowering the elemental H/C value of the residual kerogen (Hayes et al. 1983). For a population of kerogen fractions that once shared identical initial δ_{org} values but that experienced different degrees of thermal alteration, a plot of H/C versus δ_{org} can, in principal, establish the relationship between the extent of thermal alteration and the associated increase in δ_{org} of the residual kerogen (Hayes et al. 1983). If H/C values of other kerogens are then measured, this quantitative relationship could then be used to "correct" kerogen δ_{org} values for thermal alteration.

In practice, it is difficult to obtain a group of kerogens that once shared near-identical initial δ_{org} values yet have been thermally altered to varying degrees. One approach might be to analyze a set of "calibration" samples collected along a once-isotopically-uniform sedimentary bed at various distances from a contact with a thermal source (e.g. volcanic intrusive, etc.; Simoneit et al. 1981). However, such samples typically were heated at much higher temperatures and much shorter time scales than that experienced by most of the samples to which this isotopic correction would be applied. Accordingly, this particular calibration method might not accurately reflect the style of thermal alteration experienced by many of the ancient Precambrian kerogen samples. An alternative, perhaps more applicable, approach is to plot elemental H/C versus δ_{org} for a very large set of ancient kerogens that represent an interval of time over which any net long-term secular trend in δ_{org} was minor (Des Marais 1997a). Even though such a collection of kerogens may have had a considerable range of initial δ_{org} values, the isotopic variations would cancel out in a large number of samples, and any consistent net relationship between H/C and δ_{org} would become apparent.

Fortunately, the carbon isotopic composition of Proterozoic kerogens has been studied as a function of their elemental H/C values (Strauss et al. 1992a). Strauss and coworkers observed that kerogen samples that experienced various degrees of thermal alteration exhibited a correspondingly wide range of $\delta^{13}C$ values. In order to minimize long-term secular trends in δ_{org}, their data were subdivided into the following two age groups: 2.1 to 1.0 Ga and 1.0 to 0.53 Ga. Data for kerogens older than 2.1 Ga were insufficient to construct a statistically valid plot. A plot of H/C versus δ_{org} was achieved by grouping data according to their H/C values (e.g. values between 0.1 and 0.3 were plotted as 0.2, values from 0.2 to 0.4 were plotted as 0.3, and so forth). Even though the average δ_{org} values of the two age groups differed markedly, the two groups exhibited remarkably parallel isotopic trends with respect to H/C values (Fig. 4). For both groups, δ_{org} values were approximately 2 to 2.5 permil greater for kerogens with H/C values of 0.2 than for kerogens with H/C values of 0.8. These relationships were fitted with third order polynomial expressions that can be used to correct for thermal maturation effects. For kerogens having ages in the range 2.1 to 1.0 Ga, the equation is as follows:

$$\Delta\delta_{org\,(corrected)} = \delta_{org(measured)} - (4.2 - 11.9(H/C) + 11.1(H/C)^2 - 3.4(H/C)^3) \qquad (4)$$

For kerogens in the age range 1.0 to 0.53 Ga, the relationship is as follows:

$$\Delta\delta_{org\,(corrected)} = \delta_{org(measured)} - (4.3 - 8.7(H/C) + 5.1(H/C)^2 - 0.55(H/C)^3) \qquad (5)$$

Data for kerogens older than 2.1 Ga were corrected using Equation (4). Both the uncorrected data (Strauss et al. 1992b; shown as crosses) and the corrected data (open circles) derived from Equations (4) and (5) are exhibited in Figure 5. The corrected δ_{org} values are more negative than the uncorrected values, consistent with correcting for the positive isotopic shift due to thermal alteration. Also, data points corresponding to the older kerogens generally exhibit larger shifts, reflecting the larger corrections associated with the generally lower H/C values in progressively older kerogens (Strauss et al. 1992b; Des Marais 1994). However, these corrections do not markedly alter the major patterns

Figure 4. Plot of δ_{org} versus H/C values for Proterozoic kerogens in the age range 2.1 to 1.0 Ga (filled circles) and 1.0 to 0.6 Ga (open circles). These data illustrate the magnitude (about 2.5 to 3 permil) by which δ_{org} of residual kerogen is increased when thermal alteration lowers the H/C value of the kerogen from about 0.8 to 0.1 (see text). The similar slopes of the two curves indicates that this δ_{org} shift was a reproducible phenomenon among mid-Proterozoic and Neoproterozoic kerogens (Des Marais 1997a).

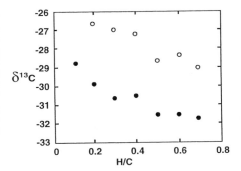

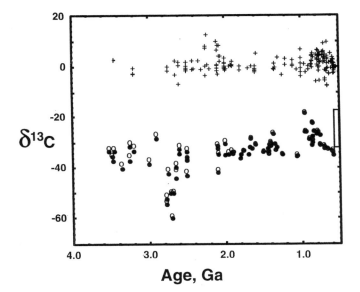

Figure 5. Plot of age versus δ_{carb} (crosses) and δ_{org} for Archean and Proterozoic kerogens. Kerogen data (filled circles) are corrected for the effects of thermal alteration (Des Marais 1997a). Uncorrected data are shown as open circles. Between 2.2 to 2.0 billion years ago, note the high δ_{carb} values and the virtual disappearance thereafter of δ_{org} values more negative than -36. Other evidence indicates that atmospheric O_2 increased substantially at this time (see text).

exhibited by the uncorrected δ_{org} values.

The observation that thermal alteration of kerogens increases δ_{org} by about 3 permil as H/C is lowered from >1 to 0.1 is also consistent with theoretical arguments (Watanabe et al. 1997), measurements of sedimentary rocks (Simoneit et al. 1981) and laboratory experiments (Lewan 1983; Peters et al. 1981).

For those ancient sedimentary rocks that have attained metamorphic grades exceeding perhaps greenschist facies, thermal alteration has proceeded beyond the stage where the original isotopic compositions can be precisely reconstructed. In addition to

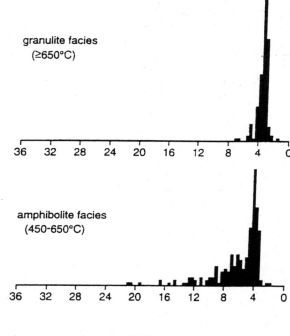

granulite facies
(≥650°C)

amphibolite facies
(450-650°C)

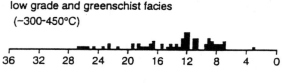

low grade and greenschist facies
(~300-450°C)

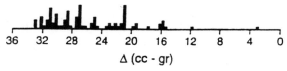

unmetamorphosed

Δ (cc - gr)

Figure 6. Distribution of measured calcite-reduced C fractionation values, Δ(cc-gr), as metamorphic grade increases (Valley and O'Neil 1981; Schidlowski 2001; Kitchen and Valley 1995; Kumar et al. 1997). The isotopic difference between carbonate and reduced C decreases at higher metamorphic grades. The narrow range of values associated with granulite facies rocks indicates that thermodynamic equilibrium between carbonate and graphite has been nearly attained.

kinetic isotope effects associated with breaking of C-C bonds, the approach to isotopic equilibrium between coexisting carbonate and reduced C phases will substantially alter values of δ_{carb} and δ_{org}, and, therefore, ε_{TOC} also. Isotopic equilibration can effectively obscure any preexisting isotopic patterns that indicate biological activity. Carbon isotopic compositions might be relatively well-preserved when kerogen-rich black shales and massive carbonates are transformed to graphitic slates and marbles, respectively. However, finely dispersed sedimentary organic carbon and coexisting carbonate can exhibit evidence of substantial isotopic exchange. The progressive obliteration of isotopic "biomarker" patterns and attainment of isotopic equilibrium between reduced carbon and carbonate is illustrated in Figure 6 (Kitchen and Valley 1995; Kumar et al. 1997). Shifts in δ_{carb} occur during thermally-driven exchange between carbonates and reduced carbon (Dunn and Valley 1992). Because graphite has eight times more carbon than calcite by weight, it only takes a small amount of graphite to shift δ_{carb} (J. Valley, pers. comm.). Calcite $\delta^{13}C$ values in Adirondack marbles declined on average about 4‰, and some samples declined as much as 8‰ just due to exchange with graphite (Kitchen and Valley 1995).

THE ARCHEAN RECORD

Planetary processes, climate and the carbon cycle

The Archean mantle influenced significantly the inventory of C in the crust, oceans and atmosphere. Higher rates of crustal production were accompanied by higher rates of mantle outgassing of C (Des Marais 1985). Subducting slabs might not have penetrated the hotter Archean mantle as deeply (McCulloch 1993), and the slabs certainly would have retained C with greater difficulty. These considerations are consistent with an early Earth in which the crustal C inventory exceeded the modern inventory (Des Marais 1985; Zhang and Zindler 1993). Alternatively, more rapid rates of mantle-crust exchange actually might have buffered Archean crustal and surface inventories of C to quantities rather similar to modern inventories (Sleep and Zahnle 2001). In either case, C cycling between the mantle and crust was more vigorous during the Archean Eon than today.

Relatively little well preserved sedimentary rock and associated C can be found in lithosphere older than 3.0 Ga. The tectonically active Archean marine basins favored rapid destruction of crust by continental collisions, partial melting and mantle/crust exchange (Windley 1984). Most of Archean continental crust had not yet become stabilized by cratonization (Rogers 1996), therefore a greater fraction of continental sediments experienced greater instability and thermal alteration. The freshest sediments occur in the 3.5 to 3.2 billion-year-old Kaapvaal Craton of South Africa and the Pilbara Block of Western Australia (Lowe 1992). These deposits are associated with episodes of greenstone activity and intrusive events that created stable microcontinents or cratons. These cratons became the nuclei of full-sized modern continents.

During the production of Archean sediments, chemical weathering was at least as effective as today. This is consistent with high CO_2 levels that compensated for a less luminous sun (Walker 1985) and thus contributed substantially to sustaining a warm climate (Lowe 1992). Biological activity might have played a major role in maintaining weathering rates (Schwartzman and Volk 1989). The crust was tectonically and magmatically less stable and produced thick first cycle sediments in the greenstone belts. The effective weathering implied by these observations is consistent both with relatively warm, moist conditions and with elevated atmospheric CO_2 concentrations (Lowe 1994).

Recently, a mineralogical study of a Precambrian paleosol (Rye et al. 1995) has indicated that, prior to 2.0 Ga, atmospheric CO_2 levels were insufficient to explain the degree of atmospheric greenhouse warming evidenced by the rock record. Accordingly, CH_4 has been proposed as an additional key greenhouse gas (e.g. Pavlov et al. 2000). Because the oxidation state of the Archean crust resembled its modern value (Delano 2001), nonbiological thermal sources of CH_4 were insufficient by themselves to maintain the requisite greenhouse warming. Accordingly, biogenic sources of CH_4 must have been substantial during the Archean Eon, consistent with the observation of extremely ^{13}C-depleted kerogens in 2.7 Ga sedimentary sequences (Hayes 1994).

Archean biosphere

Impact of life upon the carbon cycle. The consequences of the biosynthesis of abundant organic C probably represent life's most substantial impact upon the C cycle. Before life began, H_2 gas from thermal sources had escaped to the atmosphere. Some of this H_2 had even escaped to space. The earliest biosphere augmented inventories of reduced carbon by efficiently oxidizing reduced volcanic and hydrothermal species in order to convert CO_2 to organic matter. After oxygenic photosynthesis developed, the net burial in sediments of photosynthetically derived organic C allowed a stoichiometrically equivalent amount of O_2 and/or its oxidation products to accumulate slowly in the surface environment (e.g. Garrels and Perry 1974). Thus life greatly increased inventories of organic C in the HAB and SED sub-cycles, and it has simultaneously increased the oxidation state of the hydrosphere and atmosphere.

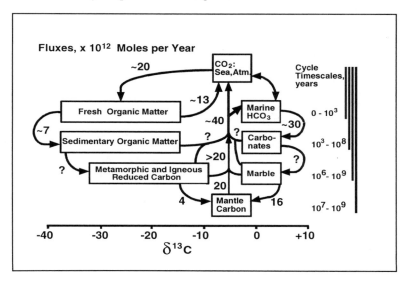

Figure 7. The biogeochemical C cycle prior to the advent of oxygenic photosynthesis, showing the much lower global primary productivity and the higher rates of thermal emanation of C (see text). Comparison with Figure 5 illustrates the enhancement of global primary productivity due to the development of oxygenic photosynthesis. Flux estimates are highly approximate, and are shown principally to illustrate the direction and magnitude of change over geologic time.

For example, Figure 7 depicts a hypothetical C cycle before the biosphere invented oxygenic photosynthesis. Unable to split H_2O to obtain H for the reduction of CO_2 to organic C, the biosphere obtained reducing power for organic biosynthesis only by utilizing reduced species derived from volcanism, hydrothermal activity and weathering. Therefore, global rates of primary productivity could not exceed the rate of thermal emanations of reduced species. To the extent that thermal reduced species were partially oxidized abiotically, biological productivity was probably less than the total thermal flux of these species. Today, the global flux of reduced species from the mid-ocean ridges is in the range 0.2 to 2×10^{12} mol yr^{-1} (Elderfield and Schultz 1996). Today, this thermal flux from oceanic crust constitutes more than 90 percent of the mantle flux (Des Marais 1985). This flux was greater in the distant past, due to higher heat flow (Turcotte 1980). Therefore, for the pre-oxygenic-photosynthetic biosphere, it is assumed that global primary productivity scaled with the square of heat flow, and thus was estimated to have been in the range (2 to 20) $\times 10^{12}$ mol yr^{-1} (Fig. 7).

When photosynthesis developed, life tapped into an energy resource that was orders of magnitude larger than the energy available from oxidation-reduction reactions associated with weathering and thermal activity. The consequences of this innovation can be quantitatively illustrated for the modern Earth. Globally, hydrothermal sources deliver $(0.13\text{-}1.1) \times 10^{12}$ mol yr^{-1} of reduced S, Fe^{2+}, Mn^{2+}, H$_2$ and CH$_4$ (Elderfield and Schultz 1996); this is estimated to sustain at most about $(0.2\text{-}2) \times 10^{12}$ mol C yr^{-1} of organic carbon production by chemoautotrophic microorganisms (Des Marais 1997b). In contrast, global photosynthetic productivity is estimated to be 9000×10^{12} mol C yr^{-1} (Field et al. 1998). Thus, even though global thermal fluxes were greater in the distant geologic past (Turcotte 1980), the onset of oxygenic photosynthesis increased global organic productivity by more than two orders of magnitude.

The microfossil record of cyanobacteria, the "inventors" of oxygenic photosynthesis, is well evident throughout the Proterozoic Eon (~2500 to 543 Ma) (Knoll 1996). For example, only cyanobacteria are known to synthesize 2-methyl bacteriohopanepolyols, and these are transformed in sediments to 2-methylhopanes, which have now been identified in rocks as old as 2500 to 2700 My (Summons et al. 1999, Brocks et al. 1999). Stromatolitic carbonates were widespread along continental margins throughout the Proterozoic (Walter 1976) and have long been associated with cyanobacterial communities. Although few stromatolites harbor diagnostic cellular fossils, the large, Paleoproterozoic stromatolitic reefs that rival modern reefs in size, architecture and extent (Grotzinger 1989) constitute firm evidence that oxygenic photosynthesis had become well established by 2500 Ma. Buick also concluded that late Archean stromatolites, observed in 2700 Ma lake deposits, required oxygenic photosynthesis in order to develop abundantly in environmental settings that lacked evidence of hydrothermal activity (Buick 1992). The discovery of sterane biomarker compounds in 2700 Ma sediments (Brocks et al. 1999), demonstrated not only that eukarya existed then, but also that free oxygen was available for sterol biosynthesis. Of course, these early eukarya might have lacked many of the distinctive features exhibited by their modern descendants. The extremely low ^{13}C/^{12}C values in kerogens 2800 Ma in age (Fig. 5) have been attributed to methanotrophic bacteria, which require both oxygen and methane (Hayes 1994).

The microfossil record of photosynthetic biota might extend further back, to 3.3 to 3.5 Ga (Schopf 1992), but these early Archean occurrences are controversial (Buick 1991). Conical and branched pseudocolumnar stromatolites occurring in 3.46 Ga carbonates and silicified carbonates of the Warrawoona Group, Western Australia were recently described (Hofman et al. 1999). Hoffman et al. concluded that microorganisms were indeed involved in the accretion of these stromatolites. They also surmised that microbial phototaxis might have played a role in shaping them, but concluded that such evidence for the role of photosynthetic biota is not yet definitive.

Archean carbon isotopic record. Sedimentary δ_{org} values ultimately become altered at elevated temperatures (Fig. 4). Alteration of δ_{org} and δ_{carb} values can be substantial if reduced and carbonate carbon can exchange isotopically (Fig. 6). However, C isotopic evidence of biological processes is far more resistant to thermal destruction than organic molecules and biological structures (e.g. cells) and textures (Hayes et al. 1983). For example, in early Archean metasedimentary rocks, Greenland, δ_{org} values ranged from -22 to -50 (Mojzsis et al. 1996; Rosing et al. 1996). Low δ_{org} values were measured for minute particles of reduced C that had been situated within grains of apatite, where the C had apparently been shielded from metamorphic alteration (Mojzsis et al. 1996). Taken together with δ_{carb} values from coeval carbonates, these yield ε_{TOC} values (see Eqn. 1) that far exceed those attained by equilibrating these carbon phases at the temperatures of metamorphism (>600°C) inferred for these rocks (Hayes et al. 1983; Rosing et al. 1996). The δ_{org} data exhibited clusters of values that spanned the range -22 to -50. Mojzsis et al.

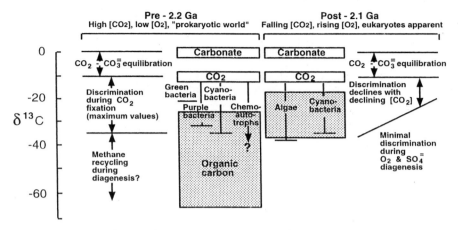

Figure 8. Range of δ_{carb} and δ_{CO2} values (open boxes) and δ_{org} values (shaded boxes), together with the processes proposed to explain their distribution prior to 2.2 Ga and subsequent to 2.1 Ga. A temperature of 15°C was assumed for the isotopic equilibrium between δ_{carb} and δ_{CO2}. The lines associated with the various groups of autotrophic bacteria and algae illustrate the maximum discrimination expected for each group. The sloped line at right depicts declining discrimination over time, perhaps in response to declining CO_2 levels.

argued that this distribution of δ_{org} values cannot be explained by Raleigh distillation during metamorphic oxidation, therefore they apparently indicate biological isotopic discrimination (Mojzsis et al. 1996; see Eiler et al. 1997 for correction).

The isotopic record of early life in the ca. 3.2-3.5 Ga volcano-sedimentary sequences in the Kaapvaal (South Africa) and Pilbara (Australia) Cratons is much more robust. For these successions, δ_{carb} values average 0±2 (Veizer et al. 1989), and δ_{org} values range from –25 to –41 (Fig. 5; Strauss et al. 1992a; Des Marais 1997a). Such δ_{org} values (25 to 41) indicate early Archean ecosystems that were driven by autotrophy. Such values are conventionally interpreted to indicate discrimination by the pentose phosphate (Calvin) cycle operating under conditions of high CO_2 that favor maximum isotopic discrimination (Fig. 8; e.g. Schidlowski 1993). However, the broad range of δ_{org} values observed for these early Archean sequences is reminiscent of the wide range of discrimination exhibited during autotrophic C assimilation by diverse microorganisms, anaerobes in particular (see Hayes, this volume). Isotopic discrimination associated with carbon assimilation by the reverse tricarboxylic acid cycle, 3-hydroxypropionate cycle, pentose phosphate (Calvin) cycle, and acetyl-CoA cycle can range from <10 to >40 permil (e.g. Preuss et al. 1989 Schidlowski et al. 1983). Consequently, the carbon isotopic record of early Archean carbonates and kerogen is consistent with, but not yet compelling for, the existence of oxygenic photosynthesis at that time. Such isotopic patterns are also consistent with isotopic discrimination by chemoautotrophic microorganisms, including methanogens, and anoxygenic photoautotrophic bacteria (Schidlowski et al. 1983).

The post-depositional diagenetic alteration of organic carbon also can contribute to the diversity of δ_{org} and δ_{carb} values (e.g. Hayes et al. 1989). Diagenetically imposed isotopic changes that are associated with anaerobic, fermentative processes can be substantially greater than those associated with respiratory processes (Fig. 8; e.g. Hayes et al. 1999). Thus the nature of the sedimentary depositional environment and the associated microbial populations can influence strongly the isotopic patterns observed.

THE LATE ARCHEAN TO PROTEROZOIC TRANSITION

Changes in the carbon cycle

The evolution of Earth's mantle and crust during the Late Archean and Paleoproterozoic (early Proterozoic) substantially affected the C cycle. Following the inevitable decay of radioactive nuclides in the mantle, the heat flow from Earth's interior declined (Turcotte 1980). This decreased the rates of both sea floor hydrothermal circulation and volcanic outgassing of reduced species. The tectonic reworking of Archean continental crust also had important consequences for the C cycle. Through a process termed 'internal differentiation' (Dewey and Windley 1981), preexisting crust may have become vertically zoned into granitic upper and granulitic lower parts. Crustal evolution during the Late Archean and Paleoproterozoic involved the modification, rearrangement and thickening (over- and underplating) of preexisting crust (Lowman 1989). New and extensive stable shallow water platforms became sites for productive benthic microbial communities and also for the deposition and long-term preservation of carbonates (Grotzinger 1989) and organic C (Des Marais 1994).

More extensive, stable continental crust would have enhanced weathering and the release of nutrients. Soil biota likely accelerated the weathering process on land, drawing CO_2 levels down (Lovelock and Whitfield 1982; Schwartzman and Volk 1989). A greater nutrient flux into coastal waters would have enhanced biological productivity (Betts and Holland 1991), which would have lowered CO_2 concentrations in surface seawater. In any case, a declining atmospheric CO_2 inventory would have contributed to falling global temperatures during the late Archean. The first well recorded glaciations occurred in the Late Archean (von Brunn and Gold 1993) and Paleoproterozoic (Harland 1983). Perhaps these events represent, at least in part, consequences of declining atmospheric CO_2 levels.

Patterns of carbonate deposition, as well as the presence/absence of gypsum/anhydrite in associated evaporites, indicate that seawater concentrations of HCO_3^- and $CO_3^=$ have declined, and $SO_4^=$ has increased, since the late Archean (Grotzinger and Kasting 1993). Late Archean and Paleoproterozoic platform sequences include relatively abundant evidence of abiotic carbonate precipitation as tidal flat tufas and marine cements. Evaporite sequences often proceed directly from carbonate to halite deposition, thus excluding gypsum/anhydrite deposition. The range of $\delta^{34}S$ values observed in Archean sedimentary sulfides is much narrower than in younger rocks (Canfield and Raiswell 1999). These observations are consistent with significantly lower seawater $SO_4^=$ concentrations and/or considerably greater HCO_3^- concentrations. In either case, the ratio of HCO_3^- to Ca^{2+} was large enough to prevent deposition of gypsum/anhydrite in marine or marginal-marine environments. These observations are consistent with the view that inorganic C reservoirs within the HAB sub-cycle were much higher during the Archean and Paleoproterozoic (Walker 1985).

Marine sedimentation on the Kaapvaal Craton: a glimpse of carbon isotopic patterns at the dawn (2.5-2.3 Ga) of the Proterozoic Eon

The Kaapvaal Craton, South Africa has preserved a remarkably complete sediment package that records coastal environments seaward to deeper water facies (e.g Klein and Beukes 1989). For example, the Kaapvaal includes the earliest known example of an extensive carbonate platform that resembles modern platforms in remarkable detail (Grotzinger and Kasting 1993). The extent and diversity of associated stromatolitic carbonate reefs are fully consistent with the existence at that time of highly productive, cyanobacterial (oxygenic photosynthetic) communities (Grotzinger 1989; Des Marais 2000).

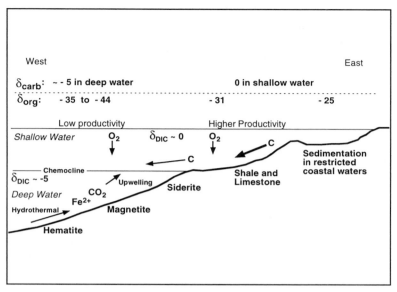

Figure 9. Schematic depositional environments for coastal facies, carbonate slope and iron formations in a stratified marine water column (adapted from Beukes et al. 1990). Distributions of minerals and trace species indicate that the deep waters were anoxic and the shallow water column was at least mildly oxidizing. The thick and thin arrows associated with C indicate higher and lower organic productivity, respectively. The regional distribution of δ_{carb} and δ_{org} values indicate that δ values of dissolved inorganic C were more negative in deep water, relative to shallow water, perhaps reflecting hydrothermal inputs. The δ_{org} values decrease offshore, indicating that ε_{TOC} increased, perhaps in response to increasingly important C cycling by chemoautotrophs and anaerobic microorganisms.

Even with the attendant uncertainties about the specific origins of biological isotopic discrimination, isotopic patterns offer insights about paleoenvironments, with implications for their long-term evolution. It is instructive to examine Kaapvaal Craton sedimentary sequences for systematic relationships between δ_{carb} and δ_{org} values and the nature of the paleoenvironment. Limestone (Campbellrand carbonates), shale, siderite and oxide banded iron formation (Kuruman Iron Formation) facies were deposited from east (landward) to west (seaward) on the submerged margin of the craton (Fig. 9; Klein and Beukes 1989). Taking into account the potential effects of post-depositional alteration upon δ_{carb}, a difference remains in the primary δ_{carb} of calcmicrosparite in the limestone (average $\delta_{carb} = -1.3$), that was deposited closer to shore, versus that of siderite microsparite in the iron formation (average $\delta_{carb} = -5.3$) that was deposited in deeper water (Beukes et al. 1990). Because both of these carbonates formed in the same basin, the water column must have been isotopically stratified with respect total dissolved inorganic C. This isotopic stratification, together with other geochemical evidence from trace and rare earth elements (Klein and Beukes 1989), indicates that deeper waters had been substantially influenced by hydrothermal inputs. Furthermore, the deep water must have been anoxic to allow Fe^{2+} to be transported in solution to the site of deposition of iron formation.

Values of δ_{org} display a pronounced negative regional trend from east to west (Fig. 9), if the transect is extended to the east to include δ_{org} values for restricted basins and coastal facies (Watanabe et al. 1997), in addition to the δ_{org} data for carbonate, shale and iron formations further offshore to the west (Strauss and Beukes 1996). Rocks that formed perhaps in restricted water bodies (evaporative lakes?) to the east have δ_{org} values ca. -25; whereas rocks from shallow subtidal environments in the middle of the transect have values ca. -31. Deeper water limestones and shales to the west have δ_{org} values ca. -35 to -44. Thus, after accounting for changes in δ_{carb}, ε_{TOC} increased from east to west. Values of δ_{org} from the shallower environments to the east are consistent with photosynthetic production and sedimentation in oxygenated waters (Watanabe et al. 1997). The more negative δ_{org} values in the deeper water facies to the west strongly indicate that organic diagenesis by methanogens and other anaerobic biota was substantial. This pattern is consistent with an anoxic deeper water environment (Watanabe et al. 1997, Hayes et al. 1999, see discussion below) that probably also had very low sulfate abundances. Methane recycling has been specifically invoked to explain the extremely negative δ_{org} values observed in late Archean kerogens (~2.8 to 2.6 Ga; Hayes 1983; Hayes 1994, see below).

Isotopic change from late Archean to Mesoproterozoic (2.7-1.7 Ga)

Several isotopic features and trends during this time interval are both robust and significant (Fig. 5). First, δ_{org} scatter widely in kerogens older than 2.1 Ga,, ranging from the mid -20 permil range to values as low as -65, with many values more negative than -35. In contrast, kerogens younger than 2.0 Ga have δ_{org} values in the narrower range -20 to -35, with virtually no values more negative than -36. To the extent that marine kerogen δ_{org} values <-35 require an anoxic, low-sulfate deep water column for their formation, the disappearance of these low δ_{org} values after 2.1 Ga indicates that the deep ocean had become more oxidized, either by oxygen or by sulfate (Fig. 8). The proposal (Canfield 1998) that sulfate levels in the deep ocean became substantial during the Mesoproterozoic is consistent with the carbon isotopic evidence.

The Paleoproterozoic interval witnessed large global excursions in δ_{carb} values (Baker and Fallick 1989; Karhu and Holland 1996; Fig. 10). Each of the positive and negative isotopic excursions between 2.3 and 2.0 Ga are documented within multiple basins globally and thus very likely represent widespread events (Melezhik et al. 1999). Additional very positive δ_{carb} values in the intervals 2.44 to 2.39 Ga and 1.92 to 1.97 Ga each represent only single sedimentary basins, therefore it is not yet established that they record global-scale phenomena (Melezhik et al. 1999). Those positive δ_{carb} values that truly reflect global isotopic excursions during the Paleoproterozoic indicate that the fraction of C buried as organic matter (f_{org}) varied repeatedly from less than 20% to more than 50% of the global carbon flux (Eqn. 3; Karhu and Holland 1996). However, at least some of the most positive δ_{carb} values apparently developed within large hypersaline, restricted basins that also sustained abundant stromatolite growth (Melezhik et al. 1999). Such organic burial events released equivalent amounts of O_2, which in turn reacted to increase the reservoirs of Fe^{3+} and $SO_4^=$. The impact of this release upon atmospheric composition must have been substantial.

The above observations indicate a consistent relationship between changing C isotopic patterns and an evolving global environment and biosphere (Fig. 8; Des Marais 1997a). The observations favor a scenario where the Archean atmosphere had a relatively low O_2 content, and its CO_2 inventory was perhaps 2 or more orders of magnitude greater than today (Kasting 1993). Given large excesses of CO_2, autotrophic microorganisms (e.g. cyanobacteria, anoxygenic photosynthetic bacteria and chemoautotrophs) assimilate ^{12}C preferentially over ^{13}C to the maximum extent possible (Fig. 8; Des Marais et al. 1989; Guy et al. 1993; Ruby et al. 1987). A CO_2 source having a $\delta^{13}C$ value near -10 (expected for CO_2 in isotopic equilibrium with marine carbonates at ~15°C, depicted in

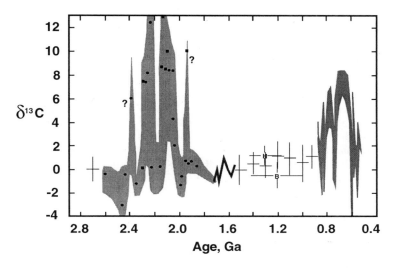

Figure 10. Carbon isotopic composition (δ_{carb}) of Proterozoic carbonates versus age. Data were taken from the following sources: each data point within the shaded area between 2.7 and 1.65 Ga represents an average of multiple analyses, see reviews by Karhu and Holland (1996 and Melezhik, et al. (1999; unmarked crosses both at 2.7 Ga and between 1.7 and 0.9 Ga are 100 Ma running averages of data compiled by Des Marais (1997a); heavy broken line between 1.7 and 1.5 Ga is from Brasier and Lindsay (1998; data labeled "H" and "B" are from Hall and Veizer (1996) and Buick et al. (1995), respectively; shaded field of data between 0.5 and 0.85 Ga are from a review by Kaufman and Knoll (1995). Question marks ("?") highlight data points that represent single basins only and therefore might reflect only regional, rather than global, isotopic excursions (see Melezhik et al. 1999).

Fig. 8), many autotrophic prokaryotes, cyanobacteria in particular, will synthesize organic C having δ_{org} values near -30. Subsequent anaerobic processing of this organic C by fermenting, methanogenic and methanotrophic (CH_4-utilizing) microorganisms can create organic C having substantially lower δ_{org} values (Hayes 1994; Hayes et al. 1999). By utilizing either acetate or CO_2 that has already been ^{13}C-depleted by primary producers, methanogens can create organic C and CH_4 that are even more ^{13}C-depleted. Furthermore, methylotrophic bacteria also prefer ^{12}C as they synthesize organic C from this highly ^{13}C-depleted CH_4 (Summons et al. 1994). Because methanogens are obligate anaerobes and methylotrophs are microaerophiles, a global environment having low but nonzero atmospheric O_2 levels would seem optimal for sustaining a globally prominent methanogen-methylotroph cycle (Hayes 1994). Indeed, using a model based upon observations of paleosols and other redox indicators, Holland estimates that pre-2.2 Ga atmospheric O_2 levels were no more than 1 to 2% of the present atmospheric level (or PAL; Holland 1994). Thus, the prominence of δ_{org} values <-35 in kerogens older than 2.2 Ga fits expectations based upon independent assessments of the paleoenvironment and the paleoecology.

Circumstances apparently changed markedly during the interval 2.2 to 1.9 Ga. Atmospheric O_2 levels rose to more than 15% PAL, as indicated by the extensive oxidation of paleosols, a severe decline in the abundance of banded iron formations and the oxidation of heavy metals such as U and Ce (Holland 1994; Knoll and Holland 1995).

Sedimentary $\delta_{sulfide}$ values became much more variable and included substantially more negative values (Cameron 1982; Canfield, this volume), indicating that marine $[SO_4^=]$ had increased significantly. Microfossils (Han and Runnegar 1992) and organic biomarker compounds (Summons and Walter 1990) constitute evidence that O_2-requiring eukaryotes existed by that time. By virtually all accounts, the surface environment had become oxidized by 2.0 Ga.

The record of δ_{carb} and δ_{org} values also is fully consistent with the oxidation of the Proterozoic environment between 2.3 and 2.0 Ga (Fig. 5). The large positive δ_{carb} excursion between 2.2 and 2.06 Ga (Baker and Fallick 1989; Karhu and Holland 1996) indicates that the relative rate of organic burial increased (f_{org} increased, see Eqn. 3). Increased net organic burial led to increased $[O_2]$, $[SO_4^=]$ and sedimentary Fe^{3+}. These oxidants probably drove the once-globally-pervasive methanogen-methylotroph cycle into much more restricted sedimentary and deep-basin enclaves. Oxic respiration and bacterial sulfate reduction became, as they are today, globally dominant pathways for organic utilization and decomposition. Notably, these pathways express minimal C isotopic discrimination (e.g. Blair et al. 1985; Kaplan and Rittenberg 1964), thus they apparently created few opportunities to form sedimentary organic C having δ_{org} values <-35, consistent with the post-2.1 Ga δ_{org} record (Fig. 8). In the post-1.9 Ga world, isotopic discrimination associated with biological CO_2-assimilation became the dominant mechanism controlling the magnitude of ε_{TOC}.

THE MESOPROTEROZOIC RECORD (1.8-1.0 Ga)

The Mesoproterozoic represents a remarkably long interval of relative stability in crustal dynamics, climate and oxidation state of the surface environment (Buick et al. 1995; Brasier and Lindsay 1998). This stability is indicated by the relatively constant δ_{carb} values (Fig. 10) that persist for 100's Ma. It has been proposed that this interval witnessed profound nutrient limitations on primary productivity; such limitations might even have been a driving force to create photosymbiotic associations in eukaryotic cells (Brasier and Lindsay 1998).

THE NEOPROTEROZOIC RECORD (0.8-0.54 Ga)

The Neoproterozoic interval witnessed supercontinent formation and breakup and profound global-scale glaciations that persisted for perhaps millions of years (Knoll 1991). It therefore is not surprising that this interval experienced substantial isotopic excursions reminiscent of those during the Paleoproterozoic (Fig. 10). Eukaryotic lineages that would lead directly to plants and animals had developed by this time. It is generally believed that the events associated with tectonics, climate and the biogeochemical cycles played important roles in the enormous diversification of plant and animal life at the dawn of the Phanerozoic Eon.

The carbon isotopic record in rocks younger than 800 Ma has been examined in much greater detail than the older record, in part because these younger successions are more numerous and better-preserved. Also, they are better understood because they are periodically punctuated with globally-pervasive glaciogenic rocks, and they are decorated with fossils that permit stratigraphic correlations. Whereas δ_{carb} values in late Mesoproterozoic and early Neoproterozoic strata vary moderately (ca. −1 to +4, Kah et al. 1999), δ_{carb} values assume exceptionally positive values after about 800 Ma (Fig. 10). These positive δ_{carb} values (+5 to +10) were interspersed several times by intervals of markedly negative excursions in δ_{carb} that attained values in the range −2 to −6 (Knoll et al. 1986). These negative excursions correlate with multiple global ice ages (Kaufman and Knoll 1995). The most negative δ_{carb} values are measured in immediately post-glacial rocks that include massive carbonates that cap the glaciogenic sequences ("cap" carbonates).

The greater abundance of analyses has allowed patterns of net biological isotopic discrimination to be assessed using the parameter ε_{TOC} (Eqn. 1). Values of ε_{TOC} offer more direct and reliable assessments of discrimination than do δ_{org} values, because δ_{org} also responds to environmentally driven variations in δ_{CO2} (compare curves a and b in Fig. 11, Hayes et al. 1999).

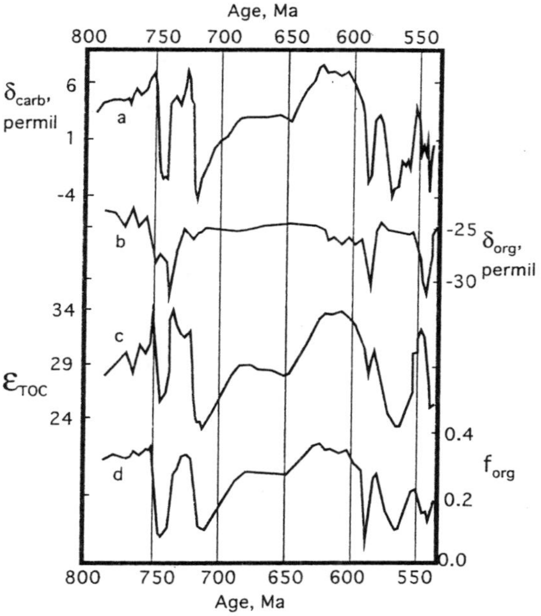

Figure 11. Neoproterozoic records of δ_{carb} (curve a) and δ_{org} (curve b) values (Hayes et al. 1999). Corresponding values for isotopic fractionation, ε_{TOC} and f_{org} during Neoproterozoic time are given by curves c and d, respectively. The periodic negative excursions are typically associated with glacial intervals (see text). Figure modified from Hayes et al. (1999).

Isotopic discrimination has varied widely during the Neoproterozoic (Fig. 11, curve c). For example, both ε_{TOC} and f_{org} (burial of organic C, relative to total C burial, Eqn. 3) experienced significant declines at the times of the first and second Sturtian glaciations, the second Varanger glaciation, and at the Precambrian-Cambrian boundary (Fig. 11, curves c and d; Hayes et al. 1999). At other times, f_{org} was significantly higher than it was during virtually the entirety of the Mesoproterozoic and the Phanerozoic. These extended intervals of high f_{org} indicate that oxygen was being added to the atmosphere. The cause of these high values may be tectonic, perhaps reflecting organic deposition in narrow, rapidly subsiding basins that formed as continents collided and then fragmented (Knoll 1992). Very positive δ_{carb} values at the onset of the glaciations is consistent with the idea that drawdown of atmospheric CO_2 due to biological productivity and organic burial might have played a role in perturbing the climate. Several aspects of the carbon isotopic record still await definitive interpretations, yet this record should help to describe those biogeochemical events that played key roles in the immense radiation of life that marked the end of the Precambrian.

THE C CYCLE, O_2 AND THE EVOLUTION OF EUKARYA

Because eukarya (i.e. the eukaryotes; e.g. algae, fungi, ultimately also plants, animals, etc.) offer a vivid example of biological evolution during the Proterozoic Eon, their potential relationship to changes in the C cycle deserves mention. The oldest known eukaryotic body fossils are spirally coiled, megascopic fossils in 2.1 Ga shales (Han and Runnegar 1992), whereas sterane biomarkers have been identified in late Archean-age rocks (Brocks et al. 1999). A major apparent increase in the diversity of fossil algae and other unicellular eukarya is recorded in 1.0-1.2 Ga rocks (Knoll 1996). The terminal Proterozoic diversification that led to plants and animals occurred between 0.53 and 0.59 billion years ago.

The evolutionary history of eukarya might be linked to the C cycle through O_2 (e.g. Knoll and Holland 1995). Molecular oxygen is required by virtually all eukarya having mitochondria and/or chloroplasts (e.g. all algae, fungi, plants and animals have mitochondria; plants also have chloroplasts). To the extent that certain subgroups of eukarya require specific minimum O_2 levels, specific groups perhaps appeared soon after their required O_2 levels were first attained. The earliest fossils of eukarya appeared during the Paleoproterozoic "C isotope" event between 2.2 and 2.0 Ga (see above). The C isotope event at the end of the Proterozoic accompanied the earliest well-documented occurrence of multicellular life (Kaufman and Knoll 1995). The coincidence in timing between these C isotope events, the various lines of evidence for atmospheric O_2 increases, and the evolutionary events for eukarya all indicate that an important biogeochemical linkage might exist (Knoll and Holland 1995). However, the final proof of this linkage remains to be established.

FUTURE WORK

Although the preceding discussion indicates that key interactions between the C cycle, atmosphere, hydrosphere and biosphere have occurred, a definitive proof of a "cause-and-effect" relationship between changes in the C cycle and biological evolution has not yet been fully achieved. What else is required? Paleoproterozoic paleontology must establish more convincingly the time of first appearance of O_2-requiring eukarya (Knoll and Holland 1995). The causes of the truly remarkable positive excursions in δ_{carb} during the Paleoproterozoic and the Neoproterozoic (Fig. 10) must be understood. Mechanistic relationships must be clarified that link the growth of the $SO_4^=$ and Fe^{3+} reservoirs to the coeval tectonic and geothermal processes. Evidence for changes in the reservoirs of O_2, $SO_4^=$, and Fe^{3+} during the Neoproterozoic should be better quantified. A more precise reconstruction of such biogeochemical phenomena promises us a deeper understanding, both of the C cycle and of our own origins.

ACKNOWLEDGMENTS

The author thanks J. Valley, D. Canfield and A. Knoll for very helpful reviews of the manuscript. This review was supported by grants from NASA's Exobiology Program and the NASA Astrobiology Institute.

REFERENCES

Baker AJ, Fallick AE (1989) Evidence from Lewisian limestones for isotopically heavy carbon in two-thousand-million-year-old-sea water. Nature 337:352-354

Barker WW, Welch SA, Banfield JF (1997) Biogeochemical weathering of silicate materials. Rev Mineral 35:391-428

Berner RA, Canfield DE (1989) A new model for atmospheric oxygen over Phanerozoic time. Am J Sci 289:333-361

Betts JN, Holland HD (1991) The oxygen content of ocean bottom waters, the burial efficiency of organic carbon, and the regulation of atmospheric oxygen. Palaeogeogr Palaeoclimatol Palaeoecol 97:5-18

Beukes NJ, Klein C, Kaufman AJ, Hayes JM (1990) Carbonate petrography, kerogen distribution, and carbon and oxygen isotope variations in an early Proterozoic transition from limestone to iron-formation deposition, Transvaal Supergroup, South Africa. Bull Soc Econ Geol 85:663-690

Blair N, Leu A, Munoz E, Olson J, Des Marais DJ (1985) Carbon isotopic fractionation in heterotrophic microbial metabolism. Appl Microbiol 50:996-1001

Brasier MD, Lindsay JF (1998) A billion years of environmental stability and the emergence of eukaryotes: new data from northern Australia. Geology 26:555-558

Brocks JJ, Logan GA, Buick R, Summons RE (1999) Archean molecular fossils and the early rise of eukaryotes. Science 285:1033-1036

Broecker WS, Peng T-H (1982) Tracers in the Sea. Lamont-Doherty Geological Observatory, Palisades, NY

Buick R (1991) Microfossil recognition in Archean rocks: an appraisal of spheroids and filaments from a 3000 M.Y. old chert-barite unit at North Pole, Western Australia. Palaios 5:441-459

Buick R (1992) The antiquity of oxygenic photosynthesis: Evidence from stromatolites in sulphate-deficient Archaean lakes. Science 255:74-77

Buick R, Des Marais DJ, Knoll AH (1995) Stable isotope compositions of carbonates from the Mesoproterozoic Bangemall Group, Northwestern Australia. Chem Geol 123:153-171

Cameron EM (1982) Sulphate and sulphate reduction in early Precambrian oceans. Nature 296:145-148

Canfield DE (1998) A new model for Proterozoic ocean chemistry. Nature 396:450-453

Canfield DE, Raiswell R (1999) The evolution of the sulfur cycle. Am J Sci 299:697-723

Deines P (1980) The isotopic composition of reduced organic carbon. *In* Fritz P, Fontes JC (eds) HEIG, Elsevier, Amsterdam, p 329-406

Delano JW (2001) Redox history of the Earth's interior since ~3900 Ma: implications for prebiotic molecules. Orig Life Evol Biosph (in press)

Derry LA, Kaufman AJ, Jacobsen SB (1992) Sedimentary cycling and environmental change in the Late Proterozoic: Evidence from stable and radiogenic isotopes. Geochim Cosmochim Acta 56:1317-1329

Des Marais DJ (1985) Carbon exchange between the mantle and crust and its effect upon the atmosphere: today compared to Archean time. *In* Sundquist ET, Broecker WS (eds) The Carbon Cycle and Atmospheric CO_2: Natural Variations Archean to Present. American Geophysical Union, Washington, DC, p 602-611

Des Marais DJ (1994) Tectonic control of the crustal organic carbon reservoir during the Precambrian. Chem Geol 114:303-314

Des Marais DJ (1995) The biogeochemistry of subtidal marine hypersaline microbial mats, Guerrero Negro, Baja California Sur, Mexico. *In* Jones JG (ed) Advances in Microbial Ecology. Plenum, New York, p 251-274

Des Marais DJ (1997a) Isotopic evolution of the biogeochemical carbon cycle during the Proterozoic Eon. Organic Geochem 27:185-193

Des Marais DJ (1997b) Long-term evolution of the biogeochemical carbon cycle. Rev Mineral 35:429-448

Des Marais DJ (2000) When did photosynthesis emerge on Earth? Science 289:1703-1705

Des Marais DJ, Cohen Y, Nguyen H, Cheatham M, Cheatham T, Munoz E (1989) Carbon isotopic trends in the hypersaline ponds and microbial mats at Guerrero Negro, Baja California Sur, Mexico: Implications for Precambrian stromatolites. *In* Cohen Y, Rosenberg E (eds) Microbial Mats: Physiological Ecology of Benthic Microbial Communities. American Society for Microbiology, Washington, DC, p 191-205

Dewey JF, Windley BF (1981) Growth and differentiation of the continental crust. Phil Trans R Soc London A 301:189-206

Dunn SR, Valley JW (1992) Calcite-graphite isotope thermometry: a test for polymetamorphism in marble, Tudor gabbro aureole, Ontario, Canada. J Metamorph Geol 10:487-501

Eiler JM, Mojzsis SJ, Arrhenius G (1997) Carbon isotope evidence for early life. Nature 396:665

Elderfield H, Schultz A (1996) Mid-ocean ridge hydrothermal fluxes and the chemical composition of the ocean. Ann Rev Earth Planetary Sci 24:191-224

Ferry JM (1991) Dehydration and decarbonation reactions as a record of fluid infiltration. Rev Mineral 26:351-394

Field CB, Behrenfeld MJ, Randerson JT, Falkowski P (1998) Primary production of the Biosphere: Integrating Terrestial and Oceanic Components. Science 281:237-240

Garrels RM, Perry EA Jr. (1974) Cycling of carbon, sulfur, and oxygen through geologic time. *In* Goldberg ED (ed) The Sea. John Wiley & Sons, New York, p 303-336

Grotzinger JP (1989) Facies and evolution of Precambrian carbonate depositional systems: emergence of the modern platform archetype. *In* Crevello PD, Wilson JL, Sarg JF, Read JF (eds) Controls on carbonate platform and basin development. Society of Economic Paleontologists and Mineralogists, Tulsa, OK, p 79-106

Grotzinger JP, Kasting JF (1993) New constraints on Precambrian ocean composition. J Geol 101:235-243

Guy RD, Fogel ML, Berry JA (1993) Photosynthetic fractionation of the stable isotopes of oxygen and carbon. Plant Physiol 101:37-47

Hall SM, Veizer J (1996) Geochemistry of Precambrian carbonates VII. Belt Supergroup, Montana and Idaho, USA. Geochim Cosmochim Acta 60:667-677

Han TM, Runnegar B (1992) Megascopic eukaryotic algae from the 2.1-billion-year-old Negaunee Iron-Formation, Michigan. Science 257:232-235

Harland WB (1983) The Proterozoic glacial record. Mem Geol Soc Am 161:279-288

Hayes JM (1983) Geochemical evidence bearing on the origin of aerobiosis, a speculative hypothesis. In Schopf JW (ed) Earth's Earliest Biosphere. Princeton University Press, Princeton, p 291-301

Hayes JM (1994) Global methanotrophy at the Archean-Proterozoic transition. In Bengtson S (ed) Early Life on Earth Nobel Symposium 84, p 220-236 Columbia Univ. Press, New York

Hayes JM, Kaplan IR, Wedeking KW (1983) Precambrian organic geochemistry, preservation of the record. In:. In Schopf JW (ed) Earth's Earliest Biosphere. Princeton University Press, Princeton, NJ, p 93-134

Hayes JM, Popp BN, Takigiku R, Johnson MW (1989) An isotopic study of biogeochemical relationships between carbonates and organic carbon in the Greenhorn Formation. Geochim Cosmochim Acta 53:2961-2972

Hayes JM, Strauss H, Kaufman AJ (1999) The abundance of ^{13}C in marine organic matter and isotopic fractionation in the global biogeochemical cycle of carbon during the past 800 Ma. Chem Geol 161:103-125

Hofman HJ, Grey K, Hickman AH, Thorpe RI (1999) Origin of 3.45 Ga coniform stromatolites in Warrawoona Group, Western Australia. Geol Soc Am Bull 111:1256-1262

Holland HD (1984) The Chemical Evolution of the Atmosphere and Oceans. Princeton University Press, Princeton, NJ

Holland HD (1994) Early Proterozoic atmospheric change. In Bengtson S (ed) Early Life on Earth. Columbia University Press, New York, p 237-244

Holser WT, Schidlowski M, Mackenzie FT, Maynard JB (1988) Geochemical cycles of carbon and sulfur. In Gregor CB, Garrels RM, Mackenzie FT and Maynard JB (eds) Chemical Cycles in the Evolution of the Earth. John Wiley & Sons, New York, p 105-173

Hunt JM (1972) Distribution of carbon in crust of earth. Am Assoc Petrol Geol Bull 56:2273-2277

Hunt JM (1979) Petroleum Geochemistry and Geology. W. H. Freeman, San Francisco

Jannasch HW, Wirsen CO (1979) Chemosynthetic primary production at East Pacific sea floor spreading centers. BioSci 29:592-598

Kah LC, Sherman AG, Narbonne GM, Knoll AH, Kaufman AJ (1999) δ^{13}C stratigraphy of the Proterozoic Bylot Supergroup, Baffin Island, Canada: Implications for regional lithostratigraphic correlations. Can J Earth Sci 36:313-332

Kaplan IR, Rittenberg SC (1964) Carbon isotope fractionation during metabolism of lactate by Desulfovibrio desulfuricans. J Gen Microbiol 34:213-217

Karhu JA, Holland HD (1996) Carbon isotopes and the rise of atmospheric oxygen. Geology 24:867-870

Kasting JF (1993) Earth's early atmosphere. Science 259:920-926

Kaufman AJ, Knoll AH (1995) Neoproterozoic variations in the C-isotopic composition of seawater: stratigraphic and biogeochemical implications. Precamb Res 73:27-49

Kitchen NE, Valley JW (1995) Carbon isotopic thermometry in marbles of the Adirondack Mountains, New York. J Metamorph Geol 13:577-594

Klein C, Beukes NJ (1989) Geochemistry and sedimentology of a facies transition from limestone to iron formation deposition in the early Proterozoic Transvaal Supergroup, South Africa. Econ Geol 84:1733-1774

Knoll AH (1991) End of the Proterozoic Eon. Sci Am 265:64-73

Knoll AH (1992) Biological and biogeochemical preludes to the Edaicaran radiation. In Lipps JH, Signor PW (eds) Origin and Early Evolution of the Metazoa. Plenum, New York, p 53-84

Knoll AH (1996) Archean and proterozoic paleontology. In Jansonius J, McGregor, D. C. (eds) Palynology: principles and applications. Am Assoc Stratigraphic Palynologists Foundation, p 51-80

Knoll AH, Hayes JM, Kaufman AJ, Swett K, Lambert IB (1986) Secular variation in carbon isotope ratios from Upper Proterozoic successions of Svalbard and East Greenland. Nature 321:832-838

Knoll AH, Holland HD (1995) Oxygen and Proterozoic evolution: an update. In Stanley S (ed) Effects of Past Global Change on Life. National Academy Press, Washington, DC, p 21-33

Kroopnick P (1985) The distribution of ^{13}C in ΣCO_2 in the world oceans. Deep Sea Research 32:57-84

Kumar MS, Santosh M, Wada H (1997) Carbon isotope thermometry in marbles of Ambasamudram, Kerala Khondalite Belt, Southern India. J Geol Soc India 49:523-532

Lambert RSJ (1976) Archean thermal regimes, crustal and upper mantle temperatures, and a progressive evolutionary model for the Earth. In Windley BF (ed) The Early History of the Earth. John Wiley, New York, p 363-373

Lewan MD (1983) Effects of thermal maturation on stable organic carbon isotopes as determined by hydrous pyrolysis of Woodford Shale. Geochim Cosmochim Acta 47:1471-1479

Lovelock JE, Whitfield M (1982) Lifespan of the biosphere. Nature 296:561-563

Lowe DR (1992) Major events in the geological development of the Precambrian Earth. *In* Schopf JW, Klein C (eds) The Proterozoic Biosphere: a multidisciplinary study. Cambridge University Press, New York, p 67-76

Lowe DR (1994) Early environments: constraints and opportunities for early evolution. *In* Bengtson S (ed) Early Life on Earth Nobel Symp 84. Columbia University Press, New York, p 24-35

Lowman PDJ (1989) Comparative planetology and the origin of continental crust. Precamb Res 44:171-195

Martin JH, Knauer GA, Karl DM, Broenkow WW (1987) VERTEX: carbon cycling in the northeast Pacific. Deep Sea Res 34:267-285

Marty B, Jambon A (1987) C/³He in volatile fluxes from the solid Earth: implicatons for carbon geodynamics. Earth Planet Sci Letters 83:16-26

McCulloch MT (1993) The role of subducted slabs in an evolving earth. Earth Planet Sci Letters 115:89-100

Melezhik VA, Fallick AE, Medvedev PV, Makarikhin VV (1999) Extreme $^{13}C_{carb}$ enrichment in ca. 2.0 Ga magnesite-stromatolite-dolomite-'red beds' association in a global context: a case for the world-wide signal enhanced by a local environment. Earth Sci Rev 48:71-120

Mojzsis SJ, Arrhenius G, McKeegan KD, Harrison TM, Nutman AP, Friend RL (1996) Evidence for life on Earth before 3800 million years ago. Nature 384:55-59

Mopper K, Degens ET (1979) Organic carbon in the ocean: nature and cycling. *In* Bolin B (ed) The Global Carbon Cycle. Wiley, New York, p 293-316

Olson JS, Garrels RM, Berner RA, Armentano TV, Dyer MI, Taalon DH (1985) The natural carbon cycle. *In* Trabalka JR (ed) Atmospheric Carbon Dioxide and the Global Carbon Cycle. US Department of Energy, Washington, DC, p 175-213

Pavlov AA, Kasting JF, Brown LL, Rages KA, Freedman R (2000) Greenhouse warming by CH_4 in the atmosphere of early Earth. J Geophy Res–Planets 105:11,981-11,990

Pawley AR, Holloway JR, McMillan PF (1992) The effect of oxygen fugacity on the solubility of carbon-oxygen fluids in basaltic melt. Earth Planet Sci Letters 110:213-225

Peters KE, Rohrback BG, Kaplan IR (1981) Geochemistry of artificially heated humic and sapropelic sediments I: Protokerogen. Bull Am Assoc Petrol Geol 65:688-705

Preuss A, Schauder R, Fuchs G, Stichler W (1989) Carbon isotope fractionation by autotrophic bacteria with three different CO_2 fixation pathways. Z Naturforschung 44c:397-402

Rau GH, Takahashi T, Des Marais DJ (1989) Latitudinal variations in plankton $\delta^{13}C$: implications for CO_2 and productivity in past oceans. Nature 341:516-518

Rogers JJW (1996) A history of continents in the past three billion years. J Geol 104:91-107

Ronov AB (1980) Osadochnaja Oblochka Zemli (Sedimentary Layer of the Earth). Nauka, Moscow

Rosing MT, Rose NM, Bridgwater D, Thompson HS (1996) Earliest proof of earth's stratigraphic record: a reappraisal of the >3.7 Ga Isua (Greenland) supracrustal sequence. Geology 24:43-46

Ruby EG, Jannasch HW, Deuser WG (1987) Fractionation of stable carbon isotopes during chemoauto-trophic growth of sulfur-oxidizing bacteria. Appl Microbiol 53:1940-1943

Rye R, Kuo PH, Holland HD (1995) Atmospheric carbon dioxide concentrations before 2.2 billion years ago. Nature 378:603-605

Schidlowski M (1993) The initiation of biological processes on Earth; Summary of the empirical evidence. In. *In* Engel MH, Macko SA (eds) Organic Geochemistry. Plenum, New York, p 639-655

Schidlowski M (2001) Carbon Isotopes as biogeochemical recorders of life over 3.8 Ga of Earth history: evolution of a concept. Precamb Res 106:117-134

Schidlowski M, Hayes JM, Kaplan IR (1983) Earth's Earliest Biosphere. *In* Schopf JW (ed) Princeton University Press, Princeton, NJ, p 149-186

Schopf JW (1992) Paleobiology of the Archean. *In* Schopf JW, Klein C (eds) The Proterozoic Biosphere. Cambridge University Press, New York, p 25-39

Schwartzman DW, Volk T (1989) Biotic enhancement of weathering and the habitability of Earth. Nature 336:457-460

Simoneit BRT, Brenner S, Peters KE, Kaplan IR (1981) Thermal alteration of Cretaceous black shale by diabase intrusions in the Eastern Atlantic: II Effects on bitumen and kerogen. Geochim Cosmochim Acta 45:1581-1602

Sleep NH, Zahnle K (2001) Carbon dioxide cycling and implications for climate on ancient Earth. J Geophys Res 106:1373-1390

Staudigel H, Hart SR, Schmincke H-U, Smith BM (1989) Cretaceous ocean crust at DSDP sites 417 and 418: carbon uptake from weathering versus loss by magmatic outgassing. Geochim Cosmochim Acta 53:3091-3094

Strauss H, Beukes NJ (1996) Carbon and sulfur isotopic compositions of organic carbon and pyrite in sediments from the Transvaal Supergroup, South Africa. Precamb Res 79:57-71

Strauss H, Des Marais DJ, Summons RE, Hayes JM (1992a) The carbon isotopic record. *In* Schopf JW, Klein C (eds) The Porterozoic Biosphere: A Multidisciplinary Study. Cambridge University Press, New York, p 117-127

Strauss H, Des Marais DJ, Summons RE, Hayes JM (1992b) Concentrations of organic carbon and maturities and elemental compositions of kerogens. *In* Schopf JW, Klein C (eds) The Proterozoic Biosphere: a multidisciplinary study. Cambridge University Press, New York, p 95-100

Summons RE, Jahnke LL, Hope JM, Logan GA (1999) 2-Methylhopanoids as biomarkers for cyanobacterial oxygenic photosynthesis. Nature 400:554-557

Summons RE, Jahnke LL, Roksandic Z (1994) Carbon isotopic fractionation in lipids from methanotrophic bacteria: Relevance for interpretation of the geochemical record of biomarkers. Geochim Cosmochim Acta 58:2853-2863

Summons RE, Walter MR (1990) Molecular fossils and microfossils of prokaryotes and protists from Proterozoic sediments. Am J Sci 290-A:212-244

Turcotte DL (1980) On the thermal evolution of the Earth. Earth Planet Sci Letters 48:53-58

Valley JW, O'Neil JR (1981) $^{13}C/^{12}C$ exchange between calcite and graphite: a possible thermometer in Grenville marbles. Geochim Cosmochim Acta 45:411-419

Veizer J, Hoefs J, Lowe DR, Thurston PC (1989) Geochemistry of Precambrian carbonates: II. Archean greenstone belts and Archean sea water. Geochim Cosmochim Acta 53:859-871

von Brunn V, Gold DJC (1993) Diamictite in the Archaean Pongola Sequence of southern Africa. J Afr Earth Sci 16:367-374

Walker JCG (1985) Carbon dioxide on the early Earth. Orig Life 16:117-127

Walter MR, ed (1976) Stromatolites. Elsevier, New York

Watanabe Y, Naraoka H, Wronkiewicz DJ, Condie KC, Ohmoto H (1997) Carbon, nitrogen, and sulfur geochemistry of Archean and Proterozoic shales from the Kaapvaal Craton, South Africa. Geochim Cosmochim Acta 61:3441-3459

Weber JN (1967) Possible changes in the isotopic composition of the oceanic and atmospheric carbon reservoir over geologic time. Geochim Cosmochim Acta 31:2343-2351

Windley BF (1984) The Evolving Continents. John Wiley & Sons, New York

Zhang Y, Zindler A (1993) Distribution and evolution of carbon and nitrogen in Earth. Earth Planet Sci Letters 117:331-345

11   Isotopic Biogeochemistry of Marine Organic Carbon

Katherine H. Freeman

Department of Geosciences
The Pennsylvania State University
University Park, Pennsylvania 16802

INTRODUCTION

The ocean accounts for over 90% of the active pools of carbon on the Earth's surface, with over 95% of marine carbon in the form of dissolved inorganic carbon (DIC) (Hedges and Keil 1995). Organic carbon dissolved in the ocean, suspended as particles or cells, and accumulating in sediments together constitute the other significant fractions of marine carbon, with organic carbon in the water column similar in quantity to the current atmospheric inventory of carbon dioxide. Isotopic partitioning among various inorganic and organic carbon phases reflects biological, physical and chemical processes, and the resulting fractionations are important tools in the study of modern and ancient carbon cycling.

The focus of this review is on the isotopic geochemistry of marine organic carbon. It will begin by setting the stage with the isotopic patterns of DIC in the modern oceans. As will be discussed below, the distribution of inorganic carbon and related nutrient concentrations as well as DIC isotopic compositions are important influences on the quantity and isotopic character of organic carbon produced in marine surface waters. The remainder of the review will discuss isotope fractionation associated with the production and preservation of marine organic carbon. The combination of organic matter composition and ^{13}C content is a potentially powerful approach for addressing the nature and pace of ecological and environmental change both in the modern and ancient ocean. This work reviews biogeochemical processes that generate, transform and ultimately preserve such signatures in marine sediments.

MARINE INORGANIC CARBON

The first truly global measurements of DIC $\delta^{13}C$ values were produced as a result of the Geochemical Ocean Sections Study (GEOSECS). The resulting data were published in a series of articles by Kroopnick and colleagues (Kroopnick et al. 1977, Kroopnick 1980, 1985) and summarized by Takahashi et al. (1980, 1981). A striking outcome of this effort is the observation of a narrow range in isotopic values for DIC on a global basis. Values in ocean surface waters do not vary more than ±0.8‰, with the global average close to 1.5‰ on the PDB scale (Kroopnik 1985). The data from the GEOSECS expeditions include uncertainties of about 0.1‰ and corrections that range up to 0.7‰ (Kroopnick 1980), precluding resolution of finer isotopic patterns. This extensive data set, until recently, served as the principal source of modern DIC isotopic data for carbon cycling studies. The GEOSECS program also evaluated inventories of ^{14}C and documented the uptake of bomb-derived radiocarbon. The program was instrumental in defining the age of ocean turnover, and the concept of deep-ocean circulation and the oceanic "conveyor belt."

The Carbon Dioxide Program at the Scripps Institution of Oceanography provides a compilation of data for marine surface water samples covering the past two decades. Samples were collected at depths to 50 m in all major ocean basins between 1978 and

1529-6466/00/0043-0011$05.00

1997. The results of analyses for total inorganic carbon (DIC), alkalinity (Alk) and $\delta^{13}C$ values of the DIC (Lueker 1998, 2000) are summarized by Gruber et al. (1999) and the data will be made available through the Carbon Dioxide Information and Analysis Center (CDIAC) at the Oak Ridge National Laboratory. This new compilation (Fig. 1) provides high-quality data with global coverage revealing latitudinal variations and recent decadal-scale changes in surface-water DIC isotopic values. Subsets of the data represent time series collected from the North Atlantic and Pacific Oceans, providing records of seasonal variations in DIC $\delta^{13}C$ values. Combined errors for isotopic analyses are 0.04‰ (Gruber et al. 1999). The very high quality of data from this global inventory and other recent regional studies (e.g. Quay et al. 1992, Gruber et al. 1998) make it possible to

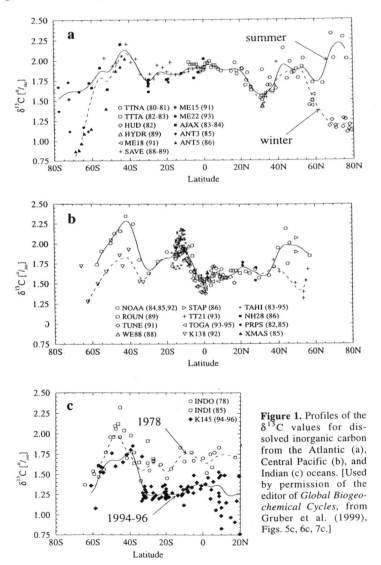

Figure 1. Profiles of the $\delta^{13}C$ values for dissolved inorganic carbon from the Atlantic (a), Central Pacific (b), and Indian (c) oceans. [Used by permission of the editor of *Global Biogeochemical Cycles*, from Gruber et al. (1999), Figs. 5c, 6c, 7c.]

evaluate processes that result in relatively small isotopic variations, which are key to understanding the influence of human-induced changes in the global carbon cycle.

Both biological and chemical processes influence the $\delta^{13}C$ of DIC in the global ocean, although the result is relatively small variations across the major ocean basins (Fig. 1). As documented by Kroopnick (1980) and Kroopnick et al. (1985) and by Gruber and coworkers (1999), DIC isotopic values show the least variation across the equatorial regions of the Atlantic and Indian, and are approximately 1.9‰ and 1.3‰ respectively. Values for DIC near the Pacific equator range from 1.5 to ~2.0‰. The higher-latitude regions generally exhibit a greater range in values, with maximum values of 2.2, 2.3 and 1.75‰ in the Atlantic, Pacific and Indian Oceans, respectively, associated with the Subantarctic Front at about 45°S latitude. Values in the Southern Ocean and Northern Atlantic Ocean reach the lowest observed values in the winter months, with observed values below 0.8 and 1.2‰, respectively. These regions also exhibit the greatest seasonal variations; with the summer months recording values over 2.25‰ in the North Atlantic, reflecting the uptake of ^{13}C depleted inorganic carbon by phytoplankton during times of high productivity.

Ocean margins and continental-type seaways can exhibit much greater ranges in DIC ^{13}C contents (Holmden et al. 1998). For example, in basins with restricted circulation, a significant quantity of inorganic carbon derived from respiration of organic carbon can cause a decrease in DIC $\delta^{13}C$ values. The inorganic carbon in marine waters near the continents can be influenced by inputs from ground waters and surface waters that contain significant quantities of nutrients and terrestrial carbon. The inorganic carbon from these sources is typically depleted in ^{13}C relative to open marine carbon. For example, in the Florida Keys, freshwater sources of inorganic carbon lower DIC $\delta^{13}C$ values by up to 4‰ relative to inorganic carbon in the Caribbean region (Patterson and Walter 1994).

A spate of publications uses DIC isotopic data (and other tracers) for evaluating the invasion of anthropogenic carbon dioxide into the ocean. These efforts are enhanced by improved precision and accuracy in available isotopic data, and they exploit the fact that atmospheric carbon dioxide is becoming depleted in ^{13}C relative to preindustrial values. This so-called "^{13}C Suess effect" is named in analogy to the depletion of global CO_2 in ^{14}C due to inputs derived from combustion of fossil carbon (Keeling 1979). The ^{13}C Suess effect is observed in the atmosphere, where global CO_2 has dropped from a preindustrial $\delta^{13}C$ value of -6.3‰ to about -7.8‰ by the end of the 20th century (Friedli et al. 1986, Francey et al. 1998, Keeling et al. 1995). The invasion of isotopically depleted carbon into surface waters of the world's oceans has become an area of active research, with a growing number of works documenting uptake of anthropogenic carbon in the marine reservoir (e.g. Quay et al. 1992, Tans et al. 1993, Heinmann and Maier-Reimer 1996, Bacastow et al. 1996, Joos and Bruno 1998, Gruber et al. 1998, Sonnerup et al. 1999, 2000; Lerperger et al. 2000, Ortiz et al. 2000, Takahashi et al. 2000, Murnane and Sarmiento 2000, Caldeira and Duffy 2000). The estimated rate of isotopic change in marine surface water DIC varies with latitude; and generally ranges from -0.1 to -0.25‰/yr.

In the modern ocean, the distribution of ^{13}C in inorganic carbon is controlled by a combination of biological and thermodynamic processes. Biological uptake of CO_2 and subsequent conversion into organic matter selectively removes carbon depleted in ^{13}C, resulting in an isotopic enrichment of residual inorganic carbon. Organic matter produced in marine surface waters has a range of isotopic values, as will be discussed below (and by Hayes, this volume), with most organic matter ranging between -20 and -30‰ (Goericke and Fry 1994). In turn, the oxidation or remineralization of organic matter releases ^{13}C-depleted carbon back into the inorganic reservoir. A much smaller fractionation is associated with biological precipitation of calcium carbonate, and this has a negligible influence on DIC ^{13}C contents (Spero et al. 1997). As organic matter is created in surface waters, and oxidized throughout the water column, but especially in deeper waters, the 'biological pump' creates an isotopic gradient with ^{13}C-enriched waters at the surface and isotopically depleted and CO_2-charged waters at depth. Nutrients are taken up and released in association with organic matter production and remineralization, respectively, such that ^{13}C depletion in inorganic carbon is associated with nutrient-rich waters, as observed in environments where deeper waters are physically advected to the surface in upwelling regions.

Phytoplankton take up and heterotrophs remineralize carbon and the nutrients nitrogen and phosphorus at nearly constant atomic ratios, when considered on a broad scale (Redfield 1934, Redfield et al. 1963). The atomic ratio for C:N:P was originally documented at 106:16:1 (Redfield 1934), with subsequent work showing variations among different organisms and growth conditions (Redfield et al. 1963, Fraga et al. 1998, 1999; Fraga et al. 1998, Rios et al. 1998) and regional variations reflecting differing influences of circulation, oxygen, microbial processes such as nitrogen fixation and denitrification (e.g. Maier-Reimer 1996, Morrison et al. 2000, Lenton and Watson 2000), and human-induced changes in nitrous oxides inputs (Pahlow and Riebesell 2000). Changes in global carbon dioxide concentrations potentially also influence the C:P and C:N ratios of marine phytoplankton (Wolf-Gladrow et al. 1999, Burkhardt and Riebesell 1997). The cycling of nitrogen, phosphorous and carbon in marine waters are unquestionably linked by photosynthesis and decay, and there is a general constancy of their atomic ratios on a coarse scale. However, such efforts suggest reconsideration of the overall constancy of the Redfield ratio through time (Pahlow and Riebesell 2000, Archer et al. 2000).

Biological-pump related surface-to-deep-water gradients in the $\delta^{13}C$ of DIC are well documented and employed in the study of modern and ancient carbon cycling related to ocean productivity and nutrient distributions (e.g. Zahn and Keir 1994, Berger et al. 1994). Nonetheless, on a finer scale, the isotopic composition of inorganic carbon does not neatly track the distribution of N and P in surface waters (Boyle 1994). As biological and physical processes influence the local manifestation of the Redfield ratio, such processes also influence the quantity, character and isotopic composition of organic carbon that is produced and remineralized. For example, variations in the lipid to carbohydrate content of cells are associated with changes in nutrient fluxes due to varying upwelling conditions (Fraga et al. 1998, 1999). As differing isotopic partitioning of carbon among biochemical compound classes is well known to occur within cells (e.g. Deines 1980, Sakata et al. 1997), it is reasonable to expect that regional variations in ^{13}C-distributions will be associated with the production and decay of cells of different chemical composition. More significantly, variations in isotopic fractionation associated with the uptake of carbon during photosynthesis (see below) further complicate any potential simple relationship between $\delta^{13}C$ of DIC and CO_2 and nutrient concentrations.

In addition to the biological factors noted above, the isotopic composition of inorganic carbon is influenced by the exchange of carbon between surface waters and the atmosphere. Carbon isotopes are fractionated with the transfer of carbon between water and the atmosphere (Siegenthaler and Munnich 1981, Zhang et al. 1995), with equilibrium fractionation resulting in atmospheric carbon dioxide about 8‰ depleted relative to the ocean. This effect is temperature-dependent, with a change in fractionation of approximately -0.1‰ per K (Mook 1986). Thus, at equilibrium, DIC in colder waters is enriched in ^{13}C relative to warmer waters. In natural waters, the time required for isotopic equilibration is slow relative to the residence time of carbon in surface waters, (Broecker and Peng 1974, Broecker and Maier-Reimer 1992), and there are additional complexities associated with regional variations in the net uptake or evasion of CO_2 (Lynch-Stieglitz et al. 1995). As a result, DIC $\delta^{13}C$ values in ocean surface waters approach to various degrees, but generally never reach isotopic equilibrium with CO_2 in the atmosphere.

PRODUCTION OF MARINE ORGANIC MATTER

The photosynthetic isotope effect

The isotopic composition of photosynthetic plankton reflects the isotopic composition of DIC and is influenced by both properties of the growth environment and cell physiology. The photosynthetic isotope effect (ε_p) is defined by the isotopic difference between dissolved CO_2 (δ_d) and photosynthetic biomass (δ_p):

$$\varepsilon_p \equiv 10^3 \left[\delta_d + 1000) / (\delta_p + 1000) - 1 \right] \tag{1}$$

which is approximated as the isotopic difference between biomass and dissolved CO_2. This formulation requires that the substrate for carbon fixation be dissolved CO_2, as is true for many marine phytoplankton and higher plants (Cooper et al. 1969). Since the late 1980s, considerable effort has focused on understanding controls on fractionation and establishing quantitative relationships between ε_p and variables that describe growth conditions. Popp et al. (1989), building on earlier work (e.g. Dean et al. 1986), identified a relationship between ε_p and evidence for variations in atmospheric CO_2 over geologic time. They reviewed arguments, based on work with modern plants, that if aquatic algae assimilate carbon by diffusional transport across cell membranes, ε_p should track dissolved carbon dioxide concentrations. The prospect for a paleo-pCO_2 barometer thus sparked efforts by many researchers to quantify the relationship based on field and culture studies (Freeman and Hayes 1992, Hinga et al. 1994, Goericke et al. 1994, Raven 1997, Wolf-Gladrow and Riebesell 1997, Rau et al. 1997). Hayes (this volume) provides a detailed discussion of the derivation of quantitative expressions that describe biological and environmental controls on ε_p. A summary of recent advances is provided below.

Using nitrate-limited growth in a chemostat culture system, Laws et al. (1995) quantified the effects of both carbon supply (dissolved CO_2 concentrations, c_e) and carbon demand (cell specific growth rate, μ) on ε_p values for the diatom *Phaeodactylum tricornutum*. The nature of this relationship was generally consistent with field observations, including those based on isotopic analyses of alkenones, which are biomarker compounds for some haptophyte algae (Bidigare et al. 1997, 1999). In samples collected mostly from the Pacific Ocean, ε_p values based on alkenone analyses varied with both the concentration of dissolved phosphate and $1/c_e$. The concentrations of phosphate observed in this study are not considered limiting to the growth haptophyte algae, and the authors suggested that a trace nutrient that tracks phosphate concentrations was responsible for the empirical growth rate-ε_p-CO_2 relationship (Bidigare et al. 1997).

The size of cells can influence the surface area across which CO_2 is transported, and carbon content of a cell tracks its volume. Thus relationships between cell geometries, growth conditions and ε_p have been observed in field studies (Pancost et al. 1997). Popp et al. (1998), again using chemostat cultures, quantified the influence of cell geometry on the relationship between ε_p and μ and c_e that was defined by Laws et al. (1995):

$$\varepsilon_p = -182 \left(\frac{\mu V}{c_e A} \right) + 25.3 \tag{2}$$

In the above equation, μ represents the specific growth rate (d^{-1}), c_e is external CO_2 (aq) concentration ($\mu M/kg$), V is cell volume (μm^3) and A is the cell surface area (μm^2). Popp et al. (1998) estimated surface area to volume ratios for cells of different geometry, with cell dimensions ranging from a radius of 0.68 μm (*Synechococcus sp.*) to 14 μm (*Porosira glacialis*). Isotopic fractionation for the prokaryotic species (*Synechococcus sp.*) employed in this study exhibited little variation with growth conditions, and did not conform to the relationship observed for the eukaryotic species. Thus, *Synechococcus* was not included in the development of Equation (2); data from this study are shown in Figure 2.

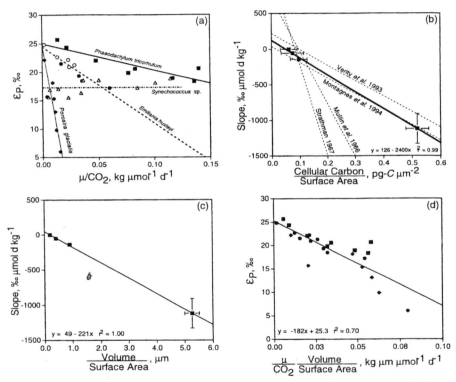

Figure 2. Chemostat experimental results representing controls on the isotopic fractionation by marine algae. (a) Plot of isotopic fractionation (ε_p) as a function of growth rate (μ) divided by the concentration of dissolved CO_2 for selected marine organisms. (b) Plot of the slopes of the relationships represented in (a) as a function of the ratio of cellular carbon to cell surface area. Dashed lines represent relationships estimated from other studies. (c) Plot of the slopes from (a) as a function of observed cell volume to surface area ratios. (d) Plot of isotopic fractionation (ε_p) as a function of the product of the growth rate divided by CO_2 concentrations and the volume-to-surface area ratios for the eukaryotic species in (2a). [Used by permission of the editor of *Geochimica et Cosmochimica Acta*, from Popp et al. (1998), Fig. 2.]

Equation (2) is applicable only under circumstances where carbon is not limiting for growth. For example, a number of authors have identified evidence for and the isotopic consequences of carbon concentrating mechanisms (CCM) in various marine algae, especially diatoms (Fry and Wainwright 1991, Laws et al. 1997, 1998; Raven 1997, Korb et al. 1998, Fielding et al. 1998, Erez et al. 1998, Burkhardt et al. 1999a,b; Clark and

Flynn 2000, Tortell et al. 2000, Rau et al. 2001). Work by Keller and Morel (1999) and Reinfelder et al. (2000) indicates that the mechanism in some diatoms is similar to the carbon concentrating mechanisms in vascular C-4 plants. Other work has explored the influence of nitrate-rich versus nitrate-limiting (i.e. chemostat) conditions on ε_p and its relationship to CO_2 and other variables, especially light-limited growth rates (Korb et al. 1996, Eek et al. 1999, Burkhardt et al. 1999, Riebesell et al. 2000a,b). Other workers have found evidence that oxygen concentration can also affect ε_p values (Berner et al. 2000), and this may prove important on timescales ranging from diurnal to geologic (Beerling 1999). Thus Equation (2) clearly is not applicable for all algae or under all marine conditions, and it is very likely to be refined in the future.

The overall approach of Popp et al. (1998) and Bidigare et al. (1997, 1999) provide laboratory data that are reinforced by field observations and, more recently, by a growing number of computer model studies. For example, Hofmann et al. (2000) in a model-based study found that CO_2 concentrations could account for most of the variation in $\delta^{13}C$ values for organic carbon in marine surface waters. The exception to this is high-latitude regions with strong seasonal variations in productivity, where growth effects are more important (Hofmann et al. 2000).

THE PHOTOSYNTHETIC ISOTOPE EFFECT AND STUDIES OF PALEOENVIRONMENTAL CHANGE

Quantitative relationships between ε_p and $\delta^{13}C$ values of organic matter that combine growth rates and carbon dioxide concentrations are being increasingly employed to evaluate sedimentary records of ε_p values (e.g. Bentaleb 1996, Popp et al. 1997, Kump and Arthur 1999, Hofmann et al. 1999, Hayes et al. 1999, Rosenthal et al. 2000). Chlorophyll (e.g. Sachs et al. 1999) and in ancient sediments, porphyrins (Hayes et al. 1989) provide direct isotopic markers for phytoplankton and avoid inputs from other members of the ecosystem. More specific biomarker-based estimates of ε_p, especially when the compound is derived from a restricted taxonomic source, help to constrain cell geometry of the source organism(s) and permit differentiation of different inputs in ancient sediments (Huang et al. 1999, Pancost et al. 1999). This is the case for alkenones (haptophyte biomarkers), which are increasingly being used to generate ε_p records that are interpreted in the context of changes in CO_2 and nutrient concentrations through time (Jasper and Hayes 1990, Jasper et al. 1994, Bidigare et al. 2000, Pagani et al. 1999a,b; 2000).

For example, using the calibration in Bidigare et al. (1997, 1999), Pagani et al. (1999a,b) produced high-resolution isotopic analyses of alkenones from multiple DSDP cores. Alkenones and planktonic foraminifera isotopic data were used to calculate ε_p records from the Oligocene/Miocene boundary to about 5 Ma. The ε_p records were employed to calculate carbon dioxide levels, assuming a range of phosphate concentrations that are similar to the low values observed in the modern ocean at the sample sites. The data indicate that CO_2 levels were low throughout the Miocene, with variations tracking glacial and related carbon-burial events. The overall low CO_2 estimates (they range ca. 320-180 ppm), are generally consistent with recent estimates based on boron isotopic analyses (Pearson and Palmer 2000). The record by Pagani et al. (1999a; Fig. 3) suggests that CO_2 did not drop dramatically during the Monterey event, at the time of East-Antarctic ice sheet expansion (EAIS), or prior to the global expansion of C-4 plant grasslands at 6-8 Ma (Pagani et al. 1999b). Pagani and coauthors suggest these events reflect ocean-circulation or hydrological shifts rather than dramatic changes in greenhouse gas concentrations.

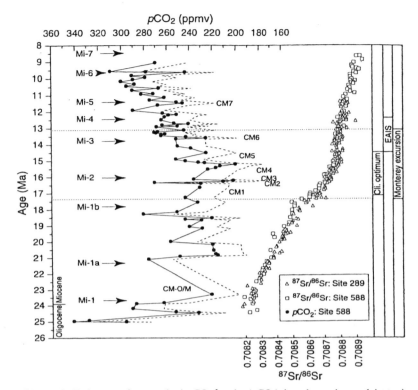

Figure 3. Estimates of atmospheric CO_2 levels (pCO_2) based on observed isotopic fractionation (ε_p) as determined by the $\delta^{13}C$ content of individual alkenone compounds from DSDP site 588. Estimates are based on a phosphate concentration of 0.3 µM and fractionation associated with enzymatic fixation of 27‰. Carbon burial events inferred from the inorganic carbon record are noted with CM, and Mi represents glacial events inferred from the oxygen isotopic record. Data for Sr isotope ratios are from sites 588 (open squares) and 289 (open triangle). Cli. Optimum is climate optimum, EAIS represents the time of the expansion of the East Antarctic Ice Sheet. Error estimates described in the original text represent 15% uncertainty of the calculated pCO_2 values. [Used by permission of the editor of *Paleoceanography,* from Pagani et al. (1999), Fig. 14.]

The interpretation that changes in sedimentary records of photosynthetic fractionation reflect past CO_2 and/or nutrient changes assumes that phytoplankton responded to their environment in a manner similar to what is observed in the modern ocean. For Paleozoic and older sedimentary records, this assumption is difficult to evaluate, as most of the modern phytoplankton flora evolved in younger eras. Laboratory and field studies of phytoplankton that have long geological records, such as cyanobacteria and green algae, are essential to strengthen the interpretation of variations in fractionation (ε_p) on long timescales. The influence on ε_p from changes in nutrients, cell geometry, or carbon dioxide concentrations will be muted or not expressed at very high CO_2 levels, as carbon supply no longer limits cell growth, and maximum fractionation is expressed. Thus even in the absence of a clear delineation of the specific environmental cause, if variations in ε_p are observed in sedimentary records, this can constrain the upper range of CO_2 concentrations. For example, if the relationship by Popp et al. (1998) holds for older time periods, the upper limit of sensitivity of this method is approximately 8-10 times the present level of atmospheric CO_2.

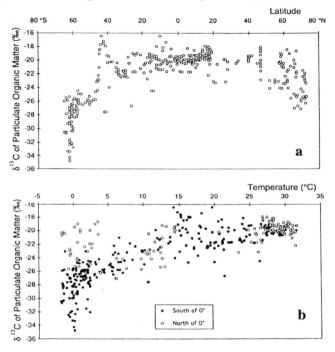

Figure 4. Surface water (mixed layer) particular organic carbon $\delta^{13}C$ values plotted as a function of (a) latitude and (b) sea surface temperature. [Used by permission of the editor of *Global Biogeochemical Cycles*, from Goericke et al. (1994), Fig. 1.]

ORGANIC CARBON IN MARINE SURFACE WATERS

In general, the properties of nutrient conditions, CO_2 abundance and phytoplankton diversity can be employed to understand the range of $\delta^{13}C$ values of organic carbon in marine surface waters. As shown by Goericke and Fry (1994) (Fig. 4), there is a significant latitudinal variation in organic matter $\delta^{13}C$ values of marine organic matter (Sackett et al. 1965, 1974; Rau et al. 1989, 1991, 1996; Fischer et al. 1991, 1997; Freeman and Hayes 1992, Francois et al. 1993, Goericke and Fry 1994, Bentaleb et al. 1998, Popp et al. 1999) and ancient marine oils (Andrusevich et al. 2000). The lower isotopic values recorded at higher latitudes reflect in part the greater concentrations of $CO_2(aq)$ associated with the colder temperatures of this region (Freeman and Hayes 1992, Goericke and Fry 1994, Hofman et al. 2000). However, variations in isotope values associated with seasonal shifts in nutrient availability in high latitude waters can be quite pronounced, as documented during the North Atlantic spring bloom (Rau et al. 1992), in the Ross Sea (Villinski et al. 2000) and in the northeast Pacific (Wu et al. 1999).

The circulation of major ocean currents can cause pronounced changes in nutrient and CO_2 concentrations, which are accompanied by major shifts in phytoplankton ecology. For example, across the polar front in the South Atlantic and Indian Ocean, a number of authors have observed strong variations in the $\delta^{13}C$ values of surface water particulate organic matter (e.g. Francois et al. 1993, Fischer et al. 1991, 1997; Dehairs et al. 1997, Popp et al. 1999). The polar front is a complex region of ocean and wind circulation that is characterized by a sharp change in temperatures, salinity and the concentrations of $CO_2(aq)$ and nutrients. It occurs between the northern extent of the Antarctic Circumpolar Current to the north and the Antarctic Divergence to the south. The Antarctic Divergence represents the region where winds change from prevailing westerlies to easterlies, as well as the upwelling of Antarctic deep water. This region is marked by changes in phytoplankton ecology, with shifts in dominant species that vary significantly in cell geometry. Thus Popp et al. (1999) used microscopic measurements of cell properties along with inorganic, bulk organic and biomarker isotopic data, to show that the isotopic composition of organic matter in this region is influenced by variations in carbon dioxide concentrations, phytoplankton taxonomy as well as nutrient-influenced changes in cell growth rates.

In polar regions, phytoplankton organic matter $\delta^{13}C$ values can also be influenced by the contributions of organic carbon derived from sea-ice algae. These organisms grow within sea-ice brines and are significantly enriched in ^{13}C relative to marine plankton found in high-latitude waters due to carbon limitation and physiological properties such as carbon concentrating mechanisms and/or the uptake of bicarbonate (Gleitz et al. 1996, Gibson et al. 1999, Villinski et al. 2000). Thus isotopic variations in circum-polar sedimentary organic carbon could reflect both variations in the contributions from ice-dwelling organisms as well as variations in CO_2 or nutrient levels influencing non-ice-hosted species (Gibson et al. 1999, Villinski et al. 2000).

The ^{13}C content of surface water particulate organic carbon (POC) shows some evidence for a decline in accord with the ^{13}C Suess effect, although this influence in not yet recorded in sedimentary organic matter. Fisher et al. (1997, 1998) suggested this anthropogenic influence accounted for the differences between the $\delta^{13}C$ of organic matter in the surface waters and sediments of the Indian, Atlantic and Southern Oceans. They suggest that surface sediment organic matter reflects preindustrial values, while POC shows anthropogenic influences, especially in colder waters, which are subject to greater CO_2 invasion. Recent ecological or nutrient changes in the surface waters are possible alternate interpretations, along with sedimentary diagenetic effects, and these require careful study. The use of molecular markers that are not subject to trophic effects or decay-related isotopic shifts (Harvey and Macko 1997b; discussed below) is a promising approach for evaluating recent anthropogenic influences on organic matter $\delta^{13}C$ values in marine waters and sediments.

Trophic effects and ecosystem studies

A large number of published studies employ carbon isotope analyses to evaluate carbon flow in marine ecosystems and food webs. Fry and Sherr (1989) provide an excellent review of the principles of this approach and its application. In addition, Gannes et al. (1997) and Hayes (1993; this volume) review the nature of isotopic shifts associated with the uptake of carbon substrates by heterotrophic organisms. In general, small isotope effects are observed between the isotopic composition of a whole organism and its food source. Fry and Sherr (1989) summarize results from 83 examples and demonstrate that the consumer value is within ±2‰ of its carbon source. In general, this agrees with laboratory and field studies that suggest the isotope effect associated with respiration of isotopically depleted CO_2 is small (DeNiro and Epstein 1978, Fry et al. 1984, Fry and Sherr 1989), such that the isotopic value of the consumer is within one permil of its food source. Nonetheless, marine field data suggest that isotopic enrichment occurs with the transfer of carbon from one trophic level to the next. Data compiled in Fry and Sherr

(1989) demonstrate consistent enrichment with trophic level in a number of marine localities, averaging around 1‰ per trophic level. The magnitude of trophic-level isotope shifts varies for different coastal and estuarine settings and the generalization of one-permil-shift-per-trophic level should be considered an approximation (Fry and Sherr 1989, del Giorgio and France 1996, Deegan and Garrit 1997).

Trophic enrichment likely reflects, in addition to respiratory effects, selective consumption of isotopically enriched dietary components (Fry and Sherr 1989, Gannes et al. 1997), such as amino acids or other biochemicals (Macko et al. 1987) or isotopically enriched tissues. An organism can partition dietary carbon into different tissues that are isotopically different from each other and from the carbon source (Macko et al. 1982, 1986). Typically, lipids are depleted in ^{13}C relative to the whole organism (DeNiro and Epstein 1977, Monson and Hayes 1982, Hayes 1993, Schouten et al. 1998). Proteins are generally enriched in ^{13}C (Deines 1980, Koch et al. 1994), although individual amino acids can exhibit a wide range of isotopic values, reflecting different synthetic pathways (Macko et al. 1987) and thus their relative distributions can influence whole-tissue δ^{13}C values (Hare et al. 1991). Hayes (1993; this volume) provides a detailed consideration of carbon flow, consumer efficiencies, and isotopic partitioning within organisms and between organisms and their carbon source.

Additional physiological effects can complicate simple interpretations of trophic effects. For example, consuming organisms do not always homogenize the isotopic values of their different dietary components, and isotopically distinct inputs can be partitioned into separate tissues (Krueger and Sullivan 1984, Schwarcz 1991, Tieszen and Farge 1993). Nitrogen exhibits a greater magnitude of isotopic enrichment with trophic level than carbon, averaging +2 to +5‰ per level (DeNiro and Epstein 1981, Ambrose and DeNiro 1986), and ^{15}N is often employed as a companion isotope to ^{13}C in food-web and ecosystem studies. In addition to the influences of dietary heterogeneity and partitioning of N among tissues, the N isotope composition of animals can also be influenced by body condition (Gannes et al. 1997). Starving animals recycle proteins from their own tissues, and the recycled nitrogen is subject to isotope effects associated with protein metabolism (Ambrose and DeNiro 1986, Macko et al. 1986, 1987). The loss of ^{15}N-depleted protein is not replaced by diet, and thus starving animals become progressively enriched in ^{15}N (Hobson et al. 1993, Hobson and Schell 1998).

Carbon and nitrogen isotopic values can be used to study the transfer of carbon within an ecosystem, although the above considerations advise caution. As discussed in detail by Gannes et al. (1997), isotopic signatures of animals can be linked to diet sources, provided metabolic processes associated with heterotrophy and body condition are taken into account. Simple differences in ^{13}C or ^{15}N contents between two animals does not necessarily indicate different diets, nor does it necessarily indicate trophic shifts or, in the case of nitrogen isotopes, starvation. Detailed understanding of the physiology and metabolism of the species is required to interpret isotopic differences among components of an ecosystem (Gannes et al. 1997). Such studies are significantly enhanced by companion laboratory studies, as well as integration of isotopic data with other chemical tracers.

Applications of isotopic data to aquatic ecosystem studies have largely focused on coastal, estuarine or lake environments, with the goal of delineating terrestrial, marine or other carbon sources for land and marine animals and birds (for recent examples, see: Weinstein et al. 2000, Gould et al. 2000, Cederholm et al. 1999, Rolf and Elmgren 2000, Hsieh et al. 2000, MacAvoy et al. 2000, Pinnegar and Polumin 2000). If food substrates are regionally distributed, with isotopically distinct patterns, organism carbon and nitrogen isotope values can reveal migration patterns (Hobson 1999). For example, Gould et al. (2000) distinguished dietary habits of seabirds in the *Puffinus* genus and used isotopic data to delineate migration patterns in the Northern Pacific Ocean.

Isotopic patterns in the keratin of baleen plates from Bowhead whales record their seasonal migration between the Bering Sea in the winter and the Beaufort and Chukchi Seas in the summer (Schell et al. 1989). Through recent and archeological samples, Schell (2000) presents a continuous Bowhead baleen isotopic record from 1947-1997 (Fig. 5). The animals consume enormous quantities of zooplankton annually (~100 metric tons; Thomson 1987) which have, in turn, preyed upon an order of magnitude more phytoplankton mass. Thus Schell (2000) argues that whale baleen is an excellent recorder of integrated isotope values for phytoplankton. The baleen records reveal a consistent decrease in the $\delta^{13}C$ values (over 2.7‰) for the Bering and Chukchi phytoplankton, which Schell suggests reflects a decrease in phytoplankton growth rates. The decline in phytoplankton growth rates (and by extension, phytoplankton productivity) possibly reflects a shallowing of mixed-layer depths and the consequent reduction in nitrate supply to North Pacific Ocean surface waters (Freeland et al. 1997). Thus Schell (2000) suggests that the whale baleen is recording a decline in phytoplankton production over the second half of the 20th century, and, if so, that this indicates a decline in ecosystem carrying capacity. He proposes that this contributes (in addition to the stress of heavy fishing of important fish stocks) to the decline in marine mammals in the Bering Sea (Loughlin 1998).

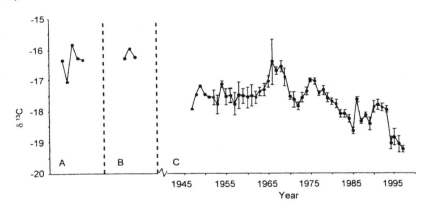

Figure 5. Average annual Bering/Chukchi carbon isotope ratios in whale baleen from (A) partial baleen plate from Punuk Island (2200 yr B.P.); (B) partial plate from St. Lawrence Island (ca. 1870 A.D.) and (C) from 37 whales taken during the 1990s. [Used by permission of the editor of *Limnology and Oceanography*, from Schell (2000), Fig. 2.]

MARINE SEDIMENTARY ORGANIC MATTER

Alteration of primary isotopic signatures

Organic matter in the ocean is delivered to sediments in the form of particles that sink from the surface to the deep ocean. The transfer of carbon can be quite effective in cases where particle diameters and densities permit rapid sinking velocities (McCave 1975), as is the case in the formation of fecal matter produced by grazing organisms. This transfer process is subject to the influences of additional grazing, currents and the systematic decay of particulate organic carbon (POC) throughout its time in the water column (Seuss 1980). A review of methods for the collection and characterization of marine particles is provided by Hurd and Spencer (1991). In the open ocean, sedimentary organic matter is largely derived from sinking particles, and it can reflect significant seasonal bias, especially in regions with seasonally elevated rates of productivity and, in particular, carbon export (Buessler 2000). Additional contributions can come from wind-blown dust particles, terrestrial sources (especially on or near continental shelves), and bacterial biomass.

In general, there are few reports of significant isotope shifts with the transfer of carbon compounds from surface waters to sediments. However, Fisher (1991) observed ^{13}C isotopic enrichment in sinking particles relative to phytoplankton, which may reflect selective grazing or loss of ^{13}C-depleted carbon associated with methane production by prokaryotes within zooplankton digestive tracts (Hayes 1993). However, the δ^{13}C values of individual lipid compounds from phytoplankton are not altered as they are packaged into fecal pellets from grazing herbivoires (Grice et al. 1998). Further, recent sediments show little evidence for significant isotopic shifts in bulk organic matter associated with heterotrophic activities (Blair et al. 1985). Meyers (1994) found little change in the δ^{13}C of sedimentary organic matter with rather extensive carbon loss in lake sediments. Based on studies of the cycling of carbon in sediment pore waters, numerical models of marine diagenesis commonly incorporate no isotopic fractionation associated with the oxic heterotrophic release of CO_2 from the decay of sedimentary organic matter (e.g. McCorkle et al. 1985, Williams and Druffel 1987, Sayles and Curry 1988, Martin and McCorkle 1993, Jahnke et al. 1997, Gehlen et al. 1999).

In contrast, the ancient record provides molecular evidence for isotopic differences between phytoplankton and heterotrophic inputs. For example, numerous studies report isotopic enrichment (up to several ‰) between molecular markers for phytoplankton (such as pristane and phytane) and compounds that are associated with grazing organisms (various *n*-alkanes) (Hayes et al. 1989, Kenig et al. 1994). In sedimentary units from the Precambrian, the isotopic difference between lipids and kerogen can be significantly greater (Hoering 1965, 1966). Logan et al. (1995) have argued that the lack of fecal-pellet-forming metazoans resulted in extended heterotrophic reworking of organic matter in surface waters before transfer to sediments. Thus isotopic enrichment of Neoproterozoic kerogens relative to extractable lipids is proposed to result from extensive heterotrophy, and that these effects are attenuated in shallow-water deposits and in Early Cambrian and younger sediments (Logan et al. 1997).

Changes in modern sedimentary bulk organic matter δ^{13}C values can be associated with the preferential loss of labile algal biomass and a relative enrichment of more refractory terrestrial-derived carbon (Prahl et al. 1997, Böttcher et al. 2000). The selective loss of compound classes with different reactivities (Westrich and Berner 1984) and δ^{13}C values (Wang et al. 1998) could account for minor shifts in marine sediments, but for algal-dominated organic matter, it is unlikely that these would be greater than 1‰ (Dean et al. 1986). Spiker and Hatcher (1984) suggested the preferential loss of an isotopically enriched carbohydrate fraction accounted for a 2-4‰ decline in the δ^{13}C of bulk organic matter in sediments from Mangrove Lake (Bermuda). However, this shift tracks changes in the δ^{13}C values of inorganic carbon phases (Dean et al. 1986), and is also associated with increased contributions from terrestrial plants relative to aquatic algae in the core (Meyers 1994).

Selective preservation of carbohydrates through sulfidization was discussed by Van Kaam-Peters et al. (1998) as a means for accounting for isotopically enrichment of 6‰ in ancient sedimentary organic matter with a high organic-sulfur content. This process has also been simulated in the laboratory with phytoplankton biomass (Sininghe Damste et al. 1998, Gelin et al. 1998) and observed for other sulfur-rich kerogens (Riboulleau et al. 2000) but it is not clear if this mechanism is widespread in normal marine sediments.

The bulk of sedimentary organic matter: an unresolved issue

The largest fraction of organic matter in marine sediments remains poorly understood in terms of both its structure and biogeochemical origin. In the absence of this understanding, it is exceptionally difficult to evaluate specific mechanisms or isotope fractionation associated with its synthesis or degradation. This "molecularly uncharacterized component" (MUC) was the topic of an in-depth review by Hedges et al. (2000), who discuss possible mechanisms for its formation, and a number of new analytical approaches to characterizing this material.

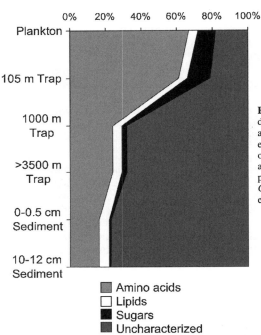

Percent of Organic Carbon

Figure 6. Cumulative biochemical class distributions (as a percent of total carbon), averaged for surveys at 9°N, 5°N and the equator. Residual "uncharacterized" carbon was obtained as the difference between total carbon and the sum of the other fractions. [Used by permission of the editor of *Geochimica et Cosmochimica Acta*, modified from Wakeham et al. (1997), Fig. 2.]

The quantitative significance of MUC in marine environments increases as materials are transported from surface waters into the sedimentary environment (Fig. 6; Wakeham et al. 1997), during bacterial decay as documented by incubation studies (Harvey and Macko 1997a) and with depth in the sediment (e.g. Parkes et al. 1993, Hedges et al. 2000). Possible mechanisms for MUC formation include:

- Polymerization of unrelated biochemicals during or following sedimentation (Tissote and Welte 1978), possibly associated with the formation of polymers from carbohydrates and proteins (Hoering 1973, Zegouagh et al. 1999) or cross-linked lipids (Harvey et al 1983).
- Natural vulcanization, or reaction of organic compounds with sulfur compounds, which renders the sulfur-linked materials less available for degradation (e.g. Kohnen et al. 1992). However, this process is not likely to be significant in many marine environments as it is requires low iron availability and abundant reduced-sulfur species.
- Selective preservation of biopolymers that are resistant to degradation reactions. These materials are widespread among plants and algae (Largeau et al. 1986, Derenne

et al. 1991; reviewed by deLeeuw and Largeau 1993), and thus may constitute a large component of organic matter in sediments.

As documented in recent studies, interactions between inorganic materials, including mineral surfaces, and organic compounds can enhance the preservation of labile materials (Mayer 1994) while also rendering it difficult to characterize structurally. The preservation of organic materials through a physical or chemical association with mineral surfaces and/or with aggregates of clays and oxides is an area of active research interest (Hedges et al. 2000).

Microbial influences on marine sedimentary organic matter $\delta^{13}C$ values

The relative importance of prokaryotic contributions to marine sedimentary organic matter, especially to the uncharacterizable fraction has been debated. Parkes et al. (1993) quantified bacterial activities, various organic fractions and the uncharacteriziable fraction in sediments from the Peru Margin. The loss of labile organic fractions (carbohydrates and proteins) clearly fueled bacterial activity in the upper sediments, with quantitative relationships observed between carbon loss and bacterial activity. However, in deeper sediments, where microbial activities and labile carbon concentrations were lower, metabolic activity increasingly consumed the uncharacterizable organic fraction. As the fraction of MUC increased with depth, microbial activity decreased, and Parkes et al. (1993) suggested this reflects an increased contribution from the cells of dead bacteria.

Gong and Hollander (1997) concluded from isotopic analyses of bulk organic matter and individual lipids that microbial biomass contributions increased under anaerobic conditions. This process could account for enhanced preservation of organic matter in the anoxic sediments relative to oxic sediments within the Santa Monica Basin, California. In this example, $\delta^{13}C$ values of organic matter at depth in the anoxic sediments were similar to the oxic sediments. However, at the surface of the anoxic sediments, organic carbon was 2‰ lower than throughout the oxic sediments, reflecting the expansion of anoxia over this interval of deposition (Hollander et al. 1997) and the contribution of isotopically depleted microbial biomass associated with chemoautotrophy in stratified basins (Freeman et al. 1994, Gong and Hollander 1997).

Hartgers et al. (1994) and Sininghe Damste and Schouten (1997) discuss observations of some striking differences between biomarker distributions and $\delta^{13}C$ values for individual biomarker compounds (both of which often indicate strong bacterial signals), and the isotopic properties of the bulk organic fraction. They suggested that biomarkers over-represent bacterial inputs, and that in fact bacterial contributions to marine organic-rich sediments are minor (Sininghe Damste and Schouten 1997). In mixed-population incubation studies, Harvey and Macko (1997a) found quantities of bacterial fatty acids did not track bacterial cell abundance, and they concurred that lipids may be poor quantitiative indicators of microbial contributions to sedimentary organic matter. Nonetheless, microbial lipids have long been employed in order to qualitatively characterize microbial contributions to a wide array of sedimentary environments (White 1983), including as symbionts within marine organisms (Abrajano et al. 1994, Jahnke et al. 1995).

Numerous studies have attempted to sort out the nature of relationships among the $\delta^{13}C$ values of microbial heterotrophic carbon substrates, biomass and individual biomarker compounds. For example, Abraham et al. (2000) evaluated the ^{13}C contents of fatty acids derived from different lipid classes within prokaryotic and fungal heterotrophs that were grown on glycerol, glucose, lactose or a complex medium. Consistent with previous work (e.g. Monson and Hayes 1980, 1982a,b; Zyakun et al. 1998), biomass and lipids in this study were isotopically depleted relative to the substrate for organisms

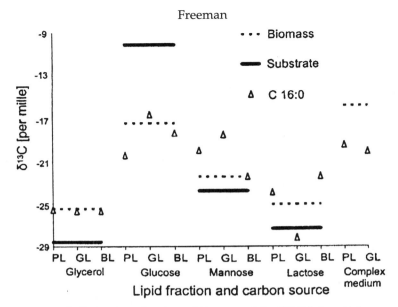

Figure 7. Carbon isotopic values of biomass and the C16:0 fatty acid from different fractions of polar and bound lipids of *Fusarium solani* compared with values for the corresponding substrate. PL, polar lipids; GL, glycolipids; BL, bound lipids. [Used by permission of the editor of *Applied Environmental Microbiology*, from Abraham et al. (1998), Fig. 3.]

grown on glucose. However, other substrates yield an array of isotopic relationships between substrate, lipids and biomass (Fig. 7), demonstrating that careful understanding of isotopic fractionation associated with biosynthetic processes is important in application of lipid isotopic data to sediment studies (Hayes, this volume). Incubation and field studies report at most minor (less than 2‰) influence of bacterial processes on sedimentary organic matter. Studies report slight ^{13}C enrichment, depletion or little change among bacterial lipids (Harvey and Macko 1997a,b) or sedimentary organic matter (Blair et al. 1985) during microbial heterotrophic degradation of marine organic matter.

Other modes of microbial metabolism besides heterotrophy can be recorded in marine sedimentary organic matter. Particularly dramatic examples are associated with microbial cycling of methane, which is strongly depleted in ^{13}C relative to other carbon substrates (Whiticar 1999). Isotopic analyses of archaeal membrane lipids reveal signatures associated with anaerobic oxidation of methane (Elvert et al. 1999, Hinrichs et al. 1999, Hinrichs et al. 2000, Pancost et al. 2000). Organisms that are responsible for the anaerobic oxidation of methane generate organic lipids (Hinrichs et al. 2000, Bian et al. 2001), and microbial biomass (C. House, pers. comm.), that are exceptionally depleted in ^{13}C, with observed values below -100‰. Field geochemical and microbiological observations suggest methane oxidation results from a syntrophic partnership between sulfate-reducing bacteria and members of the Archaea domain, possibly methanogens (Hoehler et al. 1994, Hinrichs et al. 1999, Boetius et al. 2000). Bacterial methane oxidation contributes isotopically depleted biomass to sediments, whether the process is anaerobic or aerobic (i.e. Summons et al. 1994, Jahnke et al. 1999). These signatures provide a potential record of times of elevated atmospheric methane levels and have been invoked to explain kerogen $\delta^{13}C$ values as low as -60‰ in the late Archean (Hayes 1983, 1994). Investigators have also suggested that rapid and pronounced ^{13}C-depletions in marine organic matter tracks catastrophic methane release associated with methane clathrate disruption (e.g. Dickens et al. 1995, 1997; Cramer et al. 1999, Hesselbo et al. 2000, Kennett and Fackler-Adams 2000).

Individual lipids may not be reliable quantitative indicators of bacterial inputs, and isotopic relationships between substrate, biomass and lipid reflect a rich diversity in microbial anabolic and catabolic processes. Further, the bulk of organic matter is difficult (at best) to characterize in terms of structure or its biological origin. Yet, new biochemical approaches may provide greater insights and serve as important addition to the use of lipid biomarkers and other organic geochemical methods. In particular, the extraction, characterization and isotopic analysis of genetic materials (Coffin et al. 1990, Creach et al. 1999, Radajewski et al. 2000), genetic sequencing and taxon-specific genetic probes such as fluorescent in-situ hybridization (FISH) and other methods (for example, see Murrell et al. 1998 and Bourne et al. 2000) provide unprecedented potential to evaluate microbial ^{13}C contents, activity and diversity in sediments. Studies that bridge geochemistry and microbiology promise far greater understanding of microbial influence on sedimentary materials. Interdisciplinary approaches will be key to understanding isotope fractionation associated with microbial processes in sediments and the resulting variations in the $\delta^{13}C$ of sedimentary organic materials.

SEDIMENTARY RECORDS OF PAST ENVIRONMENTS

Isotopic analyses of organic matter in marine sediments are commonly employed in the study of past oceans, climates and carbon cycling. As noted in the preceding sections, variations in the $\delta^{13}C$ of organic matter can reflect a variety of processes, which range from global to microbiological in scale. Marine organic matter isotopic data can reveal changes in carbon cycling over a range of timescales (millions of years or longer) through isotopic mass-balance calculations (reviewed in Kump and Arthur 2000, DesMarais, this volume; Ripperdan, this volume) or as a recorder of general trends in relative carbon dioxide concentrations (Hayes et al. 1999). As discussed in a previous section, high-resolution isotopic records based on specific biomarker compounds are proving a powerful means to track climatic and oceanographic change with on fine timescales and with greater precision.

Investigators must seek robust indicators of specific sources, in order to evaluate changes in the relative amounts and isotopic character of inputs from various phytoplankton, grazer, bacterial and terrestrial sources. The choice of material or molecule requires understanding of influences of diagenetic processes on its preserved $\delta^{13}C$ value, which is apparently minor, but particularly challenging, given the limitations in our understanding in the modern environment. Selective use of organic compounds with known biological sources is a welcome tool in this regard, and it can be applied to evaluate biological inputs in the context of changing ecosystems, changes in physical circulation and transport or shifts in preservation conditions and mechanisms. In environments that receive significant inputs from terrestrial materials, authors have used isotopic data for both bulk organic carbon and individual plant-derived molecules can be used to interpret shifts in the $\delta^{13}C$ values of organic materials as reflecting either changes in the inputs of terrestrial plants relative to algae (Jasper and Hayes 1993, Schlunz et al. 1999, Volkman et al. 2000) or changes in the dominance of C-3 versus C-4 plant materials within the terrestrial fraction (France-Lanord and Derry 1994, Pagani et al. 1999c, Huang et al. 2000, Freeman and Colarusso 2001).

In applications that seek to interpret sedimentary isotopic signatures in the context of global-scale processes, it is essential that the influence of the local depositional environment be considered carefully (Dean et al. 1986). The problem of global interpretation from a local sedimentary signature serves as a contrast to studies that employ computational general circulation models to estimate climatic change on a global scale. In global modeling efforts, the challenge is to extrapolate broad, regional patterns into local conditions. In the context of organic matter isotopic signatures, circulation changes may result in a change in $\delta^{13}C$ of inorganic carbon within a local marine basin, and the inorganic signal is ultimately reflected in the bulk organic matter composition (Patzkowsky et al. 1997). Alternatively, regional changes in phytoplankton population can also lead to changes in sedimentary bulk organic $\delta^{13}C$ values, and may or may not reflect widespread oceanographic or climatic change (Pancost et al. 1998, 1999; Fig. 8). These problems are especially acute in continental seas and restricted marine basins, although similar influences are also observed in open ocean localities, especially in association with changes in nutrient fluxes resulting from shifting currents and upwelling patterns.

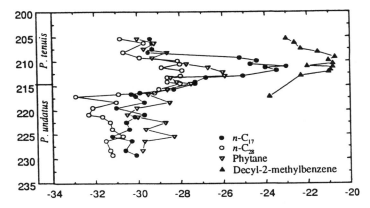

Figure 8. Molecular isotopic data from Late Ordovician epicontinental marine deposits in the U.S. mid-continent. Inputs from *Gloeocapsomorpha prisca* are presented by decyl-2-methylbenzene and inputs from other algae are recorded by C_{28} n-alkanes. C_{17} n-alkane and phytane, like the bulk organic carbon, reflect mixed contributions from both sources. [Used by permission of the editor of *Geology*, from Pancost et al. (1999), Fig. 3.]

For example, inputs from the marine alga, *Gloeocapsamorpha prisca* are represented by the biomarker decyl-2-methylbenzene, inputs from other algae are recorded by the n-alkanes C_{17} and C_{18}, and the compound phytane represents mixed contributions from both sources (Fig. 9). In Late Ordovician carbonate rocks of the US midcontinent, the isotopic values for these compounds track an isotopic excursion of approximately 3‰ in the inorganic carbon. However, the relative proportion of contributions from the two biological sources shifts across the excursion, with an increased input from *G. prisca* marking the event. The isotopic values of compounds of the compounds from distinct sources are nearly 10‰ apart. Thus the increased inputs from *G. prisca* are recorded by an increase in the isotopic composition of n-C_{17} and phytane, as well as that of total organic carbon (Pancost et al. 1999).

Thus isotopic records of marine carbon, especially organic carbon, can be influenced by regional phenomena, and potential influences must be accounted for before global interpretations are possible. Widespread or global significance can be tested by evaluation of facies-related isotopic patterns for synchronously deposited units and from globally distributed sample localities. Microbial influences on preserved $\delta^{13}C$ signatures in bulk organic carbon, although apparently minor, are not well understood. Molecular analyses can help to refine the potential for mixed biological inputs, changes in the dominant source of organic matter and possible diagenetic influences during preservation. With such tools and approaches, the application of isotopic analyses to sedimentary organic matter can offer great insight to understanding past environmental changes over a broad range of spatial and temporal scales.

ACKNOWLEDGMENTS

I thank the editors for the opportunity to contribute to this volume. D. Lambert provided invaluable assistance in the preparation of this manuscript. I am grateful for the constructive reviews provided by P. Meyers and D. Schell.

REFERENCES

Abraham WR, Hesse C, Pelz O (1998) Ratios of carbon isotopes in microbial lipids as an indicator of substrate usage. Appl Environ Microbiol 64:4202-4209

Abrajano Jr. TA, Murphy DE, Fang J, Comet P, Brooks JM (1994) $^{13}C/^{12}C$ ratios in individual fatty acids of marine mytilids with and without bacterial symbionts. Organic Geochem 12:611-617

Ambrose SH, DeNiro MJ (1986) The isotopic ecology of East African mammals. Oceologia 69:395-406

Anderson LA (1994) Redfield ratios of remineralization determined by nutrient data analysis. Global Biogeochem Cycle 8:65-80

Andrusevich VE, Engel MH, Zumberge JE (2000) Effects of paleolatitude on the stable carbon isotope composition of crude oils. Geology 28:847-850

Archer D, Winguth A, Lea D, Mahowald N (2000) What caused the glacial/interglacial atmospheric pCO_2 cycles? Rev Geophys 38:159-189

Bacastow RB, Keeling CD, Lueker TJ, Wahlen M, Mook WG (1996) The $\delta^{13}C$ Suess effect in the world surface oceans and its implications for oceanic uptake of CO_2: analysis of observations at Bermuda. Global Biogeochem Cycles 10:335-346

Beerling DJ (1999) Quantitative estimates of changes in marine and terrestrial primary productivity over the past 300 million years. Proc Royal Soc London Series B-Biol Sci 266:1821-1827

Bentaleb I, Fontugne M (1998) The role of the southern Indian Ocean in the glacial to interglacial atmospheric CO_2 change: organic carbon isotope evidences. Global Planetary Change 17:25-36

Bentaleb I, Fontugne M, Descolas-Gros MC, Girardin C, Mariotti A, Pierre C, Brunet C, Poisson A (1996) Organic carbon isotopic composition of phytoplankton and sea-surface pCO_2 reconstructions in the Southern Indian Ocean during the last 50,000 years. Organic Geochem 24:399-410

Bentaleb I, Fontugne M, Descolas-Gros MC, Girardin C, Mariotti A, Pierre C, Brunet C, Poisson A (1998) Carbon isotopic fractionation by plankton in the Southern Indian Ocean: relationship between delta C-13 of particulate organic carbon and dissolved carbon dioxide. J Marine Systems 17:39-58

Berger WH, Herguera JC, Lange CB, Schneider R (1994) Paleoproductivity: Flux proxies versus nutrient proxies and other problems concerning the quaternary productivity record. *In* R Zahn, TF Pedersen, MA Kaminski, L Labeyrie (eds) Carbon Cycling in the Glacial Ocean: Constraints on the Ocean's Role in Global Change. Springer-Verlag, Berlin-Heidelberg, p 385-412

Berner RA (1999) Atmospheric oxygen over Phanerozoic time. Proc National Acad Sci U S A 96: 10,955-10,957

Berner RA, Petsch, ST, Lake JA, Beerling DJ, Popp BN, Lane RS, Laws EA, Westley MB, Cassar N, Woodward FI, Quick WP (2000) Isotope fractionation and atmospheric oxygen: Implications for phanerozoic O_2 evolution. Science 287:1630-1633

Bian L, Hinrichs K, Xie T, Brassell SC, Iversen N, Fossing H, Jorgensen BB, Hayes JM (2001) Algal and archaeal polyisoprenoids in a recent marine sediment: Molecular isotopic evidence for anaerobic oxidation of methane. Geochem Geophys Geosystems 2, manuscript 2000GC000112

Bidigare RR, Fluegge A, Freeman KH, Hanson KL, Hayes JM, Hollander D, Jasper JP, King, LL, Laws EA, Milder J, Millero FJ, Pancost R, Popp BN, Steinberg PA, Wakeham SG (1997) Consistent fractionation of C-13 in nature and in the laboratory: Growth-rate effects in some haptophyte algae. Global Biogeochem Cycle 11:279-292

Bidigare RR, Buessler KO, Freeman KH, Hanson KL, Hayes JM, Laws EA, Milder J, Millero FJ, Pancost R, Popp BN, Steinberg PA, Wakeham SG (1999a) Correction: Consistent fractionation of C-13 in nature and in the laboratory: Growth-rate effects in some haptophyte algae (vol 11, p 279, 1997). Global Biogeochem Cycle 13:251-252

Bidigare RR, Fluegge A, Freeman KH, Hanson KL, Hayes JM, Hollander D, Jasper JP, King LL, Laws EA, Milder J, Millero FJ, Pancost R, Popp BN, Steinberg PA, Wakeham SG (1999b) Iron-stimulated changes in C-13 fractionation and export by equatorial Pacific phytoplankton: Toward a paleogrowth rate proxy. Paleoceanogr 14:589-595

Blair N, Leu A, Munoz E, Olsen J, Kwong E, DesMarais D (1985) Carbon isotopic fractionation in heterotrophic microbial metabolism. Appl Environ Microbiol 50:996-1001

Boetius A, Ravenschlag K, Schubert CJ, Rickert D, Widdel F, Gieseke A, Amann R,Jorgensen BB, Witte U, Pfannkuche O (2000) A marine microbial consortium apparently mediating anaerobic oxidation of methane. Nature 407:623-626

Böttcher ME, Hespenheide B, Llobet-Brossa E, Beardsley C, Larsen O, Schramm A, Wieland A, Bottcher G, Berninger UG, Amann R (2000) The biogeochemistry, stable isotope geochemistry, and microbial community structure of a temperate intertidal mudflat: an integrated study. Cont Shelf Res 20: 1749-1769

Bourne DG, Holmes AJ, Niels I, Murrell JC (1999) Fluorescent oligonucleotide rDNA probes for specific detection of methane oxidising bacteria. FEMS Microbiology Ecology 31:29-38

Boyle EA (1994) A comparison of carbon isotopes and cadmium in the modern and glacial maximum ocean: Can we account for the discrepancies. In R Zahn, TF Pedersen, MA Kaminski, L Labeyrie (eds) Carbon Cycling in the Glacial Ocean: Constraints on the Ocean's Role in Global Change. Springer-Verlag, Berlin-Heidelberg, p 167-193

Broecker WS, Maier-Reimer E (1992) The influence of air and sea exchange on the carbon isotope distribution in the sea. Global Biogeochem Cycles 6:315-320

Broecker WS, Peng TH (1974) Gas exchange rates between air and sea. Tellus 26:21-35

Buesseler KO (1998) The decoupling of production and particulate export in the surface ocean. Global Biogeochem Cycle 12:297-310

Burkhardt S, Riebesell U (1997) CO_2 availability affect elemental composition (C:N:P) of the marine diatom Skeletonema costatum. Mar Ecol-Prog Ser 67-76:10

Burkhardt S, Riebesell U, Zondervan I (1999a) Effects of growth rate, CO_2 concentration, and cell size on the stable carbon isotope fractionation in marine phytoplankton. Geochim Cosmochim Acta 62: 3729-3741

Burkhardt S, Riebesell U, Zondervan I (1999b) Stable carbon isotope fractionation by marine phytoplankton in response to daylength, growth rate, and CO_2 availability. Marine Ecology-Progress Ser 184:31-41

Caldeira K, Duffy PB (2000) The role of the Southern Ocean in uptake and storage of anthropogenic carbon dioxide. Science 287:620-622

Cederholm CJ, Kunze MD, Murota T, Sibatani A (1999) Pacific salmon carcasses: Essential contributions of nutrients and energy for aquatic and terrestrial ecosystems. Fisheries 24:6-15

Clark DR, Flynn KJ (2000) The relationship between the dissolved inorganic carbon concentration and growth rate in marine phytoplankton. Proc Royal Soc London Series B-Biol Sci 267:953-959

Coffin R, Devereux R, Price W, Cifuentes L (1990) Analyses of stable isotopes of nucleic acids to trace sources of dissolved substrates used by estumarine bacteria. Appl Environ Microbiol 66:2012-2020

Cooper TG, Filmer D, Wishnick M, Lane MD (1969) The active species of "CO_2" utilized by ribulose diphosphate carboxylase. J Biol Chem 244:1081-1083

Cramer BS, Aubry MP, Miller KG, Olsson RK, Wright JD, Kent DV (1999) An exceptional chronologic, isotopic, and clay mineralogic record of the latest Paleocene thermal maximum, Bass River, NJ, ODP 174AX. Bull Soc geol France 170:883-897

Creach V, Lucas F, Deleu C, Bertru G, Mariotti A (1999) Combination of biomolecular and stable isotope techniques to determine the origin of organic matter used by bacterial communities: application to sediments. J Microbiol Methods 38:43-52

De Leeuw JW, Largeau C (1993) A review of macromolecular organic compounds that comprise living organsims and their role in kerogen, coal and petroleum formation. In Engel MH, Macko SA (eds) Organic Geochemistry Plenum Press, p 23-72

Dean WE, Arthur MA, Claypool GE (1986) Depletion of ^{13}C in Cretaceous marine organic matter: Source, diagenetic, or environmental signal? Marine Geology 70:119-157

Dehairs F, Kopczynska E, Nielsen P, Lancelot C, Bakker DCE, Koeve W, Goeyens L (1997) partial derivative C-13 of Southern Ocean suspended organic matter during spring and early summer: Regional and temporal variability. Deep-Sea Res Part II-Top Stud Oceanogr 44:129-142

Deines P (1980) The isotopic composition of reduced organic carbon. In Fritz P, Fontes JC (eds) Handbook of Environmental Geochemistry: The Terrestrial Environment. Elsevier, p 329-406

del Giorgio PA, France RL (1996) Ecosystem-specific patterns in the relationship between zooplankton and POM or microplankton delta C-13. Limnology Oceanogr 41:359-365

DeNiro MJ, Epstein S (1977) Mechanism of carbon isotope fractionation associated with lipid synthesis. Science 197:261-263

DeNiro MJ, Epstein S (1978) Influence of diet on the distribution of carbon isotopes in animals. Geochim Cosmochim Acta 42:495-506

DeNiro MJ, Epstein S (1981) Influence of diet on the distribution of nitrogen isotopes in animals. Geochim Cosmochim Acta 45:341-351

Derenne S, Largeau C, Casadevall E, Berkaloff C, Rous-seau B (1991) Chemical evidence of kerogen formation in source rocks and oil shales via selective preservation of thin resistant outer walls of microalgae: origin of ultralaminae. Geochim Cosmochim Acta 55:1041-1050

Dickens GR, Castillo MM, Walker JGC (1997) A blast of gas from the latest Paleocene: Simulating first-order effects of massive dissociation of oceanic methane hydrate. Geology 25:259-262

Dickens GR, O'Neil JR, Rea DK, Owen RM (1995) Dissociation of oceanic methane hydrate as a cause of the carbon isotope excursion at the end of the Paleocene. Paleoceanogr 10:965-971

Eek MK, Whiticar MJ, Bishop JKB, Wong CS (1999) Influence of nutrients on carbon isotope fractionation by natural populations of Prymnesiophyte algae in NE Pacific. Deep-Sea Res Part II—Topical Studies Oceanogr 46:2863-2876

Elvert M, Suess E, Whiticar MJ (1999) Anaerobic methane oxidation associated with marine gas hydrates: Superlight C-isotopes from saturated and unsaturated C_{20} and C_{25} irregular isoprenoids. Naturwiss 86

Erez J, Bouevitch A, Kaplan A (1998) Carbon isotope fractionation by photosynthetic aquatic mircro-organisms: experiments with Synechococcus PCC7942, and a simple carbon flux model. Canadian J Botany 76:1109-1118

Fielding AS, Turpin DH, Guy RD, Calvert SE, Crawford DW, Harrison PJ (1998) Influence of the carbon concentrating mechanism on carbon stable isotope discrimination by the marine diatom Thalassiosira pseudonana. Canadian J Botany 76:1098-1103

Fischer G (1991) Stable carbon isotope ratios of plankton carbon and sinking organic matter from the Atlantic sector of the Southern Ocean. Mar Chem 35:581-596

Fischer G, Schneider R, Muller PJ, Wefer G (1997) Anthropogenic CO_2 in Southern Ocean surface waters: evidence from stable organic carbon isotopes. Terra Nova 9:153-157

Fraga F, Rios AF, Perez FF, Estrada M, Marrase C (1999) Effect of upwelling pulses on excess carbohydrate synthesis as deduced from nutrient, carbon dioxide and oxygen profiles. Mar Ecol Prog Ser 189:65-75

Fraga F, Rios AF, Perez FF, Figueiras FG (1998) Theoretical limits of oxygen: carbon and oxygen: nitrogen ratios during photosynthesis and mineralisation of organic matter in the sea. Sci Mar 62:161-168

France-Lanord C, Derry LA (1994) Delta C-13 of organic-carbon in the bengal fan-source evolution and transport of C3 and C4 plant carbon to marine-sediments. Geochim Cosmochim Acta 58:4809-4814

Francey RJ, Steele LP, Langenfelds RL, Allison CE, Cooper LN, Dunse BL, Bell BG, Murray TD, Tait HS, Thompson L, Masarie KA (1998) Atmospheric carbon dioxide and its stable isotope ratios, methane, carbon monoxide, nitrous oxide and hydrogen from Shetland Isles. Atmospheric Environ 32:3331-3338

Francois R, Altabet MA, Goericke R, McCorkle DC, Brunet C, Poisson A (1993) Changes in the $\delta^{13}C$ surface water particulate organic matter across the subtropical convergence in the S.W. Indian Ocean. Global Biogeochem Cycle 7:627-644

Freeland H, Denman K, Wong CS, Whitney F, Jacques R (1997) Evidence of change in the winter mixed layer in the Northeast Pacific Ocean. Deep-Sea Res Part I-Oceanogr Res Pap 44:2117-2129

Freeman KH, Hayes JM (1992) Fractionation of carbon isotopes by phytoplankton and estimates of ancient CO_2 levels. Global Biogeochem Cycles 6:185-198

Freeman KH, Colarusso LA (2001) Molecular and isotopic records of C4 grassland expansion in the Late Miocene. Geochim Cosmochim Acta 65:1439-1454

Freeman KH, Wakeham SG, Hayes JM (1994) Predictive isotopic biogeochemistry: hydrocarbons from anoxic marine basins. Organic Geochem 21:629-644

Fry B, Sherr EB (1984) $\delta^{13}C$ measurements as indicators of carbon flow in marine and freshwater ecosystems. Contrib Marine Science 27:13-47

Fry B, Sherr EB (1988) $\delta^{13}C$ Measurements as Indicators of Carbon Flow in Marine and Freshwater Ecosystems. *In* Rundel PW, Ehleringer JR, Nagy KA (eds) Stable Isotopes in Ecological Research. Springer-Verlag, New York, p 196-229

Fry B, Wainright SC (1991) Diatom sources of ^{13}C-rich carbon in marine food webs. Mar Ecol Prog Ser 76:149-157

Gannes LZ, O'Brien DM, del Rio CM (1997) Stable isotopes in animal ecology: assumptions, caveats, and a call for more laboratory experiments. Ecology 78:1271-1276

Gehlen M, Mucci A, Boudreau B (1999) Modeling the distribution of stable carbon isotopes in porewaters of deep-sea sediments. Geochim Cosmochim Acta 63:2763-2773

Gelin F, Kok MD, De Leeuw JW, Damste JSS (1998) Laboratory sulfurisation of the marine microalga Nannochloropsis salina. Organic Geochem 29:1837-1848

Gibson JAE, Trull T, Nichols PD, Summons RE, McMinn A (1999) Sedimentation of C-13-rich organic matter from Antarctic sea-ice algae: A potential indicator of past sea-ice extent. Geology 27:331-334

Gleitz M, Kukert H, Riebesell U, Dieckmann GS (1996) Carbon acquisition and growth of Antarctic sea ice diatoms in closed bottle incubations. Marine Ecology—Progr Ser 135:169-177

Goericke R (1994) Variations of marine plankton $\delta^{13}C$ with latitude, temperature, and dissolved CO_2 in the world ocean. Global Biogeochem Cycle 8:85-90

Goericke R, Montoya JP, Fry B (1994) Physiology of isotope fractionation in algae and cyanobacteria. In Lajtha K, Michener B (eds) Stable Isotopes in Ecology. Blackwell Scientific, Boston, p 199-233

Gong C, Hollander DJ (1997) Differential contribution of bacteria to sedimentary organic matter in oxic and anoxic environments, Santa Monica Basin, California. Organic Geochem 26:545-563

Gould P, Ostrom P, Walker W (2000) Food, trophic relationships, and migration of Sooty and Short-tailed Shearwaters associated with squid and large-mesh driftnet fisheries in the North Pacific Ocean. Waterbirds 23:165-186

Grice K, Breteler WCMK, Schouten S, Grossi V, De Leeuw JW, Damste JSS (1998) Effects of zooplankton herbivory on biomarker proxy records. Paleoceanogr13:686-693

Gruber N (1998) Anthropogenic CO_2 in the Atlantic Ocean. Global Biogeochem Cycle 12:165-191

Gruber N, Keeling CD, Bacastow RB, Guenther PR, Lueker TJ, Wahlen M, Meijer HAJ, Mook WG, Stocker TF (1999) Spatiotemporal patterns of carbon-13 in the global surface oceans and the oceanic Suess effect. Global Biogeochem Cycles 13:307-335

Gruber N, Keeling CD, Stocker TF (1998) Carbon-13 constraints on the seasonal inorganic carbon budget at the BATS site in the northwestern Sargasso Sea. Deep-Sea Res Part I-Oceanogr Res Pap 45:673-717

Hare PE, Fogel ML, Stafford JTW, Mitchell AD, Hoering TC (1991) The isotopic composition of carbon and nitrogen in individual amino acids isolated from modern and fossil proteins. J Archaeological Sci 18:277-292

Hartgers ES, Damste JSS, Requejo AG, Allan J, Hayes JM, De Leeuw JW (1994a) Evidence for a small bacterial contribution to sedimentary organic carbon. Nature 369:224-227

Harvey GR, Boran DA, Chesal LA, Tokar JM (1983) The structure of marine fulvic and humic acids. Marine Chem 12:119-132

Harvey HR, Macko SA (1997) Kinetics of phytoplankton decay during simulated sedimentation: changes in lipids under oxic and anoxic conditions. Organic Geochem 27:129-140

Harvey HR, Macko SA (1997a) Catalysts or contributors? Tracking bacterial mediation of early diagenesis in the marine water column. Organic Geochem 26:531-544

Hayes JM (1993) Factors controlling ^{13}C contents of sedimentary organic compounds: Principles and evidence. Marine Geol 113:111-125

Hayes JM, Popp BN, Takigiku R, Johnson MW (1989) An isotopic study of biogeochemical relationships between carbonates and organic carbon in the Greenhorn Formation. Geochim Cosmochim Acta 53:2961-2972

Hayes JM, Strauss H, Kaufman AJ (1999) The abundance of C-13 in marine organic matter and isotopic fractionation in the global biogeochemical cycle of carbon during the past 800 Ma. Chem Geol 161:103-125

Hedges JI, Eglinton G, Hatcher PG, Kirchman DL, Arnosti C, Derenne S, Evershed RP, Kogel-Knabner I, De Leeuw JW, Littke R, MichaelisW, Rullkotter J (2000) The molecularly-uncharacterized component of nonliving organic matter in natural environments. Organic Geochem 31:945-958

Hedges JI, Keil RG (1995a) Sedimentary organic matter preservation: An assessment and speculative synthesis. Mar Chem 49:81-115

Hedges JI, Keil RG (1995b) Sedimentary organic-matter preservation—An assessment and speculative synthesis—closing comment. Mar Chem 49:137-139

Hesselbo SP, Grocke DR, Jenkyns HC, Bjerrum CJ, Farrimond P, Bell HSM, Green OR (2000) Massive dissociation of gas hydrate during a Jurassic oceanic anoxic event. Nature 406:392-357

Hinrichs K, Hayes JM, Sylva SP, Brewer PG, DeLong EF (1999) Methane-consuming archaebacteria in marine sediments. Nature 398:802-805

Hinrichs K, Summons RE, Orphan V, Sylva SP, Hayes JM (2000) Molecular and isotopic analyses of anaerobic methane-oxidizing communities in marine sediments. Organic Geochem 31:1685-1701

Hobson KA (1999) Tracing origins and migration of wildlife using stable isotopes: a review. Oceologia 120:314-326

Hobson KA, Alisauskas RT, Clark RG (1993) Stable-nitrogen isotope enrichment in avian tissues due to fasting and nutritional stress: implications for isotopic analysis of diet. Condor 95:388-394

Hobson KA, Schell DM (1998) Stable carbon and nitrogen isotope patterns in baleen from eastern Arctic bowhead whales (Balaena mysticetus). Can J Fish Aquat Sci 55:2601-2607

Hoehler TM, Alperin MJ, Albert DB, Martens CS (1994) Field and laboratory studies of methane oxidation in an anoxic marine sediment: Evidence for a methanogen-sulfate reducer consortium. Global Biogeochem Cycle 8:451-463

Hoering TC (1965) The extractable organic matter in Precambrian rocks and the problem of contamination. Carnegie Inst Wash Yearbook 64:215-218

Hoering TC (1966) Criteria for suitable rocks in Precambrian organic geochemistry. Carnegie Inst Wash Yearb 65:365-372

Hoering TC (1973) A comparison of melanoidin and humic acid. Carnegie Inst Wash Yearbook 72:682-690

Hofmann M, Broecker WS, Lynch-Stieglitz J (1999) Influence of a $[CO_2(aq)]$-dependent biological C-isotope fractionation on glacial $^{13}C/^{12}C$ ratios in the ocean. Global Biogeochem Cycles 13:873-883

Hofmann M, Wolf-Gladrow DA, Takashashi T, Sutherland SC, Six KD, Maier-Reimer E (2000) Stable carbon isotope distribution of particulate organic matter in the ocean: a model study. Mar Chem 72:131-150

Holmden C, Creaser RA, Muehlenbachs K, Leslie SA, Bergstrom SM (1998) Isotopic evidence for geochemical decoupling between ancient epeiric seas and bordering oceans: Implications for secular curves. Geology 26:567-570

Hsieh HL, Kao WY, Chen CP, Liu PJ (2000) Detrital flows through the feeding pathway of the oyster (Crassostrea gigas) in a tropical shallow lagoon: delta C-13 signals. Marine Biol 136:677-684

Huang YS, Freeman KH, Eglinton G, Street-Perrott FA (1999) delta C-13 analyses of individual lignin phenols in Quaternary lake sediments: A novel proxy for deciphering past terrestrial vegetation changes. Geology 27:471-474

Hurd DC, Spencer DW (1991) Marine Particles: Analysis and Characterization. American Geophysical Union, Washington, DC

Jahnke RA, Craven DB, McCorkle DC, Reimers CE (1997) $CaCO_3$ dissolution in California continental margin sediments: The influence of organic matter remineralization. Geochim Cosmochim Acta 61:3587-3604

Jahnke RA, Summons RE, Dowling LM, Zahiralis KD (1996) Identification of methanotrophic lipid biomarkers in cold-seep Mussel gills: chemical and isotopic analysis. Appl Environ Microbiol 61:676-682

Jasper JP, Hayes JM (1990) A carbon-isotopic record of CO_2 levels during the Later Quaternary. Nature 347:462-464

Jasper JP, Hayes JM (1993) Refined estimation of marine and terrigenous contributions to sedimentary organic-carbon. Global Biogeochem Cycle 7:451-461

Jasper JP, Hayes JM, Mix AC, Prahl FG (1994) Photosynthetic fractionation of C-13 and concentrations of dissolved CO_2 in the central equatorial Pacific. Paleoceanogr 9:781-798

Joos F, Bruno M (1998) Long-term variability of the terrestrial and oceanic carbon sinks and the budgets of the carbon isotopes C-13 and C-14. Global Biogeochem Cycle 12:277-295

Keller K, Morel FMM (1999) A model of carbon isotopic fractionation and active carbon uptake in phytoplankton. Marine Ecology—Progr Ser 182:295-298

Kenig F, Damste JSS, Deleeuw JW, Hayes JM (1994) Molecular Paleontological Evidence for Food-Web Relationships. Naturwiss 81:128-130

Kennett JP, Fackler-Adams BN (2000) Relationship of clathrate instability to sediment deformation in the upper Neogene of California. Geology 28:215-218

Koch PL, Fogel ML, Tuross N (1994) Tracing the diet of fossil animals using stable isotopes. *In* Lajtha K, Michener RH (eds) Stable Isotopes in Ecology and Environmental Science. Blackwell Scientific, New York, p 63-92

Kohnen M, Schouten S, Damste JSS, Deleeuw JW, Merritt DA, Hayes JM (1992) Recognition of Paleobiochemicals by a combined molecular sulfur and isotope geochemical approach. Science 256:358-360

Korb RE, Raven JA, Johnston AM (1998) Relationship between aqueous CO_2 concentration and stable carbon isotope discrimination in the diatoms Chaetoceros calcitrans and Ditylum brightwellii. Marine Ecology-Progress Series 171:303-305

Korb RE, Raven JA, Johnston AM, Leftley JW (1996) Effects of cell size and specific growth rate on stable carbon isotope discrimination by two species of marine diatom. Marine Ecology—Progr Ser 143:283-288

Krueger HW, Sullivan CH (1984) Models for carbon isotope fractionation between diet and bone. *In* Turland T, Johnson PE (eds) Stable Isotopes in Nutrition. American Chemical Society, Washington, DC

Kump LR, Arthur MA (1999) Interpreting carbon-isotope excursions: carbonates and organic matter. Chem Geol 161:181-198

Largeau C, Derenne S, Casadevall E, Kadouri A, Sellier N (1986) Pyrolysis of immature torbanite and of the resistant biopolymer (PRB A) from extant alga. Organic Geochem 10:1023-1032

Laws EA, Bidigare RR, Popp BN (1997) Effect of growth rate and CO_2 concentration on carbon isotopic fractionation by the marine diatom Phaeodactylum tricornutum. Limnology Oceanogr 42:1552-1560

Laws EA, Popp BN, Bidigare RR, Kennicutt MC, Macko SA (1995) Dependence of phytoplankton carbon isotopic composition on growth-rate and $[CO_2](AQ)$—Theoretical considerations and experimental results. Geochim Cosmochim Acta 59:1131-1138

Laws EA, Thompson PA, Popp BN, Bidigare RR (1998) Sources of inorganic carbon for marine microalgal photosynthesis: A reassessment of delta C-13 data from culture studies of Thalassiosira pseudonana and Emiliania huxleyi. Limnology Oceanogr 43:136-142

Lenton TM, Watson AJ (2000) Redfield revisited 1. Regulation of nitrate, phosphate, and oxygen in the ocean. Global Biogeochem Cycles 14:225-248

Lerperger M, McNichol AP, Peden J, Gagon AR, Elder KL, Kutschera W, Rom W, Steier P (2000) Oceanic uptake of CO_2 re-estimated through delta C-13 in WOCE samples. Nucl Instrum Methods Phys Res Sect B-Beam Interact Mater Atoms 172:501-512

Logan GA, Hayes JM, Hieshima GB, Summons RE (1995) Terminal Proterozoic reorganization of biogeochemical cycles. Nature 376:53-56

Logan GA, Summons RE, Hayes JM (1997) An isotopic biogeochemical study of Neoproterozoic and Early Cambrian sediments from the Centralian Superbasin, Australia. Geochim Cosmochim Acta 61:5391-5409

Loughlin TR (1998) The Steller sea lion: A declining species. Biosphere Conserv 1:91-98

Lueker TJ, Dickson AG, Keeling CD (2000) Ocean pCO_2 calculated from dissolved inorganic carbon, alkalinity, and equations for K-1 and K-2: validation based on laboratory measurements of CO_2 in gas and seawater at equilibrium. Marine Chem 70:105-119

MacAvoy SE, Macko SA, McIninch SP (2000) Marine nutrient contributions to freshwater apex predators. Oecologia 122:568-573

Macko SA, Fogel-Estep ML, Engel MH, Hare PE (1986) Kinetic fractionation of nitrogen isotopes during aminio acid transamination. Geochim Cosmochim Acta 50:2143-2146

Macko SA, Fogel-Estep ML, Engel MH, Hare PE (1987) Isotopic fractionation of nitrogen and carbon in the synthesis of aminio acids by microorganisms. Chem Geol 65:79-92

Macko SA, Lee WY, Parkere PL (1982) Nitrogen and carbon fractionation by two species of marine amphipods: laboratory and field studies. J Experimental Marine Biology Ecology 63:145-149

MaierReimer E (1996) Dynamic vs apparent Redfield ratio in the oceans: A case for 3-D models. J Mar Syst 9:113-120

Martin WR, McCorkle DC (1993) dissolved organic carbon concentrations in marine porewaters determined by high-temperature oxidation. Limnology Oceanogr 38:1464-1480

McCave IN (1975) Vertical flux of particles in the ocean. Deep-sea Res 22:491-502

McCorkle DC, Emerson SR, Quay PD (1985) Stable carbon isotopes in marine porewaters. Earth Planet Sci Letters 74:13-26

Meyers PA (1994) Preservation of elemental and isotopic source identification of sedimentary organic matter. Chem Geol 114:289-302

Monson KD, Hayes JM (1980) Biosynthetic control of the natural abundance of carbon-13 at specific positions within fatty acids in Escherichia coli. J Biol Chem 255:11435-11441

Monson KD, Hayes JM (1982a) Carbon isotopic fractionation in the biosynthesis of bacterial fatty acids. Ozonolysis of unsaturated fatty acids as a means of determining the intramolecular distribution of carbon isotopes. Geochim Cosmochim Acta 46:139-149

Monson KD, Hayes JM (1982b) Biosynthetic control of carbon-13 at specific positions within fatty acids in Saccharomyces cerevisiae. Isotope fractionations in lipid synthesis as evidence for peroxisomal regulation. J Biol Chem 257:5568-5575

Morrison JM, Codispoti LA, Gaurin S, Jones B, Manghnani V, Zheng Z (1998) Seasonal variation of hydrographic and nutrient fields during the US JGOFS Arabian Sea Process Study. Deep-Sea Res Part II-Top Stud Oceanogr 45:2053-2101

Murnane RJ, Sarmiento JL (2000) Roles of biology and gas exchange in determining the delta C-13 distribution in the ocean and the preindustrial gradient in atmospheric delta C-13. Global Biogeochem Cycle 14:389-405

Murrell JC, McDonald IR, Bourne DG (1998) Molecular methods for the study of methanotroph ecology. FEMS Microbiol Ecology 27:103-114

Ortiz JD (2000) Anthropogenic CO_2 invasion into the northeast Pacific based on concurrent delta C-13 (DIC) and nutrient profiles from the California Current. Global Biogeochem Cycles 14:917-929

Pagani M, Arthur MA, Freeman KH (1999a) Miocene evolution of atmospheric carbon dioxide. Paleoceanogr 14:273-292

Pagani M, Arthur MA, Freeman KH (2000a) Variations in Miocene phytoplankton growth rates in the southwest Atlantic: Evidence for changes in ocean circulation. Paleoceanogr 15:486-496

Pagani M, Freeman KH, Arthur MA (1999b) Late Miocene atmospheric CO_2 concentrations and the expansion of C-4 grasses. Science 285:876-879

Pagani M, Freeman KH, Arthur MA (2000b) Isotope analyses of molecular and total organic carbon from Miocene sediments. Geochim Cosmochim Acta 64:37-49

Pahlow M, Riebesell U (2000) Temporal trends in deep ocean Redfield ratios. Science 287:831-833

Pancost R, Damste JSS, de Lint S, van der Maarel EC, Gottschal JC, Party MS (2000) Biomarker evidence for widespread anaerobic methane oxidation in Mediterranean sediments by a consortium of methanogenic archaea and bacteria. Appl Environ Microbiol 66:1126-1132

Pancost R, Freeman KH, Patzkowsky ME (1999) Organic-matter source variation and the expression of a late Middle Ordovician carbon isotope excursion. Geology 27:1015-1018

Pancost R, Freeman KH, Patzkowsky ME, Wavrek DA, Collister JW (1998) Molecular indicators of redox and marine photoautotroph composition in the late Middle Ordovician of Iowa, U.S.A. Organic Geochem 29:1649-1662

Parkes RJ, Cragg BA, Getliff JM, Harvey SM, Fry JC, Lewis CA, Rowland SJ (1993) A quantitative study of microbial decomposition of biopolymers in Recent sediments from the Peru Margin. Marine Geol 113:55-66

Patterson WP, Walter WL (1994) Depletion of C-13 in seawater sigma-CO_2 on modern carbonate platforms—significance for the carbon isotopic record of carbonates. Geology 22:885-888

Patzkowsky ME, Slupik LM, Arthur MA, Pancost R, Freeman KH (1997) Late Middle Ordovician environmental change and extinction: Harbinger of the Late Ordovician or continuation of Cambrian patterns? Geology 25:911-914

Pearson PN, Palmer MR (2000) Atmospheric carbon dioxide concentrations over the past 60 million years. Nature 406:695-699

Pinnegar JK, Polunin NVC (2000) Contributions of stable-isotope data to elucidating food webs of Mediterranean rocky littoral fishes. Oecologia 122:399-409

Popp BN, Laws EA, Bidigare RR, Dore JE, Hanson KL, Wakeham SG (1998) Effect of phytoplankton cell geometry on carbon isotopic fractionation. Geochim Cosmochim Acta 62:69-77

Popp BN, Parekh NR, Tilbrook B, Bidigare RR, Laws EA (1997) Organic carbon delta C-13 variations in sedimentary rocks and chemostratigraphic and paleoenvironmental tools. Palaeogeogr Palaeoclimatol Palaeoecology 132:119-132

Popp BN, Takigiku R, Hayes JM, Louda JW, Baker EW (1989) The post-Paleozoic chronology and mechanism of ^{13}C depletion in primary marine organic matter. Am J Sci 289:436-454

Popp BN, Trull T, Kenig F, Wakeham SG, Rust TM, Tilbrook B, Griffiths FB, Wright SW, Marchant HJ, Bidigare RR, Laws EA (1999) Controls on the carbon isotopic composition of Southern Ocean phytoplankton. Global Biogeochem Cycles 13:827-843

Prahl FG, De Lange GJ, Scholten S, Cowie GL (1997) A case of post-depositional aerobic degradation of terrestrial organic matter in turbidite deposits from the Madeira Abyssal Plain. Organic Geochem 27:141-152

Radajewski S, Ineson P, Parekh NR, Murrell JC (2000) Stable-isotope probing as a tool in microbial ecology. Nature 403:646-649

Rau GH (1994) Variations in sedimentary organic $\delta^{13}C$ as a proxy for past changes in ocean and atmospheric CO_2 concentrations. *In* R Zahn, TF Pedersen, MA Kaminski, L Labeyrie (eds) Carbon Cycling in the Glacial Ocean: Constraints on the Ocean's Role in Global Change. Springer-Verlag, Berlin-Heidelberg, p 307-321

Rau GH, Chavez FP, Friederich GE (2001) Plankton $^{13}C/^{12}C$ variations in Monterey Bay, California: evidence of non-diffusive inorganic carbon uptake by phytoplankton in an upwelling environment. Deep-Sea Res Part I-Oceanog Res Papers 48:79-94

Rau GH, Froelich PN, Takashashi T, Des-Marais DJD (1991) Does sedimentary organic $\delta^{13}C$ record variation in quaternary ocean [CO_2(aq)]? Paleoceanogr 6:335-347

Rau GH, Riebesell U, Wolf-Gladrow DA (1996) A model of photosynthetic C-13 fractionation by marine phytoplankton based on diffusive molecular CO_2 uptake. Marine Ecology—Progr Ser 133:275-285

Rau GH, Riebesell U, Wolf-Gladrow DA (1997) CO_2(aq)-dependent photosynthetic C-13 fractionation in the ocean: A model versus measurements. Global Biogeochem Cycle 11:267-278

Rau GH, Takashashi T, Des-Marais DJD (1989) Latitudinal variations in plankton $\delta^{13}C$: implications for CO_2 and productivity in past oceans. Nature 341:518-519

Rau GH, Takashashi T, Marais DJD, Repeta DJ, Martin J (1992) The relationship between organic matter $\delta^{13}C$ and [CO_2(aq)] in ocean surface water: Data from JGOFS site in Northeast Atlantic Ocean and a Model. Geochim Cosmochim Acta 56:1413-1419

Raven JA (1997) Inorganic carbon acquisition by marine autotrophs. Adv Botan Res 27:85-209

Redfield A, Ketchum BH, Richards FA (1963) The influence of organisms on the composition of seawater. *In* Hill MN (ed) The Sea. Interscience, New York, p 1-34

Redfield AC (1934) On the proportions of organic derivatives in seawater and their relation to the composition of plankton. James Johnson Memorial Volume, p 176-192

Reinfelder JR, Kraepiel AML, Morel FMM (2000) Unicellular C-4 photosynthesis in a marine diatom. Nature 407:996-999

Riboulleau A, Derenne S, Sarret G, Baudin F, Connan J (2000) Pyrolytic and spectroscopic study of a sulphur-rich kerogen from the 'Kashpir oil shales' (Upper Jurassic, Russian platform). Organic Geochem 31:1641-1661

Riebesell U, Burkhardt S, Dauelsberg A, Kroon B (2000a) Carbon isotope fractionation by a marine diatom: dependence on the growth-rate-limiting resource. Marine Ecology—Progr Ser 193:295-303

Riebesell U, Revill AT, Holdsworth DG, Volkman JK (2000b) The effects of varying CO_2 concentration on lipid composition and carbon isotope fractionation in Emiliania huxleyi. Geochim Cosmochim Acta 64:4179-4192

Riebesell U, Zondervan I, Rost B, Tortell PD, Zeebe RE, Morel FMM (2000c) Reduced calcification of marine plankton in response to increased atmospheric CO_2. Nature 407:364-367

Rios AF, Fraga F, Figueiras FG, Perez FF (1998) A modelling approach to the Redfield ratio deviations in the ocean. Sci Mar 62:169-176

Rosenthal Y, Dahan M, Shemesh A (2000) Southern Ocean contributions to glacial-interglacial changes of atmospheric pCO(2): An assessment of carbon isotope records in diatoms. Paleoceanogr 15:65-75

Sachs JP, Repeta DJ, Goericke R (1999) Nitrogen and carbon isotopic ratios of chlorophyll from marine phytoplankton. Geochim Cosmochim Acta 63:1431-1441

Sackett WM, Eckelmann WR, Bender ML, Be AWH (1965) Temperature-dependence of carbon isotope composition in marine plankton and sediments. Science 148:235-237

Sakata S, Hayes JM, McTaggart AR, Evans RA, Leckrone KJ, Togasaki RK (1997) Carbon isotopic fractionation associated with lipid biosynthesis by a cyanobacterium: Relevance for interpretation of biomarker records. Geochim Cosmochim Acta 61:5379-5389

Sayles FL, Curry WB (1988) $\delta^{13}C$, TCO_2, and the metabolism of organism carbon in deep-sea sediments. Geochim Cosmochim Acta 52:2963-2978

Schell DM (2000) Declining carrying capacity in the Bering Sea: Isotopic evidence from whale baleen. Limnology and Oceanography 45:459-462

Schell DM, Saupe SM, Haubenstock N (1989) Bowhead whale (*Balaena mysticetus*) growth and feeding as estimated by $\delta^{13}C$ techniques. Mar Biol 103:433-443

Schlunz B, Schneider RR, Muller PJ, Showers WJ, Wefer G (1999) Terrestrial organic carbon accumulation on the Amazon deep sea fan during the last glacial sea level low stand. Chem Geol 159:263-281

Schouten S, Breteler WCMK, Blokker P, Schogt N, Rijpstra WIC, Grice K, Baas M, Damste JSS (1998) Biosynthetic effects on the stable carbon isotopic compositions of algal lipids: Implications for deciphering the carbon isotopic biomarker record. Geochim Cosmoschim Acta 62:1397-1406

Schwarcz HP (1991) Some theoretical aspects of isotope paleodiet studies. J Archaeol Sci 18:261-275

Sininghe-Damste JSS, Kok MD, Koster J, Schouten S (1998) Sulfurized carbohydrates: an important sedimentary sink. Earth Planet Sci Letters 164:7-13

Sininghe-Damste JSS, Schouten S (1997) Is there evidence for a substantial contribution of prokaryotic biomass to organic carbon in Phanerozoic carbonaceous sediments? Organic Geochem 26:517-530

Sonnerup RE, Quay PD, McNichol AP (2000) The Indian Ocean C-13 Suess effect. Global Biogeochem Cycles 14:903-916

Sonnerup RE, Quay PD, McNichol AP, Bullister JL, Westby TA, Anderson HL (1999) Reconstructing the oceanic ^{13}C Suess effect. Global Biogeochem Cycles 13:857-872

Spiker EC, Hatcher PG (1984) Carbon isotope fractionation of sapropelic organic matter during early diagenesis. Organic Geochem 5:283-290

Suess E (1980) Particulate organic carbon flux in the oceans—surface productivity and oxygen utilization. Nature 288:260-263

Takashashi Y, Matsumoto E, Watanabe YW (2000) The distribution of delta C-13 in total dissolved inorganic carbon in the central North Pacific Ocean along 175 degrees E and implications of anthropogenic CO_2 penetration. Mar Chem 69:237-251

Thomson DH (1987) Energetics of bowheads. *In* Richardson J (ed) Importance of the eastern Alaskan Beaufort Sea to feeding bowhead whales. U S Minerals Management Service, p 417-448

Tieszen LL, Fagre T (1993) Effect of diet quality and composition on the isotopic composition of respiratory CO_2, bone collagen, bioapatite, and soft tissues. *In* Lambert J, Grupe G (eds) Molecular Archaeology of Prehistoric Human Bone. Springer-Verlag, Berlin, p 123-135

Tissot BP, Welte DH (1978) Petroleum Formation and Occurence. Springer-Verlag, New York

Tortell PD, Rau GH, Morel FMM (2000) Inorganic carbon acquisition in coastal Pacific phytoplankton communities. Limnology Oceanogr 45:1485-1500

Van Kaam-Peters HME, Schouten S, Koster J, Damste JSS (1998) Controls on the molecular and carbon isotopic composition of organic matter deposited in a Kimmeridgian euxinic shelf sea: Evidence for preservation of carbohydrates through sulfurisation. Geochim Cosmoschim Acta 62:3259-3283

Villinski JC, Dunbar RB, Mucciarone DA (2000) Carbon-13 Carbon-12 ratios of sedimentary organic matter from the Ross Sea, Antarctica: A record of phytoplankton bloom dynamics. J Geophys Res—Oceans 105:14163-14172

Volkman JK, Rohjans D, Rullkotter J, Scholz-Bottcher BM, Liebezeit G (2000) Sources and diagenesis of organic matter in tidal flat sediments from the German Wadden Sea. Continental Shelf Res 20: 1139-1158

Wakeham SG, Lee C, Hedges JI, Hernes PJ, Peterson ML (1997) Molecular indicators of diagenetic status in marine organic matter. Geochim Cosmoschim Acta 61:5363-5369

Wang XC, Druffel ERM, Griffin S, Lee C, Kashgarian M (1998) Radiocarbon studies of organic compound classes in plankton and sediment of the northeastern Pacific Ocean. Geochim Cosmoschim Acta 62:1365-1378

Weinstein MP, Litvin SY, Bosley KL, Fuller CM, Wainright SC (2000) The role of tidal salt marsh as an energy source for marine transient and resident finfishes: A stable isotope approach. Trans Am Fish Soc 129:797-810

Westrich JT, Berner RA (1984) The role of sedimentary organic matter in bacterial sulfate reduction: The *G* model tested. Limnology Oceanogr 29:236-249

White DC (1983) Analysis of microorganisms in terms of quantity and activity in natural environments. Symp Soc Gen Microbiol 34:37-66

Whiticar MJ (1999) Carbon and hydrogen isotope systematics of bacterial formation and oxidation of methane. Chem Geol 161:291-314

Wolf-Gladrow DA, Riebesell U (1997) Diffusion and reactions in the vicinity of plankton: A refined model for inorganic carbon transport. Mar Chem 59:17-34

Wolf-Gladrow DA, Riebesell U, Burkhardt S, Bijma J (1999) Direct effects of CO_2 concentration on growth and isotopic composition of marine plankton. Tellus 51b:461-476

Wu JP, Calvert SE, Wong CS, Whitney F (1999) Carbon and nitrogen isotopic composition of sedimenting particulate material at Station Papa in the subarctic northeast Pacific. Deep-Sea Res Part II-Top Stud Oceanogr 46:2793-2832

Zahn R, Keir R (1994) Tracer-nutrient correlations in the upper ocean: Observational and box model constraints on the use of Benthic foraminiferal $\delta^{13}C$ and Cd/Ca as Paleo-proxies for the intermediate-depth ocean. *In* R Zahn, TF Pedersen, MA Kaminski, L Labeyrie (ed) Carbon Cycling in the Glacial Ocean: Constraints on the Ocean's Role in Global Change. Springer-Verlag, Berlin-Heidelberg, p 195-221

Zegouagh Y, Derenne S, Largeau C, Bertrand P, Sicre MA, Saliot A (1999) Refractory organic matter in sediments from the North-West African upwelling system: abundance, chemical structure and origin. Organic Geochem 30:101-117

Zyakun AM, Zakharchenko VN, Ivanovskii RN, Keppen OI, Kudryavtseva AI, Mashkina LP, Peshenko VP (1998) Fractionation of carbon isotopes by *Ectothiorhodospira shaposhnikovii* during photomixotrophic growth. Microbiol 67:1-6

Biogeochemistry of Sulfur Isotopes

D. E. Canfield

Danish Center for Earth System Science (DCESS)
and Institute of Biology
Odense University, SDU, Campusvej 55
5230 Odense M, Denmark

INTRODUCTION

Sulfur, with an atomic weight of 32.06, has four stable isotopes. By far the most abundant is ^{32}S, representing around 95% of the total sulfur on Earth. The next most abundant isotope is ^{34}S, followed by ^{33}S, and finally ^{36}S is the least abundant contributing only 0.0136% to the total (Table 1). The natural abundances of sulfur isotopes, however, vary from these values as a result of biological and inorganic reactions involving the chemical transformation of sulfur compounds. For thermodynamic reasons, the relative abundance of sulfur isotopes can vary between coexisting sulfur phases. This is because lighter masses partition more of the total bond energy into vibrational rather than translational modes. Bonds with a higher vibrational energy are also more easily broken which is why lighter isotopes are generally enriched in the reaction products in chemical reactions with associated fractionation. Thus, for a nonreversible chemical reaction, as often occurs in biological systems, independent reactions may be written for the transformation of the light, L, and heavy, H, isotopes of reactant, A, to product, B (Eqns. 1 and 2).

$$A_H \overset{k_H}{\rightarrow} B_H \tag{1}$$

$$A_L \overset{k_L}{\rightarrow} B_L \tag{2}$$

Each of these reactions has associated rate constants, k_H and k_L, and as described above, k_H is generally less than k_L, yielding an enrichment of the lighter isotope in the product. Fractionations associated with a unidirectional process are referred to as kinetic fractionations.

Fractionations can also occur between two chemical species at equilibrium. The basis for equilibrium fractionations is thermodynamic and, as with kinetic fractionations, is related to mass-dependent differences in bond energies between light and heavy isotopes. The generalized isotope equilibrium between two chemical species is presented in Equation (3).

$$x\,A_L + y\,B_H \leftrightarrow x\,A_H + y\,B_L \tag{3}$$

Table 1. Natural abundance of stable sulfur isotopes[a]

Isotope	Abundance (%)
^{34}S	4.22
^{32}S	95.02
^{33}S	0.760
^{36}S	0.0136

[a] for Cañon Diablo troilite

1529-6466/00/0043-0012$05.00

From Equation (3) an equilibrium constant, K_{eq}, may be defined as:

$$K_{eq} = (A_H/A_L)^x/(B_H/B_L)^y \qquad (4)$$

If x and y are unity, then K_{eq} is identical to the fractionation factor $\alpha_{(A-B)}$, which is also defined as:

$$\alpha_{(A-B)} = R_A/R_B \qquad (5)$$

where $R_A = A_H/A_L$, and $R_B = B_H/B_L$. For sulfur, ratios are defined relative to ^{32}S, which is also the most abundant isotope (Table 1). Normally, ^{34}S is the heavy isotope ratioed against ^{32}S, so that, in the above example, $R_A = (^{34}S/^{32}S)_A$ and $R_B = (^{34}S/^{32}S)_B$. Isotope ratios, however, can also be expressed with any of the other heavy, minor, isotopes.

Isotopic compositions are expressed as per mil (‰) differences relative to the isotopic composition of the Cañon Diablo Troilite standard, and are presented in standard δ notation.

$$\delta^{34}S = [\{(^{34}S/^{32}S)_{sam}/(^{34}S/^{32}S)_{std}\} - 1] \times 1000 \qquad (6)$$

If heavy isotopes of other than ^{34}S are considered, then ^{34}S is replaced by the isotope of interest in Equation (6).

Fractionations are expressed, exactly, in terms of ε, with units of ‰:

$$\varepsilon_{A-B} = 1000 \times (\alpha_{(A-B)} - 1) \qquad (7)$$

For convenience, fractionations are also often expressed simply as the isotope difference, Δ, between two compounds of interest (Eqn. 8). Fractionations expressed in this way are approximately equivalent to ε as defined in Equation (7).

$$\Delta_{A-B} = \delta^{34}S_A - \delta^{34}S_B \cong \varepsilon_{A-B} \qquad (8)$$

In all of the subsequent discussion, A will be taken as the reactant species, and B as the product, in a given chemical reaction.

Microorganisms have long been known to impart fractionations during their sulfur metabolism, particularly during, but not restricted to, the process of sulfate reduction (Thode et al. 1951; Jones and Starkey 1957; Harrison and Thode 1958; Kaplan and Rittenberg 1964). Therefore, sulfur isotope studies have been useful in unraveling the processes of sulfur cycling in anoxic sediments (e.g. Kaplan et al. 1963; Goldhaber and Kaplan 1980; Habicht and Canfield, 2001). Furthermore, sulfur isotopes have been invaluable in exploring the early evolution of sulfur metabolism on Earth (e.g. Schidlowski et al. 1983; Shen et al. 2001), as well as the early history of Earth surface oxidation (Cameron 1982; Canfield et al. 2000).

As the cycling of sulfur is involved in atmospheric oxygen regulation, sulfur isotopes have also proven important in understanding the Phanerozoic history of atmospheric oxygen (Holland 1972; Garrels and Perry 1977; Berner et al. 2000). Sulfur mineralization to economic-grade ore deposits may have a biological component and sulfur isotopes provided important tool to understand the processes of sulfide-ore formation (e.g. Ohmoto and Goldhaber 1997; Shanks, this volume). Recently, a consideration of the minor isotopes of sulfur, ^{33}S and ^{36}S, has prompted new insights into the possible importance of gas-phase reactions in sulfur cycling on the early Earth, as well as in the Martian atmosphere (Farquhar et al. 2000a,b). This chapter will explore the biogeochemistry of sulfur isotopes, highlighting new advances in our understanding of the biological processes imparting fractionation and the parameters controlling the extent of fractionation. A few examples will also be provided demonstrating how sulfur isotopes can be used to unravel aspects of sulfur biogeochemical cycling.

MICROBIAL PROCESSES

Assimilatory sulfate reduction

Sulfur is an essential component of living cells, constituting around 0.5 to 1.0 % of the dry weight of prokaryotic organisms (Zehnder and Zinder 1986), being most abundant in the reduced form in the principal amino acids cysteine and methionine, key building blocks of proteins. It is also found in numerous coenzymes, vitamins and electron carriers (cofactors), crucial to cellular metabolism. Oxidized sulfur is found as sulfate esters and sulfonates in cell walls, photosynthetic membranes, and in the connective tissues of plants and animals. The sulfur found in cellular components is assimilated from the environment, most commonly as sulfate, but sometimes as sulfide, or even thiosulfate, polythionates and elemental sulfur (Le Faou et al. 1990; Thauer and Kunow 1995). When sulfate is assimilated, it must be reduced to sulfide, in a process known as assimilatory sulfate reduction, for incorporation into the principal organic sulfur compounds within cells. Assimilatory sulfate reduction is an energy-requiring process.

Assimilatory sulfate reduction begins with the active uptake of sulfate into the cell by specific sulfate-binding proteins. This processes is unidirectional and requires energy (Cypionka 1995). Following this, there are two possible pathways of sulfate reduction (Fig. 1). The APS (adenosine-5'-phosphosulfate) pathway is utilized by oxygen-producing phototrophic eukaryotes, as well as some cyanobacteria (Fig. 1). In the initial step ATP is combined with sulfate to form the intermediate APS. From here, two different routes are possible, one leading to the formation of sulfate esters and the other leading to the formation of reduced sulfur compounds. The PAPS (phosphoadenosine-5'-phosphosulfate) pathway (Fig. 1) is utilized by some by cyanobacteria, but mostly by non oxygen-producing microorganisms. In this pathway, APS formation is preceded by the formation of PAPS, leading next to sulfite, which is further reduced to sulfide in a reaction catalyzed by the enzyme sulfite reductase.

The fractionations associated with assimilatory sulfate reduction are small, with Δ values ($\delta^{34}S_{sulfate} - \delta^{34}S_{organic-S}$) for *E. coli*, the yeast, *S. cerevisiae*, and the green alga, *Ankistrodesmus* sp. in the range of 0.9 to 2.8 ‰, (Kaplan and Rittenberg 1964). In a large survey of sulfur isotope analyses of higher plants, Trust and Fry (1992) also documented small fractionations ($\delta^{34}S_{sulfate} - \delta^{34}S_{organic-S}$) of around 1.5 ‰. Significant scatter, however, of about 4 to 5 ‰ around this average was observed.

The sulfite-reductase enzyme that catalyzes assimilatory sulfate reduction is expected to have a significant isotope effect. Indeed, Kaplan and Rittenberg (1964) observed fractionations ($^{34}S_{sulfite} - {}^{34}S_{sulfide}$) of 30 to 40 ‰ during the reduction of sulfite to sulfide by *S. cerevisiae*. Presumably the small fractionations generally associated with assimilatory sulfate reduction result from unidirectional transport of sulfate into the cell. Even if internal cellular processes impart fractionation, without exchange between the internal and external sulfate pools, no net fractionation will be observed (for example, Rees 1973). Unfortunately, compound-specific sulfur isotope analytical techniques have not yet been developed. This might allow the exploration of isotope heterogeneity among specific sulfur compounds within an organism, perhaps shedding light on the internal sulfur dynamics accompanying the assimilatory sulfate reduction process.

In some instances, marsh plants and mangroves, with roots extending into sulfide-

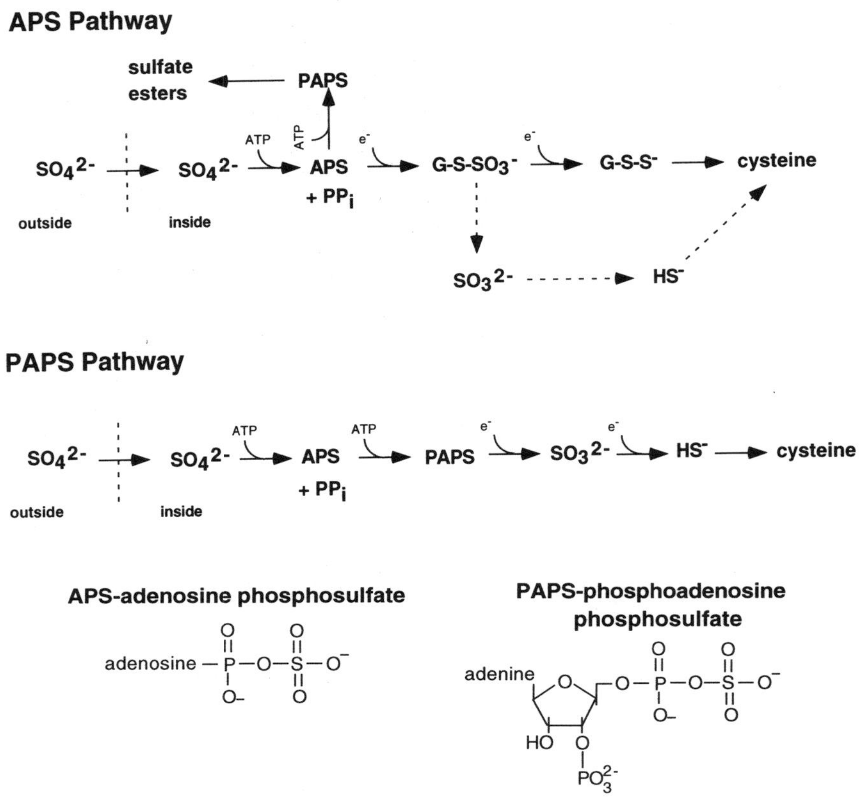

APS Pathway

PAPS Pathway

Figure 1. The most important pathways of assimilatory sulfate reduction are shown. The APS pathway is utilized by oxygen-producing, phototrophic eukaryotes and some cyanobacteria. In this pathway a bifurcation occurs where, for APS, sulfur can either be reduced to the oxidation level of cysteine or converted to sulfate-ester compounds. In the PAPS pathway, utilized by some cyanobacteria, but mostly by anoxygenic microorganisms, APS formation leads, through sulfite, to the production of reduced organic sulfur compounds.

containing sediment pore waters, contain organic sulfur with tissues considerably depleted in [34]S (see Trust and Fry 1992). For example, Stribling et al. (1998) observed *S. Alterniflora* shoots with organic sulfur δ^{34}S values as low as -5.6 ‰ , while Canfield et al. (1998b) measured mangrove leaves from Mangrove Lake, Bermuda, with a δ^{34}S of -10 ‰ . These organic sulfur compounds are 25 to 30 ‰ depleted in [34]S compared to coexisting pore water sulfate, suggesting that sulfide was a significant source of sulfur to the plants. Whether the sulfide is directly taken up by the plants, or oxidized first to sulfate in the root zone and subsequently taken up, is unclear (Trust and Fry 1992).

DISSIMILATORY SULFATE REDUCTION

Ecological and phylogenetic diversity of sulfate reducers

Dissimilatory sulfate reduction is conducted by a specialized group of prokaryotes, who gain energy for their growth by catalyzing exergonic chemical reactions in which organic carbon or H_2 (gas) is oxidized, while sulfate is reduced (Eqns. 9 and 10).

$$SO_4^{2-} + 2\ CH_2O \rightarrow H_2S + 2\ HCO_3^- \tag{9}$$

$$2\ H^+ + SO_4^{2-} + 4\ H_2 \rightarrow H_2S + 4\ H_2O \tag{10}$$

Sulfate-reducing organisms are widely distributed in anoxic environments containing sulfate and, as a group, they have a broad ecological tolerance. For example, sulfate reducers have been found active in temperatures ranging from -1.5°C (Sagemann et al. 1998), to just over 100°C (Jørgensen et al. 1990). Furthermore, sulfate reduction may be found in salinities ranging from freshwater to near halite saturation (Brandt et al. 2001).

Sulfate reducers are also phylogenetically diverse, found within several major lineages within the Bacterial Domain in the Tree of Life, but most abundant among the δ-subdivision of the proteobacteria (also known as the purple bacteria), and to a lesser extent the gram-positive bacteria (Stackebrandt et al. 1995; Widdel 1988; Castro et al. 2000). The deepest branching sulfate reducers from the Bacterial Domain are from the genus *Thermodesulfobacterium*, with optimum growth temperatures around 80°C. Within the Domain Archaea are found several members of the hyperthermophilic sulfate reducer *Archaeoglobus* sp., with maximum growth temperatures also around 80°C (Stackebrandt et al. 1995).

Principles governing the extent of fractionation

Isotope fractionation during sulfate reduction has been extensively studied, especially among mesophilic sulfate reducers (optimal growth temperatures between 20 and 40°C), and particularly for the species *Desulfovibrio desulfuricans* (for example, Harrison and Thode 1958; Kaplan and Rittenberg 1964; Kemp and Thode 1968; Chambers et al. 1975). Four long-standing observations from work on *D. desulfuricans* form the basis for our models describing the controls on isotope fractionation during sulfate reduction.

(1) When organisms utilize organic electron donors, the extent of isotope fractionation tends to respond to the specific rate of sulfate reduction (mol cell^{-1} time^{-1}), with lower specific rates generally providing higher fractionations (Fig. 2).

(2) Lower fractionations are found with H_2 as an electron donor, particularly at low specific rates of sulfate reduction (Fig. 2).

(3) Fractionations become significantly suppressed when sulfate concentrations drop below around 1 mM (Fig. 3), although the data base is small.

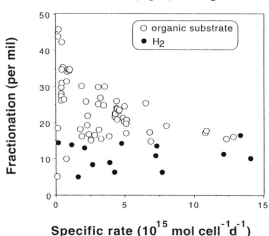

Figure 2. Fractionations are shown between sulfate and sulfide during sulfate reduction by *Desulfovibrio desulfuricans* as a function of the specific rates of sulfate reduction. Both organic substrates and H_2 are indicated as electron donors for sulfate reduction. The results are from Kaplan and Rittenberg (1964) and Chambers et al. (1975).

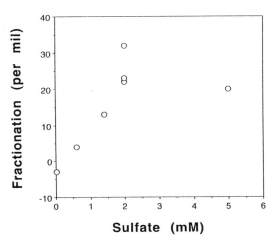

Figure 3. The extent of isotope fractionation between sulfate and sulfide during sulfate reduction at low sulfate concentrations is shown. The results are from Harrison and Thode (1958), Canfield (2001) and Canfield et al. (2000).

(4) With abundant sulfate (>1mM), fractionations range between 3 and 46 ‰ (Fig. 2), with an average around 18 ‰ (Canfield and Teske 1996).

These observations are explained by considering the basic biochemistry of the sulfate-reduction process (Eqn. 11). Sulfate reduction begins with the transport of sulfate across the cell membrane and into the cell cytoplasm, an energy-requiring process (step 1). Being a charged species, sulfate does not passively diffuse into the cell, but rather is transported together with cations to balance the charge, typically H^+ for freshwater species, and Na^+ for marine species (Cypionka 1995). Sulfate transport across the cell membrane is reversible, although considerable sulfate may concentrate within the cell, with up to 1000 times ambient sulfate levels found in freshwater sulfate reducers in a low sulfate environment (Cypionka 1995). Little isotope effect accompanies sulfate transport into the cell, and while a fractionation ($\delta^{34}S_{sulfate-out} - \delta^{34}S_{sulfate-in}$) of -3 ‰ has nominally been placed on this process (Harrison and Thode 1958; Rees 1973), a lower value is also possible.

$$\text{SO}_4^{2-} \text{ (out)} \underset{\text{Step 1}}{\rightleftarrows} \text{SO}_4^{2-} \text{ (in)} \xrightarrow[\text{Step 2}]{\text{ATP}} \text{APS} \xrightarrow[\text{Step 3}]{e^-} \text{SO}_3^{2-} \xrightarrow[\text{Step 4}]{e^-} \text{H}_2\text{S} \qquad (11)$$

Within the cell cytoplasm, sulfate is activated with ATP (adenosine triphosphate) to form the intermediate compound APS. This is a reversible reaction, as is the subsequent reduction of APS to sulfite (Postgate 1988, Akagi 1995). No isotope fractionation should accompany the activation of sulfate to form APS, however, fractionations do accompany the subsequent steps (Harrison and Thode 1958; Kaplan and Rittenberg 1964; Kemp and Thode 1968). The magnitudes of these have been explored in the classic original literature (see above references) on isotope fractionation associated with sulfate reduction. Thus, the reduction of sulfate to sulfite involves the breaking of S-O bonds and, when the reduction is conducted in the lab as an inorganic chemical process, sulfite is depleted in ^{34}S by around 22 ‰ (Harrison and Thode 1957). The possible variability in this kinetic fractionation factor and whether this result is applicable to organisms, however, remain unclear. The theoretical equilibrium fractionation between sulfate and sulfite at 20°C is of a similar magnitude, at 24 ‰ (Harrison 1957, as quoted in Kemp and Thode 1968).

The reduction of sulfite to sulfide can occur in a single step with the enzyme sulfite reductase (Eqn. 11), although indirect reduction pathways involving the intermediates thiosulfate ($S_2O_3^{2-}$) and trithionate ($S_3O_6^{2-}$) may also occur (see review in Akagi 1995). The general significance of these indirect pathways is, however, uncertain. Several studies have addressed the magnitude of isotope fractionation during sulfite reduction in resting cell cultures (Kaplan and Rittenberg 1964; Kemp and Thode 1968; Harrison and Thode 1958) or in batch cultures (Habicht et al. 1998), and in cell-free extracts of sulfate reducers (Kemp and Thode 1968). In the latter case, the cell cytoplasm is separated from the cell membrane and cell wall, and the enzymes promoting sulfite reduction are activated with ATP, with H_2 serving as an electron donor. The important point is that the organism is no longer alive or intact, and one observed the isotope effects associated specifically with the sulfite reductase enzyme (under the conditions of the experiment).

A large range of fractionations has been observed during sulfite reduction, from small fractionations ($\Delta_{sulfite-sulfide} = \delta^{34}S_{sulfite} - \delta^{34}S_{sulfide}$) of 6 ‰ with *D. salexigens* to large fractionations of 37.2 ‰ by the yeast *S. cerevisiae* (Table 2). For the yeast, however, an assimilatory sulfite reductase is used, as compared to sulfate reducers who use dissimilatory sulfite reductase (Wagner et al. 1998), and these fractionations are probably not directly relevant to isotope fractionation during dissimilatory sulfate

Table 2. Fractionation[a] associated with various steps during sulfate reduction

Step	Fractionation (‰)	reference
sulfate diffusion into cell[b]	-3 to 0	1
$SO_4^{2-} \rightarrow$ APS	0	2
$SO_4^{2-} \rightarrow SO_3^{2-}$ (experimental)	22	3
$SO_4^{2-} \leftrightarrow SO_3^{2-}$ (theoretical)	24	4
$SO_3^{2-} \rightarrow H_2S$	18.2 ± 8.2 (resting cells)	2
	9.3 ±0.8 (resting cells)	5
	11.4 ± 1.1 (resting cells)	6
	6.0 (growing cells)	7
	37.2 ± 2.5 (yeast)	5
	18.4 ± 2.2 (cell free extracts)	2

[a]fractionations are presented both as ε_{A-B} and Δ_{A-B}, where A is the reactant and B is the product.

[b]fractionation presented as Δ_{A-B}, where A is outside the cell and B is inside the cell.

1 Rees (1973)	5 Kaplan and Rittenberg (1964)
2 Kemp and Thode (1968)	6 Harrison and Thode (1958)
3 Harrison and Thode (1957)	7 Habicht et al. (1998)
4 Harrison (1957)	

reduction. Perhaps most relevant to understanding the fractionations associated with the intercellular process of sulfite reduction are the experiments on cell-free extracts, which, with a very limited data set, generate a range of fractionation ($\Delta_{sulfite-sulfide}$) with an average of around 18 ‰ (Kemp and Thode 1968; Table 2). Fractionations of 25 ‰, somewhat higher than this average, have usually been ascribed to the sulfite reduction step (Eqn. 11; step 4) (Kemp and Thode 1968; Rees 1973). Strain specific differences in the fractionation associated the sulfite reduction could occur, and determination of the magnitude of the isotope effect associated with various sulfite reductase enzymes could be a high priority item for future research.

Magnitude of fractionation during sulfate reduction by pure cultures

The maximum fractionation observed during sulfate reduction should be the sum of the maximum fractionations during the individual steps in the sulfate reduction process. Therefore, it may be no coincidence that the highest fractionation of 46 ‰ measured during sulfate reduction is comparable to the 45-50 ‰ maximum fractionation obtained by summing the likely fractionations associated with the individual sulfate-reduction steps (see Table 2). Deviations from these higher fractionations, which are common, depend, at least in part, on which step becomes rate limiting during sulfate reduction. Therefore, as observed in the original literature on fractionation during sulfate reduction (Jones and Starkey 1957; Harrison and Thode 1958; Kaplan and Rittenberg 1964), little or no fractionation will be expressed if the entry of sulfate into the cell becomes rate limiting. In this circumstance, all or most of the sulfate entering the cell will be reduced to sulfide. Thus, even with internal isotope effects, mass balance requires that they will not be expressed in the sulfide produced during sulfate reduction.

At low sulfate concentrations, the rate at which sulfate enters the cell is reduced, reducing also the rate of sulfate reduction. Sulfate in limiting supply is less exchanged back out of the cell, explaining why low sulfate concentrations reduce isotope fractionation so severely. However, the nature of the influence of sulfate concentrations on fractionation probably varies among species, particularly among freshwater and marine sulfate reducers, which have different sulfate-transport enzymes and different adaptations to low sulfate concentrations (Ingvorsen and Jørgensen 1984). The relationship between sulfate concentrations and isotope fractionation, excepting the early results of Harrison and Thode (1958), is virtually unexplored. Sulfate may also become somewhat limiting at very high specific rates of sulfate reduction if the electron donor is available in abundance. This is why reduced fractionations are also observed at high specific rates of sulfate reduction, as highlighted in the original literature (see above) on isotope fractionation during sulfate reduction.

Maximum expression of isotope fractionation will occur when the internal and external sulfate pools are in exchange equilibrium, as well as the reversible steps in the sulfate reduction process (steps 2 and 3, Eqn. 11). Under these circumstances the final step (step 4, Eqn. 11) will be rate limiting. Exchange equilibrium will be best realized when cellular metabolism is slow, or in other words, at low specific rates of sulfate reduction. This explains why high fractionations are observed at low specific rates (Fig. 2).

The generally low fractionations observed during hydrogen metabolism probably result from the ready availability of electrons through the operation of an hydrogenase enzyme. Thus, the first electron-requiring step (Eqn. 10; step 3) could be so fast that the conversion of $SO_4^{2-}{}_{in}$ to APS (Eqn. 11, step 2) becomes rate-limiting. This would tend to suppress the exchange between sulfate and sulfite, reducing fractionations both during the reduction of APS to sulfite and the subsequent reduction of sulfite to sulfide. However, experiments on isotope fractionation during sulfate reduction with hydrogen have been conducted with atmospheric pressures of hydrogen. Unknown is the extent of fractionation under environmental conditions when hydrogen is also a limiting substrate (Hoehler et al. 1998).

The principles outlined above provide a guide for understanding the controls on isotope fractionation during sulfate reduction, but they do not provide a precise means by which fractionations can be predicted. For example, a given specific rate of sulfate reduction corresponds to a broad range in fractionation, not a single value. Some of this variability could result from the metabolic state, or "well-being" of the organism. For example, when *Desulfovibrio desulfuricans* metabolizes at temperatures of less than 10°C, well outside of its optimal growth temperature of 30 to 35°C, the specific rate of sulfate reduction is greatly reduced, as would be predicted. However, fractionations are also greatly reduced, counter to predictions (Kaplan and Rittenberg 1964; Kemp and Thode 1968). At low temperatures, a stiffening of the cell membrane occurs, probably leading to reduced sulfate exchange and subsequently reduced fractionations (Canfield 2001). Furthermore, unexplained variability in fractionation occurs under what are apparently identical experimental conditions (Kaplan and Rittenberg 1964; Chambers et al. 1975). The roots of such variability are not certain, but might be ascribed to subtle differences in the exchange properties across the cell membranes, for example, or to some other unknown factor.

Also, over 100 species of sulfate reducing organisms are known, yet most of our understanding of the magnitude of isotope fractionation during sulfate reduction, and of the factors controlling it, comes from studies on a single species, *Desulfovibrio desulfuricans*. Different strains of *Desulfovibrio desulfuricans* can provide different extents of fractionation under identical circumstances (Kemp and Thode 1968) and probable, but as yet largely unexplored, differences between phylogenetically diverse species of sulfate reducers are likely. To begin to address this, Detmers et al. (2001) have measured isotope fractionation by 32 different sulfate-reducing organisms representing 28 different genera. The organisms were grown in batch cultures under optimal growth conditions. Overall, a range in fractionation from 42 ‰ to 2 ‰ was found and, generally, sulfate reducers able to completely oxidize the organic substrate (complete oxidizers) provided higher fractionations than incomplete oxidizers. *Desulfovibrio desulfuricans* is an example of an incomplete oxidizer which can oxidize lactate only to CO_2 and acetate, but not completely to CO_2 (Widdel 1988). These results also provided a few surprises. Whereas *Desulfovibrio sp.* have produced very high fractionations in resting cell suspensions and in chemostat experiments (Fig. 2), in the batch culture experiments of Detmers et al. (2001) only small fractionations of 2 to 5 ‰ were found. These differences could represent true strain specific differences in fractionation. Alternatively, they could reflect different growth conditions in the various types of experiments whose fractionation has been explored. These questions will be resolved with a systematic study of the factors influencing isotope fractionation with several phylogenetically diverse species of sulfate reducers. This should receive high priority.

Fractionation at high temperatures

Other than the suppression of isotope fractionation at low temperatures for mesophilic sulfate reducers, as discussed above, temperature has played a prominent role in the discussion of the factors influencing fractionation. Thus, for *Desulfovibrio desulfuricans,* specific rates of sulfate reduction increase with increasing temperature, and reduced fractionations down to 16-17 ‰ have been observed between 40 to 45°C (Kaplan and Rittenberg 1964). These results have been extrapolated to suggest higher specific rates should accompany fractionations reduced to near 0 ‰ at even higher temperatures (Ohmoto and Felder 1987; Ohmoto et al. 1993; Watanabe 1997). This, however, appears not to be the case. Chambers et al. (1976) observed fractionations of 14 ‰ at 55°C for the thermophilic gram positive sulfate reducer *Desulfotomaculum nigrificans*, and at 60°C a gram negative sulfate reducer was found to fractionate around 18 ‰ (Böttcher et al. 1999).

Furthermore, Canfield et al. (2000) measured the isotope fractionation associated with sulfate reduction by natural populations of sulfate reducers, metabolizing with amended organic substrate, between the temperatures of 50°C and 88°C. The natural populations were contained within sediments collected by the deep-submersible Alvin from a hydrothermal vent area in the Guaymas Basin. Fractionations generally above 10 ‰, and up to 28 ‰ , were measured (Fig. 4), with no tendency towards decreasing fractionations, except above 85°C, where reduced rates of sulfate reduction suggest temperature stress on the natural population (Canfield et al. 2000). In general, individual species of sulfate reducers can only metabolize within a relatively narrow temperature range of 20-40°C (e.g. Stetter et al. 1993; Widdel 1988; Knoblauch et al. 1999). The wide spectrum of temperatures over which sulfate reduction occurs in nature results from a variety of different organisms with overlapping temperature adaptations. From indications so far, organisms with different temperature adaptations fractionate similarly.

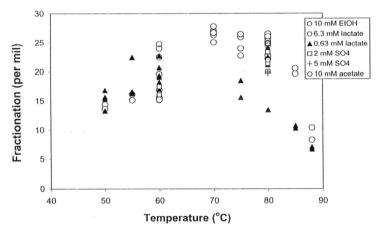

Figure 4. Shown here is the extent of isotope fractionation between sulfate and sulfide during sulfate reduction by natural populations of sulfate reducers from the Guaymas basin, Gulf of California, at high temperatures. Indicated is the type and concentration of organic substrate utilized to fuel sulfate reduction, as well as instances when sulfate concentrations were below 28 mM. Data from Canfield et al. (2000).

Isotope fractionation during sulfate reduction in nature

Anoxic environments house a variety of different sulfate reducers whose relative abundances and specific activities are difficult if not impossible to constrain. Therefore, there is no easy way to predict the expected magnitude of isotope fractionation in nature from pure culture studies. To accommodate this concern, some recent effort has been invested in exploring the isotope fractionation by natural populations of sulfate-reducing organisms metabolizing at *in situ* rates with naturally available organic substrate. In these experiments, minimally disturbed sediment samples are collected and the natural population of sulfate reducers is allowed to metabolize with the indigenous organic substrate. The sulfide produced is collected and analyzed for its isotopic composition

(Habicht and Canfield 1996, 1997, 2001; Canfield 2001). In most situations some degree of sulfate depletion occurs in the experiments, and true fractionations are obtained after correcting the observed values with a model expressing how the isotopic composition of sulfur species develops as sulfate becomes depleted. The model used depends on the type of experiment that was conducted (see below).

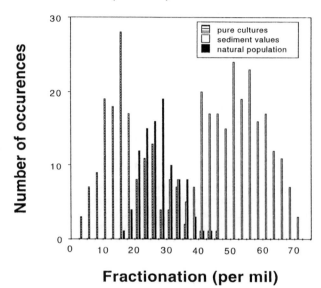

Fractionation (per mil)

Figure 5. Shown here is a compilation of the available data on the extent of fractionation during sulfate reduction by both pure cultures of sulfate-reducing organisms and natural populations metabolizing indigenous organic substrates. Also shown as a compilation of the available data on the depletion of ^{34}S ($\delta^{34}S_{seawater\ sulfate} - \delta^{34}S_{pyrite}$) into pyrite in marine sediments. Redrawn from Habicht and Canfield (1997) with additional data from Canfield (2001), and Habicht and Canfield (2001).

The results obtained so far, representing a wide spectrum of environments from rapidly metabolizing microbial mats to slowly metabolizing coastal sediments, demonstrate maximum fractionations within the same range as observed in the pure culture studies (Fig. 5). Whether this apparent upper limit to fractionation of around 45 ‰ represents a true upper limit for sulfate-reducing organisms is uncertain, and some modeling studies suggest even higher fractionations might be found in nature (Goldhaber and Kaplan 1980). However, the determination of isotope fractionations from diagenetic modeling are indirect, and studies of pure cultures and natural populations have provided no direct evidence for fractionations higher than 45 ‰.

The low fractionations frequently observed in pure bacterial cultures have not yet been seen in natural populations metabolizing under *in situ* conditions. This difference probably relates to the generally lower specific rates of sulfate reduction, with correspondingly higher fractionations, for natural populations metabolizing with substrate limitation, compared to experiments with pure bacterial cultures where organic substrate is often supplied in abundance. Indeed, Canfield (2001) found that fractionations imposed by natural populations of sulfate-reducing organisms decreased when the population was supplied with excess amounts of pure organic substrates. High rates of sulfate reduction accompanied metabolism with the amended substrate, presumably increasing the specific rates of sulfate reduction and reducing the fractionation. When the natural population metabolized pure organic substrates at limiting concentrations, rates of sulfate reduction dropped and much higher fractionations were found, approaching those observed under *in situ* conditions. In this case it is apparent that the high fractionations observed under *in situ* conditions accompanied organic matter limitation and correspondingly low specific rates of sulfate reduction.

Curiously, substrate limitation had only a limited influence on the extent of isotope fractionation during sulfate reduction by thermophilic and hyperthermophilic (optimal growth >80°C for hyperthermophiles and 45°C to around 75°C for thermophiles) populations of sulfate reducers from the Guaymas Basin (Canfield et al. 2000). Presumably these organisms metabolizing at high temperatures of 55 to 85°C (Fig. 4) had a different relationship between isotope fractionation and specific rate of sulfate reduction than observed for sulfate reducers metabolizing in the mesophilic temperature range as does, for example, *Desulfovibrio desulfuricans*. This proposition awaits experimental exploration.

SULFIDE OXIDATION

Sulfide oxidation is pervasive in marine and lacustrine environments supporting sulfate reduction. For example, in marine coastal sediments typically 90 percent or more of the sulfide produced during sulfate reduction is reoxidized (Jørgensen 1982; Canfield and Teske 1996). In some cases sulfide-oxidizing organisms might be quite conspicuous, such as the impressive mats of *Thioploca sp.* which populate low-oxygen regions of the Peruvian and Chilean Coast (Fossing et al. 1995). Also, active anoxygenic phototrophs that utilize sulfide as an electron donor, and which produce oxidized products ranging from sulfur to sulfate, may be encountered as red, purple, or green layered communities in microbial mats (Jørgensen and Des Marais 1986). They may also be found as surface films on anoxic sulfidic sediments, or as dense bacterial plates near the oxygen-sulfide interface in productive meromictic lakes (e.g. Overmann et al. 1991).

The pathways of sulfide oxidation in nature are varied, and in fact poorly known, but include: (1) the inorganic oxidation of sulfide to sulfate, elemental sulfur, and other intermediate sulfur compounds, (2) the nonphototrophic, biologically-mediated oxidation of sulfide (and elemental sulfur), (3) the phototrophic oxidation of reduced sulfur compounds by a variety of different anoxygenic phototrophic bacteria, and (4) the disproportionation of sulfur compounds with intermediate oxidation states. The first three of these are true sulfide-oxidation pathways requiring either the introduction of an electron acceptor (e.g. O_2 and NO_3^-), or, in the case of phototrophic pathways, the fixation of organic carbon from CO_2 to balance the sulfide oxidation. The disproportionation of sulfur intermediate compounds requires no external electron donor or electron acceptor and balances the production of sulfate by the production of sulfide. This process will be taken up in detail in a later section. A cartoon depicting some of the possible steps in the oxidative sulfur cycle is shown in Figure 6.

Fractionations during sulfide oxidation

A summary of the available results on the extent of isotope fractionation during sulfide oxidation is summarized in Table 3. The phototrophic oxidations of sulfide to elemental sulfur and of elemental sulfur to sulfate yield only small or negligible fractionations. Small fractionations also accompany the non-phototrophic, biologically-mediated, oxidation of sulfide to elemental sulfur, as well as the oxidation of sulfur intermediate compounds to sulfate (Table 3). However, significant depletion of sulfate in ^{34}S, and enrichments of polythionates ($S_xO_6^{2-}$) in ^{34}S, are associated with sulfide

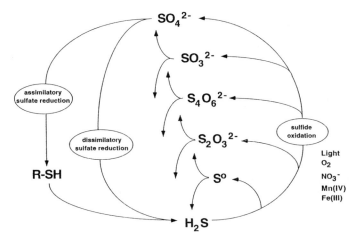

Figure 6. Outlined here are the principal pathways of sulfur-compound transformations, of interest for isotope studies in the environment. Note the numerous possible pathways of sulfide oxidation and the numerous intermediate compounds the can be formed. Each of these intermediates has a variety of possible fates including oxidation, reduction and disproportionation. Redrafted after a figure kindly made available to the author by H. Fossing.

Table 3. Fractionations during the oxidation of sulfur compounds.

Process	Generalized reaction			Fractionation[a] ‰	Reference(s)
phototrophic					
H_2S	H_2S	→	$S°$	-2 to 0	1,2
$S°$	$S°$	→	SO_4^{2-}	0	3
nonphototrophic					
H_2S	H_2S	→	$S°$	-1 to 1	4
	H_2S	→	$S_xO_6^{2-b}$	0 to -19	4
	H_2S	→	SO_4^{2-c}	10 to 18	4
$S°$	$S°$	→	SO_4^{2-}	0	1
$S_2O_3^{2-}$	$S_2O_3^{2-}$	→	SO_4^{2-}	-0.4	1
SO_3^{2-}	SO_3^{2-}	→	SO_4^{2-}	0	1
nonbiological					
H_2S	H_2S	→	$S°, S_2O_3^{2-}, SO_4^{2-}$	4 to 5	5
SO_3^{2-}	SO_3^{2-}	→	SO_4^{2-}	0.4	6

[a]fractionations are presented both as ε_{A-B} and Δ_{A-B}, where A is the reactant and B is the product.
[b] $S_xO_6^{2-}$ is a minor reaction product
[c] SO_4^{2-} is a minor reaction product

1 Fry et al. (1986)   4 Kaplan and Rittenberg (1964)
2 Fry et al . (1984)   5 Fry et al. (1988a)
3 Fry et al. (1988b)   6 Fry et al. (1985)

oxidation to elemental sulfur when these oxidized species are minor reaction products. The elemental sulfur formed, however, shows little fractionation compared to the original sulfide. It might also be expected that extensive non-phototrophic oxidation of sulfide to sulfate might produce small fractionations, although these experiments have apparently not yet been conducted.

Studies of isotope fractionation during sulfide oxidation are rather scant and important sulfide-oxidizing organisms like, for example, *Beggiatoa* sp. and *Thioploca* sp., have yet to be studied. Nevertheless, the available results suggest that biologically mediated oxidation of sulfate to elemental sulfur and sulfate lead to only minimal isotope fractionation.

Disproportionation of sulfur intermediate compounds

As mentioned above, most of the sulfide formed by sulfate reduction in the environment is ultimately reoxidized to sulfate, often through compounds in which sulfur has intermediate oxidation states (Jørgensen 1990). Examples of sulfur intermediate compounds include sulfite (SO_3^{2-}), elemental sulfur (S^o), and thiosulfate ($S_2O_3^{2-}$) (e.g. Troelsen and Jørgensen 1982; Thamdrup et al. 1994). These compounds do not generally accumulate in the environment and are, therefore, readily transformed. Until relatively recently it was assumed that oxidation and reduction were the only possible transformation pathways for these intermediates. It was quite a shock when Bak and Pfennig (1987) described a new fate for sulfur intermediates, namely disproportionation. Through a combination of good fortune and clever scientific reasoning a strain of sulfate reducing bacteria was found to disproportionate thiosulfate to sulfide and sulfate (Eqn. 12) with no external electron donor or electron acceptor required.

$$S_2O_3^{2-} + H_2O \rightarrow H_2S + SO_4^{2-} \tag{12}$$

This process has sometimes been called an "inorganic fermentation" although such a usage is potentially misleading as "fermentation" is generally associated with substrate-level phosphorylation which is probably not the process of ATP generation utilized during the disproportionation of sulfur compounds.

The original strain found to disproportionate thiosulfate, *Desulfovibrio sulfo-dismutans*, could also disproportionate sulfite to sulfide and sulfate (Eqn. 13), and later, both enrichment and pure bacterial cultures were found capable of dispropor-tionating elemental sulfur (Eqn. 14) (Thamdrup et al. 1993; Janssen et al. 1996; Finster et al. 1998).

$$4\ SO_3^{2-} + 2\ H^+ \rightarrow H_2S + 3\ SO_4^{2-} \tag{13}$$

$$4\ S^o + 4\ H_2O \rightarrow 3\ H_2S + SO_4^{2-} + 2\ H^+ \tag{14}$$

Some organisms capable of disproportionating sulfur compounds have quite general physiologies. *Desulfobulbus propionicus*, for example, is principally known as a sulfate reducer, but can also disproportionate elemental sulfur, thiosulfate, or even use oxygen as an electron acceptor (Krämer and Cypionka 1989). This organism does not grow well through sulfur compound disproportionation, but other organisms, such as *Desulfocapsa sulfoexigens*, a strict anaerobe but poor sulfate reducer, grow rapidly during the disproportionation of elemental sulfur, thiosulfate, and sulfite (Finster et al. 1998). The disproportionation of sulfur intermediates is, therefore, conducted by a diverse group of organisms with variable association to the disproportionation process (Canfield et al. 1998a). Thus, for *Desulfocapsa sulfoexigens*, the disproportionation of sulfur inter-mediates is a necessary way of life, while for *Desulfobulbus propionicus* dispro-portionation may only be employed as a maintenance strategy under microoxic conditions when sulfate reduction is not favored (Canfield et al. 1998a).

Isotope fractionation during disproportionation

The fractionation associated with the disproportionation of elemental sulfur has been explored for a variety of pure and enrichment cultures of organisms derived from both freshwater and marine environments (Canfield and Thamdrup 1994; Canfield et al. 1998a; Habicht et al. 1998). For thermodynamic reasons sulfide concentration must be kept low (<1 mM; Thamdrup et al. 1993) for elemental sulfur disproportionation to yield sufficient energy to support growth of the organism. Accordingly, cultures are generally enriched in iron oxides which buffer the sulfide concentration to low levels. A summary of the available results is presented in Table 4.

Table 4. Isotope fractionation (‰) during elemental sulfur (S^0)
disproportionation by pure and enrichment cultures.

Culture	*freshwater/ marine*	$\Delta_{(S^0-AVS)}{}^a$ *measured*	$\Delta_{(S^0-SO4)}$ *measured*	$\Delta_{(S^0-AVS)}$ *cellular*	$\Delta_{(S^0-SO4)}$ *cellular*
Desulfocapsa thiozymogenes	fresh	5.9	-17.3	5.8	-17.4
Desulfobulbus propionicus	fresh	15.5	-30.9	11.3	-33.9
Desulfocapsa sulfoexigens	marine	5.8	-16	5.5	-16.4
Dangast 1	marine	6.6	-18.2	6.2	-18.7
Dangast 2	marine	6.5	-16.7	5.8	-17.4
Weddewarden	marine	6.2	-17.9	6.9	-20.6
Gulfo Dulce S1	marine	7.9	-19.7	6.9	-20.6
Gulfo Dulce S160	marine	8.0	-17.1	6.2	-18.5
Teich 1st	fresh	6.4	-16.4	5.7	-17.1
Teich 2nd	fresh	6.2	-22	6.2	-18.6
Kugraben	fresh	7.0	-19.9	6.7	-20.2

[a]AVS = acid volatile sulfide.   Data from Canfield et al. (1998)

Due to the presence of iron oxides in the experiments, the observed fractionations do not necessarily represent the true cellular level fractionations characteristic of these organisms. This is because additional elemental sulfur may form as a result of the reaction of iron oxides with the sulfide formed during disproportionation. The observed isotopic compositions of the sulfide and sulfate depend on the extent to which this newly formed elemental sulfur either accumulates during the experiment or is re-utilized by the organism. Therefore, the experimental results have been corrected to reconstruct the cellular level fractionations during the disproportionation process (for details see Canfield et al. 1998a). In fact, in most cases the correction is quite small (Table 4).

Ignoring *Desulfobulbus propionicus* for the moment, the results summarized in Table 4 demonstrate a narrow range in fractionation where sulfide is depleted in [34]S by 6.1 ± 0.4 ‰ , and sulfate is enriched in [34]S by 18.3 ± 1.3 ‰ . There is no significant difference between the fractionations produced by either marine or freshwater organisms, and there is also no relationship between the extent of fractionation and the specific rates of disproportionation (Canfield et al. 1998a). Furthermore, the isotopic composition of elemental sulfur during a time course experiment in which most of the elemental sulfur was utilized did not change (Canfield and Thamdrup 1994), indicating that there is no

preferential uptake of ^{32}S or ^{34}S elemental sulfur by organisms during the disproportionation process.

Desulfobulbus propionicus produces fractionations nearly twice as large as the other
organisms studied so far (Table 4). This organism also differs from the others in that it
grows only poorly, if at all, while disproportionating sulfur. Perhaps *Db. propionicus*
utilizes a disproportionation pathway different from the other organisms explored. This
proposition, as well as an understanding of the factors leading to isotope fractionation
during elemental sulfur disproportionation, must await future exploration of the
biochemistry of the disproportionation process.

Table 5. Isotope fractionation (‰)
during the disproportionation of sulfite

Culture	$\Delta_{(SO_3^{2-}-H_2S)}$	$\Delta_{(SO_3^{2-}-SO_4^{2-})}$
Desulfovibrio sulfodimutans		
I	31.5±1.3	-9.2±0.6
II	37.0±1.4	-12.0±1.1
Desulfocapsa thiozymogenes		
I	20.5±1.3	-6.9±0.3
II	24.0±0.6	-8.5±0.2

Large depletions of sulfide in ^{34}S have been observed during the disproportionation
of sulfite (Habicht et al. 1998; Table 5). Thus, for *Desulfocapsa thiozymogenes*, sulfide
was depleted in ^{34}S by 21-24 ‰ while sulfate was enriched in ^{34}S by 7 to 9 ‰. For
Desulfovibrio sulfodismutans, sulfide was depleted in ^{34}S by 32-37 ‰ with enrichments
of ^{34}S into sulfate of 9 to 12 ‰. The nearly 3 to 1 difference between fractionations into
sulfide and sulfate is consistent with the stoichiometry of the sulfite disproportionation
pathway (Eqn. 13). Also, with the limited experimental database available, Habicht et al.
(1998) suggested that the magnitude of fractionation during disproportionation may be
influenced by the specific rate of disproportionation, with the highest fractionations
observed at the lowest specific rates. As discussed above, a similar trend is observed
during sulfate reduction, but not during elemental-sulfur disproportionation, where no
relationship between fractionation and specific rate has been found (see above). As with
elemental-sulfur disproportionation, the biochemistry of sulfite disproportionation has
not been explored, and therefore, we cannot yet account for the magnitude of the
fractionation observed or the apparent influence of specific rate on the extent of
fractionation.

Thiosulfate $(S_2O_3^{2-})$ is composed of an inner sulfonate sulfur $(-SO_3^-)$ and an outer
sulfane sulfur $(-S^-)$, and by expectation, the sulfate produced during disproportionation
(Eqn. 12) should be derived from the sulfonate sulfur, whereas sulfide should be derived
from the sulfane sulfur. Isotope fractionation during thiosulfate disproportionation has
been explored with pure bacterial cultures under two contrasting experimental
circumstances. In one case, sulfide, a byproduct of the disproportionation process, was
allowed to accumulate within the experimental system (Habicht et al. 1998) whereas in
the other case sulfide was actively purged from the experimental system and never
reached a concentration greater than 0.4 mM (Cypionka et al. 1999).

In all of the experiments, the sulfide produced was depleted in ^{34}S compared to the sulfane sulfur, although the magnitude of the depletion varied considerably, and was greatest in the experiment in which sulfide was purged from the system (Table 6). In the experiments where sulfide accumulated, the sulfate produced was [34]S-enriched compared to the original sulfonate sulfur. By contrast, when sulfide was actively purged from the system, the sulfate produced was [34]S-depleted compared to the initial sulfonate sulfur. In the case where sulfide accumulated, the increasing isotope difference between the sulfane and the sulfonate sulfur is consistent with isotope exchange between those two sulfur forms in the thiosulfate (Habicht et al. 1998). This apparent isotope exchange was the most important process leading to fractionation. In experiments were sulfide did not accumulate (Cypionka et al. 1999) isotope exchange was not obvious and kinetic fractionations were observed.

The impact of thiosulfate disproportionation on sedimentary sulfur systematics (further discussed below) will depend on the magnitude of the fractionation in nature and whether it is influenced by isotope exchange as some of the experimental results would indicate. Of critical importance will also be the isotope difference between sulfane and sulfonate atoms of naturally formed thiosulfate in the environment, for which no information is currently available. Our understanding of the impact of thiosulfate disproportionation in the environment will, therefore, depend on our resolving what factors influence the extent of fractionation, the degree of isotope exchange, and the isotopic composition of thiosulfate in nature.

ISOTOPIC COMPOSITION OF SEDIMENTARY SULFIDES

Sulfate reduction is the principal pathway of sulfide formation in marine sedimentary environments. Therefore, the isotopic composition of sulfide preserved in sediments

Table 6. Isotope fractionation (‰) during the disproportionation
of thiosulfate by pure and enrichment cultures

Culture	$\Delta S_{(sulfane-H_2S)}$[a]	$\Delta S_{(sulfonate-SO_4)}$[b]	reference
Desulfovibrio sulfodismutans			
I	0.7	-1.1	1
II	1.6	-0.6	1
III	2.1	-2.0	1
Desulfocapsa thiozymogenes			
I	0.7	-0.9	1
II	3.2	-2.5	1
Desulfovibrio desulfuricans	20	4	2
Løgten Lagoon	11.4	-11.3	1
Solar Lake	11.8	-12.7	1
Weddewarden	4.7	-5.0	1

[a]calculated as the isotope difference between the initial sulfane sulfur and the H_2S produced during disproportionation
[b]calculated as the isotope difference between the initial sulfonate sulfur and the SO_4^{2-} produced during disproportionation

1 Habicht et al (1998)   2 Cypionka et al. (1999)

might reasonably be expected to reflect that of the sulfide produced during sulfate reduction. However, sulfides are typically far more depleted in [34]S than can be explained by fractionations demonstrated for sulfate reducers (Fig. 5). It has sometimes been suggested that sulfate reducers in nature metabolize at much lower specific rates than yet accomplished in the laboratory fractionation experiments, thereby possibly producing higher fractionations than the experimental results would indicate. However, natural populations of sulfate reducers utilizing indigenous organic substrates produce fractionations in the same range as the pure cultures. These are not high enough to explain the depletion of [34]S in most sedimentary sulfides (Fig. 5). Furthermore, in a recent series of experiments, Habicht and Canfield (2001) have compared the fractionations produced by natural populations of sulfate reducers (ε_{SR}) with the magnitude of [34]S depletion in solid phase sulfides (mostly pyrite) relative to seawater sulfate ($\delta^{34}S_{sulfate} - \delta^{34}S_{pyrite}$). This is the first time such comparison has been made with sediments from a broad range of different environments. The results (Fig. 7) provide direct evidence that sulfate reducers do not, in most cases, fractionate sufficiently to explain the isotopic composition of sedimentary sulfides.

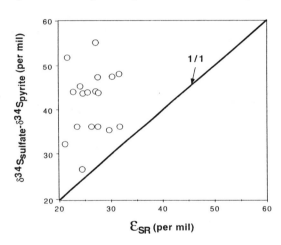

Figure 7. Shown here is the depletion of [34]S into pyrite ($\delta^{34}S_{sulfate} - \delta^{34}S_{pyrite}$) from a variety of different marine sediments, compared to the fractionations produced by sulfate-reducing organisms (ε_{SR}) in the same sediments. Also shown is the one-to-one relationship expected if sulfate reduction was the only process regulating the isotopic composition of pyrite in the sediments. Plotted from data presented in Habicht and Canfield (2001).

As the fractionation associated with pyrite formation from dissolved sulfide is less than 1 ‰ (Price and Shieh 1979), the explanation for this discrepancy must rest elsewhere, most likely isotope fractionations imparted during sulfide oxidation. The direct oxidation of sulfide to sulfate, however, even through intermediate compounds, and by a variety of different sulfide oxidation processes (see above), provides only minimum fractionation (Table 3), and is probably insufficient to explain the isotopic composition of sedimentary sulfides. By contrast, considerable fractionation accompanies the disproportionation of sulfur intermediate compounds (Tables 4, 5 and 6) and disproportionation processes probably account for the highly [34]S-depleted sulfides found in marine sediments (Jørgensen 1990; Canfield and Thamdrup 1994; Canfield and Teske 1996). One can imagine the generation of highly [34]S-depleted sulfides through several cycles of sulfide oxidation to sulfur intermediate compounds, followed by the generation of even more [34]S-depleted sulfide by disproportionation, and the oxidation of this sulfide to sulfur intermediate compounds, and so on (Canfield and Thamdrup 1994). With elemental sulfur disproportionation as an example, a cartoon representing how this process can work is shown in Figure 8.

Early evidence suggested that sedimentary sulfides became more [34]S-depleted as

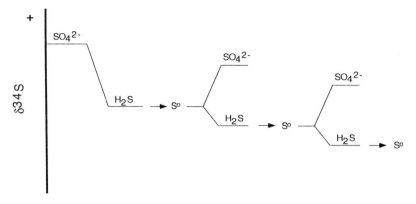

Figure 8. This cartoon shows how the isotopic composition of sedimentary sulfides can be established through an initial fractionation during sulfate reduction followed by the oxidation of sulfide to sulfur intermediate compounds and the subsequent disproportionation of these compounds to sulfate and sulfide. Further oxidation of sulfide to intermediate compounds followed by disproportionation can create quite [34]S-depleted sulfides. Figure is redrafted after Canfield and Thamdrup (1994).

more of the sulfide produced by sulfate reduction became reoxidized (Canfield and Thamdrup 1994; Canfield and Teske 1996). This relationship was interpreted to indicate that the oxidative sulfur cycle, and specifically the disproportionation of sulfur-intermediate compounds, significantly controlled the isotopic composition of sedimentary sulfides. The results of Habicht and Canfield (2001) are particularly relevant as, for the first time, the extent of sulfide oxidation could be directly compared to the amount of [34]S-depletion in sedimentary pyrites not explained by sulfate reduction. This isotope discrepancy is represented as $\varepsilon_{SR} - \delta^{34}S_{pyrite}$, and is related to the percent of the original sulfide produced by sulfate reduction that becomes reoxidized in Figure 9a. Thus while sulfides are clearly more [34]S depleted than can be explained by sulfate reduction alone, $\varepsilon_{SR} - \delta^{34}S_{pyrite}$ does not increase with the percent of sulfide oxidation.

Habicht and Canfield (2001) generated a simple mass balance model describing how the isotopic composition of sulfide should develop if sulfide oxidation proceeded through elemental sulfur, sulfite, or thiosulfate, with sulfate forming by the subsequent disproportionation of these intermediate compounds. The isotope effects associated with the disproportionation reactions were taken from experimental determinations (Tables 4-6). Thus, different disproportionation pathways generate variable extents of [34]S depletion in sulfide as the percent of sulfide oxidation increases (Fig. 9a). When these model results are superimposed against the sediment observations, it is clear that no single pathway of sulfide oxidation can explain the isotope discrepancy between the isotopic composition of sedimentary pyrites and the fractionation during sulfate reduction (Fig. 9a). It seems plausible that multiple and variable pathways of sulfide oxidation and disproportionation occur in sediments yielding no strict relationship between the extent of sulfide oxidation and the degree of [34]S depletion in sedimentary pyrites.

Habicht and Canfield (2001) could, however, show that the discrepancy between the isotopic composition of sulfide produced during sulfate reduction, and the extent of [34]S depletion in pyrites, was generally smaller at high rates of sulfate reduction, becoming greater as sulfate reduction rates decreased (Fig. 9b). It appears that at high rates of sulfate reduction a greater proportion of the sulfide produced is oxidized directly to sulfate with minimal associated fractionation (see above). By contrast, at lower rates of

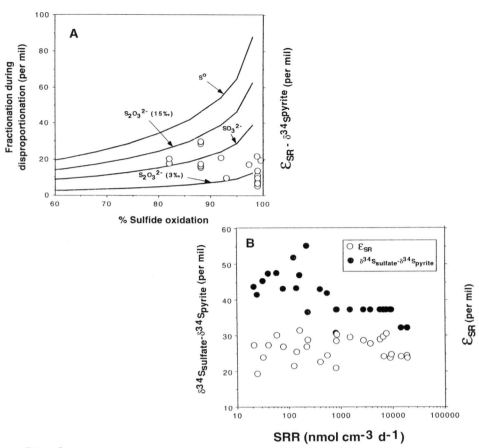

Figure 9.

(A) The isotopic composition of pyrites in the sediments studied by Habicht and Canfield (2001), were, in all instances more ^{34}S-depleted than could be accounted for by the sulfate reducing organisms in the same sediments (see Fig. 7). This extra depletion of ^{34}S in the pyrites is expressed as $\varepsilon_{SR} - \delta^{34}S_{pyrite}$, where ε_{SR} is the fractionation associated with sulfate reduction (see text). In this figure the parameter $\varepsilon_{SR} - \delta^{34}S_{pyrite}$ is plotted against the % of the sulfide produced by sulfate reduction in the sediments that becomes reoxidized. Also shown is a series of model calculations showing how the isotopic composition of sulfide would develop if sulfide oxidation was channeled through $S°$, SO_3^{2-} or $S_2O_3^{2-}$ with subsequent disproportionation of these compounds to sulfate and sulfide, followed by sulfide reoxidation to the sulfur intermediate compound, etc. (see text and Habicht and Canfield 2001). The fractionation associated with the disproportionation of $S°$ (Δ_{So-H2S}) was taken as 7 ‰, the fractionation ($\Delta_{SO3-H2S}$) for SO_3^{2-} disproportionation as 28 ‰, and two fractionations, 3 ‰ and 13 ‰ were used for thiosulfate disproportionation ($\Delta_{sulfane-H2S}$). This figure is redrafted after Habicht and Canfield (2001).

(B) This figure compares the fractionations imparted by sulfate reduction (ε_{SR}) with with depletion of ^{34}S into pyrite ($\delta^{34}S_{sulfate} - \delta^{34}S_{pyrite}$) as a function of sulfate reduction rate (SRR). As sulfate reduction rates increase, the difference between the isotopic composition of pyrite and the fractionation imposed by sulfate-reducing organisms becomes smaller. This relationship probably reflects a higher proportion of sulfide oxidation through disproportionation pathways at low sulfate reduction rates, and more direct sulfide oxidation to sulfate at high rates of sulfate reduction. Replotted after Habicht and Canfield (2001).

sulfate reduction, a higher proportion of the sulfide produced is apparently channeled through sulfur intermediate compounds followed by subsequent disproportionation reactions.

Minor sulfur isotopes

The minor isotopes of sulfur, ^{33}S and ^{36}S (Table 1), have generally not been analyzed in sulfur isotopic studies. This is, in part, because the natural abundance of the minor isotopes is very small and their analysis is technically difficult, as the sulfur must first be converted to SF_6. Also, they are expected to experience mass-dependent fractionations with ^{33}S fractionating about half as much (0.515 times) as ^{34}S, and ^{36}S fractionating about twice as much (1.91 times) as ^{34}S, compared to ^{32}S (Hulston and Thode 1965). Indeed, sedimentary sulfides and sulfates for samples younger than 2.4 billion years show mass-dependent behavior for the minor sulfur isotopes (Farquhar et al. 2000a). Therefore, for these sulfur species, the combination of processes acting to influence their isotopic composition apparently produced mass-dependent fractionations.

By contrast, mass-independent fractionations are observed in sedimentary sulfur compounds more than 2.4 billion years old (Farquhar et al. 2000a). This implies different, or additional, (bio)geochemical processes influenced the isotopic composition of sedimentary sulfides in the Archean. Farquhar et al. (2000a) suggested that atmospheric reactions involving sulfur gases, and having mass-independent isotope effects, may have been more important in the Archean, thereby explaining the isotope signal of Archean sulfur compounds. These results, therefore, imply significant amounts of sulfur cycling through the atmosphere, and show that a consideration of the minor sulfur isotopes can provide much additional information about processes of sulfur cycling. However, very little information exists as to which processes impart mass-independent fractionations, and biological processes have not yet been explicitly examined for their fractionation behavior for the minor sulfur isotopes.

FRACTIONATION CALCULATIONS

To conduct isotope fractionation calculations from experimental and environmental isotope data one must consider first whether or not the system of interest is open or closed. Mass and energy cross the boundary of an open system, whereas only energy crosses the boundary of a closed system. Examples of open systems include the oceans, lakes, sediments and cells, whereas closed systems might include parcels of water or atmosphere circulating with limited exchange, and experimental systems closed to exchange. The isotopic compositions of sulfur compounds in a closed system, whose consumption or production is associated with isotope fractionation, and in the absence of isotope exchange, develop as in a Rayleigh distillation model. No isotope exchange, for example, has been demonstrated between sulfate and sulfide. Rayleigh distillation is not, however, unique to closed systems. Thus, Rayleigh distillation applies to open systems in isotopic equilibrium where removal of a component of the system occurs in infinitely small aliquots (Valley 1986).

In a Rayleigh distillation model the isotopic ratio of a chemical species (R) is related to its isotopic ratio before any is transformed (Ro), to the fractionation factor associated with the transformation (α_{A-B}), and to the fraction of the original chemical species remaining at any particular extent of evolution of the system (f) (Eqn. 15):

$$R/R_o = f^{(1 - \alpha_{A-B})} \tag{15}$$

For example, the isotopic composition of sulfate develops during progressive consumption by sulfate reduction in a closed container in a manner consistent with this Rayleigh distillation model (Fig. 10). In this example sulfide accumulates quantitatively with the amount of sulfate depletion.

Figure 10. This graph shows how the isotopic composition of sulfate and sulfide evolve in a closed system where sulfide is produced by sulfate reduction with an α of 1.040 and an initial isotopic composition of sulfate of 20 ‰. Sulfide is assumed to accumulate quantitatively as sulfate becomes depleted. The parameter f_{SO_4} expresses the remaining fraction of the original sulfate in the system. A Rayleigh distillation model was used to calculate these results. See text for details.

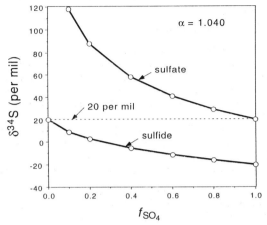

The isotopic composition of sulfate and sulfide may also develop in sediments with isotopic compositions resembling Rayleigh distillation (Fig. 11). However, sediments are not closed systems. They are influenced by mass exchange across the sediment-water interface, and furthermore, any parcel of sediment pore water is in dynamic diffusive exchange with adjacent parcels of pore water. This means that, despite appearances, the isotopic compositions of sulfate and sulfide in sediment pore waters do not develop as predicted by the Rayleigh distillation model, and such a model is inappropriate for determining isotope fractionations from sediment pore water data. More complex diagenetic models taking into account the diffusion, advection, and reaction of pore water constituents are required (Jørgensen 1979; Goldhaber and Kaplan 1980). It is recommended here that the practice of describing sediments as "closed" or "semiclosed"

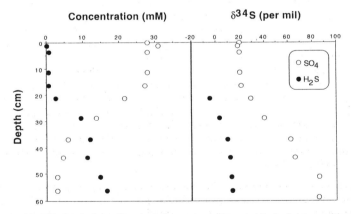

Figure 11. This graph shows how both the concentrations and isotopic compositions of sulfate and sulfide evolve with sediment depth in the sapropelic sediments of Mangrove Lake, Bermuda. Whereas the isotopic compositions of sulfate and sulfide appeared to evolve as in a Rayleigh distillation model (see Fig. 10), such a model is inappropriate for determining fractionations. This is because marine sediments are open with respect to the exchange of chemical species. See text for details. Data are replotted from Canfield et al. (1998b).

in discussing sulfur isotope dynamics is abandoned. More appropriate, and less confusing, is to discuss the extent of sulfate depletion in the sediment. Thus, the sediment experienced, for example, "large," or in another case, "limited," extents of sulfate depletion.

In a well mixed, open system, free of isotope exchange, the fractionation between coexisting chemical species may be expressed as:

$$\alpha_{(A-B)} = [\delta^{34}S_A + 1000] / [\delta^{34}S_B + 1000] \tag{16}$$

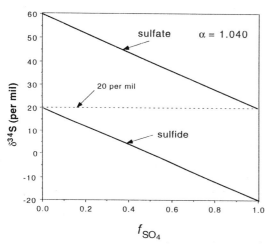

Figure 12. Indicates how the isotopic composition of sulfate and sulfide evolve in an open, well mixed, system with the same fractionation and initial isotopic composition of sulfate as in Figure 10. The parameter f_{SO4} expresses the remaining fraction of the original sulfate in the system. See text for details.

The isotopic composition of chemical species in such a system evolves quite differently than in a closed system experiencing Rayleigh fractionation. For example, the progressive reduction of sulfate, with an associated fractionation of 40 ‰, produces a linear relationship between the isotopic composition of the sulfate and the extent of the original sulfate remaining (Fig. 12). The equation expressing relationship in Figure 12 (Eqn. 19) is derived from Equation (16) with the additional mass balance constraints (Eqns. 17 and 18):

$$f_{SO_4} + f_{H_2S} = 1 \tag{17}$$

$$\delta^{34}S_{SO_4}f_{SO_4} + \delta^{34}S_{H_2S}f_{H_2S} = \delta^{34}S_{SO_{4o}} \tag{18}$$

$$\delta^{34}S_{SO_4} = \{[\delta^{34}S_{SO_{4o}}/(1 - f_{SO_4})] + [(\alpha_{A-B} - 1)/\alpha]10^3\}/[(1/\alpha_{A-B}) + f_{SO_4}/(1 - f_{SO_4})] \tag{19}$$

In addition to the symbols already defined, $\delta^{34}S_{SO_{4o}}$ is the initial isotopic composition of sulfate. A relationship as shown in Figure 12, however, is not unique to open systems, as the isotopic composition of chemical species in isotope equilibrium in a closed system develop in a similar manner (Valley 1986).

A FEW SPECIFIC APPLICATIONS

The isotope record of sedimentary sulfides

Critical aspects of the history of the sulfur cycle on Earth are potentially retained within the isotope record of sedimentary sulfides through geologic time (Fig. 13). A consideration of this record in detail would warrant a chapter in itself, and therefor, only salient points will be offered here relating to the principal factors governing the extent of

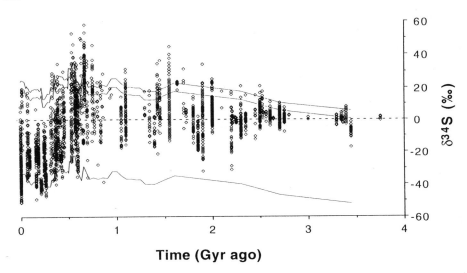

Figure 13. A compilation of the isotopic composition of sedimentary sulfides (triangles) and seawater sulfate (bar with a width of 5 ‰) over geologic time. Also shown is a line representing the isotopic composition of sulfate displaced by 55 ‰. Replotted after Shen et al. (2001).

isotope fractionation. During the Archean (<3.5 Ga) the isotope record of sedimentary sulfides (Fig. 13) generally shows $\delta^{34}S$ values within 10 ‰ of the mantle value of 0 ‰. Fractionations of <15 ‰, compared to seawater sulfate, are also indicated, although the isotope record of seawater sulfate in the Archean is scant.

One exception is the sedimentary barite deposits, once gypsum, found in the Dresser Formation, Warrawoona Group, Pilbara Block of northwestern Australia (Shen et al. 2001). These barites contain pyrites up to 20 ‰ depleted in ^{34}S compared to coexisting sulfate, and provide compelling evidence for the activities of sulfate-reducing organisms, and their early evolution. The environment depositing the original gypsum is envisioned as an evaporative lagoon accumulating high concentrations of sulfate, possibly derived from the local phototrophic oxidation of volcanogenic sulfide (Shen et al. 2001). By contrast, the much smaller fractionations ($\delta^{34}S_{sulfate} - \delta^{34}S_{pyrite}$) preserved in nearly contemporaneous sediment deposits, and other deposits throughout the Archean, probably represent sulfide formation by sulfate reduction in a sulfate-poor ocean. The concentration of sulfate was probably <1 mM, and possibly much less than this.

A large increase in fractionation ($\delta^{34}S_{sulfate} - \delta^{34}S_{pyrite}$) occurs around 2.3 Ga and probably represents an increase in seawater sulfate concentrations at this time to around 1 mM. From 2.3 Ga to about 1.0 Ga the isotopic composition of sedimentary sulfides varies widely but the maximum fractionations are of the same magnitude as observed for modern sulfate reducers with nonlimiting sulfate concentrations. The next major increase in fractionation occurs in the Neoproterozoic around 700-800 billion years ago. The maximum fractionations observed from this point in time, to the present day, exceed the demonstrated ability of sulfate reducers to fractionate during sulfate reduction. It has, therefore, been suggested that around 700-800 million years ago the isotopic consequences of the disproportionation of sulfur intermediate compounds first became expressed (Canfield and Teske 1996). The expression of disproportionation reactions has been attributed to an increase in atmospheric oxygen, allowing the oxygenation of the coastal seafloor, and the initiation of modern style sulfur cycling (Canfield and Teske 1996).

Biogenic sulfides and microscale isotope analysis

As discussed extensively in this review, sulfides depleted in [34]S are a hallmark of dissimilatory sulfate reduction. Also, the original fractionation imparted by sulfate reduction may be extensively modified through subsequent sulfide oxidation and disproportionation of sulfur intermediate compounds. The extents of [34]S depletion resulting from both sulfate reduction and the oxidative sulfur cycle are potentially variable in time and space. Furthermore, the sulfides formed in sediment horizons experiencing significant sulfate depletion may be [34]S-enriched compared to surface sediments where sulfate depletion is minimal. This is because as sulfate becomes depleted, its isotopic composition, as well as the isotopic composition of co-occurring dissolve sulfide, becomes enriched in [34]S (Fig. 11). Overall, for sulfides originating ultimately from dissimilatory sulfate reduction, significant isotope differences might be expected between sulfides in a given sediment strata, or even coexisting sulfides in the same sediment horizon. Indeed, Canfield et al. (1992) have documented isotope differences of nearly 40 ‰ between coexisting pyrites extracted from a coastal marine sediment experiencing significant sulfate depletion.

Implied from this discussion is that sulfides originating from dissimilatory sulfate reduction might, at least in many instances, exhibit significant variability in isotopic composition over short distances. Such variability might, furthermore, be best observed with microscale analytical techniques. The finest scale resolution is obtained with secondary ionization mass spectrometry (SIMS, more commonly referred to as ion microprobe) with spatial resolution of typically 15-30 µm (burning a hole 2-5 µm deep). An excellent review of the collection and interpretation of ion microprobe sulfur isotope data is provided by McKibben and Riciputi (1998). Laser ablation is also used to collect sulfur for isotope analysis, utilizing a greater amount of sample (burning a hole 100 µm in diameter and 50 µm deep), but still providing sufficient resolution to observe small scale differences in sulfur isotopic composition in many cases (Ohmoto and Goldhaber 1997).

As expected, significant sulfur isotope heterogeneity may be observed in small scale (up to 105 ‰ within 200 µm; McKibben and Riciputi 1998) in sediment-hosted sulfides. When taken together with considerations of the depositional setting, such heterogeneity can be strongly indicative of a biogenic source of the sulfide. According to McKibben and Riciputi (1998) strongly zoned pyrite nodules and clusters, with δ^{34}S values increasing from center to edge, are also indicative of dissimilatory sulfate reduction. This zonation of isotopic composition is believed to arise from the progressive precipitation of sulfide onto the growing nodule or cluster from pore solutions progressively depleted in sulfate, providing more [34]S-enriched sulfide (McKibben and Riciputi 1998).

Isotope heterogeneity is not, however, always the product of biological sulfur processing. For example, δ^{34}S values for sulfides from mantle-derived peridotites and pyroxenites have yielded δ^{34}S values from -5 ‰ to 4.5 ‰, with up to 6 ‰ differences observed on individual grains (Chaussidon and Lorand 1990). Up to 5 ‰ differences in δ^{34}S have also been measured for pyrrhotites within the Kaidun meteorite, where isotope fractionation is presumed to have occurred when the original troilite was re-crystalized to pyrrhotite (McSween et al. 1997). Furthermore, a relatively large zonation of 20 ‰ has been observed in the δ^{34}S of hydrothermal pyrite crystals in silicified rhyolite tuffs where a biogenic source of sulfide is not obvious (McKibben and Riciputi 1998). Taken together, the observation of small scale sulfur isotope variability is insufficient to conclude a sulfur source from dissimilatory sulfate reduction. Independent sedimento-logical or geochemical observations must be weighed before a decision on the biogenicity of the sulfur can be made.

Organic sulfur

Most organic sulfur and living organisms is derived, ultimately, from assimilatory sulfate reduction with an isotopic composition reflecting the sulfate source (see above). However, during the early diagenesis of organic matter in sedimentary environments, substantial additions of sulfur into the organic matter may occur, aiding in the stabilization of organic compounds towards further decomposition (Sinninghe Damsté et al. 1989). In principal, both the timing and the magnitude of organic sulfur formation in sediments can be elucidated from sulfur isotope studies. For example, in a broad survey of modern and ancient marine sediments, Anderson and Pratt (1995) have documented a relationship between the isotopic composition of organic sulfur and pyrite, with the organic sulfur typically 10-15 ‰ enriched in ^{34}S compared to the pyrite. Mass balance suggests that, on average, around 20 percent of the sedimentary organic sulfur is original biosynthetic sulfur, whereas the rest is a product of sulfur addition during early diagenesis.

Canfield et al. (1998b) have explored the dynamics of organic sulfur diagenesis in the modern sapropels depositing in Mangrove Lake, Bermuda. Here, sulfur isotope determinations revealed an immediate incorporation of sulfide, derived from sulfate reduction, into the organic matter just below the sediment-water interface. In the surface sediments where sulfate reduction was active, but the sulfide did not accumulate, between 40 to 70 percent of the total pool of organic sulfur was diagenetically derived. The amount of diagenetically derived organic sulfur was determined from a mass balance relying on a combination of sulfur isotope analyses and sulfur pool concentration determinations. Curiously, some of this early-formed organic sulfur was lost in the sulfide-containing pore waters below. But still, down to a depth of 60 cm, between 30 and 50 percent of the total pool of organic sulfur was a product of early diagenetic sulfur addition. Continued studies of the isotopic composition of organic sulfur compounds will shed further light on the timing, magnitude and pathways of organic sulfur formation in sedimentary environments.

CONCLUDING REMARKS

The classic original literature from the 1950s and 1960s on isotope fractionation during sulfur compound metabolism, particularly sulfate reduction, laid much of the groundwork by which we understand the magnitudes of fractionations imposed by the biological processing of sulfur. Most of these early concepts are still valid today. Nevertheless, the last fifteen years have also provided numerous critical advances. For example, during this period the first determinations of the extent of fractionation by natural populations of sulfate-reducing organisms have been made, placing probable limits on the extent of isotope fractionation during sulfate reduction in nature. Furthermore, significant fractionations have been identified during the disproportionation of sulfur intermediate compounds, providing a likely explanation for the isotopic compositions of sulfides in modern and ancient marine sediments. There also has been expanded work on the isotopic fractionations imposed during the phototrophic oxidation of sulfide, as well as a broader survey of fractionations imposed during sulfate reduction by pure cultures of organisms. These recent advances have been highlighted in the present chapter.

Much remains to be done, and possible future research areas have also been high-lighted in the chapter. Thus, we still do not have a predictive model for the extent of isotope fractionation during sulfate reduction. Such a model probably hinges on our understanding of the processes controlling solute exchange across cell membranes. Whether or not this is reproducible or predictable is currently unknown. Except for *Desulfovibrio desulfuricans* there has been no systematic study of the controls on isotope fractionation by a sulfate-reducing organism. As over 100 sulfate reducers, from a wide range of environmental conditions have now been isolated and identified, such an in-depth study on a few well-chosen organisms should be a high priority. We also have virtually no information on possible differences between organisms in the extent of fractionation imposed by different sulfite-reductase enzymes during sulfate reduction. Nor do we have a good model for the biochemistry of sulfur-compound dispropor-tionation processes, or for factors regulating the extent of fractionation during disproportionation. Also, very few modeling efforts have yet been explored how the extent of evolution of the sulfur system in sediments can influence the final isotopic composition of the sulfide preserved. Additional insights into our understanding of the processes imparting isotope fractionation of sulfur in nature should provide the backbone for increasing utilization of sulfur isotope studies to unravel problems of sulfur biogeochemical cycling. The next ten years should be an exciting period indeed with respect to our understanding of the biogeochemistry of sulfur and its isotopes.

ACKNOWLEDGMENTS

The author acknowledges generous support from the Danish Research Foundation (Grundforskningsfond) as well as the expert technical assistance of Mette Andersen in preparing the manuscript. Very helpful reviews were provided by John Hayes and John Valley. In addition, John Hayes is gratefully acknowledged for his tutorage on the principles of isotope fractionation calculations, although the author takes full responsi-bility for any mistakes or misrepresentations.

REFERENCES

Akagi JM (1995) Respiratory sulfate reduction. *In* Sulfate-Reducing Bacteria. LL Barton (ed) Plenum Press, New York, p 89-109

Anderson TF, Pratt LM (1995) Isotopic evidence for the origin of organic sulfur and elemental sulfur in marine sediments. *In* Geochemical Transformations of Sedimentary Sulfur. MA Vairavamurthy, MAA Schoonen (eds) American Chemical Society, Washington DC, p 378-396

Bak F, Pfennig N (1987) Chemolithotropic growth of *Desulfovibrio sulfodismutans* sp. nov. by dispro-portionation of inorganic sulfur compounds. Arch Microbiol 147:184-189

Berner RA,. Petsch ST, Lake JA, Beerling DJ, Popp BN, Lane RS, Laws EA, Westley MB, Cassar N, Woodward FI, Quick, WP (2000) Isotope fractionation and atmospheric oxygen: Implications for Phanerozoic O_2 evolution. Science 287:1630-1633

Böttcher M, Sievert S, Kuever J (1999) Fractionation of sulfur isotopes during dissimilatory reduction of sulfate by a thermophilic gram-negative bacterium at 60°C. Arch Microbiol 172:125-128

Brandt KK, Vester F, Jensen AN, Ingvorsen K (2001) Sulfate reduction dynamics and enumeration of sulfate-reducing bacteria in hypersaline sediments of the Great Salt Lake (Utah, USA). Microbial Ecol 41:1-11

Cameron EM (1982) Sulphate and sulphate reduction in early Precambrian oceans. Nature 296:145-148.

Canfield DE (2001) Isotope fractionation by natural populations of sulfate-reducing bacteria. Geochim Cosmochim Acta 65:1117-1124

Canfield DE, Habicht KS, Thamdrup B (2000) The Archean sulfur cycle and the early history of atmospheric oxygen. Science 288:658-661

Canfield DE, Thamdrup B, Fleisher S (1998a) Isotope fractionation and sulfur metabolism by pure and enrichment cultures of elemental sulfur-disproportionating bacteria. Limnol Oceanogr 43:253-264

Canfield DE, Boudreau BP, Mucci A, Gundersen JK (1998b) The early diagenetic formation of organic sulfur in the sediments of Mangrove Lake, Bermuda. Geochim Cosmochim Acta 62:767-781

Canfield DE, Teske A (1996) Late Proterozoic rise in atmospheric oxygen concentration inferred from phylogenetic and sulphur-isotope studies. Nature 382:127-132

Canfield DE, Thamdrup B (1994) The production of ^{34}S-depleted sulfide during bacterial disproprtionation of elemental sulfur. Science 266:1973-1975

Canfield DE, Raiswell R (1992) The reactivity of sedimentary iron minerals toward sulfide. Am J Sci 292:659-683

Castro HF, Williams NH, Ogram, A (2000) Phylogeny of sulfate-reducing bacteria. FEMS Microbiol Ecol 31:1-9

Chambers LA, Trudinger PA, Smith JW, Burns MS (1975) Fractionation of sulfur isotopes by continuous cultures of *Desulfovibrio desulfuricans*. Can J Microbiol 21:1602-1607

Chambers LA, Trudinger PA, Smith JW, Burns MS (1976) A possible boundary condition in bacterial sulfur isotope fractionation. Geochim Cosmochim Acta 46:721-728

Chaussidon M, Lorand J-P (1990) Sulphur isotope composition of orogenic spinel lherzolite massifs from Ariege (north-eastern Pyrenees, France): An ion microprobe study. Geochim Cosmochim Acta 54: 2835-2846

Cypionka H (1995) Solute transport and cell energetics. *In* Sulfate-Reducing Bacteria. LL Barton (ed) Plenum Press, New York, p 151-184

Cypionka H, Smock AM, Bottcher ME (1998) A combined pathway of sulfur compound disproportionation in *Desulfovivrio desulfuricans*. FEMS Microbiol Lett 166:181-186.

Detmers J, Brüchert V, Habicht KS, Kuever J (2001) Diversity of sulfur isotope fractionations by sulfate-reducing prokaryotes. Appl Environ Microbiol 67:888-894

Farquhar J, Bao HM, Thiemens M (2000a) Atmospheric influence of Earth's earliest sulfur cycle. Science 289:756-758.

Farquhar J, Savarino J, Jackson TL, Thiemens MH (2000b) Evidence of atmospheric sulphur in the martian regolith from sulphur isotopes in meteorites. Nature 404:50-52

Finster K, Liesack W, Thamdrup B (1998) Elemental sulfur and thiosulfate disproportionation by *desulfocapsa sulfoexigens* sp.nov., a new anaerobic bacterium isolated from marine surface sediment. Appl Environ Microbiol 64:119-125

Fossing H, Gallardo VA, Jørgensen BB, Hüttel M, Nielsen LP, Schulz H, Canfield DE, Forster S, Glud RN, Gundersen JK, Küver J, Ramsing NB, Teske A, Thamdrup B, Ulloa O (1995) Concentration and transport of nitrate by the mat-forming sulphur bacterium *Thioploca*. Nature 374:713-715

Fry B, Cox J, Gest H, Hayes JM (1986) Discrimination between ^{34}S and ^{32}S during bacterial metabolism of inorganic sulfur compounds. J Bact 165:328-330

Fry B, Gest H, Hayes JM (1984) Isotope effects associated with the anaerobic oxidation of sulfide by the purple photosynthetic bacterium, *Chromatium vinosum*. FEMS Microbiol Lett 22:283-287

Fry B, Gest H, Hayes JM (1985) Isotope effects associated with the anaerobic oxidation of sulfite and thiosulfate by the photosynthetic bacterium, *Chromatium vinosum*. FEMS Microbiol Lett 27:227-232

Fry B, Gest H, Hayes JM (1988a) ^{34}S/^{32}S fractionation in sulfur cycles catalyzed by anaerobic bacteria. Appl Environ Microbiol 54:250-256

Fry B, Ruf W, Gest H, Hayes JM (1988b) Sulfur isotope effects associated with oxidation of sulfide by O_2 in aqueous solution. Chem Geol (Isotope Geosci Section) 73:205-210

Garrels RM, Perry EA (1977) Cycling of carbon, sulfur, and oxygen through geologic time. *In* The Sea. Ideas and Observations on Progress in the Study of the Seas. ED Goldberg (ed) John Wiley and Sons, New York, p 303-336

Goldhaber MB, Kaplan IR (1980) Mechanisms of sulfur incorporation and isotope fractionation during early diagenesis in sediments of the Gulf of California. Marine Chem 9:95-143

Habicht KS, Canfield DE (1996) Sulphur isotope fractionation in modern microbial mats and the evolution of the sulphur cycle. Nature 382:342-343

Habicht KS, Canfield DE (1997) Sulfur isotope fractionation during bacterial sulfate reduction in organic-rich sediments. Geochim Cosmochim Acta 61:5351-5361

Habicht KS, Canfield DE, Rethmeier J (1998) Sulfur isotope fractionation during bacterial reduction and disproportionation of thiosulfate and sulfite. Geochim Cosmochim Acta 62:2585-2595

Habicht KS, Canfield DE (2001) Isotope fractionation by sulfate-reducing natural populations and the isotopic composition of sulfide in marine sediments. Geology 29:555-558

Harrison, AG (1957) Isotope effects in the reduction of sulfphur compounds. PhD dissertation, McMaster University, Hamilton, Ontario, Canada

Harrison AG, Thode HG (1957) The kinetic isotope effect in the chemical reduction of sulfate. Trans Frarday Soc 53:1648-1651

Harrison AG, Thode HG (1958) Mechanisms of the bacterial reduction of sulfate from isotope fractiontion studies. Trans Faraday Soc 53:84-92

Hoehler TM, Alperin MJ, Albert DB, Martens CS (1998) Thermodynamic control on hydrogen concentrations in anoxic sediments. Geochim Cosmochim Acta 62:1745-1756

Holland HD (1972) The geologic history of sea water- an attempt to solve the problem. Geochim Cosmochim Acta 36:637-651

Hulston JR, Thode HG (1965) Variations in the S^{33}, S^{34}, S^{36} contents of meteorites and their relation to chemical and nuclear effects. J Geophys Res 70:3475-3484

Ingvorsen K, Jørgensen BB (1984) Kinetics of sulfate uptake by freshwater and marine species of *Desulfovibrio*. Arch Microbiol 139:61-66

Janssen PH, Schuhmann A, Bak F, Liesack W (1996) Fermentation of inorganic sulfur compounds by the sulfate-reducing bacterium *Desulfocapsa thiozymogenes* gen. nov., sp. nov. Arch Microbiol 166: 184-192

Jones GE, Starkey RJ (1957) Fractionation of isotopes of sulfur by micro-organisms and their role in deposition of native sulfur. J Appl Microbiol 5:111-115

Jørgensen BB (1979) A theoretical model of the sulfur isotope distribution in marine sediments. Geochim Cosmochim Acta 43:363-374

Jørgensen BB (1982) Mineralization of organic matter in the sea bed-the role of sulphate reduction. Nature 296:643-645

Jørgensen BB (1990) A thiosulfate shunt in the sulfur cycle of marine sediments. Science 249:152-154.

Jørgensen BB, Des Marais DJ (1986) Competition for sulfide among colorless and purple sulfur bacteria in cyanobacterial mats. FEMS Microbiol Ecol 38:179-186

Jørgensen BB, Zawacki LX, Jannasch HW (1990) Thermophilic bacterial sulfate reduction in deep-sea sediments at the Guaymas Basin hydrothermal vent site. Deep-sea Res 37:695-710

Kaplan IR, Emery KO, Rittenberg SC (1963) The distribution and isotopic abundance of sulphur in recent marine sediments off southern California. Geochim Cosmochim Acta 27:297-331

Kaplan IR, Rittenberg SC (1964) Microbiological fractionation of sulphur isotopes. J Gen Microbiol 34:195-212

Kemp ALW, Thode HG (1968) The mechanism of the bacterial reduction of sulphate and of sulphite from isotope fractionation studies. Geochim Cosmochim Acta 32:71-91

Krämer M, Cypionka H (1989) Sulfate formation via ATP sulfurylase in thiosulfate- and sulfite-disproportionating bacteria. Arch Microbiology 151:232-237

Le Faou A, Rajagopal BS, Daniels L, Fauque G (1990) Thiosulfate, polythionates and elemental sulfur assimilation and reduction in the bacterial world. FEMS Microbiol Rev 75:351-382

McKibben MA, Riciputi LR (1998) Sulfur isotopes by ion microprobe. Applications of Microanalytical Techniques to Understanding Mineralizing Processes, MA McKibbin, WC Shanks III, WI Ridley (eds) Society of Economic Geologists, Inc., p 121-139

McSween HY, Riciputi LR (1997) Fractionated sulfur isotopes in sulfides of the Kaidun meteorite. Meteoritics Planetary Sci 32:51-54

Ohmoto H, Felder RP (1987) Bacterial activity in the warmer, sulphate-bearing, Archaean oceans. Nature 328:244-246

Ohmoto H, Goldhaber MB (1997) Sulfur and carbon isotopes. *In* Geochemistry of Hydrothermal Ore Deposits. HL Barnes (ed) John Wiley & Sons, New York

Ohmoto H, Kakegawa T, Lowe DR (1993) 3.4-billion-year-old biogenic pyrites from Barberton, South Africa: Sulfur isotope evidence. Science 262:555-557

Overmann J, Beatty JT, Hall KJ, Pfennig N, Northcote TG (1991) Characterization of a dense, purple sulfur bacterial layer in a meromictic salt lake. Limnology Oceanogr 36:846-859

Price FT, Shieh YN (1979) Fractionation of sulfur isotopes during laboratory synthesis of pyrite at low temperatures. Chem Geol 27:245-253

Rees CE (1973) A steady-state model for sulphur isotope fractionation in bacterial reduction processes. Geochim Cosmochim Acta 37:1141-1162

Sagemann J, Jørgensen BB, Greeff O (1998) Temperature dependence and rates of sulfate reduction in cold sediments of Svalbard, Arctic Ocean. Geomicrobiol J 15:85-100

Schidlowski M, Hayes JM, Kaplan, IR (1983) Isotopic inferences of ancient biogeochemistries: carbon, sulfur, hydrogen, and nitrogen. *In* Earth's Earliest Biosphere: It's Origin and Evolution. JW Schopf (ed) Princeton University Press, Princeton, NJ, p 149-186

Shen Y, Buick R, Canfield DE (2001) Isotopic evidence for microbial sulphate reduction in the early Archean era. Nature 410:77-81

Sinninghe Damsté JS, Eglinton TI, de Leeuw JW, Schenck PA (1989) Organic sulphur in macromolecular sedimentary organic matter: I. Structure and origin of sulphur-containing moieties in kerogen, asphaltenes and coal as revealed by flash pyrolysis. Geochim Cosmochim Acta 53:873-889

Stackebrandt E, Stahl DA, Devereux R (1995) Taxonomic relationships. *In* Sulfate-Reducing Bacteria. LL Barton (ed) Plenum, New York, p 49-87

Stribling JM, Cornwell JC, Currin C (1998) Variability of sulfur isotopic ratios in *Spartina alterniflora*. Mar Ecol Prog Ser 166:73-81

Thamdrup B, Finster K, Fossing H, Hansen JW, Jørgensen, BB (1994) Thiosulfate and sulfite distribution in porewater marine sediments related to manganese, iron, and sulfur geochemistry. Geochim Cosmochim Acta 58:67-73

Thamdrup B, Finster K, Hansen JW, Bak F (1993) Bacterial disproportionation of elemental sulfur coupled to chemical reduction of iron or manganese. Appl Environ Microbiol 59:101-108

Thauer RK, Kunow J (1995) Sulfate-Reducing Archaea. *In* Sulfate-Reducing Bacteria. LL Barton (ed) Plenum Press, New York, p 33-48

Thode HG, Kleerekoper H, McElcheran DE (1951) Sulphur isotope fractionation in the bacterial reduction of sulphate. Research London 4:581-582

Troelsen H, Jørgensen BB (1982) Seasonal dynamics of elemental sulfur in two coastal sediments. Estuarine. Coastal and Shelf Science 15:255-266

Trust BA, Fry B (1992) Sulphur isotopes in plants: a review. Plant, Cell and Environment 15:1105-1110

Valley JW (1986) Stable isotope geochemistry of metamorphic rocks. Rev Mineral 16:445-489

Watanabe Y, Naraoka H, Wronkiewicz DJ, Condie KE, Ohmoto H (1997) Carbon, nitrogen and sulfur geochemistry of Archaean and Proterozoic shales from the Kaapvaal Craton, South Africa. Geochim Cosmochim Acta 61:3441-3459

Widdel F (1988) Microbiology and ecology of sulfate-and sulfur-reducing bacteria. Biology of Anaerobic Organisns. AJB Zehnder (ed) John Wiley, New York, p 469-585

Zehnder AJB, Zinder SH (1980) The sulfur cycle. *In* The Handbook of Environmental Chemistry. O Hutzinger (ed) Springer Verlag, Heidelberg, p 105-145

13   Stratigraphic Variation in Marine Carbonate Carbon Isotope Ratios

Robert L. Ripperdan

Department of Geology
University of Puerto Rico- Mayagüez
Mayagüez, Puerto Rico 00681

INTRODUCTION

Stratigraphic variation in the carbon isotopic (δ^{13}C) value of marine carbonate and organic matter preserved within it has become a popular tool for supporting paleoclimatic hypotheses and refining stratigraphic correlation. Virtually every important biostratigraphic and sequence stratigraphic event of the last 800 million years has had a δ^{13}C dataset collected to aid in interpreting the causal environmental factors that contributed to the event. The stratigraphic resolution obtainable using δ^{13}C has yielded critical temporal constraints on the interplay between carbon cycle variation and environmental events such as mass extinction and eustasy. Although the non-uniqueness of the carbon isotopic response to different environmental stimuli has frustrated the identification of a unified mechanism linking marine δ^{13}C variation with distinct environmental forcing functions, the interpretation of δ^{13}C variation can nonetheless provide important insights into processes that influence major global climatic and surficial systems.

Early studies of marine carbonate minerals suggested that the δ^{13}C value of ancient oceans was essentially 0‰ versus the Peedee belemnite standard (PDB) (Clayton and Degens 1959; Schidlowski et al. 1975; see also Keith and Weber 1964). It was later recognized, however, that departures from the PDB standard represented secular variation in the δ^{13}C values of the marine environment rather than analytical noise, that the record was potentially rich with stratigraphic and paleoenvironmental information (Scholle and Arthur 1980; Veizer et al. 1980; Wadleigh and Veizer 1982; Arthur et al. 1985; Holser et al. 1986; Zachos and Arthur 1986) and that variation was intimately linked to the global carbon, oxygen, and sulfur cycles (Kump and Garrels 1986; Holser et al. 1988). It is now well established that carbonate δ^{13}C values vary within ±3‰ from the PDB standard for most of the last 3.5 billion years (Veizer and Hoefs 1976; Schidlowski et al. 1983) (Fig. 1). Larger variation in carbonate δ^{13}C has been found in conjunction with major events in the stratigraphic record, including the Permo-Triassic mass extinction event (Chen et al. 1984; Holser et al. 1986; Baud et al. 1989; Holser et al. 1989; Oberhänsli et al. 1989), the Cretaceous-Tertiary mass extinction interval (Zachos and Arthur 1986; Shackleton 1987; Zachos et al. 1989), the Ordovician-Silurian glacial epoch and mass extinction interval (Marshall and Middleton 1990; Middleton et al. 1991; Wang et al. 1993; Brenchley et al. 1994) and the Precambrian-Cambrian boundary interval (Magaritz et al. 1986; Knoll et al. 1986; Brasier et al. 1990).

The frequent coincidence of carbonate δ^{13}C variation with global biological events and/or oceanographic events such as sea level change led to several different hypotheses linking the phenomena, most prominently that reduced biological productivity during mass extinction events was accompanied by negative marine δ^{13}C excursions (Magaritz 1989, 1991) and that enhanced organic carbon production and/or preservation was accompanied by positive marine δ^{13}C excursions (many authors). Changing organic carbon burial rates and deep ocean anoxia resulting from eustasy-driven shifts in ocean

1529-6466/00/0043-0013$05.00

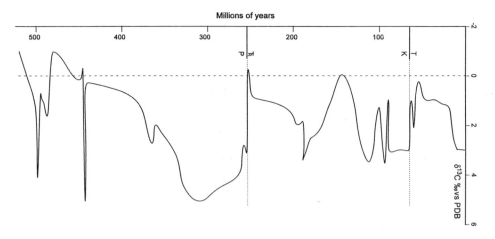

Figure 1. Secular variation in carbonate $\delta^{13}C$ values during the Phanerozoic. Adapted from Holser et al. (1995) and Walter et al. (2000).

circulation patterns and deep ocean ventilation rates have also been linked to $\delta^{13}C$ variation (i.e. Arthur and Dean 1998). More recently, change in continental weathering patterns (Kump et al. 1999) and the formation/destruction of metastable marine organic complexes such as methane clathrates (Dickens et al. 1997; Kennett et al. 2000) have been implicated in carbonate $\delta^{13}C$ variations.

The diversity of hypotheses linking causal factors to the stratigraphic variation of marine carbonate $\delta^{13}C$ values directly reflects the complexity of isotopic variation within the marine carbon cycle (Fig. 2). The stable isotopic value of any given element within its global geochemical cycle is controlled by a combination of bulk transfer and reactive modification processes. For most, either reactive modification or bulk transfer is dominant because either the kinetic and/or equilibrium isotopic fractionations are minor (especially true of heavier elements and higher temperatures), or because the proportion fractionated through reactive processes is miniscule relative to the size of the reservoir. Carbon is unique amongst elements that are abundant in common marine minerals in that it is strongly affected by both groups of processes. It has major reservoirs that (1) are readily modified by both abiological and biological reactions, and (2) are isotopically well-mixed on geologically short timescales due to the relatively large size of the atmospheric reservoir and the short residence time of marine carbon ($\sim 10^5$ y).

Variation in the carbon isotopic composition of the surface ocean—from which nearly all sedimentary carbonate rock is derived—is driven by a wide variety of mechanisms that affect both the short- and long-term carbon cycles. These mechanisms are controlled by a plethora of environmental factors and feedback systems, many of which are involved in (or modulated by) other geochemical cycles including sulfur, oxygen, phosphorus, and iron (see Berner 1999). Models that link carbon isotopic variations to causal mechanisms must therefore also rely on additional geochemical proxies and environmental analysis for refinement. However, significant insight into global processes can be obtained through relatively straightforward examination of the long-term carbon cycle. Unpromising scenarios, in particular, can often be identified solely on the basis of stratigraphic variation in marine carbonate $\delta^{13}C$ values.

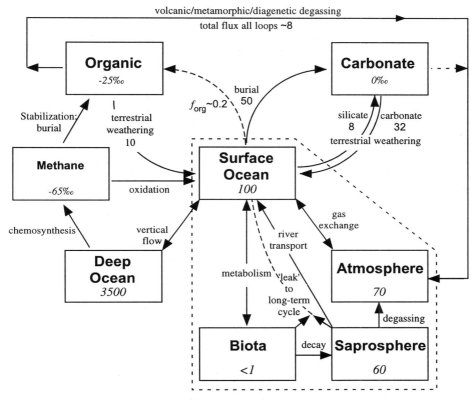

Figure 2. The marine carbon cycle. Reservoir sizes, flux estimates, and isotopic values are presumed to be representative of the Phanerozoic. The short-term carbon cycle is enclosed within the dashed box. The dashed line from the short-term carbon cycle to the organic reservoir represents the 'slow leak' of biosphere organic carbon that supplies the long-term organic carbon reservoir. Reservoir sizes in Gmoles; fluxes in Gmole/ky.

THE MARINE CARBON CYCLE

The carbon cycle is perhaps the dominant geochemical cycle on Earth. It supports the existence of life through the creation of nutritional compounds from carbon dioxide and sunlight, through the modulation of an equitable climate that permits water to remain in the liquid state, and through the provision of free oxygen to the atmosphere. The reactivity of carbon in a wide range of oxidation states make carbon a component of virtually every conceivable geological environment. The formation and weathering of carbon-bearing rocks and sediments constitute major geologic processes that are critical for maintaining a stable geochemical environment that has allowed life to persist in the oceans for at least 3 billion years.

The vast majority of carbon is found in sedimentary rocks as inorganic carbonate minerals and solid organic compounds, and in the deep interior as carbon dioxide and methane. The global carbon cycle can therefore be envisioned as the long-term transfer of deep crustal and mantle carbon to shallow crustal sediment reservoirs via the ocean/atmosphere system, where carbon is parsed into several reservoirs through reactive pathways that modify the initial carbon isotopic value of the outgassed carbon. The resulting variation in carbon isotope composition is greater than 130‰, with natural carbonates >20‰ and methane <-110‰ reported in the literature (see Hoefs 1997). Variation in the relative influence of these reactive pathways is dependent on a wide range of environmental parameters that operate on timescales ranging from sub-annual to hundreds of millions of years.

The bulk of carbon in ocean/atmosphere reservoirs occurs as inorganic bicarbonate ion (HCO_3^-). The concentration of marine HCO_3^- ($[HCO_3^-]$) is kept relatively constant on multimillenial scales by the solubility product of carbonate minerals, which balances the time-variant transfer of carbon from sources (such as volcanoes and terrestrial weathering) to sinks (marine sediment). The proportion of HCO_3^- relative to other dissolved species in the carbonate system is kept approximately constant by the enormous buffer capacity of the global ocean. The atmospheric concentration of CO_2 gas is kept near Henry's Law equilibrium with surface ocean $[HCO_3^-]$ through gas exchange during turbulent mixing. Mixing also helps maintain relatively homogeneous $[HCO_3^-]$ concentrations throughout the global surface ocean, and rapidly distributes local variation in $\delta^{13}C$ values. This strengthens the utility of stratigraphic carbonate $\delta^{13}C$ variations as globally-significant temporal markers.

Carbonate minerals form from dissolved HCO_3^- with only a minor fractionation that is relatively insensitive to temperature. The two most important calcium carbonate minerals—calcite and aragonite—typically have values that are 0.9‰ and 2.7‰ enriched in $\delta^{13}C$ relative to seawater bicarbonate at 25°C (Rubinson and Clayton 1969). The assumption that fractionation factors between seawater HCO_3^- and calcium carbonate minerals are invariant through time permits the use of stratigraphic variations in carbonate $\delta^{13}C$ values as a proxy for secular changes in seawater $\delta^{13}C$ values.

The short-term carbon cycle

The high rate of reactivity that occurs through many different carbon cycle pathways—and the large isotopic fractionations that occur at some steps within these pathways—permits the (geologically) rapid isotopic modification of the marine carbon reservoir. By far the most important of these processes is photosynthesis. Atmospheric and marine CO_2 is entrained into biological compounds with a strong discrimination against heavier isotopes during biosynthesis (see Deines 1980), thereby elevating the isotopic signature of the residual CO_2 and HCO_3^- reservoirs. An overwhelming proportion of the biologically-fixed carbon is recycled back to the surface ocean after death through degassing (via oxidative degradation) and riverine transport, greatly damping the net isotopic change experienced by the HCO_3^- reservoir. Concentration heterogeneities in atmospheric and surface ocean CO_2 developed by differential productivity rates are rapidly diminished on decadal timescales through turbulence-aided gas exchange. Together, these processes constitute the short-term carbon cycle (Fig. 2).

Several pathways within the short-term carbon cycle are associated with significant isotopic fractionations. As noted above, the fixation of CO_2 into reduced carbon compounds during photosynthesis occurs with a strong preference for incorporation of lighter isotopes into the photosynthetic products. The extent of this discrimination is often referred to as the primary photosynthetic isotope effect (ε_p) and is a function of taxonomic level, temperature, growth rate, and environmental CO_2 concentrations (O'Leary 1981; Rau et al. 1992; Hinga et al. 1994; Bidigare et al. 1997). The value of ε_p can range from -17% to -40% or more (O'Leary 1981). The more familiar fractionation factor Δ_B is a measure of the net fractionation of carbon isotopes between marine HCO_3^- and photosynthetic products, and is therefore a combination of ε_p and the ~8‰ equilibrium fractionation between dissolved CO_2 and HCO_3^-.

The degradation of biological organic matter occurs through oxidative and reducing pathways, including methanogenesis. Selective reaction of organic carbon in the marine saprosphere and terrestrial soils, selective stabilization of organic carbon compounds by chelation or mineralization, and selective postmortem transfer of biological carbon are just a few of the ways that the transfer of carbon can be modified within the short-term carbon cycle. Since biologically-produced organic compounds typically have very low $\delta^{13}C$ values (see Hoefs 1997), selective decomposition of even small fractions of biological carbon can influence the $\delta^{13}C$ value of the marine HCO_3^- reservoir.

The relative rates of mass transfer through pathways of the short-term carbon cycle are influenced by environmental factors that operate on sub-annual to millenial scales. Except under unusual circumstances, the isotopic signature of the global marine HCO_3^- reservoir is virtually unaffected by sub-millenial variation within these pathways due to the small percentage of primary productivity relative to the marine bicarbonate ion reservoir at any given moment (~0.1%; Sundquist 1985), and to the nearly conservative recycling of biologically-fixed carbon by oxidative degradation ($\geq 99\%$; Sundquist 1985; others). However, the cumulative effects of longer-ranging trends can be significant. Long-term change in biological productivity influences marine bicarbonate ion $\delta^{13}C$ values through its impact on the balance between organic and carbonate carbon burial. As will be shown below, this balance exerts significant control over the long-term isotopic evolution of the marine carbon reservoir. Conversion of organic carbon to methane or methanogenic compounds—which have extremely low $\delta^{13}C$ values—may have a profound effect on the isotopic value of the global HCO_3^- reservoir, especially during short-lived transient events where the flux of methane to or from sediment is high. Recent studies suggest that the rapid destabilization and oxidation of marine methane compounds during climate change may contribute to marine $\delta^{13}C$ variation (Kennett et al. 2000) and provide an important climate feedback via increased levels of atmospheric methane (i.e. Dickens 2000).

The long-term carbon cycle

Whereas the movement of carbon through the short-term carbon cycle is chiefly governed by interaction between the biosphere, hydrosphere, and atmosphere, the distribution of carbon on geologic timescales is strongly influenced by the interaction between rocks and the short-term carbon cycle. The long-term carbon cycle can be simplistically summarized as the removal of sedimentary carbon compounds from the surface ocean to balance the introduction of HCO_3^-, CO_2, and methane from weathering and volcanic, metamorphic, and diagenetic outgassing (Fig. 2). The movement of carbon between reservoirs is kept close to steady-state on a multimillion-year scale by the solubility products of carbonate minerals and an array of feedback mechanisms. The balancing of mass fluxes within the long-term carbon cycle has exerted a dominant control on the evolution of atmospheric oxygen and carbon dioxide concentrations over geologic time (Holland 1978).

Variation in the carbon isotopic composition of the surface ocean is driven by a combination of processes that include: changing biological productivity and shifting preservation of sinking organic carbon, sorption of atmospheric carbon dioxide by continental weathering processes during continental uplifts, infusion/removal of CO_2 into surface ocean (and atmospheric) reservoirs by changing sea surface temperatures, increased/decreased rates of vertical marine circulation, the addition of CO_2 by outgassing, and removal of carbon from active surface reservoirs by sedimentary burial. Although the kinetic isotopic fractionation of carbon in CO_2 during photosynthesis (and its perpetuation in sedimentary organic carbon) is the ultimate source of isotopic variability within the long-term carbon cycle, the degree to which the photosynthetic fractionation (ε_p) is propagated through the cycle by variation in mass transfer fluxes between reservoirs plays the leading role in controlling the $\delta^{13}C$ value of the surface ocean.

A highly reductionist argument can be made that tectonic processes are the ultimate source of all long-term carbon isotopic variation in marine HCO_3^-. Many of the geological and climatological changes that result from both short and long-term tectonic processes exert a strong influence on mass transfer functions through a diverse set of controls that include global temperature and rainfall distributions, rock exposure patterns, volcanic/metamorphic CO_2 emissions and ocean circulation. Long-term feedback systems (see Berner 1999) further modulate the responses of reservoirs and mass fluxes to environmental change.

This greatly complicates the isolation of causal mechanisms when interpreting carbonate $\delta^{13}C$ variations, and a comprehensive analysis of the role of specific geologic and climatological processes in controlling marine $\delta^{13}C$ values is beyond the scope of this paper. However, a general inspection of the role these processes play can be made through relatively straightforward mathematical expressions applied to individual pathways within the long-term carbon cycle. The development of these expressions is simplified by assuming that the net transfer of carbon into or out of the long-term carbon cycle is zero; i.e. that the carbon cycle remains at steady state. Although the reduction of atmospheric CO_2 from very high levels in the Archean (Walker et al. 1983; Kasting 1987; many others) to the present indicate a net removal of CO_2 into sediment reservoirs, the steady-state assumption is generally valid for the Phanerozoic on timescales that exceed the residence time of marine carbon ($\sim 10^5$ y).

ISOTOPIC VARIATION WITHIN THE LONG-TERM MARINE CARBON CYCLE

General principles

The sources of variation in marine $\delta^{13}C$ values can be conceptually described through (deceptively) simple models. As outlined above, the isotopic ratio of marine carbon in its active reservoirs is principally governed by removal of surface ocean carbon via biological and abiological reduction or mineralization of HCO_3^- (or CO_2), and replenishment of surface ocean carbon through a variety of restorative mechanisms. The behavior of this simple conceptual model can be expressed as a Rayleigh condensation (adapted from Rayleigh 1896):

$$\frac{R_c}{R_{DIC}} = (1 - f_o)^{[\alpha_B - 1]} \tag{1}$$

where R_{DIC} is the initial bulk isotope ratio (C^{13}/C^{12}) of dissolved inorganic carbon (dominantly in the form HCO_3^-); R_c is the instantaneous carbon isotope ratio of inorganic carbonate (which is approximately the instantaneous value of R_{DIC} due to the minimal carbon isotopic fractionation between inorganic mineral carbonate and HCO_3^-); α_B is the globally-averaged metabolic (kinetic) fractionation between organic carbon and marine HCO_3^- (note that $\Delta_B \approx 1000\ln\alpha_B$); and f_o is the proportion of marine carbon occurring as the product of metabolic fractionation. The importance of f_o and α_B as factors controlling the $\delta^{13}C$ value of marine carbonate is readily seen in this formulation (Fig. 3). Increased production of organic carbon and lower α_B values (i.e. higher metabolic fractionation constants) lead to higher $\delta^{13}C$ values in the residual HCO_3^- reservoir.

Fluctuations in α_B and f_o are accomplished by change in several components of the global marine carbon cycle, including atmospheric CO_2 concentrations, sea surface

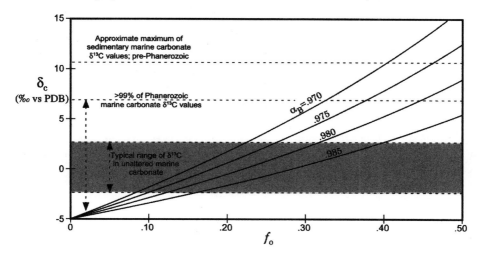

Figure 3. Predicted carbonate $\delta^{13}C$ values at different f_o and α_B using the Rayleigh approximation (Eqn. 1).

temperatures, biological productivity, and mass transfer functions including weathering, ocean circulation, and volcanic emission. Variations in these mass balance components can be broadly characterized as steady-state or transient based on the ability of restorative mechanisms to dissipate the induced variation in net carbon cycle mass flux on a given timescale (often held to be a function of the ocean residence time of carbon, ~50 ky). Carbonate carbon isotopic variation due to steady-state processes has been conventionally attributed to change in the net rate of organic carbon export to sediment reservoirs (i.e. Kump and Garrels 1986), while transient $\delta^{13}C$ variation of up to ~2% has been ascribed to change in surface ocean bioproductivity (Vincent and Berger 1985; Zachos et al. 1989; Kump 1991; Holser et al. 1995). These attributions provide suitable working hypotheses for many instances of stratigraphic variation in marine $\delta^{13}C$ values and can be easily envisioned using Equation (1). However, change in mass transfer rates through other carbon cycle pathways can also induce significant variation in ocean bicarbonate $\delta^{13}C$ values. These pathways are generally embedded within the term f_o in Equation (1), and cannot be easily extracted to evaluate isotopic variation due to change in mass transfer within the carbon cycle. This greatly diminishes the utility of the Rayleigh approximation, except under transient conditions where the rate of variation in the sequestered mass of organic carbon greatly exceeds the rate of carbon redistribution by other pathways of the carbon cycle.

The general mass balance model for marine carbon

Although Equation (1) provides an appealing form for visualizing the impact of biosynthetic reactions on the overall $\delta^{13}C$ signature of the marine HCO_3^- reservoir, the influence of mass transfer function variation is more practically evaluated using the general mass balance equation for the isotopic composition of inorganic carbon in the ocean/atmosphere system (after Kump and Arthur 1999):

$$\frac{dM_T\delta_T}{dt} = F_I\delta_I - F_B\delta_B \tag{2}$$

where M_T is the amount of inorganic carbon in the ocean and atmosphere; F_I and F_B are the integrated carbon input and burial fluxes, respectively; and δ_T, δ_I, and δ_B are the carbon isotopic signatures of the relevant mass balance components. By assuming (for the moment) that the fluxes of weathered carbon species (F_w) and volcanic/metamorphic emissions (F_v) dominate the input function of ocean carbon species ($F_w + F_v \approx F_I$), and that burial of sedimentary carbonate (B_c) and organic carbon species (B_o) dominate the export function of marine carbon ($B_o + B_c \approx F_B$), Equation (2) can be recast as:

$$\frac{dM_T\delta_T}{dt} = [F_w\delta_w + F_v\delta_v] - [B_o\delta_o + B_c\delta_c] \tag{3}$$

where δ_w, δ_v, δ_o, and δ_c are the isotopic compositions of the relevant input and output components (Kump 1991; Kump and Arthur 1999). Setting $F_T = F_I$ and $B_T = F_B$ for convenience, and defining $\Delta_B = \delta_c - \delta_o$ gives:

$$\frac{d\delta_T}{dt} = \frac{F_T[f_w(\delta_w - \delta_v) + \delta_v] - B_T[\delta_c - f_o\Delta_B]}{M_T} \tag{4}$$

where

$$f_w = \frac{F_w}{F_T} \quad \text{and} \quad f_o = \frac{B_o}{B_T} \tag{5}$$

Assuming steady state, $F_T = B_T$ and Equation (4) becomes:

$$\frac{d\delta_T}{dt} = K_T[f_w\delta_w + (1 - f_w)\delta_v + f_o\Delta_B - \delta_c] \tag{6}$$

where the factor

$$K_T = \frac{F_T}{M_T} \tag{7}$$

provides a measure of the magnitude per unit time at which carbon isotopic variations can be propagated through the carbon cycle.

Equation (6) can be further simplified to a time-independent form where

$$f_w\delta_w + (1 - f_w)\delta_v + f_o\Delta_B = \delta_c \tag{8}$$

In this form, it can be readily seen that the principal variables controlling the isotopic composition of the ocean system are the carbon isotopic value of weathering products and synthesized organic matter, the relative contributions of weathering and organic carbon to total carbon input and output, respectively, and the net fractionation between HCO_3^- and organic carbon during biosynthesis. Setting $\delta_w \approx \delta_v$ (after Garrels and Lerman 1984; Holser et al. 1988) and rearranging Equation (8) gives:

$$f_o = \frac{\delta_c - \delta_w}{\Delta_B} \tag{9}$$

which is the 'classic' form of the steady state mass balance equation (Holland 1984; Kump and Garrels 1986).

Although Equations (8) and (9) fail to provide timescales for transformation of δ_c and require that the oceans are well-mixed, they accurately provide the *sense* (positive or negative) of long-term temporal trends in δ_c due to variation in other factors, and this is often the more interesting result. First-order approximations of the expected isotopic response of marine $\delta^{13}C$ values to change in carbon input and output functions can be examined by isolated variation of individual components within Equations (6) and (8).

Variation in the burial flux of marine organic carbon

Change in the rate of organic carbon burial has been a primary hypothesis for explaining stratigraphic variation in carbonate $\delta^{13}C$ value (Holland 1984; many others). The motivation for this model can easily be seen within the modified Rayleigh equation (Eqn. 1) and the simplified steady-state mass balance formulation (Eqn. 8): in both expressions, changing f_o leads directly to different δ_c values. Interpretation of the causal mechanisms for several episodes of highly elevated $\delta^{13}C$ values—most notably during the Neoproterozoic, Early Paleozoic, and latest Permian—focused on variation in the burial (or isolation) rate of organic carbon, since until recently, no other individual factor in Equations (6) or (8) was believed to possess a range of variability sufficient to account for the >5‰ shifts seen during these episodes. Model calculations demonstrate that rapid (~10^5 yr) $\delta^{13}C$ increases of >4‰ can be accomplished with an instantaneous doubling of f_o from its presumed normal value of ~0.2 (Kump et al. 1999; see also Kump and Garrels 1986).

Unfortunately, change in f_o often defies direct attribution to an individual causal mechanism. The relative rate of organic carbon burial (f_o) is a function of the mass of organic carbon produced as primary productivity, the mass of carbonate produced, and the fraction of organic carbon production that survives oxidative degradation and is deposited as shallow or deep ocean sediment. In this context it can be observed that seemingly disparate phenomena such as biological productivity and anoxia are functionally equivalent: all else being equal, an increase in either leads to an increase in f_o.

The impact of productivity and anoxia on organic carbon burial may have several mechanistic links through the marine oxygen cycle and global climate. Oxygen generated by photosynthesis is the primary oxidant for decomposition of organic matter, and over 99% of primary productivity is rapidly oxidized in the surface ocean (Sundquist 1985). Organic matter that survives postmortem decomposition and sinks below the well-mixed surface ocean is still subject to oxidation, but in a part of the ocean where oxygen replenishment occurs primarily through vertical ocean circulation rather than photosynthetic reaction. In principal, a higher rate of primary productivity with the same fraction of postmortem organic decomposition will lead to a greater net export of organic carbon to sediment reservoirs and the deep ocean. Higher rates of carbon export from the surface ocean lead to accelerated consumption of available oxygen at the depositional interface and in the deep ocean, decreasing the oxidative potential of these environments (in the absence of accelerated vertical ocean circulation). In this example, the value of f_o would rise as a result of higher productivity *and* enhanced preservation.

The value of f_o is, in fact, dependent on many factors that influence the rate of biological productivity in the surface oceans, the concentration of dissolved oxygen in surface and deep waters, the level of degradatory microbial activity at the sediment water interface, and to a lesser extent, the absolute rate of inorganic carbonate burial. Each of these factors is, in turn, dependent on several *other* factors, many of which are also variable on geologically significant timescales including temperature, water depth, upwelling rate, and the surface area of highly productive continental shelf regions. To further complicate matters, the rate of organic matter production and export in the marine environment is subject to numerous feedbacks, including some that strongly influence primary productivity.

Many of the factors that affect rates of primary productivity and levels of ocean anoxia lack reliable geochemical or sedimentological proxies in ancient rocks. However, their impact on the $\delta^{13}C$ value of marine HCO_3^- can be qualitatively evaluated through their effect on the bioproduction of organic carbon and its preservation in sediment. A summary of selected factors and their influence on $\delta^{13}C$ is given in Figure 4.

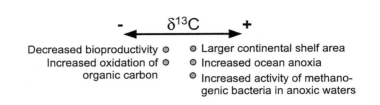

Figure 4. Some of the environmental factors that can modulate the influence of organic burial on carbonate $\delta^{13}C$ values.

Continental weathering of carbon

Carbon weathering in continental environments liberates carbonate and organic carbon from rocks, sediments, and soils. Weathering products are transported by rivers to the ocean as particulates and dissolved inorganic or organic species, or are directly released to the atmosphere as CO_2. This process—which is strongly dependent on geological and climatological factors—constitutes the major mass input function of the long-term carbon cycle (see Fig. 2).

The influence of continental carbon weathering on marine HCO_3^- $\delta^{13}C$ values can be evaluated using Equations (3) and (8). The overall carbon isotopic value of weathering products can be estimated using the relationship:

$$\delta_w = \delta_s - \varphi_k(\delta_s - \delta_k) \tag{10}$$

where δ_s and δ_k are the $\delta^{13}C$ values of eroded carbonate and kerogenic carbon, respectively, and φ_k is the fractional contribution of kerogenic carbon to the total weathered carbon flux. Substitution into Equation (3), setting $\Delta_w =_s \delta_k$ and again assuming that weathering and volcanic emissions are the dominant carbon inputs gives:

$$\frac{d\delta_T}{dt} = \frac{F_T\{\delta_v(1 - f_w) + f_w[\delta_s - \varphi_k\Delta_w]\} - B_T\{\delta_c - f_o\Delta_B\}}{M_T} \tag{11}$$

which can be simplified by assuming steady-state, so that:

$$\delta_c = [f_w(\delta_s - \delta_v - \varphi_k\Delta_w)] + f_o\Delta_B + \delta_v \tag{12}$$

Equation (12) can be further simplified by presuming, for the moment, that $f_w \approx 1$, which eliminates the δ_v terms and gives:

$$\delta_c - \delta_s = f_o\Delta_B - \varphi_k\Delta_w \tag{13}$$

Adding the term $\varphi_k\Delta_B$ to both sides of Equation (13) allows rearrangement to give:

$$\delta_c - \delta_s = [\Delta_B(f_o - \varphi_k)] - [\varphi_k(\Delta_w - \Delta_B)] \tag{14}$$

which permits an easy visualization of the expected value of $\delta^{13}C$ in carbonate minerals formed under different conditions. According to Equation (14), when the difference in absolute $\delta^{13}C$ values between organic matter and inorganic carbonate minerals is the same in the products of weathering as in the surface ocean, the $\delta^{13}C$ value of inorganic carbonate minerals being formed relative to the $\delta^{13}C$ of weathered carbonate rock is a function of the difference between the fractional rate of marine organic carbon deposition and the relative contribution of weathering-derived kerogen to ocean input, i.e.

$$\delta_c - \delta_s = [\Delta_B(f_o - \varphi_k)] \tag{15}$$

As predicted by the classical equation, intervals when organic carbon burial exceeds sedimentary organic carbon erosion (i.e. when $f_o > \varphi_k$) would have higher $\delta^{13}C$ values preserved in carbonate minerals. The prediction goes further, however. Assuming f_o and φ_k are equal, intervals where $\Delta_B > \Delta_w$ would also have elevated $\delta^{13}C$ values preserved in carbonate minerals according to the expression:

$$\delta_c - \delta_s = \varphi_k(\Delta_B - \Delta_w) \tag{16}$$

Since Δ_B is dependent on the photosynthetic fractionation factor ε_p, which is in turn dependent on the atmospheric concentration of CO_2, Equation (16) predicts the somewhat nonintuitive result that the $\delta^{13}C$ value of the surface ocean can rise or fall by variation of atmospheric pCO_2 without changing the net flux of carbon to the dissolved inorganic carbon reservoir. This mechanism for elevating surface ocean $\delta^{13}C$ values has been suggested by Kump et al. (1999) as an explanation for the very high carbonate $\delta^{13}C$ values >6 ‰ (vs PDB) found at the end of the Ordovician.

The expressions developed for describing the impact of carbonate weathering on δ_c point to several key variables, the most influential of which are related to φ_k and f_o. The overall *rate* of carbonate weathering is also important (especially as it relates to silicate weathering, described below), but the normal dominance of f_w requires extraordinary variation in other input functions to express a significant numerical difference in Equation (11). This relatively weak dependence on the value of f_w can be demonstrated by revisiting the assumption that $f_w \approx 1$ made in Equation (13). Substituting realistic values of $f_w = 0.8$, $f_o = 0.2$, $\varphi_k = 0.15$, $\Delta_B = \Delta_w = 30$, $\delta_v = -5‰$, and $\delta_s = 1‰$ into Equation (12) gives $\delta_c = 2.2‰$. Substituting the relevant values into Equation (14) gives $\delta_c = 2.5‰$. The difference between these solutions is minor compared to other sources of variation.

The percentage of weathering product derived from organic carbon—usually as kerogen—is dominantly controlled by two factors: differential exposure of organic-rich rocks (usually marine shales and mudstones), and the amenability of the liberated sedimentary kerogen to oxidation. The former can be reasonably approximated through paleogeographic and paleoenvironmental reconstruction. This approximation is deceptively complex, however, because it requires that all kerogenic carbon included in φ_k be fully oxidized: kerogenic carbon that passes unreacted and/or undissolved from point of (terrestrial) erosion into marine sediment has no influence on mass balance or δ_c. This complication—which is dependent on a diversity of variables including paleo-latitude and the concentration of atmospheric O_2—may decouple the effective value of φ_k from the paleogeographic proxy. This problem is further compounded by the difficulty in obtaining reliable estimates for the value of Δ_w at other times in Earth's history.

A brief summary of the effect of several carbon weathering-related factors on the $\delta^{13}C$ value of the marine HCO_3^- reservoir is given in Figure 5.

Silicate weathering

It has long been known that the long-term carbon cycle is linked to the weathering of silicate rocks (for example, see Berner 1992a, 1995). The influence of silicate weathering on the global carbon cycle has been the focus of intense discussion because of its potential impact on the concentration of atmospheric CO_2 through the reaction:

$$CO_2 + CaSiO_3 \rightarrow CaCO_3 + SiO_2 \qquad\qquad (17)$$

This mechanism has been proposed as a linkage between Himalayan uplift and the progressive cooling of the Cenozoic climate (Ruddiman et al. 1989; Raymo and Ruddiman 1992; Raymo 1994; see also Ruddiman 1997).

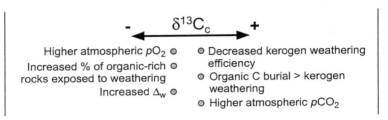

Figure 5. Some of the environmental factors that can modulate the influence of carbon weathering on carbonate $\delta^{13}C$ values.

The potential change in δ_c due to silicate weathering is substantial through its impact on atmospheric CO_2 concentrations, and by extension, on Δ_B. The effect of silicate weathering can be examined through the simplified mass balance expression in Equation (8), i.e.

$$f_w \delta_w + (1 - f_w)\delta_v + f_o\Delta_B = \delta_c \qquad\qquad (18)$$

An increase in the rate of silicate increases the riverine flux of Ca^{2+} to the oceans, which requires a compensatory deposition of carbonate to maintain the equilibrium solubility product of carbonate minerals. This serves to raise the value of B_c (Eqn. 3), which depresses the value of f_o according to Equation (8). The effect on δ_c is relatively minor (~1‰ or less) because the value of f_o is generally small. Although an increase in F_v in Equation (3) may ultimately be required to dampen fluctuations in atmospheric pCO_2 on timescales $>10^4$ years (Gibbs and Kump 1994), the net effect on δ_c from increased CO_2 outgassing is also minor except under conditions of very large increases in F_v (see next section).

A reduction in atmospheric pCO_2 due to silicate weathering provides greatest potential influence on δ_c through its impact on the photosynthetic discrimination factor, ε_p. The value of ε_p decreases with decreasing aqueous CO_2 concentrations (Arthur et al. 1985; Freeman and Hayes 1992; Rau et al. 1992; Bidigare et al. 1997). Accordingly, increased rates of silicate weathering should decrease the value of δ_c through diminution of the term $f_o\Delta_B$. Assuming a value of $f_o \approx 0.2$ (from Kump and Arthur 1999), and using the generic response curve of Kump and Arthur (1999) for the value of Δ_B at different pCO_2 values, a 10× decrease in atmospheric pCO_2 to present-day levels would lead to a reduction in δ_c of ~1.5‰. Though minor in comparison to the potential range of δ_c values that can be generated by shifting organic carbon burial rates, the potential δ_c variation due to the combination of increased carbonate production and pCO_2 reduction is nonetheless significant in comparison to the ±3‰ carbonate $\delta^{13}C$ variation observed throughout much of the geologic record (Fig. 1). It is interesting to note that a 10% increase in the fractional rate of carbonate production and a 6‰ decrease in Δ_B reasonably reproduces the observed ~2‰ fall in δ_c values during a $\sim6\times$ to 8× reduction in atmospheric pCO_2 since the Cretaceous.

The principal factors that affect silicate weathering rates are temperature, rainfall amount, lithology, topography, and vegetation. On a geologic scale, variation in these factors is primarily generated by climate change (including that generated by shifting paleogeographies), regional uplift and orogeny. Periods where major super-continents—and their dry internal regions—are prevalent are predicted to have low net rates of silicate weathering due to the lack of runoff and moisture available for fueling weathering reactions, while intervals of dispersed continents and wet climates are expected to have higher rates of silicate erosion (Gibbs et al. 1999).

A first-order prediction of stratigraphic variation in δ_c using these silicate weathering relationships is not borne out by the existing $\delta^{13}C$ dataset on timescales >10 My. Modeled bicarbonate fluxes from silicate weathering are low for the pre-Cretaceous Mesozoic, followed by a broad maxima from the Late Cretaceous through Eocene and a decline to early Meozoic values in the Quaternary (Gibbs et al. 1999). Measured $\delta^{13}C$ values have local minima in the earliest Triassic (~0‰), Late Jurassic (~0‰), and Quaternary (0.5‰), and have maxima in the Early/Middle Jurassic (~3.5‰) and Cretaceous/Paleocene (~2.5‰). The general coincidence of low silicate weathering fluxes with lower $\delta^{13}C$ values is precisely opposite that predicted by Equation (18), with only the Early/Middle Jurassic giving the anticipated correlation of higher $\delta^{13}C$ values with lower silicate weathering fluxes. Interestingly, the Early/Middle Jurassic is an interval where total (carbonate+silicate) modeled weathering fluxes are at their lowest since the early Triassic. Additional work is needed to explore the implications of these results, but they suggest that interaction with other pathways in the carbon cyle may often mask δ_c variation induced by variation in silicate weathering rates.

Volcanic and metamorphic emissions of CO_2

The CO_2 flux from volcanic emission and metamorphic outgassing is an important component of the long-term carbon cycle through direct impact on the burial rate of carbonate. Increased CO_2 emissions must be compensated for by increased carbonate burial in order to satisfy the mass balance expression (Eqn. 3), affecting the relative input and burial proportionality constants. Increased CO_2 emissions also promote a (much smaller) net increase in atmospheric pCO_2 through the consumption of Ca^{2+} during compensatory carbonate formation, which permits an increase in marine $[HCO_3^-]$ to maintain the equilibrium carbonate mineral solubility product. This, in turn, influences the value of ε_p.

The effects of increased volcanic/metamorphic CO_2 emissions on the isotopic value of marine bicarbonate can be evaluated by inserting the typically referenced value for δ_v of -5 ‰ into Equation (8), which gives:

$$[f_w\delta_w + f_o\Delta_B] - 5(1 - f_w) = \delta_c \qquad (19)$$

Increasing f_v (= $(1 - f_w)$) influences δ_c in three ways: through its impact on f_w, f_o, and pCO_2, which in turn influences Δ_B. Assuming the delivery of bicarbonate from weathering remains constant, increasing volcanic and metamorphic CO_2 emissions leads to a net decrease in f_o in accordance with Equation (5), and in conjunction with decreased f_w, gently 'pushes' the value of δ_c towards the presumed δ_v value of -5‰. Although this solution can be arrived at by intuition, Equation (19) also accurately predicts an expected *rise* in δ_c if $\delta_w < -5$‰ and f_o remains constant, and a rise in δ_c if sustained high CO_2 fluxes from emisssion sources significantly raise atmospheric pCO_2.

The rate of CO_2 release by volcanic and metamorphic processes is more directly tied to long-term tectonic processes than any other flux within the global carbon cycle, and has been approximated by rates of seafloor spreading and subduction. Estimates of global ocean crustal production for the Cenozoic imply a roughly 30% decrease in the volcanic release of CO_2 (Kump and Arthur 1997). If we assign an initial δ_w value of -7‰, allow f_w to increase from 0.75 to 0.8, allow f_o to increase from 0.2 to 0.21 and incorporate the empirically-determined decrease in Δ_B from 28% to 22% (Kump and Arthur 1997), the predicted effect would be a net decrease in δ_c of ~1‰, with more than 90% of the variation due to the change in Δ_B. This example suggests that change in the flux of volcanic/metamorphic CO_2 release primarily influences the carbon isotopic value of HCO_3^- through modification of atmospheric pCO_2. This is not entirely unexpected; since the average value of δ_w is often presumed to be $\approx \delta_v$ (Kump and Garrels 1986; Holser et al. 1988), only major changes in f_w would have an impact on δ_c, and since the value of f_w is typically >0.8, only major *increases* in the volcanic release of CO_2 can substantially alter f_w. It is likely that changing fluxes of outgassed CO_2 have not provided a direct source of variation in the stratigraphic record of marine $\delta^{13}C$ except under extraordinary circumstances.

Clathrate hydrates and sedimentary methane

The recognition that the oceans and permafrost contain enormous volumes of 'frozen' methane (>10^{16} kg; Buffett 2000) within a temperature-sensitive structural framework, that methane can escape from this structure (i.e. Dickens et al. 1997; Blunier 2000), and the knowledge that methane is a powerful greenhouse gas have led to suggestions that methane has played an important role in short- and long-term climate change throughout most of Earth's history (Sloan et al. 1992; Raynaud et al. 1997; Dickens 2000; Kennett et al. 2000; Pavlov et al. 2000). Several recent papers attribute carbonate $\delta^{13}C$ variation to change in the rate of deep ocean ventilation and/or the development of significant quantities of clathrate hydrate in anoxic waters. The models have wide appeal because they can account for very large variations in $\delta^{13}C$ found at several important levels within the geologic record.

The impact of methane clathrate formation or release on the $\delta^{13}C$ value of the marine HCO_3^- can be evaluated using a modification of Equation (8). Biologically-produced methane typically has values between -110 to -50‰ vs PDB (Schoell 1988), with a typical value of ~-65‰ (Cicerone and Oremland 1988). This is substantially lower than the average $\delta^{13}C$ value of organic carbon produced by photosynthesis, which is ~-25‰. Incorporating methane burial into Equation (8) therefore requires modification of the term Δ_B. This can be done through the simple expression:

$$\Delta_O = \Delta_B + \varphi_m(\Delta_m - \Delta_B) \tag{20}$$

where Δ_O is the net difference between the $\delta^{13}C$ values of seawater bicarbonate and the integrated organic carbon flux, Δ_m is the average $\delta^{13}C$ value of methane transferred to marine sediment, and φ_m is the fraction of biologically-produced carbon that is ultimately converted to methane. Substituting this expression into Equation (8) gives:

$$f_w\delta_w + (1 - f_w)\delta_v + f_o\Delta_O = \delta_c \tag{21}$$

It can be easily seen that even minor variation in the value of φ_m can generate significant variation in δ_c.

The formation or release of methane from clathrate hydrates can have a profound impact on the isotopic value of surface ocean HCO_3^-. As outlined above, methane formation impacts the value of Δ_O. The rapid oxidation of methane released from sediment reservoirs influences atmospheric pCO_2 in a manner similar to volcanic emissions, except that the $\delta^{13}C$ value of CO_2 converted to methane is much lower than the typical $\delta^{13}C$ value of volcanic or metamorphic CO_2. Intuitively, the $\delta^{13}C$ value of the surface ocean HCO_3^- will be lowered by the release of sediment methane. A more formal expression can be developed by modifying Equation (19) to

$$f_w \delta_w + f_o \Delta_O - \delta_g (1 - f_w) = \delta_c \qquad (22)$$

where δ_g is the average $\delta^{13}C$ value of CO_2 from volcanic sources and methane release. (Note that this expression does not require methane to vent to the atmosphere before oxidation.)

The transient impact of methane release on surface ocean $\delta^{13}C$ values is undoubtedly greater than predicted by Equation (22). Equation (22) is based on two assumptions: that the ocean carbon cycle is at steady state, and that the oceans are well-mixed. Methane release can occur at a rate that violates both of those assumptions. Kennett et al (2000) correlated millenial-scale negative excursions in foraminiferal $\delta^{13}C$ values with massive releases of methane from basinal sediments. Oscillations of up to 6‰ were identified in benthic foraminiferal species, and up to 3‰ in planktonic species. The temporal scale of the $\delta^{13}C$ variations—and therefore, the inferred maximum rate of methane release—is $<10^{-2}$ the residence time of marine carbon, which temporarily allows the isotopic composition of methane to disproportionately influence $\delta^{13}C$ values.

Methane release from terrestrial and ocean sediment reservoirs is strongly dependent on environmental variables, especially temperature. Large oscillations in methane concentrations within polar ice cores are associated with Quaternary climate cycles on orbital, millenial, and decadal time scales (Raynaud et al. 1998; others). On these timescales, the strong greenhouse properties of methane gas subject the release of sedimentary methane to temperature-dependent feedback systems. Longer-term variation may be driven by changes in ocean circulation patterns, thermal maturation of ocean basins, and levels of anoxia. Although the relatively poor geologic record of these longer-term variables—especially in the pre-Cenozoic geologic record—makes it difficult to empirically demonstrate the link between carbonate $\delta^{13}C$ variation and sedimentary methane, the potential magnitude of $\delta^{13}C$ variation from sedimentary methane cycling—and its climatic significance—suggest a bright future for this avenue of study.

Stratigraphic variation by other means: confounding factors

Stratigraphic variation in the $\delta^{13}C$ values of marine carbonates can occur through several processes other than secular variation in the marine carbon cycle, including (1) non-equilibrium processes at the formational or depositional interfaces; (2) shifting preservational bias of carbonate components with different fidelities in their record of marine HCO_3^- $\delta^{13}C$ values, including equilibrium differences in $\delta^{13}C$ between calcite and aragonite; (3) local $\delta^{13}C$ heterogeneities generated by oxidation of organic material at the sediment interface; and (4) post-depositional modification during diagenesis.

The potential $\delta^{13}C$ variation generated through these processes can be large. Biogenic carbonate formed by many calcareous species can have $\delta^{13}C$ values as much as 10‰ or more lower than the $\delta^{13}C$ value of HCO_3^- (Wefer and Berger 1991; Grant 1992). This 'vital effect' isotopic disequilibrium is particularly strong for scleractinia corals, cephalopods, echinoderms, and algae (see Hoefs 1997). Stratigraphic variation in the proportional contribution of 'vital-effected' biogenic carbonate can differentially bias whole rock $\delta^{13}C$ values away from the $\delta^{13}C$ value of seawater HCO_3^-. Proportional differences in the amount of aragonite or calcite can similarly induce stratigraphic differences of up to 2‰ due to the different $\delta^{13}C$ enrichment factors for calcite and aragonite relative to bicarbonate (0.9‰ and 2.7‰, respectively; Rubinson and Clayton 1969).

Carbonate minerals often form—or are stabilized—within pore water environments that are geochemically distinct from seawater. Local departures from seawater $\delta^{13}C$ values due to the release of decomposing organic matter are relatively common, particularly in porewaters within anoxic or near anoxic sediments (Nissenbaum et al 1972; McCorkle et al 1985). The situation is made even more complex by the activity of methanogenic bacteria, which serve to *raise* porewater $\delta^{13}C$ values through the production of isotopically light methane. Evaluating the contribution of seawater vs. porewater carbon to carbonate $\delta^{13}C$ values can be a significant challenge in studies that seek to determine secular trends in seawater $\delta^{13}C$ values.

Typically, samples are subjected to diagnostic macroscopic tests—including optical petrography, elemental ratio determinations, and X-ray diffractometry (XRD)—to determine their potential for recording secular $\delta^{13}C$ variation. Optical petrography is the method of choice for identifying individual carbonate phases deemed most appropriate for the intended application. Elemental ratio determinations—usually employing calcium, strontium, manganese, and/or iron—provide useful information about diagenetic histories, since Sr is preferentially removed during recrystallization of original metastable phases while Mn and Fe become enriched during the formation of late-stage ferroan cements (Al-Aasm and Veizer 1986; others). X-ray diffractometry is used primarily to determine the amount of dolomite, calcite, and aragonite present, since dolomite often has a different paragenetic history from coeval carbonate phases and requires a different methodology during phosphoric acid extraction of CO_2 gas.

$\delta^{18}O$ values can provide insight into the diagenetic history of a sample. Whereas $\delta^{13}C$ values are relatively robust, $\delta^{18}O$ values are sensitive to post-depositional diagenetic processes because of the extremely high elemental abundance of exchangeable oxygen (relative to carbon) in water (Magaritz 1983; Banner and Hanson 1990). This property can be used as a test of the reliability of $\delta^{13}C$ values in a given carbonate rock as a proxy for seawater $\delta^{13}C$. A linear relationship between $\delta^{13}C$ and $\delta^{18}O$ values provides evidence that the preserved $\delta^{13}C$ values are not a reliable proxy for secular variation in marine $\delta^{13}C$ values at the time of deposition. On the other hand, the lack of linear covariance between $\delta^{13}C$ and $\delta^{18}O$ values support the inference that the carbonate $\delta^{13}C$ values may retain a record of contemporaneous seawater $\delta^{13}C$.

There are two competing approaches for determining the carbon isotope value of marine carbonates: analysis of individual carbonate components and bulk analysis of all carbonate phases present in a given carbonate rock. Each methodology has both philosophical and practical advantages and disadvantages that can bias the reliability of the recovered $\delta^{13}C$ profile as a proxy for the marine HCO_3^- reservoir. The analysis of individual carbonate components focuses on obtaining the best possible record of local bicarbonate $\delta^{13}C$ values by eliminating spurious $\delta^{13}C$ values from non-equilibrium carbonate phases (i.e. some biological carbonates with potentially strong vital effects) or phases that cannot dependably preserve primary isotopic signatures (i.e. recrystallized phases, diagenetic cements). Whole rock analysis obtains an integrated $\delta^{13}C$ signal from the entire mixture of carbonate phases present in the sample material. While judicious field selection and sample preparation can eliminate the inclusion of visibly post-depositional carbonate phases and potentially diminish the heterogeneity of analyte carbonate phases (by utilizing lithologies with low initial porosities and low amounts of allochemical material), it is intuitive that component analysis is superior in its theoretical potential to delineate local secular variations in seawater $\delta^{13}C$ values. This superiority is muted, however, by the facts that local $\delta^{13}C$ profiles derive their utility as global seawater $\delta^{13}C$ proxies only through concordance with a contemporaneous global dataset, and that component analysis is inherently more time-consuming than whole rock analysis.

For many applications, the intersectional reproducibility obtained through whole rock methods is nearly equal to that of component analysis, and in some instances (notably when micrites or fine-grained lime mudstones are employed) the reproducibility of $\delta^{13}C$ stratigraphic profiles obtained through whole rock analysis may be higher due to the ease of obtaining replicate measurements. Where carbonate lithologies undergo rapid facies changes, where the proportion of allochemical or 'vital effect' carbonate varies rapidly, or where diagenetic alteration is both pervasive and extreme, whole rock $\delta^{13}C$ profiles are often unreliable and require support or replacement by $\delta^{13}C$ profiles obtained through component analysis. The presence of mixed shale and coarse-grained carbonate lithologies, in particular, can severely compromise the utility of local $\delta^{13}C$ records as a proxy for global events.

A DETAILED LOOK AT A LARGE TRANSIENT $\delta^{13}C$ EVENT: THE LATE ORDOVICIAN GLACIATION AND MASS EXTINCTION

The stratigraphic record of $\delta^{13}C$ variation during the Late Ordovician illustrates the potential insights and ambiguities that can be gained from carbonate $\delta^{13}C$ values. The Late Ordovician "Hirnantian" glaciation was one of the most profound global events of the Phanerozoic. The development of a major ice sheet in central Gondwana has been linked to eustatic sea level fall and a >50% reduction in global diversity of biological genera (Brenchley 1989). The event includes two discrete episodes of faunal turnover (Finney et al. 1999) and a >5‰ positive excursion in marine $\delta^{13}C$ values (Marshall and Middleton 1990; Middleton et al. 1991; Brenchley et al. 1994; others) within an interval of ~0.5-1 m.y. (Harland et al. 1990). The coincidence of these major phenomena has prompted models linking them to causal mechanisms, including climate forcing by a change in atmospheric CO_2 due to enhanced burial of organic carbon (Brenchley et al. 1994; Finney et al. 1999) or a change in rates of terrestrial weathering (Kump et al. 1999). The differences in these models—which invoke either lower pCO_2 or higher pCO_2 during the glacial event—illustrate the fundamental difficulties still encountered in modeling the relationship between $\delta^{13}C$, atmospheric pCO_2, and global climate.

Variation in atmospheric pCO_2 has been implicated as a direct stimulant for the development and destruction of continental ice sheets over the past ~350 ky (e.g. Shackleton and Pisias 1985; Crowley 2000; many others). Direct extrapolation of this observation to the Late Ordovician glacial episode is made difficult by the observation that Late Ordovician atmospheric pCO_2 may have been 15-20× higher than present-day levels (e.g. Berner 1990, 1992b, 1994; Crowley and Baum 1991; Gibbs et al. 1997). A sharp increase in the $\delta^{13}C$ values found at many Upper Ordovician localities (Orth et al. 1986; Marshall and Middleton 1990; Middleton et al. 1991; Long 1993; Brenchley et al. 1994; Marshall et al. 1997; Ripperdan et al. 1998; Kump et al. 1999) suggests that glaciation was accompanied by a massive disruption of the marine carbon cycle. The dramatic rise and fall of Late Ordovician marine $\delta^{13}C$ values is unparalleled in Cenozoic strata, and requires accommodation in models that link ice sheet development to a change in the marine carbon cycle and atmospheric pCO_2 levels.

Until recently, a detailed understanding of the links between glaciation, biodiversity, and the marine carbon cycle during the Late Ordovician had been hampered by cryptic evidence of sea level change and poor biostratigraphic control of isotopic samples. Detailed study of several uppermost Ordovician sections in central Nevada has resolved some of these problems (Finney et al. 1997; Finney et al. 1999), and in particular, has more clearly delineated stratigraphic relationships between sedimentological evidence for sea level change and secular variation in marine $\delta^{13}C$ values. Based on this, it can be shown that the initial rise in $\delta^{13}C$ values coincided with glacioeustatic sea level fall. After rapidly reaching an isotopic steady state at elevated $\delta^{13}C$ values, the marine carbon cycle was perturbed again and $\delta^{13}C$ values began a rapid fall prior to maximum glacioeustatic lowstand. Thus initial ice sheet expansion coincided with a positive $\delta^{13}C$ excursion, but $\delta^{13}C$ values were changed in the opposite sense during later expansion.

Late Ordovician $\delta^{13}C$ variation in the Hanson Creek Formation, central Nevada

The Upper Ordovician-aged Hanson Creek Formation is well exposed in two overlapping sections in the eastern Monitor Range. The formation represents deposition at moderate water depths along the edge of an embayed carbonate platform margin (Dunham 1977). The lower 125 m of section (exposed along Martin Ridge) consist almost entirely of monotonous dark gray to brown-gray, thin-bedded lime mudstones containing a diverse open marine fauna that includes graptolites, conodonts, chitinozoans, and radiolarians. The base of a cliff-forming, chert-bearing unit at 126 m is interpreted to represent the initiation of sea level shallowing on the basis of increased bedding thickness and carbonate abundance. This 20 m-thick chert-bearing unit is readily identified in both Monitor Range sections, and provides an unambiguous marker for offsetting from the highly fractured Martin Ridge outcrops to the relatively undisturbed Copenhagen Canyon section.

The base of the chert-bearing carbonate bed at 126 m coincides with several important biostratigraphic changes. Diagnostic *Paraorthograptus pacificus* zone graptolites disappear, then return briefly between 146-152 m during a short-lived deepening event evidenced by an increase in the abundance of shale and the cessation of chert formation. Basin-dwelling conodont assemblages are replaced at 126 m by those of shelf and slope environments.

Conodonts, *P. pacificus* zone graptolites, chitinozoans, and radiolarians disappear at 152 m, coincident with a dramatic shallowing phase expressed by rapid vertical facies changes that culminate in several meters of light-gray, cross stratified grainstone capped by quartz sandy, oolitic dologgrainstone (Finney et al. 1997). A change in siliceous sponge spicule assemblages indicates a transformation to shallow subtidal conditions. The oolitic grainstone is sharply overlain by a corroded, angular 1-6 cm thick veneer of brown, fine quartz arenite at 187 m. This exposure surface (correlative to the lower *Normalograptus persculptus* graptolite zone) signals the maximum sea-level lowstand produced by Late Ordovician glaciation (Finney et al 1999).

The thin quartz arenite bed is overlain by several meters of medium-gray wackestone representing shallow subtidal conditions. This facies is abruptly followed at 198 m by a thick succession of deeper marine, dark grey limestone with abundant chert that closely resembles the chert-bearing bed at the base of the Copenhagen Canyon section. The conodont and chitinozoan faunas that reappear in the lower portion of this upper chert-bearing unit are considered typical of the Early Silurian (Finney et al. 1999). A sparse graptolite fauna at 203 m represents the upper part of the *N. persculptus* zone (Berry 1986). The upper chert unit is overlain disconformably by the Silurian-aged Roberts Mountain Formation.

The carbon isotope stratigraphic profile from the Hanson Creek Formation at Copenhagen Canyon (Monitor Range, Nevada) is displayed in Figure 6. The profile is dominated by a large positive excursion in $\delta^{13}C$ values through the middle 40 m of the section. $\delta^{13}C$ values steadily fall by ~1‰ through the lowermost chert-bearing unit (Ka), then undergo a stronger drop in the overlying 1 meter of chert-free grainstones and shales. $\delta^{13}C$ values recover within the next half meter and remain stable at ~1.6‰ (vs. PDB) for the next ~5 m, then rise by more than 5‰ within the next 10 m, reaching a maximum of 7.3‰ (vs. PDB). $\delta^{13}C$ values fall gently by ~1‰ over the next ~14 m, decrease another 1‰ through the next 10 m, then fall ~2‰ through the next 3 m. $\delta^{13}C$

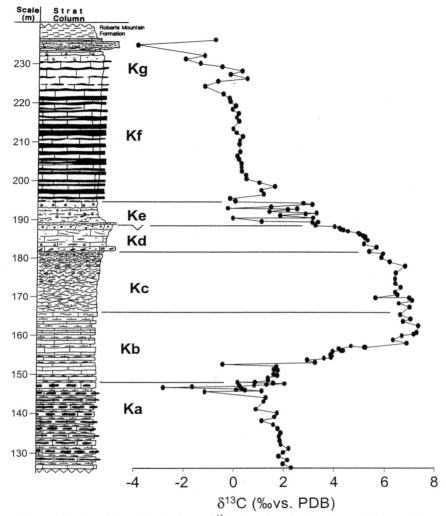

Figure 6. Stratigraphic profile of carbonate $\delta^{13}C$ values from the Copenhagen Canyon section, Monitor Range, Nevada.

values vary considerably through the next 6 meters within an interval of open-marine wackestones. After an abrupt irregularity in $\delta^{13}C$ values at 196 m—just below the base of a deep marine lime mudstone with abundant chert—$\delta^{13}C$ values resume a gentle fall from 0.5‰ to -0.2‰ through the next 25 m, to the top of the upper chert-bearing limestone.

As noted earlier, the reliability of whole rock carbonate $\delta^{13}C$ values as a record of global marine secular $\delta^{13}C$ variation is best assessed through comparison with $\delta^{13}C$ profiles from contemporaneous strata at other localities. Profiles from sections in central Sweden, the Baltic Platform, and on Anticosti Island all show a strong increase in $\delta^{13}C$ values in uppermost Ordovician strata, reaching values as high as +7‰ (Orth et al. 1986; Marshall and Middleton 1990; Middleton et al. 1991; Long 1993; Brenchley et al. 1994; Marshall et al. 1997), as does a parallel study from a nearby section in Copenhagen

Canyon (Kump et al. 1999). Faunal data confirm a direct correlation between these sections and the large positive excursion at Copenhagen Canyon.

Although major shifts in the trend of $\delta^{13}C$ variation are virtually coincident with sedimentological change at Copenhagen Canyon (Fig. 6), detailed inspection of key stratigraphic intervals reveals important discrepancies. Interpretation of these intervals is facilitated by the relatively high rates of deposition and the rapidity of facies and $\delta^{13}C$ variations that occur throughout the Copenhagen Canyon succession, permitting the accurate resolution of temporal relationships between sea level and shifts in $\delta^{13}C$.

The end of chert preservation and thinning of carbonate beds marking the top of unit Ka is followed by a ~6 m interval (Kb_1) of rapidly-changing sedimentary facies that juxtapose shales with thinly-bedded coarse grainstones (Fig. 7). Finney et al. (1999) interpreted the Kb_1 interval as a short-lived deepening event based on the brief (and final) reappearance of *P. pacificus* zone graptolites and an increase in the abundance of shale. Evidence of rapid sea level shallowing and the beginning of a sharp rise in $\delta^{13}C$ values coincides with the end of the Kb_1 interval, with $\delta^{13}C$ values reaching a maximum of +7.3 ‰ within the next 10 m.

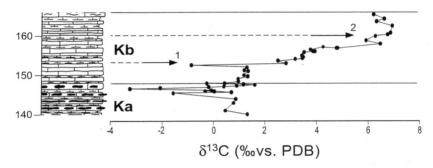

Figure 7. Carbonate $\delta^{13}C$ values from the Ka-Kb interval at Copenhagen Canyon. Sedimentological evidence suggests rapid sea-level fall beginning at the arrow labeled "1".

The Kb_1 interval is critical to understanding the possible role of marine carbon in stimulating the Hirnantian glacial event, and illustrates the more general problem of identifying the causal mechanisms that stimulated carbon isotope excursion. $\delta^{13}C$ values from most of Kb_1 were relatively uniform at ~1.6‰, after an ~2‰ negative excursion in $\delta^{13}C$ values within the meter overlying the last chert-bearing bed in unit Ka. $\delta^{13}C$ values only begin to rise rapidly *after* sedimentological evidence of sea-level shallowing at the top of the Kb_1 subunit. This suggests that the strong marine carbon cycle perturbation evidenced by >+7‰ $\delta^{13}C$ values within the next 10 meters was a *result* of sea level shallowing (and by extension, ice cap development) rather than a driving mechanism, or that it was simply another effect of the ultimate cause of sea level change.

Evidence from other Upper Ordovician localities is compatible with the observation that sea level fell prior to a substantial shift in marine $\delta^{13}C$ values. In the composite $\delta^{13}C$ profile developed from Estonia and Latvia, $\delta^{13}C$ values >+7‰ are restricted to the Porkuni Formation (Marshall et al. 1997). In central Sweden, a rise in $\delta^{13}C$ values has been identified only in the coquinas at the top of the Boda Formation (Marshall and Middleton 1990). A smaller but significant rise in $\delta^{13}C$ values found on Anticosti Island is restricted to carbonate bioherms capping the Ellis Bay Formation (Orth et al. 1986; Long 1993). Although none of these data sets contains a detailed set of transitional $\delta^{13}C$

values that could corroborate the observation that sea level fell prior to (or coincident with) the onset of rising $\delta^{13}C$ values, they all have in common the restriction that elevated $\delta^{13}C$ values only occur after a significant change in depositional environment. A rise in $\delta^{13}C$ values in the nearby Vinini Creek section (Roberts Mountains, Nevada) occurs after a gradational change from black, organic-rich siliceous mudstone to brown mudstone to light gray lime mudstone (Finney et al. 1999).

The extended interval of slowly-falling high $\delta^{13}C$ values through units Kb and Kc is terminated by a more rapid fall in $\delta^{13}C$ through units Kd, Ke and the upper 2 m of unit Kc, reaching pre-excursion values just above the base of unit Kf (Fig. 6). The slight acceleration of $\delta^{13}C$ fall through the upper part of unit Kc is most likely an artifact of compaction (Finney et al. 1997). However, the lithologies in units Kd and Ke suggest relatively rapid deposition, implying that secular $\delta^{13}C$ variation was accelerated during deposition of these units.

Models for the Late Ordovician positive $\delta^{13}C$ excursion

The close temporal relationship between the large $\delta^{13}C$ excursion, the Late Ordovician extinction interval, and the Hirnantian glacial has invited a range of hypotheses. Brenchley et al. (1994) suggested that enhanced upwelling during the glacial phase provided additional nutrients, promoting an increase in bioproductivity (and organic carbon burial) that was reflected as a sharp rise in surface ocean $\delta^{13}C$ values. Kump et al. (1999) argued that a dramatic increase in the transport of weathered carbonate relative to kerogenic carbon occurred as the result of differential continental weathering during sea level lowstand. Ripperdan (in press) argued that the scale and pattern of $\delta^{13}C$ variation during the excursion suggested the development of a transient marine reservoir of organic carbon—possibly methane clathrate.

The breadth of these hypotheses illustrates the difficulty in reliance on $\delta^{13}C$ variations for paleoenvironmental interpretation. Each is based on processes that can account for the observed >5‰ variation in carbonate $\delta^{13}C$, but are only testable via proxies that currently have large errors or are nonexistent. The productivity and organic reservoir hypotheses are separable from the weathering on the basis of their influence on atmospheric pCO_2. Enhanced organic burial leads to a net reduction in atmospheric pCO_2, while the weathering hypothesis requires increased pCO_2 and a shift in differential weathering patterns to fully account for the increased $\delta^{13}C$ values. Analysis of coeval organic material and inorganic carbonate would, in theory, help resolve this dilemma by refining time-dependent knowledge of Δ_B. Unfortunately, the results from Late Ordovician strata in central Nevada have been ambiguous (Kump et al. 1999), although the absence of a reduction in Δ_B suggests that pCO_2 did not *fall* during the glacial epoch. The development of a metastable deep ocean carbon reservoir has the advantage of requiring less net carbon isolation (and therefore, less impact on atmospheric pCO_2) because of the greater isotopic discrimination (as much as -60‰) accompanying bacterial methanogenesis. However, it requires the temporally fortuitous development of a volumetrically significant, temperature-sensitive reservoir in conjunction with a single phase of a glacial epoch that is potentially multiphasic in character. Whereas this may, in fact, be plausible, the hypothesis currently cannot be directly tested by isotopic or environmental proxies, since it relies on the formation and complete dissipation of a labile reservoir with no known environmental record except for its effect on marine carbonate $\delta^{13}C$ values.

Despite these problems, stratigraphic variation in $\delta^{13}C$ values within carbonate rocks remains one of the most powerful tools available for paleoenvironmental interpretation. Although it can be difficult or impossible to discriminate between several causal mechanisms of directional $\delta^{13}C$ change, it *is* possible to eliminate linkages between causal mechanisms from further consideration. Early studies of major biostratigraphic horizons identified a strong temporal relationship between biological extinction and negative cycles in $\delta^{13}C$, and linked the two phenomena via a reduction in productivity (i.e. Magaritz 1991). Results from the Late Ordovician glacial epoch—which hosted the second largest mass extinction on record—indicate precisely the opposite relationship between extinction and $\delta^{13}C$ (Finney et al. 1999). While this does not necessarily rule out productivity change as a factor in mass extinction, it eliminates the possibility of a durable mechanistic link between mass extinction, productivity, and other components of the long-term carbon cycle.

SUMMARY

Stratigraphic variation in carbonate $\delta^{13}C$ values occurs as the result of change in equilibrium and kinetic isotopic fractionation processes, flux relationships between carbon reservoirs within the short- and long-term carbon cycles, and other fractionation mechanisms that are unrelated to the $\delta^{13}C$ value of seawater bicarbonate. In general, processes that add or remove carbon to the marine environment via abiological mechanisms have limited effect on carbon isotopic values except through their influence on atmospheric pCO_2 or during intervals of anomalous differential weathering of carbon-bearing lithologies. Biological productivity, pCO_2, and processes that affect the preservation of organic product from point of formation to sedimentary burial potentially have a stronger impact on ocean HCO_3^- $\delta^{13}C$ values because of the large isotopic discrimination that occur during photosynthesis. Methane may play a role in modulating seawater $\delta^{13}C$ values on long and short timescales because of the extremely low $\delta^{13}C$ value of biologically-produced methane and the temperature sensitivity of marine clathrate hydrates.

Large-scale carbonate $\delta^{13}C$ variation during the Late Ordovician points to several possible sources of variation in the marine carbon cycle. A rise in carbonate $\delta^{13}C$ to >7‰ could result from a strong increase in the net transfer of organic carbon to sediment reservoirs, possibly including enhanced storage of methane. Preferential weathering of carbonate rocks over carbon-bearing siliceous rocks can also account for the observed $\delta^{13}C$ variation. The viability of these alternatives illustrates the difficulty in linking carbonate $\delta^{13}C$ variation to individual causal mechanisms.

ACKNOWLEDGMENTS

Thanks are extended to Ken MacLeod, Mark Gibbs, and Frank Corsetti for their helpful and constructive reviews. Partial support for this work was provided by the College of Arts and Sciences at the University of Puerto Rico, Mayagüez.

REFERENCES

Al-Aasm IS, Veizer J (1986) Diagenetic stabilization of aragonite and low-Mg calcite; I. Trace elements in rudists. J Sediment Petrol 56:138-152

Arthur MA, Dean WE (1998) Organic-matter production and preservation and evolution of anoxia in the Holocene Black Sea. Paleoceanogr 13:395-411

Arthur MA, Dean WE, Schlanger SO (1985) Variations in the global carbon cycle during the Cretaceous related to climate, volcanism, and changes in atmospheric CO_2. Am Geophys Union Monogr 32: 504-529

Banner JL, Hanson GN (1990) Calculation of simultaneous isotopic and trace element variations during water-rock interaction with applications to carbonate diagenesis. Geochim Cosmochim Acta 54: 3123-3137

Baud A, Magaritz M, Holser WT (1989) Permian-Triassic of the Tethys: Carbon isotope studies. Geol Rundsch 78:649-677

Berner RA (1990) Atmospheric carbon dioxide levels over Phanerozoic time. Science 249:1382-1386

Berner RA (1992a) Weathering, plants, and the long-term carbon cycle. Geochim Cosmochim Acta 56:3225-3231

Berner RA (1992b) Palaeo-CO_2 and climate. Nature 358:114

Berner RA (1994) GEOCARB II: a revised model of atmospheric CO_2 over geologic time. Am J Sci 294:56-91

Berner RA (1995) Chemical weathering and its effect on atmospheric CO_2 and climate. Rev Mineral 31:565-583

Berner RA (1999) A new look at the long-term carbon cycle. Geol Soc Am Today 9:1-6

Berry WBN (1986) Stratigraphic significance of Glyptograptus persculptus group graptolites in central Nevada, U.S.A. Hughes CP, Rickards RB, Chapman AJ (eds) Palaeoecology and biostratigraphy of graptolites. Geol Soc Spec Pub 20:135-143

Bidigare RR, Fluegge A, Freeman KH, Hanson KL, Hayes HM, Hollander D, Jasper JP, King LL, Laws, EA, Milder J, Millero FJ, Pancost R, Popp BN, Steinberg PA, Wakeham SG (1997) Consistent fractionation of ^{13}C in nature and in the laboratory: growth-rate effects in some haptophyte algae. Global Biogeochem Cycles 11:279-292

Blunier T (2000) "Frozen" methane escapes from the sea floor. Science 288:68-69.

Brasier MD, Magaritz M, Corfield R, Luo H, Wu X, Ouyang L, Jiang Z, Hambdi B, He T, Fraser AG (1990) The carbon- and oxygen-isotope record of the Precambrian-Cambrian boundary interval in China and Iran and their correlation. Geol Mag 127:319-332

Brenchley PJ, Marshall JD, Carden GAF, Robertson DBR, Long DGF, Meidla T, Hints L, Anderson TF (1994) Bathymetric and isotopic evidence for a short-lived Late Ordovician glaciation in a greenhouse period. Geology 22:295-298

Buffett BA (2000) Clathrate hydrates. Ann Rev Earth Planet Sci 28:477-507

Chen JS, Shao MR, Huo WG, Yao YY (1984) Carbon isotopes of carbonate strata at Permian-Triassic boundary in Changxing, Zhejiang. Scient Geol Sinica 1984:88-93

Cicerone RJ, Oremland RS (1988) Biogeochemcal aspects of atmospheric methane. Global Biogeochem Cycles 2:299-327

Clayton RN, Degens ET (1959) Use of C isotope analysis for differentiating freshwater and marine sediments. AAPG Bull 42:890-897

Crowley TJ (2000) Carbon dioxide and Phanerozoic climate. *In* Huber BT, MacLeod KG (eds) Warm Climates in Earth History. University of Cambridge Press, Cambridge, UK, p 425-444

Crowley TJ, Baum SK (1991) Towards reconciliation of Late Ordovician (~440 Ma) glaciation with very high CO_2 levels. J Geophys Res 96:22,597-22,610

Deines P (1980) The isotopic composition of reduced organic carbon. *In* Fritz P, Fontes JC (eds) Handbook of Environmental Geochemistry, Vol. 1. Elsevier, New York, p 239-406

Dickens GR (2000) Methane oxidation during the late Palaeocene thermal maximum. Bull Soc Geol France 171:37-49

Dickens GR, Castillo MM, Walker JCG (1997) A blast of gas in the latest Paleocene; simulating first-order effects of massive dissociation of oceanic methane hydrate. Geology 25:259-262

Dunham JB (1977) Depositional environments and paleogeography of the Upper Ordovician, Lower Silurian carbonate platform of central Nevada. *In* Stewart JH, Stevens CH, Fritsche AE (eds) Paleozoic Paleogeography of the Western United States. Pac Sect Soc Econ Paleo Mineral 7:157-164.

Finney SC, Cooper JD, Berry WBN (1997) Late Ordovician mass extinction: sedimentologic, cyclostratigraphic, and biostratigraphic records from platform and basin successions, central Nevada. Brigham Young University Geological Studies 42:79-103

Finney SC, Berry WBN, Cooper JD, Ripperdan RL, Sweet WC, Jacobson SR, Soufiane A, Achab A, Noble PJ (1999) Late Ordovician mass extinction: A new perspective from stratigraphic sections in central Nevada. Geology 27:215-218

Freeman KH, Hayes JM (1992) Fractionation of carbon isotopes by phytoplankton and estimates of ancient CO_2 levels. Global Biogeochem Cycles 6:185-198

Gibbs MT, Barron EJ, Kump LR (1997) An atmospheric pCO_2 threshold for the late Ordovician. Geology 25:447-450

Gibbs MT, Bluth GJS, Fawcett PJ, Kump LR (1999) Global chemical erosion over the last 250 my; variations due to changes in paleogeography, paleoclimate, and paleogeology. Am J Sci 299:611-651

Gibbs MT, Kump LR (1994) Global chemical erosion during the last glacial maximum and the present; sensitivity to changes in lithology and hydrology. Paleoceanogr 9:529-543

Grant SWF (1992) Carbon isotopic vital effects and organic diagenesis, Lower Cambrian Forteau formation, northwest Newfoundland: implications for $\delta^{13}C$ chemostratigraphy. Geology 20:243-246

Hinga KR, Arthur MA, Pilson MEQ, Whitaker D (1994) Carbon isotope fractionation by marine phytoplankton in culture:the effects of CO_2 concentration, pH, temperature, and species. Global Biogeochem Cycles 8:91-102

Hoefs J (1997) Stable Isotope Geochemistry. Springer-Verlag, Berlin-Heidelberg-New York, 201 p

Holland HD (1978) The Chemistry of the Atmosphere and Oceans. Wiley Interscience, New York, 351 p

Holland HD (1984) The Chemical Evolution of the Atmosphere and Oceans. Princeton University Press, Princeton, NJ, 582 p

Holser WT, Magaritz M, Ripperdan RL (1995) Global isotopic events. In Walliser OH (ed) Global Events and Event Stratigraphy in the Phanerozoic. Springer-Verlag, Berlin-Heidelberg-New York, p 63-88

Holser WT, Magaritz M, Wright J (1986) Chemical and isotopic variations in the world ocean during Phanerozoic time. In Walliser OH (ed) Global Bio-events. Springer-Verlag, Berlin, p 63-74

Holser WT, Schidlowski M, Mackenzie FT, Maynard JB (1988) Geochemical cycles of carbon and sulfur. In Gregor CB, Garrels RM, Mackenzie FT, Maynard JB (eds) Chemical Cycles in the Evolution of the Earth. Wiley and Sons, New York, p 105-173

Holser WT, Schönlaub HP, Attrep M Jr, Boeckelmann K, Klein P, Magaritz M, Orth CJ, Fenninger A, Jenny-Deshusses D, Kralik M, Mauritsch H, Pak E, Schramm J-M, Stattegger K, Schmöller R (1989) A unique geochemical record at the Permian/Triassic boundary. Nature 337:39-44

Kasting JF (1987) Theoretical constraints on oxygen and carbon dioxide concentrations in the Precambrian atmosphere. Precam Res 34:205-229

Keith ML, Weber JN (1964) Carbon and oxygen isotopic compositions of selected limestones and fossils. Geochim Cosmochim Acta 28:1787-1816

Kennett JP, Cannariato KG, Hendy IL, Behl RJ (2000) Carbon isotopic evidence for methane hydrate instability during Quaternary interstadials. Science 288:128-133

Knoll AH, Hayes JM, Kaufman AJ, Swett F, Lambert TB (1986) Secular variations in carbon isotope ratios from Upper Proterozoic successions of Svalbard and East Greenland. Nature 321:832-838

Kump LR (1991) Interpreting carbon-isotope excursions: Strangelove oceans. Geology 19:299-302

Kump LR, Arthur MA (1997) Global chemical erosion during the Cenozoic: weatherability balances the budget. In Ruddiman W (ed) Tectonic Uplift and Climate Change. Plenum Press, New York, p 399-426

Kump LR, Arthur MA (1999) Interpreting carbon-isotope excursions: carbonates and organic matter. Chem Geol 161:181-198

Kump LR, Arthur MA, Pazkowsky ME, Gibbs MT, Pinkus DS, Sheehan PM (1999) A weathering hypothesis for glaciation at high atmospheric pCO_2 during the Late Ordovician. Paleogeogr Paleoclimat Palaeoecol 152:173-187

Kump LR, Garrels RM (1986) Modeling atmospheric O_2 in the global sedimentary redox cycle. Am J Sci 286:337-360

Magaritz M (1983) Carbon and oxygen isotope content of Recent and ancient coated grains. In Peryt TM (ed) Coated Grains. Springer-Verlag, Berlin, p 27-37

Magaritz M (1989) ^{13}C minima follow extinction events: a clue to faunal radiation. Geology 17:337-340

Magaritz M (1991) Carbon isotopes, time boundaries, and evolution. Terra Nova 3:251-256

Magaritz M, Holser WT, Kirschvink JL (1986) Carbon-isotope events across the Precambrian/Cambrian boundary on the Siberian Platform. Nature 320:258-259

Marshall JD, Brenchley PJ, Mason P, Wolff GA, Astini RA, Hints L, Meidla T (1997) Global carbon isotopic events associated with mass extinction and glaciation in the late Ordovician. Palaeogeogr Palaeoclimat Palaeoecol 132:195-210

Marshall JD, Middleton PD (1990) Changes in marine isotopic composition and Late Ordovician glaciation Jour Geol Soc London 147:1-4

McCorkle DC, Emerson SR, Quay P (1985) Carbon isotopes in marine porewaters. Earth Planet Sci Lett 74:13-26

Middleton PD, Marshall JD, Brenchley PJ (1991) Evidence for isotopic change associated with Late Ordovician glaciation, from brachiopods and marine cement of central Sweden. Geol Surv Canada Papers 90-9:313-321

Nissenbaum A, Presley BJ, Kaplan IR (1972) Early diagenesis in a reducing Fjord, Saanich Inlet, British Columbia I: Chemical and isotopic changes in major components of interstitial water. Geochim Cosmochim Acta 36:1007-1027.

Oberhänsli H, Hsü KJ, Piasecki S, Weissert H (1989) Permian-Triassic carbon isotope anomaly in Greenland and in the southern Alps. Hist Biol 2:37-49

O'Leary MH (1981) Carbon isotope fractionation in plants. Phytochem 20:553-567

Orth CJ, Gilmore JS, Quintana LR, Sheehan PM (1986) Terminal Ordovician extinction: geochemical analysis of the Ordovician/Silurian boundary, Anticosti Island, Quebec. Geology 14:433-436

Pavlov AA, Kasting JF, Brown LL, Rages KA, Freedman R (2000) Greenhouse warming by CH_4 in the atmosphere of early Earth. J Geophys Res E 105:11,981-11,990

Rau GH, Takahashi T, Des Marais DJ, Repeta DJ, Martin, JH (1992) The relationship between $\delta^{13}C$ of organic matter and $[CO_2,$ aq] in ocean surface water: Data from a JGOFS site in the northeast Atlantic Ocean and a model. Geochim Cosmochim Acta 56:1413-1419

Rayleigh JWS (1896) Theoretical considerations respecting the separation of gases by diffusion and similar processes. Philos Mag 42:493

Raymo ME (1994) The Himalayas, organic carbon burial, and climate in the Miocene. Paleoceanogr 9: 399-404

Raymo ME, Ruddiman WF (1992) Tectonic forcing of late Cenozoic climate. Nature 359:117-122

Raynaud D, Chappellaz J, Blunier T (1998) Ice-core record of atmospheric methane changes; relevance to climatic changes and possible gas hydrate sources. In Henriet JP, Mienert J (eds) Gas hydrates; relevance to world margin stability and climate change. Geol Soc Spec Publ 137:327-331

Ripperdan RL, Cooper JD, Finney SR (1998) High-resolution $\delta^{13}C$ and lithostratigraphic profiles from Copenhagen Canyon, Nevada: clues to the behavior of ocean carbon during the Late Ordovician global crisis. Mineral Mag 62A:1279-1280

Rubinson M, Clayton RN (1969) Carbon-13 fractionation between aragonite and calcite. Geochim Cosmochim Acta 33:997-1002

Ruddiman WF (1997) Tectonic Uplift and Climate Change. Plenum, New York, 535 p

Ruddiman WF, Prell WL, Raymo ME (1989) Late Cenozoic uplift in southern Asia and the American West; rationale for general circulatino modeling experiments. J Geophys Res D 94:18,379-18,391

Schidlowski M, Eichman R, Junge CE (1975) Precambrian sedimentary carbonates: carbon and oxygen isotope geochemistry and implications for the terrestrial oxygen budget. Precam Res 2:1-69

Schidlowski M, Hayes JM, Kaplan IR (1983) Isotopic inferences of ancient biochemistries: carbon sulfur, hydrogen, and nitrogen. *In* Schopf JW (ed) Earth's Earliest Biosphere: Its Origin and Evolution. Princeton University Press, Princeton, NJ, p 149-186

Schoell M (1988) Multiple origins of methane in the earth. Chem Geol 71:1-10

Scholle PA, Arthur MA (1980) Carbon isotopic fluctuations in Cretaceous pelagic limestones: potential stratigraphic and petroleum exploration tool. AAPG Bull 64:67-87

Shackleton NJ (1987) The carbon isotope record of the Cenozoic history of organic carbon burial and of oxygen in the ocean and atmosphere. Geol Soc Am Spec Publ 26:423-444

Shackleton NJ, Pisias NG (1985) Atmospheric carbon dioxide, orbital forcing, and climate. *In* Sundquist ET, Broecker WE (eds) The Carbon Cycle and Atmospheric CO_2: Natural Variations Archean to Present. Am Geophys Union Monogr 32:303-317

Sloan LC, Walker JCG, Moore TC Jr, Rea DK, Zachos JC (1992) Possible methane-induced warming in the early Eocene. Nature 357:320-322

Sundquist ET (1985) Geological perspectives on carbon dioxide and the carbon cycle. *In* Sundquist ET, Broecker WS (eds) The Carbon Cycle and Atmospheric CO_2: Natural Variations Archean to Present. Am Geophys Union Monogr 32:5-59

Veizer J, Hoefs J (1976) The nature of $^{18}O/^{16}O$ and $^{13}C/^{12}C$ secular trends in sedimentary carbonate rocks. Geochim Cosmochim Acta 40:1387-1395

Veizer J, Holser WT, Wilgus CK (1980) Correlation of $^{13}C/^{12}C$ and $^{34}S/^{32}S$ secular variations. Geochim Cosmochim Acta 50:1679-1696

Vincent E, Berger, WH (1985) Carbon dioxide and polar cooling in the Miocene: the Monterey hypothesis. *In* Sundquist, ET, Broecker WS (eds) The Carbon Cycle and Atmospheric CO_2: Natural Variations Archean to Present. Am Geophys Union Monogr 32:455-468

Wadleigh MA, Veizer J (1982) $^{18}O/^{16}O$ and $^{13}C/^{12}C$ in lower Paleozoic articulate brachiopods: implications for the isotopic composition of seawater. Geochim Cosmochim Acta 56:431-443

Walker JCG, Klein C, Schidlowski M, Schopf JW, Stevenson DJ, Walter MR (1983) Environmental evolution of the Archean-Early Proterozoic Earth. *In* Schopf JW (ed) Earth's Earliest Biosphere: Its Origin and Evolution. Princeton University Press, Princeton, NJ, p 260-290

Walter MR, Veevers JJ, Calver CR, Gorjan P, Hill AC (2000) Dating the 840-544 Ma Neoproterozoic interval by isotopes of strontium, carbon, sulfur in seawater, and some interpretive models. Precamb Res 100:371-433

Wang K, Orth CJ, Attrep M Jr, Chatterton BDE, Wang X, Li J-J (1993) The great latest Ordovician extinction on the South China Plate: Chemostratigraphic studies of the Ordovician-Silurian boundary interval on the Yangtse Platform. Palaeogeogr Palaeoclimat Palaeoecol 104:61-79

Wefer G, Berger WH (1991) Isotope paleontology: growth and composition of extant calcareous species. Mar Geol 100:207-248

Zachos JC, Arthur MA (1986) Paleoceanography of the Cretaceous/Tertiary boundary event; inferences from stable isotopic and other data. Paleoceanogr 1:5-26

Zachos JC, Arthur MA, Dean WE (1989) Geochemical evidence for suppression of pelagic marine productivity at the Cretaceous/Tertiary boundary. Nature 337:61-64